WORLDMARK
CHRONOLOGY
of the Nations

Volume 3 – Asia

Timothy L. Gall and Susan B. Gall, Editors

Detroit
San Francisco
London
Boston
Woodbridge, CT

The Gale Group

Editorial
Shelly Dickey, Project Editor
Matthew May, Assistant Editor
With contributions from the Cultures and Customs Team
Rita Runchock, Managing Editor
Production
Mary Beth Trimper, Production Director
Evi Seoud, Production Manager
Wendy Blurton, Senior Buyer
Product Design
Cynthia Baldwin, Production Design Manager
Michelle DiMercurio, Senior Art Director
Graphic Services
Barbara J. Yarrow, Graphic Services Director
Randy Bassett, Image Database Supervisor
Pamela R. Reed, Imaging Coordinator
Permissions
Maria Franklin, Permissions Manager
Margaret Chamberlain, Permissions Specialist

Library of Congress Cataloging-in-Publication Data

Worldmark chronology of the nations / Timothy L. Gall and Susan Bevan Gall, editors.
p. cm.
Includes bibliographical references and index.
v. 1. Africa -- v. 2. Americas -- v. 3. Asia -- v. 4. Europe.
ISBN 0-7876-0521-2 (set) -- ISBN 0-7876-0522-0 (v. 1) -- ISBN 0-7876-0523-9 (v. 2)
-- ISBN 0-7876-0524-7 (v. 3) -- ISBN 0-7876-0525-5 (v. 4)
1. Geography. I. Gall, Timothy L. II. Gall, Susan B.
G133.W93 1999
910 21--dc21

99-044217
CIP

The paper used in this publication meets the minimum requirements of American National Standard for Information Sciences—Permanence Paper for Printed Library Materials, ANSI Z39,478-1984.

Contents

Contributors

Editors: Timothy L. Gall and Susan Bevan Gall
Associate Editors: Daniel M. Lucas, Eleftherios Netos
Editor, Photo Research: Michael Cikraji
Typesetting and Design: Bridgette Nadzam
Graphics: Rebecca Kimble, Hannah Lissauer
Editorial Assistants: Jill Coppola, Rebecca Kimble, Susan Stern, Jennifer Wallace

ADVISORS

JANE L. THOMAS. Librarian, McNeil High School, Round Rock Independent School District, Austin, Texas
WENDI GRANT. Teacher, Geography, McNeil High School Library, Round Rock Independent School District, Austin, Texas
FLO RANKIN. Library Media Specialist, Hoover High School, North Canton, Ohio
MARK KURT. Teacher, Lake Forest High School, Lake Forest, Illinois
KEITH PLATTE. Teacher, Social Studies, Kalamazoo Central High School, Kalamazoo, Michigan

CONTRIBUTORS

OLUFEMI A. AKINOLA, Ph.D. W.E.B. DuBois Institute, Harvard University
LESLIE ASHBAUGH, Ph.D. Department of Sociology, Seattle University
VICTORIA J. BAKER. Department of Anthropology, Eckerd College
IRAJ BASHIRI. Professor of Central Asian Studies, Department of Slavic and Central Asian Languages and Literature, University of Minnesota
THEA BECKER. Researcher/Writer, Cleveland, Ohio
HEATHER BOWEN. Researcher/Writer, Washington, D.C.
GABOR BRACHNA. Researcher/Writer, Cleveland, Ohio
SALVADOR GARCIA CASTANEDA. Department of Spanish and Portuguese, The Ohio State University
ERIK CHING. History Department, Furman University
FRANCESCA COLECCHIA. Modern Languages Department, Duquesne University
LEAH ERMARTH. Worldspace Foundation, Washington, D.C.
JENNIFER FORSTER. Researcher/Writer, Kent, Ohio
ALLEN J. FRANK, Ph.D., Researcher/Writer
DIDIER GONDOLA. History Department, Macalester College, St. Paul, Minnesota
ROBERT GROELSEMA. Ph.D. African Bureau, U.S. Agency for International Development (USAID)
HIMANEE GUPTA. Researcher/Writer, Honolulu, Hawaii
BRUCE HEILMAN. University of Dar es Salaam in Tanzania
ROSE M. KADENDE-KAISER. Director, Women's Studies Program, Mississippi State University
EZEKIEL KALIPENI. Department of Geography, University of Illinois at Urbana-Champaign
RICHARD A. LOBBAN, Jr., Ph.D. Department of Anthropology, Rhode Island College
IGNACIO LOBOS. Journalist, Honolulu, Hawaii
DERYCK O. LODRICK. Visiting Scholar, Center for South Asian Studies, University of California, Berkeley
DANIEL M. LUCAS. Researcher/Writer, Cleveland, Ohio
PATRIZIA C. MCBRIDE. Department of German, Scandinavian, and Dutch, University of Minnesota
MEGAN MENTREK. Researcher/Writer, Cleveland, Ohio
WILLIAM MILES. Department of Political Science, Northeastern University
EDITH T. MIRANTE. Project Maje, Portland, Oregon
CAROL MORTLAND. Crate's Point, The Dalles, Oregon
NYAGA MWANIKI. Department of Anthropology and Sociology, Western Carolina University
ELEFTHERIOS NETOS. Ph.D. Candidate, Kent State University
BRUCE D. ROBERTS, Department of Anthropology and Sociology, University of Southern Mississippi
GAIL ROSEWATER. Researcher/Writer, Cleveland, Ohio
JENNIFER SPENCER. Researcher/Writer, Columbus, Ohio
JEANNE-MARIE STUMPF. Researcher/Writer, Cleveland, Ohio.
CARMEN URDANETA. M.A. Researcher/Writer, Boston, Massachusetts
KIMBERLY VARGO. Researcher/Writer, Cleveland, Ohio
JEFF WASICK. Kent State University.
GERHARD H. WEISS. Department of German, Scandinavian, and Dutch, University of Minnesota
ROSALIE WIEDER. Researcher/Writer, Cleveland, Ohio
JEFFREY WILLIAMS, Ph.D. Cleveland State University

Preface

Worldmark Chronology of the Nations contains entries on 192 countries of the world. Arranged in four volumes—Africa, Americas, Asia, and Europe—*Worldmark Chronology of the Nations* follows the organizations of its sister sets, *Worldmark Encyclopedia of the Nations* and *Worldmark Encyclopedia of Cultures and Daily Life*. Within each volume, entries are arranged alphabetically.

Each volume begins with a general timeline of world history. This timeline provides a history of the world from prehistoric times to the present, lending a context to the entries on individual nations.

The nations profiled are those that exist at the end of the twentieth century. Profiles cover the territory within the modern-day geographic borders. History of earlier nation-states and empires is included in entries for each modern-day nation that has evolved within their region. Emphasis is on social and cultural history, helping the user of *Worldmark Chronology of the Nations* gain an understanding of a country's history beyond its succession of political leaders and military conflicts.

Each entry begins with an overview essay. This essay provides a general introduction to the country, its history, and its people from the beginning of time to the present. Trends in all aspects of a nation's development—politics, the military, literature, art, industry, religion, society, and relations with its neighbors and the other nations of the world—are included.

Following the introduction is a timeline of history, with dated entries describing the key people and events that shaped the nation. Notable people whose achievements helped shape the society and who contributed to world history are profiled. Entries may cover an era, a decade, a range of years, a specific year, a period of months, a specific month, or a specific date. A heading summarizes the event's significance, while the paragraphs that follow provide further details.

A current map accompanies each entry; in addition, over 100 historic maps illustrate the location of earlier empires, kingdoms, and depict transfers of territory between nation-states. Sidebars provide context for events, empires, people, and organizations of significance, such as the Cold War, Ottoman Empire, and League of Nations.

A comprehensive bibliography following the organization of the work (by continent and country), glossary of terms, and comprehensive index appear at the back of each volume.

Over fifty university professors, professional writers, journalists, and expert reviewers contributed to the preparation of this *Worldmark Chronology of the Nations*. Many have carried out historical research in the country about which they wrote. All are skilled researchers with expertise in their chosen area of study.

Acknowledgments

The editors express appreciation to staff of The Gale Group who were involved at various stages of the project: Linda Irvin and Kelle Sisung, who assisted with development of the initial concept of the work; Shelly Dickey, Project Editor, Matthew May, Assistant Editor, who guided the editorial development of the entries with contributions from the Cultures and Customs Team, Rita Runchock, Managing Editor; Cynthia Baldwin and Michelle DiMercurio, who were responsible for the design of the volumes' covers; and Mary Beth Trimper, Evi Seoud, and Wendy Blurton, who supervised the printing and binding process. In addition, the editors express appreciation to William Becker, archivist of the Cleveland State University Photo Archive, for his assistance with photo research. Also helping with images were members of the Gale Group Graphic Services department, Barbara J. Yarrow, Randy Bassett, and Pamela R. Reed; and Permissions department, Maria Franklin and Margaret Chamberlain.

SUGGESTIONS ARE WELCOME: The first edition of a work of this size and scope is an ambitious undertaking. We look forward to receiving suggestions from users on ways to enhance and improve future editions. Please send comments to:

Editors
Worldmark Chronology of the Nations
The Gale Group
27500 Drake Road
Farmington Hills, MI 48331–3535
(248) 699-4253

Timeline of World History

c. 2,600,000 B.C. *Homo Australopithecus* walks the earth

Homo Australopithecus, the earliest ancestor of *Homo Sapiens Sapiens* (present-day human beings) lives in sub-Saharan Africa. *Homo australopithecans* are hunter-gatherer nomads who fashion stone tools.

c. 100,000 B.C. Emergence of *Homo Sapiens Sapiens* in Africa

The earliest ancestors of the modern human species, *Homo Sapiens Sapiens,* emerges in Africa. They have larger brains than their predecessors and are hunter-gatherers who communicate by means of sophisticated language. By 30,000 B.C. they are the only human-like hominid (two-legged primate) left.

c. 8000 B.C. Agricultural revolution

Humans begin raising crops and domesticating animals. This results in the greatest change in human lifestyle to this point; humans start settling in fertile river valleys. This change in life patterns leads to the development of the first civilizations.

c. 3200 B.C. Sumerians establish first civilization

The first civilization emerges in Sumer, in Mesopotamia, along the Euphrates River. The establishment of civilization brings with it government, religion, urbanization, and a specialized economy (including trade between cities). Significantly, the Sumerians establish a pictographic system of writing called Cuneiform. A writing system allows for record keeping as well as communication, particularly for trading purposes.

c. 2850 B.C. Old Kingdom established in Egypt

During the period of the Old Kingdom, the lands along the Nile River in northern Africa are the second region of the world to form a civilization. Arguably the most impressive achievement of the Old Kingdom are the pyramids which are built as the tombs of pharaohs (kings). The Egyptians of the Old Kingdom also establish a calendar based on the annual flooding of the Nile and develop a form of writing known as hieroglyphics which they often write on papyrus scrolls (early paper).

2200 B.C. Emergence of Chinese civilization

Chinese civilization emerges along the Yellow River with the establishment of the Xia (Hsia) dynasty. This is the first step in the formation of a Chinese empire. To this day, Chinese civilization is the oldest in existence.

1728–1686 B.C. Code of Hammurabi

King Hammurabi of Babylonia writes a law code for his subjects. The laws, which seek to maintain stability in the kingdom, pertain to topics as diverse as marriage, taxes, and business contracts. The Code of Hammurabi is the oldest law code in the world.

c. 1500–1000 B.C. Emergence of Indian civilization, basis of Hinduism

Indian civilization emerges along the Indus River valley. An early achievement of the Indus civilization is the compilation of the first *Vedas*, spiritual texts which form the basis of Hinduism.

c. 1200 B.C. Emergence of Hebrew civilization, Judaism

The emergence of the Hebrew civilization gives rise to Judaism, a monotheistic religion. Although the Hebrews subsequently succumb to internal division and foreign invasion, their religion is one of the most influential in the world. Both Christianity and Islam are offshoots of Judaism.

c. 1000 B.C. Emergence of the Kingdom of Kush

The Kingdom of Kush emerges in present-day Sudan. Centered around the city of Meroe, the Kingdom of Kush is the first sub-Saharan African civilization.

c. 1000 B.C. Emergence of first American civilizations in Mexico and Peru

Civilization emerges in the Americas in the areas of present-day Mexico and Peru.

8th century B.C. **Emergence of Classical Greek civilization, *Iliad* and the *Odyssey* compiled**

Classical Greek civilization arises out of a dark age. The Greeks organize themselves into city-states both on the Greek mainland as well as on hundreds of the islands that make up the Aegean and Ionian seas. Great seafarers, the Greeks establish colonies and trade throughout the Mediterranean and Black Seas.

Greek civilization is also well-known for its many cultural achievements. Homer's compilation of the *Iliad* and the *Odyssey* are among the oldest works of Western literature. Greek civilization reaches its height during the Athenian Golden Age (see 5th century B.C.).

563–483 B.C. **Life of Buddha**

Siddharta Gautama, an Indian prince, seeks eternal happiness as a way of escaping suffering in the world. After a personal odyssey, he adopts the name Buddha (enlightened one) and forms a belief system that becomes known as Buddhism. In subsequent centuries the Buddhist faith spreads throughout much of Asia including Japan, Korea, and China (where it becomes the dominant form of religious belief).

551–479 B.C. **Life of Confucius**

Chinese philosopher Confucius advocates human behavior based upon morality and justice. His writings, compiled in the *Analects,* are part of a program of political reform.

509 B.C. **Roman Republic founded**

The Roman monarchy is overthrown and replaced by a republican system of government. Under this system, citizens with voting rights (landholding males) elect representatives to legislate for them. The foundation of the republic leads to stronger and more effective government which eventually unites all of the Italian peninsula. The republican system of government breaks down when Rome subsequently expands to encompass all of the Mediterranean world as well as much of southern and western Europe. The Roman Republic becomes the Roman Empire.

480–431 B.C. **Athenian Golden Age**

During the Athenian Golden Age, Athens becomes the center of Greek politics and culture. The city's democratic government allows all of its citizens (roughly a third of its population) to participate in government at some point in their lives. Art, philosophy, and theater flourish. The Parthenon is erected, Socrates (469–399) debates pupils and populace, and Euripides (485?–406?) writes plays. The Golden Age ends with the outbreak of the Peloponnesian War (431–404) against Sparta, which ends in an Athenian defeat.

c. 4th century B.C. **Emergence of Mayan civilization**

Mayan civilization emerges in the area of present-day Mexico. The Maya are the first civilization to emerge in the Americas and are best-known for their sophisticated mathematics which they apply to the study of astronomy and which enables them to devise a highly accurate calendar.

356–323 B.C. **Life of Alexander the Great**

In only thirteen years King Alexander III (the Great) of Macedon creates an empire that extends from Greece to the Indus River. His conquests spread Greek civilization throughout the Middle East and into Asia.

4th–2nd centuries B.C. **Hellenistic Era**

In the wake of Alexander's death, his empire fragments as his successors quarrel for the spoils. Nevertheless, a new cosmopolitan culture emerges that combines Greek and Eastern influences.

3rd century B.C. **Erection of the Great Wall of China**

In an effort to keep foreign invaders at bay, the Chinese emperor constructs a wall stretching across the northern frontier of China. Known as the Great Wall, this fortification stretches for over a thousand miles (1,620 kilometers) and becomes one of the Seven Wonders of the World.

A.D. 1st century **Axumite Kingdom established**

The Axium Kingdom emerges in Ethiopia. This state soon adopts Christianity and becomes a Christian stronghold in sub-Saharan Africa as it has remained to the present.

5 B.C.–A.D. 29 **Life of Jesus, Rise of Christianity**

In Roman Judea, Jesus of Nazareth claims to be the Jewish messiah (savior). Although he gains a following among those who embrace his message, the Roman authorities view him as an obstacle to their rule and he is sentenced to death. After his crucifixion, his followers claim that he is resurrected. His life is the basis of a new religion, Christianity, that comes to dominate the Roman Empire by the fourth century. It subsequently spreads throughout Europe into Asia, Africa, and, ultimately, the Americas and Australia. It is presently the largest faith in the world with over one billion followers.

5th century **Middle Ages, rise of feudalism**

The Middle Ages follow the fall of the Roman Empire in Western Europe. This period is characterized by the establishment of small kingdoms based upon the economic system of feudalism. Under this regime, peasant serfs work for lords who provide protection from invaders. In turn, the lords provide services for a king in return for land.

440–61 **Reign of Pope Leo the Great**

Under Pope Leo the Great, Rome increases its power over the other four patriarchates (the five original seats of church government: Rome, Constantinople, Antioch, Jerusalem, Alexandria) of Christianity. Subsequently, the pope becomes the leader of the Roman Catholic branch of Christianity.

528 **Justinian's Code**

The Byzantine (Eastern Roman) emperor Justinian, (483–565) establishes a code of laws for his empire called the *Corpus Juris Civilis*, which forms the foundation of much of European law today.

7th–8th centuries **Rise and spread of Islam**

Under the direction of Mohammed (570–632), the Islamic religion emerges among the Arabs. An offshoot of Judaism and Christianity, Mohammed claims that he is the last prophet of God. After Mohammed's death, the Muslims (followers of Islam) create an Islamic empire that stretches from the North African coast and Spain as far east as Persia. Islam later spreads throughout much of sub-Saharan Africa, and central and southeast Asia. It is presently the second-largest religion in the world.

8th century **Emergence of Ghana**

The African kingdom of Ghana emerges in the northwest part of Africa and is the first sub-Saharan African civilization. The kingdom establishes trade across the Sahara with Morocco. Although Ghana is an advanced civilization it falls victim to invasion and disintegrates in the eleventh century.

8th–9th centuries **Carolingian Renaissance**

The rise of the Carolingian Empire in western Europe, results in a renewed stress on learning. Newly-established monasteries become centers of education and preservation where monks study Greek and Roman classics (as well as the scriptures) while also copying and translating the old texts.

10th century **Rise of Russia**

The Varangians establish the first Russian state centered in Novgorod. This later becomes the foundation of the Russian Empire that stretches across Siberia.

1054 **East-West split in Christianity**

Christianity splits into two branches: Roman Catholicism, centered in Rome; and Eastern Orthodoxy, centered in Constantinople, the capital of the Byzantine Empire. The split is of vital importance over time as the two halves of Europe undergo significantly different historical development.

11th–13th centuries **Crusades foster greater East-West contact**

In an effort to oust the Muslims from the Holy Land, Christian warriors from western Europe wage a series of holy wars (crusades) against the Muslim Turks (who control the Middle East). All but the first crusade are military defeats for the Europeans. (Indeed, the Fourth Crusade, rather than attacking the Holy Land, captures Constantinople in 1204.) Nonetheless, they help bring western Europe out of its isolation during the Middle Ages.

1192 **Feudal Japan emerges**

Feudalism emerges in Japan. Under this system, the emperor becomes a figurehead and political power rests in the hands of the *Shogun*, the chief warlord. Japan remains under this system until the Meiji Restoration in 1868.

c. 1200 **Height of Inca civilization**

The Inca establish an empire in the Quechua mountains along the Pacific coast in South America. The Inca government is a theocracy in which the emperor holds absolute authority. Among the Inca achievements are the creation of thousands of suspension bridges across rivers and numerous religious sculptures.

12th century **Rise of the Mongol Empire**

Under the leadership of Genghis Khan (1162–1227), the Mongols establish an empire that stretches from China, through central Asia, into Europe.

c. 1300 **Aztec civilization**

The Aztec civilization in present-day Mexico reaches its height in the fourteenth century. The Aztecs have an advanced civilization that includes sophisticated cities and government. Along with the great pyramids and temples they build, they are known for their bloody ritual human sacrifices meant to please their gods.

1347–50 **Black Death sweeps through Europe**

Brought in from Asia, Bubonic Plague (known as the Black Death) spreads across Europe and kills approximately half its population.

14th–16th centuries **Era of the Renaissance**

A period of intellectual and cultural rebirth, the Renaissance, begins in northern Italy and later takes root in northern Europe, particularly Holland. The central characteristic of the Italian Renaissance is the emphasis on classical (ancient Greek and Roman) civilization. Renaissance artists include Leonardo da Vinci (1452–1519), Michaelangelo (1475–1564), and Raphael (1483–1520). Much of their work graces churches and cathedrals in Rome, most notably St. Peter's.

1445 Gutenberg invents printing press

Johann Gutenberg (1400–67) invents the printing press which allows for the widespread publication of books. Mass production of books leads to widespread access to ideas and is instrumental in bringing people closer together.

1453 Constantinople falls to the Ottoman Turks

Constantinople, the capital of the Byzantine Empire, falls to the Ottoman Turks. The Ottomans, a Muslim people, establish an empire that, at its height, includes all of Asia Minor, all of the Balkans and central Europe as far as Vienna, the North African coast, the Fertile Crescent, and the Caucasus. The empire lasts until 1923.

Late 15th century–19th century African slave trade

The African slave trade begins with the Portuguese. Over the course of four centuries approximately twenty-two million black Africans are uprooted forcibly. The half who survive the voyage are sent primarily to the Americas where they primarily work in agriculture.

1492 Columbus makes first voyage to the Americas

While searching for a shorter route to Asia by sailing west, Christopher Columbus (an Italian in the service of the Spanish) instead reaches the West Indies and the South American coast. However, not recognizing his mistake, he embarks on three return journeys over the next decade. His voyages promote further European exploration that results in the eventual colonization of North and South America.

Early 16th century Protestant Reformation

A revolt against certain practices of the Catholic Church led by German monk Martin Luther (1483–1546) results in a full-scale revolt against the Papacy in western Europe. Religious wars ensue between the Catholics and Protestants (those who split from Rome) and leave much of western Europe politically and economically devastated.

1500–1700 European exploration and colonization of the Americas

European explorers make a series of voyages to the Americas. Although their early journeys focus on discovery, they soon turn to expansion. Spain and Portugal conquer most of the Americas (although England, France, and the Netherlands stake claims as well) and establish vast empires in the "New World." The Europeans soon begin exploiting their new territories for mineral and agricultural wealth (often through slave labor). Most of the native population dies of disease or warfare while the Europeans begin establishing their territories.

1519–21 Magellan circumnavigates the globe

An expedition led by Portuguese explorer Ferdinand Magellan (1480–1521) sets sail to circumnavigate the globe. Although Magellan and most of his crew die, the survivors return to Portugal in 1521 and prove that the world is round.

1519–25 Spanish conquest of the Aztecs

The Spanish, under Hernan Cortez (1485–1547) conquer the Aztec Empire in Mexico. The rapid conquest of the Aztecs illustrates the advantages held by Europeans over non-Europeans in weapons technology.

1564–1616 Life of William Shakespeare

English playwright and poet William Shakespeare creates his artistic and literary masterpieces. Not only do his works become widely-acclaimed as the best writings in the English language, they receive international recognition, are translated into dozens of languages, and his plays are performed throughout the world.

1581 Beginning of Russian expansion into Siberia

Russian traders begin expanding their activities eastward into Siberia, incorporating the land into the Russian Empire as they travel. This begins one of the greatest expansions in history as the Russian Empire eventually grows to encompass one-sixth of the entire land area of the world.

17th–18th centuries Age of Reason, Enlightenment

The principles of liberal political thought emerge during the Enlightenment. Writers such as John Locke (1632–1704), Voltaire (1694–1778), and Jean-Jacques Rousseau (1712–78) stress the importance of individual liberty based on private property, religious toleration, and the social contract between a government and its people. These ideas become the key ideology behind the American and French Revolutions and remain an important part of Western political thought through the end of the twentieth century.

1618–48 Thirty Years War

When Catholic Habsburg prince Ferdinand II becomes king of Protestant Bohemia, Protestant and Catholic states in Germany, and eventually throughout central Europe, fight one another for political and religious dominance.

c. 1750 Industrial Revolution

Industrialization begins to develop in England. Characterized by steam-powered manufacturing and mass production, the industrial revolution sets the stage for the modern age. The development of industry soon spreads to western Europe and North America and propels them to the forefront of world economic, political, and military power. With the West dominant throughout the globe, the rest of the world attempts to industrialize as well.

1770 Cook charts the east coast of Australia

English Captain James Cook (1728–79) charts the east coast of Australia while on a South Sea expedition. His discovery paves the way for the eventual European colonization of the island-continent.

1776 Adam Smith writes *The Wealth of Nations*

British economist Adam Smith writes *The Wealth of Nations*. This work stresses that the key to expanding a nation's wealth lies in the establishment of a capitalist economic system characterized by free enterprise, private property, and an absence of government interference in the economy.

1776–83 American War of Independence

In the name of Enlightenment principles of political freedom, Great Britain's thirteen American colonies declare their independence and, by 1783, emerge victorious as the United States of America. The new country comes to dominate the North American continent and by the twentieth century becomes the most powerful nation in the world.

1789 French Revolution

With its goals of *Liberté*, *Egalité*, and *Fraternité* (Liberty, Equality, and Fraternity), revolution breaks out in France. By 1792 it overthrows the monarchy and institutes a republic. Although the revolution ultimately degenerates into the imperial dictatorship of Napoleon Bonaparte (1769–1821) it serves as a model for subsequent liberal nationalist revolutions throughout the world.

Early 1800s Latin American revolts against Spain

Spain's colonies in Latin America rise in revolt against Spain. By the mid-1820s, the once vast Spanish Empire in South America is reduced to minor island possessions.

1839–42 Opium War

The Chinese forbid the importation of opium into their ports by the the British, who ship it from India. The British wage war on China forcing them to reopen their ports. The Chinese defeat is a severe setback to the forces that oppose Western imperialism.

1848 Revolutions in Europe

Liberal nationalist revolutions sweep across the European continent. In the short-term, the revolutions fail, although their long-term impact is profound as they form the basis of future liberal nationalist political programs that play an important role in European politics in the late nineteenth century.

1848 Karl Marx publishes the *Communist Manifesto*

In the wake of revolutions that sweep across Europe, Karl Marx publishes his *Communist Manifesto*, which advocates the overthrow of capitalism in favor of socialist societies in which all property is communal and workers rule. Although his work has no immediate impact, his ideas form the basis of future Communist revolutions in the twentieth century.

1868 Meiji Restoration

In an effort to combat Western imperialism, Japanese patriots stage a revolution that "restores" the emperor's authority. As a result of this successful revolt, Japan embarks upon a policy of rapid modernization that makes it a leading power by the first decade of the twentieth century. Its defeat of Russia in the Russo-Japanese War (1904–05) is the first time that a non-European power defeats a European power in modern times.

1885 Karl Benz designs first automobile

The German Karl Benz designs the first automobile. Within decades, his invention becomes the primary means of ground transportation in the industrialized world.

1903 Wright brothers make first powered flight

The American Wright brothers make the first powered flight in a light airplane in Kitty Hawk, North Carolina. By the late twentieth century, air travel becomes one of the safest and primary means of long-distance transportation.

1911–49 Chinese Revolution

In 1911, the Manchu dynasty falls and is replaced by a republic led by Sun Zhongshan (Sun Yat-Sen, 1866–1925). The revolution is aimed at modernizing China and throwing off Western imperialism. Sun is succeeded by Jiang Jieshi (Chiang Kai-Shek, 1887–1975). By the late 1920s, however, a Communist movement led by Mao Zedong (Mao Tse-Tung, 1893–1976) challenges Jiang for control and a bloody civil war ensues. By 1949 the Communists are victorious and drive Jiang's forces off the mainland to the island of Formosa (Taiwan).

1914–18 World War I

World War I pits two great alliances against each other: the Triple Alliance (Germany, Austria-Hungary, the Ottoman Empire, and, later, Bulgaria) versus the Triple Entente (Great Britain, France, Russia, later joined by the United States) and their associates. The war results in an Entente victory but only after most of Europe is destroyed. The war kills around nine million people. Political instability follows the war and leads to World War II in 1939.

1917 Russian Revolution

The monarchical tsarist regime falls in Russia. After a brief period in which a provisional government takes over, Russia is governed by the Bolsheviks, who establish the first Communist government in the world. The Communists seek to create a socialist state run by the working class. In practice, however, their one-party regime relies upon force and widespread suppression of human rights to remain in power.

1939–45 World War II

World War II pits the expansionist Axis (Germany, Italy, and Japan) and its associates against the Allies (Great Britain, France, China, the United States, and the Soviet Union) and its partners. By 1945 the Allies are victorious. Their victory, however, comes at great cost as most of Europe and Asia are destroyed. The war kills around sixty million people, including twenty-seven million Soviet citizens and six million European Jews. Jews are singled out for extermination by Nazi Germany in a policy known as the Holocaust.

1947–91 Cold War

The wartime alliance between the United States and the Soviet Union (known as the superpowers) breaks down into a relationship of mutual distrust and hostility. The Cold War is the commonly used name for the prolonged rivalry and tension between the United States and the Soviet Union which lasts from the end of World War II to the break-up of the U.S.S.R. in 1991. The Cold War encompasses the predominantly democratic and capitalist nations of the West, which are allied with the U.S., and the Soviet-dominated nations of eastern Europe, where Communist regimes are imposed by the U.S.S.R. in the late 1940s. Although the United States and the Soviet Union never go to war with each other, the Cold War results in conflicts elsewhere in the world, such as Korea, Vietnam, and Afghanistan, where the superpowers fight wars either by proxy or with their own troops against allies of their superpower rival.

1950s–60s Decolonization

Throughout the world European colonial empires collapse and previously subjugated peoples receive their independence. Freedom from colonial rule presents many challenges to the newly-independent states. Economic underdevelopment, widespread poverty, and tenuous political stability all plague these new countries.

1969 Man walks on the moon

On July 20, 1969, U.S. astronauts Neil Armstrong (b. 1930) and Edwin "Buzz" Aldrin (b. 1930) become the first men to walk on the moon. This marks the first time human beings have ever set foot upon another celestial body. It is only sixty-six years since the Wright brothers' first flight and only eight years since Soviet cosmonaut Yuri Gagarin made the first manned space flight.

1980–Present Information revolution

The increasing importance of computers in daily life—particularly for the transmission of information results in the so-called "Information Revolution."

1989 Collapse of Communism in Eastern Europe

As part of the liberalizing policies of Soviet leader Mikhail Gorbachev (b. 1930), the U.S.S.R. allows its East European satellites to go their own way. As a result, Communist regimes throughout Eastern Europe collapse.

1991 Collapse of the Soviet Union

Gorbachev's liberalization policies lead to a rise in nationalism among the peoples of the Soviet Union. When his reforms prove unable to rescue Communism, the Communist Soviet Union disintegrates into its constituent republics. Russia, the chief republic of the old U.S.S.R. is a mere shadow of its predecessor and is plagued by political and economic instability for the rest of the decade.

1992 European Union formed

The nations of the European Community (Belgium, Denmark, France, Germany, Greece, Ireland, Italy, Luxembourg, Netherlands, Portugal, Spain, and the United Kingdom) form the European Union. More than an economic union (as its predecessor was), the European Union seeks to establish a common currency, defense and foreign policy.

1997: Summer Asian economic crisis

Beginning in Thailand, a financial crisis strikes East Asia. Caused by over-investment and currency devaluation, the crisis brings economic growth in East Asia to a halt and threatens global prosperity. Prior to this economic reverse East Asian economies are the fastest-growing in the world and comprise one-third of the world economy.

Afghanistan

Introduction

Afghanistan, the "Land of the Afghans," is located in the mountains of South Asia's northwestern borderlands. Until the early twentieth century, 'Afghan' was used to refer only to Pashto-speakers although now the term encompasses the many ethnic groups that inhabit the region. Afghanistan emerged as an independent state in 1747 when Ahmad Shah Abdali (Durrani) was crowned king by an assembly of Pashtun (also Pakhtun, Pathan) chieftains. Rallying the Pashtun tribes to his banner, Ahmad Shah carved out an empire that extended from eastern Persia to the plains of northern India, from the Amu Darya to the Indian Ocean. In subsequent centuries, peoples in outlying provinces broke away from Afghan rule, leaving the territorial outlines of the modern state.

Location, geography, and the character of the tribes that inhabit Afghanistan hold the key to understanding the region's history. The mountains and deserts of Afghanistan lack the resource base to support large populations and powerful political states. The Afghani kingdoms that have arisen in the past invariably have based their power on the wealth of the surrounding lowlands. However, Afghanistan lies in close proximity to regions which have given rise to one of the world's first civilizations (the Indus valley) and to centers of great historical empires (Persia, northern India). To the north lie the steppes and deserts of Central Asia, the source of a continuing stream of tribal peoples (Aryans, Kushans, Mongols) that have passed through or settled the area. Afghanistan also lies on the great Silk Road and the trans-Asian trade routes linking India, China, and the Mediterranean. The history of Afghanistan thus extends back far before 1747, and—despite the region's relative geographic isolation—is one of continuing cultural, political and economic contacts with the surrounding lands.

The modern state of Afghanistan has an area of 250,001 square miles (647,500 square kilometers), containing a population estimated at 22.1 million people. The country is completely landlocked, sharing borders with Iran, Pakistan, the People's Republic of China, and three former Soviet Republics, Tajikistan, Uzbekistan and Turkmenistan.

The mountainous core of the country, the Central Highlands, is formed by the Hindu Kush ("Killer of Hindus"), a mountain system whose eastern peaks exceed 20,000 feet (6,100 meters). The Hindu Kush originates in the Pamir Knot in Central Asia, and trends south and west across Afghanistan towards the plateaus of Iran. The mountains spread out into subsidiary ranges such as the Koh-i-Baba, the Feroz Koh, and the Kasa Murgh ranges towards central and western Afghanistan. Climate is severe at higher elevations where alpine conditions prevail, and the region is generally arid. Settlement and agricultural activities are largely confined to the valleys of rivers such as the Kabul, Marghab, and Hari Rud. Pastoralism, based on camels, goats, sheep, and yaks, is common in the region.

North of the mountainous interior, the land descends to the plains of the Amu Darya (the ancient Oxus River). The river itself forms the international boundary between Afghanistan and the Central Asian states of Uzbekistan and Tajikistan. The only extensive lowlands in Afghanistan, the Northern Plains, represent a continuation of the steppes of Central Asia. Its fertile soils are irrigated and produce a considerable proportion of the country's food. Wheat is the staple crop, along with maize, barley and rice. Nuts and fruits, vegetables, legumes, and cotton are also cultivated.

Southwestern Afghanistan consists of high plateaus and deserts extending west to Iran and south to Pakistan. Although the region is crossed by the Helmand River, it is essentially arid and includes the sandy Registan Desert and the salt flats of the Dasht-i-Margo.

Afghanistan's rugged environments are matched by the variety and individualism of its peoples. The country is populated by a mosaic of ethnic groups, each with its own dialect or language, customs, and cultural traditions. Afghanistan's diversity is a legacy of the country's tumultuous past. Throughout history, it has served as a crossroads for countless invaders, many of whom have left an indelible imprint on the country's character. The earliest recorded settlements in modern-day Afghanistan date at least 6000 years to the Harappan civilization of the Indus River valley. The first significant foreign influence came with the arrival of the Aryan invaders around 2000 B.C. The Persian and Greek conquests brought the region into contact with major empires. By the third century B.C., however, most of Afghanistan was part of the Mauryan Empire, centered in India. Indian and Greek influence

predominated until the first century A.D. Afghanistan's position as a crossroads of trade received a boost during the period of Kushite rule (c. A.D. 50–225). Under the Kushites, Afghanistan prospered not only as a commercial center, but also as a center of Mahayana Buddhist culture. Kushite influence ended with the rise of the Persian Sassanid Empire, which conquered most of Afghanistan in the mid-third century.

The next major conquest of Afghanistan proved the most significant. Arab invasions during the seventh century introduced Islam to the country and established a cultural legacy that lasts to the present. In subsequent centuries, Afghanistan would fall to further foreign conquerors, including the Mongols, Tamerlane, and the Mughals. Not until 1747 did Afghanistan emerge as a state following the election of Ahmad Khan Abdali (1724–73) as *amir* (king) by Afghan tribal chiefs. Yet the emergence of statehood did not eliminate foreign interference in Afghan affairs, a circumstance that stemmed from the state's continued importance as a strategic crossroads. By the nineteenth century, the United Kingdom and Russia made Afghanistan the center of the rivalry in central Asia. As a result, a series of wars broke out between Afghanistan and Britain, culminating in the 1919 Treaty of

Rawalpindi, by which Britain recognized Afghan independence.

For much of the twentieth century, the predominant theme in Afghanistan has been the struggle between modernization and tradition with foreign meddling thrown into the mix. Efforts by kings Nadir Shah (d.1933) and Zahir Shah (1914–) at modernization were thwarted by political repression, increasing radicalism among the youth. Although Zahir Shah's 1964 constitution established a constitutional monarchy with fundamental freedoms, continuing political polarization and droughts in the early 1970s culminated in the monarchy's overthrow in 1973.

A communist coup in 1978 led to further instability as the leaders of the revolt fell out and the Soviet Union mounted an invasion in December of 1979 to prop up the tottering regime. As popular resistance to the communist regime intensified, Afghanistan became the site of a major civil war. Heavy losses led to a Soviet withdrawal in 1989 and to the defeat of the communist regime in 1992 at the hands of the *Mujahidin* (anti-Soviet Islamic resistance). Less than two years later, however, the country plunged into yet another civil war as the Islamic fundamentalist *Taliban* (seekers) began a campaign that resulted in their effective takeover of the country by 1996. Through its imposition of Islamic law, the *Taliban* victory reversed many of the westernizing reforms that had been instituted by the Zahir Shah and his successors. Education was reformed to emphasize Islam, and women's rights were sharply curtailed. As of 1999, Afghanistan was largely isolated diplomatically and tensions with Iran threatened to erupt into full-scale war.

Afghans tend to identify with their own community rather than with any abstract concept of an "Afghan" nation. Pashtuns and Tajiks, for example, speak languages (Pashto and Tajiki, a dialect of Dari [Farsi]) belonging to the Iranian branch of the Indo-European language family. In northern Afghanistan, however, the language of the Turkomans and Uzbeks belong to the Turkic branch of the Ural-Altaic language family. Hazaras, like the Tajiks, speak a dialect of Dari or Persian, but their physical appearance indicates a Mongolian ancestry. Aimaqs, Baluchis, and Nuristanis are but a few of the other ethnic groups present in Afghanistan.

Pashtuns make up nearly one half of Afghanistan's population. They, in fact, comprise a group of tribes (e.g. the Durrani and the Ghilzai) who speak the Pashto language. Though some Pashtuns have settled in cities, many continue as agriculturalists and nomadic herdsmen. Prior to the Soviet invasion of 1979, over 300,000 Pashtun nomads crossed annually from Afghanistan into Pakistan. The life of the Pashtun is governed by *Pashhtunwali,* or the Pashtun code of honor. This is based on concepts such as hospitality (*melmastia*), the right of asylum (*nanawati*), bravery (*tureh*), defense of the honor of one's women (*mamus*), and the blood feud (*badal*).

A tribal social structure, family loyalty, and concepts of proper conduct characterize not only Pashtuns, but virtually all the peoples of Afghanistan. These values, combined with traditional means of livelihood and a deep-rooted belief in Islam, help define the nature of Afghan society. They are its strength, but at the same time outline its weaknesses. The fighting spirit of the tribal warrior has helped build Afghan empires. Afghanis defeated British colonial armies in the nineteenth century, and threw back the mechanized divisions of the Soviet army in the 1980s. Yet, with independence secure, the country has embarked on a civil war that continues to tear the country apart. In a very real sense, Afghanistan is a still a country of feuding tribes and political factions seeking to forge a national identity and move into the twentieth century.

Timeline

Pre-8000 B.C. Old Stone Age peoples

Old Stone Age (Paleolithic) peoples inhabit Afghanistan as early as 100,000 B.C.

c. 8000–4000 B.C. Early settlements

Neolithic communities living in caves near Bamiyan (Chihil-Sutoon) and Mazar-i-Sharif (Aaq Kaprak) are domesticating plants and animals. Agricultural villages such as Mundigak exist near Kandahar by the end of this period.

c. 3000–2000 B.C. Harappan sites

Mundigak flourishes as the westernmost city of the Harappan (or Indus valley) civilization. Shortughai (perhaps a trading outpost) in northern Afghanistan suggests that Harappan influence extends north of the Hindu Kush mountains.

2000 B.C. Aryan migrations begin

The first wave of nomadic Aryan tribes enters the region. Aryan migrations into Afghanistan continue over a period of a thousand years.

c. 1500 B.C. Aria, land of the Arians (Aryans)

The *Avesta*, the sacred book of the Zoroastrian religion, mentions that the region west of Herat is called Aria or the land of the Aryans.

c. 550–331 B.C. Achaemenid empire

Darius (r. 544–486 B.C.), ruler of Persia's Achaemenid empire, conquers the lands that now make up Afghanistan. Aria, Bactria, and Gandhara are among the satrapies (provinces) that send tribute to the Persian king. Articles of tribute include leather goods, shields and lances, bulls, and the two-humped Bactrian camels.

330–327 B.C. Alexander conquers Afghanistan

Alexander the Great defeats the Persian king Darius III (r. 336–330 B.C.) in 331 B.C. He subdues the eastern satrapies of Drangania, Arachosia (both in southern Afghanistan), and Bactria, before crossing the Hindu Kush into India.

305 B.C. Chandragupta defeats the Seleucid Greeks

The Indian king Chandragupta Maurya defeats Seleucus Nikator, the Greek general who acquires Alexander's eastern empire. Seleucus cedes extensive territories to Chandragupta in return for 500 war elephants. The Hindu Mauryas rule virtually all of Afghanistan except for Bactria.

261 B.C. Asoka converts to Buddhism

Asoka, the third Mauryan emperor (reigned c. 265–238 B.C. or c. 273–232 B.C.), adopts Buddhism as his state religion.

c. 250 B.C. Independent Greek kingdom in Bactria

The Greek satrap of Bactria declares his independence and establishes a kingdom at Balkh on the northern plains of Afghanistan. Under Demetrius I (reigned c. 190–167), the Greco-Bactrians push south, conquering much of modern Afghanistan.

170 B.C. Indo-Greek kingdom at Taxila (Taksasila)

Bactrian Greeks establish a kingdom at Taxila in Gandhara (now in Pakistan). Under King Menander (r. 155–130 B.C.), Hellenistic and Indic elements combine to create a distinctive "Indo-Greek" culture. During the later years of his reign, Menander annexes the lands of northeastern Afghanistan.

c. 128 B.C. Decline of Bactria

Bactria falls to the Scythians, who in turn are displaced by the Yüeh-Chih, a nomadic people originating in central China. The Yüeh-Chih create five principalities south of the Oxus River in the uplands east of the region.

c. 87 B.C. The Indo-Parthians

Displaced by the Yüeh-Chih, the Scythian tribes move south and west, and settle in southwestern Afghanistan around Zaranj. The Indo-Parthians are descendants of Scythians and the local populations.

The Indo-Parthians become independent from the Parthian (Persian) empire around 87 B.C. Within a few decades, they conquer the Indo-Greek principalities in northeastern Afghanistan. Only Bactria and Aria do not fall under Indo-Parthian rule.

c. 50–225 A.D. The Kushans

Kajula Kadphises, prince of Kuei-Shang (hence "Kushan"), unites the Yüeh-Chih principalities. He crosses the Hindu Kush to conquer the Indo-Parthian kingdoms of northeastern and central Afghanistan. One of his successors, Kanishka (r. 120–162), rules over one of the four great empires of his day. The Kushan kingdom, located astride the trade routes of Asia, acquires considerable wealth from the trade that passes between India, China, and Rome.

The Kushans establish capitals near Kabul and at Peshawar, near the heart of the former Indo-Greek kingdom of Gandhara. This area is strongly Buddhist and Kanishka converts to Mahayana Buddhism (the northern form of the religion). Buddhist culture and art flourish under his patronage. The Gandharan school of art, which blends Greco-Roman and Kushan influences with Indian themes, enters one its most creative phases.

c. 241 Sassanids gain control of Afghanistan

The Sassanid dynasty of Persia conquers most of the Kushan lands in Afghanistan. For the next three and a half centuries, the Sassanids exert varying degrees of political control over the region.

370–c. 530 Huna kingdoms in northern Afghanistan

The Hunas (also called Hepthalites), a nomadic tribe from Central Asia, establish an independent kingdom in Bactria in 370. They defeat the Sassanids in 454 and move south to conquer the Kabul valley and Gandhara. A second Huna kingdom emerges in the region of modern Ghazni in the middle of the fifth century.

630 Hsüan-tsang reaches Kabul

The Chinese pilgrim Hsüan-tsang (c. 600–664) reaches Kabul on his journey from China to South Asia. One of the most remarkable travelers of all time, Hsüan-tsang spends the next fourteen years traversing South Asia recording the details of Buddhist life in the Middle Ages. His travels in Afghanistan take him to Balkh, Bamiyan, Kunduz, and Gandhara.

c. 650–700 Arab invasions of the seventh century

Following the death of the prophet Muhammad in A.D. 632, Islam spreads through the Middle East carried on a tide of Arab expansion. The Arabs overthrow the Sassanid empire and within a few years control much of the former Sassanid lands in western Afghanistan. From there, the Arab caliphate mounts repeated invasions to capture Kabul, Herat, and Balkh during the remaining decades of the seventh century.

870 Arabs capture Kabul

Yaqub ibn Lais, a general of the Saffarid dynasty, captures Kabul and brings Islam to the very heart of Afghanistan.

The closing decades of the ninth century find most of Afghanistan under the rule of Islamic dynasties which have broken away from the Arab caliphate. The Saffarids supplant the Taharids of Khorasan (the region around Herat) in the northwest and extend their lands to include Balkh, Bamiyan, and Kandahar. Only the lands east of the Hindu Kush, where the Hindu Shahis seize power around 890, are not under Muslim rule.

962 Rise of the Ghaznavids

Alptegin (d. 963), a Turkish slave and general of the Samanid dynasty of eastern Persia rebels against his masters and establishes himself at Ghazna. His successors found the Ghaznavid dynasty (977–1186) and build the first great Islamic empire in Afghanistan.

998–1030 Mahmud of Ghazna

Mahumd of Ghazna (971–1030) rules over an empire that stretches from Mesopotamia to the Ganges River. He is viewed with fear and loathing in Hindu India, where he loots and pillages the wealth of the plains and destroys the temples of the "idol-worshippers" (Hindus).

Mahmud's capital at Ghazna becomes a center of Islamic culture and learning. The poets and scholars attracted to his capital include Firdausi (934?–1020?), author of the *Shahnama* (Book of Kings), a collection of legends and history of Persia and Afghanistan, the scholar Al-Biruni (973–1048), the historian Abu'l Fazl Baihaqi (995–1077), and the philosopher Al-Farabi (d. 1077).

1150–1217 The Ghurid dynasty

The Ghurids, from a mountainous region in central Afghanistan (the modern province of Ghowr), rise to power in the mid-twelfth century. They sack Ghazna in 1151 and quickly conquer the rest of the country.

Muhammad Ghuri (more correctly, Mu'izz-ud-Din Muhammad ibn Sam) begins raids into India in 1175. He defeats a coalition of Hindu Rajput kings at the Second Battle of Tarain (1192) and captures Delhi (1193). His conquests bring the north Indian plains, as far as the Ganges delta, into the Ghurid empire.

Qutb-ud-Din Aybak (d. 1210), Muhammad Ghuri's viceroy in Delhi, declares his independence from the Ghurids in 1206.

1207 Jalaluddin Rumi Balkhi is born

Jalaluddin Rumi is born in Balkh, and is therefore called Balkhi by Afghanis.
Jalaluddin is acclaimed as the most eminent Sufi (Muslim mystic) poet writing in Persian. He founds the Sufi order of "whirling dervishes," known for the whirling dance that forms part of the order's rituals.

1214 Khwarazm Shahs take Ghazna

The Khwarazm Shahs of Khiva (in Uzbekistan) rise to prominence in the mid-twelfth century. They extend their control into eastern Persia and southwards into Afghanistan, capturing Ghazna and absorbing the Ghurid lands.

1220–1370 Mongol rule

Genghis (Chingez) Khan (1162–1227) and his Mongols invade Afghanistan from Central Asia, conquering the northern areas of Herat, Balkh, Bamiyan and Kabul.

Genghis Khan's successors found the Il-Khan dynasty (1256–1353), whose empire includes Persia, Mesopotomia, and all of Afghanistan. The peoples of the Il-Khan lands are Muslims, and the one of the Il-Khan rulers, Mahmud Ghazan (r. 1295–1304) becomes a Sunni (Orthodox) Muslim.

1271–95 Marco Polo's travels

Marco Polo (c. 1254–1324), the Venetian merchant, stays in Badakshan in northeastern Afghanistan for around a year on his journey to the court of the Great Khan in China. He describes his journey in *Il milione* (translated as the *Travels of Marco Polo*).

c. 1333 Ibn Batutah passes through Afghanistan

Ibn Batutah (1304–1368 or 1369) visits Herat and Kabul en route to India and China. He describes his travels in his *Rehla* (see *Ibn Batuta*, trans. H. A. R. Gibb., London, 1924).

1370 Timur becomes king of Balkh

Mongolian conquerer, Timur (1336–1405), known in the West as Tamerlane (a corruption of Timur-i Lang or "Timur the Lame"), becomes king at Balkh.

Within two decades, Timur subdues virtually all of Afghanistan and Persia and eventually pushes the borders of his empire to the shores of the Mediterranean. In 1398, he invades northwestern India from Afghanistan, and conquers Multan and the Punjab. Timur utterly devastates the city of Delhi.

1370–1504 Cultural revival under the Timurids

Timur's successors are responsible for a revival of artistic and intellectual life. Herat emerges as an important center of Persian culture under the patronage of the royal court. Schools of miniature painting flourish at Herat, Shiraz and Tabriz, with the artist Behzad (died c. 1525) at Herat being unsurpassed as a painter of illuminated manuscripts. Writers at the Herat court include the last great poet of classical Persian, Jami (Mulla Nur-ud-Din Abdur Rahman Jami [1414–1492]). The Timurids develop a distinctive architecture, using blue and turquoise tiles to decorate their buildings. Timur's mausoleum at Samarkand is the most notable example of this style. Herat and Samarkand are major commercial centers, known for their carpets, ivory ware, and other goods.

1504 Babur captures Kabul

Babur (1483–1530) seizes Kabul.

Babur, whose actual name is Zahir-ud-Din Muhammad, inherits the Central Asian kingdom of Ferghana (Fergana) in

1494. Attacks by Uzbeks force him to flee south into Afghanistan, where he occupies Balkh, Kabul, and Ghazna. He begins raids into India in 1519, and in 1526 wins a decisive victory over the Afghan sultan of Delhi (Ibrahim Lodi) at Panipat.

Babur becomes the first of India's Mughal emperors ("Mughal" is a corruption of the word Mongol, though Babur actually is a Turk).

1510 Safavids take Herat

The Safavid dynasty of Persia (1501–1629) captures Herat and brings areas of western Afghanistan under its control.

1526–1747 Mughal-Persian rivalries in Afghanistan

For over two centuries, Afghanistan is the scene of a struggle for power between competing empires. Eastern Afghanistan forms the Kabul *subah* (province) of the Mughal empire, although Mughal authority is restricted mainly to the valleys and plains. The Safavids rule western areas, and periodically wrest Kandahar from the Mughals. The northern areas of Balkh and Badakhshan come under Uzbek control.

1709 Ghilzais revolt against the Safavids

The Hotaki Ghilzais, under Mir Wais Khan (r. 1709–1715), rise up against their Persian masters. The Ghilzais invade Persia in 1716 and capture Isfahan in 1722. This brings an end to Safavid rule in Afghanistan and creates a short-lived Afghan empire that includes Persia and much of Afghanistan. The Ghilzai Shahs are driven out of Persia by Nadir Quli Beg in 1729.

1732 Nadir Shah invades Afghanistan

Nadir Quli Beg, who becomes Shah of Persia in 1736 as Nadir Shah, captures Herat. He seizes Kandahar in 1738, and proceeds north to take the Mughal outposts of Kabul, Ghazni and Jalalabad. He sacks the Mughal capital of Delhi the following year, adding Sind and all the Mughal lands west of the Indus to his empire. One of Nadir Shah's own tribesmen assassinates him in 1747.

Afghanistan as a State

1747: October Ahmad Shah Durrani elected king

A council of tribal chiefs elects Ahmad Khan Abdali (1724–73), commander of Nadir Shah's Afghan bodyguard, *amir,* or king. Ahmad Khan adopts the title *Durr-i Durran* "Pearl of Pearls", and is more commonly known as Ahmad Shah Durrani. He is crowned *amir* of the Afghan tribes.

1747–72 Ahmad Shah Durrani's reign

Ahmad Shah creates the last great Afghan empire, with its capital at Kandahar. He defeats the Mughals, annexes the Mughal provinces of Multan, Lahore, and Kashmir, and sacks the great Mughal cities of Delhi and Agra.

In addition to empire-building, Ahmad Shah lays the foundations of the state of Afghanistan and establishes the Durrani dynasty (1747–1973) as its ruling house. He sets up the administrative structures of a central government, ruling with the aid of a prime minister and a council of advisers drawn from the leading Pashtun tribes.

1747–1818 Afghanistan's Sadozai Durrani rulers

Ahmad Shah belongs to the Sadozai lineage of the Durrani tribe. His successors come from the same lineage, although they face continual rebellion, intrigue, usurpers and betrayal. Timur Shah (r. 1773–93) moves his capital to Kabul in 1776 because of opposition from local chieftains around Kandahar. The Sadozai Durrani dynasty is severely weakened by squabbles over the succession to the throne among Timur's twenty-three sons after his death.

1809: June 7 Anglo-Afghan treaty of friendship

The British, concerned with the threat of a Franco-Russian invasion of India, send Mountstuart Elphinstone to meet with Shah Shoja (r. 1803–09; 1839–42) in Peshawar. Afghanistan and Britain sign a treaty of friendship, but Shah Shoja is deposed by his brother Shah Mahmud (r. 1800-03; 1809–18) before he can return to his capital. This is the first direct contact between Afghanistan and Britain.

1818 Civil war in Afghanistan

Dost Muhammad (1793–1863), a member of the Barakzai (or Muhammadzai) lineage of the Durrani tribe, seizes Kashmir. He captures Kabul in 1826, and assumes the title of amir. The lands he controls, however, are much reduced from the empire ruled by first amir, Ahmad Shah. Balkh lies in the hands of the Uzbeks, Afghan vassals in Baluchistan and Sind assert their independence, and the Sikhs in the Punjab hold Peshawar.

1837: September British and Russian envoys arrive in Kabul

The British send Alexander Burnes to negotiate an alliance with Dost Muhammad and arrange a peace between the Afghans and the Sikhs, who have occupied Kashmir and captured Peshawar. Burnes turns down Dost Muhammad's request for help in regaining Peshawar. He returns to India when a Russian emissary, Ivan Vitkevitch, arrives in Kabul.

1839–42 First Anglo-Afghan War

Fearing an Afghan alliance with the Russians, the British seek to restore Shah Shoja to the throne. A British army invades Afghanistan and installs Shah Shoja as king.

The Afghan tribes resent the foreign occupation of their land and uprisings break out all over the country. Following the assassination of two of their envoys in Kabul, Burnes and Sir William MacNaghten, the British decide to withdraw to India. What follows is one of the worst military disasters to befall the British in South Asia. Some 4,500 troops and 12,000 camp followers march out of Kabul on January 6, 1942. A sole survivor, Dr. Bryden, escapes to India to tell the story of the retreat. Shah Shoja is killed by Afghans after the British depart Kabul.

After a British force retakes Kabul in the summer of 1842 and destroys much of the city, the British government decides on the complete evacuation of Afghanistan.

1843–63 Dost Muhammad consolidates his rule

Dost Muhammad returns to the throne, and consolidates his rule in Afghanistan, suppressing tribal rebellions and re-acquiring Kandahar (1855), Balkh and the northern Afghan lands (1859), and Herat (1863).

Concerned with Russian and Persian interest in India, Britain reopens diplomatic relations with Afghanistan (Treaty of Peshawar, 1855). Despite considerable pressure from his people, Dost Muhammad refuses to involve the Afghans in the Indian mutiny against the British in 1857.

1863 Shir Ali becomes Amir

Shir Ali (r. 1863–66; 1868–79), Dost Muhammad's third son, ascends the Afghan throne. He puts down several revolts by relatives, but is deposed briefly as amir by two elder half-brothers Muhammad Afzal (r. 1866–67) and Muhammad Azim (1867–68).

1878: July 22 Shir Ali receives Russian envoy

The Russian General Stolietoff arrives in Kabul empowered by the Russian government to draft an alliance with the Afghan ruler.

1878: November 22 Second Anglo-Afghan War begins

Alarmed at the Russians' presence in Kabul, Britain sends a military mission to Afghanistan. When Shir Ali refuses to allow the mission to cross the Afghan border, Britain invades Afghanistan. Shir Ali flees north seeking Russian help, but dies in Mazar-i-Sharif on February 21, 1879.

1879: May 26 Treaty of Gandamak

The Treaty of Gandamak ends the war and makes Afghanistan a virtual British protectorate. Britain recognizes Yaqub Khan, Shir Ali's son, as amir. The Afghan ruler remains independent in his internal affairs but the treaty requires him to follow British advice in his relations with foreign powers. The Afghan amir also cedes to the British certain strategic territory, including the Khyber Pass.

1879: September 3 British Residency at Kabul stormed

Afghans attack the British Residency at Kabul and kill the resident, Luis Cavagnari. The British invade Afghanistan again, but their attempts to pacify the country end with their defeat at Maiwand (July 27, 1880).

1880: July 22 British recognize Abdur Rahman as Amir

The British, fearing a repeat of the disaster of the First Anglo-Afghan War, recognize Abdur Rahman Khan (r. 1880–1901) as ruler of Afghanistan and withdraw their forces.

1880–1901 Abdur Rahman modernizes Afghanistan

Abdur Rahman, a cousin of Shir Ali, returns to Afghanistan from exile in Central Asia. He gathers an army around him and proclaims himself amir (Yaqub Khan abdicates in 1879). He gains recognition from the British and a free hand in Afghanistan's internal affairs at the cost of reaffirming the terms of the Treaty of Gandamak. Britain grants the amir a subsidy of twelve lakhs (1.2 million) rupees.

Abdur Rahman captures Kandahar and Herat, puts down numerous rebellions, and sets about destroying the power of the tribal chiefs and religious leaders. He strengthens the institutions of central government and establishes a standing army and police force.

1887 Afghanistan's northern borders are fixed

Russian advances south into Afghan-held territory lead to the involvement of Britain, which has guaranteed Abdur Rahman protection against Russian aggression. The two great powers edge towards war, but eventually reach a compromise and define the border between Afghanistan and Russian- occupied lands to the north.

1893: November 12 Durand Agreement signed

Abdur Rahman and Sir Henry Mortimer Durand sign an agreement in Kabul that defines the boundary between Afghanistan and British India. The Durand Line, as it is called, passes through the middle of the Pashtun tribal lands, dividing the Pashtun tribes between Afghanistan and India.

1901: October 3 Habibullah proclaimed amir

Habibullah Khan, eldest son of Abdur Rahman, succeeds to the throne after his father dies on October 1. As Habibullah I (r. 1901-19), he continues his father's policies of modernization.

1903 Amir founds Habibia School

Habibullah establishes Afghanistan's first modern school in Kabul.

Education in Afghanistan at this time is strictly in the hands of the mullahs (a Muslin teacher of the sacred law), who run the mosque schools (*maktabs*) and secondary schools (*madrasahs*). Leading Muslim clerics object to the founding of Habibia School, but the amir presses on with his plans for modernizing education.

1905 British reaffirm treaties

The British reaffirm their 1880 and 1893 agreements made with Abdur Rahman.

1907: August 31 Anglo-Russian Convention

Britain and Russia sign the Anglo-Russian Convention of St. Petersburg defining the countries' relations in Central Asia. Habibullah refuses to ratify the convention because Afghanistan is not included in the negotiations.

1910 First telephone line in Afghanistan

The first telephone line in Afghanistan links Kabul and Jalalabad.

1911 Mahmud Tarzi begins publishing *Siraj al-Akbar*

Mahmud Tarzi (1865–1933) begins publishing the biweekly newspaper *Siraj al-Akbar* (Torch of the News) in Kabul. Mahmud Tarzi is a prominent Afghan nationalist who espouses pan-Islamic and modernist ideas.

1914–18 World War I

Germany sends the Niedermayer-Hentig mission (1915) to Kabul to gain Afghanistan's support for military action against British India. Despite this and pressure from nationalists to side with Germany, Habibullah keeps Afghanistan neutral during the conflict.

1918 Kabul Museum opens

The Kabul Museum opens. The museum houses important collections of archeological and ethnographic materials from the Hellenistic, Greco-Buddhist, Ghaznavid, and other periods of Afghan's past.

1919: February 20 Habibullah is assassinated

Resentment at Habibullah's stance during World War I leads to his assassination by enemies associated with the anti-British movement in Afghanistan.

1919: February 25 Amanullah seizes power

Amanullah Khan (r. 1919–29), a younger son of Habibullah, seizes the throne from Nasrullah Khan who rules for five days after his brother's death.

Amanullah assumes the title King (*Padshah*) in 1923. He creates the first Afghan parliament, builds a new capital (Darulaman), and provides for the separation of religious and secular authority. He enlists Soviets and Turks to train his armed forces. His social reforms include founding modern schools and relaxing the traditional restrictions Islam places on women.

1919: May 3 Third Anglo-Afghan War begins

When Britain refuses to renegotiate the existing Anglo-Afghan treaties, Amanullah declares Afghanistan independent and launches attacks on British forces along the North West Frontier. The tribes of the region are in ferment, and a war-weary Britain enters negotiations with Amanullah.

1919: June Afghanistan recognizes the Soviet Union

Wanting to demonstrate his independence by establishing diplomatic relations with European powers, Amanullah recognizes the Soviet Union.

1919: August 8 Treaty of Rawalpindi signed

The Treaty of Rawalpindi formally ends the Third Afghan War. Britain recognizes Afghanistan's independence. The Anglo-Afghan Treaty of 1921 (Treaty of Kabul) normalizes relations between the two countries.

1921: February 28 Soviet-Afghan treaty

Amanullah signs a treaty of friendship with the new government in Russia initiating a special relationship between Afghanistan and the Soviet Union. The treaty provides for Soviet technical and financial aid to Afghanistan.

1922: September 9 French archeologists to excavate in Afghanistan

Afghanistan and France sign an accord giving France rights to conduct archeological excavations in Afghanistan.

1924: May Khowst rebellion

Resentful of the king's reforms, Afghans rise up against Amanullah's rule. The rebellion, led by the Mangal tribe and based at Khowst, is put down by the king with Soviet aid.

1926 Afghani becomes new monetary unit

The Afghani (Af.) replaces the Afghan *ropia* as the country's currency.

1927–28 The king visits Europe, Turkey, and Persia

Impressed by what he sees on his foreign tour, King Amanullah renews his efforts at social reforms. He orders the emancipation of women, outlaws polygamy, and decrees compulsory education for both sexes. A new dress code in Kabul allows women in Kabul to go unveiled, and requires Afghans to wear Western dress.

Nadir Khan returns a salute during the review of troops at the celebration of Independence Day in Kabul, the capital. (EPD Photos/CSU Archives)

1929: January 14 Amanullah abdicates

Amanullah's reforms lead to an uprising of Shinwari Pashtuns and other tribes, led by the country's mullahs. Amanullah abdicates in favor of his younger brother Enyatullah. One of the leaders of the rebellion, a Tajik bandit from Kohistan, seizes the throne and rules as Habibullah II. He garners little support among the Afghan tribes, but his opponents are too disunited to oppose him.

1929: October 17 Muhammad Nadir Shah proclaimed king

Sardar Muhammad Nadir Khan becomes king and rules as Muhammad Nadir Shah.

Nadir Khan (1883–1933), a great grandson of one of Dost Muhammad's brothers, is in France when Amanullah abdicates. He returns to Afghanistan, gathers support among the tribes, and defeats Habibullah.

Nadir Khan continues the process of modernization, although at a slower pace. He placates the mullahs and religious conservatives, and repeals some of Amanullah's more radical reforms. He returns to some of the tribal traditions abandoned by his predecessors, and recognizes the role of the national gathering of tribal chiefs, the *Loya Jirgah,* in Afghanistan's political life.

1931: October 31 Nadir Shah confirms new constitution

Nadir Shah drafts a new constitution which establishes Sunni Islam as Afghanistan's official religion, requires that the king be a Sunni, and declares the Hanafi School of Islamic Law to be supreme in the country.

1932 School of medicine opens

The school of medicine opens in Kabul with French and Turkish professors.

1933: November 8 Nadir Shah assassinated

Nadir Shah is assassinated and succeeded by his 19-year old son Muhammad Zahir Shah (r. 1933–73). The new king proceeds cautiously with economic and social development.

1936 Pashto becomes the national language

Pashto is proclaimed the national language of Afghanistan.

1937 Pashto Academy founded

The Pashto Academy is founded to promote the Pashto language and Pashtun literature.

1939–45 World War II

Afghanistan remains neutral during World War II.

1946 Kabul University inaugurated

The school of medicine in Kabul is combined with the schools of law, political science, and natural sciences to form Kabul University.

1946: November 9 Afghanistan joins United Nations

The United Nations General Assembly approves Afghanistan's entry into the UN.

1950: July 11 UNICEF funds to lower infant mortality

UNICEF (United Nations Children's Fund) provides Afghanistan with $100,000 to lower infant mortality rates. Afghanistan's demographic characteristics are typical of a developing country, and infant mortality is among the highest in the world.

1953: September Daud Khan seizes power

Lt.-General Muhammad Daud Khan becomes prime minister, introducing far-reaching educational, social and economic reforms. He allows the voluntary unveiling of women, abolishes the traditional Muslim custom of purdah, and generally improves the status of women. He initiates state-planning and turns to the USSR for economic aid and assistance in training and arming the military. The regime remains politically repressive, however, with little political opposition tolerated.

1955: December 21 U.S. offers to mediate Pashtunistan dispute

The United States offers to mediate the dispute between Afghanistan and Pakistan over the creation of a homeland for Pashto-speaking tribes along Pakistan's border with Afghanistan.

Daud Khan takes a strong stance on the "Pashtunistan" issue, which periodically causes strained relations between Afghanistan and Pakistan.

1956 First five-year plan begins

Daud Khan begins a program of economic modernization. The first of a series of five-year plans (1956–61), financed by the Soviet Union and the United States, has as its aim the creation of the communications infrastructure necessary for economic growth and development.

The Afghan economy is essentially agricultural, with its main items of export being karakul (a breed of sheep) skins, fresh and dried fruits, and carpets.

1959 Helmand valley project is completed

A United States company completes work on the Helmand valley project.

The Afghan government creates the Helmand Valley Authority in 1947 to oversee an irrigation and electrification project in the Helmand valley in southwestern Afghanistan. It is patterned after the Tennessee Valley Authority in the United States. Poor planning and minimal benefits make the project a costly failure.

1963: March Daud Khan resigns

Daud Khan resigns to ease tensions with Pakistan over the Pashtunistan issue, and is replaced by Dr. Muhammad Yusuf Khan. Pakistan's closing of the border following unrest in tribal areas and military clashes in 1961 disrupts Afghanistan trade and causes economic instability.

1964: October 1 New constitution

Zahir Shah proclaims a new constitution that makes Afghanistan a constitutional monarchy. The constitution guarantees Afghan citizens rights to property, the freedom of assembly, a free press, the right to form political parties and certain other freedoms associated with democratic forms of government.

1964: October 1 Official languages

The new constitution designates Pashto and Dari as Afghanistan's official languages.

1964–73 The constitutional period

Although elections are held in 1965 and 1969, many of the reforms envisaged in the new constitution never materialize. Fighting between governments and the legislature lead to a stalemate in parliament, and the country has five prime ministers (Yusuf Khan, Muhammad Haishim Maiwandal, Nur Ahmad Etemadi, Abd az-Zahir, and Muhammad Musa Shafiq) in less than ten years. Politics in the country become polarized, with students and the radical left confronting Afghanistan's traditional élites.

Socioeconomic conditions during this period also slow the pace of political reforms. Afghanistan experiences severe droughts in the early 1970s, with that in 1971 being the worst in recorded history. Widespread starvation and loss of livestock cause hardships across the country, along with resentment at the government's handling of the crisis.

1973: July 17 Daud Khan deposes King Zahir Shah

Former prime minister Daud Khan deposes the king in a bloodless coup. He abolishes the monarchy and creates a republic with himself as president.

Daud Khan initiates a program of rapid economic development, tightening state controls over the economy and nationalizing several industries. Although still maintaining good relations with the Soviet Union and the United States, he courts Iran and the oil-rich states of the Gulf region.

1978: April 27 Communist coup

Marxist military commanders in Kabul stage a bloody coup, killing Daud, his family, and most of the prominent leaders in the government. The rebels release members of the People's Democratic Party of Afghanistan (PDPA) imprisoned by Daud Khan, who form a new government. Nur Muhammad Taraki becomes head of the "Revolutionary Council" and president of the new Democratic Republic of Afghanistan.

1978–79 Afghanistan under Taraki and Amin

Taraki's government announces reforms based on traditional Marxist-Leninist ideology—state-based socialism, land reforms, equality of women, abolition of dowries, and other progressive programs. The traditional Afghan flag is replaced by an all red flag, Soviet advisors enter the country, and a policy of repression against opponents of the regime begins.

Taraki's reforms strike at the very heart of traditional Afghan culture and society. Within three months, anti-Communist uprisings occur in Nuristan and Kunar, and quickly spread to other parts of the country.

Infighting between the People's (Kalq) and Banner (Parcham) factions of the PDPA quickly causes trouble within the new government. Hafizullah Amin (1929–79) seizes power and kills Taraki in September 1979. The Afghan army collapses as a result of purges, mutinies, and desertions, and Amin asks for more Soviet military aid.

1978: February 14 U.S. ambassador killed

Adolph Dubs, United States ambassador to Afghanistan, is killed in a rescue attempt by government security forces following his kidnapping.

1979: December 27 Soviet invasion

Soviet forces invade Afghanistan and overthrow and execute President Hafizullah Amin. Babrak Karmal, leader of the Banner faction of the PDPA, becomes president. The Soviets justify their actions by claiming Amin's government has been overthrown by an internal revolution and that military support has been requested by the new regime.

1980–86 The Karmal regime

Karmal continues with the "revolution," although he initiates policies involving conciliation, respect for tribal and religious customs, and creation of a broad-based national government. However, he faces a civil war, widespread economic chaos, and a disintegrating army and bureaucracy. Karmal becomes increasingly dependent on Soviet military support.

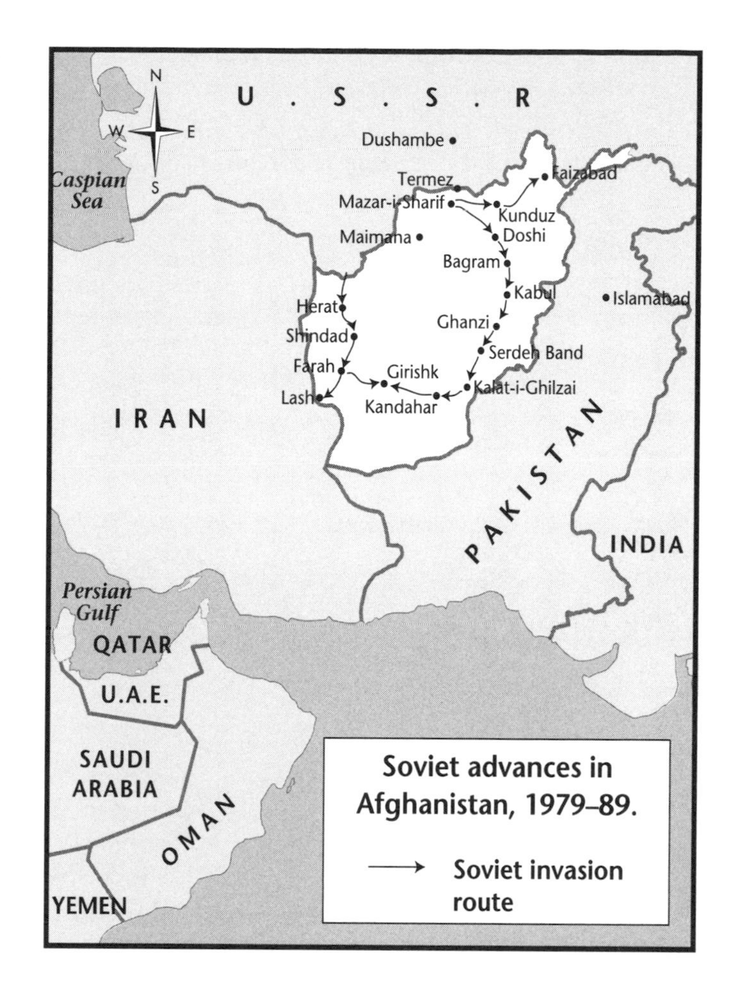

Soviet advances in Afghanistan, 1979–89.

1981: May 9 Two million refugees in Pakistan

Pakistan officials estimate two million Afghan refugees from the civil war in Afghanistan are in Pakistan. Before the end of the war, refugee totals reach more than three million in Pakistan and over two million in Iran.

1981: August 22 Mujahidin form an alliance

Five Afghan resistance groups enter an alliance and create a fifty-member advisory council.

Afghan resistance fighters adopt the name *mujahidin,* fighters in a holy war *(jihad)* against an "infidel" government and its Soviet allies. Poorly armed and uncoordinated, anti-government guerrillas attack roads and bridges, government installations, and harass Afghan and Soviet army units. Many mujahidin groups operate out of the refugee camps around Peshawar in Pakistan.

1984: July 26 U.S. Congress announces aid to Afghan resistance

Intelligence sources report that the United States House of Representatives approves $50 million in covert aid for the Afghan resistance. This increases to $280 million in 1995. Saudi Arabia, Israel, and China also assist the mujahidin.

1985: May **Islamic Union of Afghan Mujahidin (IUAM) formed**

The seven principal Peshawar-based mujahidin groups form an alliance to coordinate political and military operations.

1986: April 2 **U.S. to supply missiles**

The United States agrees to supply the mujahidin with the *Stinger* missile. This shoulder-fired ground to air weapon helps turn the tide against Afghan and Soviet bombers and particularly against the Red Army's helicopter gunships. Terror-bombing of villages suspected of harboring rebels is a common tactic of government and Soviet troops.

1986: May 4 **Najibullah replaces Karmal**

Dr. Najibullah, former head of Afghanistan's secret police, replaces Karmal as General Secretary of the PDPA. He becomes president in September.

Najibullah continues existing government policy, although with a greater emphasis on "national reconciliation." He legalizes political parties, announces his willingness to form a coalition government with opposition parties, and introduces a new constitution. Najibullah's policies gain little support from mujahidin groups in elections held in 1988.

1987: October 24 **Shia groups form new alliance in Iran**

Shia resistance groups based in Iran form a group called the Islamic Unity Party (IUP).

1988: April 14 **Agreement on Soviet withdrawal signed**

UN sponsored talks, ongoing since 1981, culminate in the Geneva Accords. Under the agreement, signed by Afghanistan, Pakistan, the Soviet Union, and the United States, Soviet troops will withdraw from Afghanistan within nine months. An estimated 105,000–115,000 Soviets are in Afghanistan at the height of the conflict.

1988 **Civil war continues**

Despite the Soviet withdrawal and increased mujahidin activity against the Kabul government, Najibullah survives and appears to be strengthening his hold on power. He survives several coup attempts. Supplies continue to flow to both sides in the conflict in violation of the Geneva accords.

1989: February 23 **Mujahidin form government-in-exile**

The IUAM meets in Rawalpindi, Pakistan, and forms an interim government-in-exile known as the Afghan Interim Government (AIG). The AIG gains little support from guerrilla commanders and tribal leaders inside Afghanistan.

1991: May 21 **UN peace proposal**

UN Secretary-General Javier Pérez de Cuéllar issues a five-point proposal for a settlement in Afghanistan. The governments of Afghanistan and Pakistan welcome the peace proposal, but the AIG rejects it.

1992: April 16 **Najibullah resigns**

Mujahidin successes force Najibullah' resignation.

The northern town of Mazar-i-Sharif falls to the mujahidin after General Abdul Rashid Dostam's Uzbek militia mutiny. The Jam'iat-i Islami, a largely non-Pashtun guerrilla force led by the Tajik commander Ahmad Shah Masoud, captures the strategic airbase of Bagram, north of Kabul.

1992: April 25 **Kabul falls to mujahidin**

Government forces surrender Kabul to Masoud and Gulbuddin Hekmatyar's Pashtun-dominated Hizb-i Islami. Fighting immediately breaks out in Kabul between rival mujahidin factions.

An interim Islamic Jihad Council is formed to establish a new government, which quickly gains recognition by the international community.

1992: June 28 **Rabbani government**

Burhanuddin Rabbani, leader of the Jam'iat-i Islami, becomes president of an interim government.

Heavy fighting continues in Kabul and other provinces between Hekmatyar's forces and other mujahidin groups. Conflicts also involve confrontations between Shias and Sunnis.

1993: March 7 **Peace accord**

An agreement signed in Islamabad, Pakistan between competing factions calls for an immediate cease-fire and the formation of an interim government with Rabbani as president and Hekmatyar as prime minister.

1993–94 **Civil war continues**

The relations between Rabbani and Hekmatyar remain contentious, with the two leaders opposing each other in a range of policies. Fighting continues in Kabul and other areas, the opposing groups often being defined along ethnic or sectarian lines as well as political. The civil war continues, despite international efforts to negotiate a settlement.

1994: August–September **Emergence of the Taliban**

The Taliban appears on the scene and begin operations in southern Afghanistan.

The Taliban are supported and trained by Pakistan. "Talib" means "seeker of religious truth," and the nucleus of the groups is provided by Afghan refugees who are students at madrasahs (Islamic schools) all over Pakistan. The Taliban are puritanical in their outlook, espousing fundamental

Islamic beliefs, and seek to replace corrupt mujahidin leaders and eliminate modern Western influences from Afghani society. The head of the Taliban movement is Maulvi Muhammad Omar Akhund, although like all Afghan groups the Taliban has a tendency to split into competing factions.

1994: September Land mines problem highlighted

An international conference on land mines meeting in Oslo addresses the worldwide problem of land mines. The UN estimates there are some ten million mines in Afghanistan, a result of the war with the Soviet Union and the ongoing civil war. Between twenty and twenty-five Afghans, most of them civilians and many of them children, are injured by mines every day.

1995: February–March Taliban successes in south

The Taliban rapidly gain control of ten provinces, mostly in southern and southern-eastern Afghanistan.

The Taliban reportedly work in the interests of the common people—executing drug traffickers, disarming local militia, suppressing banditry, and restoring law and order to a war-weary countryside.

1996: September 27 Taliban forces capture Kabul

After some reverses, the Taliban captures Kabul and overthrows Rabbani's government. The new government is recognized by Pakistan, but few other countries afford it official status.

The Taliban declares Afghanistan to be an Islamic state, and imposes strict Islamic codes. Women cannot work or be educated beyond the age of ten, and must be in purdah. Television, non-religious music, gambling, and alcohol are banned. Traditional Islamic punishments such as amputations and stoning are enforced, and all males are required to attend mosques.

1997: June 26 Taliban denies involvement in drug trafficking

The Taliban government, in a broadcast on Radio Voice of Sharia from Kabul, denies any involvement in drug production and trafficking.

However, opium production in areas of southern Afghanistan controlled by the Taliban is up ten percent over the previous year. The opium is transported to Pakistan where it is processed into heroin and smuggled to Western countries through Central Asia. UN officials claim drugs is Afghanistan's largest export, with all Afghani factions being involved in the trade and using the cash generated to purchase arms.

1998: February 8 Major earthquake in Afghanistan

A severe earthquake hits the Takhar province of northern Afghanistan. At least 4,500 people are dead and many thousands more injured or homeless.

1998: August 9 Taliban takes Mazar-i-Sharif

Taliban forces capture the city of Mazar-i-Sharif from the Northern Alliance, a coalition of opposition forces including General Dostam, Masoud, and Shia factions. The fall of the city gives the Taliban virtual control of the entire country. The Taliban kill several diplomats from Iran, which supports anti-Taliban forces in Afghanistan, in the fighting. Iran and Afghanistan move close to war.

1998: August 20 U.S. missiles strike terrorist camp in Afghanistan

Cruise missiles hit a terrorist camp near Khowst, in eastern Afghanistan. The attack is in response to the August 7 bombing of the United States embassies in Nairobi, Kenya and Dar es Salaam, Tanzania by terrorists believed to be associated with the dissident Saudi, Osmana bin Laden.

1998: October 8 Taliban-Iranian battle

Unconfirmed reports from Tehran indicate that Taliban fighters and Iranian troops fight a three-hour battle along the Afghanistan-Iran border northwest of Herat.

Bibliography

Adamec, Ludwig W. *Historical Dictionary of Afghanistan.* Metuchen, NJ: Scarecrow Press, 1991.

Afghanistan, Past and Present. Moscow: USSR Academy of Sciences, 1982.

Amin, Hamidullah. *A Geography of Afghanistan.* Omaha, NE: Center for Afghanistan Studies, 1976.

Barfield, Thomas. "The Afghan Morass." *Current History,* January 1996, vol. 95, no. 597, pp. 38+.

Clifford, M. L. *The Land and People of Afghanistan.* Philadelphia: Lippincott, 1973.

Dupree, Louis. *Afghanistan.* Princeton, NJ: Princeton University Press, 1980.

Edwards, David B. *Heroes of the Age: Moral Fault Lines on the Afghan Frontier.* Berkeley: University of California Press, 1996.

Fredericksen, Birthe. *Caravans and Trade in Afghanistan: the Changing Life of the Nomadic Hazarbuz.* New York: Thames and Hudson, 1996.

Giradet, Edward. *Afghanistan: The Soviet War.* London: Croom Helm, 1985

Griffiths, John Charles. *Afghanistan: Key To A Continent.* Boulder, CO: Westview Press, 1981.

Noelle, Christine. *State and Tribe in Nineteenth-century Afghanistan: The Reign of Amir Dost Muhammad Khan (1826–1863).* Great Britain: Curzon Press, 1997.

Nyrop, Richard F. and Donald M. Seekins, eds. *Afghanistan: A Country Study.* 5th ed. Washington, DC: U.S. Government Printing Office, 1986.

Olesen, Asta. *Afghan Craftsmen: the Cultures of Three Itinerant Communities.* New York: Thames and Hudson, 1994.

Spain, J. W. *The Way of the Pathans.* London: Hale, 1962.

Australia

Introduction

Both a country and a continent, Australia lies southeast of Asia with the Pacific Ocean to its east and the Indian Ocean to the west. New Zealand is roughly 1,200 miles (1,930 kilometers) away, beyond the Tasman Sea. With a total area of 2,968,000 square miles (7,687,000 square kilometers), Australia is almost as large as the United States. Six states make up the Commonwealth of Australia: New South Wales, Victoria, South Australia, Western Australia, and Queensland, all on the mainland, and the island state of Tasmania. Australia also encompasses two territories: the Northern Territory and the Australian Capital Territory where Canberra, the seat of government, is located. Although Australia is known for its vast, sparsely inhabited interior, or "outback," four-fifths of its estimated population of 18.3 million live in urban areas or suburbs. In addition to large-scale urbanization, another significant population trend has been the growth of ethnic and racial diversity. The land that placed strict immigration limits on nonwhites for much of the twentieth century today prides itself on the diversity of its residents, one-fourth of whom were born abroad.

Thought to be the world's oldest continent, Australia can be divided geographically into four major regions: the eastern highlands, which stretch from Cape York Peninsula south to Tasmania; the eastern lowlands, a sandy coastal plain; the central plains, which include the Great Artesian Basin, the world's largest internal drainage area; and the western plateau, consisting of vast deserts and plains. Because of Australia's geographical isolation from other continents, it is populated by many unique animal species, notably marsupials (animals that carry their young in a pouch after birth) such as the kangaroo, wallaby, and koala.

Early History of Australia

The recorded history of Australia began in the seventeenth century with the discovery of the continent by the colonial powers of Europe. In 1606, Spaniard Luis Vaez de Torres sailed through the strait that was to be named for him, and in the same year sailors aboard a Dutch ship sighted what was to become the Cape York Peninsula. The first Dutchman reached Australian land in 1616; further discoveries followed throughout the century. The English first explored the continent in 1688. Nearly one hundred years later, Captain James Cook led an expedition to the east coast of Australia, named the land New South Wales, and claimed it for England.

Having just lost its colonies in North America in the American Revolution, the British were eager to find another place where they could send convicts in order to relieve the overcrowding of their prison system. In 1788, the first English penal colony was established at Botany Bay (present-day Sydney). The practice of shipping convicts to Australia, called "transportation," was to last well into the next century, with the various colonies ending it on their own between 1840 and 1868. Altogether, about 161,000 English convicts were brought to Australia, and they made up the great majority (some 90 percent) of the new land's inhabitants. Not long after transportation began, settlers who were not convicts also began arriving in Australia (1793). In the first decades of settlement by the British, the continent's native aboriginal population was nearly wiped out from the combined effects of violent confrontations, disease, and alcohol consumption.

The discovery of a ridge route across the Blue Mountains in 1813 opened up the interior of the continent to settlement, providing land for growing wheat and other crops and grazing space for sheep and cattle, leading to the growth of the wool and meat industries By 1829, Britain had claimed all of Australia. The discovery of gold in the 1850s, first in New South Wales and then in Victoria, changed life in Australia in a number of ways. It attracted newcomers from various countries, many of whom stayed on even after the initial "gold fever" had worn off, permanently changing the demographic makeup of the land. The bustle of mining activity spurred the economy, as demand for food, housing, supplies, and equipment rose. Within a decade, rail and telegraph service had been introduced, and Australia's population had nearly tripled.

As modernization came to Australia, cultural pursuits flourished. In the closing decades of the century, the first art museum and art school were opened, an architectural institute was formed, the first symphony orchestra was organized, and landscape painting reached new levels of artistry, gaining the

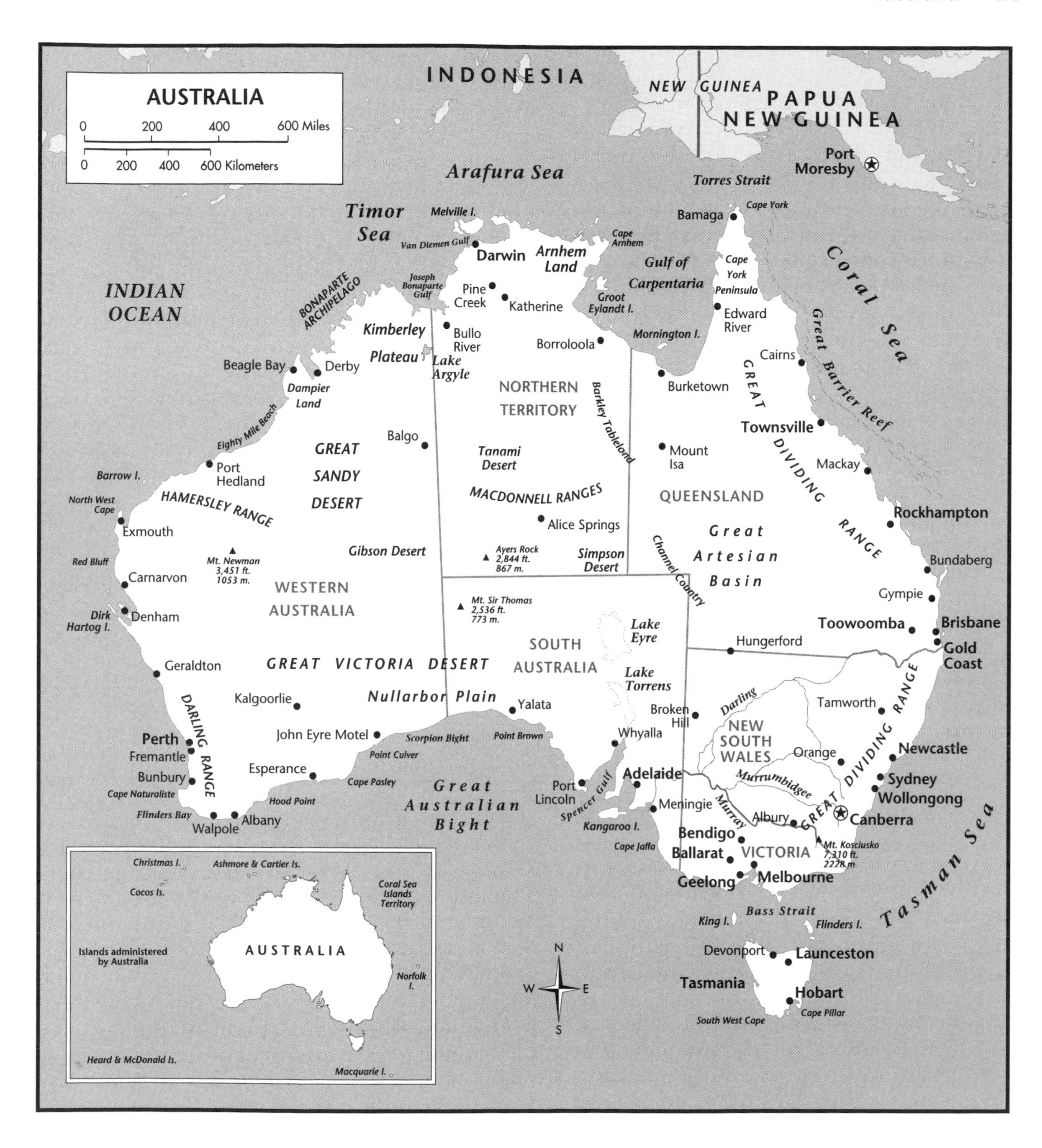

first international recognition of Australian artists. In the 1890s, Australia moved toward independent nationhood, beginning with the drafting of a federal constitution in 1891 and ending with Queen Victoria's approval of independence in 1900 and the proclamation of the Commonwealth of Australia on January 1, 1901.

Twentieth-century Australia

The new nation remained part of the Commonwealth of Nations and was thus automatically committed to supporting Britain's declaration of war on Germany in 1914, a commitment it honored with distinction at Gallipoli and other battles. At the conclusion of the war, Australia became a member of the League of Nations and acquired a mandate to govern the part of New Guinea previously controlled by Germany.

In 1927, the Australian capital was moved to Canberra, a site chosen and newly developed in accordance with the constitutional provision that the newly federated states have a new capital. By the early 1930s, Australia was feeling the effects of the global depression then under way, and its government took austerity measures to deal with the downturn. In 1939, Australia once again supported Britain in fighting a world war. Expeditionary forces were sent to the Middle East and North Africa and, following the Japanese attack on Pearl Harbor in 1941, Australia declared war on Japan. Thousands of Australians were either killed or captured by the Japanese as prisoners of war.

After the war came a long period of rule (1949–72) by the Liberal-Country coalition, led until 1966 by R.G. Menzies, the longest-serving prime minister in Australian history. The postwar period was an era of rapid population growth, industrial expansion, and rising foreign investment. Diplomatically, Australia's strategic alliance with the United States was the cornerstone of its foreign policy, although this relationship was called into question during the Vietnam war when Australia sent over 4,000 troops to support the battle against communist forces in the north of the country. Growing dissent at home forced the withdrawal of all Australian combat troops by 1971. In the early 1970s, Australia's government also came under fire for its economic policies as the country struggled with the effects of worldwide recession and inflation. After a brief period of labor rule (1972–75), the Liberal and National coalition was returned to power. However, it was the centrist Labor government of Prime Minister Robert Hawke, coming to power in 1983, that restored the economic health of the nation through a program of privatization and other pro-business policies. Hawke's Labor government was reelected to an unprecedented four terms, although its final term (1990–93) was marred by the onset of another recession. In March, 1986, Australia severed its ties with the British Commonwealth.

Contemporary Australia

In recent decades, a significant social trend has been the effort by Aboriginals to assure that discrimination does not deprive them of their full civil liberties, and to reclaim control of their traditional tribal lands. The first land-rights protest by Aboriginals took place in 1962, when the Yirrkala Aborigines claimed that bauxite mining on their ancestral lands at Gove in the Arnhem Land reserve was desecrating sacred sites. Although the original suit was denied, as was an appeal of that verdict, the Aboriginal Land Rights Act of 1977 gave Aborigines control of reserves and other lands totaling roughly 18 percent of the Northern Territory. They also won the right to approve or deny the leasing of these lands for mining purposes, with the proceeds of leases going to benefit the Aboriginal community. In addition, government programs have been instituted to improve education, health care, housing, and employment opportunities for Aborigines.

In 1988, the bicentennial of Australia's first European settlement, Port Jackson, was celebrated through a variety of projects and activities, including a race by tall ships from thirty different countries, the unveiling of the Bicentennial Science Centre in the capital city of Canberra, and the Australian Bicentennial Exhibition, which traveled to locations throughout the country. It is indicative of today's emphasis on Australia as a multiracial, multiethnic nation that although the bicentennial commemorated European settlement, its theme was "living together."

Timeline

c. 38,000 B.C. First Aboriginals believed to have inhabited Australia

Archaeological evidence suggests the arrival of the first aboriginal people.

European Exploration

1606 Torres Strait is discovered

Explorer Luis Vaez de Torres (fl. 1605–13) sails through the Torres Strait.

Luis Vaez de Torres is believed to be Spanish, although some historical records indicate that he may have been born in Brittany, in present-day France. He commands one of the three ships in the expedition of Pedro Fernandez de Quiros of Portugal. When the crew mutinies, Quiros abandons Torres. Torres sails west through a strait, now known as the Torres Strait, between New Guinea and Australia. He sights the southern tip of Australia during this journey, but he mistakes it for another island.

1606 First recorded sighting of Queensland

Dutch sailors aboard the ship *Duyfken* sight the west coast of what becomes known as the Cape York Peninsula.

1616 Dutch explorer arrives on western coast of island

Dutch explorer Dirk Hartog lands on an island in present-day Shark Bay in Western Australia.

1642: November 24 Tasmania is discovered

A Dutchman, Abel Janszoon Tasman (1603–c. 1659), discovers the island that is now named for him. He initially calls it Van Diemen's Land.

Tasman is born near Groningen in the Netherlands. He is sent by Antony Van Diemen (1593–1645), governor general

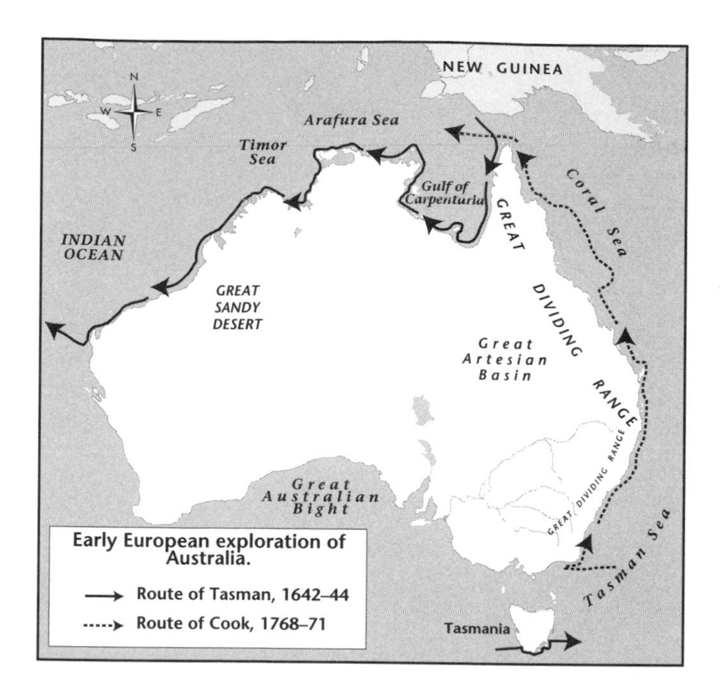

of Batavia, to explore the southern hemisphere. It is during his first expedition in 1643 that he discovers Tasmania, a large island off the coast of mainland Australia, and New Zealand. On the same voyage, he discovers Tonga and Fiji. He also makes a second voyage to the northwest coast of Australia (see 1644).

1644 Tasman explores northern Australia

Abel Tasman (see 1642) explores and charts most of the north coast of Australia, but fails to find the passage to the Pacific Ocean (the Torres Strait).

1688 Englishman visits western coast

British subject William Dampier (1652–1715) explores the western coast of Australia aboard the *Cygnet,* landing at King Sound.

Dampier is born in Somerset, England. In 1679, he joins a band of pirates who sail to Central America and plunder the coastline communities. In 1683, he seizes a ship near Sierra Leone on the west coast of Africa and sails to the Philippines, China, and Australia. He returns to England in 1691, and publishes an accound of his travels titled *New Voyage Round the World* (1697). In 1699–1700, he explores the west coast of Australia.

1770 Cook explores Australia

An expedition by the Royal Society to the Pacific led by English navigator Captain James Cook (1828–79) reaches the east coast of Australia; Cook sails the *Endeavor* from present-day Victoria to the Torres Strait. Naming the land New South Wales, Cook claims it for King George III of England.

English Settlement

1788: January 26 First convicts transported to Australia

Since the American Revolution has deprived the English penal system of a haven for unwanted criminals, the government shifts these transports to New South Wales to relieve overcrowding in English prisons. The first convicts are transported to Botany Bay (present-day Sydney), and a penal colony is established. Between 1788 and the end of the transportation system in the mid-nineteenth century, approximately 161,000 convicts are transported from England, making up 90 percent of all Australian settlers.

1788: February 9 First music is performed in Australia

The first recorded performance of a musical piece is the playing of "The Rogue's March" at the disciplining of a sailor caught in the tents of women convicts.

1789 First play is performed

The first performance of a play is given by a group of convicts, who perform the comedy *The Recruiting Officer* by George Farquhar.

1793 Free settlers reach New South Wales

The first group of settlers who are not convicts reaches New South Wales. Among these are John Macarthur (1767–1834) and his wife, Elizabeth Veale Macarther (1766–1850). The couple name their land Elizabeth Farm. They prosper there, despite John's many absences (see 1808). Elizabeth introduces merino sheep to Elizabeth Farm, and successfully breeds them. By around 1816, the Macarthurs have launched the Australian wool industry.

19th century Aboriginal population is decimated by British settlement

In the first decades of British settlement, the aboriginal population suffers a steep drop from attacks, disease, and alcohol consumption. The aboriginal population of Tasmania falls from an estimated 4,000 to 500 and is eventually eradicated altogether. In Victoria a population of over 10,000 falls to 2,000 within thirty years of the colony's establishment at mid-century.

1801–03 First circumnavigation of Australia

Englishman Matthew Flinders (1774–1814) sails his ship, the *Investigator,* all the way around Australia, becoming the first to circumnavigate the continent. Flinders produces a map of Australia and recommends that the name "Australia" be given to the entire continent. A river in Queensland is named for him as are mountains in south Australia, the Flinders Range.

1803 First newspaper is founded

Australia's first newspaper, the *Sydney Gazette and New South Wales Advertiser*, is established.

1803 Convict settlement founded in Tasmania

A second penal colony is established in Tasmania (formerly Van Diemen's Land).

1808 Bligh is ousted as governor

Captain William Bligh (1754–c.1817; the object of the mutiny on HMS *Bounty*) is unseated after serving two years as governor of New South Wales, in an episode that becomes known as the Rum Rebellion. The rebellion was led by John Macarthur (1767–1834; see 1793) and was comprised of settlers and British soldiers. Bligh was freed in 1810, and Macarthur was sent back to England.

1813 Ridge across the Dividing Range is discovered

Gregory Blaxland, William Lawson, and William Wentworth discover a ridge route that extends most of the way across the Blue Mountains, opening the interior of the continent to settlement.

1817 First bank is founded

Australia's first bank, the Bank of New South Wales, is established.

1824–25 Penal colony formed in Queensland

A settlement for transported convicts from England is established in Queensland.

1824 Bathurst and Melville islands are annexed

Fearing French encroachments on territory claimed for Britain, the British annex Bathurst and Melville islands.

1825 Tasmania becomes a separate colony

Tasmania (then known as Van Dieman's Land) is made a separate colony.

1826 Colonial art school is founded

The colonial government's first attempt at providing art education is the establishment of the Van Dieman's Land Mechanics School of Art in Hobart, offering classes to working men in reading and writing as well as art.

1829 Britain expands claim to Australia

Britain expands its claim on Australia to include the entire continent.

1831 Land grant system is ended

Under the Ripon Regulations, Crown lands must be disposed of at auction; in addition, the former system of land grants is terminated.

1833 Theatre Royal is opened

The Theatre Royal is established in Sydney by Barnett Levey for both musical and theatrical performances.

1836 First bishop is appointed

Australia's first bishop from the Church of England is installed.

1836 First piece is published by an Australian composer

"The Minstrel Waltz" by Thomas Stubbs purports to be the first musical composition published by an Australian-born composer.

1840–68 Transportation of convicts ends

One by one, the Australian colonies stop receiving convicts transported from England, beginning with New South Wales (1840) and ending with Western Australia (1868).

1842 New South Wales is granted a legislative council

Moving toward a greater degree of colonial autonomy, Britain gives the colony of New South Wales a thirty-six-member legislative council. Freed convicts can vote, but property qualifications are established.

1850s First labor organizing takes place

Workers begin joining forces to agitate for an eight-hour day and other labor reforms.

1850 Colonies win the right to draft constitutions

The Australian Colonies Government Act gives the colonies permission to adopt constitutions of their own.

1850 The University of Sydney is founded

Australia's first university, the University of Sydney, is established.

1851 Gold rush spurs economy and population growth

Gold is discovered in New South Wales and Victoria, drawing hundreds of prospectors to the site, boosting foreign immigration, and producing dramatic economic growth in the southeastern colonies. New towns spring up, and the population of Australia nearly triples in a decade. Other sectors of the economy, including agriculture, manufacturing, and construction are stimulated as the demand for food, housing, mining equipment, and other commodities grows. The gold rush also changes the ethnic composition of the population, drawing immigrants from all over the world, many of whom stay and settle in Australia. Gold is later struck in Queensland and Western Australia as well.

1852 Steamship service to Australia is launched

The *Chusan* becomes the first steamship to operate regular service to Australia.

1853 Founding of the University of Melbourne

The University of Melbourne becomes Australia's second university

1854 First railway opens

Australia's first railroad, connecting Melbourne and Port Melbourne, begins operations.

1854: December 3 Eureka stockade riots

An armed miners' revolt at the gold fields at Ballarat (Victoria) is crushed by government troops. The miners' discontent centers on a required monthly licensing fee and the brutal tactics often employed by the special police forces used to collect the fee and check licenses. When the colonial government fails to make a conviction in a case involving the murder of a miner named James Scobie, the Ballarat Reform League is created by local miners, five hundred of whom proclaim their own republic and build a stockade in the Eureka district.

On December 3, the stockade is overrun by a government force of three hundred. An estimated twenty miners are killed, and thirteen others are charged with sedition but are later released. The major demands of the League—revocation of the costly mining licenses and universal manhood suffrage—are later met by the government.

1855 The name "Tasmania" is officially adopted

The name "Tasmania" is officially adopted for what was formerly called Van Dieman's Land. It is named for Abel Tasman (See 1642: November 24.)

1858 Telegraph service begins

Telegraph service is introduced to Australia, linking Adelaide, Melbourne, and Sydney.

1859: May Rules are developed for Australian football

The first set of rules for what will become Australian Rules Football, based on the game of football (soccer) as played at English public schools, is drawn up by seven members of the Melbourne Cricket Club. The ball can be either kicked or passed by hand. Two eighteen-member teams compete on an oval field with four goal posts at each end.

1860 First south-north crossing of the continent

Robert O'Hara Burke (1820–61) and William John Wills (1834–61) are the first to cross Australia from south to north, journeying from Melbourne to the Gulf of Carpentaria with a band of eighteen men. Wills is a surveyor who arrives in Australia in 1852 and is hired by the new observatory in Melbourne. He joins Burke's expedition to cross the continent. The men reach the Flinders River but run out of food. They eat their camels and continue on foot. Only Burke, Wills, and two others complete the journey and reach the Gulf. They are the first Europeans to cross Australia from north to south. On their return trip, three die of starvation when they find their camp abandoned. The remaining explorer, John King (1838–72) survives and is found and aided by aborigines. After three months, he is rescued by a search party.

1861 Birth of soprano Nelli Melba

Helen Porter Mitchell, later known as the operatic soprano Nellie Melba (1861–1931), is born in suburban Melbourne. She studies voice in Australia and London, and makes her operatic debut, using the stage name Nellis Melba in honor of her hometown of Melbourne, as Gilda in a production of Guiseppe Verdi's *Rigoletto* in Brussels, Belgium, in October, 1887. Melba becomes well-known internationally, singing with such well-known opera stars as Italy's Enrico Caruso. She and Caruso are pioneering recording artists, at first on cylinders and later on 78 rpm record disks. Melba is so popular that two culinary dishes–Melba toast, a cracker-like bread, and Peach Melba, a dessert featuring peaches and raspberries–are created in her honor.

1861 First art museum opens

Australia's first public museum devoted totally to artworks (later named the National Gallery of Victoria) is established in Melbourne, where it occupies part of the public library. Beginning with a collection of sculptural casts, it expands by acquiring paintings (mostly British). The museum is separated from the library in 1944, and a separate building is constructed for it in the Victorian Arts Centre in 1968. The museum's dominant position in Australian art is reinforced by the establishment of its art school (see 1870).

1864 Rugby is introduced to Australia

Rugby Union (amateur rugby) is first played by teams of students at the University of Sydney and crews of visiting ships from Britain.

1867 Artist Arthur Streeton is born

Sir Arthur Ernest Streeton (1867–1943) is renowned for making landscape painting a legitimate art form in Australia. Streeton is born at Mt. Duneed, Victoria, and studies art at the Melbourne National Gallery School. In the 1880s, he becomes closely associated with contemporary artists, including Tom Roberts and Frederick McCubbin (see 1889). Streeton becomes known for his use of bright colors and light tones and his square brush strokes. In 1890, he becomes the first Australian artist to have his work displayed in a public collection when the Art Gallery of New South Wales acquires his painting *Still Glides the Stream.*

In the 1890s, Streeton and his colleagues establish several artists' camps in Little Sirius Cove, Mosman. During this period, his subjects shift from seascapes to scenes from the interior, including the Blue Mountains. A well-known work completed at this time is *The Purple Noon's Transparent Might* (1896). In the last decades of his life, Streeton is widely regarded as the "grand old man" of Australian painting, and he is knighted in 1937. He dies in 1943.

1870 Art school opens in Victoria

The National Gallery of Victoria Art School launches a program of professional training modeled on those in Britain, France, and Germany, becoming one of the foremost influences in shaping the direction of Australian art. The school's first director is Eugene von Guérard.

1870 Birth of composer Arthur Hill

Arthur Francis Hill, a seminal figure in the development of Australian composition, is born in Melbourne and raised in Wellington, New Zealand. At the age of fifteen he travels abroad to study at the Leipzig Conservatorium in Germany, winning a prize for violin performance upon graduation. He returns to New Zealand in 1891, becoming the conductor of the Wellington Orchestra Society. Six years later, he moves to Australia, where he is a founder of the New South Wales Conservatorium of Music in Sydney and serves as its first professor of composition and music theory. Hill opens a music school of his own in 1935. His compositions, which include symphonies, choral and chamber works, operas, and solo instrumental music, reflect both his European training and his interest in folk traditions, including Celtic, Maori, and Aboriginal music.

1870: January 3 Birth of author Henry Handel Richardson

Henry Handel Richardson (1870–1946), the pseudonym of Ethel Florence Lindesay Robertson, is raised in Melbourne and attends the Presbyterian Ladies' College. In 1888, she leaves Australia to study music in Leipzig and lives abroad permanently from that time on, settling in England six years later. She is best known for her trilogy *The Fortunes of Richard Mahony* (1930). Considered the finest Australian fiction published by that date, it consists of the novels *Australia Felix* (1917), *The Way Home* (1925), and *Ultima Thule* (1929). Among Richardson's works with a musical theme are her first novel, *Maurice Guest* (1908), and her last, *The Young Cosima* (1939), which is based on the private lives of the composer Richard Wagner and his wife. Richardson dies in 1946.

1871 The Victorian Institute of Architects is formed

Australia's first professional association for architects is founded. Its mission includes the supervision of architectural training programs.

1873 Birth of artist George Washington Lambert

George Washington Lambert (1873–1930), the first Australian member of London's Royal Academy, is born in Russia to American parents who emigrate to Australia in 1887. He begins his career as a newspaper illustrator, later becoming a protégé of Julian Ashton. Lambert's first public success comes with the painting *Across the Black Soil Plains* (1899), now exhibited in the Art Gallery of New South Wales. In 1900, he is awarded the Society of Artists' Travelling Scholarship, which enables him to journey to Paris, where he spends the next twenty years. In 1921, Lambert returns to Sydney, where he is particularly acclaimed as a portrait painter. His emphasis on draftsmanship has a major influence on art education in Australia. He dies in New South Wales in 1930.

1874 Publication of *For the Term of His Natural Life*

The first distinguished Australian novel, *His Natural Life,* written by Marcus Clarke (1846–81), paints a vivid portrait of the convict settlement in Tasmania. (It is published in revised form with the title *For the Term of His Natural Life* in 1882; this version is produced as a film.)

Clarke is born and educated in London, England. He is sent to Australia when his family suffers financial difficulty. He works as a bank clerk and in sheep farming while pursuing his interest in writing. His first novel, *Long Odds,* is published first in installments. In 1890, a collection of his works is published under the title, *The Selected Works of Marcus Clarke.*

1877 First football association is formed

The Victorian Football Association (VFA), which consists of eight member clubs, is formed to promote and regulate the emerging sport of Australian Rules Football.

1880s and 1890s Period of artistic achievement

Australian art flourishes, as the first graduates of native art schools form professional networks, show their works in exhibits, and open studios in which to show and sell them. Artistic standards are expected to be as high as those in Europe and works with specifically Australian subject matter are encouraged. Outdoor (*plein* air) painting is introduced. The first art purchases are made by individuals and institutions, and art criticism begins to appear. At first, the center of artistic activity is Melbourne, but it shifts to Sydney after the early 1890s.

1880 Refrigeration expands the meat industry

The meat industry receives a boost from the introduction of refrigerated cargo ships. The first shipment of refrigerated Australian meat arrives in London.

1881 Women win right to divorce adulterous spouses

The New South Wales Matrimonial Causes Amendment Act enables women to sue for divorce solely on grounds of adultery.

1882: July 8 Birth of composer Percy Grainger

In the opinion of many, Percy Grainger (1882–1961) is the most important composer Australia has produced. Grainger is born in Melbourne and receives his early musical education from his mother. He leaves for Germany at the age of thirteen to pursue advanced studies with teachers including Busoni. In 1901, he settles in London and begins a performing career as a pianist. By 1905, he has become interested in folk music and undertakes the project of collecting and transcribing British folk songs, becoming the first to use a phonograph for this type of work. Influenced by his friendships with the composers Edvard Grieg and Frederick Delius, Grainger begins integrating folk themes into original compositions of his own and wins recognition as a composer by 1912.

Although Grainger later settles in the United States, becoming a U.S. citizen in 1914, he retains his identity as an Australian composer. He spends time in Australia in the 1920s and then returns during the following decade to oversee the construction of the Grainger Museum at the University of Melbourne (see 1935). Although famed for his composition of the popular "Country Gardens" and other short pieces, Grainger composes, arranges, and edits a varied musical output of some 400 works. Grainger becomes head of the music department of New York University in 1932. The composer dies in 1961.

1884 McKay Harvester is developed

Hugh McKay of South Australia contributes to the development of Australian agriculture by developing a combine harvester that bags threshed grain.

1888 First professional symphony orchestra is formed

Australia's first professional orchestra is brought together under the direction of Frederick Cowen during the Melbourne Centennial Exhibition.

1889: August First avant-garde exhibition is held

Australia's first avant-garde (experimental) art show, the *9 by 5 Impression Exhibition,* is presented in Melbourne by a circle of artists known as the Heidelberg School (named for a Melbourne suburb). It features sketches of rural and city scenes drawn on cigar-box lids. Artists of the Heidelberg School include Tom Roberts, Arthur Streeton, and Frederick McCubbin (see 1867).

1890 Women's suffrage league is formed

Established in Sydney by Rose Scott, the Australian Woman's Suffrage League lobbies for the vote and for reforms in other areas, including labor.

1890 Maritime strike

A maritime strike is the first in a series of major strikes in the 1890s.

1890–92 Strikes fail to win advances for labor

A series of strikes fail to achieve advantages for laborers.

1891 First federal constitution is drafted

A convention produces the first draft of a constitution for a proposed federation of the Australian colonies.

1892 Women's grounds for divorce are expanded

Grounds on which women may seek divorce are expanded to include drunkenness, cruelty, and desertion. Divorced mothers also win the legal right to gain custody of their children.

1893 Married Women's Property Act is passed

Married women win the right to own property separately from their husbands.

1894 South Australia grants women's suffrage

South Australia becomes the first colony to give women the vote. Over the next fourteen years, women's suffrage is instituted in Western Australia (1899), New South Wales (1902), Tasmania (1903), Queensland (1905), and Victoria (1908).

1894 Women win the vote in South Australia

South Australia is the first Australian colony to grant female suffrage.

1898 Second federal convention completes constitution

A second convention finishes the constitutional draft produced at the first one (see 1891).

1900: July 9 Queen Victoria approves Australian federation

Queen Victoria gives her assent to the joining of the six Australian colonies in a federation.

1901 Immigration Restriction Act is passed

The Immigration Restriction Act limits immigration by Asians and Pacific Islanders while encouraging immigration from Great Britain. Nonwhite immigrants are screened out by being required to take written dictation tests in English. Racial preferences in immigration remain in effect until the 1960s.

The Commonwealth of Australia

1901: January 1 Commonwealth of Australia is formed

Persuaded by the advantages of cooperation in areas including defense, irrigation, and trade, Australia's six self-governing colonies form the Commonwealth of Australia. The first prime minister of the commonwealth—a federation with its own constitution—is Edmund Barton, and a ceremonial opening of the first parliament by the Duke of York takes place.

1902 Women's suffrage granted on national level

Women's suffrage is mandated for all elections on the national level, at a time when it is not yet the law in all of the six member nations of the Commonwealth.

1902 National library is founded

The National Library of Australia is established in Canberra. Originally linked with the Commonwealth Parliamentary Library, it is later separated. In the 1990s, it houses over 4.7 million volumes.

1902: July 17 Birth of novelist Christina Stead

Acclaimed novelist Christina Stead is best known for *The Man Who Loved Children* (1940), her novel about a fragmented family in turmoil. A widely traveled author, she spends time living in London, Paris, and the United States, and writes for the motion picture industry for a brief period in the 1940s. Stead's work receives renewed interest with the rebirth of feminism in the 1960s and the advent of feminist literary criticism. However, Stead disavows the label "feminist author." Her novels include *The Beauties and the Furies* (1936), *For Love Alone* (1944), *The People with the Dogs* (1952), *Dark Places of the Heart* (1966), and *Miss Herbert* (1976). Stead dies in Sydney in 1983.

1903 Publication of novel *Such Is Life*

Such Is Life by Joseph Murphy (writing under the pseudonym Tom Collins), details rural life in Victoria and is noted for its narrative technique.

1906 Papua becomes an Australian territory

Fearing that the island of New Guinea may be annexed by Germany, Australia has claimed the southeast portion—known as British New Guinea, or Papua—as a protectorate. Under the Papua Act, Papua becomes a federal territory, and the eastern half of New Guinea is also brought under control by Australia. Under the leadership of the first lieutenant governor, Hubert Murray, all of Papua is explored and divided into administrative units.

1907 First professional rugby games

The first professional rugby (Rugby League) matches are played between an Australian team and a team of New Zealanders on its way to England.

1909 Federal government assumes states debts

Australia's federal government assumes all remaining debts of the individual states.

1910 First airplane flight is made

The first powered air flight is made by J.R. Duigan in Victoria in a craft of his own design.

1911 The Northern Territory becomes a separate entity

The Northern Territory is separated from South Australia.

1912: May 28 Birth of author Patrick White

Nobel-prize winning author Patrick White (1912–90) is born in London to Australian parents and raised in Australia. He returns to England for his secondary and university education, which includes studies at King's College, Cambridge. White serves in the Royal Air Force in World War II. His first novel, *Happy Days,* is published in 1939. Later novels include *The Tree of Man* (1955), *Riders in the Chariot* (1961), *The Eye of the Storm* (1973), *The Twyborn Affair* (1979), and *Memoirs of Many in One* (1986). White's novels are known for their use of allegory and myth. While distinctly Australian, they transcend nationality in their concern with basic human problems, including isolation and the search for meaning. In addition to novels, White also publishes plays, short stories, poetry, an autobiographical work, and essays. In 1973, he is awarded the Nobel Prize for Literature.

White dies in Sydney in 1990.

1914 World War I begins

Great Britain declares war on Germany, automatically committing its possessions, including Australia, to war. Australia's government vows strong support for Britain and occupies German New Guinea.

1914 Opera league presents performances

For two seasons, the Australian Opera League stages productions of operas by Australian composers.

1915 Gallipoli campaign

Australian troops distinguish themselves in the campaign to take the Gallipoli peninsula in Turkey, giving Allied forces access to the Dardanelles and the Sea of Marmara in order to attack Istanbul, the Turkish capital. The initial invasion is marred when the Turks gain advance warning of it. The Allied invasion force gains a foothold but is unable to accomplish its military objective, and a stalemate ensues from April to December, when Allied troops are evacuated. Altogether, some 8,500 Australians die at Gallipoli, and the invasion date, April 25, becomes an annual day of commemoration known as Anzac (Australian and New Zealand Corps) Day.

1915 Australian wins Nobel Prize for Physics

William L. Bragg shares the Nobel Prize for Physics with his father, William H. Bragg, becoming the first Australian-born person to win a Nobel Prize.

1915 Birth of artist James Gleeson

James Timothy Gleeson, the foremost advocate of surrealism in Australian art, is born in Sydney and studies art at the East Sydney Technical College and later in the United States. He begins to experiment with surrealism by the late 1930s, exhibiting his first surrealist painting in 1939 at the Contemporary Art Society Exhibition in Sydney. He plays a prominent role in the public debate over the merits of surrealism launched by the showing of Salvador Dali's work *L'Homme Fleur* in Australia the same year. In addition to his own paintings, Gleeson also supports the surrealist cause through articles and lectures. Also a distinguished art critic, Gleeson is appointed a trustee of the National Gallery in Canberra.

1916 First conservatory is established

The New South Wales State Conservatorium of Music is founded.

1917 Transcontinental railroad opens

Construction is completed on the Trans-Australian line linking the Western Australia and South Australia rail lines. Due to changes in engineering personnel when these lines were constructed, the gauge of the lines is not uniform, and there are five breaks of gauge in the new transcontinental line.

1919 Australia joins League of Nations

Following World War I, Australian representative take part in the Paris Peace Conference, and Australia becomes a member of the League of Nations. The League mandates Australian control of German New Guinea, seized by Australia during the war.

1920 Communist and Country parties are formed

The Communist Party of Australia and the federal Country Party are established.

1922 Australian airline is launched

Australian airline Qantas begins regular flights. In its first year of operation, it carries 1,100 passengers.

1926 Scientific research organization is formed

The Australian government establishes the Commonwealth Scientific and Industrial Research Organization for specialized research.

1927 Capital is moved to Canberra

The Nationalist-Country Coalition government moves the Australian capital to Canberra.

1929–31 Australia racked by depression

Australia feels the effect of falling world prices as the result of the global economic depression. The government takes austerity measures to counter the effects of the downturn.

1929 Birth of composer Peter Sculthorpe

One of Australia's most acclaimed postwar composers, Sculthorpe is born in Launceston, Tasmania, and demonstrates talent in both music and painting. He studies at the Melbourne University Conservatorium and later teaches at the University of Sydney, where he influences many younger Australian composers. He is probably best known for his *Sun Music,* a series of works evoking the Australian landscape. Other works include *Irkanda IV* and *Sixth String Quartet.* Most of his music is written for a modest number of performers, but he has also composed the large-scale theatrical work *Rites of Passage.* Sculthorpe's music is characterized by its accessibility and, often, its melancholy and slow-moving character. Sculthorpe, who has achieved an international reputation, works throughout his career to further the concept of a distinctly Australian music.

1930 Death of artist George Washington Lambert

Lambert, one of the foremost painters of his era and a member of the Royal Academy, dies at Cobbity, New South Wales. (See 1873.)

1932 Australian Broadcasting Commission is formed

The Australian Broadcasting Commission is set up. In addition to nationalizing radio stations throughout the country, it plays a central role in the advancement of classical music in Australia by establishing studio orchestras in all the states, and sponsoring and broadcasting recitals and subscription concerts. Its first studio ensembles are formed in Sydney and Melbourne, followed by smaller groups in Adelaide, Brisbane, Perth, and Hobart.

1933 Australia claims Antarctic territory

Australia claims an extensive region of Antarctica as the Australian Antarctic Territory.

1935 Construction of the Grainger Museum

The Grainger Museum is built on the grounds of Melbourne University by composer Percy Grainger to house his work, including published music as well as manuscripts and correspondence. (See 1882.)

1935 Broken Hill gains iron and steel monopoly

Broken Hill Proprietary Company wins a nationwide monopoly on iron and steel production when it acquires Australian Iron and Steel.

1935: October 7 Birth of author Thomas Keneally

Acclaimed for his historical fiction, Thomas Keneally is born in Sydney and spends time in a Roman Catholic seminary, although he leaves without being ordained a priest. His novels range widely in subject matter, from *Bring Larks and Heroes* (1967), which is about the early years when Australia was a penal colony, to *Gossip from the Forest* (1975), set at the end of World War I, to *Confederates* (1979), set during the American Civil War. *The Chant of Jimmie Blacksmith* (1972), about a notorious murder spree by a half-caste aborigine, wins international attention for Keneally and is made into a successful film in 1980. Keneally's most famous work, *Schindler's List* (first published in 1982 as *Schindler's Art*), about a German entrepreneur who saves over 1,000 Jewish lives during World War II, is made into an Academy Award-winning film by American filmmaker Steven Spielberg in 1993.

1939 Worst loss of life to bushfires in Australian history

The most damaging bushfires in Australia's history rage in Victoria, causing 71 deaths.

1939: September 3 Australia declares war on Germany

Australia joins "the Motherland," Great Britain, in declaring war on Nazi Germany. The Australian government organizes an expeditionary force, which takes part in campaigns in the Middle East and North Africa.

1940 Australian troops arrive in the Middle East

The first units of the Australian Imperial Force (AIF) sail for the Middle East, where they see action in Libya within the year.

1941: December 8 Australia declares war on Japan

Following Japanese attacks the day before on both Pearl Harbor and Malaya (where Australian troops are stationed) Australia declares war on Japan.

1942: April MacArthur arrives in Australia

U.S. General Douglas MacArthur arrives in Australia after being routed from the Philippines by Japanese troops. As commander in chief of Allied forces in the southwest Pacific area, he marshals his forces to mount a counteroffensive against the Japanese. Some 15,000 Australians are among the Allied prisoners captured by the Japanese when they take Singapore.

1943 Council is founded to support the arts

The Australia Council is created to support the performing arts by sponsoring dance, music, and drama events.

1943 Artist Arthur Streeton dies

Acclaimed landscape artist Sir Arthur Streeton dies at Olinda, Victoria. (See 1867.)

1944 Liberal Party is founded

Robert Menzies (1894–1978) founds the Liberal Party.

Born in Victoria, Sir Robert Gordon Menzies begins his career as a barrister before entering politics. He becomes a member of the Victoria parliament in 1928, goes to the federal house of repesentatives (1934), and is attorney general for the Commonwealth of Australia (1935–39). He becomes prime minister (1939–41) and leads the opposition party (1944–49). In 1949, he becomes premier as leader of a coalition. (See 1949–66). He holds office for sixteen years, a time during which Australia's economy thrives. In 1956, he heads the Five Nations Committee created to negotiate a settlement over the Suez Canal. He retires from Parliament in 1966.

1945 Australia joins the United Nations

Australia becomes a founding member of the United Nations.

1945 Musica Viva is founded

Musica Viva is established to promote chamber music performances. For five years, it maintains its own ensemble of chamber players. Later, it organizes festivals, manages tours by visiting artists, and sponsors tours abroad by Australian groups.

1946–49 Social welfare legislation is passed

The Australian government enacts legislation creating a broad program of social welfare benefits, including unemployment, health, old-age, and disability benefits.

1946: March 20 Death of author Henry Handel Richardson

Novelist Henry Handel Richardson (pseudonym of Ethel Florence Lindesay Robertson) dies in Sussex, England. (See 1870.)

1947 First professorship in fine arts is established

The University of Melbourne establishes the country's first academic position devoted solely to the fine arts.

1949–66 Long period of Liberal-Country party rule

The Liberal-Country Party is in control of the government for twenty-three years, sixteen of them under the leadership of R.G. Menzies (see 1944). It is a period of rapid industrial and population growth, suburban expansion, and rising foreign investment.

1949 Rural electrification agency is formed

The Snowy Mountains Authority is created to institute a program that will irrigate the dry interior region of southeastern Australia and generate electricity inexpensively.

1949 Labor tensions cause economic and political turmoil

Union conflict in the coal industry touches off anti-Communist sentiment and criticism of the ruling government. Legal remedies and government troops are used to end a national coal strike.

1949: December Election launches era of Liberal-Country party rule

The Liberal-Country party, headed by R.G. Menzies (see 1944), returns to power, inaugurating a twenty-three year period of uninterrupted rule.

1950–53 Australian troops fight in the Korean War

Australian troops support the U.S. by fighting Chinese and Korean communist forces in the Korean War.

1951 ANZUS defense pact is formed

Australia, New Zealand, and the United States sign the ANZUS treaty, which serves as a base for Australia's defense policy through the 1980s.

1954 Australia joins SEATO

Australia joins the Southeast Asia Treaty Organization.

1954 Theater foundation is founded

The Elizabethan Theatre Trust is formed to provide financial support for theater, ballet, and opera.

1955 Australian forces are stationed in Malaya

Australia sends troops to Malaya to assist in British anti-Communist operations there.

1956 First televisions broadcasts are aired

Television is introduced to Australia.

1956 Olympics are held in Melbourne

The Olympic games are held in Melbourne, Australia. Local participants garner thirteen gold, eight silver, and thirteen bronze medals. The country's overall ranking is surpassed only by those of the United States and the U.S.S.R.

One of only three nations to have participated in every Olympic Games since the modern games were instituted in 1896, Australia is later selected to host the summer Olympic Games for a second time in 2000.

1958 Council for the Advancement of Aborigines is formed

A multiracial federal council is formed to work for the rights of the nation's aboriginal population.

1960 Iron ore deposits found at Pilbara

Extensive deposits of iron ore are discovered at Pilbara in Western Australia. It is developed at sites including Mt. Newman, Mt. Tom Price, Mt. Goldsworthy, and Robe River.

1960 Heather McKay wins first of fourteen squash championships

Heather Pamela Blundell McKay (b. 1941) wins the first of fourteen Australian squash championship titles (1960–73). She wins the British Open squash tournament for sixteen years straight (1962–77) and is world champions twice, in 1976 and 1979. From 1962–80, she is undefeated. Her husband, Brian McKay, is also a champion squash player. The couple moves to Canada in the mid-1970s.

1960: March Adelaide Festival of Arts is inaugurated

Australia's most prominent arts festival, held biennially for three weeks, is launched. It encompasses a wide range of activities, including concerts, recitals, and exhibitions, and features popular as well as classical music. A Writers' Week program is also associated with the festival, as well as a number of community events. Over the years, a number of notable Australian works are premiered at the festival, including a play by Patrick White (see 1912: May 28) titled *Night on Bald Mountain* and ballets by Sir Robert Helpmann (1909–86).

1960 Australian wins Nobel Prize for medicine

The Novel Prize for Physiology or Medicine is awarded to Sir Macfarlane (Frank) Burnet (1899–1985) for his work in the field of immunization.

Born in eastern Victoria, Burnet studies medicine at Melbourne University and moves to London for further study. He returns to Melbourne in 1928 to join the staff of the Walter and Eliza Hall Institute for Medical Research and becomes director there in 1944. Burnet's work improves the understanding of how the body produces antibodies to fight invasion of foreign substances. He shares the Nobel Prize with Peter Medawar, whose experiments support Burnet's theory. Burnet is knighted in 1951.

1961 Death of composer Percy Grainger

Internationally recognized Australian composer Percy Grainger dies. (See 1882.)

1961 Ballet company is formed

The Australian Ballet is launched. Under the direction of Dame Peggy (Margaret) Van Praagh (1910–90) and Sir Robert Murray Helpmann (1909–86), the company establishes an international reputation and tours widely overseas. It is known for incorporating distinctly Australian elements into its ballets, including Australian themes and music, choreography, and set design by Australians.

1961 Rod Laver wins Wimbledon tennis championship

Tennis player Rod Laver (b. 1938) wins the singles tennis championship at the prestigious Wimbledon tournament in England. He wins it again in 1962, the same year he wins what is referred to as the Grand Slam—championships at Wimbledon, the U.S. Open, French Open, and the Australian Open. In 1962, he gives up his amateur status. Between 1964–70, Laver is the professional world singles title holder five times.

Until 1968, participants in the Wimbledon tennis tournament are limited to amateurs. That year, when professionals are first allowed to play, Laver wins the championship. He wins again at Wimbledon in 1969 in the course of winning his second Grand Slam.

1962 First Aboriginal land rights protest

The Yirrkala Aborigines send a petition painted on bark to the federal government protesting bauxite mining on their ancestral lands at Gove in the Arnhem Land reserve. They report that their sacred sites are being desecrated and destroyed. The case is appealed all the way to the Supreme Court in Darwin (See 1971.)

1964 The *Australian* is founded

The national newspaper, the *Australian,* begins publication. In the late 1990s, it is one of only two national newspapers in circulation.

1964 First elections are held in Papua New Guinea

The territory of Papua New Guinea elects its first House of Assembly representatives.

1965 Australia sends troops to Vietnam

Supporting the policies of its ally and defense partner, the United States, Australia sends 800 troops to assist U.S. and South Vietnamese forces in battling the North Vietnamese communists.

1966 Racially based immigration policy is ended

The White Australia policy, in effect since the turn of the century (see 1901), is terminated, setting off a boom in immigration from Asia and the Middle East.

1966 Prime Minister Menzies retires

Prime Minister R.G. Menzies (see 1944) retires from office after serving the longest continuous term in Australian history.

1966 Aboriginal livestock herders strike for equal pay

Aboriginal workers in cattle stations in the Northern Territory go on strike demanding pay equal to that received by their white peers. They later occupy a tract of their ancestral land at Wattie Creek for eight years and are eventually granted a lease rather than the land rights they are seeking.

1966: April Forces in Vietnam are increased

Sir Harold Edward Holt (1908–67), Menzies's successor as prime minister, increases the Australian troop commitment in Vietnam to 4,500. Following the death of the first Australian in combat, opposition to the country's military involvement sparks a long-running national debate over Australia's support for U.S. foreign policy. Holt strongly supports the United States with the slogan "All the way with LBJ," referring to U.S. president Lyndon Baines Johnson. Holt dies in 1967 while swimming near Melbourne.

1969 Oil and gas production begins at Bass Strait

Production is begun at the Barracouta field, off the Gippsland coast, which is the site of the first oil and gas reserves discovered in the area. Later discoveries include the Marlin, Halibut, and Kingfish fields.

1970s and 1980s Film industry wins recognition

Australia's filmmakers gain international recognition. Among the most prominent film directors are Peter Weir (b. 1944) and Bruce Beresford. Two of Weir's early short works—*Michael* (1970) and *Homesdale* (1971)—win the Australian Film Institute Grand Prix. His first feature film is *The Cars That Ate Paris* (1974), followed by *Picnic at Hanging Rock* (1975), *Gallipoli* (1980), *The Year of Living Dangerously* (1982), *Witness* (1985), *Dead Poets Society* (1989), *Fearless* (1993), and *Crocodile Dundee* (1986). Beresford's films include *The Getting of Wisdom* (1977); *Breaker Morant* (1980), *Driving Miss Daisy* (1989), *Tender Mercies* (1993). His 1998 film, *The Truman Show* received a nomination for an Academy Award

1970s Country struggles with economic ills

Global recession and inflation affect the Australian economy and result in criticism of the Labor government's policies, including tariff cuts, wage concessions, and expensive social programs.

1971 Combat troops come home from Vietnam

After mass demonstrations protesting Australia's involvement in the Vietnam war, all Australian combat troops are withdraws from Vietnam.

1971 Darwin court decides against Aboriginals in land case

The Supreme Court upholds a lower court ruling in a land rights case protesting bauxite mining on ancestral Aboriginal lands in the Arnhem Land reserve. (See 1962.)

1971 First Aboriginal wins parliamentary seat

Neville Bonner becomes the first Aboriginal to win a seat in Parliament.

1972: December Labor party is returned to office

National elections return the Labor party to office for the first time since 1949 under the leadership of Gough Whitlam. The Labor government ends the military draft, pardons draft resisters jailed during the Vietnam era, and establishes diplomatic relations with the People's Republic of China.. Social welfare and education programs are expanded, and universal health insurance is introduced.

1973 Voting age is lowered

The legal voting age is lowered to eighteen for federal elections and elections in all states that have not yet lowered it.

1973 Patrick White wins Nobel Prize for Literature

Author Patrick White (1912–90) is awarded the Nobel Prize for Literature. (See 1912.)

1974: December 25 Darwin destroyed by a cyclone

A cyclone strikes Darwin, in the Northern Territory, creating Australia's most costly natural disaster in terms of property damage and one of the worst in terms of lives lost. More than half the town's residences are destroyed, and sixty-six people are killed. Wind speeds of 217 kilometers per hour are registered before the gauge breaks, and all communications with the outside are severed. An evacuation operation moves 35,000 people out of the area. Reconstruction work is not completed until 1977.

1975 Liberal and National coalition is returned to power

Rocked by economic woes, the country returns the Liberal-National (formerly Liberal-Country) coalition to power under Prime Minister J. Malcolm Fraser.

1975 Independence is granted to Papua New Guinea

The territories of Papua and New Guinea become the independent state of Papua New Guinea.

1975: November Budget controversy sparks constitutional crisis

When the Labor party budget is stalled in the Senate and a government shutdown is threatened, Governor-General Sir John Kerr dissolves Parliament and calls for new elections. Gough Whitlam is dismissed as prime minister and elections are scheduled for December.

1975: December Liberal-National party is returned to power

Labor is ousted by the Liberal-National Party, which wins majorities in both houses of Parliament. Malcolm Fraser becomes the new prime minister.

1977: January 18 Worst rail accident in Australian history

In Australia's worst rail disaster, a commuter train derails after leaving Granville, New South Wales, and hits the pylons supporting a four-lane overpass, which collapses onto the train, killing eighty-three and injuring 213. An investigation lays the blame on the condition of the tracks rather than human error, and the government launches a $200 million program rail improvement program in New South Wales.

1978–79 Vietnamese "boat people" arrive in large numbers

"Boat people," refugees leaving Vietnam by boat, begin arriving in Australia. Almost all are allowed to remain as permanent residents. In June of 1979, the government agrees to accept 32,00 refugees and accepts an additional 14,000 in 1979–80.

Dancers rehearse for a taping of "The Carol Burnett Show," a U.S. television program. Carol Burnett was invited to be the first entertainer to tape a television program at the newly opened Sydney Opera House. (EPD Photos/CSU Archives)

1979 Award-winning film, *My Brilliant Career,* stars Judy Davis

My Brilliant Career (1979), directed by Gillian Armstrong and starring Judy Davis, wins international recognition. The film tells the story of a young Australian woman at the end of the nineteenth century who pursues opportunities for education and independence, defying social pressures to conform to traditional roles. The film wins the Australian Film Institute "Best Film" award (1979) and is nominated for an Academy Award for Costume Design (1980). Davis wins the British Academy Award for Best Actress (1980).

1980 Government supports U.S. boycott of Olympic games

The Australian government supports the United States call for a boycott of the Moscow Olympics to protest the U.S.S.R. invasion of Afghanistan. However, most Australian athletes disregard the government's request and compete in the games.

1980–83 Severe drought affects eastern Australia

Drought spreads from New South Wales to affect Victoria, Queensland, and South Australia as well. Some areas report an all-time record low rainfall for the second half of 1982.

1980 Aboriginal women are injected with contraceptive hormone

Through a government program, young aboriginal women are injected with the synthetic hormone Depo-Provera without their knowledge in a measure designed to reduce the birthrate of the aboriginal population.

1980 National Museum of Australia is established

The National Museum of Australia is founded in Canberra.

1982: October The Australian National Gallery opens

The Australian National Gallery opens in Canberra.

1983–91 Labor government in power

The centrist Labor government of Robert Hawke restores the health of the Australian economy through pro-business policies that include privatization of publicly owned enterprises.

1983 Australia wins America's Cup yacht race

The yachting team aboard the *Australia II* defeats the U.S. vessel *Liberty* in the America's Cup contest at Newport, Rhode Island.

1983: March 31 Death of author Christina Stead

Novelist Christina Stead dies in Sydney. (See 1902.)

1984 Cocos Islands become part of the Australian Commonwealth

The Cocos (Keeling) Islands vote to join the Australian Commonwealth.

1985–86 Legal ties with Great Britain end

Australia and Great Britain agree to terminate all remaining legal connections.

1985 The first Australia Games are held

The Australia Games are inaugurated in Melbourne, with more than 1,600 entrants, including athletes from thirty foreign countries.

1986: March 2 Separation from Britain becomes official

Queen Elizabeth II formally approves the separation of Australia from the British Commonwealth under the UK Australia Act.

1987 Labor government wins third consecutive national election

Robert Hawke's centrist Labor government is elected to a third term, becoming the first to serve more than two consecutive terms.

1987: October U.S. Stock market crash affects Australia

A period of frenzied speculation in securities ends when the Australian stock market crashes.

1988 Bicentennial of European settlement is feted

Australians commemorate the bicentennial of the continent's first European settlement, Port Jackson. Federally funded projects at the state, local, and territorial level focus on the theme of "living together." Included are the Australian Bicentennial Exhibition, which travels to roughly fifty locations throughout the country; the inauguration of the Bicentennial Science Centre in Canberra; and a tall ship race from Hobart to Sydney, with entrants from thirty countries invited to participate.

1988 All levels of government enjoy balanced budgets

For the first time since 1938, all levels of Australian government—territorial, state, and federal—are balanced.

1990 New regulatory body is appointed for securities

In the wake of the stock market crash in 1987, a new securities commission is created.

1990: March 24 Labor government is elected to a fourth term

The Labor government of Robert Hawke is elected for the fourth time, but this short-lived triumph is soon followed by an economic recession.

1990: September 30 Author Patrick White dies

Nobel Prize-winning novelist Patrick White dies in Sydney. (See 1912.)

1992 Unemployment hits sixty-year high

With the Australian economy in recession, unemployment soars to eleven percent (from six percent in 1989)—the highest percentage since the Great Depression of the 1930s.

1993 Labor wins another federal election

The Australian Labor Party (ALP) maintains its parliamentary majority in national elections.

1995: January Foreign affairs minister visits Cuba

Foreign affairs minister Gareth Evans becomes the first Australian minister to visit Cuba, where he meets with Fidel Castro. Both nations seek to bolster their political and economic ties.

1995: January 30 Howard becomes head of the Liberal Party

Alexander Downer steps down as head of the opposition Liberal Party and is replaced by John Howard, who headed the party between 1985 and 1989.

1996: February 13 Victoria port to be privatized

In the first privatization of an entire Australian port, the government of Victoria sells the port of Portland to a group of investors for $22 million (U.S. dollars).

1996: March 2 Labor is unseated after thirteen years

After a record five terms in power, the Labor party loses its majority to a Liberal Party-National Party coalition led by John Howard, who becomes the new prime minister. The Liberal and National parties win ninety-four out of 148 seats in the House of Representatives, compared to forty-seven for Labor. The focal issue of the campaign had been the economy, with the Liberal-National side challenging the Labor government's claims about the economic health of the nation.

1996: April 28 Tasmania gunman unleashes Australia's worst massacre

Twenty-nine-year-old Martin Bryant, wielding a semiautomatic rifle, opens fire on crowds of tourists in Port Arthur, Tasmania, killing thirty-five and injuring eighteen. The dead include three hostages whom Bryant holds in a guest cottage before setting fire to it and fleeing. Bryant, the perpetrator of the worst massacre ever recorded in Australia, is found to have a history of mental problems. He is later indicted on thirty-five counts of murder.

1996: May 17 First Aboriginal judge is sworn in

Bob Bellear becomes Australia's first Aboriginal judge.

1998: October Liberal-national government wins reelection

The Liberal-National government of John Howard is elected to a second term. Howard vows that his second administration will seek reconciliation with the Aboriginal population.

Bibliography

Atkinson, Alan. *The Muddle-headed Republic.* New York: Oxford University Press, 1993.

Bassett, Jan. *The Oxford Illustrated Dictionary of Australian History.* New York: Oxford University Press, 1993.

Berndt, Ronald Murray. *The Speaking Land: Myth and Story in Aboriginal Australia.* Rochester, Vt.: Ineer Traditions International, 1994.

Blainey, Geoffrey. *The Tyranny of Distance: How Distance Shaped Australia's History.* South Melbourne: Macmillan, 1982.

Bolton, Geoffrey, ed. *The Oxford History of Australia.* New York: Oxford University Press, 1986–90.

Dyster, B. *Australia in the International Economy in the Twentieth Century.* New York: Cambridge University Press, 1990.

Gunther, John. *John Gunther's Inside Australia.* New York: Harper & Row, 1972.

Hancock, Keith, ed. *Australian Society.* New York: Cambridge University Press, 1989.

Heathcote, R.L. *Australia.* London: Longman, Scientific & Technical, 1994.

London, H.I. *Non-White Immigration and the "White Australia" Policy.* New York: New York University Press, 1970.

Lourandos, Harry. *Continent of Hunters-Gatherers: New Perspectives in Australian Prehistory.* Cambridge: Cambridge University Press, 1987.

Moore, Andrew. *The Right Road?: A History of Right-Wing Politics in Australia.* Melbourne: Oxford University Press, 1995.

Rickard, John. *Australia, A Cultural History.* London: Longman, 1996.

Azerbaijan

Introduction

The Republic of Azerbaijan is a former Soviet republic in the Caucasus Mountains region located on the southwestern shore of the Caspian Sea. Azerbaijan is the eastern part of a larger historical region known as Transcaucasia, that also includes Armenia and Georgia. The Republic of Azerbaijan is the nation state of the Azerbaijanis, or Azeris, a Turkic people. Historically the territory of Azerbaijan comprised not only the territory of the modern Republic of Azerbaijan, but also a significant portion of northwestern Iran, and today at least as many Azerbaijanis live in Iran as live in the Republic of Azerbaijan. Since the seventh century A.D. Azerbaijan has been part of the Islamic world, and for much of that time has been under the strong cultural influence of Persia. However, after becoming part of the Soviet Union in 1920, it became firmly integrated into the Soviet system.

The origins of the name Azerbaijan are unclear, but it has been used to refer to the region since the Middle Ages. There are a number of folk etymologies for the name Azerbaijan, but some linguists have suggested that the toponym originated from "Atropaten," the name of an Achaemenid satrap who ruled what is today Southern Azerbaijan in the fifth or fourth century B.C. and who eventually gave his name to the region.

The territory of the Republic of Azerbaijan consists of 86,200 square kilometers. Roughly half of the republic's territory is mountainous, while the other half consists of alluvial plains. These plains are for the most part located along the Caspian seacoast and in the center of the country, especially along the Kura River. The highest mountains are in the north of the country, the Caucasus range, which reaches heights of over 13,000 feet. The republic is bordered in the north by Russia, that is, by the Russian republic of Daghestan, to the northwest by the Republic of Georgia, to the east by Armenia, and to the south by Iran. Azerbaijan's neighbors across the Caspian Sea on its eastern shore are the Central Asian republics of Kazakstan and Turkmenistan.

Azerbaijan's territory is not contiguous. The province of Nakhichevan is separated from the Azerbaijani "mainland" by a strip of Armenian territory and is sandwiched between Armenia and Iran. Within Azerbaijan's territory is the disputed enclave of Nagorno-Karabakh. The enclave's population is overwhelmingly Armenian, and in 1988 the enclave's leadership sought to secede from Azerbaijan, sparking a war that intensified after the collapse of the Soviet Union and eventually drew in the republic of Armenia itself. As a result of defeats at the hands of the Armenians, Azerbaijan lost its control over Nagorno-Karabakh, as well as much of its western territory bordering Armenia.

The central alluvial plain is especially fertile, and is well suited for cereal cultivation and stockbreeding. The more mountainous regions are also used extensively for stockbreeding, and also for grape and other fruit cultivation. Azerbaijan has had a large oil industry since it was first established in 1872, and was one of the major oil producing regions of both imperial Russia and the Soviet Union. The oil industry was intensively developed during the Soviet era, and in the post-Soviet era Azerbaijan's enormous off-shore reserves have attracted the attention of the international oil industry.

The capital of Azerbaijan is the city of Baku, a major port city on the Caspian Sea, and the cultural, political, and economic center of the country. Until the early 1990's the city was ethnically mixed, with Azeris, Armenians, and Russians living in mixed neighborhoods, but has since become a much more Azeri city. The city is especially prominent as the economic center of the oil-rich Caspian sea region.

Azerbaijan's population numbers about 7.7 million, according to a 1994 estimate. According to the 1989 Soviet census the majority of the population at that time were Azeris, making up 82.7 percent of the total. The Azeris speak a language known as Azeri or Azeri Turkish, a language mutually intelligible with Turkish spoken in the Republic of Turkey. Azeri is the republic's official language, but Russian is still widely used in commerce and politics. Since 1991 Azeri has been the official language of Azerbaijan. Until 1920 the language was written in the Arabic script, and was for all intents and purposes indistinguishable from the literary Ottoman Turkish language of the Ottoman Empire. The Latin script began to be introduced in Azerbaijan in 1922, and in 1924 became the official script for the language. The Latin alphabet proved more suitable to the Azeri language than Arabic. In the late 1930s the Soviet authorities decreed that the script be changed to a modified Cyrillic script. In 1991 the Azerbaijani authorities declared a shift back to the Latin script.

Under Soviet rule the largest minorities were Russians (5.6 percent) and Armenians (also 5.6 percent). However since the collapse of the Soviet Union and the war with Armenia, many Russians and Armenians have fled the country, substantially reducing their numbers. The exodus of Armenians was especially accelerated by anti-Armenian pogroms (organized persecutions) that took place in 1990 and 1991. In addition to these groups, there is also a large community of Lezgins, a Daghestani nationality, making up 3.2 percent of the overall population. Smaller minorities include Georgians, Kurds, Jews, Talysh, and Tats, although the number of Jews has declined since the fall of the Soviet Union as a result of emigration to Israel and the United States.

The Azeris are traditionally a Muslim people, and since the sixteenth century have belonged to the Shia branch of Islam. The Azeris were the only major Muslim nationality in the Russian empire or the Soviet Union to be Shias, the minority branch of Islam. Although seventy years of Soviet anti-religious policy undermined and eliminated many of the Azeris' Islamic institutions, the collapse of the Soviet Union and the declaration of religious freedom has resulted in a new interest in and revival of Islam and Islamic institutions.

Historically, Azerbaijan occupied a much larger territory that included not only the territory of the modern Republic of Azerbaijan, but also much of northwestern Iran. This territory, centered around the city of Tabriz, continues to be known as Iranian Azerbaijan, and whereas there are nearly five million Azeris in the republic of Azerbaijan, there are at least as many ethnic Azeris in Iran. The territory of historical Azerbaijan was partitioned between Persia and Russia in 1813, when the Persians and Russians signed the treaty of Gulistan, in which Persia ceded to Russia that part of Azerbaijan that lay north of the Araks River and north of the Talysh mountains, the landmarks that still mark the boundary between Iran and the Republic of Azerbaijan.

Although the territory of Azerbaijan has been home to human civilization since the Paleolithic era, and had a developed political and social culture already in antiquity, the ethnic history of the Azeris begins in the eleventh century, when the region was conquered by the powerful Seljuk empire, and was settled by waves of Muslim Turkic tribesmen from Central Asia. These nomads soon abandoned nomadism for agriculture and a more sedentary life. Traditionally the Azeris lived in agricultural villages, as well as in cities, and in terms of material culture and diet differed little from their Turkish and Iranian neighbors. The Azeri diet featured various sorts of fruits and vegetables, rice, bread, and meat.

Under Russian rule most of Azerbaijan was included within the Province (*guberniya*) of Baku and became one of the most economically advanced and industrialized Muslim regions of the Russian empire, and it was in Azerbaijan that

Pan-Turkic and later Azeri nationalism, as well as Islamic modernism, found especially fertile soil.

Although Azeri nationalists established a short-lived state in 1918, the territory of Azerbaijan was soon conquered by the Red Army and integrated into the Soviet Union. Soviet policies, especially collectivization, the Cultural Revolution, the purges of the 1930s, and anti-religious measures radically transformed Azeri society, economically and socially. Furthermore, Azeri cultural life was severely restricted to fit Soviet norms and goals. At the same time, the country's political elites, such as the then head of the Azerbaijani Communist Party, Geidar Aliev, and to a large degree, cultural, elites by the late 1980s were in large measure products of the Soviet system, and were for the most part committed to its preservation. With the fall of the Soviet Union in 1991, and Azerbaijan's formal declaration of independence in that year, power soon went to the nationalist forces, who sought to break with the Soviet past, while establishing an Azeri nation state and prosecuting the war with Armenia over Nagorno-Karabakh. By 1992, defeat in the war with Armenia, and the major economic and social dislocation caused by the war, encouraged the former Communist leader Aliev to overthrow the elected nationalist government, and in the following year Aliev was formally elected president of Azerbaijan. Under Aliev, Azerbaijan's situation has stabilized to some degree. Although the Nagorno-Karabakh conflict remains unresolved, a cease-fire with the Armenians has brought a degree of calm, and the country's potential as an exporter of oil and natural gas has attracted substantial foreign investment.

Timeline

c. 8000 B.C. First rock drawings appear

The oldest rock drawing in the territory of Azerbaijan dates from around this time, located near the site of Gobustan. Other later rock drawings are found near the sites of Ashperon, Giamigai, and Nakhichevan, and suggest that those who made them were familiar with seafaring. Today the Gobustan Museum preserves prehistoric artifacts and cave paintings.

c. 5500 B.C. First permanent human settlements appear

First permanent human settlements founded. These earliest settlements appear to be agricultural settlements, and their inhabitants are also involved in pottery production.

c. 3500 B.C. Copper Age gives way to Bronze Age

Settlements that use and produce copper tools become widespread in northern Azerbaijan. At this time, to about 1000 B.C. bronze tools are in wide use. Settlements at this time multiply along major rivers and in upland valleys

c. 1500 B.C. Hurrian state emerges

The Hurrian state is centered in the area between Lakes Van and Urmiya, and includes among its territories part of western Azerbaijan. The Hurrians, an Indo-European people, create monumental works of pictorial art.

720 B.C. Sakas enter Azerbaijan

The Sakas, a branch of the nomadic Scythian people who populate the Eurasian steppe, according to the Greek historian Herodotus, cross into Azerbaijan in pursuit of an enemy people. Of these, a group known as the Massagetae establish themselves in Northern Azerbaijan, as well as east of the Caspian Sea.

715 B.C. Establishment of the Medean Empire

The Medes, who inhabit much of Southern Azerbaijan, establish a powerful state that expands to the west and south. Medes are instrumental in keeping the Assyrian empire out of Azerbaijan.

625–585 B.C. Cyrus II conquers Medea and creates Persian Empire

The Achaemenids, Persian vassals of the Medes, conquer Medea and annex it to their own empire.

c. 340 B.C. Satrapy of Atropates created

Atropates rules Southern and Northern Azerbaijan from the town of Frasp to the south of modern Tabriz. Atropates rules the Medes, Albanoi (no relation to modern Albanians), Cadusians, and Sacesinians. The latter three peoples live in the territory of Northern Azerbaijan. Modern Azerbaijani historians consider Atropates' satrapy (ancient Persian province) to be the nucleus of the first Azerbaijani state, and they believe the name Azerbaijan to be derived from the name Atropates.

331 B.C. Satrapy of Atropates absorbed into Macedonian Empire

Darius III, the Achaemenid ruler of the Persian Empire is decisively defeated by the Macedonians of Alexander the Great at the Battle of Gaugamela. Atropates, who fought under Darius, pledges his allegiance to Alexander, and rules his satrapy as a Macedonian vassal.

320 B.C. Kingdoms of Atropatene and Albania established

After the Death of Alexander the Great the Macedonian Empire in the East divides. The Atropatene Kingdom, which is centered in Southern Azerbaijan, finds itself eventually in the sphere of influence of the Seleucid dynast, who rules most of Iran. The Albanian Kingdom (no relation to modern Albania) corresponds fairly closely to the territory of the modern Republic of Azerbaijan, that is, it is located in the Kura Valley north of the Araks River. The economies of these states are based primarily on agriculture and sedentary stockbreeding. Various crafts are also widespread, including glass-making.

20 B.C. Atropatene and Albanian Kingdoms annexed by Parthia

The Atropatene Kingdom, which conducts relations with the Roman Empire, is annexed to the Parthian Kingdom, based in Eastern Iran, as is Albania.

A.D. 226 Parthian dynasty overthrown by Sassanids

Atropatene and Albanian Kingdoms come under the rule of the Iranian Sassanid dynasty and are ruled by Sassanid satraps.

c. A.D. 300 Christianity becomes state religion in Albania

Albanian rulers make Christianity the official religion, supplanting Zoroastrianism and other local religions. Albania thus becomes one of the first officially Christian states, and brings itself closer into the sphere of the Byzantine Empire, the Sassanids chief rival.

637–40 Arabs defeat Sassanids and occupy Albania

As a result of the rapid expansion of the Arab Caliphate and Muslim power, the Arab armies under the caliph Umar I decisively defeat the Sassanid armies and occupy the Iranian Capital of Ctehesiphon. By 640 the Arab armies have reached the Caucasus Mountains.

640–900 Islamization and Rebellion in Azerbaijan

When the Arab armies occupy Azerbaijan they encounter communities that are largely Christian. Although conversion to Islam under the Caliphate is not forcible, certain official policies, especially concerning taxation, make conversion to Islam attractive to many local communities. A number of anti-Arab rebellions also take place at this time, centered in Azerbaijan. The largest of these rebellions is the Hurramite rebellion, which lasts off and on from 780 until 880.

700–900 Arab-Khazar rivalry in Azerbaijan

Beginning early in the eighth century the Khazars, a powerful nomadic empire based along the lower Volga River, form an alliance with the Byzantine Empire against the Arabs. The Khazars are able to stop the momentum of the Arab advance, and hold the Arabs south of the Caucasus Mountains. The battleground in this conflict is mainly Azerbaijan, with the local Albanian rulers occasionally rebelling against the Arabs. By 750 the territory of Azerbaijan becomes annexed to the Arab Caliphate.

950–1050 Local Muslim rulers in Azerbaijan

By 950 the Arab Caliphate surrenders much of its central authority to local Muslim dynasties. These include the Mazyadids centered near Bailakan, the Derbent and Sheka emirates, the Sheddadids centered in the important city of Ganja, and the Sajids in Tabriz.

1054 Rulers of Azerbaijan submit to Seljuk rule

In the middle of the eleventh century the Seljuk dynasty, supported by Muslim Oghuz Turks from Central Asia, conquer Iran, Iraq, Syria, and Asia Minor. As a result of these large-scale victories, the local dynasties in Azerbaijan submit to Seljuk rule.

1050–1150 Oghuz settlement of Azerbaijan

One of the consequences of the Seljuk triumphs is the large scale settlement of Iran, Iraq, Azerbaijan, and Anatolia by Oghuz tribesmen, known as Turkmens. In ethnic terms the modern Azeri nation is descended from these Turkmen settlers. It is unclear to what extent Turkic speaking peoples were already established in Azerbaijan, but there is no question that the medieval Turkmens form the ethnic foundation of the modern Turkmen nation.

1050–1200 Cultural Blossoming under the Seljuks

The cultural flowering in the Azerbaijani cities of Ganja, Tabriz, Shirvan and others takes place under Seljuk rule. In the twelfth century Azerbaijan especially excelled in the creation of literary works, written in Arabic, Turkic, and especially Persian; the most well-known of these poets is Nizami Ganjavi (1141?–1209?). His most celebrated work is *Khamsa,* comprised of five epic poems.

Together with the written literary works, the most important work of Azerbaijani oral literature also dates from this period. This work, *Dedem Qorqut Kitabi* (The Book of Dede Korkut), is the oral epic of the Turkmen nomads.

1120 Ajami Nakhichevani is born

Ajami, son of Abubekr, known as Ajemi Nakhichevani, is born. He works as an architect in Nakhichevan in western Azerbaijan, creating spectacular mausoleums. He employs intricate brickwork with carved glazes to create a uniquely decorative structure. During his lifetime, Nakhichevan becomes the center of Azerbaijani architecture.

1231 The Mongol conquest

Following their conquest of China and Central Asia, the Mongols move to permanently conquer and occupy western Iran and the Transcausasus. In 1231 the Mongols, under their leader Genghis (Chingis) Khan (1162–1227), conquer all of these areas. Mongol troops had already passed through Azerbaijan in 1225, during an earlier campaign. For the most part, the immediate aftermath of the Mongol conquest is wholesale destruction, especially of Azerbaijan's urban centers.

1256–1335 Ilkhan dynasty

In 1256 Hülegü, Genghis (Chingis) Khan's grandson, establishes himself as ruler of Iran and Azerbaijan, and makes Tabriz his capital. In the thirteenth and fourteenth centuries Azerbaijan becomes a battleground between the Ilkhans and the Khans of the Golden Horde, centered to the north along the lower Volga River. The last of the Ilkhanid rulers of Azerbaijan is Abu Sa'id, who dies in 1335.

1388–1503 Qaraqoyunlu and Aqqoyunlu

The Qaraqoyunlu (Black Sheep) and Aqqoyunlu (White Sheep) are local Turkic dynasties based in Azerbaijan and northern Iraq that develop a sharp rivalry and compete for influence between themselves and against the Ottoman Empire in Turkey. The Qaraqoyunlu are Shiites (Shia branch of Islam) based mainly in Azerbaijan, while the Aqqoyunlu are based further to the west in eastern Anatolia. A powerful local Azerbaijani dynasty at this time are the Shirvanshahs, centered in Ardebil and Baku.

1438 Birth of poet Muhammed Fizuli

Muhammed Fizuli (1438–1556) is born. He becomes an important cultural influence with his poetry, which draws on folk themes and traditional folktales.

1456 The "Blue Mosque" built in Tabriz

The Blue Mosque, located in Tabriz, is one of the outstanding monuments of Islamic architecture in Azerbaijan. The mosque is remarkable for its marble walls and engraved designs.

1470–1502 Flowering of Azerbaijani literature

The period of Qaraqoyunlu and Aqqoyunlu dominance corresponds to a period of substantial development in Azerbaijani literature, written primarily in Persian, but also in Turkic. Of particular prominence is the poet Imadeddin Nasimi; other notable poets include Tabrizi and Khatai Tabrizi, author of the poem *Yusuf and Zulayha*. The Qaraqoyunlu ruler Jahanshah and the Aqqoyunlu ruler Yaqub-padishah are themselves accomplished poets as well, producing both Persian and Turkic verse works. The fields of historiography and Islamic philosophy also develop at this time. One prominent philosopher is Muhammad ibn Musa who wrote the work *Hikmet al-Ayn*.

The Modern Era

1500–03 Safavid conquest of Azerbaijan

The Safavids are a Shiite order that is initially based in Ardebil, and who are the followers of a local thirteenth-century Sufi *shaykh* (Arab chief) named Safiaddin. In 1499 their 13-year-old Shah, Ismail, had begun to expand his power in the Ardebil region. In 1500 he moves against the Shirvanshahs of Northern Azerbaijan, and captures their city of Shemakha. Then Ismail expands his realms, conquering Nakhichevan, and defeating the now-weak Aqqoyunlu confederation.

c. 1505 Shah Ismail makes Shiism official religion of Iran

Under the Safavids Shiism becomes the official religion of Iran, including Azerbaijan, establishing the dominance of Shiism in Azerbaijani religious life.

1508 Shah Ismail conquers wide territory

Ismail has conquered all of Iran, Iraq, and Armenia, and much of eastern Anatolia and Central Asia. The capital of the Safavids becomes Tabriz, and Azerbaijan again serves as the political center of a dynasty ruling all of Iran.

1514: August 23 Safavid defeat at Battle of Chaldiran

In this battle Shah Ismail's army is decisively defeated by the Ottoman Turks, stopping Shiite and Safavid expansion to the west, and effectively establishing the western limits of Azerbaijan. Intermittent wars and raiding between the Ottomans and Safavids are to last another hundred years.

1589 Caucasus region occupied by Ottomans

Nearly all of the southern Caucasus region, including Azerbaijan, is under Ottoman occupation, and the Safavids themselves move their capital from the Azerbaijani city of Tabriz, to Isfahan, in central Iran.

1612 Peace treaty between Safavids and Ottomans

In the face of pressure from the Safavid ruler Shah Abbas, the Ottomans cede Azerbaijan back to Iran.

1722–24 Russian troops overrun Caspian coastal cities

A Russian fleet sent from Astrakhan on the orders of tsar Peter the Great occupies Derbent in spring of 1722. In June of the following year Russian troops occupy Baku. That summer Russian troops further occupy the mouth of the Kura River and Salyan.

1735: March 10 Treaty of Ganja

Under pressure from the new Iranian ruler Nadir Shah, Russia cedes back Baku and Derbent.

1747 Nadir Shah dies

After the death of the Persian ruler Nadir Shah in 1747 minor local rulers in Azerbaijan begin to establish their independent realms.

1748–1800 Era of minor Azerbaijani principalities

After the death of Nadir Shah, Hajji Chelebi establishes his own small state in Sheka and Shirvan, Panah-Ali Khan establishes the khanate of Karabakh, and Husayn Ali Shah establishes his own realm in Kuba. In the eighteenth century these local rulers try to play off the major powers—Russia, Turkey, and Iran—which are competing to dominate Azerbaijan. As the eighteenth century progresses however these small states fragment into ever smaller city-states.

1803–13 First Russo-Iranian War

The Russians defeat Iran in the first Russo-Iranian war.

1813: October 13 Treaty of Gulistan

Following the Iranian defeat, Iran is forced to cede all of Azerbaijan north of the Araks River to Russia. Northern Azerbaijan becomes part of the Russian empire.

1828: February 10 Treaty of Turkmenchay

Following the Iranian defeat in the Second Russo-Iranian War, Nakhichevan passes over to permanent Russian control.

1830–40 Growth of silk industry

Under Russian rule silk begins to develop into a major local industry, and is especially developed in and around the town of Shemakha.

1832 First Azerbaijani newspaper appears

In Tiflis (modern-day Tblisi) the first Azerbaijani newspaper, called *Tatar News* (until 1917 Azerbaijanis were known to Russians as "Transcaucasus Tatars") begins publication.

1840 Administrative division of Transcaucasia

In an effort to provide more efficient administration, the larger region of Transcaucasia is divided by the Russians into two large administrative units. In the west they create the Tiflis *guberniia* (province), centered in the city of Tiflis and in the east they create the Caucasus *oblast* (district) which is centered in the Azerbaijani city of Shemakha and comprises much of the territory of modern-day Azerbaijan.

1846: December 6 Law on the rights of *beks* and *agas*

The Azerbaijani land-owning nobility are known as *beks* and *agas*. The Russian authorities seek to administer Azerbaijan at the local level using the local nobility. As a result they pass a law confirming land-owning and other civil privileges of the Azerbaijani elite.

1861 Most serfs in Russia freed

As part of a program of modernization, Tsar Alexander II frees Russia's serfs, and those serfs working in Russia's colonial territories are freed. This action makes him known as the "Tsar Liberator".

1870: May 14 Abolition of serfdom

Russian authorities pass a law allowing all serfs to buy their lands from their lords outright; however, since few peasants possess the means to do so, they are allowed to make payments to the state. Although the serfs of Russia proper had been freed about a decade earlier (see 1861), in some of Russia's colonial territories the process has not been completed, and many serfs are still working under indentured arrangements.

1872 Beginning of oil industry in Baku

Petroleum had long been known to exist in Azerbaijan, but it only begins to be commercially developed after 1872. The center of the oil industry is the town of Baku, which as a result of Russian and Armenian immigration, soon becomes the industrial and commercial center of Azerbaijan. At this time oil is shipped from Baku to the industrial centers of Russia via the Caspian Sea and then up the Volga River. The development of large-scale industrial activity leads to the further development of shipping in the Caspian Sea, and Baku becomes the chief port. Another consequence of the commercial and industrial development of Azerbaijan is the creation of a large western-oriented Azerbaijani bourgeoisie and the creation of Azerbaijani national identity.

1883 Opening of the Batumi-Baku Railroad

This railroad connects Baku with the Black Sea port of Batumi, linking Azerbaijan with the export markets of Europe and the Mediterranean. In 1900 the South Caucasus railroad links Baku with the North Caucasus city of Vladikavkaz and the inner provinces of Russia proper.

1885 Birth of composer Uzeyir Hajibeyov

Uzeyir Hajibeyov (1885–1948) is born. He writes the first national Azerbaijani opera, and also composes the national anthem of Azerbaijan. He founds the Azerbaijan Symphonic Orchestra, and is considered the leading modern composer of Azerbaijan.

1899 Social-Democrats begin organizing among Baku workers

The Social Democrats and other socialist-oriented parties find fertile soil among the multi-national industrial workers in Baku. The Bolshevik faction of this party are already able to organize a general strike in Baku in 1904. One of the chief Social Democrat activists in Baku is Joseph Stalin.

1905–07 Revolutionary violence in the Transcaucasus

The outbreak of revolutionary violence across Russia in 1905 following its defeats in the Russo-Japanese War is strongly felt in Azerbaijan. Although much of the violence is primarily a result of conflicts between revolutionary groups and the Russian authorities, some of the violence takes on an ethnic character as well, especially between Armenians and Azerbaijanis, and there are many cases of massacres and other violence perpetrated by one group against another.

1909 Roman Catholic Cathedral built in Baku

A Roman Catholic Cathedral is constructed in Baku. It is seriously damaged during World War II (1939–45).

1911 Creation of Musavat Party

The Musavat Party is organized by the Pan-Turkic nationalist Mehmed Emin bey Resul Zadeh. Resul Zadeh had been involved in Bolshevik agitation with Stalin and later went to Iran to take part in revolutionary activities there. In 1910 he returned to Baku and founded the nationalist party Musavat (Equality), which was concerned with the political unification of all Turkic peoples. However, while the Musavat Party is mainly pro-Turkish, it meets opposition from the Shiite clergy, who are suspicious of Sunnite Turkey.

1914 World War I Begins

During World War I (1914–19) the Russian Empire, including Azerbaijan, finds itself at war with Germany, Austria and Turkey. On the whole, Azerbaijanis are loyal to Russia, and in the Russian army Azerbaijanis serve in their own separate units as privates and as generals, displaying the weakness of pan-Turkic ideas among the population.

1917: March 14 Tsar Nicholas II abdicates

With the abdication of the Russian tsar, central authority in Russia gradually weakens, encouraging a period of intense political debate among Azerbaijanis as to whether Azerbaijan should remain an autonomous part of the new federation, or whether the Transcaucasus region as a whole should become independent of Russia.

1917: November 7 Bolshevik seizure of power

With the overthrow of the Provisional Government by the Bolsheviks. The Bolshevik seizure of power reflects a larger current of radicalization in the Russian empire, which is also felt in Azerbaijan. Following the Bolshevik coup nationalist forces in Azerbaijan seek to detach the country from Russia, while Bolsheviks, based primarily in Baku, seek to keep the region within Russia.

1918: April 22 Transcaucasus Federation declares independence from Russia

Under pressure from Germany and Turkey, which are still waging war against Russia, the provisional Transcaucasus Federation, which initially refuses to recognize Bolshevik authority, declares its independence. The Federation consists of Armenia, Georgia, and Azerbaijan.

1918: May 28 Azerbaijani Republic declares independence

Two days after the defection of Georgia from the Transcaucasus Federation, Azerbaijani nationalists declare their country independent, with Resul Zadeh as president. This action is partly the result of Turkish pressure, and partly out of the need to acquire Turkish military support in the struggle against the Bolsheviks in Baku. At this time the provisional capital of the Azerbaijani republic moves from Tiflis to Ganja.

1918: June 27 to July 1 Turkish occupation of Azerbaijan

Under pressure from a Bolshevik military campaign against Ganja, the nationalist forces, in alliance with the Turkish Army, retake Baku and by the end of July they have destroyed the Bolshevik organization in Azerbaijan.

1918: November 17 British occupation of Azerbaijan

After the surrender of Turkey to the allied powers, Turkish troops leave Azerbaijan and are replaced by British troops from Iran.

1918: December 26 British troops leave Azerbaijan

Recognizing the government of the Democratic Republic of Azerbaijan, British troops leave its territory.

1918–20 Conflicts with Armenia

Ethnic conflict in Nakhichevan and Nagorno-Karabakh results in a prolonged state of armed conflict between the Armenian Dashnak government and Azerbaijan, precipitating ethnic violence and the migration of refugees.

1919 Baku State University founded

The Baku State University is founded. It offers degrees in mathematics, physics, chemistry, biology, geology, and geography.

1920 Museum of the History of Azerbaijan is founded

The Museum of the History of Azerbaijan of the Academy of Sciences of Azerbaijan is founded. It houses extensive collections of artifacts, especially weapons, vessels, and implements used in daily living.

1920 Azerbaijan Technical University founded in Baku

The Azerbaijan Technical University is founded in Baku to offer training in automation and computer technology, electrical engineering, auto mechanics, metallurgy, robotics, and transportation technology.

1920: March Denikin's White Army invades Azerbaijan

Denikin's defeated White Army invades northern Azerbaijan in the name of an "indivisible Russia," but also to escape the pursuing Red Army. The Azerbaijani government enters into an alliance with the Bolsheviks against the Whites.

1920: April 27 Bolsheviks seize power in Azerbaijan

Red Army forces converging from the north and Communist organizations in Baku overthrow the Democratic Republic of Azerbaijan and declare an Independent Soviet Azerbaijan.

1922: March 12 Soviets create the Transcaucasian Soviet Federative Socialist Republic

With the blessings of the Russian communist leadership in Moscow, the communist leadership in Baku relinquishes Azerbaijan's independence. Azerbaijan joins with Armenia and Georgia to form the Transcaucasian Soviet Federative Socialist Republic which becomes part of the Union of Soviet Socialist Republics (U.S.S.R). (See map. Dashed line indicates border of Transcaucasian Soviet Federative Socialist Republic.)

1923: July 7 Nagorno-Karabakh Autonomous District (AO) is formed

Soviet leaders make this overwhelmingly Armenian area into a nominally "autonomous" district within the republic of Azerbaijan. The area had been fought over between Azerbaijanis and Armenians before the Bolshevik seizure of the region.

1929–31 Collectivization of Azerbaijani peasantry

Under orders from Stalin, private land-holding is effectively banned throughout the Soviet Union. Azerbaijani landowners and prosperous peasants are massacred or shipped *en masse* to prison camps and poorer peasants are made to join collective farms (*kolkhozes*) or state farms (*sovkhozes*). These actions cause widespread social upheaval in the countryside. At this time all independent religious organizations, including mosques and schools, are permanently closed and much of the Islamic clergy is exterminated, effectively ending traditional religious life in Azerbaijan.

1936: December Creation of Azerbaijani Soviet Socialist Republic (SSR)

The modern borders of Azerbaijan are established with the creation of the Azerbaijani SSR, a union republic of the USSR. The Azerbaijani SSR also includes the Nakhichevan Autonomous Soviet Socialist Republic (ASSR), which was created in 1924 and is separated from the Azerbaijani "mainland" by a strip of Armenian territory.

1937 Purges of Azerbaijani Communist Party

Although tens of thousands of Azerbaijanis had perished or been sent to labor camps (the Gulag) by 1937, in this year Stalin begins arresting and executing members of the Azerbaijani Communist Party, essentially turning over the membership and creating a new, more malleable organization.

1939–45 Azerbaijan in World War II

Nazi Germany and the Soviet Union sign the Nazi-Soviet pact. Two years later, Nazi Germany attacks the Soviet Union. Azerbaijan becomes crucial to the Soviet war effort, meeting much of the Soviet Union's fuel needs and supplying hundreds of thousands of men to the Red Army.

1940 Birth of musician Vagif Mustafa Zadeh

Vagif Mustafa Zadeh (1940–79) is born. He creates a musical movement in the 1960s that combines elements of jazz composition with traditional Azerbaijani folk music known as *mugam*.

1943–46 Soviet troops occupy Iranian Azerbaijan

In 1943 the British and Soviets occupy Iran to forestall a pro-Axis coup in that country. Following the end of the war the Soviet forces do not evacuate their zone, and install the pro-Soviet "Autonomous Government of Azerbaijan" in Iranian Southern Azerbaijan. In April 1946 however, bowing to American pressure, the Soviet return Southern Azerbaijan to Iranian control.

1959 Khrushchev purges Azerbaijani Communist Party

Soviet leader Nikita Khrushchev, seeking to reduce opposition elements in the Azerbaijani Communist Party (ACP), expels much of the membership, replacing it with members more loyal to him.

1969 Heydar Aliyev heads Azerbaijani Communist Party

Heydar Aliyev (Geidar Aliev) is the dominant figure in Azerbaijani politics throughout the late Soviet period. A Brezhnev loyalist, he becomes a full member of the Politburo of the Communist Party of the Soviet Union (CPSU) in 1982.

1969 Museum of the History of Azerbaijan begins undersea archaeological exploration

The Museum of the History of Azerbaijan of the Academy of Sciences of Azerbaijan begins underwater archaeological exploration in the Caspian Sea. Extensive discoveries provide evidence of ancient people living in the area around the sea.

1985 Mikhail Gorbachev comes to power

Mikhail Gorbachev becomes Fist Secretary of the CPSU and subsequently introduces the policies of *glasnost* (openness) and *perestroika* (reorganization) which encourage more open debate and more criticism of Soviet politics and life. The policies soon result in a growing demoralization of Communist Party elites throughout the Soviet Union and the emboldening of emerging nationalist forces.

1988: February 20 Nagorno-Karabakh assembly (Soviet) votes to secede from Azerbaijan

The council of predominantly Armenian Nagorno-Karabakh votes to secede from the Azerbaijani SSR and join the Armenian SSR. Eight days later this action sparks anti-Armenian pogroms and riots in the Azerbaijani port city of Sumgait. In an attempt to calm the situation, the Soviet authorities in January of 1989 replace local authorities in Nagorno-Karabakh with a special commission under neither Azerbaijani or local Armenian control.

1989: July 16 Azerbaijani National Congress holds constituent meeting

The nationalist and anti-Communist Azerbaijani National Congress vows to challenge the Armenians and later in the summer organizes a blockade of Nagorno-Karabakh.

1989: September 23 The Supreme Soviet of Azerbaijan declares sovereignty

Under pressure from the Azerbaijani National Congress, the Supreme Soviet declares sovereignty in a symbolic move.

1990: January Widespread anti-Armenian violence in Azerbaijan

After widespread pogroms against Armenians in Baku, Moscow authorities declare a state of emergency in Azerbaijan. Following a massive government protest in Baku, Soviet army and naval units break the blockade of Baku's port and raid the offices of the popular front.

1991: April Fighting in Nagorno-Karabakh escalates

As Soviet and Azerbaijani troops join forces against Armenians in Nagorno-Karabakh, fighting escalates to the point of becoming an undeclared war.

1991: October 18 Azerbaijani Parliament declares full independence

Following the support of Communist Party leader Ayaz Mutalibov for the anti-Gorbachev coup in August of 1991, Communist authority dissolves and the National Front encourages a declaration of independence.

1992: February 29 Russian forces leave Nagorno Karabakh

Following a series of Armenian successes, Russian forces leave the prosecution of the war to the Azerbaijanis, who themselves are gradually pushed out of the region by Armenian troops during the spring of 1992.

1992: March 2 Azerbaijan admitted to the United Nations

Azerbaijan is admitted to the United Nations. Other nations admitted at the same time are Armenia, Bosnia and Herzegovina, Croatia, Georgia, Kazakstan, and the Kyrgyz Republic (later Kyrgyzstan).

1992: June 7 Ebulfaz Elchibey becomes president

Ebulfaz Elchibey, leader of the Popular Front, is elected president in national elections, forming Azerbaijan's first post-communist government.

1993: June Elchibey overthrown in coup

Unable to put down a mutiny by dissident troops, Elchibey is soon overthrown by members of the formerly defunct Supreme Soviet, led by Geidar Aliev, the former Communist chief of Azerbaijan.

1993: October 3 Aliev formalized as president

Under questionable polling conditions, Aliev wins elections against two other candidates with 98.8 percent of the vote (97.6 percent turnout reported).

1994 War between the Russian army and Chechnyan separatists

Separatist agitation by Chechnyan separatists in the Russian republic of Chechnya unleashes full-scale warfare against the Russian army. In a war that lasts over one year, 30,000 people lose their lives before a shaky cease-fire takes effect. The outbreak of fighting in Chechnya results in renewed apprehension among other leaders in the Caucasus who fear a new round of fighting in their own republics. Separatists in Abkhazia demand to break away from Georgia while the conflict over Nagorno-Karabakh is still fresh.

1996: June 3 Aliyev–Yeltsin meeting

Aliyev meets with Russian President Boris Yeltsin to discuss the recent conflicts in the Caucasus region. Also attending are the presidents of Armenia, Georgia, and high-ranking local officials from the region of the northern Caucasus within Russia. Yeltsin says Russia would continue to honor the cease-fire in Chechnya despite recent assaults on Russian military personnel there. He also pledges Russia's efforts to resolve—and not exploit—the ethnic conflicts in the Caucasus.

1997 Over 600,000 Azeri refugees

Over 600,000 Azeri refugees remain from the war in Nagorno-Karabakh.

Bibliography

Altstadt, Audrey. *The Azerbaijani Turks.* Stanford, Calif.: Hoover Institution Press, 1992.

Bennigsen, Alexandre & S. Enders Wimbush. *Muslims of the Soviet Empire: A Guide.* Bloomington & Indianapolis: Indiana University Press, 1986.

Hunter, Shireen. "Azerbaijan: Searching for New Neighbors." in *New States, New Politics: Building the Post-Soviet Nations*, edited by Ian Bremmer and Ray Taras, Cambridge & New York: Cambridge University Press, 1997, pp. 437–470.

Ismail, Makhmud. *Istoriia Azerbaidzhana.* Baku: Azerbaidzhanskoe gosudarstvennoe izdatel'stvo, 1995.

Nichol, James. "Azerbaijan." in *Armenia, Azerbaijan and Georgia.* Area Handbook Series, Washington, DC: Government Printing Office, 1995, pp. 81–148.

Bahrain

Introduction

Bahrain, surrounded by the rich waters of the Persian Gulf, built its early fortunes from the sea, its divers bringing to the surface translucent pearls that captivated people from faraway lands. And then, from the depths of Bahrain's harsh landscape, oil gushed to the surface, lifting the nation from poverty.

Travel, fishing, and trade on the sea became instrumental to Bahrain's existence and growth. With no nails available, early vessels were stitched together with huge needles and cord. Although appearing fragile, the vessels withstood a great deal of punishment from winds and rough seas.

Bahrain became a great trading empire more than 4,000 years ago, when it was known as Dilmun. The island became a must stopping point for traders moving cargo between India and Mesopotamia. For centuries, it was known as the "Island of a million palm trees," which once flourished thanks to the abundant water springs.

In ancient mythology, Bahrain was a sacred island, a place where wise men and heroes came to live in eternal bliss. The Sumerian *Epic of Gilgamesh,* the oldest poetic saga in the world, describes Dilmun as a paradise, an elusive place hidden from mere mortals. In the epic, King Gilgamesh spends most of his time searching for the island.

Yet this is a country firmly rooted in reality. As it moves into the twenty-first century, Bahrain faces a future without oil and has looked back to its trading past to keep the nation moving.

Located in the western Persian Gulf, eighteen miles north of Qatar, the State of Bahrain consists of thirty-three islands with a total area of 239 square miles (620 square kilometers), roughly 3.5 times larger than Washington, D.C. Only six of the islands are inhabited, with most of the population living in Bahrain, the main island.

Bahrain is connected by causeways and bridges to Muharraq and Sitra islands and to Saudi Arabia. The islands are essentially desert. There are natural springs and artesian wells along a narrow strip of land on the north coast. The landscape consists of low rolling hills with rocky cliffs. The highest point is Jabal ad-Dukhan, a rocky, steep hill that rises to 450 feet.

Modern Bahrain was shaped in the mid-1800s, when members of the Al-Khalifa family fought each other for control of the islands for nearly three decades. The violent struggle was settled with British intervention in 1869, when Isa Bin Ali Khalifa assumed the throne. Khalifa cleared up the line of succession, naming his son as the state's next emir. Members of the Al-Khalifa family continue to rule the nation.

Because of its important geographical position, European powers had long sought to establish themselves in the Gulf. The Ottomans tried many times to gain a foothold. Late in the twentieth century, Iran and Iraq had claims to small Gulf states.

By the eighteenth and nineteenth centuries, the British managed to drive away all European competitors who wanted a piece of the Persian Gulf. They struck favorable agreements that gave England the right to conduct foreign affairs and prevented the Arab states from striking agreements with other nations. The Arab states, often weakened by internal struggles, were in no position to reject British offers.

Under British control, Bahrain remained a small, impoverished, and isolated state. Other than pearls, it had little to offer the world. But oil changed the course of this small nation in the 1930s. With oil revenues, the country changed dramatically, becoming essentially a welfare state that generally improved the lives of its citizens.

Bahrain's population grew by nearly forty-five percent between 1981 and 1991, from 350,000 people to more than 508,000. Migration of foreign workers was responsible for the dramatic increase in population. By 1990, there were more than 171,000 temporary workers from Arab countries, India, Pakistan, Iran and the Republic of Korea, including many skilled workers from Europe.

About sixty-seven percent of the population are indigenous Bahrainis, who are mostly of Arab descent. Iranians make up about twenty percent of the population, followed by other Arabs, Pakistanis and Indians.

Bahrain, much like its Arab neighbors, is a conservative country, with little interaction between men and women in public places. Revealing clothes almost always are insulting and etiquette is very important to Arab people. It is considered impolite to refuse tea or coffee in any social or business gathering.

Most Bahrainis, men and women, still wear traditional clothes, although more are wearing Western-style clothes. The traditional dress for men is a floor-length shirt-dress, which in Bahrain is called a *throbe.* They are usually white, although the colors change in winter to black, brown or blue. The reason for this change in color has little to do with tradition, and more to do with the practical usage of lighter colors in summer, (to keep cooler) and darker colors in the winter (to keep warmer). Men also wear a loose headscarf called a *gutra*. It is usually white, with a black rope made of wool and called *agal*. Some women, especially in small villages, wear the traditional black *abba*, a cloak that covers their entire bodies. A thin black gauze over the face permits the woman to see.

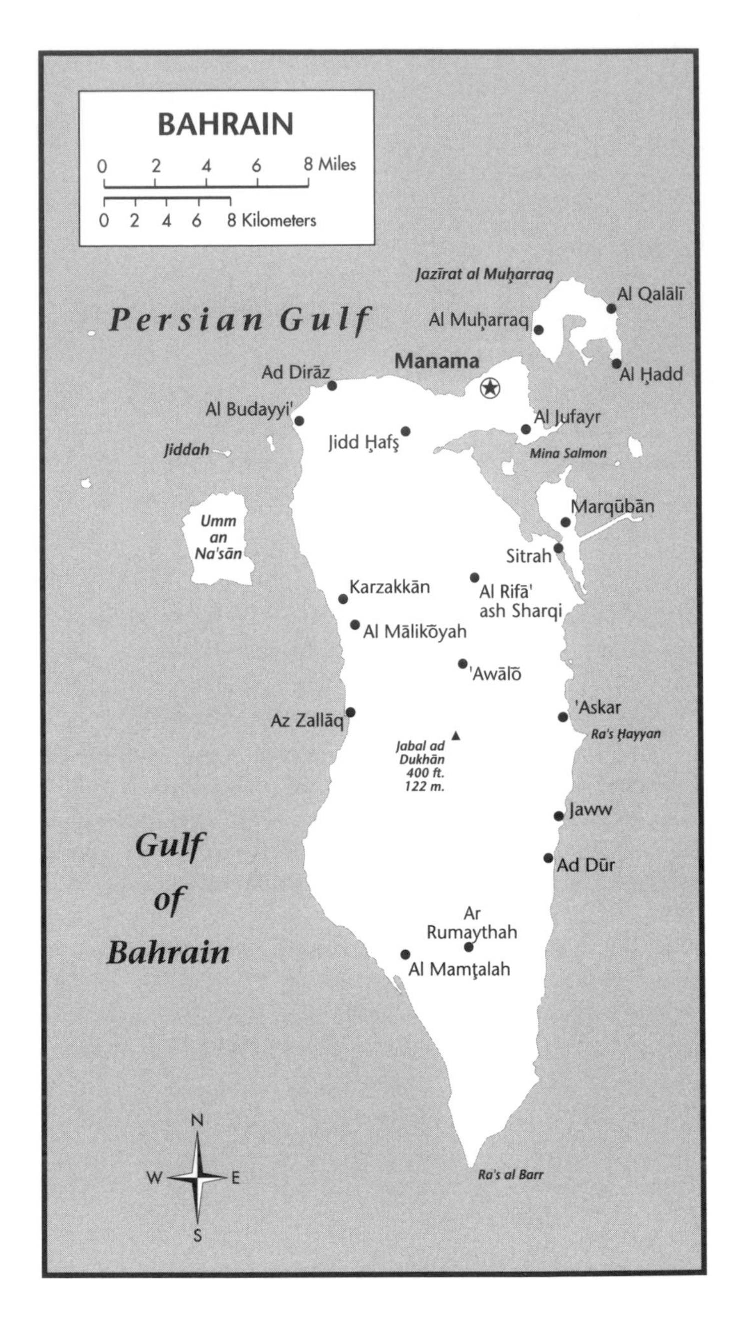

Timeline

6000 B.C. Bahrain once part of Arabic mainland

It is believed that the main island of Bahrain breaks away from the Arabian mainland around 6000 B.C. The islands play host to some of civilization's earliest inhabitants, but little is known about them.

2000 B.C. A trading center thrives in the Persian Gulf

Bahrain becomes a great trading empire around 2000 B.C., when it is known as Dilmun.

Dilmun appears on some of the earliest written documents in the world: tablets written in the Sumerian language. The Sumerians, one of the great ancient civilizations, settle in what is now southern Iraq and develop a writing system before 3000 B.C.

Dilmun thrives because of its underground water and strategic location along the trade routes between Mesopotamia and India. In Sumerian mythology, it is a sacred island.

The mythological Dilmun is described in the Sumerian *Epic of Gilgamesh,* the oldest poetic saga in the world. Dilmun is a paradise, home to wise men and heroes where they may enjoy eternal life. King Gilgamesh spends most of his time searching for the island. When he finds it, he seeks the help of Ziusudra, the lone survivor of a flood. Ziusudra (the Sumerian equivalent of the biblical Noah) helps Gilgamesh find the flower of eternal youth, but a serpent eats the flower and Gilgamesh returns to his kingdom, a little wiser but still mortal.

The real Dilmun is blessed with ample underground water and its position quickly establishes the islands as an important trade center.

600 B.C. Babylonians claim Dilmun

Dilmun goes into decline and is conquered by the Assyrians, an ancient people who settle in northern Mesopotamia as early as 3000 B.C. By 600 B.C., the Babylonians take control of Dilmun.

For the next several hundred years, the islands of present-day Bahrain fall to different empires that value their geographic position. Connected to the rest of the world only by sea, the Bahrainis build and sail their own boats.

Trade and the pearl industry increase the need for vessels, and Bahrain begins to import wood from the Indian subcontinent.

A.D. 640 Bahrain adopts Islam

The emergence of Islam in present-day Saudi Arabia under the Prophet Muhammed (570-632) has a profound influence on the countries of the Middle East. Islam, like Christianity and Judaism, is a monotheistic religion that acknowledges the absolute sovereignty of God. Conquering Arabs quickly spread in all directions, taking Islam with them.

Bahrain becomes one of the first territories outside the Arabian mainland to adopt Islam when the Prophet Muhammed sends a letter to the ruler of Bahrain inviting him to adopt the religion.

A.D. 900–1200 Muslim dynasties rule Bahrain

Two Muslim dynasties, the Umayyad caliphs of Damascus and the Abbasid caliphs of Baghdad, rule Bahrain and Eastern Arabia for more than 300 years.

Historically, the term caliph means "successor" to the prophet Muhammed and temporal and spiritual head of Islam. The caliph's main duty is to guard the faith, execute, and preserve justice and uphold the rights of his subjects. The caliph must also wage *jihad* (holy war) and defend the frontiers of Islam. The caliph can be deposed if these duties are not carried out.

1154 Historian describes Bahrainis happy with their leader

The Arab/Spanish geographer Al-Idrisi visits Bahrain and writes that its people are happy with their leader, apparently an independent king who owes nominal allegiance to the Caliph of Baghdad.

1485 Portuguese explorer visits Bahrain

Portuguese explorer Duarte Barbosa visits the islands and writes about the quantity and quality of pearls found there.

The pearl industry remains important to Bahrain and other Gulf States for many centuries. In Bahrain, it is the oldest industry, with references dating back to 200 B.C. An Assyrian inscription from that time describes a "parcel of fish eyes from Dilmun." In A.D. 100, a Roman visitor describes the area as famous for its pearls.

Pearling is a difficult and often brutal job, with boats constantly at sea from May to October. The work is divided between divers and pullers. The divers only carry a nose clip, leather guards to protect their hands, a knife to cut open the shells and a bag for the pearls. They usually tie rocks to their feet to sink to the bottom, where they stay for up to a minute. Painful stings from Jellyfish are a constant hazard. When they need air, they tug on the rope and they are pulled to the surface.

While pearling makes the boat owners rich, divers and pullers barely manage on subsistence wages. They often borrow money from the boat owners before the pearling season starts and can never pay it back. If a diver dies, his son must assume the debt and work until it is paid off.

1487 Omanis invade Bahrain

The growing Omani Empire conquers Bahrain and builds Arad Fort on the island of Muharraq.

1522–1602 Portuguese occupy Bahrain

Portugal becomes the first European power to take an interest in the Persian Gulf. In 1498, the famous explorer Vasco de Gama visits Oman's northern coast, the Strait of Hormuz and the sheikdom of Julfar, which now is the United Arab Emirates.

The Portuguese consider the Gulf important because they are trying to establish an empire in India. From the Gulf, they can control the sea routes linking Bombay and Goa in India to Lisbon. In the next twenty years, the Portuguese control much of the lower Gulf and later extend their power as far as Bahrain. In the meantime, the Ottoman Empire expands south, but is repelled in the Gulf by the Portuguese. In time, the Portuguese and the Ottomans lose influence in the Gulf to England, which strikes protection agreements with Arab states.

1602 Portuguese give up Bahrain

The Portuguese governor executes the brother of Rukn El-Din, one of Bahrain's wealthiest traders. Rukn El-Din leads an uprising against the Portuguese and drives them off the island. He quickly asks Persia, now Iran, for protection. Persia remains in control through the end of the seventeenth century.

1782 Bahrain's ruling family makes its way from Kuwait

The Al-Khalifa, Bahrain's ruling family, travel south from Kuwait sometime in the mid-eighteenth century. The Al-Khalifa family help their distant relatives, the Al-Sabah, establish rule in Kuwait before they move to present-day Qatar. There, they settle in Zubara, apparently to get involved in the pearling business. Zubara quickly grows into a large town.

In 1782 or 1783, Sheikh Ahmed Al-Fatih, a member of the Al-Khalifa family, drives the Persian garrison out of Bahrain. He appears to receive some help from the Al-Sabah family in Kuwait. The islands provide protection for the Al-Khalifa family, who are constantly defending themselves against Wahhabi raids in Zubara. The Wahhabis are ancestors of the present Saudi royal family and at one point manage to briefly hold on to Bahrain.

1796 Bahrain's ruler dies, Oman invades

Sheikh Ahmed Al-Khalifa dies, leaving his two sons, Abdullah and Salman, as joint rulers. Three years later, the Omanis again invade Bahrain. Abdullah escapes to the Arabian mainland and Salman retreats to Zubara.

1820 Abdullah and Salman regain the island

The brothers retake the island in 1820 and sign a protection treaty with the British, who over the past 200 years have slowly built up their presence in the region.

The British, like the Portuguese, also want control of the sea routes between the Gulf and India. During the seventeenth and eighteenth centuries, the British drive French and Dutch competition out of the region. The English sign treaties with the small Arab states, today known as the nations of Kuwait, Oman, Qatar, Bahrain, and the United Arab Emirates, as well as present-day Saudi Arabia.

Known as Exclusive Agreements, the documents give England control of foreign affairs, forcing small states to seek British permission to make treaties with any nations or to cede land. In exchange, the British promise to protect them from Turkey and Persia or any other foreign threat, including the Russians and Germans, who are interested in the area.

The British administer the Gulf from India and the Indian rupee becomes the common currency of the Arab states. With India's independence in 1948, the Indian rupee is replaced by the Gulf rupee, which remain in circulation until 1971 in some Arab states.

1840 Family struggles for power lead to turbulent era

In 1825, Salman dies, leaving his son Khalifa to rule jointly with his uncle Abdullah. When Khalifa dies in 1834, Abdullah rules Bahrain alone for several years.

Khalifa's son, Muhammed Bin Khalifa, begins to challenge Adbdullah's rule. Sometime in 1840, he sets himself up as co-ruler and rival of Abdullah. He rules from Muharraq, a neighboring island. In 1843, he deposes Abdullah, who dies in exile in 1848.

During his rule, Muhammed signs a Treaty of Perpetual Peace and Friendship with the British. Other agreements give England more control over Bahrain.

Khalifa's reign is anything but peaceful. Soon after taking power, his cousin Muhammed Bin Abdullah (the son of former ruler Abdullah) begins raids on Bahrain from the Arabian mainland. At the same time, Bahrain and neighboring Qatar go to war. The war ends in 1868, when Muhammed Bin Khalifa flees to Qatar. His brother Ali proclaims himself ruler of Bahrain.

But Muhammed Bin Khalifa is not done. In Qatar, he rebuilds his forces, attacks Bahrain and kills his brother Ali in 1869. Khalifa forgives his cousin, Muhammed Bin Abdullah, who returns to Bahrain and assumes a role in the government.

Abdullah, whose father was exiled by Khalifa, quickly takes revenge by deposing and imprisoning Khalifa.

1869 The British intervene in family politics

The British, who have remained out of the family struggle, send their Royal Navy to Bahrain and deport Khalifa and Abdullah to Bombay.

Isa Bin Ali Khalifa, the son of the murdered Ali, takes power under British supervision. The British force the twenty-one-year-old emir to clear up the line of succession and appoint a crown prince.

Early 1900s The pinnacle of the pearl industry

At the beginning of the twentieth century, more than 900 boats are engaged in the pearl industry. The first type of boat built specially for pearling is the sambouq, later replaced by the jalibut.

1919 Bahrain introduces free education system

The government becomes the first in the Gulf to provide free education for all children. In time, Bahrain reaches the highest literacy rate among Arabian Gulf states, including the highest rate among women.

1923 Isa's rule punctuated by little change

Isa becomes notorious for his conservatism and opposing even modest modernization reforms, yet under his rule, Bahrain enjoys a peaceful time. In 1923, the British force Isa, who is in his mid-70s, to hand over the day-to-day operations of the kingdom to his son, Hamad.

1932: June 1 Oil discovery launches Bahrain into new era

Hamad's father, Isa, dies in 1932, the same year oil is discovered. With oil revenues, Hamad, the country's new emir, begins to modernize Bahrain, providing electrical power to most of the country and building roads, mosques, schools and hospitals.

Long before the oil industrial age and the birth or the automobile and airplane, oil is mentioned in ancient texts and is found in natural pools or oozing from the ground in small quantities. Arabs, including those living in Bahrain, use the oil to light lamps and waterproof their boats.

The British become aware of oil in Bahrain early in the twentieth century but do little to exploit it, believing there is not much there. But oil discoveries in nearby Iran spark interest in Bahrain.

Explorers discover oil in commercial quantities in 1932 and exports begin in 1936, with the opening of a refinery. Bahrain becomes the first country in the Gulf to export oil, reaping the benefits well ahead of its neighbors. The discovery comes at a good time. Its pearl industry, which drives the economy, comes to a halt in the 1930s when the Japanese begin to cultivate pearls.

Bahrain's oil reserves are quite small when compared to some of its neighbors. The country does not enjoy the conspicuous consumption of Kuwait and follows a more fiscally conservative development plan.

1939 Bahrain sides with England during World War II

Bahrain has one of three refineries operating in the Middle East during World War II (1939–45). Bahrain is mostly quiet during the war, with a failed Italian bombing attempt of its refinery in October 1940.

1942 Hamad's son continues country's development

Hamad dies in 1942 and is succeeded by his son, Shaikh Salman bin Hamad Al-Khalifa, who continues to develop the country with oil revenues. Salman dies in 1961 and is succeeded by his son, Isa bin Salman Al-Khalifa.

1952 Bahrainis demand government reforms

The Middle East in the 1950s is punctuated by growing resentment against the British. Egyptian leader Gamal Abdel Nasser's speeches against British colonial privileges strike a chord with people of the Middle East.

In Bahrain, members from the Shiite and Sunni communities demand a more open political system, including the creation of a parliament, Western-style trade unions, and the removal of the British adviser to the emir.

1956 English dignitary receives cold welcome

Bahrain sees its first signs of unrest when stones are thrown at the British Foreign Secretary, Selwyn Lloyd, during his visit to Bahrain. In response, the emir's government deports several government opponents. In November, several Bahrainis lose their lives in anti-British riots. The British send troops to protect the oil fields.

Saudi Arabia, which refines some of its crude oil at a Bahrain facility, cuts its supply to punish the small state, which is perceived to be too close to the British.

1958 New labor laws are enacted

The government creates an eight-hour day, forty-eight-hour work week, sets minimum wage, overtime and severance pay, paid vacations and sick and maternity leave. It also sets up regulations to control the growing foreign workforce.

1968–71 Road to independence

In 1968, the British announce they plan to withdraw from the Gulf by 1971, leaving the small emirates to decide their fate.

Between 1968 and 1971, the nine emirates of the southern Gulf consider forming a federation, but talks break down when Bahrain demands greater representation because of its larger population and more advanced economy.

In 1971, Iran gives up efforts to claim Bahrain after the United Nations reports that Bahrainis don't want to become part of that nation.

1971: August 14 Bahrain declares independence

Sheikh Isa Bin Salman Al-Khalifa declares independence after Gulf states fail to agree on a confederation. That decision also prompts neighbor Qatar to declare independence.

1971: September 21 Bahrain admitted to the United Nations

Bahrain is admitted to the United Nations (UN). Other countries admitted the same year are Bhutan, Oman, Qatar, and United Arab Emirates.

1972 Assembly formed to draft a new constitution

A constituent assembly in charge of drafting a constitution is elected.

1973: May–June New constitution announced

The Emir announces the constitution and calls for elections for a new National Assembly (see 1973: December), which convenes in December. The constitution is ratified in June.

Under the constitution, Bahrain is a constitutional monarchy headed by the emir. A council of ministers works under a prime minister who is appointed by the emir. Members of the ruling family hold most of the important government posts, including the ministries of foreign affairs, defense, justice, and interior. The brother of the emir becomes the prime minister. The constitution also calls for a National Assembly composed of thirty elected members and fourteen cabinet ministers.

1973: December National Assembly convenes

After elections were called for in May, the newly elected National Assembly convenes.

1974 State Security Act passed

Trade unions and political parties are repressed by the government. A new State Secutiry Act is passed that allows detainees to be held for up to three years without a trial.

1975: August Emir dissolves assembly

The Emir dissolves the assembly amid charges of Communist influence. The Emir claims that radical assembly members make it impossible for the executive branch to function.

The Emir rules without challenge and bans all trade unions and strikes, even though the constitution allows unions. Political parties are illegal but several anti-government groups remain active, including pro-Iran supporters who are repressed by the government.

1977 U.S. vessels allowed in Bahrain

Bahrain and the United States sign a formal agreement allowing U.S. Navy ships to visit the islands. The agreement includes security cooperation.

1978 University of Bahrain in Manama Library is founded

The University of Bahrain in Manama establishes a library. It becomes known as one of the largest collections in Bahrain, with approximately 70,000 volumes by the late 1990s.

1980 Arabian Gulf University founded

The Arabian Gulf University is founded to provide opportunities for study in science, engineering, and medicine.

1981 Bahrain claims attempted coup against government

Following an alleged coup attempt in Bahrain in 1981, many people suspected of having links with Iran are arrested, convicted and sentenced to long-term imprisonment. Others are held without trials or exiled. Bahrain claims members of the Shiite community support Iran and wants to install a similar type of government in the islands.

1986 New road connects Bahrain to Saudi Arabia

Bahrain, with Saudi money, builds a four-lane highway on top of a fifteen-mile causeway, physically and symbolically connecting the island-nation to Saudi Arabia. The new road gives a boost to business and tourism.

1986 University of Bahrain is formed

Two institutions—University College and Gulf Polytchnic—merge to form the University of Bahrain.

1990 Bahrain opens a handicrafts center

A new handicrafts center shows off the country's traditional crafts and provides a place for artisans to sell their work. This center is important in bridging the old and new aspects of Bahrainian society by providing a modern structure for the selling of tradtional crafts.

Among the crafts present at the new center, cotton weaving, embroidery, Sitra crafts, and carpentry is represented. Cotton weaving, mostly done in small villages on the western coast, is seen throughout the center. Embroidery, a flourishing craft with women from several villages working intricate patterns onto traditional garments is also present. Sitra craftsmen, who make mats from grass that grows along saline coastal soils also make baskets out of dried palm leaves that can be seen. Traditional carpentry that is accounted for includes curved doors, lattice panels and furniture, all made by hand.

1991 Bahrain, U.S. become closer allies

The United States and Bahrain strengthen security agreements after Iraq invades Kuwait in 1990. The United States becomes one of the main suppliers of weapons to Bahrain, which allows American forces to store military material on the island for possible future conflict. Both countries sign a treaty that gives the United States access to facilities on the island. The country is home to the U.S. Navy's Fifth Fleet.

1992 Emir sets up Consultative Council

The emir appoints a thirty-member Consultative Council, but it has no legislative powers. All laws are dictated by the emir or introduced by him and approved by the cabinet.

1994–96 Civil unrest shakes the island

Bahrain, like its traditional neighbors in the Gulf, begins to experience severe civil disturbances from a Shiite-led resistance opposed to the ruling family. The same group claims it wants to create an Islamic democracy in the country.

Troubles begin when the emir refuses to accept a petition, reportedly signed by 25,000 Bahrainis, calling for democratic reforms. One of the main demands is restoration of the National Assembly and broader distribution of the country's wealth.

In 1996, the government arrests forty-four people, claiming they have been planning to overthrow the ruling family. The emirate breaks relations with Iran, accusing that country of fomenting civil disturbances, which claim twenty-five lives in two years.

Bahrain is an Islamic country and about seventy percent of it indigenous population is Shiite Muslim and thirty percent are Sunni Muslims.

The role of Ali, the cousin of the Prophet Muhammed, has divided the two sects since the seventh century. Sunni, who make up about eighty-five percent of all Muslims worldwide, regard Ali as the fourth successor of Muhammed, while the Shiites say the right of succession belongs to him. Iran is a Shiite nation. Shiites are a majority in Iraq, with a large Sunni minority.

1994 Special Security Court established

A special Security Court is established to handle anti-government cases. It does not allow for even basic human rights, however.

1994 Government encourages more fishing

There are more than 300 species of fish found in Bahraini waters, but fishing has declined because of heavy industrial pollution, mainly from oil production. The fish catch totals 7,629 tons in 1994.

The government encourages traditional angling and is attempting to increase the annual catch. It improves fishing and freezing equipment and establishes cooperatives.

Bahrainis continue to use traditional fishing methods, with a history that stretches back many centuries. Most fish are caught in 'hadra' wooden traps, which from the air look like giant arrowheads pointing to the sea. The long sides of the arrows are wooden fences, which guide the fish moving along the shore into a trap at the head of the arrow. Fishermen collect their catch at low tide. Each trap is known by a particular name and fishing sites are registered. Poaching is considered a serious offense.

Fishermen also use less elaborate traps that look like large lobster pots. They fill them with ground bait to entice the fish through a narrow opening. They also use a poison named *sim* to stun the fish. Shrimp fishing has been practiced for centuries.

1995 Gas reserves down to twenty-two years

Bahrain's gas reserves are estimated at 100,000 billion cubic meters, enough to last for twenty-two years. But oil is another story. Bahrain has one diminishing oil field expected to run out before the turn of the century. While its economy has been driven by oil for more than six decades, its development has been tempered by limited reserves.

The country appears to be going back to its roots as a trade center for the Gulf region. It provides plenty of storage space for goods in transit and drydock facilities for marine engine and ship repairs. By 1995, the service industry accounts for eighteen percent of all economic activity in the country. Despite a change in strategy, the country suffers through negative growth.

1996 Country changes laws to assure jobs for its citizens

The government announces that private firms must have at least one Bahraini worker. Firms with more than ten workers are required to increase their Bahraini workforce by 5 percent each year until reaching fifty percent. All new enterprises must have a workforce that is twenty percent Bahraini.

The government is concerned it has too many foreign workers, who make up sixty-seven percent of the labor workforce by the mid-1990s. Among Bahrainis, about half of the estimated 80,000 workers are employed by the government.

1997 Planned attack on U.S. forces revealed

The United States claims it has uncovered a plot to attack its military bases stationed in the country.

1997: March 27 Human rights organization calls for fair trials

Amnesty International, a human rights organization based in London, asks the Bahraini government to stop unfair State Security Court trials, which sentences fifteen of its citizens to long jail terms on this day.

In a press release, Amnesty International claims some of the fifty-nine defendants accused of committing crimes against the state and planning a coup were tortured to extract confessions. Their attorneys are not given enough time to prepare a defense and had no access to their clients until the start of the trials, the human rights organization says.

Amnesty International has repeatedly criticized Bahrain for the treatment of its citizens. The country, it says, has engaged in a consistent pattern of human rights violations since the early 1980s. One of their targets is the Shiite population, whose members are accused of supporting Iran.

Bibliography

Al-Khalifa, Hamad Bin Isa. *First Light: Modern Bahrain and its Heritage.* New York: Columbia University Press, 1994.

Al-Tajir, Mahdi Abdalla. *Bahrain, 1920–1945: Britain, The Shaikh and the Administration.* London and New York: Croom Helm, 1987.

Ali, Wijdan. *Modern Islamic Art: Development and Continuity.* Gainesville, FL: University Press of Florida, 1997.

Crawford, Harriet. *Dilmun and its Gulf Neighbors.* Cambridge: Cambridge University Press, 1998.

Fakhro, Munira A. (Munira Ahmed). *Women at Work in the Gulf: A Case Study of Bahrain Fakrhro.* London and New York: Kegan Paul International, 1990.

Nugent, Jeffrey B., and Theodore Thomas (eds.) Bahrain and the Gulf: Past Perspectives and Alternate Futures. New York: St. Martin's Press, 1985.

Robison, Gordon. *Arab Gulf States.* Hawthorn, Aus.: Lonely Planet Publications, 1996.

Wheatcroft, Andrew. *The Life and Times of Shaikh Salman Bin Hamad Al-Khalifa (Ruler of Bahrain, 1942–1961).* New York: Columbia University Press, 1995.

Bangladesh

Introduction

Bangladesh is a modern state, born in the throes of a bloody civil war in 1971. The events of that year, however, were a culmination of complex historical and cultural forces that may be traced back nearly three thousand years. Bangladesh literally means the "land of the Banglas (or Bengalis)," early settlers who gave their name to a geographical and, later, administrative region of the Indian subcontinent. Over the centuries, the peoples of this area developed a unique cultured and regional identity. Although Bangladesh exists today largely because of irreconcilable differences between religious communities in the subcontinent, the country embodies in a very real way the history and cultural values of the Bengali people.

The Republic of Bangladesh lies at the head of the Bay of Bengal in the eastern part of the Indian subcontinent. Virtually all of the country's territory (56, 977 square miles) is situated on the low-lying alluvial plains and delta of the Ganges and Brahmaputra Rivers. Crossed by some 700 rivers fed by monsoon rains and melted snow from the Himalayas, the well-watered, fertile region supports a population of 126 million people. Annual rainfall reaches 200 inches (5000 mm) in the northeast, most of this being associated with the summer monsoon (May to October). Temperatures reach between 90ºF and 100ºF (32ºC to 38ºC) in April, the hottest month of the year. Geography also creates serious problems for Bangladesh. The numerous rivers make surface transport difficult, the low-lying terrain is subject to frequent flooding, and tropical cyclones (hurricanes) from the Bay of Bengal cause extensive economic damage and loss of life.

The modern state of Bangladesh encompasses the eastern section of Bengal, a region that for much of the past has been, isolated from the main currents of Indian history. It was often "frontier" territory, located on—or even beyond—the margins of the subcontinent's great empires. States that grew up in Bengal frequently maintained an independent, or sometimes tributary (money paid for political independence) political status. The region had Hindu kingdoms and shared in the religious and social structures of the Hindu religion. It formed part of a Buddhist empire in the third century B.C. By the early thirteenth century, the rulers of Bengal were Muslim and remained so for over 500 years. It was during this period that many Bengalis converted to Islam, thereby providing the underlying rationale for Bangladesh's existence today. It was in Bengal that the British first established their political authority in India during the late seventeenth century, and from Bengal that they carved out their Indian empire. And it was, indirectly, the actions of the British in leaving their empire that contributed to the creation of the state of Bangladesh.

Muslims in the Indian subcontinent—concerned that they would be a minority in a Hindu-dominated state when the region became independent in 1947—demanded their own country. All efforts at a compromise failed and when the British left India, Muslim majorities in the northwest and in eastern Bengal were separated from India to form Pakistan. Although both areas shared a common religion, the population in the eastern region was culturally quite different from that in the west. East Pakistan resented its second-class status within Pakistan. Bengalis constituted a majority of Pakistan's population yet were hardly represented in the upper levels of political and military leadership. In another instance, West Pakistan tried, and failed, to impose its Urdu language (essentially a dialect of Hindi) upon the Bengalis. Pakistan's central government imposed martial law on East Pakistan several times. Matters came to a head in 1970–71 following elections for the Pakistani national assembly when the East Pakistani Awami League, campaigning on a platform of Bengali autonomy, won a majority of the seats. Pakistan's President, General Yahya Khan, refused to accept Bengali demands for autonomy, annulled the election results, and postponed elections indefinitely. Bengalis immediately embarked upon a campaign of civil disobedience which turned into civil war in March 1971 when West Pakistan authorized its military forces in East Pakistan to crack down on the demonstrators.

Independence

With the outbreak of violence in East Pakistan, Bengalis proclaimed independence on March 26, 1971, and called their new state "Bangladesh." Bengali resistance, bolstered by aid from Pakistan's arch-foe India, held out against a determined

West Pakistani effort to crush the rebellion. Thousands of Bengalis died and many more fled, particularly to India. By early December, India had massed troops along its border with East Pakistan. Yahya Khan, fearing an Indian intervention in East Pakistan on the side of the Bengalis, authorized preemptive air strikes against Indian airfields on December 3. The following day, India and Pakistan were at war for the third time in twenty-four years. It was no contest. India invaded East Pakistan and in twelve days Pakistani troops surrendered. Bangladeshi independence was recognized and Sheik Mujibir Rahman (Mujib) became the new state's first prime minister.

The years since independence have been turbulent for Bangladesh. The state is extremely poor and among the most densely populated in the world. The 1971 war destroyed much of the economic progress that had occurred since independence. Political instability also beset the country at regular intervals. Mujib became president in 1975 but fell to an assassin's bullet in a military coup that same year. The coup leader, Major-General Ziaur Rahman (Zia) became president in 1977 and ruled until he was assassinated in a 1981 coup. Another coup followed in 1982 and martial law remained in force until 1986. Since then Bangladesh has struggled with the democratic process but has not returned to martial law.

Bangladeshis are Bengali in culture. Although they may be Muslim by religion, and follow Muslim religious practices and social patterns, they share Bengali cultural traditions that extend back countless centuries. Like Bengalis, Bangladeshis speak Bengali (also, Bangla), an Indo-Aryan language related to the Hindi and the other languages spoken in northern India. Bengali is written in its own script which, again, is related to the Devanagari used for Sanskrit, Hindi, and other Indo-Aryan languages. The Bengali literary tradition include both Hindu and Muslim writings, and authors such as Rabindranath Tagore (1861–1941) are revered by Hindu and Muslim alike. Muslim contributions to Bengali literature include the Sufi (a mystical sect of Islam) devotional works of the thirteenth and fourteenth centuries and secular Muslim writings such as the mid-seventeenth century *Padmavati* of Alaol. Dhaka, the capital of Bangladesh, has emerged as a modern literary center. Writers such as Shamsur Rahman, Hasan Azizul Haq and Selina Hossain are among the forefront of modern Bangladeshi writers.

Dance in Bangladesh is essentially of Hindu (e.g. *kathakali, bharata-natya*) or folk (*dhali, manipuri*) origin, because Islam does not feature this art form. Music, however, embraces both Hindu and Muslim traditions. The devotional songs of Hinduism (*dhrupad, kirtan*) have their counterparts in the *qawwali* of the Muslims. It is the area of folk culture, however, that Bangladesh has evolved its own character. The boatmen's songs (*battiali*), *bhawaiya, and baul,* for example, are unique expressions of a particular way of life and folk culture that crosses any specific religious affiliations. Underneath the formal Muslim culture of Bangladeshis is a commonality of rural life that unites all people living in Bengal, regardless of their modern political labels.

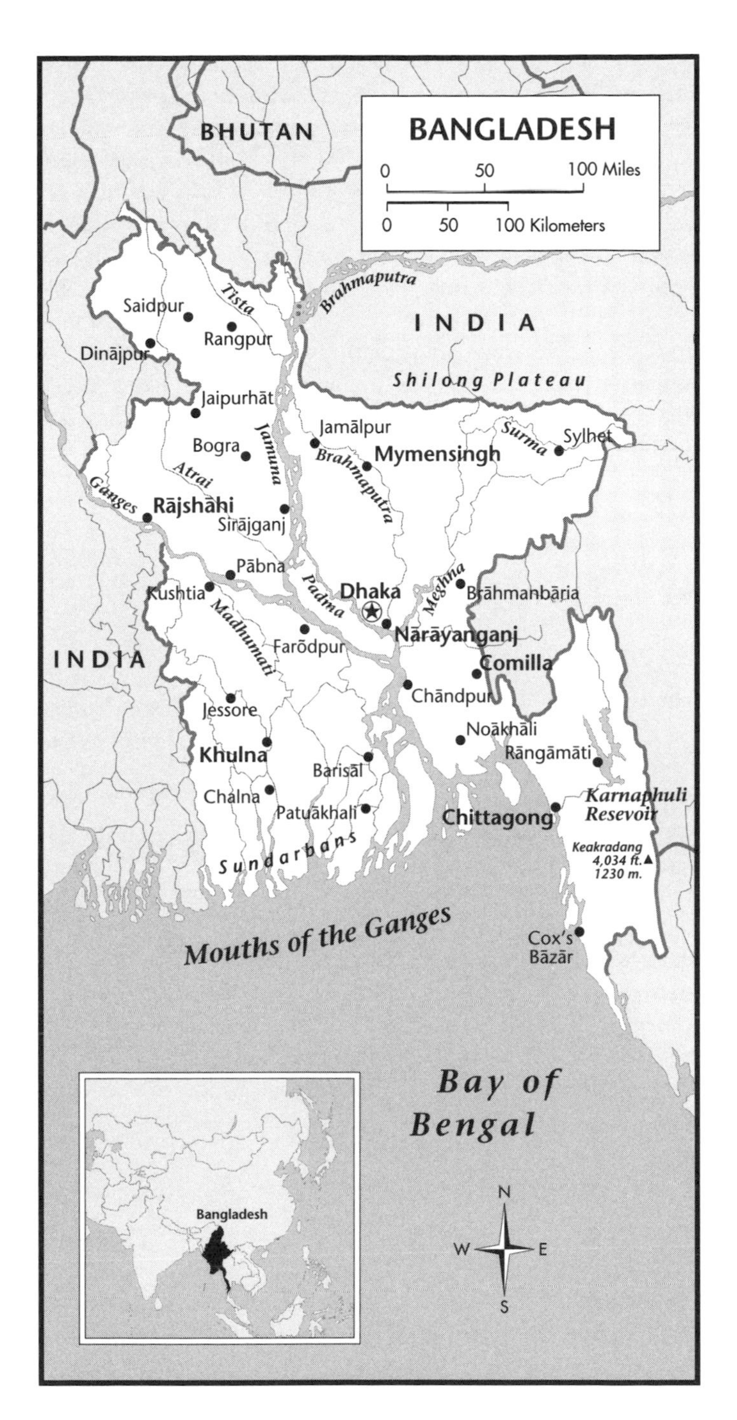

Bangladesh is a country of villages, of rural peasants engaged in subsistence agriculture. Rice is the major food crop, supplemented by oilseeds, pulses, grains, and vegetables. Bangladeshis eat fish and chicken; those who can afford it eat beef. The country is not self-sufficient agriculturally, and has to import food grains. Nutritional levels are low, and malnutrition and disease form a cycle that grips much of the population. Attempts to raise standards of living are hindered by high rates of population growth, low levels of education, a restricted mineral resource base, and widespread poverty.

Economic development in Bangladesh has been hindered by political instability and social unrest during the nearly three decades since the country's birth. A series of military coups, periods of martial law, and intermittent democratic governments have resulted in a lack of any continuity in economic policies. They have also created an environment that generates little confidence for foreign investment. The general level of development and political situation have not been helped by the burden of the frequent natural disasters affecting the region. Bangladesh, today, faces challenges as great as any in its short history.

Timeline

c. 1000 B.C. Early settlement in Bengal

The Bang tribe settles in the region of the Ganges delta.

Aryan-speaking invaders, who enter the Indian subcontinent from the northwest during the second millennium B.C., push southeastwards into the upper Ganges valley, forcing many groups to flee their homelands. The Bang, who speak a Dravidian (indigenous to the Indian subcontinent) language and represent a pre-Aryan element in the population of the Indian subcontinent, are among the many peoples in northern India who are displaced at this time. They migrate southeastwards down the Ganges valley and the lands where they eventually settle come to be known as "Banga." The many names by which this region is known throughout history—Banga, Bangala, Bangal, Bengal—all hark back to its early inhabitants.

c. 800–600 B.C. References in early Sanskrit literature

The Ganges delta is peripheral to the early centers of Aryan culture in India. However, the Vedic literature and the Indian epics, the *Mahabharata* and the *Ramayana,* identify regions such as Vanga (in Sanskrit, the classical language of India, the 'b' in Banga becomes a 'v') and Pundra in the area of modern Bengal.

321–185 B.C. The Mauryan empire

Vanga and Pundra form the eastern extremity of the first great empire in India's history, the Mauryan empire. The *Arthasastra,* a text attributed to Kautiliya, prime minister and chief counselor to the Mauryan emperor Chandragupta (r. 321–296 B.C.), relates that the region produces cotton, silk, and woolen fabrics.

By the time of Ashoka (reigned c. 265–238 B.C. or c. 273–232 B.C.), the Mauryan empire extends over virtually all of the Indian subcontinent. Ashoka adopts Buddhism as the state religion throughout his lands. Archeological evidence,

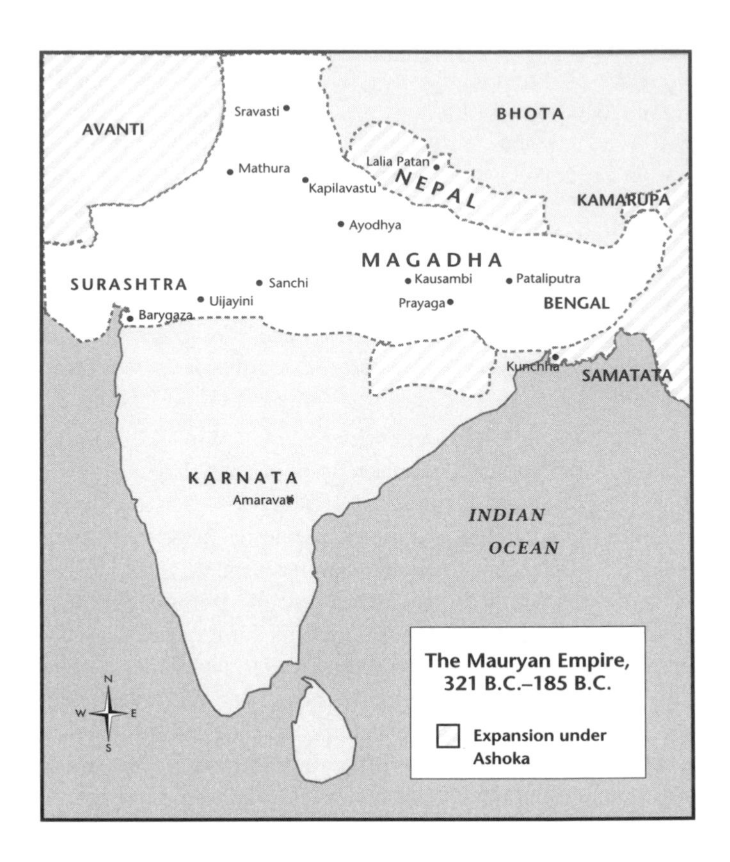

The Mauryan Empire, 321 B.C.–185 B.C.

including a Mauryan inscription, reveals that Mahasthangarh in what is now the Bogra District of Bangladesh is an important Buddhist center.

Buddhism originates in the Ganges valley some 500 miles to the northwest of Bengal during the fifth century B.C. The religion is based on the teachings of Siddhartha Gautama (c. 563–c. 483 B.C), a Hindu prince who comes to be called the Buddha (the Enlightened One or the Wise One). Although Buddhism retains many philosophical concepts of Hinduism, its teachings are, in effect, a reaction against Hindu social and religious practices. Buddhism, in its original form, has no gods, no priests, no social classes. Salvation is within the reach of anyone who accepts the Buddhist *Dharma* (Law) and the beliefs embodied in the Four Noble Truths. These four basic tenets are: (1) suffering exists, (2) suffering is caused by ignorance, (3) ignorance and desire must be eliminated to achieve Nirvana (the ultimate state), and (4) one can achieve Nirvana by following the eightfold path. By contrast, Hindus worship many gods, have a priestly caste (Brahmans), and are born into a rigid class structure.

185 B.C.–A.D. 320 Post-Mauryan era

In the years following the decline of the Mauryas, the kingdoms of the Bengal region are left very much to their own devices. Some, at times, fall under regional powers such as the Shungas (r. 185–28 B.C.), while others maintain a virtually independent existence. Mahasthangarh (or Pundravardhana) remains an important Buddhist religious site.

A.D. 320–510 Gupta dynasty

Vanga and Pundra form the easternmost territories of the Gupta empire. Further to the east, the kingdom of Samatata is independent, but pays tribute to the Guptas.

606–647 The reign of Harsha

Vanga falls within the territory of the short-lived Harsha empire.

c. 750–1150 The Pala dynasty

In the confusion that follows the disintegration of the Harsha empire, the people of Bengal (Vanga and neighboring areas) elect Gopala to be king. Gopala (ruled c. 750-770) establishes the Pala dynasty, and his successors rule over a stable and prosperous kingdom. The Pala state emerges as the dominant political force in eastern India, expanding to include most of the areas of modern Bengal and Bihar. In addition to controlling the rich agricultural lands of the lower Ganges valley, the Palas derive a substantial income from commercial dealings with southeast Asia. Staunchly Buddhist, the Pala kings built monasteries, universities, and monuments throughout their lands. The ancient Buddhist city of Mahasthangarh flourishes as a center of religion, learning, and the arts.

1150 Senas overthrow the Palas

The Senas, tributary chieftains of the Palas, invade the Pala kingdom from the south and seize the territories of their former overlords. As orthodox Vaishnava Hindus (i.e. they worship the god Vishnu), the Senas are ardently anti-Buddhist in their outlook. They reintroduce Hinduism, along with its religious and social practices, into the region.

Hindu society is divided into four main groups (*varna*) which are sometimes loosely called castes. These are hereditary in nature, and determine one's standing—and often occupation—in society. They are also ranked, with the Brahman castes occupying the highest position by virtue of their control of ritual and religious practices. Beneath the Brahmans, in status if not in terms of actual power and wealth, are *ksatriyas*, the warrior elites which provide the kings and rulers of the country. The *vaisyas*, the third category, are the mercantile class, while peasants and artisans fall into the last group, the *sudras*.

1202 Muslims seize the Sena capital

Nadia (in West Bengal), the capital of the Sena kingdom, falls to the Muslim armies of Bakhtiyar Khalji, general of the Ghurid dynasty that rules in Delhi. The Hindu Sena ruler, Lakshmanasena, retreats to eastern Bengal, where his successors retain power until c. 1245.

Muslim ideals of equality and social justice find popular appeal among Buddhists and lower caste Hindus who resent the Brahmanical structures imposed on them by the Senas. Mass conversions to Islam, particularly in eastern Bengal, begin in the thirteenth century and continue through several hundred years of Muslim rule. Sufi saints, who follow a mystical tradition of Islam, play an particularly important role in these conversions.

Muslims in Bengal, who are mostly Sunnis (the majority sect of Islam), develop specific social structures which survive in modern Bangladesh. The basic social unit is the family (*paribar* or *gushti*), consisting of an extended family living in a household (*chula*) residing in a homestead (*bari*). Beyond the circle of immediate relatives is 'the society' (*samaj*), a voluntary association concerning itself with village issues such as supporting the local mosque and mullah (priest), or settling village disputes. In matters of marriage, divorce, inheritance and related social practices, Bangladeshis follow normal Islamic patterns.

1280 Delhi reasserts authority over West Bengal

Following a period when West Bengal is virtually independent under Muslim governors of the Delhi sultanate, the Slave Dynasty reasserts Delhi's control over West Bengal.

1318 Delhi sultanate extends authority to East Bengal

East Bengal becomes a province of the Delhi sultanate.

1338 East Bengal breaks away from Delhi

East Bengal revolts against the Tughluqs, who now rule in Delhi.

1346–1490 Ilyas Shahi dynasty

West Bengal breaks away from the Delhi sultanate under Shamsuddin Ilyas Shah (r. 1346–58), who annexes the east in 1352. The Ilyas Shahi dynasty rules an independent Bengal for almost a century and a half.

During this period there are times of misrule and many sultans of Bengal die violent deaths. Some rulers, however, bring peace and prosperity. Sikander Shah (1358–89) is a patron of learning, law and justice, and under Jalal-al-Din Mohammed Shah (1415–31), Bengal experiences, peace, and prosperity. Trade increases and architecture flourishes. The Adina Mosque at Pandua and the Shait Gumbad Mosque at Bagerhat date from this period.

1490 Sayyid Dynasty

Alauddin Husain Shah overthrows the ruling sultan and founds the Sayyid dynasty of Bengal.

1517 Portuguese arrive at Chittagong

The first organized Portuguese expedition to Bengal arrives at Chittagong under the command of João de Silveira.

1536 The first European trading posts

The territorial governor, Mahmud Shah, gives the Portuguese permission to set up the first European trading posts (known in South Asia as "factories") in the region at Chittagong, now in Bangladesh, and Satgaon, on the Hooghly River in West Bengal.

Over the next one hundred and fifty years, the Dutch, French, British, and even the Danes, establish factories to exploit the resources of Bengal. The main items of trade include saltpeter (used in making gunpowder), silk, cotton, sugar, muslins, opium and various spices.

1539–76 Afghan kings rule Bengal

The Afghan Sher Shah Suri seizes Bengal from the Mughals, following its capture by Humayan's troops in 1538. The Suris and their successors, the Afghan Karranis, remain independent of the ruling power in Delhi.

The Mughal dynasty, founded in 1526, rules extensive areas of India from its capital at Delhi (and also Agra) between the early sixteenth and mid-eighteenth centuries. Akbar (r. 1556–1605) and his successors create a great empire.

1576 Akbar conquer Bengal

Daud Shah of Bengal invades Mughal territory, but is defeated and killed by Akbar's generals. The Mughals annex Bengal, installing their own people as governors rather than entrusting this task to local chieftains. They set about integrating the new province (*subah*) into the Mughal empire, building a road to Delhi, establishing a postal service, and introducing other administrative changes. Bengal becomes one of the most important economic provinces of the empire providing the Mughals with items ranging from war elephants to cloth.

1608 Dhaka becomes provincial capital

Dhaka becomes the capital of the Mughal province of Bengal.

1632 Shahan Jehan destroys Hooghly

The imperial army of emperor Shah Jehan destroys the Portuguese factory at Hooghly, causing the Portuguese to leave—and opening the way for the British in Bengal.

1650 British arrive in Bengal

The British East India Company, founded by royal charter in 1600 and given the monopoly of Britain's trade with Asia, receives Mughal permission to establish a trading post at Hooghly.

Although it has factories on India's west and southeast coast (e.g. at Surat and Madras), Hooghly is the Company's first settlement in the Ganges delta.

1686 British at Calcutta

Armed conflict between the British and the Nawab of Dhaka, the Mughal governor, leads to the Hooghly factory being evacuated. Job Charnock, the British leader, sets up his headquarters at a village named Kalikata, on a swampy, malarial site on the east bank of the Hooghly River.

Although the British withdraw from Bengal for a short period, the Mughals permit them to return. In 1690, Charnock comes back to Kalikata where he builds Fort William. From such humble beginnings rises the city of Calcutta, which grows to be Bengal's dominant city and second in importance only to London in the British Empire.

1756: June 20 The Black Hole of Calcutta

Britons die in the Black Hole of Calcutta, in one of the most infamous, but perhaps exaggerated, incidents in India's colonial history.

The Nawab of Bengal, Siraj-ud-Daulah, angered at the military fortification of Calcutta, attacks and captures the British settlement. He places all of his prisoners, thought to number 146 people (the exact number is a matter of controversy) in the local jail, a room roughly eighteen by fourteen feet in dimensions. With no water and a lack of ventilation, only 23 people survive until the morning.

The incident results in a British expedition commanded by Robert Clive (1725–74) to retake Calcutta, and indirectly triggers events that lead to British rule in Bengal.

1757: June 23 Battle of Plassey

A British force, under Clive, defeats Siraj-ud-Daulah and his French allies at a battle near Plassey (Palashi), a village roughly 100 miles north of Calcutta and just west of the modern India-Bangladesh border. The British capture and execute Siraj-ud-Daulah, and appoint their own candidate, Mir Jafar, governor of Bengal.

1765 The Bengal *Diwani*

The Mughal emperor, defeated by Company forces at the Battle of Buxar (1764), grants the British the Bengal *diwani* (right to collect revenues). The British thus become the effective—if not titular—ruler in Bengal, Bihar, and Orissa.

1793 Lord Cornwallis and the Permanent Settlement

Lord Charles Cornwallis introduces the Permanent Settlement of the Bengal Revenue Act. (In the late eighteenth century, Bengal refers not only to the geographic region but also the Bengal Presidency, an administrative unit of the British that extends from the Bay of Bengal to beyond Delhi). Cornwallis, the British commander who capitulated at Yorktown during the American Revolution, is Governor-General of India between 1786–93.

Although aimed at reforming the tax system, the Bengal Revenue Act has one unanticipated effect. Under the

Mughals, peasants paid taxes to hereditary *zemindars* or tax collectors who kept some of the revenues and remitted the rest to the government. The new system recognizes the zemindars as owners of the land from which they collects taxes and the peasants become their tenants. In eastern Bengal, the zemindars are for the most part Hindus, who emerge as the elite at the expense of Muslims. Many also become absentee landlords, moving to Calcutta and siphoning off the resources of their lands. This system of land tenure, the *zemindari* system, is not abolished in eastern Bengal until after independence in 1947.

1857: May 10 The Indian Mutiny breaks out

Indian sepoys (troops) in the East India Company's Bengal army turn on their British officers at Meerut, a town near Delhi, in a mutiny (for Indian nationalists, a war of independence) that spreads across northern India and shakes British rule to its very core.

The mutiny is put down by 1858, but has consequences for Bengal and later, Bangladesh. The British believe Bengalis are a major element among the mutineers, and cease recruiting them into the military. As a consequence, Bengalis are poorly represented in Pakistan's armed forces when the country becomes independent nearly a century later.

1858: September 1 Company rule ends

As a direct consequence of the mutiny, the British Crown assumes control of all the East India Company's possessions in India. Henceforth, the Governor-General is replaced by a Viceroy as head of the British government in India.

1885: December 28 Indian National Congress founded

The first president of the Indian National Congress is a Bengali Hindu. Few Muslims join this national movement, partly because of the "Two Nation Theory" of Sir Syed Ahmad Khan (1817–98). Khan, a leading Muslim educationalist, argues that differences between Hindus and Muslims are so significant they should be considered separate nations even though they inhabitant the same territory. This theory underlies later developments, such as separate electorates for Hindus and Muslims, and the eventual creation of a Muslim state on the Indian subcontinent.

1905 Partition of Bengal

Lord Curzon, Viceroy of India (1899–1905), partitions Bengal for administrative reasons. Eastern Bengal (roughly the area of Bangladesh today) is joined with Assam, while western Bengal, Bihar, and Orissa form the new Bengal Province.

Hindus oppose the change because they are now a minority in Bengal Province, outnumbered by the people in Bihar and Orissa. Muslim Bengalis, however, support the new administrative system, which creates a Muslim majority in the eastern province. Dhaka becomes the provincial capital of Eastern Bengal and Assam.

1906: December Muslim League founded

Leading Muslims meet in Dhaka, where they found the All-India Muslim League. The goal of the organization is to promote the interests of Muslims in India.

1911 Partition annulled

The British government reunites the eastern and western areas of Bengal to create a new province. However, it also announces the moving of India's capital from Calcutta to New Delhi.

1913 Tagore wins Nobel Prize

Rabrindranath Tagore, Bengal's greatest literary figure, wins the Nobel Prize for Literature for *Gitanjali,* a collection of poems.

Tagore (1861–1941), who writes in Bengali, is a gifted poet, novelist, story writer, essayist and composer. His works range from poems and songs celebrating rural life to social and political writing. He is influential in bringing Indian culture to the West, where he has a considerable following in the early twentieth century.

Tagore's work "My Golden Bengal" is the National Anthem of Bangladesh.

1940: March 23 Muslim League adopts Lahore Resolution

At its annual session in Lahore, the Muslim League resolves that no independence plan that does not include independent autonomous states for Muslim majority areas in northwest and northeast India is acceptable to Muslims.

1943 Bengal Famine

Up to three million people die in Bengal in the worst famine in modern times.

Famine is common in India, usually reflecting the failure of the monsoon rains. The 1943 famine, however, is a man-made rather than climatic disaster. Concerned with the rapid military advance of the Japanese towards India, the British government seeks to deny them transportation by banning the use of "country" (native) boats in Bengal. A poor monsoon, food shortages, and the disruption of the food distribution systems lead to extensive loss of life in the region.

1946: August 16 Calcutta riots

Bloody riots break out between Muslims and Hindus in Calcutta.

Muslims under Husan Shahid Suhrawardy (1893–1963), leader of the Muslim League, seek a united Bengal, separate from India and ruled by a Hindu-Muslim coalition. This position is unacceptable to the Indian National Congress, the Hindu-dominated nationalist movement in India, which does not want to give up control of Calcutta with its Hindu majority.

The riots are a precursor of communal relations in the Indian subcontinent during the months leading up to independence. Massacres of Hindus and Muslims in cities across Bengal lead to some two and a half millions Hindus fleeing East Bengal. An estimated three-quarters of a million Muslims depart Hindu-majority areas for the refuge of East Bengal.

Pakistan and Bangladesh

1947: August 14 Pakistan gains its independence

At midnight, Pakistan becomes an independent nation.

Beyond the logistical problems raised by two territorial units, West Pakistan and East Bengal, separated by 1,000 miles (1,600 km) of hostile Indian territory, the new country faces numerous problems. In one of the greatest migrations in history, an estimated 15 million cross the newly-created international frontiers between India and Pakistan. Refugees create a significant economic burden on Pakistan and India. The migrations have a significant impact in East Bengal, where the majority of government and administrative officials are Hindus who leave for India. East Bengal suffers from a serious shortage of administrators to run the country.

1947 Economic consequences of partition

Beyond the immediate economic upheaval and loss of life associated with the partition of India, the separation of East Bengal has long-term economic implications. The boundaries determined by the Radcliffe Commission (the British commission that draws the boundaries of the new states) not only disrupt the region's transportation system, they destroy long-standing economic relationships.

Historically, East Bengal forms the economic hinterland of Calcutta, but the new international borders place the city in India. Though India and Pakistan have an agreement to allow free trade and the free movement of goods, people, and capital, this arrangement soon breaks down. This hits East Bengal, one of the world's major jute-producing regions, particularly hard. Jute is the fiber of either of two native East Indian plants (*Corchorus olitorius* and *C. capsularis*) belonging to the linden family that is used to produce rope, twine, sacking, and burlap. Jute is the main cash crop of the East Bengal farmer and the mainstay of the region's economy, but the main processing plants are located in Calcutta. In 1949, the Bank of Pakistan has to purchase the entire jute crop for the year to prevent economic disaster in East Bengal.

1949: June Founding of Awami League

Husan Shahid Suhrawardy founds the Awami ("People's") Muslim League as an alternative party to the Muslim League. He soon drops the word "Muslim" from the name, stating that party membership is open to non-Muslims as well. The new party gains little strength in West Pakistan, but emerges as the strongest party in East Bengal.

1952: February 21 Martyrs' Day

Students and others demonstrate in Dhaka demanding equal status for the Bengali language. Police open fire on the marchers, killing several students. The date is remembered annually in Bangladesh as Martyrs' Day and a monument is later erected on the spot where the students are killed.

Language illustrates that the government of Pakistan in Karachi is separated from East Bengal by more than just distance. Pakistan adopts Urdu as a national language. Urdu evolves in the sixteenth and seventeenth centuries from the mix of languages spoken by Muslim soldiers in India and the local speech. Its grammatical construction and basic vocabulary are similar to Hindi, but uses the Perso-Arabic script. Urdu is the language spoken by Muslims in the north Indian plains, rather than in Bengal. The attempt to impose Urdu on East Bengal generates considerable resentment among the populace, and Pakistan eventually allows Bengali equal status as an official language.

1952 East Bengali grievances

In addition to the language problems, several contentious issues arise between East Bengal and West Pakistan. The central government of Pakistan has its capital at Karachi and is dominated by West Pakistanis, although East Bengalis make up sixty percent of Pakistan's population. West Pakistanis support a strong central government, whereas East Bengalis favor regional autonomy. The Pakistani military is made up mostly of West Pakistanis, who resist the inclusion of Bengalis in the armed services. West Pakistanis—regarded by East Bengalis as outside masters—are brought as to fill the gaps left by Hindu administrators and civil servants who have fled to India. Few Bengalis rise to position of importance in the government, even in East Bengal. East Bengalis also resent the draining of economic resources from the east to the west wings of Pakistan.

1955 East Pakistan

East Bengal is renamed East Pakistan.

1958: October 7 Martial Law

Pakistan's president Iskander Mirza declares martial law in Pakistan.

1958: October 28 Ayub assumes power

General Mohammed Ayub Khan dismisses Mirza and assumes the presidency of Pakistan.

1963 Mujib becomes leader of Awami League

Sheikh Mujibur Rahman (1921–75), widely known as Mujib, becomes leader of the Awami League.

Mujib begins his political career as a student in Calcutta in 1940 with the Muslim Students' Federation, an arm of the Muslim League. He is a founding member of the Awami League, and later its principle organizer in East Bengal, and later East Pakistan. Known for his organizational abilities rather than his administrative skill, he emerges as an important figure in the fight for Bangaldeshi independence from Pakistan.

1963 Louis Kahn designs new Parliament building

Pakistan commissions Louis Kahn, the renowned American architect, to design a regional capital for East Pakistan. Kahn's National Assembly building in Dhaka opens in 1982.

1965 War with India

Border clashes between Pakistani and Indian troops in Kashmir and the Rann of Kutch escalate to a full scale war. Although the main engagements take place in the northwest, Pakistani planes based in East Pakistan bomb targets in West Bengal and Assam. Ground skirmishes occur along the East Pakistan-India border. A cease-fire is reached on September 23, and hostilities formally ended by an agreement signed at Tashkent in the Soviet Union in January 1966.

1970: November Killer cyclone

A cyclone sweeps out of the Bay of Bengal and hits the low-lying delta region. Winds exceed 100 mph (160 kph) and a storm surge (a wall of water pushed ahead of the storm) of eighteen feet hits the coast, inundating coastal lands that are little more than four to ten feet above sea level. Lack of an early warning system and difficulties in transportation make evacuation impossible. Although the exact numbers are not known, estimates put the loss of human life between 300,000 and 600,000.

1970: December 7 Elections in Pakistan

In elections, Mujib and the Awami League win a majority of the seats in the national assembly. However, all these are in East Pakistan and they win none of the seats contested in West Pakistan. Mujib claims that his overall majority gives him the right to frame a constitution incorporating the Awami League's Six-Point Plan which is based on the principle of provincial autonomy. This is unacceptable to General Yahya Khan, Pakistan's president, and Zulfiqar Ali Bhutto, whose Pakistan People's Party gains a majority in West Pakistan.

Near Dacca, women participate in mass military training spurred by the tensions between Pakistan and India. (EPD Photos/CSU Archives)

1971: March 1 Yahya Khan postpones convening of Parliament

Following the breakdown of talks between Yahya Khan, Bhutto and Mujib to resolve the constitutional crisis, Yahya Khan dissolves his civilian cabinet and postpones indefinitely the meeting of the newly elected legislature.

The population in East Pakistan reacts with outrage, engaging in strikes, demonstrations, and civil disobedience, and virtually bringing government and business to a standstill. Militant Bengali nationalists organized as the Mukti Bahini ("Freedom Fighters") begin terrorist attacks on central government offices. General law and order begins to break down throughout the province.

1971: March 25-26 West Pakistanis crack down

The Pakistani army, led by Lieutenant-General Tikka Khan and composed primarily of troops from West Pakistan, launches an offensive against the Bengali nationalists. Although Mujib is arrested and jailed in West Pakistan, most of the Awami League leadership escapes.

1971: March 26 Independence proclaimed

Bengalis proclaim the "independent, sovereign republic of Bangladesh." This day is celebrated in Bangladesh as Independence Day.

1971: April-November **Civil War in East Pakistan**

Pakistani regulars extend their operations throughout the country, attempting to stamp out the nationalist insurgency. The Mukti Bahini finds an ample supply of recruits among the local population and, reinforced by army units such as the East Bengal Rifles which join the nationalist cause, survives the Pakistani offensive. India supports the guerrillas, offering safe bases along the border, training and equipment. (India sees it in her own interest to remove the danger of a hostile Pakistan threatening her on two flanks.)

The civilian population of East Bengal suffers grievously during the conflict. Anyone thought to be remotely associated with the nationalists is ruthlessly hunted down by the West Pakistanis. Bengalis claim that the army indiscriminately kills teachers, students, the intelligentsia, or any others who might be seen as providing leadership cadres for the nationalists. Some three million civilians are killed and another ten million refugees flee to India.

1971: December 3 **Pakistani air strikes on India**

In response to India's military buildup and incursions along the border of East Pakistan, the Pakistani Air Force launches preemptive air strikes against Indian airfields.

1971: December 4 **Pakistan and India declare war**

Fighting a holding action in the northwest, India invades East Pakistan. The Indian Air Force destroys the Pakistanis' air capability in the east. Indian troops, accompanied by Mukti Bahini forces, launch a five-pronged ground attack aimed at Dhaka. The Pakistani army fights hard, but is in a militarily untenable situation.

1971: December 16 **Pakistani forces capitulate**

Lieutenant-General A. A. K. Niazi and some 75,000 Pakistani troops surrender to the Indian commander, Lieutenant-General J. S. Aurora.

1971: December 20 **Yahya Khan resigns as president of Pakistan**

In the face of demonstrations in West Pakistan against the military government, General Yahya Khan resigns. He is replaced by Bhutto, who promises a return to constitutional government.

1972: January 11 **Mujib returns to Bangladesh**

Mujib returns to Dhaka to take up the reins of government as Bangladesh's prime minister. Under its new constitution, Bangladesh becomes a parliamentary democracy, with a single-house legislature elected by universal suffrage. The new nation, though populated mostly by Muslims, is officially a secular state and committed to the principles of nationalism, socialism, and democracy.

1972 **Indo-Bangladesh Joint Rivers Commission set up**

Bangladesh and India set up a joint commission to consider problems arising from the use of waters from rivers such as the Ganges and Brahmaputra which flow through both countries.

1973: March 7 **Bangladesh's first elections**

The Awami League wins 292 out of 300 seats contested for the new national assembly.

The new government pursues its socialist policies by nationalizing banks, insurance companies, and other industries. However, despite huge amounts of foreign aid (US $2.5 billion between 1971–74), Bangladesh drifts towards economic and political chaos.

1974 **India completes construction of the Farakka Dam**

India builds a dam to divert the waters of the Ganges River into the Hooghly River to reduce silting (the buildup of earthly sediment) at the port of Calcutta. The site of the dam is Farakka, some eleven miles upstream from the point where the Ganges enters Bangladesh.

India's use of Ganges water and the broader issue of water rights is a major political irritant between India and Bangladesh. Diversion of water from the Ganges before it flows into Bangladesh has serious consequences downstream. It reduces water available to Bangladesh for irrigation and other uses, prevents renewal of agricultural soils by the annual deposition of silt, increases silting in the country's waterways, allows salt water incursions in the delta, and upsets the delicate ecology of the region.

1975: January 25 **Mujib becomes president**

In the face of deteriorating conditions, Mujib amends the constitution to create a presidential system in Bangladesh. Mujib becomes president with power concentration in his own hands. He makes Bangladesh a one-party state, eliminating all parties except the Awami League (now renamed the Bangladesh Krishak Srimak Awami League or BAKSAL for short).

1975: August 15 **Mujib assassinated**

Mujib and several members of his family are murdered in a coup staged by relatively junior officers in the army. An abortive counter-coup by senior army officers favoring Mujib's policies is suppressed by regular units of the army supporting Major-General Ziaur Rahman (Zia).

1976: November 30 **Zia becomes Chief Martial Law Administrator**

Zia assumes the post of Chief Martial Law Administrator in Bangladesh under President A. S. M. Sayem.

1976 Grameen Bank of Bangladesh

Mohammed Yunus, a Bangladeshi professor of economics, sets up the Grameen ("Rural") Bank. The bank lends small amounts of money to the poor. The initial loan is usually no more than US $20 per household, and is made without collateral. The heart of the bank's operations is a "lending circle," a group of five women (usually) who collectively manage and guarantee their loans. Once the loan is repaid, a borrower becomes eligible for larger loans. The bank has some two million borrowers, of whom ninety percent are illiterate, and ninety-four percent are women. Grameen Bank pioneers "micro-lending" and becomes a model for banks in other developing nations.

1977: April Zia becomes president

Zia replaces Sayem as president, and subsequently his political party (which later comes to be called the Bangladesh Nationalist Party) obtains impressive majorities in elections held in 1978 and 1979.

Zia develops into a charismatic leader, expounding the policy of self-reliance. His programs emphasize family planning and increased food production as a basis for economic development, a withdrawal from the socialist policies of Mujib, and regional cooperation among the countries of South Asia. He also abandons secularism, making Islam a basic principle of the state.

1977: May 29 Shanti Bahini launches armed offensive

The Shanti Bahini "Peace Fighters," an organization of militant tribal separatists, launches its first offensive against government forces in the Chittagong Hill Tracts of southeastern Bangladesh.

Roughly one percent of Bangladesh's population is of tribal origin. Most, unlike Bengalis, are of Tibeto-Burman descent, they are Mongoloid in appearance, and non-Islamic in religion and culture. Tribal groups such as the Chakma, Marmas, Mros, and Tipperas are concentrated in the hills of southeastern Bangladesh known as the Chittagong Hill Tracts. Resentful of incursions of Bangladeshis from the plains into their lands and by their treatment as second-class citizens by the Bangladeshi government, tribal insurgents take up arms in the cause of regional autonomy.

1981: May 31 Zia assassinated

During his years in power, Zia survives about twenty coup attempts, but is finally assassinated by a group of disaffected army officers. His vice-president Abdus Sattar replaces him as president.

1982: March 24 Military coup overthrows Sattar

The Chief of Army Staff, Lieutenant-General H. M. Ershad assumes power in a bloodless coup, on the grounds of political corruption and economic mismanagement. Ershad rules under martial law, consolidating his power by keeping the army content and appointing many senior officers to political office.

1983: December 11 Ershad assumes presidency

Ershad becomes president and retains this office through several elections that are either rigged or boycotted by the opposition.

1985: December SAARC

A summit meeting at Dhaka ratifies the formation of the South Asian Association for Regional Cooperation (SAARC). This is an association of seven Asian countries (Bangladesh, Bhutan, India, Maldives, Nepal, Pakistan, and Sri Lanka) which agree to mutual cooperate on regional economic issues.

1986: November Martial law repealed

Ershad repeals martial law and restores the 1972 constitution.

1987: July-August Ershad legislation opposed

Wide spread public opposition, including strikes and violent demonstrations, greet an Ershad plan to appoint army representatives to elected local district councils.

1987: August-September Floods

Extensive flooding creates widespread devastation in Bangladesh and overshadows political events.

1988: June Islam becomes the state religion

The Eighth Amendment to the constitution makes Islam Bangladesh's state religion. This has little effect on the lives of the country's eighty-five percent Muslim population, but raises concerns among the Hindus, tribal groups, and other religious minorities in the country.

1988 Political unrest and more floods

Political unrest due to economic devastation, accompanied by violence and numerous arrests, continue though 1988. Bangladesh experiences the worst flooding in recorded history, with waves of floods in August, December, and January 1989, and a severe cyclone in late November.

1990: December 4 Ershad resigns

Opposition groups step up anti-government activities, staging strikes and demonstrations across Bangladesh in late 1990. Unrest is made worse by communal Muslim-Hindu violence following the destruction of a mosque at Ayodhya in India by Hindu militants. Ershad declares a state of emergency, but eventually resigns in the face of mounting opposition in the country.

1991: February 27 Democratic elections in Bangladesh

Parliamentary elections bring victory to the BNP, and Begum Khaleda Zia (Zia's widow) becomes prime minister. The president's role in government is reduced to titular head of state.

1991: May Cyclone causes destruction

A devastating cyclone hits Bangladesh, causing up to 250,000 deaths and massive economic damage.

1993: June 4 Feminist author persecuted

The *Bangladesh Times* reprints an article by Taslima Nasreen, a feminist author, in which she says the *Koran* should be revised and brought up to date to reflect the modern world. Fundamentalist Muslims are outraged by this and her other allegedly anti-Islamic writings. Demonstrators demand her arrest and execution, and a price is put on her head. Nasreen goes into hiding, eventually escaping Bangladesh and finding political asylum in Sweden.

1994 Bridges over the Meghna and Jamuna Rivers

The Japanese Friendship Bridge across the Meghna River, built with foreign aid, completes the direct road link between Dhaka and Chittagong. Work starts on the multipurpose Jamuna Bridge, which will link Dhaka with western Bangladesh when completed in 1998 (in Bangladesh the Brahmaputra is called the Jamuna River). Although funded largely by the World Bank and Japan, this is the most expensive project (US $700 million) ever undertaken by the government. Critics of the bridge project foresee serious environmental and social complications.

1995 U.S. aid totals $94 million

U.S. aid to Bangladesh during the year totals $94 million. Total aid from the U.S. since the time of independence amounts to some $3.4 billion. This started as emergency relief following the 1971 war, but now has as its aims stabilizing population growth, protecting health, encouraging broad based economic growth, and promoting democracy. Other foreign aid programs include the multi-billion dollar Flood Action Plan (FAP), sparked by the catastrophic flooding of the late 1980s.

1995: July 4 Child Labor

The Bangladesh Garment Manufacturers and Exporters Association (BGMEA), the United Nations Children's Fund (UNICEF), and the International Labor Office (ILO) sign a memorandum of understanding to remove children under fourteen years of age from the Association's more than 2,000 factories. Under the program funded by UNICEF and ILO and supported by the Bangladeshi government, the children are to be placed in schools. As many as 50,000 child workers (some estimates place the number at 300,000) are affected. Bangladesh is the fifth largest supplier of cotton apparel to the U.S.

This move is part of an international effort to eliminate the problem of child labor in developing countries.

1996: June Awami League wins elections

Facing widespread unrest in the country, Zia holds elections in February and wins by a landslide. Opposition parties boycott the elections and force Zia's resignation. In new polls the Awami League is victorious. Sheikh Hasina Wajed, leader of the Awami League and daughter of Mujib, becomes prime minister.

1996: July 1 University of Dhaka celebrates seventy-five years

The University of Dhaka celebrates the seventy-fifth anniversary of its founding.

Apart from its brief years as capital of Mughal Bengal, Dhaka lags behind Calcutta in areas of culture, learning and the arts. Following Pakistan's independence, however, Dhaka emerges as an important center of education and intellectual activity. It is the site of several universities, including the University of Dhaka and the Bangladesh University of Engineering and Technology. The dancer and writer Rashid Ahmad "BulBul" Choudhury founds the Bulbul Fine Arts Academy in Dhaka in 1955. The Nazrul Academy promotes the works and ideals of the Muslim poet Kazi Nazrul Islam (1898–1976), one of the giants of Bengali literature.

1996: December 12 Ganges water treaty signed

Bangladesh and India sign a treaty by which the countries agree on an allocation of water from the Ganges. This replaces the expired treaty signed in 1977.

1997: December 2 Treaty ends tribal uprising

Representatives of the government and tribal insurgents sign a treaty ending the two-decade long tribal rebellion in the Chittagong Hill Tracts.

1998: June 25 Indian company wins natural gas contract

The government announces that the publicly-owned Indian Oil and Natural Gas Commission (ONGC) is awarded a US $15 million contact to drill natural gas wells in Bangladesh. Although Bangladesh is poor in oil and petroleum resources, it has promising natural gas reserves.

1998: July 25 Floods affect Bangladesh

Monsoon rains create nation-wide floods that affect two-thirds of Bangladesh. An estimated 10 million people are displaced or isolated.

1998: August **Economic policies liberalized**

Commerce minister Tofail Ahmed announces economic policies aimed at reducing restrictions on imports, lowering tariffs, and accelerating exports from Bangladesh. He proposes development in areas such as agro-processing, software exports, and the leather industry to expand the country's narrow export base which now rests on fish exports and garment production.

1998: August 11 **Mass opposition protests**

Opposition groups announce mass protests against the Awami League government. Among their demands are adequate relief for flood victims, the restoring of law and order, an end to political repression, and economic policies to better the common people.

The politics of confrontation continue in Bangladesh. The main political figures, prime minister Sheikh Hasina and opposition leader Khaleda Zia, have little liking for each other. They each believe the other was associated in some way with the death of their relatives, Mujib and Zia. Given the volatility of the political scene, the tradition of protest and often violence in politics, widespread poverty among the masses, and the frequency of major natural disasters, Bangladesh faces an uncertain and troubled future.

Bibliography

Afsaruddin, Mohammad. 1990. *Society and Culture in Bangladesh.* Dhaka: Book House.

Ahmed, Nazimuddin. *Islamic Heritage of Bangladesh.* Dacca: Government of the People's Republic of Bangladesh, 1980.

Baxter, Craig. *Bangladesh: A New Nation in an Old Setting.* Boulder and London: Westview Press, 1984.

Bigelow, Elaine. Bangladesh: the Guide. Dhaka: AB Publishers, 1995.

Eaton, Richard M. *The Rise of Islam and the Bengal Frontier, 1204-1760.* Berkeley: University of California Press, 1993.

Gardner, Brian. *The East India Company.* New York: Dorset Press, 1971.

Heitzman, James, and Worden, Robert L., eds. *Bangladesh, a Country Study.* 2nd ed. Washington, DC: Federal Research Division, Library of Congress, 1989

Johnson, B. L. C. 1982. *Bangladesh.* 2nd ed. London: Heinemann Educational Books.

Newton, Alex. *Bangladesh: A Lonely Planet Travel Survival Kit.* Hawthorne, Victoria, Australia: Lonely Planet, 1996.

O'Donnell, Charles Peter. *Bangladesh: Biography of a Muslim Nation.* Boulder and London: Westview Press, 1984.

Bhutan

Introduction

Bhutan is a small Buddhist kingdom located on the northern mountain rim of South Asia. The name Bhutan is of Indian derivation and has its origins in a word meaning the borderland of Bhot or Tibet. The people of the country, however, call their land "Druk Yul" (Land of the Thunder Dragon) and refer to themselves as Drukpa. The reigning monarch of Bhutan is known as the *Druk Gyalpo* or Dragon King.

Bhutan is one of the least-densely populated countries in South Asia. A 1995 government census put the population at 816,000, but this figure excluded large numbers of immigrants of Nepalese origin. By some estimates, the total population may exceed 1.8 million people. Bhutan has an area of 18,217 square miles (47,182 square kilometers), roughly one-third the size of New York State.

Bhutan is a mountainous country. The Great Himalayas, with peaks rising above 23,000 feet (7000 meters), lie in the north of the country. Mount Kula Kangri, on the border with Tibet, reaches an elevation of 24,784 feet (7554 meters). Central Bhutan is situated in the Inner Himalayas. In this region, mountains of lesser elevation (but still exceeding 16,500 feet [c. 5000 meters] in places) separate north-south running valleys of rivers flowing towards the Indian plains. Population is concentrated along the valley floors, which lie at elevations between 3,500–9,800 feet (1000—3000 meters). The Ha, Paro, Thimphu and Punakha Valleys are among the more important valleys of central Bhutan. Southern Bhutan consists of the foothills of the Himalayas which range from 2,000–5,000 feet (600–1500 meters) in elevation. The narrow belt of lowlands bordering the foothills is known as the Duars (literally "doors"), because it provides access to the strategic central valleys of Bhutan. Southern Bhutan has a sub-tropical monsoon climate, with excessive rainfall and high summer temperatures. Climate becomes progressively cooler as one moves northwards towards the alpine regions of the Great Himalayas.

History

Geography has greatly influenced Bhutan's history and culture. Mountainous terrain and dense forests have isolated Bhutan from direct contact with the outside world, and also hindered internal communications. Yet Bhutan lies between India and Tibet, and throughout its history the country has been exposed to cultural and political influences from these ancient centers of civilization in Asia.

The Bhotia, the dominant ethnic group in Bhutan, are a Mongoloid people of Tibetan origin. They speak Dzongkha, a Tibetan dialect, follow Tibetan Buddhism, and in their dress and other aspects of culture reflect traditional Tibetan patterns. It is the Bhotias, who are concentrated in the western valleys of Bhutan, who are the country's political and social elites. The Sharchops of eastern Bhutan, like the Bhotia, are a Mongoloid people and practice Tibetan Buddhism. However, they speak Tsangla (a Tibeto-Burman language) and show cultural similarities with the peoples of India's eastern borderlands rather than with Tibet. The third major ethnic group, the Nepalis, are relative late-comers to the region. Physically and culturally related to the peoples of northern India, Nepalis began settling in southern Bhutan during the late nineteenth century. Unlike the Bhotia and the Sharchops, Nepalis follow the Hindu religion.

Buddhism has been an integral part of Bhutan's history and culture for nearly fifteen hundred years. From the seventh century on, Tibetan influences extended into the region, with a form of Tibetan Buddhism eventually emerging as the dominant religion. Between 1651–1907, Bhutan was a theocracy, i.e. a state governed by religious leaders, and virtually all leading political figures during this time were Buddhist monks.

The traditional centers of power in Bhutan were the *dzongs,* the fortress-monasteries built at strategic locations throughout the country to control its valleys and routes of communication. The dzongs in major valleys such as Paro, Thimphu, and Tashigang frequently provided regional power bases from which local lords attempted to extend their control over the country. Even today, the seat of Bhutan's government is the Tashichho dzong in the capital Thimphu.

As religious institutions, dzongs and monasteries (*lakhangs*) were centers of Buddhist learning and education. They kept libraries of traditional Buddhist scriptures, and produced Bhutan's own religious literature and histories such as the *Namthars*. They were also centers of art, because almost all Bhutanese art (e.g. *thankas*, *mandalas* (circular designs of geometric form), frescoes, and illuminated manuscripts) and sculpture are religious in nature. Even Bhutan's music and famous masked dances are usually religious in nature and performed by monks at religious festivals. All along the roads in Bhutan, colorful prayer flags are mounted on poles by prayerful Buddhist travelers.

The Bhutanese are mostly engaged in traditional subsistence agriculture. The society is matrilineal, which means the Bhutanese pass on their family land through daughters. When a couple marries, the man moves into his new wife's home. Rice is the staple food, supplemented by maize and, more recently, potatoes. Herding of sheep and yak is thc only activity possible in the northern regions. Despite their Buddhist heritage, the Bhutanese eat meat and a typical meal might consist of *thugpa*, a meat soup, served with rice and curried dishes. Tea, taken with salt and butter, and beer (*chang*) are popular. Salt and tea were important items in Bhutan's historic trade with India and Tibet.

The modernization of Bhutan's economy began only in the 1960s. This required the creation of roads, communications, transportation, banks—virtually an entire economic infrastructure. Obviously an expensive undertaking, this has been accomplished with considerable foreign aid, especially from India. Bhutan has limited natural resources to support industrialization, although its forests and hydro-electric potential are important assets. Tourism, although strictly controlled by the government, is another developing industry which is a major earner of foreign currency. Development plans have also placed considerable emphasis on improving social conditions by expanding scrvices such as health care and education.

Modernization in Bhutan, however, has created problems. By lowering barriers to the outside world, Bhutan's monarchy exposed its subjects to alien ideas such as concepts of political freedom and popular democracy. The ethnic unrest that has afflicted southern Bhutan since the mid-1980s is as much a struggle between the old and new political orders as it is a conflict between competing Bhutanese and Nepalese cultures.

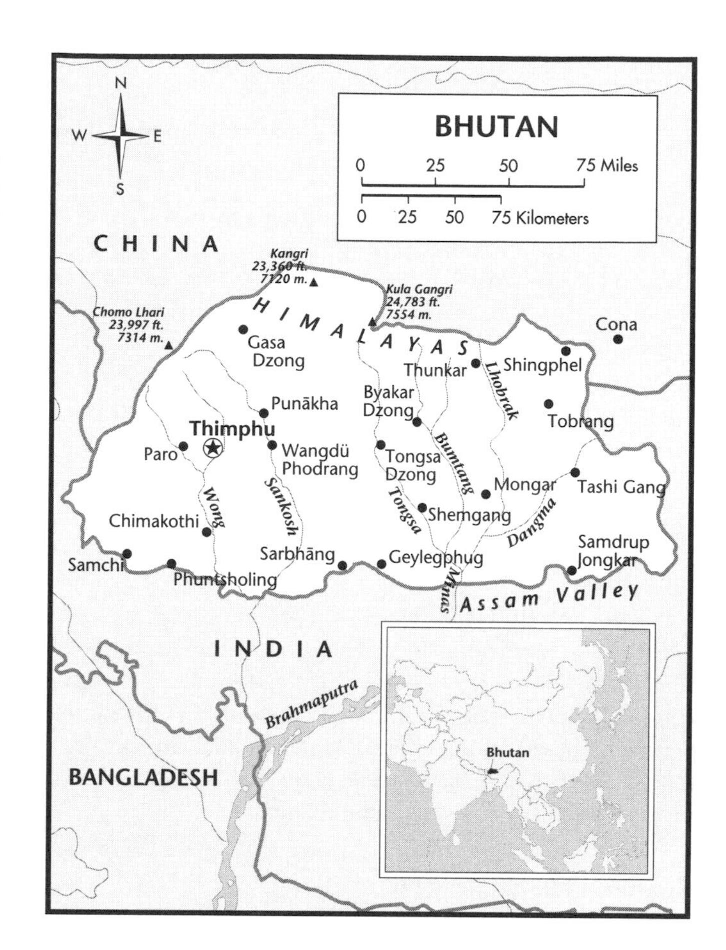

Timeline

c. 2000 B.C. Evidence of early populations

The prehistory of Bhutan is shrouded in the mists of time, although stone tools, weapons, and stone megaliths (pillars) suggest the region is inhabited at this time.

c. 500 B.C.–A.D. 500 Lho Mon, or the southern country

Early Bhutanese historical works tell us that the land is called Lho Mon or the "southern country," because the region lies to the south of Lhasa in Tibet. Sometimes it is called Lho-Mon Tsenden Jong, or "southern sandalwood country," because of its abundance of this precious, aromatic wood which is not available in Tibet.

A later Bhutanese chronicle tells that the people of Lho Mon stay in one place, that there is neither trade, nor any communication, nor education. Each village has its own dialect, customs, and cultural practices. There are no laws, taxes, government, or formal religion.

There are, however, legends concerning the introduction of the Bon religion into Lho Mon. Bon, which is widespread in Tibet at this time, involves superstitious practices, the worship of nature gods (sun, moon, hills, trees, rivers, etc.), good and evil spirits, sorcery, magic and witchcraft. One tradition says that Lhase Tsangma of Tibet, who followed the Bon religion, was exiled from his country. He settled in Tashigang in eastern Bhutan, and spread Bon throughout the region. Elements of the Bon religion survive in Buddhist practices in Bhutan even to this day.

c. A.D. 640 Tibet's king establishes monasteries in Bhutan

The renowned Tibetan king, Srong-brt-san-sgampo (c. 608–50), establishes Buddhist monasteries at Paro and Bumtang.

747 Guru Rimpoche visits Bhutan

The Buddhist monk Padmasambhava, also known as Guru Rimpoche (Precious Teacher), visits Bhutan.

Buddhism is one of the world's great universal religions, i.e. it is not restricted to peoples of any specific ethnic group or culture. The religion is based on the teachings of a Hindu prince who was born in northern India during the sixth century B.C. Siddhartha Gautama (c. 563–c. 483 B.C.) came to be known as the Buddha, the Enlightened One, and after his death his disciples spread the new religion across the Indian subcontinent and beyond.

Although Buddhists have been present in the region for nearly two centuries, Guru Rimpoche firmly establishes Mahayana (Northern) Buddhism as the dominant religion in Bhutan. Bhutan's religious, social, and political institutions are firmly based on three fundamental Buddhist principles—faith in the Buddha, the *Dharma* (Law), and the *Sangha*, the order of monks founded by the Buddha himself.

816–36 Tibetans settle in Bhutan

Tibetan records and Bhutanese *Namthars* (biographies of important saints and rulers) refer to the incursion of Tibetan troops into Bhutan during the reign of the Tibetan king, Ralpachen. These invaders refer to themselves as the *Milongs* ("those who will not return") and make their homes in the valleys of northern and central Bhutan.

836–42 Buddhists flee Tibet

The Tibetan king Glang-dar-ma (r. 836–42) outlaws Buddhism in Tibet and reinstates Bon as the state religion. He persecutes Buddhists, who flee southwards into the Himalayas. Many Buddhist monks settle in Bhutan, strengthening the position of Buddhism in the region.

1220 The Drukpa sect comes to Bhutan

Lama Phajo Drugom Shigpo arrives in Bhutan, bringing with him the teachings of the Drukpa Kagyupa (*'Brugpa Bka'-brgyud-pa*) school of Buddhism. A "lama" is a monk in Tibetan Buddhism.

During the rebirth of Buddhism in Tibet in the eleventh century, numerous sects with their own interpretations of Buddhist doctrine, methods of worship, and monasteries appear across the country. Drukpa, which derives its name from the monastery in Tibet where it originated, is one of several schools of thought belonging to the Kagyupa (or Red Hat) sect of Tibetan Buddhism.

c. 1433 Bridge-building in Bhutan

A Tibetan lama, Thangton Gyalpo (1385–1464), arrives in Bhutan. He is famous for building bridges, some of which are supposedly still in use, all over Tibet and Bhutan. These take the form of iron chains, slung across gorges and ravines, from which rope walkways are suspended. These structures stimulate trade and cultural contact between the valleys, and contribute to the unification of the country during the seventeenth century. Similar bridges are common throughout the Himalayas today.

1616 Ngawang Namgyal arrives in Bhutan

Ngawang Namgyal (1594–c. 1651) flees to Bhutan from Tibet.

Ngawang Namgyal is an important lama and a prince who is in the line of succession to become head of the Drukpa Kagyupa sect. He is driven out of Ralung, the sect's main monastic complex in Tibet, by a rival claimant who has the support of the ruler of Tsang, a kingdom in eastern Tibet.

1616–c. 1651 Ngawang Namgyal unifies Bhutan

Ngawang Namgyal is the person most responsible for unifying Bhutan. Prior to his arrival, the region is a patchwork of minor kingdoms and independent monastic estates, peopled by different ethnic groups with differing laws and customs. Using the monasteries of his own sect as a power base, Ngawang Namgyal extends his control over the western valleys of Bhutan. He codifies laws, and establishes a stable form of government based on Buddhist principles. As a leading member of the Drukpa Kagyupa sect, he ensures that this school emerges as the dominant form of Buddhism in Bhutan. It is from this sect that the country takes its name Druk Yul.

1617 Ngawang Namgyal defeats the Tibetans

The Bhutanese defeat an invading Tsang army at the battle of Densa Drug Chhoeding in the Paro Valley. Local tradition, preserved in the Namthars and oral histories, tells that Ngawang Namgyal uses magical powers to destroy his enemy.

1627 First Europeans visit Bhutan

Two Jesuit priests, Estevão Cacella and João Cabral, visit Bhutan. They are the first Europeans known to have visited the country.

1629 Nwagang Namgyal builds the Simtokha Dzong

Nwagang Namgyal lays the foundation of the Simtokha Dzong, which is built on a strategic ridge overlooking and guarding the Thimphu Valley. By building dzongs all over the country to protect settlements and communication routes, Nwagang Namgyal strengthens Bhutan's defenses against foreign aggression, curbs local feudal lords, and centralizes power in his own hands.

1629–39 Tsangs invade Bhutan

Three times during this period (in 1629, 1634, and 1638), Tsang armies from Tibet invade Bhutan. On each occasion they are driven back. After his last victory over the Tsangs, Nwagang Namgyal assumes the title Shabdrung Rimpoche, becoming both the spiritual and temporal (i.e. non-religious) leader of Bhutan.

1643 Gushri Khan invades Bhutan

Gushri Khan, the Mongol who now controls Tibet, enters Bhutan with a mixed force of Mongols and Tibetans. The invading armies are defeated by Shabdrung Nwagang Namgyal and flee the battlefield. A similar fate befalls a Tibetan invasion of Bhutan in 1647.

1650 The Code (*Tsa Yig*) of Shabdrung Nwagang Namgyal

The Shabdrung codifies the customary laws and regulations of Bhutan. The new legal code, which remains in force until the 1960s, is based on the Buddhist dharma (law).

The legal code also bans the sale and purchase of slaves. The institution of slavery exists in Bhutan, but slaves are usually prisoners of war or their descendants. No Bhutanese can be sold into slavery unless he is a criminal. Slaves are the property of the state and are not owned by individuals. Slaves are allowed to marry, own houses (but not land) and the state provides them with food, shelter and clothing for services rendered.

1651 Bhutan adopts "dual" system of government

Shabdrung Nwagang Namgyal establishes the form of government outlined in his legal code. The Shabdrung Rimpoche becomes the head of state and the supreme authority in all religious and civil matters. Under him, the functions of government are divided between a spiritual and a temporal (i.e. responsible for non-religious affairs) leader. From this time forward, women of Bhutan adhrer to certain conventions in style of dress and hairstyle, cropped to ear length. The garment they wear is a *kira,* an ankle-length dress made of handwoven fabric, tied with a sash. Each family has a unique design for their woven kira cloth.

The *Deb* (or *Druk Desi*), an elected official, looks after the general administration, revenues, and military affairs of the country. The Shabdrung nominates Umzey Tenzin Drugyel (r. 1651–65) for this position and he becomes the first Deb. The Shabdrung also appoints a *Je Khenpo* or Head Abbot to lead the Central Monk Body (the community of Buddhist monks supported by the state) and manage the religious institutions of the country. Neten Pekar Jungney (r. 1651–73) takes office as the first Je Khenpo.

In addition, the Shabdrung sets up a regional administration, with a hierarchy of officials having clearly defined duties and responsibilities. He also appoints a *Penlop*, or territorial governor, to administer important areas such as Thimphu, Paro, and Tongsa.

c. 1651 Shabdrung Nwagang Namgyal dies

The Shabdrung dies around this time, but his death is kept a secret until 1705 to prevent Bhutan from disintegrating. The pretense is kept up that he is alive and in spiritual retreat in Punakha Dzong.

Bhutan's leaders eventually adopt the principle of reincarnation to determine the succession to the Shabdrung, in the same way Tibetans chose a Dalai Lama. The soul of the dead Shabdrung enters the world in a new body, and makes itself known by displaying intimate knowledge of its past life. A common test, for example, requires the candidate—often an infant or child—to identify personal items of the dead monk. This method of selection frequently results in lengthy periods of time without a Shabdrung, or with the chosen successor being a minor. The British come to call the Shabdrung's successors the *Dharma Rajas*.

1657 Tibetan invasion

Tibet invades Bhutan for the third time. By the mid-seventeenth century, Tibet is under the control of the Yellow Hat sect of Tibetan Buddhism (the Gelugpa sect) and its spiritual and temporal ruler, the fifth Dalai Lama. The Dalai Lama attempts to exert his authority over the rival Red Hats in Bhutan but his armies are turned back by Tenzin Drukda (r. 1656–68), the second Deb. Bhutan preserves its sovereignty and its independence from Tibet.

1676–1700 Bhutan invades Sikkim

Internal opposition to Bhutan's central government results in its opponents seeking help from Tibet and Sikkim, the kingdom immediately to the west of Bhutan. Bhutanese troops enter Sikkim in pursuit of a local lord who is in rebellion against the government. Bhutan again invades Sikkim in 1700, and its ruler (the Chogyal) flees to Tibet.

1714 Tibet invades Bhutan

Lajang Khan, the last of the Mongol rulers of Tibet, invades Bhutan in retaliation for Bhutanese pressure on the Gelugpa sect's stronghold at Tawang monastery. Bhutanese forces rout the combined Tibetan-Mongol armies.

1728–72 Civil wars threaten Bhutan's stability

Internal dissension, which has threatened Bhutan's stability since the death of the Shabdrung, erupts into a series of civil wars that last for nearly fifty years. Some conflicts originate in the tangled web of succession to the Shabdrung, while others reflect regional struggles or local revolts against the authority of the central government.

1731 Tibet imposes suzerainty on Bhutan

Taking advantage of civil conflict in the west, King Pho-lha of Tibet mounts an invasion of Bhutan and succeeds in imposing Tibetan suzerainty (overlordship) on the country. This relationship formally passes on to Tibet's own overlord, the Ching dynasty of China. However, Bhutan soon repudiates its subordinate political status.

1772 Bhutan occupies Cooch Behar

Bhutan overruns Cooch Behar, a Hindu kingdom lying to the south on the plains of Bengal.

Bhutan's relations with Cooch Behar date to 1730, when the country's raja asks for Bhutanese help against the Mughals, the Muslim rulers of India. By the 1770s, Bhutan views Cooch Behar as a dependency and has a small garrison of troops stationed there. In 1772, Bhutan intervenes in a dispute over succession, drives out Khagendra Narayan, a claimant to the throne, and installs its own candidate as raja of Cooch Behar.

1773: April 5 Cooch Behar becomes a British dependency

Khagendra Narayan signs an agreement by which the East India Company is to assist him in becoming raja of Cooch Behar. This agreement with the East India Company, which governs British territory in India, makes Cooch Behar a virtual dependency of Britain. A small British force enters Cooch Behar, drives the Bhutanese out of the country, and pursues them into Bhutan.

1774: April 25 Anglo-Bhutanese treaty is signed

Unnerved by their reverses, the Bhutanese ask the Panchen Lama (the second-ranking lama in Tibet) to intercede with the British. Warren Hastings, the Governor-General of India, sees this as opportunity to establish friendly relations with Tibet and to open up trade relations with Tibet and Bhutan. A peace treaty is signed between Britain and Bhutan.

At this time in the eighteenth century, trade caravans leave Bhutan every year carrying goods to Rangpur in Bengal. They arrive in February and March, bringing China silks, yak-tails, ponies, wax, walnuts, musk, lac, madder (a plant used to make red dye), blankets, cloth, and silver. On their return to Bhutan, they carry indigo, cloves, cardamom, camphor, sugar, various types of cloth, copper, dried fish, and tobacco.

1774–83 British missions to Bhutan

The British send several missions to Bhutan and Tibet with the objective of improving political relations and encouraging trade. George Bogle and Alexander Hamilton visit Bhutan in 1774, Hamilton makes two more journeys to the country in 1776 and 1777, and Captain Samuel Turner leads a party to Bhutan and Tibet in 1783. The reports and journals of these visitors provide the earliest detailed European accounts of Bhutan and the Bhutanese. Turner, who witnesses a civil war in Bhutan, describes his travels in *An Account of an Embassy to the Court of Tseshu Lama and Narrative of a Journey through Bhutan.*

1783–1864 Border disputes with India

Disputes along Bhutan's southern border with India remain a constant source of discord between the two countries. Problems range from Bhutanese raids into British territory to obtain slaves or recapture Bhutanese refugees to non-payment of tribute for subject lands in the Duars. India and Bhutan exchange numerous missions to resolve these questions, but they achieve no lasting solution.

1841: September 6 India annexes the Assam Duars

The Governor-General of India issues the order for the annexation of the Assam Duars.

The Assam Duars form the eastern section of the narrow strip of plains that extends along the base of the Himalayan foothills south of Bhutan. Bhutan controls these lands, having secured them from a weak Assam government nearly a century earlier for payment of an annual tribute. When Britain acquires Assam in 1826, it inherits this payment of tribute. By 1841, Britain claims Bhutan is in arrears in its payments. Britain uses this, along with a series of border incidents in 1834 and 1835, as an excuse to take the Assam Duars from Bhutan.

1842–64 The Bengal Duars and the Eden mission

Border problems continue in the Bengal Duars, the western portion of the Duars. Bhutan's control is less secure in these areas, Bhutanese and British territories are intermingled in a confused manner, and boundaries are poorly defined. Internal disturbances and cross-border incidents, culminating in a series of armed Bhutanese incursions into British territory in 1862, lead the British to send a mission to Bhutan to resolve issues between the two governments.

The mission, headed by Ashley Eden, reaches Punakha in 1864. Bhutanese officials reject the terms of the peace treaty offered by the British, humiliate the British envoy, and send him away empty-handed.

1864: November 12 Britain declares war

Britain declares war on Bhutan and annexes the Bengal Duars in what is called the 'Duar War.'

1865: November 11 The Treaty of Sinchula

In the treaty that ends the Duar War, Bhutan cedes Britain both the Assam and Bengal Duars, and certain other territories. In return, Britain agrees to pay Bhutan an annual subsidy of 50,000 Indian rupees. The treaty also provides for the release of all subjects of Britain, Sikkim, and Cooch Behar held in Bhutan, mutual extradition of criminals, and maintenance of free trade. Britain establishes a quasi-protectorate over Bhutan.

1865–85 Civil wars in Bhutan

Following the its defeat in the Duar War, Bhutan undergoes a series of civil wars in which regional rivals compete for power. These struggles usually pit factions which are pro-British against those which are pro-Tibetan.

1882 Ugyen Wangchuk is appointed Tongsa Penlop

Ugyen Wangchuk is appointed to the position of Tongsa Penlop. He defeats his rivals, puts down several insurrections, and ends the periodic civil wars in Bhutan with his resounding victory over rebels at Changlingmithang, near Thimphu, in 1885. He appoints his own nominee to be Deb, and emerges as the most powerful man in the country.

Bhutan in the Twentieth Century

1903 The last Deb assumes power

Chhogley Tuelku Yeshe Ngoedub, the fifty-fourth in the line of temporal rulers established in 1651, is appointed as Deb. During his rule, he falls into disfavor with the public as well as with government ministers and powerful men such as the Tongsa Penlop. His arbitrary actions and apparent disregard for the welfare of the people lead to his removal in 1907.

1903–04 Ugyen Wangchuk assists the Younghusband expedition

Already the power behind the throne, Ugyen Wangchuk solidifies his relationship with the British by accompanying a mission led by diplomat and explorer Sir Francis Younghusband to Lhasa and mediating the 1904 Anglo-Tibetan Treaty. He is later knighted by the British for his services.

1907: December 17 Ugyen Wangchuk is elected king

Ugyen Wangchuk (r. 1907–26), the Tongsa Penlop, is unanimously elected hereditary king by the lamas, State Councilors, regional governors, and representatives of the people of Bhutan.

Over two and a half centuries of government under the dual system of government introduced by Shabdrung Nwagang Namgyal thus comes to an end. The kingdom is unified in the person of its new sovereign, a firm, powerful, and shrewd monarch who is well-equipped to lead the country into the new century.

1910: January 8 The Treaty of Punakha is signed

Well aware of the threat posed by China on its northern border, the king seeks to strengthen his ties with Britain by revising the terms of the 1865 Treaty of Sinchula. Under the terms of the new Anglo-Bhutanese treaty, Britain increases its allowance to Bhutan and undertakes not to interfere in the country's internal affairs. Bhutan, on its part, agrees to be guided by the British government in its external affairs. This effectively counters the growth of China's influence in Bhutan.

1911 The first modern school opens in Bhutan

The first modern school opens in Bumtang in central Bhutan. It is attended by fourteen boys, including Gyalsay (Prince) Jigme Wangchuk, the son of the king. Ugyen Wangchuk's internal reforms include the introduction of the western system of education, improving internal communications, and encouraging trade and commerce with India. He also builds and repairs many temples and monasteries, and promotes Buddhist education throughout the kingdom.

1926: August 26 King Ugyen Wangchuk dies

The king of Bhutan dies, and is succeeded as Druk Gyalpo by his twenty-one year old son Jigme Wangchuk (r. 1926–52).

1949: August 8 Indo-Bhutanese Treaty

Bhutan signs a treaty with newly-independent India, by which India assumes the former British protectorate over Bhutan. The terms of the treaty are essentially similar to those of the 1910 Treaty of Punakha. India undertakes not to interfere in Bhutan's internal affairs, while Bhutan agrees to be guided by India in its external relations. India increases the amount of the subsidy provided to Bhutan, and returns former Bhutanese territory seized by Britain during the Duar War. The two countries agree to free trade and commerce between their territories, to the mutual extradition of criminals, and India allows Bhutan to import arms and war materials across Indian territory.

1952 Jigme Dorji Wangchuk becomes king

King Jigme Wangchuk dies and is succeeded on the throne by Jigme Dorji Wangchuk (r. 1952–72).

Jigme Dorji Wangchuk is often called the father of modern Bhutan. During his twenty-year reign, he does much to modernize the country. He abandons the traditional policy of national isolation and opens up the country to the outside world. He introduces democratic political institutions, undertakes social and economic reforms, abolishes slavery, and promotes economic development.

1953 Constitutional reforms

The king establishes the National Assembly of Bhutan (*Tshogdu*) composed of indirectly-elected representatives of the people, members nominated by various bodies of monks, and officials selected by the king. The assembly, which holds office for a three-year period is both a legislative and an advisory body. The king constitutes a Royal Advisory Council (*Lodyo Tshogdu*) in 1963.

1958 Bhutan grants Nepalis citizenship

A provision of the Nationality Law of Bhutan, 1958 makes foreigners (i.e. Nepalis) living in Bhutan for ten years eligible for citizenship, if they own land. This makes Bhutanese citizenship available to Nepalis for the first time.

At the same time, Bhutan places restrictions against further Nepali immigration. Subsequent laws prohibit Bhutanese of Nepali origin, even though citizens, from leaving the southern part of the country to settle elsewhere in Bhutan. This causes bitter resentment among the Nepali population.

1959 Bhutan grants asylum to Tibetan refugees

Following the Chinese army's occupation of Tibet, Bhutan accepts some 6,000 refugees from across its northern border. The involvement of some refugees in domestic affairs and their unwillingness to accept governmental authority leads Bhutan to require, in 1979, that they accept Bhutanese citizenship or be expelled to India.

In the wake of events in Tibet, Bhutan aligns itself with India against China, and closes its northern border with Tibet. Bhutan also begins a program of modernization designed to offset the possibility of Chinese encroachment into Bhutan.

1961 Modernization begins

Modernization in Bhutan really begins with the implementation of the First Five-Year Development Plan (1961–66). India provides the planning and the finances for the scheme, which emphasizes creating the basic economic infrastructure of a road system, transportation, communications, and power supply. In addition, the plan seeks to improve agriculture, promote industrial growth, and develop educational and health facilities.

1962 The first vehicles in Thimphu and Paro

The first motor vehicles appear in Thimphu and Paro. Road construction is given the highest priority in the First Five-Year Plan. Indian engineers build roads from India into Bhutan and construct east-west routes in Bhutan itself. This road system serves strategic as well as economic purposes. Although India has no formal responsibility for Bhutan's defense, it takes the position that any attack on Bhutan is an attack on India.

Bhutan introduces a modern postal service.

1964: April Prime minister is assassinated

A disaffected soldier assassinates the Prime Minister, the king's brother-in-law Jigme Palden Dorji. The conspirators are mostly members of the Royal Bhutan Army. The incident arises from factional differences between Wangchuk loyalists and "modernist" Dorji supporters over the extent of Indian involvement in Bhutan's affairs.

1965: July The king survives an attempt on his life

The tense political situation in Bhutan continues with an unsuccessful attempt on the king's life. The Dorji family is not implicated and the would-be assassins are pardoned by the king.

1968: May Bhutan becomes a constitutional monarchy

King Jigme Dorji Wangchuk voluntarily surrenders his veto power over the National Assembly and announces Bhutan is to be a constitutional monarchy.

1971: September 21 Bhutan joins the United Nations

India sponsors Bhutan's membership in the United Nations.

1972: July 21 King Jigme Dorji Wangchuk dies

King Jigme Dorji Wangchuk dies and is succeeded by his sixteen-year old son, Jigme Singye Wangchuk.

1974: June 4 Jigme Singye Wangchuk is crowned king

Jigme Singye Wangchuk is enthroned as the fourth hereditary king of Bhutan. The young ruler, who has received a traditional Buddhist education as well as western training at schools in India and London, continues the policies of modernization and development begun by his father.

1975: April Sikkim becomes a state of India

Sikkim, a protectorate of India, is a monarchy that has existed for over three hundred years. However, Nepali settlement in Sikkim results in a large immigrant population that grows to outnumber the native Sikkimese. In a plebiscite to determine the country's form of government, the Nepali majority vote to become the twenty-second state of India.

This event in neighboring Sikkim has clear implications for Bhutan's internal policies. It provides impetus to speed the pace of reform and modernization, so that the monarchy is not rejected by the populace. Yet is also raises fears that the Bhutanese and their culture will be swamped by growing numbers of Nepalis. This eventually leads to policies that discriminate against peoples of Nepali origin.

1985 The king abandons modernization

The king and his advisors reach the conclusion that modernization has failed. All opposition to the monarchy is suppressed, Bhutan is cut off from the outside world, tourism is curtailed, and television antennas are dismantled so that programs from neighboring countries cannot be received.

1985: June 10 Bhutan Citizenship Act

A revised nationality act removes land ownership by Nepalis as a prerequisite for Bhutanese citizenship. Nepalis resident in Bhutan for twenty years, speaking Dzongkha, familiar with Bhutanese culture, and prepared to swear an oath of loyalty, can apply for citizenship. This appears to enfranchise large numbers of Nepalis living in Bhutan. However, the requirement that the applicant's presence in Bhutan during this time have been registered with the Bhutanese authorities effectively denies most Nepalis citizenship.

1986 The Code of Cultural Correctness (*Driglam Namsha*)

The government embarks on a policy of Bhutanization, forcing Nepalis (now known as Lhotshampas) to adopt Bhutanese culture. The seventeenth-century Shabdrang's Driglam Namsha is imposed on Nepalis, who are required to wear traditional Bhotia clothes and speak Dzongkha in public places—despite the fact that most don't speak the language. The practice of any religion other than Mahayana Buddhism is prohibited.

Nepalis see these restrictions as violations of their human rights, and openly defy them. Bhotias, in turn, view such Nepali actions as a clear case of rebellion against the monarch and the state.

1986 Bhutan's media

Kuensel, founded as an internal government bulletin, becomes the country's only newspaper.

With low literacy rates, circulation of print media in Bhutan is limited. Oral tradition is very strong, however, and the Bhutan Broadcasting Service (established 1973) is widely listened to. There are no television stations in Bhutan, although in some areas transmissions from India and Bangladesh can be received.

1988 King Jigme Singye Wangchuk marries

Although he privately married four sisters in 1979, Bhutan's king marries his wives again in a public ceremony. This is to legitimize the eventual succession to the throne of his eldest son Jigme Gesar Namgyal Wangchuk.

Bhutanese law allows a man to have more than one wife, although the practice is not widespread in modern times. Women enjoy considerable equality, own land and property, and can sue for divorce. However, women are poorly represented in political life.

1988: October Chukha hydro-electric project is opened

The Chukha hydro-electric project, funded largely by India, is officially opened by the President of India. With a capacity of 336 MW (megawatts), Chukha provides electricity to most of western Bhutan. India buys the plant's surplus power.

1988–89 Unrest spreads in southern Bhutan

A government census and checking of citizenship registration in 1988 reveal large numbers of illegal Nepali immigrants in southern Bhutan. Mostly post-1961 arrivals brought to Bhutan by labor contractors or attracted by Bhutan's free health care and education, many have married locally and raised families. Resentment among Nepalis at government immigration policies and attempts to impose Bhutanese culture on them lead to ethnic strife between Nepalis and Bhutanese. Pro-democracy activists agitating against the government stir things up even further. Refugees begin to arrive in Nepal claiming they have been expelled from Bhutan.

In mid-1989, anti-government literature is distributed in Bhutan. The government cracks down on "anti-national" activities, arresting suspected ring-leaders.

1990–96 The uprising in southern Bhutan

Sporadic violence against government forces and large-scale anti-government demonstrations take place in southern Bhutan in 1990. Terrorists continue their activities for the next several years. They attack schools and bridges, kill government officials, and threaten and rob government supporters. Security forces policing southern Bhutan are charged by Nepalis with beatings, rapes, robberies, and forcible evictions.

Nepalis flee Bhutan in large numbers, most finding their way to refugee camps in eastern Nepal. By 1996, Nepal is sheltering an estimated 100,000 refugees.

1995 Dzongkha becomes official language

The government declares Dzongkha the medium for official communication, continuing its policy of Bhutanization.

Dzongkha (the language of the Dzong) has developed since the seventeenth century and is a Tibetan dialect based on the speech of the Punakha Valley. It is written in a script adapted from chhokey, the classical language derived from the Tibetan Buddhist scriptures.

1996–97 Exhibitions by the National Museum

The National Museum of Bhutan organizes a series of exhibitions devoted to the early and medieval history of Bhutan.

1996: March Chukha II and Chukha III hydro project agreement

Bhutan and India sign an agreement to jointly construct the second and third phases of the Chukha hydroelectric project at Tala and the Wangchu reservoir. The planned increase in capacity is 1,620 MW.

1996: April Refugee problem remains unresolved

The Joint Ministerial Committee set up by Bhutan and Nepal to deal with the refugee problem meets for the seventh time. Failure to reach an agreement reflects Nepal's position that the refugees be resettled in Bhutan, and Bhutan's claim that the refugees are not Bhutanese citizens.

1996: September 10 Bhutan votes against U.N. ban on nuclear tests

Bhutan joins India and Libya as the only countries voting against the Comprehensive Test Ban Treaty (CBCT) approved by the United Nations.

1997–98 Security problems along India's border

Indian separatist groups seek refuge from Indian security forces in Bhutanese territory along the border with Assam.

1997: July 1 Eighth Five-Year Plan begins

The Eighth Five-Year Plan (1997-2002) commences, with "national security," i.e. internal security, as its primary goal. Expenditures during the plan period are targeted at US $845 million.

1998: January Amnesty International charges human rights violations

Amnesty International, a human rights organization, issues a report alleging human rights violations by government forces against anti-government supporters, mostly Sharchops, in eastern Bhutan.

1998: June 29 The king relinquishes more power

The National Assembly meets to debate reforms outlined in a kasho or royal edict issued by the king. The proposals call for an elected National Assembly and the provision that the National Assembly can force the king to abdicate by a vote of no-confidence. Supporters of the monarchy claim this shows a willingness to bring Bhutan into the modern age and to deal with the country's ethnic problems. Government opponents argue this is merely an attempt to improve Bhutan's image in the international community.

1999 Dress code still in effect

The Bhutanese adhere to a mandated dress code. The men are required by law to wear ghos: calf-length robes made from solid, colored, striped, or plaid fabrics. A violation of the dress code results in a fine of about $10, approximately one-week's earnings for the average Bhutanese man. He may also be sentenced to serve one night in jail.

1999 Television started

Bhutan begins television transmission in the capital city of Thimpu in June.

Bibliography

Aris, Michael. *Bhutan: The Early History of a Himalayan Kingdom.* Warminster, England: Aris & Phillips, 1979.

Chakravarti, B. *A Cultural History of Bhutan.* 2 vols. Chittaranjan, India: Hilltop Publishers, 1979.

Crossette, Barbara. *So Close to Heaven: The Vanishing Buddhist Kingdoms of the Himalayas.* New York: A. A. Knopf, 1995.

Regional Briefing. *Far Eastern Economic Review.* 6 May 1999, 16.

Hasrat, Bikrama Jit. *History of Bhutan: Land of the Peaceful Dragon.* Thimphu, Bhutan: Education Department, Royal Government of Bhutan, 1980.

Karan, P. P. *The Himalayan Kingdoms: Bhutan, Sikkim, and Nepal.* Princeton, NJ: Van Nostrand, 1963.

Matles, Andrea, ed. *Nepal and Bhutan: Country Studies.* Washington, DC: Federal Research Division, Library of Congress, 1993.

Pommaret-Imaeda, Françoise. *Bhutan.* Lincolnwood, Ill.: Passport Books, 1991.

The Southern Bhutan Problem. Available at http://www.bhutan-info.org/index.htm

Strydonck, Guy van. *Bhutan: A Kingdom in the Eastern Himalayas.* Boston: Shambala, 1985.

World Bank Country Study. *Bhutan, Development in a Himalayan Kingdom.* Washington, DC: World Bank, 1984.

Brunei Darussalam

Introduction

The Sultanate of Brunei, *Negara Brunei Darussalam* (Nation of Brunei, Abode of Peace) is located on the northeastern tip of the island of Borneo, the third-largest island in the world. One of the smallest nations in the world, 2,227 square miles (5,769 square kilometers), Brunei is only slightly smaller than Delaware, yet it is one of the wealthiest countries in the world. Brunei is the sole surviving ancient Malay Muslim kingdom in Southeast Asia.

Borneo is part of the Greater Sunda Group of the Malay Archipelago or group of islands. The Malay Archipelago, also known as the "East Indies," extends along the Equator for more than 3,800 miles. Borneo is subdivided into three nations, Malaysia, Brunei, and Indonesia, surrounded by the South China Sea to the north and northwest, the Java Sea to the south, the Celebes Sea to its east, and the Sulu Sea to the northeast. The two small enclaves of Brunei each border on the South China Sea. Brunei is situated in an equatorial climate consisting of high temperatures (73–89°F) with high humidity and high annual rainfall. However, typhoons, earthquakes and severe flooding are extremely rare. Its terrain in the east is flat coastal plains with beaches; the west is hilly with a few mountain ridges.

In 1997 Malays comprised sixty-four percent of the population and Chinese twenty percent. Indigenous ethnic groups are Dusans, Kedeyans (Kadazan), Iban (Sea Dayak), Bisaya (Bisayah), and Maruts. The capital is Bandar Seri Begawan in and around which nearly two-thirds of the population is found. The official language is Malay, but English and Chinese are spoken as well, reflecting Brunei's commercial and ethnic ties to China and its historical relationship with Great Britain. Muslim (Sunnite) is the official religion practiced by sixty-three percent of the population. Buddhism is practiced by fourteen percent of the people, mainly Chinese.

Although Brunei is a constitutional *sultanate,* the sultan dominates the country's political process, ruling by decree under a state of emergency first imposed in 1962 following a rebellion. It has been renewed every two years ever since, and no legislative elections have been held during this time. The sultan, a traditional Islamic monarch, is both chief of state and head of the government. A Council of Cabinet Ministers is appointed and presided over by the sultan. Several councils, Religious (religious matters), Privy (constitutional matters), and Succession (issues of succession to the throne) are appointed by the sultan. The Supreme Court, chief justice and judges, are sworn in by the sultan for three-year terms.

Both daily life and culture are dictated by the national ideology of *Melayu Islam Beraja* (MIB, or Malay Muslim Monarchy). Formulated by the country's current sultan, Hassanal Bolkiah, it integrates traditional Muslim values and practices with loyalty to the monarchy. Buoyed by oil wealth, Brunei's standard of living is second only to Japan's for the Southeast Asia region. The government controls the ownership of automobiles, and only the Sultan can own a Rolls Royce. There is no taxation on personal income.

History

There is evidence of trade between Brunei and China as early as the sixth century A.D. Contact between Brunei and India brought Buddhism and Hinduism to Brunei by the sixth and seventh centuries. Beginning in the seventh century, Brunei was dominated by two powerful kingdoms in succession. The Sumatran Srivijaya kingdom, based in southern Sumatra, rose to dominance in the seventh century and prevailed for hundreds of years. It was followed by the Indonesian Hindu kingdom of Majapahit (c. 1200-1400). As Majapahit declined in the late fourteenth and early fifteenth centuries, an Islamic sultanate emerged in Brunei. At the height of its power it ruled over an area encompassing most of northern Borneo and extending to the Philippines. By the end of the sixteenth century, however, the power of the sultanate began to decline due to internal unrest. In the seventeenth century Dutch expansion weakened the sultanate's influence even further. By the beginning of the nineteenth century, the territory ruled by Brunei's sultan included only present-day Brunei itself, Sarawak, and North Borneo (now called Sabah).

From the middle of the nineteenth century, British influence in the region grew. A treaty of friendship was signed with Brunei's sultan in 1847. By 1888 Brunei was declared a British protectorate, retaining internal autonomy while Great Britain took charge of its foreign affairs, keeping rival European powers at bay. Early in the twentieth century, a resident British commissioner was installed in the country, although his responsibilities were still confined to defense and foreign relations. At the end of the 1920s, large reserves of oil and natural gas were discovered, a discovery that was to transform the future nation into one of the wealthiest states in the region. (In addition, offshore reserves were discovered in the 1960s.)

Brunei officially attained full internal autonomy in 1959, and its first scheduled elections were held in 1962. They were won by an Indonesian-backed nationalist party, the Brunei People's Party, but the sultan prevented the newly elected government from taking office. In response, the nationalists staged a rebellion against Sultan Omar Ali Saifuddin, but it was put down with help from Britain. The sultan then declared a state of emergency and gave himself the right to rule by decree. From that time on, Brunei's sultans have renewed the emergency decree every two years, and the country's political process remains in suspension to the present day. There are no elections, political parties are outlawed, and dissidents are detained without trial. The legal system is governed solely by Islamic law. 1967 Omar Ali Saifuddin abdicated; his eldest son, Muda Hassanal Bolkiah ascended the throne the following year.

Brunei became a fully independent member of the British Commenwealth in 1984. In 1992, the nation celebrated Sultan Hassanal Bolkiah's twenty-fifth year on the throne. One of the world's wealthiest men, he is also an extremely popular and accessible ruler, traveling widely among his people without bodyguards—an unusual practice for any head of state.

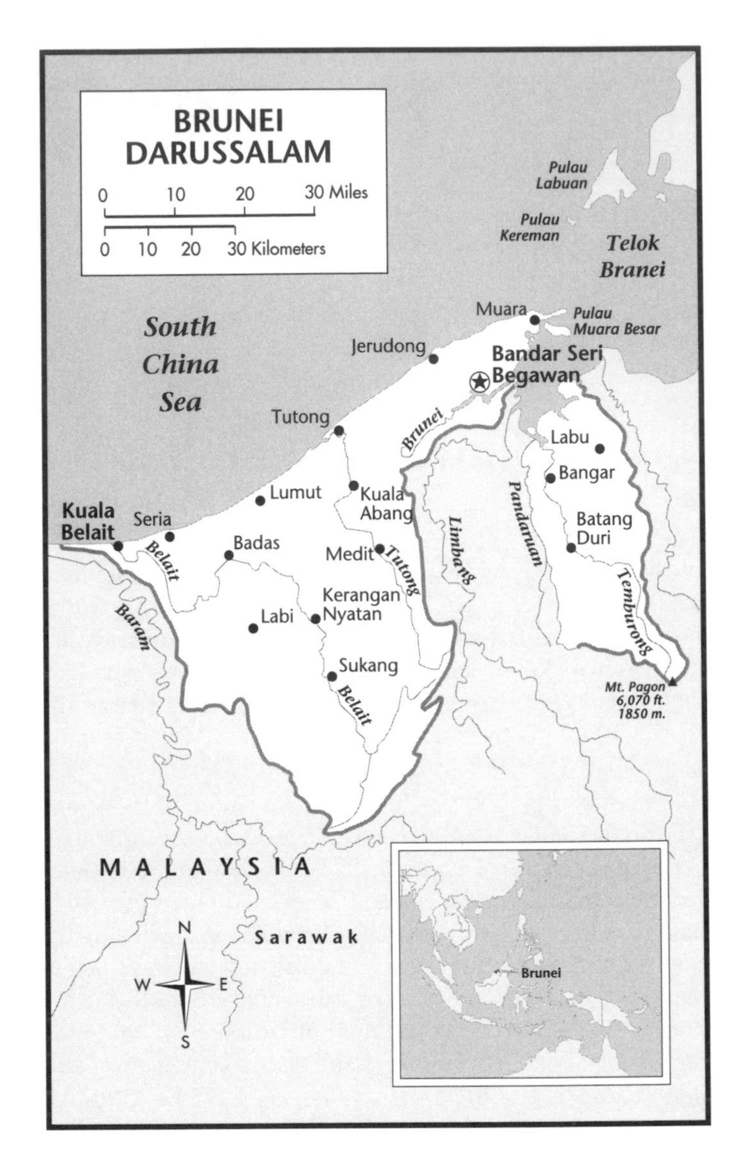

Timeline

40,000–38,000 B.C. Occupation of the Niah Caves near Niri, Sarawak

Archaeological evidence—the discovery of a human skull—suggests that early humans occupied the Niah Caves during this period. Rock paintings from around A.D. 700 depicting funerary rituals are also found within the cave system.

Early Regional Developments in the Malay Archipelago

6th–7th centuries A.D. Contact between India and Brunei

Hindu and Buddhist statues dating from the sixth and seventh centuries indicate contact between Brunei and India.

6th–7th centuries Contact between China and Brunei

Chinese records mention the export of gold and silver from the Sarawak region in the beginning of the seventh century. There are also records of tribute (gifts of goods or money for which favors are expected in return) paid by the king of Brunei to the Emperor of China. Brunei is called variously *Puni* or *Po-ni*, *Puli*,or *Polo*.

950 Sribuza, ancient Brunei

An Arab sea captain writes of a place located on Brunei Bay. It is ruled by members of the royal family of Funan (present-day Cambodia). Funanese are seafarers. Other sources mention the existence of a trading center simultaneous with the existence of Sribuza.

7th–14th centuries The Sumatran kingdom, Srivijaya, dominates the region

The kingdom of Srivijaya is near Palembang on the southern end of Sumatra. Dominance of the Strait of Malacca controls

Indian and Chinese trade, as well as that of Sumatra, Java and the Malay peninsula. Thirteenth-century Chinese records list some products: resins, camphor, sandalwood, cloves, cardamom, elephant tusks, and pearls. India's influence spreads further through Hinduism and Buddhism. As a stronghold of Mahayana Buddhism, Srivijaya attracts scholars and pilgrims. When Marco Polo lands in North Sumatra on his return to Venice from China in 1292, he writes of an Islamic town.

1200s–1400s Majapahit dominates the region

In the early 1200s Majapahit, a Hindu kingdom in East Java, rises to prominence. Its sphere of influence includes Sumatra, Brunei and Bali. A famous chief minister, Gadjah Mada, is highly regarded in Indonesia. He is virtual ruler from 1331 to his death in 1364. He expands the empire, codifies law, and commissions the epic poem, *Nagarakertagama*. However, conditions deteriorate in the late fourteenth century, and civil war breaks out by the beginning of the fifteenth century. Hostility among Islamic and Hindu-Buddhist traditions hastens Majapahit's end.

1215 Written account of P'o-ni by Chao Ju-kua

Chao describes P'o-ni as a settlement of 10,000 people in fourteen districts. According to Chao's account, the king's palace is built of wood and thatched with palm. The native inhabitants worship two sacred pearls enshrined in a temple.

The Emergence of Brunei

1360s Brunei's Islamic sultanate is founded

Brunei's ruler, Alak Betatar, converts to Islam and takes the name Sultan Muhammad Shah, thus creating a sultanate (kingdom ruled by a sultan), which grows stronger in the early fifteenth century, as Majapahit's power wanes.

1375 Chinese dignitary Wang Sen-ping governs Sabah region

The region comes under the control of Chinese dignitary Wang Sen-ping when his daughter marries the sultan of Brunei. Wang Sen-ping's granddaughter eventually marries an Arab, who ascends to the throne as Sultan Ali Sharif.

1400–1511 Malacca empire emerges

Around 1400 the port city, Malacca, is founded by Parameswara, a descendant of the royal house of Srivijaya. He flees Palembang after an attack by the Majapahit forces. Once a follower of Hinduism and Buddhism, he converts to Islam and takes the name Iskander Syah. His successors are Islamic. Malacca dominates trade during the fifteenth century thus promoting the spread of Islam in the archipelago.

1408–26 Sultanate of Sultan Ahmad

The second sultan of Brunei, Sultan Ahmad, rules from 1408–26. He marries a Chinese woman, and their daughter marries Sharif Ali who succeeds his father-in-law as sultan.

1426–32 Sultan Sharif Ali, third Sultan of Brunei

Sultan Sharif Ali, an Arab, becomes the third sultan of Brunei. He receives the right to the sultancy by marrying the daughter of Sultan Ahmad (see 1408–26).

1432–85 Reign of Sultan Sulaiman

Sultan Sulaiman reigns as the fourth sultan of Brunei.

1485–1524 Rule of Sultan Bolkiah

Sultan Bolkiah ascends the throne as the fifth Sultan of Brunei. During his rule, Brunei experiences regional growth in power which constitutes a "golden age" for Brunei. Sultan Bolkiah is a popular sultan, remembered as a great warrior and navigator.

1511 Portuguese capture Melaka

The Portuguese capture Melaka (Malacca) and take control of trade there. Displaced Indian-Arab Muslim merchants switch their trade to Brunei.

1521: July 8 Magellan's crew lands in Brunei

Ferdinand Magellan and his crew are given a lavish reception by the king of Brunei. The city is built over the water on stilts.

1530 The Portuguese visit Brunei

Portuguese explorers and merchants, operating from their regional base on neighboring Melaka, visit Brunei.

1578: March 14 Spanish naval expedition against Brunei

Under the Captain-General of the Philippines, Francisco de Sande, the Spanish send a naval expedition to Brunei. They demand tribute (forced payment of money or goods) and order the sultan to become a vassal of the King of Spain. The Spanish also demand that Catholicism replace Islam as the religion of Brunei. The sultan's refusal to honor these demands triggers a battle that ends in a stalemate: the Bruneians lose, but the Spanish also retreat due to illness. The following year the second Spanish attempt to gain control over Brunei is defeated in a battle off Maura.

1641 Dutch capture Melaka

The Dutch capture Melaka from the Portuguese. They then divert trade from smaller ports like Brunei, to Batavia (Jakarta), Indonesia.

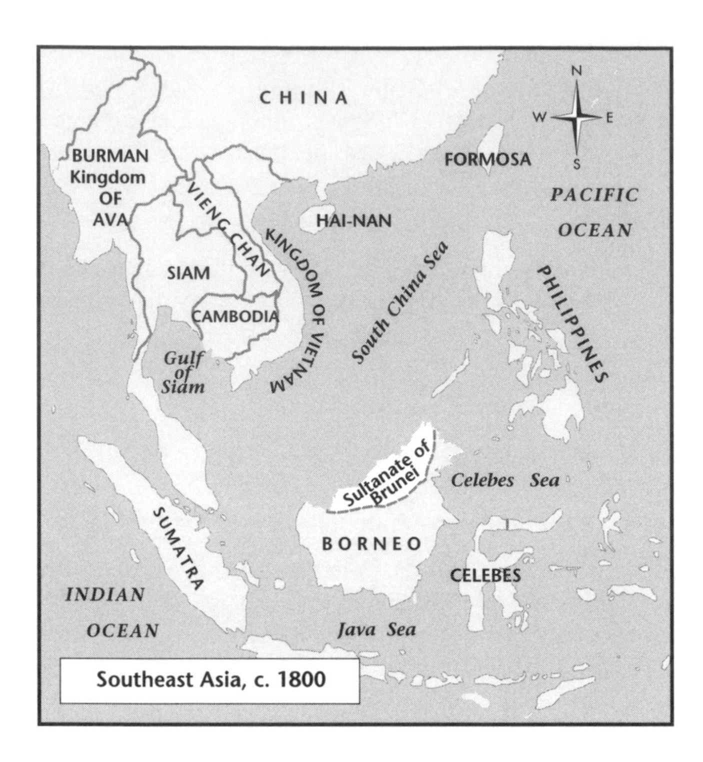

Southeast Asia, c. 1800

1700s European influence grows as sultanate's power declines

The decline in Brunei's power, begun by the end of the sixteenth century, continues, as European dominance in the region grows. From the seventeenth century through the first part of the eighteenth, Dutch power is paramount. Then the Dutch are gradually eclipsed by the British in the region.

1800s England becomes dominant in the region

By the beginning of the nineteenth century, the territory governed by the sultanate of Brunei has been greatly diminished from its one-time expanse. The sultan controls only Sarawak, part of present-day Sabah (then called North Borneo), and the area that currently constitutes the state of Brunei.

1828–52 Reign of Sultan Omar Ali Saifuddin II, twenty-third Sultan of Brunei

Omar Ali's mother prevents her father from being apppointed *Yang Di Pertuan* i.e., installed as king. Sultan Muhammad Alam's son, Raji Api, fights for the throne within his family, but Omar Ali Saifuddin II is installed. Raja Api's brother, Raja Muda Hassim, turns to Englishman James Brooke. This sequence of events is the ruin of Raja Muda Hassim and his family.

1841 James Brooke made Governor of the territory of Sarawak

Brooke forges a political alliance with Raja Muda Hassim. The Raja requests Brooke's help in suppressing a rebellion. Brooke's success later results in Raja Muda Hassim's support of his governorship of Sarawak. This alliance is the beginning of a dynasty of independent "white rajahs" who govern Sarawak until 1946. Sir Charles Anthony Johnson Brooke, Sir James' nephew, succeeds him in 1868. In 1917 his eldest son, Charles Vyner de Windt Brooke, succeeds him. Vyner Brooke attempts to establish self-government in Sarawak. After the Japanese occupation in World War II, he ends the family's rule on July 1, 1946, but not without a bitter family feud.

1846 James Brooke made Rajah (king) of the territory of Sarawak

The Sultan gives Brooke Sarawak in perpetuity. Brooke declares Sarawak independent. He proclaims himself an independent rajah of Sarawak.

1846 Massacre of entire Raja Muda Hassim family

Nobles unhappy with Brooke's measures, arrange to have Muda Hassim's family killed.

1847: May 27 Treaty with Great Britain

Treaty of Friendship and Commerce is signed between Her Majesty and the Sultan of Borneo [Brunei], Sultan Omar Ali Saifadeen. The sultan gains British assistance for the suppression of piracy and continued commercial relations. The British acquire rights to the island of Labuan in Brunei Bay.

1852–85 Reign of 24th Sultan, Sultan Abdul Mumin

Sultan Abdul Mumin is not aligned with the two major families whose succession has been disputed. His reign is of great importance because during this time both Sarawak and the British North Borneo Company encroach on Brunei territories. From his weak position he gathers support by having his chief minister make an *amanah* (sacred oath) not to divide Bruneian territory any further.

1863 Treaty with Sarawak

A treaty is made with Sarawak ceding the Bintulu and Mukah territories.

1877 North Borneo ceded to British merchants

The sultan of Brunei leases North Borneo to British merchants who later establish the British North Company. (See 1881.)

1881 British North Borneo Company (BNBC) receives royal charter

The British North Borneo Company is officially chartered. (In the twentieth century the territories it controls are known as Sabah.)

1885–1906 Sultan Hashim Jalilul Alam

Sultan Hashim Jalilul Alam reigns as the twenty-fifth sultan.

1888: September 17 British-Brunei [Protectorate] Treaty

Under Sultan Hashim Jalilul Alam Akamadin, Brunei becomes a British protectorate. Sarawak and North Borneo also become protectorates. While Brunei retains internal independence, foreign relations are conducted by Britain.

1899 Revolt

Two Bruneian districts, wanting to be ruled by Sarawak, stage a revolt.

1899 First oil exploration

A well is drilled near Bandar Seri Begawan, the capital, in the first attempt to find oil in Brunei. Although the site does not yield a well suitable for commercial production, it triggers interest in oil prospecting.

The Residential System

1905: December 5 and January 2, 1906 British-Brunei [Protectorate] Document

A Resident British commissioner is established. Further encroachment on Bruneian territory is halted. Britain has the right to administer Brunei's internal and external affairs, except in regard to Malay custom and Islamic religious affairs.

1906–24 Sultan Muhammad Jamalul Alam II

Sultan Muhammad Jamalul Alam II, the twenty-sixth Sultan, reigns for eighteen years.

1908–09 The introduction of rubber in Malaysia and Brunei

The native home of the rubber tree is the Amazon basin of Brazil. Vast rubber estates (plantations) owned by British companies in Malaysia and Brunei are worked by people from Tamil Nadu, India.

1914 The first Malay school opens

A school offering instruction in the Malay language opens in Brunei Town, the first to open in Brunei.

1914 First well strikes oil

Anglo Saxon Petroleum Company (later Brunei Shell Petroleum Company) strikes oil at Belait-w. Although the well is not commercially viable, Anglo Saxon oil does apply for and subsequently acquire all oil rights in Brunei.

1916 The first Chinese school opens

The first school to offer instruction in the Chinese language opens in Brunei Town.

1924–50 Reign of the twenty-seventh Sultan, Sultan Ahmad Tajuddin II

The twenty-seventh Sultan, Sultan H. H. Ahmend Tajuddin Akhazul Khairi Wad-din, ascends the throne at eleven years of age due to his father's death from malaria at age thirty-five.

1929 Discovery of an oil field at Seria

The first commercially viable oil well is drilled at the large oil field at Seria, the first successful well in Brunei. By 1931 oil production transforms Brunei's economy.

1931 An English school opens

A school opens in Kuala Belait, offering instruction in English.

1934 Palace of the sultan completed

The largest residential palace in the world, the home for the sultan, is completed at an exorbitant cost. The royal family and their staff occupy the nearly 2,000 rooms reached by over forty staircases. The banquet hall accommodates 4,000 guests.

1939 Brunei Youth Front is formed

The Brunei Youth Front, considered the first organization with poltical aims, is formed. It establishes the begining of nationalism in Brunei.

1941: December 16 Japanese forces attack Brunei

The Japanese, engaged in World War II (1939–45), stage a successful attack on Brunei.

1942–45 Brunei under Japanese control

Brunei falls to Japan for the duration of World War II.

1945: June 10 Allied Forces land in Brunei

Brunei is liberated from Japan when the Allies land in Brunei.

1946: July British civil administration is restored

Both Sabah (British North Borneo Company) and Sarawak, under Rajah Vyner Brooke, surrender to the British government and become Crown Colonies.

1950–67 Reign of the twenty-eighth Sultan, Sultan Omar Ali Saifuddin III

Al Marhum Sultan Haji Omar Alu Saiduddin III is hailed as first absolute monarch and the father or architect of modern Brunei.

1954 District Advisory Councils (DACs) are established

Set up in all four districts of Brunei, Councils have ninety-one members. One third are chosen by the people and the remainder are appointed by the government.

1958 Construction of mosque completed

The Munshi Ahmed Omar Ali Saifuddiend Mosque, one of the most spectacular mosques in Asia, is constructed.

1959: September 29 1959 Brunei Agreement

The 1959 Brunei Agreement between the United Kingdom and Brunei on Defence and External Affairs revokes the 1906 Agreement and grants Brunei internal self-government with Britain retaining responsibility for defense and foreign affairs.

1960: May 29 Formation of the Barisan Buruh Bersatu Brunei

The Brunei Labour Front (BULF), a trade union organization, becomes involved in politics. It is later outlawed after the 1962 rebellion (see 1962: December 8).

1962: August Brunei's first elections

The Brunei People's Party (Parti Rakyat Brunei, PRB) is victorious. The new Legislative Council is not convened, so these militant nationalists stage an armed uprising.

1962: December 8 Brunei Rebellion

Comprised of a small group who want to do away with the monarchy, the victorious nationalist PRB are prevented from taking office. They stage a revolt, with Indonesian backing, against Sultan Omar Ali Saifuddin and the Malaysia proposal. They are quickly put down by the sultan, who has British assistance, but the sultan decides against federation in any case. From this time on, the sultanate is ruled by decree under a national state of emergency. The legislature never convenes, and many sections of the constitution are ignored. From this point forward, the ruling sultan renews the "emergency" every two years.

1963: January Brunei Alliance Party (BAP) is formed

Two smaller pro-Malaysia parties merge. The BAP supports the Malaysia proposal and Sultan Omar Ali Saifuddien's policy. When the Sultan decides to keep Brunei out of Malaysia, the BAP continues to support Malaysia. The party comes to nothing.

1965 Hassanal Bolkiah marries

The oldest son of the sultan, Hassanal Bolkiah, marries his cousin, Princess Saleha. Marriage between cousins is common in Brunei; the sultan's wife is also his cousin.

1966: August Political groups merge

The Parti Barisan Kemerdekaan Rakyat/PNKR (Brunei's People's Independence Party) forms. The PNKR requests of Britain swift independence. Britain is willing to grant independence but Sultan Omar Ali Saifuddien III opposes Britain's stance. (See 1967: October 4.)

1967: October 4 Sultan Omar abdicates

Omar Ali Saifuddien III abdicates rather than weaken the monarchy by complying with Britain's wishes to grant independence to Brunei. (Some observers speculate that the sultan is encouraged by the British to retire.) The sultan's eldest son, Muda Hassanal Bolkiah, becomes Sultan Hassanal Bolkiah, the twenty-ninth Sultan of Brunei.

1968: August 1 Coronation of Hassanal Bolkiah

Twenty-two-year-old Hassanal Bolkiah is officially crowned the twenty-ninth Sultan of Brunei.

1975: May 14 Royal Brunei Airlines

Royal Brunei Airlines takes its inaugural flight.

1979: January 7 Brunei renegotiates treaty with the United Kingdom

A new treaty, the Treaty of Friendship and Cooperation with Britain, provides for independence within five years.

1981: October 28 Sultan Hassanal Bolkiah marries his second wife

Mariam (or Miriam) binti Abdul Aziz, a former Royal Brunei airline stewardess, becomes the sultan's second wife. The sultan's father and younger brother, Mohamed, both disapprove of this marriage to a commoner and refuse to recognize the marriage. The sultan's other brother, Jefri, approves of the marriage. The marriage is not made public until one year later.

1984: January 1 Brunei attains full independence from the United Kingdom

Brunei becomes the one hundred and fifty-ninth sovereign nation in the world. It also becomes a member of the British Commonwealth. The United States opens an embassy in Bandar Seri Begawan on this same date. The monarchy is buttressed by a revised constitution that consolidates power within the family. Later that year Brunei joins the Association of South-East Asian Nations (ASEAN), the Organization of Islamic Countries (OIC), opens an embassy in Washington D.C., and is admitted to the UN.

1984 Sultan declares *Melayu Islam Beraja* the state philosophy

Melayu Islam Beraja (Malay Muslim Monarchy—MIB) is declared the official state philosophy by the sultan. By emphasizing traditional values MIB is interpreted as an attempt to counter Western values and modernization.

1985 Education reforms

Brunei schools begin an era of reform. Bilingual education, with instruction in both Malay and English, is begun in government schools.

1985 Brunei National Democratic Party forms

The Brunei National Democratic Party (BNDP) is formed, made up predominantly of businessmen loyal to the Sultan.

1986 Brunei National Solidarity Party

This party's membership is open to Chinese, unlike the BNUP (see 1986). Abdul Latif bin Chuchu is elected president but resigns a few months later.

1986 Brunei National United Party forms

An offshoot of the BNDP (see 1985), the Brunei National United Party (BNUP) emphasizes greater cooperation with the Government. The BNUP's membership is open to Muslim and non-Muslim, but still excludes Chinese.

1986: September Death of Sultan Omar

Sultan Omar, father of Sultan Hassanal Bolkiah, dies. After his abdication (see 1967: October 4), Sultan Omar served as Defense Minister and assumed the royal title of Seri Begawan.

1986–87 U.S. Government solicits Brunei for funds to aid the Nicaraguan Contras

The $10 million donation is credited to the wrong bank account and never reaches its destination. It is traced and returned to Brunei with interest.

1988 Top leaders arrested

The two top leaders of the BNDP, President Haji Abdul Latif bin Abdul Hamid and Secretary-General Haji Abdul Latif bin Chuchu, are arrested as they are about to fly to Australia. Held under the Internal Security Act, which allows detention for up to two years without charges being filed, they are detained until 1990.

1990: May Haji Abdul Latif bin Abdul Hamid dies

Haji Abdul latif bin Abdul Hamid, President of the BNDP, dies.

1990 A national ideology is promoted by Sultan Hassanal

Melayu Islam Beraja (Malay Muslim Monarchy—MIB) is reaffirmed as a state ideology promoted by the sultan. It affirms traditional values by stressing Malay culture, Islam, and the monarchy.

1991 Sale of alcohol is banned and stricter dress codes are introduced

The sultan's emphasis on traditional values extends to a ban on the import of alcohol, a stricter dress code, and a ban on the public celebration of Christmas.

1992 Malay Muslim Monarchy taught in schools

Malay Muslim Monarchy (MIB) is made a compulsory subject in the schools.

1992: October Sultan celebrates his Silver Jubilee

His Majesty Sultan Haji Hassanal Bolkiah Mu'izzaddin Waddaulah celebrates his Silver Jubilee, marking twenty-five years on the throne.

1993 Brunei establishes diplomatic relations

Brunei establishes diplomatic relations for the first time with China, Vietnam, and Laos.

1994: March The East Asian Growth Area (EAGA) is created

Brunei, the Philippines, Malaysia and Indonesia create the EAGA. Four sectors, tourism, air transport, shipping, and fisheries, are earmarked for expanded trade among the members. The Muara Export Zone opens as a regional transshipment center.

1994: November 29 Defense cooperation

Brunei and the United States sign a memorandum of understanding on defense cooperation.

1995 Sultan extends emergency rule

Although emergency rule is routinely extended every two years (see 1962), this is the first time the sultan publicizes his decision to extend it.

1995 Brunei ends appeals to Privy Council in London

The right to appeal to the Privy Council in London ends for criminal cases. Only civil cases may be appealed to the Privy Council.

1995 Tariffs on imported cars enacted

Despite no automobile manufacturing in Brunei, almost 150,000 private cars are in the country. Tariffs are increased from a flat rate of twenty percent to rates ranging from forty to two hundred percent depending on the car model. Thus the sultan attempts to control private car ownership, since the small country's system of roads suffers from so many vehicles.

1995: February The Brunei Solidarity National Party holds inaugural assembly

The party supports the Sultan despite being dormant for nearly ten years.

1996 Michael Jackson performs for Sultan's birthday

U.S. pop superstar Michael Jackson performs for the Sultan's fiftieth birthday.

1997 Sultan is richest man in the world

Fortune and *Forbes* magazines estimate the Sultan of Brunei's personal wealth at US $38 billion, a figure that makes him one of the richest men in the world.

1997 Royal scandal develops over claim of imprisonment in harem

A former Miss USA, Shannon Marketic, files a lawsuit against the sultan's brother, Prince Jefri Bolkiah. She claims that she was hired to work as a model in Brunei but then imprisoned in Prince Jefri's harem, which contains some sixty women. The case is thrown out of a U.S. federal court because the sultan and his brother, Prince Jefri, enjoy diplomatic immunity.

1997 Relaxing the rules to promote tourism

In a continuing effort to promote tourism, alcohol is served as "tea" (actually beer poured from teapots), women are unveiled in public, and choices for entertainment are expanded, via satellite, video, and movies.

1998: August Sultan Hassanal Bolkiah selects son as heir apparent

Prince al-Muhtadee Billah, age twenty-four, is proclaimed heir to the Islamic Sultanate of Brunei. The oldest son of the sultan by his first wife, Queen Saleha, Billah is crowned heir in a lavish ceremony.

1998: August Prince Jefri of Brunei leaves the country

Reports in the international media describe Prince Jefri, the brother of the Sultan of Brunei, as having vanished. He may have embezzled $8 billion from the Brunei treasury, through his company, Amedeo. Media reports claim his whereabouts are unknown, while Prince Jefri has in fact, traveled to England to escape controversy over his finances.

Bibliography

Bartholomew, James. *The Richest Man in the World: The Sultan of Brunei.* London: Penguin Group, 1990.

Chalfont, Lord Alun. *By God's Will: A Portrait of the Sultan of Brunei.* London: Weidenfeld and Nicolson, 1989.

Cleary, Mark and Shuang Yann Wong. *Oil, Economic Development and Diversification in Brunei Darussalam.* New York: St. Martin's Press, Inc., 1994.

"Fairy Tale's Over for the Kingdom of Brunei." *Fortune,* February 1, 1999, 90.

Leake, David, Jr. *Brunei: The Modern Southeast-Asian Islamic Sultanate.* Jefferson, N.C.: McFarland & Company, Inc. 1989.

Major, John S. *The Land and People of Malaysia & Brunei.* New York: HarperCollins 1991.

Pigafetta, Antonio. *Magellan's Voyage.* Trans. and ed. by R. A. Skelton. New Haven: Yale University Press, 1969.

Singh, D.S. Ranjit. *Brunei 1839–1983: The Problems of Polticical Survival.* London: Oxford University Press, 1984.

Vreeland, N. et al. *Malaysia: A Country Study.* Area Handbook Series. Fourth edition. Washington, D.C.: Department of the Army, 1984.

Cambodia

Introduction

Cambodia is a small, rural, agricultural country in Southeast Asia. About the size of the state of Oregon, Cambodia is ringed by rugged forested mountains on its borders with Thailand to the north and west, Laos to the northeast, and Vietnam to the east. Its southwestern border is the Gulf of Thailand. Cambodia's history includes five centuries as the greatest empire in Southeast Asia followed by centuries of decline at the hands of its neighbors Vietnam and Thailand. This was followed by almost a century as a French colony, four years as a concentration camp and killing field, and a current reputation as a hotspot always ready to explode into violence and war. Through time, Cambodian leaders have had to alternatively compromise and acquiesce to the demands of Thailand, Vietnam, France, and the United States in order to maintain a semblance of independence.

Cambodia lies completely within the tropics. Its climate is monsoonal, with markedly wet and dry seasons of approximately equal length. Rain falls between April and September, and the dry season lasts the other six months. Temperatures and humidity are high throughout the year. Forests, which previously covered about two-thirds of the country, are being cut down at an incredible rate; some scientists say Cambodia's forests will be gone soon unless the government takes steps to halt or slow this process.

The Kingdom of Funan, located in the lower Mekong River delta region and founded in the first century A.D., is generally believed to be the first Khmer (Cambodian) kingdom. By the fifth century, Funan controlled much of present-day Cambodia, with its population concentrated on the lower Mekong River and the Tonle Sap River below the Great Lake. Around 500 A.D., the Kingdom of Chenla attacked Funan and brought an end to its dominance. They moved the capital to Angkor Borei.

Cambodia's primary resource has always been its land, and agricultural development has been the cornerstone of every Khmer administration since the third century, except for Lon Nol's rule from 1970 to 1975 when the United States supported the country. The staple food of Cambodia is rice, followed closely by fish. Cambodians grow rice in the lowlands, principally around the two great rivers of Cambodia: the Mekong and the Tonle Sap. Because this is where the rice is located, the population also is concentrated primarily on the cultivated lands along the Mekong and Tonle Rivers and around the Tonle Sap Lake. The greatest concentration is in the southeast, along the lower Mekong River and the Tonle Basak Rivers. Most of the farming population clusters on elevations safe from the annual floods. Fish are abundant in the interior fresh lakes of rivers, especially in the Tonle Sap Lake in the middle of the country and in the Tonle Sap and Mekong Rivers.

Cambodian Society

More than eighty-five percent of the population are ethnic lowland Khmer. Other minority groups include the Chinese, Vietnamese, and Cham, who also live in the lowlands. Most Chinese live in urban areas, where they dominate commerce and finance. Many Vietnamese are fishermen or small businessmen in towns or in Phnom Penh. The Cham usually live in separate villages, where the vast majority also cultivate rice and fish.

To talk about Cambodia is to talk sooner or later of the Khmer Rouge years from 1975 to 1979, when Communist Cambodians ("Red Cambodians" or "Khmer Rouge" in French) ran the country. In an effort to return Cambodia to its glory of a thousand years before, its leaders attempted to return it to an agricultural state without western influence. After the destruction and devastation of the civil war in the early 1970s, the Khmer Rouge regime in the late 1970s, and the return to civil war since, Cambodia remains primarily a peasant society, in many ways worse off now than during the 1960s.

Cambodian villages are distinguished by being situated in rows, either along rivers or, more recently, along roads. Today, as for centuries, a typical Cambodian village is comprised of cultivators of wet rice who live in houses set among palm trees, fruit trees, gardens, and wild flora. Many village houses are built on stilts to protect their inhabitants from the rising waters. The rural homes of those better off are made of wood while those with fewer resources build their houses from thatch. Bamboo floors hold little furniture but allow air

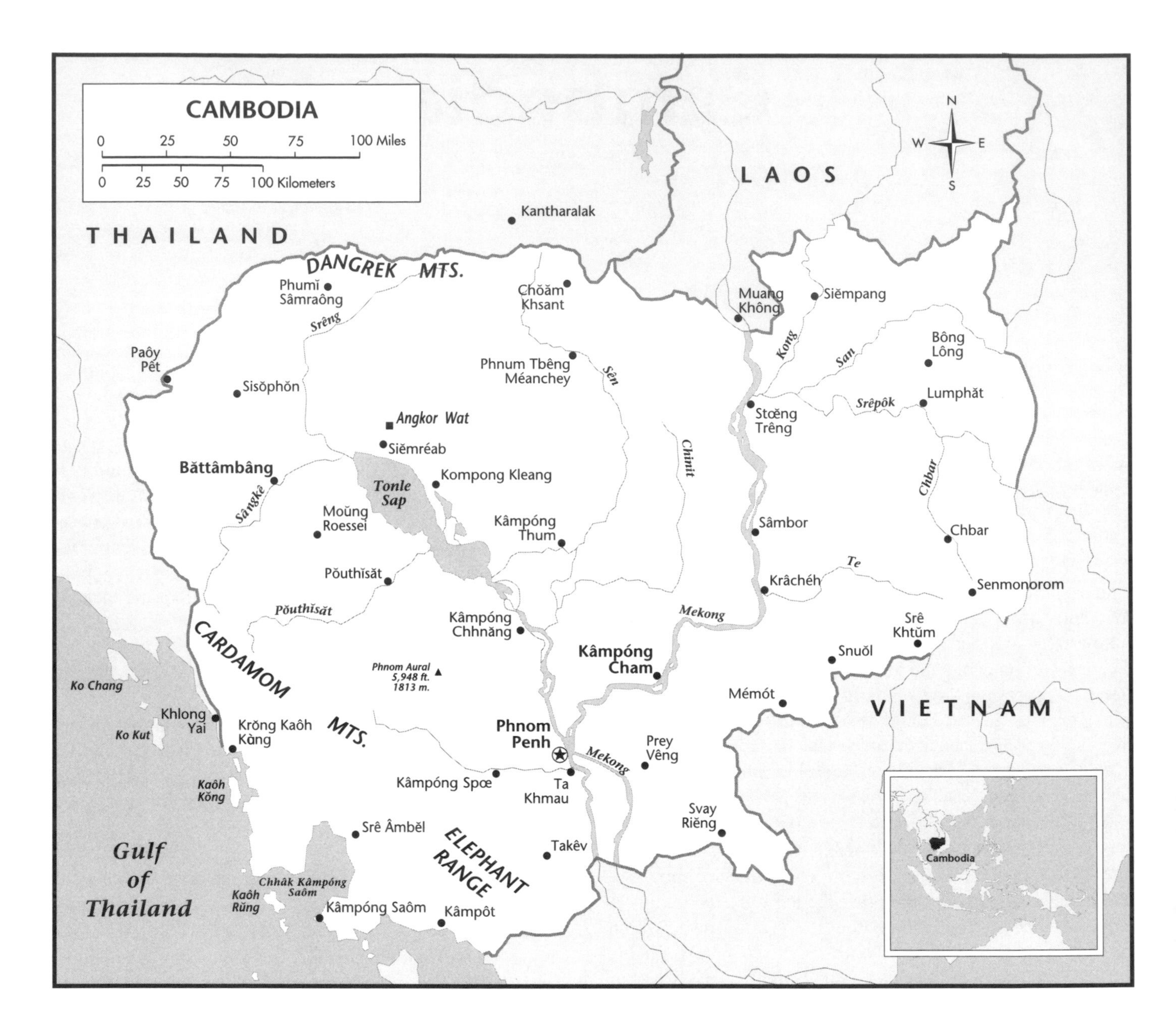

circulation and an easy cleaning system (food can be shoved through slits in the floor to chickens, pigs, and dogs waiting below). The village itself sits amidst a patchwork quilt of ricefields. Most rice cultivation is done by draft animals, either oxen or water buffalo, and by plow.

Most Cambodians live in households containing nuclear families, sometimes with extended relatives also in the household. Each household is the primary unit of production and consumption, with most households raising enough food to support the family and bring in a little extra money. Families and neighbors do much of their own labor, although cooperative work parties made up of extended family members and fellow villagers are common during heavy work periods. Rice cultivation involves heavy work at various times in the planting season: plowing the fields to prepare them for the seed and seedlings, transplanting the seedlings to the ricefields, harvesting, and threshing.

As primarily subsistence cultivators, Cambodians also cultivate their own gardens and fruit trees, fish during the rainy season, and produce additional household items such as mats, basket, and twine. Most families earn extra money by raising pigs and chickens, making palm sugar, selling their surplus crops, selling cooked food, and working as laborers during the slack rice cultivation season.

Villagers are joined as kinfolk, friends, and neighbors. Villages are also the lowest level of an administrative hierarchy that ascends to the national level. The political turmoil of the past three decades has caused each government—of whatever stripe—to emphasize Cambodian national identity. Thus, despite the tensions among them, the ongoing threat of war, and a continuing concern for obtaining the necessities of life, Cambodians are tied together by identity as fellow sufferers of the Khmer Rouge period and as residents of the country of Cambodia.

Cambodians are also tied together by religion. Virtually every village in Cambodia houses a Theravada Buddhist temple complex. The temple serves as a moral, social, and educational center for villagers. Monks are accorded the highest respect as living embodiments of the Buddha and moral virtues. Cambodians believe that human beings go through a cycle of reincarnations and that meritorious deeds are rewarded in the next life by an improvement in one's life. In contrast, being reborn in a worse position will punish evil deeds. Cambodians try to earn merit by becoming monks; by supporting the monks and the temple by giving money, food, and service; by attending religious services and rituals; and by following Buddhist rules of conduct. These include, most importantly, not killing living creatures, not stealing, not lying, not engaging in improper sexual relationships, and not drinking alcohol.

Cambodians's major annual festivals include the New Year, Prachum Bon or Festival of the Dead in which Cambodians honor their ancestors, and Katun, during which Cambodians give gifts to the monks. In addition to earning merit at their festivals and ceremonies, Cambodians socialize with their relatives, friends, and fellow villages.

During the Khmer Rouge years, all semblance of regular life disappeared. Families were separated, children became mobile work teams, and villagers were dispersed and recombined in ways that destroyed old relationships. Work was done in cooperative groups under the command of Khmer Rouge soldiers. There was no leisure time, socializing, traveling, or celebrating. Religion was banned, as were all types of rituals to mark the normal passages of time, such as weddings and funerals. Healing rituals were forbidden. Villagers did not follow the rhythms of the land, seasons, family, or village. Rather, they were subject to the capriciousness of young soldiers.

Since 1979, when the Vietnamese pushed the Khmer Rouge from power, life has returned to some normalcy. Despite continued civil war and disputes over control of the country, many Cambodians have returned to the rhythms of peasant life. They center on their families, practice their religion, struggle to feed their families, and dream of better lives for their children.

Cambodia's urban areas are located along navigable rivers. The capital Phnom Penh is located at the confluence of the Mekong and Tonle Sap Rivers. City folk have always lived different lives from lowland rice farmers. The lives of urban dwellers center on their various occupations: soldier, policeman, government official, small businessman, bicycle repairman, hair dresser. Urban Cambodians live in villas, apartments, or shacks, depending on their financial resources. The greater one's financial resources, the greater access one has to education, health, and other opportunities. Since those with more resources live primarily in the cities, more urban Cambodians have access to services and resources than do most rural Cambodians.

The hill people of Cambodia live differently from any of the lowland Cambodians, either urban or rural. Upland Cambodians are tribal peoples who speak their own languages, practice their own religion and customs, and support themselves by slash-and-burn farming. They cut and burn trees, plant their crops in the newly prepared ground for several years, then move on to prepare more fertile grounds elsewhere. Isolated from lowland Cambodians by mountainous terrain, distance, and a climate that creates formidable travel obstacles much of the year, highland Cambodians are often looked down upon by other Cambodians. Highland Cambodians were a source of troops for Communists rebels in the late 1960s and early 1970s and continue to be viewed by many Cambodians today as strange and non-Khmer.

Cambodia today includes not only the Cambodians within its borders but hundreds of thousands of Cambodians resettled in other lands. They can vote in Cambodia's elections and maintain intense interest in Cambodia's affairs. Over 170,000 Cambodians now live in the United States.

Timeline

7000 B.C. Hunters and gatherers in Cambodia

Hunters and gatherers occupy Cambodia.

2000 B.C.–A.D. 1 Neolithic occupation

In the first and second millennia, according to archaeological evidence, Neolithic peoples occupy Cambodia. These people may be Malayo-Polynesians, ancestors to the people who later occupy insular Southeast Asia and many of the Pacific Ocean islands. These groups probably mix with the Mon Khmer, who arrive from the north. The Khmer of Cambodia probably migrate from southeast China before the beginning of the Christian Era and before the arrival of neighboring Lao, Siamese, and Vietnamese.

These Neolithic groups are defined as such because they worked metal, including iron and bronze.

By the beginning of the Christian Era, these inhabitants live in organized societies located primarily along the coast and in the lower Mekong River Valley and delta areas, close to water and transportation routes. There they cultivate irrigated wet rice and utilize domesticated animals.

A.D.100s–1,000 Development of Khmer civilization

Subjected to influence from India, the region becomes one of several "Indianized" states on the Southeast Asian peninsula.

Indian influence includes philosophy, literature, art, architecture, social organization, law, the monarchy, and religion.

A.D.100s Kingdom of Funan

Funan is one of the earliest and most significant of the Indianized states. Funan is located in the lower delta areas of the Mekong River created, according to legend, from the union of an Indian Brahman named Kaundinya to the daughter of the Naga snake god, a king who lived and ruled over water and the underworld. The father of the young prince Kaundinya gave his son and daughter-in-law the land of Cambodia as a wedding gift.

The Funan Empire is supported by agriculture and fishing. Its expansion is probably due to intensive rice cultivation made possible by an extensive irrigation system (the existence of this system was confirmed by archaeological work in 1995 and 1996). Maritime trade is also important in the development of Funan, and traffic and communication with neighboring communities and kingdoms are primarily done by water throughout the region.

200s Chinese delegation visits Funan

The first historical reference to Funan is by a Chinese delegation that visits it. The name "Funan" may come from Chinese. Some modern scholars believe it to be a Chinese version of the Old Khmer word *bnam*, the modern form of which is *phnom*, which means mountain.

Early Cambodians vie for dominance over the rich and fertile lands of the lower Mekong River Valley and the central plains around the Tonle Sap, or Great Lake, in northwestern present-day Cambodia.

400s Expansion of Funan

Funan reaches its zenith with the accession of Kaundinya II to the throne. He has extensive control over his population and good relations with neighboring powers. Trade with India and China increases. King Kaundinya II sends two tributary missions to China, and relations with that powerful country improve.

By the fifth century, Funan controls much of present-day Cambodia, with its population concentrated on the lower Mekong River and the Tonle Sap River below the Great Lake. It also dominates neighboring regions of northern Cambodia, southern Laos, southern Siam, and areas of northern Malay Peninsula, all of which have to pay tribute to Funan.

Indian merchants, sailors, diplomats, and migrants continue Indian influence over the population of Funan. The elite use the Sanskrit language; in addition, an Indic alphabet is introduced and the Indian legal code is in effect.

The Khmer royal court and the elite adopt the Indian Brahmanic cult, which centers on a god-king. The Mahayana Buddhist religion is introduced into Cambodia from India.

514 Rudravarman ascends the throne

Rudravarman is the last king of an independent and autonomous Funan.

Early 500s Conflict in Funan

Dynastic strife and frequent civil wars destabilize the Kingdom of Funan. Its neighbors increasingly threaten the government and people.

Late 500s Decline of Funan

Formerly subject to it, the Kingdom of Chenla attacks Funan from the north. Chenla defeats Funan and the king flees. The capital of the country is changed to Angkor Borei in southern Cambodia.

500s–800 Expansion of Chenla

The people of Chenla are Khmer, who expand Chenla in all directions. In the early 600s, Chenla consolidates its power, becomes the dominant force in the area, and makes Funan a subject nation.

Over the next 300 years, the Kingdom of Chenla gains dominance over the areas of southern Vietnam and central and upper Laos and direct control of the populations located in western Cambodia and southern Siam. The first Chenla king to rule over the expanded territory is Bhavavarman, probably the grandson of Rudravarman, last king of an independent Funan.

700s Division into Land Chenla and Water Chenla

Members of the royal family engage in factional disputes over leadership, leading to a split in the kingdom that divides the country into two rival powers. The north becomes the Land Chenla, and the south becomes the Water Chenla.

A period of turmoil follows. The northern Land Chenla, or upper Chenla, maintains a much more stable existence, while the southern Water, or lower, Chenla, engages in considerable conflict. Lower Chenla is torn as rival dynasties fight for power.

Late 700s Attacks on Water Chenla

Water, or lower, Chenla is attacked by Indonesian pirates while the country is apparently a subject country under the Sailendra dynasty in Java.

790 End of Chenla

A ruler from Java, whom he offended, kills the last Water Chenla ruler.

802 Beginning of the Angkor period

After surmounting his rivals, the first ruler of a new Khmer state takes the throne. This prince, who has been ruling a minor state just north of the delta area, takes the name Jayavarman II. Before being named king of Water Chenla by the Javanese king, he may have lived temporarily in Java as an exile or hostage. Thus begins the Kambuja or Angkor era in Cambodian history.

802–1432 Angkorean period

Cambodia now develops into the dominant nation and culture in Southeast Asia. The Angkor period endures from the beginning of the ninth century to the middle of the fifteenth century and is known as the Classical Era in Cambodian history. India continues to spread its influence over the country. A literature in Sanskrit flourishes and the arts and architecture expand exponentially. The country reaches its zenith in size, dominance, and culture.

The Angkorean kings construct large barays, or manmade lakes, and a sophisticated system of canals and dikes which support an elaborate system of intensive agriculture with two, even three, rice crops a year.

Early 800s Expansion of the Angkor Kingdom

King Jayavarman II frees the country of Kambuja from Javanese suzerainty (overlordship) and begins unifying the various factions of the nation. He begins a campaign of conquest, first consolidating his power in the area around the great lake of Tonle Sap. He establishes his capital, Hariharalaya, near the site where Angkor is eventually constructed. Jayavarman II restores the cult of the devaraja, or god-king, in which the king is viewed by his people as a god.

944–68 Reign of Jendravarman II

During the reign of Jendravarman II, Upper (Land) and Lower (Water) Chenla are reunited. The kingdom expands in size and the state increasingly becomes an imperial theocracy. As the god-king, the ruler is the protector of religion and sacred law and the owner of the land, the kingdom, and its people. Officials number approximately 4,000 in the tenth century and are appointed by the court. They swear loyalty oaths to the king and give their daughters to the royal harem. Administration is in the hands of royal ministers of state, who control police, justice, and forced labor. These provincial administrators act as rulers of their vassal states, each with varying degrees of local autonomy from the king.

1112–50 The Great Suryavarman II

The kingdom again expands during the reign of Suryavarman II, this time by a series of wars. The Kingdom of Champa in southern Vietnam, the Annamese in northern Vietnam, and the peoples residing as far west as the Irrawaddy River in Burma all yield to the superior force of the Khmer king.

Suryavarman II builds Angkor Wat, the largest religious building in the world, which is considered by most to be the most impressive architectural monument in Southeast Asia. Suryavarman II also establishes a network of roads and an irrigation and reservoir system.

1112–50 Peasant revolts

The magnificent building and irrigation projects and the expansionist wars draw heavily upon the population, who pay heavy taxes and are forced into labor. In the middle of the twelfth century, the peasants finally begin to revolt.

1150 Dynastic upheaval

The successful reign of Suryavarman II is followed by thirty years of dynastic upheavals.

1177 Destruction of the City of Angkor

Maurading Cham from the Kingdom of Cham in Vietnam successfully invade Cambodia and destroy the city of Angkor.

1181–1215 Reign of Jayavarman VII

Jayavarman VII successfully repels the Cham invaders. He then goes on to expand the Kingdom of Angkor to its greatest extent. He builds the great capital city complex of Angkor Thom, which includes ten miles of walls and the beautiful temple of Bayon.

1200–1400 Angkor loses ground to its neighbors

The Siamese people to the west increasingly challenge the authority of the Khmer rulers over the Chao Phraya River Basin and the upper Mekong River, where numerous Siamese reside. Siamese kingdoms are established on Khmer land, and the Khmer are unable to reclaim the areas.

1215 Decline of Angkor

The death of Jayavarman VII ushers in a period of decline that leads to the eventual disintegration of the Kingdom of Angkor. Peasant revolts against court demands for money, food, forced labor, and military service. Neglect of the hydraulic works—irrigation canals and reservoirs—leads to erosion in the effectiveness of rice cultivation.

Theravada Buddhism spreads among the common people, bringing a message of salvation through individual effort and eroding support for an extravagant aristocracy feeding off an enslaved people.

1353 Wars with Siam and Laos

A Siamese army captain captures Angkor. Although recaptured by the Khmer, Angkor's wars against the Siamese continue. The Siamese loot Angkor numerous times, carrying off thousands of scholars, craftsmen, and artists into slavery in Siam. At the same time, Khmer lose territory to the Lao kingdom of Lan Xang along the present-day border between Cambodia and Laos.

1431 End of the Angkor Period

Warfare has been constant between the Khmer and Siamese through the latter part of the fourteenth century and the early fifteenth century. In 1431, the Siamese attack, capture, and sack Angkor Thom, the grand city of the Angkorean kings. The capital is then abandoned. The sack of Angkor Thom marks the end of the Angkor Era and its production of architecturally magnificent monuments and temples, its sculptures, decorations, and inscriptions, and its system of irrigation canals and reservoirs, roads, hospitals, and administrative centers.

At the end of the Angkorean Period, Cambodians are divided into clearly delineated classes. At the top are the king and a hereditary aristocracy that includes the royal family, nobles, and court Brahmans, religious men serving the court. Next are the civil servants appointed by the king, who receive a share of the taxes, revenues, and forced labor collected to run the government. These civil servants are non-hereditary and primarily commoners. Next are freemen, primarily farmers, followed by slaves. In addition, there is the clergy, held in high esteem. Highlanders, ethnic groups who inhabit the mountains of Cambodia, are held in low esteem and, when they are lucky, are ignored by lowland Cambodians. The Cham, refugees from the Champa Empire in central Vietnam, leave descendants who practice Islam and live in separate villages. By and large, the Cham are on friendly terms with the majority Khmer.

1434 Capital moved to Phnom Penh

The capital of Cambodia is moved to the present-day site of Phnom Penh, where the capital remains. There then follows a period of relative peace and calm.

1473 The Siamese attack the Khmer

The Siamese again attack the Khmer, but the Khmer manage to repel them three years later.

1516 Ang Chan takes the throne

From the late 1400s to the early 1500s, there is considerable dynastic upheaval within the country. This ends when a member of the royalty, Ang Chan, takes the throne. Ang Chan rules for half a century, moving the capital north of Phnom Penh to Lovek on the Tonle Sap River. The new capital contains a number of Buddhist temples within its compound.

1531 Invasion of Siam

Ang Chan invades Siam and successfully repels their counterattack. Several decades later, Siam itself is weakened by attacks from Burma and internal dynastic disputes of its own.

1564 Another invasion of Siam

After repeating its invasion of Siam, Ang Chan reaches the capital of Ayutha, which is occupied by the Burmese.

Mid 1500s Arrival of Europeans in Cambodia

European contact with Cambodia is limited primarily to missionaries and Portuguese and Spanish explorers and adventurers. Buddhist opposition keeps them from establishing a foothold in the country.

1580s Strong new Siamese ruler

In the midst of repeated attacks on Siam by the Khmer, Siam gains a strong new ruler, Phra Naret. He allies his country with Cambodia against the Burmese. When Cambodia and Siam have a dispute, the Siamese again attack Cambodia.

1590s Spanish presence in Cambodia

Spanish adventurers become active in the politics of Cambodia. When the Siamese advance toward the capital city, Lovek, Khmer King Sattha asks Spanish soldiers to request aid from the Spanish governor in the Philippines. The Spanish expedition arrives too late to help, but their presence influences the choice of the next king, Barom Rachea II, on whom the Spaniards continue to have enormous influence. The Spaniards push the Khmer king for a Spanish Protectorate over Cambodia.

1594 Siamese capture Lovek

On yet another attack on Cambodia, the Siamese capture the capital at Lovek. The Siamese place their own government in the capital city, marking the first time that a foreign power has control over the Khmer kingdom.

Foreign traders, primarily from Southeast Asia and China, come into Cambodia to do commerce, since Khmer are primarily rice cultivators and fishermen supporting their own families and conducting little trade beyond the village level.

1599 Spanish presence ends

A dispute between Spaniards and Malay in Phnom Penh occurs in the last year of the 1500s, resulting in the deaths of the Spanish there and the end of their influence in Cambodia.

1600s–1864 Siam and Vietnam struggle for control

Siam and Vietnam struggle for control over Cambodia. Cambodian kings seek the support first of Vietnam, then of Siam, and always with an eye toward maintaining independence from either neighbor to the east and west.

Early 1600s Chinese establish themselves in Cambodia

By the early 1600s, Chinese traders establish themselves in Phnom Penh where they comprise one-seventh of the population. Residing in separate quarters of the city, they govern themselves under the ultimate authority of the Khmer.

Dutch traders also establish themselves in Phnom Penh in the early 1600s. The Khmer capital is moved from Lovek to Odong.

1609 Chinese in Cambodia

A Portuguese traveler says there are 3,000 Chinese in Phnom Penh among the city's 20,000 inhabitants.

1620 Alliance with Vietnam through marriage

Cambodia is allied with Vietnam through the marriage of King Jayajettha to a Vietnamese princess, the daughter of King Sai Vuong of the Nguyen dynasty, which controls southern Vietnam. Although the alliance allows Cambodia to resist Siamese pressure, the consequences are long-term.

1623 Vietnamese migration into Khmer territory

The Cambodia monarchy agrees to a Vietnamese request to establish a customhouse at Prei Kor, the site today of Ho Chi Minh City. This facilitates the settlement of Vietnamese migrants into the delta controlled by Cambodia.

1640s Hostility toward the Dutch

Cambodians are hostile toward the Dutch traders, who capture and kill some of the residents in Phnom Penh.

1655 Disappearance of European presence

The Khmer and Dutch traders sign a treaty, but it fails to give the Dutch the trading monopoly they want in Cambodia. Their influence in the country gradually decreases, disappearing by the end of the century along with all European influence.

Late 1700s Vietnam controls the delta

By the end of the eighteenth century, Vietnamese settlers occupy the delta and Vietnam extends its control over the delta to its present-day boundaries.

1795 Ang Eng becomes king

The Siamese install young Ang Eng, who has taken refuge in the Siamese capital of Bangkok, as king in the Khmer capital of Odong. The Siamese then annex three Cambodian provinces in the north.

Early 1800s Vietnam dominates Cambodia

Conflict between Khmer and Siamese results in Vietnamese domination over Cambodia, which revolts. When Siam comes in on the side of the Khmer, the Vietnamese and Siamese negotiate and join in suzerainty over Cambodia.

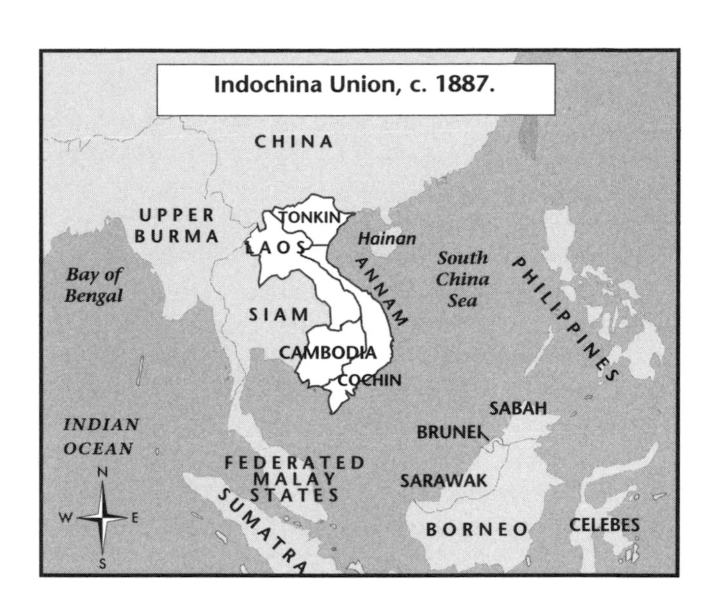

Indochina Union, c. 1887.

1846–59 Ang Duong becomes king

Ang Duong, Ang Eng's grandson, becomes king, but is subject to both Vietnam and Siam. Ang Duong fears Cambodia's neighbors will take over his country, so he appeals unsuccessfully to Napoleon III (1808–73) of France for assistance.

Cambodians say that modern Khmer literature begins with the poetry of Ang Duong.

1863–64 France takes control of Cambodia

King Norodom signs treaties beginning the era of the French protectorate over Cambodia, which lasts until 1953. France offers to protect the country from both Vietnam and Siam. France's control over Cambodia at the expense of Siam is formalized by the Protectorate Treaty of 1863. The French demand the return of Cambodia's royal regalia, the symbols of Khmer royal power, from Siam. They crown Norodom king—over his brother—on July 3, 1864. As a result, Cambodia is protected from both its neighbors. France gains the right to settle in the country, exploit Cambodia's resources, import duty-free goods, and establish courts for itself. France also has exclusive control of Cambodia's foreign affairs and the obligation to defend Cambodia against her enemies.

Late 1880s–Early 1990s French Protectorate

France does little to change Cambodia economically, since it is a country comprised of self-sufficient subsistence households, each supplying its own needs. The French do not reshape the economy but add rubber and corn as cash crops. They also do not contribute much infrastructure, such as railroads or roads. The French put little emphasis on industry, and what little there is is primarily for processing local materials for local use.

Vietnamese migrate into Cambodia, working as rubber plantation laborers and as civil servants for the French administrators. As their numbers grow, they increasingly take over fishing jobs and operate small businesses. The Chinese increase in number, both from migration and natural increase. Fleeing poor conditions in China and looking for economic opportunities, large numbers of Chinese migrate into Cambodia. They dominate commerce, small industry, and handicrafts. Chinese manage much of Cambodia's commercial life, especially in the capital of Phnom Penh and the towns.

Although not many French migrate to Cambodia, their political influence is great. Their social influence is less so. Their plantations employ primarily Vietnamese. Cities expand, some schools are built, goods increase, and communication improves. Phnom Penh especially sees a new look as the French establish businesses and other modern buildings, villas, and wide boulevards. However, most Khmer, and especially those in rural areas, are little influenced by these changes.

1860s–70s Cambodians revolt against French rule

The Cambodians conduct a number of minor rebellions against the French soon after the Protectorate is installed, but the French are able to put them down.

1864–1904 King Norodom's rule

King Norodom's reign is peaceful, but he is powerless. A French governor general has control over the country and lives in Phnom Penh near the king. The governor general is the real ruler of Cambodia. The governor general rules through local governors located throughout the country, while the king is the national symbol and religious leader of the Khmer. A hierarchical, centralized Khmer bureaucracy remains, but has little power.

France's primary interest in Cambodia is in its location next to Vietnam, allowing France to more easily exploit the resources of Vietnam. Cambodia also serves as a buffer state against Siam and England, with interests farther east in Burma and India. France is primarily concerned with maintaining peace and order.

1880s King Norodom begins rebelling

While France is preoccupied with gaining control of central Vietnam, the Khmer king tries to assert more independence. In 1884, he refuses France's demand to join a customs union with south and central Vietnam under French direction. Angered, France moves soldiers and gunboats up the Tonle Sap River to the capital of Phnom Penh to threaten the king. Norodom yields, signing a treaty that makes Cambodia a de facto French colony. The Cambodians now revolt against the new treaty, led by the king's brother Si Votha. It takes the French over a year and numerous losses to regain full control of the country.

1884 France abolishes debt slavery

Although France's main goal during the eighty years of Cambodia's protectorate was to protect French interests in Vietnam, her presence in Cambodia has several results. Among them, France abolishes debt slavery, making it illegal for a debtor to become a slave to the person he owes.

1887 Cambodia becomes part of the Indochina Union

Cambodia becomes part of the Indochina Union, which includes Annam (central Vietnam), Tonkin (north Vietnam), Cochin (south Vietnam, Laos, and Chan-chiang on the coast of the South China Sea).

1887 Succession to King Norodom

At the death of King Norodom, the French, passing over Norodom's sons, appoint Sisowath the next king in hopes that he will be easier to control.

1904–27 King Sisowath rules

Sisowath rules with as little power as his brother. His son, Sisowath Monivong, who rules until 1941, succeeds him.

1911 Establishment of French schools

Until now, Cambodian education remains traditional: some boys are taught in the temple by monks, and education primarily emphasizes religious texts and matters. In 1911, King Sisowath decrees mandatory education for all boys: beginning at age eight, they are to be educated in the traditional way at temple schools. This decree is basically ignored. French efforts to establish schools for Cambodian children largely fail. The Cambodians prefer traditional education for their children, while the French are not really interested in educating the Cambodians, thinking that to do so might encourage the Khmer to question the French administration.

1930 Beginning of Buddhist Institute

French scholar Suzanne Karpeles establishes the Buddhist Institute in Phnom Penh. This institute nurtures the beginnings of a Cambodian independence movement and provides a refuge for Son Ngoc Thanh, one of Cambodia's first nationalists.

Ho Chi Minh (1892–1969) founds the Indochinese Communist Party for Vietnam, Cambodia, and Laos.

1937 Discovery of Vishnu statue at Angkor Wat

A Cambodian has a dream and, obeying it, begins to dig in the lake west of Angkor Wat; the largest religious structure in the world, it had been built by Suryavarman II in the twelfth century as a temple-tomb. The Cambodian digger finds a large bronze head from a reclining Vishnu, the central deity of Angkor Wat. It is magnificent and impressive. Later, more

Sisowath Monivong is crowned king. He wears the golden headdress of the Sisowaths, and sits on a jeweled throne in the palace at Pnom Penh. (EPD Photos/CSU Archives)

fragments of its legs, arms, and chest are found in the same location.

1940 Japanese take over Southeast Asia

The Japanese take over control of Southeast Asia, allowing the French to rule Cambodia.

1941 French select Sihanouk to be king

Considering Monivong's son, Monireth, to be too independent and thus difficult to control, the Vichy French (French collaborators with France's enemies during World War II—in this case, Japan) select a great-grandson of King Norodom's, Prince Norodom Sihanouk (b. 1922) to be the next king. The selection is made by pressuring the Cambodian Council of Ministers. During his reign, Sihanouk unites both branches of the royal family. A mere high school student, he is viewed by the French as passive and malleable.

The Japanese march into Phnom Penh.

1941 Thailand takes two provinces from Cambodia

After a series of border incidents, Thailand (Siam, which changed its name in 1939) invades Cambodia. Japan intervenes in disputes between France and Thailand and forces France to allow Thailand to seize the Cambodian provinces of Battambang and Siem Reap, the latter the home of the famous Angkorean temple ruins.

1942 First anti-French demonstrations

Son Ngoc Thanh organizes the first anti-French demonstrations in Phnom Penh in support of Buddhist nationalists. In Thailand, the Cambodians organize Freedom committees (called Issarak in Cambodia) against the French.

1943 French try to replace Khmer alphabet

The French administration says it will replace the Khmer alphabet of forty-five letters with the Roman alphabet, which it sees as a move toward modernization. Cambodians, especially the clergy, oppose this reform and it eventually dies.

1945: March 12 Sihanouk declares independence

As the war nears an end, the Japanese remove the Vichy French from power in Cambodia and encourage Cambodia (and Vietnam and Laos) to declare independence from France, within the Japanese-directed Greater East Asia Co-Prosperity Sphere. Proving false the earlier French expectations of his character, King Sihanouk declares independence for his country.

1945 Nationalist leader becomes Prime Minister

World War II ends, and Sihanouk asks the French to return to Cambodia.

Son Ngoc Thanh, a nationalist who fled Cambodia early in the war (see 1930), returns from Japan to Cambodia and becomes Prime Minister in May. In August, he sets up his own anti-French government, still acting as Prime Minister.

1945: October Son Ngoc Thanh is jailed

Allied troops occupy Phnom Penh and arrest Son Ngoc Thanh on charges of collaborating with the Japanese. He is sent to France where he is placed under house arrest. Some followers flee to areas controlled by Thailand in northwest Cambodia where they form the Khmer Issarak, meaning Free Khmer. Their goal is to gain full independence from France.

1945–46 Return of the French

Yielding to the inevitable, King Sihanouk sends delegates to Vietnam to negotiate a new arrangement for France's return to Cambodia. The agreement is signed in January 1946 and states that Cambodia is an autonomous state within the French Union. As the Cambodians realize, however, the French remain in control of Cambodia. Cambodians are left to decide minor matters.

1946: September Electing a Consultative Assembly

Two parties compete for representation at the Consultative Assembly, which is meeting to decide on a constitution. One of the parties is the Liberal Party which, despite its name, is conservative and connected to the court. It consists primarily of people with land, government position, education, and status in Khmer society. The other party, the Democratic Party, consists of Cambodians of lower education, occupation, and economic standing. It includes Khmer Issarak, the Free Khmer. The two parties dispute over continuing French rule and the role of the monarchy in a new government. The Liberal Party wants French control and a strong monarchy, while the Democratic Party favors an end to colonialism and unlimited royal power. The Democratic Party wins, rejects the constitution draft, and sends a new draft to King Sihanouk

1946: November Return of Battambang and Siem Reap

Thailand returns to Cambodia the two provinces it was ceded in 1941.

1946 First Indochina War

The First Indochina War begins as Ho Chi Minh declares Vietnam's independence from the French. Viet Minh and French armies begin fighting.

1947: May 6 New constitution

King Sihanouk signs the constitution draft although he is unhappy since it limits the monarchy. Its vagueness, however, will aid Sihanouk later. The constitution recognizes Sihanouk as the spiritual head of the country.

1947: December Electing a National Assembly

The first representatives are elected to Cambodia's National Assembly, with the Democratic Party gaining a considerable majority. The Democratic Party is now dominated by the Khmer Issarak. The Democrats are opposed to the royalists, with the Democrats supporting full independence and the royalists, more pragmatically, willing to accept some French involvement.

1949 Saloth Sar goes to France

Saloth Sar (1926–98), the future Pol Pot, goes to Paris to begin his studies. He remains there until 1953, when he returns to Cambodia.

1949: November France grants Cambodia qualified independence

A treaty with France grants independence to Cambodia, with qualifications. France continues to oversee foreign affairs and remains involved in some judicial and economic matters. Thus, Cambodians control the new armed forces, while France controls wartime military operations. King Sihanouk rules through a provisional government, justifying his control with the threat to the government by the Khmer Issarak.

Late 1940s Growth of the Viet Minh

King Sihanouk is also concerned about the Viet Minh, a Communist group seeking independence for Southeast Asia from France.

1951 Democratic Party wins elections

In September, the Democrats win the election for seats in the National Assembly but fail to develop strong programs. Tension between the Democratic Party and King Sihanouk intensifies. Sihanouk asks France to release Son Ngoc Thanh from prison. Son Ngoc Thanh returns to Cambodia in October.

The Khmer People's Revolutionary Party is created from the Indochinese Community Party.

1951–52 Son Ngoc Thanh's return to Cambodia

After Son Ngoc Thanh's return to Cambodia, he publishes a newspaper demanding the withdrawal of French troops from Cambodia. King Sihanouk closes down his newspaper and Son Ngoc Thanh flees the capital to join the Khmer Issarak in rebelling against the government. While Cambodian factions quarrel, demonstrations against the French continue.

1952–55 Sihanouk seizes direct rule of Cambodia

Sihanouk assumes power over the country, announcing himself as Prime Minister and dismissing the cabinet. He says he will reestablish control and security throughout the country and seek full independence from France. Sihanouk dismisses the National Assembly and institutes martial law throughout the country. He travels through Europe and America, seeking support for Cambodia's independence.

1953–54 Cambodia gains full independence

In the face of increased rebellion, France agrees to make Cambodia independent. Cambodia pushes for complete control over its own defense, courts, and money. In August, France transfers control over the police and judiciary. In September and October, France transfers control of the Khmer armed forces.

At the time of its independence, Cambodia's economy is underdeveloped in Western terms. There is little development capital, an untrained civil service, low per capita income, and inadequate infrastructure such as transportation, power, and communication facilities. However, there is also an abundance of cultivable land, a surplus of agricultural products for export, considerable resource possibilities, and relatively equitable land distribution with no traditional landed aristocracy.

1953: November 9 Independence Day

Cambodia celebrates its Independence Day.

Early 1950s Viet Minh intrude on new government

The Viet Minh, a Communist group, soon threatens the newly independent government by seeking to overthrow the monarchy and install a Communist government. Viet Minh activities relate as much to Vietnam as to Cambodia, as the Communists try to influence the outcome of the upcoming Geneva Conference.

1954 Geneva Conference

The Geneva Conference is held to address issues in Southeast Asia. North Vietnam, Cambodia, Laos, and South Vietnam are joined by France, the United States, the Soviet Union, China, and Great Britain. The participants agree that hostilities in Indochina, primarily between North and South Vietnam, will cease. In Cambodia, all Viet Minh forces will withdraw within ninety days and other Khmer resistance groups will be demobilized in thirty days. By October, 1954, most Viet Minh forces withdraw from Cambodia. Son Ngoc Thanh—in exile in Thailand—agrees to submit to the government, but King Sihanouk will not see him. Cambodian Communist leader Son Ngoc Minh and approximately half the Cambodian Communist movement go into exile in Vietnam. Sihanouk gains complete control over Cambodia.

1955 Sihanouk fails to abolish political parties

Following the requirements of the Geneva Agreement, Sihanouk calls for a national referendum. Saying that the king has attained independence and security for his country, the referendum receives 99.8 percent of the vote. The king then decides to conduct another referendum, this one asking his countrymen to change the constitution in order to dissolve the political party system, which leaves power solely in the hands of the king. In the face of considerable opposition, King Sihanouk announces that elections will be held, as the existing constitution requires.

1955: March 2 Sihanouk gives up the throne to become a politician

Sihanouk abdicates in favor of his father King Suramarit so he can freely enter political life outside constitutional restrictions on the monarch. Assisted greatly by his royal standing, which he retains as a prince, Sihanouk founds the People's Socialist Community, called the Sangkum.

1955: September Sihanouk wins national elections

The Democratic Party, the small pro-Communist party called the People's Party, and Sihanouk's Sangkum Party meet in national elections. Sihanouk's party wins eighty-three percent of the vote and all of the seats in the National Assembly. Sihanouk becomes Prime Minister and pursues a neutralist stance for his country.

1955–58 Political instability in the government

Political stability is not easily achieved. Nine governments rule between the 1955 and 1958 elections. The various interests of the members of Sihanouk's party make it unstable. However, opposition parties are unable to benefit from this factionalism. Sihanouk alternates between holding office as Prime Minister and retiring, always returning to resume power and restore stability.

Late 1950s Economic development

With financial aid for other countries, including the United States, a harbor is constructed in Sihanoukville, now Kampong Som. The harbor allows Cambodia access to the sea without having to go through Saigon in Vietnam.

1955–60 Growth of the Cambodian Communists

Cambodian Communists concentrate on developing their power in the rural areas as well as in the capital of Phnom Penh. In 1959, the Communist leader within the country of Cambodia publicly defects to Sihanouk's party.

Late 1950s–Early 1960s Conflict with Thailand and South Vietnam

Sihanouk's neutralist policies offend his neighbors. Relations with Thailand are often broken and border incidents are common. Relations with South Vietnam are even worse, with constant arguments over the treatment of Cambodians in South Vietnam by the Vietnamese. South Vietnam accuses Cambodia of allowing Vietnamese Communists to seek sanctuary along the Vietnamese-Cambodian border and allowing supplies to come down from North Vietnam through Cambodia to South Vietnam. Cambodia, in turn, accuses South Vietnam of giving sanctuary to the Khmer resistance fighters, called the Khmer Serei (formerly the Khmer Issarak).

Late 1950s–1960s Sihanouk period

Cambodia's neighbors, Vietnam's war, and Thailand's ties to the United States threaten Cambodia's stability and neutrality. Despite war, political unrest, opposition to his rule, and economic difficulties, Sihanouk succeeds in keeping Cambodia out of the Vietnam War. Cambodia remains primarily agricultural, traditional, Buddhist, and isolated from other countries.

1960s Growth of education

King Sihanouk increases the number of public secondary schools in the country from nine in 1953 to 110 in 1966. Sihanouk's emphasis on education also spurs the construction of elementary schools, private schools, vocational schools, and universities, including the University of Phnom Penh. Khmer schools increasingly emphasize Cambodian language and history and vocational training rather than French language and classics.

1960s Cambodia breaks with the United States

Accusing the United States of aiding a plot against Sihanouk by Son Ngoc Thanh and his followers, Sihanouk breaks off relations with the United States.

1960 Sihanouk becomes head of state

Sihanouk's father, King Suramarit, dies. Sihanouk takes power as head of state, remains the leader of his Sangkum party, and becomes supreme commander of the armed forces. His mother, Queen Kossomak, remains at the palace as a symbol of the monarchy. There is no longer any effective opposition to Sihanouk's rule. Despite his autocratic way of reigning, Sihanouk continues to dominate Khmer politics through the 1960s.

The Cambodian Communists hold a congress in Phnom Penh and found the Cambodian Marxist-Leninist party, called the Workers Party of Kampuchea.

1963 Saloth Sar becomes head of Communists

Cambodian Communists hold their second congress and Saloth Sar becomes head of the party. In the face of growing oppression from the government, most of the Communist party leaders flee into the countryside.

1965 Sihanouk breaks with the United States

Partly in reaction to the United States sending troops into Vietnam, Sihanouk breaks off diplomatic relations with America.

1966 Beginning of the Khmer Rouge

A group of radical, primarily French-educated young people change the name of the Cambodian communist group from the Vietnamese Workers' Party of Kampuchea (another name for Cambodia) to the Communist Party of Kampuchea. This group of Cambodian Communists, later named the Community Party of Cambodia, is also known as the Red Cambodians, as Prince Sihanouk named them, or the Khmer Rouge in French. At the same time, they begin to split from the Vietnamese Communists, who dominated them previously. The new Communist party members begin an uprising against Prince Sihanouk's neutralist government, although the Cambodian Communists feel somewhat abandoned by the Vietnamese Communists.

Sihanouk allows the Vietnamese Communists to use Cambodia's border areas with Vietnam and allows materials and goods to go to the Vietnamese Communists through its seaport.

1966 Peasant rebellion

An insurrection against a rice tax is instigated by peasants in Samlaut in northwest Cambodia. This peasant rebellion encourages the Cambodian Communists, or Khmer Rouge, to actively wage an armed struggle against Sihanouk and his government.

1966–69 Sihanouk wages war against the Khmer Rouge

Prince Sihanouk responds to the Khmer Rouge—only 5,000 in number in 1966—by waging war against them.

1969 America begins bombing Cambodia

The United States begins secretly bombing Vietnamese targets inside Cambodia.

1970: March 18 Coup against Prince Sihanouk

While Prince Sihanouk is out of the country, his government is overthrown in a coup led by one of his top generals, Marshal Lon Nol (1913–85). Others, including Sirik Matak and Chang Heng, assist in the coup. With this coup, 2,000 years of uninterrupted government by a monarchy comes to an end. Opposition to Sihanouk's capriciousness, cruelty, and arrogance is accompanied by dissatisfaction over the war and the poorly managed economy.

1970: March 23 Sihanouk forms a resistance group

Sihanouk responds by forming the National United Front of Kampuchea, a resistance group that includes the Khmer Rouge, against whom he has been fighting since the late 1960s.

1970: May Kent State killings

Protesting the Nixon Administration's invasion of Cambodia, four Kent State University students are killed by the United States National Guard. The shootings are the most serious results of a series of demonstrations by Americans against the Vietnamese War.

1970: October 9 Founding of Khmer Republic

Lon Nol abolishes the monarchy and Cambodia becomes the Khmer Republic.

1970–75 Increasing war

Lon Nol's regime is tainted by corruption and incompetence. As American aid pours into the country to assist the American war effort in South Vietnam, corruption increases a hundredfold. The authoritarian government is unable to accept the reality of its military situation.

The spread of the Vietnam War into Cambodia comes with American and South Vietnamese raids. Khmer agriculture is destroyed, industry declines, and a third of the population becomes refugees. Lon Nol's government fights a losing battle with the Khmer Rouge.

1971 North Vietnamese fight Cambodian army

The North Vietnamese Communist army continues fighting against the national army of Cambodia while the Khmer Rouge Communists build their forces. In December, North Vietnamese troops defeat the Cambodian army in a major battle. This is the beginning of the end for the Cambodian armed forces.

1972: May 10 Cambodia gets a new constitution

The constitution establishes an independent, democratic, and social republic. Lon Nol is elected president of the Khmer Republic.

1973 Heavy American bombing of Cambodia

The Cambodian Communists refuse to negotiate at the Paris peace talks and are not involved when the Paris Agreement ending the war in Vietnam is signed in Paris. The Khmer Rouge take over much of the fighting in Cambodia from their Communist brothers, the North Vietnamese. Within the zones they have conquered, the Khmer Rouge instigate new rules, including forming the farmers into cooperatives.

The Nixon Administration drops more bombs on Cambodia in the first six months of the year than were dropped on Japan during all of World War II. The United States Congress forces an end to the bombing of Cambodia on August 15.

1974 Khmer Rouge gain in strength

The Khmer Rouge take over much of the territory of Cambodia, causing hundreds of thousands of Cambodians to flee into towns and into the capital of Phnom Penh. The capital grows to three and four times its normal size. The city cannot take in enough food to feed its population. The Cambodian army loses more battles. The Khmer Rouge win control of the resistance, purging Communists who have returned from North Vietnam, ethnic Vietnamese residing in Cambodia, and Sihanoukists.

1975: January Final Khmer Rouge push

The Khmer Rouge begin their final offensive on New Year's Day.

1975: April 17 Khmer Rouge take Cambodia

The Khmer Rouge take over the city of Phnom Penh, completing their conquest of Cambodia. They evacuate the cities, telling inhabitants that the Americans are coming to bomb. The cities and towns are emptied, a process that takes weeks in some areas. Tens of thousands die.

The Khmer Rouge close off the country to all foreigners, taking Westerners to the Thai border by truck and crossing all entry points into the country. The Khmer Rouge now put their plans into effect to return all Cambodia to an agrarian way of life, one that is self-sufficient, Communist, and purified of all Western influence. China is their primary ally. The Khmer Rouge establish the state of Democratic Kampuchea. They then concentrate on establishing the entire population in rural cooperatives where they work from before sunup to after sundown in slave-like conditions. Only a few light industries remain in the once active cities and towns. By July, they Khmer Rouge have begun setting up party and government offices in Phnom Penh.

1975: April 30 Vietnam War ends

North Vietnamese troops take Saigon, bringing an end to the Vietnam War.

1975: May 12 *Mayaguez* incident

The merchant ship *Mayaguez* is captured by Khmer Rouge sailors off Cambodia, leading to the United States briefly bombing Cambodia.

1975: December Additional evacuations and migrations

By the end of the year, the Khmer Rouge order a second evacuation of city residents from the southwest of Cambodia to the north and northwest.

1975–79 Khmer Rouge regime

Democratic Kampuchea lasts for four years. During this time, the deaths of between 1 and 2 million Cambodians take place from execution, starvation, illness, and neglect. The country is ruled by a Standing Committee of the Central Committee of the Communist Party of Kampuchea.

1976 The revolution expands

At a party congress, Khmer Rouge leaders decide to increase the speed and expanse of their revolution. They write a constitution. The government of Democratic Kampuchea is announced for the first time and Pol Pot is named Prime Minister. Sihanouk, who has been brought back as a figurehead, resigns as head of state. The Khmer Rouge begins executions of party leaders who have ties to the Vietnamese Communists. They also complete a purge of their own cadre in northern Cambodia.

1977 Purges and skirmishes

The Khmer Rouge completes purges of pre-1975 leaders and continues purges of its own members. Disputes between Vietnam and Cambodia and Thailand and Cambodia break out over their borders, leaders, and various other issues. The border disputes with Vietnam are the most serious.

1977: September Pol Pot breaks his silence

Pol Pot announces that his country is Communist. He then travels to China and North Korea, where he announces that he is the former Saloth Sar.

1977: December Cambodia breaks with Vietnam

Border disputes now reach crisis level, Democratic Kampuchea suspends relations with Vietnam.

1978 Growing internal and external difficulties

The Vietnamese withdraw their troops from the Cambodian border, leaving the Khmer Rouge to claim victory in their border war with Vietnam. In the spring, Vietnam's leaders decide to invade Cambodia and overthrow the Pol Pot government.

The Khmer Rouge continue their purges, now in the northwest and the east. The Cambodians hold another congress, which reveals plans to increase industrialization in Cambodia.

1978: December Formation of a government to take over Cambodia

Vietnam radio announces the formation of a Khmer national salvation front that is opposed to the Pol Pot government.

1978: December 22 Vietnam invades Cambodia.

Vietnamese troops invade Cambodia.

1979: January 7 Vietnamese troops capture Phnom Penh

Vietnamese troops enter Phnom Penh, with the Khmer Rouge fleeing just before them. The next day, the Vietnamese install a new government, headed by former Khmer Rouge junior officers, Heng Samrin (b. 1934) and Hun Sen. This new government, the People's Revolutionary Council, is run by Khmer Rouge defectors and advised by Vietnamese officials. Over 150,000 Vietnamese troops occupy Cambodia, now called the People's Republic of Kampuchea. Khmer Rouge troops are pushed to the Cambodia border with Thailand.

The Khmer population immediately goes on the move, returning to home villages looking for relatives, food, information, and relief from the Khmer Rouge. Others are forced to accompany Khmer Rouge soldiers to the Thai border, where the Khmer Rouge have settled into border camps.

1979–80 Refugee camps on the Cambodia-Thai border

Seeking food and shelter, waves of Cambodians arrive at the border between Cambodia and Thailand. Famine conditions are created by massive movements of people within Cambodia, fighting between the invading Vietnamese and retreating Khmer Rouge, disruption of the agricultural cycle, and a lack of draft animals and seed. These conditions are relieved through international aid. The international community provides more aid to Cambodian refugees on the Cambodian-Thai border than ever provided to any group before.

Eventually, Cambodians come to the refugee camps on the border and in Thailand seeking resettlement in third countries. Tens of thousands of Cambodians are resettled in third countries, causing a serious drain on Cambodia's few remaining skilled laborers and educated classes. Over 400,000 Cambodians who have fled the country return home from refuge in Thailand, Laos, and Vietnam.

1979–89 Return to normalcy

Over the next decade, Cambodia slowly rebuilds itself. Buddhism, family life, farming, and marketplaces are reestablished despite enormous obstacles, including a shortage of men, poor health as a result of years of malnutrition, and illness. Rebuilding the country is also handicapped by an international blockade which prevents economic development, financial assistance, and international aid from coming into Cambodia—seemingly the Western world's way of punishing the country for Vietnam's invasion.

Another major obstacle is the development of resistance groups on the Thai-Cambodian border that continue to fight against the Vietnamese-led government. These groups include the Khmer Rouge; Prince Sihanouk's followers, led by his son Prince Rannariddh; and a non-Communist group led by former prime minister Son Sann. The groups obtain fighters out of the refugee camps, where they receive support, and send them back into Cambodia to fight the occupying Vietnamese troops, who are supported by Cambodian soldiers.

1982 Coalition resistance government

While in exile, ousted Prince Sihanouk and former Prim Minister Son Sann form a coalition government with the Khmer Rouge. This coalition government and army is furnished money and weapons by the Association of Southeast Asian Nations (ASEAN), who oppose growing influence in the area by the Soviet Union and Vietnam. The coalition government, now including the Khmer Rouge, holds Cambodia's seat in the United Nations until 1990 and is recognized as the legal government of Cambodia by the United States.

The Khmer Rouge launch guerrilla resistance against the Vietnamese-run Cambodian government with weapons secured from China.

1989 Cambodia's population

Cambodia's population reaches 8.4 million, much less than the 11.5 million that would be expected from Cambodia's usual growth rate. Females make up 60-65 percent of the population and almost half the population is under fifteen years of age. These statistics are the results of two decades of war.

1989: May Cambodia gets new constitution and name

Cambodia's constitution is revised, restoring citizens's rights to private property and granting them freedom of religion. The country is renamed the State of Cambodia.

1989: September Vietnamese withdraw their troops

The Vietnamese withdraw virtually all of their troops from Cambodia.

1990: July International community slows support to Khmer Rouge

The United States withdraws diplomatic recognition of the government in exile. It continues to support the non-Communist factions until 1991. China agrees to stop supplying the Khmer Rouge with financial and military support.

1991 United Nations announces a peace plan

The United Nations announces a plan for peace in Cambodia in which the United Nations will play a key role.

1991 Cambodian refugees remain outside the homeland

Over 16,000 Khmer refugees remain in Vietnam and over 350,000 remain in Thailand. Meanwhile, the four warring factions—the three resistance groups on the border and the State of Cambodia—sign a peace agreement in Paris.

1991: November The prince returns to Cambodia

Norodom Sihanouk returns to Cambodia as head of the Supreme National Council, a twelve-member body comprised of representatives from all four factions. This council acts in an advisory capacity to the United Nations.

1991–93 The U. N. Peace Plan

The 2 billion dollar Cambodian peace plan instituted by the United Nations is the largest and most expensive United Nations effort ever. The United Nations Transitional Authority in Cambodia (UNTAC) deploys over 22,000 personnel, including 15,000 soldiers, to oversee the cease fire among the factions and to organize and supervise the 1993 elections. By organizing and managing a national election, the United Nations hopes Cambodia can elect a national assembly and government.

UNTAC monitors the daily operations of five key ministries of the existing State of Cambodia government run by Hun Sen (b. 1952): defense, interior, finance, foreign affairs, and information. UNTAC oversees disarming each faction's soldiers.

The Khmer Rouge end up refusing to participate, and their forces remain armed. They attempt to disrupt UNTAC operations, attack United Nations personnel several times, and massacre ethnic Vietnamese living in Cambodia. Because the Khmer Rouge do not cooperate, UNTAC is limited in obtaining compliance from three other factions. This period sees human rights abuses, harassment of opposition parties, and occasional murders.

1992: March UNHCR repatriates Cambodians from Thailand

The United Nations High Commissioner for Refugees (UNHCR) begins repatriating 370,000 Khmer refugees from Thai refugee camps to Cambodia. More than 25,000 Cambodians live in Vietnam, 152,000 in the United States and 90,000 in over twenty other countries.

1993: May Cambodia holds elections

Internationally supervised elections are held, with more than ninety percent of registered voters going to the polls. More than forty-two percent of registered voters vote on the first day despite Khmer Rouge propaganda threatening death to anyone who goes to the polls. Prince Sihanouk's royalist party wins forty-five percent of the vote.

Unwilling to turn over complete power to the royalist party, Hun Sen threatens war if he is not included in the new government. Sihanouk's party forms a coalition government with the State of Cambodia administration, who wins thirty-eight percent. Sihanouk's son, Prince Norodom Rannariddh, becomes First Prime Minister. Hun Sen becomes Second Prime Minister.

1993: September The Prince becomes King again

The new government declares itself a monarchy, becomes the Kingdom of Cambodia, and declares Sihanouk king again. Prince Sihanouk assumes the throne, fifty-two years after he was first crowned king of Cambodia. He remains a symbol of national unity to his people and a stabilizing force in the continuing turmoil of Cambodia.

1995 Thieves are stealing Cambodia's heritage

Gangs of Cambodians and occasional foreigners are stealing Cambodia's treasures, including ceramics, statues, and bas-reliefs located in temples throughout the county, but primarily in the north and west. French scholars estimate that ninety percent of Cambodia's cultural heritage has been destroyed during two decades of war.

1997: March **Stolen art returns to Cambodia**

New York's Metropolitan Museum of Art returns two priceless artifacts to Cambodia that were stolen from ancient Cambodian temples. The first head was stolen from a temple in Cambodia in 1940; the second was taken in 1985. The return was done in accordance with new international protocol about returning goods stolen from other countries.

1997: June **Khmer art exhibition in America**

An exhibition of 100 Cambodian works of art tours the United States. The sculpture is from ancient Cambodia and Angkor between the sixth and sixteenth centuries.

1997: July 5 **Coup overthrows first prime minister**

Second Prime Minister Hun Sen overthrows First Prime Minister Rannariddh in a bloody coup. Rannariddh flees the country, leaving Hun Sen in complete control.

1997: July **Pol Pot on television**

For the first time in twenty years, the Western world sees Pol Pot, broadcast on television from his hut on the Thai-Cambodian border. He is put on trial by his comrades, who fear that he may try to have them killed or have them take the blame for the Khmer Rouge atrocities now that much of the world is calling for an international trial of Khmer Rouge leaders. His comrades say that Pol Pot is imprisoned for life.

1998: April 15 **Pol Pot dies**

Nineteen and a half years after he was pushed from power by a Vietnamese invasion of Cambodia, the leader of the Khmer Rouge dies of an apparent heart attack. Since losing the country in 1978, Pol Pot has spent much of his time on the Thai-Cambodian border, directing a resistance by forces that at one time totaled 50,000 soldiers but in recent years had been reduced to a few thousand. One-legged Ta Mok, Pol Pot's sadistic "butcher," and Nuon Chea take over the leadership role for the few remaining forces.

1998: July 26 **National elections**

Cambodia holds its first national elections since the United Nations-sponsored elections of May 1993 and the coup that overthrew First Prime Minister Rannariddh the year before. Supported by international financial assistance and numerous international observers, Cambodians go to the polls despite threats of—and actual—violence.

Some human rights workers claim that more than 100 opposition party members, primarily members of the royalist party of ousted First Prime Minister Rannariddh, have been killed by members of the Second Prime Minister's party. Hun Sen's powerful Cambodian People's Party is joined in the race by Rannariddh's royalist party and former finance minister Sam Rainsy's party.

The election sees a high voter turnout despite a Khmer Rouge attack that kills ten people on election morning.

1998: August **Hun Sen claims victory**

Hun Sen claims victory in the national elections held in July.

Bibliography

Barron, John and Anthony Paul. *Murder of a Gentle Land. The Untold Story of Communist Genocide in Cambodia*. New York: Reader's Digest Press. 1977.

Becker, Elizabeth. *When the War Was Over: The Voices of Cambodia's Revolution and Its People*. New York: Simon and Schuster. 1986.

Chanda, Nayan. *Brother Enemy. The War after the War.* New York: Harcourt, Brace, Jovanovich. 1986.

Chandler, David P. *A History of Cambodia*. Boulder, CO: Westview Press. 1983.

Chandler, David P. *The Tragedy of Cambodian History. Politics, War, and Revolution since 1945*. New Haven, CT: Yale University Press. 1991.

Ebihara, M. M., Carol A. Mortland, and Judy Ledgerwood, eds. *Cambodian Culture since 1975. Homeland and Exile*. Ithaca, NY: Cornell University Press.

Kiernan, Ben, ed. *Genocide and Democracy in Cambodia. The Khmer Rouge, the United Nations and the International Community.* New Haven, CT: Yale University Southeast Asian Studies. 1993.

Kiernan, Ben. *How Pol Pot Came to Power.* London: Verso. 1985.

MacDonald, Malcolm. *Angkor*. London: Jonathan Cape. 1958.

Ponchaud, Francois. *Cambodia Year Zero*. London: Allen Lane. 1978.

Shawcross, William. *Sideshow: Nixon, Kissinger and the Destruction of Cambodia*. London: Andre Deutsch. 1979.

China

Introduction

China, the world's most populous country, also has the most ancient roots of any civilization still in existence—its prehistory dates back some 600,000 years, and its oldest dynasty some 4,000 years. Its long and complicated relationship with the West has included periods of isolation, interaction, defiance, and emulation of Western ways (including the import of Communism). As the twentieth century nears it end, China, the world's only remaining Communist superpower since the dissolution of the U.S.S.R., is working to assure its place among the world's industrial leaders through economic modernization and reform and to maintain an active role diplomatically in the community of nations.

Located in southern Asia, China is the world's third-largest country, occupying an area slightly larger than the United States. Its 3,646, 448 square miles (9,596,960 square kilometers)—figures that include Taiwan, which is claimed as a province—are surpassed only by the areas of the Russian Federation and Canada. China shares borders with more than a dozen different countries, including Russia, India, Pakistan, Afghanistan, and Vietnam. About 90 percent of China's population lives in the eastern part of the country, where lowlands form the setting for both agriculture and industry. The sparsely populated west is composed largely of mountains and high plateaus. China's population is estimated at 1.23 billion. Its capital city of Beijing (Peking) has a population of 14 million.

Early History

The history of human habitation in China goes back at least 35,000 years to the first appearance of hunter-gatherers in the Yellow River basin. By 6500 B.C. Neolithic (late Stone Age) agrarian cultures were beginning to develop, with the help of irrigation techniques. China's earliest dynasty, the Xia (Hsia) flourished in the second millennium B.C., with advances that included bronze tools and vessels, boat building, calendars, and the production of silk cloth. The first Chinese culture to leave written records was the Shang dynasty (1554–1045 B.C.), which produced books using wooden tablets and strips of bamboo. A feudal system was introduced by the Zhou dynasty (1122–221 B.C.), China's longest period of rule by one family. This period also marks the true beginning of the Chinese literary tradition, with the creation of five classic works that included the *I Ching* (Book of Changes); *Shu Ching* (Book of History); and *Ch'un Ch'iu* (Spring and Autumn Annals). The Eastern Zhou period, in particular (771–221 B.C.), was a great period of cultural flowering, during which the philosophers Confucius (551–479) and Laozi (Lao Tzu, c. 604–531) laid the foundations for two of China's major belief systems (Confucianism and Daoism).

China Unified

The decline of the Zhou dynasty was followed by the chaotic Period of Warring States, which ended with the ascent to power of the Qin dynasty, which unified all of China under one government for the first time. The long feudal period ended, a strong centralized government was established, an efficient system of taxation was implemented, and construction was begun on the Great Wall, a fortification stretching for roughly 1,500 miles and considered one of the Seven Wonders of the World. The Qin dynasty ruled China for only fifteen years (221–207 B.C.) but introduced a system of government administration that lasted for centuries. The civil service examinations that staffed these governments were introduced under the Han dynasty that followed (202 B.C.–A.D. 220). In this era paper was invented, the silk trade developed, and Buddhism reached China from India. The Tang (618–907) and Song (960–1279) were periods of political unity and cultural achievement. China's two greatest poets, Li Bo and Fu Du, lived during the Tang period, and movable type was developed in the Song period. The art of landscape painting reached new heights in both eras. Chinese culture was also enriched by contact with India and the Middle East. With the Mongol invasion in the thirteenth century, China experienced its first period of foreign rule. However, the Yuan dynasty wisely retained most aspects of native Chinese culture while improving the country's infrastructure. A Chinese dynasty came to power once again in 1368 with the beginning of the Ming period, which ushered in a long period of stability.

One unfortunate aspect of China's history has been the prevalence of natural disasters. The two most deadly earthquakes in recorded history occurred in China. The first, in 1556, killed more than 800,000 people; the second, in 1976, killed roughly 700,000. Many thousands of Chinese have lost their lives to famine caused by crops ruined either by drought or flooding. The world's worst recorded famine killed as many as thirteen million people in China between 1876 and 1878.

China's last dynasty placed the country once again under rule by outsiders—this time the Manchus (who came from Manchuria, to the north). It was the Manchus who required the shaved heads and long pigtails that Westerners long associated with the Chinese. By the time of Manchu rule, Western powers were starting to open China to trade. The British and Dutch had both secured trading rights by 1700, but the Manchu government kept these rights strictly limited through the eighteenth century. However, the West successfully challenged Chinese power in the nineteenth century with the two Opium wars and won growing concessions, and China was significantly weakened in the Sino-Japanese War in 1894–95.

In addition to widespread anti-foreign sentiment, another growing tendency by the 1890s was opposition to the Manchu government, which was overthrown in a 1911 rebellion. The revolutionary leaders formed a republic, headed at first by Sun Zhongshan (Sun Yat-sen). By 1917, however, rival warlords had seized power and remained in control until they were overthrown by the combined forces of the nationalist

Guomindang (Kuomintang) and the Communists. However, the Nationalists started attacking the Communists, beginning a twenty-year battle for supremacy that ended only with the formation of the People's Republic of China in 1949. The internal struggles between the two groups were even maintained during the Japanese occupation of China during the second Sino-Japanese War, which began in 1937 and lasted throughout World War II. With the Communist victory, Nationalist leader Jiang Jieshi (Chiang Kai-shek) fled to Taiwan with his supporters and established the Republic of China there.

People's Republic of China

The People's Republic of China, under the leadership of Mao Zedong (Mao Tse-tung), aligned itself with the Soviet Union and, like other Communist nations in the postwar period, implemented ambitious programs to collectivize agriculture and promote rapid development of heavy industry. Collectivization peaked with the Great Leap Forward in the late 1950s and the formation of massive agricultural communes comprising as many as 20,000 people.

As in other Communist societies, individual freedoms were limited. Mao's one attempt to encourage freedom of expression and open debate, the "Hundred Flowers" campaign of 1956, resulted in so much criticism of the government that it was followed by a strict crackdown, and freedom of speech became even more limited than before. However, the height of repression was reached during the Cultural Revolution of the late 1960s, when thousands of intellectuals and others accused of spreading capitalism and other aspects of Western culture were persecuted. Some were uprooted from their homes and jobs and moved to reeducation camps; others were tortured or killed. All told, an estimated 400,000 Chinese lost their lives for political reasons during this period. Internationally, a rift developed between China and the Soviet Union during the 1950s, and by 1960 the Soviets had withdrawn technical and economic aid from China. By the early 1970s, Chinese relations with the United States (which still recognized the Republic of China on Taiwan as the only Chinese government) began to thaw. In 1972 President Richard M. Nixon made a historic trip to China, which was followed by the expansion of trade and cultural ties between the two nations. In 1979 the U.S. established formal diplomatic ties with the People's Republic.

In the 1980s, under the leadership of Deng Xiaoping, China's economic policies were liberalized to allow for a greater degree of private enterprise. A movement away from farming in large collectives, with a new program called "individual responsibility," allowed farmers to retain a share of agricultural profits. China established ties with the World Bank and the International Monetary Fund. Joint ventures with foreign companies were encouraged, and by 1991 the People's Republic even had a stock exchange. However, human rights policies were not liberalized, and by the mid-1980s a pro-democracy movement had sprung up, primarily among Chinese students. Agitation for democratic freedoms reached a peak in the spring of 1989, following the death of Hu Yaobang, a popular government figure perceived as sympathetic to democratic reforms.

Between April and June, growing crowds demonstrated in Beijing and other Chinese cities, and the student demonstrators were, increasingly, joined by workers. Some 2,000 hunger strikers occupied Tienanmen Square from the middle of May, on the eve of an official visit by Soviet premier, Mikhail Gorbachev. On May 17 a crowd of over a million protesters filled Tienanmen Square, virtually paralyzing the city, and the government declared martial law. The standoff between the government and the demonstrators was finally ended on June 4, when armed troops and tanks converged on Tienanmen Square and assaulted the crowds. An estimated 1,000 people were killed, thousands more were injured, and China's action drew condemnation from foreign governments.

In the 1990s China remained an active member of the international community. In 1990 the United States extended China's most-favored-nation trade status, in spite of domestic opposition due to the Tienanmen massacre. The following year, China and Vietnam resolved their dispute over Cambodia, and diplomatic relations between the two nations were normalized. In 1992, the border between India and Tibet was opened for the first time in more than thirty years. In 1996 China joined the United States, Russia, France, and Great Britain in signing the Comprehensive Test Ban Treaty (CNTBT). In a historic ceremony on July 1, 1997, the British turned control of their crown colony Hong Kong back to China after 156 years. It was renamed the Hong Kong Special Administrative Region of China.

In the summer of 1998, President Bill Clinton traveled to China for summit talks, the first since the Tienanmen massacre in 1989.

Timeline

c. 600,000 B.C. Early tool-making species lives in Lanthian

Based on fossils found in 1963, an early proto-human who uses tools is thought to have existed in the Lanthian (Lan-t'ien) region.

c. 500,000–210,000 B.C. Upright hominid inhabits Beijing (Peking) region

A near-human who walks upright and uses fire, today called Beijing Man (*homo erectus pekinensis*), lives in limestone caves near present-day Beijing during the Middle Pleistocene period. Remains are first discovered at Zhoukoudian in the twentieth century (see 1923). His brain is close to the size of modern humans.

c. 50,000–35,000 B.C. Homo sapiens appears

Homo sapiens, or Neanderthal man, appears in the Yellow River region, living as a hunter-gatherer.

c. 6500 B.C. Neolithic cultures develop

Farming, weaving, and pottery are introduced simultaneously in various parts of China, and metal tools replace stone tools. Development of irrigation methods leads to expansion of agriculture.

5000–3000 B.C. Agriculture becomes diversified by region

Millet growing is introduced to the northern regions, and rice growing is begun in the south.

c. 5000 B.C. Yangshao culture flourishes

The Yangshao, an advanced Neolithic culture seen as a precursor to classic Chinese civilization, flourishes. Distinctive signs of Yangshao culture are painted pottery, rice cultivation, and the use of certain weapons and instruments.

c. 3000–2000 B.C. Longshan culture

Longshan culture (whose first remains are found in Shandong Province) is characterized by towns with a class system, earthen houses, domesticated horses and sheep, use of a potter's wheel, and a primitive form of writing.

Early Dynasties

c. 2205–1766 B.C. Xia (Hsia) dynasty flourishes

China's first dynasty, with its center in present-day Shanxi province, is said to have been founded by the legendary Yu the Great. Thought to have included parts of Henan province as well, this civilization, which uses bronze tools and vessels, is one of the transitional societies between China's prehistoric and classical urban civilizations. It has extensive cultural influence throughout China. Other distinguishing features include calendar-making, the use of silk cloth, and the construction of boats and timber houses.

1554–1045 B.C. Shang Dynasty

A fairly advanced civilization develops in the Yellow River Valley during the Shang Dynasty, the first Chinese culture to leave written records. Centered in Henan, it extends over a large area. Walled cities are built, and a royal and priestly ruling class emerges. Altogether there are about thirty rulers between the beginning and end of the dynasty. Wheels with spokes are used for chariots, weapons and vessels are of bronze, and books are made from wooden tablets and strips of bamboo. Trade expands and money consisting of cowrie shells is introduced. Ancestor worship is practiced, and warfare with neighboring peoples is frequent.

c. 1150 B.C. Wen Wang reigns over Zhou (Chou) state

Legendary monarch Wen Wang reigns over the Zhou kingdom in western China. Under his rule, Zhou territory is greatly expanded. Later, emperors of the Zhou dynasty consider him to be an exemplary ruler. He is imprisoned for three years by the last Shang emperor. His son, Wu Wang, founds the Zhou dynasty.

The Zhou Dynasty

1122–221 B.C. Zhou dynasty rules China

China's longest-lived dynasty comes to power, aided by a slave rebellion within the Shang dynasty. The Zhou, already rulers of a kingdom to the west of the Shang empire, establish a feudal system of fiefdoms ruled by relatively independent lords who pay taxes to the emperor and take part in ceremonial events demonstrating their allegiance to the throne.

The Zhou dynasty is the historical period of the five great classics that begin China's literary tradition: the *I Ching* (Book of Changes); *Shu Ching* (Book of History); *Shih Ching* (Book of Songs); *Li Ching* (Book of Rites); and *Ch'un Ch'iu* (Spring and Autumn Annals of the State of Lu).

771–221 B.C. Eastern Zhou period

The western Zhou capital of Hao is abandoned due to local unrest and invasion from abroad, beginning the decline of the Zhou dynasty. The capital is moved eastward to Luoyang and the Eastern Zhou dynasty is established. Culture flourishes during this period, with the works of the great philosophers Confucius (551–479), Lao Tzu (c. 604–531), and Mencius (c. 371–c. 288). Confucianism and Taoism begin to supplant older religions. Military developments include introduction of the crossbow and iron weaponry. The Eastern Zhou period is commonly divided into the Spring, Autumn, and Warring States periods.

c. 604–531 B.C. Life of Lao Tzu

Revered philosopher Lao Tzu founds Daoism (Taoism) and writes the *Tao-te Ching.*

551–479 B.C. Life of Confucius

The teachings of this philosopher dominate Chinese culture until the modern era. Confucius (the Latin term for *K'ung Fu Tzu,* or Master Kung) is born to a noble but impoverished family. His magistrate father dies when he is a child. After beginning a career as a government official, he resigns and becomes a teacher who attracts many followers. At the age of fifty Confucius reenters public life, becoming the prime minister of his province but resigns because of political clashes and is exiled for thirteen years, which he spends traveling and teaching.

Confucius is generally regarded as the author of the *Spring and Autumn Annals* and editor of the *Shih Ching* (Book of Songs) and *Shu Ching* (Book of History). The wisdom of Confucius is first compiled by his pupils in the *Analects*, which describes the actions and words of their master in a number of different situations. After his exile, the philosopher returns to his home province of Lu, where he dies. His disciples eventually found China's first university at the site of his tomb.

453–221 B.C. Warring States Era

The Chou dynasty, under attack by tribes originating in northern and western China, also battles independent armies raised by rebellious feudal lords who fight among themselves as well.

475–221 B.C. Rise of the Qin (Ch'in) dynasty

The Qin dynasty emerges and gains power, ultimately unifying all of China.

c. 371–c. 288 B.C. Life of Mencius

An important figure in the development of Confucianism, Mencius authors the most important Confucian text, the *Book of Mencius.*

c. 298–230 B.C. Life of philosopher Xunzi

Xunzi organizes Confucius's teachings into a coherent system, thus helping found the tradition of Confucianism in spite of his own disagreement with its principles.

Unification and Expansion

221–207 B.C. Qin (Ch'in) ruler unifies China

Shi Huangdi, the ruler of the Qin kingdom in the northwest, prevails over the other warring states to unify all of China under one government for the first time, triumphing over the Zhou dynasty and ending its feudal system. A centralized government rules over counties and provinces, instituting a tightly controlled system of taxation and requiring the peasantry to serve in the army and on crews constructing public works. Construction is begun on the Great Wall connecting the fortifications of the former feudal states to protect the kingdom from invading foreign tribes. Within its borders, the parts of the country are unified by a network of roads, and a standard system of weights and measures is adopted.

213 B.C. Book burning ordered by Qin emperor

Qin Shi Huangdi orders the burning of thousands of books in an attempt to suppress ideas he considers dangerous to the perpetuation of his power.

206–202 B.C. Peasant revolt leads to end of Qin rule

In response to their virtual enslavement in the service of the Qin army and work crews, the Chinese peasants revolt against the emperor, ultimately bringing down his dynasty after only fifteen years in power. However, the system of government administration introduced by Qin Shi Huangdi endures for centuries, adopted by subsequent dynasties and even copied abroad in other Asian countries, including Japan.

The Han Dynasty

202 B.C.–A.D. 220 Han dynasty rules

The Han dynasty is founded by a leader of the peasant rebellion against the Qin dynasty. Order is restored and the empire expands westward. Civil service examinations are introduced to government administration, and paper is invented. Irrigation is improved, agriculture is developed further, and the silk trade develops. Buddhism is introduced from India, and Confucianism spreads. Eventually wars with the Huns drain the empire's wealth and its peasant population, many of whom are drafted to serve in the army. Deteriorating conditions lead to peasant revolts, and the empire is further weakened by growing divisiveness in the imperial court and in the military.

145–90 B.C. Life of historian Sima Qian

Sima Qian, China's first acknowledged historian, writes *Shiji.*

Six Dynasties Period

A.D. 220–589 Six dynasties period

A period of chaos follows the collapse of the Han dynasty, as small kingdoms battle each other in multiple civil wars. A number of different dynasties are founded and collapse during this period, all with their capitals at Nanjing.

A.D. 220–264 Three kingdoms period

The Wu, Wei, and Shu Han kingdoms divide China among themselves in the period immediately following the rule of the Han dynasty. The exploits of warriors of this period are later idealized and celebrated in Chinese literature.

265–316 Jin dynasty restores order

China is briefly reunified under General Sima Yan, ruler of the Jin dynasty. However, the dynasty's rule fragments after his death in A.D. 290.

317 The Jin court moves to Nanjing (Eastern Jin period)

Invasions force the Jin dynasty to move its capital south to Nanjing, and the political unity of China gives way to fragmentation into separate dynasties.

589–618 Sui dynasty rules China

The Sui dynasty reunites China and begins to build the Grand Canal linking the rivers of the north and south. The system of civil service examinations is refined.

The Flowering of Chinese Culture

618–907 Tang dynasty period

The Tang dynasty, established by Li Yuan, rules China. The period of Tang rule is one of power, prosperity, cultural achievement, and territorial expansion. China reclaims the regions to its west, as well as Korea and Annam (present-day Vietnam), giving it an even larger territory than that of the Han empire that preceded it. The Tang dynasty also maintains friendly relations with foreign powers, opening China to cultural influences from India and the Middle East. Buddhism flourishes. A new legal system is created, based on Confucian principles, and land reform is instituted, limiting the number of large estates. Chinese art and poetry reach a new pinnacle. Drama evolves as a training academy for singers and dancers, the Li Yuan (Pear Garden), is established. Landscapes, known as *shan-shui* "mountain-water," become the predominant mode of painting. With the development of block printing, written literature becomes accessible to a much wider public.

655 China conquers Korea

Korea is conquered during the reign of Tang empress Wu Zhao.

c. 700–62 Life of poet Li Bo

Li Bo's work, known for its romanticism, is part of the great cultural flowering of the Tang period. He and Du Fu (see 712–70) are commonly considered China's two greatest poets. Li Bo spends several periods of his adult life traveling. In between, he is a court poet at Changan and for a time serves as poet laureate on a military expedition led by Prince Lin. When the Prince falls into disfavor and is executed, Li Bo is arrested and jailed twice. He is supposed to be exiled to Yelang but benefits from a general amnesty and lives out the rest of his life in eastern China. Famous as one of China's great drinkers, as well as one of its greatest poets, Li Bo writes poems celebrating this pastime, as well as poems on themes including mortality, solitude, nature, and friendship.

712–70 Life of poet Du Fu

Du Fu is one of China's most revered poets. He receives a Confucian education but, unlike most scholars, does not obtain a civil service position because of his failure to pass the government examination. He travels extensively in his youth, writing poems inspired by the beauty of the natural landscape. Eventually he obtains a low-level position at court. However, security eludes him, and he undergoes periods of hardship and destitution, the most severe occurring as a result of the An Lushan Rebellion of 755. He spends parts of the 760s traveling and, according to legend, dies from overeating following a ten-day fast. Many of Du Fu's later poems describe the plight of the country's poor people and the horrors of war. His poems are famed for the intricacy of their classical allusions and use of language and his command of poetic technique, especially in the genre of poetry known as *lüshi.*

751 Military defeat signals decline of the Tang dynasty

The Chinese defeat by Arab forces at Talas marks the end of the Tang dynasty's greatest period and the beginning of a long military decline.

768–824 Life of writer Han Yu

Han Yu, known as the "Prince of Letters," is one of the greatest literary figures of the Tang period. He advocates Confucianism over the more popular teachings of Daoism and Buddhism. Although he also writes poetry, he is known primarily for his essays, which demonstrate the simple prose style he espouses. These essays, which are still admired today, include "On the Way," "On Man," and "On Spirits."

907–960 Downfall of the Tang dynasty is followed by fragmentation

The Tang dynasty is overthrown by invaders from the north, inaugurating a period of chaos and division in which the empire fragments into five northern dynasties and ten kingdoms in the south.

927–78 Life of emperor and poet Li Yu

Li Yu, the last Tang emperor, is also known as a poetic master of the *ci* form. He succeeds to the throne in 961, taking over from his father, Li Jing, who is also a poet. In 974, Taizu, founder of the Song dynasty, captures the Tang capital of Jinling. Li Yu is taken to the Song capital of Bian, where Taizu's brother, Taizong, has him poisoned after Taizu's death in 976. Although Li Yu's poems cover a range of subjects, he is most famous for the late poems lamenting his fate following the downfall of the Tang dynasty and his capture by the Song.

960–1279 Song (Sung) dynasty

A period of unity and cultural achievement is inaugurated with the assumption of power by the Song dynasty, whose rule is customarily divided into two periods (northern and southern). Under the Song, political power becomes more centralized. Major urban centers of commerce develop with

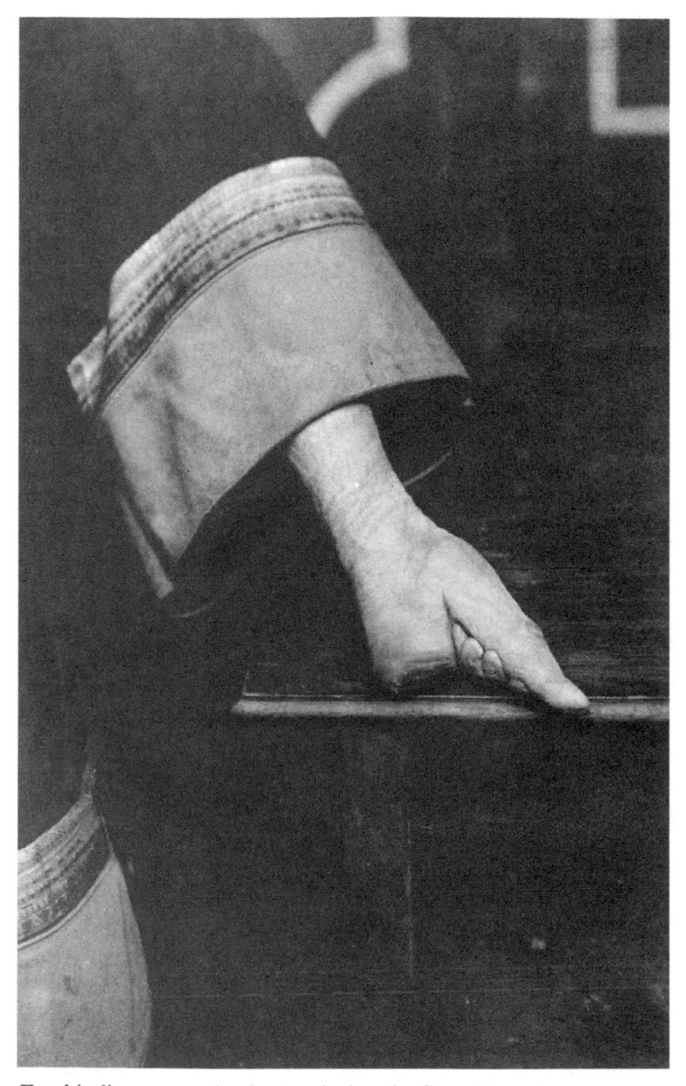

Footbinding, a practice begun during the Song Dynasty, was popular among the upper class as a symbol of wealth and subserviance. (EPD Photos/CSU Archives)

the rise of a new mercantile class. Footbinding, a practice designed to enhance the desirability of women, becomes widespread among the upper class. Cultural advances include the invention of movable type. In poetry, the *ci*, a genre with roots in popular song, reaches its fullest development. Landscape painting becomes more sophisticated and refined, with the use of blurred outlines and other impressionistic techniques.

960–1127 Northern Song period

During the first part of the Song period, known as the Northern Song, the capital, established by Taizu, the founder of the dynasty, is located at Kaifeng.

1081 Birth of poet Li Qingzhao

Li, considered by many to be China's greatest woman poet, is known for her *ci* poetry (a popular lyrical genre of the Song period). Only fragments of her work, originally composing six volumes of poetry and seven volumes of essays, survive into the modern era.

1127–1279 Southern Song

The Song empire pulls back from northern China because of invasions by nomadic tribes, moving its capital to Hangzhou.

c. 1180–24 Life of painter Xia Gui

Xia Gui is renowned for his landscape paintings. He and his contemporary, Ma Yuan, are the leading painters of the Southern Song period.

1275–92 Marco Polo's stay in China

In the service of the Mongol emperor, Kublai Khan, Marco Polo visits China as part of his voyage around the world.

Mongol Rule

1279–1368 Mongols overrun China and establish the Yuan dynasty

Mongols led by Kublai Khan overthrow the Song dynasty and seize control of China. Kublai Kahn inaugurates the Yuan dynasty—the first foreign dynasty to rule China—as its first ruler. China under the Yuan dynasty benefits from cultural exchanges with West Asia and Europe, which bring such items as cloisonné (an enameling technique) and certain Western musical instruments to China. The city of Beijing undergoes extensive renovation, as does the Grand Canal. Water works and roads are improved, and granaries are built throughout the country.

The Yuan period marks the formal beginning of China's drama tradition, the forerunner of the famed Beijing opera. Since many government positions are filled by foreigners, many Chinese scholars who formerly would have been civil servants turn to literature, producing plays and novels. The Beijing opera combines song, instrumental music, mime, dancing, and poetry to present narratives about legendary Chinese heroes.

c. 1330–1400 First Chinese novels are written

The development of the Chinese novel begins with the two early works, *San Guo Zhi* (The Romance of the Three Kingdoms) and *Shui Hu Zhuan* (The Water Margin), both by Luo Guanzhong. Both are long, loosely plotted works that collect and rework earlier material from the Chinese oral tradition of travel and adventure tales.

Restoration of Chinese Rule

1368–1644 The Ming dynasty

China's Mongol rulers are succeeded by the Ming dynasty, founded by a Buddhist monk turned rebel leader. China enjoys a long period of stability during the Ming period, which is an age of exploration for the Chinese, who complete expeditions to the Indian Ocean and the South China Sea. Eventually, though, the dynasty is weakened by Mongol attacks and by having to defend Korea against invading Japanese.

1405–33 Great sea voyages are undertaken

China launches seven ambitious maritime expeditions, commanded by an admiral named Zheng He. Most take place during the reign of Yong Le. In junks (ships) as long as 400 feet, the Chinese seamen visit destinations including Java, Sumatra, Ceylon, the east coast of Africa, Aden, and the Red Sea.

1406–27 Vietnam is annexed

During the reign of Yong Le, China annexes Vietnam (known as Annam).

1420 Capital is moved to Beijing

With the completion of the Imperial Palace, China's capital is moved from Nanjing to Beijing. Altogether, the city will be home to twenty-four Chinese emperors.

1421 Construction of the Forbidden City begins

The Forbidden City, a complex of residences, royal halls, pavilions, and artificial lakes, is one of China's great architectural achievements. Although located on the site of the previous Mongol capital of Cambaluc, it is an entirely new city, with a nine-gate 40-foot-high wall extending for 14 miles.

16th century *Journey to the West* is published

Journey to the West (Xiyou ji), China's most famous comic novel, is published sometime between 1500 and 1582 by Wu Chengen. It is based on an actual event—a pilgrimage by the Buddhist monk Xuanzang (602–64) to India in search of religious texts. The events of this journey have already passed into Chinese folklore, and Wu adapts them to create a classic comic novel. The book's 100 chapters detail eighty-one adventures that befall the protagonist, Tripitaka, who is accompanied by three animal spirits. In addition to its effectiveness as an adventure story, *Xiyou ji* is also prized for its satire on Chinese society of its time. An English translation of the novel by Arthur Waley, called *Monkey,* is published in 1942.

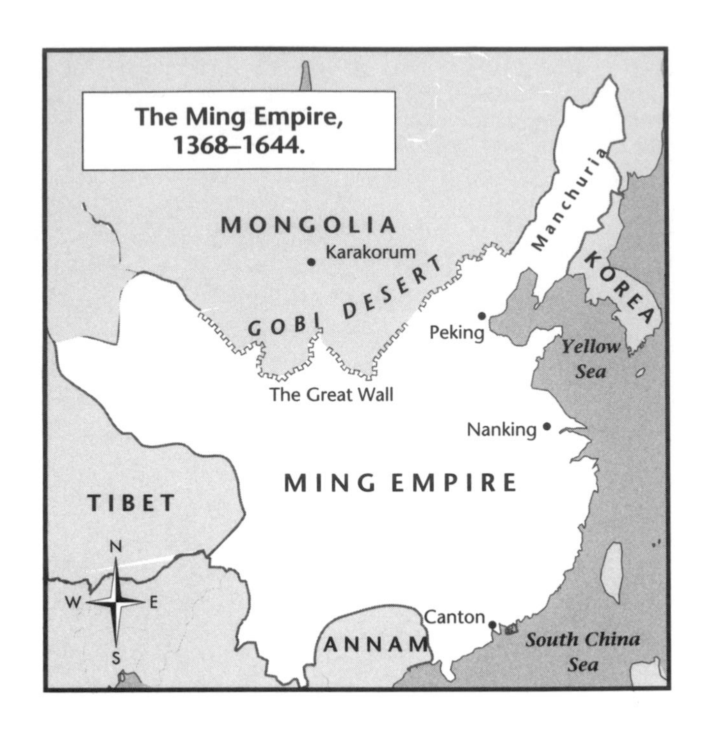

1516 Portuguese arrive in China

Portuguese explorers become the first Westerners to arrive on Chinese soil.

1520 First Western envoy reaches China

The Portuguese envoy, Tomé Pires, arrives in China. The emperor refuses to receive him.

1556: February 2 Deadliest earthquake in recorded history kills over 800,000

The most lethal earthquake ever recorded takes over 800,000 lives along the Wei-ho River valley in central China.

1557 The Portuguese settle Macao

The Portuguese establish a settlement at Macao.

1592 Japan attempts invasion of China

The Japanese emperor Toyotomi Hideyoshi invades Korea as part of a projected attack on Japan (it is never carried out).

Late 16th century First realistic novel about Chinese society appears

Jin ping mei (The Plum in the Golden Vase) by Xiao-xiao Sheng is the first Chinese novel based on realistic details of contemporary society rather than on historical legends. One of the most widely read Chinese novels of all time, it portrays the family of a wealthy but unscrupulous businessman, Qing Ximen, and has raised controversy because of erotic passages describing the protagonist's licentious activities.

1637 English traders reach China

The first English traders arrive in China.

Manchu Rule

1644–1912 Qing (Manchu) dynasty

The Qings (also known as Manchus because their homeland is Manchuria) capture Beijing and found the Qing dynasty, China's last imperial line. Although a clear division is maintained between the conquering Manchus and the majority Han Chinese, the Qing rulers retain China's Confucian system of government administration and preserve most of its cultural elements. The Qing dynasty protects its territory by enlarging its boundaries, and the Chinese empire eventually grows to cover its greatest area in history, including Outer Mongolia, Taiwan, and part of Central Asia. The greatest challenge to Qing rule is posed by the arrival of traders and other visitors from the West. The Qing are responsible for the classic appearance that Westerners associate with the Chinese: the shaved head and *queue*, or pigtail.

1655 Dutch secure trading rights

The Dutch win limited trading rights with China.

1679 Famous painting manual published

Jieziyuan Huazhuan (Manual of the Mustard Seed Garden), a classic five-volume painting manual that is still in use, is published.

1689 Treaty of Nerchinsk signed with Russia

The Treaty of Nerchinsk, China's first treaty with a Western power, ends border skirmishes by establishing the border between the Russian region of Siberia and Manchuria in northern China.

1699 British East India Company establishes a presence in China

The British East India Company sets up a trading post, known as a "factory," in Canton, which becomes a trade center for porcelain, silk, lacquer, musk, and other commodities, most notably tea, as the British demand for the beverage grows.

1724 Spread of Christianity is banned

Teaching Christianity is declared illegal and most Christian missionaries are deported.

1728 Massive encyclopedia is published

The encyclopedia, *Tu Shu Ji Cheng*, begun in the previous century, is published.

1760 "Guangzhou system" for trading is established

To retain control over the activities of foreign traders, China restricts trade to Guangzhou (present-day Canton).

1792 *Dream of the Red Chamber* is published

Hong Lou Meng (The Dream of the Red Chamber) by Cao Xueqin, commonly regarded as the greatest Chinese novel, is published after circulating for years in manuscript form. The semiautobiographical work tells the story of the decline and subsequent recovery of a family belonging to the gentry class. It is noted for its complex plot and rich character development.

1793 British request for trade concessions is rebuffed

King George III sends Lord Macartney as an official envoy to the Chinese emperor Qian Long to request the opening of additional ports at Tianjin (Tientsin), as well as other trade concessions. Requested to prostrate himself before the emperor in the full Chinese bow (*kowtow*), Macartney refuses and instead is allowed to go down on one knee, the traditional bow before the British sovereign. The trade concessions are refused.

1796 Chinese ban the use of opium

The rapid spread of the opium trade creates widespread opium addiction in China, which persists in spite of an official ban on the use of the drug.

1807 Missionary arrives in China

Robert Morrison, the first Protestant missionary to come to China, arrives in Guangzhou.

The Empire Declines

1839–42 Opium War

China and Britain go to war over Chinese trade restrictions in general, and, in particular, their attempts to halt the opium trade, which include destroying supplies of British opium stockpiled in Canton Harbor. The Chinese are easily overpowered by British forces using modern firearms, and they are forced to sign the Treaty of Nanjing, granting the British numerous concessions. Five ports are opened to British trade, the island of Hong Kong is ceded to the British, tariffs are limited, and Britain gains most-favored-nation trade status. British nationals are also granted extraterritoriality (exemption from Chinese laws).

1843: June 25 Hong Kong becomes a British possession

Hong Kong is declared a crown colony of Great Britain.

1850–64 Taiping Rebellion

China's greatest popular uprising in modern times, the Taiping Rebellion, is led by a village teacher, Hong Xiuquan (1814–64), who proclaims his own state as an alternative to Manchu rule and gathers an army of thousands, capitalizing on economic problems, widespread discontent with Manchu rule, and antagonism toward aggression by Western powers. The rebellion is defeated, but thirty million lives are lost over a fourteen-year period, and the Qing dynasty is significantly weakened. The Taiping Rebellion serves as a model for nationalist and communist rebels in the twentieth century.

1856–60 Second Opium War

A second war with the British (and, this time, the French, also) is triggered with the arrests of Chinese aboard a British ship sailing from Hong Kong. The Chinese are defeated and forced to allow new trade concessions, including the opening of its borders to all Westerners.

1858: June 26 The Chinese accept the Treaty of Tianjin (Tientsin)

The Chinese are forced to sign the Treaty of Tianjin which provides for the Yangtze River and five new ports to be opened to trade.

1859 Opium War is resumed

After the Chinese violate provisions of the Treaty of Tianjin (Tientsin), fighting in the second Opium War against the British is resumed, as Chinese troops attempt to bar the entry of foreign diplomats into Beijing.

1859: October British occupy Beijing

British forces occupy Beijing.

1859: October 18 Imperial palace is burned down

British and French forces set fire to the imperial Summer Palace.

1860 Russia acquires Chinese territories

Russia gains control of parts of Manchuria.

c. 1861–74 Self-strengthening movement

Chinese officials and scholars attempt to strengthen their country by studying Western ideas and models and applying them to their own political and economic practices. The goal is not modernization but the preservation of the existing Chinese culture and way of life by all available means, and a prominent motto is "Chinese learning for essence [substance]; Western learning for use."

1862–74 Tongzhi Restoration

The Westernization efforts of the Self-strengthening Movement are counteracted by the conservative agenda adopted during the reign of the Tongzhi Emperor and largely implemented by his mother, the empress dowager Cixi (Tz'u Hsi).

1866 Birth of Sun Zhongshan

Honored as the hero of the Chinese revolution that overthrows imperial rule, Sun Zhongshan (Sun Yat-sen) is born in Canton. He studies in Honolulu and Hong Kong and is trained as a physician. He also converts to Christianity. By the 1890s, Sun begins to work for the overthrow of the Manchu government and has to leave the country in 1895 after a failed revolt in Canton. He lives in Japan and then London, gaining international support for his political goals, and organizes a Chinese revolutionary league in Tokyo in 1905.

Sun returns to China in 1911, after revolutionaries, aided by the military, force the last Manchu emperor to abdicate. He is elected president of the Chinese Republic, but control of the country passes first to military leader, Yuan Shikai, and, later, to warlords. In 1923 he organizes the Guomindang, a nationalist party, with support from Russia. The following year he takes part in negotiations with warlords in northern China. Sun Zhongshan dies of cancer in 1925.

1866 Scientific publisher founded

Jiangnan Arsenal, which publishes translations of scientific works, is established in Shanghai.

1870: June 21 Tianjin (Tientsin) Massacre

In an outbreak of anti-foreign feeling, an angry mob kills twenty-one French nationals following rumors of abuses at a Catholic orphanage.

1876–78 Famine ravages northern China

A three-year absence of the normal annual monsoons results in drought, crop destruction, and, finally, the world's worst recorded famine. As many as thirteen million people are thought to have died as a result of the disaster, which leads to widespread starvation, banditry, and cannibalism. The Manchu government adds to the tragedy by keeping it secret from the rest of the world for two years instead of asking for relief aid, which is finally sought by the British when they become aware of the situation.

1876 First railway is completed

Construction of China's first railroad, the British-owned twelve-mile line running from Shanghai to Woosung, is completed.

1880 First telegraph line is completed

China's first telegraph line, running from Shanghai to Tianjin, becomes operational.

1884–85 Sino-French War

China aids its possession, Vietnam, in fighting the French. The Chinese are defeated, and France gains control of Chinese lands in Southeast Asia, which become known as French Indochina.

1887 Yellow River flood kills 1.5 million

The Yellow (Huang Ho) River breaks through a dike, flooding a massive area and resulting in the deaths of an estimated 1.5 million people.

1887: October 31 Jiang Jieshi is born

Jiang Jieshi (Chiang Kai-shek), future leader of the Chinese Nationalist government, is born in the province of Zhejiang. He embarks on a military career at the age of 18, receiving his training at the Baoding Military Academy and a Japanese military academy in Tokyo. While in Japan, he meets Chinese nationalist, Sun Zhongshan, who is gathering support for the overthrow of the Manchu dynasty. Chiang takes part in a variety of revolts between 1911 and 1925, when he takes charge of the Guomindang (Nationalist) forces in southern China following Sun's death.

After briefly allying his forces with the Communists, Chiang turns on them in 1927, carrying out a massacre of Communists in Shanghai and pursuing a civil war against them for two decades, through the 1930s, World War II, and beyond. By May 1949 the Communists are victorious, driving the Nationalists out of mainland China to Taiwan, where Chiang establishes the Republic of China, recognized by the United States as the official Chinese government until 1979. Chiang is elected president of the Republic of China for five successive terms, serving in that position until his death in 1975.

1893: December 26 Birth of Mao Zedong (Mao Tse-tung)

Mao Zedong, leader of China's Communist Revolution, is born in Hunan Province to a middle-class peasant family. He establishes a pattern of rebellion at a young age, defying the harsh discipline of both his father and his schoolteachers. In 1918 he moves to Beijing, where he is first exposed to Marxism. He is one of the founding members of the Chinese Communist Party in 1921. During the 1920s Mao organizes peasants in the countryside, authoring a legendary article on the subject in 1927. When the Communists are routed from their stronghold, Jianxi, in southern China, Mao leads the year-long 6,000-mile Long March to a new base in the north, becoming the uncontested leader of the Communist party. After World War II, Mao leads the successful Communist Revolution which results in the creation of the People's Republic of China in 1949, and he is named chief of state as well as party chairman.

In the 1950s Mao presides over plans to collectivize agriculture and raise both agricultural and industrial production. In 1957 he launches the Hundred Flowers movement, an experiment in tolerance that backfires when the level of political criticism is so great that the government is threatened by it and cracks down again. In the 1960s, Mao presides over China's rift with the Soviet Union and leads the Cultural Revolution, a purge of alleged capitalist elements in Chinese society spearheaded by newly organized youth groups, the Red Guards. After the death of Mao's main political rival, Lin Baio, in 1971, Premier Zhou Enlai takes over most political leadership duties from Mao, whose health is declining. Thought to be suffering from the effects of either Parkinson's disease or a stroke, Mao dies on September 9, 1976 in Beijing.

1894–95 First Sino-Japanese War

China is defeated by Japan in the first Sino-Japanese War. The Treaty of Shimonoseki is signed, and China is required to pay a large sum of money to the Japanese. Japan obtains control of Taiwan and other concessions. This Chinese defeat leads to a "scramble for concessions" by other nations, including Great Britain, Germany, Russia, and France, in the following years.

1896 Motion pictures are introduced

The first motion pictures are shown in China.

1896 Postal service begins

China's postal service is inaugurated.

1896 Birth of poet Xu Zhimo

One of China's most distinguished modern poets, Xu Zhimo is an innovator who introduces Western influences and Chinese vernacular into his work. He is educated at Beijing University and in the United States, where he studies political science and economics. In 1920 he moves to England, where he studies at Cambridge and is inspired to embark on a literary career by his exposure to the English Romantic poets. Returning to China, Xu begins writing poetry using the vernacular. He later works as an interpreter for the Indian poet Rabindranath Tagore, who serves as yet another poetic influence.

Xu teaches poetry at several Chinese universities and edits the literary supplement of the major newspaper *Chen bao* (Morning Star). He is also a founder of the literary journal *Xinyue shudian* (Crescent Moon Book Company), that publishes Western literary works. Xu Zhimo publishes four poetry collections in his lifetime. Xu dies on November 31, 1931.

1898 Great Britain leases Hong Kong

The New Territories of Hong Kong and Weihei in Shandong province are leased for 100 years by Britain.

1898 Birth of statesman Zhou Enlai (Chou En-lai)

Zhou Enlai is born in Jiangsu (Kiangsu) Province to a wealthy family. As a youth he pursues studies in France, Germany, and Japan. In the 1920s he is converted to Marxism and leads an uprising by workers and peasants in Shanghai in 1927. He is a leader of the Long March in 1934–35. He holds the positions of premier and foreign minister in the People's Republic of China and often represents his country abroad. He is also instrumental in the normalization of relations between China and the United States. Under attack from the Gang of Four in the 1970s, Zhou becomes ill in 1974 and dies two years later, on January 8, 1976. More than a million Chinese fill Tienanmen Square to honor him when his death is announced.

1899–01 Open Door Policy instituted by the U.S.

The United States implements the Open Door Policy to equalize trading privileges among Western powers in China and protect Chinese sovereignty.

1899: February 3 Birth of author Lao She

Lao She (pen name of Shu Sheyou) is born in Beijing. After serving as a school principal, he travels to England, where he teaches Chinese and becomes familiar with the works of Charles Dickens, which inspire his first novel. His early novels are humorous and satirical. Eventually he abandons his early belief in the power of the individual to affect society, a change that is evident in his most celebrated work, *Luotuo Xiangzi* (Camel Xiangzi; 1936), the tragic story of a Beijing rickshaw puller. This work, in a pirated version with an altered ending, achieves great success in the United States, where it appears as *Rickshaw Boy* (1945). Starting with the Sino-Japanese war, Lao's works become patriotic and propagandistic in nature, and decline from their earlier quality. Lao dies on October 24, 1966.

1900–01 Boxer Rebellion

A secret society called Ho Ch'uan (literally, "righteous, harmonious fists," loosely translated as "Boxers") becomes a focus for anti-foreign sentiment among the Chinese. Claiming some 140,000 members by the end of the nineteenth century, the group stages a rebellion, supported by the Manchu court of the dowager empress Cixi. Raids are carried out against foreign missions and other property in the northern provinces of Chihi, Shansi, and Shantung. Railroads and other installations are attacked, and buildings are burned down. Beijing is occupied for two months. Much of the Boxers' bitterest hostility is directed against Chinese Christians. Ultimately, 18,000 troops from Britain, France, the U.S., and other Western powers join forces to defeat the revolt, and China is required to pay $333 million in reparations.

1904: August 22 Birth of statesman Deng Xiaoping

Deng Xiaoping, one of the major Chinese leaders of the twentieth century, is born Deng Xixian, in a village in Sichuan province. The son of a well-to-do family, he leaves for France in 1920 to work and study. While there he joins the Communist Youth League. Deng also studies briefly in the Soviet Union before returning to China, where he becomes active in the Communist Party, changing his name to Deng Xiaoping. He leads troops in the Communist rebellion against Jiang Jieshi's nationalist Guomindang party and joins forces with Mao Zedong in 1931. In 1934–35 Deng participates in the historic Long March by the Communist forces.

Once the People's Republic is formed, Deng rises rapidly in the Communist hierarchy, becoming secretary general of the Communist Party by 1956. By the 1960s, however, Deng's fortunes begin to falter, as he is purged in the Cultural Revolution and goes into internal exile in the south. After being rehabilitated in 1973, Deng falls out of favor again in 1976 but makes a second comeback by the following year. By 1978, Deng's plans for economic liberalization are being adopted, and under his guidance China assumes full diplomatic relations with the United States. In 1989, Deng's reputation suffers when the Tienanmen demonstrations are violently suppressed. The same year, Deng gives up all official government positions, but he remains a dominant political figure into the 90s. In his final years, he suffers from Parkinson's Disease and respiratory ailments. He dies on February 19, 1997.

1905 Civil service examinations are abolished

The civil service examination system, a mainstay of traditional Chinese society for 2,000 years, is dismantled.

1905 Sun Zhongshan forms the United League

Tong Men Hui (the United League), a Chinese revolutionary group, is formed in Tokyo by Sun Tay-sen.

1906: September 18 Typhoon kills thousands in Hong Kong

A typhoon with 100-mile-per-hour winds devastates Hong Kong. The huge death toll of 10,000 is blamed partially on the delay in firing the "typhoon gun," a cannon in the harbor that is fired only twenty minutes before the arrival of the storm, giving residents little time to flee to safety. The British, well protected in their brick houses, suffer few losses, with all but twenty deaths occurring among the Chinese.

1909 National Library of Beijing is founded

The National Library of Beijing, China's largest library, is established. At the end of the twentieth century it contains nearly sixteen million works.

1910 China seizes Tibet; Dalai Lama flees

The Chinese gain control of the government of Tibet; the Dalai Lama flees.

1911: September Flooding of Yangtze River kills over 200,000

The Yangtze River bursts its banks, flooding the provinces of Nganhwei, Hupei, Hunan, and Ichang. Crops and livestock are completely destroyed, and 100,000 people are drowned. Another 100,000 die from the ensuing famine. Over half a million people are left homeless and migrate to Manchuria and Mongolia.

The Republic

1911: October–December Chinese revolution overthrows imperial line

Beginning with a military uprising in Wuhan, a revolution led by nationalist leader Sun Zhongshan overthrows the Manchu government, ending China's last imperial dynasty. Sun Zhongshan becomes the president of China's first republic.

1912 Women's suffrage is enacted

Women gain the right to vote on the same basis as men.

1912: January 1 The Republic of China is formed

The Republic of China is established.

1912: February 7 Last Manchu ruler abdicates

Two millennia of imperial rule in China end with the forced abdication of the infant emperor Puyi (P'u-yi), implemented by the dowager empress Cixi (Tz'u Hsi).

1912: February 15 Yuan is chosen as new leader of the Chinese republic

Sun Zhongshan turns over leadership of the Chinese republic to Yuan Shikai (Yuän Shih-kai), whose rule soon becomes repressive, as the new republic is torn by divisions between political factions.

1912: August Guomindang (Kuomintang) party is organized

Sun Zhongshan founds the *Guomindang* (People's National Party).

1912: December Yuan declares himself emperor

Yuan Shikai, the successor to Sun Zhongshan, declares himself emperor of China.

1913 Artists begin using nude models

A major and controversial sign of Western influence on traditional Chinese culture is the use of nude models by Chinese artists, which remains a continuing source of debate on both moral and political grounds.

1915 China concedes to Japan's Twenty-One demands

China, confronting from a weak position Japan's growing militarism, is forced to accept most of the conditions in the Twenty-One Demands presented by Japan.

1917–26 Period of rule by warlords

Following the death of Yuan Shikai and his failure to establish a new dynasty, China enters a period of rule by rival warlords.

1917–23 New Culture Movement

A period of protest against China's warlords by students and other intellectuals leads to the May Fourth Movement (see 1919: May).

1917: August 14 China sides with the Allies in World War I

China enters World War I, declaring war on Germany.

1918 Beiping Academy of Art is founded

The Beiping Academy of Art in Beijing is the most important of the new art schools established in the first two decades after the creation of the Chinese republic.

1919: May May Fourth Movement

At the Versailles Peace Conference following World War I, Germany's former possessions in China's Shandong province are awarded to Japan, triggering mass student demonstrations. Due to this pressure from the people, the Chinese government does not sign the treaty.

1920 China joins the League of Nations

1921 Chinese Communist party is formed

With aid from Russia's Communists, the Chinese form their own Communist party. Russia further allies itself with China by renouncing all claims to Chinese territory and opposing those of other foreign powers.

1924 Whampoa Military Academy is founded

The Whampoa Military Academy, founded by Sun Zhongshan, plays a central role in training the nationalist troops that defy the warlords and unify China. Its first director is Jiang Jieshi, who oversees the training of the first three graduating classes. At the school's inception, Nationalists and Communists cooperate in running the academy and teaching its classes. Zhou Enlai is one of its early political instructors.

1925: March 12 Sun Zhongshan dies

Guomindang leader Sun Zhongshan dies (see 1866). Leadership of the party is taken over by Jiang Jieshi.

1927: March Guomindang defeats warlords and captures Shanghai

Nationalist Guomindang forces capture Shanghai. Jiang Jieshi begins outright attacks on the Communists, with whom the Guomindang had joined forces in 1923. Chiang's forces launch attacks on Communists in Shanghai, and they are also driven out of other areas controlled by the Guomindang, including Wuhan.

1928: October Jiang Jieshi forms a government in Nanjing (Nanking)

Jiang Jieshi, leader of the nationalist Guomindang, forms a national government that is officially recognized internationally and receives aid from other nations. Chiang continues to wage war against the Communists.

1929–31 Famine afflicts the north

Northern China suffers from a disastrous famine.

1930 Socialist writers' group is formed

The League of Left-Wing Writers is instrumental in the development of social realism in Chinese literature. Social realist works portray realistically problems and events in contemporary society and promote leftist philosophies.

1931: February 12 Mine explosion in Manchuria kills 3,000

An explosion in the Fushun coal mines east of Mukden traps 3,000 miners. The Chinese and Japanese issue conflicting accounts of the incident, for which no official cause is ever found.

1931: November 31 Poet Xu Zhimo dies

Innovative and influential poet Xu Zhimo dies. (See 1896.)

1932: February The Japanese occupy Manchuria

The Japanese invasion of Manchuria, begun the previous year, is completed.

1934–40 Major plays of Cao Yu are written

Best-known of the new playwrights who achieve fame following the May Fourth Movement, Cao Yu (b. 1910) demonstrates the influence of Western plays on Chinese drama. His major plays, which include "Thunderstorm," "Sunrise," Wilderness," and "Peking Man," win a wide audience.

1934–35 Long March by the Communists

Jiang Jieshi's Nationalist government continues its war on the Communists following the Japanese invasion of Manchuria. Blockaded by Nationalist troops in the south of the country, a force of over 100,000 Communists breaks through and undertakes the famous, year-long 6,000-mile (9,700-kilometer) Long March on foot to Shaanxi Province in the northwest, establishing their base there. Mao Zedong becomes the unchallenged leader of the Communists.

1935: July 4 Yellow River flooding kills 30,000

The North China Plain is flooded when the Yellow River breaks through a levee in Shantung province following heavy rains. Five million people lose their homes and 30,000 lose their lives in the disaster.

1937: July 7 Second Sino-Japanese War

The second Sino-Japanese War begins with hostilities at the Marco Polo Bridge. The Japanese ultimately occupy many of China's major population centers and continue their occupation through World War II. Nanjing, the capital, is captured and looted, and some 40,000 civilians are slain. The Guomindang retreats to Sichuan and Yunan provinces.

1939–45 World War II

Japan occupies China throughout World War II, and among the Chinese themselves Communist and nationalist Guomindang forces battle each other to unseat the Japanese and control the country. Japan retains firm control of urban areas, while Communist power spreads in northern rural areas, where land reforms are carried out and local militias are raised to fight the Japanese. The Nationalist government, led by Jiang Jieshi, is weak and plagued with corruption.

1939: September–November Famine kills 200,000 in Hopei province

Crop-destroying floods in the Yellow River basin cause a famine that kills 200,000 people. The same autumn, an estimated twenty-five million bushels of rice are destroyed in three neighboring provinces due to fear of glutting the market after a bumper harvest. The occupying Japanese intercept international relief sent to the famine-stricken area.

1942–43 Famine kills almost three million in Honan province

Failure of all three of the province's crops—rice, millet, and corn—in the summer of 1942, after poor harvests the preceding years, leads to one of the worst famines in history. The tragedy is compounded by government neglect and corruption, as relief funds are siphoned off by bankers and tax collectors. The peasants are reduced to eating tree bark and to cannibalism. Conditions in the province are documented by American journalists Theodore H. White and Annalee Jacoby in the spring of 1943 in their book, *Thunder Out of China.*

1942 Mao gives influential lecture on literature and art

Mao's influential "Yenan Talks on Literature and Art" become a cornerstone in the promotion of social realism in the arts. His main focus is the subordination of art to politics and its use in educating and uniting the people. He also calls for the creation of literature written by common working people, including soldiers and peasants.

The People's Republic of China

1945–49 China's civil war continues

The struggle between the Nationalists and the Communists continues following the surrender of Japan and the end of World War II.

1946 Marshall arranges cease-fire in civil war

General George C. Marshall of the United States Army, negotiates a cease-fire between the Guomindang and the Communists, but it does not last.

1949 Women's right to vote upheld

The new People's Republic of China upholds women's right to vote.

1949 People's Daily newspaper founded

The *People's Daily,* which becomes China's leading newspaper, is founded. By the end of the twentieth century, it has a circulation of three million.

1949 First National Congress of Writers and Artists

The First National Congress of Writers and Artists meets to encourage the type of social realist literature advocated in Mao Zedong's "Yenan Talks on Literature and Art." Prominent examples of early social realism in the People's Republic of China are the land reform novels of Ting Ling; works by Chao Shu-li and Chou Li-Po; the plays of Lao She; and the play *The White-Haired Girl* by Ho Ching-chih.

1949: October 1 People's Republic of China is formed

China's Communist forces take control of the Chinese mainland, establishing the People's Republic of China, with its capital at Beijing. The Nationalist Guomindang, under the leadership of Jiang Jieshi, retreats to Taiwan, where they establish a Nationalist government. Mao Zedong is named chairman of the newly formed People's Republic of China, and Zhou Enlai becomes the premier. The new state is immediately recognized by the Soviet Union. The United States does not grant diplomatic recognition to the People's Republic and continues to regard the Nationalist government on Taiwan as the official Chinese government.

1949: November 1 Chinese Academy of Sciences founded

China's research institutes are combined into a single organization.

1950: February 14 Sino-Soviet Friendship Pact is signed

Mao visits the Soviet Union and persuades Soviet leader Josef Stalin (1879–1953) to formalize a political and economic alliance with the People's Republic of China. The agreement guarantees financial and technological aid to China.

1950: June 28 Agrarian Reform Law is adopted

The first stage of China's land-reform program is carried out. Land is redistributed and given to the peasants to farm collectively in groups of thirty to forty families, who provide each other with mutual assistance and share the income from their crops.

1950: October China enters the Korean War

When the United Nations forces enter Korea to aid the Republic of Korea in the south, China sends troops to help the Communist Democratic People's Republic of Korea in the north and to protect Manchuria against the perceived threat posed by the U.N. forces, which includes troops from the United States.

1951 China takes over Tibet

China captures Tibet in what is announced as a "peaceful liberation."

1953–57 First Five-Year Plan

The Chinese government makes a major effort to industrialize and socialize the country in its first Five-Year Plan. Production of coal, iron, and steel rises rapidly, as does construction of power facilities, dams, and railways. Collective farms, modeled after those in the Soviet Union, are established.

1954: August Yangtze and Hwai rivers flood, killing thousands

Flood waters of the Yangtze and Hwai rivers rise to a record ninety-six feet following what is called the heaviest rainfall in a century. The Yangtze bursts through a dam designed by a Chinese architect working from Soviet models. The flood is a significant setback for the Communist government's first Five-Year Plan.

1954: September First National People's Congress is held

The First National People's Congress meets and adopts a new constitution. Mao and Zhou Enlai remain in charge of the government and are joined by Liu Shaoqui.

1956: May The Hundred Flowers campaign

Mao undertakes a campaign to expand freedom of expression and his government's heavy censorship. Its name is based on an ancient Chinese saying, "Let a hundred flowers bloom, and a hundred schools contend." Political debate and criticism of the government are allowed and even encouraged. Within a year, however, the idea backfires, as the criticism of Communist rule grows into a storm of protest and demands for greater freedom spread throughout the country.

1957: June Anti-Rightist campaign silences protests

In response to the unexpected dissent unleashed by the Hundred Flowers movement, a harsh government crackdown silences criticism and dissent. Hundreds of thousands of people are banished to labor camps, high officials are dismissed, and public confessions of wrong-doing are demanded from many.

1958–60 The Great Leap Forward

Following the success of the first Five-Year Plan, a second plan is created, even more ambitious than the first, to expand dramatically both agricultural and industrial production. Farms are organized into massive collective communes containing as many as 20,000 people and all private ownership is abolished.

1958–60 Classical ballet is introduced to China

During the Great Leap Forward, classical Western ballet is introduced to China by eminent dancers from the Russian Bolshoi and Kirov companies.

1959–61 The "Three Bad Years"

Instead of the improvement anticipated from the Great Leap Forward, China suffers a period of decline and deprivation as the new plan fails to produce the desired results, and a series of bad harvests and natural disasters, including drought, flooding, and locusts, adds to the difficulties.

1959 Oil reserves are discovered in northeast China

Oil deposits are discovered in the Sungari River basin. China's first oil field begins operations a year later.

1959: March Tibetan revolt is crushed

A revolt against Chinese rule is defeated, and the Dalai Lama (the Tibetan spiritual leader) takes refuge in India.

1959: April Second National People's Congress

At the Second National People's Congress, Liu Shaoqui is named to replace Mao Zedong as chairman of the People's Republic, but Mao still remains chairman of the Communist Party.

1960 Sino-Soviet split

Clashes and ideological disagreements with Soviet premier Nikita Khrushchev throughout the 1950s, following the death of Stalin, lead to a growing rift between China and the Soviet Union. Krushchev withdraws promised technical and economic aid from China and helps India, a country with which the Chinese are engaged in a border dispute.

1962: October–November Border dispute with India escalates to open war

Chinese troops attack Indian forces on the Aksai-Chin plateau. Hostilities end by November.

1964: October China joins the "nuclear club"

China carries out its first test of an atomic bomb. With this test, China becomes the world's fifth nuclear power, joining the United States, Soviet Union, Britain, and France.

1965–69 The Cultural Revolution

Chairman Mao launches the Great Proletarian Cultural Revolution to strengthen his control over the country and root out any perceived sympathy for capitalism or Western influences. China's young people, organized into units of the Red Guards, are encouraged to turn against older Chinese who they feel have strayed from the strict Communist ideology of the Revolution. Many older leaders, including People's Republic chairman Liu Shao-ch'i, are purged, and thousands of educated urban Chinese are sentenced to hard labor or killed. It is estimated that as many as 400,000 die because of activities during the Cultural Revolution.

1966–70 Universities close

China's universities are closed during the Cultural Revolution. When they reopen they are placed under the direction of Revolutionary Committees who make all admissions and curriculum decisions. Students' political views are closely monitored, and all students are required to spend two years doing agricultural labor. Academic degrees are abolished as signs of

After the Cultural Revolution, agriculture is carried out in communal farms like this one, where tea and grain are grown in terraces on rocky slopes. (EPD Photos/CSU Archives)

unequal status and not reinstated in Chinese universities until 1980.

1966: October 24 Novelist Lao She dies

Satirical novelist Lao She dies. (See 1899.)

1967: June First hydrogen bomb is tested

China explodes its first hydrogen bomb. A hydrogen bomb is several times more powerful than an atomic bomb.

1969 Chinese and Russian troops clash

Border clashes between China and Russia take place near the Ussuri River. Diplomatic talks fail to produce a resolution, and both countries continue to maintain troops in the region, although the fighting ends. Russia and China compete with each other for political, economic, and military influence abroad.

1970 Beijing subway opens

China's first subway, in Beijing, opens to the public.

1970 First space satellite is launched

China launches its first space satellite.

1971: April U.S. table tennis team visits China

China paves the way for improved relations with the United States by inviting a U.S. table tennis team to visit. The United States reciprocates by sending presidential advisor Henry Kissinger (b. 1923) to China, where he makes plans for a presidential visit the following year by Richard Nixon (1913–94).

1971: October 25 China is admitted to the United Nations

The People's Republic of China wins the Chinese seat in the UN by a vote of 76–35 in the General Assembly (17 abstentions), replacing the government of Taiwan, which is expelled from the organization. In a break with its previous policy, the United States votes in favor of its admission to the UN, although it continues to maintain diplomatic relations with Taiwan until January 1, 1979.

1972: February 21–28 President Nixon visits China

Long-time Communist opponent, President Richard M. Nixon makes a historic visit to the People's Republic of China as an introductory step toward normalizing relations between the two countries. He holds talks with Communist party chairman Mao Zedong and Premier Zhou En-lai. The

Shanghai Communiqué is issued, outlining the positions of the two nations. Following the meeting, trade between the two nations expands, and cultural exchanges are implemented.

1973: May U.S. and China establish liaison offices

As a step toward full diplomatic relations, the People's Republic of China and the United States set up liaison offices in each other's capitals.

Mid-1970s China dominates the sport of table tennis

China fields three-time world champion Zhuang Zedong (b. 1942), as well as runner-up Li Furong (b. 1943). Other top players include Zhang Xielin, Hsu Shao-fa, and Huang Liang.

1975 Gang of Four opposes modernization plans

A radical political faction consisting of Mao Zedong's fourth wife, Jiang Qing (Chiang Ch'ing), and three colleagues (Zhang Chunqiao, Yao Wenyuan, and Wang Hongwen) opposes Zhou Enlai's plans for modernization in industry, agriculture, technology, and other areas. They eventually become known as the Gang of Four. Their opposition to Deng Xiaoping, widely considered the front-runner to succeed Mao as party chairman, results in his removal from politics for a period of time.

1975: April 5 Jiang Jieshi dies

Chinese nationalist leader Jiang Jieshi dies in Taipei, Taiwan. (See 1887.)

1976: January 8 China mourns death of Zhou Enlai

Premier Zhou Enlai dies. A crowd of over a million Chinese gathers in Tienanmen Square to mourn his loss. He is replaced by Hua Guofeng (Hua Kuo-feng). (See 1898.)

1976: April 5 Memorial service for Zhou ends in riots

Government attempts to stop a Beijing demonstration honoring Zhou Enlai result in widespread rioting. The Gang of Four blames the incident on Deng Xiaoping.

1976: July 28 Earthquake near Beijing kills 700,000

The worst earthquake the world has seen in over 400 years results in an estimated 700,000 deaths and virtually levels the city of Tangshan. Nearly 800,000 more people are injured from the quake, whose shocks radiate for almost 100 miles. Property damage amounts to the equivalent of hundreds of millions of dollars.

1976: September 9 Mao Zedong dies

Communist Party chairman and China's leading statesman Mao Zedong dies at the age of 82. (See 1893.)

1976: October 7 Gang of Four is arrested

Jiang Qing and the other members of the Gang of Four are arrested on the orders of premier Hua Guofeng.

1976: October 24 Hua becomes party chairman

Hua Guofeng officially succeeds Mao Zedong as chairman of China's Communist Party, while simultaneously serving as premier and chief of the armed forces.

1977: February Cultural Revolution is blamed on the Gang of Four

Hua Guofeng blames the Cultural Revolution on the Gang of Four and proclaims that it is officially over.

1977: July Deng Xiaoping is rehabilitated

Deng Xiaoping is returned to power after falling out of favor for the second time.

1978: February 26–March 5 Fifth National People's Congress

A new constitution is adopted at the Fifth National People's Congress, as well as an ambitious ten-year modernization plan. Deng Xiaoping, who has been fully rehabilitated politically, is a leading figure in the modernization effort. Deng's program for opening China's economic markets is adopted later in the year.

1978–79 Democracy Wall movement

Pro-democracy activists call for major democratic reforms to accompany economic modernization, communicating criticism of the Gang of Four, and other political opinions in essays written on posters and mounted on a wall in the Xidan district of Beijing. Once the posters begin to criticize not only past governments and movements but also the current Chinese government, the wall is closed down.

1979 The Cultural Revolution is repudiated

The Chinese government formally denounces the Cultural Revolution and begins a reassessment of Mao Zedong aimed at ending his being worshipped as a cult figure. Some of his policies are criticized, especially those toward the end of his career, and his image is removed from public places.

1979 Dissident writer Liu Binyan is released

Writer Liu Binyan, imprisoned since 1958, is released by the government.

1979: January 1 China and the U.S. establish diplomatic relations

Full diplomatic relations are established between China and the United States. In the same month, Deng Xiaoping becomes the first Chinese Communist leader to visit the U.S.

1980s Deng oversees economic reforms

China's economy is liberalized by reforms including the reintroduction of individual farming in place of collective farms, in a system known as "individual responsibility," by which farmers are allowed to keep a share of the profits from agricultural production. Also adopted are such free-market institutions as private enterprise, stock markets, and joint ventures with foreign investors. Banking and taxation are reformed, as well as the systems for rationing, and wage and price controls.

1980 China joins international financial community

The People's Republic is admitted to the World Bank and the International Monetary Fund.

1980 Hua is replaced as party chairman

The political position of Hua Guofeng, the successor chosen by Mao, weakens as Deng Xiaoping becomes more powerful. Hua is replaced as party chairman by Hu Yaobang and as premier by Zhao Ziyang.

1980: November Gang of Four trial opens

The Gang of Four goes on trial for their role in the Cultural Revolution, in proceedings that last over a year.

1981: January 25 Gang of Four is sentenced

After more than a year, the trial of the Gang of Four ends and its members are sentenced. Mao's widow, Jiang Qing, is sentenced to death, but her sentence is later commuted to life imprisonment. She later commits suicide.

1982 "Fifth Generation" Chinese filmmakers comes of age

The first class graduates from the Beijing Film Academy, reopened in 1977 after being shut down during the Cultural Revolution. This group, which is known as the "Fifth Generation" of Chinese filmmakers, includes leading directors Chen Kaige (*Farewell My Concubine;* 1993); Tian Zhuangzhuang (*The Blue Kite;* 1993); and Zhang Yimou, China's most internationally recognized director (*Red Sorghum,* 1987; *Ju Dou,* 1990; *Raise the Red Lantern,* 1992; and *To Live,* 1994).

1984: January China joins the IAEA

China joins the International Atomic Energy Agency.

1984: August Reagan visits China

U.S. President Ronald Reagan (b. 1911) makes an official visit to the People's Republic of China.

1984: September 26 Agreement is signed for return of Hong Kong

China and Great Britain sign an agreement providing for the return of Hong Kong to China in mid-1997.

1985 Historic Catholic church is reopened

Beitang Catholic Church, which has been closed since 1858, is reopened after undergoing restoration.

1986–87 Students begin pro-democracy demonstrations

Thousands of students take part in demonstrations demanding greater personal freedom that draw crowds of as many as 50,000 in Beijing, Shanghai, and Nanjing.

1986: October 12–18 Queen Elizabeth visits China

Queen Elizabeth II of Great Britain pays a state visit to China.

1987 Li Peng becomes new premier

Li Peng takes office as China's premier.

1987 New National Library building is completed

Construction of a new building to house the National Library is completed in the suburbs of Beijing. It is one of the largest library buildings in the world.

1987: April 13 Portugal agrees to the return of Macao

An agreement is signed providing for the return of Macao, a Portuguese colony since the sixteenth century, to China in 1999.

1987: September 27 Tibetan monks lead mass demonstrations

Monks lead mass protests against Chinese rule of Tibet. Rioting and violent clashes with police are reported.

1988 China competes in the Seoul Olympics

Chinese athletes compete in the summer Olympic Games held in Seoul, South Korea, where the men's and women's swim teams compile impressive records in the diving events, both platform and springboard.

1989: February Avant-garde art exhibit opens

An exhibition of *avant-garde* (experimental) art opens in Beijing's National Gallery. It is soon closed by government authorities.

1989: March Martial law is imposed in Tibet

In response to violent demonstrations in the preceding two years, martial law is imposed in Tibet.

1989: April 15 Death of Hu Yaobang sparks demonstrations

Former Communist party general secretary Hu Yaobang, dismissed for failing to repress student demonstrations in 1986 and 1987, dies of a heart attack. His death sparks new marches in Beijing by students honoring his memory and demanding democratic freedoms. The demonstrations spread to other cities.

1989: April 21 Government bans demonstrations

The government, tolerant of the initial demonstrations following the death of Hu Yaobang, becomes nervous as the protests grow in size and spread throughout the country. A ban is issued on all protests, but it is defied as 100,000 students gather in Beijing's Tienanmen Square. They are joined by increasing numbers of workers discontented due to government corruption, high inflation, and other problems.

1989: May 13 Hunger strike begins on eve of Gorbachev visit

Some 2,000 students begin a hunger strike in Tienanmen Square as mass demonstrations threaten to embarrass the government during the planned visit of Russian leader Mikhail Gorbachev. A scheduled ceremony in Tienanmen Square is moved to the airport because the demonstrators cannot be dispersed.

1989: May 15 Russian leader Mikhail Gorbachev visits China

Mikhail Gorbachev (b. 1931) visits China, becoming the first Russian leader to meet with top Chinese officials in thirty years. Full diplomatic ties are resumed between the two countries.

1989: May 17 One million gather in Tienanmen Square

A crowd of over a million gathers in Tienanmen Square, paralyzing normal operations in the capital and elsewhere. Salaried workers, journalists, small businessmen, and others join in the protests. Unconscious hunger strikers are taken to hospitals.

1989: May 19 Government imposes martial law

Martial law is imposed and 1,000 troops are sent to dismiss the crowds now encamped in Tienanmen Square. Communist Party chairman Zhao is criticized as being too soft on the demonstrators, and his position is given to Premier Li Peng. Drinking fountains in Tienanmen Square are shut off. Thousands of people crowd Beijing, defying government troops and halting their progress. Television broadcasts by foreign reporters are cut off.

Activities in Xian are brought to a standstill by 300,000 demonstrators, and 500,000 march in Hong Kong.

1989: June 4 Violent troops assault Tienanmen Square demonstrators

Thousands of troops supported by tanks and armored personnel carriers converge on Tienanmen Square in the predawn hours, trapping the crowds there by blocking off the square with armored vehicles. They then advance on the protesters, beating and opening fire on students and other unarmed civilians. It is estimated that over 1,000 are killed and over 10,000 injured in one of the world's worst peacetime massacres in the modern era. Further protests are sparked throughout the country, and thousands are arrested. The events in China also draw international censure.

1989: June 24 Jiang Zemin becomes new party secretary

Following the brutal repression of the Tienanmen Square protests, Jiang Zemin, mayor of Shanghai, is named as the new general secretary of the Communist Party, replacing Zhou Ziyang.

1990 U.S. extends most-favored-nation status

The United States extends most-favored-nation trade status for China in spite of opposition due to the Tienanmen massacre.

1990: April Martial law in Tibet ends

Martial law is lifted in Tibet.

1991 Normalization of relations with Vietnam

Diplomatic ties with Vietnam are normalized after the two countries resolve their dispute over Cambodia.

1991 China's first stock exchange opens

The Shanghai stock exchange begins operations.

1992 China explodes its largest nuclear weapon ever

China detonates a 1,000-kiloton nuclear bomb in underground testing.

1992 Tibet-India border is reopened

The border between India and Chinese-controlled Tibet is opened for the first time in over thirty years.

1992 China wins women's table tennis honors at the Olympics

Deng Yaping (b. 1973), two-time world champion, wins the Olympic gold medal in women's singles table tennis.

1993–94 Peasant protests and riots take place

Peasants protest local government corruption and payment for produce with IOUs instead of cash.

1993 Chinese film wins top honors at Cannes

Farewell My Concubine, directed by Chen Kaige, shares the highest honors at the Cannes Film Festival. It also wins the U.S. Golden Globe award for Best Foreign Film and is nominated for an Academy Award in the same category. The film, which revolves around the Beijing Opera, covers over sixty years of Chinese history, from the early years of the republic to the end of the Cultural Revolution.

1993: May U.S. renews most-favored-nation trade status

President Bill Clinton (b. 1946) announces that the United States will extend most-favored-nation trade status for China for another year in spite of previous ultimatums regarding improvement of China's human rights record.

1994: June 6 Jetliner crash is China's worst air disaster

The crash of a China Northwest Airlines jet near Xian is the worst aviation disaster in Chinese history, killing all 160 passengers and crew aboard the aircraft.

1994: August 23 Typhoon Fred causes over 700 deaths

Zhejiang province is struck by a typhoon that kills over 700 people and results in $1.16 billion worth of property damage.

1995 Successor chosen by Tibet's Dalai Lama is detained

The six-year-old boy named by the Tibet's exiled Dalai Lama as his preordained successor by virtue of reincarnation is detained by Chinese authorities, who substitute another child in the boy's place.

1995 Chinese return to supremacy in table tennis

Chinese athletes win all seven world championship table tennis titles, including singles, doubles, and team titles. Leading players include Wang Tao (b. 1967) and Lu Lin (b. 1970).

1995: September 4–15 UN Conference on Women is held in Beijing

The fourth United Nations World Conference on Women, attended by delegates from over 180 countries, is held in Beijing. The gathering endorses a "Platform for Action" containing measures to improve women's rights. The scheduling of a UN-sponsored forum on women in Huairou at the same time is criticized as a stratagem to limit the Chinese public's awareness of women's rights issues.

1996: September 24 China signs nuclear test ban treaty

China signs the Comprehensive Test Ban Treaty (CNTBT), along with the United States, Russia, France, and Great Britain. The treaty is signed at the United Nations headquarters in New York during the opening session of the fifty-first General Assembly.

1996: December 11 New Hong Kong governor is named

Chinese businessman Tung Chee-hwa is named to become the new governor of Hong Kong when the British crown colony reverts to Chinese rule in 1997. Tung, who is selected by a specially appointed 400-member election committee, will take over from the British-appointed governor, Christopher Patten.

1997: February Muslim revolt is stopped with force

The Chinese military crushes a revolt by Muslims in Xinjiang, near Kazakhstan, killing over 100 people.

1997: February 19 Death of Deng Xiaoping

Longtime political leader Deng Xiaoping dies of respiratory failure at the age of 92. An official mourning period of six days is declared. Deng is hailed as the architect of China's free-market reforms, but he is also remembered for his role in repressing the Tienanmen Square protests in which hundreds of protesters were killed by police. (See 1904.)

1997: March 20,000 workers protest in Nanchong

A labor protest in Nanchong involving 20,000 workers is China's largest since the Communist revolution of 1949.

1997: July 1 Hong Kong is returned to China

The British depart from Hong Kong, returning the island to Chinese rule after 156 years. The territory is renamed the Hong Kong Special Administrative Region of China. A formal handover ceremony is held, and Tung Chee-hwa becomes Hong Kong's first chief executive under Chinese rule.

1997: September 12–18 Fifteenth Communist Party congress convenes

The Chinese Communist Party holds its fifteenth five-year congress in the Great Hall of the People in Beijing. Over 2,000 delegates attend the gathering, which is considered an occasion for President Jiang Zemin to consolidate his political power following the death of Deng Xiaoping earlier in the year. Jiang's dramatic program to privatize most of China's state-owned enterprises is approved.

1997: October 29–November 3 Jiang Zemin visits the U.S

President Jiang Zemin travels to Washington, D.C. to meet with U.S. president Bill Clinton, in the first U.S.-China summit meeting since 1989. The U.S. agrees to export nuclear power technology to China, progress is made on other trade, and environmental issues are discussed. However, groups of demonstrators protesting human rights abuses in China greet Jiang at various stops on his visit.

1998: May 4 Beijing University observes one hundredth anniversary

Beijing University marks its one hundredth year in existence with special ceremonies that include the opening of a new library and an appearance by President Jiang Zemin.

1998: June 25–July 3 President Clinton visits China for summit talks

President Bill Clinton becomes the first U.S. president to visit China for summit talks since the 1989 government crackdown on demonstrators in Tienanmen Square. Clinton and Chinese president Jiang Zemin have an unprecedented televised exchange over human rights that is broadcast live to the Chinese public.

Bibliography

Bailey, Paul. *China in the Twentieth Century.* New York: B. Blackwell, 1988.

Cotterell, Arthur. *China: A Cultural History.* New York: New American Library, 1988.

Creel, Herrlee Glessner. *Chinese Thought from Confucius to Mao Tse-Tung.* Chicago: University of Chicago Press, 1992.

Ebrey, Patricia B. *The Cambridge Illustrated History of China.* New York: Cambridge University Press, 1996.

Evans, Richard. *Deng Xiaoping and the Making of Modern China.* New York: Viking, 1994.

Gray, Jack. *Rebellions and Revolutions: China from the 1800s to the 1980s.* New York: Oxford University Press, 1990.

Howland, Douglas. *Borders of Chinese Civilization: Geography and History at Empire's End.* Durham, N.C.: Duke University Press, 1996.

Hsu, Immanuel Chung-yueh. *The Rise of Modern China.* 5th ed. New York: Oxford University Press, 1995.

Hsu, Kai-yo. *Literature of the People's Republic of China.* Bloomington: Indiana University Press, 1980.

Lin, Bih-jaw, ed. *The Aftermath of the 1989 Tienanmen Crisis in Mainland China.* Boulder, Colo.: Westview Press, 1992.

Link, Perry, Richard Madsen, and Paul G. Pickowicz, eds. *Unofficial China: Popular Culture and Thought in the People's Republic.* Boulder, Colo.: Westview Press, 1989.

Ogden, Suzanne. *China's Unresolved Issues: Politics, Development, and Culture.* Englewood Cliffs, N.J.: Prentice Hall, 1995.

Salisbury, Harrison E. *The Long March: The Untold Story.* New York: Harper & Row, 1985.

Schrecker, John E. *The Chinese Revolution in Historical Perspective.* New York: Greenwood Press, 1991.

Snow, Edgar. *Red Star over China.* New York: Bantam, 1978.

Worden, Robert L., Andrea Matles Savada, and Ronald E. Dolan, eds. *China, a Country Study.* 4th ed. Washington, D.C.: Library of Congress, 1988.

Cyprus

Introduction

The turbulent history of Cyprus, an island nation in the Mediterranean, has created a deep ethnic and religious division between its two major ethnic groups, the Greek and Turkish Cypriots. Since 1974 the two communities have been geographically and politically separated. In spite of international efforts to bring the two sides together, the political future of the island remains uncertain.

With an area of 3,571 square miles (9,250 square kilometers), Cyprus is the third-largest island in the Mediterranean Sea. Its landscape is dominated by two mountain ranges, the Troodos Massif in the southwest and the Kyrenia Mountains in the north. Between lies the Mesaoria, a low, fertile plain. Since 1974 the northern third of the island has been under the control of the Turkish Cypriot community, first as the Turkish Cypriot Federated State and, since 1983, as the Turkish Republic of Northern Cyprus. The remainder of Cyprus is administered by the Greek Cypriot-controlled Republic of Cyprus. Cyprus had a total population of 658,000 according to a 1997 census. Its capital, Nicosia, is located in the northern part of the island.

The name "Cyprus" is derived from *kypros,* the Greek word for the island's major natural resource, copper, which was discovered as early as 3000 B.C. and became an early factor in drawing traders from nearby areas. In addition, its strategic location in the Mediterranean made it both attractive and vulnerable to neighboring populations from an early date. In the last 1,500 years before the start of the Christian era, Cyprus was controlled by the Egyptians, Greeks, Phoenicians, Assyrians, Persians, and Romans, who annexed it in 58 B.C. and presided over a long period of peace and commercial expansion that encompassed the introduction of Christianity in A.D. 45. With the division of Rome's territories in 324, Cyprus became part of the Byzantine Empire, and the Greek Orthodox Church grew to be a formidable force on the island, both spiritually and politically.

The seventh century began a difficult period of Arab invasions, which continued intermittently for three centuries, causing massive suffering and destruction on the island before the invaders were finally driven away decisively in 965 under the leadership of a Greek general. In 1191 Cyprus came under the control of the English king Richard I, when a shipwreck forced his party of Crusaders to land there. Richard's stake in Cyprus was eventually transferred to Guy de Lusignan, who instituted a Frankish dynasty that ruled the island for three hundred years. The Frankish rulers brought with them France's language and aspects of its culture, but these remained largely confined to the nobility because a rigid class system separated the aristocratic newcomers from the native Cypriot population, whose culture and daily lives were largely unchanged. Cyprus's wealth was also concentrated in the hands of another group of foreigners—Italian merchants from Genoa and Venice, who eventually won control of the island when its aristocratic line declined in the fifteenth century.

In 1571 Cyprus fell to the Turkish Ottoman empire (see sidebar, Timeline: 1571), which dismantled the feudal system imposed by the Frankish monarchs and administered Cyprus through a system of religious communities, or *millets.* Thus the island's Greek Orthodox leaders also became its civic leaders, further expanding the role of the Orthodox church and the unity of the Greek community. The native Greek Cypriots' resentment toward an indifferent and corrupt Ottoman administration gave rise to a growing sense of mistrust and antipathy between the islands' Greeks and Turkish populations, a breach which would have lasting political consequences.

As the power of the Ottoman empire declined, the Turks sought British aid to fend off Russian aggression in areas they controlled. In 1878 Britain agreed to take over the administration of Cyprus in return for military protection against Russia. The British modernized the government bureaucracy and improved education and medical care on the island. When the outbreak of World War I found Britain and Turkey on opposite sides of the conflict, Britain annexed Cyprus, and in 1925 it became a British crown colony. British rule became repressive in the 1930s following protests over taxation. Nonetheless, Cypriots served with the British in World War II to support the Allies. After the war, British rule was liberalized, but Cypriot public opinion demanded a change. However, its two ethnic groups wanted different changes. The majority Greeks maintained their tradition demand for *enosis* (union with Greece), while the Turkish Cypriots favored *taksim* (partition between Greece and Turkey).

In 1955 a Greek Cypriot underground organization, EOKA (National Organization of Cypriot Fighters), launched an armed campaign to end British rule over Cyprus. By 1958, violence on the island had reached the level of civil war. Early in 1959 an agreement reached by the British and both groups of Cypriots (as well as both Greece and Turkey) provided for the establishment of an independent republic, with power shared by the Greek and Turkish populations. The Republic of Cyprus was proclaimed on August 16, 1960. Greek Cypriot leader Archbishop Makarios III was elected president. By the end of 1963, though, tensions between Greek and Turkish Cypriots had erupted into violence. The Turkish community withdrew from the Makarios government, which imposed a trade embargo on the Turkish enclaves. In March 1964 United Nations peacekeeping forces arrived on the island, and an uneasy peace was eventually achieved. However, the Greek and Turkish Cypriot communities maintained their separation, and the Republic of Cyprus essentially became a Greek Cypriot state. In the late 1960s, however, the Makarios came into conflict with the military government that had come to power in Greece in 1967. By 1968, many Greek Cypriots began to abandon *enosis* as a feasible solution to the Cyprus problem and advocated an independent, neutral Cyprus with a strong, centralized government as envisioned under the 1960 constitution. (The abandonment of *enosis* put Greek Cypriots in conflict with the Greek dictatorship which viewed Greece as the "center of Hellenism.") Also in 1968, Greek and Turkish Cypriot leaders met for talks, but the two sides still maintained conflicting demands—a unitary state for the Greeks and partition for the Turks—and a political stalemate resulted.

In July 1974, President Makarios was overthrown and forced into exile by Cypriot National Guard members cooperating with the military dictatorship in Greece. Fearing that Greece would now take over Cyprus, Turkey invaded the island and occupied the northern third. The following year, the Turks proclaimed a separate, federated state, and, in 1983, an independent republic, the Turkish Republic of Northern Cyprus (TRNC). The only foreign country to recognize the northern republic was Turkey. Nevertheless, the island has remained divided. Turkish leader Rauf Denktash, the president of the TRNC, was reelected to a second term in 1990. In the late 1990s, tensions on Cyprus, and between Greece and Turkey, were heightened by Cyprus's pending admission to the European Union, which would recognize only the southern, Greek-controlled portion as the legal government of the entire island.

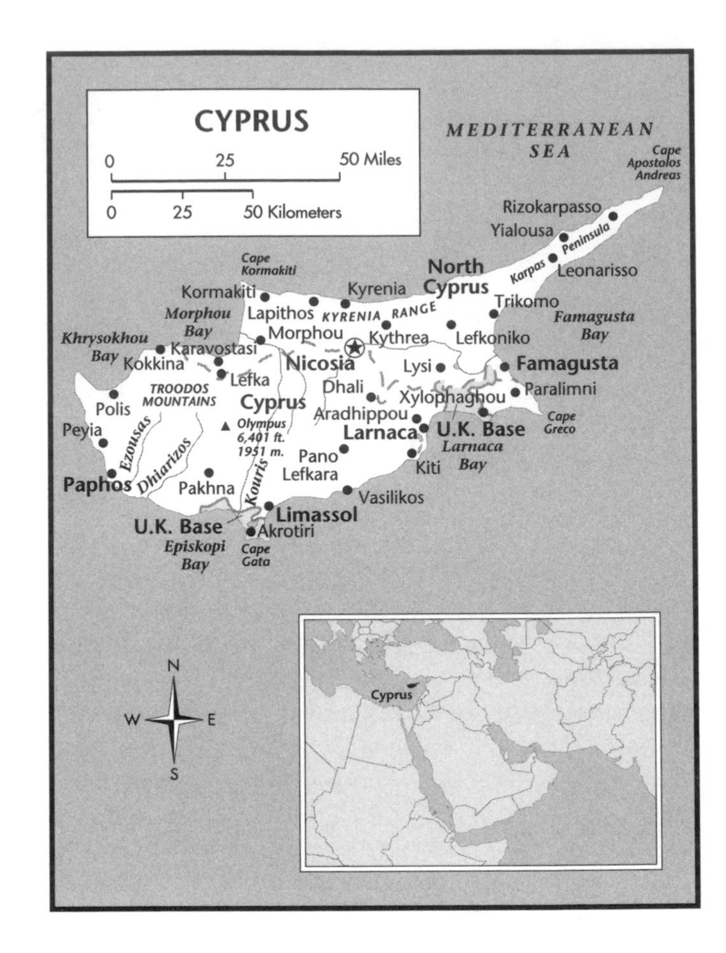

Timeline

5800 B.C. Neolithic era

Neolithic civilization exists on Cyprus. Crops are grown and livestock is raised. Distinctive pottery and figurines of fertility goddesses are produced.

3000 B.C. Copper is discovered

Copper, Cyprus's major natural resource, is discovered during the Chalcolithic period, making the island attractive to traders from nearby areas. The name *Cyprus* is either derived from or serves as the derivation for the Greek term for copper (kypros). Copper from Cyprus becomes additionally valuable when it begins to be used in the production of bronze.

Foreign Invasion

1500 B.C. Egyptians control Cyprus

Cyprus comes under the control of Egypt during the reign of Thutmose III, who promises to protect the island from invasion by other powers in return for obtaining a trade monopoly on Cyprus's copper and other commodities.

1400–1250 B.C. Copper trade with Greeks grows

Traders from the Greek Mycenaean and Minoan civilizations inaugurate a brisk copper trade with Cyprus, breaking the monopoly imposed by Egypt. Greeks begin to settle on Cyprus.

1184 B.C. The end of the Trojan Wars speed Greek settlement

Increased numbers of Greeks flock to Cyprus as the Trojan Wars conclude, in search of a peaceful homeland. They bring the Greek language and firmly establish their culture on the island, influencing local art by introducing the potter's wheel and building temples to honor their deities.

800 B.C. Phoenicians settle on Cyprus

The ship-building Phoenicians, who have traded with Cyprus since the third millennium B.C., begin to settle there, mostly at Kition (present-day Larnaca), to the southeast. They share political power with the Greeks already on the island.

708 B.C. Cyprus is conquered by the Assyrians

Assyrians, under the leadership of Sargon II, seize control of Cyprus, maintaining their dominance for roughly a century. However, the native Cypriot rulers still wield a fair degree of independence, especially economically, and their already considerable wealth continues to grow. Cyprus's seven city-states grow to ten during the period of Assyrian rule. Their kings are military, religious, and political leaders.

600 B.C. Egypt regains control of Cyprus

Maritime strength enables the Egyptians to overpower the Phoenicians and Cypriots and retake Cyprus. The local kings must pledge allegiance to the Egyptian pharaohs, but they retain autonomy within their own jurisdictions. As in the previous period of Egyptian control, Egypt is primarily concerned with establishing a monopoly over Cypriot trade.

500 B.C. Persians conquer Egypt and control Cyprus

King Darius of Persia becomes the ruler of Cyprus after conquering the Egyptians. The foremost Cypriot city-state at this time is Salamis, whose king, Onesilos, leads a revolt against the Persians.

333 B.C. Alexander the Great conquers the Persians

With the conquest of Persia by Alexander the Great, Cyprus experiences a brief period of independence, which ends with the death of Alexander ten years later.

294–58 B.C. Ptolemaic period

Ptolemy, one of Alexander's generals, gains control of Egypt and Cyprus. He establishes a central government for Cyprus, terminating its network of city-states.

58 B.C. Roman annexation

The Romans seize control of Cyprus, inaugurating a long period of peace after centuries of invasions. They develop mining and commerce on the island and build up its infrastructure, dividing it into four administrative districts: Amathus, Lapithos, Paphos, and Salamis. Salamis becomes the major commercial center, while the government is headquartered in Paphos.

15 B.C. Earthquake levels Salamis

The city of Salamis is shattered by an earthquake but rebuilt by the emperor Augustus Caesar.

A.D. 45 Introduction of Christianity

Christianity is brought to Cyprus during the reign of Claudius I, supposedly by the apostle Paul and Cyprus native Barnabas. Paul and Barnabas convert the proconsul Sergius Paulus, making him the first Roman government official to become a Christian.

The Byzantine Period

324 Establishment of the Byzantine Empire

Cyprus becomes part of the Byzantine Empire established by Constantine in the reorganization that divides the Latin- and Greek-speaking portions of Roman territory. A strong centralized government is maintained on Cyprus, as well as a rigidly defined class system, with landowners and merchants among its upper ranks. Tenant farmers are reduced to the status of vassals and forbidden to leave their land. The principal conflict is a religious one between the Cypriot church, which struggles to retain its independence from the powerful churches of the surrounding areas.

332 and 342 Earthquakes shatter Cyprus's cities

Salamis and other cities are destroyed by earthquakes. Salamis is rebuilt and renamed Constantia, and the capital is moved to Paphos.

488 Church of Cyprus attains religious autonomy

The Byzantine emperor Zeno grants special privileges to Cypriot archbishop Anthemius of Constantia, showing that the Church of Cyprus is considered autocephalous (independent of other religious jurisdictions, or patriarchates, including Antioch). This status increases the political power of the church and its leaders, both now and in future historical periods.

6th century Silk production flourishes

Silkworms are brought to the Byzantine Empire from East Asia, and silk production thrives on Cyprus.

647–965 Arab invasions decimate Cyprus

During the reign of Constans II, Arab raiding parties begin landing on Cyprus, and the attacks continue intermittently for three centuries, killing and enslaving thousands and destroying many cities and towns, as well as all the churches on the island. Arab aggression in the region is finally ended by military resistance under the leadership of General Nicephorus Phocas.

1184 Isaac Comnenos seizes power

Isaac Comnenos proclaims himself king of Cyprus and inaugurates a period of harsh dictatorial rule.

European Rule

1191: April Richard I takes control of Cyprus

Richard I ("Richard the Lion-Hearted"), en route to the Third Crusade to recapture Jerusalem, and his fleet are shipwrecked off the coast of Cyprus. Maltreatment at the hands of its ruler, Isaac Comnenos, prompts Richard to overthrow him and capture the island. Deciding not to administer it himself, however, Richard sells it to the military order called the Knights Templar.

1192 Cypriot rule passes to Guy de Lusignan

Following a Cypriot revolt, the Knights Templar turn the island over to Guy de Lusignan, the Christian ruler of Jerusalem, ending the era of Byzantine rule. Lusignan dies shortly afterwards, but his successors retain power for over three hundred years, establishing an independent kingdom on Cyprus. During this period the island thrives culturally and commercially. A feudal system similar to those in Europe is instituted, with native Cypriots at the bottom and foreigners (the ruling Franks of the nobility and an Italian merchants class) at the top. The Roman Catholic Church becomes the official church of Cyprus, but the majority of the people retain their allegiance to the Orthodox Church of the Byzantine era. Traders from Genoa and Venice gain increasing power on the island.

Early 14th century Medieval polyphony flourishes on Cyprus

During the reign of King Janus (1398–1432), court composers, possibly brought from France by Janus's wife, Charlotte de Bourbon, produce one of the most significant bodies of late medieval polyphony, from which survive a manuscript containing Latin and French motets, polyphonic mass movements (Glorias and Credos), and short instrumental pieces, as well as six mass cycles in plainchant.

1347 Bubonic plague reaches Cyprus

The Bubonic plague, or Black Death, spreads to Cyprus, and from there to Florence, where one of the very worst outbreaks occurs.

1489 Venice gains control of Cyprus

The independent kingdom on Cyprus declines in the early fifteenth century, and the island comes under Venetian control following an alliance contracted through the marriage of a Cypriot king to the daughter of a Venetian noble. Like the Frankish Lusignan dynasty before them, the Venetians gain as much wealth as possible from the island while keeping its population subservient. They expand trade of commodities including sugar, cotton, saffron, and hemp, and mine salt at Larnaca.

1562 Cypriots revolt against Venetian rule

Cypriots launch an unsuccessful revolt against the exploitative Venetian regime. Within a decade, however, the Venetians are ousted by the Turkish Ottoman Empire.

The Ottoman Period

1571: August Cyprus falls to the Ottomans

After a nearly year-long siege, Famagusta falls to the Turks, beginning three centuries of Ottoman rule. Abolishing the feudal system adopted by the Frankish rulers, the Ottomans rule through semi-autonomous *millets,* or religious communities. This arrangement has the effect of uniting the island's Greek Orthodox community, which forms the basis of these administrative units. Declining trade due to competition from Atlantic routes, Ottoman mismanagement, and a series of natural disasters set off an economic decline on the island.

Three circumstances contribute to the growth of Greek nationalism among native Cypriots: religious loyalty, resentment of Ottoman rulers for their inefficiency, and widespread poverty. By the end of the Ottoman era, the goal of *enosis,* or union with Greece, is already being spread among educated Cypriots.

1575 Orthodox bishops are reinstalled

The Ottomans allow the archbishop of the Church of Cyprus and three of its bishops to return to their religious jurisdictions.

17th century Population drops due to natural disasters and violence

The population of Cyprus hits an all-time low in the seventeenth century, due partly to a series of natural disasters (famine, drought, locusts, and plague) and partly to pirate attacks. Corrupt administration by the Ottoman government exacerbates the hardships suffered by Cypriots.

The Ottoman Period

The Ottoman Empire began in western Turkey (Anatolia) in the thirteenth century under the leadership of Osman (also spelled Othman) I. Born in Bithynia in 1259, Osman began the conquest of neighboring countries and began one of the most politically influential reigns that lasted into the twentieth century. The rise of the Ottoman Empire was directly connected to the rise of Islam. Many of the battles fought were for religious reasons as well as for territory.

The Ottoman Empire, soon after its inception, became a great threat to the crumbling Byzantine Empire. Constantinople, the jewel in the Byzantine crown, resisted conquest many times. Finally, under the leadership of Sultan Mehmed (1451–1481), Constantinople fell and became the capital of the Ottoman Empire.

Religious and political life under the Ottoman were one and the same. The Sultan was the supreme ruler. He was also the head of Islam. The crown passed from father to son. However, the firstborn son was not automatically entitled to be the next leader. With the death of the Sultan, and often before, there was wholesale bloodshed to eliminate all rivals, including brothers and nephews. This ensured that there would be no attempts at a coup d'état. This system was revised at times to the simple imprisonment of rivals.

The main military units of the Ottomans were the Janissaries. These fighting men were taken from their families as young children. Often these were Christian children who were now educated in the ways of Islam. At times the Janissaries became too powerful and had to be put down by the ruling Sultan.

The greatest ruler of the Ottoman Empire was Suleyman the Magnificent who ruled from 1520–1566. Under his reign the Ottoman Empire extended into the present day Balkan countries and as far north as Vienna, Austria. The Europeans were horrified by this expansion and declared war on the Ottomans and defeated them in the naval battle of Lepanto in 1571. Finally the Austrian Habsburg rulers were able to contain the Ottomans and expand their empire into the Balkans.

At the end of the nineteenth century, the Greeks and the Serbs had obtained virtual independence from the Ottomans, and the end of a once great empire was in sight. While Europe had undergone the Renaissance, the Enlightenment, and the Industrial Revolution, the Ottoman Empire rejected these influences as being too radical for their people. They restricted the flow of information and chose to maintain strict religious and governmental control. The end of Ottoman control in Europe came in the First Balkan War (1912–13) in which Greece, Montenegro, Serbia, and Bulgaria joined forces to defeat the Ottomans.

During the First World War, (1914–18), the Ottoman Empire allied with the Central Powers and suffered a humiliating defeat. In the Treaty of Sévres, the Ottoman Empire lost all of its territory in the Middle East, and much of its territory in Asia Minor. The disaster of the First World War signaled the end of the Ottoman Empire and the beginning of modern-day Turkey. Although the Ottoman sultan remained in Constantinople (renamed in 1930, "Istanbul"), Turkish nationalists under the leadership of Mustafa Kemal (Ataturk) (1881–1938) challenged his authority and, in 1922, repulsed Greek forces that had occupied parts of Asia Minor under the Treaty of Sévres, overthrew the sultan, declared the Ottoman Empire dissolved, and proclaimed a new Republic of Turkey. That following year, Kemal succeeded in overturning the 1919 peace settlement with the signing of a new treaty in Lausanne, Switzerland. Under the terms of this new treaty, Turkey reacquired much of the territory—particularly, in Asia Minor—that it had lost at Sévres. Kemal is better known by his adopted name "Ataturk", which means "father of the Turks". Ataturk, who ruled from 1923 to 1938, outlawed the existence of a religious state and brought Turkey a more western type of government. He changed from the Arabic alphabet to Roman letters and established new civil and penal codes.

1804 Uprising in Nicosia

Turkish Cypriots in Nicosia and nearby villages stage a rebellion against their Ottoman governor. Troops are brought in from Asia Minor to stop the revolt.

1821: July 9 Mass execution of Cypriot religious leaders

Fearing that the war for independence in Greece will carry over to Cyprus, the Ottoman rulers arrest the entire leadership of Cyprus's religious community—its archbishop, Kyprianos, as well as its bishops, priests, and even laymen who play a prominent role in the church. All are executed, either by hanging or decapitation.

1865 First major archaeological digs

The first extensive archaeological excavations take place.

1874 First antiquities law is adopted

Cyprus's Ottoman rulers adopt an antiquities law specifying that one-third of the items excavated by archaeologists go to the government. The Cyprus Museum is established in Nicosia to house these objects.

Rule by Britain

1878: July 13 Britain takes over administration of Cyprus

The Ottoman Turks turn over administration of Cyprus to the British in an attempt to ward off aggression by Russia. The arrangement is formalized in an agreement known as the Cyprus Convention. Sir Garnet Wolseley is appointed High Commissioner for Cyprus. British administration begins to bring Cyprus into the modern world after centuries of isolation under the Ottomans. The government bureaucracy is modernized, an educational system is established, with separate facilities for Christians and Muslims, and Cyprus's first hospital is built. Other public health measures include controlling malaria by draining swamps near Larnaca.

1881 First census is taken

The first census taken on Cyprus shows a total population of 186,173, of which 140,793 are Greek Cypriots and 42,638 are of Turkish origin.

1882 A constitution is adopted

Under British administration, Cyprus adopts a constitution establishing an eighteen-member Legislative Council made up of both elected and appointed officials. Both Muslim and non-Muslim representation on the council is mandated.

1895 British pass education law

The Education Law enables local governments to finance schools by levying taxes. Within twenty years the number of schools on Cyprus rises from 76 to 179.

1900: January 23 Birth of painter Adamantios Diamantis

Diamantis, one of Cyprus's most influential painters and art teachers of the twentieth century, studies art in London in the early 1920s and then returns to Cyprus to begin a career as a painter and teacher. Early in his career, Cubism and other modernist styles play an important role in his work, although his subject matter focuses on the people and landscapes of his native land. From the 1960s on after the middle period of greater realism, his work is characterized by abstract forms and freer expression. His well-known works include *Coachmen of Asmaalti* (1943), *At the Festival of Our Lady of Araka* (1942), and the masterpiece, *World of Cyprus* (1967–72), composed of eleven acrylic panels portraying the land and people of Cyprus. Diamantis dies in Nicosia on April 28, 1994.

1909 Agricultural cooperative movement is founded

Following a tour of cooperatives in Germany and Britain, farmers belonging to a village society found Cyprus's agricultural cooperative movement.

1910: February 24 Birth of painter Telemachos Kanthos

Painter and engraver, Telemachos Kanthos, studies art in Athens, where he is influenced by the work of Paul Cézanne. When he returns to Cyprus, he settles in Nicosia, where he paints and teaches, producing vibrant landscapes and scenes from Cypriot village life. He is also known for his simply composed but powerful engravings, including the series *Hard Times,* reflecting the political events of 1974. Kanthos dies in November 1993.

1913: August 13 Birth of Archbishop Makarios III

Cyprus's foremost twentieth-century leader, Michael Christendous Mouskos, is born in the Paphos district. When he becomes a monk, he takes the name *Makarios,* which means "blessed." He pursues religious studies in Athens, where he enters the priesthood in 1946, and later at Boston University. Makarios is elected bishop of the Cypriot See (religious jurisdiction) of Kition in 1948 and Archbishop of Cyprus in 1950. This position also makes him the ethnarch (political leader) of the Greek Cypriot community, and he vigorously promotes union with Greece (*enosis*).

In 1956 Makarios is deported by the British, who accuse him of supporting the terrorist activities of EOKA (National Organization of Cypriot Fighters), the Greek Cypriot resistance organization. From exile in Athens, he coordinates efforts to win enosis, as civil war is waged in Cyprus. In February 1959 he participates in talks with the prime ministers of Greece, Turkey, and the United Kingdom that result in the establishment of the Republic of Cyprus. In December he is elected its first president. After two extensions of his initial term, Makarios is reelected in 1968 and 1973. He is deposed and exiled in 1974 by pro-enosis extremists allied with the military junta in Greece but is later reinstated. Archbishop Makarios dies on August 3, 1977, with Cyprus divided into separate Greek and Turkish zones following the Turkish invasion of 1974.

1914 Britain annexes Cyprus

When World War I breaks out, Turkey sides with Germany, and Britain then annexes Cyprus. Britain offers the island to Greece in return for its military participation, but Greece's King Constantine declines, preferring to maintain a neutral position.

1920–40 Cypriot artists integrate European modernist styles

Many painters and architects, studying abroad during the British colonial period, are influenced by the avant-garde movements occurring in Europe. Upon returning to Cyprus, they integrate these new ideas with elements of traditional Byzantine religious art and Cypriot folk art to produce works in distinctive new styles. Telemachos Kanthos (see 1910) produces impressionistic landscape paintings. Yeoryios Yeoryiou (1901–72) merges native Cypriot traditions with the techniques of Mannerism and Expressionism. Adamantios Diamantis (see 1900) captures the spirit of the Cypriot people in his *World of Cyprus* (1967–72).

1923 Britain's wartime annexation becomes permanent

Turkey recognizes the permanent annexation of Cyprus by Britain in the Treaty of Lausanne.

1924: June 8 Birth of painter Christoforos Savva

Savva, a major innovator in postwar Cypriot art, studies in London and Paris. Cubism is an early influecne. He later begins experimenting with a variety of media, including wire and appliqué, as well as sculpture. His work attains an increasing degree of abstraction toward the end of his career, which is cut short by his early death in 1968. Savva's works include *Nude* (1957), *Composition with Two Circles* (1967), and *Sun* (1968).

1925 Cyprus becomes a British crown colony

Cyprus becomes a crown colony of Britain, administered by a governor. The Legislative Council is expanded to twenty-four members.

1925 Agricultural Bank is founded

Working through cooperative societies, Cyprus's Agricultural Bank provides farmers with medium- and long-term loans.

1925 First trade unions are founded

Cyprus's first trade unions are established. The Pancyprian Federation of Labor (PEO) is the oldest and most prominent labor organization.

1930–35 Tsangarides influences architecture of Nicosia

Odysseus Tsangarides (1907–74) lays the foundation for modern urban planning in Nicosia during his tenure as municipal architect of the island's capital.

1931: October Civil unrest over tribute payments brings on British repression

A proposed increase in taxes heightens existing discontent over the Cyprus Tribute, a payment exacted since the Cyprus Convention (see 1878) to cover Britain's administration costs for Cyprus. Leftover funds intended to be turned over to the Turkish sultan (before annexation) are instead used to pay Crimean War loans on which Turkey defaulted. This practice enrages Cypriots, whose funds are thus being used to pay a debt they have not incurred. In spite of modifications in the twentieth century, the Cyprus Tribute remains an incendiary issue.

When Britain raises taxes further because of deficits brought on by the Great Depression, Cypriot anger erupts into violence. Six persons are killed and many are injured in rioting, and the British governor's residence in Nicosia is burned down. In response, the British institute a harsh crackdown on both Greek and Turkish Cypriots, suspending the constitution and normal political activity, imposing censorship, and exiling the religious leaders of the Greek Orthodox Church. More than 2,000 Greek Cypriots are arrested. Elections are suspended, and appointed officials fill all elective posts. All displays of either Greek or Turkish nationalism are suppressed, and a ban is placed on teaching the language or history of either country.

1936 Folk art collection is inaugurated

The most significant collection of Cypriot folk art is launched by the Society of Cypriot Studies. Now gathered in the Old Archbishopric in Nicosia, it includes 4,000 photographs. Its archives date back to the 1800s.

1937 Britain issues laws governing the Cypriot Church

The British colonial administration further antagonizes Cypriots by adopting laws affecting the internal affairs of the Greek Orthodox Church. These include a provision requiring the British governor's approval of any archbishop elected by the church.

1939–45 World War II

Cypriots support the Allied cause during the Second World War, sending 6,000 volunteers to fight in the Greek campaign under British command. Altogether, some 30,000 Cypriots serve with the British in the course of the war. Cyprus itself becomes a strategic naval base and a center for training and supplies, and the large number of Allied troops stationed on the island boosts local economy. No attempt is made to invade the island, but it undergoes air raids.

1943 Municipal elections are reinstituted

The first municipal elections are held since the disturbances of 1931. A Communist group, the Progressive Party of the Working People (AKEL), wins control of the major cities of Limassol and Famagusta.

1946 Britain liberalizes rule

The British take measures to relax the harshness that characterized its prewar rule. It allows a legislative body to form and draft a new constitution and lets exiled religious leaders back on to the island. However, the Greek Cypriot desire for enosis is undiminished.

1946 British change land ownership laws

Britain ends the communal ownership system for grazing land that has been in effect since the Ottoman period. All restrictions on private ownership of land are removed.

1950–74 Postwar generation of artists favors experimental approaches

Cypriot artists trained in Europe just before and after World War II establish a vigorous tradition of experiment and innovation, introducing into their work elements of such movements as Fauvism, Cubism, Constructivism, and Surrealism. Notable artists of this period include Christoforos Savva (1924–68), Stass Paraskos (b. 1933), Stelios Votsis (b. 1929), Vera Hadjida (b. 1936), and George Skoteinos (b. 1937).

Cyprus's first distinguished sculptors are active during this period. They include Andreas Savvides (b. 1930), known for monumental sculptures and innovative combinations of materials, and Andy Adamos (1936–90), whose work reflects African influences from time he spent living in South Africa.

1952 Electricity Authority of Cyprus is founded

The Electricity Authority of Cyprus (EAC) is established as a public corporation to generate, transmit, and distribute electric power.

1954 Turkish Cypriot labor federation is formed

Turkish Cypriots found their own labor federation, called Turk-Sen.

1954: December Greece seeks union with Cyprus

Greece petitions the UN to respond to the Greek Cypriots' demands for independence from the UK and *enosis*. The appeal is ignored.

1955: April 1 Greek Cypriots launch underground movement

With a series of bombings in Nicosia, EOKA, led by Colonel George Grivas (1898–1974), begins an armed guerrilla rebellion to win independence from Britain and union with Greece. In the next four years, hundreds of people die in armed clashes, and hundreds are imprisoned in jails and camps. The unity of NATO (the North Atlantic Treaty Alliance), of which Greece, Turkey, and the UK are all members, is seriously threatened.

1956: March 9 Archbishop Makarios is exiled

The UK exiles Archbishop Makarios II Mouskos, the head of the Cypriot Church, to the Seychelles Islands, accusing him of supporting the EOKA rebellion.

1958: June British "partnership plan" is refused

The UK offers Cyprus a "partnership plan" that would give the Greek and Turkish Cypriot communities greater internal autonomy while maintaining British rule. Turkey approves the plan, but Greece and the Greek Cypriots reject it.

1958: June 7 Nicosia bombing intensifies hostilities

The bombing of the Turkish press office in Nicosia raises the level of hostilities between Greek and Turkish Cypriots to civil war. Churches are burned down, looting is widespread, and many Cypriots leave their homes to migrate to areas in which their own group is the majority, turning Cyprus's cities into collections of ethnic enclaves.

1959: February 19 "London agreement" is signed

An agreement is reached by Greece, Turkey, Britain, and the Greek and Turkish Cypriot communities, providing for the establishment of an independent republic on Cyprus, with fair representation for both the Greek and Turkish Cypriots. The UK will be allowed to maintain two military bases on the island, Greek Cypriots agree to set aside their demand for *enosis*, and Turkish Cypriots agree to abandon *taksim* (partition). Britain, Greece, and Turkey are to act as guarantors of the island's independence and have the right to station limited military contingents on Cyprus, and retain the right to military intervention in order to uphold Cypriot independence.

Turkish Cypriot women and children huddle around a radio, listening for news about the battles between Greek and Turkish factions. (EPD Photos/CSU Archives)

The Republic of Cyprus

1960: August 16 Cyprus gains its independence

Cyprus becomes an independent republic and is admitted to the United Nations the same year. The following year it becomes a member of the British Commonwealth. The Greek Cypriot religious leader, Archbishop Makarios, is elected president, continuing the long-standing tradition of political rule by the island's religious leader, or ethnarch. To maintain an ethnic balance, the vice president, Fazil Kücük, is elected by the Turkish Cypriot minority. Because the government as a whole is similarly divided into officials serving two opposed constituencies, compromise and cooperation are difficult, and conflict predominates. Greek Cypriot nationalists continue their age-old agitation for *enosis*, while many Turks call for *taksim*, or partition of the island between Greece and Turkey.

1963: December Armed clashes between Greek and Turkish Cypriots

Following disagreements over proposed constitutional amendments that threaten the rights of Turkish Cypriots, hostilities break out between the Greek and Turkish communities, starting in Nicosia and spreading to the rest of the island. The government institutes emergency measures. Turkish government officials, led by Vice President Kücük, remain in the Turkish quarter of Nicosia and withdraw their participation in the Makarios government, ruling over the island's Turkish enclaves. Their isolation is increased by a trade embargo imposed by the Greek Cypriot government.

1964: March 4 UN sends troops to Cyprus

The United Nations Security Council votes to send peacekeeping forces to Cyprus to quell violence between Greeks and Turks. Nevertheless, 500 people die in fighting the following summer.

1964: August 10 Cease-fire agreement is reached

Greek and Turkish Cypriots agree to a UN-sponsored cease-fire.

1964: December Cease-fire ends

Fighting erupts once again between Greek and Turkish Cypriots.

1965: December UN Cyprus resolution seeks to prevent intervention

A UN resolution on Cyprus warns against intervention by Greece of Turkey (which nearly go to war over Cyprus two years later).

1967 Crisis on Cyprus threatens to turn into a Greek-Turkish war

Following skirmkishes between Greek and Turkish Cypriots, Turkey threatens to invade Cyprus and thus ignite a Greek-Turkish war. Stormy seas in the eastern Mediterranean allow U.S. envoy Cyrus Vance enough time to broker an agreement and avoid a Turkish invasion. Under the terms of the deal, Turkey promises not to invade the island and Greece promises to withdraw its military forces (estimated at over 10,000 troops) in excess of those allowed under the terms of guarantee first outlined in the London Agreement of 1959.

1968 Greek and Turkish Cypriots hold talks

Political tensions between the Greek and Turkish communities begin to relax. By now, many Greek Cypriots, including Makarios, are wary of *enosis* (partly due to opposition to the military government that came to power in Greece in April 1967). The Greek-controlled government lifts its trade embargo on the Turkish sector, and representatives of the two communities hold talks. The Greeks maintain their demand for a strong, unitary state, while the Turks want a looser, federal form of government and equal status, rather than the minority status offered by the Greeks.

1968: July 13 Death of painter Christoforos Savva

Painter and sculptor Christoforos Savva dies at the age of forty-four. (See 1924.)

1970 Major copper producer scales down production

With the decline in Cyprus's copper reserves, the American Cyprus Mines Corporation, the island's oldest and largest copper producer, begins to cut back its operations.

1973 Makarios is reelected president

Archbishop Makarios is reelected to the presidency.

1973: June 1 Trade agreement with EC goes into effect

Cyprus's "association agreement" with the European Economic Community (EC) goes into effect. It provides for a mutual reduction in import taxes and restrictions, to be gradually instituted over a five-year period.

1974: July 15 Makarios is overthrown by Greek military officers

After Makarios calls for the removal of Greek military officers from Cyprus's National Guard, the National Guard, acting in concert with Greece's military dictatorship (which desires enosis rather than an independent Cyprus), overthrows Cyprus's president, who flees the island. The military appoints Nikos Sampson president.

1974: July 20 Turkey invades Cyprus

Fearing a Greek takeover, Turkish Cypriots request help from Turkey, which invades Cyprus. The United Nations rapidly brokers a cease-fire. On July 23, Sampson resigns and Glafcos Clerides, speaker of the Cypriot House of Representatives, becomes acting president. On that same day, unable to counter the Turkish invasion force, the Greek military regime falls and is replaced by a civilian government. Later that month peace talks begin in Geneva. Despite these developments, Turkey continues building up its forces in Cyprus.

1974: August 14 Turkey launches a second attack and occupies the north

After the breakdown of talks in Geneva, Turkey initiates a full-scale invasion, seizing control of the northern thirty-seven to thirty-eight percent of the island, which is now divided into Greek and Turkish sectors along the Attila Line, while the Green Line divides the capital city of Nicosia. The Greek Cypriot-controlled Republic of Cyprus retains the districts of Limassol and Paphos, as well as most of Larnaca and Nicosia. The Turkish Cypriots take control of most of Famagusta and Kyrenia. Thousands of Greeks in the north flee to the south, while Turks in the Greek-controlled southern portion flee northward.

1974: December Makarios returns

Archbishop Makarios returns to Cyprus and resumes the presidency.

1975–1990s Cypriot art is influenced by political division

A new generation of artists comes of age during the political upheaval of the 1960s and 1970s that results in the division of the island into Greek and Turkish zones (and, eventually, two separate republics). The search for identity and the role of the artist become key themes in Cypriot art of this period. One of the most famous works produced is Cyprus's most outstanding sculpture, the 10-meter-high statue of Archbishop Makarios by Nikos Kodjiamani (b. 1946) erected in the capital city of Nicosia, just outside the Archbishop's Palace. Other prominent artists of this period include Andreas Ladammatos (b. 1940), Emin Cizenal (b. 1949), and Andreas Charalambous (b. 1947).

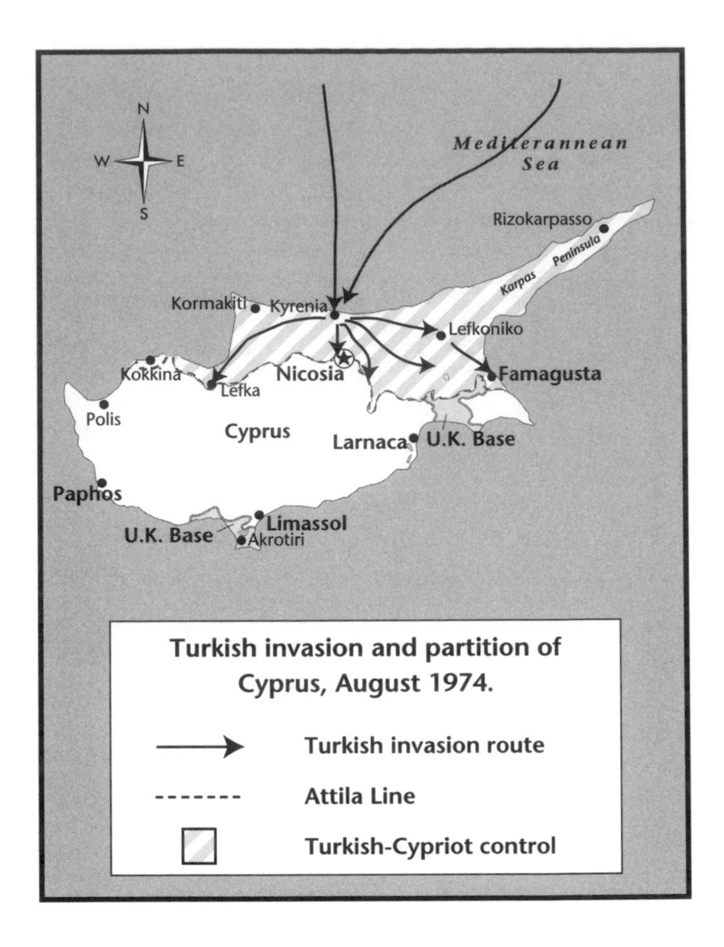

Turkish invasion and partition of Cyprus, August 1974.

1975–76 Emergency economic plan is implemented

To cope with the economic problems caused by the splitting off of the Turkish section in the north, the Republic of Cyprus adopts the Emergency Action Plan, designed to foster the development of labor-intensive industries including textiles and shoes.

1975: February 13 Turkish Cypriots proclaim a federated state

The Turkish Federated State of Cyprus is proclaimed. Former Vice President Rauf Denktash is named president. The Turkish action is declared illegal by the Makarios government.

1977: August 3 Archbishop Makarios dies

The Republic of Cyprus's first president dies of a heart attack. (See 1913.)

1978 Spyros Kyprianou is elected president

Following the death of Archbishop Makarios (see 1977), Spyros Kyprianos is elected the second president of the Republic of Cyprus.

1979 Eastern Mediterranean University is founded

Eastern Mediterranean University is established by Turkish Cypriots.

1982 The first offshore banking license is granted

The Central Bank of Cyprus sets rules for offshore banks, which can serve only non-Cypriots and do business only overseas. In the 1980s offshore banking grows into an important sector of the financial industry, and offshore firms of other kinds transacting business with countries in the Near East prosper as well. By the mid-1980s, almost 3,000 offshore firms are registered with the Republic of Cyprus, including insurance businesses and real estate companies, architects, and consultants.

1983 Kyprianou is reelected

Spyros Kyprianou is elected to a second term as president of the Republic of Cyprus.

Two Republics

1983: November The Turkish Republic of Northern Cyprus is proclaimed

Turkish Cypriot leader Rauf Denktash proclaims the northern, Turkish area of Cyprus a separate republic. The action is repeatedly condemned by the United Nations, and the only country to recognize the new republic is Turkey. The UN attempts to effect a reconciliation between the two sides, but the presence of 25,000 Turkish troops in the north is a barrier to negotiations.

1985 The Turkish Republic adopts a constitution

The government of the Turkish Republic of Northern Cyprus approves its own constitution.

1988: February Vassiliou is elected president of the Republic of Cyprus

Leftist Georghios Vassiliou defeats two-term incumbent Spyros Kyprianou in the election for the presidency of the Republic of Cyprus.

1989 Historical museum is established

The Leventis Municipal museum, the first historical museum on Cyprus, is founded.

1989: March New coastal hotel licenses are banned

The growth of tourism in the 1980s prompts measures to avoid overdevelopment of coastal areas, which has begun to cause water shortages and traffic jams and strain existing waste disposal systems beyond their capacity.

1989: May Troops withdraw from Nicosia borderline

Troops withdraw from twenty-four posts along the line that divides the Greek and Turkish sections of Cyprus's capital, Nicosia. However, over 2,000 UN peacekeeping troops and 20,000 Turkish troops are still stationed on Cyprus.

1990 Denktash is reelected in northern Cyprus

Rauf Denktash is elected to his second term as president of the Turkish Republic of Northern Cyprus. Observers fear that the island may remain politically and ethnically divided permanently.

1990: July 4 Cyprus applies for admission to the European Union

The Republic of Cyprus applies for membership in the European Union (EU).

1992 University of Cyprus opens

The University of Cyprus is established in the capital city of Nicosia.

1993: February Glafcos Clerides is elected president of the Republic of Cyprus

Glafcos Clerides, a conservative, is elected president of the Republic of Cyprus, replacing Georghios Vassiliou.

1993: October European Union announces intent to admit Cyprus

The European Union confirms its intention to grant membership to Cyprus.

1993: November 18 Death of painter Telemachos Kanthos

Painter and engraver Telemachos Kanthos dies. (See 1910.)

1994: April 28 Death of painter Adamantios Diamantis

Adamantios Diamantis, one of the foremost Cypriot painters of the twentieth century, dies in Nicosia. (See 1900.)

1996: August Worst border violence in two decades occurs

Two Greek Cypriots are killed and over fifty are hurt when Greek Cypriot demonstrators breach military barriers and clash with Turkish troops in the buffer zone between the Greek and Turkish sections.

1997: January Clerides buys Russian surface-to-air missiles; Turkey threatens air strikes

President Clerides announces his government's decision to buy Russian S-300 surface-to-air missiles. Deployment is orginally scheduled for the summer of 1998, but is put back to the spring of 1999. These missiles, which can target Turkish aircraft as far away as the Turkish mainland are intended to provide air cover for the Greek Cypriot forces which have no air force of their own. In response to the Greek Cypriot action, the Turkish government threatens to launch air strikes to destroy the missiles before they can become operational.

The missiles are to be installed at a military airfield and a naval base, both of which are under construction and will be used by the Greek Air Force and Navy, respectively. Given the fact that Greek Cypriot forces are vastly outnumbered and outgunned and that Greece is too far away to provide effective aid in the event of renewed fighting, many international observers believe that Clerides's decision to buy the missiles is really intended to put the Cyprus issue back on the international agenda. Instead, the missiles, rather than the Cyprus problem, become the issue and the purchase is roundly condemned in Western circles.

1997: November 3 Leaders of Greece and Turkey sign cooperation pact

Following weeks of rising tensions between the two nations, the Greek and Turkish prime ministers, Costas Simitis and Mesut Yilmaz, agree to cooperate in improving relations between their countries, which are strained over the division of Cyprus into Greek and Turkish sectors.

1998: February 15 Greek Cypriot President is reelected

Glafcos Clerides is elected to a second five-year term as president of the Republic of Cyprus, narrowly defeating independent candidate, George Iacovou, in a second round of elections.

1998: March 15 Compromise permits scheduling of EU admission process

A compromise is reached on conditions for the proposed admission of Cyprus to the European Union. France had threatened to block membership for Cyprus because only the Greek Cypriots would be represented. The new agreement specifies that progress on reunification of the Greek and Turkish sectors will be a condition of admittance to the EU.

1998: March EU membership talks begin

The European Union begins talks with Cyprus and five other prospective member nations. The proposed admission of Cyprus to the EU has heightened tensions among the island's two ethnic communities, as well as Greece and Turkey because only the Greek Cypriot Republic of Cyprus will be considered for admission. The Turkish Cypriot republic in the north is not recognized by any members of the international community except Turkey.

1999: January Clerides announces S-300s will not be deployed on Cyprus

Under pressure from the United States and the European Union, and fearing Turkish retaliation, President Clerides announces that he will not deploy the S-300 surface-to-air missile system on Cyprus. Instead the missiles are to be deployed on the Greek island of Crete in April.

Bibliography

Crawshaw, Nancy. *The Cyprus Revolt: The Origins, Development, and Aftermath of an International Dispute.* Winchester, Mass.: Allen & Unwin, 1978.

Denktash, Rauf. *The Cyprus Triangle.* Winchester, Mass: Allen & Unwin, 1982.

Foley, Charles, and W.I. Scobie. *The Struggle for Cyprus.* Stanford, Calif.: Hoover Institution Press, 1975.

Hart, Parker T. *Two NATO Allies at the Threshold of War: Cyprus, a Firsthand Account of Crisis Management, 1965–1968.* Durham, N.C.: Duke University Press, 1990.

Jennings, Ronald C. *Christians and Muslims in Ottoman Cyprus and the Mediterranean World, 1571–1640.* New York: New York University Press, 1993.

Koumoulides, John T.A., ed. *Cyprus in Transition, 1960–1985.* London: Trigraph, 1986.

Polyviou, Polyvios G. *Cyprus: Conflict and Negotiation, 1960–1980.* New York: Holmes & Meier, 1981.

Salem, Norma, ed. *Cyprus: A Regional Conflict and its Resolution.* New York: St. Martin's Press, 1992.

Solsten, Eric, ed. *Cyprus, a Country Study.* 4th ed. Washington, D.C. Government Printing Office, 1993.

Tatton-Brown, Veronica. *Ancient Cyprus.* Cambridge, Mass.: Harvard University Press, 1988.

Federated States of Micronesia

Introduction

The Federated States of Micronesia (FSM), a group of 607 islands in the Pacific Ocean, located just north of the equator, cover 702 square kilometers of land, with an estimated population of 125,377 in 1996. The land mass of the FSM is about four times that of Washington, D.C.

There are four states or island groups within the Federated States of Micronesia: Chuuk, Kosrae, Pohnpei, and Yap. These are the newly adopted names for the states which when they were districts within the Trust Territory of the Pacific Islands, were called by their Westernized spellings: Truk, Kusaie, Ponape, and Yap.

Chuuk is the largest and most populous, having almost half the total population. Kosrae has the fewest inhabitants and also the smallest land area.

There are nine Micronesian ethnic groups represented. Among these, Polynesians are represented by the populations of Nukuoro and Kapingamarangi. There are an estimated 900 Filipinos as well as around 1,000 Asians, mostly from China and India, in the FSM. Approximately 400 Americans reside permanently in the Federated States of Micronesia.

The tropical climate brings heavy, year-round rainfall to the region, especially in the eastern islands of Pohnpei and Kosrae. The heaviest rainfall is found in Pohnpei which receives about 3,330 inches per annum, making it one of the wettest places on earth. Moreover, the islands lie on the southern edge of the typhoon belt. In the typhoon season from June until December, typhoons cause occasional severe damage.

Geologically, islands range from high, volcanic mountains to low coral atolls. The highest point is Mt. Nahna Laud on Pohnpei which rises to 798 meters above sea level.

Among the Federated States' limited natural resources are forests, marine products, and some deep-sea bed minerals. Tuna exports to Japan have become a major source of revenue for the islands, along with scuba and deep sea tours of numerous World War II shipwrecks on reefs and in lagoons.

History

In prehistory, waves of migrants reached the area from various parts of the Pacific. A number of extensive and complex pre-contact trading networks also linked the current states with islands beyond their present political structure. Yap engaged in long-distance trade across hundreds of kilometers with the island of Palau for calcium carbonate discs called *raay*. Better known as "giant disc money" of Yap, a raay measured about a half meter in diameter with a central hole. After the nineteenth century when larger ships were available to the Palauans and Yapese, colossal discs which measured several meters across became popular.

Archaeologists have excavated early construction of low platforms and islet fills at Nan Madol on Pohnpei dating back to about 1,000 years ago. Settlement at the site dates back 1,900 years. Houses were likely built out over the reefs. Archaeologists speculate that flooding and subsidence caused the inhabitants of Nan Madol to begin creating artificial islets. A similar site has been excavated at Lelu on Kosrae dating back 2,000 years. The populations of the major islands increased gradually based on trade and exchange between island centers on Yap, Chuuk and Palau. Archaeologists and demographers estimate that the region reached its maximum population between 700 and 500 years ago.

Division of labor and ways of determining wealth between men and women varied from island to island, and in some cases from region to region on the same island. On Chuuk, the men of the lagoon region were responsible for gardening while on the western atolls of the region, gardening was considered women's work. In traditional Micronesian societies, wealth also differed between men and women. On Pohnpei, items that defined a male's wealth included giant yams, pigs, and kava while items that defined a woman's wealth consisted of cloth goods, sugarcane, and coconut oil.

European contacts with the previously isolated Micronesians had devastating effects on the population. It has been estimated that at least thirty percent of the regional population was decimated by European-borne diseases. Chuuk was one of the least affected islands because the population consistently repelled European advances.

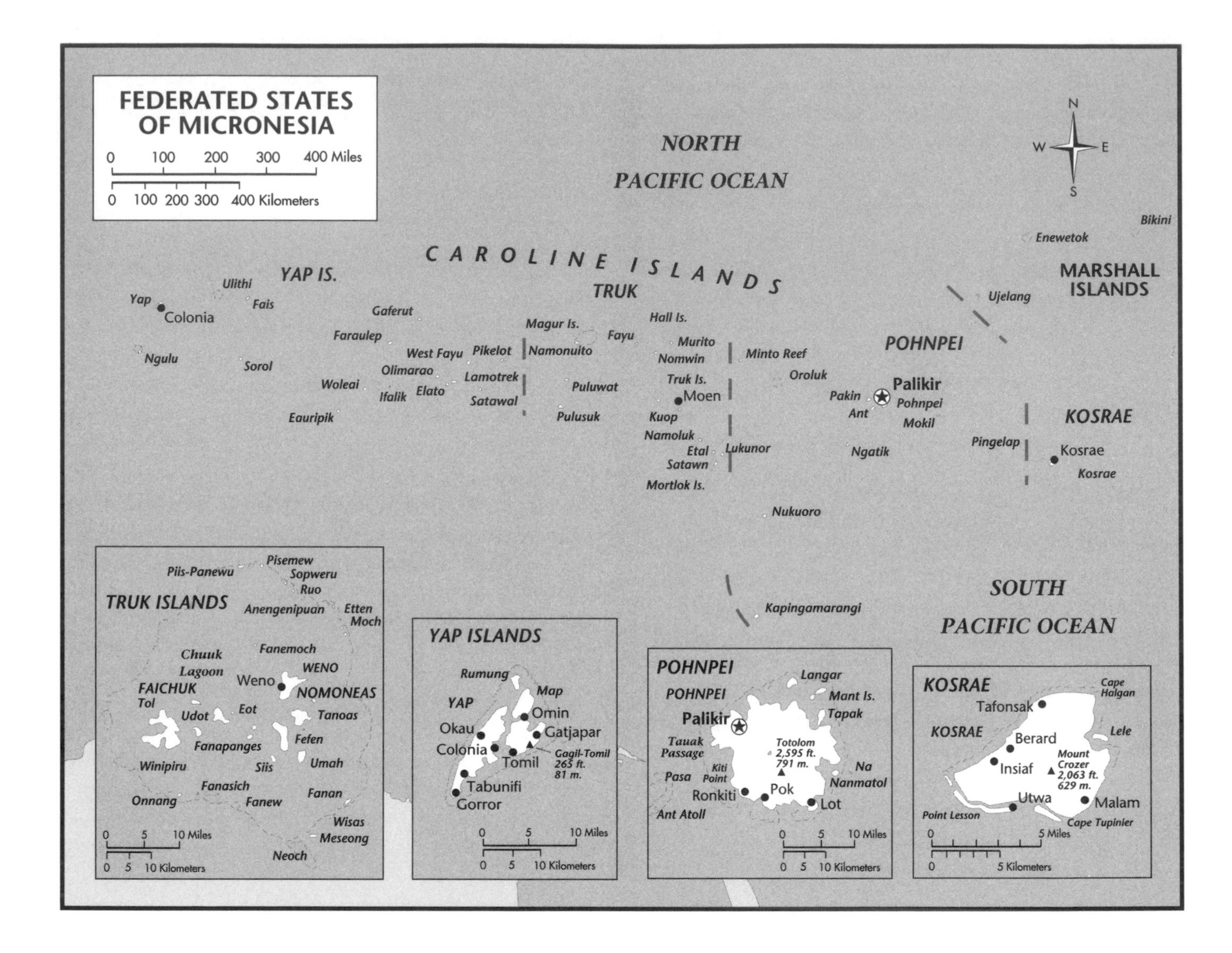

World War II had dramatic consequences for the islands currently in the Federated States of Micronesia. Chuuk, then known as Truk, was Japan's offshore naval headquarters as Pearl Harbor was for the United States. Admiral Yamamoto anchored his combined fleet in Chuuk Lagoon. Americans bombed Truk Lagoon on February 17 and 18, 1944 leaving ten warships and thirty-one merchant vessels sunk in Chuuk Lagoon or on the offshore reefs.

The constitutional government of the Federated States of Micronesia is in free association with the United States through a compact which provides for a level of financial and developmental support as well as protection during the formative years of the new nation. While the previous trusteeship only allowed limited movement to the United States for employment, the current Compact, which expires on January 1, 2001, allows free entry for Micronesian residents to the United States.

The executive branch of government represented by the President and Vice-President, follows the U.S. system in that the President is both the chief of state and the head of government. Unlike the American system of government, however, the President and Vice-President are elected by Congress from the four senators-at-large. The President and Vice-President cannot both be from the same state. Both the President and the Vice-President are elected to serve four-year terms with elections taking place in May.

The legislative branch consists of a unicameral congress with fourteen members elected by popular vote. Four congressional members, one from each state, serve four-year terms. The other ten congressional members, who serve two-year terms, are elected from single-member districts delineated by population: five from Chuuk, three from Pohnpei, and one each from Yap and Kosrae. There are no formal political parties as such in this structure.

The judicial branch, the Supreme Court, is headed by the Chief Justice. There are also up to five Associate Justices who are appointed by the President with advice and consent of Congress. Justices of the Supreme Court serve for life.

Although the Japanese attempted to imbue the Micronesians with the ideals and rituals of Shinto, partly though the construction of Shinto shrines, their efforts did not hold with the indigenous population. Most of the inhabitants of the Federated States are Christians: about fifty percent Roman Catholics, and forty-seven percent Protestant. The remaining three percent claim no religious affiliation.

English is the official and common language of the Federated States of Micronesia. There are also eight major indigenous languages spoken: Yapese, Ulithian, Woleaian, Chuukese, Pohnpeian, Kosraean, Nukuoro, and Kapingamarangi. Nukuoro and Kapingamarangi are Polynesian Outlier languages while the other six are Micronesian languages which belong to the Oceanic branch of the Austronesian language family.

Literacy is high in the FSM with around ninety percent of the population over the age of fifteen able to read and write English.

The primary source of revenue for the FSM is aid from the United States. In the period between 1986 and 2001 when the Compact expires, $1.3 billion will have been contributed by the United States to the FSM.

Defense of the Federated States of Micronesia is the responsibility of the United States.

Timeline

c. A.D. 1000 Early people develop stone money

The people on the island of Yap develop stone money, sculpted discs of stone. Some stone money discs are as large as twelve feet (four meters) in diameter.

1250 Kosrae's city of Lelu thrives

Kosrae's ancient stone city is on the tiny island of Lelu (Lay-luh). The people living here are ruled by royalty. The ruins of the stone city, including what are believed to be king's palaces, consist of over one hundred structures. The population is divided into clans, and members of the same clan do not marry each other.

1527–29 Spanish explorer sails past Pohnpei

The Spanish explorer, Alvaro de Saavedra Ceron, sails through a number of Micronesian islands and atolls, including Pohnpei and possibly many other of the Caroline Islands.

1686 Francisco Lezcano names Yap island, *La Carolina*

While sailing in the region of Micronesia, Spanish Captain Francisco Lezcano sights the island of Yap and names it, *La Carolina*, after King Charles II of Spain.

1820s First European ships arrive at Kosrae

The first European trading vessels begin to call at Kosrae. The estimated population of Kosrae is 3,000 when the first Europeans arrive. The diseases Europeans introduce diminish the indigenous population to approximately 300 by the 1880s.

1830s Men of Ngatik Island are exterminated

Ngatik Island, part of the present-day state of Pohnpei, is over-run by British pirates. The men of Ngatik Island are systematically exterminated to enable the pirates to have unimpeded access to Ngatik women.

1852 Missionaries arrive on Kosrae

European missionaries arrive on the island and convert nearly all islanders to Christianity. Local customs such as singing are preserved in Christian church services.

1869 American missionary, Benjamin Snow, introduces new government structure

Civil unrest, based on depopulation and hostilities between followers of traditional beliefs and newly converted-Christians, besets Kosrae. In an attempt to restore order, the paramount chief tries to revive pre-Christian rituals and ban Sabbath observance. The sudden death of the paramount chief of Kosrae weakens the power base of the lesser chiefs. An American missionary stationed on Kosrae, Benjamin Snow, urges the new paramount chief to accept a new government structure for the island. A council is established which contains seven of the island's original, titled chiefs and another seven regional representatives as well as the paramount chief. This arrangement enables Snow to get Christian converts into the political arena.

1874 First Christian King of Kosrae is installed

The Council of Kosrae, established by Benjamin Snow (see 1869) only five years earlier, ousts the non-Christian paramount chief and installs the first Christian "king" of Kosrae.

1885 German warship lands on Pohnpei and demands annexation

A German warship lands on Pohnpei with a company of marines and demands that the five high chiefs of the island sign a treaty of annexation. Although the chiefs sign the treaty, Germany does not establish a presence on the island.

1885: August German naval contingent seizes Yap

Spain had done little to assert its claims to the various island groups in the north Pacific Ocean, including the Caroline Islands. German traders had asserted trading rights in the neighboring Marshall Islands in 1878. A German naval contingent decides to seize the island of Yap in August, 1885, provoking a violent Spanish reaction in which Spain threatens war against Germany.

1885: September Possession of Caroline Islands submitted to Pope Leo XII

In an attempt to avert overt hostilities, the Spanish claim to the Caroline Islands is submitted to Pope Leo XII (1810–1903) for arbitration.

1885: December Pope Leo XII settles dispute over Caroline Islands

Pope Leo XII upholds Spain's claim to the Caroline Islands, contingent upon Germany's right to trade and to establish coaling stations for its ships in the Pacific.

1886 Spanish arrive to fend off Germans

Having previously ignored their possession, Pohnpei, the Spanish arrive in 1886 in response to German attempts to take over the island. Pohnpeian resistance to the Spanish occupation is protracted and intense. The following year, the Pohnpeians assassinate the Spanish governor and force the Spanish colonists from the island. The Spanish return about four months later but spend the short duration of their rule in protected fortifications.

1899 Spain sells Caroline Islands to Germany

After its loss of the Spanish-American War to the United States in 1898, the Spanish government is weakened both militarily and economically. Germany offers Spain the equivalent of U.S. $4.5 million to buy Spain's remaining possessions in Micronesia. Spain accepts. This enables Germany to become the dominant European force in Micronesia.

1910 Sokehs Rebellion on Pohnpei

As they had done previously with the Spanish in their country, the Pohnpeians once again take up arms and assassinate the German Governor and several other German officials. The Sokehs Rebellion is crushed by German authorities after six months of violence. The leaders of the rebellion are all executed.

1914 Japanese take control of Micronesia

With the outbreak of World War I (1913–18), Japan takes advantage of Germany's focus on war in Europe and occupies all German-held islands until the end of the war.

1917 German cruiser sinks

The SMS *Cormoran*, a 335-foot-long World War I German auxiliary cruiser, is deliberately sunk as the conflict decreases.

1919 Japan gets mandate over Micronesia

Although the Japanese push for complete sovereignty over Micronesia after the end of World War I, a mandate is issued under the League of Nations for Micronesia, then Japan develops the region as though it had been granted total domination over Micronesia. Micronesia becomes an integral part of the Japanese Empire after the Japanese walk out of the League of Nations in 1933.

1944: February 17–18 Americans bomb Chuuk

During World War II U.S. military aircraft bomb the Japanese air and naval forces stationed at Chuuk. The U.S. destroys 270 aircraft and sinks an estimated sixty warships and merchant vessels harbored in Chuuk Lagoon. No U.S. forces land on Chuuk.

Chuuk Lagoon is altered forever by this military activity. When ships sink, they create a new platform for marine life on the silty bottom of the lagoon. Many varieties of coral, anemones, and clams thrive in this new environment. The Tokai Maru, a 465-foot Japanese freighter, lies in 138 feet of water with the mast coming within thirty-five feet of the surface.

1946 New York diocese takes over administraiton of the Caroline and Marshall Islands

Francis A. McQuade, S.J., the superior of the New York Province of the Roman Catholic Church, adds Micronesia to New York's list of overseas mission for which they are responsible.

1947 The United Nations creates the Trust Territory of the Pacific Islands

The United Nations unites Ponape, which includes Kusaie, Truk, Yap, Palau, the Marshall Islands, and the Northern Mariana Islands, into a territory called the Trust Territory of the Pacific Islands. This area is administered by the United States Department of the Navy until 1951 when the authority over the region is passed to the Department of the Interior.

1959 Alcohol is legalized on Chuuk

Reversing a prohibition established under early colonial regimes, Chuuk legalizes the sale and consumption of alcohol in 1959.

1962 Headquarters of the Trust Territory of the Pacific Islands is transferred to Saipan

The headquarters of the Trust Territory of the Pacific Islands, originally located in Hawai'i, then in Guam, is transferred to Saipan in the Mariana Islands. At this time, the TTPI have six district components: the Mariana Islands, the Marshall Islands, Ponape, Palau, Truk, and Yap.

1965 Pohnpei Agriculture and Trade School is founded

The Pohnpei Agriculture and Trade School (PATS) is founded by Father Costigan to provide training for islanders. This school is the first post-secondary school in Micronesia.

1965: July 12 First Congress of Micronesia of the TTPI

In the first meeting of the Congress of Micronesia of the Trust Territory of the Pacific Islands (TTPI), the new Congress adopts the official flag of Micronesia.

1970 Neylon is named Bishop of Micronesia

Martin J. Neylon (b. 1940), a Jesuit priest, is named bishop of the Diocese of the Carolines, a post he holds for twenty-five years (see 1995).

1972 United States negotiates with trust territories

In developing an instrument to achieve independence for the administrative districts within the Trust Territory, a compact of free association is envisioned that will allow for self-government, with the United States retaining authority over foreign affairs and providing for defense. Delegates from the Northern Mariana Islands want a closer administrative relationship with the United States, although the United States tries to negotiate equally with each district. The United States decides not to press for political unity for the districts and opens separate negotiations with the Northern Mariana Islands. This action leads to the establishment of the Commonwealth of the Northern Mariana Islands in 1976. Following suit, Palau and the Marshall Islands opt to negotiate separate agreements from the rest of the districts of the TTPI.

1977 Prohibition laws for Moen, Chuuk

The urban center of Moen is placed under prohibition (alcohol is forbidden) due to the unwavering efforts of church women and their male allies. Although there are at least three attempts to repeal this prohibition, the law remains in effect in Moen, the only urban center on Chuuk, as of 1999.

1978: July 12 Federation of states within Micronesia is formed

The four districts of the former Trust Territory votes in a referendum to form a Federation under the Constitution of the Federated States of Micronesia. The United Nations recognizes the Federation as a legitimate act of self-determination.

1979: May 10 Constitution of the FSM is implemented

The First Congress of the Federated States of Micronesia convenes in Congressional Chambers in Kolonia, Pohnpei state thereby implementing the Constitution.

1979: May 11 Congress elects first President and Vice President of the Federated States of Micronesia

The Congress elects Tosiwo Nakayama of Chuuk and Petrus Tun of Yap as president and vice president respectively of the newly formed nation.

1982 Office of Public Auditor (OPA) is established

The Office of Public Auditor (OPA) is established, which takes its authority from the constitution, and is charged with auditing the financial accounts of all government-funded agencies and organizations.

1986: January 14 U.S. Congress approves Compact of Free Association

The Compact of Free Association, known as Public Law 99-239, passes in the U.S. Congress (see 1986: November 3).

1986: November 3 Compact of Free Association takes effect

The Compact of Free Association between the United States and the Federated States of Micronesia takes effect.

The Federated States of Micronesia begins to receive cash payments and other benefits designed to support FSM's transition to a self-sufficient economy from the United States. Included among these benefits is availability of Pell Grants for students enrolling in college.

1987 Watershed protection legislation is introduced on Pohnpei

In response to a study done in 1983 which found Pohnpei forests reduced to only fifty-five percent of the island's area, watershed protection legislation is introduced to save them.

1990: December 22 Official termination of trusteeship

The United Nations Security Council officially terminates the trusteeship over the Federated States of Micronesia, although the Compact of Free Association entered into in 1986 already declared the trusteeship terminated.

1991: September 17 Federated States of Micronesia becomes a full member of the United Nations

Along with Marshall Islands, Democratic People's Republic of Korea, Latvia, Estonia, Lithuania, and the Republic of Korea, the Federated States of Micronesia becomes a member of the United Nations (UN).

1995: May 11 Bailey Olter wins elections

Bailey Olter is elected to a second term as President.

1995: July 31 "Fish Air" flies first round trip

Air Micronesia begins to transport fresh fish to Guam. At Guam, the cargo is tranferred to other airlines for transport to

Japan. Known as "Fish Air," the airline flies aircraft that have the capacity to carry 35,000 pounds (16,000 kilograms). Fish Air plans to fly round trips between Micronesia and Guam daily.

1995 Scuba divers report increase in manta ray population

Yap Island, a popular destination among scuba divers, witnesses an increase in the manta ray population. The creatures are identified using photos taken by divers. Each manta ray has a unique pattern of black spots on its underside. Estimates of the number of manta rays counted range from 45 to 125.

1995: February Bishop Amando Samo is appointed

Replacing retiring Bishop Neylon, Bishop Amando Samo becomes bishop of the Diocese of the Carolines (see 1970). Most of the island parishes are led by priests who are island natives; Bishop Samo is the first islander to head the diocese.

1996 Micronesia economy is based on U.S. aid

A large portion of the gross domestic product of the Federated States of Micronesia is comprised of U.S. financial assistance.

1996: July President of Micronesia suffers stroke

Only a little over a year into his second term, President Bailey Olter suffers a severe stroke. Vice-president Jacob Nena is appointed Acting President for a period of 180 days.

1996: November President Bailey Olter declared incapacitated

Congress declares President Bailey Olter incapacitated. The Constitution of the Federated States of Micronesia calls for a 180 day waiting period if the President suffers an illness or injury that prevents him/her from performing the duties of office. During that period, the Vice President serves as Acting President. At the end of the 180 period, the President either resumes duties or is declared incapacitated by the Congress. In November of 1996, the Congress of the Federated States of Micronesia declares President Olter incapacitated. Vice-president Jacob Nena is slated to be President.

1997: May 9 Jacob Nena takes office as President

Jacob Nena who had been vice president to president Bailey Olter, takes the oath of office as president of the Federated States of Micronesia. Nena will serve the remaining two years of Olter's term.

2001: January 1 Compact of Free Association expires

The Compact of Free Association between the United States and the Federated States of Micronesia is scheduled to expire on January 1, 2001. There is a provision in the Compact for its renewal.

Bibliography

Denoon, Donald, ed. *The Cambridge History of the Pacific Islanders.* Cambridge: Cambridge University Press, 1997.

Hanlon, David. *Upon a Stone Altar: A History of the Island of Pohnpei to 1890.* Honolulu: University of Hawai'i Press, 1988.

Hezel, Francis. *The First Taint of Civilization: A History of the Caroline and Marshall Islands in Pre-Colonial Days, 1521–1885.* Honolulu: University of Hawai'i Press, 1983.

Levesque, Rodrigu, ed. *History of Micronesia.* Honolulu: University of Hawaii Press, 1994.

Ritter, Philip. "The population of Kosrae at contact," Micronesia 17 (nos. 1-2), 1981.

Scarr, Deryck. *History of the Pacific Islands: Kingdom of the Reefs.* Sydney: Macmillan Australia, 1990.

Sinoto, Yosihiko, ed. *Caroline Islands Archaeology.* Bernice P. Bishop Museum Pacific Anthropological Records 35, 1984.

Smith, Gary. *Micronesia: Decolonization and U.S. Military Interests in the Trust Territory of the Pacific Islands.* Canberra: Australian National University Press, 1991.

Spate, Oskar. *The Pacific Since Magellan, vol. 1: The Spanish Lake.* Canberra: Australian National University Press, 1979.

Fiji

Introduction

The republic of Fiji , formerly known as Viti, an archipelago comprised of more than 300 islands, is located in the southwestern Pacific, about 500 miles (800 kilometers) southwest of Samoa and 1,100 miles (1,800 kilometers) north of New Zealand. The total land mass of Fiji is 7,055 square miles (18,272 square kilometers). Only about 106 islands in the group are inhabited. The port city, Suva, is located on the island of Viti Levu, which, in 1999 had a population of approximately 200,000 people. The other large island is Venua Levu.

The larger islands are volcanic in origin, and some rise to over 3,000 feet (9,000 meters). Fiji has a tropical climate with two distinct seasons. From November to April the weather is hot and wet, while a cooler and drier period lasts from May through October.

Crops that are tropical in nature, such as sugarcane, coconut, ginger, cocoa, and rice, are the chief exports of this island nation. Fishing is also important as a subsistence food and as an export. Newer government-supported industries, such as pine timber, beef cattle, and dairy products, are helping to expand Fiji's economy. Gold and copper are also mined. By far the largest industry of the twentieth century is tourism. The beautiful beaches and coral atolls have attracted vacationers interested in snorkeling and SCUBA diving. The beach area on the south coast of Viti Levu is attractive to residents of Australia, New Zealand , and North America. Vacationers can arrive at the airport at Nadi or by ship at the port city of Suva, which can accommodate large cruise ships.

History

Fiji is believed to have been inhabited as early as 1500 B.C. by nearby Polynesians, a broad-nosed, light-skinned people. Archaeologists have discovered horizontal-lined and geometric-shaped pottery, called *lapita,* which dates from this period. Later, about 500 B.C., Melanesian peoples arrived and intermarried with native Fijians. This blend of peoples is what as referred to now as the Native Fijians. Fijian authority was determined by family descent. The chief or *turaga* was the living representative of god, the *vu*. Feuds occurred between many clans or families of Fiji. The male, or chief of the clan, would cannibalize the losers, often drinking the ceremonial kava from the victim's skull and eating the bodies also. Even though the early Fijians were talented artists and boat-builders, they soon became known for their strange practices, and thus early European discoverers of Fiji called these islands the Cannibal Isles.

In 1643 the first documented European visitor to Fiji, Abel Jansen Tasman, was looking for a new southern continent and instead discovered Fiji, Tasmania, and New Zealand. The first non-native to come ashore in Fiji was Captain James Cook who landed on Vatua, which he named Turtle Island, in 1779. However, the credit for actual charting of the islands is given to William Bligh, captain of the ship *Bounty.* In May 1789 Captain Bligh and his crew were chased by Fijian warriors, and they barely escaped with their lives. Bligh returned to Fiji in 1792 but did not leave his ship.

Most people had heard tales of the cannibalistic Fijian warriors and, therefore, avoided exploration of Fiji. However, traders had heard of the abundance of sandalwood along the coasts, and Fiji was from time to time inhabited by traders and beachcombers. Other non-native inhabitants, escaped convicts from Australian penal colonies, took advantage of the wars between Fijian tribes and instructed the Fijians in the use of guns. As early as the 1830s many European and American beachcombers and traders were living in colonies at Levuka. Also at this time Methodist missionaries from the nearby island of Tonga converted the Fijians to Christianity. They accomplished this by giving the Fijians their first written language.

While foreigners made a few friendships among some tribes, they were considered enemies of others. When Swedish adventurer Charles Savage tried to befriend one warring tribe, he was taken captive and eaten. With European influence waning, the cannibal chief Naulivou and his brother Tanoa became the undisputed rulers of Fiji. In the 1840s, *Ratu* (chief) Seru Cakobau, son of Tanoa proclaimed himself *Tui Viti* which means King of Fiji. Other explorers and adventurers continued coming to Fiji for trade in sandalwood and *beche-de mer* (a sea slug), but most met unfortunate ends.

In 1847, rivalries and fights occurred between the Tongan chief and Cakobau, ruler of the Fijians. With both parties struggling for power, Cakobau enlisted British aid and finally on October 10, 1874, ceded Fiji to Great Britain. The British governor of Fiji, Sir Arthur Gordon, and his successor, Sir John Thurston, began transforming Fiji into a Western-style civilization.

The British colonial period on Fiji was marked by the development of the sugarcane plantations, the importation of laborors from India as indentured servants, and the continued struggle for control and ownership of land. Asian Indians came to Fiji in search of a better life. They agreed to work for ten years in return for food and shelter. They were able to lease small plots of land from the government for farming and raising livestock. Many decided to remain even after their term of indenture had ended. This period lasted from about 1879 to 1916 when the practice of indenture officially ended. The racial animosity that grew up between the Native Fijians and the Indians continued into the twentieth century causing uprisings, terrorist attacks, and coups.

In 1904, a legislative council was formed, beginning a form of representative government in Fiji. The Council consisted of six elected Europeans, and two Fijians nominated by the Chiefs. In 1916, an Indian was appointed to the Council. By 1929 the Council had five representatives from each community. However, this body did not have governmental control; only the British-appointed governor could make laws. The Europeans and Fijians usually sided with each other, and the Indians were generally left out. Finally on October 10, 1970, the Fijians voted to become independent of Great Britain. The first Fijian governor was Ratu Sir George Cakobau, great-grandson of the original Cakobau.

Although Fiji was then an independent nation, the population differences between Fijians and Indians continued to influence the country's direction. By law the land could only be owned outright by native Fijians. Two major political parties, the Alliance Party (Fijian) and the Fiji Indian Alliance (Indian), were formed, based on race. Not until 1985 did the Fiji Labor Party, started by Dr. Timoci Bavadra, hope to unite all Fijians under an economic identity instead of a race-based policy. In 1986 the Labor Party joined with the Fijian Nationalist Party to form a Coalition government. After the election, the *Taukei* (landowners) Fiji-for-Fijians started a terrorist campaign designed to unseat the government. They barricaded highways and organized protest marches to gain support for the Fijians. On May 14, 1987 Lieutenant Colonel Sitiveni Rabuka staged a mititary coup of Bavadra's government. No one was killed, but suddenly Rabuka was in charge of a new, but not elected government. Many countries, including England, refused to acknowledge Rabuka as the legitimate leader of Fiji, and refused to see Rabuka's personally appointed foreign minister Ratu Mara. However, President Reagan of the United States agreed to aid Fiji in exchange for permitting United States nuclear warships to visit ports in

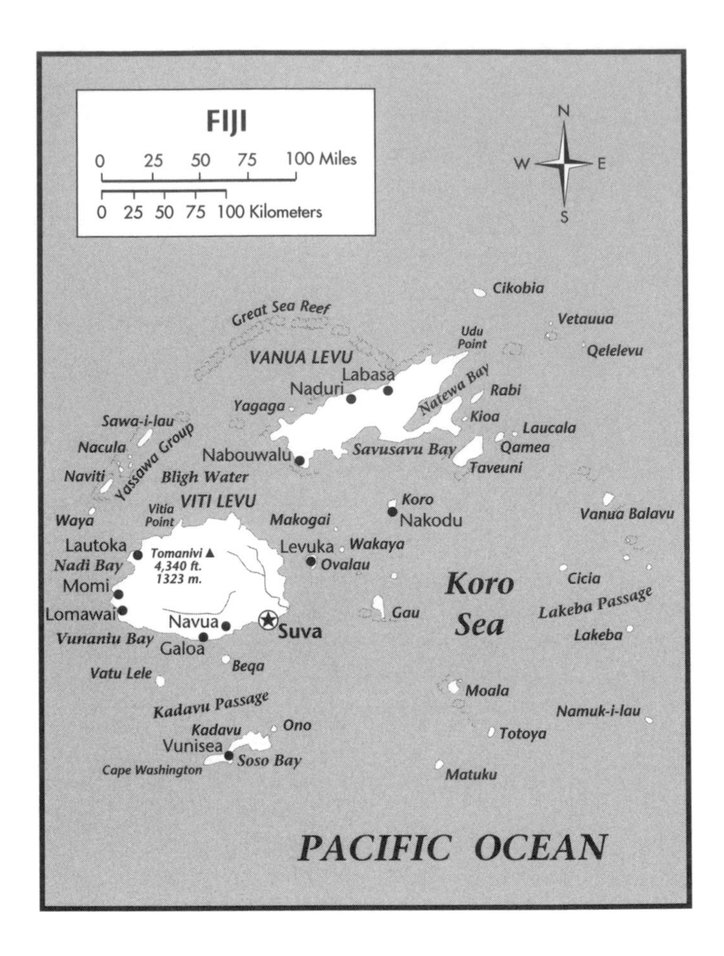

Fiji. This agreement was highly controversial because many accused the United States of being involved in Rabuka's coup, a suspicion not substantiated in fact.

As of the late 1990s, Fiji remains an independent republic. A constitution ratified in 1990 still favors native Fijians, but amendments in 1997 made laws fairer to all Fiji inhabitants. However, two important sociological problems face the nation of Fiji in the 1990s; continued racial hostilities between the Fijians and the Indian population, and the struggle of the Fijians to accept a government that is in direct contradiction to the rule of deity-kings. Regarding the first problem, the Prime Minister, Sitiveni Rabuka, continues to disrupt efforts at equality. His two successful coups have left the government worried about future overthrows. He has faithful followers who might carry out acts of violence. The Indian population, subject to threats of violence and lacking a voice in Parliament, has left Fiji in large numbers, taking with them successful professionals such as doctors, lawyers, and businessmen. This emigration leaves Fiji without a successful middle class.

The second problem concerns basic religious beliefs. Historically, the islanders' deities and the worship of them, including cannibalistic ceremonies, blended with Methodism making a third religion that some anthropologists explain by suggesting that the Fijians accepted Christian beliefs and transformed their native ones in order to get goods and services from European explorers. Cakobau, the Fijian king, stated that he embraced both religions in order to satisify all the gods.

The native Fijians traditionally accepted the power of their king transferred through his sons, not by elections. The government of Fiji in the 1990s is the result of British colonial rule and the parliamentary system of government does not appeal to many Fijians who wish to return to the old values and desire the return of all lands to native Fijians. The result of Fiji's discovery and colonization by the western world has resulted in a nation caught in the middle of conflicting values and trying to define itself in the twenty-first century.

Timeline

1500 B.C. Beginnings of civilization

The first people, most likely of Polynesian descent, arrive on what is known today as Fiji. Most are casual voyagers to the islands. Fiji, being a few days' sail from Samoa and Tonga, receives many such visits before any permanent settlements are established.

1290 B.C. Some permanent settlers arrive

The first permanent settlers to arrive are broad-nosed, light-skinned, Austronesian-speaking peoples. They arrive from nearby islands.

Proof of their civilization exists in their unique *lapita*-style pottery which is characterized by geometric horizontal bands. They eat pigs, chicken, rats, and dogs. They fish the waters. The islands are rich in timber for building shelter and bowls. There is no evidence of large-scale farming, but rather evidence of soil erosion from overfarming and frequent migration. Evidence of these communities can be found in the sand dunes near Singatoka.

500 B.C. Melanesian peoples arrive

People from other Polynesian areas arrive to settle permanently on Viti (Fiji). They are of a slightly different racial mix, but they intermarry and create what is recognized as the modern-day Fijian people. They adopt the surrounding areas' social structure based on male descent and the power of the chief who is called *turaga*. The chief is the physical representation of the spirit of their god or *vu*. They arrange their society into groups called *vanua* who are almost continually at war with one another.

These tribes practice cannibalism and often use the skulls of new victims to drink a ceremonial beverage they call kava or *yanggona*. They also often cook and eat other body parts of their enemies. Other rituals include the sacrifice of young men and women when a new house is constructed or a new boat is launched. Women are regarded as prizes or spoils of war.

The Fiji people are artistic and expressive. They sing, dance, and create works of art. They trade with nearby Tonga and Samoa.

A.D. 1643 Tasman discovers Fiji

Abel Tasman (1603–59), a Dutch sailor, circumnavigates these islands but does not actually land. Looking in vain for a large southern continent, he discovers that these islands are not attached to any other land mass. Tasman also discovers New Zealand and Tasmania, which is named for him.

1779 Captain Cook lands on Fiji

Captain James Cook (1728–79), an English mariner and explorer, actually drops anchor on Fiji near Vatoa, which he names Turtle Island. He does not venture further inland.

1789 Captain William Bligh chased by Fijian warriors

Captain William Bligh, of the ship *Bounty*, makes contact with the people but is chased and almost captured by the Fijians. In making his escape, he travels between the two main islands of Fiji for which route he is credited as being the first European explorer of Fiji. The section of water between the two islands becomes Bligh Water. Captain Bligh's crew later mutinies and sets him adrift in open waters. He survives and establishes trade in the area.

1792 Bligh returns to Fiji

Captain Bligh returns to Fiji, but he is afraid of the cannibals and does not leave his ship.

1804 American discovers sandalwood on Fiji

An American sailor from the wrecked ship *Argo* discovers the vast forests of sandalwood. Word of this discovery spreads, and beachcombers and traders begin to arrive on Fiji, eager to export and sell the fragrant sandalwood in Europe. Sandalwood is also prized by the Chinese for use in making incense sticks.

1814 Sandalwood trade declines

Within a decade, traders deplete the sandalwood forests and trade with Fiji declines.

1827–50 New trade in Beche-de-Mer revives interest in Fiji

The discovery of the *beche-de-mer,* an edible sea slug, revives interest in trade. Europeans enjoy beche-de-mer smoked and consider it a delicacy. Because the beche-de-mer has to be processed prior to shipment, traders set up permanent factories and become long-term residents of Fiji. Some Europeans

A traditional ceremony, Vilavila-I-Revo or "jumping into the fire" is still performed centuries later. Here, firewalkers walk on white-hot stones with bare feet. The firewalkers do not suffer burns on their feet. (EPD Photos/CSU Archives)

intermarry with Fijians. They are the founders of the Euro-Fijian race that remains today.

1830s–40s First missionaries arrive from Tonga

Methodist missionaries arrive in Fiji. They make some progress converting natives and are the first to translate the native Fijian language into writing. Some of the spellings they give are not exactly phonetic. Many missionaries are also victims of cannibalism.

1847 Tongans invade Fiji

Enele Ma'afu, of a royal Tongan family, exerts influence in Fiji. He challenges the Fijian ruler, Cakobau, by claiming that Christian power gives him the right to rule Fiji.

1850s Cakobau declares himself Tui Viti (king) of Fiji

Ratu (Chief) Seru Cakobau (b. 1811) becomes the main ruler of Fiji by allying with Enele Ma'afu. Together, they defeat the warring Fijian tribes.

1860s Cakobau puts down revolt

King Cakobau enlists the aid of approximately two thousand Europeans living in Fiji to help him win the support of Britain's Queen Victoria to assert his authority, and put down all rebellion.

1860s Cotton Becomes Major Crop in Fiji

When the U.S. Civil War (1861–65) brings cotton production in the United States to a halt, Europeans see an opportunity to establish cotton plantations on Fiji, from which they hope to make large profits. When the U.S. Civil War ends and cotton production is resumed in the U.S., the profitability of cotton enterprises in Fiji drops. When native Fijians lose their jobs on the cotton plantations, they begin warring and King Cakobau's reign is jeopardized.

1874: October 10 Cakobau cedes Fiji to Great Britain

Ratu Seru Cakobau, seeing that his nation is in danger of collapse, appeals to Britain to take Fiji into the Commonwealth.

(Commonwealth nations are independent countries that are tied economically to Britain.)

1875 Cakobau brings measles epidemic to Fiji

Cakobau and his brother visit Australia, where they contract measles which they bring back to Fiji. Having no immunity, the Fijians lose one-third of their population in this epidemic.

1877 First British governor arrives in Fiji

Sir Arthur Gordon (1829–1912) becomes the first governor of Fiji. He establishes his influence among the warring tribes. He passes a law that Fijians no longer have to work on cotton plantations. Gordon suggests the importation of Indians (from India) as laborers on newly formed sugarcane plantations.

1879 First indentured servants arrive from India

Many Indians begin arriving on Fiji and start working as indentured servants on sugarcane plantations. They agree to work five years in exchange for their passage from India. After their term of labor is over, many Indians stay on Fiji. This is the beginning of racial difficulties that develop between native Fijians and Indians which continue into the 1990s.

1904 First representative government on Fiji

For the first time, Fijians have a voice in their government. A Legislative Council is created consisting of Europeans and Fijians nominated by the Council of Chiefs.

1916 Indians are admitted to Legislative Council

Fijians of Indian descent are admitted for the first time to the Legislative Council.

1929 Reform of Legislative Council

A reform grants five seats on the Legislative Council to each community. However, the Legislative Council is an advisory body only; the governor still has complete control of government.

1944 Fijians insist on special government treatment

Difficulties between native Fijians and Indians intensify. Ratu Sir Lala Sukuna creates a separate administration for Fijians and fights for special treatment and access to land for native Fijians.

1941–1945 Fijians help Allies in World War II

Fijian troops fight on the Allied side (United States, Britain, and France) in the battles on Solomon Islands in the Pacific Ocean during World War II (1939–1945). Suva, the capital city, is the headquarters for the British Imperial Administration in the South Pacific.

1954 Ratu Sukuna forms Fijian Association

Sukuna asks the British government to support him in his demands agains the Indians. He does not want to grant equality to Indian residents of Fiji.

1954 Racial hatred between Fijians and Indians dominates politics

Many of the Indians who came to Fiji as indentured workers are now free. Most elect to remain on Fiji, and thus they become the dominant population of Fiji. The native Fijians resent this domination by people they view as outsiders. Fijians insist that Fiji should be governed by and for the Fijians.

The Fijians' traditional religious ideas are based on the land and on inhabitants' connection to their lands. These ideas reinforce beliefs that Indians should not have equal rights, since they are not connected by ancestry to the land of Fiji. Nevertheless, the majority population of Indians continues to push for more representation in government.

1963 Suffrage extended

For the first time, all women and native Fijians are given the vote.

1955 Dovi is first woman in parliament

Adi Losalini Dovi becomes the first woman member of parliament. She serves from 1966–70.

1970 Fiji Independence Medal awarded to Dovi

Adi Losalini Dovi is awarded the Fiji Independence Medal.

1970: October 10 Ratu Mara leads Fiji to independence

On the anniversary of the day that Cakobau ceded Fiji to the British, Ratu Mara declares Fiji an independent nation. The government becomes a republic.

1970s–80s Coalition governments control racial politics

Racial politics continue to dominate the Fijian government. The only way to maintain order is by a coalition government.

1985 Bavadra forms the Labor Party

Dr. Timoci Bavadra forms a new political party, the Labor Party, whose membership cuts across racial and religious lines.

1987 Coalition government wins re-election

Coalition has broad-based support. Most in the Labor Party want Fiji to become a multi-racial society in which everyone is equal and everyone is called a Fijian.

1987 Terrorist movement threatens peace

The extremist Fiji-for-Fijians *Taukei* (landowners) begin terrorist activities in protest of the coalition government. They barricade streets and threaten civilians. Also they organize protest marches and fire-bomb buildings.

1987: May 14 Coup d'etat ends coalition government

Lieutenant Sitiveni Rabuka enters the Parliament with armed troops and stages a coup d'etat ousting Prime Minister Bavadra. There is no bloodshed and Rabuka takes control of the government. Other nations refuse to do business with Rabuka. Publication of daily newspapers is suspended, and journalists are removed from their offices.

1987: May 14 United States suspected of being part of coup

Journalists and other observers accuse the United States of supporting the coup because the U.S. wants to maintain nuclear bases on Fiji. The United States Government denies any involvement and continues to maintain relations with Rabuka.

1987: July–August Fijians review the constitution

Prior to 1987 the Fijian government is composed of two houses of Parliament, with representation selected by native Fijians, Indians, Euro-Chinese, and the general population. In 1987 a constitutional review committee suggests a *unicameral* (one house) parliament with a Fijian as prime minister. The country is divided on accepting the new constitution. There are terrorist threats. The government, hoping to unite the people by establishing a Government of National Unity, prepares for new elections.

1987: September 26 Rabuka stages second coup

Fearful of losing his job, Rabuka proclaims himself head of state. He arrests hundreds of people opposing him. He closes the newspapers, imposes a curfew, and has many people tortured.

1987: October 7 Rabuka declares Fiji a republic

Rabuka has considerable support among native Fijians who detest Western involvement and who dislike the Indian majority. Rabuka backs Christian fundamentalism and the return to traditional Fijian beliefs. He bans all Sunday activities, including transportation and trading, and demands the conversion of all Muslims and Hindus.

1987: October 16 Fiji expelled from Commonwealth

Sensing the end of representative government, Great Britain expells Fiji from the Commonwealth of Nations. The Fijian governor general, Ratu Ganilau, becomes disgusted with Rabuka's abuse of power. He accuses Rabuka of terminating the rights of Indians and other non-Fijians.

1987: December 5 Rabuka appoints Ganilau and Ratu Mara

Rabuka sees that the terrorist government ruled by the Taukei is not working. He appoints the popular Ratu Ganilau as president and Ratu Mara as prime minister. They set a two-year guideline to return Fiji to a true representative government. The censorship and oppression of Rabuka's regime remain. Citizens fear that if Rabuka's wishes are not followed, he will stage a third coup.

1988 More oppression

The censorship of the press continues. People are blacklisted, and critics are prohibited from leaving the island.

1988 Coups ruin Fijian economy

Unrest in Fiji keeps tourism down. Foreign diplomats are expelled. Many Indian professionals emigrate to Canada, the United States, Australia, and New Zealand. The government issues a security decree allowing arrest and detention for any reason, the imposition of curfew laws, and even the right to shoot on sight any suspected person. In short, all the hallmarks of a democracy are gone.

1997 Government as of 1997

As of 1997, the Fijian government has a chief of state who is Ratu Sir Kamiese Mara. The actual head of the government is Prime Minister Sitiveni Rabuka who has engineered two governmental coups. The government is still advised by the Council of Chiefs and the cabinet, appointed by the prime minister. There is only one non-elective body of government, the Senate, with thirty-four seats. Twenty-four are reserved for ethnic Fijians, nine for Indians and others, one for the island of Rotuma. These are appointed by the president.

1997: August Fiji asks re-admittance to Commonwealth of Nations

Both the Prime Minister, Sitiveni Rakuba, and the National Federation Party leader, Jai Ram Reddy, petition Great Britain for readmission to the Commonwealth. Australia, New Zealand, and other Pacific Commonwealth nations agree that Fiji's constitutional reforms are good enough to justify readmitting Fiji to the Commonwealth.

Late 1990s Jai Ram Reddy Head of National Federation Party (Indian)

The two principal leaders of Fiji in the late 1990s, Sitiveni Rabuka and Jai Ram Reddy, represent the major political factions dominating Fiji. The struggle for dominance continues between native Fijians and Indians.

1997 Tourism is Fiji's main attraction

Tourism is Fiji's main industry. The warm climate and ocean reefs make it attractive to divers from all over the world. Even though Fiji is still struggling for a true democracy, its economy is increasing due to tourism. Turtle Island, first discovered by Captain James Cook, is now a luxury vacation resort.

1998: August 17 Vijay Singh wins golf tournament

Vijay Singh, a thirty-five-year-old Fijian of Indian descent, wins the Professional Golfers' Association (PGA) Championship at Sahalee Country Club near Seattle, Washington. Singh wins $540,000 in his first major tournament win.

Bibliography

Crozier, Brian. "Of Fiji, Race, and All That," *National Review,* December 4, 1987, 26.

"Fiji," *The World Almanac 1998,* Mahwah, NJ: World Almanac, pp. 763–64.

"Fiji Ends its Isolation," *The Economist,* May 17, 1997, 48.

Kaplan, Martha: *Neither Cargo Nor Cult*. Durham, NC: Duke University Press, 1995.

Lal, Brij V. *Broken Waves.* Honolulu: University of Hawaii Press, 1992.

Lal, Victor. *Fiji: Coups in Paradise.* London: Zed Books Ltd., 1990.

Ogden, Michael R. "Republic of Fiji," *World Encyclopedia of Political Systems,* 3rd edition, New York: Facts on File, 1999.

"Shameless in Fiji," *The Economist,* December 1, 1990, 34.

"Sorry About That: Fiji," *The Economist,* August 16, 1997, 33.

Stanley, David. *Fiji Islands Handbook.* Los Angeles: Moon Publications, Inc., 1985.

Vaughn, Roger. "The Two Worlds of Fiji," *National Geographic,* October l995, 114.

India

Introduction

The modern state of India came into being on August 14, 1947, "...at the stroke of the midnight hour...," in the words of its first prime minister, Jawaharlal Nehru (1889–1964). Britain was withdrawing from its colonial Indian empire, and Muslim demands for a separate state led to the partition of the Indian subcontinent between India and Pakistan. This change resulted in the Indus valley and the northwestern mountain region, as well as East Bengal, being assigned to the new state of Pakistan. The remaining lands, along with some former princely states and Portuguese and French colonies, make up modern India. Lakshadweep and the Andaman and Nicobar Islands, island groups lying in the Indian Ocean, are also part of Indian territory.

The Republic of India, with an area of 1,222,559 square miles (3,166,414 square kilometers), is just over one-third the size of the United States. Its population, however, is nearly four times that of the United States. With 952 million (1997 estimate) people, India has the world's second largest population, after China.

By the end of the twentieth century, the term "India" identified the modern political state. In the past, however, the name was used to refer to the entire Indian subcontinent. India is derived from "Sindhu," the local name for the Indus River (now in Pakistan). The Persians called the river Hindu, which passed to the ancient Greeks as Indos, which became India in Western usage. Here, the name India is used in the historical and geographical sense of the Indian subcontinent, unless reference is made to events after 1947 or indicated otherwise.

India is the largest state on the Indian subcontinent. (The subcontinent includes India, Bhutan, Nepal, Pakistan, and Bangladesh.) It stretches 1,900 miles (about 3,000 kilometers) from its mountainous border with China in the north to Cape Comorin, at the southern tip of the Indian peninsula. On the west lies the Islamic Republic of Pakistan, while some 1,800 miles (2,900 kilometers) to the east, India shares borders with China and Myanmar (Burma). India also has borders with Bangladesh and the Himalayan kingdoms of Nepal and Bhutan.

Geographically, India falls into three broad physical regions. The Himalayas, with many of the world's highest peaks (including Mount Everest at 29,028 feet [8,848 m]), create a formidable barrier in the north. They separate India from the Tibetan plateau, which forms a transitional zone where the cultures of South Asia and Central Asia meet. South of the mountains lie the alluvial plains of the Brahmaputra, Ganges and Indus Rivers. Except for the Rajasthan (or Thar) Desert in the west, the plains are densely populated and support the bulk of India's agriculture. The third region is formed by the uplands that constitute the Indian peninsula. The southern part of this region is known as the Deccan (literally "the South"), which is culturally quite distinct from the rest of the country. The Eastern and Western Ghats (mountains) separate the interior plateaus from the narrow coastal plains that fringe the peninsula.

Virtually every aspect of life in the region is influenced by the South Asian monsoon. The subcontinent experiences three distinct seasons. Winters are generally cool and dry, with bright, sunny skies and pleasant temperatures. From late February, however, temperatures rise steadily as the hot season approaches. Maximum temperatures in May and June may exceed 115°F (46°C) in the northwestern plains. Monsoon rains reach southwest India in late June and move northwards bringing relief from the high temperatures. Cherrapunji, in the northeast, with the distinction of being the wettest place on earth, averages nearly 450 inches (11,500 millimeters) of rain annually. The onset of the rains marks the beginning of the summer (*kharif*) growing season. Water is abundant, and the land comes alive with plants and animals. A bad monsoon can mean a poor harvest, food shortages, and hardships for many. The monsoon begins to retreat south towards the end of September, the rains die out, and temperatures drop as the cold season nears.

Although geography defines the Indian subcontinent as a broad physical region, the question is often raised about an underlying rationale for the post-1947 India to exist as a single political unit. For only a few times in the past, e.g. during the Mauryan or Mughal periods, has India existed as a single political entity. The subcontinent has more commonly been occupied by numerous regional kingdoms, each with its geographical base, unique cultural characteristics, and often a history of competing for political dominance. Even today, people tend to identify as much with Bengali, Tamil, Punjabi, or other regional cultural traditions as with an "Indian" one. The modern state of India faces political dynamics similar to those of past great empires—a central power attempting to exert its control over diverse peoples, cultures and regions that show a tendency to break away from the whole.

History

India's history has been essentially inward-looking, one of continually absorbing and assimilating foreign peoples and influences into a uniquely Indian culture. Peoples speaking Dravidian languages who entered the subcontinent in early times probably developed the Harappan civilization of the Indus valley. In modern times, these people are associated with the Dravidian languages and Dravidian culture of South India. Later, invading Aryan tribes settled northern India and are responsible for the dominant Aryan culture of the northern regions. The subsequent history of India is one of waves of invaders sweeping through the northwestern mountain passes onto the plains of India. The Persians, Greeks, Parthians, Kushans, and White Huns were among those who entered the subcontinent. Some groups left little imprint on the region. Others, such as the Muslim conquerors who entered India at the beginning of the eleventh century, added new dimensions to Indian civilization.

The Europeans reached the Indian subcontinent by sea, motivated by the rewards of trade with Asia. When the British East India Company gained the upper hand over its European competitors, another chapter opened in the history of South Asia. While the nature of British colonial rule continues to be the focus of much debate, the legacy of Britain is clear, from the widespread use of English to the development of transportation networks to a legal system based on British common law. Of no less significance are the democratic principles and political structures of British India's successor states, as well as their territorial outlines and the political problems arising out of the partition of the British Indian empire.

Society

There are, nonetheless, certain overarching structures that define Indian civilization. The most important of these is Hinduism, its beliefs, customs, and traditions. Some eighty percent of Indians are Hindus. Hindus worship many gods but believe in the supremacy of Shiva or Vishnu. A class of priests, the Brahmans, exists to perform rituals and the necessary rites of passage. Worship at temples and pilgrimage to holy places, such as Varanasi or the sacred Ganges River, is an important aspect of Hindu life. Hindu philosophy shares many concepts with Buddhism and Jainism, the other indigenous religions. Hindus believe the soul (*atman*) is caught in an endless cycle of rebirths (*samsara*). Actions in one's last existence determine the nature of the next incarnation (the law of *karma*). Righteous behavior (*dharma*) causes the soul to move through levels of existence towards salvation or total release (*moksa*) from the physical world. Practicing non-injury to living things (*ahimsa*), and especially the cow, which most Hindus view as sacred, is a means to this end. As a result of this philosophy, most Hindus are vegetarians.

Central to Hinduism is a social system that had its origins among the Aryan tribes who entered India during the second millennium B.C. Social distinctions already existed among the Aryans, but new structures evolved which integrated indigenous peoples into Aryan society. The resulting system is described in the *Veda* (literally "knowledge"), the collection of hymns and sacred works composed by the Aryan tribal priests beginning sometime between 1500 and 1200 B.C. Society was divided into four classes known as *varna*, the Sanskrit word for color. The highest ranked class were priests (*brahmana*), followed by warriors (*ksatriya*), peasant-farmers (*vaisya*), and serfs (*sudra*). Varnas form the basis of the caste system in India today, and many Hindus see caste as divinely ordained because of its Vedic origins.

Caste influences every aspect of a Hindu's life. One is born into a caste and, no matter what one's occupation or achievements, caste determines one's ritual status in traditional Hindu society. It defines social relationships, whom one marries, what foods one eats and (in the past) where one lived in a community and from which well one drank water. A fifth category was added to the four varnas—the "untouchables," people whose ritual standing is so low that physical contact defiles and requires purificatory rites by high caste Hindus. India's 1950 constitution legally abolished untouchability and instituted measures to eliminate caste inequities, but caste remains a significant factor in Hindu and, indeed, Indian society. Caste is so pervasive that even Muslims and Christians, whose religions espouse social equality, adopted some aspects of the caste system.

In addition to Hinduism and the caste system, Indian society is defined by its villages. Despite the presence of huge cities, such as Calcutta and Bombay, nearly three-quarters (73.2 percent) of Indians live in rural areas. In fact, because it is an occupational as well as a social system, the caste system is especially important in villages. Traditionally, individual castes performed specific tasks in the village economy, serving a *jajman* or landowner in a hereditary relationship that passed down through the generations. This jajmani system, an exchange of services for a share of the harvest, is being replaced by a cash economy, but is still significant in many rural areas. Most villagers, outside the service castes, derive their livelihood from agriculture. Wheat and other cereals are the staple crops in the drier northern and western areas, while rice dominates in the east and south. Legumes, pulses (edible seeds of plants with pods), oilseeds and vegetables are other important food crops. While beef is not eaten, cattle are kept for milk and draught purposes. Only non-Hindu groups such as Muslims, Christians, or tribals will eat chicken, goat, and other meat.

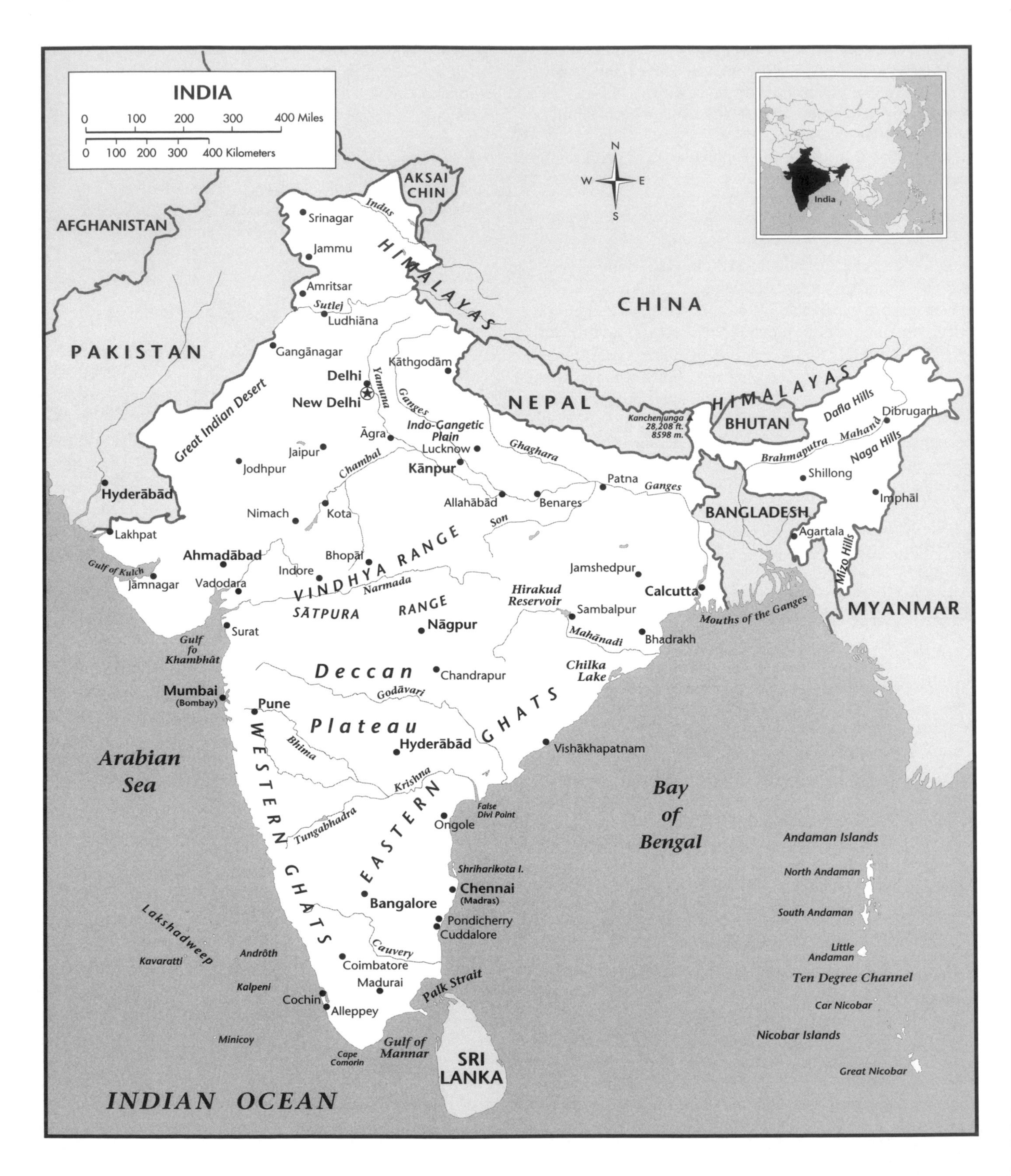

With a largely agricultural population, Indian demographic characteristics are typical of a developing country. By the 1990s the annual rate of population increase, though declining, was 1.9 percent. This means roughly 18 million new mouths to feed each year. The per capita Gross Domestic Product (GDP) is $350. Despite an emerging middle class, estimates of people living below the poverty line run as high as 40 percent. Inadequate sewage disposal, contaminated drinking water, and poor nutrition contribute to poor health. Infant mortality rates are high, numbering 78.4 deaths per 1,000 live births. Literacy stands at 48 percent, a figure that drops significantly among females and certain disadvantaged groups in rural areas.

Improving living conditions is only one of the complex tasks facing the Indian government today. India has a good mineral resource base and a thriving industrial sector. The sheer size of its population, however, and (until recently) restrictive government economic policies have impacted the rate of economic development. The problems of poverty and a sizable class of underprivileged citizens are compounded by periodic social unrest, including caste divisions, sectarian violence, and a tendency for some groups, such as the Sikhs, Tamils, and Gorkhas, to seek independence. Until the early 1990s, strong leadership in New Delhi succeeded in preserving India's political unity. Since that time, however, the weaknesses inherent in government by coalition resulted in an era of political flux. Issues such as a rising tide of Hindu fundamentalism, troubles in Kashmir, and the potential of a nuclear face-off with Pakistan make central government effectiveness even more significant for India's future.

Timeline

Pre-8000 B.C. Human evolution

The Indian subcontinent plays a yet undefined role in the biological evolution of humans. Bone fragments of *Ramapithecus,* a prehominid ape thought to be the oldest direct ancestor of modern man, indicate the presence some fifteen million years ago of this genus in the Siwalik Hills in what is now Pakistan.

Stone tools and artifacts reveal Paleolithic ("Old Stone Age") peoples living in the subcontinent as far back as perhaps 500,000–300,000 years ago. Archeologists distinguish between a northern Soan culture and the southern Madras culture of the peninsula.

Peoples and cultures evolve over the next several hundred thousand years. By around 10,000 B.C., humans little different from modern populations are found throughout the subcontinent.

c. 8000 B.C. Beginnings of agriculture

Environmental changes at the end of the Pleistocene Ice Age coincide with cultural advances that lead to the first great technological revolution in human history, the development of agriculture. Agriculture first appears in the Fertile Crescent, the band of hills bordering the Tigris-Euphrates valley, and extending into the mountains of Iran. From here, it is possible that ideas and practices associated with agriculture spread eastwards into the western borderlands of India.

6000–4000 B.C. Early farming communities in India

Agricultural settlements are well established in the uplands of Baluchistan that lie to the west of the Indus valley. Mehrgah, on the Bolan River, is the earliest agricultural village as yet discovered in South Asia and dates to perhaps 7,000 B.C. Its inhabitants live in mud-brick houses, cultivate cereals such as barley, and herd domestic sheep, goats, and cattle. Farming villages, cultivating rice, appear in the Vindhyan Hills, just south of the Ganges valley towards the end of this period.

c. 4000 B.C. Pre-Harappan settlements

Population moves from the uplands of Baluchistan onto the Indus valley plains, where pre-Harappan settlements such as Amri and Kot Diji appear.

c. 3200 B.C. Beginnings of urbanization

The first cities appear in the Indus valley, with sites like Amri and Kot Diji showing features associated with Harappan urban civilization.

c. 2500–2000 B.C. Harappan cities

Mohenjo-Daro and Harappa are the dominant cities of the Indus valley, though lesser towns, such as Chanhu-Daro and Kalibangan (in modern Rajasthan), share many of their characteristics. Urban centers are planned (and sometimes fortified), houses are constructed of fired bricks, and cities have public sewage systems. Granaries attest to agricultural surpluses, which in turn suggest an evolving social structure.

The discovery of what has been identified as a dock at Lothal, in Gujarat, indicates the importance of trade in Harappan culture. The Harappans have a standardized system of weights and measures. Seals, small squares of soapstone typically inscribed with pictures of animals and the yet-undeciphered Indus script, may be used for identifying packages and goods being traded. Some scenes depicted on seals appear to be of ritual significance and suggest religious dimensions to life in the Indus valley cities.

c. 2000–1500 B.C. Aryans invasions

Several waves of nomadic tribes enter India from Central Asia through the mountain passes of northwestern India. The invaders speak Aryan languages, they are pastoralists, and they introduce the horse and two-wheeled chariot to South Asia.

c. 1750 B.C. Harappan cities decline

Harappan civilization begins to break down around 1750 B.C. Cities decline in importance and eventually disappear altogether. Although the invading Aryans are blamed for this, the actual reason for this decline is more likely environmental factors that disrupt agricultural production and commerce.

c. 1500–600 B.C. The Vedic age

The nomadic Aryan tribes occupy the Punjab, the plains of northwestern India drained by the Indus and its five tributaries. Around 1000 B.C., they make the transition from pastoralism to settled agriculture and in the following centuries push their settlements southeastward toward the Ganges River. By the end of this period Aryan settlements exist in the central and eastern Gangetic plain.

Extant Vedic literature from this period preserves the songs and hymns composed by Aryan priests and handed down through generations by word of mouth. The earliest of the Vedas, the *Rigveda*, probably dates to around 1200 B.C. to a time when the Aryans were still centered in the Punjab. Other texts, such as the *Brahmanas, Upanisads,* and *Sutras*, date from the Late Vedic age and reveal social, economic, and political transformations in Aryan society during this time.

The Vedas, sacred books of Hinduism, describe religious beliefs, rituals, and the social structure that come to underlie the very fabric of Hindu civilization.

c. 600 B.C. Tribal kingdoms in the Ganges valley

Numerous small tribal kingdoms in the Ganges valley coalesce into sixteen major states (*mahajanapadas*). Urbanization begins along the Ganges valley.

c. 563–c. 483 B.C. Life of Buddha

Siddhartha, the founder of Buddhism, is born in Lumbini, a town (now in Nepal) on the central Ganges plain some 100 miles (180 kilometers) northeast of Varanasi. Son of a tribal king, Siddhartha belongs to the ksatriya or ruling warrior caste of Hindu society. Siddhartha leaves his home and family at the age of 29 years and embarks on the search for supreme knowledge. He finds Enlightenment at Bodh Gaya (in Bihar state) and is subsequently honored as the Buddha, the Wise or Enlightened One.

c. 517 B.C. Achaemenid empire

Darius, ruler of Persia, conquers Gandhara and subsequently pushes the eastern borders of the Achaemenid empire to the Indus River.

c. 500 B.C. Rise of Magadha

The kingdom of Magadha, with its capital at Pataliputra (the modern Patna), emerges as the dominant power in the Ganges plains. It subjugates its neighbors and eventually comes to rule most of northern India. Magadha reaches its greatest extent under the Nanda kings (361–321 B.C.), who conquer areas of central India and the eastern coastal plain (Kalinga).

?468–467? B.C. Mahavira attains *nirvana*

Mahavira, the great reformer of the Jain religion, dies (attains nirvana or salvation).

In Jain tradition, Mahavira is the twenty-fourth and most important in the line of *Tirthankaras* (teachers). Jains are strong believers in the concept of *ahimsa*, or non-violence to living creatures.

c. 400 B.C. Indian epics composed

Earliest probable date for the composition of the Indian epic poems *Mahabharata* and *Ramayana.*

327–325 B.C. Alexander invades India

Alexander the Great of Macedonia (356–323 B.C.) invades northwestern India with a small army of Greeks. He crosses the Indus River and defeats the Indian King Porus at the Battle of Hydaspes (Jhelum) in July 326 B.C.

321 B.C. Chandragupta founds Maurya dynasty

Chandragupta Maurya (or Sandracottus, c. 350–c. 250 B.C; r. 321–297 B.C.) seizes the throne of Magadha and founds the Maurya dynasty (321–185 B.C.).

The Mauryan kings rule over the first great empire in India's history. Chandragupta conquers the lands west of the Indus River ruled by the Greeks and their vassals and extends his territory as far as the borders of modern Iran.

c. 302 B.C. Megasthenes in India

Megasthenes (fl. 300 B.C.) is sent by Macedonian general Seleucus I (c. 358–281 B.C.) as ambassador to Chandragupta, emperor at the Mauryan court. Megasthenes's work, *Indica*, is the most reliable account of India known to the Greek world.

c. 290? B.C. Kautilya's *Arthasastra*

Kautilya, prime minister and advisor to Chandragupta, authors the *Arthasastra*, a famous treatise on political economy and statecraft.

c. 273 B.C. Asoka ascends throne

Asoka brings all but the extreme south of the subcontinent under Mauryan rule. No agreement on the exact dates of Asoka's era of rule exist. They are recorded by different sources as c. 265–238 B.C., c. 269–232 B.C., or c. 273–232 B.C. Asoka makes Buddhism the state religion and proclaims the principles of right conduct in edicts carved on rocks and pillars in strategic locations throughout the empire. He sends missionaries to the eastern Mediterranean, southern India, and Ceylon (Sri Lanka).

c. 250 B.C. Bactrian Greeks

Greeks establish an independent kingdom in Bactria on the plains of the Oxus River (the modern Amu Darya) north of the Hindu Kush mountains. The Bactrian Greeks cross the

Hindu Kush under Demetrius I (reigned c. 190–167) and establish an empire in northwestern India.

c. 185 B.C Shungas replace Mauryas

Pushmitriya Shunga, a Mauryan general, assassinates the last Mauryan ruler and founds the Shunga dynasty.

155–130 B.C. King Menander

Menander rules a powerful Indo-Greek kingdom with its capital at Taxila (Taksasila) in Gandhara (now in Pakistan). Menander is commemorated in Indian literature and his conversion to Buddhism is the subject of the Indian text *Milinda-pañha* or the "Questions of Milinda [Menander]".

c. 100 B.C. Scythian invasions

Scythians, nomadic peoples from Central Asia, invade the Indian subcontinent and conquer the Indo-Greek kingdoms in the northwest. They are known as Sakas in India but adopt Indo-Greek culture, using Greek names of the months and issuing coins in the Indo-Greek style.

1st century B.C. Rise of the central and eastern kingdoms

The Andhras (also called Satvahanas) of central India and the kings of Kalinga in eastern India challenge the political dominance of the northern Indian states.

A.D. 20 Parthians

Parthians (Persians) conquer the Saka lands in northwestern India and establish a short-lived kingdom. Gondophernes (reigned c. A.D. 20–46) is the most famous of the Parthian rulers. Tradition is that St. Thomas the Apostle lives at Gondophernes' court, and that the king becomes Kaspar, one of the three kings of the east, in Christian mythology.

c. 50–225 The Kushans

The Yüeh-Chih, known in India as the Kushans, are a nomadic people originating in China. They establish themselves in Bactria during the second century B.C. and from there conquer the Indo-Parthian kingdoms of Afghanistan and northwestern India. Further Kushan conquests create one of the greatest empires of its time, and Kanishka (r. 120–162) is regarded as a Buddhist ruler comparable with Asoka.

?100–300? Date of the *Natya-sastra*

Muni Bharata is the semi-legendary author of the *Natya-sastra,* the earliest Indian treatise on theater, dance, and music. The meaning of the word, Natya, combines dance and drama, since the two are inseparable in most eastern societies including India. Some scholars place the writing of this treatise as late as A.D. 1000, but the Natya rules he describes there are practiced for hundreds of years, perhaps since as early as 1500 B.C.

c. 150 Gandharan school of art

Buddhist culture and art flourish under Kanishka's patronage and the Gandharan school of art enters one its most creative phases. Gandharan art blends Indian Buddhist themes with Greco-Roman and Kushan influences reflecting facial structure, dress, and decorative motifs.

c. 174–203 Andhras pre-eminent in the Deccan

The Andhra kingdom reaches its zenith under Yajna, joining the Sakas and Kushans as the leading powers in India.

c. 255–385 Vakatakas supplant Andhras

The Vakataka dynasty rises to prominence in the Deccan. It blocks Gupta expansion south of the Vindhya Mountains but maintains cordial relations with the Gupta emperors. At times linked to the Guptas through marriage, the Vakatakas aid in the spread of Gupta culture in central and southern India.

320–510 Gupta dynasty

Chandragupta I (reigned 320–c. 330), a relatively unknown local chief in the kingdom of Magadha, establishes the Gupta dynasty. His son, Samudragupta (reigned c. 330–c. 380), earns a reputation as one of the great conquerors in Indian history. Samudragupta expands the Gupta domains to include all of the Ganges valley and subjugates the independent republics of the Indus plains. He campaigns successfully in southern India, making lands his tributaries as far south as the Pallava kingdom at Kanchipuram. At its greatest, the Gupta empire includes all the lands from Assam to Kashmir and northern Afghanistan as well as Samudragupta's southern conquests.

320–510 The Golden Age of the Guptas

Gupta achievements go far beyond the political integration of northern India. The Gupta emperors, and in particular Chandragupta II (reigned c. 380–415), preside over a flowering of Indian science, art and culture. Astronomy and mathematics reach new heights, and the decimal system of notation begins.

The Gupta age is the classical period of Sanskrit literature, the age of the dramatist and poet Kalidasa (?350–00?) and other writers. The *Puranas* (works belonging to the category of non-Vedic Hindu scriptures) begin to take shape at this time. They present traditional, orthodox Hindu concepts through myth, symbol, and philosophical dialogues. Hindu architecture and sculpture assume new directions, with the defining achievement of Gupta architecture being the development of the Hindu temple.

402 Fa-hsien enters India

Fa-hsien (or Fa Xian, fl. 400 A.D.), a Chinese Buddhist monk, crosses the Pamirs into India on a pilgrimage to the sacred places of Buddhism. His *Fo Kuo Chi* ("Record of Buddhist Kingdoms") is a valuable source of information on India in the fifth century.

455–c. 530 Huna conquests

The Hunas, or White Huns, invade the Punjab but are driven back west of the Indus by Skandagupta (r. 455–67). Hunas continue to mount incursions into India and eventually succeed in making the Guptas their vassals (c. 510). Although Hindu armies drive the Hunas from India in 528, the Guptas never effectively recover from the Huna conquests.

543–66 Rise of the Southern kingdoms

The mid-sixth century sees the emergence of the Chalukya kingdom with its capital at Badami (Vatapi) in Karnataka. At the same time, the Pallavas rise to prominence on the coastal plains south of Madras in modern Tamil Nadu. Their capital is Kanchipuram.

606–47 Harsha's empire in northern India

Harsha, also called Harsha-vardhana, subjugates the warring kingdoms that appear in the Ganges valley following the disintegration of the Gupta empire. His empire extends from the Ganges delta to the northern Indus plains, although the Chalukyas stop his expansion southward. Harsha is one of the great monarchs of his age. He is highly cultured, and patronizes Buddhism, promoting art, learning, and religion. Harsha's empire disintegrates rapidly after his death.

629–45 Hsüan-tsang's pilgrimage to India

The Chinese Buddhist monk, Hsüan-tsang (Xuan Zang, A.D. 602–64), embarks on a pilgrimage to the holy Buddhist sites in India. He reaches the Ganges valley in 633 and spends the next sixteen years traveling 40,000 miles to the most distant parts of the subcontinent. His travels become the subject of a novel *Xiyouzhi* (Monkey, 1593; English translation, 1942).

c. 630 Chalukyas defeat Harsha

The Chalukya defeat of Harsha confirms their position as the most powerful kingdom in the Deccan. During the late seventh and early eighth centuries, the Chalukyas are constantly at war with the Pallavas. The Chalukyas are orthodox Hindus, and during their reign Buddhism declines and is replaced by Brahmanical religious practices.

642–720 Pallava achievements

Pallavas achieve their political and artistic zenith. King Narasimhavarman captures the Chalukya capital of Badami in 642. He and his successors are patrons of the arts, and the famous Rock Temples and the relief, "Descent of the Ganges," at Mahabalipuram are completed during his reign. Narasimhavarman II (c. 680–720) builds the Shore Temple at Mahabalipuram. The distinctive southern style of temple tower (*puram*) is perfected.

711 Arabs conquer Sind

Arabs conquer Sind and the southern Punjab and annex these lands in the Indus valley to the empire of the Umayyad caliphs. These events provide Islam with a political base in India for the first time in the history of the subcontinent.

c. 750–70 Rise of the Palas of Bengal

Gopala (r. c. 750-70) establishes the Pala dynasty in Bengal. The Pala kingdom is stable and prosperous and emerges as a dominant political force in eastern India. The Pala kings are Buddhists, and monasteries, universities, and Buddhist culture flourish under their rule.

752–56 Rashtrakutas overthrow Chalukyas

Feudatories (vassals who hold land granted in return for services) of the Chalukyas rebel against their overlords and establish a new kingdom in the Deccan. The Rashtrakutas dominate central India for over two centuries.

788–836 Life of Shankara

Shankara, also called Shankaracharya, is the leading Hindu philosopher of medieval times. His teaching leads to a resurgence of Brahmanical Hinduism in India at the expense of Buddhism. He founds four great *maths* or monasteries at the limits of India, which continue to be important Hindu religious institutions.

c. 783 Rise of the Rajputs

The Rajputs clans rise to prominence in Rajasthan and northwestern India.

One consequence of the Huna invasions of the fifth and sixth centuries is the continuing migration of Central Asian tribes into India. Gurjara and Rajput tribes are among these groups. The Pratihara clan of the Gurjara tribe of Rajputs establishes itself at Ujjain, just north of the Vindhya Ranges in what is now western Madhya Pradesh State. Within a hundred years, the Gurjara Pratiharas rule a powerful confederacy of Rajput principalities in northern India.

c. 839–910 Gurjara Pratihara empire

Under Bhoja (reigned c. 836–885) and Mahendrapala (r. 885–910), the Gurjara Pratihara domains reach from the Arabian Sea to the mouths of the Ganges and from the Himalayas to the Vindhyas. The Rajputs of northwestern India block Arab expansion into India for several centuries.

The Rajputs illustrate the process by which India absorbs and assimilates alien peoples. They conquer the indigenous populations, establish kingdoms, and legitimize their position as rulers in Hindu society by adopting ksatriya status. All Rajput clan genealogies trace their ancestry back to Solar and Lunar lineages of ancient India or a legendary purification by fire on Mount Abu in Rajasthan. Specialized castes of bards and genealogists preserve the lineages of the Rajputs. Rajputs develop a culture based on valor, chivalry, and defense of the Hindu religion.

c. 897 Chola dynasty founded

Under King Aditya, the Cholas of the Kaveri (Cauvery) valley in southern India throw off Pallava rule and establish an independent Chola kingdom. The Cholas conquer the Pandya kingdom to the south, but expansion northward is halted by the Rashtrakutas.

939–68 Rashtrakuta dominance in the Deccan

The Rashtrakuta kingdom of the central Deccan is at its most powerful under Krishna III. The Rashtrakutas are the first southern power to rule virtually all of peninsula India and also extend their control north of the Vindhya mountains.

973 Later Chalukyas

Taila seizes the Rashtrakuta throne. He claims descent from the Chalukyas of Badami and founds a dynasty, the Chalukyas of Kalyani.

985–1044 Chola supremacy in the south

The Cholas attain imperial status, becoming the dominant power in south India. Rajaraja I (r. 985–1014) conquers the lands south of the Kaveri River, captures Ceylon, and even mounts a naval expedition to the Maldive islands in the Indian Ocean. Rajendra I (1014–44) gains fame by successfully campaigning as far north as the Ganges Delta. The Cholas are one of the few Indian powers to undertake maritime expansion, conquering the Shrivijaya empire in Southeast Asia in 1025.

Chola rule is a period of great prosperity for southern India and sees the crystallization of Tamil culture. Economic and social institutions, such as the famous temple towns, the trade guilds, Chola administrative systems, religious developments, temple art, and architecture, all come to be seen as the standard of south Indian culture.

1001–26 Mahmud of Ghazna invades northwest India

Mahmud of Ghazna (971–1030), raids far and wide on the north Indian plains, devastating Mathura, Kanauj, and other Indian cities. He sacks the famous Shiva temple at Somnath in Gujarat in 1024 and reportedly carries off 6.5 tons of gold. To Hindus, Mahmud is the very embodiment of wanton destruction and Muslim fanaticism.

Terra Cotta Horses

Terra-cotta horse shrines are frequently seen in Hindu villages. The horses may be three inches tall (as seen in neighboring Nepal) or up to twelve feet or more in the rural shrines of the Indian state of Tamil Nadu, where the sculptures may qualify as the largest found anywhere on earth.

In Tamil Nadu, the horse provides transportation for the god Ayanar, who is believed to represent two main Hindu gods. Ayanar, who is half Shiva and half Vishnu, is created when Shiva is visited by Vishnu in the form of an incredibly beautiful woman named Mohini. Shiva is so excited by Mohini's beauty that he spills his seed. From that seed grows Ayanar. There are hundreds of Ayanar shrines throughout Tamil Nadu, usually some distance outside the village in a grove of trees. The shrines vary, but many consist of a terra cotta horse and a plain stone or small terra-cotta figure representing Ayanar.

c. 1100 Ramanuja and the *bhakti* revival

The teachings of the Tamil philosopher, Ramanuja, lead to a revival of Vaishnavism, the sect of Hinduism that worships the god Vishnu. This period also sees a resurgence of the *bhakti* cults, a devotional movement dedicated to the worship of Krishna (an incarnation of Vishnu).

1179–1205 Last Hindu king of Bengal

Lakshmana Sena is the last Hindu ruler of Bengal.

1186–1206 Ghurid conquest of north India

Muhammad Ghuri defeats the Ghaznavids, captures the Punjab, and advances towards the Ganges. He defeats a confederacy of Hindu Rajputs under Prithviraj Chauhan at the second battle of Tarain (1192). He captures Delhi the following year and quickly extends Afghan rule over much of northern India and the central Ganges valley. Muhammad Ghuri is assassinated in 1206.

1206 The Delhi sultanate

Qutb-ud-Din Aybak (d. 1210), Muhammad Ghuri's viceroy in India, declares his independence. He found the Mamluk or Slave King dynasty (1206–90) of the Delhi sultanate. A Muslim power based in the Delhi area is to rule in northern India for the next six and a half centuries.

1221–22 Mongol incursions

Genghis Khan and his Mongols pursue a fugitive prince into the northwestern plains and advance to the Indus before withdrawing. The Mongols continue to mount raids deep into India during the remainder of the thirteenth century.

1266–86 Delhi sultanate under Balban

Balban's court is a refuge for Muslim rulers, scholars, saints, and artists displaced by the ravages of the Mongols in Persia, Iraq and Syria.

1293 Marco Polo

The Venetian traveler Marco Polo (1254–1324) visits India on his return journey by sea from China. He calls at the ports of Kovalam, Calicut and Cambay.

1296–1306 Mongol attacks repulsed

Jalal-ud-Din (r. 1290–96) and Ala-ud-Din (r. 1296–1316), the first two rulers in the Khalji dynasty of Delhi sultans (1290–1320), repel several Mongol invasions.

Ala-ud-Din continues the Muslim campaigns against the Hindu Rajputs of Rajasthan, capturing the Rajput strongholds of Ranthambhor (1301) and Chitor (1303).

1309–11 Ala-ud-Din conquers south India

Ala-ud-Din embarks on the conquest of south India and brings virtually all of the Indian peninsula under Khalji control.

1320–88 Tughluq dynasty of Delhi sultans

Tughluq rulers replace the Khaljis as Delhi sultans. Muhammad ibn Tughluq (r. 1325–51) builds a new joint capital at Daulatabad in central India so as to better control his south Indian lands. When he returns from Daulatabad to Delhi again, his empire begins to disintegrate. The sultanate of Malabar (1334), Bengal (1337), the Vijayanagar kingdom of south central India (1346), and the Bahmanis all break away from Tughluq rule.

Firoz Shah (r. 1351–88) is the last important Tughluq sultan.

1334 Ibn Batutah arrives in Delhi

Ibn Batutah (1304–68 or 1369), the Muslim traveler from Tangier, arrives at the Tughluq court at Delhi. He spends several years at Delhi, before being appointed the sultan's envoy to China in 1342.

1398–99 Timur invades India

Timur (1336–1405), also known as Tamerlane, invades the Indian subcontinent from his base in Afghanistan. He conquers Multan and the Punjab and destroys Delhi. The destruction wrought in Delhi is so devastating that the city remains unoccupied for many years.

1440–1518 Mystical poet Kabir

Kabir, a mystic and a poet, attempts to bridge Hindu and Muslim thought, preaching the unity of all religions and the equality of all men. His teachings give rise to the Kabirpanth and other religious sects.

1450 Lodi dynasty established

The Lodis (1450–1526), of Afghan origins, replace the Sayyids (1414–50) as Delhi sultans and begin the process of restoring sultanate power and prestige. The Delhi sultanate emerges as the most powerful state in north India in the late fifteenth and early sixteenth centuries. Sikander Lodi (1489–1505) moves his capital from Delhi to Agra.

1469 Birth of Nanak, founder of Sikhism

Nanak (1469–1538) is born in the Punjab. He founds the Sikh religion, becoming the first of ten *gurus* (teachers) revered by the Sikhs (*Sikh* comes from the Sanskrit word for "disciple"). Nanak's teachings combine elements of Hinduism and Islam. While they accept many basic concepts of Hindu philosophy, they are monotheistic, deny the role of priests, and advocate a classless society. Sikhs are concentrated in the Punjab region.

1498 The Portuguese land in India

The Portuguese explorer, Vasco da Gama (1460–1524), arrives at Calicut, on India's southwest coast, the first European to reach India by sea.

1505–29 Vijayanagar dominates the south

Under King Krishnadeva Raya, Vijayanagar rises to a position of pre-eminence in south India. Krishnadeva Raya subdues the Deccan sultans and acquires the coastal lands of Andhra from the Hindu Gajapatis. By the mid-sixteenth century, virtually the entire south is under Vijayanagar hegemony.

1510 Portuguese conquer Goa

Afonso de Albuquerque captures Goa, whose natural harbors and location on the sea routes to the Far East make it of great strategic significance to the Portuguese.

1526: April 21 Babur establishes Mughal empire

Babur (Zahir-ud-din Muhammad, 1483–30) defeats and kills Ibrahim Lodi, the Sultan of Delhi, at the traditional Indian battle field of Panipat.

Babur, a descendant of both Genghis Khan and Tamerlane, is forced to flee from his kingdom of Ferghana (Fergana) in Central Asia. He establishes himself in Afghanistan and from there invades India. Although Babur is vastly outnumbered at Panipat, his use of artillery wins the day. After his victory, he assumes the throne of Delhi and becomes the first of the Mughal (a corruption of "Mongol") emperors (r. 1526–30).

In addition to founding one of the greatest empires in Indian history, Babur is an accomplished poet and man of letters. His legacy includes an autobiography, *Babur-nameh* (Memoirs of Babur).

1540 Sher Shah defeats Humayun

Sher Shah, an Afghan adventurer, defeats Babur's son Humayun (r. 1530–40; 1555–56) in battle and founds the Sur dynasty. Sher Shah dies in 1545, however, and squabbles over the succession weaken his dynasty.

1555: July Humayun regains his throne after years in exile

Humayan spends fifteen years in exile before the Persian Shah grants him military aid to regain his throne. He returns to India to defeat Sher Shah's successors. Humayan dies in Delhi in January 1556, after falling down the staircase of his library.

1556–1605 The Mughal empire under Akbar

During Akbar's reign, his conquests bring all of northern India under Mughal rule. He reduces Rajput opposition through marriage alliances and military conquests, until only Mewar State in Rajasthan opposes Mughal rule. Many Rajputs serve with distinction in the Mughal armies. Akbar conquers the Mirza sultans of Gujarat and extends Mughal control into eastern Afghanistan. Bengal falls in 1576, and the southern boundary of the empire is pushed south to the Godaveri River.

Akbar's principal achievement is the creation of the empire's administrative structure. He reforms the revenue system, basing new assessments on actual productivity, and puts in place an efficient mechanism for collecting taxes. He reorganizes the nobility and the army, with every Mughal official holding a *mansab* (rank) which determines pay and the extent of his military obligation to the emperor. Akbar follows a policy of religious toleration and even attempts (unsuccessfully) to create a new religion. He enlists Persians, Indian Muslims, and Hindu Rajputs into the imperial service. Akbar abolishes the *jizya*, a tax imposed by Islamic rulers on non-Muslim subjects.

Akbar is also a patron of the arts. He maintains a large studio at court where painters from Persia train Indo-Muslim and Hindu artists. Mughal painting is an important tradition in the art of India.

1569–84 Fatehpur Sikri

Akbar builds a new capital west of Agra near Sikri. The city has been described as one of the boldest achievements in the history of world architecture. Fatehpur Sikri is abandoned in 1584, probably because of a lack of water, and today stands deserted.

1600: December 31 British East India Company receives royal charter

Elizabeth I, Queen of England, signs a royal charter founding the East India Company. The purpose of the company is to exploit the lucrative spice trade of the East Indies (the islands of Southeast Asia). Thwarted in Southeast Asia by the strong Dutch presence, the East India Company subsequently turns it attention to South Asia.

1605–27 Jahangir's reign

Akbar's son, Prince Salim, assumes the throne under the name Nur-ud-Din Jahangir. Jahangir ably manages the affairs of the empire. He puts down a rebellion by his son, subdues an Afghan uprising in Bengal, and finally defeats the Rajput Rana of Mewar. His armies continue to fight in the Deccan.

Jahangir marries a Persian who introduces Persian culture and Persian artists to the court. Under Jahangir's patronage, court artists produce portraits and naturalistic studies of birds, animals and flowers that are among the masterpieces of Mughal art.

1612 British base at Surat

The British establish their first foothold on the Indian subcontinent at the port of Surat, in Gujarat State.

1616–18 Embassy of Sir Thomas Roe

Sir Thomas Roe, ambassador to the court of Jahangir, obtains official permission for the British East India Company to trade in the Mughal empire.

1627–58 Shah Jahan

Shah Jahan continues the tradition of the Great Mughals, campaigning in the Deccan and subduing Golconda and Bijapur. His efforts to retake Babur's old lands in Central Asia are unsuccessful, and future Mughal expansion is directed toward the south.

1643: February Taj Mahal completed

Twenty thousand artisans spend eleven years on the banks of the Yamuna River at Agra completing the mausoleum Shah Jahan builds for his beloved wife Mumtaz Mahal. Built of white marble and decorated with an inlay of semi-precious stones, the Taj stands as the masterpiece of Mughal architecture.

Other structures built by Shah Jahan include the Jama Masjid and Red Fort at Delhi and the Pearl Mosque at Agra.

1646–80 Shivaji, the Maratha leader

Shivaji (1627–80), a Maratha chief, establishes himself in hill forts in the Western Ghats in the Konkan (the coastal and western areas of Maharashtra State). From here, he challenges the rule of the Mughals, harassing imperial forces and striking out to sack cities such as Surat (in 1666 and again in 1670). Shivaji lays the foundations for a Hindu state in western India.

By 1674, Shivaji is powerful enough to have himself crowned king with all the ritual and ceremony of a great Hindu ruler. The Marathas are in constant conflict with the Mughals, and Maratha light cavalry raids as far south as Tanjore. Maratha horsemen come to be feared across India by Hindu and Muslim alike.

1658–1707 The empire under Aurangzeb

Aurangzeb (1618–1707) extends the empire to its furthest limits. He conquers the sultans of the Deccan and overruns virtually the entire India peninsula. At the time of his death, Aurangzeb's lands stretch from Afghanistan to Bengal, and only Assam and the southern tip of India lie outside his domains.

The Mughal empire, however, is so vast as to be almost unmanageable. Aurangzeb defeats the Rajputs who are in revolt again and puts down an uprising among the Sikhs, but internal disturbances continue to strain Aurangzeb's resources. The Marathas, under Shivaji, remain a festering problem. Aurangzeb's own policies also contribute to the empire's downfall. He abandons the religious tolerance of his predecessors and attempts to enforce orthodox Islamic rule. He destroys Hindu temples despite Hindu opposition, favors Muslims over Hindus, and reimposes the jizya tax on non-Muslims.

1690 British at Calcutta

The British build Fort William near the village of Kalikata on the east bank of the Hooghly River in Bengal. The settlement founded at this swampy, malarial site grows into the city of Calcutta, second in commercial importance only to London in the British Empire.

1699 Sikhs organize as militaristic sect

Gobind Rai (1666–1708), the tenth Sikh Guru, organizes the Sikhs into a fighting fraternity known as the *Khalsa.* Aurangzeb's persecution of the Sikhs at the end of the seventeenth century leads to a lasting enmity between Sikhs and Muslims.

1707–19 Decline of the Mughals

Following Aurangzeb's death, a series of weak, short-lived emperors preside over the disintegration of the Mughal empire.

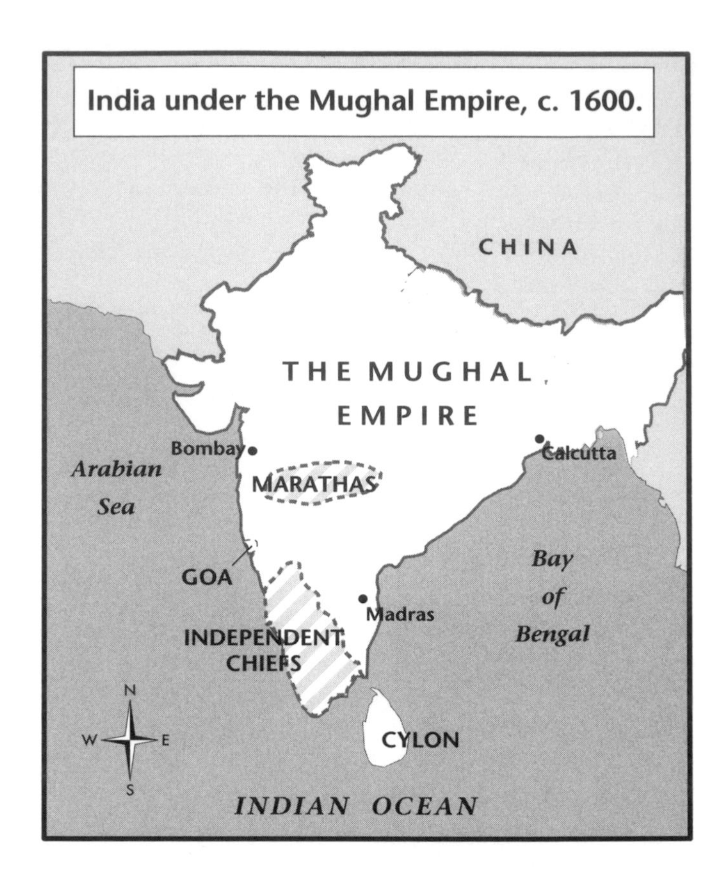

1720–40 Peshwas extend Maratha rule over north India

The Peshwa (chief minister) emerges as the supreme political and military leader of the Marathas. By the mid-eighteenth century, a powerful Maratha Confederacy rules or holds as vassals nearly all the lands of central India. Marathas capture Delhi briefly and continue raiding as far afield as the Punjab, the Ganges delta, and the southern peninsula.

1737: October 11 Cyclone ravages Calcutta

A cyclone strikes Calcutta along the banks of the Ganges River. Powerful winds and floodwater wash away ships, buildings, and an estimated 3,000 people.

1738–39 Nadir Shah of Iran invades subcontinent

The Shah of Iran invades India, overruns the Punjab and sacks Delhi. He carries off the Peacock Throne of the Mughal emperors and the famous Koh-i-Nur diamond.

1742–63 Carnatic campaigns

The Carnatic, the southeast coastal region of India, becomes a battleground between the French and British East India Companies. Competition for commercial domination leads to complicated and constantly changing political alliances between Europeans and local rulers. The French and British attack each other's settlements, with varied success. The French, under Joseph François Dupleix, capture Madras in 1746 (though it is restored to the British in 1748). Robert

Bibi-ka-Muqhara is built as a mausoleum for Rugia Daurani, the wife of Aurangazer. The white-domed marble edifice is built near Aurangabad, India, to rival the Taj Mahal at Agra. (EPD Photos/CSU Archives)

Clive seizes Arcot, capital of Nawab of Carnatic (an ally of the French) and successfully withstands a lengthy siege. The British defeat of the French at the Battle of Wandiwash in 1760 ends the French dream of an Indian empire. The French retain their colony of Pondicherry, south of Madras.

1747–62 Afghan invasions

Afghans exert pressure on the Mughals' northwestern frontier, with Ahmad Shah Abdali (Durrani) seizing the provinces of Kabul, Lahore, and Multan. The Afghans inflict a serious defeat on the Marathas in 1761 at the Third Battle of Panipat near Delhi, halting Maratha expansion to the northwest. They then proceed to sack the Mughal cities of Delhi and Agra.

1756: June 20 Black Hole Incident

The Nawab of Bengal captures Calcutta, and confines his British prisoners in the local jail. Because of overcrowding, no water, and a lack of ventilation, many prisoners die by the morning. The East India Company sends Robert Clive (1725–74) from Madras to retake Calcutta.

1757: June 23 Battle of Plassey

A British force, under Clive, defeats the Nawab of Bengal and his French allies at the Battle of Plassey. This battle traditionally marks the beginning of British rule in India.

1761 The rise of Mysore

Haider Ali seizes the throne of Mysore and within a short time rules all of southern India except for the Carnatic. Haider Ali is a formidable leader and challenges the British, fighting several wars with them. Continuing conflict with the Marathas to the north restricts his ability to inflict a decisive defeat on the British in the south.

1765 The Bengal *Diwani*

Company forces defeat a Mughal army at the Battle of Buxar (1764). The following year, the Mughal emperor grants the

British the Bengal *diwani* (finance ministry). This gives the British the permanent right to collect taxes and establishes them as the rulers of Bengal, Bihar, and Orissa.

1770 Bengal famine

Contemporary estimates suggest that up to ten million people die in a famine in Bengal.

1772–1833 Life of Ram Mohun Roy

Ram Mohun Roy is an important social reformer from Bengal who challenges aspects of traditional Hinduism such as idol-worship, the caste system, and widow-burning. He proposes a modern, Western curriculum be taught in schools to prepare Indian students for modern life. He founds the Brahmo Samaj (Society of Brahma), a Hindu reform sect, in 1828.

1773 Warren Hastings becomes Governor-General

Warren Hastings becomes the first Governor-General of Bengal. Although he is impeached by his critics in the British Parliament in 1785, during his tenure Hastings lays the foundations of British dominion in India.

1793 Permanent Settlement of the Bengal Revenue Act

Lord Charles Cornwallis reforms the tax system in the Bengal Presidency. This continues the system introduced by the Mughals but recognizes the hereditary tax collectors (*zemindars*) as actual owners of the land from which they collect taxes. The *zemindari* system of north India contrasts with the *ryotwari* system of southern areas, in which peasants pay taxes directly to the state.

1799 Final defeat of Tipu Sultan

The British defeat and kill Haider Ali's son, Tipu Sultan, at Seringapatam. Areas of Mysore State are annexed by the British and the Nizam of Hyderabad. This marks the beginning of British territorial dominion in south India.

1801 Ranjit Singh establishes Sikh kingdom

Ranjit Singh (1780–1839) declares himself Maharaja in the Punjab, creating a powerful Sikh state that extends from the Sutlej River to Kashmir.

1813 Missionaries enter India

A new East India Company charter permits Christian missionaries to enter India. Before this time, the Company had barred missionaries from the territories it controls.

1814–16 Anglo-Gorkha War

Gorkha incursions into India from Nepal, in the Himalayas, lead to an invasion of Nepal by British forces. After a hard-fought conflict, the Gorkhas sue for peace and cede large areas of their territory to Britain.

1818 Final defeat of Marathas

Following the First (1775–83) and Second (1803–05) Maratha Wars, the British finally succeed in subduing the Marathas. The Third Maratha War (1817–18) ends with the British deposing the Peshwa and annexing his territories.

1829 Law prohibiting *sati*

The Indian government passes a law banning the practice of *sati*. This custom, prevalent among the ksatriya castes, requires a widow to burn herself alive on the funeral pyre of her husband.

1839–42 First Anglo-Afghan War

Fearing an Afghan alliance with the Russians, the British invade Afghanistan and install their candidate, Shah Shoja, as *amir.* The Afghan tribes rise up in rebellion, and the disastrous British retreat from Kabul in 1842 counts among the worst military disasters in British military history.

Britain and Afghanistan confront each other again in the Second Anglo-Afghan War (1878–80) and in the Third Anglo-Afghan War (1919), King Amanullah declares his independence.

1843–49 Britain annexes Sind and Punjab

Britain extends the borders of its Indian empire west of the Indus by conquering first Sind (1843) and then the Sikh state of Punjab, following the First (1845–46) and Second (1848–49) Sikh Wars.

1853 The first railroad opens

The first stretch of railroad line opens between Bombay and Thane. The British subsequently build an extensive rail system in the subcontinent, for strategic as well as economic reasons. Into the 1990s, the Indian rail system is one of the most heavily traveled in the world.

1857: May 10 Indian Mutiny

Indian troops at Meerut kill their British officers and march to Delhi, where they proclaim Bahadur Shah, the last of the Mughals, as emperor. *Sepoys* (native soldiers) throughout the Bengal army, with some Muslim and Hindu princes, rise up against the British. Europeans in cities across the Ganges plains flee to safety or are killed. Lucknow withstands a prolonged siege and is finally relieved by British forces.

1858 End of "John Company"

British forces retake Delhi, bringing the mutiny to a close, and exile Bahadur Shah to Burma. The events of 1857–58 bring about changes in British rule in India, not the least of which is that the British Crown takes direct control of the country's government. This ends over 250 years of the British East India Company's involvement in India.

1867 Press and Registration of Books Act

According to the terms of a new law, the Press and Registration of Books Act of 1867, every printer in India is required to send copies of all new books to the government. This effort begins the establishment of national libraries.

1869: November Suez Canal opens

The Suez Canal, linking the Mediterranean and Red seas, officially opens. The new route shortens the sea voyage between Britain and India by roughly 5000 miles (8000 kilometers).

1872 Census established

The first census enumerates population in British India. Censuses are later carried out in some of the princely states and eventually throughout the entire subcontinent. The decennial census of India is an important source of demographic and socio-cultural information on the peoples of India.

1875 Aligarh College founded

Sir Syed Ahmad Khan (1817–98), a leading Muslim educationalist, founds the Muhammadan Anglo-Oriental College at Aligarh in Uttar Pradesh. It later becomes Aligarh Muslim University.

1882 Theosophical Society's headquarters in India

Russian Madame Helena Blavatsky (1831–91) and Henry S. Olcott (1832–1907) establish the headquarters of the Theosophical Society near Madras. Theosophy, a pantheistic philosophical-religious system blending Eastern and Western ideas, attracts a strong following in India. Annie Besant (1847–1933), a British social reformer who becomes a leader of the Indian independence movement (see 1917), is the society's president from 1907 to 1933.

1885 Indian National Congress founded

The Indian National Congress meets for the first time in Bombay. Initially a forum for Indians to express political ideas, Congress evolves into the major vehicle for Indian nationalism and, after independence, is India's dominant political party.

1889 Ahmadiya sect founded

Mirza Gulam Ahmad (1835–1908) founds the modern Ahmadiya sect of Islam in the Punjab. At the end of the twentieth century, members of the sect are found mostly in Pakistan. The term "Ahmadiya" also designates various Sufi orders of considerable antiquity.

1905 Partition of Bengal

The British government partitions Bengal. The new Bengal Province consists of western Bengal, Bihar, and Orissa, while Eastern Bengal and Assam merge to form a separate province.

1906: December 30 Muslim League founded

Muslims meet in Dhaka in Eastern Bengal and found the All-India Muslim League, the first India-wide Muslim political organization in the subcontinent.

1909 Morley-Minto reforms

Separate electorates for Muslims is a leading principle of the constitutional reforms put in place by the Indian Councils Act (the "Morley-Minto reforms"). These reforms conferred greater power upon the councils.

1910 Aurobindo establishes ashram

Aurobindo Ghose (1872–1950), seer, poet, nationalist, and philosopher, flees British India for Pondicherry. He establishes an *ashram* (retreat) where he teaches his Vedantist philosophy and attracts an international following.

1911: December Delhi Durbar

King George V (1865–1936) of England visits India and is crowned King Emperor of India at an imperial *durbar* (reception) held just north of Delhi. The king announces that India's capital is to be moved from Calcutta to Delhi.

1912–29 Building of New Delhi

Sir Edwin Lutyens (1869–1944) and Herbert Baker (1862–1946) plan and design the new capital at Delhi, which is conceived as a monument to British imperialism in India.

1912 First silent films produced

French filmmaker and inventor, Charles Pathe (1863–1957), brings his newsreel cameras to India in the early 1900s. The first Indian films are made, including the famous *Raja Harishchandra* by Dhundiraj Phalke. Since filmmaking technology is in its infancy, Phalke looks at individual frames of his films by holding the film up to a window between his fingers.

1913 Tagore wins Nobel Prize

Rabrindranath Tagore (1861–1941), the Bengali poet and writer, is the first Asian writer to win the Nobel Prize for Literature.

Tagore is a member of a wealthy Hindu family, and manages his family estates for nearly twenty years. During this time, he collects Hindu legends that he later uses in his writing. His first work, *A Poet's Tale*, is published in 1878 when Tagore is seventeen. He receives a knighthood in 1915 but resigns it four years later in protest of British policies in India. Among his notable works are *Binodini,* published in 1902 and considered the first modern Indian novel; *Gitanjali*, a volume of verse; *Chitra*, a play; *The Religion of Man* (1931); and *Farewell My Friend* (1940).

1914–18 World War I

Over one million Indian troops fight for Britain in Europe and Mesopotamia. At first, Indians show support and loyalty for Britain in its war effort. As the war progresses, however, a wave of riots and revolutionary activity leads to the repressive measures of the Defence of India Act of 1915.

1916: December Lucknow Pact

Mohammed Ali Jinnah (1875–1948), head of the Muslim League, and Bal Gangadhar Tilak (1856–1920), the Congress leader, arrange an agreement for Hindu-Muslim unity in the struggle for independence.

1917 Annie Besant elected president of Indian National Congress

English theosophist and activist, Annie Besant (1847–1933; see 1882), is elected president of the Indian National Congress, a post she holds until 1923.

1918 Ramanujan elected to the Royal Society of London

Srinivasa Ramanujan (1887–1920), recognized by many as a mathematical genius, becomes the first Indian elected to the prestigious Royal Society of London.

1918 Influenza deaths

More than seventeen million Indians die in the influenza pandemic. Most regions of the subcontinent experience a decline in population between 1911 and 1921.

1919: April 13 Massacre at Amritsar

British troops, under General Richard E. H. Dyer, open fire on 10,000 Indian demonstrators in Amritsar, in the Punjab. They are trapped in an area known as Jallianwalla Bagh, which has only one exit. Official British reports estimate the death toll at 329, with 1,200 wounded.

1919 The Khilafat movement

Muslims in India start the Khilafat movement. Its purpose is to restore the Turkish sultan to power after Turkey's defeat in World War I. The Turkish sultan has been the Caliph, the leader of the world's Sunni Muslims, for over two centuries.

1920: April 7 Sitar master Ravi Shankar is born

Rabendra (Ravi) Shankar is born in Benares, the youngest of five sons of Dr. Shyama Shankar, a Bengali Brahman, and Hemangini Shankar. Ravi and his brother, Uday, are influential in creating international exposure for the music of India.

In 1947, three of the five Shankar brothers—Ravi, Rajendra, and Debendra—form a cultural group called the India Renaissance Artists (IRA). During his long career, Shankar works with jazz saxophonist, Bud Shank; violinist, Sir Yehudi Menuhin; flutist, Jean-Pierre Rampal; minimalist composer, Philip Glass; and conductors, Zubin Mehta and Andre Previn. He teaches former Beatle George Harrison to play the sitar. Harrison incorporates what he learns into the songs he writes for the Beatles and for his solo performances. In 1967 Shankar performs for an audience of about 50,000 at the Monterey Pop Festival in Monterey, California. George Harrison calls Shankar "the Godfather of World Music."

1920: May–June Gandhi outlines his non-cooperation movement

Mahatama Gandhi (1869–1948) begins his campaign of non-cooperation with the British in India.

Mohandas Karamchand Gandhi, also called the Mahatma "Great Soul," is born October 2, 1869, in Porbander, in Gujarat. He studies law in England and goes to work in South Africa. He challenges the South African government's racial policies by openly disobeying laws that enforce discrimination. He coins the term, *satyagraha* ("holding on to the truth"), to describe his tactics of passive resistance to the authorities. Gandhi returns to India in 1915 and emerges as the head of the Indian Congress. His policy of non-violent civil disobedience becomes the hallmark of the nationalist campaign for India's independence.

1921 Birth of filmmaker Satyajit Ray

Satyajit Ray (1921–92) is born in Calcutta. He graduates from Santiniketan University and begins a career as an artist at an advertising agency. With the support of the government, he completes his first film in 1955. Set in rural India, *Pather Panchali* (On the Road), wins acclaim at the Cannes Film Festival in France that year and becomes the first of the so-called Apu trilogy that also includes *Aparajito* (The Unvanquised, 1956), and *Apu Sansar* (The World of Apu, 1959).

During the 1960s, his efforts include documentaries and film adaptations of folktales; in the 1970s he turns his attention to political issues. He depicts famine in *Ashanti Sanket* (Distant Thunder, 1973), and lack of ethics in business in *Jana-Arnaya* (The Middle Man, 1975). Two other works are *Ghare-Baire* (The Home and the World, 1984) and *Agantuk* (The Stranger, 1992). He dies in 1992.

1922: March 10 Gandhi is arrested

Gandhi (see 1920: May–June) is arrested, tried for sedition, and sentenced to six years in jail. The British release him on grounds of ill-health after he serves two years.

1930: March Gandhi's Salt March

The Indian government has a monopoly on the production and sale of salt, which it taxes heavily. Gandhi (see 1920: May–June) organizes a well-publicized protest march from

Ahmedabad to the beach at Dandi, in Gujarat. There he picks a grain of salt, a simple but symbolic act that defies the laws of the colonial government.

1930 C. V. Raman wins Nobel Prize for Physics

Sir Chandrasekhara Venkata Raman (1888–1970) wins the Nobel Prize for Physics for his discovery that light changes wave lengths when it passes through transparent materials. The process bears his name: the Raman effect. His other works include an investigation of the physics of musical instruments.

1931 First Indian film made

The first Indian talking film is made in Bombay. The Indian movie industry, based in Bombay, grows to be the largest (in terms of output) in the world. The bulk of the films are popular movies made in Hindi, although regional film industries flourish elsewhere in India. Satyajit Ray (1921–92; see 1921), the Bengali filmmaker, wins several international awards for his films.

1935 Government of India Act

The Government of India Act establishes India as a federation, comprising the provinces and the princely states. The central government, i.e. the British government, remains in overall control. The act also separates Burma from British India.

1936 Kabaddi demonstrated at Olympics

The Indian game, *kabaddi,* is demonstrated at the 1936 Olympics in London. The game is played by two seven-person teams. One team sends a player to invade the other team's space. The invader constantly chants the word, "kabaddi," derived from a Hindi word that means "holding breath." The invader tries to accomplish his mission in one breath. The Olympic committee did not elect to add the game to its official competitions.

1937: February Congress wins elections

In the first provincial elections held under the Government of India Act, 1935, the Congress Party forms governments in seven out of eleven provinces.

1938 Uday Shankar founds India Cultural Center

The dancer and choreographer, Uday Shankar, founder of modern Indian ballet, establishes the India Cultural Center in Almora, Uttar Pradesh. Shankar's adaptations of western theatrical technique to traditional Indian dance forms help popularize Indian dance on the international stage. Shankar earns an international reputation producing Hindu plays and ballets and assists the Russian ballerina, Anna Pavlova, in developing and performing her ballet, "Radha-Krishna." He tours, visiting London, England; Vienna, Austria; Berlin, Germany; Budapest, Hungary; and Geneva, Switzerland. In 1931, he organizes a dance troupe in India and tours the United States.

1939–45 World War II

The armed forces of India play an important role in Britain's fight against the Axis powers. Indian troops serve in North Africa and Southeast Asia.

1939: October Congress governments resign

Congress provincial governments resign in protest because Indians are not consulted before British India declares war against Germany.

1940: March 23 Lahore Resolution

Jinnah asserts at the Muslim League's meeting at Lahore that the Muslims of India are a separate nation and demands a separate Muslim state.

1942: August Quit India movement

Following the failure of the Cripps mission to persuade Congress politicians to support the British war effort, Congress launches its "Quit India" movement. The British government arrests and jails Mahatma Gandhi (1869–1948), Jawaharlal Nehru (1889–1964), Vallabhai Patel (1875–1950), and others in the Congress leadership. Many younger nationalists undertake violent, subversive activities against the government.

1943 Bengal famine

As many as three million people die in Bengal in the worst famine in modern times. The famine is dramatically presented on the screen in the evocative film, *Ashanti Sanket* (Distant Thunder), directed by Satyajit Ray (see 1921) in 1973.

1943: October Bose organizes Indian National Army

Subhas Chandra Bose (1897–1945), also known as *Netaji* (Respected Leader), organizes Indian troops held prisoner of war by the Japanese to fight against the British. The Indian National Army (INA) sees action in Burma.

1946: August 16 Hindu-Muslim riots

A day of protest (Direct Action Day) called by the Muslim League is marred by communal violence, with particularly bloody killings in Calcutta. Hindu-Muslim riots spread to other regions of the Indian subcontinent and continue until independence.

1947: February 20 Mountbatten becomes Viceroy

Clement Atlee announces in the British House of Commons that Lord Louis Mountbatten is to become Viceroy of India.

He also states the government's intention to hand over power in India no later than June, 1948.

1947: June 3 Partition announced

The British government announces that India will be partitioned between Muslims and Hindus.

Independent India

1947: August 15 Inauguration of India as a Dominion

Celebrations mark the inauguration of the independent Dominion of India. Mountbatten remains as Governor-General. Nehru continues as prime minister and leader of the new government of India.

Jawaharlal Nehru (1889–1964), also known as *Pandit* (Teacher) Nehru, comes from a prominent Kashmiri Brahmin family. He receives his education in England, where he attends Harrow School and Trinity College, Cambridge, before qualifying as a barrister at London's Inner Temple. On his return to India, Nehru meets Gandhi and becomes active in the Congress Party. Gandhi appoints him president of the Congress, and by the early 1930s he is seen as Gandhi's successor. Nehru spends more than nine years in jail as a result of his activities as a leader of the nationalist movement. Nehru is strongly committed to socialist principles and a policy of Indian neutrality in world affairs.

1947: August Radcliffe Awards

A storm of violence sweeps the country, and particularly the Punjab, when the Radcliffe Awards, announcing the new Indo-Pakistan frontiers, are made known. The historical and cultural regions of Punjab in the northwest and Bengal in the east are partitioned between India and Pakistan. Sikhs in the Punjab, in particular, are outraged at the dismembering of their homeland. In one of the largest migrations in history, anywhere from twelve to fifteen million people flee their homes seeking safety in Muslim Pakistan or Hindu India. Estimates place the number of deaths during this period of communal violence as high as one million.

1947: August Accession of princely states

The India Independence Act of 1947 requires the rulers of over 600 princely states in British India to accede to either Pakistan or India at the time of independence. By and large, this transition occurs with little problem. However, India has to annex the states of Junagadh (November 1947) and Hyderabad (August 1948) by "police actions."

1947: October 26 Kashmir accedes to India

The Mahajara of Kashmir formally accedes to India.

Kashmir, one of the most important states in India, is ruled by a Hindu, but its population is largely Muslim. Both Pakistan and India see its accession as vital because of its strategic location in the Himalayas. Hoping to retain his independence, the Maharaja of Kashmir initially declines to join either Pakistan or India. Pathan tribesmen from Pakistan invade Kashmir to free their Muslim "brothers." The Mahajara seeks help from India, and accession is the price of Indian military support. Pakistan refuses to accept the accession, claiming that Kashmir's population is largely Muslim, and sends troops to confront the Indian forces in the state.

1948: January 30 Gandhi assassinated

Nathuram Godse, a young Hindu radical, shoots and kills Mahatma Gandhi (1869–1948; see 1920: May) in Delhi. Godse is unhappy at Gandhi's position on turning over Pakistan's share of the treasury of British India.

1948 India wins gold at London Olympics

The Indian men's field hockey team wins the first of three consecutive gold medals at the Olympic Games in London, England. Field hockey, cricket, and soccer are popular sports in India.

1949: January 1 Kashmir cease-fire

The United Nations arranges a cease-fire in Kashmir. A proposed plebiscite to allow Kashmiris to decide their future never takes place. Kashmir remains divided between India and Pakistan, a flash-point in relations between the two countries.

1950: January 26 Republic of India

India's constitution comes into effect, creating a democratic, federal republic with a parliamentary form of government. India is instituted as a secular state. Rajendra Prasad becomes India's president and Nehru continues as prime minister.

The constitution is based to a large part on the 1935 Government of India Act, yet the framers of the constitution draw freely on the examples of other countries. A preamble, a list of fundamental rights, and the powers and function of the Supreme Court all mirror the United States constitution; India's federal system draws on the experience of Canada and Australia, and Ireland provides the model for a list of directive principles of national policy.

1950 Union and state governments

The constitution clearly defines the new federal structure of India.

The Union, or central government, consists of the Executive (headed by the president), the Judiciary, and the Legislature. Parliament comprises the president and two legislative houses, the *Lok Sabha* and *Rajya Sabha.* The Lok Sabha, the 542-member lower house, is elected on the basis of universal adult suffrage in the states and union territories (plus two nominated representatives of the Anglo-Indian community).

The Rajya Sabha or Council of States consists of members elected by state legislatures as well as some appointed by the president.

The state governmental structure mirrors the central government, although most states have only a unicameral (single) legislative body. The state chief minister fulfills the role of prime minister at the center. The state is headed by a governor, with broad discretionary powers, who is appointed by the president of India.

The central government exercises a dominating role in its relationship with the states. This role is best exemplified in the "President's Rule," which gives the central government powers to dissolve legally-elected state governments. Some prime ministers, especially Indira Gandhi, use President's Rule for party politics rather than constitutional reasons.

1950 National Planning Commission established

Nehru establishes the National Planning Commission to direct the course of India's economic growth. India develops a "mixed economy," with the state controlling certain industries and other activities being left in the hands of the private sector. The first five-year plan, based on the Soviet model, is implemented in 1952–56.

1950 Mother Teresa begins her work with the poor

Agnes Gonxha Bojaxhiu (August 26, 1910–September 5, 1997), an Albanian nun, founds the Missionaries of Charity in Calcutta. Taking the name of a sixteenth-century nun, Mother Teresa devotes the rest of her life to ministering to the poor, sick, and dying in the slums of India's largest city. Mother Teresa becomes a figure of international standing and is awarded the Nobel Peace Prize in 1979.

1950s–60s Indian music popularized

Tours by sitarist Ravi Shankar and Ali Akbar Khan, who plays the *sarod,* popularize Indian music around the world.

1952 First general elections

Congress wins an overwhelming victory in the first national elections held in an independent India, confirming Nehru's supremacy on the Indian political scene.

1952 National Forest Policy enacted

The National Forest Policy, India's first environmental preservation action, is enacted. The policy is supported by India's first prime minister, Jawaharlal Nehru, who has sought to preserve India's dramatic natural environment.

1955 Ali Akbar Khan tours with violinist Yehudi Menuhin

Ali Akbar Khan, who plays the sarod, tours the United States and United Kingdom with violin virtuoso Yehudi Menuhin. Khan appears on U.S. television and records "Music Of India: Morning and Evening Ragas," the first album of Indian classical music.

1955–56 The "Hindu Code"

The government passes a series of acts reforming laws governing Hindu marriage, succession, guardianship, and adoption. This legislation reflects Nehru's interests in social reforms and modernizing traditional Hindu law.

1956 First nuclear reactor

The first nuclear reactor in South Asia begins operating at Trombay, just outside Bombay.

1956 Buddhist conversions

Led by Dr. Bhimrao Ranji Ambedkar (1893–1956), an untouchable and former cabinet minister, 200,000 untouchables convert to Buddhism in Nagpur in central India.

Born near Bombay, Ambedkar receives his education in Bombay and at Columbia University in New York and the London School of Economics in England. He becomes a member of the Bombay Legislative Assembly and leads the 60 million untouchables of India in efforts to improve their living conditions. Ambedkar is appointed Law Minister of India in 1947 and is the principal author of the Constitution. Among his writings is *Annihilation of Caste,* published in 1937.

1956–66 States reorganization

State boundaries are reorganized to more closely reflect linguistic patterns. Boundaries of states in the south are adjusted in 1956, Bombay province is split into Gujarat and Maharashtra states in 1960, and Haryana is separated from Punjab in 1966.

1957: October Wildlife week inaugurated

The Indian government steps up efforts to preserve the environment and its native plants and animals. An annual wildlife week is inaugurated to call attention to environment issues, especially related to endangered species.

1959: March Dalai Lama flees to India

The Dalai Lama, along with 100,000 followers, seeks asylum in India following an abortive Tibetan revolt against Chinese communists. The Dalai Lama is the head of the Yellow Hat (Gelugpa) sect of Tibetan Buddhism, and spiritual and temporal leader of Tibet. This leads to worsening of already poor relations between India and China.

1959 First TV in Delhi

The first television transmissions in India are broadcast in Delhi. Doordarshan (literally "seeing from afar"), the Indian television company, is controlled by the Indian government which strictly regulates programming.

1960: September Indus Waters Treaty

India and Pakistan sign the Indus Waters Treaty which resolves the dispute over the use of the waters of the Indus tributaries caused by location of the international boundary in 1947. Supported by considerable international financial assistance, the two countries reach agreement only after prolonged negotiations mediated by the World Bank.

1961: December Goa liberated

Indian troops invade and occupy Goa, the Portuguese enclave on India's west coast. Goa becomes a state of India in 1962.

1962 Border war with China

Increasing tensions along India's border with China, along with Chinese construction of a road through the Indian territory of Aksai Chin in the northwest, leads to war with China. India is ill-prepared for the conflict and is humiliated by China, whose forces penetrate to the plains of Assam before withdrawing.

1963 Indianization of Indian press

An Indian group buys *The Statesman,* thus completing the Indianization of the Indian daily press.

1964: May 27 Nehru dies

Nehru dies and is succeeded as prime minister by Lal Bahadur Shastri (1904–66).

Shastri is born in Benares and joins Gandhi's independence movement at the age of sixteen. He is imprisioned seven times by the British for his pro-independence activities. He becomes a minister for the railways in Nehru's government in 1952, minister of transport in 1957, minister of commerce in 1958, and home secretary in 1960. He dies while in the Soviet Union for discussions on ways to mediate tensions with Pakistan (see 1966: January 11).

1965: September War with Pakistan

Pakistan launches Operation Grand Slam on September 1, aimed at capturing Kashmir. India counters by crossing the Punjab border on September 6, inflicting heavy casualties on Pakistani armor and occupying several areas of Pakistani territory. The United Nations (UN) arranges a cease-fire on September 23.

1966: January 11 Shastri dies and Indira Gandhi comes to power

Prime Minister Shastri dies in the Soviet Union the day after signing the Tashkent Accord with Ayub Khan, ending the war over Kashmir.

The Congress old guard chooses Indira Gandhi (1917–84), Nehru's daughter, to succeed Shastri. They see her as a weak candidate and select her to forestall the prime ministerial ambitions of Morarji Desai (1896–1995).

1967 Language Bill continues government use of English

National legislation continues the use of English in government in areas where Hindi is not spoken.

Linguistic diversity is a problem in India, where there are some fifty major regional languages, and the census identifies a total of 1,653 languages, dialects, and their variants. Attempts to impose Hindi as a national language face opposition from non-Hindi speakers. The Indian constitution recognizes fifteen official languages (Assamese, Bengali, Gujarati, Hindi, Kannada, Kashmiri, Malayalam, Marathi, Oriya, Punjabi, Sanskrit, Sindhi, Tamil, Telugu, and Urdu).

c. 1968 Green Revolution begins

The central government introduces high-yielding, hybrid varieties of wheat, rice, and other grains that produce dramatic changes in the country's agriculture. Dubbed the "Green Revolution," this expansion more than doubles India's food production between 1965–66 and 1984–85 and makes India self-sufficient in food.

1969 Congress split

Mrs. Gandhi is expelled from the Congress Party but breaks away from the more conservative group (now called Congress [O]) to form her own branch of Congress. She calls national elections in 1971 and her Congress (R) party wins a two-thirds majority in the Lok Sabha.

1970 Sulabh Movement founded to improve sanitation

Sociologist Bindeshwar Pathak founds the Sulabh movement to promote the goal of installing toilets for 100 percent of India's population. Only about 30 percent of those living in urban areas have access to toilets. Almost no sanitation facilities exist in rural areas. The Sulabh Movement recruits volunteers to install toilets. By the late 1990s, they have installed about 700,000 private pour-flush toilets. In addition, they install public toilets which cost people one rupee to use.

1971 Princes' privy purses abolished

Mrs. Gandhi eliminates the government subsidies, privileges and formal status which the rulers of India's princely states received when they merged their territories with India at the time of independence.

Every six years, devout Hindus engage in a mass bathing ceremony in the waters of the Ganges River, considered holy. (EPD Photos/CSU Archives)

1971: August 4 Indo-Soviet Friendship Treaty

India and the Soviet Union sign a Treaty of Peace, Friendship and Cooperation. India develops close relations with the Soviet Union as a balance to Pakistan's accord with China.

1971: December 4 Bangladesh's liberation begins

India intervenes in Pakistan's civil war, invading East Pakistan after the Pakistani Air Force launches pre-emptive airstrikes against Indian airfields. Fighting a holding action against the Pakistanis in the northwest, the Indian army advances on Dhaka and on December 16 forces the surrender of its Pakistani defenders. East Pakistan becomes the independent nation of Bangladesh.

1972 Men's field hockey team wins Olympic medal

At the summer Olympics, the Indian men's field hockey team wins the bronze medal.

1972 Sanctuaries created to protect animal and bird species

The government declares certain animals and birds, including tigers and elephants, protected. "Project Tiger" creates eleven sanctuaries, forests, and national parks for tigers exclusively.

1974: May 5 India explodes nuclear device

India joins the nuclear club when a nuclear device explodes in the Rajasthan desert.

1975: June 26 Mrs. Gandhi declares an emergency

Mrs. Gandhi moves to end challenges to her authority by requesting the president of India to declare an emergency.

The energy crisis, a series of bad harvests, high inflation, and a national strike of railroad workers in 1974 result in a serious domestic crisis in India. In 1975, a court declares invalid Mrs. Gandhi's election to the Lok Sabha in 1971 because she uses government facilities in her campaign. Mrs.

Gandhi uses the economic crisis as a rationale for imposing the Emergency, which is the only constitutional way she can remain in office.

Opposition to Mrs. Gandhi and the Emergency grows, led by such venerable political figures as Jayaprakash Narayan. Mrs. Gandhi counters this resistance by imprisoning opposition leaders.

1977: March Gandhi defeated

Opposition parties form a coalition and campaign as the *Janata* (People's) Party to contest national elections called by Mrs. Gandhi. The Janata Party wins a majority in the Lok Sabha by putting up only one candidate to challenge the Congress candidate. Morarji Desai becomes prime minister.

1980: January Mrs. Gandhi's return to power

Against all odds, Mrs. Gandhi wins the elections and returns to power. She now heads her own branch of the Congress Party, the Congress (Indira) Party.

1980: June Sanjay Gandhi dies

Sanjay Gandhi (b. 1946), Mrs. Gandhi's younger son who is being groomed to succeed his mother, dies when the stunt plane he is piloting crashes on the outskirts of the capital. Rajiv Gandhi (1944–92), the prime minister's elder son, enters politics and subsequently wins in a landslide election victory in Sanjay's constituency.

1982–84 Regional unrest

The Indian government faces widespread disturbances, protests, communal violence and terrorism in many parts of the country. In Assam, local people protest against Bengali immigrants; Kashmir faces disturbances linked to local elections; caste violence breaks out in Bihar and Gujarat; and violence between Sikhs and Hindus continue in the Punjab, where Sikh extremists are demanding a separate Sikh state (Khalistan).

1984: June Golden Temple action

Mrs. Gandhi orders the Indian army to storm the Golden Temple in Amritsar, where Sikh extremists led by Jarnail Singh Bhindranwale are entrenched. Bhindranwale and many others are killed in the fierce fighting that follows.

The Golden Temple is the most sacred place in the Sikh religion, and many Sikhs are outraged at what they see as desecration of the temple.

1984: October 31 Mrs. Gandhi assassinated

Two Sikhs of Mrs. Gandhi's own bodyguard shoot and kill the prime minister in her garden in revenge for the attack on the Golden Temple. The killing is followed by attacks against Sikhs in Delhi and other north Indian cities.

President Giani Zail Singh, himself a Sikh, sets aside all parliamentary conventions and immediately installs Rajiv Gandhi as prime minister. He knows that no other Congress leader can stand as a national symbol at this critical time. Gandhi later is elected leader of the Congress Party.

1984: December Congress wins decisive victory at polls

General elections to the Lok Sabha gives Rajiv Gandhi and the Congress the largest parliamentary majority in Indian history, with 403 out of the 513 contested seats.

1985: January New economic policy

Rajiv Gandhi announces major changes in economic policy. The government is to loosen controls on foreign trade to stimulate exports, lessen the protection of domestic industries from foreign competition, reform the management of public sector enterprises, and target corruption and inefficiency in the bureaucracy.

1986: June Mizos sign peace accord

Leaders of the Mizo National Front (MNF) sign an agreement with the central government ending the twenty-five-year Mizos rebellion in northeastern India.

1987: February New states

Mizoram and Arunachal Pradesh, in the northeast, are officially admitted as the twenty-third and twenty-fourth states of India. The Union Territory of Goa becomes India's twenty-fifth state in May.

1987: July India, Sri Lanka sign accord

India and Sri Lanka sign an accord by which an Indian Peace Keeping Force (IPKF) is to disarm Tamil guerillas in northern Sri Lanka who are fighting for independence from the Sinhalese dominated government. What is meant to be a short police action turns into a protracted war that costs India its men, money, and reputation. The last Indian forces withdraw from Sri Lanka by May, 1990.

1989: November 22–26 Congress lose elections

Congress loses its overall majority in the Lok Sabha. Rajiv Gandhi resigns as prime minister.

1989: December 2 V. P. Singh forms minority government

Vishwanath Pratap Singh (b. 1931), former Congress (I) minister and leader of the coalition National Front, forms a minority government with the parliamentary support of the Communist parties and the Bharatiya Janata Party (BJP).

1990: August–October Mandal Commission riots

Violent demonstrations occur across northern India in protest against the government's decision to implement the recommendations of the Mandal Commission. The Commission proposes raising from 22.5 percent to 49.5 percent the quota of government and public-sector jobs reserved for deprived communities and the lower castes.

1990: November Chandra Shekar becomes prime minister

The government loses a vote of confidence in parliament, and V. P. Singh resigns. Chandra Shekar, leader of a breakaway socialist faction of the Janata Dal party known as Janata Dal (S), forms a minority government with the support of Congress (I).

1991: May 21 Bomb kills Rajiv Gandhi

A suicide bomber kills Rajiv Gandhi while he is campaigning at a political rally for parliamentary elections near Madras in Tamil Nadu state. The woman who carries the bomb is apparently associated with a Tamil separatist group.

1991 P. V. Narsimha Rao forms Congress government

Congress wins a convincing victory in national elections. P. V. Narsimha Rao becomes the first Congress prime minister in twenty-five years who is not a member of the Gandhi family.

1992: October Babri Masjid mosque destroyed

In Ayodhya, Uttar Pradesh, Hindu fanatics tear down the Babri Masjid mosque, believed to be built on the birthplace of the Hindu god Ram. This incidents sets off riots in cities across northern India. Followers of the BJP, a fundamentalist Hindu party, are prominent in inciting Hindu animosity toward the Muslims.

1996 India participates in Olympic games

At the summer Olympics, Indian tennis player, Leander Paes, wins the bronze medal in singles. This is the first Olympic medal for an individual competition won by an Indian athlete since 1952. Leander's father had been a member of the Indian men's field hockey team that won a bronze medal in the 1972 Olympic games.

At the Winter Olympics in Nagano, Japan, India's only competitor is Shiva Keshavan, age sixteen. Shiva has been competing in the luge for just one year. In 1995, the International Luge Federation recruits athletes from Asia, and Shiva is among them. Since he was born near the Himalayas, Shiva is an accomplished skier as well.

1996: May–1997: March Three coalition governments

Following general elections the BJP is the largest single party in the Lok Sabha. The president asks Atal Behari Vajpayee, the BJP leader, to form a government. Vajpayee, however, lacks support in parliament, and his government resigns after only thirteen days in office. No party wishes to see another round of elections, so a fourteen-party coalition led by the Janata Dal forms the United Front government under H. D. Deve Gowda. Within a year, Congress withdraws its support and brings about the resignation of the Gowda government. In March 1997 a sixteen-party United Front coalition forms a government with Inder Kumar Gujral as prime minister.

1996: November Two airplanes collide near New Delhi

Over 300 people die when a Saudi Arabian Airlines Boeing 747-100 airplane and a Kazakstan airline aircraft collide in midair near New Delhi. The cause of the crash is believed to be navigational or air traffic control error.

1997: April 21 Gujral assumes leadership of coalition

Inder Kumar Gujral becomes the twelfth in the country's history to lead the government. He heads a coalition government that replaces the previous coalition government that lost support of the All India Congress Party. Gujral pledges stability in the government and expresses commitment to speeding up economic reforms.

Inder Kumar Gujral was born and educated in Pakistan, migrating to India in 1948. He served as information and broadcast minister and envoy to Moscow in the government of Indira Gandhi. He resigned from both government service and the Congress Party in 1980.

1997: May Narayanan becomes president

The first president of India to come from a low caste, Kocheril Raman Narayanan, is elected by members of the national and state legislatures. Narayanan belongs to the Dalit community, representing the lowest rank in the Hindu social hierarchy.

1997: September 5 Mother Teresa dies

Mother Teresa (1910–97) dies in Calcutta. Contrary to her wishes, the Indian government gives her a state funeral that is televised around the world. (See 1950.)

1997: September 29 India launches space rocket

For the first time in the history of its space program, India places a satellite in orbit atop an Indian-built rocket, the PSLV-C1. The rocket lifts off from the Shah Launch Centre at Shriharikota in Andhra Pradesh State.

1997: December **Narayanan dissolves parliament**

Just seven months after taking office, president Kocheril Raman Narayanan dissolves parliament and calls for an election in February 1998. This action is triggered by the Congress Party's withdrawal of support of the coalition, United Front.

1998: March 20 **BJP government at the helm**

Once again, Congress withdraws it support from the United Front government. As a result of national elections in February, the BJP emerges as the largest single party in the Lok Sabha, though without an absolute majority. The BJP forms a coalition government again led by Vajpayee.

1998: May 11–13 **India explodes nuclear devices**

India explodes a series of nuclear devices, a response to Pakistan's successful tests of the Ghauri intermediate range missile. Despite appeals by the international community, Pakistan conducts a series of nuclear tests of its own, thus becoming a member of the world's nuclear club. These events cause great concern, given the enmity between India and Pakistan, for the potential for nuclear conflict in the Indian subcontinent.

1998: May 13 **United States sanctions**

U.S. president Bill Clinton announces economic sanctions against India. These sanctions are required by a U.S. law prohibiting U. S. economic and military aid to countries developing nuclear weapons.

1998: June **Cyclone hits Gujurat state**

A cyclone hits Gujurat state in northwestern India, killing more than 1,000 people and affecting more than 5 million. Power is lost in over eighty percent of Gujurat. The hardest-hit area is the port city of Kandla on the western coast of India, just north of Bombay. Buildings, machinery, ships, cargo, pipelines, and railway tracks are destroyed and trees and utility poles are uprooted there. The cylcone drives waves over nearby salt pans and shanty settlements, destroying them and washing them into the ocean. (Salt pans are shallow pools where sea water is collected and evaporated to produce salt.)

1998 **Violence against Christians on the rise**

United Christian Forum for Human Rights, a human rights group, reports ninety incidents of violence against Christians, including rape, Bible-burning, and church vandalism.

1998: October 20 **Sitarist Anoushka Shankar releases first album**

Anoushka Shankar, daughter of internationally renowned sitarist Ravi Shankar, releases her first album featuring five tracks of classical sitar music composed by her father.

1998: December **State elections result in Congress Party gains**

The Congress Party wins victories three key state elections—Madhya Pradesh, the largest Indian state; Delhi; and Rajasthan. Sonia Gandhi, widow of Rajiv Gandhi, leads the Congress Party and may become the next prime minister. The Hindu nationalist Bharatiya Janata Party (BJP) loses voter support in part because of the rising cost of food, notably the staple of the Indian diet, onions.

Bibliography

Allchin, Bridget and Raymond. *The Birth of Indian Civilization: India and Pakistan before 500 B.C.* Harmondsworth, England: Penguin Books, 1968.

Basham, A. L., ed. *A Cultural History of India.* Oxford: Clarendon Press, 1975.

Craven, Roy. C. *A Concise History of Indian Art.* New York: Praeger, 1976.

Gardner, Brian. *The East India Company.* New York: Dorset Press, 1971.

Heitzen, James and Robert L. Worden, eds. *India: A Country Study.* Washington, DC: Federal Research Division, Library of Congress, 1996.

Kulke, Hermann and Dietmer Rothermund. *A History of India.* London and New York: Routledge, 1986.

Naipaul, V. S. *India: A Wounded Civilization.* New York: Knopf, 1977.

Robinson, Francis, ed. *The Cambridge Encyclopedia of India, Pakistan, Bangladesh, Sri Lanka, Nepal, Bhutan and the Maldives.* Cambridge: Cambridge University Press. 1989.

Schwartzberg, Joseph E., ed. *A Historical Atlas of South Asia.* 2nd impression. New York and Oxford: Oxford University Press, 1992.

"She Knows Her Onions. (Sonia Gandhi leads India's Congress Party to victories in state elections)." *Time International,* December 14, 1998, vol. 152, issue 23, p. 14.

Spate, O. H. K. and Learmonth, A. T. A. I*ndia and Pakistan: A General and Regional Geography.* 3rd ed. rev. London: Methuen, 1967.

Spear, Percival. *A History of India, Vol. 2.* Harmondsworth, England: Penguin Books, 1965.

Thapar, Romilla. *A History of India, Vol. 1.* Harmondsworth, England: Penguin Books, 1966.

Watters, Thomas, ed., *On Yuan Chwang's Travels in India (A.D. 629–645).* New Delhi: Munshiram, 1973.

Wolpert, Stanley. *India.* Berkeley: University of California Press, 1991.

Indonesia

Introduction

A nation since gaining independence from Dutch colonization in 1945, Indonesia was at one time a source of inspiration to many in the postcolonial world. However, inspiration soon gave way to an often brutal authoritarianism and notorious corruption. Since the ouster of longtime leader President Suharto (b. 1921) in 1998, Indonesia faces the dual tasks of finding its way internationally in the post-Cold War age, and recovering domestically from the ravages of the Southeast Asian economic crisis of the late 1990s.

Indonesia is situated on an archipelago, a chain of as many as 17,000 islands curving between the Indian Ocean and the Pacific Ocean. It is spread over 3.1 million square miles (7.9 million square kilometers) of land and sea. In terms of population, it is the fourth largest nation in the world, with some 200 million people. Despite advances in voluntary family planning, population density in many areas, especially on the island of Java, remains extremely high, and Indonesian cities are some of the most crowded on earth.

The Indonesian Archipelago is part of the Pacific Rim's "Ring of Fire," prone to volcanic eruptions and earthquakes. Volcanoes have wreaked havoc on Indonesia, but they also fertilized the soil for rich farmland. Java alone has twenty volcanoes, including the particularly active Mt. Merpati, and other islands are formed of volcanic peaks and crater lakes. Indonesian rain forests possess extraordinary biodiversity which have been under siege from logging companies and oil palm plantation ventures intent on selling off the valuable trees and replacing them with vast palm groves. Massive forest fires have been a particular risk of such operations, along with the displacement of tribal people from their traditional way of life.

International tourism has created pressures on Indonesians, as Australian surfers flocked to Bali for the waves and Balinese culture became commercialized to accord with foreigners' preconceptions of the exotic. Burial rites on Sulawesi have sometimes been overwhelmed by busloads of European onlookers. Other pressures on traditional life have included government resettlement and religious conversion, displacing forest peoples such as the Mentawi of Siberut island, to a precarious life in towns. The "transmigration" resettlement program pursued by the regime of former president Sukarno (1901-1970) sent people from highly populated regions to settle more remote areas like Kalimantan or Irian Jaya, putting them in conflict with the indigenous inhabitants who perceived them as unwelcome exploiters.

History

Starting in the first century A.D., Java and Sumatra were strongly influenced by Indian civilization, which brought both the Hindu and Buddhist religions to the region. By the seventh century, independent kingdoms sprang up on both islands. The last and most famous of these was the Hindu kingdom of Majapahit on Java, which rose to prominence in the late thirteenth century and flourished for two hundred years.

The first Europeans to arrive in present-day Indonesia were the Portuguese, who seized control of the trading center of Malacca in 1511. By the end of the sixteenth century, the Dutch and English had followed Portugal's lead, and by 1610 the Dutch drove out the Portuguese and established themselves as the reigning power, referring to Dutch-controlled parts of the Indonesian Archipelago as the "Dutch East Indies." Dutch colonization, which lasted 350 years (with a brief interruption during the Napoleonic era), resulted in the systematic exploitation of Indonesia's people and land. The Dutch monopolized the spice trade and, in the nineteenth century, forced a shift from the farming of subsistence crops to export crops grown and exported for the profit of the Netherlands. The initiation of this policy in 1830 led to rice shortages, famine, and mass migration from Java. Some enlightened Europeans attempted to reform the system but, for the most part, the colonial days were bitter for the Indonesians. Resistance fighters challenged foreign domination in major revolts in the nineteenth and at the beginning of the twentieth centuries.

In the early twentieth century, the Dutch instituted reforms under what was known as the Ethical Policy, but by then a nationalist movement was under way. In 1927, nationalist leader Ahmed Sukarno (1901–70) founded the Nationalist Party of Indonesia (PNI), which worked for Indonesian

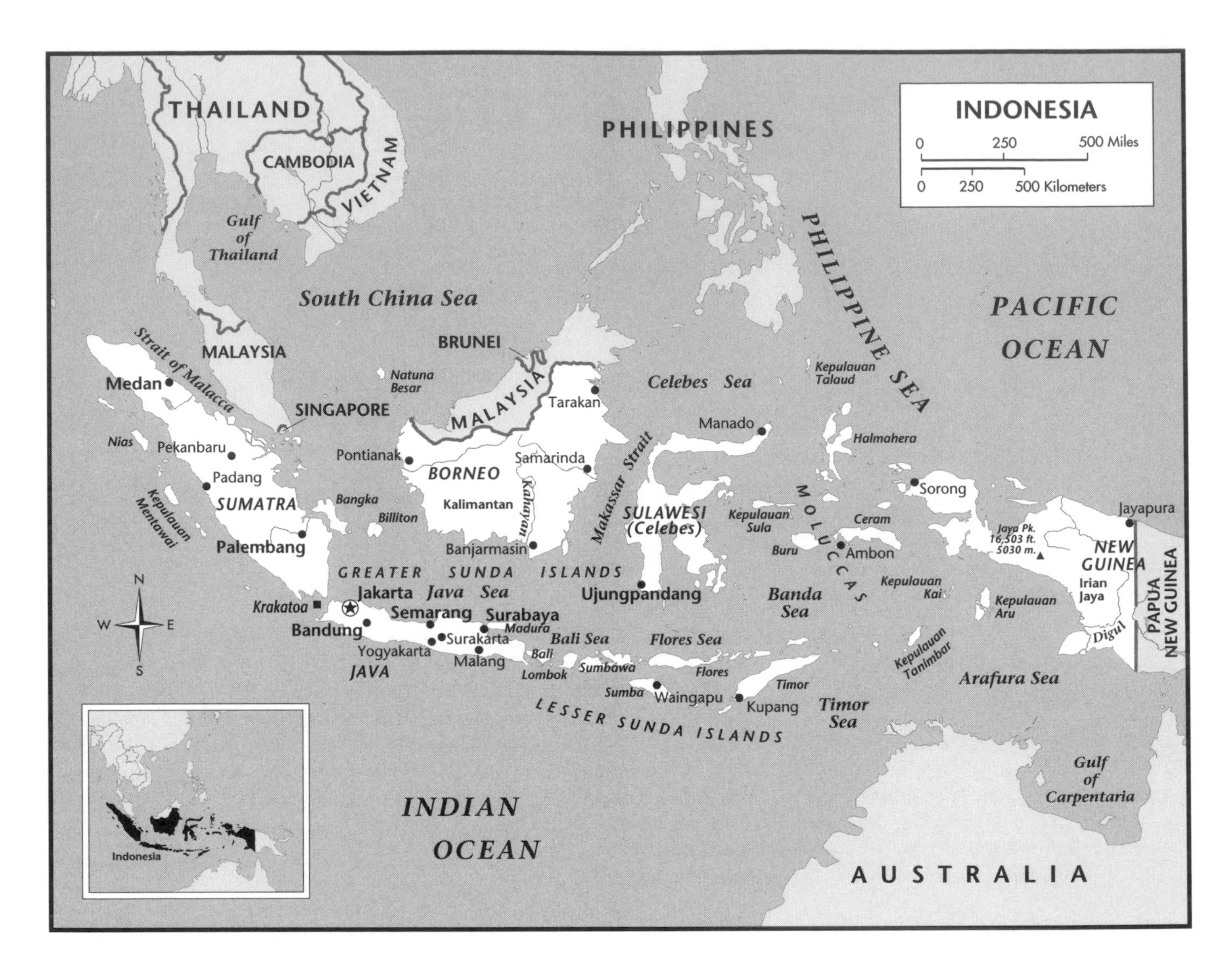

independence from colonial rule. The move toward independence received additional impetus during World War II, when Dutch rule was replaced by Japanese occupation. Although the Japanese treated the Indonesians harshly, they supported independence as preferable to a return of the Dutch presence in the region. Following the Japanese surrender in August 1945, Sukarno and his fellow nationalist leader, Muhammad Hatta, declared Indonesian independence. A four-year war with the Dutch followed. Finally, with help from the United Nations, an agreement was reached in late 1949. The Dutch officially transferred sovereignty over Indonesia to the newly formed Republic of the United States of Indonesia, a parliamentary democracy with Sukarno as president and Hatta as vice president.

The Sukarno government proved to be inefficient and corrupt, and as opposition to its activities grew, Sukarno's leadership became increasingly authoritarian. Dissent in Indonesia was weeded out and self-expression stifled. Intellectuals and young people demanding democracy or labor rights were called communists and imprisoned for years. The press and the arts were subjected to censorship. The military became ever more powerful in Indonesian life, setting policy and entering into big business ventures. Sukarno proclaimed martial law in 1957, and by 1959, following uprisings in Sumatra and Sulawesi, restored the 1945 constitution which concentrated power in the hands of the executive branch. Calling the new political system "Guided Democracy," he dissolved parliament. At the same time, his administration weakened the nation's economy by rapidly nationalizing Dutch companies and expelling Dutch citizens from Indonesia. By the fall of 1965, members of the military staged a coup d'état, but it was defeated. A top military official, General Thojib Suharto, was put in charge of restoring order to the country, and his influence rapidly grew, as that of Sukarno waned. Meanwhile, civil unrest continued, with several weeks of anti-Communist and anti-Chinese violence on Java and Sumatra that killed as many as half a million people.

In March 1966, Sukarno formally transferred power to General Suharto, who became the nation's acting president. Suharto reversed the government's pro-Communist foreign relations stance under Sukarno and made overtures to the West, seeking foreign aid and investment to further economic growth in Indonesia. Under Suharto, Indonesia formally annexed Netherlands New Guinea, which had been under United Nations administration since 1962, renaming it West Irian (later changed to Irian Jaya). In 1967 Indonesia became a founding member of the Association of Southeast Asian Nations (ASEAN). Domestically, Suharto instituted the policy of "transmigration," moving impoverished farmers from the overpopulated islands of Java and Bali to more sparsely settled areas including Sumatra and Kalimantan.

Exploiting Indonesia's natural resources of oil, gas, and timber, Suharto achieved rapid economic growth. However, excessive logging, which also increased the incidence of forest fires, caused irreparable harm to the nation's rain forests, giving Indonesia Asia's highest deforestation rate. Suharto himself became one of the richest men in the world, and his family was allowed to involve itself in numerous lucrative business deals. An entire system of business-government links grew up, making Indonesia notoriously corrupt. While Indonesia's economy exhibited an impressive growth rate up until the Southeast Asian currency crisis, much of the wealth was siphoned off by corruption. Multinational investors spurred the "Tiger" economy, but much of their involvement was based on extracting resources such as oil, mining, and timber. International manufacturers became controversial in the 1990s when they were accused of exploiting cheap subminimum wage labor.

In 1975 Portugal withdrew from East Timor, which had been its colonial possession, and the left-leaning Fretelin, a pro-independence group, stepped up its activities. In response, Indonesian troops invaded and occupied East Timor in December 1975. In their attempts to destroy Fretelin, they victimized the entire population of the former colony, which had been declared a province of Indonesia. With food supplies withheld and agriculture sabotages, between 100,000 and 250,000 East Timorese died between 1976 and 1980. In spite of its human rights abuses in East Timor, however, Indonesia, as a Cold War ally, continued receiving military aid from the United States throughout the 1980s. In 1991 Indonesian troops fired into a crowd at a funeral for an East Timorese activist. An estimated one hundred people were killed in the massacre, which was witnessed by foreign journalists and drew world attention to the Indonesian occupation of the Timorese. The United States and other Western governments began threatening to withhold aid until human rights abuses in East Timor ended, and the United Nations placed Indonesia on a human rights "watch" list in 1993. In 1996 two East Timorese activists won the Nobel Peace Prize for their efforts toward a solution to the continuing crisis in their homeland.

However, unrest continued in East Timor, and ethnic violence in Kalimantan killed hundreds in 1997. In the second half of that year, the Southeast Asian economic crisis spread to Indonesia, causing the national currency, the rupiah, to lose about half its value between August and October, and created a banking crisis that forced the closing of sixteen banks by the government. The austerity measures required for receipt of a $23 billion IMF bailout created widespread popular discontent and civil unrest. Two months after being elected to his seventh term as president, Sukarto was forced to resign, as protesters occupied the parliament building and even his own Golkar party called for the seventy-six-year-old leader to step down. He was succeeded by his vice president, B.J. Habibie, and the nation's political situation stabilized.

Today, Indonesia faces a new stage in its history. However, economic meltdown and a legacy of ethnic violence have yet to be overcome. With its immense territory, population and resources, Indonesia will undoubtedly continue to have a major influence in Asia and beyond. There is great hope that the forces of openness, tolerance and democratization may create a better future for its people.

Timeline

63 million years ago The Indonesian Archipelago is formed

The Archipelago, the chain of islands that would be known as Indonesia, is formed during the end of the Cretaceous Period, when undersea shelves are compressed and folded into immense ridges: the Sunda Shelf (part of the Asian land-mass) and the Sahul Shelf (part of the Australian land-mass) are separated by seas more than three miles deep and dotted with volcanoes.

1.8 million to 500,000 years ago Early humans live on Java

Early humans, forms of Homo erectus (an ape-like human who walks upright), live on Java, part of the Sunda shelf.

100,000 to 50,000 years ago Homo solonesis evolves

Homo solonesis, an early form of Homo sapiens (the modern human, capable of reason) or a late form of Homo erectus with a large brain, evolves on Java.

40,000 years ago Early humans use tools

Early humans living on Java make and use simple tools, such as flaked-pebble choppers and stone hand-axes.

20,000 B.C. Australo-Melanesians settle the island of Papua

The huge island of Papua, to the east of the Archipelago, is settled by Australo-Melanesian people, who speak languages related to those of present-day Pacific Islanders.

13,000–6000 B.C. The Indonesian islands are separated

As the sea level rises, the large islands of Java, Sumatra, and Kalimantan separate from the Asian landmass, and Papua separates from Australia. Plant and animal species remain distinct in the two areas, which also feature numerous smaller islands.

c. 7000 B.C. Agriculture develops on Papua

Australo-Melanesian people develop agriculture on Papua. They drain swamps to plant crops, possibly taro, a starchy root vegetable.

c. 3000–1500 B.C. Australo-Melanesians and Austronesians migrate to the Indonesian islands

Australo-Melanesian people migrate from The Philippines and Taiwan southeast to the Indonesian islands, and Austronesian people related to the present-day Malays migrate there too, inhabiting islands including Java, Sumatra, and Bali. The inhabitants of the region use stone tools and create pottery. They build wooden canoes and houses, and domesticate animals such as dogs and cattle. They live by farming (including rice-growing) and fishing. The use of shadow puppets may have appeared in this time, as a way to "act out" storytelling for gatherings of people.

c. 1000 B.C. Bronze tools and gongs are in use on Papua

The people of Papua use bronze tools and musical gongs, probably brought there from Vietnam. The gongs, suspended to be played, are ancestors of Indonesian *gamelon* orchestras.

c. 500 B.C. Indonesian islands trade with Southeast Asia, and then India

Seaborne trade between the Indonesian islands and the Southeast Asian mainland to the north becomes a way of life for many. Metal goods and ceramics are particularly important trade goods. By the first centuries A.D., the trade routes expand to the west and Indian cultural influences come to the islands of the Archipelago, particularly Java. The Hindu and Buddhist religions find their way to Java from India and are adopted by royal courts of small states on the island, whose rulers become literate in the Indian language, Sanskrit. Hindu epic poems are portrayed in the shadow puppet theater called *wayang*, and in dance.

5th century A.D. Seafarers from Indonesia settle off the coast of Africa

The island of Madagascar off the east coast of Africa is settled by Austronesian seafarers who sail all the way from the Indonesian Archipelago, bringing Asian food crops and agricultural techniques, such as terraced rice fields, with them.

7th century The Sumatran kingdom of Srivijaya thrives

Srivijaya, a wealthy kingdom of eastern Sumatra thrives through trade and conquers much of Sumatra, Java, and the Malay Peninsula to the north. The kingdom dominates the trading routes amid the islands.

7th–10th centuries Indian-influenced kingdoms build temples and monuments

On Java, the Hindu kingdom of Mataram gives way to the kingdom of Sailendra, which is Buddhist. Borobudur, a massive stone structure representing a mythical peak, Mount Meru, with hundreds of carved Buddha images, is built by the Sailendra kingdom, as is a huge stone Hindu temple nearby, the Prambanan. The island of Bali adopts Hinduism. Stone temples are built there, where the people make offerings of fruit and flowers to the Hindu gods.

10th century The *Gamelan* musical form develops

Orchestral ensembles known as *gamelan*, featuring bronze gongs and other percussion instruments, become an important musical form. The *gamelan* is associated with aristocratic court occasions on Java, and with more popular village life on Bali.

11th century Buddhist scholars visit Srivijaya

Buddhist scholars Atishar and Yijing (a Chinese monk) visit Srivijaya. They write down their observations of the prosperous Sumatran kingdom.

1049 The Kediri kingdom begins

Airlangga (991–1046), a powerful Javanese-Balinese ruler in east Java, steps down from his throne to become a holy man and divides his kingdom for his two sons. One rules Kediri, which grows rich from the trade in spices from other islands.

1290s Marco Polo visits Sumatra

Returning to Europe from China (journey 1292–95), the Italian adventurer Marco Polo stops in Sumatra, which he describes as a vast island, wealthy in spices.

1292 King Keranagara is killed and a new dynasty begins on Java

King Keranagara (1268–92), monarch of eastern Java, is killed in 1292 by a rival following an invasion by Kublai Khan's Mongols. Prince Vijaya (d. 1309), Keranagara's son-in-law, begins a new dynasty based at Majapahit, and the Mongols are forced to withdraw from Java. From 1293 well into the fourteenth century, Majapahit becomes a great Hindu state, enjoying a "Golden Age" by trading and commanding tribute from far-off islands. A Javanese epic by the court poet Prapanca, *Nagara-krtagama,* tells the saga of the Majapahit empire's expansion.

Late 13th century Islam takes hold in Sumatra

Spread by Arab and Indian traders and holy men, the religion of Islam converts the ruling aristocracy and many of the common people in much of Sumatra. Islam later is adopted by people of the islands of Sulawesi and Java.

14th century *Sarungs* are worn by everyone

As cotton cloth is imported to Java, Sumatra, and other islands, the wrap-around skirt called the *sarung* is worn by men and women, rich and poor alike.

1330–64 Prime Minister Gajah Mada wields power

Gajah Mada, prime minister to King Hayam Wuruk and Queen Tribuwana of Majapahit, extends and consolidates the empire throughout the Archipelago through use of the navy and political maneuvering.

1345–46 ibn-Battuta visits Sumatra

Arab author Muhammad ibn-'Abdullah ibn-Battuta visits Sumatra during his travels in Asia and observes the Islamic way of life there.

15th century Islamic states grow

The Malay Peninsula city-state of Malacca becomes a seafaring power and makes converts to Islam, as does Gowa, a state in Sulawesi.

1405 A Chinese fleet enters the region

Chinese Admiral and Grand Eunuch Zheng He (1371–1435) brings his massive fleet of ocean-going war junks to ports in Java and Sumatra, suppressing the activities of Chinese pirates off the Sumatran coast.

1520s Majapahit empire disintegrates

Malay trading states to the north, as well as China, become powerful enough to take commercial privileges and power away from Majapahit, and the dynasty ends with a bitter conflict over who has the right to be the next ruler.

1596 Catholic missionaries gain a foothold

With Portuguese traders based on the Malay Peninsula making spice-trade forays to the Malaku Islands, a Spanish Jesuit, St. Francis Xavier founds a mission there, at Ambon, converting local people to Catholicism.

17th century Batik becomes a textile art of Java

Batik, a method of textile design using beeswax to form decorative patterns with dyeing, becomes a popular art form in Java and spreads to other areas of Indonesia. Clothing made of batik fabrics remains popular internationally to the present.

1602 United East India Company is formed

The Dutch parliament issues a charter to form the United East India Company, which will, in addition to its commercial interests in trading for spices and other commodities, have military, diplomatic, and governmental powers in the Archipelago. The Dutch refer to the parts of the Indonesian Archipelago under their control as "the Dutch East Indies" or "the Netherlands Indies."

1609 Dutch forces use brutal tactics to control spice trade

The United East India Company monopolizes the source of nutmeg, the Banda Islands in the Malaku region, by preventing the island from trading with the English. Dutch forces kill as many as 15,000 people there and deport the rest of the inhabitants of the Banda Islands, replacing them with indentured servants and slave laborers. Dutch forces also attempt to eliminate the indigenous people of the island of Ceram in the Malaku region in order to control clove production in 1656. In 1659 the Dutch company burns the old royal capital of Palembang on Sumatra, in a fight to control that area's trade in pepper.

1613–46 The powerful state of Mataram rises on Java

Sultan Agung (r. 1613–46) expands his power from Mataram, his state on Java, to Madura and Surabaya on the island of Sulawesi, but is unsuccessful in attempts to take over Jakarta and Bali. His relentless military campaigns impoverish the Javanese people and hurt the interisland trade.

1619 Coen is named Governor General

The United East India Company devises the post of governor general to administer its holdings from a base in Indonesia, and Jan Pieterszoon Coen (1587–1629) takes to the task with great aggressiveness. He serves from 1619–1623 and from 1627–1629. His troops seize Jakarta, a west Java port, and make it his capital, which is renamed Batavia. From Batavia, Coen seeks to control the entire spice trade, often in competition with England's own East India Company.

1637 A seven-volume book, *"The Garden of Kings"* is commenced

Bustan as-Salatin (The Garden of Kings) is written by Nuruddin ar-Riniri, from Gujerat, India, under the patronage of Sultan Iskandar Thani of Aceh, the state in northern Sumatra. The seven-volume book, partly based on Persian writings, contains stories of Moslem rulers and information about science and medicine. Following its completion, the book is translated into several Indonesian languages.

1671 Mataram weakens and begins concessions to the Dutch

Amangkurat I (r. 1646–77), the successor to Sultan Agung as ruler of Mataram, seeks protection from the Dutch during a rebellion in 1671 and has to take refuge in Dutch territory in 1677. The next ruler of Mataram, Amangkurat II (r. 1677–1703), cedes territory and several trade monopolies to the Dutch.

1696 Coffee comes to Java

United East India Company traders bring coffee plants from South America to Java for plantations. The first coffee from Java is sold in The Netherlands in 1712, and the drink eventually becomes so identified with the Indonesian island that "Java" becomes a lasting international slang word for coffee.

1704–08 The First Javanese War of Succession

Conflict among the aristocracy of Mataram over who will succeed as the next ruler leads to more concessions to the United East Indies Company. Supported by the Dutch, a new ruler, Pakubuwona I (r. 1705–19), allows them increased military presence on Java as well as naval dominance.

1719–23 The Second Javanese War of Succession

With his kingdom further weakened by its own rivalries in a war of succession, King Amangkurat IV (r. 1719–26) of Mataram hands over more power to the United East Indies Company.

1740 Chinese settlers are massacred

In the Dutch capital, Batavia (their name for Jakarta), approximately 10,000 rebellious Chinese settlers are killed, with the apparent approval of the United East Indies Company's governor-general.

1746–55 The Third Javanese War of Succession

Continued turmoil in Mataram leads to the division of the state into the small sultanates of Yogyakarta and Surakarta.

1780–84 War between England and The Netherlands

United East India Company trading in Indonesia is disrupted by war in Europe, as the British navy harasses the Dutch merchants' ships at sea.

1799 The United East India Company is disbanded

While a protectorate of Napoleon's France, the government of The Netherlands discovers corruption and debt of the United East India Company and lets the company's charter lapse in order to disband it. The Dutch government takes over the company's territories.

1808 A reformist governor general takes office

The Dutchman Herman Willem Daendels (1762–1818) is made governor-general of the Indonesian islands by the French king of the Netherlands. Influenced by the concepts of the French Revolution, Daendels begins to institute reforms in Java's political system, which causes great resentment in the Javanese aristocracy.

1811: August The British take over Java

British forces take possession of Java from the Dutch/French, and Thomas Stamford Raffles (1811–16) is put in charge as lieutenant governor by the British East India Company. Raffles travels widely in the Archipelago, researching local cultures and natural history. Reforms he tries to institute include ends to crop production quotas, forced labor, and slave trading. Raffles's administration ends with the return of Dutch control to Java in 1816.

1815: April Gunung Tambora erupts, causing immense destruction

Gunung Tambora, a volcano on Sumbawa island, south of Bali, erupts, killing as many as 92,000 people. Its clouds of ash cause "a year without a summer" as far away as Europe.

1817 Botanical Gardens are founded

The Botanical Gardens are founded at Bogor, on Java. The gardens contain a huge variety of tropical plants. Conservation efforts for plants throughout the Archipelago are planned and administered from the Bogor Gardens offices.

1817 Rebellion breaks out in the Spice Islands

In the Malaku region, known as the Spice Islands, rebellion against Dutch rule breaks out. Pattimura (?1783–1817), also called Thomas Matulessy, an indigenous person who had been a sargent major in the British army, leads a rebel force, occupying a Dutch fort on Saparua Island. He is captured by the Dutch and executed in 1817. Pattimura's last words to the Dutch colonials are an ironic "Have a pleasant stay here, gentlemen." Another indigenous rebel leader is Martha Christina Tiahaha, who fights the Dutch on Nusa Laut island until they capture her. She starves herself to death while being taken away to exile.

1821–38 The Padri War leads to Dutch control in Sumatra

Dutch colonial forces fight against the followers of the *padri* (Islamic religious leaders) in rural Sumatra. By backing the more secular *adat* factions, the Dutch expand their power and agricultural control on the large island.

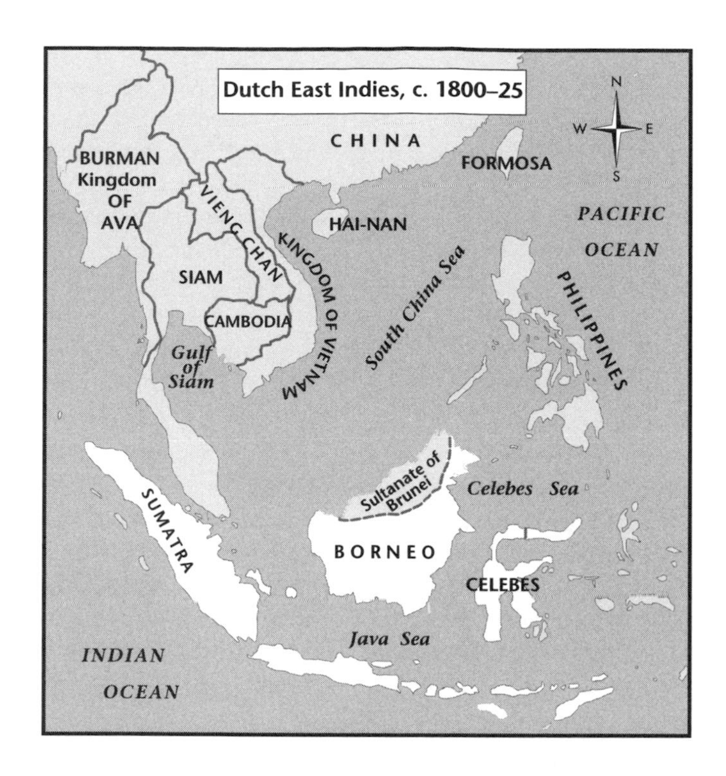

1824 The Anglo-Dutch Treaty defines colonial domains

The Anglo-Dutch Treaty gives British traders and colonists dominion over the Malay Peninsula. The Dutch control territories known as the Dutch East Indies, including Sumatra (except for Aceh, which retains its independence) and the islands to the south, including Java.

1825–30 Aristocrats rebel against the Dutch in the Java War

A Dutch plan to build a road across sacred land sparks a rebellion of the aristocracy, which becomes a long guerrilla conflict. Pengeran Diponegoro (c. 1785–1855), the son of a Yogyakarta sultan, leads the Javanese. With a mystical sense of destiny, Diponegoro inspires the common people to rise up in revolt, until he is arrested and exiled in 1830. Dutch forces extend their military presence throughout the island, and some 200,000 Javanese perish during the war.

1830 The Cultivation System takes hold over Java

The Dutch begin enforcing their Cultivation System which compels Javanese villages to produce export crops for the profit of the colonial elite. The exploitive excesses of the Cultivation System lead to shortages of rice, famine, and the migration of impoverished villagers from Java. Huge profits from Java's exports bolster the economy of The Netherlands, which dominanates world trade in tropical products. The Indonesian colony is administered by Dutch officials using Chinese commercial middlemen and lower bureaucrats from the Javanese aristocracy.

1854 Cinchona, a plant used against malaria, is brought to Java

Seeds of the cinchona plant from South America are brought to Java for plantations. Quinine, an effective antimalarial medicine, is extracted from cinchona's bark. Java becomes the world's leading quinine producer.

1859 East Timor is established as a Portuguese colony

While the Dutch hold power over most of the Archipelago, Portugal administers the eastern part of one of the southernmost islands, Timor. The Portuguese convert most of the population to Roman Catholicism and export aromatic sandalwood from East Timor.

1860 *Max Havelaar*, a protest novel, inspires reforms

An exposé of Dutch colonists' exploitation of Java's people through their Cultivation System is published. Written by former Dutch official Eduard Douwes Dekker (1820–87) under the pseudonym Multatuli, *Max Havelaar: or the Coffee Auctions of the Dutch Trading Company* shows the evils of the colonial system to readers back in The Netherlands. With popular indignation creating pressure for change, the Dutch government begins to ease its economic demands on Java. In the new "Liberal Policy," trade and investment in the Archipelago become less centralized and more privatized. Trading routes between the colonies and Europe are made more efficient by the opening of the Suez Canal in 1869, which opens a passageway through Egypt into the Mediterranean.

1863 The Wallace Line defines a convergence of ecosystems

A British naturalist, Alfred Russel Wallace (1823–1913), roams the islands of the Archipelago, studying plant, animal, and insect species. In 1863 he publishes a paper with a red line marking the convergence of Southeast Asian and Australian ecosystems, west of the island of Sulawesi. His findings have a major influence on the study of species' migration and their evolution in their habitats.

1873–1903 The Dutch attack Aceh and a long conflict follows

To end independent Aceh's diplomatic contacts with other nations and to gain control of its pepper trade, Dutch forces attack Aceh's capital, Banda Aceh. With Islam as its inspiration, Aceh's guerrilla war of resistance is tenacious.

1880 Workers are restricted by the "Coolie Ordinance" on Sumatra

On Sumatra's plantations, laborers (mostly from Java or China) can be jailed under the Dutch "Coolie Ordinance" if they break their contracts. The ordinance, opposed by the United States as a form of unfair competition in the tobacco trade through virtual slave labor, stays in place until 1936.

1883: May 20 and August 26–28 A massive volcano, Krakatau (Krakatoa), erupts

Krakatau, a volcano located between Java and Sumatra, erupts, hurling six cubic miles of land into the atmosphere and causing a tidal wave that kills as many as 36,000 people in western Java.

1890 The Indonesian Archipelago's oil industry begins

Following extraction of oil in Sumatra, the Royal Dutch Company for Exploration of Petroleum Sources in the Netherlands Indies is formed.

1899 "A Debt of Honor," an essay, calls for fair return

Conrad Theodoor van Deventer, a Dutch lawyer, publishes "A Debt of Honor," an essay calling for The Netherlands to return the profits from the Indonesian Archipelago's cash crops to the indigenous farmers in the form of social welfare programs and education.

1900 A rural passive resistance campaign opposes the Dutch

Javanese followers of Surantika Samin, head of a religion called The Science of the Prophet Adam, participate in a tax resistance and labor refusal campaign against the Dutch colonists.

1901 The Ethical Policy reforms the Dutch colonial administration

Influenced by van Deventer's ideas, the government of the Dutch East Indies begins to implement a set of reforms called the Ethical Policy, including education for the people of the Archipelago. A program of transmigration of people from densely populated Java is commenced. This controversial policy of offering poor farmers land on the other islands continues long after Indonesia's independence in the twentieth century.

1904–08 Balinese aristocrats oppose the Dutch through suicide

The Dutch colonial administration strengthens and expands its hold on the islands through military power. Opposed by the aristocracy in regions of Bali, the Dutch forces launch a military campaign which is met by mass suicide. The Balinese rulers and their families march into Dutch gunfire.

1905–09 The first wildlife protection laws are enacted

The Dutch administration enacts its first conservation laws, designed to protect orangutans, rhinoceros, birds of paradise, and elephants in the Archipelago. Many of the islands' species have become endangered when they are hunted for products such as feathers and ivory.

1907 A new petroleum company is formed

The Dutch petroleum exploration company (see 1890) merges with a British firm, Shell Transport and Trading Company, to form Royal Dutch Shell, which will come to control the vast majority of the Archipelago's petroleum. With automobiles becoming an important new industry in the world, export of petroleum becomes more and more lucrative. Rubber plantations are established to provide rubber for car tires.

1908 Nationalism stirs medical students, who form a political group

As the people of the Dutch-occupied islands become more educated, their resentment of foreign control and exploitation is heightened by international movements, including Islamic reformism and Russia's Bolshevik revolution (1917). Students at Batavia's School for Training Native Doctors form *Budi Utomo* (Noble Endeavor), a political organization. The group uses Malay (rather than Javanese) as its language, as a way to link the people of all the islands.

1910 A nationalist party is formed

The Indies Party, a nationalist group calling for self-government in the Archipelago, is founded by E.F.E. Douwes Dekker (1880–1950), a relative of the author of *Max Havelaar* (see 1860).

1911 Kartini's book, *From Darkness to Light* is published

The writings of Raden Ajeng Kartini (1879–1905), a Western-educated Javanese princess, are published in a book called From Darkness to Light, after her death in childbirth. Kartini calls for rights for the women of Indonesia, particularly for their education. Her advocacy inspires the founding of "Kartini schools" for girls on Java.

1912 An Islamic political organization is established

Islamic merchants form the Sarekat Islam (Islamic Union), which becomes an increasingly popular force of opposition to colonial rule when people from all social classes and religions join it.

1913: July The Native Committee is formed to promote self-rule

In Bandung, Java, E.F.E. Douwes Dekker (see 1910) and other nationalists form the Native Committee. The intellectuals plan to petition The Netherlands for a parliament for Java and the other islands. They publish a pamphlet of their ideas, titled "If I Were to be a Dutchman." The author of the pamphlet, Dr. R.M. Suwardi Surjaningrat, as well as Douwes Dekker, and Dr. Tjipto Mangunkusumo, another founder of the Native Committee, are exiled to The Netherlands as punishment for subversion (attempting to overthrow a government). In Europe they work to promote nationalism for the islands of the Archipelago, using the name "Indonesia" for their homeland.

1914: May **A left-wing group is organized**

The Indies Social Democratic Association is organized in 1914. It evolves into the Communist Association of the Indies, and then in 1924 becomes the *Partai Komunis Indonesia* (PKI, Indonesian Communist Party).

1918: May **The People's Council, an advisory body, meets**

The People's Council is formed with the stated purpose of advising the governor general, but it has no real legislative powers. Most of the original thirty-nine members of the council are Dutch, although there are some Indonesian appointees.

1921 **A Javan rhinoceros reserve is established**

A nature reserve is established at Ujung Kulon by the colonial administration for one of the world's rarest mammals, the Javan rhinoceros. The reserve, including an island formed by the eruption of the volcano Krakatau, becomes an Indonesian national park in 1980.

1926: November **Indonesian Communists attempt a revolt**

A Communist revolt is repelled by Dutch forces in the cities and countryside of Java and Sumatra. The colonial administration arrests some 13,000 people in suppressing the uprising, imprisoning and exiling thousands of them.

1927 **Films are produced by and for Indonesians**

The first Indonesian movies are made for the local market, often featuring *kroncong* music. Influenced by Portuguese folksongs, *kroncong* tunes featuring guitar and ukulele become popular with the poor, and the sentimental songs take on nationalistic meanings for Indonesians.

1927: June 4 **Sukarno participates in founding a nationalist party**

A charismatic Javanese architectural engineer, Ahmed Sukarno (1901–70) becomes influential in the nationalist movement.

Sukarno is born in Surabaya on eastern Java and is instrumental in the Indonesian movement for independence. He participates in the founding of the Nationalist Party of Indonesia, becoming its chairman. The party is intended to organize the public and agitate aggressively for Indonesia's independence from colonial rule. He becomes the first president of the independent nation (see 1945) and serves for over twenty years (see 1965: October 2 and 1966: March 11). He formally retires in 1968, and dies two years later (see 1970: June 21).

1929: December 29 **Sukarno is arrested by the colonial authorities**

The Dutch governor general has the leaders of the Nationalist Party of Indonesia arrested and tried for sedition (stirring up rebellion), and the party banned. Sukarno is given a four-year prison sentence. Another nationalist leader, Dr. Mohammed Hatta (1920–80), then forms the Pendidikan National Indonesia (PNI, Indonesian Educational Union) to replace or revive Sukarno's banished party.

1930 **Clove cigarettes become popular with Indonesians**

In 1930, Ong Hok Liong, a Chinese trader, begins to manufacture and sell cigarettes called *kreteks* on Java. Filled with a high-tar-content mixture of tobacco and sweet-scented cloves, *kreteks* are smoked by men throughout the Archipelago.

1933 **Sukarno makes dramatic speeches**

An extremely gifted dramatic orator, Sukarno expounds on various themes in his message of Indonesian nationalism. After he is released from arrest, he makes speeches to audiences around Java. In 1933 he promotes *"Marhaenism"*, his term for a self-sufficient peasant society that rejects foreign exploitation. He also draws on Islamic and Marxist concepts and makes references to characters from traditional shadow-puppet plays in his speeches.

1934 **Nationalist leaders are arrested and exiled**

Two prominent nationalist leaders, Dr. Mohammed Hatta (see 1929: December 29) and Sultan Sjahrir (1909–66), organizers of small pro-independence groups, are arrested by the Dutch authorities and sent to a penal colony in West Papua. Even while imprisoned there, Hatta and Sjahrir continue to plan for Indonesia's overthrow of foreign domination.

1940–41 **World War II (1939–45) disrupts Dutch colonial power**

As The Netherlands is occupied by Nazi Germany, Japan demands resources from the Dutch Indies administration which is cut off from the Dutch government.

1942: February 27–28 **Battle of the Java Sea takes place**

In an immense naval battle off the coast of Java, the Battle of the Java Sea, a fleet of American, British, Dutch, and Australian war-ships is defeated by the Japanese navy, leaving Java vulnerable to Japanese occupation.

1942: March 9 **The Dutch Indies are surrendered to Japan**

Without fighting on land, the Dutch Indies administration surrenders to the Japanese forces, and most of the Archipelago is placed under Japanese military control. Japanese troops often abuse the Indonesian people, using an estimated 4 to 10 million as forced laborers, and confiscating food crops. On Java, the Japanese sixteenth Army exhibits some tolerance for local political and Islamic movements and promotes Indonesian cultural themes in an effort to wipe out Dutch colonialism. The Japanese also encourage the use of Bahasa Indonesia, based on the Malay trading language, as a unifying language for the Archipelago. They change the capital's name from Batavia back to Jakarta. Europeans are imprisoned by the Japanese in camps, where many die during the war.

1943 Japan consolidates Islamic groups

The Japanese occupiers force Islamic organizations to form a single group called Masyumi.

1944: September 7 Japan favors Indonesian independence

Facing defeat in the war, Indonesia's Japanese occupiers become more active in promoting the independence movement, hoping to forestall Dutch reoccupation of the Archipelago. On September 7, Japanese prime minister General Koiso makes a promise of independence for Indonesia.

1945: March A committee plans for an independent Indonesia

With delegates from throughout the Archipelago, the Investigating Committee for Preparatory Work for Indonesian Independence is organized. Masyumi, the Islamic group's goal of Islam becoming the state religion, is thwarted through a compromise in which the constitution would simply call for a belief in one god. Sukarno (see 1927: June 4) is named president, and Hatta (see 1929: December 29) vice president. The delegates plan for Indonesia to include the Portuguese colony of East Timor and the British territories on Borneo and the Malay Peninsula, as well as all of the Dutch Indies.

1945: June 1 Sukarno defines the *Pancasila,* five principles of the new nation

In a speech to the constitution drafting committee, Sukarno defines five guiding principles, called the *Pancasila*, for independent Indonesia: national unity, internationalism or humanitarianism, representative government, social justice, and belief in God.

1945: August 17 Indonesia declares its independence

Three days after Japan's surrender, Indonesia's independence is declared by Sukarno and Hatta. The nation, the Republic of Indonesia, is divided into eight provinces, and governors are appointed by the new Central Indonesian National Committee.

1945: September Allied troops land in Java

Allied troops land to take charge of Indonesia after the Japanese surrender. The British commander of Southeast Asia, Admiral Earl Louis Mountbatten, is in charge of administering Indonesia. Turmoil follows, as local forces resist a return to Dutch rule.

1945: October 28 British forces clash with Indonesians

In the city of Surabaya, on Java, British troops attempt to pacify armed young nationalists, and hundreds of the British soldiers are killed. From November 9 to 24, the British bring in more troops, aerial bombardment, and battleships to suppress the Surabaya uprising. Thousands of Indonesians die, but the fierce resistance convinces the Allies that the independence movement is a serious force.

1947: May 25 The Dutch partially recognize the Indonesian Republic

The Netherlands recognizes the Republic of Indonesia by signing the Linggajati Agreement. Brokered by the British, the agreement recognizes Java and Sumatra as the Republic of Indonesia, but the rest of the Archipelago is to remain under Dutch auspices.

1947: July 21 Dutch troops invade and then retake much of the Republic

The Netherlands invades Java and Sumatra in what it terms a "police action," retaking nearly all of the land from the Republic. Only the Yogyakarta region of Java remains in the Republic's hands, though it is under a Dutch blockade. The United Nations undertakes sponsorship of negotiations between the Dutch and the Indonesians.

1948: January 17 The Renville Agreement recognizes Dutch control

The United Nations agreement is ratified by the Dutch and Indonesian governments. It recognizes Dutch control of nearly the entire Archipelago, although it does call for referendums on self-rule.

1948: September Conflict breaks out between Communists and the Republic

The Communist Party of Indonesia (PKI) fights with forces of the Republic on Java and retreats to the city of Madiun, where the rebellion is eventually suppressed by troops loyal to the Republic. This gives the Republic anti-Communist credentials valued by the West during the Cold War.

1948: December A second "police action" expands Dutch control

Dutch forces retake Yogyakarta in their second "police action," which puts them back in control of all of Java. They arrest Sukarno and Hatta. The Dutch bombing of Yogyakarta and suppression of the Republic provokes a strong negative reaction in the United States, Britain, and much of Asia. On January 28, 1949, the United Nations Security Council adopts a resolution calling on The Netherlands to reinstate the Republic's government and hand over all of Indonesia to the Republic by July 1, 1950.

1949: December 27 The Netherlands transfers sovereignty to Indonesia

Following a Round Table Conference held from August to November 1949, The Netherlands decides on troop withdrawal and formally hands over sovereignty of the Archipelago (except for West Papua) to become a federal state, now called the Republic of the United States of Indonesia (RUSI).

1950: March 31 A national airline is founded

Indonesia's national airline, Garuda (named after a mythical Hindu bird-king), is established as a joint venture with the Dutch airline KLM. It eventually becomes an international airline on its own.

1950: April Rebellion flares in South Malaku

Rebels against the central government declare the Republic of South Malaku at Ambon on Ceram island. When the insurrection is suppressed, 12,000 of the rebel soldiers, who had previously served with the Dutch army, go into exile in The Netherlands with their families.

1950: May Indonesia evolves into a unitary state

The many small states of the RUSI are absorbed into a unitary state, the Republic of Indonesia, by May 1950. It is governed from the national capital, Jakarta.

1950: August 14 A provisional constitution institutes parliamentary democracy

A provisional constitution establishes a parliamentary democracy for the Republic of Indonesia, with Sukarno as President.

1955: April 18–24 The Asia-Africa Conference meets in Indonesia

In the Javanese university city of Bandung, Sukarno convenes a meeting of Asian and African leaders from twenty-nine countries. China's Zhou Enlai, India's Nehru, and Egypt's Nasser are among the heads of state attending. The conference is a show of solidarity for the post-colonial world, and it adds to Indonesia's prestige as a newly independent nation.

1955: September The first general election is held in Indonesia

Indonesia holds its first general election, in which nearly 38 million men and women vote. The Indonesian Nationalist Union (Partai Nasional Indonesia, PNI), led by Sukarno, wins the largest percentage of the vote.

1956: October Disgruntled military officers attempt a coup d'état

Military officers attempt a coup d'état (sudden overthrow of a government) because of the government's efforts to stifle corruption among officers on Sumatra and other islands. Some military commanders remain defiant even after the coup is unsuccessful.

1956: December 1 Hatta resigns as vice president

Independence leader Mohammed Hatta resigns his post as vice president, disappointed with the Indonesian Republic's centralized power and persistant corruption.

1957: March 14 Sukarno proclaims martial law in Indonesia

With continued instability from military factions and Islamic and regional groups, Sukarno solidifies his power by discarding the parliamentary system. He backs his growing authoritarianism with support from both the PKI and mainstream military officers.

1957: December Dutch companies are put under military management

Sukarno's government nationalizes Dutch companies in Indonesia, including Royal Dutch Shell. A new national oil company, Permina, is formed. Military officers are involved in running the nationalized companies, particularly in the oil industry.

1957: December 5 Dutch citizens are expelled from Indonesia

Continuing his campaign to remove the remnants of colonialism, Sukarno has the Justice Ministry order the expulsion of tens of thousands of Dutch nationals from Indonesia.

1958: February 10 Regional rebellions break out

Insurrection occurs in Sumatra and then in Sulawesi. Eventually the rebellions are suppressed by Sukarno's military. Discovery of United States covert aid to the rebels alienates Sukarno from the West and drives him closer to Communist Bloc alliances.

1959: July 5 Sukarno restores the 1945 Constitution

To launch the period of "Guided Democracy," Sukarno returns Indonesia to the 1945 Constitution and dissolves the House of Representatives. "Guided Democracy" is promoted by Sukarno as his alternative to the checks and balances of a parliamentary system (see 1945: June 1).

1959: November **Restrictions are placed on Chinese traders in Indonesia**

With the military maneuvering against the Communist China-backed PKI, the Indonesian government places restrictions on Chinese residents. A decree forbidding Chinese from trading in rural areas is enforced, and tens of thousands of Chinese are expelled from Indonesia.

1960: March **A new legislature is formed**

The House of People's Representatives is formed as a new legislature which includes military and Communist representatives. Sukarno balances those factions and appeals to the populace through his inspirational speeches. He builds expensive monuments in Jakarta for the country's prestige, but ordinary people suffer from inflation and poverty.

1961: December 19 **Indonesia intends to seize West New Guinea**

Sukarno announces that he intends to use military force to seize West New Guinea from Dutch occupation. Following mediation by the United States, The Netherlands hands its authority over West New Guinea to the United Nations on August 15, 1962.

1963 **Sukarno is named President for Life**

Indonesia's Provisional People's Deliberative Assembly, a Guided Democracy equivalent of a parliament, confers the title "President for Life" on Sukarno.

1963: March **Sukarno reveals "Guided Economy"**

To complement his authoritarian "Guided Democracy," Sukarno promotes a "Guided Economy," emphasizing national self-sufficiency, including Indonesian industries functioning without foreign investments.

1963: May 1 **Indonesia takes over West New Guinea**

Fulfilling a United Nations settlement, The Netherlands turns over the responsibility of governing West Papua to Indonesia (see 1961: December 19).

1963: September 25 **Indonesia is hostile to newly independent Malaysia**

As Malaysia becomes independent from Britain, Sukarno views it as a colonial puppet state and announces that Indonesia will "gobble Malaysia raw." Some Indonesian troops are sent into Malaysia in what is called the *Konfrontasi* (Confrontation). A cease-fire between Indonesia and Malaysia is arranged by the United States in January 1964.

Late 1963 **Communists begin forcible land reform**

The PKI, extremely popular in rural Indonesia, with over two million members, begins forcibly taking land away from landlords and distributing it to peasants. The well-organized PKI is affiliated with trade unions and has direct ties to Communist China.

1964: March **United States aid is rejected**

Wary of American interference in Indonesia, Sukarno declares that the United States can "go to hell with its aid."

1964: August 17 **Sukarno declares "A Year of Living Dangerously"**

In a speech marking Indonesia's independence anniversary, Sukarno declares "A Year of Living Dangerously," speaking of "the Romanticism of Struggle" and warning his opponents that his revolutionary struggle will be victorious.

1964: December **Indonesia withdraws from the United Nations**

In protest of the United Nations' acceptance of Malaysia as a member, Indonesia withdraws its own membership.

1965 **Tensions rise between Communists and military**

With Sukarno's government close to Communist China, and the PKI being offered arms by Chinese premier Zhou Enlai for workers and peasants, the military is on edge. Sukarno is rumored to be in ill health.

1965: September 30 **A coup d'état fails**

A group of seemingly pro-Communist military officers, calling itself the September 30 Movement, attempts to seize power in a coup d'état. They execute five Indonesian army generals. The next day they take over government radio stations, announcing that they are averting a right-wing coup backed by America's Central Intelligence Agency. The involvement of the United States, China, and the PKI in the September 30 coup are a matter of lasting controversy, as are the roles of military rivals and Sukarno himself.

1965: October 2 **General Suharto rises to power**

General Thojib Suharto (b. 1921), a career army officer with experience in military campaigns on Java, is given the task of restoring law and order in the wake of the September coup attempt. Sukarno's power disappears, and Suharto's increases.

Suharto is born on Java and serves in the Dutch colonial army. In 1965 he becomes the army chief of staff and assumes executive power of Indonesia in 1967. He becomes president in 1968, a post to which he wins re-election six times, serving for thirty-two years. (See 1998: June.)

1965: December Massive violence kills hundreds of thousands of Indonesians

The coup attempt having failed, mob violence directed at PKI members takes place on a huge scale on Java, Sumatra, and Bali. With some military instigation, suspected Communists and their families are murdered by villagers. Chinese people are particularly victimized in the mass murder frenzy. Estimates of those killed in the weeks of the bloodbath range from 200,000 to half a million or more.

1966: March 11 Power is formally transferred to Suharto

Sukarno signs an executive order which transfers power in Indonesia to General Suharto, and the next day Suharto becomes acting president. He sets in process the "New Order," in which economic growth is emphasized by seeking foreign aid and investment. Suharto's economic planning is influenced by Indonesian academics known as the "Berkeley Mafia" because some had studied at the University of California at Berkeley.

1966: August 11 Indonesia transforms its foreign relations

In a complete change from Sukarno's adversarial foreign policy, Suharto formally ends the *Konfrontasi* (Confrontation) with Malaysia on August 11, 1966, and Indonesia rejoins the United Nations the next month. Indonesia moves from its Communist Bloc ties to closer relations with Western nations.

1967: August Association of Southeast Asian Nations is formed

A regional organization, the Association of Southeast Asian Nations (ASEAN) is formed, mainly as a non-Communist trading body, with Indonesia one of its charter members—along with Malaysia, Thailand, Singapore, and the Philippines.

1968 The military uses its power against rivals

Suharto's military hunts down people it considers to have been involved in the 1965 coup attempt, arresting some 200,000 and executing many. Some remain in custody until the 1980s and even 1990s. Suharto's principle of *"Dwifungsi"* (dual functions) gives the military a crucial place in social, governmental, and economic affairs.

1969 A referendum joins West Papua to Indonesia

In a referendum called the "Act of Free Choice," conducted with United Nations assistance, West Papua is said to give its assent to becoming part of Indonesia. The legitimacy of the referendum is questioned, as the voting is limited to certain community leaders, who simply give their consent to the territory's absorption. The Organisasi Papua Merdeka (OPM, Free Papua Movement) fights against joining Indonesia, and the Indonesian government launches a counterinsurgency campaign in West Papua.

1969 Five-year economic plans begin

Suharto's "technocrats" (fast-development economic advisors) devise a series of five-year plans for Indonesia's economic progress. The National Development Planning Council sets goals and programs for agriculture and infrastructure improvement, then for industry, transport, and communications. Foreign aid and investments fuel these efforts.

1970s Bottled tea becomes a popular drink

Sweetened cold tea in bottles, called *the botol*, becomes a popular soft drink in Indonesia, a concept that will later spread to other parts of the world.

1970s East Timor explores options regarding Indonesia

Different political factions in East Timor favor different solutions for the post-colonial future. Some favor integration with Indonesia; others, like the Revolutionary Front for an Independent East Timor (Fretilin), desire independence.

1970: June 21 Sukarno dies

Isolated from power, Sukarno falls ill and dies. The burial site requested by two of his five wives is denied, being too close to the capital, Jakarta, for the comfort of his successor, Suharto.

1971: July 3 A general election is won by Golkar

Golangan Karya (Functional Groups), known as Golkar, a party formed and backed by the military, becomes the most powerful political force in Indonesia. Under Suharto, other political parties decrease and are absorbed, leaving Golkar as a massive power-base. In 1971 Golkar wins the general election with 62.8 percent; it also wins over 60 percent of the vote in the 1977 and 1982 elections.

1971: November 6 A scientist arrives to study orangutans

Birute Galdikas (b. 1946), a Canadian scientist, arrives in Kalimantan's forest to begin her study of orangutans in their natural habitat. While studying the habits of the wild apes, she also becomes an advocate for their protection. Orangutans are threatened by the loss of their forests to logging and plantations. Adult orangutans are often killed so that the babies can be taken away and sold as pets. In 1982 the area of Galdikas's research, Tanjung Puting, is declared a national park by the Indonesian government.

1972 West Papua is renamed

The Indonesian government renames West Papua, calling it "Irian Jaya" (Victorious Irian). The island as a whole is known as New Guinea, and the independent eastern half is the nation of Papua New Guinea.

1972 *Dangdut* music rocks Indonesia

A new kind of folk-rock music, called *dangdut* after the sound of the beat of its Indian *tabla* drumming, sweeps the islands. Influenced by Middle Eastern music, Indian film soundtracks, old *kroncong* folk tunes, and Western rock, *dangdut* uses electric guitars and a heavy beat. Played in night markets, it becomes featured in Indonesian movies. Rhom Irama is the most famous *dangdut* star, and his songs have political and Islamic (against alcohol and corruption) content.

1974 The last Chinese-language school in Indonesia is closed

As part of an effort to force Indonesia's Chinese minority population to assimilate into the majority culture, Chinese-language schools are phased out. Only one (government controlled) Chinese-language newspaper is allowed; Chinese people are encouraged to take Indonesian names; and Chinese signs are forbidden on buildings.

1974: January 15 Riots break out in Jakarta

Unemployed youth and students riot in the streets of Jakarta during a visit by the Japanese prime minister. Intended as a protest against foreign domination of the Indonesian economy, the riots also express discontent against the authoritarian government and resentment of Chinese businessmen. The unrest is known as the Malari Affair (from an abbreviation of *Malapetaka 15 Januari*—disaster of January 15th).

1975: March The national oil company, Pertamina, is caught in scandal

While oil revenues help to grow Indonesia's economy, the national oil company, Pertamina (formed from Permina [see 1957: December] and other firms) becomes extremely corrupt. The extent of mismanagement by Pertamina's boss, General Ibnu Sutowo is revealed, and he is forced to resign with the company $10 billion in debt.

1975: August 26 East Timor is no longer administered by Portugal

While Portugal disengages itself from its colonial possession of East Timor, Australia seems to support the island territory's integration into Indonesia. The pro-independence group Fretilin (see 1970s) becomes aggressive, seizing arms and controlling much of East Timor. It is viewed as a left-wing threat by Indonesia, which backs integrationist factions against it.

1975: December 7 Indonesia mounts an invasion of East Timor

Dili, the capital of the former Portuguese colony which Fretilin had declared the Democratic Republic of East Timor, is invaded by some 30,000 Indonesian troops. On July 15, 1976, East Timor is declared a province of Indonesia.

1976–80 The death toll rises from Indonesia's occupation of East Timor

The Indonesian military occupation forces attempt to suppress continued Fretilin guerrilla resistance. Their counterinsurgency campaign tries to keep food supplies from the Fretilin forces and damages local agriculture, resulting in mass civilian starvation and disease. Of an East Timorese population of about 650,000, some 100,000 to 250,000 die during the first four years of Indonesian occupation.

1977: May 23 Exiles from Malaku hijack a Dutch train

Young activists from Malaku, protesting Indonesia's hold on their islands, hijack a commuter train and take a primary school hostage in the Netherlands to bring attention to their cause. After two weeks, Dutch troops take back the train, with several hijackers and a hostage killed in the raid.

1979 Novelist Pramoedya Ananta Toer is released from prison camp

Pramoedya Ananta Toer (b. 1925), Indonesia's foremost novelist, is released from arrest after fourteen years. Most of the time he had been held in a prison camp on Buru, a remote Malaku island. While held there because of his political views, Toer completes a series of historical novels, including *This Earth of Mankind* and *Child of all Nations*. Published overseas, the books win praise for their expressive portrayal of Indonesian history and aspirations, but all of Toer's works are banned in Indonesia.

1980s–90s Indonesia is a close ally of the United States

As an anti-Communist state, Indonesia becomes a major recipient of military aid from the United States and a strategic ally. However, the Indonesian government is often criticized in the United States Congress for its repression of East Timor and its lack of political freedom.

1980 Suharto's business ties become controversial

In 1980 controversy arises over Suharto's links with Liem Sioe Liong, an extraordinarily wealthy Chinese-Indonesian businessman whose conglomerates invariably win favorable contracts. Despite constraints on Indonesia's press, foreign reporters also expose the questionable business dealings of Suharto's children and other relatives.

1980: May 5 **A prestigious group calls for political freedom**

The Petition of Fifty, a group of influential academics, politicians, and retired military officers, calls for the government to grant Indonesia more political freedom. These demands have no effect on government policy.

1983–85 **A crackdown on crime goes outside the law**

In several Indonesian cities, the police and military crack down on crime by secretly executing as many as 5,000 suspected criminals, then dumping their bodies in public to serve as a warning to others.

1983 **Excessive logging and fires destroy rainforests in Indonesia**

With the government promoting export of rainforest timber from Kalimantan (Indonesian Borneo), Sumatra, and other regions, Indonesia achieves the highest deforestation rate in Asia, losing 700,000 to 1 million hectares (1,729,000 to 2,470,000 acres) each year. In 1983 massive forest fires in Kalimantan burn approximately 3 million hectares (7,410,000 acres) of rainforest and wetlands. Rich in plant and animal species, and home to indigenous people, the rainforests are logged and made vulnerable to fire by companies exporting tropical hardwoods to Japan and elsewhere. In many areas the natural forests are replaced by oil palm plantations.

1984: September 12 **The military shoots into crowd of Moslem opponents**

Moslem groups show signs of discontent, and a crowd of Moslems is shot into by government troops in Tanjung Priok, Jakarta's port. A pamphlet called "The White Paper" is distributed, disputing the government's account of the shooting. There are Moslem riots and bombing incidents in Jakarta, as religious groups oppose the government's forcing all political groups to avow support of *Pancasila* (see 1945: June 1).

1988 **Indonesia takes on a mediating role in Cambodian civil war**

Working through ASEAN, (see 1967: August) Indonesia mediates between warring Cambodian factions. Indonesian-brokered peace conferences in Jakarta and Paris result in a partial settlement to Cambodia's civil war in 1991.

1990–91 **Rebellion stirs in Aceh**

Fighting for land rights and Islamic identity, a separatist movement rises in Aceh, in northern Sumatra. More than one thousand Acehnese die during the rebellion.

1990: August 8 **Diplomatic relations are restored with China**

After decades of mistrust, Indonesia restores its diplomatic relations with China, with an interest in improved economic ties.

1991: September **Army suppresses strikes over subminimum wages**

A series of strikes over payment of less than minimum wages is suppressed by the army on Java. The idea that Indonesia's economic development relies on extremely low wages to attract foreign manufacturers draws attention overseas, resulting in an international boycott of Nike sports shoes in protest of low wages and poor conditions for Indonesian shoe factory workers.

1991: November 12 **Indonesian troops open fire on a procession in East Timor**

Indonesian troops open fire on a funeral procession for a slain activist in Dili, East Timor, killing an estimated one hundred people. Witnessed by foreign journalists, the event focuses world attention on the occupation of East Timor.

1992 **Sukarno's daughter, Megawati, becomes politically prominent**

Megawati Sukarnoputri (b. 1947), the oldest daughter of Sukarno, wins the chairmanship of the Democratic Party, the smallest of the three political parties still allowed in Indonesia.

1992: March **Indonesia rejects Dutch aid because of human rights criticisms**

Indonesia decides to reject aid from The Netherlands due to that country's continued criticisms of human rights violations in Indonesia. A foreign aid donor consultation group that had been led by the Dutch is disbanded, and a new group is formed without them, with World Bank support.

1992: December 12 **A huge earthquake shakes the island of Flores**

An earthquake, 6.8 on the Richter scale, devastates the island of Flores, south of Bali, killing more than 2,500 inhabitants.

1993: March 10 **Suharto is re-elected President for the sixth time**

For the sixth time, Suharto wins a five-year term as president. There is no obvious choice for his successor in the event that he might not run for a seventh term.

1993: May 8 Marsinah, a labor activist, is found murdered

Marsinah, a twenty-five year old east Java watch factory worker earning about $1.00 a day, who had been a workers' representative, is arrested, raped, and stabbed to death. Her death becomes an international symbol of Indonesia's violations of workers' rights. The U. S. government threatens Indonesia with sanctions on trading privileges unless it stops the military suppression of labor union activities.

1994 Indonesia hosts Asia-Pacific Economic Cooperation summit

Heads of state from around the Pacific Rim meet at Bogor, on Java, for a summit on Asia-Pacific Economic Cooperation (APEC). Attendees including China's President Jiang Zemin, and the United States president Bill Clinton, express their commitment to free trade and praise increased investment opportunities throughout the region.

1994: April 14 Anti-Chinese rioting occurs on Sumatra

Beginning as labor unrest by workers who are not getting paid the minimum wage, a disturbance in Medan, a Sumatran city, becomes a riot aimed at Chinese-owned businesses.

1994: June 21 The government curtails the press

Three weekly magazines, *Tempo*, *DeTik,* and *Editor*, are banned by the government for violating the ethics code for journalists and endangering national security. They had reported on conflicts between B.J. Habibie (b. 1936), the minister for research and technology, and the military over his arms purchasing methods. In June and August the military and police break up protests by hundreds of journalists.

1996: January–March Unrest shows discontent on Irian Jaya

In January, Free Papua Movement rebels kidnap several scientists on Irian Jaya, and two are killed in a military rescue operation. In early March indigenous people riot near a massive copper and gold mine operated by Freeport Minerals, an American company. The local people demand part of the environmentally controversial mine's profits, and Indonesian troops are called in to pacify them. Also in March, a riot breaks out in the Irian Jayan city of Jayapura, demonstrating local anger at resettlement of people from other parts of Indonesia there through the transmigration program.

1996: June 20–22 Opposition hopes focus on Megawati

While Megawati Sukarnoputri becomes a leading opposition hope, the government takes steps to unseat her as chairwoman of the Indonesian Democratic Party. In protest, her supporters occupy the Democratic Party headquarters in Jakarta. When rivals attack the building on July 27, rioters in the city rampage against government buildings and businesses. A government crackdown on activists follows.

1996: October 11 The Nobel Peace Prize is awarded to East Timorese activists

The Nobel Peace Prize is awarded to two East Timorese activists: Roman Catholic bishop Carlos Felipe Ximens Belo (b. 1948) and Jose Ramos-Horta (b. 1950), an exiled former Fretilin leader and journalist. The Peace Prize honors their "work towards a just and peaceful solution" to the conflict over East Timor.

1997: February Ethnic violence sweeps Kalimantan

In Pontiak, a city in West Kalimantan, indigenous Dayak people, angered by an attack on schoolgirls, massacre hundreds of people from Madura, an island off Java, who were settled on Kalimantan through the transmigration program.

1997: May 29 Golkar wins another general election

Golkar again wins the general election, this time with 74.5 percent of the vote. Megawati and her supporters are not allowed to run in the election, and there are accusations of fraud, as well as election-related rioting in Java and Kalimantan.

1997: July The Indonesian currency plunges, and the economy goes into sharp decline

Asian currencies fall in value against the U. S. dollar, as international investors lose confidence in the "tiger economies" of Southeast Asia, due to thier overspending on extravagant infrastructure projects and their ingrained corruption. The Indonesian currency, the *rupiah,* falls hard, losing some fifty percent of its value between July and October. Food shortages and massive unemployment are predicted for Indonesia.

1997: August Haze from Indonesian forest fires blankets much of Southeast Asia

With El Niño drought conditions prevailing, huge forest fires, caused by oil palm plantations' land-clearing and logging, spread across much of Kalimantan and Sumatra. In addition to the devastation of at least 800,000 hectares (nearly 2 million acres) of rainforest, the fires cause a haze of air pollution which spreads north to Malaysia and Singapore. The fire settles into peat bogs, smoldering uncontrollably. On September 26, a Garuda airliner crashes in Sumatra, where forest fire smoke limits visibility, and all 234 passengers are killed. Collisions between ships in the region are also blamed on low visibility due to the haze.

1997: November The Indonesian government takes measures to shore up the economy

The Indonesian government liquidates sixteen private banks which have run out of funds, including two owned by

Suharto's own relatives. The nation is revealed to have an immense, U.S.$117 billion foreign debt. Other countries come to Indonesia's aid with credit, including Japan, Singapore, the United States, Brunei, Malaysia, and Australia. The World Bank offers a U.S.$23 billion "bail-out" package, but it includes difficult financial constraints.

1998: February Drought and fires cause hardships

Irian Jaya suffers from severe drought, and hundreds of people there die from famine and disease. The haze of forest fires returns to the region, with the smoky air pollution from more than five hundred fires in Borneo and Sumatra reaching Singapore.

1998: March 10 Suharto wins seventh term

The National Assembly chooses Suharto to be president of Indonesia for his seventh consecutive term, and B.J. Habibie becomes vice president. Student demonstrations occur throughout Indonesia. In many urban areas, Chinese-owned shops and houses are burned by rioters, Chinese women are gang-raped, and many people of Chinese ethnicity flee the country in terror. Human rights groups charge that the government is scapegoating the Chinese minority to deflect the blame for rising prices of consumer goods.

1998: May 14 Rioting and anti-Suharto protests inflame Jakarta

Reacting to Suharto's continued rule and economic difficulties, thousands of rioters in the streets of Jakarta burn and loot office buildings, shops, and other businesses. Police and the military attempt to maintain order with troops and tanks. Many establishments attacked are linked to Suharto's family, but even more simply belong to resented ethnic Chinese. The outbreak of rioting follows Suharto's introduction of financial austerity measures, causing increases of food and other commodities, to follow the requirements of a U.S.$43 million rescue for Indonesia's economy from the International Monetary Fund. The Jakarta rampages are sparked when police shoot six student protesters. An estimated five hundred people die during the riots.

1998: June Suharto resigns as president after 32 years in power

As student protesters occupy the Parliament building in Jakarta, opposition leaders and legislators from Suharto's own Golkar party call on the seventy-six-year old president of Indonesia to step down. While the military encircles the students, causing expectations of a brutal crackdown on their demonstration, Suharto announces his resignation. The presidency is handed over to his vice president, B.J. Habibie (b. 1936), a devout Moslem "technocrat" with a fondness for grand schemes and little support within the army. Student leaders and other oppositionists, such as Moslem activist Amien Rais, appear to accept the transition while calling for widespread reforms, particularly multiparty elections. In the first weeks following the transition, some political prisoners are freed, constraints on the press are lifted, and a new openness sheds light on the Suharto family's economic empire, as well as environmental and labor issues.

1998: August 6 Troop withdrawal from Aceh is announced

Indonesia's armed forces chief general Wiranto announces that all troops will be withdrawn from Aceh, northern Sumatra, where they had been in occupation since an attempted uprising in the early 1990s. New openness following Suharto's resignation allows citizens to make open allegations of human rights abuse by the Indonesian military in Aceh, including revelations of mass graves of civilian victims. Smaller-scale troop withdrawals are also announced for East Timor after Habibie takes office.

Bibliography

Amnesty International. *Power and Impunity: Human Rights Under the New Order.* New York: Amnesty International, 1994.

Bellwood, Peter S. *Prehistory of the Indo-Malaysian Archipelago.* Honolulu, Hawaii: University of Hawaii Press, 1997.

Blair, Lawrence and Lorne Blair. *Ring of Fire.* New York: Bantam Press, 1988.

Broughton, Simon, ed. *World Music: The Rough Guide.* London: The Rough Guides Ltd., 1994.

Cribb, Robert. *Historical Dictionary of Indonesia.* Metuchen, N.J.: Scarecrow Press, 1992.

Far Eastern Economic Review. *Asia 1998 Yearbook: A Review of the Events of 1997.* Hong Kong: Review Publishing Company, 1998.

Fischer, Louis. *The Story of Indonesia.* New York: Harper & Brothers, 1959.

Fraser-Lu, Sylvia. *Indonesian Batik: Processes, Patterns and Places.* Singapore: Oxford University Press, 1986.

Frederick, William H., ed. *Indonesia: A Country Study.* Washington, D.C.: Library of Congress, 1993.

Holt, Claire. *Art in Indonesia: Continuity and Change.* Ithaca, N.Y.: Cornell University Press, 1967.

Koch, Christopher. *The Year of Living Dangerously.* New York: St. Martin's Press, 1979.

Legge, J. D. *Indonesia.* Englewood Cliffs, N.J.: Prentice-Hall, 1964.

———*Sukarno: A Political Biography.* New York: Praeger, 1972.

Lewis, Norman. *An Empire of the East: Travels in Indonesia.* New York: Henry Holt, 1994.

Lindsay, Jennifer. *Javanese Gamelan: Traditional Orchestra of Indonesia.* Singapore: Oxford University Press, 1992.

Lingard, Jeanette, trans. *Diverse Lives: Contemporary Stories from Indonesia.* Oxford: Oxford University Press, 1995.

Muller, Kal. *Spice Islands.* Lincolnwood, Ill: Passport Books, 1997.

Neill, Wilfred T. *Twentieth-Century Indonesia.* New York: Columbia University Press, 1973.

Pinto, Constancio and Matthew Jardine. *East Timor's Unfinished Struggle: Inside the Timorese Resistance.* Boston: Southend Press, 1997.

Ricklefs, M. C. *A History of Modern Indonesia.* Bloomington, Ind.: Indiana University Press, 1981.

Schwarz, Adam. *A Nation in Waiting: Indonesia in the 1990s.* Boulder, Colo.: Westview Press, 1994.

Stommel, Henry and Elizabeth Stommel. *Volcano Weather: The Story of 1816, the Year Without a Summer.* Newport, R.I.: Seven Seas Press, 1983.

Tsing, Anna Lowenhaupt. *In the Realm of the Diamond Queen: Marginality in an Out-of-the-way Place.* Princeton, N.J.: Princeton University Press, 1993.

Toer, Pramoedya Ananta. *This Earth of Mankind.* New York: William Morrow & Co., 1975.

Wallace, Alfred Russel. *The Malay Archipelago.*Reprint, Singapore: Oxford University Press, 1989.

Iran

Introduction

Located in southwest Asia, Iran, with an area of 636,296 square miles (1,648,000 square kilometers), is bound to the north by Azerbaijan, the Caspian Sea and Turkmenistan; to the west by Turkey and Iraq; to the east by Afghanistan and Pakistan; and to the south by the Persian Gulf and the Sea of Oman. Iran's major geographical features are the Elburz mountains (18,000 feet or 5,800 meters in elevation), which stretch from east to west in the north of the plateau and the Zagros range (13,000 feet or 4,200 meters in elevation), which stretches from north to south in the west. Two vast deserts, Loot and Kavir, form the plateau's center. The climate, moderate in the north, becomes increasingly warm as one travels south.

Iran has a population of 65.6 million with a population density of 103 per square mile (fifty-seven percent urban). Iran's ethnic mix includes fifty-one percent Persian, twenty-four percent Azerbaijani Turk, and seven percent Kurd. The principal languages spoken are Persian, Turkish, Kurdish, and Luri. The population is concentrated along the rivers Karun, Atrak, and Sefid, whose sources flow from the higher mountain elevations. Iran's population is more than ninety-five percent Shi'ite Muslim; the remaining population is comprised of 30,000 Zoroastrians, 330,000 Christians, and 32,000 Jews.

History

Archaeologists believe that as early as 6000 B.C., and well before the arrival and development of the Indo-European Iranians, the plateau was occupied by a primitive people with only rudimentary systems of governance and trade. Later, around 3000 B.C., an advanced kingdom known as Elam dominated the area.

Around 1500 B.C., Iranians (people of Indo-European descent) migrated from the north. By the seventh century B.C. they were monotheists who believed in the deity Ahura Mazda and a host of lower gods. Man, they reasoned, a creation of Ahura Mazda, struggled with an invisible evil intruder trapped in his domain. Guided by Ahura Mazda and aided by *farr,* humanity could be led by Zoroaster to a fulfilling and prosperous life on the earthplane.

With the rise of the Achaemenid ruler Cyrus the Great around 500 B.C., the Persian Empire was born. The Achaemenids overtook areas inhabited by the Medes, Babylonians, Lydians and Egyptians. Under kings Darius and Xerxes I, the empire extended as far as India and Greece.

The Achaemenians' later attempt at annexation of the Greek city-states led to an invasion by Alexander the Great of Persia. Alexander defeated the Achaemenid forces, and a century of Greek rule followed. Iranians' strong ties to their land and divinely ordained kings helped them overcome Hellenic domination. Building on the reconstruction efforts of the Parthians who moved into the area in the third century, the Sassanians, a priestly caste, reestablished the divine (and the secular) hierarchies that reflected the will of Ahura Mazda. The Sassanian era continued until the seventh century.

The inflexibility of the Greeks prevented the incorporation of Iranian culture into the Greek world. However, the Arab culture that overtook the Sassanian Empire in the seventh century exerted its influence on the Iranians, replacing Zoroastrianism with Islam.

The Safavid dynasty of Iranian rulers (1502–1736) made Shi'ism (branch of Islam) the official religion and permitted a degree of European thought and lifestyle. The fourth Safavid ruler, Shah Abbas (r. 1587–1628) moved the capital to Esfahan. The turquoise domes of the mosques in cities like Esfahan reflect the thriving Muslim culture, and contrast with the steeples of Christian churches also constructed during this era.

The Qajars, a Turkish tribe from northern Iran, gained power in the late eighteenth century. Qajar rule (1795–1925) was characterized by the intrusion of British and Russian political and economic domination and by a steady rise of nationalism. Twice defeated, Iran ceded the regions west (1813) and east (1825) of the Caspian to Russia. Debts owed Britain (incurred to pay war indemnities and royal expenses) led to additional concessions. These concessions helped push the Iranian economy to the verge of collapse.

Recognizing the shortcomings of the ruling elite, progressive individuals in the military, at court, and among the clerics called on patriotic Iranian intellectuals to foster Ira-

nian nationalism. Their goal was to use education, adoption of modern technology, and a new constitution to unify the people, reform the government, and overthrow foreign domination. A constitution was granted in 1906.

The Soviets engineered the creation of a Social Democratic Party in Baku, Azerbaijan, with land distribution forming a major plank in its platform. This led, in 1920, to the Jangali Movement and the formation of the Communist Party of Iran. The movement was defeated in 1922, and the Party was dissolved in 1934.

The era known as Pahlavi Iran (1925–79) began when Reza Shah, the first of the Pahlavi dynasty of rulers, took power. Pahlavi Iran, a feudal society, was divided along regional and tribal lines and dominated by foreign ideological and economic interests. The interests of Britain and the Pahlavis converged when unification of the country as a sizable market for manufactured goods, suppression of political consciousness through denial of education, and division of profits resulting from unencumbered access to natural and human resources was concerned. They differed where Iranian nationalism was concerned. In 1953 Iranian nationalism put an end to overt British dominance.

Using a program of secular nationalism, the Pahlavis made two attempts at reform. The first, a progressive reform devised to decrease political and economic dependence, ushered Iran into the new age. The second, known as the "White Revolution," failed for a number of reasons including flawed planning and disregard for the rights, let alone the sentiments, of Iran's growing middle class. Rather than adjusting Western progress to the needs of Iran, it sought to reshape Iran to the needs of the West.

Throwing off the British yoke was a national goal. Germany was approached twice to supply Iran with assistance, such as construction of a steel mill, to make Iran technologically self-sufficient. World War II (1939–45) prevented Iran from achieving this goal, however. In the 1950s, oil and the control of oil rights, began to dominate Iran's politics and economy. Muhammad Mosaddeq (Mussadiq), leader of the National Front, attempted to win assistance from the United States to prevent the Soviet Union from gaining control of oil in northern Iran. When the United States declined to assist and Britain imposed an embargo on Iranian oil, Mosaddeq used nationalistic fervor to win support for the expulsion of both the shah and British interests. The shah, supported by the United States and Britain, returned after just four days, and the United States gave financial assistance to help stabilize the situation. However, the shah could not end Iranian nationalism.

Ironically, the shah's return dealt a mortal blow to secularism, the very ideal on which the government, ruled by his family, was based. In the 1960s, the question was not how, but when, foreign interests would be overthrown, and who would oversee the oil money flooding the country. The secularists, the shah among them, opted for a strong military, modern urban centers, and the building of dams to generate hydroelectric energy. The clergy, representing the anti-secular point of view, demanded a more equitable sharing of wealth and power, a better judiciary, a less Western society, and most importantly, governance by an assembly of Islamic jurists. To suppress these demands, the shah resorted to repression. Using the secret police force, SAVAK, originally created to safeguard against Communist infiltration, he quelled all opposition and continued to Westernize the country. By 1963, the division between the secularists and the clerics was great.

The Communist Party was renamed the Tudeh Party in 1942. By 1944, claiming 25,000 members, it participated in the Fourteenth Majlis (National Assembly), capturing eight seats. By 1946, it created two pro-Soviet independent republics, Azerbaijan and Mahabad, in northern Iran. In 1949, blamed for an unsuccessful attempt on the life of the shah, Tudeh was outlawed; surveilance of its clandestine activities was assigned to SAVAK.

In 1965, the Tudeh was internationalized. Furthermore, its main objections to a secular-nationalist Iran having been neutralized by the White Revolution (the shah's 1962–63 liberalizing reforms), it made common cause with Ayatollah Ruhollah Khomeini (ayatollah is the highest rank in Shia Islam). Together they sought to rescue Iran from a total cultural, socio-economic, and technological dependence on the West. Henceforth, Iran was to defend its own interests against both the East and the West.

As a theocracy (religious government), Iran's affairs are administered by a *Faqih* (supreme religious leader), a president, and a cabinet. The president and the cabinet do not have independent decision-making powers. They answer to the Faqih and a group of religious authorities appointed by the Faqih. The Iranian constitution codifies Islamic principles of government for a unicameral legislative body, the Islamic Consultative Assembly. Judicial affairs are administered by the Supreme Court.

Timeline

5000 B.C. Early humans in Iran

Sialk, a settlement to the south of present-day Tehran, is ruled by women who serve as creators of agriculture and a repository of knowledge about edible roots. Skilled hunters, gatherers, and potters, women also serve as guardians of the fire. Similar settlements exist at Susa, Tepe Hisar, Tepe Gujan, Zuriyeh, and Hasanlu.

The dead are buried under the floor. Shells from the Persian Gulf indicate familiarity with trade, while carved bones, hammered objects, stone tools, knife blades, sickle blades, axes, and scrapers bespeak the existence of a rudimentary civilization. Pottery, although coarse, is painted with primitive designs mimicking everyday objects. A primitive textile industry uses stone spindles. *Pisé* housing protects them against the elements.

4000 B.C. Pre-Elamite civilization

The settlers barter among themselves: wheat and barley for furs and arrows. Communal labor is employed for building houses, cleaning ground, and providing irrigation. Use of the horse facilitates transportation and travel, while discovery of precious stones leads to the creation of jewelry. Metal and pottery works display realistic designs of ibexes, birds, and boars on bowls, utensils, and jars. Women serve as commanders of the army. Next-of-kin marriage is practiced.

3000 B.C. The Elamite invasion

Signatures on cylinder seals identify ownership or mark prices. Pottery, produced on the wheel, carries more sophisticated designs; jewelry is refined. Bricks, made in molds, improve housing.

2000 B.C. The Aryan invasion

The Mede and Pars tribes invade the plateau through the Caucasus. The Medes settle the region south of the Caspian Sea, assume leadership, and create an empire; the Pars move south to Parsamush.

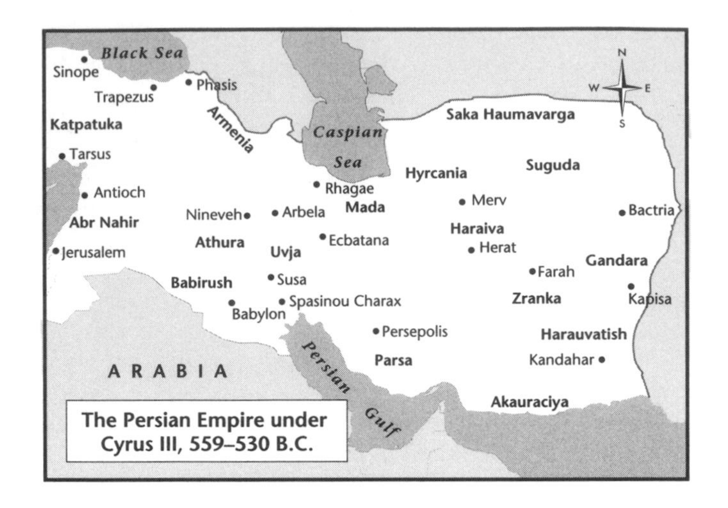

The Persian Empire under Cyrus III, 559–530 B.C.

1000 B.C. The Iranian world order

The newcomers establish an *Ahuric* (divine) order comprising the king of kings, who occupies the uppermost level of the hierarchy, and lower kings who, in turn, are supported by princes, nobles and free men. Similarly, a hierarchy of priests, landowners, craftsmen, peasants, and slaves appears that caters to the spiritual and material needs of the ruling class.

614 B.C. Cyaxeres captures Ninevah

The Medean chief, Cyaxeres (Uvkhstra), captures the Assyrian capital, Nineveh; Parsamush becomes a vassal (feudal state) of the Medes.

559 Cyrus the Great

Cyrus III (r. 559–530 B.C.) declares Anshan independent, captures Ecbatana (Medean capital), and incorporates Lydia and Babylonia into the Achaemenian Empire.

Founder of the Persian Empire, Cyrus is a great military commander and a generous and broad-minded individual. He does not impose his will or his form of government on his subjects nor does he keep captives. Indeed, he frees some 40,000 Jews held in captivity in Babylonia (605–562 B.C.) and assists their return to Jerusalem.

530 B.C. Cambyses III

Before he marches on Egypt, Cyrus's eldest son, Cambyses III (r. 530–522 B.C.) secretly kills his own brother. While he is in Egypt, a Magian usurper, pretending to be the slain bother, ascends the throne. On his way back to the heartland, Cambyses dies in Egbatana, Syria.

522 B.C. Darius the Great

Darius I (r. 522–486 B.C.) overthrows the usurper Gaumata, assumes the kingship, and creates a powerful, centralized state. During his first year, he fights nineteen battles, captures nine kings, and annexes Egypt. He replaces the feudal system that divided Iran with a system of *satrapies* (administrative units) each consisting of a network of governors, military generals, secretaries, tax collectors, and spies. He crushes the revolts of the Greek cities of Asia Minor (see 499 B.C.) but is defeated at Marathon (see 490 B.C.).

Darius is a warrior, an administrator, and a builder. He orders the Temple at Jerusalem to be rebuilt (519 B.C.) and expands the Persian Empire only after he becomes convinced that he rules over a prosperous people. His palace at Persepolis still stands. His *Chapar* (a system of communication), consisting of 1,677 miles of roads and 111 stations, later serves as a model for the United States' Pony Express.

519 B.C. Temple is rebuilt

The Jewish Temple at Jerusalem is rebuilt at Darius's behest.

499 B.C. Darius crushes a revolt

Darius crushes a revolt by the Greek cities of Asia Minor.

490 B.C. Athenians defeat Darius

Athenians defeat Darius at Marathon and his empire shrinks to the limits of Asia Minor.

486 B.C. Xerxes I

When Darius dies, his son, Xerxes I (r. 486–465 B.C.), viceroy of Babylon, ascends the throne. Furthering his father's expansionist policy, Xerxes defeats the Spartans at Thermapolae but is defeated at Salamis (see 480 B.C.). After Salamis, Xerxes returns to Iran, gives up desire for conquest, and becomes involved in court affairs. He is assassinated by his uncle.

484 B.C. Egypt and Babylonia become part of the Empire

Under Xerxes I (see 486 B.C.), Egypt and Babylonia become integral parts of the Empire of the Medes and Persians.

480 B.C. Xerxes is defeated

Xerxes defeats the Spartans at Thermapolae but is defeated at Salamis. The Persian defeat at Salamis turns the tide in the war against the Greek city-states. After the Persian defeat at Mycale, the following year, Xerxes's forces retreat back to Persia.

465 B.C. Artaxerxes I

During the rule of Artaxerxes I (r .465–424 B.C.), Athens, Greece, keeps cultural contact with Iran. At the same time, Artaxerxes secretly concludes the pact of Delos with other Greek states to invade Iran.

424 B.C. Iran's role in the Peloponnesian War

Darius II (r .404–358 B.C.), son of Artaxerxes I, brings about the Peloponessian War by making gold an issue in the conflict between Athens and Sparta.

404 B.C. Rule of Artaxerxes II

Artaxerxes II (r. 404–358 B.C.), son of Darius II (see 424 B.C.), becomes king and pursues an expansionist policy, using gold rather than might to bring Athens and Sparta under Iranian control. In the process, Artaxerxes neglects his own provinces; his unhappy *satraps* (viceroys) destroy his grandiose plan (see 400 B.C.).

401 B.C. Artaxerxes defeats Cyrus

Artaxerxes defeats Cyrus, his brother, at Cunaxa.

400 B.C. Xenophon leads Greeks home

Xenophon leads ten thousand Greeks home, safely, from Cunaxa after the death of Cyrus.

383 B.C. Birth of Philip II of Macedonia

As king and supreme commander of the League of Corinth, Philip II of Macedonia (383–336 B.C.) revolutionizes the art of warfare and builds a powerful state with the ultimate goal of crushing Persia. He is assassinated, however, before he can carry out his planned invasion of Persia. He is succeeded by his son, Alexander the Great (see 336 B.C.).

358 B.C. Rule of Artaxerxes III

Strong but brutal Artaxerxes III (r. 358–338 B.C.), son of Artaxerxes II, reconquers Egypt and reestablishes unity in the Empire. After his assassination, his son, Arses (r. 338-335 B.C.) rules the Empire.

338 B.C. Rule of Arses

Arses (r. 338–335 B.C.), son or Artaxerxes III (see 358 B.C.), becomes king after his father is assassinated.

336–323 B.C. The Age of Alexander the Great

Alexander (356–323 B.C; r. 336–323 B.C.), son of Philip II of Macedonia (see 383 B.C.), takes over his father's leadership when Philip II is assassinated. Alexander frees the Greek states of Asia Minor. Alexander then occupies Phoenicia, captures Egypt, and defeats the Persians at Granicus, Issus, and Gaugamela. The Persians' four capitals and their territories in Central Asia become parts of Alexander's empire.

Alexander owes his success to his father and Aristotle. Philip II taught Alexander the art of war. Aristotle, the founder of the scientific method, instructed Alexander in the finer art of building cultures. Using this dual approach, Alexander intends to unite the Greeks and the Persians and inaugurate a new era of trade, enhanced civilizations, and commonality of cultures.

335 B.C. The Last Achaemenian monarch

Although courageous, Darius III (r.335–330 B.C.) is fooled by the might of his own army. Defeated by Alexander, Darius III escapes to Ecbatana and the northern provinces.

330 B.C. Bessus kills Darius III

Bessus, satrap (viceroy) of Bactria, kills Darius III for his jewelry, ending the Achaemenian empire.

323 B.C. Death of Alexander the Great

Alexander (see 336–323 B.C.) dies in Babylon.

312 B.C. Seleucid Empire established

Selecus I Nicator (356–280 B.C.), the satrap of Babylonia, inherits most of Alexander's conquests in the east. A brilliant Mesopotamian general, Selecus I builds Seleucia, and through courage, common sense, and fair government, establishes the Seleucid empire.

250 B.C. Bactria and Parthia break away

Bactria and Parthia declare independence from Seleucid rule. The Bactrians populate what is present-day northern Afghanistan. Alexander chooses Bactria as one of the main centers for the development of his new Greco-Persian alliance of cultures. The Parthians are a nomadic people of Iranian extraction, located to the east of the Caspian Sea. Under the Arsacid dynasty, they rule Persia for several centuries.

246 B.C. Ptolemy III rules

Ptolomy III (d. 222 B.C.) occupies Mesopotamia.

242 B.C. Antionchus III the Great

Antiochus III (242–187 B.C.) reestablishes Seleucid rule over Bactria and Parthia (217 B.C.) and extends Seleucid domain as far as present-day Kabul, Afghanistan. He is the commander that goes the farthest to reunite the divided empire of Alexander and unify its disparate cultures.

190 B.C. Antiochus defeated

Antiochus III is defeated by Rome at Magnesia, Lydia.

175 B.C. Antiochus IV faces challenges

Antiochus IV, Epiphanus (175–164 B.C.), son of Antiochus III, becomes king. He faces a Jewish rebellion under the Maccabees and a quickly disintegrating empire. His efforts to defeat the Parthians and restore the empire to its former grandeur fail.

171–138 B.C. Mithradates I

Mithridates I (d. 138 B.C.; r. 171–137 B.C.), taking advantage of the struggles between the Seleucids and Rome, incorporates Media and Mesopotamia into independent Parthia and expands his kingdom to the former Achaemenian limits. To accommodate his Greek subjects, he establishes good relations with Greece. It is the Greeks, however, that revolt against him.

129 B.C. Ctesiphon is built

The city of Ctesiphon is built on the east bank of the Tigris River, facing Seleucia. The city serves as the capital of the Parthians and their Sassanian successors.

57 B.C. Orodes I comes to power

Orodes I (56–36 B.C.) becomes Emperor of Iran.

53 B.C. Orodes I defeats Crassus

Orodes I defeats Crassus at Carrhae, captures Syria and Judea, and invades Asia Minor. Ventidius restores lost eastern provinces to Rome.

20 B.C. Augustus establishes the Roman Empire

Through political skill and intelligence, Gaius Octavius Augustus (63 B.C.–AD14) transforms the post-Caesar chaos in Rome to his own advantage and establishes himself as Rome's first emperor. He regains Armenia and Transcaucasia and restores the Roman standards captured at Carrhae.

AD 51–77 Reign of Vologeses I

Vologeses I (AD 51–77) and Roman general Corbulo make Armenia a buffer between Parthia and Rome. Vologeses makes peace with Nero in exchange for granting his brother, Tiridates, the rulership of Armenia. The de-*Hellennization* process that begins under Volgeses leads to the revival of *Zoroastrianism* under the later Parthians and Sassanians.

224 Reign of Ardashir I

Founder of the Sassanian Empire, Ardashir I (r. 224–241) expands the Iranian frontiers in all directions, reduces the power of local kings, and centralizes the government. Furthermore, he replaces the liberal Seleucid and Parthian practices with his own imperial brand of forced conformity. His doctrine of the inseparability of state and religion guides Iran for four centuries under his successors and continues to influence political and religious thought in Iran.

241–272 Reign of Shapur I

A great soldier and empire-builder, Shapur I (r. 241–72) defeats and captures the Roman Emperor Valerianus at Edessa. He keeps the Emperor Valerianus prisoner for the rest of his life and humiliates him by having him kneel before his horse every time he mounts. Shapur I commands a wide range of interests as evidenced by his orders to translate Greek and Indian works into Middle Persian and by his support of Mani, the founder of *Manicheism.* Manicheism is a dualistic (worships both good and evil) religion that borrows some of its doctrines from Zoroastrianism.

309 Birth of Shapur II

Shapur II (309–79) becomes ruler of Iran at birth. His rule is marked by both cultural expansion and political triumph.

313 Constantine the Great legalizes Christianity

Roman emperor Constantine (280?–337) the Great legalizes Christianity with the Edict of Milan. This act places great strain on the political relations between Iran and Rome; Iranians view their Christian compatriots as Roman spies. He converts to Christianity on his deathbed in 337.

459 Hephthalites dominate Iranian affairs

During the rule of King Firuz (r. 459–484), Iran is plagued with a seven-year drought, an invasion by the Hephthalites, and religious conflicts among its prominent Christian sects. Defeated and unable to pay war indemnities, Firuz leaves his son, Qubad, as a hostage with the nomads. From this time until the time of Khusrau I (see 531), the Hephthalites interfere in almost all Iranian affairs, including succession.

484 Firuz is killed by the Hephthalites

Hephthalites kill Firuz, throwing Iranian affairs into chaos.

488–96 First rule of Qubad

Qubad rules twice. This first rule is marred by the interference of the Hephthalites and by the emergence of Mazdak, a reformer who preaches a new, dualistic religion imbued with communistic ideas. The nobles, whose interests are threatened, try and imprison Qubad.

499–531 Second rule of Qubad

During his second term, Qubad restores stability, discredits Mazdak, and persecutes his followers.

531 Rule of Khusrau I Anushirawan

Qubad undermines Mazdak, and the Mazdakite movement ends. Known as the Just and the *Anushirawan* (the immortal soul), Khusrau I (531–78) welcomes nine exiled Athenian philosophers to his court and, with their help, establishes Gundishapur University. Their legacy remains as a unique model of learning even after they return to Greece (see 549). Khusrai I also settles the conflict brought about by the Mazdakite movement between the lower-classes and the aristocrats and reforms the administration, the military, and the government.

549 Athenians return to Greece

Athenian philosophers (see 531), who establish a model for learning in Iran, return to Greece.

570 Prophet Muhammad is born

Muhammad (570–632), an ordinary man from the town of Mecca in present-day Arabia, is born and lives as a trustworthy merchant. According to Islamic belief, Muhammad receives Allah's command while living in Mecca. Accounts of Muhammad's words and actions are incorporated in the body of the *Shari'a* or Islamic Law.

575 Yemen is captured

Persians capture Yemen and expel the Abyssinians.

590–628 Rule of Khusrai II Parviz

Unlike his grandfather Khusrau I, Khusrau II (d. AD 628; r. 590–628) is cruel and selfish. Interested in luxury, he lives in an opulent palace surrounded by the most precious treasure chests known in the entire history of Iran. Expanding the empire, he captures Hira and Jerusalem; takes the "True Cross" to Ctesiphon and lays siege to Constantinople. Defeated by Emperor Heraclius (627–28) near Ctesiphon, Khusrau II is murdered by his son.

590 Bahram Chubin takes the throne

Bahram Chubin (590–91) usurps the throne.

591 Khusrau II returns

Khusrau II (see 590–628) returns and expands the Empire.

622 The *Hijra,* Muhammad's flight from Mecca

In Mecca, discussion of Muhammad's prophecy leads to plots to assassinate him. Muhammad flees to Medina from Mecca. Usually referred to as the *Hijra,* this flight marks the beginning of the Islamic lunar calendar. In Medina, Muhammad serves as judge, law-giver, and commander.

627–28 Khusrau II defeated

Emperor Heraclius defeats Khusrau II near Ctesiphon.

627–33 Rule of several kings

Qubad II, Ardashir III, and several kings rule; Muhammad's message arrives at the Sassanid court.

632 Muhammad dies

Muhammad dies, leaving no successor. Debates over succession divide the community into rival Sunni and Shi'ite sects. The *Sunnis,* following the Arab tribal rule, elect Abu Bakr through consensus. The *Shi'ites* opt for Ali, the Prophet's cousin and son-in-law. The Shi'ites believe in the supremacy of bloodline. They also believe that people of certain bloodlines have access to knowledge and wisdom that is unperceived by ordinary individuals.

633–51 Last Sassanian monarchs

A dozen kings succeed one another before Yazdagird III (r. 633–651), the last *scion* (son of a prominent figure) of Khusrau II, is discovered in Istakhr where he has been hiding from king-makers. During his rule, Iranians are defeated at al-Qadisiyya and Nihavand (642).

642 Defeat at Nihavand

Iranians are defeated at Nihavand.

651 End of Sassanian empire

Yazdagird III is assassinated at Merv; Sassanian empire ends.

661 The Sunni/Shi'i split

Kharijite Ibn-i Muljim's assassination of Ali results in the division of Islam into two sects: Sunni and Shi'i. Ali forms the foundation of the Shi'ite doctrine of the *Imamate* and serves as the first *Imam.* His descendants, emphasizing the significance of connection with the "knowledge" of the Prophet, promote Shi'ite thought in secret. The Sunnis follow the *Qur'an* and the four schools of law.

762 Establishment of the *Ja'fari Doctrine*

Imam Ja'far al-Sadiq establishes the *Ja'fari Doctrine,* the official religious doctrine of Iran which states that Ali and his successors are divinely ordained as sole authority to reveal the true meaning of the *Qur'an.*

765 Death of Ja'Far al-Sadiq

Ja'far al-Sadiq, scholar and sixth Shi'ite Imam, dies.

813–27 Ma'mun tries to unite sects

Upon becoming *caliph* (religious leader of all Islam) in 813, Ma'mun, using Mu'tazilite principles, attempts to bridge the gap between the Sunni and the Shi'i factions; he even nominates Imam Ali al-Riza to be his heir apparent (817) and proclaims that the *Qur'an* is created (827). Ma'mun's contributions to the development of Islamic culture include support for native scholars and translations of major Greek works into Arabic.

821–872 Rise of Iranian cultural awareness

Ma'mun's efforts to unite Islam are supported by the Taherids and Samanids, who are awarded the eastern lands and who pay nominal allegiance to Baghdad. The Samanids (850–999) are of special importance for their role in the revival of ancient Iranian traditions and for their promotion of Persian culture, especially the Persian language.

873 First disappearance of the Twelfth Imam

The Eleventh Shi'ite Imam dies. His son, whose existence is often in dispute and who would be the Twelfth Imam, disappears from 873–940. This period is referred to as the *Lesser Occultation*. Shi'ites believe that the Twelfth Imam will remain in hiding until the end of time.

922 al-Hallaj executed

Celebrated Sufi al-Hallaj (858–922) is executed by orthodox religious authorities because he undergoes moments of extreme ecstasy during one of which he utters *ana al-Haqq* (I am the Truth).

923 al-Tabari dies

Abu Ja'far Muhammad ibn Jarir al-Tabari, Muslim historian and commentator famous for his *Annals of Apostles and Kings,* dies.

925 Muhammad Zakariyya al-Razi dies

Outstanding Muslim physician al-Razi (b. c.841) dies. An encyclopedic scholar interested in philosophy, alchemy, mathematics, and medicine, al-Razi recognized smallpox and measles as two distinct diseases that plague children.

932 Buyid ruler takes control

Buyid ruler, Mu'izz al-Dawlah, assumes control of northern Iran, makes the Abbasid Caliph a figurehead.

940 Second and last disappearance of the Twelfth Imam

The Fourth Representative of the Hidden Imam refuses to name a successor. The "Great Occultation" begins and will last until the coming of the *Mahdi* (a Muslim messianic leader). Meanwhile, the affairs of the community remain in the hands of the deputies of the Hidden Imam. Ayatollah Ruhollah Khomeini's claim to the rulership of Iran in the twentieth century draws on this source of power.

941 Poet Rudaki dies

Abu 'Abd Allah Ja'far Rudaki (858–941), Central Asia's celebrated bard, dies. Born in the district of Rudak near the city of Samarqand, he memorizes the entire text of the *Qur'an* by the time he is eight years old. A special ward of Nasr ibn Ahmad (r. 913–42), he becomes both rich and famous. Rudaki is known mostly for his *Ju-yi-Muliyan* (The Muliyan Brook) ode and his simple style that reflects the charm of the pre-Islamic literature of Iran.

962–63 The Shi'ite ascendancy

Buyids institute the ritual of public mourning for al-Hussein and his followers to be held on the tenth of *Muharram* (a month of the Islamic calendar). Younger son of Ali and grandson of the Prophet, al-Hussein is killed in Karbala by the forces of the Sunni Caliph, Mu'aviyyeh. The way is paved by the Buyid ruler, Mu'izz al-Dawlah, who assumes control of northern Iran in 932 and makes the Abbasid Caliph a figurehead.

980 Great epic poem, *Shahnama* (Book of Kings), is written

Firdawsi (also known as Firdowsi [Abdul Qasim Hassan ibn-I-Ishaq ibn-I Sharafshah], 940?–1020) writes his major work, *Shahnama* or *Shahname* (Book of Kings), around this time. Firdawsi, born to a family of landed gentry in the town of Tus, grows up in a rural culture suffused with the words and deeds of legendary heroes. At the age of forty, he sets himself the task of collecting, organizing, and writing in verse the most cherished stories and myths about his homeland. The resulting epic work, *Shahnama,* is 60,000 couplets in length and forms the foundation of Iranian national identity.

1020 Death of the poet, Firdawsi

Firdawsi (935–1020), author of the *Shahnama* (Book of Kings), dies. (See 980.)

1037 Abu Ali Ibn-i Sina dies

Physician and philosopher Ibn-i Sina (980–1037) dies. Usually cited alongside Hippocrates and Galenus, Ibn-i Sina is known in the East as *Shaykh al-Ra'is* and in the West as the "Prince of the Physicians." He leaves between 276 and 456 manuscripts, contributions to the fields of medicine, phenomenology, philosophy, alchemy, mineralogy, mathematics, literature, astronomy, and music. Of these contributions, between forty-four and fifty-nine books and articles are devoted to medicine. The contents of nine of these books, written in Arabic and Perso-Tajik languages, are in poetry; the rest are in prose.

1040 Beginnings of Turkish ascendancy

The defeat of the Ghaznavids by the Seljuqs at Dandanqan marks the beginning of Turkish domination of Iranian lands. All hope of possible revival of Samanid power dissipates as regions as far away as Constantinople fall into Turkish hands.

1047 Abu Rayhan al-Biruni

Abu Rayhan al-Biruni (973–1047), scientist, philosopher, and scholar, dies. Biruni studies mathematics and the theory of numbers and leaves copious publications including *al-Tafhim.* This most important work, written in both Arabic and Persian, deals with astronomy, mathematics, geometry, and geography. al-Biruni's most remarkable work, however, is his calculation of the circumference of the earth, twelve miles short of twentieth century calculations.

1070 Saljuqs defeat Byzantines

The Saljuqs of Rum defeat the Byzantines at Manzikert and gain control of eastern and central Anatolia.

1090 Hassan Sabbah establishes a reign of terror

Hassan Sabbah seizes Alamut and exerts control through violent responses to acts against him. During this time, the Nizari branch of the Isma'ili Shi'ites begins.

1092 Khawja Nizam al-Mulk assassinated

Prime Minister Nizam al-Mulk (1017–92), famous for his *Siasatname* (Book of Government) in which he guides the king through the tortuous path to just government, is assassinated. His adherence to the strict Sunni rules alienates the Isma'ili minority, the assassins who murder him.

1122 Omar Khayyam dies

Omar Khayyam (1021–1122), Persian philosopher, mathematician, and poet, best-known for his work *Ruba'iyyat* (Quatrains) that was translated into English by Edward Fitzgerald, dies. Khayyam finds life at once bitter and sweet; bitter because of the depth of the vacuum in which it rests and sweet because of the delight it imparts to plants, animals, and man. His *Jalali Calendar,* completed at the behest of Seljuq Malik Shah (1079), bespeaks the depth of his knowledge in mathematics and astronomy.

1140 Birth of Nizami of Ganja

Persian poet Nizami Ganjavi (c. 1140–1203 or 1217), well-versed in mathematics, astronomy, medicine, jurisprudence, music, and the arts, is born. Nizami incorporates the knowledge of his time into his poetry. His *Khamsa* (Quintet) consists of *Makhzan al-Asrar* (the Treasury of Mysteries) about esoteric subjects; *Khusrau va Shirin* (Khusrau and Shirin) about the power of love; *Layla va Majnun* (Leyli and Majnun) about the object of love; *Haft Gunbad* (Seven Princesses) about the many faces of love; and *Eskandar Nama* (The Romance of Alexander) about knowledge and might. The beauty of the *Khamsa* is unsurpassed in Persian literature.

1210 Jalal al-Din becomes a Sunni believer

Jalal al-Din, Master of Alamut and Chief of the Assassins becomes a Sunni believer.

1221 Death of Farid al-Din Attar

Persian poet Attar (1142–1221) dies. A son of a prosperous pharmacist, Attar has an excellent education, especially in medicine, Arabic, and theosophy. His major poetic works include *Asrar Nameh* (Book of Secrets) about Sufi ideas, *Elahi Nameh* (Divine Book) about asceticism, and most importantly *Manteq al-Tayr* (The Conference of the Birds) in which he crosses the seven valleys of love.

1256 Ilkhanid dynasty established

Hulagu Khan conquers Alamut, establishes Ilkhanid dynasty.

1273 Jalal al-Din Rumi dies

Sufi Jalal al-Din Rumi (1207–73) dies. Jalal al-Din's father leaves Balkh at the time of the Mongol invasion (1219–20) and settles in Konya, Turkey. In 1244, Jalal al-Din meets Shams-i Tabrizi, his teacher, whose name he chooses as a pseudonym and for whom he writes a *divan* or collection of poems containing 36,349 *distichs* (two successive lines of verse that form a unit) and 1,983 quatrains. Rumi's best-known work, however, is *Mathnavi-i Ma'navi* (devoted to the Intrinsic Meaning of all Things). Containing 7,000 couplets, the *Mathnavi* begins with *The Song of the Reed,* suggested by Attar.

1292 Poet Sa'di dies

Shaykh Muslih al-Din Sa'di (1213–92), Iran's cultural icon, dies. Sa'di travels extensively in the West as far as Mecca and, allegedly, in the East as far as Transoxania and, possibly, India. He is known for his 1257 *Bustan* (The Garden) and the 1258 *Gulistan* (The Rose Garden). *Gulistan* is the first classical Persian work to be translated into a Western tongue. The themes of this quintessential Muslim humanist include justice, love, humility, acceptance, contentment, and repentance.

1369 Tamerlane victories

Tamerlane (1336–1405) conquers Khurasan and Transoxania and establishes court at Samarqand. In a series of campaigns against Persia, Tamerlane captures Herat in present-day Afghanistan (1380). In 1393, he captures the Fars province of present-day Iran and Iraq. In 1398, he invades and conquers India. His most successful victory happens when he captures the Ottoman Sultan Bayazid in the Battle of Angora (1402). Tamerlane dies en route to China.

1389 Death of Shams al-Din Hafiz

Persian poet Hafiz (1320–89), who allegedly recites the *Qur'an* in fourteen different forms, dies. Versifying some of the main themes of the holy book is, perhaps, the reason behind the special appeal that Hafiz' *divan (collection of poems)* has for the general public. At times placed at the level of the holy *Qur'an,* some Iranians use Hafiz' divan to look into the future.

Hafiz is undeniably the master of the art of *ghazal* (sonnet), which he develops and perfects. The interpretation of Hafiz's ghazals has been highly controversial. The controversy arises from Hafiz's use of allegorical symbolism next to profane love.

1480 In the wake of Mongol Rule

Two competing Turkish dynasties, the Qaraquyunlu and the Aqquyunlu, rule in the west. In Transoxania, defeated by Russia, Uzbek descendants of Chingiz Khan's son, Juji, fill the vacuum.

Modern History

1501–24 Portuguese seek trade relations

The Portuguese, moving towards India, seek trade relations with Iran.

1502–1736 Safavid era

The marriage of Shahrbanu (daughter of Yazdigird III) and al-Hussein (son of Imam Ali), belief in the Hidden Imam's absolute right to rulership, and a need to rescue Iranian culture from absorption by Arab and Turkish Sunni cultures prompt Isma'il (see 1499) to unite the seven major Qizilbash tribes, defeat the Aqquyunlus, enter Tabriz, and crown himself king. Isma'il adopts Shi'ism as Iran's official religion even though only a quarter of the population is Shi'ite.

1524 Sir Anthony Jenkinson establishes commerce

Responding to Portuguese intentions on Iran (1501), the British send Sir Anthony Jenkinson to Iran to establish commercial relations.

1578–98 Shah Abbas the Great

Internal tribal struggles, Uzbek unrest in Khurasan, and the Ottomans' threat to Qazwin define the circumstances under which Shah Abbas I (r. 1578–98) comes to power. The killing of his meddling guardian, reduction of the Qizilbash to ceremonial guards, and involvement of a young cadre of officials in governmental affairs constitute Abbas's initial steps at reform. Helped by foreign advisors Anthony and Robert Shirley, he introduces a new military system, builds cannons and ships, and trains soldiers. A new system of education, a new approach to medicine, a flourishing trade, and a new lifestyle distinguish Abbas's reign.

1600 The British in Safavid Iran

Unable to satisfy the nobles and the clergy, the British close their silk textiles and armaments factories and leave. British military aid to Persia, however, continues (see 1602).

1602 Portuguese expelled

Aided by the British navy, Iranians expel the Portuguese from the Persian Gulf.

1640 Death of Mullah Sadra

Mullah Sadra, Persian theologian and philosopher, dies.

1722 Mahmud Afghan proclaims himself king

Mahmud Afghan takes Esfahan and proclaims himself king, ending the Safavid dynasty.

1736 Nadirquli establishes dynasty

Nadirquli (r. 1736–47) drives the Afghans out, establishes the Afsharid dynasty.

1750–95 Zand Dynasty

Karim Khan (r. 1758–79), a tribal leader from Fars province, defeats all contenders for the Iranian throne and establishes himself in Shiraz. By fostering foreign trade and encouraging European merchants to enter the Iranian market, he ushers in an era of exemplary peace and prosperity.

1792 Death of Vahid Behbahani

Vahid Behbahani, who forces the *Akhbari* school of Shi'ism out of Persia and establishes the ascendancy of the *Usuli* school, dies.

1795–1925 Qajar Dynasty

Agha Muhammad Khan (1779–97), a eunuch at the Zend court, leads the Qajar tribe to victory over the Zend, establishes the Qajar dynasty, and moves the capital to Tehran. Within a year, he is assassinated by his own servant.

1797 Gulistan and Turkmanchay Treaties

Fath Ali Shah (r. 1797–1834), Agha Muhammad's nephew, becomes king. He is known for his long beard and 2,000 children—over sixty princes waiting in the wings. Fath Ali Shah is defeated twice by Russia. After the first defeat, according to the Gulistan Treaty (1813), he loses five cities in the Caucasus, gives up Georgia and Daghistan, and the right to a navy on the Caspian Sea; after the second defeat according to the Treaty of Turkemanchay (1825), he loses the rest of the Caucasus and everything north of the Aras river. War damages imposed exceed 3,000,000 pounds sterling.

During Fath Ali Shah's reign, 'Abbas Mirza (1789–1833), Crown Prince, Governor of Azerbaijan, and Commander of the Iranian forces in the Caucausus introduces European military techniques into Iran. The ablest of Fath Ali Shah's sons, he sends groups of Iranian youths abroad to learn new military techniques, bring a printing press to Tabriz, and translate military manuals. His efforts, however, are undermined by his father who refuses to supply provisions and troops and blames him for the defeats.

1813–25 Caucasus won by Russia

Fath Ali Shah (1799–1834) loses the Caucasus to Russia.

1834–48 Rule of Muhammad Shah

Narrow-minded, superstitious, and incapable, Muhammad Shah (r. 1834–48) tries to keep the British out of Afghanistan but fails. During his rule, Babism, an indigenous reform in Shi'ite Islam, is established (1844). The reform is intimately tied to the Twelfth Imam or Mahdi who, the Babis believe, speaks to the faithful through the *bab* (gate). In subsequent years, Babism is rivaled by the more internationally-oriented Baha'ism.

1848–96 Nasir al-Din Shah and the Great Amir

Accompanied by Mirza Taghi Khan Amir-i Kabir (1798–1852), Nasir al-Din Shah (r. 1848–96), absolute ruler, womanizer, food connoisseur, and hunter enters Tehran and coronates himself. With the Shah's consent, Amir-i Kabir, prime minister from 1848 to 1851, launches a far-reaching program of revitalization to restore Iran's former glory. The reforms adversely affect the court, the tribal hierarchy, and British and Russian legations. Sectors of society affected convince the Shah to eliminate the Amir, his supposed rival to the throne. After the murder of the Amir, Nasir al-Din Shah occupies Herat (1856). But unable to withstand British pressure, following the Treaty of Paris (1857), he agrees to evacuate Afghanistan and recognize its independence. He also faces Russian invasions which cost him Bukhara (1868), Khujand (1870), and Merv (1884).

Amir-Kabir introduces fundamental reforms leading to the westernization of Iran. His reforms include abolishing all titles as well as all bribes and moneys received by officials. Officials are assigned regular salaries commensurate with their tasks. He inspects the treasury books personally, stops all subsidies made to the court, especially to the Queen Mother, and promotes education by instituting the *Dar al-Fonun* in1850 (see 1850). He also curtails the influence of Russia and Britain in Iran.

1850 Dar al-Fonun

Dar al-Fonun is established to educate future diplomats and military generals. The execution of the Bab and the massacre of his followers leads to the Baha'i movement's beginning.

1890 Rise of political consciousness

British Major Gerald F. Talbot's company is awarded a monopoly for the production, sale, and export of all tobacco in Iran. The concession pits Iranians, especially the clergy, against the British government and the Shah, who has received 15,000 pounds sterling. The Shah and the British government retreat.

1901–06 Beginning of oil diplomacy

British discover oil in southern Iran and proceed to gain concessions for exploitation, refining, and transporting of the crude worldwide. After receiving 12,000 pounds sterling and a cut of the profits, Muzaffar al-Din Shah (1896–1907) grants the rights.

Iranians, questioning the legitimacy of the agreement, demand an assembly to examine the document. They also demand a constitution according to which similar decisions can only be reached in a public forum. When 14,000 protesters occupy the British Legation, the Muzzafar al-Din Shah grants a constitution.

1907–09 Agreement between Britain and Russia

An Anglo-Russian Convention divides Persia into British and Russian spheres of influence.

1908: June Shah attempts to disband Majles

Muhammad Ali Shah (r. 1907–11) attempts to disband the Majles and strengthen the monarchy. The Constitutionalists retaliate. The Shah, aided by Britain and Russia, shells the Majles and suspends the constitution (June, 1908) while the Anglo-Russian Convention divides Persia. Muhammad Ali is deposed by the Constitutionalists and the Bakhtiari tribe that comes to their aid.

1911 Last Qajar shah

Muhammad Ali is replaced by his eleven-year-old son, Ahmad Shah (r. 1911–25). The parliament is restored. An Anglo-Persian Oil Company is formed. Iran is divided into British (south) and Russian (north) spheres of influence. A refinery at Abadan is completed.

1916 Social-Democrat Party is formed

Social-Democrat Party of Azerbaijan is formed in Baku.

1917 End of Russian influence

Direct Russian imperialist influence in Iran ends.

1919 Majles refuses to ratify agreement

The Majles refuse to ratify the Anglo-Persian oil-exploitation agreement. Muhammad Musaddiq, a vociferous defender of Iranian rights, is expelled.

1920 Communist Party is formed

First Communist Party of Iran (CPI) is organized in Gilan.

1921: May Reza Shah the Great stages a coup

Colonel Reza Khan, a Russian-trained Cossack officer, stages a military coup and becomes Iran's minister of war. The Russo-Persian Treaty is signed. (See also 1923.)

1922 Jangala movement defeated

The Jangali movement, led by Mirza Kuchik Khan, is defeated; Gilan Republic is abolished.

Eleven-year-old Ahmad Shah (r. 1911–25) becomes shah when his father is disposed. (EPD Photos/CSU Archives)

1923 Reza Khan is prime minister

Two years after staging a coup, Colonel Reza Khan becomes Iran's prime minister.

1925–38 Reza Shah the Great

Constituent Assembly entrusts the rulership of Iran to Reza Khan Pahlavi (1878–1944) who crowns himself shah, abolishes capitulation, and cultivates amicable relations with Germany.

All employees are required to wear uniforms. A vigorous program of upgrading transportation provides roads and services and a network of grain silos, sugar factories, and national banks cooperates to stabilize Iran's economy. To satisfy the country's technical needs, a program of training Iranians, both men and women, abroad is devised. Marriages and divorces are registered by governmental officials. A teachers training college and Iran's first university become operational. In foreign correspondence, Iran's official name is changed from Persia to Iran. (In Iran, the country's name has always been Iran.)

1936 Communism enters Northern Iran

Reza Shah dissolves the Communist Party of Iran (CPI) and imprisons fifty-three of its members. Begun as the Social-Democrat Party of Azerbaijan in Baku in 1916 to facilitate direct Russian imperialist influence in Iran, the Party leads to the Jangali movement—under Mirza Kuchik Khan—in Gilan (see 1920), and to the creation of the Gilan Republic. Reza Khan abolishes the Gilan Republic in 1922.

1939–41 Large-scale immigration from Russia to Iran

Large numbers of White Russians, pursued by the Red Army, emigrate to Iran. Iran announces neutrality in World War II; Allies (Britain from the south, Soviets from the north) invade Iran.

1941 Reza Shah abdicates

Reza Shah abdicates in favor of his son, Muhammad Reza, a figurehead ruler. Reza Shah dies in exile in Johannesburg in 1945.

1941 Parvin E'tesami

Parvin E'tesami (1907–41), women's rights advocate, dies of typhoid fever. Using conventional imagery, Parvin holds logical debates among people and things and draws moral conclusions. Themes of her poetry include poverty and enlightenment of the masses, the plight of orphans, and predestination.

1942 Rise of the Tudeh Party

Following the same ideals and operating according to the same platform, the Marxist, pro-Soviet Tudeh Party rises from the ashes of the outlawed CPI.

1943–44 The Allies and Iran

The United States sends its first military mission to Iran. Not an occupation force, the mission ensures protection of U.S. interests against possible Soviet aggression. Britain's Winston Churchill, U.S. President Franklin D. Roosevelt, and the U.S.S.R.'s Joseph Stalin meet in Tehran, recognize Iran as the "bridge of victory," and pledge assistance at war's end.

1944 Tudeh gains seats in the Majles

The Communist Party, Tudeh, claiming 25,000 members, occupies eight seats in the Fourteenth Majles. Five hundred thousand participate in a rally organized by the Party. The Soviet Red Flag is hoisted at Sepah Square.

Tudeh Party, founded by four of the fifty-three CPI members incarcerated by Reza Shah in 1936, forms out of the outlawed CPI in 1942 to continue its ideals and strategy.

1945 Reza Shah dies

Reza Shah dies in exile in Johannesburg, South Africa.

1945 Soviet influence

The USSR foreign minister arrives in Tehran to negotiate northern Iranian oil concessions. To remain in Iran after World War II, the Soviets proclaim two Autonomous Republics in Azerbaijan and Kurdistan.

1946 Iran asks UN for help

Iran appeals to the United Nations (UN) Security Council against Soviet aggression in Azerbaijan and Kurdistan. The UN agrees to protect Iran's rights.

1946 Sayyid Ahmad Kasravi assassinated

Ahmad Kasravi (1890–1946) is assassinated in Tehran. Kasravi intends to change Iran by attacking and debunking its constitutional monarchy and clerical hierarchy. He writes books, delivers lectures, and gathers followers, some of whom continue his *unadulterated* way to this day.

1947 Majles reject oil concession

Majles rejects the Iran-Soviet oil concession.

1949 Assassination attempt blamed on Tudeh

An unsuccessful attempt to assassinate the Shah, blamed on the Tudeh, results in persecution of the Tudeh (Soviet-affiliated party).

1950–51 The Musaddiq era

Muhammad Musaddiq (1880–1967) opposes the Iranian monarchy (see 1908). Arrested, he is exiled to Europe where he studies economics and political science. He enters the Majles in Iran in 1915 and, by 1918, becomes minister of finance. He avoids politics throughout Reza Shah's reign.

In 1949, to prevent the shah from signing an oil treaty with Britain, Mussaddiq leads the social-democratic National Front into power and, as the Chair of the Committee on British Concession, nationalizes the oil industry (1951). Musaddiq introduces a bill to create SAVAK, the shah's secret police force.

Musaddiq is designated Man of the Year by the U.S. news magazine, *Time*. The Majles impose Musaddiq on the shah as prime minister. The shah accepts his nomination—the first such prime minister in the country's history. A liberal democrat, Musaddiq demands that the shah must reign, but not rule. He also suggests that the prime minister should be in charge of all governmental affairs; members of the royal house must not interfere in governmental affairs; the Queen Mother and the shah's twin sister should leave the country; and the army should be placed under the prime minister. The shah refuses to sign the legislation that would make these ideas law.

1951 Sadeq Hedayat commits suicide

Sadeq Hedayat (1903–51), founder of modern Persian fiction, dies. Hedayat is known for his *Blind Owl,* a Buddhaesque sojourn into the underworld mapped out into a Kafkaesque existence in the ruins of the ancient city of Ray in the south of present-day Tehran. The motives for his suicide—suicidal tendency, depression, and drug addiction—are questioned by Iraj Bashiri (b. 1940). In his 1984 *Fiction of Sadeq Hedayat,* Bashiri cites Hedayat's political satires, especially his *Pearl Cannon,* as grounds that political intrigue may have affected his life.

1952 Negative equilibrium policy

Musaddiq's policy of Negative Equilibrium, i.e., independence from both super powers, does not sit well with U.S. President Dwight Eisenhower. Eisenhower fears that the weakening Iranian economy will lead Iran to increasing reliance on the Tudeh and, eventually, on communism.

1953 CIA-backed coup

Musaddiq is encouraged to resign by the shah and the United States. Musaddiq's government, plagued by economic blockade, frozen Iranian assets, and threats of military action, succumbs to a coup backed by the U.S. Central Intelligence Agency (CIA). The shah flees while his generals and the CIA (under the direction of Allen Dulles, Kermit Roosevelt, H. Norman Schwartzkopf) carry out a coup in his favor. Within three days, the shah returns and bans political opposition.

Musaddiq is brought before a military tribunal and sentenced to three years of solitary confinement for high treason. After his release in 1956, he remains under house arrest until his death.

1954 Iran gains minimal control of its oil fields

Iran receives aid and nominal control of its oil fields. Management and development rights remain with the Anglo-American consortium.

1955 End of martial law

Martial law in Tehran, in effect since 1951, ends.

1956 Tudeh becomes Islamic

Tudeh, now part of the international community, is grafted to Islam.

1957 SAVAK established

Sazman-i Amniyyat va Edari-i Keshvar (SAVAK), Iran's secret police is formed. The bill for the formation of SAVAK was sent to the Majles by Muhammad Musaddiq in 1950 but is not enacted until now. The intention behind formulating the bill is containment of Soviet activities in Iran through its Tudeh network.

1960 Khomeini Succeeds Borujerdi

Ayatullah Muhammad Hussein Borujerdi, a moderate cleric dies. The vacuum is filled by a comparatively junior cleric, Ruhullah Khomeini (1902–89) but no single *marja'*, or point of emulation, emerges. A radical *mullah* (teacher of Islamic law), Khomeini opposes the shah's reforms and calls for an Islamic form of government. Indirectly, Khomeini also opposes two other groups of *ayatullahs* (a Shi'ite Muslim religious leader). The conservative group headed by Ayatullah Khoi, which believes the clergy should stay out of politics altogether, and the moderate group headed by Ayatullah Golpayegani, which believes in a genuine constitutional monarchy in which an assembly of five Islamic jurists makes sure that the bills passed by the parliament are in agreement with the *Shari'a* law.

1960: January Death of poet Nima Yushij

Nima Yushij (1895–1960), the father of modern or new Persian poetry or *she'r-i now,* dies. Nima changes the tradition of Persian poetry with the notion that poetry must serve the people. His predecessors write for the court and the nobles. Nima's poetry has a new meter and rhythm, uses daring imagery bordering on surrealism, and adds local color.

1961 Spade work for the White Revolution

The shah abandons the Arab League for relations with Israel, secures his wealth in the Pahlavi Foundation, dissolves the Majles, and begins Land Reform. These rapid changes infuriate the clergy, especially Khomeini.

1961 Iran Freedom Party formed

Mehdi Bazargan (b. 1906) forms the Iran Freedom Party as a counterweight against the Tudeh Party.

1962–63 U.S. interest grows

The National Front gains in strength. The shah sends in a request to the U.S. Congress for additional assistance. U.S. President John F. Kennedy urges the shah to gear his economic reforms to American interests. Islamic clergy protest the shah's dictatorial rule.

1963: January 26 The White Revolution

The White Revolution, calling for land reform, literacy, and women's rights, is launched. Taking place in two phases, the White Revolution changes Iranian life drastically. The major planks in its reform bill are abolition of peasant/landlord tenure system; sale of industrial shares to former landlords; female suffrage; formation of education, health, and agricultural corps; and establishment of village courts. These reforms take the power away from the clerics and place it in the hands of bureaucrats and foreign agribusiness ventures. Companies such as Dow Chemical, Bank of America, John Deere, and Royal Dutch Shell operate freely in Iran's economic sectors. IT opens way for foreign agribusiness ventures. Tudeh, which has consistently asked for similar reforms, is helpless.

1963: March Khomeini speaks out

Khomeini denounces the White Revolution and calls for a more Islam-friendly government. The shah's commandos attack the Faiziyeh Theological College and beat up Khomeini's followers. In subsequent weeks, large numbers of unarmed people are murdered in the streets and bazaars of Tehran, Qom, and Shiraz. Khomeini is temporarily imprisoned then released.

1963: August Immunity for U.S. military

Military advisors from the United States are given diplomatic immunity.

1964 Arrest of Dr. Ali Shair'ati

Dr. Ali Shari'ati is arrested upon arriving in Iran from Paris where he has been studying.

1964: November 4 Khomeini exiled

Khomeini is exiled to Bursa, Turkey. He returns to Najaf, Iraq in October, 1965.

1965–67 Shah purchases fighter planes

Shah buys two squadrons of F-14 fighter-bombers from the United States and uses them against the rebellious Qashqai tribes in the south; SAVAK is accused of intensifying repression.

1967: February Farrokhzad dies in a car crash

Forugh Farrokhzad (1935–67) is killed in a car accident. She is one of only two well-known modern Iranian poets, Farrokhzad stands in direct contrast to Parvin E'tesami. While E'tesami follows the old masters, the energetic and highly controversial Farrokhzad follows the footsteps of Nima.

1968 British military plans to leave region

Britain finalizes plans to pull its military out of the Persian Gulf.

1969: January Artists and writers jailed

Fourteen artists and writers are incarcerated and tortured by a military tribunal; the Iranian embassy in Rome is occupied as part of a worldwide Iranian student protest against the shah; SAVAK operates inside the United States.

1969 Jalal Al-i Ahmad dies

Jalal Al-i Ahmad (1923–69), stylist, short-story writer, essayist, and one of the architects of the 1979 Islamic Revolution, dies. Al-i Ahmad is known for his 1963 *Gharbzadagi* (Westoxication), a series of essays condemning Iran's rapid incorporation of Western ideas into its traditional Islamic society.

1970 Amnesty International reports on Iran

Taking into consideration the case of fourteen artists and writers incarcerated and tortured by military tribunal in Iran (see January, 1969), Amnesty International and French news journal *Le Monde* report mass jailings, arbitrary arrests, and torture inside Iranian prisons.

1971 Fedayeen and Mujahidin

The Fedayeen (anti-Soviet Marxist-Leninist) and the Mujahidin (radical Islamic nationalist with Marxist tendencies) become active and remain so until 1976–77 when they splinter into factions; the British military leaves the Middle East.

1971: October 2,500 years after Cyrus the Great

At the cost of more than $60 million at Persepolis, the shah celebrates the 2,500th anniversary of the founding of the Persian Empire. Food and water are flown in from Paris.

1972: May Nixon visits Tehran

At Tehran University, Richard Nixon is met with a silent protest staged by students. The Shah agrees to protect U.S. interests in the Middle East in exchange for arms; Iranian military purchases reach $519 million.

1973 Iran does not join OPEC oil boycott

Iran does not participate in the oil boycott and raises the price of its crude. Former CIA head, Richard Helms, becomes the U.S. ambassador to Iran. A twenty-year oil agreement with the consortium of Western firms is concluded.

1974 Oil revenues reach $18 billion

Iranian oil revenues reach $18 billion. The French government contracts to build two nuclear plants at $2 billion each.

1974: May Amnesty International cites worst human rights record

Amnesty International singles out Iran as the country with the worst record on human rights.

1975 Iran becomes a one-party state

In reaction to the Amnesty International report, Iran seeks to suppress international pressure and civil unrest by becoming a one-party state; riots follow government-sponsored price increases.

1975: March 6 Iran settles dispute with Iraq

Iran and Iraq settle the Shat al-Arab dispute by Iran's withdrawing support for Iraqi Kurds.

1976 Testimony before International Commisison of Jurists

In Geneva, Switzerland, prisoners who have been subjected to SAVAK torture, speak to the International Commission of Jurists.

1977 U.S. advises Iran on policy matters

U.S. President Jimmy Carter, keeping human rights in the forefront of his policy agenda, presses the shah to moderate his stance. William L. Sullivan, the new U.S. ambassador to Iran, repeatedly meets with the shah to discuss strategies whereby the Iranian government can suppress opposition.

1977: April Military trial open to the press

After Amnesty International singles out Iran as the country with the worst record of human rights (see 1974: May), the military trial of dissidents is opened to the press for the first time in more than five years.

1977: June Ali Shari'ati is murdered

Ali Shari'ati (1933–77), former University of Mashad teacher and a victim of SAVAK torture (1972–75), dies under mysterious circumstances in London, England. Along with Jalal Al-i Ahmad, Shari'ati is one of the foremost exponents of the Islamic Revolution that sweeps the shah from power.

1977: June 12 Plea for free elections

In an open letter, moderate leaders of the National Front plead with the shah to discontinue despotism and allow free elections. The shah appoints Jamshid Amouzegar, a liberal, prime minister and frees 572 political prisoners.

1977: December Iran–U.S. relations at peak

Iran's military purchases reaches $5.8 billion. Over 7,000 Americans work in more than thirty para-military organizations in Iran. Bell International, Hughes Aircraft, Computer Sciences Corporation, Harsco, TRW, Rockwell International, GTE, Lockheed, and Harris Corporation top the list.

The stakes are high, yet the U.S. Government views the Iran–U.S. relations with optimism. Even though 8,500 Iranians, defying mace and tear-gas, demonstrate outside the White House in which Carter entertains the shah (November 15). U.S. President Carter visits Iran in December and calls Iran an island of stability in a sea of troubles. Toasting the shah on New Year's Eve, he says, "Your Majesty, your view of human rights and mine are the same."

1978: January 7–16 Khomeini on the periphery

Islamic clergy stage massive rallies in Qom, protesting land reform policies, the ban of the veil, and an attack on Khomeini in a Tehran newspaper. Police open fire on unarmed demonstrators, killing dozens. Forty-day cycles of killings and mourning begin. Khomeini becomes the focus of the opposition.

1978: February 17–19 Demonstrations result in violence

To mourn the Qom deaths, religious leaders call for business shutdowns. In Tabriz and other cities, demonstrations are interrupted by army tanks, and scores are killed.

1978: February 28 International pressure on the Shah

Amnesty International accuses Iran of denying defenders the right to a fair trial. This statement follows a protest by the International League for Human Rights on June 23, 1977, and an exposure of SAVAK atrocities to the International Commission of Jurists in Geneva by prisoners subjected to SAVAK torture (1976).

1978: May 9–10 Tensions escalate

Army, police, and SAVAK step up activity. Police kill two Shi'ite clergy at the home of Ayatollah Shari'at-Madari.

1978: August 18–20 Movie houses set on fire

Movie houses in Meshed, Reza'iyyeh, and Shiraz are set on fire, and 430 die in a movie theater fire in Abadan. SAVAK is blamed for setting the fires. Anti-shah riots break out in Abadan.

1978: August 27 New prime minister takes action

Sharif-Emami becomes prime minister, abolishes the portfolio of the minister of state for women's affairs, closes the casinos, and resumes the use of the traditional calendar. These actions do little to quell the unrest. People riot, shut down the oil industry, and continue work stoppage in many service sectors.

1978: September 7–9 Martial law declared

Martial law is declared in Tehran and eleven other cities. Troops open fire on unarmed demonstrators in Jaleh Square, killing 3,000, including 700 women. U.S. President Jimmy Carter reaffirms U.S. support for the shah.

1978: September 24–October 4 Oil workers strike

Ten thousand oil workers go on strike; industrial actions against the government lead to strikes in all service sectors.

1978: October 6 Khomeini travels to Paris

Khomeini, expelled from Iraq, joins Abolhassan Bani-Sadr and Sadeq Ghotbzadeh in Paris.

1978: October 31 Oil industry shutdown

The oil industry is shut down. Speaking with Crown Prince Reza, U.S. President Jimmy Carter reaffirms U.S. support for the shah.

1978: November 1 Anniversary of Khomeini exile

The sixteenth anniversary of Khomeini's exile is commemorated.

1978: November 3–5 Khomeini gains ally

Sanjabi joins forces with Khomeini; Prime Minister Sharif-Emami resigns.

1978: November 6–8 Shah admits mistakes

The shah admits mistakes and appoints a military government led by General Azhari. He defines his intention to reign rather than rule Iran. In addition, he orders the arrest of former prime minister Hoveida, SAVAK Chief General Ne'matollah Nassiri, and six former cabinet members. Khomeini and Sanjabi rule out cooperation with the shah.

The U.S. State Department confirms that the U.S. is supplying Iran with tear gas, shields, helmets, and batons.

1978: November 13–14 Oil workers return to work

Oil workers return to work in response to government threats. Two hundred workers, suspected of organizing strikes in the oil fields, are arrested. Oil workers stage a work slowdown in Ahwaz.

1978: November 19–20 Shah frees political prisoners

The shah frees 210 political prisoners. At the same time, the United States denies Soviet allegations of interference in Iranian military affairs.

1978: November 28 Government bans processions

Government bans processions during the month of Muharram, announces a revision of laws to conform with Islamic principles.

1978: November 30 Pay raise for civil servants enacted

A bill granting a twenty-five percent increase in pay for 70,000 civil servants is approved.

1978: December 1–5 Demonstrators defy procession ban

Demonstrators defy ban and stage processions at the cost of 700 lives. President Carter appoints George Ball to study the Gulf crisis.

1978: December 6–18 Khomeini calls for a general strike

Khomeini calls for a national day of mourning and a general strike.

1978: December 13–15 Proposals over role of Shah

Former prime minister Ali Amini suggests the establishment of a Regency Council. Sanjabi suggests that the shah relinquish power to the Regency Council. Khomeini insists on a referendum on the monarchy.

1978: December 29 Bakhtiar government

The shah asks Shapur Bakhtiar (1916–91) to form a civilian government. Paris-educated Bakhtiar is an opponent of both the shah and Khomeini. Bakhtiar is the second man in Iran's National Front, a monarchist, and former deputy minister of labor in Mosaddeq's cabinet. He intends to phase out martial law, return the military to the barracks, put an end to heavy military spending, abolish the SAVAK (except for the intelligence department), and stop sale of oil to Israel and South Africa. Khomeini can return to Iran from exile provided he does not interfere in Iranian politics.

1978: December 30–31 Bakhtiar denounced by Khomeini

Khomeini and the National Front denounce Bakhtiar's appointment; Americans are advised to leave Iran; General Azhari resigns.

1979: January 1 Khomeini makes proposal about U.S. relations

Khomeini declares willingness to work with the United States, provided the shah leaves Iran, and the United States ends its interference in Iranian affairs.

1979: January 3 Bakhtiar tries to maintain control

General Robert E. Huyser arrives in Iran to redirect the military to support Bakhtiar and to prepare a contingency plan in case the Bakhtiar government collapses. Bakhtiar halts the sale of oil to South Africa and Israel and announces the shah's decision to leave the country temporarily.

1979: January 5 Dual government

A dual government rules, one under Bakhtiar (who is criticized for capitulating to the shah and the West) and the other under Khomeini (who directs the oil workers to produce enough oil for domestic consumption only).

1979: January 8–11 Bakhtiar makes further proposals

Bakhtiar presents his cabinet to the Majles and proposes a seventeen-point program, including dissolution of SAVAK, greater political rights, and a greater role for the religious leaders in government.

1979: January 13 Khomeini moves to center

Government announces the formation of a nine-man regency council; Khomeini appoints a Provisional Revolutionary Council to displace Bakhtiar's "illegal" government and introduces a provisional government to oversee elections and an assembly to write a constitution.

Khomeini (1902–89) studied Shi'i traditions under a student of the political activist Mirza Hassan Shirazi. In 1941, in Qom, he opposed the reforms of Reza Shah in a religious tract called *Kashf al-Asrar* (Disclosing Mysteries) and supported the candidacy and, later, the anti-Pahlavi activities of Ayatullah Borujerdi. When Borujerdi died in 1962, Khomeini, filling the vacuum, vehemently opposed the changes in the constitution, i.e., the right of women to vote, their equality with men, and modification of the oath taken by officials. In 1963, he denounced the shah's "White Revolution," calling it a western ploy for the disestablishment of the Islamic culture of Iran. Khomeini's subsequent activities resulted in the downfall of the Pahlavi dynasty.

1979: January Shah flees

Having transferred some $2–4 billion to his accounts abroad, the shah flies to Egypt; Khomeini calls on members of the government and the Regency Council to resign. Khomeini refuses to meet with Bakhtiar in Paris, unless Bakhtiar resigns. The government allows Khomeini to return from exile, and nonessential U.S. government personnel leave Iran.

1979: February 10–16 Khomeini return marked by celebrations

Three million people take to the streets to celebrate Khomeini's return. Khomeini appoints Mehdi Bazargan prime minister in a provisional government. Mass desertions in the army and a rout of the elite Imperial Guard by Khomeini supporters prevent the United States from implementing a coup. Bakhtiar's government falls, and Tudeh expresses support as Khomeini takes control of Iran. SAVAK Chief General Ne'matollah Nassiri is executed. The 30,000-man Imperial Guard is dissolved.

1979: March 7 People's Revolutionary Council created

People's Revolutionary Councils become operational.

1979: March 8–12 Women protest dress code

Tehran women protest the required Islamic dress code.

1979: March 12–20 Voices of moderation

Shari'at-Madari and Bazargan convince Khomeini to place a ban on the Revolutionary Council's recent trials. They cannot convince him, however, to give people a wider spectrum of choices by stopping the projected referendum on the monarchy. Kurdish tribesmen and government officials in Sanandaj fight over the Kurds' demand for autonomy. The government announces its intention to grant limited autonomy to the Kurds and appoints a Kurd, Ibrahim Yunesi, governor of Kurdistan province.

Turkmen tribesmen and government officials in Gunbad Qabus fight over autonomy; unrest is reported among Baluchi tribesmen.

1979: March 30 Shah exiled

The shah flies to the Bahamas.

1979: March 31 National referendum

Voting begins to establish a democratic republic with a freely-elected parliament—a theocracy. In it, the Ayatollah acts as a *Faqih* (high counselor). The republic intends to break ties with Israel in support of the Palestinian cause, stop shipment of oil to Israel and South Africa, break ties with monarchical states like Saudi Arabia and Persian Gulf States, and export Shi'ism to other Middle Eastern nations. It intends to engage in a radical redistribution of personal income, turn most companies over to the people, establish foreign cooperatives, and modernize agriculture. It plans to create a people's army, censor books, movies, and all things Western, and treat women with full respect.

1979: April 1 Islamic Republic of Iran established

Khomeini announces the establishment of the Islamic Republic of Iran.

1979: April 5–13 Former officials executed

Thirty-five high-ranking former officials, including former prime minister Hoveida, are executed.

1979: April 21–26 Deadly ethnic conflicts continue

Hostilities between the Kurds and the Turks in Naghdah, Azerbaijan, result in 1,000 deaths.

1979: May 1 Ayatullah Mottaheri assassinated

Ayatullah Morteza Mottaheri, important Revolutionary Council member and ideologue, is assassinated. A student of Khomeini, Mottaheri is the founder of Husseinie-i Ershad and author of a number of books on why Islam is relevant to modern living. The Husseinie plays a major role in familiarizing the people with the practical and theoretical aspects of the revolution.

1979: May 9 Executions continue

The government begins the execution of private individuals on charges of corruption.

1979: June 9–July 6 Government nationalizes industry

Government nationalizes thirty-seven banks, all insurance companies, and large-scale industries. Bazargan's government is advised to share power with the Revolutionary Council. Council members serve as ministers while cabinet members sit on the Council.

1979: June 10 Shah moves to Mexico

The shah, living in exile, leaves the Bahamas for Mexico.

1979: July 19 Government advised to share power

Bazargan's government is advised to share power with the Revolutionary Council. Council members serve as ministers while cabinet members sit on the Council. Khomeini bans playing music on radio and television stations.

1979: August 14–September 6 Kurd rebellion crushed

The Kurdish rebellion that begins in March 1978 is finally crushed. Kurdish tribesmen have presented a list of demands for autonomy. Government denial of their demands leads to battles in Sanandaj (March 18–21) and Naghdah (April 21–26, 1979). After 1,000 Kurds are killed, the government promises limited autonomy.

1979: September 10 Ayatullah Taleghani dies

Ayatullah Mahmud Taleghani (1910–79) dies. The main cleric in Tehran to support Musaddiq, Taleghani goes into retirement in 1953 to study constitutional issues. He emerges as a major force against Khomeini in the codification of the constitution and in relation to the position of the Faqih. These differences, however, are reconciled (April 18–20, 1979). Taleghani also signs an accord with the Kurds to settle their differences. Khomeini rejects the accord. Two days after Taleghani's death, the Constituent Assembly grants supreme power to the Faqih.

1979: October 19–22

The United States informs Iran of its intention to allow the shah to be admitted to the United States temporarily to receive medical attention. Iran responds with a warning of consequences if the former monarch is granted admission to the United States. The shah flies to New York City.

1979: November 4 Militant students storm the U.S. embassy in Tehran

Iranian militants storm U.S. embassy, seize ninety hostages, including sixty-two Americans, and demand the return of the shah. The United States refuses the demands. Instead, President Carter freezes official Iranian assets and suspends U.S. imports from Iran.

1979: November 6 Revolutionary Council assumes power

The Bazargan government resigns; the Revolutionary Council takes over its duties.

1979: November 19–20 A few hostages are released

Thirteen hostages—five women and eight men—are released.

1979: November 26–27 United Nations appeals for settlement of hostage crisis

Shari'at-Madari denounces the taking of the hostages. The United Nations Security Council appeals to both countries for moderation.

1979: December 2–3 New Islamic constitution

New Islamic Constitution is approved. The referendum is boycotted in Azerbaijan, Kurdistan, and Baluchistan. The shah, having been refused admission by Mexico, is moved to Lakeland Air Force Base in the United States.

1979: December International relations with Iran worsen

One hundred eighty-three Iranians are given notice to leave the United States within five days. To prevent any Iranian design on Afghanistan and its own Islamic republics in Central Asia, the Soviet Union invades Afghanistan (December 24).

Shah arrives in Panama. International Court of Justice orders Iran to release the hostages.

1979: January 1–3 Waldheim in Tehran

United Nations Secretary General Kurt Waldheim arrives in Tehran but cannot mediate or negotiate with the captors. Khomeini, the main figure deciding the fate of the hostages, refuses to meet with Waldheim even after the latter pledges to investigate the human rights abuses that have occurred during the shah's rule.

1980: January 25–30 Bani-Sadr elected president

Bani-Sadr is elected Iran's first president.

1980: March 14 Islamic Republic Party strong in elections

In the first election for the National Assembly, Islamic Republican Party, with strong fundamentalist leanings, wins plurality.

1980: March 23 Shah moves to Egypt

The shah, seeking a place to live, returns to Egypt.

1980: April 7 Khomeini supports militant holding hostages

Khomeini rules the hostages must remain in the hands of the militants; President Carter breaks off diplomatic relations with Iran and expels the remaining five Iranian diplomats.

1980: April 23 Trade agreements with Communist bloc

Iran announces new trade accords with the Soviet Union, Romania, and East Germany.

1980: April 24–29 Failed hostage rescue

An attempt by the U.S. military to rescue the hostages fails, and eight U.S. servicemen die. Iran moves the hostages to various locations to foil further rescue attempts. U.S. Senator Edmund Muskie replaces Secretary of State Cyrus Vance, who resigns in disagreement over the failed rescue mission.

1980: June 2–5 Conference in Tehran

Sixty countries participate in the Conference on U.S. Intervention in Iran, held in Tehran. Americans participating include former U.S. attorney general Ramsey Clark.

1980: July 10 Coup attempt fails

An attempted coup, led by Iranian Air Force officers, is foiled.

1980: July 27–29 Last shah dies

Muhammad Reza Pahlavi (1919–80) dies in exile in Egypt. He ascended the throne in 1941 at the age of twenty-one. Confronted by Musaddiq on his support of foreign oil interests, he fled the country for the first time in 1953. Returned to power with the support of the U.S. Central Intelligence Agency (CIA), he became an autocrat, using his formidable SAVAK to keep the antagonized middle class and religious leaders in line. In 1979, he was challenged once again by Ayatollah Khomeini who established an Islamic Republic in Iran. A king without a country, the former shah visits Egypt, the Bahamas, Mexico, United States, and Panama before he dies in Egypt. He is buried in a Cairo mosque. Former U.S. president Richard Nixon attends the funeral. Ronald Reagan praises the shah as a "loyal and valued friend."

1980: August 9–10 New nominee for prime minister

Mohammad Ali Rajaie is nominated prime minister.

1980: September 12–1981: January 20 Final hostage negotiations

Warren Christopher and Behzad Nabavi sign the final agreement. The hostages are released. Khomeini's conditions for ending the crisis are the return of the shah's wealth, cancellation of U.S. claims against Iran, unfreezing of Iranian assets, and U.S. guarantee not to interfere in Iranian affairs. Demand for an apology is dropped.

1980: September 19–22 Iran-Iraq war begins

Border skirmishes end in large-scale military conflict between Iraq and Iran. Oil refineries in both nations are bombed.

Border disputes between the two countries have been going on for centuries. Prior skirmishes were settled in the seventeenth century and again in 1937. In 1975, the shah forced the Algiers Treaty on Iraq by agitating and arming Iraq's Kurdish population. The median line in the Shat al-'Arab became the boundary line.

The conflict flares up again in 1979, when Khomeini begins exporting his revolution into Iraq, his place of exile for fourteen years. Recognizing the Shi'ite threat, Iraqi leader Saddam Hussein (b. 1937) invades Iran, captures fifty square kilometers of Iranian territory, and lays siege to Khorramshahr and Abadan. Iran liberates Khorramshahr (May, 1982) but becomes bogged down in the marshes around Huvayzah. Iraq sues for peace.

The economic, political, and ideological dynamics of the war harden each country's position. Iraq regards the 1975 Algiers Treaty as nonbinding, asks for the removal of the Iranian army from its territory, and payment of damages. In addition, Iran must recognize Iraq's full sovereignty over the Shatt al-Arab, relinquish Little and Big Tunb and Abu Musa to the United Arab Emirates. Iran considers Iraq the aggressor and asks for payment of $150 billion for damages to its petrochemical industries, removal of Saddam Hussein, his trial, and installation of a Shi'ite ruler.

1980: September 28 United Nations call for fighting to stop

The UN Security Council calls for a halt to the fighting between Iran and Iraq.

1981 Sa'id Soltan-pour is executed

Modern Iranian revolutionary poet Sa'id Soltan-pour is executed by firing squad. Besides being a poet, Soltan-pour is a teacher, actor, and director. As an intellectual he was harassed by the SAVAK and incarcerated and tortured at Qasr and Evin prisons.

1981: June 6 Khomeini becomes army commander

Khomeini replaces Bani-Sadr as army commander.

1981: June 21–24 Bani-Sadr ousted

Receiving the Iranian Parliament's overwhelming vote of no confidence, Bani-Sadr, Iran's first president, is dismissed by Khomeini.

Bani-Sadr helped Khomeini in Paris (1978). Later, he became a member of the Revolutionary Council. As Iran's acting foreign minister (November 11, 1979), he called for the return of the shah and his assets. He was elected Iran's first president in January, 1980. Mohammad Ali Rajaie becomes president following Bani-Sadr's ouster.

1981: June 28 Bomb blast kills Islamic leader

Ayatullah Mohammad Beheshti, head of the Islamic Republican Party and Iran's Chief Justice, dies in a bomb blast that kills seventy-four. Beheshti, a student and staunch supporter of Khomeini, studied languages in Europe and headed the mosque at Hamburgh. On June 21, 1981, as a member of the parliament and Revolutionary Council, he won an overwhelming vote of no confidence against Bani-Sadr.

1981: August 30 President Rajaie is killed

Mohammad Ali Rajaie, Iran's president since June 21, 1981, is killed by a bomb. Rajaie, a simple teacher, was nominated prime minister in August 9, 1980. In October, he told a New York news conference that the release of the hostages was "not far away."

1981: October 2 Khamenei elected president

Mohammad Ali Khamenei is elected president.

1982: March Iran gains territory from Iraq

Iran launches an offensive, regains border areas occupied by Iraq.

1982: September 22 Sadeq Qutbzadeh is executed

Sadeq Qotbzadeh, former Khomeini aide, member of the Revolutionary Council and Foreign Minister, is executed by firing squad on charges of treason. Qotbzadeh, a U.S.-educated Iranian, helped Khomeini in Paris (1978). Later he became the Director of the National Iranian Radio and Television. A hardliner, he replaced moderate Bani-Sadr as foreign minister (November, 1978). In 1982, he is implicated in a plot against the clerical government.

1983: March–April Iran alleged responsibility for embassy bombing

The U.S. embassy in Beirut, Lebanon, is destroyed, and Iran is blamed. Four Mujahidin hideouts are discovered and destroyed.

1983: April–May Tudeh outlawed

Tudeh is outlawed, and members are given one month to report. Eighty-four Russian diplomats are expelled on charges of interfering in Iran's affairs and having links to the Tudeh.

1983: May–June Soviet Union expels diplomats

The Soviet Union expels three Iranian diplomats, and an exhibition sponsored by the Hujatiyyeh is forced to close.

1983: June–July France leases fighter planes to Iraq

France leases fighter planes equipped with Exocet missiles to Iraq. Iran closes the French consulate in Esfahan and the cultural institute in Tehran. Khomeini submits his will to the Assembly of Experts.

1983: August–September Members of Baha'i faith accused of spying

Baha'is, accused of spying, close their six hundred assemblies in Iran.

1983: November–December Trial of Tudeh military

The entire military wing of Tudeh (about three hundred people) is put on trial. The trial reveals that F-14 manuals, information on spare parts, and numbers of Iranian planes have been compromised.

1984 Moderate Ayatollah dies

Moderate Grand Ayatollah Kazem Shari'atmadari dies of cancer at the age of 86. As the senior theologian in Qom, the leading Azeri Mujtahid (jurist), and the spokesman of the Azeri clergy, Shari'atmadari exerts great influence on the course of the Islamic Revolution. Since his 1982 alleged involvement (with Sadiq Qotbzadeh) in a plot against the clerical government, he has been under virtual house arrest.

1984 Government encourages women's rights

Women are encouraged by the government to become involved in education. The government also makes it legal for wives to sue for divorce.

1984 Death of Ayatollah Shirazi

Ayatollah Haj Seyyed Abdullah Shirazi (95) dies.

1984: January–March Tudeh trial verdicts and war intensifies

Tudeh officers convicted as Party members and spies receive thirty-year to life sentences; ten officers, referred to the High Judiciary Council, are executed. The U.S. Government lists Iran among the nations supporting terrorism.

The war between Iraq and Iran intensifies. Baghdad, Karkuk, and the Faw peninsula on the Iraqi side and Dezful, Kharg, Hoveyzeh, and the oil-rich artificial Majnoon Island on the Iranian side are repeatedly bombed. Amid tanker wars, the U.S., UN, and Iran condemn Iraq for its alleged use of chemical weapons.

1984: April–May Spy for Soviets confesses

Ehsan Tabari confesses to espionage on behalf of the Soviet Union.

1985–86 Arms for hostages negotiations

Senior U.S. officials negotiate an exchange of the U.S. hostages for arms. (This becomes known as the Iran-Contra affair).

1985: March–April Fighting between Iran and Iraq continues

Twenty-four Iranian cities are hit by Iraqi rockets. Iran devastates Karkuk and the center of Baghdad. Iraq intensifies bombing of Iranian cities, especially Dezful. Iran accuses Iraq of using chemical weapons.

1985: June–July Soviet Union withdraws personnel from Iran

The Soviet Union removes its technical personnel from Iran; Khoiniha, leader of the students who seized the American embassy, becomes Prosecutor General.

1985: August–September Economy suffers

Iraqi bombs damage ten of fourteen tanker berths at Kharg. Iran's economy suffers from a sharp decline in agriculture and industrial production.

1985: November–December Panel nominates Khomeini's successor

Assembly of experts nominates Ayatollah Hosseinali Muntazeri as Khomeini's successor.

1985: December–1986: January Conflicts continue

Baghdad attacks Hoveyzeh and oil-rich, artificial Majnoon Island; Turkey and Iran sign a bilateral trade agreement worth $3 billion.

1986: January–February Iran captures Iraqi territory

Iran captures Iraq's Faw peninsula plus 800 square kilometers of Iraqi territory. Iraq damages Iran's Ganaveh pipelines, reducing production by one-half; Ayatollah Muhammad Reza Golpayegani views the war simultaneously as a great loss and an agent of national solidarity.

1986: February–March Women asked to support national defense

Khomeini and Rafsanjani ask women to become the army's scientific and cultural bulwark against the West. The United

States condemns Iraq for its alleged use of chemical weapons, and Iran demands that Iraq be recognized as the aggressor.

1986: June–July Ban on luxury items

To stem a domestic economic crisis and to trim Iran's imports to conserve foreign currency, more items are added to the list of rationed goods and fewer luxury items are imported.

1986: September–October Iran at the Asian Games

The Iranian team sent to the Tenth Asian Games in Seoul, Korea, wins six gold medals including one in bicycling and one in tae-kwan-do. Four Iranian athletes, including one bronze-medal winner, defect and seek asylum at the Iraqi consulate.

1986: November 6 Newspaper reports on U.S. arms sale to Iran

Lebanese newspaper *Al-Shiraa* reports on the United States's secret arms sales to Iran. U.S. president Ronald Reagan acknowledges the sale but denies the hostage connection.

1987: February–March Crime and punishment in Islamic Iran

According to a 1987 Iranian Department of Justice report, 6,467 individuals receive bodily punishment for crimes. Of these, 1,096 are charged either with drinking alcoholic beverages or with fornication. There are several instances of cutting fingers and stoning. Iran has 2,564 judges, six hundred of whom are members of the clergy.

1987: April 29 Ihsan Tabari dies

Ihsan Tabari (1917–87), Tudeh theoretician and founder, dies. In 1947, when Tudeh was outlawed, Tabari fled to Moscow where he supervised the Persian section of Radio Moscow. In 1984, he confessed to espionage on behalf of the Soviet Union.

1987: November–December Khomeini changes his will

Five years after issuing his will (see 1983: June), Khomeini submits a new will to the Assembly of Experts and the religious archives in Meshed. The new will is prompted by the changes that have taken place in the country.

1988: July 3 U.S. shoots down commercial airliner by mistake

A U.S. navy warship, mistaking an Iranian commercial airliner for an F-14 fighter jet, shoots it down. All 290 people aboard die.

1988: July 18 Cease-fire agreement reached

Khomeini agrees to cease-fire negotiations with Iraq.

1988: December 21 Iran offers incentives for expatriates to return

Iran's Labor Ministry facilitates repatriation of Iranian professionals by providing travel expenses and housing.

1989: February 8 Islamic Revolution accomplishments celebrated

Tenth year of Islamic Revolution is celebrated; Khamenei distinguishes pride and national integrity, rather than functional party systems, as the main achievement of the revolution.

1989: February 14–18 Khomeini targets author Salman Rushdie with death threat

Khomeini issues a *fatwa* (religious ruling) that Salman Rushdie, British author of *Satanic Verses,* should be executed for blasphemy.

1989: March Montazeri is dismissed as Khomeini's successor

The overt criticism of the regime by moderate Ayatullah Husseinali Montazeri (b. 1922) causes him to resign as Khomeini's successor. Like his former teacher, Montazeri is an uncompromising opponent of the former regime. In his letter to Khomeini, Montazeri admits to a lack of preparation to undertake the burdensome task of leading the nation. Khomeini agrees with Montazeri and thanks him for his honesty.

Montazeri is nominated by the Assembly of Experts to succeed Khomeini (December, 1985) as the Supreme Ruler and *Marja'*. He had been elected to the position in August, 1986. The Supreme ruler watches over the activities of the president and has the power to dismiss him.

1989: April 7 Oil flows again at Abadan

After eight years, oil flows through the pipes of the gigantic Abadan refinery.

1989: April 28 Constitutional review undertaken

Khomeini empowers a twenty-member commission to review the constitution.

1989: May 5 Iran and Japan conclude agreement

Iran and Japan sign a $1.9 billion agreement for an oil refinery at Arak.

1989: May 12 Majles speaker advocates retaliation

Majles speaker Rafsanjani advocates the killing of five American, British, and French citizens in retaliation for every Palestinian killed.

1989: May 19 Iran and China reach agreement

Iran and China sign economic and cultural cooperation agreement.

1989: June 2 Prime minister position eliminated

Post of prime minister is eliminated; all power is concentrated in the presidency.

1989: June 3 The Khomeini legacy

Ayatollah Ruhullah Khomeini (1902–89), frail, old leader who inspires as much fear as he inspires hate for the Iranian monarchy and the West, dies. His ten-year legacy includes Shi'ite activism, Islamic fundamentalism, hostage-taking, the Iran-Contra affair, the Rushdie *fatwa,* and the Iran-Iraq war. *Time Magazine*'s 1979 Man of the Year, Khomeini restores Iranian history to its sixteenth-century course. Khomeini's twenty-nine-page will does not name a successor.

1989: June 4 Khamenei nominated

Two hours after the death of Khomeini, his former student, Ali Khamenei (b. 1940) is confirmed by the Assembly of Experts as the nation's spiritual leader or *faqih*. The experts choose Khamenei, who has served as a Revolutionary Committee member (1979), a Central Committee of the Islamic Republican Party member, and as president (1981–89).

1989: June 23 Araki succeeds Khomeini

Ayatollah Mohammad Ali Araki succeeds Khomeini as *marja'* (the world Shi'ite religious leader).

1989: June 30–July 7 Rafsanjani visits Moscow

Rafsanjani's visit to Moscow results in the sale of gas to the Soviets and the purchase of arms from them. The deal is worth $6 billion. Rafsanjani also signs an economic agreement worth $15 billion.

1989: July 28 Rafsanjani becomes president

Majles speaker (1980–89), former acting commander of the armed forces (1988–89), and Deputy Chair of the Council of Experts, Ali Akbar Hashemi Rafsanjani (b. 1934), becomes president. Moderate and pragmatic, he begins to rebuild Iran through a program of economic liberalization.

1989: August 6 Khamenei becomes spiritual leader

Khomeini's student, Ali Hoseini Khamenei (b. 1940) is confirmed by the Assembly of Experts as the nation's Spiritual Leader. Revolutionary Committee member (1979) and a Central Committee of the Islamic Republican Party member, Khamenei served two-terms as president (1981–89).

1990: April 5 Agreement with Italy

Italy and Iran sign a $13 billion agreement to cooperate in the fields of steel, gas, petrochemical, and industrial productions.

1990: April 8 Cooperation between Iran and the USSR

Iran begins exporting natural gas to the Soviet Union after eleven years. The agreement, according to which Iranian gas is transferred via a pipeline to Astara, yields $300 million for Iran. It also serves as a foundation for future cooperation.

1990: May 13–15 Open letter to Rafsanjani

Ninety liberals send an open letter to Rafsanjani complaining about the economy, lack of freedom, and political and judicial insecurity. Eighteen of them are arrested and accused of pandering to the whims of foreign colonialists.

1990: June 21 Devastating earthquake

An earthquake registering 7.3 on the Richter scale shakes northern Iran: 45,000 are killed, 100,000 are injured, and 400,000 are left homeless.

1990: July 3–August 21 Iran-Iraq war ends

Iran and Iraq normalize relations, and the last Iraqi soldier leaves Iran. Iran and Iraq exchange POWs.

1990: August Rafsanjani supports *sigha*

Rafsanjani's support of *sigha* (temporary marriage) angers feminists. To stem prostitution, Shi'ite law allows temporary marriages between willing partners of the opposite sex. After a set period, the two separate with no obligation to each other. Critics claim *sigha* itself is prostitution.

1990: August 27 Iran and Britain normalize relations

Iran and the United Kingdom normalize relations.

1991 Rafsanjani loosens controls

Rafsanjani decentralizes the military and government command system and introduces free-market mechanisms.

1991: February Refugees flood into Iran

One million Kurdish refugees flee Iraq and enter Iran.

1991: May 3 Police step up enforcement of Islamic codes

Police detain 800 women for violations of the Islamic dress code (appearing in public in "bad *hijab*") and close fifty shops for "neglecting Islamic codes."

1991: August 6 Shapur Bakhtiar assassinated

In Paris, three Iranians knife to death former prime minister Shapur Bakhtiar and his aide. Denying any involvement, the

Iranian government attributes the deed to perennial struggle for power among Iranian factions abroad.

1991: October 4 Islamic bridal dress established by law

The government establishes a law making it obligatory for brides to wear Islamic bridal dress.

1991: October 18 Segregation of men and women while swimming

As part of a continuing effort to isolate women, the Islamic government builds thirty walls along the shoreline of the Caspian Sea to ensure segregation of bathers. Women's continuous defiance of restraints express their frustration with these measures.

1991: December 6 United States and Iran settle claims

The United States pays $278 million to settle Iran's claim regarding military equipment left in the United States for repair.

1991: December 27 Government allows private investment

The government permits private sector investment in petrochemical industry, and Rafsanjani urges against "anti-Western sentiments."

1992: January 10 Iran's damages from war are assessed

The United Nations assesses Iran's wartime damage at $97.2 billion, but Iran claims $1000 billion.

1992: July 10 Agassi wins at Wimbledon

Iranian-American tennis champion Andre Agassi (b. 1970) wins at Wimbledon. Agassi is the winner of many international tennis championships, including the U.S. Open and Australian Open. In 1996, he wins a gold medal in the Summer Olympics.

1992: July 31 Iran buys fighter planes from Russia

Iran buys 110 combat aircraft from Russia.

1992: August 8 Ayatullah Khoi dies

Ayatullah Sayyid Abolqassem Khoi (92) dies; Golpayegani, a moderate ayatullah, becomes the Shi'ite world's only *marja'*. Unlike the Sunnis who follow the four schools of law, the Shi'ites believe in the primacy of knowledge. This knowledge, which theoretically remains in the family of the Prophet while the Imams live, is transferred to the most pious Shi'ite Muslim, the point of emulation or *marja'*.

1992: October 23 Debate over Iranian assets abroad

Ayatullah Ahmad Jannati warns against the return of Iranian exiles who, along with their power, bring corrupt Western values detrimental to Muslim artists trying to express themselves. Jannati reacts to the persuasive measures adopted by the government to bring lost Iranian assets back to boost the country's ailing economy.

1992: November 20 Iran threatens to block Straight of Hormuz

Iran threatens to sink its only submarine in the middle of the Straight of Hormuz.

1992: November 27 VCR ban lifted

The government lifts its ban on videocassette recorders (VCRs).

1992: December 4 Iran buys submarines from Russia

Iran buys three submarines from Russia at $4.5 million.

1992: December 18–25 Divorce rights amended

Men lose the right to unilateral divorce. Under the shah, women had gained a number of rights, including the right to divorce. After the revolution these rights were lost. To compensate for the loss, the government agrees to send an expert woman lawyer into divorce courts to assure that women's rights are respected.

1993: January 22 Iran and the West

Iran condemns the United States' attack on Iraq. Some Iranians cite the wars in Bosnia, Palestine, Somalia, India, Tajikistan, and the Caucasus as part of a Western wave of anti-Islamism.

1993: February 12 Majles indicts Iran TV for Western programming

Majles' investigative committee report indicts Iran TV for showing too many American westerns and for allowing its female staff to appear on the air improperly dressed.

1993: July 9 Lifestyle of clergy subject of protest

Iranians protest relatively luxurious lifestyle of the clergy and its disparity with the living conditions in the rest of society. One of the major goals of the Islamic revolution was the creation of an equitable society in which past titles and class distinctions play no role. This ideal society, however, has not materialized.

1993: July 23 Islamic dress code enforced

Vice squads reward women who abide by the strict Islamic dress code and fine transgressors. One hundred seventy-eight shops are closed for selling alcohol, pornography, and unacceptable music.

1993: October 8 Majles defeat women's council proposal

Majles (261 males, 9 females) defeat a proposal to create a committee on women's issues.

1993: November 26 Treatment of the Baha'is

The U.S. senate reaffirms its 1982, 1984, 1988, 1990, and 1992 resolutions condemning Iran's treatment of its religious minority group, the Baha'is. The Baha'is are relatively apolitical people who strive for a progressive society with international dimensions. They are the first to accept westernization as a means of introducing change into an otherwise medieval Iran.

1993: December 3 Ban on courses for women lifted

Restrictions on what women can study are lifted. The Islamic government of Iran contends that women, involved in taking care of their husbands and children, have little need for mathematics and physics. As a result, women could take only half of the math and a quarter of the humanities and experimental courses offered.

1993: December 17–24 The politics of the Marja'

Moderate Grand Ayatullah Muhammad Reza Golpayegani, age 96, world-Shi'ite marja', dies of a heart attack. Khamenei, Araki, and Sistani are nominated as marja'. Khamenei's nomination creates a conflict between the religious and political strands of the republic and opposes the Shi'ite principle of unencumbered decision making. Sistani's appointment requires that the marja' be moved from the city of Qom, Iran, to Najaf, Iraq.

1994: January 7 Araki named marja'

Khamenei names Araki as marja'. Efforts are expended on the creation of a single marja' for all Shi'ites.

1994: February 22 New culture minister calls for purge of Western influences

Mustafa Mirsalim becomes culture minister and calls for a purge of Western influences.

1994: February 25 Korean oil tankers purchased

Iran purchases five oil tankers from South Korea; six of the fourteen units of the Arak petrochemical complex become operational.

1995: March 15 Sa'idi Sirjani arrested

Influential Iranian literary critic and author, Aliakbar Saidi Sirjani, is arrested on alleged drug, homosexuality, and treason charges. Sa'idi Sirjani has made a unique contribution to Perso-Tajik culture by editing and annotating Sadriddin Aini's *Reminiscenes,* an account of the history of Bukhara at the turn of the twentieth century. Sa'idi Sirjani later dies in prison at the age of sixty-three.

1995: March 17 Sex education banned

Iran opposes sex-education in schools.

1995: July 21 Agreement reached with Tajiks

In Tehran, Tajik President Rahmonov and UTO leader, Nuri, reach agreement to end inter-Tajik hostilities.

1995: August 23 Iran–U.S. relations re-examined

U.S. President Bill Clinton signals willingness of the United States to discuss differences with Iran. Since the hostage crisis of November 4, 1979, Iran and the United States have been at odds. Before this incident, during the era of the shah, the two countries were partners in many fronts, especially in security concerns and commerce.

1995: August 24 Trade surplus reported

Iran's balance of trade shows a $6 billion surplus.

1995: October First woman deputy health minister appointed

Ashraf-ol-Sadat Sanei becomes Iran's first woman deputy health minister.

1995: November 20 Steel mills of Iran

Rafsanjani picks a site for the Kurdistan Steel Mill. Iran's first reformer, Reza Shah, fails in supplying Iranians with a steel mill that can satisfy their technological needs, especially in making their own cars. His son remedies the need by importing engines and building some of the parts in a steel mill built near Isfahan by the Soviets. The Islamic Republic expands this to a network of steel mills, supplying Iran's increasing need for steel.

1996: January Construction project begun

Construction of a major art and culture complex comprising an amphitheater, three movie houses, an art center, and a library begins in Tehran.

1996: February 24 Caspian oil: a regional economic factor

The first Iranian-made oil platform in the Caspian Sea becomes operational. In the past, Caspian oil was a political issue among the Western nations, Iran, and the former Soviet Union. In the 1990s, however, Caspian oil emerges as an economic factor affecting not only Iran but the West, Russia, and the other countries on the Caspian shore.

1996: February 24 U.S. pays reparations

The United States agrees to pay compensation for shooting down an Iranian airbus (see 1988: July 3).

1996: March 25 Gas refinery begins operation

The Sarkhun-2 gas refinery (500 million cubic feet capacity) is inaugurated.

1996: March–April Economic sanctions expanded

At the same time that Rafsanjani calls expatriates a valuable asset for Iran, America expands its economic sanctions to cover foreign firms trading with Iran.

1996: April–May Economic and social unrest

The price of bread rises thirty to fifty percent; Hizbullah supporters invade movie theaters, break video game machines, and beat up moviegoers. Four pages of the Shahtahmasbi *Shahname* sell for $2.9 million.

1996: April 19 Oil fleet expands

Within ten months, with an addition of five 300,000 ton oil tankers, Iran's oil tanker fleet reaches 3.8 million ton capacity.

1996: May 15 Fiber optic network follows Silk Road

Twenty-one thousand kilometers of optic fiber network along the Silk Road contributes to Iran's participation in international trade; completion of Sarakhs-Tajan railroad makes the overland Silk Road operational.

1996: May 30 Declining demand for carpets

Government studies the reasons for a decline in sale of Iranian carpets abroad. Given the alarming drop since the early 1990s, carpet exporters receive more support and better facilities (August 10, 1995).

1996: December 5 Majles forbids foreign words

Majles forbids the use of foreign words by all sectors of society and empowers the Farhangestan to coin necessary terms. Foreign words for which an equivalent is not available are exempt.

1997: January 17 Tehran sewer system launched

The president inaugurates the first stage of Tehran's sewage system and announces plans for the same for ten other cities.

1997: January 22 Developing a viable tanker fleet

Iran negotiates to buy five new tankers. Japan, South Korea, China, and Croatia have been competing for the 260 million dollar project since April 19, 1996. With the addition of these five 300,000-ton oil tankers to the five purchased from South Korea (February 25, 1994), Iran's oil tanker fleet reaches a 3.8 million ton capacity.

1997: February Rushdie reward reaches $2,500,000

The reward for killing Salman Rushdie (b. 1947) reaches $2,500,000. Originally (see 1989: February), Khomeini issued a *fatwa* (religious ruling) that the Iranian-British author of *The Satanic Verses* be executed. At that time, a reward of $1,000,000 was offered.

1997: February 28–April 8 Tabriz petrochemical complex opens

Tabriz petrochemical complex becomes operational. Iran, the second largest producer of petrochemical products in the Middle East, serves 6,000 factories.

1997: April 16 Regional cooperation

Natiq Nouri proposes a regional axis encompassing Iran, Russia, China, and the newly-independent republics of Central Asia.

1997: April 21 Iran-Russian relations

Russia and Iran sign agreements on oil and gas cooperation. After the Revolution, Russia and Iran become closer commercial partners. Iran begins building six ocean liners at a Persian Gulf shipyard; Russia and Iran sign agreements on oil, gas cooperation.

1997: May 23 The Khatami Difference

Mohammad Khatami (b. 1943), a moderate cleric, wins the presidential elections (seventy percent of the vote). Candidate Khatami, outlining his political, cultural, and economic plans for the nation, states that while striving to be free, religious, and responsible individuals, Iranians must constantly be watchful of exploitative nations.

Khatami uses the Internet to bring his views to the voters. After the elections, he consults experts and professionals to form a cabinet with an original outlook and approach to the nation's needs. He chooses Dr. Masumeh Ebtekar, a woman physician with an outstanding record of service, as a vice-president.

As Minister of Culture, Khatami emphasizes the necessity of sports, music, film, and books for the development of a well-balanced society. When criticized, he resigns, citing the "stagnant and retrograde" cultural climate that "bedevils cultural activities" in Iran. Rafsanjani rejects the resignation.

Khatami calls on Iraq to pay reparations for its 1980–1988 aggression against Iran.

1997: May 23 Adherence to strict Islamic codes

Khatami chooses Dr. Ebtekar, a woman physician with an outstanding record of service, as a vice-president. This follows Ashraf-ol-Sadat Sanei becoming Iran's first woman deputy health minister in October, 1995. Women's situation, however, has been erratic due to inconsistent government policies. In 1979, women protested against the required Islamic dress code; in May 3, 1990, police detained 800 women for appearing in public in bad *hijab* and for neglecting Islamic codes. Women were forbidden from wearing "see-through" dresses and men could not wear shirts open in the front (May 13-15, 1990). By October 4, 1991, the government had even established obligatory Islamic bridal dress (October 4, 1991).

1997: June 20 Canal proposed

Experts propose connecting the Caspian and the Persian Gulf via a canal. This would increase trade between Iran and several landlocked former Soviet republics, while also affording the latter access to the high seas.

1997: July 22 Rafsanjani's legacy: a moderate Iran

Majles honors Rafsanjani. When he took over the presidency, Iran was monolithically anti-Western and staunchly anti-American. In 1997 Iran is moderate enough for Khatami to state that he is not against normalizing ties with the United States.

1997: August 7 Academics protest television broadcast

One hundred and eighteen Iranian university professors protest some British TV networks' intention to air *Not Without My Daughter,* a made-for-television movie about the daughter of an Iranian father and American mother who is not permitted to leave Iran with her mother.

1997: October 7–8 U.S. sends warship to Persian Gulf

U.S. redeploys the warship *Nimitz* in the Persian Gulf.

1997: October 31 Crackdown on smoking

Smoking and advertising tobacco products are forbidden in Iranian public places.

1997: November 6 Soccer team qualifies for World Cup

Iran's soccer team qualifies for the World Cup; riotous celebrations paralyze Tehran.

1997: November 9 M. A. Jamalzadeh dies

M. A. Jamalzadeh (c. 1895–1997) dies. Although Jamalzadeh left Iran at the age of thirteen, he never lost touch with its culture. The founder of the Iranian short-story genre, he is best known for his satiric piece, *Farsi Shekar Ast* (Persian is Sugar), depicting an innocent rural youth trying to understand Iran, a land divided linguistically, ethnically, ideologically, and militarily. Jamalzadeh returned to Iran towards the end of his life and was writing in support of the current regime.

1997: November 10 The moderates' bumpy road

Mayor Gholamhossein Karbaschi, a staunch Khatami supporter, is questioned and released on bail. Karbaschi was first temporarily imprisoned on corruption charges on April 21. Hardliners are likely to scrutinize the activities of not only Karbaschi but all those who support Khatami's relatively open administration.

1998: January 9–18 Khamenei opposes rapprochement with the U.S.

President Clinton proposes direct talks between Iranian and U.S. governments at a time when Khatami seeks avenues of dialogue with the United States. The stumbling block is Khamenei, Iran's spiritual leader who, along with his hardline supporters in the upper level of the Iranian government, reject any Iran-U.S. rapprochement.

1998: June 21 Iran beats U.S. in World Cup

The Iranian soccer team beats the United States' team in the World Cup, 2–0.

Bibliography

Albert, David H. *Tell the American People: Perspectives on the Iranian Revolution.* Philadelphia: Movement for a New Society, 1980.

Bacharach, Jere L. *A Near East Studies Handbook, 570–1974.* Seattle: University of Washington Press, 1974.

Famighetti, Robert (ed.). *The World Almanac and Book of Facts.* New York: St. Martin's Press, 1998.

Ghirshman, Roman. *Iran.* London: Penguin Books, 1965.

Iran Almanac and Book of Facts. Tehran: Echo of Man, 1971.

Nikazmerad, Nicholas M. "A Chronological Survey of the Iranian Revolution," *Iranian Studies,* vol. 8, nos. 1–4, 1980.

Nyrop, Richard F., (ed.) *Area Handbook of Iran: A Country Study.* Washington, D.C.:The American University, 1978.

Rahman, H. U. *A Chronology of Islamic History: 570–1000 CE.* London: Mansell, 1989.

Salinger, Pierre. *America Held Hostage: The Secret Negotiations.* Garden City, N.Y.: Doubleday & Company, 1981.

—I. Bashiri

Iraq

Introduction

The state of Iraq has only existed since 1920, when it was created as a British mandate, or territory given to Britain with the requirement that it establish a government there. However, the area it now occupies was home to some of mankind's earliest civilizations. Iraq was known in classical times as Mesopotamia, a name given it by the ancient Greek and meaning "the land between the two rivers" (the Tigris and the Euphrates). Some scholars and archaeologists say the biblical Garden of Eden lay in Mesopotamia. Mesopotamia made up a large part of the area known to historians as the Fertile Crescent, an arc of fertile land reaching from the Persian Gulf to the southeastern coast of the Mediterranean. Some of the world's earliest cities were built along the banks of the Tigris and the Euphrates Rivers: Eridu, Ur, Sumer, Akkad, Assyria, and Babylon. Mesopotamian civilizations gave to the world the wheel, writing, a code of law, the 60-minute hour, and the 360-degree circle.

Iraq lies in the heart of the Asian Middle East, just to the northwest of the Persian Gulf. With an area of approximately 170,000 square miles (440,300 square kilometers), about twice as large as the state of Idaho, Iraq is bounded on the south by Saudi Arabia, Kuwait, and the Persian Gulf; on the west by Jordan and Syria, on the east by Iran (formerly Persia); and on the north by Turkey.

In the northern region, Iraq's topography is mountainous, with elevations up to 7,000 feet near the Turkish border; Iraq's highest peak, Haji Ibrahim, reaches 11,811 feet. Rainfall here is adequate for growing crops. The region's major city is Mosul, near the site of the ancient city of Nineveh. This area was the first in Iraq to be developed for its oil reserves. Northern Iraq has been home to the Kurdish people since the seventh century A.D. The Kurds are an Indo-European-speaking people, neither Arab nor Persian, and are of the Sunni Muslim sect. Since the mid-twentieth century, they have fought for the establishment of a Kurdish nation in the region of northeastern Iraq, southern Turkey, and northwestern Iran known as Kurdistan. The Kurds today remain probably the world's only ethnic community of more than 15 million that has not attained independent statehood.

Central Iraq consists of a broad alluvial plain and valley formed by the Tigris and the Euphrates, where the majority of the population lives. The ancient civilizations of Sumer and Akkadia learned to build irrigation channels between the two rivers, supporting the agriculture that fed their cities. As the region was conquered by various empires, Mesopotamia's economy and standard of living depended upon whether the current ruler maintained and improved these irrigation systems or let them fall into ruin. Iraq's capital, Baghdad, lies on the Tigris River in east central Iraq, near the sites of ancient Ctesiphon and Babylon. The two great rivers that bring life to this central region have also caused death and destruction over the centuries as they flooded their banks.

In the low-lying, hot and humid southern portion of Iraq (some of the world's highest atmospheric temperatures have been recorded here) are salt marshes stretching to the Persian Gulf, and the Shatt-al-Arab, a channel formed by the confluence of the Tigris and the Euphrates. This waterway forms part of the boundary between Iraq and Iran and has been the cause of fighting between the two countries in recent years. The marshlands of southern Iraq are home to tribes of the traditionally oppressed Muslim sect, the Shia. These people (called the Marsh Arabs), like the Kurds, have risen in rebellion against the current Baath Party government under Iraqi President Saddam Hussein and have been repressed by the Iraqi military. Hussein's army even resorted to draining the salt marshes that support the Shias' agriculture, after the Persian Gulf War, in which Western nations allied in the fight against Iraq after it occupied and annexed the nation of Kuwait, just to its south.

Most of western Iraq lies in the Syrian Desert, very sparsely populated and crossed by pipelines carrying oil from the fields of northern Iraq to the Mediterranean ports of Tripoli, Lebanon, and Haifa, Israel.

About three-quarters of Iraq's 17.5 million population is Arab. Iraq has been influenced by Arab Muslims since A.D. 637, when Caliph (Islamic religious and political leader) Umar I conquered the ruling Persian dynasty. Kurds make up about 20 percent of Iraq's population, and Bedouins, Turkomans, Jews, and Yazidis account for the remainder. About 70 percent of Iraq's population lives in urban centers; many rural

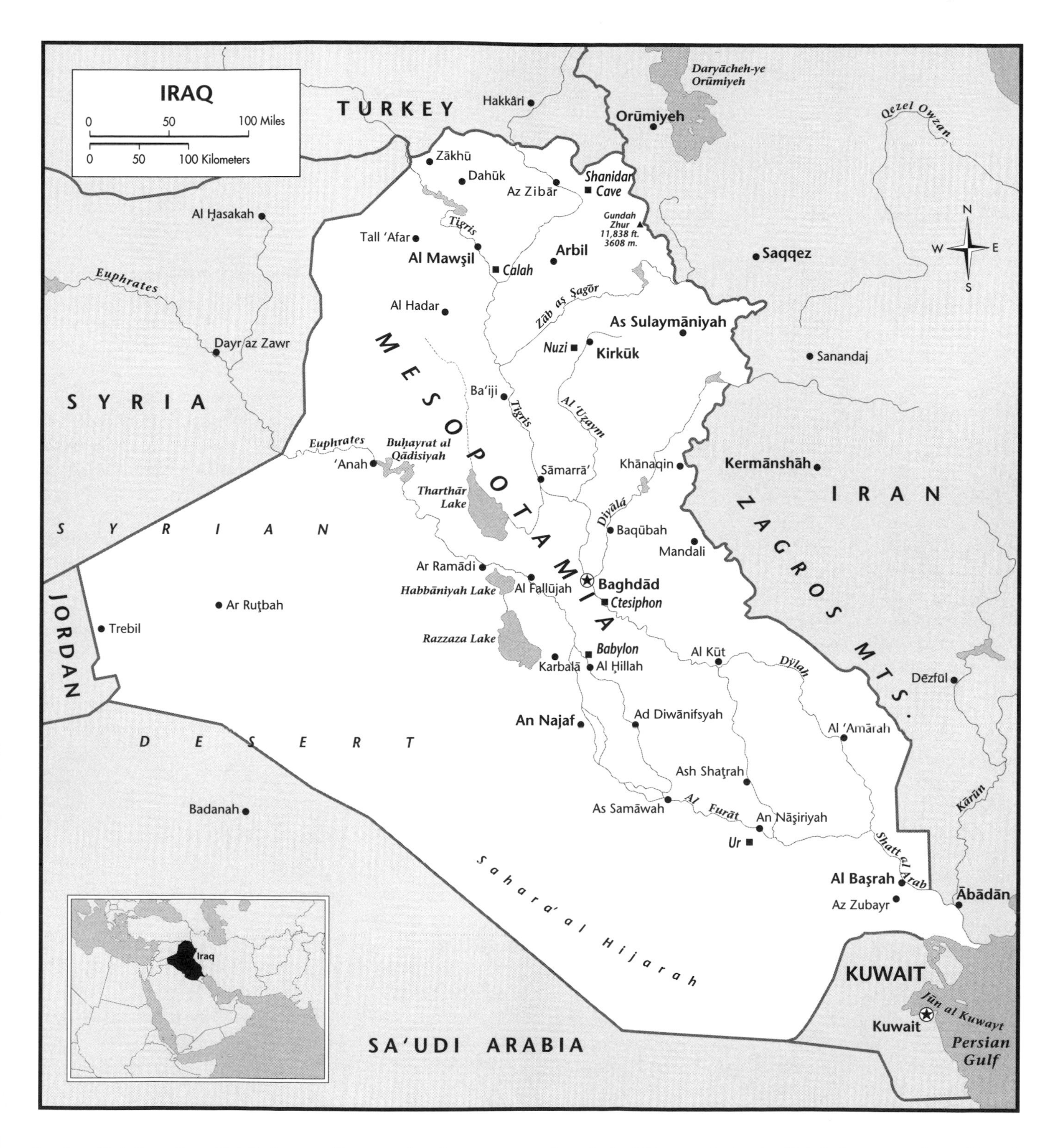

Iraqis still follow the old tribal nomadic ways, herding camels, horses, and sheep to seasonal pastures. Arabic is the country's official language, with Kurdish the major second language.

Sunni Muslims (the majority sect of Islam) have nearly all of the political and economic influence in Iraq, even though the Shia population is slightly larger, at about 60 to 65 percent of the total Muslim population of 95 percent. Small communities of Christians and Jews also thrive in Iraq.

Education in Iraq is free and compulsory through the sixth grade, although schools are not available to all children in rural areas and only about half of the population over age

15 is literate. The nation has seven universities and some 20 technical schools. All leading libraries are located in Baghdad.

History

In the mid-1600s, under Ottoman Turkish rule, Mesopotamia was divided into three major regions, or *vilayets*, that have remained into present times. These regions are based upon the major cities of Mosul in the north, Baghdad in central Iraq, and the Shatt-al-Arab port of Basra in the south. Iraq is divided into 18 political divisions, or provinces, each headed by a governor.

Iraq's history has been one of prosperity and growth or neglect and decline, depending upon the ruling regime. Over the centuries since the golden civilizations of Sumer, Akkad, Assyria, and Babylon, Mesopotamia has been ruled by the Persians, the Greeks, the Romans, the Parthians, the Arabs, the Mongols, the Turks, and finally the British, before again being brought under Arab control. The longest period of rule in Iraq was that of the Ottoman Turks, who, except for brief periods of conquest by the Persians, controlled the region from 1534 until World War I.

The Hashemite dynasty of three Arab kings, Faisal I, Ghazi I, and Faisal II, who ruled from 1921 to 1958, ended with a coup d'etat by members of the Iraqi army under General Abdul Karim Kassem, who implemented some important civil rights laws but was himself overthrown by military officers of the revolutionary Baath Party after only four years. His successor, Abd al-Salam Arif, who took control in 1963, based his policy on neutrality and cooperation with neighboring Arab nations. Arif was killed in a helicopter crash in 1966, and his brother, Abd ar-Rahman Arif, took his position. By 1968, however, the second Arif was overthrown by military Baathists under Ahmed Hassan al-Bakr, who declared himself president. In the following year, al-Bakr appointed Baathist Saddam Hussein as his second in command. The two ruled Iraq as chairman and vice chairman of the newly established Revolutionary Command Council (RCC), which remains in power in Iraq under President Saddam Hussein, who took control after al-Bakr's resignation in 1979.

In spite of numerous executions of alleged conspirators and the barbaric treatment of the Kurds and other groups in Iraq, as well as the restriction of many of the Iraqis' civil rights, a kind of personality cult has grown up around Saddam Hussein, fueled by his own propaganda, that has kept him in power for the past two decades. His apparent hunger for power engaged Iraq in two major wars between 1980 and 1991—the Iran-Iraq War and the Persian Gulf War. However, the secrecy with which Hussein has governed Iraq is being dissolved with the seizure of official documents from security headquarters after the Gulf War. Rebels obtained them in an uprising and the papers are being studied in the United States by human rights organizations.

In the aftermath of these two wars, Iraq and its neighboring nations have suffered immeasurably. As a result of massive bombings of oil fields and tankers, Iraq, Kuwait, and Saudi Arabia, as well as the shallow waters of the Persian Gulf, experienced heavy pollution from damaged oil wells and ruptured tankers spurting their contents into the air and spilling them into the water. Captain Robert E. Luchun of the U. S. Coast Guard, senior oil spill adviser in Saudi Arabia, told *National Geographic* magazine in 1991 that the best estimate of the amount of oil spilled into the Persian Gulf as a result of the Gulf War was between four and six million barrels, an amount equivalent to "a fleet of monster tankers [running] aground." Massive cleanups have been underway by international teams since 1991, but such extensive environmental damage takes time to repair.

Within Iraq itself, the massive international debt burden incurred in financing the two wars, in addition to the war damage and heavy economic sanctions imposed by the United Nations during and since the Gulf War, has taken an overwhelming toll. Iraq's war debt, including reparations, was estimated at $200 billion in 1991, and inflation rates in the country soared to 2,000 percent. As recently as mid-1997, the cost of a dozen eggs on the open market was said to be equal to half a government worker's monthly salary, according to the *Houston Chronicle*. Saddam Hussein fed his people using emergency rations rather than accept the "oil for food" plans proposed by the United Nations. In 1996, however, he accepted a plan under which Iraq is allowed to sell about one-fifth of the amount of oil it sold before the Gulf War, with earnings of $2 billion to be used for food, soap, medicines, and infrastructural improvements to better the lives of the Iraqi people.

Iraq Today

Oil, as it is for all Middle Eastern countries, is the most important economic sector for Iraq. In 1979, before the Iran-Iraq War, Iraq was the second-largest oil-producing nation in the Middle East, after Saudi Arabia. The petrochemical industry was also very large in Iraq before most plants were damaged in the wars.

Agriculture is still the main occupation of Iraqis, but the small farmer has for decades been compromised by the large landholder, just as the average wage earner has been under the thumb of the wealthy industrialist. This imbalance has prevented the Iraqi economy from stabilizing and continues to prevent a general prosperity. Cooperative rural farms are on the rise, however, and agricultural reform could improve the situation. Chief crops in Iraq are dates in the south, and wheat, barley, linseed, beans, millet, and rice elsewhere.

A major challenge for Iraq as a nation is to establish peace among warring ethnic groups, namely, the Kurds in the north, now split into three factions—the Kurdistan Democratic Party (KDP), led by Massoud Barzani; the Patriotic

Union of Kurdistan (PUK), led by Jalal Talahani; and the Maoist-inspired Kurdistan Workers Party (PKK), under the elusive and powerful guerrilla leader Ocalan. Other groups historically at odds are the Sunni and the Shia Muslims of central and south Iraq, respectively.

The United States and other Western powers, as well as Jordan's late King Hussein, have called for the ouster of Iraqi president Saddam Hussein, who has survived some 25 coup attempts. His ruthlessness as a dictator, however, might be the only force keeping the three ethnic blocs intact. A frequently cited speculation that Iraq should—and in the future might—be dissolved as a nation and partitioned into two or three different states has yet to become a reality.

About one hundred miles south of Baghdad, archaeologists find more than 900 tablets in Sumerian, and the temple for Inanna, goddess of love and war. (EPD Photos/CSU Archives)

Timeline

6500 B.C. Beginnings of civilization

Agriculture and the domestication of animals begin in what is now northern Iraq, where rainfall is adequate and cereal grains grow wild. The prehistoric agricultural site of Jarmo, a prepottery Neolithic culture near present-day Kirkuk in northeastern Iraq, dates to this period.

6000 B. C. Agricultural settlements in the Fertile Basin

Farming and stockbreeding begin in the fertile basin between the Tigris and the Euphrates Rivers when an early people migrate from the Turkish and the Iranian highlands. These agricultural settlements are the beginnings of the prehistoric Ubaid culture and later become the Sumerian city-states of Nippur, Uruk, and others.

c. 5000 B.C. Eridu is Mesopotamia's oldest city

Scientists have discovered artifacts at the site of Eridu, the southernmost city near the Euphrates and the one nearest the Persian Gulf, dating to before 5000 B.C., making it Mesopotamia's oldest city. Eridu is the center of worship of the god Ea and later becomes the first royal Sumerian city.

c. 4000 B.C. Sumerian civilization begins

The Sumerians, wandering shepherds of unknown origin, come to the southern, or lower, region of Mesopotamia. Their language is unrelated to any known family of languages. Forming stable communities, they build irrigation systems and are able to create a surplus of grain to sustain their communities. The region becomes known as Sumer.

3500 B.C. World's first cities are built

Modern archaeological discoveries have shown that the world's first cities are built around this time in ancient Sumer. These include the city-states of Eridu, Ur, Nippur, Uruk, and Lagash. Evidence has been found of a complex of thirteen Sumerian city-states reaching from what is now Baghdad to the Persian Gulf. Each city-state is ruled by a king, and daily life centers around a god and a temple. Powerful priests emerge as mediators with local dieties. Because of their food surplus, the Sumerians have time for intellectual and artistic pursuits.

3200–2850 B.C. Sumerians begin use of writing

Writing appears in Mesopotamia. The Sumerians develop a script originally based on pictographs, it later evolves into cuneiform wedge-shaped characters. They also develop the elements of modern mathematics and create the first accurate calendars and the method of telling time using sixty minutes (the number sixty is sacred to their god An) and twelve hours. They hold the world's first formal school, open only to males and called the *edubba*. The potter's wheel turned on its side becomes the first wheel used for moving objects from one place to another. This concept enables Sumerians to invent the first wheeled chariot and to use their new mobility to build the first monumental structure known as the ziggurat, a pyramid-shaped brick temple with a staircase winding around the outside and a platform for worship at the top. The Sumerians also invent the plow.

c. 2334–2108 B.C. King Sargon creates first empire

The great Akkadian king Sargon I (rules c. 2334–2279 B.C.), or, in Assyrian, Sharrukin, conquers Sumer and goes on to create the first Mesopotamian empire, with his capital at Akkad, a city-state in what is now central Iraq. Sargon (as distinguished from Sargon II, Assyrian ruler from 721–705 B.C.) unites Akkad and Sumer into one great kingdom, later to be called Babylonia. His empire reaches to northern Syria and south along the western shore of the Persian Gulf, perhaps as far as Oman. He establishes trade over a wide area, as far west as Crete and Greece and as far east as India.

Sargon's empire is passed on to his two sons, Rimush and Manishtusu, and then to his grandson Naram-Sin, who eventually builds temples in Nineveh, the capital of ancient Assyria, in what is now northern Iraq.

2108 B.C. Sargon's empire overthrown

Sargon's great-grandson Shar-kali-sharri fights to hold on to the empire but is overthrown by the Guti, a mountain people from a region east of Akkad.

2060–1950 B.C. Sumerian renaissance

After the defeat of the Guti by Sumerian king Utukhegal of Uruk, Sumer experiences a renaissance under the Third Dynasty of Ur, beginning with the reign of Urnammu of Ur (c. 2060–2043 B.C.). During this period, Sumerian literature and mathematics flourish.

2000 B.C. Ur is home of Biblical Abraham

The Mesopotamian city of Ur (referred to as Ur of the Chaldees in the Bible), in today's southeastern Iraq, is mentioned in the Bible as the birthplace and home of Abraham, the patriarch who leads his people to the land of Canaan (Israel).

2000 B.C. Sumerians begin recording history

The Sumerians/Akkadians begin recording their history as well as instruction for future generations. They are the first authors of poetry and mythology and create the first epic poem, the story of King Gilgamesh of Uruk (who ruled around 2700 B.C.), which includes an account of a great flood. Archaeological treasures of cuneiform tablets have been found in Sumerian "libraries" recovered from the ruins of Mesopotamia's great cities.

1830 B.C. First Babylonian dynasty begins

The Third Dynasty of Ur has fallen. The Amorites, a Semitic people from the west, and the Elamites, Caucasians from the east, have battled for control of Mesopotamia. Under the Amorites, Akkad and Sumer are united to form Babylonia, with its capital at Babylon (56 miles south of present-day Baghdad, Iraq), founded about 4000 B.C. and controlled by the Amorites since about 2200 B.C. The Amorite ruler Sumuabum founds the first dynasty of Babylon.

c. 1792–c. 1750 B.C. Reign of Hammurabi

The Amorite conqueror and king Hammurabi, or Hammurapi (r. 1792–1750 B.C.), rules Babylonia. He creates the first great Babylonian empire, extending north from the Persian Gulf through the Tigris-Euphrates river valleys and west to the Mediterranean Sea. Under Hammurabi commerce, the arts, and astrology flourish. Hammurabi compiles a celebrated code of laws known as the Code of Hammurabi, which commands "an eye for an eye." His code amends the common law of Babylonia, established by Urnammu of Ur, a king of Isin who ruled the land from 1889 to 1871 B.C. Hammurabi's code, like Babylonian codes before it, protects the weak and poor, especially women and children. Hammurabi's code establishes laws governing women's rights in marriage, including the right of a woman to remarry or return to her parents' home if her husband abandons her or if she "hates" him and "refuses him his conjugal rights." The code also addresses labor and wages, tenure and rent, medical procedures, public order, and criminal and civil justice.

1595–1150 B.C. Rule of the Kassites

The Kassites, a barbaric people of unknown origin from the mountains to the east, begin a 400-year rule during which little progress is made other than an alliance with Egypt.

1350 B.C.–612 B.C. Assyrians rise to power

The Assyrians, a Semitic people from Assyria, in what is now northern Iraq, in 1350 begin a rise to military power from their stronghold at Assur on the Tigris River. They establish a capital at Nineveh (near present-day Mosul), named for their god Nina. The fierce Assyrian king Tiglath-pileser I (who reigns from about 1115–1077 B.C.) brings Assyria to its greatest level of power through military campaigns in the north and west. By the end of the Assyrian dynasty in 612 B.C., the Assyrians have ruled over a vast empire extending throughout the Fertile Crescent, from the Persian Gulf to Egypt.

704–681 B.C. Arts flourish in Nineveh

During the reign of Assyrian king Sennacherib, Nineveh becomes one of the greatest of ancient cities. The arts and literature flourish. Irrigation systems are improved and the cultivation of cotton is introduced.

612 B.C. Ninevah destroyed

Nineveh is destroyed by King Nabopolassar (r. 625–605 B.C.) of Babylonia, and his allies, the Chaldeans and the Medes.

605 B.C.: September 7 Nebuchadrezzar II is crowned

Nebuchadrezzar (Nebuchadnezzar) II (c. 630–562 B.C.), son of King Nabopolassar, is crowned king of Babylon after defeating the Egyptian army under Necho II at Carchemish, a city on the western bank of the Euphrates at what is now Syria' s border with Turkey. The Babylonian victory ends Egypt's power in Asia.

605–562 B.C. Reign of Nebuchadrezzar II

The Chaldeans of Babylon under Nebuchadrezzar II inherit Assyrian lands and power and continue their western conquest into Egypt and to Jerusalem, which finally falls to Babylon in 586, thus ending the kingdom of Judah (a kingdom in southern Palestine between the Mediterranean and the Dead Sea). Many Jews are taken captive and held as slaves in Babylonia.

Nebuchadrezzar rebuilds the city of Babylon to its most magnificent, including constructing the Hanging Gardens of Babylon, which become one of the Seven Wonders of the Ancient World cited by the Greeks. He also rebuilds walls, palaces, and temples and restores other Babylonian cities. During Nebuchadrezzar's reign, Babylon is the largest city in the known world, covering some 2500 acres. A seven-storied ziggurat (temple tower) at Babylon is often linked with the biblical Tower of Babel.

562 B.C. Death of Nebuchadnezzar

At Nebuchadnezzar' s death, his son Awil-Marduk takes the throne but rules only two years before being killed. The great Chaldean Empire of Babylonia soon falls.

539 B.C.: October 29 King Cyrus takes Babylon.

King Cyrus II of Persia (c. 585–c. 529 B.C.), called Cyrus the Great, unites the Medes and Persians. He marches on Babylonia, and the gates of Babylon are opened to his general almost without a struggle. Cyrus takes the city. He frees the Jewish slaves, allowing them to return to Palestine.

539–529 B.C. Achaemenians rule Babylonia

A 208-year period of rule in Mesopotamia by the Achaemenian dynasty (named after Achaemenes, the dynasty's founder) of Persia (now Iran), of which Cyrus II is the first king begins. Cyrus is a just ruler who respects the customs of the Babylonians, requiring them only to pay him tribute and obey him.

529 B.C. Death of Cyrus II

Following Cyrus's death, there is a brief period of political instability in Babylonia.

522–486 B.C. Darius the Great assumes power

Persian king Darius I (550–486 B.C.), called Darius the Great, takes the throne. Darius is a conqueror who further extends his empire. He brings great economic prosperity to Babylonia, building roads and organizing the bureaucracy. The Babylonian irrigation systems and structures are maintained, and Sumerian customs are retained.

486 B.C. Darius I dies

After Darius's death however, Persian rule deteriorates, and by the fourth century B.C., most Babylonians oppose the Achaemenians. Cities have fallen into decline, trade is reduced, and the population has been heavily influenced by the migration of Persians (Iranians) into Mesopotamia (Iraq). The languages of Mesopotamia are replaced by Aramaic, the Achaemenian Empire's official language.

331–323 B.C. Mesopotamia under Alexander the Great

Alexander III (356–323 B.C.), king of Macedon (Macedonia), who is known as Alexander the Great, conquers Babylonia, defeating the Persians. Alexander restores Babylonian customs and worship of the local gods and plans to establish a seat of his great empire at Babylon. He becomes ill with fever there, however, and dies at the age of 32 before his plans can be carried out.

312–126 B.C. The Seleucid dynasty

Alexander the Great's general Seleucus (c. 358–281 B.C.) inherits most of the Persian Empire and founds the Hellenistic Seleucid dynasty in 312 B.C., when he becomes king of Babylon. He and his successors build great city-states, modeled on those of Greece, in the Middle East.

126 B.C.–A.D. 226 Parthians enrich Mesopotamia

The Parthians (or Arsacids, named for Arsaces, founder of the Parthian state and dynasty), a people of Scythian descent (Scythia was located in what today is part of the former USSR) who migrated from the steppes of what is now Turkistan to northeastern Iran, conquer Mesopotamia, having previously taken Persia. They retain much of the Greek language and influence, adding it to their own. Their peaceful and prosperous rule enriches Mesopotamia. The Parthian capital is at Ctesiphon, on the eastern bank of the Tigris River opposite Seleucia.

227–651 Sassanids rule unsuccessfully

Mesopotamia falls under the control of the Sassanids, the last native Persian dynasty, who struggle throughout their rule with Roman aggression and invasion. The Sassanids neglect the Mesopotamian cities and irrigation systems, allowing them to fall into ruins. The palace at Ctesiphon collapses when the Tigris River floods its banks, and soon afterward the city is taken by Bedouins, nomadic warriors from Arabia. Sumero-Akkadian civilization comes to an end.

622: July 16 Roots of Islam

The prophet Muhammad (c. 570–632) is said to have fled with a few followers from his home in the western Arabian city of Mecca to the city of Medina, to the north of Mecca. This 210-mile flight is known as the Hegira (or Hejira).

Muhammad organizes his Meccan followers and Medinan tribes, including some Jews, into the commonwealth of Islam (meaning "surrender to the will of God"). By 630, he has led his army to the border of Syria and controls all of Arabia. Muhammad dies at Mecca in 632 and his body is entombed at Medina. These two cities become, respectively, the first and second most important holy cities to Muslims, the followers of Islam.

632 Abu Bakr first caliph

Abu Bakr (c. 573–634), said to be the prophet Muhammad's first male convert and also his close companion and adviser, becomes the first caliph, or successor of Muhammad as secular and spiritual head of Islam. He begins a conquest of Mesopotamia (Iraq) and Syria for the expansion of the faith.

637 Arab Muslims take Mesopotamia

Arabian Muslims under the caliph Umar I (c. 586–644) conquer the Persian Sassanid dynasty by taking its capital at Ctesiphon after a series of successful battles against the Persians. The Muslims consider their conquests to be holy wars (called *jihads*) and spare the lives of women and children and all nonsoldiers. They also keep destruction of the cities to a minimum. Those citizens who do not wish to convert to Islam may pay a tax, or *jizya*, to live as non-Muslims. Most tribes at the time of the Muslim conquest of Mesopotamia are Christian.

Umar founds two garrisoned cities in Mesopotamia: Al-Kufah, on the western bank of the Euphrates, south of present-day Baghdad; and Basra, a seaport near the Persian Gulf.

661 The Sunni and Shia Muslim split

Two monumental events that split the Muslim world occur on Iraqi soil. The first occurs when Ali ibn Abi Talib (c. 600-661), Muhammad's cousin and a contender for the title of caliph, is murdered by a member of a dissenting group of Muslims called the Kharajites. Ali is proclaimed a saint and the true successor to Muhammad by his followers, a small group called the Shia that splits from the Sunni, or orthodox, Muslims.

Ali' s challenger, the Sunni Muslim governor of Damascus, Syria, Muawiyah I (c. 602–80), is named caliph and founds the Umayyad dynasty (its name derived from an ancestor, Umayyah), moving the Muslim capital to Damascus. Iraq continues to be ruled by Arab caliphs through the Umayyad caliphate (office or dominion of a caliph), with its capital at Damascus, until 750.

680 Husayn, Ali's son, is killed

The second event that splits the Muslim world occurs when Ali's son Husayn (c. 629–80) is killed in battle leading a small band of Shia Muslims in a revolt against the Sunni caliph Yazid I at Karbala, in central Iraq.

750–1055 Abbasid rule begins

The Abbasid caliphate begins its rule of Iraq with Abu al-Abbas as-Saffah (722–54), a descendant of Muhammad' s uncle Abbas. At this point in history, the Islamic empire is the greatest civilization west of China.

Mesopotamian jurist and theologian Abu Hanifah (699–767) analyzes existing legal doctrine and develops a structured code of laws, the first of the four orthodox Islamic schools of law.

762 Baghdad becomes the capital

The second Abbasid caliph, Abu Jafar al-Mansur (c. 709–75), builds a new Muslim capital at Baghdad, on the Tigris River in central Iraq.

786–809 Caliphate of Harun ar-Rashid

Under the fifth Abbasid caliph, Harun ar-Rashid (c. 763–809), Baghdad becomes the center of the Muslim world and of Arabic culture. He is the caliph in the famed Arabian *tales The Thousand and One Nights* (or *Arabian Nights*). Ar-Rashid is a supporter of arts and literature and becomes friends with the great Arab poets Abu Nuwas and Abu al-Atahiyah.

813–833 Caliphate of al-Mamun

On Harun ar-Rashid's death, his sons al-Amin and al-Mamun (786–833) struggle for the caliphate. Al-Mamun has the greater support and takes over as ruler. He continues to build Baghdad and to support the flowering of Arab culture, which reaches its height during his reign as caliph.

Al-Mamun establishes a House of Knowledge, where philosophical, literary, and scientific works are translated from Greek, Syriac, Persian, and Sanskrit texts. Hunayn ibn Ishaq al-Ibadi (808–73), known in the West as Johannitius, is the most famous of these Arab scholars. He is known for his Arabic translations of the works of Galen, Ptolemy, Hippocrates, Plato, and Aristotle.

Yaqub al-Kindi (died c. 870) is known as the Philosopher of the Arabs. He is one of the first Arab students of the great Greek philosophers and formulates a system that combines the views of Plato and Aristotle.

Science is also at its peak in Baghdad, with the construction of an astronomical observatory by al-Farghani (d. c. 850), whose work is carried on by the Arab astronomer and mathematician al-Battani (c. 858–929), the best known Arab astronomer of the Middle Ages. He used trigonometric methods to calculate the length of the year, equinoxes, and eclipses. Al-Mamun' s caliphate is also known for great advances in geometry and trigonometry, and the opening of medical schools and a free public hospital.

The Shia Muslims (the Abbasids belong to the Sunni sect) of Iran are unhappy with al-Mamun's choice to keep the capital at Baghdad, however, and begin to break away from the Abbasid caliphate, triggering its decline.

Separate dynasties within the empire begin to break off under their own rulers, and the Abbasid caliphate is finally overthrown in 1055 when the Seljuq Turks come to power.

1055–1258 Rule of the Seljuq Turks

Turks have been brought to Baghdad for many years as slaves (called Mamluks), but in 1055 they have risen to power, and the Turk Toghril Beg (c. 990–1063) puts an end to the brief reign of Persian tribal chiefs, founding the Seljuq dynasty (named for his grandfather, Seljuq, of the Turkish tribe of the Oguz, who lived north of the Oxus River, or Amu Darya). As Sunni Muslims, the Seljuqs rule peacefully.

1073–92 Seliq Malik Shah reigns

Iran and Iraq experience a cultural and scientific renaissance during the rule of the Seljuq Malik Shah (1055–92), who expands the Turkish empire and inspires a group of astronomers, including Omar Khayyam, to reform the calendar.

Under Malik Shah's rule, Persian vizier Nizam al-Mulk (c. 1018–92) establishes the Nizamiyah University in Baghdad. Its chief professor is the Islamic philosopher and jurist al-Ghazali (1058–1111), a Persian, who writes authoritative standard works on law, theology, and philosophy. For a time he abandons teaching to become a Sufi mystic and sets up a monastic community. One of his works relates Sufism to Islam.

After the death of Malik Shah, Turkish rule declines and ends with the coming of Mongol invaders from the Far East. The coming of the Mongols also signals the end of the nominal Abbasid caliphate.

The Mamluks will go on to establish a dynasty in Egypt and to defeat the Mongols under Hulegu Khan at Ain Jalut, Israel, in 1260.

1258–1335 Mongolian invasion

Baghdad, a city of 800,000, is sacked by an invasion of Mongols, or Tatars, led by Hulegu (c. 1217–65), grandson of the fierce Mongolian conqueror Genghis Khan (Temujin) and brother of Kublai Khan, who rules the vast Mongolian Empire from Tatu (now Beijing, China). Hulegu massacres Baghdad's scholars, religious leaders, scientists, and poets, making a pyramid of their skulls, and destroys the city's canals. The Mongols set up a capital at Tabriz in Iran, and Iraq becomes part of Hulegu's Il-khanate until the death of the last Mongol Il-khan, Abu Said, in 1335.

1401 Timur the Lame conquers Baghdad

Following a brief period of local rule, Timur the Lame, or Tamerlane (1336–1405), a Turkic conqueror, includes Iraq in his wide conquest, sacking Baghdad in 1401 and wiping out Islamic scholarship and arts except in his capital of Samarkand in what is now Uzbekistan. Iraqi cities fall into decay, agriculture wanes, and the people are forced into tribal pastoralism in the river valleys.

Modern History

1509 Rule of the Safavid Empire

The Safavid Empire of Persia (now Iran) declares Shia Islam (see Premodern Regional Chronology, A.D. 622) Persia's official religion and conquers Mesopotamia (now Iraq). The Safavids rule for a brief period, fighting the Ottoman Turks (a warrior tribe originating on the central and eastern Asian steppes) for Baghdad (see Premodern Regional Chronology, A.D. 750-1055) and deepening the conflict between the Shia and Sunni Muslims (see A.D. 661 and 680).

1534 Mesopotamia is under Ottoman rule

The Ottoman Turkish sultan Suleyman I (c. 1494–1566), called Suleyman the Magnificent, conquers Baghdad. Except for a short period of Persian control during the seventeenth century, Mesopotamia remains an Ottoman province until World War I, when it is occupied by the British. The Sunni Muslims gain political and economic power that will carry them into the twentieth century, while the Shia sect is oppressed.

1638 Mesopotamia is divided into three regions

Mesopotamia becomes an official part of the Turkish Ottoman Empire and is divided into the three regions, or *vilayets*, of Basra in the south, Baghdad in central Mesopotamia, and Mosul in the north.

1639: May 17 Borders created through temporary peace

The Ottomans and the Safavids make a temporary peace, and borders are drawn up for Persia (now Iran) and Mesopotamia (now Iraq) that remain basically the same into modern times.

1640 Kurds migrate to Iraq

Many of Persia's Kurdish tribes migrate to northern Iraq, where Kurds have lived since ancient times. The Kurds are neither Arabs nor Turks, and their language is Aryan.

1780–1802 Rule of Suleyman II

Mesopotamia is ruled by the greatest of the Mamluk Turkish (see A.D. 1055–1258) governors, Suleyman II, who abolishes executions not decreed by religious courts, outlaws confiscation, and otherwise raises the level of justice.

1831: April Bubonic plague devastates Baghdad

Bubonic plague devastates Baghdad, killing 2,000–3,000 people a day. Food supplies are exhausted, and the city collapses. Then the Tigris River floods its banks, ruining the ruling Mamluk Turks' improvements under leader Daud (governed 1816–31), who worked to modernize the city, refurbishing canals, introducing industry, and starting a printing press. The flood destroys the city, and soon afterward the Ottomans resume control.

1869 Midhat Pasa and the tanzimat

Midhat Pasa (1822–83) is appointed Ottoman governor of Baghdad. He sets out to modernize Iraq, modeling it after Western countries. His reforms come to be called the *tanzimat*. He reorganizes the army, creates codes of law, improves provincial administration, and secularizes the school system. In addition, he gives the Iraqis some control over local government. Midhat Pasa also begins publishing the first official newspaper, *al-Zawra* (a favorite name for Baghdad).

However, land reforms based on legal property rights instead of the feudal system (one marked by the domination of a privileged class) are enacted and land and agriculture fall into the hands of tribal sheikhs. The sheikhs become profit seekers, transporting crops to the West, whereas their tribesmen are reduced to the role of poor sharecroppers.

November 16 Suez Canal opens.

The Suez Canal, a ship canal across the Isthmus of Suez, Egypt, connecting the Red Sea with the Mediterranean opens, giving Iraq better access to European markets. The canal has been built by a French company.

1908 Nationalist movement begins

A new Ottoman ruling clique, the Young Turks, institutes a number of reforms after taking power in Istanbul, Turkey's capital. Bringing back the Ottoman constitution of 1876, they offer representation in the parliament at Istanbul to Iraqis and others ruled by the Ottomans. However, the Young Turks' "Turkification" policy offends the Iraqis and gives birth to a nationalist movement.

1908 Education is modernized

The free public schools started by Midhat Pasaare are improved upon. The three-year Law College of Baghdad is founded, the only institution of higher education in the country.

1914 Ottoman Empire enters World War I

The Ottoman Empire enters World War I (1914–1918) on the side of the Germans. A British expeditionary force from India lands in the Shatt-al-Arab waterway in southern Iraq and occupies part of Basra. A long campaign follows.

1918: November British control most Iraqi territory

The British take Mosul in northern Iraq, giving them control over all three Iraqi provinces (*vilayets*), except for the Kurdish highlands on the Turkish and Persian borders and land on the banks of the Euphrates south of Baghdad, including the Shia Muslim cities of Karbala and An Najaf.

1920s Women seek rights

Iraqi women seek better educational opportunities and recognition as full citizens, including freedom from the veil (according to Islamic tradition, women must veil themselves in public). Aswa Zahawi founds the Women's Rising Group, which begins publishing the journal *Leila*, advocating education and employment rights for women.

1920: April 25 Iraq is mandated to Britain

The collapse of the Ottoman Empire stimulates Iraqi hopes for independence, but on April 25, 1920, at the San Remo Conference in Italy, Mesopotamia is declared a League of Nations mandate under British administration.

1920: July Arabs rebel against British rule

Mesopotamian nationalists are angered at the British mandate, especially when British governor Colonel Arnold Talbot Wilson appoints Indians to staff government positions instead of Iraqis. For the first time, Sunni and Shia Muslims, tribal and urban Iraqis, unite to form one body against what they perceive as British colonialism. Riots and revolts break out in what comes to be known as the Great Iraqi Revolution, and the British must call in Royal Air Force bombers to restore order after three months of anarchy.

1920: October 1 State of Iraq is proclaimed under provisional government

Sir Percy Cox, British acting minister to Persia (now Iran), is named high commissioner of the new state of Iraq, and the British establish an Arab provisional government in the country. It is to be a kingdom with a council of Arab ministers but will still be under Cox's supervision and British control.

1920: December 23 British and French define regional borders

Britain and France agree on the delineation of the borders between Syria, Iraq, and Palestine.

1921: August 23 Faisal I becomes king

Following a plebiscite, with 96 percent of the people in favor, the government proclaims Faisal I (1885–1933)—an Arab and a son of Husayn ibn Ali, emir (ruler) of Mecca and king of the Arabian province of the Hejaz—king of Iraq.

King Faisal founds the Hashemite dynasty in Iraq. The British-created monarchy is alien to Iraqis, however, and the people are never fully confident of the government.

1922: June 18 **Kurdish rebellion begins**

Sheikh Mahmud Berezendji, a Kurdish chief from Sulaimaniya in Iraq's Mosul region, leads a Kurdish revolt and proclaims himself "King of Kurdistan."

1922: October **First Anglo-Iraqi treaty**

Iraq and Britain ratify the first of a series of Anglo-Iraqi treaties, which gives Britain continued power and voice in Iraqi government, keeping the country politically and economically dependent on Britain.

1923 **Mosul dispute with Turkey**

During negotiations leading up to the Treaty of Lausanne, signed on July 24, in which Turkey repudiates its claims to the non-Turkish territories lost during World War I, the issue of the Mosul region of northern Iraq, believed to be rich in oil, leads to a prolonged dispute over the question of establishing a Kurdish state in the region.

1923: April 30 **British treaty is amended**

In an amendment to the treaty with Britain of October, 1922, Iraq is to attain independence when it becomes a member of the League of Nations, or not later than four years from the conclusion of a peace agreement with Turkey. Iraq's constituent assembly ratifies the agreement in March, 1924.

1924: July 10 **Constitution is adopted**

The constitution provides for a liberal parliamentary government in which the Iraqi ministers will answer to a two-chamber parliament. All Iraqis are declared equal in the eyes of the law. Islam becomes the official state religion and Arabic the official language. The legal system is divided between Sunni and Shia Muslim courts. Women do not have the right to vote.

1925 **Iraq Petroleum Company acquires oil rights**

A concession is granted to an international oil cartel called Iraq Petroleum Company to develop the oil reserves of the Baghdad and Mosul regions.

1925: January 13 **Mandate agreement extended**

The League of Nations recommends that the alliance relationship with Britain be extended if Iraq is to be granted the Mosul region. (The city of Mosul is located in northern Iraq, on the west bank of the Tigris River.) The two countries agree to a new treaty, which extends their relationship to twenty-five years, or until Iraq joins the League of Nations. However, Iraqi nationalists continue to oppose the treaty and to demand independence from Britain.

1926: June 5 **Iraq given control of Mosul**

In an agreement between Britain and Turkey, Iraq gains most of the disputed Mosul region.

1927: October **Oil discovered**

The Iraq Petroleum Company's first big oil strike occurs near Kirkuk in northern Iraq. The well produces 80,000 barrels of oil per day. The gusher floods the countryside with oil for ten days before workers can bring it under control.

1927: November–December **Iraq seeks admission to League of Nations**

King Faisal I visits London in November to request Iraq's admission to the League of Nations. In December, Britain recognizes Iraq's independence and agrees to support Iraq's admission to the League of Nations in 1932. In return, Iraq grants permission for three new British air bases and for British officers to train the Iraqi army.

1929: August **Treaty of friendship with Persia**

A treaty of friendship is signed with Persia (modern Iran).

1930: February **Treaty of friendship with Saudi Arabia**

Iraq concludes a treaty of friendship with Saudi Arabia.

1930: June **Anglo-Iraqi treaty is signed**

King Faisal's closest adviser, Nuri as-Said (1888–1958), negotiates a new Anglo-Iraqi treaty that provides for mutual assistance between Iraq and Britain in case of war. Britain is granted the use of Iraqi military bases near Basra and at Al-Habbaniya, west of Baghdad. The treaty is to become effective when Iraq joins the League of Nations. Iraqi nationalists oppose the leasing of air bases, and Assyrians and Kurds oppose the treaty because it leaves their status ambiguous.

1930: November 16 **Parliament ratifies British treaty**

The parliament ratifies the 1927 treaty with Britain, which provides for Iraq's admission to the League of Nations in 1932. The treaty is presented to the League in January 1931.

1931: January **League of Nations membership proposed**

The treaty providing for Iraq's admission to the League of Nations is presented to the League.

1931: April **Rebellion ends**

Second Kurdish rebellion ends when Sheikh Mahmud surrenders.

1932: April–June Rebel Kurds are expelled to Turkey

Another Kurdish insurrection is ended when British planes assist the Iraqi army in driving the Kurdish rebels across the border into Turkey.

1932: October Iraq joins League of Nations.

Iraq becomes a sovereign state and is admitted to the League of Nations. This independence, however, is overshadowed by disputes over Kurdish, Assyrian, and other ethnic issues, as well as by tensions between Sunni and Shia Muslims.

1933: August Assyrians are massacred

A group of Christian Assyrians are massacred while trying to cross the border from Syria. The League of Nations tries unsuccessfully to arrange resettlement elsewhere for the Assyrians.

1933: September 8 Death of King Faisal

King Faisal I dies in Bern, Switzerland, while under medical treatment there. He is succeeded by his son Ghazi I (1912–39).

1934 Nationalism superceded by local politics

Under the monarchy, Prime Minister Nuri as-Sa'id (1888–1958) dominates Iraq with support of the upper-middle and upper classes. Tribal, religious, and local loyalties take precedence over any sense of Iraqi nationalism. King Faisal I, during his reign, considered the existence of political parties desirable for the political development of Iraq. During the decade from 1935 to 1945, however, political parties are not effective political factors.

1934: July 14 Iraq's first oil pipeline opens

An oil pipeline opens, connecting the Mosul oil-producing region with Tripoli in Syria. A second line, to Haifa, Israel, opens in January 1935.

1935: March Hashimi forms government

General Yasin Pasha el Hashimi forms a national government

1936: April 12 Treaty is concluded with Sa'udi Arabia

Iraq concludes a treaty of friendship and nonaggression with Sa'udi Arabia. The agreement becomes the first step in the creation of a general Arab alliance.

1936: October 29 Prime minister is overthrown

General Bakr Sidqi, a prominent Pan-Arab leader supported by King Ghazi, with politicians Abu Timman, a Shia Muslim, and Hikmat Sulayman, a Turkoman, carries out a successful *coup d'etat*. The parliament is abolished, and Sidqi rules as a dictator.

1937: August 11 Sidqi assassinated

General Sidqi is assassinated by a group of military officers. A series of military coups follows over the next four years.

1939: April 4 King Ghazi dies

King Ghazi is killed in an automobile accident in Baghdad. In the disturbances that follow, the British consul, rumored to have arranged the accident, is stoned to death by a mob. Ghazi is succeeded by his three-year-old son, Faisal II (1935–58), whose uncle, emir Abd al-Ilah (r. 1939-53), rules as regent until Faisal II ascends the throne in 1953.

1941: May 2 Pro-German prime minister attempts coup

At the outbreak of World War II, a pro-Axis regime led by Prime Minister Rashid Ali attempts a government take over. British forces invade the country and a rebellion breaks out.

May 31 British forces occupy Iraq

After bombing Iraqi airfields, British forces enter Baghdad, bringing the rebellion to an end. Iraqi's pro-Nazi prime minister, Rashid Ali, flees to Iran. Five days later, the British occupy Mosul. The two nations conclude a truce, and a pro-British government is installed. Iraqi nationalists, now a large and powerful group, divorce themselves from the monarchy.

1943: January 17 Iraq declares war

The Iraqi government is the first free Muslim state to declare war on Germany, Italy, and Japan during World War II (1939–45).

1945: March Arab League is founded

The Arab League is founded by Egypt, Iraq, Jordan, Lebanon, Saudi Arabia, Syria, and Yemen. Libya joins the league in 1953; Sudan in 1956; and Morocco and Tunisia in 1958.

1945: August The Iranian Kurdish Democratic Party is founded

A new political party, the Iranian Kurdish Democratic Party, forms.

1945: October 20 Arab states threaten war over creation of Jewish state

U.S. president Harry Truman requests that 100,000 Jewish refugees from Europe be immediately admitted to Palestine. Iraq joins Egypt, Lebanon, and Syria in warning the United States that the creation of a Jewish state in Palestine will lead to war.

1945: December 21 Iraq becomes a charter member of the United Nations

Iraq is among the fifty countries to sign the charter creating the United Nations.

1946: March 29 Formation of new political parties

Five new political parties are founded, including the Socialist National Democratic Party (Al-Hizb al-Watani al-Dimuqrati), the pro-Communist People's Party (Ash-Sha'b), and the National Union Party reformist (Al-Ittihad al-Watani).

1947 Treaty between Iraq and Transjordan

Iraq and Transjordan (now Jordan) conclude a treaty of alliance.

1947: November 29 United Nations votes to partition Palestine

A United Nations (UN) Special Commission on Palestine recommends the partitioning of Palestine into an Arab and a Jewish state, with the holy city of Jerusalem being international. The UN General Assembly adopts this resolution, but the Islamic Asian countries vote against the proposal. A measure to question the International Court of Justice on the United Nations' authority to force the partitioning upon the country against the majority of the people's will fails by a narrow margin. Civil war breaks out in Palestine.

1947: October 26 British withdraw most of their military from Iraq

Britain withdraws all but two air force detachments from the country.

1947: December 17 Jewish settlements in Palestine attacked

The Arab League announces its intention to prevent the partitioning of Palestine by force, and raids against Jewish settlements in Palestine begin.

1948: January 15 Portsmouth Treaty fails after popular uprising

The Iraqi government signs yet another treaty with Britain, this one called the Portsmouth Treaty (after the naval base where it is signed). The treaty calls for the withdrawal of all British troops from Iraq but for continuing British involvement in Iraqi affairs, which is bitterly opposed by the Iraqi people. Students demonstrate in Baghdad the following day, and four people are killed.

1948: January 27 Iraqis become violent

The violent Wathbah (uprising) follows, resulting in Iraq's worst clash between police and demonstrators. Some 80 people are killed and several hundred are wounded. The people call for the abolition of the treaty and the dissolution of the parliament, followed by a free election. Many members of the parliament resign, including Prime Minister Salih Jabr. The Portsmouth Treaty is repudiated by the succeeding cabinet, but the British simply reinstate the 1930 treaty, which is still in effect for ten years.

1948: May 15 Arab-Israeli War begins

The day after the state of Israel is proclaimed, the Arab League states—Egypt, Iraq, Jordan, Lebanon, Saudi Arabia, and Syria—launch an all-out armed attack against Israel. This marks the beginning of what is known as the Palestine or Arab-Israeli War (1948–49). Israeli forces have the advantage in both weapons and training and quickly establish the upper hand.

Iraq's conservative politicians use the Palestine War as a pretext for the suppression of the pro-Communist People's Party and the reformist National Union Party. Only the Socialist National Democratic Party is allowed to function without interference.

1949: February Arab-Israeli war ends

Through a series of truces between Israel and the Arab League countries, fighting ends. The armistice agreements fail to provide for a transition to peace, however, because the issue of a home for the Palestinian Arab refugees remains unresolved. A peace treaty is never signed, and sporadic Arab incursions along the borders are met with Israeli reprisals.

1949 Two new political parties are founded

Prime Minister Nuri as-Said founds the Constitutional Union Party, with a pro-Western, liberal reform platform intended to attract both the older and younger generations.

Former prime minister Salih Jabr founds a new opposition party, the Nation's Socialist Party, which advocates a democratic and nationalistic pro-Western and Pan-Arab policy.

1951: September 3 Arab League tightens blockade of Israel

The members of the Arab League are encouraged to intensify the economic blockade of Israel and completely shut of its oil supply.

1952: November 23 Rioters protest continued presence of British forces

Continued popular opposition to the presence of British forces results in anti-British and anti-United States rioting. Army chief of staff General Nur al-Din Mahmud is asked by Prime Minister Mustafa-l-Umari to form a new civilian cabinet. Mahmud declares martial law, bans all political parties and public demonstrations, and closes the major newspapers.

1953: January 17 Constitutional Union Party dominates elections

Nuri as-Said and his Constitutional Union Party are victorious in the nation's first direct parliamentary elections.

1953: May 2 Faisal becomes king

King Faisal II ascends to the throne on his eighteenth birthday.

1954 Political parties dissolved

Prime Minister as-Said dissolves all parties, including his own, on the grounds that they had resorted to violence during the elections.

1955: February 24 Iraq and Turkey conclude the Baghdad Pact

A mutual defense alliance known as the Baghdad Pact is signed between Iraq and Turkey. Britain, Iran, and Pakistan join the agreement in the same year. The Baghdad Pact presents a direct challenge to anti-Western Egypt and splits the Arab world into two camps—those favoring an alliance with the West and those who believe the Arab world should remain neutral. A cold war begins between Egypt and the northern Arab nations.

1955: November Middle East Treaty Organization founded

The prime ministers of Iraq, Iran, Pakistan, and Turkey, and the British defense minister attend the first meeting of the Baghdad Pact council in Baghdad. They establish the Middle East Treaty Organization (METO). The United States supports but does not join the alliance until 1956, when it agrees to establish ties with the organization.

1956: April 14–19 Baghdad Pact council meets in Tehran

The alliance gains an agreement from the United States to establish economic and other ties with the organization.

1956: July 26 Nassar nationalizes Suez Canal

Egypt's president Gamal Abdel Nasser announces that he is nationalizing the Suez Canal Company and creating a managing Egyptian Canal Authority.

1956: October 29–November 6 The Suez crisis

Britain, France, and Israel join forces to invade Egypt in an attempt to win back international use of the Suez Canal. They inflict heavy damages on Egypt's air force and capture Port Said, but Egypt has blocked the Suez Canal with sunken ships, preventing the Western alliance from entering the canal. The United States, other British Commonwealth nations, and the Soviet Union oppose the attack and refuse to provide help. The UN General Assembly calls for a cease-fire, and Britain and France accept it on November 6, forcing Israel to stop the attack as well. Egypt regains complete control of the Suez Canal. As a result of the nationalization of the canal, President Nasser's popularity among his people grows.

1958: January 17–30 United States promises military support

Baghdad Pact members meeting in Ankara, Turkey, obtain a pledge of military support from U.S. secretary of state John Foster Dulles. Iraq, Iran, and Pakistan also conclude a cooperative customs agreement. The United States pledges its commitment to the defense of the Baghdad Pact nations.

1958: February 14 Federation with Jordan negotiated

King Faisal II negotiates a cooperative arrangement with neighboring Jordan.

1958: May 27 Arab Union parliament opens

King Faisal II opens the first session of the parliament of the Arab federation of Iraq and Jordan (see 1958: February 14). Nuri as-Said, former Iraqi prime minister, is premier of the Arab Union.

1958: July 14 King Faisal II is assassinated in coup

In a coup d'etat led by Iraqi general Abdul Karim Kassem (1914–63), King Faisal II, Crown Prince Abd al-Ilah, and Prime Minister Nuri as-Said are assassinated. The monarchy is abolished, and Iraq is proclaimed a republic. General Najil al-Rubai heads the new cabinet.

1958: December 29 League for the Defense of Women's Rights is recognized

Under Prime Minister Kassem, the League for the Defense of Women's Rights, founded in 1952, is finally recognized. By the mid-1960s, however, the authoritarian regime's harassment and assaults forces most branches of the league to close down. Members are arrested and tortured under the Baathist regime, which begins in 1963.

1959 Naziha Dulaimi becomes first woman cabinet minister

The president of the League for the Defense of Women's Rights, Naziha Dulaimi, becomes the first woman cabinet minister in the Arab world when she is named Minister for Municipalities.

1959: January Tensions erupt with United Arab Republic

Tensions between Iraq and the United Arab Republic (UAR) develop when the UAR carries out a campaign against Syrian Communists. Iraq supports the Syrian Communist Party.

Egypt charges Iraq with attempting to create divisions in the Arab world.

1959: March Iraq withdraws from the Baghdad Pact

Following Iraq's withdrawal, the remaining members of the Baghdad Pact meet in Ankara, Turkey, in August to reorganize as the Central Treaty Organization (CENTO). The United States supports the new organization but does not join it.

1959 Colonel Shawwaf's rebellion proves unsuccessful

Colonel Abdul Wahhib Shawwaf leads a rebellion against the government with the support of the nationalists. The rebellion, which may also have support from the United Arab Republic (UAR), is unsuccessful and its leaders are executed. Following the rebellion, General Kassem conducts a roundup of Iraq's Communist leaders; many are put on trial and some are executed.

1959: June 26 Activities of Popular Resistance Front curtailed

Prime minister Kassem restricts the activities of the Popular Resistance Front, a pro-Communist group.

1959: October Assassination attempt by Baath Party

Prime Minister Kassem escapes an assassination attempt with only minor injuries. A young Saddam Hussein and other members of the Iraqi branch of the Baath Party, an underground organization of Pan-Arab intellectuals founded in 1947, are believed to have a part in the plot. Hussein flees Iraq, but a number of other Baathists are arrested and tried for treason.

1959: December Family relations law is revised

Prime Minister Kassem revises family relations law, traditionally dictated by Islamic law. Under the new code, women are given equal rights with men concerning inheritance. Polygamy is limited; the minimum marriage age is raised to 18; and women are protected from arbitrary divorce. The code draws religious opposition, however, and does not survive after Kassem's regime.

1960: September 14 OPEC holds first meeting

The Organization of Petroleum Exporting Countries (OPEC) holds its first meeting in Baghdad. Charter members are Iraq, Kuwait, Iran, Saudi Arabia, and Qatar.

1960–64 Revolution erupts in Kurdish region

Kurdish nationalists in northern Iraq renew their battle to establish the autonomy of Iraqi Kurdistan within a democratic Iraq.

1961: July I Iraq claims Kuwait

Refusing to recognize Kuwait' s claim to independence, Premier Kassem declares Kuwait (a former British protectorate at the southern tip of Iraq on the Persian Gulf) part of Iraq, saying Iraq has inherited the Ottoman claim to the territory. British troops are deployed in Kuwait to prevent a threatened Iraqi invasion. The British troops are gradually withdrawn and are replaced by Arab League troops.

1961: July 20 Arab League admits Kuwait

The Arab League votes to admit independent Kuwait to membership. The Iraqi delegation charges the league with aligning with British imperialism and walks out of the meeting.

1963: February 9 Kassem is ousted in military coup

After a bloody two-day battle, a military junta made up of Baathist officers and civilians and their Arab nationalist allies overthrows the Kassem regime and executes Prime Minister Kassem. The newly formed National Council of the Revolutionary Command (NCRC) assumes control. Colonel Abd al-Salam Arif becomes provisional president, and Ahmed Hassan al-Bakr is named premier. The new government begins an anti-Communist campaign, using the Iraqi army to round up members of the Iraqi Communist Party (ICP). The new regime follows a policy based on neutralism, cooperates with Syria and Egypt, and works to improve relations with Iran and Turkey.

1963: March 1 New government guarantees Kurdish rights

The Revolutionary Council pledges to guarantee "the rights of the Kurds." The Kurdish rebel leader, Mustafa al-Barzani, threatens a war against the new government if Kurdish autonomy is not granted.

1963: May 11 al-Bakr forms new government

The cabinet resigns, and Prime Minister Ahmed Hassan al-Bakr forms a new government. The key positions in the new government are held by members of the right wing of the Baathist movement.

1963: May 14 Iraq recognizes independent Kuwait

Kuwait becomes a member of the United Nations, and Iraq is compelled to recognize Kuwait as an independent state.

1963: June 10 Government attacks Kurds

The Iraqi government, led by Arif and al-Bakr, launch an offensive against the Kurds. Some 300 civilians are killed and are buried in a mass grave in Sulaimaniya, a Kurdish town in northeastern Iraq.

1963: November Baathist extremists attempt coup

A second coup d'etat is attempted by Baathist extremists from the left. The coup is supported by the ruling Syrian wing of the party. Provisional president Abd as-Salam Muhammad Arif and the moderates survive the coup attempt, and in the aftermath Iraq's Baathist organizations recede from prominence until the late 1960s.

1964 Kurds establish alternative government

The Kurds set up their own constitution and body of laws, giving their Revolutionary Command Council power to establish a civil administration, set up a judiciary, and impose taxes.

1964: January 13–17 Jordan River diversion opposed

Arab League opposes Israeli diversion of Jordan River. At a meeting of the heads of state of the Arab League nations, plans are formulated to oppose Israel's planned diversion of the waters of the Jordan River.

1964: February 10 Cease-fire is achieved in Kurdistan War

President Arif and the Kurdish leader, Mustafa al-Barzani, conclude a cease-fire. The Iraqi government recognizes the national rights of the Kurds, as part of the Iraqi union.

1964: September 5–11 Arab League supports Palestinians

The Arab League meets to consider ways to divert the Jordan River's waters before they reach the areas under Israeli control. League members also agree to establish a United Arab Command, which will include a Palestine army. The Palestine representative, Amhed Shukairy, is named chairman of the Palestine Liberation Organization (PLO).

1964: October 4 Council approves unification with Egypt

President Abdel-Salam Arif's plan for uniting with Egypt is approved by the National Revolutionary Council.

1965 Baath Party reorganizes

The Iraqi Baath Party overcomes its rivalries and reorganizes, with former prime minister Ahmed Hassan al-Bakr as secretary-general and Saddam Hussein (b. 1937) as his deputy.

1965: April Kurds demand autonomy

Kurdish rebels renew their revolt, demanding their own autonomous territory and support for a standing Kurdish army.

1966: March Separatist Kurds attacked

The government begins a major offensive against the separatist Kurds.

1966: April 13 President Arif is killed

President Arif dies in a helicopter accident. His brother, General Abd ar-Rahman Arif, is named to succeed him as president.

1966: June Cease-fire negotiated

The Kurds agree to a cease-fire in return for another promise of regional autonomy. The Kurdish rebels are not disarmed.

1966: August 9 New cabinet is formed

Prime Minister Abdul Rahman al-Bazzaz, appointed in 1965, resigns. A new cabinet is formed under General Naji Taleb, a radical nationalist who strongly opposes peace with the Kurds.

1966: December 10 Arab League supports Jordan against Israel

The Arab League votes to send Saudi Arabian and Iraqi troops into Jordan to defend the country against Israel.

1967: June 5–10 The Six-Day War (Arab-Israeli War)

The Israeli air force conducts surprise attacks on 25 Arab airfields, destroying nearly all the Egyptian, Jordanian, Iraqi, and Syrian planes on the ground. Simultaneously, Israeli tanks sweep over the Sinai Peninsula, reaching the Suez Canal on June 8.

Israel gains control of the Gaza Strip, the Sinai Peninsula (as far as the Suez Canal), the West Bank of the Jordan River, Jerusalem, and the Golan Heights in Syria. With the capture of Arab Jerusalem on June 7, the Israeli government unites the city but promises freedom of access for all religions. At the request of President Nasser of Egypt, and with Israeli agreement, the United Nations arranges a cease-fire on June 10. The Egyptian government charges that British and U S. planes have participated in the attack. Despite denials from Britain and the United States, most Arab states, including Iraq, break off diplomatic relations with the two countries. Israel retains control of the West Bank territory following the war. The Suez Canal is once again blocked by sunken ships.

1967: June–July Soviets pledge military aid to Arab states

In a visit to Arab states in the Middle East, Soviet president Nikolay Viktorovich Podgorny promises to replace the military equipment lost by Egypt, Iraq, and Syria.

1968: July 17 Baathists overthrow Arif

The right wing of the Baath party, in conjunction with dissatisfied military officers, overthrows President Abd ar-Rahman Arif. The Baathists establish the Revolutionary Command Council (RCC) under General Ahmed Hassan al-Bakr. Bakr is installed as both president and prime minister. Baathist military men hold all key cabinet posts, and the Tikritis—Sunni Arabs from the town of Tikrit, northwest of Baghdad—related to al-Bakr become especially prominent. Saddam Hussein, a Tikriti, plays an important role behind the scenes.

The Baathists organize as the Arab Baath Socialist Party and become the ruling political group in Iraq, espousing a Socialist, Pan-Arab platform.

1969: November 9 Hussein becomes second in command.

The Revolutionary Command Council is enlarged from five to fifteen members, and Saddam Hussein al-Tikriti is named vice chairman of the council, making him the second in command in the Baathist regime.

1970: January 21–24 Government executes alleged coup plotters

The execution of thirty-seven pro-Western, right-wing men and women accused of plotting to overthrow the government arouses worldwide condemnation. By the end of the year as many as 86 people are officially executed by the Baath. Relations are broken off with Iran, said to have supported the plot, and many Persians are deported from Iraq. Iraq's Communist Party opposes the harsh actions and by June several hundred Communists are in jail.

1970: March 11 RCC fails to grant Kurdish state

The Iraqi Revolutionary Command Council (RCC) issues a communique giving the Kurds a number of new rights, including recognition of "the existence of the Kurdish nation;" provisions for better education, with classes taught in Kurdish; support for Kurdish authors; amnesty for those participating in revolution; and participation by Kurds in Iraqi government.

1970: July Baathist constitution is published

The Baathists publish a new constitution that establishes the country's government as a People's Democratic Republic, with socialist and Pan-Arab goals. Islam is declared the state religion, but religious freedom is granted. Although the constitution states that Iraq cannot be partitioned, both Arabs and Kurds are recognized and Kurds are given certain national rights. The Iraqi president, as head of the Revolutionary Command Council (RCC), is given chief executive power in both the state and the armed forces and is responsible for the national budget. Provision is made for the election of a national assembly, but this will not come to pass until 1981.

Iraq joins other Arab states in rejecting UN cease-fire. Egyptian President Nasser accepts the U.S. proposal on July 23 after extensive conferences with the Soviet Union. Jordan and Lebanon accept on July 26, followed by Israel on July 31. Syria, Algeria, Iraq, and the Palestine Liberation Organization (PLO) refuse to participate in peace negotiations.

1972: June 1 Government nationalizes oil field

Since coming to power in 1968, the Baathist regime has focused increasingly on economic problems. In 1972, the regime nationalizes Iraq Petroleum's Kirkuk field, a major oil producer in northern Iraq. Iraq is the first Arab Persian Gulf country to take this action.

1972: December 19 General Federation of Iraqui Women is established

The Revolutionary Command Council passes a law outlining the main task of the General Federation of Iraqi Women (GFIW), the only women's organization in which Iraqi women are allowed to participate. (Membership in any other is a capital offense.) The task, according to the law, is to mobilize women "in the battle of the Arab nation against imperialism, Zionism, reactionism and backwardness. " The importance of marriage and family is emphasized, although some strides are being made in education and employment equality for women.

1973 War erupts

The Arabization program continues in northern Iraq, and war again breaks out between the Baathists and the Kurds.

1973: October 6–24 Iraq aids Syria in conflict over Golan Heights

The government of Iraq provides material aid to Syria during that country' s renewed conflict with Israel over the Golan Heights. Iraq strongly opposes the cease-fire, which ends the conflict but leaves Israel in control of the Golan Heights.

1974: March Iraq is at war with the Kurds

Kurdish insurgents, known as the Peshmerga, mount another revolt, with Iranian military support. The Iraqi army counters with a major offensive in what comes to be known as the Fifth Kurdistan War.

1975: March 6 Iran withdraws support for Kurds

Iraqi president Ahmed Hassan as-Bakr reconciles with the shah of Iran at an OPEC summit meeting in Algiers. Part of the bargain is that Iran withdraw its support for the Kurds and close its borders to Kurdish refugees. In return, Iran gets the Iran-Iraq border in the Shatt-al-Arab waterway—the joint outlet of the Tigris and the Euphrates Rivers at the Persian Gulf—redrawn at the median instead of at the eastern shore.

1976: June Guerrilla attacks erupt again

Guerrilla activity reemerges and during 1977 and 1978, the Iraqi army destroys the Kurdish villages along Iraq's border with Turkey, Iran, and Syria.

1977 Women in the military

Iraqi women are allowed to enter the military forces as officers, if they have a health-related degree, and can be appointed as warrant officers or army nurses if qualified. Most women serve in administrative or medical positions until 1981, when they begin serving in combat positions. By 1987, as a shortage of male soldiers develops, women are serving in combat roles in the air force and the Air Defense Command.

1979: March 2 Kurdish leader dies

Former Kurdish Democratic Party (KDP) leader Mustafa Barzani dies in the United States, where he has recently resettled. His son Massoud becomes leader of the KDP.

1979: July 16 Saddam Hussein becomes president

Iraqi president Ahmed Hassan al-Bakr resigns, apparently because of poor health. Saddam Hussein al-Tikriti is immediately sworn in to succeed al-Bakr as president of Iraq and chairman of the Revolutionary Command Council. Approximately three weeks later, after a series of "trials" conducted to uncover an alleged plot against the Iraqi government, 22 former officials are executed.

1980: June 7 Israel destroys nuclear reactor near Baghdad

Israeli bombers destroy the Osirak nuclear reactor near Baghdad. Israel claims the facility will be used to produce nuclear weapons, a charge Iraq denies.

1980: September Invasion of Iran leads to Iran-Iraq War (1980–88)

Kurdish and Shia Muslim issues have caused rising tensions between Iraq and Iran since the Iranian revolution of 1979, in which Islamic fundamentalists took control of Iran's government under the leadership of Shia religious leader Ayatollah Khomeini. In September 1980, President Saddam Hussein of Iraq renounces the 1975 border agreement concerning the Shatt-al-Arab waterway and invades Iran.

Iraqi forces take the Khuzestan region of southwestern Iran where Arab dissidents are in revolt against the Iranian government. They capture the major southern city of Khorramshahr and the island of Abadan, on the Shatt-al-Arab, destroying its large oil refinery.

1980: November National Assembly elected

The first National Assembly is elected by popular referendum, however, it has no real power.

1982: May–June Iran recaptures Khuzestan

Iranian forces reorganize and stage a counterattack, recapture Khuzestan, and drive Iraqi soldiers out of Abadan and Khorramshahr and from all disputed Iranian territory.

Iraq makes a bid for peace and withdraws troops from Iranian territory. Iran, ignoring peace proposals, launches major offensives on the Iraqi oil port of Basra. Iraq inflicts heavy losses on the Iranian forces.

1983 Iran-Iraq War spreads to Persian Gulf

The Iran-Iraq land war reaches a standoff with Iranian and Iraqi troops setting up an elaborate system of trenches.

The Iraqis attack Iranian oil installations in the Persian Gulf, disrupting but not stopping oil exports from the main oil terminal at Khark Island. In mid-1983, Iraq receives French jets bearing Exocet missiles. Iran threatens to close the Strait of Hormuz if Iraq uses the missiles. The United States declares the strait a vital interest and says it will use military force to keep it open because of the large volume of oil that passes through it on the way to the West.

Iraq also begins to attack civilian targets in Iran using long-range missiles. The attacks cause heavy casualties, and Iran responds by shelling Iraqi border cities.

1986: February Iranian forces gain advantage

Iranian forces capture the port of Fao, at the mouth of the Shatt-al-Arab at the southernmost tip of Iraq. The Iranians now control all of Iraq's border on the Persian Gulf and are within reach of the major Iraqi city of Basra.

1986: April Khomeini issues conditions for ending conflict with Iraq

Iran's Islamic leader, Ayatollah Khomeini, renews his demands for ending the war: Iraqi President Saddam Hussein must step down, and Iraq must admit responsibility and pay war reparations. Iran rejects all demands for a cease-fire until its demands are met.

1986: November United States provides military aid to Iran

It is revealed that the United States has supplied Iran with $30 million in spare parts and antiaircraft missiles in hopes that Iran will exert pressure on terrorist groups in Lebanon to release American hostages held there.

1987 Persian Gulf tanker war

Iranian forces seize several islands in the Shatt-al-Arab waterway opposite Basra. The war soon becomes a tanker war in the Persian Gulf, as both sides attack ships transporting oil, supplies, and arms to the enemy.

1987: May 8 U.S. ship accidently bombed

Iraqi missiles inadvertently hit a U.S. frigate, the Stark, killing thirty-seven men. President Saddam Hussein offers a public apology.

1988: March 16 Poison gas is dropped on Kurdish village

The Kurdish village of Halabja, in northeastern Iraq, is devastated by a poison gas attack from the air that kills some 5,000 civilians. Iran and Iraq blame one another for the attack, but the truth remains unknown.

1988: July 18 Iran accepts cease-fire

1988: August 20 War with Iran ends

The peace is considered little more than an armed truce because none of the issues fought over are resolved and the two countries remain enemies. The Baghdad government begins the postwar reconstruction process, which includes rearming. Iraq has suffered enormous casualties and physical damage and has acquired a massive international debt.

1990: February 18 Murder of woman accused of adultery not prosecutable

The Revolutionary Command Council decrees, as in patriarchal times, that a man cannot be prosecuted for killing his mother, daughter, sister, or any other female relative on grounds that she has committed adultery.

1990: August 2 Iraq invades and annexes Kuwait

Iraqi president Saddam Hussein accuses Kuwait of illegally drilling oil from a border oil field that the two countries share and of producing oil beyond its OPEC quota to drive down prices, costing Iraq $1 billion a month in lost profits. On August 2, Iraq invades and occupies Kuwait. Kuwaiti defense forces offer little resistance, and most senior government officials escape into exile in Saudi Arabia.

1990: August 6 Operation Desert Shield begins

Iraq rejects a United Nations (UN) resolution calling for its immediate withdrawal from Kuwait. The UN imposes extensive economic sanctions on Iraq. President Hussein's army moves toward Saudi Arabia, and world superpowers fear he will take the Saudi oil fields, giving him, with those of Iraq and Kuwait, control over half of the world's oil reserves.

King Fahd of Saudi Arabia asks for and receives military support from the United States and Britain in what becomes known as Operation Desert Shield. Hussein makes peace with Iran to help protect Iraq's eastern border. He also holds thousands of non-Iraqi men hostage in Iraq and Kuwait but releases them by the end of the year after talks with U.S. president George Bush.

1990: August 28 Iraq claims Kuwait

Iraq declares Kuwait its nineteenth province.

1990: November 29 United Nations sets deadline for withdrawal from Kuwait

The United Nations Security Council gives Iraq until January 15, 1991, to remove its forces from Kuwait.

1991: January 17 Persian Gulf War begins

A devastating air war—the heaviest bombing in history—begins with an attack on Baghdad by a combined force of twenty-nine United Nations-member countries led by the United States. The air attack on Iraq and Kuwait is followed by a ground attack.

1991: February 24 Operation Desert Storm

The ground attack led by the United States follows the air attack on Iraq and Kuwait and becomes known as Operation Desert Storm. Iran, which remains neutral during the Persian Gulf War, receives Iraqi planes that are flown across the border for safekeeping.

1991: February 28 Hussein accepts ceasefire

After forty-three days of fighting Iraqi forces are driven from Kuwait, and President Saddam Hussein is forced to accept the UN cease-fire. Iraq is defeated but not occupied. Despite vast destruction of cities, weaponry, and chemical and nuclear facilities, in addition to tens of thousands of casualties, Hussein's regime remains firmly in control.

1991: March–April Saddam Hussein besieges Kurdish and Shia minorities

The Hussein regime moves to crush uprisings by the Kurds in the north and the Shia in the south following the war. No-fly zones imposed by the West give the Kurds some safety but do little to protect the Shia Muslims in the south. Clashes occur between UN allied peacekeeping forces and Iraqi forces in both areas.

During the spring of 1991, some 1.5 million Iraqis flee the country for Turkey or Iran to escape Saddam Hussein's increasingly repressive rule. Most of the refugees are Kurds, who later resettle in areas in Iraq not controlled by the government. Allied humanitarian groups set up along the northern Iraqi border offer some 600,000 Kurds emergency medical help, food, and a safe haven until they can return home.

1993: May 27 Border between Iraq and Kuwait affirmed by UN

The United Nations (UN) reaffirms the boundary between Iraq and Kuwait. Relations with Kuwait have remained tense since the end of the Persian Gulf War. Some Baghdad officials continue to assert Iraq's claims to Kuwait. The UN Security Council reaffirms the decision of a boundary demarcation commission establishing the border between the two nations.

1993: June 26 **U.S. bombs Baghdad**

The United States bombs Iraqi intelligence headquarters in Baghdad after a report that the Iraqis had planned to assassinate former president George Bush on his visit to Kuwait in April.

1994 **UN economic sanctions**

The United Nations continues to enforce strict economic sanctions on Iraq and a ban on oil sales. It also requires that Iraq submit its nonconventional weapons renters to regular inspections. Iraq has complied with these terms. Hussein in turn strives to crush Kurdish resistance by applying an embargo in northern Iraq and continues to use military force against Shia Muslims in the south.

1994: October **Iraq troops threatening Kuwaiti border**

Iraq again begins to align troops along the Kuwaiti border. U.S. president Bill Clinton sends forces to keep him Saddam Hussein in check.

1995 **Countries break sanctions on oil exports**

During the year, international support for the sanctions erodes as France, Russia, Romania, and Jordan all seek to purchase oil.

1995: August **Intergroup conflict ends**

The warring Kurdish Democratic Party and the Patriotic Union of Kurdistan in the north agree to the end a fifteen-month conflict among the Kurds.

1995: October 15 **Saddam Hussein is reelected**

As the only candidate, Saddam Hussein is reelected to another seven-year term as president. He takes 99.96 percent of the vote.

1996: May **UN agrees to Iraqi sale of oil**

Iraq and the United Nations (UN) agree that Iraq may sell up to $2 billion worth of oil. The proceeds of the sale are to be used for food and medicine in Iraq, to pay war reparations, and to pay UN expenses.

1996: September 2–3 **U.S. attacks radar installations**

In response to Iraqi attacks on Kurdish settlements in the north the United States launches missile attacks against Iraqi radar installations in the south and extends the no-fly zone from the thirty-second to the thirty-third parallel.

1996: October **Kurdish conflict continues**

The intergroup accord fails (see 1995: August), and fighting again erupts between the Kurdish factions represented by the Patriotic Union of Kurdistan and the Kurdish Democratic Party. Attempts by the United States to arrange a cease-fire lead to an agreement, but no actual cease-fire.

1997: May 14 **Turkish soldiers cross Iraqi border**

Turkish soldiers and tanks cross the Iraqi border to fight Turkish Kurdish rebels of the Kurdistan Workers Party, or PKK, who have been using Iraqi mountain hideouts to attack Turkish towns. The PKK has also been skirmishing with Iraqi Kurds along the border, and Kurdish Democratic Party leader Massoud Barzani has asked Turkey for help in keeping the Turkish group out of Iraq. Iraqi Kurds enjoy a de facto state in northern Iraq, thanks to continuing U.S. enforcement of a no-fly zone over the region, keeping President Saddam Hussein's army from attacking.

1997: June **Oil for food plan is sponsored by UN**

A United Nations-sponsored "oil for food" plan begins. Under this arrangement, Iraq may sell about one-fifth of the amount of oil it sold before the Persian Gulf War (see 1991: January 17). About a third of the profits from this oil sale will go to pay war reparations. The remaining two-thirds buys food, medicine, and soap and detergent for the impoverished Iraqi people who have suffered as a result of war damage and economic sanctions. Money is also allocated for improvements in services such as electricity, water, agriculture, and education.

Bibliography

Bulloch, John. *Saddam's War: The Origins of the Kuwait Conflict and the International Response.* Boston: Faber and Faber, 1991.

Chaliand, Gerard, ed. *A People without a Country: The Kurds and Kurdistan.* New York: Olive Branch Press, 1993.

Farouk-Sluglett, Marion, and Peter Sluglett. *Iraq since 1958: From Revolution to Dictatorship.* New York: I. B. Tauris, 1990.

Gunter, Michael M. *The Kurds of Iraq: Tragedy and Hope.* New York: St. Martin's Press, 1992.

Helms, Christine Moss. *Iraq: Eastern Flank of the Arab World.* Washington, D.C.: The Brookings Institution, 1984.

Jehl, Douglas. "In Iraq, small things like soap make big difference." *Houston Chronicle*, Sunday, June 1, 1997, Sec. 31A.

Mansfield, Peter. *A History of the Middle East*. New York: Viking, 1991.

Marr, Phebe. *The Modern History of Iraq*. Boulder, CO: Westview Press, 1985.

Metz, Helen Chapin, ed. *Iraq: A Country Study*. 4th ed. Washington, D.C.: Library of Congress, Federal Research Division, 1990.

Onaran, Yalman. "Turks cross Iraqi border to fight Kurds." *Houston Chronicle,* Thursday, May 15, 1997, Sec. 21A.

Paxton, John. *The Statesman's Year-Book Historical Companion*. New York: St. Martin's Press, 1988.

Simons, Geoff L. *Iraq: From Sumer to Saddam*. New York: St. Martin's Press, 1994.

Welteran, Bruce. *World History: A Dictionary of Important People, Places, and Events, from Ancient Times to the Present.* New York: Henry Holt, 1994.

Israel

Introduction

The modern day State of Israel is located in the Middle East on the eastern shore of the Mediterranean Sea. Its neighbors are Lebanon to the north; Syria, West Bank, and Jordan on the east; the Gaza Strip and Egypt on the west. The West Bank and the Gaza Strip are administered by Israel, but these lands were still in dispute as of 1999. Tenuous borders have been constantly under revision since the 1948 founding of the State of Israel. Although its present-day borders have been established by many mandates and re-arranged by hostilities and treaties, the land that was once referred to as Palestine has been inhabited by civilized peoples for more than five thousand years.

As of 1999, the population was 5,644,000, with 591,400 in its capital city, Jerusalem. Approximately 4.6 million of these residents are Jewish and claim descent from the ancient Hebrews. The term *Hebrew* refers to the peoples of ancient Israel. The term *Jew* refers to people adhering to the faith of Judaism. In the modern period, it is used to describe those who adhere to Judaism as well as secular descendants of those who practiced the religion. The term *Israeli* refers to all the citizens of the modern State of Israel, whether Jewish, Muslim, or Christian. The remainder of Israel's population is overwhelmingly Palestinian-Arabs.

The only Middle East country with minimal oil reserves, Israel's chief economy comes from diamond cutting and electronics, with some exports of citrus fruits and cotton. Also profitable is collecting potash from the Dead Sea. Glass production from sand glass has become a major industry, too. Tourism is one of Israel's major sources of income. Israel, a desert country, has done a remarkable job of reclaiming the land, planting vast forests, and creating its own supply of fruits and vegetables. The *kibbutz,* or communal farmland, has helped increase productivity and immigration.

Early History

Of the three major religions (Judaism, Christianity, and Islam) that trace their roots to ancient Palestine, Judaism has the oldest roots, stretching back to the arrival of the Hebrews around 1250 B.C. The Hebrews, a nomadic Semitic tribe, arrived in the land of *Canaan* (the land west of Jordan in Palestine, so-called Promised Land). The term *Hebrew* probably comes from a Canaanite term *hapiru* meaning wanderers or nomads. The Hebrews alternately fought and lived peaceably with other tribes in this area: Babylonians, Assyrians, Hittites, Amorites, and Kassites, to name a few. All of these peoples, called the Canaanites, had a much more advanced urban civilization and were well versed in martial arts and city planning. They were the prime movers among many bands of people inhabiting the Fertile Crescent between the Tigris and Euphrates Rivers. Though there is no archeological evidence of his arrival, Abraham of the *Old Testament* was supposedly the first of the Hebrews to cross into Canaan. The Hebrews basically lived on the fringes of Canaanite society, barely eking out a subsistence living. Urban Canaanites and nomadic Hebrews were distinct groups who shared dome characteristics. While they both spoke Semitic languages, the Canaanites worshiped Baal, and the Hebrews mostly worshipped Yahweh. Another difference related to their governments, for the Canaanites had a monarchy but the Hebrews were ruled by judges.

About 1000 B.C. the Hebrews established a short-lived monarchy which had a long lasting effect. The three kings who ruled during this brief monarchy were Saul, David, and Solomon. Saul was appointed by the judge Samuel to unite the Hebrews who were facing war with the Philistines. Saul was killed in battle, and David became the next king, during whose reign (1050–920 B.C.) the tiny Hebrew kingdom became a mini-empire. David defeated the Philistines and the kingdoms of Ammon, Moab, Edom, and Damascus, and he established the city of Jerusalem. It was made to look like the cities in other kingdoms the Hebrews had become familiar with—the Egyptians and Mesopotamians. David formed a strong army, taxed his people severely, and employed slave labor. Solomon continued the tradition of David: he built a magnificent palace and a glorious Temple to honor Yahweh. Even though Solomon created a great kingdom, at his death his people revolted because of heavy taxes. The kingdom was split into two competing areas, Judah and Israel, and the Hebrews never fully controlled the land of Israel again until the twentieth century.

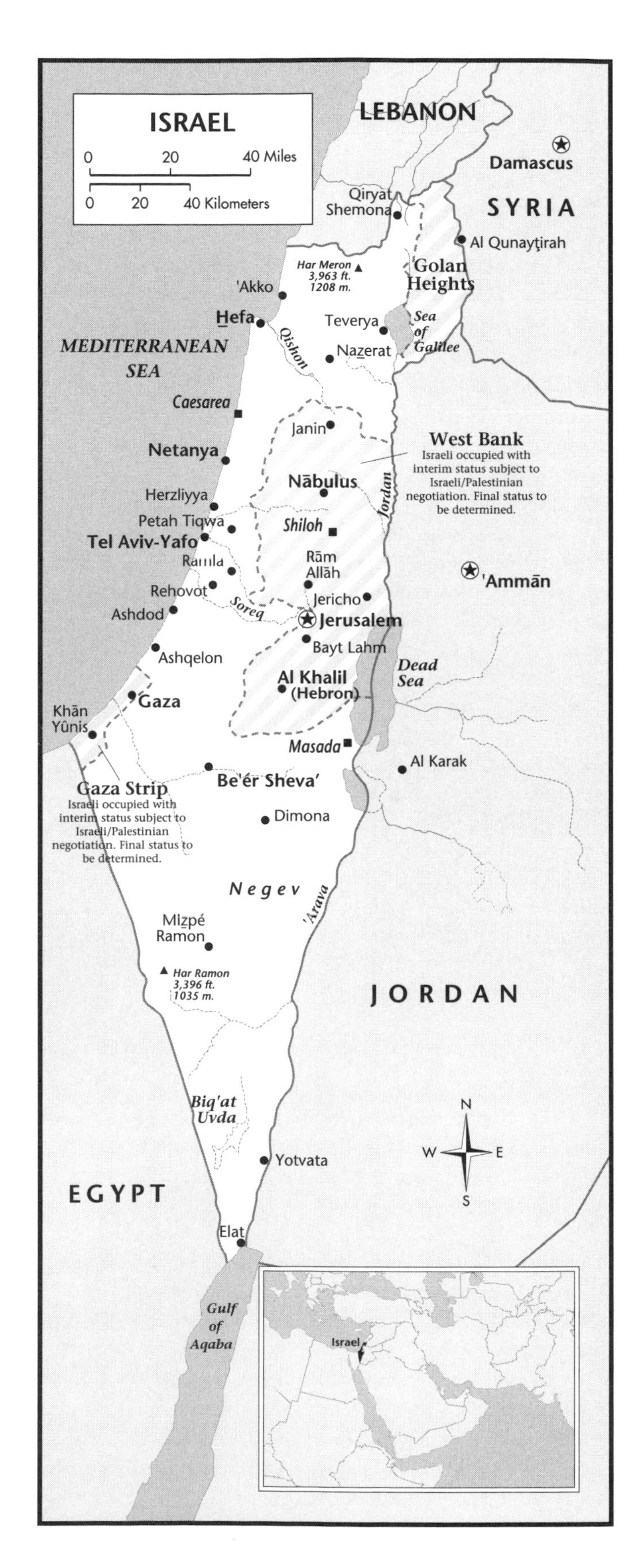

Soon other local tribes took over Judah and Israel. After Solomon died, the northern kingdom, Israel, with its capital city at Samaria, was conquered by the Assyrians. The southern kingdom, Judah, succumbed soon after. The Assyrians forced the ten tribes of Israel to scatter and no one knows what happened to them. They are often referred to as the "ten lost tribes." The two remaining tribes of the kingdom of Judah were permitted to remain together until the Babylonian conquest in the sixth century B.C., when King Nebuchadnezzar destroyed Jerusalem and exiled 10,000 Hebrews to Babylonia (called the *Babylonian Captivity*).

For centuries afterward, the Hebrews made several attempts to refound their homeland. One chance for Israel came when Cyrus the Great of Persia (600–529 B.C.) conquered the Babylonians in 538 B.C. He was impressed with the faith of the Hebrews and ordered them to return to Judah and rebuild the Temple. However, the vast majority of Hebrews chose to remain outside the Holy Land. Those who did return created a theological state, as ordered by Cyrus, dedicated to Yahweh. The Persians were largely unconcerned with the political development of Judah. This benign neglect continued until the conquest of the Persian empire by Alexander the Great (356–323 B.C.). Alexander also gave the Hebrews leniency to worship Yahweh; however, the Hebrews resisted assimilating into Greek culture.

After Alexander, Judah was conquered and re-conquered by various empires. The desecration of the Temple during the reign of Antiochus IV (d. 163 B.C.), the Syrian king, incited the Hebrew revolt led by the Maccabees (a priestly family) in 168 B.C. For a brief period, after the Maccabees' victory, Judah (also called Judea) became an independent state again. The Hebrews rededicated the Temple which gave rise to the Festival of Hannukah still celebrated today by Jews. Under the leadership of Judas Maccabaeus, the Hebrews established the Hasmonean dynasty which was equal in organization and power to the monarchies of Saul, David, and Solomon. This powerful dynasty was brought to an end by Roman rule and internal conflicts between rival brothers, Hyrcanus II and Aristobulus II.

In 67 B.C., Pompey of Rome (106–48 B.C.) invaded Judea. Hyrcanus II willingly supported Pompey but Aristobulus took refuge in the Temple and resisted. In retaliation, Pompey reduced the area controlled by Judea and placed most of Judea under Roman rule. Before Roman rule, the Hebrews had been governed by a political and religious court called the *Sanhedrin.* However, the Romans installed their own king, Herod, who was perhaps of Hebrew ancestry but was loyal to Rome. The Romans allowed the Sanhedrin to control the religious law as long as it did not interfere with Roman law. Herod remained in power for more than thirty years (37–4 B.C.), during which time Jerusalem became a famous city. Herod built fortresses, such as Masada, for his own protection and that of the city. Josephus (37 B.C.– 100 A.D.), a Jewish historian, suggests Herod was both hated and successful. The Jews hated him because he was ruthless and because he violated their laws. Josephus also states that Herod gave to the poor and at times reduced heavy taxes.

When Herod died, the new emperor, Caesar Augustus, divided Judea into several parts, making it a small province. Its most infamous administrator was Pontius Pilate (1? A.D. 50? A.D.), believed to have been involved in the crucifixion of Jesus. During Pilate's period, the Jews opposed placing Roman portraits on military and state flags because their religion forbade idolatry. They opposed the census which increased taxation. To make matters worse, Jewish factions were in conflict, and these groups were also at odds with the Roman government. In 66 A.D., a group of Jews known as *Zealots,* (fanatical, militant Jews) fought for independence from Rome. In response Emperor Vespasian laid siege to Jerusalem and destroyed the Second Temple. Most of the uprising was swiftly defeated, but a small group of Zealots remained at the fortress of Masada. When their situation became desperate, the Zealots committed suicide rather than submit to Roman rule.

At the beginning of the second century A.D., Christians were still regarded as Jews, even by Jews, but they were distinguished by their belief that Jesus was the Messiah. Though most Jews believed in the coming of the Messiah, only a small band of followers identified Jesus as such. Still, Jews and Christians were equally persecuted by the Romans. In 130 A.D., the Roman emperor Hadrian (76–138) visited Jerusalem and decided to rebuild it directly on the ruins of Solomon's Temple. Enraged Jewish citizens rebelled, staging the Second Jewish War or the Bar-Kochba Rebellion, and capturing Jerusalem for three years. Hadrian crushed the rebellion and mandated that no circumcised male (i.e. no male Jew) be allowed to enter the city of Jerusalem. This effort to separate was the beginning of the break between the new Judeo-Christians and the Jews.

Once Constantine (280–337), legalized Christianity, Jerusalem became a place of Christian pilgrimage. In fact, Palestine continued to be a holy Christian site until the advent of Islam in 632. In 638 the armed forces of Caliph Omar Ibn al-Khattab conquered Jerusalem, and Muslims controlled the city for the next 450 years. The Muslims believed that the Temple Mount of the Hebrews was also their sacred site as the prophet Mohammed (570–632) was said to have spent his last night on earth at the Temple of Solomon before entering heaven. On this site the Muslims established a mosque, called the Dome of the Rock. The Al Aqsa Mosque was also built. During the Muslim occupation, Christians either left or were forced to convert to Islam. Many Muslim factions fought amongst themselves. The Ummayeds were followed by the Abbasids (from Iraq), who in turn were conquered by the Fatimids and the Seljuk Turks. The Seljuks conquered Jerusalem in 1071 and were in turn conquered by the Christian Crusaders in 1098. The Muslim period saw Jerusalem divided into residential areas, each voluntarily occupied by one group: Christians, Jews, or Muslims.

The Crusades, which emanated from Christian Europe, at the behest of various popes, turned Palestine into a battlefield. The Crusaders established the Latin Kingdom of Jerusalem in 1100. They massacred Jews and burned their villages, but their conquest of Jerusalem was short-lived. In 1187, the Muslim leader Salah ad Din (Saladin, 1138–1193) defeated the Christians at the Battle of Hattin. The Mameluks or Mamluks, slave-traders from Syria and Egypt, held Jerusalem for the next four centuries. During that time, Europe became more interested in trading with the Muslims than fighting with them. Then, in the eleventh and twelfth centuries, Jews were expelled from England and France; they made their way to Palestine, increasing the population decimated by the Crusades.

From the eleventh century until about 1517, the Mameluks ruled Palestine. The collapse of the Byzantine Empire led, however, to the Ottoman takeover of Palestine. When the Jews were expelled from Spain and Portugal in 1492, the Ottoman Turks took them in. The Hebrew word for Spain is *Sepharad*; thus, the Jews from this area are called *Sephardim.* By contrast Jews from other parts of Europe (England, Germany, France, Poland, Hungary, Russia) are known as *Ashkenazim.* Each group claims its own social and cultural distinctions and its own language. The Spanish dialect spoken by Jews of Spain and Portugal is called *Ladino,* while the Jews of Eastern Europe speak a German dialect called *Yiddish.* The Ottoman Yutkish rule was good to the Palestine Jews and their community thrived. For example, Joseph Nasi, a Portuguese Jew, became a financial advisor to the Ottoman rulers Suleiman II (1496–1566) and Selim II (1566–74). As a reward for his services, the Ottomans gave Nasi permission to settle Jews in the area of Tiberias in Palestine. As a result, Jews from Europe began to trickle into Palestine and settle there.

The Modern Era and Zionism

Even by the early nineteenth century, though, the population of Palestine was sparse. The Jewish population numbered about 50,000. On the outskirts the Arab population, both Sunni and Shi'ite, were at war with each other. The Christian population numbered about 3,500. In 1833, Mehmet Ali (1769–1849), the Albanian viceroy of Egypt, decided to attack the Ottoman Empire in Palestine. The French helped him, but the British intervened, and with their aid, Palestine was returned to the Turks. This change benefited the Jews because previously they were not allowed to reside permanently in Palestine. With the British mandate, Jews from all over Europe began to re-settle Palestine.

The first great wave of immigrants, in the late nineteenth century, were Russians escaping anti-Semetic tsarist policies. However, arriving in Palestine, these Russian Jews found themselves stateless and at the mercy of the religious courts which were largely Sephardic. There were open, religiously based conflicts between the Russian—including Ashkenazim—and the Turkish favored Sephardim. These polarities were culturally based. The Turks shared the same political history and their dietary laws were similar to the *kosher* laws of the Sephardim. Thus, until the end of Ottoman rule in 1917, the Sephardic rabbis had the political upper hand. The advent of the Zionist movement, however, shifted the political power to the Ashkenazim.

Zionism, or the movement toward an independent Palestine, arose out of a history of persecution and displacement. Russian Jews who previously were forced to live on the outskirts of civilization, called the Pale of Settlement, were allowed in the late nineteenth century to become members of Russian society and study in public schools. In Germany, Jews were allowed to become functioning members of German society. Jews quickly adopted the language and customs which had previously been denied to them. They shed their confining religious traditions and adopted a more liberal religion called Reform Judaism. This more secular movement freed Jews from the restraints of the 613 Orthodox commandments of the Old Testament involving dress, dietary laws, and Sabbath observance. In Russia, this new found opportunity to assimilate into Russian society while maintaining a Jewish identity ended abruptly with the murder of Tsar Alexander II (1818–1881). His son Alexander III (1845–1894), sensing the liberal policies of his father were the cause of his death, created a policy of Jewish extermination or conversion by force which generated mass emigration of Russian Jews to other European cities or to Palestine.

Those who emigrated to Palestine were shocked by the primitive living conditions there. These urban Jews were now confronted with farm life and hard times in finding employment. Many of them returned home; others crowded into the Palestinian cities of Jerusalem, Hebron, Safed, and Tiberias. The events which followed these mass migrations to Palestine produced a political machine that caused the birth of the state of Israel. For example, in 1882 a young Lithuanian Jew, Eliezer Ben-Yehuda (1858–92), emigrated to Israel and set about to learn to speak the Hebrew as a daily language. Hebrew had been used by the Jews only in prayer since the Babylonian Captivity. Religious Jews thought it profane to use the sacred tongue to describe common everyday events. Eliezer Ben-Yehuda brought the language of Hebrew back home to Palestine.

Another watershed event helped to create the modern state of Israel was the Dreyfus Affair. Alfred Dreyfus (1859–1935), a Jewish officer in the French army, was falsely accused of treason. Even though he was found innocent, France became a hotbed of anti-Semitism. Theodor Herzl (1860–1904), a Jewish journalist from Vienna, covered the Dreyfus trial and began publishing articles calling for a Jewish state where Jews could live free from persecution. He did not specifically envision Palestine, but he predicted the creation of such a state. Herzl traveled the world looking for economic and political support. The bulk of economic support came from Russian Jews who had the most to gain, since they were being persecuted in their homeland. Herzl offered to pay off the Turks' national debt if they would give Palestine to the Jewish people. But Constantinople felt Herzl could not make good on his offer and they refused.

Jewish emigration to Palestine began to worry Arab residents. Loyal to Islam, they viewed any creation of a Palestinian Jewish state as an encroachment. The sophisticated organizational skills of European Jewry were anathema to tribal Arabs, who were not used to fighting in a political arena. The advent of World War I halted Jewish emigration to Palestine. The Allied victory in World War I guaranteed the partitioning of the Ottoman empire. Zionists saw an opening in world politics for the creation of a Palestinian state. However, support for a Jewish homeland came from Britain, under the urging of Chaim Weizmann (1874–1952). Weizmann implored Arthur James Balfour (1848–1930), the British foreign secretary, to write to Lord Rothschild in England, declaring Britain's commitment to a Jewish homeland, without prejudice to civil and religious practices of existing non-Jewish communities. This letter became known as the Balfour Declaration (1917).

The Zionists were disappointed because the declaration did not state specifically that Palestine would be a uniquely Jewish state; nevertheless, the League of Nations approved a British mandate over Palestine. The British were hesitant to proclaim a Jewish state in a land that had a large Arab majority. Arab leaders even staged a mock coup d'état in which they proclaimed King Faisal (1885–1933) king of Syria and Palestine. Faisal was told by the British that an independent Syria and Palestine would not be created and that Sir Herbert Samuel (1870–1963) would be in charge of Palestine under the Balfour Declaration. By 1920, Jewish settlement in Palestine had reached about 100,000. By 1935, it was 300,000. The modern city of Tel Aviv had 100,000 inhabitants. Still, until the beginning of World War II in 1939, the majority of Jews lived assimilated lives in Europe and America. Few of these Jews were convinced to give up their comfortable lifestyles to move to Palestine.

Independence

The aftermath of World War II impelled the nations of the world to recognize the need for an independent state of Israel. During the years leading up to World War II, Britain had acquiesced to Arab demands for limiting Jewish immigration to Palestine. In 1939, the *White Paper*, published in Britain, indicated that Jewish immigration in Palestine was to be limited and then curtailed after five years. This declaration was motivated by British hopes for Arab support during World War II. After the war, the fact of the 6 million Jews massacred by the Nazis turned world support in favor of creating a Jewish homeland. The United Nations, in 1947, proposed partitioning Palestine between Arabs and Jews. In May, 1948, David Ben-Gurion (b. David Gruen, 1886–1973) proclaimed the State of Israel. Ben-Gurion was its first prime minister. The United States was the first nation to recognize Israel's independence. Immediately the surrounding Arab states declared war on Israel.

The war proved disastrous for the Arabs. They lost most of the territory that had been given to them by the Balfour Declaration. Palestinians were dispersed to other Arab countries or held as refugees in Israel; Jordan, which was to be the Palestinian homeland, refused to take them in. Continued terrorist attacks kept Israel on the defensive. In 1958 the Palestinian Liberation Organization (PLO) was formed with the sole purpose of undermining Israel's statehood. Yasir Arafat (b. 1929), its leader, organized several attacks from other Arab countries. In two short but major wars, the Six-Day War in 1967 and the Yom Kippur War in 1973, Egypt and Syria united to launch attacks on Israel. In both wars Israel emerged victorious and gained more territory, a positive outcome which proved to be problematic for Israel. Palestine, which had controlled the city of Jerusalem, lost control of that city in the war.

As of 1999, the Palestinians want to establish their state in East Jerusalem, and Israel refuses to cede the conquered territory. Politics in Israel are divided on the question of how to achieve a lasting peace. The Labor Party favors ceding some land to the Palestinians for self rule. The Likud party favors keeping the land. The encroachment of Israeli settlements on the newly acquired West Bank of the Jordan River has kept this issue alive for more than twenty years. Israel maintains a well-prepared military with men and unmarried women being required to serve. The only citizens exempt from this requirement are the Ultra-Orthodox who refuse to serve on religious grounds and Arabs who find it a conflict of interest to go to war against other Arab nations. There are, however, Arabs who volunteer and who are accepted for military duty.

The foundation of a Jewish state might have reduced widespread anti-Semitism, but it did not solve internal dissension within Judaism. Various factions have struggled over issues concerning religious freedom. The Orthodox have continued to limit the granting of marriage and divorce decrees to Orthodox laws, to the dismay of Conservative and Reform Jews. Jews also cannot remarry unless granted a Jewish divorce, a *get.* Sometimes a Jewish husband, who has control over the granting of the *get*, will refuse, thus leaving the life of an undivorced woman in limbo. Sabbath laws (similar to Blue Laws) forbid certain activities on Saturday. Orthodox Jews protest, sometimes violently, against violators of these laws.

Also, since Israel's inception, the government has granted the right of citizenship under the Law of Return to Jews wishing to settle in Israel. However, the Orthodox maintain their right to decide who is a Jew. Under their strict definition, the child of a Jewish mother is declared Jewish as are individuals converted to the Orthodox tradition. The child of a Jewish father and a non-Jewish mother and converts to Reform and Conservative Judaism are not Conservative and Reform members are outraged by these distinctions. Children of *gentile* (non-Jewish) mothers in mixed marriages are considered illegitimate and have few rights. They may not marry other Jews unless properly converted under Orthodox law. The Falashas, a black tribe who came to Israel from Ethiopia, claim descent from the tribe of Dan and have completely different customs which often conflict with Orthodox traditional practices. Moreover, most Jews living in Israel today are not religious and are treated with disdain by the Orthodox. In all, many social and religious problems face Israel today.

At the close of the twentieth century, Israel is maintaining a fragile peace with her Arab neighbors. Egypt and Jordan actually recognize the existence of an Israeli state while other Arab nations still refuse to do so. Israel has proved a faithful ally of the United States and other Western countries, maintaining a strong democratic government, a self-sufficient economy, and a highly educated populace. Tourism thrives because outsiders are drawn to its holy sites. Jews and Arabs both claim rights to Israel as descendants of a common patriarch, Abraham. Christians claim the Holy Land as the birthplace of Jesus, though Eastern Orthodox Christians are in theological conflict with Roman Catholics. The main issues Israel must address are those which will lead to lasting peace in the Middle East. These issues are not completely under Israel's control, although Israel's strong government is a stabilizing factor. Access to Jerusalem, which ideally would be a shared city, may only be granted under tight security from all other nations. Jewish extremist and terrorist groups confront Arab terrorist groups constituting an impediment to peace discussions. Israel's many political parties run a coalition government; pursuing peace plans involves having to make side deals with separate groups to get a consensus. In a national election in June, 1999, Benjamin Netanyahu (b.1949) of the Likud Party was defeated by Ehud Barak of the Labor Party. This election may indicate a trend to more open negotiations with Arab nations.

Perhaps the most notable cultural event of the founding of the State of Israel is the re-birth of the Hebrew language, which had been dormant for four thousand years. Hebrew is now the primary language of the Israelis, although English is a major secondary language. No other country in the world has Hebrew as its language.

Timeline

8000 B.C. Ancient peoples inhabit the area

Most people living at this time are nomads. Some subsist as small bands of hunters and gatherers. The area between the Tigris and Euphrates rivers comes to be known as the Fertile Crescent.

7000 B.C. City of Jericho

The ancient city of Jericho has about two thousand inhabitants. The city is enclosed by massive walls.

4000 B.C. Semitic peoples enter Jericho

Bands of nomadic Semitic peoples come from Mesopotamia and Egypt causing much fighting and destruction of property.

3500 B.C. New villages emerge

Evidence survives to modern times of pottery making and copper artistic pieces being forged. The Sumerian city of Kish provides the oldest known stone tablet containing pictographs. (picture-like writing) Kish is probably the most sophisticated city outside China and Egypt.

3100–2200 B.C. Early Canaanite period

Population increases and forms into city-states. People speak a Semitic tongue which is the ancestor of languages such as Canaanite, Phoenician, Moabite, Hebrew, and Aramaic.

2600 B.C. City of Ur

Modern excavations at the city of Ur yield evidence of a palace for a king.

2370 B.C. Akkadians invade Sumeria

The Akkadian king, Sargon I, turns the Fertile Crescent into an empire.

2200–2000 Amorites come to Canaan

A Semitic-speaking people, the Amorites, invade the area and command control of all Mesopotamia (Fertile Crescent). They found their capital city, Babylon. Their most famous ruler is Hammurabi (1728–1687) who is responsible for creating a famous code of civil law. Eventually, these people come to control the area of Palestine.

1550 B.C. First Hebrews settle in Canaan

The nomadic Hebrews wander into Canaan, led it is believed by Abraham of the *Old Testament.* The Canaanite civilization is much more highly developed than that of the Hebrews. The Hebrews are not one united people but are composed of many different tribes. The Hebrews are no match for the militaristic Canaanites and Philistines. The Hebrews live on the fringes of Canaanite society.

1250–1050 B.C. Occupation of Canaan

The Hebrews continue to worship Yahweh but often worship the Canaanite Baal. They build altars to both. The Hebrews are governed by judges who decide tribal altercations. They continually battle the Philistines with varying results. The Hebrews ask the judges to choose a king for them, but the judges refuse.

1050–920 B.C. The Hebrew monarchy

In spite of the judges' warnings that a monarchy is against the wishes of Yahweh, the Hebrews ask for a king. The judges pick Saul. (This story is recounted in the *Old Testament* book, *Samuel*). Saul is good-looking and a good warrior, but he lacks administrative skills. He argues constantly with Samuel and often switches his allegiance from Yahweh when things are not going well. He is jealous of his protégé, David. Saul and his three sons die in battle against the Philistines. During his two-year reign Saul manages to unite the twelve tribes into one political unit. The next king, David, brings with him the reputation of having slain the Philistine giant, Goliath.

With Saul's heir apparent, Jonathan, killed in battle, David takes on the monarchy. He captures Jerusalem from the Jebusites. He conquers other tribes and puts down rebellions among his own people. He takes many wives and concubines. He is attributed with having written the *Book of Psalms*, perhaps the most notable being the *Twenty-Third Psalm.* David's successor, Solomon, is the son of David and Bathsheba, a Jebusite woman. Solomon is a great builder and constructs a magnificent Temple. He incurs large debts which require sending some of his people to do slave labor in other countries. Upon his death, ten of the original twelve tribes revolt and refuse to be ruled by Solomon's son, Rehoboam. These ten tribes secede, and the monarchy, along with the powerful kingdom of Solomon, is destroyed.

920–597 B.C. Two kingdoms of Judah and Israel

Divided, the kingdoms of Israel and Judah are easy prey for conquering tribes. The northern kingdom of Israel, with its ten tribes, is overcome by the Assyrians in 740 B.C. The southern kingdom of Judah succumbs shortly thereafter. The Babylonian king, Nebuchadnezzar, (also sometimes referred to as Nebuchadnezzar, 605–562 B.C.) becomes enraged with little pockets of resistance and, in revenge, destroys Solomon's Temple in 586 B.C. Also, all Hebrews are exiled from Jerusalem; many are taken to Babylon (the *Babylonian Captivity*).

539 B.C. Hebrews return to Jerusalem

Cyrus the Great of Persia conquers the Babylonians and, impressed with the faith of the Hebrews, enjoins them to return to Jerusalem to rebuild the Temple. Not many Hebrews take advantage of this return because they are living comfortably in exile in other kingdoms. Those who do return set about to rebuild the Temple.

516 B.C. Dedication of the Second Temple

The rebuilt Temple of Solomon is constructed on the same site as the first Temple.

332 B.C. Alexander the Great conquers Persian Empire

Alexander of Macedon is the new ruler of the Western world. Now the state of Judah, once under Persia, is a vassal state of Greece. The Greeks let the Jews run their own kingdom. The lifestyle of the Greeks is abhorred by the Jews who refrain from idolatry and nudity. Culturally, for the Jews, the *Torah* (*Pentateuch* i.e. first five books of the *Old Testament*) is translated into Greek and preserved in the library at Alexandria. The *Torah* now becomes the defining text for the absolute religious laws of the Jews. With this publication, Judaism becomes recognized as an established Mediterranean religion.

168 B.C. Antiochus IV (Epiphanes) desecrates the Temple

The Seleucid (Syrian) king, Antiochus IV, inherits part of Alexander's empire. He puts up idolatrous pictures on the Jewish Temple which enrage the Jews who stage a revolt. The Jews, under the leadership of the Maccabeans (Hasmonean Dynasty), manage to defeat the Syrian king and rededicate the Temple. This rededication gives rise to the celebration of *Hanukkah,* which means rededication. Judah temporarily becomes an independent state. The kingdom is again as large as it was under King David. However, now there are internal religious disputes between the Pharisees, who want to include oral law and written law into religious interpretation, and the Sadducees, who hold only to written law. One of the greatest Judaic scholars, Hillel, is a Pharisee.

64 B.C. Roman law ends Hasmonean Dynasty

The religious quarrels end when the Romans choose their own leader to rule the kingdom of Judah.

37 B.C. Herod Becomes Ruler of Judah

Herod, also known as Herod the Great, is chosen by the Romans as ruler of Syria Palestina, as the Romans call Palestine. He is loyal to Rome, but being of Jewish ancestry, considers himself to be also king of the Jews. He is a war hero (Battle of Actium—31 B.C.) and a great builder. He begins the reconstruction of the Great Temple. Herod is detested by the Jews because of his loyalty to Rome. His loyalty changes depending on whoever is in power at the time. Herod is an Idumean, recently converted to Judaism, and ambitiously ruthless. He manages to get rid of all competing members of his family, marries many women, and rebuilds the Temple of Solomon. He executes all remaining members of the Hasmonean Dynasty (Antigonus, Hyrcanus, and Aristobulus). He also executes many of the members of the Sanhedrin (Jewish high court) because of past differences. Herod mints coins with an eagle on them and likewise places his symbol of the eagle on the Temple. This is regarded as blasphemy by the Jews and riots erupt in which these symbols are removed from the Temple.

6 B.C. Birth of Jesus

Historians estimate this date as the birth of Jesus.

4 B.C. Death of Herod

Upon Herod's death, many underground groups seek to overthrow the Roman rule. One of the main groups, the Zealots, work to assassinate Roman officials. Also operating are various messianic groups who aspire to restore the kingdom of David.

25–36 A.D. Rule of Pontius Pilate

Pilate, the prefect of the Roman province of Judea, is forced to deal with Jewish plots and rebellions. Jews consider Roman symbols on public works to be idolatrous. Jews want, in fact, to subvert Roman rule. Pilate draws more ire when he uses religious funds to build an aqueduct. The Jews stage a mass rebellion and Pilate retaliates with equal violence. He is commonly believed to have been involved in the trial and crucifixion of Jesus.

27–30 Ministry of Jesus

Little is known about the life of Jesus, but he preaches to a small band of followers in Judea. Coming to Jerusalem for the celebration of Passover, Jesus is regarded by the Pharisees as dangerous because he does not adhere to the strict interpretations of the *Torah*. The Romans regard him as an enemy because they see him as an influential person who challenges authority. Passover is a crowded time in Jerusalem, and the Romans prepare to avert open acts of rebellion. Jesus is arrested and tried by a Roman court for claiming to be the Messiah, and Pontius Pilate condemns Jesus to be crucified. Claims of being the Messiah are viewed by the Romans as treasonous because of hostility to Roman rule of Judea. Even though Jesus is tried by a Roman court, his followers blame the *Sanhedrin* (Jewish court) for promoting his death.

66–67 First Jewish War

Eleazar, captain of the Temple and leader of the Zealots, proclaims in an open act of rebellion that no Roman sacrifices are to be accepted in the Temple. He occupies Herod's palace and organizes an underground revolt. When the Romans surround the palace, Eleazar pretends to surrender, and then he slaughters all the Roman soldiers.

70 Romans re-take Jerusalem

The Roman commander in Jerusalem, Vespasian, (9–79) is ordered by the emperor of Rome, Nero (37–68), to lay siege to Jerusalem. The Jews are defeated and the Second Temple is destroyed. Some of the Zealots take refuge in the fortress of Masada. There they continue to resist for three years until they are surrounded and most have succumb to starvation. The remainder commit suicide rather than submit to Roman rule. After the war, a small band of Jews (Pharisees) are allowed to settle in Jamnia, on the coast. Their leader, Yohanan ben Zakkai, establishes a settlement and a Torah school. Yohanan does not include any messianic teachings because they cause trouble with the Romans and also they are espoused by early Christians.

132–35 Second Jewish War

The Roman emperor Hadrian (76–138) rebuilds the city of Jerusalem and renames it Aelia Capitolina. Infuriated Jews stage a second rebellion. Under the leadership of Simeon Bar Kochba, the greatest rabbi of his time, the Jews are defeated, having no chance against the Roman army. Hadrian henceforth forbids all Jews (all circumcised males) from ever entering Jerusalem. Hadrian sells Jewish rebels into slavery, thus dispersing many Jews to all parts of the Roman Empire. Thus, Hadrian unknowingly helps to spread Judaism throughout the Roman Empire.

200–400 Rise of Christianity

More and more *gentiles* (non-Jews) convert to Christianity, even though they suffer persecutions. The Jews are pushed to a far corner of Tiberias but continue to build new synagogues. (The word *synagogue* is often used to describe a place of Jewish worship. Many Jews do not use the word *temple* out of reverence to Solomon's Temple.) At first the early Christians are regarded as another Jewish sect. Conversions to Jesus' ministry are confined to believing Jews. However, most Jews do not accept Jesus and the apostle Paul (b. Saul of Tarsus, d. A.D. 64) begin to convert gentiles to the new sect of Christianity.

306–37 Constantine is Roman Emperor

Constantine the Great (280–337) becomes emperor of Rome. He relocates his capital city from Rome to Byzantium which he renames Constantinople. In 313, he issues the Edict of Milan, which legalizes Christianity, while on his deathbed he himself converts to the religion, thereby giving it legitimacy.

632 Death of the Prophet Mohammed

The prophet Mohammed (570–632) founds the Islamic religion. The Qu'ran (Koran*),* its holy book, is said to have been dictated by God to Mohammed. Mohammed spends his last night on earth at the Temple Mount in Jerusalem. His followers build a mosque there they call Dome of the Rock.

637 Moslems invade and conquer Palestine

The forces of Omar Ibn al-Khattab conquer Palestine and install an Islamic leadership that lasts 450 years.

661–750 The Syrian Umayyad Dynasty

This Syrian dynasty rebuilds the area of the Temple Mount and establishes the Dome of the Rock, the al-Aqsa Mosque, and the Gate of Mercy.

750–969 The Iraqi Abbasid Dynasty

This dynasty continues Arab rule and tries to discredit the Umayyads. Their capital is in Baghdad.

969 Jerusalem falls to Egyptian rule

Under the reign of caliph al-Hakim of the Fatimids of Egypt, Jerusalem enters a period of decline and neglect. Al-Hakim destroys all synagogues and churches.

1071 The Seljuk Conquest

The Seljuks completely ransack and pillage Jerusalem. Their persecution of Christians prompts the Christians to organize the Crusades to re-take the city.

1099: July 15 The first crusade conquers Jerusalem

The *Crusaders* (warriors of the cross) conquer Jerusalem. At the urging of Pope Urban II (1042–99), Christians set off on a holy war to battle the Muslims. Urban dies in 1099 without receiving news of the Christian victory. The Crusaders build castles and establish special orders of knighthood for the protection of Christian visitors to the Holy Land. The Latin Popes of Rome stand as rivals to the bishops at Constantinople. Even though the Roman Popes do not visit the Holy Land, they are recognized by Christian Western European as the rightful leaders of the Church.

1187: October 2 Re-conquest of Jerusalem by the Muslims

Under the leadership of Saladin (1137–93), the Muslims retake the holy city of Jerusalem during the Battle of Hattin (1187). Saladin is the sultan of Egypt which in turn rules Palestine. In victory, Saladin is neither vengeful nor destructive. With the failure of the Third Crusade, Richard the Lion-Hearted of England (1157–99) and Philip II of France (1165–1223) manage a truce with Saladin by which Christians are allowed to visit holy sites in Jerusalem. Saladin is personally kind to the Jews. His personal physician is Moses Maimonides (1135–1204), a Sephardic Jew born in Muslim-ruled Cordoba, Spain. The Greek patriarchs, who are rivals to the Latin Popes, now have residency in Palestine. It is up to the Turks to manage the Roman and Eastern Orthodox religious problems.

1250–1516 Mamluks rule Palestine

The Mamluks are a slave trading dynasty from Egypt and Syria. They have little concern for the welfare of their people but manage to control their empire for about four hundred years.

1290 England expels Jews

During the Middle Ages, many countries engage in anti-Semitic activities as a result of attributing natural disasters such as bad weather and plagues to the Jews. The Jews who are expelled from England go to Palestine.

1390 France expels Jews

France expels Jews for many of the same reasons England does. French Jews also go to Palestine.

1453 Ottoman Turks capture Constantinople

The fall of Constantinople constitutes the beginning of Turkish rule for most of the Middle East.

1517 Ottomans capture Palestine

Under their leader Selim I (1467–1520), the Ottomans defeat the Mamluks and capture Palestine and add it to their growing empire. They massacre over 40,000 Shi'ites to ensure the rule of the Sunni Muslim sect. Selim's son, Suleiman the Magnificent (1496–1566), rebuilds the city of Jerusalem along the old walls. Suleiman's improvements can still be seen in modern-day Jerusalem. Palestine is administered from Damascus (Syria) until 1830. Muslims continue to enjoy tax-free benefits which encourage many to convert to Islam. The Greeks, Armenians, and Jews are regarded as a merchant second-class. Joseph Nasi, a Jewish financier, serves sultans Suleiman and Selim II (1641–91) who give him permission to settle Jews in Tiberias. Nasi is a Sephardic Jew who speaks Ladino, a dialect of Spanish written with Hebrew characters. Non-Muslims still do not have the right to purchase land in Ottoman-controlled territories (see sidebar).

1701 Polish Jews (Ashkenazim) settle Palestine

A rival group of European Jews who speak Yiddish, a German dialect with Hebrew characters, whose leader is Judah the Pious (Yehuda Ha-Chasid) settles in Jerusalem. At this time Palestine is still rather rural and newcomers have a harsh life.

1774 Catherine the Great defeats Turks

Catherine the Great of Russia (1729–96) defeats the Turks and ensures that Greek Orthodoxy will have more influence than Roman Catholicism in Palestine.

1800 Napoleonic Wars threaten Ottomans in Palestine

Napoleon Bonaparte (1769–1821), Emperor of France, tries to extend his territory into Africa and the Middle East. This leads to Great Britain's involvement in Egypt and Palestine in an attempt to halt Napoleon.

1805 Ottomans name Muhammad (Mehmet) Ali Viceroy of Egypt and Palestine

Muhammad Ali (1769–1849) is named *pasha* (Ottoman governor) of Egypt and Palestine. Muhammad Ali, an Ottoman subject born in Macedonia, is in the army which helps to defeat the French under Napoleon. As a reward, he and his son rule Palestine. Even though he has a virtually independent control of the area, he launches an attack on the Ottomans with the French as his ally. The French, under King Louis-Philippe (1773–1850) , sends military aid in an attempt to re-establish French influence in the area. Mehmet loses to the Ottoman forces, and he is forced to retreat and abandon all of Syria, Palestine, and the Sinai Desert. However, during his control of Palestine, Mehmet introduces many reforms. Non-Muslims can now settle permanently in Jerusalem, and he accepts foreign consulates. Christians are welcome and come to set up churches. Jews also begin rebuilding synagogues. Jews are now allowed to pray at the Western Wall of Solomon's Temple.

Protestants take advantage of Muhammad Ali's tolerance to try to set up missionaries in Palestine. Previously, anyone trying to proselytize Muslims were punished. The British Protestant churches also want to proselytize the Jews of Palestine because they believe proselytization of Jews to be a sign of the coming of the Messiah.

1840 Return of Ottoman Rule

With the defeat of Mohammed Ali, the Ottomans take control of Palestine once again. The Turks are openly hospitable to Jewish settlement in Palestine. Many Jews from Russia, persecuted by Tsar Nicholas I (1796–1855), begin to make permanent *aliyah* (settlement) in Palestine.

1842 Turkish government gives official status to Sephardic rabbi

The Turkish government allows the chief rabbi to arbitrate disputes between Jews. The chief rabbi now has access to Turkish high officials. Since the chief rabbi is of the Sephardic tradition, the Ashkenazim now feel threatened. They have different traditions regarding preparing *kosher* meat and burying the dead. Officials in the Turkish government talk only to the Sephardim who share many Turkish customs. The Anglican Church of England sends a bishop to Palestine to set build a church and to attract converts. This move is not viewed favorably by the Turks who put many obstacles in the way of the church's completion. Few Protestants live in Palestine.

1854–56 Crimean War

Feeling that he has a right to protect Christian churches there, Tsar Nicholas provokes the Turks by trying to conquer Constantinople. This action results in the Crimean War (1854–56). Now Britain and France are involved in the Middle East.

The Ottoman Period

The Ottoman Empire began in western Turkey (Anatolia) in the thirteenth century under the leadership of Osman (also spelled Othman) I. Born in Bithynia in 1259, Osman began the conquest of neighboring countries and began one of the most politically influential reigns that lasted into the twentieth century. The rise of the Ottoman Empire was directly connected to the rise of Islam. Many of the battles fought were for religious reasons as well as for territory.

The Ottoman Empire, soon after its inception, became a great threat to the crumbling Byzantine Empire. Constantinople, the jewel in the Byzantine crown, resisted conquest many times. Finally, under the leadership of Sultan Mehmed (1451–1481), Constantinople fell and became the capital of the Ottoman Empire.

Religious and political life under the Ottoman were one and the same. The Sultan was the supreme ruler. He was also the head of Islam. The crown passed from father to son. However, the firstborn son was not automatically entitled to be the next leader. With the death of the Sultan, and often before, there was wholesale bloodshed to eliminate all rivals, including brothers and nephews. This ensured that there would be no attempts at a coup d'état. This system was revised at times to the simple imprisonment of rivals.

The main military units of the Ottomans were the Janissaries. These fighting men were taken from their families as young children. Often these were Christian children who were now educated in the ways of Islam. At times the Janissaries became too powerful and had to be put down by the ruling Sultan.

The greatest ruler of the Ottoman Empire was Suleyman the Magnificent who ruled from 1520–1566. Under his reign the Ottoman Empire extended into the present day Balkan countries and as far north as Vienna, Austria. The Europeans were horrified by this expansion and declared war on the Ottomans and defeated them in the naval battle of Lepanto in 1571. Finally the Austrian Habsburg rulers were able to contain the Ottomans and expand their empire into the Balkans.

At the end of the nineteenth century, the Greeks and the Serbs had obtained virtual independence from the Ottomans, and the end of a once great empire was in sight. While Europe had undergone the Renaissance, the Enlightenment, and the Industrial Revolution, the Ottoman Empire rejected these influences as being too radical for their people. They restricted the flow of information and chose to maintain strict religious and governmental control. The end of Ottoman control in Europe came in the First Balkan War (1912–13) in which Greece, Montenegro, Serbia, and Bulgaria joined forces to defeat the Ottomans.

During the First World War, (1914–18), the Ottoman Empire allied with the Central Powers and suffered a humiliating defeat. In the Treaty of Sévres, the Ottoman Empire lost all of its territory in the Middle East, and much of its territory in Asia Minor. The disaster of the First World War signaled the end of the Ottoman Empire and the beginning of modern-day Turkey. Although the Ottoman sultan remained in Constantinople (renamed in 1930, "Istanbul"), Turkish nationalists under the leadership of Mustafa Kemal (Ataturk) (1881–1938) challenged his authority and, in 1922, repulsed Greek forces that had occupied parts of Asia Minor under the Treaty of Sévres, overthrew the sultan, declared the Ottoman Empire dissolved, and proclaimed a new Republic of Turkey. That following year, Kemal succeeded in overturning the 1919 peace settlement with the signing of a new treaty in Lausanne, Switzerland. Under the terms of this new treaty, Turkey reacquired much of the territory—particularly, in Asia Minor—that it had lost at Sévres. Kemal is better known by his adopted name "Ataturk", which means "father of the Turks". Ataturk, who ruled from 1923 to 1938, outlawed the existence of a religious state and brought Turkey a more western type of government. He changed from the Arabic alphabet to Roman letters and established new civil and penal codes.

They see the Russians as threatening their access to India if the Russians control Constantinople. England and France help the Ottoman Empire to defeat the Russians.

1864 Telegraph connects Palestine with Europe

The installation of telegraph lines link Palestine with Europe.

1838–80 Palestine still sparsely populated

Still sparsely populated, Palestine has by 1850 fewer than 3,500 Christians. The Jewish population is concentrated in Jerusalem, Hebron, Safed, and Tiberias. In 1860, Moses Montefiore (1784–1885), a Jewish philanthropist, establishes the first Jewish settlement outside the walls of Jerusalem. He starts a collective farm (the modern day *kibbutz).* No roads connect cities, and many people coming to live in Palestine find life surprisingly harsh. Non-Muslims still cannot buy property. Land is leased to them by Muslims. There are power struggles between Protestants, Catholics, and Eastern Orthodox to attract converts and control the holy sites. The Temple Mount is only open to Jews and Muslims.

1873–79 Ashkenazic Jews name chief rabbi

Finally, the Ashkenazim get their own chief rabbi who will be the arbiter of their religious laws. They no longer have to obey the Sephardic rabbi.

1881 Jews arrive in Palestine from Russia

Tsar Alexander II (1818–81) of Russia is assassinated. His son, Alexander III (1845–94), who becomes tsar is convinced that the Jews are to blame for the death of his father. He vows to convert one-third of the Jews, expel one-third, and starve one-third. There are many *pogroms,* or wholesale killings of Jews. Fleeing these conditions many Jews emigrate to Palestine. Moses Montefiore and the Rothschild families of Europe help finance the relocation of impoverished Russian Jews in Palestine.

1882 Beginnings of Zionism

The term *Zionism,* which comes from Zion, the Jebusite fort captured by King David, now comes to mean a homeland for the Jewish people. Jews search for a peaceful existence. At first, they seek entry to any country that will give them shelter. The BILU, an acronym for "House of Jacob come ye, and let us go," brings Jews to Palestine who attempt to survive by farming the land. A harsh life for many urban Jews, some return to Russia. In 1882, 7,000 immigrants come to Palestine. A second nationalist movement, *Hovevi Zion* (Lovers of Zion), succeeds in founding several farm communities.

1894 The Dreyfus Affair

Anti-Semitism in Europe gives rise to Zionism. In France, army general Alfred Dreyfus is falsely accused of treason. One of his defenders, among many, is Theodor Herzl, a Jewish reporter assigned to the Dreyfus case for the Vienna newspaper *Neue Freie Presse.* He publishes *Die Judenstadt (*The Jewish State) in which he predicts the necessity of a Jewish homeland for the protection of the Jewish people. He prefers Palestine but says any country will be acceptable as long as it is safe. He travels throughout Europe looking for political and financial support for his plan. He is rejected everywhere. Russia allows him to encourage Jews to emigrate to Palestine, but the Russian government will not allow speeches promoting the establishment of a Jewish state.

1901 Creation of the Jewish National Fund

Financial support comes mostly from Russian Jews who stand the most to gain from a Jewish homeland. The World Zionist Organization establishes a fund for the purpose of purchasing land in Palestine. The land would belong to the Fund but be leased to Jewish farmers. Herzl is not successful in getting large donations from wealthy Jews, but the blue boxes (like piggy banks) fill up with support from many middle and lower income Jewish families. Most of the support comes from funds collected in these little banks.

1911 Founding of the First Kibbutz

The kibbutz, or collective farm, proves to be the backbone of early Israeli economy. In 1911 four adventuresome people found Kibbutz Deganya on the shores of Galilee. The kibbutz members share all work, property, and decision-making. In modern-day Israel, fewer than three percent of the people live on a kibbutz, but its importance in cementing the lives and fortunes of the newly arriving immigrants cannot be discounted. Without the kibbutz, many new immigrants would be unable to survive.

1914–18 World War I population in Palestine

Before the advent of World War I, there are about 70,000 Jews in Palestine, a very small minority compared to the Arab population. However, there is no particular animosity or feelings of nationalism in any group. Palestine is a relatively unpopulated place, providing room to grow. World War I changes many feelings. The Turks, who allied themselves with Germany and the Austro-Hungarian Empire, are against the Russians, and Jews who support Russian immigration in Palestine are suspect. Many of them flee to America. They have a hard time persuading Russian refugees in America to come to Palestine.

1916 Sykes-Picot Accord

Mark Sykes of Britain and Charles Georges Picot of France decide to partition Palestine as part of the spoils of war. The French want to control Syria and Lebanon and cede Palestine and Iraq to the British. With the Turks on the losing end, everyone wants a piece of Palestine. Meanwhile, Britain tries to negotiate a settlement with the Zionists and the Arabs.

1917: November 2 The Balfour Declaration

Lord Arthur James Balfour (1848–1930), the British Secretary of State for foreign affairs, is an evangelical Christian who believes that all Jews must re-locate in the Holy Land before the Messiah can come. In this declaration he states that the British government favors the establishment of a Jewish homeland in Palestine "without prejudice to the civil and religious rights of non-Jews living in Palestine."

1918 Founding of Hebrew University

The Hebrew University is established on Mount Scopus. The official language of instruction is Hebrew although most students and professors are still new to using the ancient tongue.

1919 British Mandate for Palestine

The League of Nations (precursor of the United Nations) establishes a British mandate for Palestine (see sidebar).

1919–24 Periods of *Aliyot* (Immigration) to Palestine

Many people from various countries come in the hopes of creating a Jewish state.

1920 Herbert Samuel heads Palestine

Herbert Samuel (1870–1963) becomes the first high commissioner of Palestine. Although he is of the Jewish faith, he is not a Zionist and does not want to offend his Arab subjects living in Palestine. By trying to court Arab favor, he does not please the Arabs and he angers the Jews.

1924–29 Jews flee Poland

34,000 Polish Jews escape violent anti-Semitism in Poland. Many of them are small shopkeepers who cannot survive in the fragile Palestine economy. Many of them leave for other countries.

1932 The first Maccabiah Games

The Maccabiah Games, sometimes called the Israel Olympics, are held. Competitors in the Maccabiah Games are Jewish athletes from around the world. Events range from archery to water polo.

1934–39 Jews leave Germany

Jews escaping repressive measures in Germany go to many countries, including Palestine (see 1939–45).

1939 Britain limits Jewish Immigration

Great Britain, at war and needing Arab help, decides to limit over the following five years the number of Jews coming into Palestine to 75,000 and thereafter to prevent entry without

League of Nations

Formed in the wake of World War I, the League of Nations—the forerunner of the United Nations—was the world's first international organization in which the nations of the world came together to maintain world peace. Headquartered in Geneva, Switzerland, the League was officially inaugurated on January 10, 1920. During the life span of the organization, over sixty nations became members, including all the major powers except the United States. Like the United Nations, the League of Nations pursued social and humanitarian as well as diplomatic activities. Unlike the United Nations, the League was primarily oriented toward the industrialized countries of the West, while many of the regions today referred to as the Third World were still the colonial possessions of those countries.

The League's structure and operations, which were established in an official document called the *League of Nations Covenant,* resembled those adopted later for the United Nations. There was an Assembly composed of representatives of all member nations, which met annually and in special sessions; a smaller Council with both permanent and nonpermanent members; and a Secretariat that carried out administrative functions. The League's procedures for preventing warfare included arms reduction and limitation agreements; arbitration of disputes; nonaggression pledges; and the application of economic and military sanctions.

Although the League of Nations solved a number of minor disputes between nations during the period of its existence, it lost credibility as an effective peacekeeper during the 1930s, when it failed to respond to Japan's takeover of Manchuria and Italy's occupation of Ethiopia. Ultimately, the organization failed in its most important goal: the prevention of another world war. However, it did make contributions in the areas of world health, international law, finance, communication, and humanitarian activity. It also aided the efforts of other international organizations.

By 1940 the League of Nations had ceased to perform any political functions. It was formally disbanded on April 18, 1946, by which time the United Nations had already been established to replace it.

Arab consent. The document outlining this decision is known as the *White Paper.*

1939–45 World War II and the Holocaust

Over six million Jews are exterminated by the Nazi government during World War II. By 1942, most of the world is aware of the planned extermination of the Jews. Many Jews in Palestine clandestinely help German Jews enter Palestine. But those who are discovered are sent back to Germany. After the war, revelations about the concentration camps increase world sentiment for the establishment of a Jewish homeland. However, the Tory government in Britain, which manages Palestine, is solidly against this idea. Led by Ben-Gurion, the Jewish underground, known as the *Hagana,* fights the British for the liberation of Palestine.

1946 Hagana bombs King David Hotel

The elegant King David Hotel in Jerusalem, headquarters for the British administration of Palestine, is bombed. The Hagana warn workers in the hotel to evacuate and then explode an entire wing of the hotel. There are many casualties, including Jews. The Jews are mobilized to get more illegal immigrants into Palestine. American Zionists purchase an old British war vessel and re-name it the *Exodus.* The British attack the ship and send its passengers, Jews from displaced persons' camps in Europe, back to Europe. This story is a factor in ending the British mandate of Palestine.

1947 Jews and Arabs vie for control

By 1947, there are about 600,000 Jews and 1,300,000 Arabs in Palestine. The Arabs are confident that the British White Paper limiting Jewish immigration into Palestine will assure them (Arabs) of controlling Palestine. The Jews will either have to leave or submit to an Arab government. The British are openly supportive of the Arabs because they need oil.

1947: February 18 Britain renounces Palestinian mandate

The British government renounces the mandate and turns the problem of Palestine over to the United Nations (see sidebar). A special committee, United Nations Special Committee on Palestine (UNSCOP), is sent to Palestine to decide the fate of the Jews and Arabs. Committee members decide that the British mandate will be ended and that Palestine will be partitioned into an Arab state and a Jewish state.

1947: November 29 UN adopts a plan for partition

The United Nations creates a plan for partitioning Palestine into two sovereign states, one Jewish and one Arab. An international zone will include Jerusalem and Bethlehem. President Harry S. Truman of the United States (1884–1972) lobbies the United Nations for approval of the plan. He gets the British to agree because the United States can make loans to Great Britain. Britain agrees to end the mandate on May 15, 1948. The Arabs living in Israel seek political and military support from neighboring Arab nations.

1948: May 15 British mandate ends and Arabs prepare to attack

Jews announce the formation of the State of Israel as soon as the British mandate on Palestine ends. Arabs, with support of the surrounding Arab countries, make plans to attack Israel as soon as the British pull out. On May 15, a war begins. Egypt, Syria, Iraq, Lebanon, Saudi Arabia, and Jordan launch a coordinated attack on Israel.

1949 Armistice

Israel emerges victorious from the war. Israel possesses western Jerusalem and the Arabs have eastern Jerusalem. When the UN plan for an area of Arab-controlled land does not become reality, Jordan is encouraged by other Arab states to annex the region west of the Jordan River, known as the West Bank, which it does. Although there is a fragile peace, Egypt and Syria refuse to recognize Israel.

1949 Israel's first election

The *Knesset* (Parliament) elects David Ben-Gurion as Israel's first prime minister. However, he must form a coalition government and joins with the Orthodox rather than the extreme left Marxists. This causes him to grant concessions to religious Jews. All kosher laws are to be observed, no unnecessary work may be done on the Sabbath (Saturday), and the decisions of "who is a Jew" are to be determined by the Chief Orthodox Rabbi.

1949: February Israel and Egypt sign armistice

Egypt and Israel sign an armistice which leaves Egyptian forces in the Gaza Strip, an area granted to the Palestinian Arabs under the UN partition plan.

1949: March Israel and Lebanon sign armistice

Israel and Lebanon reach an armistice agreement.

1949: April Israel and Transjordan sign armistice

Israel and Transjordan (later Jordan) reach an armistice agreement that leaves Jordanian forces in occupation of most of West Bank (of the Jordan River), an area, like the Gaza Strip, that the UN partition plan gives to the Palestinian Arabs. In addition, the armistice leads to the division of Jerusalem, with the eastern half of the city under Jordanian occupation.

1949: May United Nations accepts Israel as a member state

Israel joins the United Nations.

The United Nations

The United Nations (UN) is the international organization created by the major Allied powers at the end of World War II to maintain peace and encourage cooperation among the world's sovereign nations. The UN officially came into being on October 24, 1945, when its charter was ratified by a majority of the fifty nations whose representatives had drafted the document at a conference in San Francisco the previous April. The organization replaced the older League of Nations that had been in existence between the world wars. Since its inception, the United Nations has been headquartered in the UN Building in New York City. In 1998 it had 185 members.

As specified in its charter, the UN consists of six main bodies, of which the most prominent are the General Assembly and the Security Council. All member nations send representatives to the General Assembly, which holds regular sessions annually, as well as special sessions. Each nation may send up to five delegates but still has only one vote. The General Assembly passes resolutions on issues of concern to the international community, amends the UN charter, elects judges to the International Court of Justice (described below), and handles budgetary and other matters. The Security Council is, as its name suggests, concerned with international security. It has five permanent members—the United States, Britain, Russia, China, and France—and ten nonpermanent members elected to two-year terms by the General Assembly. The nonpermanent memberships are distributed so that all the different regions of the world are represented on the Council. The Security Council can recommend ways of resolving international disputes, impose sanctions, and pass binding resolutions. It has deployed peacekeeping forces to regions around the world, including Bosnia, Haiti, Northern Ireland, and Somalia. The Security Council also nominates the UN Secretary General.

The remaining four UN bodies are the Secretariat, the UN's administrative arm; the International Court of Justice (or World Court), headquartered in the Hague, which settles disputes; the Economic and Social Council, which oversees economic, humanitarian, and cultural programs; and the Trusteeship Council. The Secretariat is headed by the UN Secretary-General. As the single most prominent UN official, the Secretary-General often acts as a UN spokesman and as a conciliator in international disputes in addition to his duties as an administrator.

The United Nations also sponsors a number of additional agencies and programs, including the United Nations Children's Fund (UNICEF); the United Nations Educational, Scientific and Cultural Organization (UNESCO); the World Health Organization (WHO); the International Monetary Fund (IMF); and the International Bank for Reconstruction and Development (the World Bank).

Twice in its history, major political developments around the world have led to "growth spurts" in UN membership. In 1960 sixteen new African nations were admitted to the world body, reflecting the end of Europe's colonial empires. In the 1990s, the break-up of the Soviet Union into separate republics and developments in other parts of Eastern Europe brought nearly two dozen new member nations into the UN, bringing the total number of members to 185.

1950: July 5 Passage of the Law of Return

The Law of Return solves the problem of Jewish immigration to Palestine. Henceforth there are no limits on immigration for Jews wishing to live in Palestine. This act creates an onslaught of refugees from Europe, Asia Minor, and those expelled from Arab states after the War of Independence.

1950: October The Maccabiah Games are played in Ramath-Gan Stadium

The third Maccabiah Games, first staged in 1932, are held in Ramath-Gan Stadium in Tel Aviv. Seating 80,000 spectators, the massive stadium was built to host the games.

1956: July 26 Egypt nationalizes the Suez Canal

The Suez Canal, which gives Israel access to the Red Sea, is nationalized by Egypt's president, Gamal Abdel Nasser (1918–70). This move threatens Israel's access to trade and water. France and Britain ask Israel to launch a surprise attack on Egypt and offer to help out. Israel, with efficient military, holds all of the Sinai Desert. The United Nations compels Israel to withdraw, but the benefit is that now there will be a peace-keeping force to secure Israel's borders with Egypt. Israeli foreign minister, Golda Meir (1898–1978) issues a warning that any country trying to block free access to the

The Ramath-Gan Stadium in Tel Aviv, the largest stadium in the Middle East with seating for 80,000 is inaugurated for the Maccabiah Games. Athletes from eighteen nations competed. (EPD Photos/CSU Archives)

Gulf of Aqaba will find itself liable to forceful withdrawal by Israeli troops. Israel gains relative peace with Egypt.

1959: October Creation of *Fatah*

Yasir Arafat creates the Palestinian terrorist group called *Fatah*. This later becomes the Palestine Liberation Organization (PLO). (See 1964: May.)

1962: May 30 Adolf Eichmann hanged

Adolf Eichmann (1906–62) is the only person to be executed in the history of Israel's statehood. Eichmann is one of the chief Nazi officers who sent millions of Jews to their deaths in German and Polish concentration camps during World War II.

1964 Problems of water rights

Both Syria and Israel begin diverting water from the Jordan River. There is much sniper fire on both sides.

1964: May Founding of the PLO

The PLO (Palestine Liberation Organization) is founded by Arabs living in Palestinian East Jerusalem. At first it has no terrorist aims. However, a Syrian group called the Fatah wants to attack Israel from Jordanian territory so as not to arouse suspicion of Syrian involvement. Also the leader of the Fatah is Yasir (also Yassir) Arafat, the pseudonym of Rahman Rauf Arafat al-Quana al-Husseini. He takes this organization to new heights of terrorism.

1967: June The Six-Day War

On May 22, Nasser informs all nations he is closing the Gulf of Aqaba to all ships delivering Israeli goods. International law says that creating a blockade is an act of war. The United States is busy with the war in Vietnam and cannot help. Acting independently, Israel launches a surprise attack on Egypt. Egypt's air force is destroyed on the ground and Israel occupies the entire Sinai Peninsula. Syria and Jordan go to Egypt's

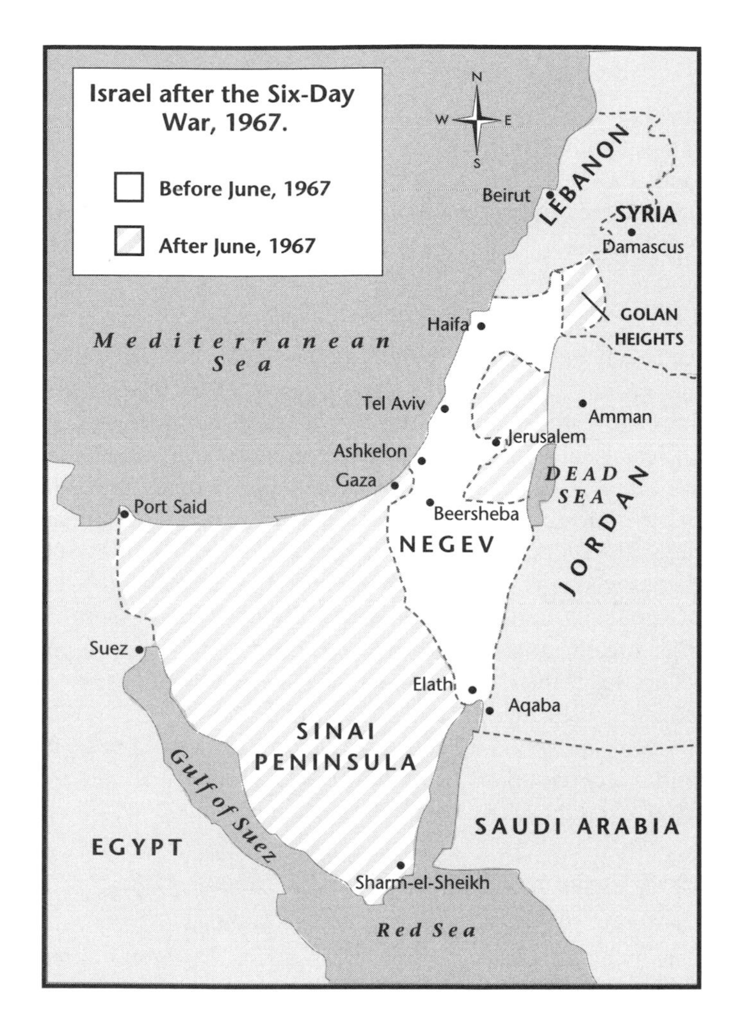

defense. The Israelis not only defeat these armies, they occupy the Golan Heights, Samaria, Judea, and all of Jerusalem. A result of the Six-Day War is that now Israel and Jordan are at peace. The Israeli government indicates that all territories gained in the war are negotiable for peace except for Jerusalem, Sharmel-Sheikh, and the Golan Heights, which is a strategic location.The Arabs reply is absolutely in the negative. For the Arabs, there is no peace, no recognition of Israel as a state, and no negotiations with Israel. Nasser keeps bombarding the Israeli border trying to force the Israelis to retreat to the pre-1967 boundaries.

1968 Amos Oz publishes *Mikhael sheli*

Amos Oz (b. Amos Klausner, 1939), writing in Hebrew, publishes his best-known novel, *Mikhael sheli* (published as *My Michael* in English in 1972). It is considered symbolic of the struggles of the cultures of Jerusalem to co-exist. Oz's other works include *Makon aher* (1966; published in English as *Elsewhere, Perhaps* in 1973); *La-ga 'ath ba-mayim, la-ga'ath ba 'ruah* (1973; published in English as *Touch the Water, Touch the Wind,* 1975), *Menuah nekhonah* (1982; published in *English as A Perfect Peace,* 1985.

1970 Jordan expels the PLO

Jordan, enjoying somewhat peaceful relations with Israel, expels the PLO. King Hussein of Jordan and Yasir Arafat sign a peace agreement.

1970 September Nasser dies

Gamal Nasser dies and Anwar Sadat (1918–81) becomes president of Egypt.

1972 The PLO murders Israeli Olympians

The PLO stages an attack at the Olympic Games in Munich, Germany. They take the Israeli team hostage, and when rescuers try to free the hostages, many are killed in the opposing gunfire. In spite of a day of mourning, the Olympic Games continue.

1973 Bethlehem University founded

Bethlehem University is founded. Tensions between Palestinians and Israelis affect the atmosphere at the university.

1973: October 6 The Yom Kippur War

Yom Kippur (The Day of Atonement) is the most solemn holiday of the Jewish faith. On this day, when many Israeli Jews are in synagogues worshipping, the Egyptians launch an all out attack on Israel. At the same time, the Syrians launch a coordinated attack on the Golan Heights. The United States quickly sends supplies to Israel. The Israeli army gets through enemy lines and Egypt is forced to surrender when Israeli troops are within fifty miles of Cairo. Furthermore, the Israeli army invades Syria and finds itself on the outskirts of Damascus. The United Nations calls for a cease-fire. The Israelis are shocked at how unprepared they are initially for the war and blame their prime minister, Golda Meir, of the Labor Party and the Minister of Defense, Moshe Dayan (1915–81). Golda Meir is forced to resign.

1974 April Israelis elect General Yitzhak Rabin

General Rabin (1922–95) is Chief of Staff during the Six-Day War. Immediately before taking office, he is Israel's ambassador the United States.

1974: October Arab Nations recognize PLO

Arab nations now recognize the PLO as the official spokesman of the Palestinian people. Yasir Arafat is still the leader. The main idea of the charter of the PLO is that Israel shall not be a state.

1975: November United Nations tries to expel Israel

The United Nations accuses Israel of using Zionism for racist reasons. The counter-charges are that Arab countries expel Jews and Christians living within their borders. The U.N. also states that there cannot be peace in the Middle East until there is a Palestinian state.

1976 Entebbe rescue

A French airliner is hijacked by PLO terrorists and forced to land at Entebbe Airport in Uganda. The terrorists release all but Israeli and Jewish passengers. The Israeli army stages a surprise raid and frees the hostages. The Israeli commander, Jonathan Netanyahu, is killed in action. (His brother Benjamin becomes prime minister in 1996).

1977 Likud Party comes to power

After a long domination by the Labor Party, the Likud Party comes to power and forms a coalition government with Menachem Begin (1913–92) as its prime minister. Begin chooses the Labor Party Defense Minister Moshe Dayan as his minister of foreign affairs, hoping to strengthen his power.

1978: September Camp David peace accords

The President of the United States, James Earl (Jimmy) Carter (b. 1924), receives Menachem Begin and Anwar Sadat, the President of Egypt, for a peace conference at his presidential retreat at Camp David, Maryland. Israel agrees to cede its possession of the Sinai Peninsula to Egypt in exchange for unfettered passage through the Suez Canal. Israel also surrenders all Egyptian artifacts it has held from the occupation of the Sinai. President Carter acknowledges the receipt of a letter from President Sadat stating that East Jerusalem shall belong to the Palestinians. He does not comment on the demand. Also, Israelis and Egyptians will have permission to visit each others' countries.

1979: March Egypt and Israel sign peace treaty

The Camp David Accords take effect as Egypt and Israel sign a peace treaty. Thus, Egypt becomes the first Arab state to establish diplomatic relations with Israel. This agreement isolates Egypt from its fellow Arab states and leads to Sadat's assassination two years later (see 1981: October).

1981 Israel bombs nuclear plant in Iraq

Israeli intelligence sources reveal that the Soviet Union is arming Iraq. The Israelis bomb a nuclear plant in Osiraq, near Baghdad. The plant is completely destroyed.

1981 October Islamic terrorists assassinate Sadat

Arab terrorists, in the belief that he is too conciliatory towards Israel, assassinate Sadat.

1982 Israel leaves Sinai Peninsula

As agreed to in the Camp David Accords, the last Israeli troops leave the Sinai Peninsula.

1982 September PLO leaves Lebanon

Yasir Arafat takes his organizational headquarters to Tunisia.

1982 September 14 Israel massacres civilians in Sabra and Chatila

Defense Minister Ariel Sharon invades Palestinian refugee camps looking for terrorists. In the process the military kills more than seven hundred civilians. Israeli citizens are horrified and the career of Menachem Begin is ruined. He retires a year later.

1987 Beginning of Palestinian *Intifada*

An Israeli tank crashes into Arab civilian cars in Gaza killing four people. This incident sets off major riots and demonstrations and a group calling itself *Intifada,* which means shaking off Israeli rule, forms. The Gaza Strip has a huge Arab refugee camp. No Arab country is willing to take the refugees in as it will decrease the chances of a demand for a Palestinian state. The Intifada is composed of men in their twenties and thirties who engage in small-time terrorism such as throwing stones at cars and disrupting events. The PLO and other terrorist organizations are worried because the Intifada will not acknowledge their authority and the older organizations are in jeopardy of losing respect. The Intifada actually makes life worse for the Arabs by such acts as forbidding them to go to work on certain days or by boycotting products. Arabs are terrorized if they refuse Intifada's demands.

1988: November PLO declares Palestine a state

The PLO has a meeting in Algiers and proclaims Palestine a state with its capital city in East Jerusalem. This brings about disunity among the Israelis. Likud, the right-wing party, does not intend to recognize the PLO. Labor, the left-wing party, wants to open negotiations with the PLO and agrees to defining boundaries which leads to a Palestinian state.

1989 Russians allow Jews to emigrate

Thousands of Russian Jews immigrate to Israel. This event results in loud protests from Arabs who object to increased Jewish population in Israel.

1990: September Extremists mar peace plans

An extremist Jewish group announces its plans to ascend the Temple Mount and lay a cornerstone for the rebuilding of the Temple of Solomon. The mayor of Jerusalem and the high court forbid this action. Arabs circulate leaflets saying that the Jews are planning an attack on El Asqa, one of Islam's most holy sites. A confrontation ensues and many Arabs are killed. Israel refuses to allow the United Nations to investigate the matter.

1991: January–February Gulf War

Iraq invades Kuwait and the United States is drawn into war to protect the Kuwaitis. Arab countries are angry with the United States for interfering, and they launch scud missiles at Israel. The United States pleads with Israel not to counterattack and Israel agrees, thus guaranteeing itself a return favor from the United States.

1991: December U.N. repeals Zionism condemnation

The United Nations repeals an earlier resolution condemning Israel for promoting racism by encouraging Jews to immigrate to Israel.

1992: June Labor Party elects Rabin as prime minister

Rabin is elected on the basis of his hard line stance on Israel's maintaining the Golan Heights and boundaries won in the Six-Day War. The coalition government wants to exclude Likud and instead join Meretz, a group which will trade land for peace. Also joining the group is a Sephardic-based religious party called Shas.

1992 Rise of Foreign Minister Shimon Peres

Shimon Peres, a political rival of Rabin, is needed to help negotiate the secret Oslo Accords. His philosophy is that a healthy economy, like that of the European Union, will make the old enemies, Israel and her neighbors, agree to peace.

1993: August Oslo Peace Accords

Under the peace agreement negotiated in Oslo, Norway, Arabs in Gaza and Jericho will have their own police force, known as the Palestinian Authority, and will be elected by Arab voters. Israeli officials will not have the right to interfere or use their security forces in these areas. Israel also agrees to cede Samaria, Judea, and Gaza. Israelis are outraged when they discover that Peres has indeed created a de facto Palestinian state within Israeli borders. Arafat has stationery designed that shows all of Israel as Palestine with Arafat its president.

1993: September Arafat disavows terrorism

Arafat is awarded the Nobel Peace Prize. One member of the committee resigns saying he refuses to give a peace prize to a known terrorist.

1993: December Vatican establishes diplomatic relations

A political milestone is reached when the Vatican recognizes the state of Israel.

1994 Purim massacre

Purim is supposed to be a joyous holiday. On this day, an Israeli doctor, Major Baruch Goldstein, while attending religious services at the Tomb of the Patriots in Hebron, hears on the loud-speaker coming from the mosque next door threats to kill Jews. He returns the next day and opens fire killing twenty-nine Arab worshippers. He is killed by return fire. This event is an embarrassment to the Israeli government which condemns terrorism.

1994: May Israel withdraws forces from Jericho and the Gaza Strip

As agreed to under the terms of the Oslo Accords, Israeli forces withdraw from the West Bank city of Jericho and the Gaza Strip. These territories now come under the control of the Palestinian Authority.

1995: November 4 Yitzhak Rabin assassinated

Rabin is assassinated by an Israeli student, discontented with Rabin's policies. Citizens are outraged. Labor, which was previously predicted to lose the election, now has the sympathy of the people. The election date is moved to May 1996.

1996: February 25 Hamas and Israeli government engage in deadly attacks

The Islamic extremist group, Hamas, launches two suicide bombings of civilians. Hezbollah, another Arab terrorist group, begins bomb attacks in northern Israel. The Israeli government retaliates with Operation Grapes of Wrath, killing ninety-one Lebanese civilians who are in the midst of Hezbollah terrorists. The United Nations condemns Israel's retaliation.

1996: May Likud elects Netanyahu

Benjamin Netanyahu (b. 1949) is elected prime minister. Netanyahu has to address the problem of the Arab-occupied Israeli territory of Hebron and must come to terms with the fact that the European Common Market recognizes the Palestinian state. Netanyahu forms his political coalition with all types of self-serving extremist groups making it virtually impossible to reach a consensus.

1996: August 2 Government ends freeze on construction

One of the biggest obstructions to achieving peace in the Middle East is that after the Six-Day War, Israel occupies many previously-owned Arab territories. Israel commences constructing housing on these sites, thus blocking future return of this land to the Arabs. Israel's previous agreement to stop further construction is defunct, and the building of new Israeli housing resumes, again igniting Arab passions.

1996: September Netanyahu openly denounces Palestinian State

The Likud Party, which is militant in its demands never to cede land for peace, has its leader, Netanyahu, oppose the creation of a Palestinian state.

1996: September 23 Israel encroaches on Arab Holy Site

Israel opens a new exit tunnel at the Western Wall. The Arabs, whose holy site is the Dome of the Rock on top of that site, say that their holy site has been violated. A new wave of Palestinian violence ensues.

1997: January 12 Jordan's King Hussein meets with Arafat and Netanyahu

In an effort to restart the peace process, Jordan's King Hussein meets with Palestinian leader Arafat and Prime Minister Netanyahu.

1997: March 13 Israeli schoolgirls killed by bomb

Israel is criticized for increasing population in disputed Arab territories. A Jordanian extremist hurls a bomb killing seven Israeli schoolgirls to protest these settlements.

1997: May Israel and Jordan sign water agreement

Israel and Jordan sign an agreement on water use between the two countries. Water supply is a major issue in Middle East politics due to its scarcity.

1998: March Israel withdraws from Lebanon

Israel agrees to the United Nations Security Council's Resolution 425 to withdraw its military troops from Lebanon.

1998: May 17 Israel celebrates fiftieth anniversary

Elaborate celebrations occur in the Jewish community.

1998: October 23 Wye River Accords

Palestinian leader Yasir Arafat, Israeli prime minister Benjamin Netanyahu, King Hussein of Jordan (who travels to the talks from a hospital where he is receiving treatment for advanced cancer), and President Bill Clinton meet at Wye River, Maryland, to discuss peace between Israel and the PLO. Israel makes concessions as to the West Bank and Gaza in exchange for Arafat's denunciation of the terrorist organizations, Hamas and the Islamic Jihad. The agreement is vaguely worded and can be interpreted in many ways in order for each country to save face. Not completely binding, the agreement depends a lot on good faith. All parties involved seem happy.

1998: November Hamas and Jihad resume terrorist activities

A Hamas sympathizer sets himself on fire showing that Hamas and the Islamic Jihad are committed to the the reconquest of the West Bank and East Jerusalem.

1999: May 8 Arafat threatens to declare Palestinian statehood

On the eve of Israeli elections, Arafat threatens to declare Palestine statehood. Netanyahu says if Arafat does, Israel will have to re-occupy the West Bank and Gaza, regions currently controlled by Palestinian security forces.

1999: May Israeli elections

In a stunning reversal of fortune, Netanyahu loses the election to Ehud Barak (b. 1942), of the Labor Party.

Barak has served in the Six-Day War and the Yom Kippur War. He is well-educated and has degrees in physics and engineering.

1999: June Barak reveals his peace plans

Barak plans to visit President Clinton in Washington, D.C. to talk about peace in the Middle East. It is no secret that the United States is happy with the election results in Israel. Barak is viewed as a warrior turned peacemaker in the style of Yitzhak Shamir. Barak indicates his willingness to communicate with Arafat. His landslide victory and strong coalition indicate that he will not have to make so many concessions to fringe political groups. The Israeli public is also more willing to support him in a land-for-peace agreement with the Palestinians.

1999: June Keys to Christian Holy Site to be shared

Since the twelfth century, the Muslims have controlled the keys to the Church of the Holy Sepulcher. After his conquest of Palestine, Saladin solved the problem of disputing Christian sects by having Muslims guard the tomb of Jesus. Now, that honor may be shared when an exit door is added. Church leaders are still divided as to who should get the key.

1999: June Hamas sees support waning

Hamas is forced to concede that the election of Barak, a moderate who may allow the creation of a Palestinian state, counters their organization which is extremely militant. Also, the Palestinians strongly support Arafat as their spokesman.

Bibliography

Blumberg, Arnold. *The History of Israel.* Westport, Conn.: Greenwood Press, 1998.

———. *Zion before Zionism.* Syracuse, NY: Syracuse University Press, 1985.

Grant, Michael. *The History of Ancient Israel.* New York: Charles Scribner's Sons, 1984.

Metz, Helen Chapin. *Israel: A Country Study.* Washington, DC : Library of Congress, 1990

Sachar, Abram Leon. *A History of the Jews.* New York: Alfred A. Knopf, 1955.

Shanks, Hershel. *Ancient Israel.* Englewood Cliffs, NJ: Prentice-Hall, 1988.

Japan

Introduction

In the mid-nineteenth century, Japan was a feudal society governed by a warrior class; today it is the world's second-greatest industrial power, combining modern Western mores with centuries-old cultural traditions nurtured and refined by 250 years of isolation from other societies. The nation's remarkable postwar economic growth has been called "the Japanese miracle." In addition to traditional industries such as steel and automobiles, Japan is also a leader in such new fields as the global communications revolution.

An archipelago in the North Pacific Ocean, Japan consists of four principal islands—Hokkaido, Honshu, Shikoku, and Kyushu—and numerous small ones. Japan's total land area is 145,883 sq mi (377,835 km), or roughly the size of the state of Montana. The largest Japanese island, Honshu, accounts for three-fifths of this area. Nearly three-fourths of Japan's land is mountainous—the highest peak is Mt. Fuji, on the island of Honshu, which rises to 12,388 ft (3,776 m). Only about one-fifth of the land is suitable for agriculture. Based on its 1990 census figures, Japan was the world's seventh most populous country, and its population is expected to reach 126,500,000 by the year 2000.

Early History

Japan is thought to have been inhabited by about 30,000 B.C., by paleolithic hunter-gatherers using stone tools. Between 10,000 and 300 B.C. the first pottery in Japan—and many think, in the world—was produced during the Jomon era, which is named for its distinctive patterns of twisted cords. Agriculture appeared by around 300 B.C., as well as bronze and iron tools. During the Kofun period (A.D. 300–701), distinguished by its majestic mounded tombs, Japan's imperial dynasty, the Yamato family—whose descendants still occupy the throne—came to power, and in 552 Buddhism was first introduced by way of Korea.

For the remainder of the first millennium (and nearly two centuries beyond), power and influence were centralized at the imperial court, which changed locations several times. The Nara period (710–784), when the capital was situated at Nara, saw a significant introduction of Chinese religious, political, and cultural ideas. Of the first two written histories of Japan, both produced in the early eighth century, one was written in Chinese. Near the end of the century, the first poetry collection, consisting mostly of the popular *waka* verse type, was written.

Heian Period

The Heian period (794–1185) is famous for the flowering of aristocratic Japanese culture. The capital was moved to the newly built city of Keiankyo (now Kyoto), which had been designed on a Chinese-style grid plan. In the mid-ninth century, *kana*, a written system for representing spoken Japanese phonetically using modified Chinese characters, was developed, furthering the creation of a Japanese literary tradition, including prose vernacular works. In the first part of the tenth century, Japan's (and, according to some, the world's) first novel, *The Tale of Genji*, was written by Lady Shikibu Murasaki. Politically, one family, the Fujiwara, wielded great influence at court during the Heian period, and the nobility's control of the lands beyond the capital was weakened as two rival clans of warriors, the Minamoto and the Taira, became increasingly powerful.

Military Rule through Shoguns

By 1185 Japan was under the control of a military government ruled by warriors known as *samurai.* The top military leader, or *shogun*, became the person who wielded the true power over the nation, rather than the emperor, although officially the imperial line continued to rule. Nevertheless it was the shogunate that controlled the landed estates (*shoen*) and their administrators (*jito*), in what was essentially a feudal system of government that lasted almost eight hundred years. Zen Buddhism, which provided the basis for many of the *samurai* ideals, flourished, and great Zen temples, as well as the classic sculpture of the Great Buddha of Kamakura, were built. Under the patronage of the *samurai*, a number of characteristic Japanese art forms flourished, including the highly stylized *Noh* drama and the comic *kyogen* theater. The tea ceremony was adopted and influenced architectural styles.

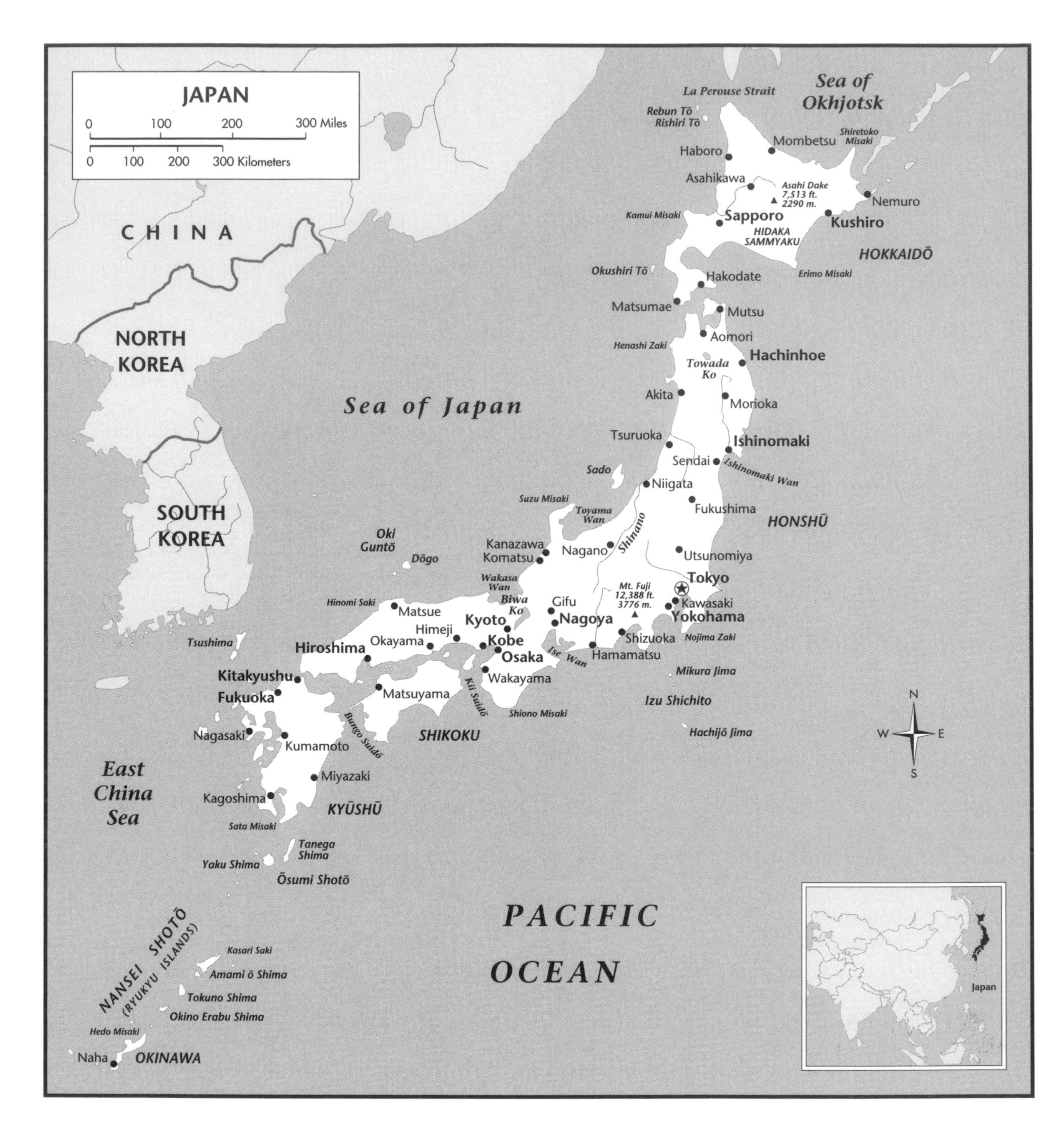

In 1603 Ieyasu Tokugawa founded the last of Japan's shogunates and moved his capital from Kyoto to Edo (present-day Tokyo), inaugurating the Edo Period, which lasted until 1867. This long period of internal peace (the longest in Japan's history) was achieved by the shogunate's nearly absolute consolidation of power. Internally, it took close control over the country's feudal lords (*daimyo*), requiring them to establish residences in the capital. Externally, it introduced an extended period of national isolation, in which virtually all contact with foreigners was banned.

Culture Flowered during Isolation

During this period of isolation, Japanese culture thrived. Early in the Tokugawa shogunate, the Kabuki theater tradition was born, first performed by women and then, after this practice was outlawed, by young men. Also in the seventeenth century, Japan's great master of haiku poetry, Matsuo Basho, flourished, bringing the haiku form to its mature form as a major poetic genre. Between the late seventeenth and mid-nineteenth centuries, Edo was also the center of an artistic tradition known as Ukiyo-e ("pictures of the floating world"), which centered on print portraits of the city's actors and the denizens of its red-light district. Associated with this movement was the refinement of the wood-block print by such masters as Hokusai and Hiroshige.

Western influence Began

Japan's isolation was forcibly ended when Commodore Matthew Perry arrived in 1853 and pressured the Japanese into opening their ports to foreigners. The already weakened shogunate, which the people held responsible for this unpopular move, lost even more ground, and by 1867 the last shogun had resigned and was replaced by a coalition government. The emperor Matsushito (at this time a mere boy called Meiji) was restored to power, beginning the period known as the Meiji Restoration (1868–1912).

With Western influences playing a central role, Japan underwent a period of modernization and growing power. By the end of the century it was pursuing a policy of expansionism abroad, winning control of Korea and Taiwan in the Sino-Japanese War (1894–95) and gaining further ground in the Russo-Japanese War (1904–1905). Japan gained more concessions from the Chinese during World War I (1914–18), in which it sided with the victorious Allies against Germany.

Political Structures Formed and Japanese Militarism

Japan's first parliamentary cabinet was formed after the war, and political parties became an important force for the first time in the 1920s, as universal male suffrage was instituted. By the 1930s, however, militaristic elements gained growing power over Japan's government. In 1931 Japan invaded Manchuria (a section of northeastern China), withdrawing from the League of Nations two years later. By 1940 Japan was expanding into Southeast Asia and had allied itself with the Axis powers of Germany and Italy. Japan entered World War II (1939–45) when it bombed U.S. military installations at Pearl Harbor on December 7, 1941. Throughout the course of the war, U.S. forces gradually reversed early military gains by the Japanese. By 1945 Japan had been overpowered militarily but still refused to surrender. On August 6 and 9 the United States released the world's first atomic bombs over the Japanese cities of Hiroshima and Nagasaki, and the Japanese surrendered on August 14.

Period of Economic Growth

From 1945 to 1952, Japan was under military occupation by the United States, which had a strong influence on its political, economic, and social policy. By the 1950s Japan had begun the economic rise that would make it one of the world's economic superpowers by the end of the decade. Exports and investment abroad grew rapidly. The country regained full sovereignty in 1952, and in 1955 the Liberals and Democrats merged to form the Liberal Democratic Party, which would govern Japan into the 1990s. In 1972 Japan established diplomatic relations with the People's Republic of China. In the same year the United States returned the island of Okinawa, occupied since World War II, to Japan.

Japan's strong economic growth continued from the 1960s through the 1980s except for a brief dip during the oil crisis of 1973–74, which provided an unwelcome reminder of Japan's dependence on foreign oil, as fuel prices rose and national income declined for a short time. In the early 1990s the "bubble economy" created by real estate and stock market speculation collapsed, and a period of economic slowdown and retrenchment or cutting expense began.

The Liberal Democratic Party (LDP) lost its majority in the government in 1993, after thirty-eight years, ceding political control to a seven-party coalition founded by Morihiro Hosokawa, who became prime minister. In the following years, a series of successive governments were voted out of office for their failure to significantly improve the nation's economic slowdown, and also because of continuing corruption scandals plaguing government officials.

Timeline

c. 28,000 B.C. Paleolithic habitation of Japan

Archaeological evidence shows that Japan is inhabited by this date (and possibly much earlier) by hunter-gatherers using stone tools. Migration from Korea and China was made possible by land bridges exposed when sea levels dropped.

c. 10,000 B.C.–300 B.C. Jomon period

Japan is inhabited by hunter-gatherers who live in settled villages. They hunt boar and deer, fish, and gather plants. They also produce what is thought to be the world's first pottery, characterized by its distinctive twisted cord markings ("Jomon" means "cord-marked"). The appearance of ceramics and polished stone tools in a non-agricultural civilization is unique.

300 B.C.– A.D. 300 Yayoi period

Metal tools and agricultural communities become common. Crops include rice, millet, barley, and wheat. Wetland rice cultivation is especially important. Bronze and iron are used for both tools and ritual objects. Social stratification and political systems develop.

Imperial Rule

A.D. 300–710 Kofun period

Government becomes more centralized, class divisions become more elaborate and important, and written records appear. The name "kofun" comes from the mounded tombs thought to have been built for the rulers of Japan's first political dynasty, the Yamato, who eventually come to rule the whole country.

6th century Chinese painting is introduced to Japan

Black-and-white drawings in the Chinese tradition, known as *suiboku*, are introduced to Japan along with Buddhism.

c. 550–710 Asuka period

The Yamato become Japan's first imperial dynasty. The capital of the Yamato court is established in the Asuka region, which becomes the site of Buddhist temples. The Yamato are supported by powerful clans, or *uji.*

552 Buddhism brought to Japan

Buddhism is introduced to Japan from Korea. Construction of religious sites shifts from tomb building to temple building. Japan's ruling families use Buddhism to reinforce their power.

Buddhism originates in the Ganges valley of India during the fifth century B.C. The religion is based on the teachings of Siddhartha Gautama (c. 563–c. 483 B.C), a Hindu prince who comes to be called the Buddha (the Enlightened One or the Wise One). Buddhism, in its original form, has no gods, no priests, no social classes. Salvation is within the reach of anyone who accepts the Buddhist *Dharma* (Law) and the beliefs embodied in the Four Noble Truths. The four basic tenets are: (1) suffering exists, (2) suffering is caused by ignorance, (3) ignorance and desire must be eliminated to achieve Nirvana (the ultimate state), and (4) one can achieve Nirvana by following the eightfold path.

593–628 Reign of Empress Suiko

Empress Suiko's court is established at the Toyoura Palace in Asuka. Buddhism is the imperial religion.

604 Royal instructions issued for government officials

Suiko's regent, Prince Shotoku, issues moral and religious directives for government officials, beginning a period of strong Chinese influence that continues with the Taika reforms of 645.

645 Taika reforms

Emperor Kotoku issues the Taika Reforms, which increase the power of the imperial line over the clans *(uji)* and establish systems for taxation and land distribution.

694 Capital built at Fujiwara

Japan's first permanent capital city, modeled after the grid plan developed by the Chinese Tang dynasty (618–907), is built at the mouth of the Asuka valley. Houses of aristocrats are built on a street grid, and the city also includes ministerial buildings, a palace, and an imperial residence.

701 Taiho law issued

Emperor Mommu decrees the Taiho laws, which further empower the emperor. Under Mommu, a system of standardized weights and measures is instituted and diplomatic ties with China are resumed after a period of rupture.

710–94 Nara period

The Nara period is inaugurated by the construction of a new capital city at Nara, called Heiankyo (site of present-day Kyoto), and based on the Chinese capital at Ch'angan. It will serve as the imperial residence until 1869. Chinese cultural influences increase, and Japan's rulers embrace Confucianism in addition to Buddhism. Political administration, including taxation and land control, becomes increasingly centralized.

712 First Japanese history, *Kojiki,* is written

Scholars complete the first chronicle of Japan, *Kojiki* (*Record of Ancient Matters*), in a written form of the Japanese language. Composed of legends, genealogies, songs, and historical records, it narrates the history of Japan from mythical times.

720 Second history, *Nihon Shoki,* is completed

A second, and more sophisticated history of Japan, written in classical Chinese, is completed.

c. 775 First Japanese poetry collection compiled

The *Man'yoshu*, Japan's first poetry anthology, is compiled. Written using Chinese characters, it is considered a cultural milestone. Most of its poetry consists of the 31-syllable *waka*, a widely used typed of verse.

794–1185 Heian period

The imperial government rules Japan from the new capital of Heiankyo. A sophisticated courtly culture is developed. The Fujiwara family becomes increasingly powerful by marrying their daughters into the imperial family, building up great landed estates (*shoen*), and through a complicated series of political intrigues. The Fujiwara are effective administrators, but the Heian period is increasingly characterized by lack of strong, centralized military power, as imperial control of the country beyond the capital becomes more and more tenuous toward the end of this period.

805 Tendai Buddhism is established

The monk Saicho returns from China and introduces an important new Buddhist sect, Tendai Buddhism, at a monastery called Enryakuji on Mount Hiei. Tendai Buddhism is an attempt to reconcile the various sects of Buddhism.

816 Shingon Buddhism is introduced

Shingon, the second of two influential Buddhist sects imported from China in the ninth century, is introduced by Kobo Daishi. It emphasizes rituals and magic and becomes popular at the imperial court.

Mid-9th century Native phonetic writing system is developed

Kana, a system for phonetically representing the spoken Japanese language, is developed, using shortened Chinese characters. A significant cultural advance, the creation of kana is a major contribution toward the development of a national literary tradition, including vernacular prose literature.

887 Ako Incident

Fujiwara no Mototsume challenges the power of Emperor Uda. Fujiwara wins almost total control of the government as regent.

10th–12th centuries Growing power of military families

Through their control of *shoen* (landed estates), two rival military families, the Minamoto and the Taira, have become powerful during the Heian era, jockeying for power with each other and with the Fujiwara nobles of the capital. Eventually they grow strong enough to challenge Japan's rulers.

Early 10th century *The Tale of Genji* is written

The Tale of Genji (*Genji Monogatari*) is written by Lady Shikibu Murasaki. One of the first novels ever produced (according to some, *the* first novel), it is a classic of world literature. Consisting of ffity-four books, it describes the elegant and refined existence of nobles at the imperial court, centering on the life and loves of Prince Genji. It is believed by some that the last fourteen books were written by a different author.

967–1068 Regency government

The Fujiwara wield continuous control of the government as regents (relatives of the emperor). All emperors during this era have Fujiwara mothers and grow up in Fujiwara households. The dominant political figure of the period is Fujiwara no Michinaga, three of whose grandsons become emperors. The regency period is known for the flowering of literature.

1068–73 Reign of Emperor Go-Sanjo

The reign of Go-Sanjo marks the end of Fujiwara domination. Political power shifts to a series of former emperors.

12th–13th centuries Popular forms of Buddhism are introduced

With the introduction of a phonetic Japanese vocabulary called kana and decreasing emphasis on scripture, new, less aristocratic forms of Buddhism become popular. These include Pure Land Buddhism, or Jodo, founded by Honen (1133–1212) and True Pure Land Buddhism (Shinsu), established by Honen's follower, Shinran (1173–1262).

1156 Hogen War

Two rival families, the Minamoto and Taira, compete for power in a conflict over the royal succession. After backing the victor, Taira no Kiyomori becomes an important force at the royal court, winning influence for his family.

Military Rule

1185–1333 Kamakura period

Military governments dominate, and the imperial court at Kyoto becomes less powerful. A new ruling military class is formed and increases its power by controlling tax-free landed estates (*shoen*) administered by stewards called *jito*.

1185 Battle of Dannoura (beginning of Kamakura period)

Minamoto no Yoritomo and his warriors overthrow the dominant Taira family and set up a military government, or *bakatu*, at Kamakura, south of present-day Tokyo, with Yorimoto as the ruler. This inaugurates the Kamakura period (see 1185–1333), which begins centuries of military rule in Japan.

1192 Kamakura shogunate established

Minamoto no Yoritomo conquers northern Honshu, uniting all of Japan for the first time. He is officially named *shogun* (top military leader) by the emperor and becomes the de facto or unofficial ruler of Japan, establishing the Kamakura shogunate. The court is still located at Heian, but a new military aristocracy actually holds the political power, taking control of the landed estates (shoen) through stewards (jito).

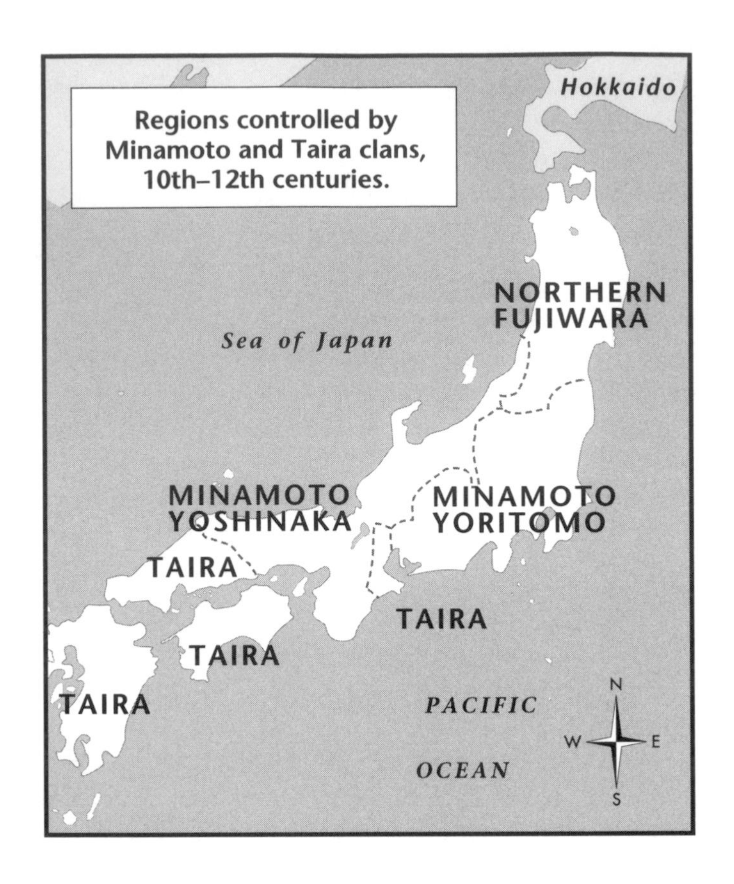

1199 Hojo family gains power

Yoritomo dies and power falls into the hands of Hojo Tokimasa, whose family retains control until the fourteenth century through regents.

1221 Revolt by Go-Toba crushed

Former emperor Go-Toba attempts to gain power when there are no more Minamoto descendants to rule the shogunate. He is defeated and banished, and Kamakura rule is reinforced.

c. 1250 Composition of the *Heiki Monogatari*

The military epic detailing the conflict between the Minamoto and Taira warrior families and the decline of the imperial line is composed and spread throughout Japanese society, chanted by traveling storytellers accompanying themselves on a lutelike stringed instrument called the *biwa.*

1274 Mongol invasion fails

After Japan's rulers refuse to recognize the sovereignty of the Mongol Emperor, Kublai Khan (1216–1294), a Mongol force invades Japan from Korea, but the 40,000-man expedition is destroyed by a fierce storm after only one day of battle.

1281 Second Mongol attack is defeated

A second Mongol invasion, by a force of 150,000, is crushed by a typhoon after two months of fighting.

1333–1568 Muromachi period

The Hojo rulers of the Kamakura shogunate are overthrown by Emperor Go-Daigo, ending a 150-year dynasty. In the aftermath, the Ashikaga establish a new shogunate (see entry: 1336 Ashikaga Takauji seizes Kyoto) and install their headquarters in the Muromachi section of Kyoto, inaugurating what becomes known as the Muromachi period.

1336 Ashikaga Takauji seizes Kyoto

The warrior Ashikaga Takauji, who had helped Go-Daigo seize the throne, deposes him. Go-Daigo flees and sets up a rival court at Yoshino, south of Nara. Civil warfare between supporters of the two factions continues until the end of the century, when a compromise is reached, and the Ashikaga shogunate is established.

1338–1568 Ashikaga shogunate

Ashikaga Takauji is appointed shogun. The Ashikaga remain in power for over two hundred years, ruling from Kyoto with the assent of the provincial military leaders (*shugo-daimyo*). These provincial lords become increasingly powerful, and central authority is weakened.

c. 1350–1400 Development of *kyogen* plays

The comic drama known as *kyogen* develops simultaneously (and complementary to) the serious Noh drama. In contrast to Noh plays, kyogen drama involves colloquial speech about everyday happenings and exposes the folly and weaknesses of the characters in a comic manner. Kyogen plays feature a number of stock characters, including the feudal lord who is outwitted by his subordinates, the newly married bridegroom, the devil who is bedeviled by human failings, and the ascetic wise man who is like ordinary mortals.

Mid-14th century Noh drama is created

Kan'ami (1333–84) originates the Noh drama by introducing a dance with a sung narrative, the *kusemai*, into the short plays of the traditional *sarugaku* variety shows. ("Noh" means "accomplishment" or "skill.") The highly stylized performances are accompanied by singing and instrumental music, and the actors wear elaborate costumes and masks. The repertoire of about 250 Noh plays are grouped into the following categories by subject matter: *nan* (man), *nyo* (woman), *shin* (*god*), *ki* (demons), and *kyo* (madness).

1364–1443 Life of Zeami, son of Noh theater originator

Zeami, the son of Kan'ami, who originated the Noh drama, continues and refines the theater tradition begun by his father. As an associate of the shogun Yoshimitsu, Zeami has the opportunity to bring Noh theater to the court, developing a classic two-part drama with standard movements and gestures, stylized performances, and a rich symbolic tradition.

The plays, for which no scenery is used, are performed by actors, a chorus, and an ensemble of four musicians. Zeami also establishes strict rules for the composition of Noh texts.

1369–95 Reign of Ashikaga Yoshimitsu

Ashikaga Yoshimitsu reigns as shogun and ends the civil war between the two rival courts in the north and south of Japan, reunifying Japan's imperial house and curbing the power of the provincial governors. He also reinstitutes trade with China after a period of isolationism.

15th–17th centuries The Kano family dominates Japanese painting

The Kano family of painters occupies a central position in Japanese art, forming the core of the Kano school of ink painters, which is centered at their studios in Kyoto. Leadership of the school and painting techniques are passed down from one generation to the next through an apprenticeship system that also allows the inclusion of talented painters who are not family members. In exceptional cases, these painters are formally adopted into the family. The apprenticeship system also gives the Kano studios enough manpower to undertake very large painting projects.

Kano painters are the official court painters for the Ashikaga shogunate (see 1333), the warlords who ruled afterward, and the Tokugawa shogunate (see 1603). In addition, they paint works commissioned by provincial lords, merchants, and Buddhist temples.

Among the better-known Kano masters are Kano Masanobu and Kano Motonobu, both famed for their ink painting in the early Muromachi period.

1467–77 Onin War and Sengoku period

The Onin War inaugurates a longer period of civil unrest, lasting into the late sixteenth century, and called the Warring States, or Sengoku, period. Kyoto is destroyed, and the power of the shoguns is drastically reduced. Local warlords gain power in the countryside, building fortified towns.

1540s Arrival of first Westerners

The first Portuguese traders arrive in Japan and take over most of the trade between Japan and China. They are accompanied by Christian missionaries who convert as many as 300,000 Japanese by the end of the century. By the beginning of the sixteenth century, the Portuguese are followed by Spanish, Dutch, and English traders.

1549 Christianity introduced to Japan

Christianity is introduced to Japan by the Spanish Jesuit St. Francis Xavier (1506–52).

1568–82 Oda Nobunaga gains control of central Japan

Nobunaga captures Kyoto and seizes control of the government, forcing out the current shogun.

1582–98 Toyotomi Hideyoshi in power

Toyotomi Hideyoshi, a general, succeeds Oda Nobunaga as ruler of Japan. By the 1590s, he has unified the entire country under his rule, the first leader in generations to do so. He sets up systems of taxation and a land survey, and gives the samurai warriors a place within his administrative bureaucracy.

1592, 1596 Hideyoshi invades Korea

Hideyoshi makes costly attempts to overrun Korea as part of an invasion of China. The first attempt fails, and the second is cut short by Hideyoshi's death in 1598.

1597 Persecution of Christians begins

Hideyoshi orders the execution of foreign priests (the "Twenty-Six Martyrs"), in the beginning of a crackdown on foreign influence in Japan.

The Edo Period, 1600–1867

1600 Edo (or Tokugawa) period begins as Tokugawa Ieyasu comes to power

At the Battle of Sekigahara, Tokugawa Ieyasu wins the struggle to succeed Hideyoshi, triumphing over supporters of Hideyoshi's heir, Hideyori.

1603 Tokugawa shogunate is founded

Tokugawa Ieyasu founds his own shogunate and moves his capital from Kyoto to Edo (present-day Tokyo). Under the Tokugawa shogunate, the shogun officially owns all land and exercises strict control over the local military lords. The country experiences over two hundred years of peace. The shogunate maintains its power by instituting a political structure known as the *Bakuhan* system under which the military lords (*daimyo*) govern their own domains (*han*), while still remaining loyal to the shogun. They are kept under the court's influence by being required to maintain residences in the capital and spend at least every other year there. Society is strictly divided into the following four classes modeled on the Chinese system: *samurai* (warriors), merchants, artisans, and peasants.

1603 The Kabuki tradition is begun

A female attendant at the Izumo Shrine, named Okuni, leads an all-female comic performance dressed as a man. The idea is copied in performances by other women, and these become known as *kabuki odori* (Kabuki dance) (also as *onna* kabuki—women's kabuki). From them evolves one of Japan's main theatrical traditions, featuring fixed, formalized dialogue, generous use of gestures, elaborate costume changes, and often, masks. Some of the plays deal with the nobility and warriors of early Japanese history, while others are about the common people.

Due to the unruly behavior of the audiences, the Tokugawa shogunate eventually forbids women to take part in kabuki plays (see 1629), and they are replaced by young men. However, the rowdiness at performances of *wakashu* (young men's) kabuki continues, and the shogunate finally bars such performances altogether and demands a reform of the kabuki genre (see 1652).

1614 Eviction of foreign missionaries is announced

The shogun Tokugawa Ieyasu announces that all foreign missionaries will be forced to leave the country.

1629 Women are banned from performing in Kabuki plays

The shogun Iemitsu bars women from participating in Kabuki performances. Even women's roles in the plays are henceforth performed by men.

1635 Long period of national seclusion begins

The Japanese shogunate begins a 200-year policy of seclusion from foreign influence in order to secure its power. Travel abroad by Japanese is outlawed.

1639 Foreigners are excluded from Japan

In response to the Shimabara Rebellion (1638) by Japanese Catholics, the shogunate bars foreigners from the country as part of its policy of national seclusion.

1644 Birth of poet Matsuo Basho

Matsuo Basho (1644–94), poet, essayist, and writer of travel sketches, is famed as the father of modern *haiku* poetry, which he is instrumental in transforming into a mature art form. *Haiku* is characterized by its three unrhymed lines of five, seven, and five syllables.

Born in Iga province, Basho begins traveling around the Kyoto area in 1666 and moves to Edo in 1672, becoming a teacher of haiku and adopting the name Tosei ("Basho," which is another literary name, is not adopted by the poet until later). His poems are characterized by their humor, earthiness, and wordplay.

After moving into a small hut in the Fukagawa district, the poet begins calling himself Basho ("banana plant"), in honor of the plant that stands near his home. He becomes interested in Zen Buddhism. In 1683 Basho and his students publish the collection *Minashiguri* ("Empty Chestnuts"). After this publication the poet undertakes trips to different areas of Japan, resulting in poetry journals that are among the finest examples of this genre. They include *Nozarashi kiko* ("The Records of a Weather-Exposed Skeleton"; 1685) and *Oi no kobumi* ("The Records of a Travel-Worn Satchel; 1690-91), both based on travels to parts of western Japan. After setting out on another journey in 1694, the poet becomes ill and dies in Osaka on November 28, 1694.

1649 Stiff new tax is placed on farmers

The Tokugawa shogunate enacts a new tax that places great financial pressure on farmers forcing many to live on millet, send their children to cities to find work, and have their daughters work as prostitutes to earn extra money.

1652 Shogunate bans kabuki performances

The Tokugawa shogunate bans kabuki performances in their present style, requiring that kabuki actors perform plays based on kyogen (see 1350). Also, the younger actors are eliminated, and the new genre that emerges is called *yaro* (men's) kabuki.

1680–1850 Flowering of the Ukiyo-e painting tradition

Probably the most characteristic art form of the Edo period was Ukiyo-e ("pictures of the floating world," an artistic tradition revolving around Tokyo's red-light district). Through ukiyo-e, wood block prints (*hanga*), become the dominant mode of Japanese art. Ukiyo-e pictures mostly depict Tokyo's brothel districts and Kabuki theaters, although some artists associated with this tradition paint landscapes. Important Ukiyo-e masters include Matabei Iwasa (1578–1650), Suzuki Harunobu (1725–70), Katsushika Hokusai (see 1760), and Ando Hiroshige (see 1797).

1680–1709 Reign of Tokugawa Tsunayoshi

Tsunayoshi engages in lavish spending and virtually bankrupts the shogunate. He is known as the "dog shogun" because of his edict against killing dogs.

1685 Collaboration begins between originators of puppet theater

Gidayu Takemoto (1652–1714) and Monzaemon Chikamatsu (1653–1724), the originators of Japan's puppet theatre, today called *bunraku*, begin their professional collaboration. Chikamatsu writes the texts, and Takemoto chants them. As the puppet theatre evolves in the first part of the eighteenth century, the puppeteers begin operating the puppets standing directly behind them, in full view of the audience, which remains a distinguishing feature of bunraku. The puppet theatre usually performs historical plays about samurai (warriors) or tragic dramas about star-crossed lovers.

1694: November 28 Poet Basho dies

Basho, considered the father of modern haiku poetry, dies in Osaka. (See 1644.)

1716–45 Reign of Tokugawa Yoshimune

Yoshimune institutes regulations to reform the government administrative and financial bureaucracy.

1720 Ban on Western literature is lifted

The shogunate ends the ban on European books.

1721 Guilds are legalized

Merchant's guilds and monopolies are legalized. Important trading houses, including Mitsui, Sumitomo, and Konoike, become increasingly powerful.

1737 Powerful earthquake creates record-breaking *tsunami*

A massive earthquake strikes Japan, killing thousands and setting off what is thought to be the largest tsunami (huge wave) in history, possibly reaching heights of two hundred feet. The coastal city of Kamaishi is completely destroyed.

1760 Birth of painter and print artist Hokusai

One of Japan's most famous artists, Hokusai (1760–1849) is born in Edo and originally named Tokitaro (he will later take multiple professional names as an artist). He begins an apprenticeship at the age of nineteen in the studio of Shunsho Katsukawa and produces his first set of portraits in 1779. Eventually he switches from his early subject—portraits of actors—to landscapes, becoming famous for print series including *Fuhaku samjurokkei* ("Thirty-six views of Mt. Fuji"; early 1830s) and *Fugaku hyakkei* (One Hundred Views of Mt. Fuji"; 1834). The artist adopts the name Hokusai in 1796. His last major print project is *Hyakunin isshu uba ga etoki* ("Illustrations of the One Hundred Poems"), intended to accompany a poetry anthology. Hokusai dies in 1849.

1782–87 Crop failure causes famine

The Temmei famine, caused by failed crops, exacerbates the existing economic problems of the shogunate.

1787 Troubles spark Edo riots

Three days of rioting in Edo are triggered by the nation's mounting social and economic troubles.

1790s Shiju school of painting is established in Kyoto

The painter Goshun Matsumura (1752–1811) establishes the Shiju school of painting, which produces scrollwork featuring domestic subjects, such as still-lifes, plants, and animals, that are popular with missle-class patrons in Kyoto and Osaka. The name "Shiju" comes from the street where Matsumura's studio is located. Although the Shiju style is eventually adopted in Edo (Tokyo), it retains a strong base in Kyoto throughout the eighteenth century.

1793 Volcano destroys Japanese island of Unsen

The island of Unsen, site of a volcano called by the same name, is destroyed when the volcano blows up during a period of earthquakes. Fissures pouring lava run from the volcano to the sea, as the entire island, with all of its 53,000 inhabitants, sinks into the ocean.

1797 Birth of artist Ando Hiroshige

One of Japan's most revered artists, Hiroshige (1797–1859) is born in Edo (Tokyo) during the period when the wood-block print is at its peak, the son of an official in the Edo fire department. (His birth name is Ando Tokutaro; he adopts Hiroshige later as a professional name.) Orphaned at the age of twelve, he enters the fire service himself soon afterwards but then opts for a career as an artist, studying under Toyohiro Utagawa. He produces his first landscapes in the late 1820s. Based on sketches made during a trip to Kyoto, the artist produces the print series *Fifty-Three Post Stations of the Tokaido*, which brings him rapid popularity and success by the early 1830s.

Other well-known print series by Hiroshige include *Noted Places of Edo, Eight Landscapes of Edo Suburbs, One Hundred Scenic Spots of Edo,* and *Eight Scenes of Omi.* (The set of eight scenes, or *hakkei*, is a common print genre consisting of a prescribed set of views based on Chinese art.) Unaffected by the Tempo Reforms (see 1841–43), which crimp the style of many fellow artists, Hiroshige continues to produce landscapes. The greatest of his late works are three triptychs (set of three hinged panels) on the themes of snow, moon, and flowers. Hiroshige dies in a cholera epidemic in October 1859, leaving a farewell poem that reads: "I have let fall my brush in the East, and depart to enjoy the wondrous scenery of the West." He is buried in the Asakusa district of Tokyo.

1841–43 Tempo Reforms are introduced

The shogunate adopts the Tempo Reforms in an attempt to restore the political, economic, and moral health of the shogunate. Guilds and monopolies are banned, and ownership of lands belonging to the provincial lords (*daimyo*) is ordered to be transferred. Strict censorship of the arts is instituted, as prints of courtesans and actors, a staple of the "floating world," are banned. Due to their unpopularity, most of the Tempo Reforms are later reversed.

1853–54 Commodore Perry demands opening of Japan's ports

A squadron commanded by U.S. commodore Matthew Perry (1794–1858) arrives in Japan. Perry forces the Japanese to open their ports, ending two centuries of isolation. The Treaty of Kanagawa is signed. Japan agrees to treat shipwrecked sailors kindly, to allow foreign ships to land for supply, and to permit American vessels to dock at two Japanese ports.

1857: March 21 Earthquake, fires decimate Tokyo

Fires following an earthquake spread through Tokyo, killing 107,000 people. Firefighters are unable to battle the blaze because the water delivery system has been destroyed by the quake.

1858: July 29 Treaty is signed with U.S. to open ports

The Harris Treaty is signed, opening five ports to trade with the United States. Treaties with other nations follow, establishing trade relations, restricting Japanese tariffs, and establishing consular provisions for foreign residents of Japan. These treaties are unpopular with many Japanese, who regard them as manifestations of the shogunate's weakness, and this disapproval ultimately helps lead to its collapse. Anti-foreign sentiments are widespread, and there are attacks on foreigners, with retaliation from the British and other Western forces.

1867 The last shogun resigns

Pressure from imperial courtiers and widespread anti-foreign sentiment force the resignation of the last shogun, Tokugawa Hitotsubashi. A coalition government takes his place.

The Meiji Period

1868–1912 Meiji restoration

Courtiers and warlords (daimyo) seize control of the imperial palace in Kyoto and restore the boy emperor, Meiji (Matsushito), to the throne to rule directly. The royal capital is officially moved from Kyoto to Edo, which is renamed Tokyo. The Meiji Restoration, referring to the period until the emperor's death, is a period of Western influence, modernization, and growing economic and political power for Japan.

1868 Government issues Charter Oath

The new imperial government adopts the Charter Oath, mandating government reforms, including the establishment of a legislative body and an end to Japan's strict centuries-old feudal system of class divisions.

1869 Power of the provincial lords is reduced

The provincial lords (daimyo) are required to accept imperial control over their domains, over which they are then named as governors within a highly centralized system of government administered by a Western-style bureaucracy.

1870s Baseball is introduced to Japan

Americans teaching English in Tokyo introduce the sport of baseball to Japan.

1873 Ban on Christianity is removed

With the opening of Japan to the West, Christianity is legalized for the first time since the early seventeenth century.

1873: May Women get the right to divorce

A new law gives women the right to divorce their husbands and decrees that a woman who has neither a husband nor a son may be officially considered the head of a household (*koshu*). This status is only temporary, however, until she remarries or until any of her daughters marry.

1885 Cabinet is appointed

The Grand Council of State is replaced by a cabinet headed by a prime minister.

1889: February Constitution adopted

The emperor adopts a new constitution establishing a constitutional monarchy with a bicameral or two-part legislature, including an elected house of representatives. Most of the power is wielded by the appointed cabinet, and the emperor retains full sovereignty over the nation. The constitution also recognizes the right of citizens to own property and other basic rights. However, women are barred from serving as empress (as they have in the past) or from participation in politics on any other level.

1894 Aoki-Kimberley Treaty with Britain

The British sign a pact agreeing to give up the rights gained through earlier treaties by 1899, including most-favored-nation trade status. As Japan becomes more powerful, similar agreements are signed with other nations.

1894–95 Sino-Japanese War

The rival territorial ambitions of Japan and China lead to war between the two countries. China is forced to give up control of Korea and Taiwan.

1896: June 15 Tsunami kills 28,000 on northeast coast

An earthquake emanating from the Tuscarora Deep, a crater underneath the ocean, trigger a series of tsunamis that wash up on 100 miles of coastline, destroying villages along Japan's northeast coast. With heights between 10 and 110 feet, their speed is estimated at 500 miles per hour.

1899: June 11 Birth of novelist Yasunari Kawabata

Kawabata (1899–1972), one of the foremost modern Japanese literary figures, and the first to win a Nobel Prize, is born in Osaka. He is orphaned at a young age, and his other close relatives also die during his childhood. After attending Tokyo University, Kawabata makes his literary debut with *Izu no odoriko* (*The Izu Dancer*; 1926), a semiautobiographical novel that centers on a visit to a traveler's lodge. He is best known for his postwar novels *Yukiguni* (*Snow Country*; 1948), *Sembazuru* (*Thousand Cranes*; 1959), and *Yama no oto* (*The Sound of the Mountain*; 1970). He also writes many short stories, and specializes in a type of extremely short fiction he calls *tanagokoro*, or "palm-of-the-hand story."

Hallmarks of Kawabata's style include structural looseness and flexibility, juxtaposition of sharply contrasting elements, and thematic emphasis on traveling, isolation, and death. In 1968 Kawabata becomes the first Japanese author to be awarded the Nobel Prize for Literature (there has been only one other since then, Kenzaburo Oe in 1994; see 1994: October 13). Kawabata's death in 1972 is generally thought to have been a suicide and is often linked to the suicide of his friend and colleague Yukio Mishima (see 1925: Janurary 14).

1901: April 29 Birth of Emperor Hirohito

Hirohito (1901–89), the 124th Japanese emperor, is born in Tokyo. He becomes heir apparent to the throne upon the death of his grandfather, Emperor Matsushito, in 1912. After serving in both the army and navy, Hirohito travels extensively in Asia and Europe, becoming the first Japanese prince to tour the West. He marries Princess Nagako Kuni on January 26, 1924.

Hirohito becomes emperor upon the death of his father, Yoshihito (1879–1926), on December 25, 1926, and is formally installed in November 1928. Although the official name for Hirohito's reign (Showa) means "enlightened peace," this period is one of growing Japanese militarism leading up to World War II and climaxing in the atomic bomb attacks on Hiroshima and Nagasaki. Hirohito is allowed to remain on the throne after Japan's defeat in the war. On January 1, 1946, he officially renounces the imperial claim to divinity. The royal family's increasing identification with its people is evident in the marriage of Akihito's son, Crown Prince Akihito, to a commoner, Michiko Shoda, rather than a member of the nobility.

In Hirohito's later years he pursues a longtime interest in marine biology, publishing a book on the subject in 1962. His 1971 visit to Europe is the first such journey by a Japanese emperor. On the same trip, Hirohito travels to Alaska, where he becomes the first Japanese emperor to meet a U.S. president (Richard M. Nixon). Hirohito dies in the imperial palace on January 7, 1989.

1904–05 Russo-Japanese War

Threats of Russian expansion into Korea trigger a new war. Japan is victorious, and Russia is forced to recognize Japanese supremacy in Korea, as well as ceding its rights to Manchuria and the southern Sakhalin Islands. Japan becomes the dominant power in East Asia, as well as the first Asian nation to inflict a military defeat on a Western power.

1905: September 5 Treaty of Portsmouth

The Russo-Japanese War is officially ended by the Treaty of Portsmouth, signed at Portsmouth, New Hampshire.

1907 "Gentleman's Agreement" with U.S. on immigration

In response to anti-immigration sentiment in the United States, Japan agrees to restrict immigration to the United States.

1907 Government sponsors major art exhibition

The Meiji government sponsors an art exhibition, the *Bunten*, modeled on those of the Paris salons. It includes both Western- and traditional Japanese-style artworks.

1910 Birth of film director Akira Kurosawa

Japan's most internationally acclaimed director, Kurosawa begins working in films in the mid-1930s after starting out to be a painter. He achieves professional and commercial success with his first film, *Sugata Sanshiro* (1943), the story of a young student of the judo, a Japanese martial art. His films in the immediate postwar period include the notable antiwar film *Waga seishun ni kui nashi* (*No Regrets for our Youth*; 1946). Others, such as *Yoidore tenshi* (*Drunken Angel*; 1948), portray the devastation of postwar Tokyo.

Roshomon (1950), one of the director's most famous films, wins the grand prize at the 1951 Venice Film Festival. Set in medieval Japan, it tells the story of a murder through the differing perspectives of four different people, including the murderer and the victim. *Shichinin no samurai* (*The Seven Samurai*; 1954) is another of the director's best-known works, and possibly the most widely seen Japanese movie. Later notable films include *Kumonosujo* (*Throne of Blood*; 1957), a version of Shakespeare's *Macbeth*; *Dersu Uzala* (1975), filmed in the Soviet Union; *Kagemusha* (1980); and *Ran* (1985), which is based on *King Lear*. Kurosawa received a special honorary Oscar award at the Academy Awards ceremony in 1990.

1910: August Annexation of Korea

Japan formally annexes Korea, a Japanese protectorate since 1905.

The Modern Era

1912 End of the Meiji period

Emperor Matsushito dies and is succeeded by Yoshihito, inaugurating the Taisho period. (See 1901: April 29)

1914–18 World War I

With the outbreak of World War I, Japan joins the Allies and declares war on Germany, seizing control of the Marshall, Mariana, and Caroline Islands in the North Pacific and attacking German bases in Shandong Province (China). At the close of the war, Japan emerges as a world power and major participant in the Versailles Peace Conference, which awarded it control of the aforementioned former German possessions in the Pacific and a permanent seat on the League of Nations Council.

1915 The Twenty-One Demands

Japan presents China with set of ultimatums known as the Twenty-One Demands, guaranteeing Japan privileges on Chinese territory and arousing widespread antagonism among the Chinese.

1920s Political parties gain power

For the first time, political parties become an important force in Japanese politics. Two major parties, the Seiyukai and Minseito, are regularly alternating power by the end of the decade. A number of left-wing and other radical parties are also formed. The introduction of universal (male) suffrage further democratizes Japanese politics and expands the power of the urban proletariat.

1920s Sosaku hanga movement begins

A new movement called *sosaku hanga* ("creative prints") is introduced among modern printmakers, led by Kanae Yamamoto (1882–1946) and his colleagues. A central tenet is that the artist should be responsible for the complete production of a print, including not only designing it but also carving the wood block from which the print is made and doing the actual printing. The *sosaku hanga* movement continues to gain strength in the following decades and becomes very popular following World War II.

1920 Feminist coalition is formed

Raicho Hiratsuka and Fusae Ichikawa form the feminist Coalition of New Women (*Shinfujin Kyokai*), which publishes the newspaper *Josei Domei* (Women's League) and petitions the parliament for female suffrage and the elimination of laws placing other types of limits on women's participation in politics.

1922 Formation of Japanese Communist Party

The Japan Communist Party is established and promptly outlawed by the government.

1922 *Karate* begins to gain popularity

The martial art of *karate* becomes widely known following a demonstration in Tokyo by karate master Gichin Funakoshi.

1923 Tokyo is decimated by an earthquake

The massive Kwanto earthquake levels the city of Tokyo. Many of the thousands of deaths are from fires caused by the quake. Even those who flee to bodies of water are still not safe. When the Standard Oil facilities explode, large quantities of oil are spilled into Yokohama Bay and ignite around those who have sought safety there. In other cases, fireballs crash into ponds and pools, heating them to the boiling point.

In Yokohama, the earthquake also sets off unquenchable flames, as well as gigantic landslides that push an entire village and its inhabitants into the sea. The combined death toll in Tokyo and Yokohama reaches 143,000, with 200,000 injured and 500,000 made homeless.

1925: January 14 Birth of novelist Yukio Mishima

One of the major Japanese authors of the twentieth century, Mishima is born in Tokyo, the son of a government official. After studying law at the University of Tokyo, Mishima begins his literary career, publishing his first novel, *Kamen no kokyhaku* (*Confessions of a Mask*) in 1949, to immediate acclaim. Later novels include *Ai no Kanaki* (*Thirst for Love*: 1950), *Kinjiki* (*Forbidden Colors*; 1954), and *Kinkajuki* (*the Temple of the Golden Pavilion*; 1956). He also writes a number of Noh plays that are modern reworkings of traditional plays. Mishima's final work, the four-part *Hojo no umi* (*Sea of Fertility)*, serialized between 1965 and 1970, is widely considered to be his masterpiece.

Although Mishima had a broad knowledge of Western literature, he was highly critical of Western influences on Japan in the postwar period. In his own life, he sought a return to the Japanese traditions of the past, including militarism, forming a controversial eighty-member private army, the *Tate no Kai* (Shield Society). On November 25, 1970, together with several other members of this group, Mishima seized and occupied the office of a top general at Japan's military headquarters in Toyko, where he committed suicide by ritual disembowelment (*seppuku*).

1926: December 25 Emperor Taisho dies

Emperor Taisho (the "reign name" for Yoshihito) dies and is succeeded by Hirohito, whose reign is known as the Showa period. (See 1901: April 29.)

1930: November Prime minister assassinated

Prime Minister Hamaguchi Yuko is assassinated as part of a growing right-wing effort to place the military in control of the government through a program of assassinations and coups d'état.

1931: September Japan invades Manchuria

Following a deliberately created provocation, Japan overruns the Chinese region of Manchuria and turns it into a puppet state, renaming it Manchukuo. This action marks the growing power of the military over the civilian government.

1932: May Second prime minister is assassinated

Prime Minister Inukai Ki becomes the second prime minister to be assassinated within two years. The assassination marks the rise of military power in Japan.

1933 Japan withdraws from League of Nations

Following international condemnation of its takeover of Manchuria, Japan withdraws from the League of Nations.

1934: September 21 Typhoon kills 4,000 in Osaka

A typhoon, or violent tropical cyclone, sweeps through Osaka at over 125 miles per hour, killing thousands of residents and destroying schools, hospitals, and other public buildings, as well as over 3,000 textile factories, the core of the city's economic base.

1935 Government seeks control of the arts

The Minister of Education, Matsuda Genji, issues an order reorganizing the Imperial Fine Arts Academy with the goal of establishing government control of artists, who are expected to produce patriotic artworks supporting the government's militarism.

1935: January 31 Birth of novelist Kenzaburo Oe

Nobel Prize-winning novelist Kenzaburo Oe (b. 1935) is born on the island of Shikoku. By the time of his graduation from Toyko University in 1959, he has won the prestigious Akutagawa Prize for his story "Shiiku" ("The Catch"), published in a collection by the same name the preceding year, and published a well-received first novel, *Memushiri kouchi* (*Pluck the Buds, Shoot the Kids*). Oe goes on to distinguished himself from many of his contemporaries through his involvement in politics and social issues, which color his writing, provoking bitter attacks from members of right-wing groups. His social concerns are expanded in the 1960s by two personal experiences—a visit to Hiroshima and the birth of a brain-damaged child. Later works by Oe include *Pinchi ranna chosho* (*The Pinch Runner's Record*; 1976); *Atarashii hito yo meza meyo* (*Awake, New Man*; 1983); and *Shizuka na seikatsu* (*A Quiet Life*; 1990). Oe also publishes an autobiographical work, *Boku ga honto ni wakakatta koro* (*When I was Really Young*; 1992).

In 1994 Kenzaburo Oe was awarded the Nobel Prize for Literature, becoming the second Japanese novelist so honored (the first was Yasunari Kawabata; see 1899: June 11).

1936 Professional baseball association is formed

Seven baseball organizations form the Japan Professional Baseball Association.

1936 Anti-Comintern Pact

Japan and Germany sign a formal agreement to band together in opposing the power of the Soviet Union.

1937–45 Second Sino-Japanese War

Following a clash between Chinese and Japanese troops at the Marco Polo Bridge near Beijing, Japan mounts a full-scale invasion of China. This incident begins an eight-year war that results in the eventual occupation of most of the Chinese coast (which holds over half of China's population) by Japanese troops. The Japanese armed forces are ruthless in their conquest and commit large-scale attrocities—particularly against civilians—of which the "Rape of Nanaking" is the most publicized.

The Sino-Japanese War isolates Japan even further diplomatically and puts Japan on a collision course with Western powers, especially the United States, which oppose Japanese expansion. By 1941, American opposition to Japanese expansion in Southeast Asia leads to an oil embargo against Japan. With no substantial domestic oil supply from which to draw, and unwilling to relinquish its territorial gains, Japan decides upon a war of conquest throughout Southeast Asia and the Pacific (see World War II).

1940 Political parties are dissolved

Japan's political parties are dissolved and replaced by the Imperial Rule Assistance Association. By now the militarists' domination of Japanese politics is complete.

1940: September Japan enters into joint pact with Germany and Italy

Japan approves the Tripartite Pact authorizing mutual assistance between itself and Germany and Italy.

1941–44 Hideki Tojo serves as prime minister

General Hidecki Tojo (1884–1948) is in office as both prime minister and war minister, making military control of Japan's government absolute.

World War II

1941: December 7 Japanese bomb Pearl Harbor

After negotiations with the United States fail, Japan enters World War II with the attack on American military installations at Pearl Harbor on Oahu, Hawaii. The bombing is intended to eliminate U.S. interference in Japanese plans to expand into Southeast Asia (Greater East Asia Co-Prosperity Sphere).

In addition to the attack against Pearl Harbor, the Japanese strike against American, British, and Dutch colonies throughout Asia and the South Pacific. Japanese forces attack and conquer Burma, the Dutch East Indies, New Guinea, the Philippines, Singapore, and the Solomon Islands. These conquests are to serve not only as sources for the raw materials the Japanese economy sorely needs, but also as Japan's defense perimeter which shields the Japanese home islands from direct attack.

1942: April The U.S. begins bombing Tokyo

The first U.S. bombing raids on Tokyo are carried out.

1942: June Japanese are stopped at Midway

The Japanese navy is blocked at Midway, tiny islands at the tip of the Hawaiian chain. Initial Japanese military campaigns in China, Southeast Asia, and the Pacific are successful. However, counterattacks by the United States slowly rescind Japan's early gains.

1943 Nimitz leads successful U.S. advance

Under Admiral Chester Nimitz (1885–1966), U.S. forces advance against the Japanese in the Central Pacific.

1944 U.S. bombing of Japan intensifies

The United States bombs the Japanese mainland with B-29s and undertakes daylight bombing of Tokyo.

1944 Mariana Islands are taken

The U.S. captures the Mariana Islands, located in the western Pacific but east of the Philippines, by midsummer. Prime minister Tojo is forced to resign.

1945: March Japanese population is mobilized

All Japanese over the age of six are ordered to perform war-related service in preparation for an expected U.S. invasion.

1945: June Okinawa is captured

Okinawa, south of the Japanese mainland, is captured by the Allies, and Burma, a country in Southeast Asia, is retaken.

1945: July Japan refuses to surrender

Japan rejects an ultimatum of surrender from the United States, becoming the last Axis power to continue fighting.

1945: August 6 U.S. drops atomic bomb on Hiroshima

The United States releases the world's first atomic bomb over Hiroshima, a city of over 400,000. The blast kills between 75,000 and 80,000 people and destroys some 62,000 buildings, rendering virtually all the rest uninhabitable.

1945: August 9 Atomic bomb dropped on Nagasaki

A second atomic bomb is dropped by the United States, this one over the city of Nagasaki. Damage from this blast is more localized, as parts of the city are protected by hills. All structures in the area closest to the center of the explosion, however, are almost totally destroyed. An estimated 35,000 people are killed, and many more are injured.

1945: August 14 Japan surrenders

Japan surrenders, cedes control over its outer islands, and agrees to military occupation by the United States. A seven-year military occupation of Japan by the United States begins, with goals of demilitarization and democratization. It is headed by General Douglas MacArthur (1880–1964).

The Postwar Period

1946: April 10 Women vote for the first time

Japanese women vote in their first election, having been granted the vote a year earlier. Thirty-nine women win seats in parliament.

1946: May Japan becomes a constitutional monarchy

After rescinding the centuries-old doctrine of imperial divinity, Emperor Hirohito agrees to the adoption of a new constitution, making Japan a constitutional monarchy.

1947: February U.S. blocks general strike

A planned general strike by labor unions is blocked by the U.S. occupation authority.

1947: May 5 Equal Rights Amendment expands women's rights

Under pressure from the U.S. occupation authority, an Equal Rights Amendment is adopted that enables women to sue for discrimination in economic, social, and political matters. The law also provides for establishment of a Department of Women's and Children's Affairs.

1948 War crimes trials end

Trials of Japanese accused of crimes in World War II conclude. Wartime leader Tojo is executed.

1949 Hideki Yukawa wins Nobel Prize in physics

The Nobel Prize in physics is won by Japanese physicist Hideki Yukawa (1907–81). Yukawa is known for his discoveries in the fields of particle physics, including his theory of nonlocal fields and meson theory. At the invitation of J. Robert Oppenheimer (1904–67), Yukawa accepts a visiting professorship at Princeton University between 1948 and 1953. Afterward, he returns to Japan, where he teaches, serves in government posts, and lobbies for the peaceful use of nuclear energy.

1949 Judo federation is established

The All-Japan Judo Federation (Zen Nihon judo Remmei) is organized.

Musicians in traditional costume and playing traditional instruments perform in the ornate music chamber of the Imperial Palace in Tokyo. This performance was the first to include members of the general public—in this case, music students invited for the occasion—in the audience. (EPD Photos/CSU Archives)

1949: November **Music room at Imperial Palace opens to the public**

Musicians perform for the first time before an audience that includes members of the general public. Until now, only members of the royal family were permitted to attend performances in the palace.

1950s **Popularity of baseball grows**

Baseball becomes Japan's most popular spectator sport. Total annual game attendance grows from 2.5 million at the beginning of the decade to 9 million by the end.

1950–53 **Korean War**

The Korean War benefits the Japanese economy by the United States' demand for war material and other goods. Japan's growth rate remains at about nine percent throughout the 1950s.

1951: September **U.S.-Japan peace treaty signed**

The treaty formally ending hostilities between Japan and the United States is signed in San Francisco. The United States continues to occupy Okinawa and the Bonin Islands.

1952 **Japan and U.S. sign security treaty**

A separate security agreement is signed by Japan and the United States allowing U.S. troops to be stationed in Japan.

1952: April 28 **Japan regains full sovereignty**

Japan regains full sovereignty as the United States occupation ends.

1953 **First televised baseball game**

A new era in Japanese baseball is launched with the first televised game between the Hanshin Tigers and the Yomiuri Giants.

1955 Liberal Democratic Party is formed

The Liberals and Democrats merge to form the Liberal Democratic Party, which remains in power through the early 1990s.

1956 Japan hosts judo championship

The first world judo championship is hosted in Tokyo.

1956 Japan joins the United Nations

Japan becomes a member of the United Nations.

1964 Karate federation is established

A federation of Japan's various karate organizations is founded.

1964: July 18–19 Earthquake and massive flooding hit coastal areas

A minor earthquake touches off severe floods in coastal areas along the Sea of Japan, leading to 108 deaths, as well as injuring 233 and rendering 44,000 homeless. Both natural and manmade structures weakened by the quake give way, adding to the destruction, as hillsides are washed away in landslides, and bridges and dams collapse.

1964: October Japan hosts the Olympics

Japan hosts the Olympic Games. In preparation, the high-speed "Bullet Train" railway between Tokyo and Osaka is constructed, and construction is begun on an expressway system for autos.

Judo is a formal entry in the Olympics for the first time.

1965 Japan and South Korea sign a peace agreement

A peace agreement between Japan and South Korea is signed.

1968 Japan reclaims Bonin Islands

The Bonin Islands, occupied by the United States since World War II, are restored to Japan. (See 1951: September.)

1968 Yasunari Kawabata wins Nobel Prize

The Nobel Prize for Literature is awarded to Yasunari Kawabata (1899–1972), renowned Japanese novelist. (See 1899: June 11.)

1970: November Death of novelist Yukio Mishima

Mishima, one of Japan's foremost authors, commits suicide. (See 1925: January 14.)

1972 Diplomatic ties with China are resumed

The state of war between China and Japan, officially maintained since 1937, is ended by a joint agreement, and diplomatic relations are restored. Japan breaks diplomatic ties with Taiwan.

1972: April 16 Death of novelist Yasunari Kawabata

Kawabata, Japan's first Nobel Prize-winning literary figure, dies, most likely at his own hand. (See 1899: June 11.)

1972: May 15 Okinawa is returned to Japan

Japan takes possession of Okinawa from the United States.

1973–74 Japanese economy threatened by OPEC actions

Rising oil prices by Middle East suppliers (Organization of Petroleum Eporting Countries, or OPEC) hurt the economy of Japan, which imports all its oil. Prices for all types of fuel rise, and the country's national income declines in 1974 for the first time during the postwar era.

1974 Eisaku Sato receives Nobel Peace Prize

The Nobel Peace Prize is awarded to former prime minister Eisaku Sato (1901–75). During his years in office (1964–72), Sato presided over the normalization of relations with Korea in 1965, signed the nuclear nonproliferation treaty in 1970, and negotiated the return of Okinawa to Japanese administration following U.S. occupation in 1972.

1974 Tanaka resigns over Lockheed scandal

Prime Minister Kakuei Tanaka is arrested and imprisoned after revelations that he accepted bribes from the Lockheed Corporation. He resigns the prime ministership but retains his seat in the legislature and his leadership in the Liberal Democratic Party (LDP). Although voters in Tanaka's district continue to support him, nationwide support for the LDP begins to erode, and the party loses its majority in the lower legislative house.

1978 Peace treaty is signed with China

Japan and the People's Republic of China sign a peace and friendship treaty.

1978 New international airport is opened

The New Tokyo International Airport in Narita on Chiba peninsula opens after delays caused by opposition from local farmers.

1981: March 8 Radioactive leak contaminates nuclear power plant

Leaking radioactive waste contaminates the nuclear power plant at Tsuruga on Japan's western coast. It is later found to have been caused by an employee who neglected to turn off a valve, allowing water to fill a sludge tank, which then overflowed. Over fifty workers assigned to clean up the spill are exposed to radiation, and possible widespread contamination of the Sea of Japan is feared.

1985: August 12 Worst airline crash in Japanese history

A Japan Air Lines Boeing 747 crashes into a mountainside, killing 475 people in the world's worst crash involving a single plane, and Japan's worst air disaster. The plane, flying from Tokyo, was carrying 524 people, many flying home to honor their ancestors at the midsummer festival of Obon. Due to an improperly repaired bulkhead, the pilot lost control of the plane shortly after takeoff, and it was not regained for the remaining half hour before the plane crashed into Mount Ogura and exploded. Acres of mountainside forest caught fire, and all but forty-nine of the plane's passengers were killed.

1986 Takako Doi becomes the first Japanese woman to head a political party

Takako Doi takes over leadership of the Socialist party, becoming the first woman ever to head a Japanese political party. Doi will later become the first female speaker of the House (see 1989: August).

1988 Recruit Cosmos scandal rocks Japan

Top Liberal Democratic Party leaders, including Prime Minister Noboru Takeshita and former prime minister Yasuhiro Nakasone, are involved in a scandal involving insider stock trading of shares of the Recruit Cosmos Company in 1984–86.

1989: January 7 Emperor Hirohito dies

After the longest reign of any Japanese emperor, Hirohito dies at the age of eighty-seven after a long illness. He is succeeded by his son, Crown Prince Akihito, who adopts the "reign name" of *Heisei*, meaning "peace and prosperity." (See 1901: April 29.)

1989: April Prime Minister Takeshita resigns

Prime Minister Noboru Takeshita is forced to resign over the Recruit Cosmos stock trading scandal. Charges against Takeshita include receiving over 150 million yen from Recruit Cosmos. Former prime minister Yasuhiro Nakasone admits profiting from the sale of shares in the company and gives up his positions with the LDP but keeps his seat in the legislature. A Nakasone political associate, Uno Sasuke, becomes prime minister.

1989: May Government official indicted on Recruit Cosmos charges

Eighteen government officials are indicted in connection with the Recruit Cosmos stock trading scandal.

1989: July LDP loses ground in elections

An unpopular consumption tax enacted under Takeshita and sex scandals surrounding the new prime minister, Uno Sasuke, weaken support for the LDP. Consequently the party loses municipal elections in Tokyo early in July and suffers its first parliamentary defeat ever in the midterm Senate elections at the end of the month, when the Socialist party wins a major victory. Uno resigns the prime ministership after only two months.

1989: August Choice for prime minister is split for first time

For the first time in the postwar era, the upper and lower houses of the Japanese legislature (the Diet) choose different candidates to succeed Uno Sasake as prime minister. The Senate chooses Takako Doi (see 1986), the female leader of the Socialist party, while the lower house picks LDP candidate Toshiki Kaifu. According to Japanese law, the lower house choice—Kaifu—becomes the new prime minister.

1990: May Japan apologizes for Korean aggression

Japan issues a formal apology to Korea for its past aggression against that nation.

1990: October Japanese stock market crashes

Anxiety over a possible bank crisis and high interest rates triggers a steep drop in the Tokyo Nikkei Index to a low of 20221.86, representing a forty-eight percent drop from its level at the end of 1989. This crash signals the end of Japan's "bubble economy," built on speculation in stocks, real estate, and investments. Japan's long period of sustained growth comes to a halt, giving way to a period of economic slowdown and retrenchment or reduction of expenses.

1991: June Prime Minister Kaifu resigns

After Toshiki Kaifu's proposed electoral reforms are rejected by his own party (the LDP), he announces that he will resign once a new prime minister can be chosen.

1991: November Kiichi Miyazawa becomes prime minister

Kiichi Miyazawa, former finance minister and LDP member who had been forced to resign in the Recruit Cosmos scandal (see 1988), becomes prime minister, succeeding Toshiki Kaifu.

1992 Government tries to halt economic downturn with spending measure

The Japanese government implements an $83 million spending program in an attempt to head off an economic slump stemming from the decline of the real estate and stock markets, as the Nikkei Index hits its lowest level in over six years.

1992: January Miyazawa apologizes for wartime abuse of women

Prime minister Kiichi Miyazawa issues a formal apology for Japan's abuse of Korean women during its occupation of Korea during World War II, when women and girls were forced into prostitution by the military.

1992: February New LDP scandals revolve around Kanemaru

Shin Kanemaru, a top LDP political figure, becomes embroiled in a scandal involving illegal contributions from a trucking company supported by organized crime (*yakuza*).

1992: August Kanemaru resigns from LDP due to financial scandal

Key LDP figure Shin Kanemaru resigns from the beleaguered LDP after being fined for accepting $4 million in contributions from the mob-supported Sagawa Kyubin trucking company. Kanemaru's downfall, capped by arrest for tax evasion, paves the way for the end of the LDP's thirty-eight years of political dominance.

1993: July LDP toppled in general election

Nearly forty years of LDP political rule ends with the victory of the New Party, a seven-party coalition founded by Morihiro Hosokawa, who becomes prime minister. Socialist leader Doi Takako (see 1986 and 1989: August) is elected speaker of the house of representatives, becoming the first woman to hold that post.

1993: December Lower house passes electoral reform bill

An electoral reform bill sponsored by Prime Minister Morihiro Hosokawa is passed by the lower house of the Japanese Diet (legislature). The bill, aimed at fighting the widespread corruption that afflicts Japanese politics, bans direct campaign contributions to individual candidates by corporations and changes the system that assigns multiple representatives from a single district to the lower house. In addition, public funding for campaigns is increased, and electoral districts are reshaped.

1994: January 21 Electoral reform bill is rejected by upper house

The upper house of the Japanese Diet defeats the electoral reform bill that is the centerpiece of Prime Minister Morihiro Hosokawa's political program, following approval of the legislation in the lower house (see 1993: December). Rejection of the bill threatens the stability of the coalition that Hosukawa put together to defeat the ruling Liberal Democratic Party in the July 1993 elections. The proposed legislation would place new restrictions on corporate campaign contributions and reapportion representation in the lower house. The lower house does not have the two-thirds majority needed to overturn the upper-house vote, but a compromise bill is passed later in the month.

1994: April 8 Prime Minister Hosukawa resigns

Prime Minister Morihiro Hosukawa announces he will step down in response to allegations of financial wrongdoing resulting from a 1982 loan he received from the notorious Sagawa Kyubin trucking company, which is rumored to have underworld connections. The resignation of Hosukawa after only eight months in office threatens the survival of the seven-party coalition that brought him to office in 1993, ending the continuous thirty-eight-year dominance of the Liberal Democratic Party.

1994: April 25 Hata is elected prime minister

Tsutomu Hata is elected by the Diet to be Japan's new prime minister, succeeding Morihiro Hosukawa. The following day, the Social Democratic Party withdraws from the governing coalition, presenting Hata with his first political challenge.

1994: April 26 Taiwanese jetliner crash lands at Nagoya Airport

In Japan's second-worst air disaster, a Taiwanese jet crash lands and catches fire at Nagoya Airport west of Tokyo. Almost all passengers and crew die: 264 are killed, with only 7 survivors. According to police reports, alcohol consumption by the pilot and co-pilot is implicated in the disaster, although engine failure and an incorrect landing approach are also considered possible factors.

1994: June Release of nerve gas causes injury and death

The nerve gas sarin is released in the city of Matsumoto in central Japan, killing seven people and injuring more than two hundred. The person or persons responsible are never found.

1994: June 25 Hata resigns as prime minister

Tsutomu Hata resigns from the post of prime minister after only two months in office. Hata's resignation comes on the eve of an anticipated no-confidence vote by the lower house of the legislature. Hata's attempts to continue the political reforms of his predecessor, Morihiro Hosukawa, alienated his political supporters, and both the Liberal Democratic Party and the Social Democratic Party aligned themselves against his coalition government, which gave it little chance of survival. Hata's government is Japan's third in one year.

1994: June 29 Murayama becomes new prime minister

The lower house of parliament elects Tomiichi Murayama to be Japan's fourth prime minister in one year, succeeding Tsutomu Hata, whose government only lasted for two months.

Murayama, leader of the Social Democratic Party, becomes Japan's first Socialist prime minister since 1948. He will head a coalition government formed by an alliance between the Social Democrats and the Liberal Democratic Party, both of whom had opposed the policies of his predecessor.

1994: October 13 Nobel Prize awarded to Kenzaburo Oe

Kenzaburo Oe, author of more than two dozen novels, is awarded the Nobel Prize for Literature. (See 1935: January 31)

1995: January 17 Major earthquake strikes Kobe

The region around the city of Kobe is struck by an earthquake measuring 7.2 on the Richter scale and triggering hundreds of fires. Preliminary estimates place the death toll at over 4,000 and it is thought likely to top 5,000. It is estimated that over 20,000 people are injured, and 275,000 are made homeless and forced to take refuge in emergency shelters. Hundreds of thousands of people are without electrical power or gas, and as many as a million are cut off from a water supply. It is expected that property damage could total billions of dollars and require two to three years of rebuilding.

1995: March 20 Nerve gas attack contaminates Tokyo subway system

Open cans of sarin, a deadly nerve gas, are placed on three cars in the Tokyo subway system during the morning rush hour. As the gas evaporates and spreads through cars and tunnels, 10 people are killed, some 5,000 become ill, and the subway system is forced to shut down. Foremost among those suspected are members of Aum Shinrikyo, a religious sect with branches throughout the country. Large amounts of chemicals are found in police raids on the group's offices.

1995: November 2 Major Japanese bank indicted for trading activities in U.S.

The Daiwa Bank is indicted on fraud and conspiracy charges by the U.S. Department of Justice in connection with illegal bond trading in its New York office. The Federal Reserve orders Daiwa to close all its United States operations by February 1996.

1996: January 5 Prime minister Murayama resigns

Tomiichi Murayama steps down as prime minister after eighteen months in office. While admired for his character, he has largely been viewed as a weak leader, and compromises between different factions of his coalition have weakened its ability to govern effectively. (See 1994: June 29.)

1996: January 11 Hashimoto is elected prime minister

Ryutaro Hashimoto (b. 1938) of the Liberal Democratic Party is elected by the lower house of parliament to succeed Tomiichi Murayama as the nation's prime minister. Hashimoto, who is fifty-eight, has served as minister of finance and trade in previous governments. As minister of trade in 1995 he was acclaimed for his tough stance in negotiations with the U.S. over automobile exports. Following this episode, he became leader of the LDP in September 1995.

1996: October 20 LDP nears majority in general elections

The Liberal Democratic Party comes close to winning a majority in Japan's first elections since 1993, when the LDP failed to get a parliamentary majority for the first time since 1955. It is expected that the party will once again form a ruling coalition with other parties and reelect Prime Minister Ryutaro Hashimoto.

1997: April 22 Siege at Peruvian embassy ends

Peruvian troops storm the Japanese embassy in Lima, Peru, ending the 126-day-long siege by leftist guerrillas, who seized the building during a Christmas party in December, taking the hundreds of guests, including numerous Japanese nationals, hostage. The seventy-two remaining hostages who are freed include Japan's ambassador to Peru and the brother of Peruvian president Alberto Fujimori.

1997: November 24 Major brokerage firm announces it will close

Yamaichi Securities, one of Japan's four major brokerage houses, announces it will close, following revelations that it had concealed the true extent of its debt, which is even larger than the approximately three trillion yen previously reported. The failure of Yamaichi heightens the already growing concern about the state of Japan's financial industry. It follows the closing of a bank and another securities company earlier the same month.

1998: January 28 Finance minister resigns due to scandal

Japan's finance minister, Hiroshi Mitsuzaka, resigns following revelations that two of his officials accepted lavish bribes from four banking companies between 1994 and 1997. Prime Minister Ryutaro Hashimoto takes on the post of finance minister until a replacement can be named.

1998: February 7–22 Winter Olympics are held in Nagano

The eighteenth Winter Olympics are held in Nagano. Competitors include 2,450 athletes from 72 countries. Japan's Emperor Akihito and Empress Michiko attend the opening ceremonies, at which figure skater Midori Ito lights the Olympic flame. Japanese athletes win a total of ten medals in the sixteen-day competition, and Japanese organizers of the games are praised for their efficiency in planning and running the event. However, bad weather forces postponements in almost all the skiing events.

1998: July 13 Ryutaro Hashimoto resigns as prime minister

Prime Minister Ryutaro Hashimoto resigns following a poor showing for his party, the LDP, in elections to the upper house of the Diet. The vote is attributed to the failure of the LDP under Hashimoto to reverse Japan's economic downturn, which has slid into recession. In his resignation, Hashimoto takes personal responsibility for the defeat.

1998: July 24 Keizo Obuchi is named to head the LDP

Foreign Minister Keizo Obuchi is named the head of the Liberal Democratic Party, an appointment that virtually guarantees that he will be appointed prime minister by the legislature.

1998: July 30 Keizo Obuchi is named prime minister

Newly elected LDP party leader Keizo Obuchi is formally chosen as Japan's new prime minister by parliament. He appoints former prime minister Kiichi Miyazawa as finance minister (See 1991: November and 1992: January).

Bibliography

Beasley, W. G. *Japanese Imperialism, 1894–1945.* New York: Oxford University Press, 1987.

Cortazzi, High. *Modern Japan: A Concise Survey.* New York: St. Martin's Press, 1993.

Demente, Boye Lafayette. *Japan Encyclopedia.* Lincolnwood, Ill.: NTC Publishing, 1995.

Dolan, Ronald E., and Robert L. Dolan, eds. *Japan, a Country Study.* 5th ed. Washington, D.C.: Library of Congress, 1992.

Fukutake, Tadashi. *The Japanese Social Structure: Its Evolution in the Modern Century.* 2d ed. Tokyo: University of Tokyo Press, 1989.

Hane, Mikiso. *Modern Japan: A Historical Survey.* 2d ed. Boulder, Col.: Westview Press, 1992.

Masumi, Junnosuke. *Contemporary Politics in Japan.* Berkeley: University of California Press, 1995.

Morton, W. Scott. *Japan: Its History and Culture.* 3d ed. New York: McGraw-Hill, 1994.

Packard, Jerrold M. *Sons of Heaven: A Portrait of the Japanese Monarchy.* New York: Scribner, 1987.

Perkins, Dorothy. *Enclyclopedia of Japan: Japanese History and Culture, from Abacus to Zori.* New York: Facts on File, 1991.

Reischauer, Edwin O. *Japan: The Story of a Nation.* 4th ed. New York: Knopf, 1989.

Richardson, Bradley M. *Japanese Democracy: Power, Coordination, and Performance.* New Haven, Conn.: Yale University Press, 1997.

Thomas, J. E. *Modern Japan: A Social History Since 1868.* New York: Longman, 1996.

Tsuru, Shigeto. *The Economic Development of Modern Japan.* Brookfield, Vt.: E. Elgar, 1995.

Waswo, Ann. *Modern Japanese Society, 1868–1994.* New York: St. Martin's Press, 1993.

Jordan

Introduction

NBC correspondent Hanson Hosein traveled in 1998 from his residence in Tel Aviv to Israel's border crossing with Jordan. As he crossed into Jordan, he marveled at the beauty of the Jordan River valley, the country's pastoral communities, its farms and the glimpses that he caught of the Dead Sea.

"It amazed me how much Jordan resembled Israel: verdant plains, hills covered in pines, and the desert," recalled Hosein in a personal dispatch sent to acquaintances. "Of course," Hosein added, "it shouldn't have come as that much of a surprise; the border that created Israel and Transjordan in the early half of this century was merely splitting one land into two."

Such an observation underscores many of the difficulties that Jordan experiences in the 1990s. Although it lies in a land with a long and varied history, the modern nation-state of Jordan is relatively new. It was created by Great Britain after World War I and its borders did create an artificial barrier that, as Hosein noted, split one land into two. Perhaps the most visible reminder of this is the West Bank, a butterfly-shaped piece of land that lies on the western banks of the Jordan River and contains some of the most sacred cities and monuments of the Jewish, Christian and Muslim faiths. Jordan and Israel have fought for control of the West Bank almost as long as they have existed as nation-states. Another reminder is the fact that more than half of Jordan's current population consists of Palestinians who fled from Israel in 1948 after Zionist nationalists established a Jewish regime.

Jordan today consists of 32,175 square miles (83,335 square kilometers), and is bordered by Syria and Iraq to the north, Saudi Arabia to the east and south, and Israel to the west. The Gulf of Aqaba separates the southwestern edge of Jordan from Egypt. The country is slightly smaller than the state of Indiana and consists largely of desert. Although some semitropical vegetation exists in the Jordan Valley, less than 1 percent of the country is forested. The desert regions receive less than eight inches (20 centimeters) of rain annually, while other parts of Jordan get up to 23 inches (58 centimeters) of rainfall in a typical year.

History

Bedouin tribes and other nomadic groups historically have populated Jordan, and traces of their influence exist in Jordan's population. However, about seventy-one percent of the country's 4.7 million residents live in urban areas. About twenty-seven to twenty-eight percent of the population lives in villages, and about two to three percent continue to lead nomadic or semi-nomadic lives. Most Jordanians (nearly ninety-two percent) are Sunni Muslims, and the population is virtually all Arab. However, traces of Greek, Egyptian, Persian, European and Negroid influences linger in the population and the country's cultural heritage.

As part of the Fertile Crescent, an area of land that connects Africa to Asia, Jordan was a major transit zone for many civilizations. Over the centuries Egyptians, Semites, Greeks, Seleucids, and other groups ruled the area. It became part of the Roman Empire about 100 B.C. and was predominantly Christian for about 300 years. After Muslim invaders entered the area in the seventh century A.D., Islam quickly replaced Christianity as the main faith. Despite a century of domination by the Crusaders, Islam has remained the area's primary religion.

The story of modern Jordan begins with a hazy incident known as the Hussein-McMahon correspondence. While Jordan was under the rule of the Ottoman Empire, a British leader promised to give Sharif Hussein—whom Ottoman Turks had established as an emir of Mecca—an independent Arab state if Hussein helped the British overthrow the Ottoman Empire. Hussein enlisted the help of his sons, Faisal and Abdullah, who fought against the Ottoman Turks with British scholar-adventurer Thomas Edward Lawrence, whose life is depicted in the movie, *Lawrence of Arabia*. After the battle, Palestine, Syria and Lebanon were freed from Turkish forces. Faisal received the land that is known as Iraq, while Abdullah was awarded the emirate of Transjordan.

Great Britain created Transjordan (later renamed Jordan) by dividing the ancient land of Palestine, which the British controlled after World War I, to satisfy two conflicting promises: the promise of an Arab state to Hussein and a separate agreement to establish a Jewish homeland in Palestine. A series of negotiations granted Transjordan its independence in

1946. Israel received its independence two years later and was immediately invaded by several Arab nations.

During this 1948 war, Abdullah annexed the West Bank, claiming it was part of the land that had been promised to his father. Israeli occupied the territory in 1967, and the sovereignty of the area became a hotly contested issue until 1988 when Jordan formally renounced its claim to the territory. A 1994 agreement in the ongoing Middle East peace process designated the West Bank as part of the future Palestinian state.

King Hussein I (1935–99), ruled Jordan from 1953 to 1999. Hussein was the great-grandson of Sharif Hussein and a *Hashemite*, a term that refers to the Hashemite branch of the tribe of the Prophet Muhammed, founder of Islam. This bloodline to the Prophet gave him the ability to wield a great deal of power in Middle East affairs, and Hussein used that power to play a double role: he promoted peace between Israel and the Arab world and encouraged economic growth in Jordan on one hand but suppressed dissent and political participation among his people on the other hand.

Although Jordan's Constitution of 1952 established a democratically elected parliament, Hussein banned political parties after 1956 and exerted a near-complete level of authority beginning in the late 1960s. He insisted that these moves were necessary because of the region's ongoing tensions and promised to introduce democratic reforms once questions of Palestinian sovereignty and control of the West Bank were resolved. He kept this promise by allowing elections to take place in 1989 and by allowing political parties to reorganize in 1992. In following years, however, Hussein clamped down on dissent once again, causing many critics, such as Palestinian scholar Edward Said, to suggest that Jordan is not the stable, benign country that Hussein wanted it to appear to be.

Many of Jordan's recent acts of restriction are linked tensions in the Mideast peace process. In the early 1990s, Hussein played a key role in resolving the issue of Palestinian sovereignty with Israel. Following the assassination of Israeli Prime Minister Yitzak Rabin, a conservative leader, Benjamin Netanyahu rose to power in 1996. Netanyahu showed less willingness to support peace, and Hussein feared that criticism of the Israeli leader might undo the work he and Rabin began. Hussein realized he could not rule forever and did not want a failure in the peace agreement to tarnish his legacy. In 1997, he called for many restrictions on Jordan's mass media and gave the government the power to shut down newspapers that printed unfavorable comments about the royal family or other Arab leaders. He insisted that such rules were needed to achieve a lasting peace in the Middle East. Critics, however, contend that this may undo the reforms that began in Jordan a decade earlier. In a 1998 essay for *Newsweek* magazine, Said noted that "despite the creation in 1989 of democratic reforms ... Jordan is really just a quasi-police state presided over by a seemingly benevolent despot."

Dissent grew in the 1990s as Jordan's economy worsened. Much of Jordan's income traditionally came from workers who would travel to work in other countries, particularly the oil-rich Gulf States of the Arabian Peninsula. Before 1990, 340,000 Jordanians worked overseas; 270,000 of whom were employed in Gulf States. Gulf States also provided Jordan with lucrative subsidies to promote economic growth. These subsidies ended when the Gulf War began in 1991 and Hussein threw his support behind Iraq. The war also forced thousands of migrant workers to return to Jordan, straining social services and causing unemployment to soar. Financial difficulties forced Jordan to seek help from the International Monetary Fund, and the strict conditions imposed by that organization caused prices of food and other necessities to spiral.

These difficulties increased as many Palestinians in the country opposed Hussein's efforts to seek closer ties to Israel. This can be seen in how Ahmed Daqamseh, a Jordanian soldier, was treated after being convicted of murdering seven Israeli girls in March 1997. Residents of Jordan wrote songs about the soldier and left flowers at his jail, while lawyers fought to defend him in court.

As peace talks stalled, Hussein also began to show impatience with Israel. "The peace process is not dependent on Jordan alone," Hussein said in a 1997 news conference. "The peace process and its continuation depends on whether there is a real wish from the Israeli side at the responsible level to progress towards peace."

The Khazhan (treasury), built into the sandstone cliffs of the ancient city of Petra, survives into the twentieth century. (EPD Photos/CSU Archives)

Hussein was succeeded by his thirty-seven-year-old son, Crown Price Abdullah, who has promised to support the peace process. Many proponents of the peace talks viewed posititvely the election of Ehud Barak as Israeli prime minister in May 1999. Arabs viewed Barak's predecessor, Netanyahu, as too dependent upon the support of groups opposed to the peace process such as Israeli settlers living in the occupied West Bank, and religious Jews who favored the creation of a "Greater Israel".

Timeline

2000 B.C. Jordan River area is settled

Neolithic remains found in Jericho put the region's history as beginning around 7800 B.C. Jericho, located in what is now known as the West Bank, has the reputation of being the world's lowest city. It is 825 feet (251 meters) below sea level.

1300 B.C.-A.D. 636 Waves of invaders conquer Palestine

Egyptians conquer Palestine in 1600 B.C. Palestine, which was named Philistia after the Philistines who occupy the southern coastal areas, includes the land between the southeastern end of the Mediterranean Sea and the Lebanon Mountains.

The West Bank falls under domain the of King David, a Hebrew, about 1000 B.C. and becomes part of the Kingdom of Judah. After periodic invasions by various groups, the Macedonian king, Alexander the Great, conquers Syria and Palestine. His conquest introduces Hellenistic (a blend of Greek and Near Eastern culture) civilization to the region.

Other invasions follow, and by about 100 B.C. the area is part of the Roman Empire. A flourishing civilization develops along the East Bank of the Jordan River. Meanwhile, the Nabataean kingdom in southern Jordan forges an alliance with Rome and develops a distinctive culture which blends Arab, Roman and Greek elements. This kingdom establishes Petra as its capital and builds its structures, including the ornate Khazen or treasury, from red sandstone cliffs.

As Islam spreads, Muslims begin to invade the region. By 636, Arab rule is firmly in place, and Jordan becomes predominantly Arab and Muslim.

1100–1291 Crusaders arrive

As the influence of Islam spreads, Christians in Europe organize a series of assaults to regain control of Jerusalem, Palestine and much of the Levant (the lands along the eastern shores of the Mediterranean Sea). These military campaigns last through the fourteenth century.

Crusaders arrive in Palestine in 1095 and retake Jerusalem from Muslim rulers. Jerusalem falls to Muslims in 1187. Led by the Egyptian sultan, Saladin, Muslim forces push back the Crusaders at Qalat er-Rabad, or Ajlun Castle, an area north of Amman, the present-day capital of Jordan. By 1291, the Mamlukes, a highly-trained military group, push the Crusaders out of the Middle East.

A.D. 1517–1917 Ottoman era begins

After the Ottoman Turks gain control of the area, they fold lands east of the Jordan River into an administrative division known as a *vilayet,* based in Damascus, Syria. The West Bank becomes part of the Jerusalem *sanjak,* a smaller unit, within the vilayet of Beirut in present-day Lebanon.

The Ottoman Law of 1913 establishes Shariah courts to deal with legal matters that pertain to the Muslim community. Some of these procedures are used in present-day Jordan.

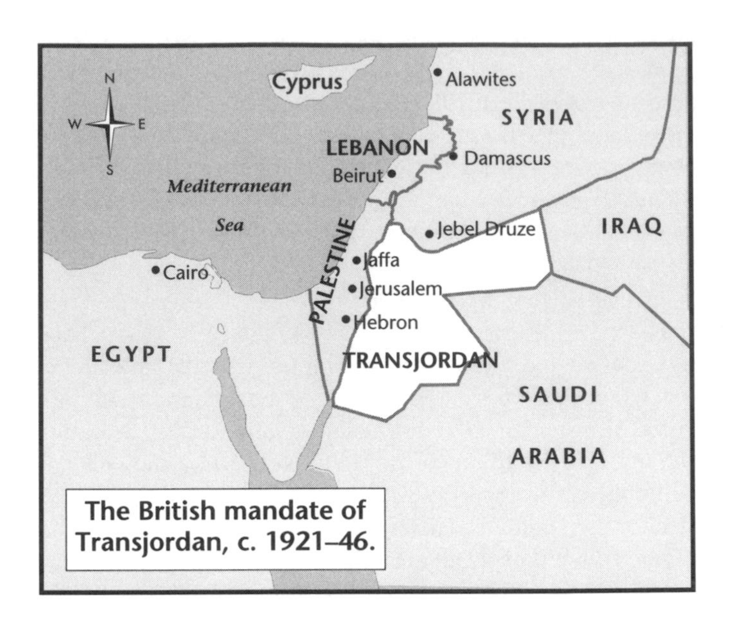

The British mandate of Transjordan, c. 1921–46.

1880 Amman is resettled

This city, the current capital and business center of Jordan, is established by Muslim Circassian refugees from Russia. Ottoman rulers send the Circassians to the region to build a buffer zone along their empire's southern border.

An earlier settlement that dates to 4000 B.C. falls into disrepair after A.D. 1200.

1916: May 16 Arabs revolt against Turks

As World War I is waged, a British leader promises to help Sharif Hussein, the great-grandfather of King Hussein I, create fully independent Arab states if Hussein helps to overthrow the Ottoman Empire. Sharif Hussein is a descendant of the Prophet Muhammad of the Banu-Hashem clan, and the Hashemite dynasty that he establishes stems from that relationship. He proclaims an Arab Revolt, and with help from the British, the revolt frees Palestine, Syria and Lebanon from Turkish forces.

1920–23 Britain gains ruling authority

When World War I ends, the League of Nations develops a mandate system in which parts of the Middle East are divided into administrative regions. Through this system, Great Britain receives a mandate to govern the area that includes Jordan, Israel, the West Bank, Gaza, and Jerusalem.

Britain separates Transjordan from Palestine and establishes a semi-autonomous region known as the Emirate of Transjordan. Britain installs Hashemite Prince Abdullah, son of Sharif Hussein, as emir of Transjordan while the governing of Palestine remains in the authority of a British high commissioner.

Abdullah settles in Amman and establishes it as the capital of Transjordan.

1930 Emir cultivates painter

Jordan's pre-modern history shows few distinctive artistic trends. This begins to change as artists migrate to Transjordan. Among these artists is Ziyaeddin Suleiman (1880–1945) who moves to Amman from Turkey. He paints Jordanian landscapes and portraits and becomes a friend of Abdullah who persuades him to stay in Jordan. Suleiman agrees to do so, and in 1938 he exhibits his work at the Philadelphia Hotel in Amman.

1946: May 25 Jordan gains independence

With its independence from Great Britain, Jordan is established as a constitutional monarchy and Abdullah is proclaimed king.

1948: May 14 Arab-Israeli war begins

Israel is established as the homeland for Jews. Transjordan joins other Arab nations in supporting Palestinians who oppose Israel. War breaks out, and Abdullah gains control of the West Bank, claiming this land was part of the territory the British had promised his father. An armistice agreement with Israel gives Transjordan control of the West Bank.

The manner in which Abdullah receives the West Bank is unclear. Some scholars, such as Palestinian scholar Edward Said, suggest that Abdullah gains the West Bank as a reward for cooperating with Zionist nationalists who help build the state of Israel. Regardless of how the deal unfolds, Abdullah's land-grab angers many Arab nations who see the move as unduly ambitious.

The war also brings nearly 1 million Palestinians into Transjordan.

1950: April 24 Country is renamed

Transjordan holds elections and is renamed the Hashemite Kingdom of Jordan. Jordan sets up three governing districts on the West Bank: Nablus, al-Khalil, and al-Quds, which includes Jerusalem. The United States recognizes Jordan's right to administer the area but insists that ultimate authority over the West Bank must result from a future agreement.

1951: July 20 Abdullah is assassinated

A Palestinian Arab kills Abdullah in Jerusalem. His son, Talal, becomes king but is declared unfit to rule because of a mental illness, and his son, Hussein I, is proclaimed Jordan's ruler. A regency is installed to rule until Hussein turns 18 in 1953.

1952: January 8 Constitution takes effect

According to Jordan's constitution, executive authority is vested in the King and a Council of Ministers. The King signs and implements all laws and may veto a measure passed by Jordan's National Assembly.

A prime minister leads the Council of Ministers, whose members are appointed by the King. Jordan's legislative body—the National Assembly—includes a Chamber of Deputies, with eighty members, and a Senate with forty members. Chamber members are elected to four-year terms. The King appoints Senators to eight-year terms.

1953: May 2 King Hussein takes office

1955 Village becomes part of urban city

As population increases in Al-Barha, this village in northern Jordan becomes part of Irbid, the country's second largest city. This development is part of a trend that reflects the urbanization of Jordanian society, but it also shows the persistence of traditional household and family patterns.

Anthropologist Lisa M. McCann's study of Al-Barha suggests that a typical household includes an elderly married couple, their unmarried children and married sons, sons' wives and sons' children—a pattern that has prevailed for more than a century. McCann, however, argues that the persistence of this household structure has more to do with economics than tradition. Because rental costs are high, families can enhance their income by forming joint households. McCann's findings suggest that this household pattern is typical in Jordan and many other Arab nations.

1955: December 14 Jordan joins United Nations

1957: April 25 Hussein abolishes political parties

After a group of militants advocating a pan-Arab nationalism allegedly attempt a coup, Hussein abolishes all political parties. Candidates are allowed to run for office only after passing through a screening procedure run by Hussein's Interior Ministry.

1958: July King Faisal II of Iraq is assassinated

A military coup leads to the death of King Faisal II of Iraq, Hussein's cousin. Hussein fears aggression from Syria and turns to the West for support. British troops arrive in Jordan from Cyprus.

1960–70 Port of Al-Aqaba is developed

Although Jordan had begun developing this port after the 1948 Arab-Israeli war, full scale work on this project starts during this period. The port on the northern edge of the Gulf of Aqaba, an arm of the Red Sea, is Jordan's only outlet to the ocean. It separates Egypt's Sinai Peninsula from northwestern Saudi Arabia and becomes a key transit point for Jordan's phosphate industry.

Aqaba dates to 1000 B.C. and was part of ancient Bedouin caravan routes. It also has a history of being a port for Muslim pilgrims who are traveling to Mecca, the birthplace of Muhammad the Prophet and Islam's most sacred city. Aqaba became less prominent in this respect, however, when the Suez Canal opened in 1869.

1966 Government encourages arts

Jordan establishes a Department of Culture and Arts to encourage the development of theater, literature and fine arts.

1967: June 5–11 Arab-Israeli war breaks out

As the Arab nations focus their attention on Israel, terrorist raids often are launched from Jordan. Israel retaliates against these actions, and the growing presence of the Palestinian Liberation Organization leads to terrorist raids within Jordan.

Jordan joins Syria, Iraq and Egypt in fighting against Israel in 1967. During the six-day war, Israel captures the West Bank, the Golan Heights in Syria, and Sinai Peninsula in Egypt.

The loss deprives Jordan of rich agricultural land and hurts its tourist industry. The war brings more Palestinian refugees to Jordan. With the refugees comes a rise in the power of Palestinian freedom fighters known as the *fedayeen.*

1968–71 Hussein tightens authority

Jordan's constitution gives the King a great deal of power, which Hussein uses to assert a near-complete level of authority. He suspends elections and dissolves the lower house of the National Assembly. He justifies these moves as necessary because of Jordan's ongoing tensions with Israel. He also contends that such control is necessary to maintain peace among increasingly hostile factions within the country.

Clashes between the heavily-armed Palestinian fedayeen and government forces escalate in 1970 into a ten-day war. Hussein's forces suppress the revolt and drive the freedom fighters out of Jordan permanently in July 1971.

1973 Wildlife law is passed

Jordan's government prohibits hunting of birds or wild animals and fishing without a license. The law also prohibits persons from cutting trees, shrubs and plants.

1973: April Women receive the right to vote

1974: October 28 Hussein supports Palestinian accord

Arab nations meet in Rabat, Morocco, and pass a resolution that recognizes the Palestinian Liberation Authority as the "sole legitimate representative" of the Palestinian people on any liberated piece of Palestinian land. Hussein endorses the measure even though it calls into question Jordan's claim to the West Bank.

1977 New Civil Code takes effect

The New Civil Code of 1977 is established to regulate civil procedures in legal matters. Jordan's judicial system consists of four levels of civil and criminal court, religious courts, and tribal courts. While religious courts deal with such personal issues as marriage, divorce, wills, testaments, and the custody of orphans, the civil courts deal with most other legal matters. Jordan's tribal courts lose importance as members of these communities start to use the civil courts.

1979 National handicrafts center opens

Craftspeople and artisans work with the government to open the El-Aydi Jordan Crafts Center, a project that helps revive such traditional crafts as weaving, glass blowing, pottery, basket weaving, embroidery, olive wood carving, and silver jewelry making. Other attempts to revive such crafts follow: the Bani Hamida project promotes carpet weaving among women in the Bani Hamida mountains; the Jerash Handicraft Society trains women to develop such skills as weaving, embroidery and carpentry, and the Salt Traditional Handicraft Center markets traditional crafts.

1979 Egypt and Israel sign peace accord

After Egypt and Israel normalize relations, other Arab nations isolate Egypt from diplomatic affairs. Hussein refuses to participate in talks between Israel and Egypt on the future of the West Bank.

1980 Fine arts gallery opens

Wijdan (b. 1939), a painter and art historian who dedicates her life to promoting art from Arab, Islamic and Third World artists, establishes in 1979 the Royal Society of Fine Arts in Amman. This organization opens the Jordan National Gallery of Fine Arts, known for being the only gallery in the world to extensively collect works by Muslim and Third World artists.

1982: June Israeli forces invade Lebanon

Israel retaliates against PLO rocket attacks from bases in Lebanon by invading that country. A truce between Israel, the PLO, and Syria results in the removal of 14,000 Palestinian and Syrian fighters from Lebanon. The explusion of these fighters brings Jordan into the debate, as Hussein begins to coordinate a peace initiative with the PLO.

1983 Jordan airport opens

Jordan opens the Queen Alia International Airport twenty-five miles south of Amman. It is named after Hussein's third wife, who had died in a plane crash seven years earlier.

1984 Palestinian woman publishes autobiography

Poet Fadwa Tuqan (b. 1917), a Palestinian woman living in the West Bank, describes in her autobiography, *Difficult Journey-Mountainous Journey*, the struggle for a Palestinian homeland and the broader fight among Arab women for reform of their society's restrictive mores.

Tuqan recalls that her father wanted her to write political poetry for Palestinians. But how, asks Tuqan in her autobiography, can a poet write about the world if she is sheltered from it, secluded within her household's walls, forbidden to participate in public life? "The poet must know the world before it can be healed through poetry," she writes. "How else can the political issues be weighed?"

1985: February Jordan and PLO announce accord

Hussein agrees to work with the PLO to develop a "peaceful and just settlement" to the issue of Palestinian sovereignty.

1985 Development plan takes effect

This five-year plan sets aside government funds to upgrade Jordan's road and transport facilities. As a result of this plan, a network of roads connects the country to Syria, Iraq and Saudi Arabia, and all of the country's roads become paved within a decade.

1988 Jordan achieves child immunization goal

Jordan manages to immunize all of its children, surpassing an average rate for the rest of the world by two years. Achievement of this goal shows dramatic improvements in the country's health care system and places Jordan as one of the world's top ten countries in reducing infant mortality.

1988 Jordan gives up claim to West Bank

After Hussein formally renounces his claim to the West Bank, Jordan retains an administrative role in the region until a final settlement on the area's status can be reached.

1989 Jordan signs agreement with IMF

A growing level of debt and steady decline in per capita income prompt Hussein to enter into an agreement with the International Monetary Fund. Jordan reschedules $573 million of debt and raises prices on several commodities.

During this period the official cost of living rises fifty-six percent. Political discontent sweeps the country, which prompts Hussein to begin reforming the country's democratic processes. The first relatively free parliamentary elections since 1956 take place later in the year, and Hussein appoints a commission to rewrite the National Charter.

1990 Parliament ratifies anti-discrimination law

The government approves the Convention for the Elimination of All Forms of Discrimination against Women. The convention and other laws in Jordan allow women the right to work, to take maternity leaves and to run for political office. Asma Khodr, a lawyer and the President of the Jordanian Woman's Union, notes in an assessment that such laws have given women a stronger legal status.

Societal restrictions, however, continue to plague Jordanian women. Khodr points out that a woman cannot travel without her husband's permission and that Arab societies rarely stand up for oppressed or abused women. She points out that Jordan laws excuse men who kill wives, sisters or daughters suspected of committing adultery.

1991 Persian Gulf War begins

Jordan throws its support behind Iraqi leader Saddam Hussein. This angers Iraq's enemies—the United States, other Western nations, and Gulf states such as Saudi Arabia. Many Gulf states cut subsidies they traditionally have provided Jordan.

As a country which depends heavily on worker remittances from Gulf States, Jordan is unable to access money that typically had flowed into the country from Kuwait. In addition, the war deprives the country of its export markets, oil supplies and other foreign aid. Nearly 300,000 Jordanian workers are forced to return to the country, mainly from Kuwait. This causes unemployment to rise and strains on the government's ability to provide basic services. Adding to the difficulties is the arrival of 120,000 Iraqi refugees.

1991 Jordan joins peace talks

Jordan, Syria, Lebanon and Palestinian representatives agree to participate in direct peace talks with Israel. The United States and Russia sponsor these talks, and Hussein's willingness to participate helps repair Jordan's relationship with Western countries.

1992: February Jordan restarts economic reforms

Jordan enters into a new agreement with the IMF and reschedules $771 million of its debt. The changes stimulate Jordan's economy and reduce inflation. Unemployment, however, hovers at twenty to twenty-five percent.

1993: November 8 Parliamentary elections take place

Twenty-two political parties participate in these elections. Parties that support Hussein's economic reforms and pro-Western policies win fifty-four of the available eighty seats. The Islamic Action Front wins eighteen seats, the largest number of any single party.

1994 West Bank's status is established

As part of the Middle East peace process, the West Bank becomes part of an area that eventually will be under Palestinian home rule. Israel and Jordan sign a treaty that ends the state of war that had existed since Jordan's early years of independence.

Hussein emerges from the talks as a leading Middle East peace negotiator, but his relationship with Israel stirs protests in Jordan. Many Jordanians, particularly those of Palestinian descent, are suspicious of Israel's motives in the peace process and fear that Hussein's sympathies may undermine their stake to the West Bank.

1996 Woman becomes first taxi driver

Muyassar Abu Hawa, 53, begins driving a taxi in Jordan after her husband dies of a heart attack, a move that breaks new ground for working women. Fewer than one out of every four women work outside the home, and when Abu Hawa first goes on the job, men tell her that "driving is a man's job." She responds that her husband left her with a family to support and debts that must be paid.

1996: August 16 Rise in bread price triggers riots

Three days after Prime Minister Abdul Karim Al Kabariti announces that the price of bread will be doubled, nearly 2,000 demonstrators clash with police in the southern town of Karak. The protesters burn banks and government buildings, and the army imposes a curfew to restore order.

In *The Middle East*, Kirk Albrecht notes that the price increase would cause many to suffer. Nearly one in three Jordanians live below the poverty line, and bread is a major part of the diet. Prices also have soared on such items as heating and cooking fuel, flour, grain and dairy products.

1998: August King has cancer

Hussein tells his subjects, via a satellite link from the Mayo Clinic in Rochester, Minn., that he has cancer. Although he is receiving treatment, the news causes many confidantes to think about how to hold Jordan together after Hussein's death. The king's brother, Prince Hassan, has been the designated heir for thirty-three years but has done little to win trust among Palestinians who comprise a majority group in Jordan. He also has little loyalty among Bedouin tribes whose members have supported Hussein throughout his reign.

1999: January 19 Hussein names Prince Abdullah heir

King Hussein returns from the United States, where he received treatment for six months for lymphatic cancer. Upon his arrival in Jordan, he annouces that his brother, Prince Hassan, 51, is no longer his designated heir. He designates his eldest son, Prince Abdullah, 37, as heir to the throne. Abdullah is the son of Hussein and his second wife, Toni Gardiner.

1999: January 20 Hussein returns to Mayo Clinic

King Hussein, suffering serious medical symptoms, returns to the Mayo Clinic in Minnesota for further treatment.

1999: February 7 King Hussein dies

King Hussein, after returning to Jordan with a diagnosis that no further treatment was recommended for his cancer, dies. Hussein is succeeded by Crown Prince Abdullah.

Hussein's funeral is attended by twenty-seven presidents, eight monarchs, eight crown princes, fourteen prime ministers, as well as other dignitaries and one million Jordanians. The United States delegation includes President William Clinton as well as former Presidents George Bush, Jimmy Carter, and Gerald Ford. The United States also expresses its support for King Abdullah by extending $300 million in aid to Jordan over the next three years.

Bibliography

"A Not So Loyal Opposition: Jordan." *The Economist*, August 30, 1997, Vol. 344, No. 8032, p. 31.

Albrecht, Kirk. "The Pains of Restructuring." *The Middle East*, October 1996, no. 260, p. 20–21.

Ali, Wijdan. *Modern Islamic Art: Development and Continuity.* University Press of Florida, Gainesville, Fla. 1997.

Badran, Margot and Miriam Cooke. *Opening the Gates: A Century of Arab Feminist Writing*, Indiana University Press, Bloomington, Ind. 1990.

Contreras, Joseph and Christopher Dickey. "The Day After: King Hussein's Second Bout of Cancer Raises Questions About Jordan's Political Future." *Newsweek*, Vol. 132, No. 6, p. 38, August 10, 1998.

Gauch, Sarah. "Stalled Peace's Victim: Free Press," *Christian Science Monitor*. April 28, 1998, p. 1.

Guide to Places of the World, John Palmer, editor. The Reader's Digest Association Ltd., London. 1995.

Hosein, Hanson. Personal correspondence. August 1998.

"Jordan: Women's Struggle for their Rights." *WIN News*, (Women's International Network) Spring 1997, vol. 23 no. 2, p. 57.

"King Hussein of Jordan, remarks following the Israeli assassination attempt in Amman, 5 October 1997." *Journal of Palestine Studies,* Winter 1998, Vol. 27, No. 2, p. 149–50.

Jehl, Douglas. "As Jordan's Troubles Grow, Uneasy Lies Its Crown," *New York Times*. March 22, 1998, p. 8.

McCann, Lisa M. "Patrilocal Co-Residential Units (PCUs) in Al-Barha: Dual Household Structure in a Provincial Town in Jordan." *Journal of Comparative Family Studies*, Summer 1997, vol. 28, no. 2, pp. 113–235.

Peretz, Martin. "King of Peace: Jordan's King Hussein," *The New Republic*, Oct 13, 1997 Vol. 217, No. 15, p. 54.

Said, Edward W. "Deconstructing King Hussein," *Newsweek*. Aug. 10, 1998, Vol. 132, No. 6, p. 40.

Shahin, Mariam. "Cracks Become Chasms," *The Middle East*. December 1997, No. 273, p. 13–14.

Kazakstan

Introduction

Kazakstan, officially the Republic of Kazakstan, is the second largest of the Commonwealth of Independent States (CIS), once part of the former Soviet Union. Only the Russian Federation is larger. With an area of 1.05 million square miles (2.7 million square kilometers), Kazakstan is twice the size of Alaska and roughly the same size as western Europe. Its boundaries reach from the Caspian Sea on the west to China on the east. It is bordered on the north by Russia and on the south by the republics of Kyrgyzstan, Uzbekistan, and Turkmenistan. Kazakstan's geography ranges from the Karagiye Depression east of the Caspian Sea at 433 feet (132 meters) below sea level to the lofty peaks of the Altai and Tien Shan mountain ranges on its border with China. Kazakstan's climate is generally dry, with very cold winters and very hot summers. Its capital city is Alma-Ata (Almaty), located in southeastern Kazakstan. Plans are to move the capital to the city of Akmola (formerly Tselinograd) in north central Kazakstan by the year 2000.

This vast desert, steppe, and mountain region of Central Asia has long been home to Turkic-speaking tribes, nomads who traversed the flat, low-lying plains and fertile grasslands on horseback, herding their sheep to summer and winter pastures and taking their homes—the sturdy felt-lined round tents called *yurts*—with them. Heavily influenced in both population and lifestyle by a long Mongol (or Tatar) rule beginning with the great conqueror Genghis Khan in the early thirteenth century A.D. and continuing until the Russian prince Ivan the Great threw off the "Tatar yoke" in 1480, Kazakstan's native people, the Kazaks, are Turkic-speaking people with Mongol features. Many even today hold to the ancient Central Asian lifestyle of sleeping in a yurt and drinking the fermented mare's milk drink called *koumiss* as a dietary staple. The newly independent nation's emblem, the *shaneraq,* is the wooden wheel holding together a yurt frame.

The influx of Russian and other settlers into Kazakstan that began in the sixteenth century continued throughout imperial Russian rule under the czars of the Romanov dynasty and on into the early twentieth century.

Beginning in 1906, the self-sufficient nature of the Kazaks caused much national resistance, first to Russian and then to Communist rule. This resistance culminated in the deaths or exile of thousands of Kazaks and the suppression of their lifestyle. In the 1950s and 1960s, Communist-leader Nikita Khrushchev's Virgin Lands program converted huge tracts of Kazakstan grazing land to wheat and cereal production and brought Russians and Ukrainians by the train loads into northern Kazakstan to farm the fertile black soils. The Kazak nomads, forced to settle on farms year-round and give a large portion of their livestock to the Soviet government, rebelled, often slaughtering their animals rather than turning them over to the government. This action led to many being exiled to China or Mongolia. Thousands of Kazaks starved to death, unwilling and unable to adapt to the new lifestyle.

Eventually, the Kazaks came to be a minority in their own homeland. Today, however, the Kazaks have regained a plurality of forty-six percent, compared with the Russian population representing thirty-eight percent of the total 16.7 million people in Kazakstan (a 1995 estimate). The remaining ethnic groups are Ukrainian, German, and more than 100 other nationalities, including Chinese. Many Kazak citizens have moved to the larger cities, causing a decline in rural population. Most heavily populated is the capital city of Alma-Ata, in southeastern Kazakstan, with about 1.1 million people. Other major cities are Karaganda (Qaraghandy), Semipalatinsk (Semey), and Pavlodar in northeastern Kazakstan, and Chimkent (Shymkent) and Dzhambul (Zhambul) in southern Kazakstan.

Although Kazak was declared the country's official language in 1995, seventy percent of the population speaks Russian and only about thirty-six percent speaks Kazak. The Kazaks are heavily "Russianized," many with Turkish, Arabic, or Persian surnames with Russian endings, and many families in Kazakstan are "international."

Kazakstan was one of the last Soviet republics to declare its independence, waiting until December 16, 1991, just before Soviet President Mikhail Gorbachev stepped down, finalizing the dissolution of the Union of Soviet Socialist Republics (USSR). Becoming a member of the Commonwealth of Independent States (CIS) on December 26, 1991, it adopted a constitution on January 28, 1993 which guarantees

the same basic freedoms that people in the United States and other democratic nations have. With universal suffrage beginning at age eighteen, Kazakstan's voters elect a president for a five-year term, who serves as head of state and appoints a prime minister and members of a council of ministers. The prime minister in turn appoints certain other ministers. Under new election rules established by Kazakstan's president Nursultan Nazarbayev (b. 1940) in 1995, the sixty-seven members of the lower house of Kazakstan's parliament are elected by the voters on a nonpartisan basis. Some of the members of the upper house are elected and some are appointed by the president. The country's ruling political party is the People's Unity Party of Kazakstan.

President Nazarbayev is the central figure in Kazakstan's politics, having been the country's leader since before the dissolution of the Communist Party and the USSR. He was elected, unopposed, in December of 1991, reelected in 1995, and again in January 1999.

After a long struggle for economic and social reform, aided by billions of foreign dollars, Kazakstan today has the highest per capita income in Central Asia, a stable political environment, and a strong industrial and agricultural base. Numerous tax breaks and other incentives have attracted U.S. and other investors in large numbers. Oil and gas are also major attractions for foreign investment. Second only to Russia among the Soviet Union's oil producing countries in 1989, Kazakstan is believed to contain the world's largest untapped oil and gas reserves (about 35 billion barrels of oil), according to President Nazarbayev. Its current annual output is about 22 million tons of oil, or about 500,000 barrels per day. Kazakstan aspires to be one of the world's leading oil producers.

Kazakstan has four basic economic regions: the eastern, with its diversified manufacturing centered around the capital city of Alma-Ata (Almaty); the desertlike, sparsely populated western region, with huge developed and undeveloped oil including the untapped Tengiz field; the central and southern regions, with sprawling cotton fields and copper, lead, and zinc mines; and the most-developed northern region, with one of the world's major wheat-producing areas as well as heavy industry, such as coal and metallurgy, and mining of gold, copper, iron ore, and other metals. Mining is an important industry for all of Kazakstan. In addition to the metals already mentioned, the country produces manganese; nickel; chromite, the chief source of chromium (Kazakstan and South Africa hold two-thirds of the world's resources of this metal); cobalt; uranium; molybdenum; bauxite; titanium; platinum; and vanadium. Kazakstan, where raising sheep has been a traditional way of life, is also one of the world's leading producers of wool.

Along with a promise of prosperity in Kazakstan's future comes the ghost of environmental problems inherited from the time of Soviet rule. Years of heavy crop irrigation have so depleted the waters of rivers (Syr Darya in Kazakstan and Amu Darya in neighboring Uzbekistan) that the surface area of the Aral Sea, an inland sea on the border between the two countries that was fed by the rivers, has shrunk by more than half. Its receding waters have left behind salt flats heavily contaminated by pesticides, defoliants, and fertilizers used in cotton fields. High winds raise salt storms that carry the toxic dust over local farms and villages, where many people are experiencing immune system disorders, respiratory diseases, throat cancer, and hepatitis. The area's drinking water is also scarce and contaminated. In the northeastern part of Kazakstan, the Soviet Union's nuclear weapons industry has left a legacy of birth defects, cancers, and other maladies caused by radioactive fallout from the Semipalatinsk (Semey) nuclear test site near the city of Kurchatov, where atomic and nuclear weapons have been detonated both above and below ground, since 1949.

Also in Kazakstan's northeast, particularly in the city of Ust Kamenogorsk (Oskemen), heavy metals contaminate the soil, water, and air, caused by years of pollution from lead and zinc, titanium and magnesium smelting (for the purpose of turning uranium into nuclear reactor fuels), and fallout from nuclear testing. Ust Kamenogorsk (Oskemen) is estimated to be the most heavily polluted city in the former Soviet Union.

Russia and the United States are both involved in helping Kazakstan clean up its nuclear waste and other pollutants. All former Soviet nuclear warheads have been dismantled and transported to Russia, and the United States recently purchased uranium from nuclear weapons for reprocessing. However, progress on cleaning up the environment in Kazakstan is slow, and in the meantime the people suffer from severe health problems.

Social services in Kazakstan are improving, including health care and nutrition for women and children, which has been well below the worldwide average in the developing country. Contraception was all but unavailable as recently as the 1980s, and abortion has been the most popular form of birth control.

Nearly 100 percent of Kazakstan's citizens are literate. Educational reform continues, with plans for building new schools, training more teachers, buying more books, and strengthening national and local educational management. Moreover, Kazakstan has over 12,000 libraries, with the Kazak National Library in Alma-Ata being the largest. It has fourteen reading rooms and can seat more than 1,300 people.

World interest in Kazakstan is growing. In 1994, sixteen different countries and eleven international organizations committed a total of $1 billion in addition to the United States' $3.96 million to help move the country forward into the twenty-first century. There are nine current U.S. economic agreements with Kazakstan and five major international trade agreements.

As Kazakstan moves from a vast, sparsely populated Soviet republic to a prosperous, democratic nation, many of its native people hold on to their history and customs. Kazaks whose forebears were exiled to Mongolia in previous years are beginning to return to their homeland, many of them settling in yurts on the steppes as they await a chance to return to raising sheep and other livestock and enjoying familiar nomadic traditions. This simple way of life is one they consider worth retaining in a fast-developing modern land.

Timeline

300,000–150,000 B.C. Migration begins

Bands of archaic Homo sapiens begin a migration from the Middle East and the Black Sea regions into the *steppes* (flat, expansive, treeless lands) and forest lands of central Asia during the middle Paleolithic Period (Old Stone Age).

200,000 B.C. Neanderthals live in Europe

Neanderthals appear in Europe. Fossil remains found by present-day archaeologists show Neanderthals migrate as far east as Teshik-Tash, a site near the Amu Darya river in what is now Uzbekistan. These early peoples live at the edge of the Eurasian ice sheets during the Pleistocene Epoch, or Ice Age, seeking shelter in caves or beneath overhanging rocks. Physically powerful because of the harsh environment in which they have evolved, Neanderthals are skilled toolmakers; hunt in groups, killing large animals at close range; have social organization and carry out rituals, including burying their dead; and probably speak a very primitive language.

30,000 B.C. Cro-Magnons replace Neanderthals

Neanderthals are replaced by anatomically modern humans, referred to by paleontologists as Cro-Magnons.

33,000–9,000 B.C. First cultural revolution

Small hunter-gatherer groups of modern humans in Europe and Asia begin to create likenesses of themselves and of animals and begin to mark the passage of time. Cave paintings, ivory carvings, elaborately designed tools, and primitive forms of jewelry are among archaeological evidence found in modern times to show that mankind experiences a cultural revolution during this period, known as the Upper Paleolithic.

A great Upper Paleolithic culture is known to have spanned the territory from what is now Austria and the former Czechoslovakia eastward into the present-day Commonwealth of Independent States (former Soviet Union). As far east as Malta in Siberia, archaeologists have discovered an ivory plaque incised with dots in a spiral design believed to indicate the passing of days. Small sculptures and carvings of the female body, often called Venus or fertility goddess, have been found throughout this territory.

21,000–18,000 B.C. The mammoth hunters

Herds of woolly mammoths, horses, and bison roam the steppes of eastern Europe and central Asia. Upper Paleolithic hunters choose the mammoth as one of their primary sources of food, clothing, and shelter.

More than thirty modern archaeological digs (beginning in 1879) revealing the hunters' way of life have been completed near the village of Kostenki, Russia, on the river Don between Kazakstan and Ukraine.

18,000 B.C. Human migration southward

The people in the region of modern-day Kazakstan move south as glaciers advance and the climate becomes more harsh. With southward movement, they experience cultural change.

9000 B.C.–5000 B.C. The Mesolithic period

During the transitional period between the Old Stone Age (Paleolithic) and the New Stone Age (Neolithic), the earth's climate warms, and the great ice sheets begin to retreat. Groups of hunter-gatherers move farther and farther north in central Asia, into forested zones. Archaeologists have found evidence of Mesolithic culture as far north as Kunda, Estonia. Estonia (formerly the Estonian Soviet Socialist Republic) borders Russia on the west and lies between Finland and Poland, on the Baltic Sea.

5000–2000 B.C. The Neolithic period

The Neolithic period brings the earliest forms of agriculture and the raising of livestock. Most groups still depend mainly on hunting, fishing, and gathering for food; however, people are beginning to settle and form communities, especially in the southern regions of central Asia, where the climate and soils are favorable to agriculture. They glean knowledge from neighbors in the Middle East, where cultivation is more advanced and sometimes transport plants and animals with which to start crops and herds of their own.

During this period, people in central Europe begin to use copper to make tools and weapons, although beads, pins, and awls were first made from copper in Iran and Turkey in about 8000 B.C.

2000–1500 B.C. Chariots may have been in use

Twentieth-century excavations east of the Ural Mountains uncover evidence of two-wheeled chariots. These may be the earliest people to make and use chariots in the world.

3000–1000 B.C. The Bronze Age and the Iron Age

Toward the end of the Neolithic period, groups of people in Europe, Asia, and the Middle East begin to use tin and copper to make bronze, which they at first use only for ornamentation but later use to make tools and weapons. The pastoral-agricultural economy in central Asia becomes well developed, and metal workers are particularly skilled in the Caucasus and Transcaucasus regions (southern Russia, Georgia, Armenia, and Azerbaijan).

1000 B.C. Various cultures begin to use iron for tools and weapons

People of the Caucasus and Transcaucasus areas are the first to develop this technology in central Asia. The use of iron spreads rapidly eastward across the continent to Siberia, encouraging trade and the further development of agriculture.

700–100 B.C. The Scythians

The Scythians, a fierce, nomadic culture with horses and a love of wine and gold, migrate from Asia into what is now western Russia, Ukraine, and eastern Europe. They speak a Persian tongue, have a special reverence for fire, and, according to the ancient Greek historian Herodotus, drink from the skulls of those they kill in battle. Archaeological evidence shows women warriors are among the ranks that conquer the area from the Danube River in present-day Romania to the Don River and the Caucasus Mountains in southern Russia. This area becomes known as Scythia.

The Scythians bury their kings and other important persons in *kurgans,* or burial mounds, along with horses, family members, servants, tools, jewelry, and provisions for the afterlife. The main tomb lies several feet beneath the mound and is often divided into chambers.

Most of the Scythian burial mounds found by archaeologists are located along the Dnieper River in southern Ukraine. Some are as high as sixty-five feet. One tomb near the

Dnieper holds treasures in gold jewelry which are discovered by archaeologists in 1971, even after robbers have taken their share. There is archaeological evidence of a Scythian city the Greeks called Neapolis ("new city") Scythica on the Crimean peninsula, to the south of the Dnieper River valley.

The Scythians trade extensively with the Greeks, especially for gold ornaments, wine, and pottery, according to archaeological finds. They provide the Greeks with wheat and a market for their goods. During the third century B.C., however, the Scythians begin to raid Greek cities.

600–200 B.C. The Pazyryk and other cultures

Grouped together under the name Scytho-Siberian, many customs found among the Scythians are shared by dozens of other tribes of nomadic horsemen and shepherds of the steppes of central Asia. They include a culture known to archaeologists as the *Pazyryk* (from a word meaning "mound"). Pazyryk burial mounds, or *kurgans,* very similar to those of the Scythians, have been studied throughout what is now the Autonomous Republic of Altai (Altay), which borders Kazakstan on the northeast. A few mounds have been uncovered in northeastern Kazakstan, near the Altai Mountains.

In 1993, a team of archaeologists discover a Pazyryk tomb dating to about 400 B.C. on the Ukok Plateau in Altai, just east of Kazakstan's border and just north of Altai's border with China. It is the tomb of a woman dubbed "the lady" by the archaeologists. Her frozen body is mummified and reveals two tattoos, one of an elaborately designed deer and the other of a dragonlike creature seen in Scythian art. Horses, food, and many other items of her everyday life are found in the tomb and give archaeologists clues to the Pazyryk culture.

Like the Scythians, the Pazyryks bury horses, food, and other provisions with their dead. They drink *koumiss,* a drink made from fermented mare's milk, and move their herds of sheep to summer and winter pastures. They live in round, felt-covered portable structures—known today as *yurts*—in summer and in log structures during winter. They use wool to make felt for clothing, carpets, and yurt coverings. Their primary foods are meat and dairy products, with some wild herbs, teas, vegetables, and perhaps some grain obtained in trade. Although bits of gold foil adorned "the lady," most Pazyryk tombs yield more treasure in wool, wood carvings, tools, and other items. Evidence shows bronze and iron are both used by the Pazyryks. Stylized drawings and carvings of animals decorate their tools, clothing, jewelry, and other items.

500 B.C. People of Europe

The first Slavonic people in central and western Europe split into eastern, western, and southern groups, with the eastern becoming the largest. These peoples are the ancestors of the groups that establish the Russian state.

300–200 B.C. Usun take over Scythians

The Scythians are gradually overrun by the Usun, allied tribes of Turkic-speaking Mongolic nomads.

108 B.C. Neapolis burns

The Scythians meet their downfall when a Greek army retaliates for Scythian raids on Greek cities, attacking and burning Neapolis.

Other tribes invade from the north, east, and west, laying claim to the lands of Scythia. The present-day Ossetians of southern Russia and Georgia say they are the descendants of the Scythians and a later tribe, the Sarmatians. The Ossetians carry on certain beliefs and customs known to be Scythian, such as reverence for horses, consumption of *koumiss,* and rituals involving fire.

Modern History

A.D. 50–900 Early agriculture

The eastern Slavs begin farming and trading along the Dnieper, Desna, Dniester, Volga, and other rivers in what are now western Russia and Ukraine. Slavonic communities are raided by a number of tribes, including the Huns, the Goths, the Khazars, and the Bulgars from the east and south and the Vikings from the north. The Scandinavians raid and settle the area more than do other tribes. The group of peoples created from intermarriage between Slavs and Scandinavians forms communities that establish a small forest princedom known as Rus.

A.D. 400–600 Altai Turks attack

Turks from the Altai Mountains, just east of today's Kazakstan, attack and diminish the Usun, allied Turkic-speaking tribes of Mongolian nomads.

739 Arab armies enter region

Arab armies cross the Syr Darya river in what is now southern Kazakstan and claim the southermnost part of the Turkic state. Uighurs and Karluks, tribes from the east (the area that is now the Sinkiang Uighur region of China) finally defeat the Turkic peoples.

766 Karluk dominance

The Karluks establish dominance over what is now eastern Kazakstan.

c. 800–4000 Arab influence

Southern portions of the region come under Arab influence, and Islam is introduced.

c. 1020 First reference to Kazaks

The first authenticated reference to the Kazaks is from a Persian explorer. The Kazaks are a branch of the Mongolic/Altai Turkic Kyrgyz people of Turkish origin.

1054 Seljug rule

Kazakstan is part of the region controlled by the Seljuqs, a Turkish dynasty ruling western Asia during the eleventh through fourteenth centuries.

c. 1162 Great conqueror Genghis Khan is born

Genghis Khan (c. 1162–1227), whose original name is Temujin, meaning "blacksmith," is born near the Onon River in northeastern Mongolia. Genghis Khan becomes not only a great warrior but an astute ruler who makes use of the talents of artisans and scholars among the peoples he conquers. He adapts a written language for the Mongols from the Uighurs, Turks of what is now western China. He is tolerant of all religious practices. The khan himself worships Tengri, "the ruler of heaven."

1180–90 Vengeance against the Tatars

In early manhood, Temujin begins to form alliances among local Mongolian clans and launches a vengeance raid on a Turkic-speaking Tatar tribe, members of which killed his father when Temujin was nine years old. Temujin and his men wipe out the Tatars, killing nearly all the men and boys and enslaving the women. The name Tatar, or Tartar in Europe, is used thereafter to refer to the Mongol invaders who conquer central Asia and eastern Europe.

1206 Temujin is proclaimed *Genghis Khan*

At a great assembly of his people, the Mongols, Temujin is proclaimed *Genghis Khan,* which means "Universal Ruler" or "Oceanic Ruler." He organizes a powerful army and begins a conquest against the Tanguts (a Tibetan people) of Hsi Hsia on the Yellow River in what is now northern China. By 1210, Genghis Khan and his army have defeated the Tanguts, and the great khan moves eastward to defeat the Jin dynasty at Zhongdu (later Beijing) by 1215.

1215–27 Mongol empire is created

Genghis Khan builds a capital city at Karakorum on the upper Orhon River in central Mongolia then moves on to take Kara-Khitai in western Mongolia in 1218. From there his army advances into the wealthy Muslim empire of Khwarizm in northern Persia (along a great trade route from China to Europe known as the Silk Road), founded by Turks of the Seljuq dynasty. Genghis Khan and his warriors—with their lightweight protective armor; *mangonels* (devices used to throw heavy stones to break down walls); clever tactics in battle; and their use of naphtha, a petroleum and sulfur mixture used to make firebombs—overrun the great cities of Utrar (on the Syr Darya in present-day southern Kazakstan), Bukhara, Samarkand, and Urgench (all in present-day Uzbekistan), and Merv (in present-day Turkmenistan). From there the great khan's army conquers territory in Pakistan, Afghanistan, Iran, Iraq, and southern Russia, then moves back eastward to the central Asian steppes and home to Mongolia.

Genghis Khan's army, 10,000 men at its height, moves in a vast and orderly procession during its conquests. With the nomadic warriors are enormous herds of pack animals, spare mounts, sheep, and goats. Approximately 40,000 women and children accompany the men, taking along their collapsible round tents called *gers* for shelter. Genghis Khan's large golden ger, however, remains assembled and is carried on a wagon pulled over the steppes by dozens of oxen. The women play an important role by milking animals five times a day (*koumiss*, a drink made from fermented mare's milk, is a staple for the Mongols), killing the wounded enemy on the battlefield, and collecting stray arrows.

1227 Death of Genghis Khan

Genghis Khan dies, either from typhus or after a fall from a horse, according to historical accounts, as a second campaign against Hsi Hsia begins. He orders the complete destruction of this kingdom from his deathbed and names Ogodei (1185–1241), his third son by his principal wife, Borte, as his successor. Ogodei takes the title *khagan,* or "khan of khans."

Genghis Khan's body is transported back to Mongolia by his soldiers, who kill everyone encountered along the way to keep his place of burial from being discovered. The great conqueror is believed to be buried near a mountain called Burkhan Khaldun near the Onon River in eastern Mongolia, but his remains are never found.

1233–67 The khanates, including the Golden Horde

After the death of Genghis Khan, his son and successor Ogodei raids and conquers the Jin dynasty of northern China in 1233. The Song dynasty to the south falls to the Mongols by 1235, and in a push to expand the empire begun by his father, Ogodei, from his throne in Karakorum, sends his armies westward across central Asia to the Volga River in the Russian Principalities. With the conquest of much of Russia and Ukraine complete by 1240, the Mongolian army, led by Ogodei's nephew Batu Khan (d. 1255), son of Genghis Khan's eldest son, Jochi (d. 1227), rides into Hungary and Poland in 1241, aided by the general Subotai. News of Ogodei's death in Karakorum stops Batu's army just short of a conquest of Vienna, Austria, in 1242.

At the death of Genghis Khan, the Mongolian empire was divided among the khan's four sons—Jochi, Chagatai (d. 1241), Ogodei, and Tolui (d. 1231)—into four khanates, or fiefdoms. By the time his grandsons rule, these khanates are much larger. Ogodei's son Guyuk (1206–48) rules the great empire in Mongolia and China until his death, when Tolui's son Mongke (r. 1251–59) takes the throne.

Chagatai is given the khanate west of Mongolia and China, from the Altai Mountains bordering Kazakstan on the northeast to the Amu Darya in present-day Uzbekistan. Southeastern Kazakstan is included in the Chagatai Khanate. Chagatai's grandson Kara Hulegu is his successor.

Tolui receives the conquered territory in western Asia and Persia. By the time his son Hulegu (c. 1217–65) proclaims himself *Il-khan,* or subordinate khan loyal to the great khan Mongke, the Ilkhanate has expanded to include the area from Pakistan to Turkey.

Batu Khan is heir to his father's portion of Genghis Khan's lands—the huge western territory of the steppes that includes land from Lake Balkhash, in present-day eastern Kazakstan, to Hungary. Batu builds a capital at Sarai (near present-day Leninsk), on the Volga River in southern Russia. His empire becomes known as the Golden Horde, so named because Batu rules from a huge, gold brocade-lined ger; "horde" comes from the Mongol word *ordu,* meaning camp or fiefdom. The Golden Horde is also known as the Kipchak Empire, Kipchak being a name for the Mongol-Turkic peoples known as the Tatars after merging with the local tribes of central Asia.

The system under which the Mongols tax the people of the Golden Horde is known to historians as the *Tatar yoke.* This yoke is finally thrown off by Russian grand prince Ivan III Vasilyevich (1440–1505), called Ivan the Great, in 1480.

Batu Khan is succeeded by his brother Berke (d. 1267), who rules the Golden Horde from 1257 to 1267 and is the first Mongol ruler to adopt the Islamic faith. When in 1260 the Egyptian Mamluk dynasty deals the Mongols their first real defeat, at Ain Jalut in what is now Israel, Berke supports the Mamluks. The native peoples of Kazakstan are Islamic largely because of the influence of the Mongols. In modern times, most Kazaks are Sunni Muslims.

1328 Moscow becomes seat of power

The Mongols assign administrators in the Russian city of Moscow, in the forested lands of western Russia, the task of collecting taxes from other principalities in the western portions of the Golden Horde. Moscow thus begins to assume a leadership role.

c. 1395–97 Timur attacks Golden Horde

The Golden Horde is routed by Turkic conqueror Timur (1336–1405), or Timur the Lame (also known in English as Tamerlane), whose capital is at Samarkand in the area known as Transoxiana. Transoxiana lies north of the Oxus River (now the Amu Darya) and northeast of northern Iran (formerly Persia), in what is now Uzbekistan. A barbaric conqueror, Timur invades territory from India to Moscow and is on his way to a conquest of China when he dies.

1440s Golden Horde is divided

The weakened Golden Horde is divided, probably by the ruling khan for his heirs, into four independent khanates: Crimea, Kazan, Astrakhan, and Siberia (or Sibir). What is now Kazakstan belongs to the khanate of Siberia.

1480–1502 Golden Horde is vanquished

The city of Moscow is the center of a Russian state that is growing more powerful.

After a number of attacks on Moscow, the Mongols are defeated by the Russian princes in 1480, after which Ivan III Vasilyevich, or Ivan the Great, denounces the Tatar yoke. By 1502, with the death of its last khan, Saiyid Ahmad, the Golden Horde comes to an end.

1538–80 Kazaks are divided

The Kazak people themselves become divided into three large territorial groups, or hordes: the Greater (or Elder) Horde of southern Kazakstan; the Central (or Middle) Horde of northern and eastern Kazakstan; and the Lesser (or Younger) Horde of western Kazakstan. Each territorial group is subdivided into races, tribes, sections, branches, and family units called *auls.* These divisions retain their influence on Kazak society into modern times. The three dialects of the Kazak language, which belongs to a Turkic branch of the Uralo-Altaic language family, correspond to the three hordes.

1552–81 Ivan the Terrible conquers territory of former Golden Horde

Ivan IV Vasilyevich (1530–84), called Ivan the Terrible, conquers the khanates of Kazan and Astrakhan (in what is now western Russia) in 1552–56. Ivan wins the khanate of Siberia in 1581 by employing as warriors the Cossacks (from the Turkish *kazak,* meaning free person).

The Cossacks, a native Russian and Ukrainian farming people settling the Don, Dnieper, and Ural River valleys in the fourteenth and fifteenth centuries, fall under the authority of the Russian *czars* (emperors from the Slavonic word for Caesar), who use Cossack men, ages 18 to 50, in a special Russian cavalry to conquer and control new lands and as frontiersmen to guide settlers into unknown areas.

The Cossacks fight the Mongols (Tatars) in the Crimean Peninsula and in the Caucasus (the region between the Black and Caspian Seas, including the Caucasus Mountains). After aiding in the conquest of Siberia, the Cossacks of the Don River valley—known as the Don Cossacks—lead expeditions to colonize Siberia. The Cossacks become a people of the steppes, living mainly in the area bounded on the west by territory north of the Black Sea and on the east by the Altai Mountains of Siberia—an area that includes northern Kazakstan.

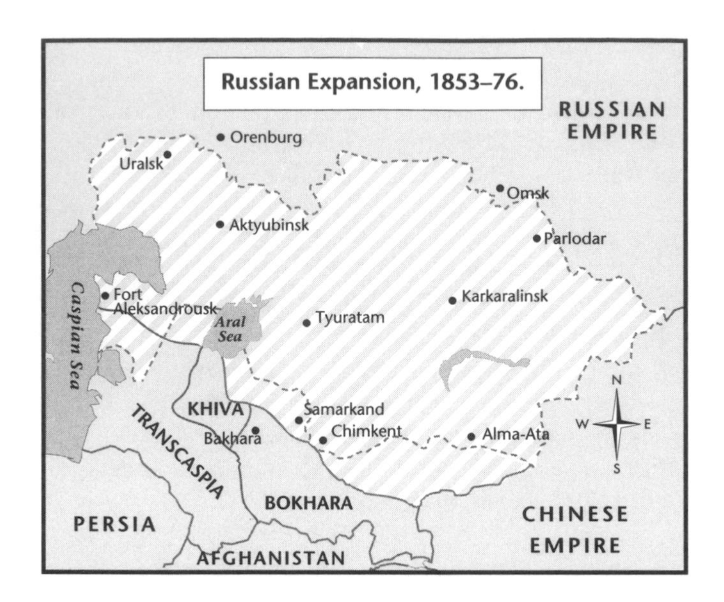

Early 1700s Industrial development begins in Russia

Industrial development begins in Russia under Peter I (Peter the Great), a czar of the Romanov dynasty which rules from 1613–1917. This development includes the building of factories in the Ural Mountains region north of Kazakstan, where iron ore is plentiful.

Mid-1700s Cossacks defend northern border

Cossack soldiers along the northern border of Kazakstan defend the Russian frontier against Kazaks and other nomadic tribes that attempt to invade Russia from the south.

1830s–1866 Moscow takes Kazakstan

The Russians of Moscow, under czars Nicholas I and Alexander II of the Romanov dynasty, begin a southward offensive, taking all of present-day Kazakstan (then part of Western, or Russian, Turkistan) by 1866. Large numbers of Cossacks and other Slavic immigrants have settled the black earth region and the steppe. The Russians begin another advance southward from the Lake Balkhash region in southeastern Kazakstan, where they build a fort on the site of Alma-Ata, now Kazakstan's capital, in 1855. From there they take the cities of Tashkent and Samarkand in what is now Uzbekistan, only to continue their expansion until all the territory up to the border of Afghanistan comes under Russian control.

1837–47 Qasimov leads resistance against Russians

The Kazaks fiercely resist Russian expansion into Central Asia. Leading the most famous military campaigns against the Russian advance is Khan Kenesary Qasimov (Kasimov), a Kazak revolutionary who later comes to be considered a national hero of Kazakstan. The Kazaks resist the Russian takeover for nearly sixty years.

1845: August 10 Poet Abay Kunanbayev is born

Abay (Abai) Kunanbayev (1845–1904) is born in Chingistan, in eastern Kazakstan. Abay, as he comes to be known, wins fame as a beloved poet and philosopher who provides both ethnic and personal inspiration to the Kazak people, even in modern times. Socially active through his art, he encourages the preservation of the Kazak culture but also, through his work as a translator, makes Russian classical literature available throughout his homeland. The son of a feudal nobleman, Abay receives an education in both Kazak and Russian in the city of Semipalatinsk. He revives poetry as an art in Kazakstan and is also a composer who often sets his poems to music. An intellectual and a humanist, Abay is considered the father of modern Kazak literature. In Kazakstan, many streets, theaters, schools, and libraries are named for Abay. He dies in Chingistan on July 6, 1904.

1855 Fort and trade center is founded

Russians found a fort and trade center in southeastern Kazakstan, called Zailiyskoe. The city that develops around it is known as Vernyi from 1855 to 1921. It becomes known as Alma-Ata (now Almaty), meaning "apple place," in the 1920s, developing into an economic center as a terminus on the Turkistan-Siberian Railroad. It becomes the capital of Kazakstan under Soviet rule in 1929.

1897: September 28 Birth of Mukhtar Auezov

Kazak novelist, folklorist, playwright, and linguist Mukhtar Auezov is born into a nomad family in Chingistan. He receives a college education, graduating in 1919 from the teachers college in Semipalatinsk and in 1928 from the Department of Oriental Studies at Leningrad University. Auezov becomes a member of the Kazak Academy of Sciences in 1946. He is most famous for the multi-volume epic poem, *Abay Golo* (The Journey of Abay; 1942–56), based on the life of well-known Kazak poet Abay Kunanbayev (see 1845). *Abay Golo* portrays steppe life in Kazakstan during the latter part of the nineteenth century. Auezov also writes stories and about twenty plays, most developed around the theme of the Kazak people's road to revolution. In addition, he translates Russian classics and writes sketches on Turkestan and India. He is awarded the State Prize in 1949 and the Lenin Prize in 1959.

1900: April 13 Kazak writer Mukanov is born

Sabit Mukanov, considered a founder of Kazak Soviet literature, is born near Tuasarsk. Although his family is poor, Mukanov studies and becomes a teacher, continuing his studies in Orenburg, Leningrad, and Moscow after the Russian revolution and later becoming a member of Kazak Academy of Sciences. His poems expand the expressiveness of Kazak poetry. He directs the Kazak Writers' Union, beginning in

1935. Mukanov's works include an autobiography, *The School of Life,* and a historical novel, *Botagos.*

1906–17 Kazaks are overrun and forced into exile

Some 140,000 Russians per year stream into Kazakstan during the period from 1906 to 1914. The Russians' agricultural way of life disrupts the Kazaks' nomadic lifestyle. Russian control creates tension that reaches a peak in 1916, when Russia issues a decree drafting Kazaks into military service during World War 1. The Kazaks raid Slavic settlements, killing thousands. To punish them, Czar Nicholas II forces some 300,000 Kazaks to leave their own lands. They flee to the neighboring Sinkiang Uighur region of China (formerly Eastern, or Chinese, Turkistan), home to other Turkic-speaking nomadic tribes. When the Kazaks return to their lands in 1917, Russian settlers kill some 80,000.

1917–18 The Russian Revolution begins

A sprawling but disadvantaged Russia with insufficient industry, agriculture, and transportation for its size has suffers many defeats during World War I. Starving workers in the capital city of St. Petersburg (renamed Petrograd) go on strike, and soldiers soon join them. Czar Nicholas II is overthrown on March 15, and on November 7 the Bolsheviks, a minority party in the soviet, the council of workers and soldiers leading the socialist parties, seize power in Russia, and move the capital back to Moscow. Their leader, Vladimir Ilich Ulyanov (1870–1924), known as Lenin—who formulates the official Communist ideology, Marxism-Leninism—makes peace with Germany in 1918, and Russia withdraws from World War I. Under the terms of the peace treaty, however, the Bolsheviks surrender some of the former Russian Empire's best agricultural lands and other natural resources and about one-third of its population to Germany.

1917 Muslim people of Central Asia form congress

Learning of the czar's abdication, the Muslim peoples of Central Asia, including Kazakstan, unite to form the All-Russian Muslim Congress.

1917 Kazakstan declares independence

Kazakstan declares its independence from Russia and establishes an autonomous republic in eastern Kazakstan in 1918, but the Bolsheviks soon acquire this territory.

1918 Asian rivers slated to irrigate cotton fields

A decree is issued that the waters of two large central Asian rivers, the Amu Darya and the Syr Darya, which flow into the Aral Sea (lying in what is now southern Kazakstan and northern Uzbekistan) should be used to irrigate millions of acres of cotton fields to help the Russian Empire become self-sufficient. This area will soon grow 90 percent of the country's cotton, but the Aral Sea will begin shrinking as the waters that feed it are diverted to the cotton fields.

1920–25 Kazakstan becomes Kirghiz A.S.S.R.

All of modern-day Kazakstan again becomes an autonomous republic, under communist control, known as the Kirghiz Autonomous Soviet Socialist Republic (A.S.S.R.), until 1925. The Kirghiz are a people closely related to the Kazaks and often confused with the Kazaks by the Russians, who use the term Kirghiz to refer either to the Kirghiz and Kazaks together or to the Kazaks alone. The communists call the Kirghiz people the Kara-Kirghiz.

1920s–30s Communism under the Bolsheviks

The Bolsheviks establish communism as a way of life, which means nationalization of farms, factories, and transportation, taking them from individual ownership to production that will benefit the people as a whole in their promised rise to world power. The communists endeavor to provide recognition of the many nationalities in Russia, including the Kazaks, but at the same time aim to crush their independence by uniting them with Russia under the Communist flag.

1920 Bigamy, bride-price, and bride-stealing made illegal

Bigamy (having more than one spouse), bride-price (payment made by the groom and his family to the family of a bride), and bride-stealing are banned by law in the Kazakh republic. The paying of *kalym,* or bride-price, has been a common cultural practice. The laws are widely ignored, especially in rural areas.

1924 Death of Lenin; Stalin's rise to power

Lenin's health fails after he suffers a stroke in 1921. He dies in January 1924, realizing he has failed to accomplish the socialist organization of the USSR he had hoped for in so short a time. The Communist Party's general secretary of the Central Committee, Joseph Stalin (1879–1953), whose birth name is Iosif Vissarionovich Dzhugashvili, assumes control after Lenin's death, establishing himself as a dictator determined to carry out Lenin's ideals. Part of his goal includes the reculturation—or political and social refinement—of the peoples of Eurasia.

1925: May 12 Kazak A.S.S.R.

The Kirghiz Autonomous Soviet Socialist Republic is renamed the Kazak Autonomous Soviet Socialist Republic (A.S.S.R.).

1929 First attempt at collectivization fails

Soviet leader Joseph Stalin outlines the goals of his first Five-Year Plan, part of which is to put the Russian people in automobiles and the peasants on tractors. Late in the year he demands the collectivization of small homesteads into large-scale farms managed by the peasants as a group. A sizeable portion of the yield is to be given to the Soviet government. The hardest-working peasants, called *kulaks,* are driven from their lands and exiled to Siberia for their capitalist tendencies. As a result of collectivization, a miserable failure, some 14.5 million people starve or are killed for objecting to the program.

Kazakstan is a primary target for Stalin's collectivization program, since it has a large area of potential farmlands. However, the forced settling of the nomadic Kazaks onto collective farms decimates the population and culture. Unable to accept this lifestyle, some 1.5 million Kazaks are killed and thousands more leave their country for China and elsewhere. Virtually all of their livestock are killed, either by the Kazaks themselves to keep them from the communists, or by guerrillas. An estimated three million Kazaks die, mostly from starvation.

1929–30 The Arabic alphabet, used for written Kazak, is replaced by Latin script

The Russian government under Stalin replaces the Arabic alphabet used for writing the Kazak language with the Latin alphabet.

1932 Karakalpak separates from Kazak republic

Karakalpak, a region southeast of the Aral Sea, is administratively separated from the Kazak Autonomous Soviet Socialist Republic and assimilated into nieghboring Uzbekistan.

1933–38 The deception of the Soviet system

Under Stalin's second Five-Year Plan, collectivization takes hold and is more successful. In 1936, a new Soviet constitution declares that socialism has been achieved and that the Soviet Union has the world's most democratic constitution. Underneath this facade is a government that uses terror to control its people. Thousands, including Soviet military officers, are executed for alleged conspiracy against Stalin. By 1938, some eight million people—including intellectuals, artists, teachers, and holy men—are held in *gulags,* or labor camps, one-fourth of them dying there. Kazakstan and Siberia are used as outlands to which suspected enemies of the Soviets are exiled.

1936 Kazakstan becomes constituent republic

Kazakstan becomes a constituent republic of the federal Union of Soviet Socialist Republics (U.S.S.R.); it is called the Kazak Soviet Socialist Republic. Armenia, Azerbaijan, and Georgia also become republics of the USSR. Ten Soviet republics are to have directly elected assemblies that vote by secret ballot. However, the Communist Party remains the only legal political organization.

1939–40 U.S.S.R. thrives

The U.S.S.R. enjoys a period of prosperity under Stalin, as industry, highly mechanized collective farms, and transportation improvements increase the standard of living and national pride. Intellectuals and artists are relatively free to pursue their work, and scientific advances are made, including the construction of an atomic bomb. Soviet athletes become known for their excellence.

1940 The Cyrillic alphabet replaces Latin

The Cyrillic alphabet replaces Latin for written languages as part of the U.S.S.R.'s Russification Program.

1941: June 22 World War II begins in the Soviet Union

World War II (1939–45) begins in the USSR. Soviet leader Joseph Stalin is stunned by the news that the Nazi German army under Adolf Hitler has attacked Russia with the intention of overcoming the Soviets and taking their country for German expansion. Stalin has maintained a successful trade relationship with Germany since 1922 and signs a pact with the Nazis in August 1939 that divides Poland, giving the USSR former Russian Empire lands in that country.

Under Stalin, the Russians retreat toward Moscow, the Ural region, Siberia, and the Central Asian region, burning their own cities and farms to slow the German advance. Germans who have settled in the Soviet Union are exiled to Siberia and Kazakstan.

During World War II in Russia the women, elderly men, and adolescents keep the coal mines, furnaces, and factories producing materials and ammunition for the war. As the male soldiers die, women join the ranks of infantry who continue to fight the Germans.

Kazaks serve in Soviet military. More than 450,000 Kazaks serve in five national military divisions during World War II, including the 96th, the 105th, and the 106th Cavalry divisions (the Soviet cavalry is crucial in view of the country's inferior weapons technology; the Kazaks, as descendents of fierce Mongolian and Turkish horsemen, and the Cossacks, known for their superior cavalries, make up the finest among these divisions) and the 100th and 101st Infantry divisions. Kazaks also serve in the 38th Infantry and the 316th, or *Panfilov,* antitank division, whose members die trying to stop the advance of German tanks. Kazaks fight in every major battle of World War II on Soviet and German soil.

1943: January Russia defeats Germans at Stalingrad

Russian soldiers defend the city of Leningrad for two years and defeat the German army at Stalingrad (now Volgograd).

1944 Stalin orders large-scale deportation

Soviet leader Joseph Stalin orders the deportation of more than a million Chechens, Volga Germans, Crimean Tatars,

Ingush, and Balkars to Kazakstan and Siberia. Many remain in Kazakstan after "rehabilitation" (see 1957).

Mid-1945 Soviet victory in WWII

By mid-1945, the Russians, with the help of the Allies, finally defeat the Germans, and the Soviet Union gains land as far west as central Europe. The Soviet Union has borne a tremendous share of destruction for its victory, however. Of the estimated total lives lost worldwide during World War II—5.5 million—the Soviet share is estimated at 49 percent.

After World War II, the USSR quickly rebuilds its factories, farms, and cities. The lands in Central Asia that became home to Soviet citizens fleeing the Germans experience greater economic development because the refugees bring technical expertise with them.

1946 USSR rises to world power

Both the USSR under Joseph Stalin and the United States rise to the rank of world power. Stalin believes another war will be waged against his country and strives to be ready for it by increasing work on the atomic bomb. His faith in communism as the best way for his people and for the world causes Sir Winston Churchill, Britain's former prime minister, to tell a U.S. audience that "An iron curtain has descended across the [European] Continent," separating communist central and eastern Europe from democratic western Europe. Thus begins the "cold war," in which east and west, communism and democracy, the USSR and the United States and their respective allies are locked in mutual distrust, hatred, and rivalry for the next three decades.

1947 City of Temirtau is founded

Karaganda Metal Works, one of the biggest steel plants in the Soviet Union, needs workers, so a settlement is established. This small city, Temirtau, attracts thousands of people from Russia, Ukraine, and other Soviet republics who move seeking work at the plant. The steelworks also provide employment for criminals when they are released from a prison camp nearby.

1949: August First atomic bomb is tested

The Soviet Union tests its first atomic bomb, in northeastern Kazakstan's Semipalatinsk region. Neighboring villages and towns are exposed to radiation fallout. The USSR also uses atomic energy to produce electric power for the large country.

1953: March 5 Death of Stalin

The great Soviet leader Joseph Stalin, "the Man of Steel," dies after suffering a stroke on March 1. Some five million people come to view his body in Moscow, many being crushed to death in the crowd. Stalin is buried next to Lenin in the mausoleum on Red Square.

1953: August Hydrogen bomb test

First H-bomb is tested. The Soviet Union tests its first hydrogen (H) bomb, only nine months after the United States tests its bomb.

1954: December Khrushchev and the Virgin Lands campaign

New First Secretary of the Communist Party, Nikita Sergeyevich Khrushchev (1894–1971), begins a campaign to grow corn and other grains to feed the Soviet people and their cattle. Ignoring warnings of insufficient rainfall, he chooses 60 million acres of the steppes of Kazakstan and southwestern Siberia—the "virgin lands" that have pastured sheep and other animals since ancient times—as prospective prime farmlands.

1957 Krushchev rehabilitates land

Krushchev sets a goal of twenty million acres of grain to be harvested during the first year of his Virgin and Idle Lands Program. Volunteers arrive by the trainload from Russia and Ukraine, bringing farm machinery with which to plow the steppes. Many of the millions of people who had been relocated by Stalin to Kazakstan remain to participate in this land rehabilitation effort.

The tractors and harvesters are too few, however, and production falls short of Khrushchev's five-year goals as unharvested and unstored grain lies decaying in the fields. Erosion impoverishes the land, and by 1964—when Khrushchev is dismissed from office—the Soviet Union must import food for its people. The Virgin Lands Program dies in disgrace.

1958 Krushchev becomes premier

Khrushchev becomes premier of the Soviet Union, a post he holds until 1964.

1961: April 12 Soviets put first man in space

The Soviet Union makes world history when astronaut Yuri Gagarin (1934–68), a major in the Soviet air force, becomes the first man to travel in space. The earth satellite, Vostok I, carries him in a single orbit around the earth, lasting one hour and forty-eight minutes. The Vostok I is launched from the Soviet Union's Baikonur Cosmodrome space facility in south central Kazakstan.

1965 Era of success for USSR

When Leonid Brezhnev (1906–82) becomes First Secretary of the Communist Party, with Aleksey Kosygin (1904–80) as premier of the Soviet Union and Nikolay Podgorny (1903–83) as president of the USSR, the nation experiences heights of success. It matches or surpasses the United States in many areas, including defense, diplomacy, sports (winning more

Olympic Games gold medals than any other country), the arts, and industry. In the following year, Kazakstan enjoys its largest grain harvest—25.5 million tons.

1965 Atomic Lake is created

An underground atomic device is detonated sixty-five miles south of the city of Semipalatinsk (now Semey), in Kazakstan. The device is detonated specifically for the peaceful purpose of creating a reservoir, which becomes known as Atomic Lake. The population is exposed to radiation fallout.

1969 Skeletal remains of fifth century human discovered

A burial site excavated east of Alma-Aty is believed to be from the fifth century. The skeletal remains are unearthed.

1970 Treaty signed

First treaty to ban nuclear weapons testing is signed. The USSR and the United States begin Strategic Arms Limitation Talks (SALT 1) in 1969 and sign a nuclear nonproliferation treaty in 1970.

1975 Aitmatov and Mukhamedzhanov publish play

Kirghiz writer Chingiz Aitmatov and his Kazak co-author Kaltai Mukhamedzhanov co-author and publish a play, *The Ascent of Mount Fuji.* The play is originally performed at the Sovremennik Theater in Moscow in the winter of 1973. It is first performed in English on June 4, 1975, in Washington, D.C. Aitmatov wins a Lenin Prize, the country's highest literary honor.

1985 Gorbachev is elected

Mikhail Gorbachev (b. 1931), from a Cossack family in the area north of the Caucasus Mountains, is elected general secretary of the Communist Party. He believes in *perestroika* (restructuring) and *glasnost* (openness).

1986 Elections

The parliamentary elections result in Kazaks holding more than fifty percent of the seats in the Kazak parliament, or Supreme Soviet.

1986: April 26 Chernobyl nuclear accident

An accident occurs involving a nuclear reactor at the Chernobyl Nuclear Power Plant in northern Ukraine. The worst nuclear disaster in history, the accident at Chernobyl kills some 5,000 people and disables another 30,000, according to estimates by a citizens' study group. It also contaminates thousands of acres of farmland and spreads radiation as far as some parts of Western Europe.

1986: December Kazaks protest appointment of Russian

Anti-Russian riots—almost unheard of in peaceful Kazakstan—break out in the capital, Alma-Ata. Kazaks are incited to riot when Mikhail Gorbachev accuses longtime Kazak Communist Party leader Dinmukhamed Kunaev of corruption and replaces him with a Russian loyalist, Gennadi Kolbin, completely unknown to the Kazaks. This unusual governmental act is reversed in 1989.

1988 Aral Sea level is critical

A portion of the Aral Sea, a large inland sea lying partly in southwestern Kazakstan and partly in Uzbekistan, separates from the main body of water. Heavy irrigation in the region has redirected the waters of the Amu Darya and the Syr Darya. These two rivers, both formerly large rivers feeding the Aral Sea, yield no flow into the sea between 1974 and 1986. Now known as the Small Aral Sea and the Large Aral Sea, the bodies of water represent only about twenty-five percent of the Aral's volume in 1960.

As a result of the significant decrease in water volume, the lake's salinity (salt concentration) triples, and all of its twenty-four species of fish die. As the water level recedes, toxic chemicals used to spray crops are exposed and deposited by evaporation on the seabed, creating a toxic dust that contaminates towns for miles around. A high infant mortality rate and rate of birth defects, immune system damage, throat cancer, hepatitis, and respiratory diseases result in the region. The drinking-water supply is critical around the Aral Sea and is often contaminated too. The Communist Party *Politburo* enacts guidelines in September for restoring water to the Aral Sea, but the environmental damage will take many years to correct.

1989: January Kazak is declared official language

The republics bordering on the Baltic Sea enact a new law requiring Russians living there to learn the native language of their republic. The Central Asian republics, including Kazakstan, follow suit. Kazak is declared the official language of Kazakstan, even though a majority of native Kazaks are Russian-speaking, and less than one percent of Russians in Kazakstan speak fluent Kazak.

1989: June Nazarbayev is appointed First Secretary

In the face of increasing anti-Russian sentiment in Kazakstan, Soviet President Mikhail Gorbachev replaces the Russian Gennadi Kolbin with ethnic Kazak Nursultan A. Nazarbayev (1940–), a metallurgist from the southern *Kazakhjuz,* or clan, as First Secretary of the Communist Party in Kazakstan.

1989: November Communism begins to collapse

The Communist regime in eastern Europe collapses after the fall of the Berlin Wall—separating East and West Germany—

on November 9. A huge public rally of support for pro-democracy, former Communist Party leader, Alexander Dubcek in Czechoslovakia follows later in November.

1990 Russian-speakers organize Unity movement

As Kazakstan struggles for independence, members of the Soviet ministries in Alma-Ata who fear losing control of the goverment organize the first Russian-speaking movement in Kazakstan, known as Unity. Supported by common people opposed to nationalism, it soon collapses because, ironically, of its diversity of interests.

1990 Pollution causes health problems

After an outbreak of nosebleeds among children in the town of Ust-Kamenogorsk (now Oskemen), one hundred miles east of the town of Semipalatinsk in northeastern Kazakstan, health officials discover the soil, air, and population are contaminated with heavy metals—lead, zinc, mercury, arsenic, and cadmium. An old lead-and-zinc smelter, a smelter for titanium and magnesium, and a plant for making reactor fuel rods from uranium—as well as radioactive dust from the nuclear test site at Semipalatinsk—are blamed for the high rates of cancer, birth defects, and immune system abnormalities found among the population here.

Soviet president Mikhail Gorbachev declares a moratorium on nuclear testing at the Semipalatinsk test site after public protest concerning radiation-caused illnesses in the area. By this time, however, 470 nuclear tests have been carried out, beginning in 1949, 116 of them in the air and the rest underground.

1990: April Alash Freedom Party is founded

Kazakstan's first political party, Alash National Freedom Party, is founded. Led by Aron Atabek and Rashid Nutushev, it is a small party whose leftist members are drawn from Kazak nationalists, Islamists, and pan-Turkic nationalists. Alash advocates a Kazakstan free from Western influence and a return to Turkic roots.

1990: August Nazarbayev elected president

Kazakstan's 510-member Supreme Council elects Nursultan Nazarbayev president of the Kazak Soviet Socialist Republic.

1991–98 Kazaks return to Kazakstan

Nearly 40,000 families return to KazakstanKazakstan. Most come from Mongolia, Iran, Turkey, China, and Afghanistan, but some also migrate from Russia and other ex-Soviet states. Those returning to their homeland hope to find housing and employment. Some are disappointed, since the government and economy are struggling. In addition, many Kazaks speak only the Kazak language, and the common language of business and government is Russian.

1991: August Six republics declare independence

Ukraine, Belorussia, Moldavia, Azerbaijan, Uzbekistan, and Kyrgyzstan declare their independence from the Soviet Union.

Following an attempt by hard-line Communists to take over the Gorbachev government in Moscow, Nursultan Nazarbayev resigns from the Communist Party. He also closes the Semipalatinsk nuclear test site in northeastern Kazakstan, in use since 1949.

1991: December Commonwealth of Independent States is formed

Negotiations held in Byelorussia (now Belarus), which borders Russia on the west, create an informal network of most of the former Soviet republics for economic and political cooperation. This network is known as the Commonwealth of Independent States (CIS). Its supporters finalize details of the agreement in Kazakstan's capital, Alma-Ata, in late December.

1991: December 1 Tereshchenko is elected prime minister

Sergei Tereshchenko, a fluent Kazak-speaking Slav, is elected prime minister of Kazakstan.

1991: December 16 Kazakstan declares independence

Kazakstan declares independence from the Soviet Union.

1991: December 26 Kazakstan is officially independent

Kazakstan becomes an officially independent state the day after Mikhail Gorbachev resigns as president of the USSR (December 25), marking the dissolution of the Soviet Union. In Kazakstan's first democratic presidential election, Nursultan Nazarbayev, running unopposed, captures ninety-five percent of the popular vote, confirming his election as president by the Supreme Council (see 1990: August).

1992 Kazakstan joins IMF, World Bank, and CSCE

Kazakstan joins the International Monetary Fund and the World Bank, which provides funding for infrastructure projects in the developing country. Kazakstan also joins the Conference on Security and Cooperation in Europe (CSCE).

1992 Health care law enacted

Kazakstan's parliament passes a new law designed to improve health care. Most of the state's hospitals are located in the cities and are dilapidated, understaffed, and poorly supplied.

Health care for women is especially inadequate, with abortion the most common method of family planning and the number of abortions equaling the number of births. Common modern methods of contraception are scarce and expensive. Most Kazak women of childbearing age are anemic, and oral iron supplements are unavailable.

1992 Kazakstan Institute of Management, Economics, and Strategic Planning founded

President Nazarbayev founds the Kazakstan Institute of Management, Economics, and Strategic Planning. The Institutes goals are to provide education and training needed to rebuild the Kazak economy, using visiting professors from universities in the West.

1992 Oil pipeline consortium is formed

Kazakstan and the Arabian nation of Oman form a consortium for the construction of an oil pipeline to supply other nations with Kazak oil.

1992: January US establishes embassy

The United States is the first country to open an embassy in the newly independent Kazakstan.

1992: March 2 Kazakstan joins the UN

Kazakstan—along with other former Soviet republics of Armenia, Azerbaijan, Georgia, Kygyz Republic, and Moldova—is admitted to the United Nations (UN).

1992: May Kazakstan signs agreement with Chevron

Kazakstan's government signs an agreement with U.S. oil company Chevron to develop one of the world's largest untapped oil reserves, the Tengiz field, in north central Kazakstan.

1992: May 18–20 Negotiations with United States unsuccessful

Kazak president Nursultan Nazarbayev makes his first official visit to Washington, D.C. His attempt to obtain guarantees of security from U.S. president George Bush in return for Kazakstan's giving up strategic nuclear weapons fails. By the middle of the year all tactical nuclear weapons have been moved to Russia for storage and destruction. Nazarbayev's government has pushed for ratification of the Strategic Arms Reduction Treaty (START).

1992: May 23 Agreement with the United States

Kazakstan and the United States sign the Lisbon protocol to the Strategic Arms Reduction Treaty (START), making Kazakstan a party to the START treaty and committing the republic to reductions in strategic nuclear weapons.

1992: June Kazak proposed as official language

Kazakstan's newly proposed draft constitution upholds the declaration of the native language, Kazak, as official language of the state. Russian is deemed an important second language.

1992: November 15 Kazakstan issues currency

Kazakstan issues its own currency, the *tenge,* replacing the former Kazak ruble, which was printed in Russia. One tenge equals 500 rubles.

1992: December Kazakstan ratifies the Nuclear Nonproliferation Treaty

Kazakstan and the United States sign an agreement concerning silo launchers of intercontinental ballistic missiles and the prevention of nuclear weapons proliferation. The United States ultimately spends several hundred million dollars in aid to Kazakstan as a result.

1993: January 28 Kazakstan adopts new constitution

Kazakstan adopts a new constitution, providing for the election of a president by the voters to a five-year term; the appointment of a prime minister and certain members of a council of ministers by the president; the appointment of other ministers by the prime minister; and the election of 177 members of the unicameral legislature to five-year terms.

The new constitution guarantees the rights of free speech, protest strikes, privacy, protection from double jeopardy, and the admission of evidence. The state retains the right to all natural resources. Mutual parent-child obligations are recognized, and political parties based on religious beliefs or loyalties are prohibited.

1993: June Peace Corps volunteers arrive in Kazakstan

Volunteer members of the U.S. agency, Peace Corps, arrive in Kazakstan for the first time. They help develop small businesses and teach English as a second language.

1994 British Council sponsors training for Kazak accountants

The British Council selects the Institute of Chartered Accountants of Scotland (ICAS) to provide training for accountants in Kazakstan. The Kazak accountants will then provide training to others in international accounting methods.

1994 Japanese loan to titanium refinery

Chori, a Japanese company, loans $3.2 million to a refinery processing titanium ore in Kazakstan.

1994 IMF supplies financial aid

The International Monetary Fund gives Kazakstan $13 billion in financial aid for economic development. Because the demise of the USSR has severed important industrial ties to

Russia and other republics, however, more than one thousand industries in Kazakstan shut down in the spring of 1994.

1994: January Miners and steelworkers protest

About two thousand workers—most of whom are miners and steelworkers—stage a protest in northeastern Kazakstan in Leninogorsk (near Oskemen). They are upset over nonpayment of salaries and the deterioration of mines and manufacturing facilities.

1994: February President Nazarbayev meets with Clinton

Kazakstan's president Nursultan Nazarbayev visits the United States, meeting with President Bill Clinton. Clinton promises $311 million in economic assistance in 1994 as a reward for Kazakstan's signing documents acceding the republic to the Nuclear Nonproliferation Treaty as a non-nuclear power. It will also receive $85 million to help dismantle its nuclear weapons. The two leaders declare their mutual support in the areas of defense and security.

1994: March 7 Elections are called unfair by Russians

President Nazarbayev, running unopposed, wins an overwhelming victory in the presidential elections. Russian legislators call parliamentary elections unfair because Kazaks will hold fifty-eight percent of the seats, yet only make up forty-two percent of the country's population. Russians and Ukrainians together hold thirty-three percent of the seats.

1994: March 28 Agreement with Russia over nuclear weapons

President Nursultan Nazarbayev of Kazakstan and President Boris Yeltsin of Russia sign an agreement according to which Russia assumes jurisdiction over nuclear weapons stored in Kazakstan. All warheads are to be removed within fourteen months, with silos and missiles to be dismantled within three years. Russia is also to control the Baikonur Cosmodrome space center complex in south central Kazakstan for twenty years at a cost of $115 million per year. The complex is nine times as large as the United States' Kennedy Space Center.

1994: May Scholars call Solzhenitsyn "an extreme chauvinist"

Scholars in Kazakstan describe Russian author Aleksandr Solzhenitsyn "an extreme Russian chauvinist" after he proposes reunification of northern Kazakstan and Russia in a Russian renewal.

1994: July The Central Asian-American Enterprise Fund is incorporated

The Central Asian-American Enterprise Fund is incorporated. It promotes private-sector economic development in Kazakstan, Uzbekistan, Turkmenistan, Kyrgyzstan, and Tajikistan. Funded by the U.S. government over a three to four-year period, it concentrates on such projects as food processing, textile manufacturing, distribution and transportation, and the production of consumer goods.

1994: July 6 Parliament votes to move capital

Kazakstan's parliament votes to move the capital from Alma-Ata, in southeastern Kazakstan, to Tselinograd, the capital of Soviet leader Nikita Khrushchev's Virgin Lands agricultural program (see 1956 and 1957). Plans are made to move the capital and rename the city Akmola, located in north central Kazakstan, by the year 2000. President Nazarbayev says Alma-Ata is overcrowded and unable to expand. Other theories about the reason for the move revolve around Kazakstan's wish to establish a firm hold on its heavily Russian northern territory and point to Alma-Ata's proximity to China.

1994: July 8 Summit meeting

In a summit meeting held in Alma-Ata, the leaders of Kazakstan, Kyrgyzstan, and Uzbekistan agree to form a common union for defense and economic purposes. The union is open to other Central Asian countries wishing to join.

1994: July 26 Nuclear inspection agreement

Kazakstan's prime minister Sergei Tereshchenko signs an agreement with the International Atomic Energy Agency (IAEA) for regular inspection guaranteeing the country's compliance with the nuclear nonproliferation treaty signed (see 1993: December). An IAEA inspection conducted at this time finds no unusual levels of radiation above ground at the Semipalatinsk nuclear test site in northeastern Kazakstan, despite longtime charges that nuclear contamination has caused high rates of cancer and birth defects in the area.

1994: August Russian troops continue to man space center

The Russian Space Agency says 28,000 Russian troops are still needed to maintain the Baikonur space center. This is a reduction from the 70,000 troops stationed there—including military construction workers—in 1990. All Russian troops and civilians stationed at Baikonur are to be under Russian law (see 1994: March 28).

1994: September 6 Disarmament agreement with Japan

Kazakstan signs an agreement with Japan in which that country will supply $11 million to help the former Soviet republic dismantle its nuclear weapons.

1994: October–November United States takes Kazak uranium

A large quantity of uranium removed from nuclear weapons is shipped to the United States for reprocessing. The uranium has been stored at Oskemen in northeastern Kazakstan. Earlier in the year Kazak officials had asked for U.S. help in disposing of the material.

1994: October 11 Tereshchenko is forced to resign

President Nazarbayev asks for the resignation of Prime Minister Sergei Tereshchenko and his cabinet for failure to implement a plan to improve Kazakstan's economy. The parliament approves Nazarbayev's nomination of Akezhan Kazhegeldin, first deputy prime minister, to replace Tereshchenko.

1994: October 19 Summit in Istanbul, Turkey

A two day summit of the leaders of the Turkic-speaking countries—Kazakstan, Uzbekistan, Turkmenistan, Azerbaijan, Kyrgyzstan, and Turkey—ends in Istanbul, Turkey, with hopes of forming closer cultural, economic, and political ties between the countries. One goal is the construction of oil and gas pipelines from Central Asia to Europe through Turkey.

1995 Diphtheria epidemic

Diphtheria, a contagious upper respiratory infection, kills at least 51 people and afflicts about 700 more in Kazakstan in the worst outbreak in decades. The large number of cases is blamed on poor sanitation, a shortage of vaccines, and the people's failure to accept vaccination as a valid method of disease control.

1995: March 6 Constitutional Court invalidates election

A crisis erupts in the government when the Constitutional Court invalidates the results of the 1994 election due to "numerous irregularities. "

1995: March 11 Call for new elections

The Kazak parliament attempts to overrule the Constitutional Court, but the court ignores its amendment as unlawful. President Nazarbayev is forced to side with the court over the opposition of the legislative majority. He dissolves the parliament and calls for new elections. Until elections can be held, Nazarbayev rules by decree.

1995: March 29 Kazakstan and Canada sign trade agreement

Kazakstan and Canada sign a trade agreement and a declaration of principles that facilitate trade and secure Canadian business interests in Kazakstan. Canada also announces a $1.1 million technical assistance package for Kazakstan that will focus on mining, oil and gas production, and agricultural projects in which Canadian companies are involved.

1995: April New tax code

A comprehensive new tax code and other economic reforms (all established by President Nazarbayev), as well as the establishment of a securities and exchange commission, are designed to encourage foreign investment in Kazakstan.

1995: April 29 Referendum confirms Nazarbayev's presidency

A national referendum confirms Nursultan Nazarbayev as president of Kazakstan until the year 2000, with ninety-five percent of the vote. Nazarbayev claims that over ninety percent of the population participated in the voting, a figure that is widely disputed. None of the embassies in the capital chooses to particpate in monitoring the voting.

1995: April 31 Last nuclear device dismantled

The last of the former-Soviet nuclear devices in Kazakstan is destroyed.

1995: August 30 New constitution approved

A new consitution is approved by eighty-nine percent of the voters in a referendum supported by President Nazarbayev. The new constitution extends Nazarbayev's presidential power, effectively enabling him to rule by decree. Kazak opposition parties boycott the event.

1995: October Nuclear test tunnel is sealed

The United States and Kazakstan sign an agreement to seal permanently Kazakstan's nuclear test tunnel, guaranteeing that the facility will never be used again.

1996 Subway system planned

A contract to build a $600 million subway system in Alma-Ata is awarded to a Turkish firm and a Canadian firm. The project is expected to be completed by the year 2000.

1996 Interest in Islamic traditions grows

During the Soviet era, Islamic traditions were suppressed. Kazak men, particularly those with newly acquired wealth, seek to restore traditional lifestyle practices such as bigamy (having more than one wife). A percentage of these wealthy Muslims, referred to by some as *Kazanovi,* regard a second wife as a status symbol. Women over twenty-five may be willing to become second wives rather than face remaining single. In rural areas, laws against bride-stealing (taking a bride against her will) are widely ignored, and stolen brides rarely report their husbands, since they have little chance of finding another husband.

1996 Birth rate declining

The Kazak government statistics agency reports a decline in the birth rate from 25.1 per 1,000 inhabitants to 15.2 per 1,000 inhabitants in the period from 1985 to 1996. One government official, predicting serious underpopulation of the country if the trend continues, blames feminism for encouraging women to neglect their childbearing duties.

1996 Aid for Institute of Management, Economics, and Strategic Planning dwindles

The Kazakstan Institute of Management, Economics, and Strategic Planning, founded in 1992, faces a crisis as international aid diminishes. The Institute has been providing training since its founding (see 1992) and has been recognized as a leader in supporting the development of the Kazak economy.

1996 January New tax code encourages growth of Akmola

Tax code amendments exempt foreign and local construction companies from corporate income, land, and property taxes. These amendments are designed to encourage construction in Akmola, named Kazakstan's new capital. The Kazak government is scheduled to begin moving to Akmola in 1997.

1996: February Involvement in World Trade Organization

Kazakstan gains observer status in the World Trade Organization and applies for full membership.

1996: March Import and export duties lifted

The government lowers or eliminates export and import duties on many goods, including new cars and agricultural products. Imports must meet a certain hygiene standard.

1996: June First HIV infection in Temirtau

The central-Kazakstan town of Temirtau registers its first HIV infection (the virus that causes AIDS). Within two years, it is known as AIDS City with 631 or 795 people officially registered as HIV-positive in Kazakstan living there. Over ninety percent of them use intravenous drugs; the city has high unemployement since the steelmaking plant closed.

1996: June 28 Infrastructure improvement to help agriculture announced

Kazakstan's Ministry of Agriculture outlines a proposed water resources management and land improvement project that will increase family farm production and income. The project will focus on infrastructure to improve irrigation systems and farm roads and upgrade government agencies involved in coordinating these services. It will encourage and train women in managing these family farms.

1997 Move of capital begins

President Nazarbayev begins action to move the capital from Alma-Ata to Akmola. Seventy percent of the Akmola population is Russian, Ukrainian, or German, and just thirty percent are Kazaks. By moving the capital to Akmola, Nazabayev hopes to defuse claims that Kazakstan should be annexed to Russia.

1997 Railroad construction project

An extensive railway expansion and improvement project is undertaken by the Kawasaki Steel Corporation of Japan. The project is scheduled to be completed by December 2000.

1998: June 10 New capital officially opens

The new capital, known during the Soviet era as Tselinograd (Virgin Lands City), is renamed Akmola ("white tomb") in the early days of Kazakstan. It is now known as Astana (capital).The climate in the new capital challenges government workers and visitors. Winter temperatures are commonly -40°F, and summer temperatures soar to nearly 100°F. In addition, the city is plagued by mosquitoes.

The population of the new capital is about 300,000 and climbing. Company investing there receives tax advantages from the Kazak government. The new telecommunications system is the most-sophisticated in Central Asia.

1999: January 10 Nazarbayev wins re-election

Nursultan Nazarbayev is re-elected president for another seven years, winning with over eighty percent of the vote, the first contested presidential election in Kazakstan. The second-place candidate is Communist Party candidate Serikbolsyn Abdildin, who receives just twelve percent of the vote. Nazarbayev has headed the government in Kazakstan for ten years (see 1989).

Bibliography

Conflict in the Soviet Union: The Untold Story of the Clashes in Kazakstan. New York: Human Rights Watch, 1990.

Edwards-Jones, Imogen. *The Taming of Eagles: Exploring the New Russia.* London: Weidenfeld & Nicolson, 1992.

Kalyuzhnova, Yelena. *Kazakstani Economy: Independence and Transition.* New York: St. Martin's Press, 1998.

Olcott, Martha Brill. *The Kazakhs.* Stanford, CA: Hoover Institution Press, Stanford University, 1987.

World Bank. *Kazakstan: The Transition to a Market Economy.* Washington, DC: World Bank, 1993.

Kiribati

Introduction

When one thinks about the thirty-three islands that make up Kiribati, the image of "deserted islands" might come to mind. Kiribati, which is pronounced "Kiribass," is not only one of the world's smallest and least-known nations, but also one of its youngest and most underdeveloped. Located at the point where the International Date Line crosses the Equator in the Pacific Ocean, Kiribati spans an area of more than 2 million square miles. It includes three island chains: the Gilbert Islands, the Phoenix Islands, and the Line Islands. Its total landmass, however, is about four times the size of Washington, D.C., and its population is about 78,000. It became an independent nation on July 12, 1979.

All of the islands except for Banaba are atolls, which means that each consists mainly of a ring of land surrounding a lagoon. Few of the islands rise more than thirteen feet above sea level, which makes a possible rise in the ocean level from global warming a threat to their very existence. In recent years, the use of carbon dioxide has created a greenhouse effect in the earth's atmosphere, causing the polar ice caps to melt. This melting has caused many island nations in the Pacific to try to pressure more industrialized societies to reduce their emission levels of carbon dioxide and other gases. However, their pleas have largely gone unheard.

Pre-European History

Like many islands in the Pacific, much of Kiribati's early history can be traced through a tradition of storytelling. Stories about the islands' creations and the ancestors of their indigenous people were passed from generation to generation. Certain families on each island were known as storytellers. Because Kiribati is so spread out geographically, its oral histories vary widely from island to island, which has made it virtually impossible for historians to find a single way to depict the pre-European history of the region. Upon the nation's independence from Great Britain in 1979, however, the new government of Kiribati sponsored a workshop in which a group of historians from the country's islands gathered to produce what is regarded as an authoritative history. In this text, which is titled *Kiribati: Aspects of History*, a creation story based on the traditions of Beru Island is considered representative of the islands' overall history.

Archaeological evidence indicates that the islands were settled around 1000 B.C. by Micronesians who had migrated from Southeast Asia. Samoans, Tongans, and other Polynesian peoples also migrated to the islands, and over time a distinct cultural group evolved from a blend of the Micronesian and Polynesian influences.

Arrival of Europeans

European explorers first began sailing into the area in the sixteenth century. By the early nineteenth century, the islands had become known as the Gilbert Islands, while the nearby islands that comprise present-day Tuvalu were identified as the Ellice Islands. Many well-known sea captains recorded the location of the islands on their atlases and occasionally dropped anchor, but sustained European contact did not begin until the mid-nineteenth century.

In the nineteenth century, Europeans took advantage of the islanders' innocence. One early visitor taught the islanders how to make sour toddy, a fermented beverage, that soon left many inhabitants in a frequent state of intoxication. Travelers on merchant and whaling ships introduced tobacco, and soon many became addicted to it. The desire for tobacco was so strong that many families willingly traded the women in their family for the drug. These women then were taken aboard the whaling ships as prostitutes. When missionaries arrived in the islands in the latter half of the nineteenth century, they were shocked. Many reported widespread drunkenness, prostitution, and nudity. Scholars who study Kiribati have criticized Europeans for taking advantage of the islanders' lack of knowledge about the outside world. At the same time, however, scholars note that the islanders voluntarily participated in the activities that the outsiders introduced and didn't accept all of the new beliefs as inherently correct.

The manner in which the islanders greeted Christianity illustrates this point. Although nearly every resident practices Christianity today, missionaries in the nineteenth century experienced great difficulties in trying to convert the inhabitants. When the first groups arrived in the early 1850s, the islanders were deeply committed to their own cosmology, which involved a relationship between themselves and their spirit-ancestors, and many gods. Many greeted the idea of Christianity's one god with disinterest. "A new ideology could not be accepted in isolation," according to historian Barrie MacDonald, who added that it would have to make sense in the people's terms before it would find a place in their society.

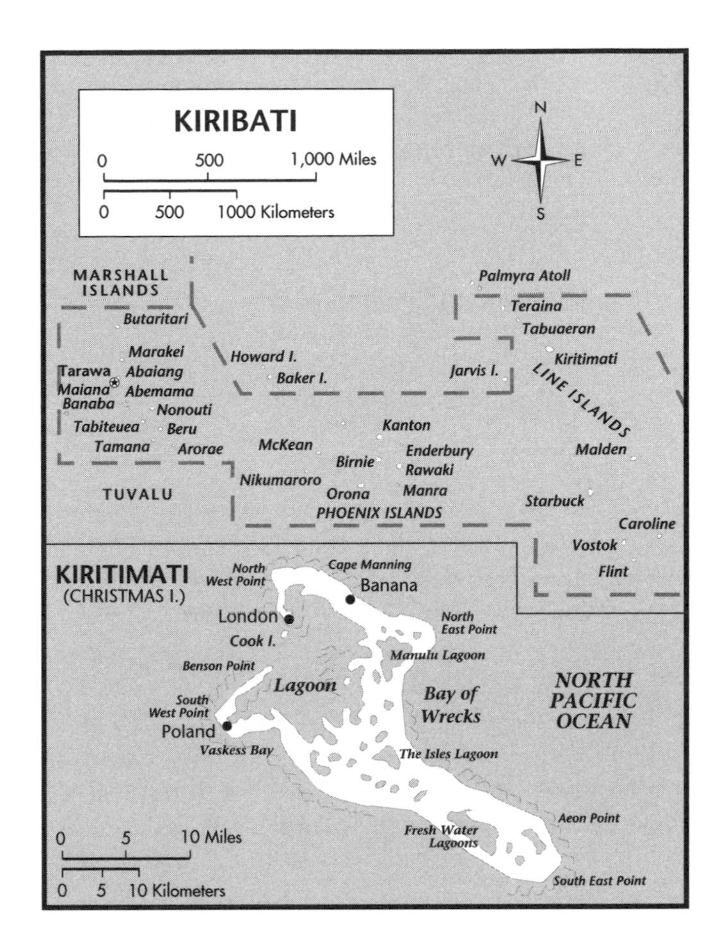

Islands as a British Protectorate and WWII

Historians such as MacDonald note that while the arrival of Europeans did affect the islanders, their customs and way of life were not greatly touched. This changed after 1892 when Capt. E. H. M. Davis declared the islands to be a British protectorate. In the following years British administrators brought together a series of culturally-distinct island groups under one government and established the Gilbert and Ellice Island Colony. During this time, phosphate deposits were discovered on the island of Banaba, and much of the British administration's efforts became focused on mining this valuable mineral.

The colony became a critical World War II (1939–45) battleground after the Japanese attacked Banaba in 1941. Japanese troops occupied the islands and remained there until Americans drove them out in 1943. World War II devastated the islands. Many homes were ruined, and many traditional food supplies, such as coconut trees and banana plants, were destroyed. After U.S. troops left in 1946, British rule was restored. But the British Empire had been financially ruined, and many of the modern buildings that had been built before the war were either left to decay or were replaced with such enterprises as coconut plantations. The phosphate industry was revitalized, but the rulers realized that the deposits would soon be depleted.

Independence

Because British rulers brought the Gilbert and Ellice chains together, many have thought of these two islands groups as interchangeable. In reality, however, the two chains are quite different culturally and socially. These differences became more apparent as nationalists in both chains began to lay the groundwork for establishing their own independent nations. By the early 1970s, the differences became irreconcilable, and residents of the Ellice group voted to separate from the Gilbert group, establishing themselves as the present-day nation of Tuvalu.

After the islands gained their independence in 1979, phosphate production on Banaba ceased. The deposits had been exhausted. Since then, the country has relied on fishing, subsistence agriculture, and exports of copra (dried coconut from which coconut oil is produced) for its livelihood. It also has depended on aid from Great Britain, New Zealand, Japan, and Australia. Although Japan, Australia, and the United States have made some investments, Kiribati is likely to remain economically troubled in the near future.

Despite its problems, Kiribati continues to be something of a paradise, untouched by modern life. An article by J. Maarten Troost offers a sense of this. Troost, a former resident of Washington, D.C., wrote in *The Washington Post* that he lived on "what could best be described as a sandbar in the Pacific, an island 200 yards wide, part of an obscure Micronesian nation." There were no televisions, no Internet connections, and no news magazines. He got his news from two newspapers and a radio station, which mostly reported on boats and canoes lost at sea. He rarely shaved or used such toiletries as deodorant, which he said costs $13 on Tarawa. His hair was cut by a barber at a beach, and his clothing consisted mainly of the colorful *lava-lava* cloth and flip flops.

While this makes Kiribati appear like a pleasant getaway, the island faces severe health and environmental threats. One American couple, Roger and Marilyn Young, discovered this while on a yacht trip to Fanning Island. During the trip, Marilyn Young stubbed her toe on the deck. The cut became infected, she developed the flu, and her temperature shot up to 102 degrees. Fanning Island had no doctor, and only a handful of nurses were available. The Youngs wound up traveling to Christmas Island to find a doctor. On the way, the U.S. Coast Guard intercepted them. Knowing that the Coast Guard was available reassured them, but it also gave them a sense of how difficult it would be to receive medical treatment in an emergency.

Timeline

1000 B.C. Migrants settle in Kiribati

Archaeological evidence indicates the islands in present-day Kiribati have been inhabited for at least 3,000 years. The earliest peoples were migrants from Southeast Asia. Languages in these islands resemble those from an Austronesian family which evolve in Southeast Asia and begin to spread throughout the Pacific Island basin around 3,000 B.C.

Over the next several hundred years, other travelers from surrounding Pacific Islands visit and settle in Kiribati.

Many plants that grow in Kiribati also come from Southeast Asia. The coconut and breadfruit, for instance, are believed to be native to that region. So is pandanus, a tree that produces a long, spindle-like leaf, known as lauhala, that is used for weaving mats, hats, and clothing. It also is used as roofing material.

500–400 B.C. Samoans arrive

Oral histories depict stories of a Samoan migration about five to six hundred years after the islands were settled by Southeast Asians. This argument fits well with an overall Polynesian belief that the peoples of Samoa migrated to islands throughout the Pacific.

However, much of Kiribati's traditional way of life comes to be shaped in a pattern that resembles a Micronesian culture rather than a purely Polynesian one. The *kaainga* system of settlement is one example of this. In this patriarchal system, families stay together from one generation to the next, occupying and controlling the resources of particular areas of land. When an existing *kaainga* grows too large, members of an existing group leave to set up a new *kaainga*. The amount of land that a particular kaainga occupies usually is regarded as a sign of its wealth and prestige.

Most *kaainga* depend on the sea and land for sustenance. Families tend to be large, and many have several adopted children. Elderly people are highly respected. It is believed that when people die, they are being punished by the gods. So those who survive to old age are seen as being particularly loyal to the gods. Upon death, it is believed that the spirit leaves the body and moves north to where the god Nakaa waits, making nets. Spirits that are not trapped in the nets return to hover near their ancestral lands.

A.D. 1300 Second wave of Samoan migration occurs

Some oral histories report that inhabitants of present-day Kiribati and Tuvalu travel to Samoa, and that their descendents return to the Micronesian islands several generations later. Other histories describe a wave of Samoan migration as a major event in the area's history. Although the accounts vary, it appears that people from Samoa did begin to settle in the Gilbert and Ellice island chains between the early fourteenth and mid-seventeenth centuries.

For these islanders, life was quite difficult. Unlike Samoa, the islands in Kiribati and Tuvalu lie only a few feet above sea level. Soil is sandy, and rainfall is sparse, which produces a shortage of food and fresh water. Foods that thrive elsewhere in the Pacific, such as banana and the taro root, fare poorly in these islands.

Inhabitants, over time, cultivate *babai*, a root plant similar to taro, and coconuts. Many islanders use the sap from the coconut tree to make a milky drink known as toddy.

1537 Spanish explorers sight Christmas Island

Spanish explorers in the early sixteenth century begin sailing into the South Pacific in search of *Terra Australis Incognito*, a mythical southern continent that was supposed to be filled with wealth. On one such expedition, the crew of a ship kills its captain, Hernando de Grijalva, and sets sail for the East Indies. Only seven crew members survive, but they record sighting the islands of Nonouti and Kiritimati, or Christmas Island.

1765 HMS Dolphin drops anchor

In what is believed to be the first personal contact between the islanders and Europeans, British Commodore John Byron of the *HMS Dolphin* stops in the area. Byron hopes to trade goods with the islanders. More than sixty canoes visit the ships, but the islanders traveling in the canoes show little interest in trade. Rather, they help themselves to coconuts on the ship that had come from nearby Tokelau and head back to their homes.

1777 Capt. Cook names Christmas Island

British explorer Capt. James Cook arrives on what is now known as Christmas Island (or Kiritimati) on December 24, 1777. He spends the following day on the island and names it Christmas Island in honor of the occasion.

The world's largest atoll, Christmas Island is part of the Line Island chain, one of the three major island chains that comprise present-day Kiribati. It becomes a British territory in the early twentieth century and is folded into the Gilbert-Ellice Islands Colony in 1916. During World War II (1939–45), it serves as a garrison for soldiers from the United States and New Zealand. After the war, the British use the island as a testing site for nuclear explosions, a move that generates a strong anti-European sentiment among many inhabitants.

1820–1840 Russian map maker names islands

In the late eighteenth and early nineteenth centuries, Europeans slowly gather knowledge about Oceania (the region of the south Pacific consisting of Australia, New Zealand, and the groups of islands to the north and east, including the Philippines and the Indies), as more islands are spotted, identified, and placed on world maps. Eventually, Adam von Krusenstern (1770–1846), a Russian explorer and cartographer, puts together a comprehensive atlas. He gives the Gilbert Islands their name after the sea captain Thomas Gilbert, who had commanded convict ships with co-captain John Marshall. The Marshall Islands, north of present-day Kiribati, are named after him.

1835 "Baoba" lands on Butaritari

As European ships begin to sail through the Pacific more frequently, contact between the islanders and outsiders increases. Outsiders rarely spend much time in the islands, however. They are mostly merchants who drop anchor briefly and exchange such goods as iron hoops and beads for coconuts.

One early exception to this trend is Robert Wood, nicknamed "Baoba" and "Grey." He is dropped off by a ship and is adopted by a *kaainga* on Butaritari. He learns the local island customs and participates in dancing. The family that adopts him even tattoos him with *tekitoko*, the insignia used to identify dancers.

"Baoba," in turn, teaches islanders to use coconut oil for lighting lamps at night. Before his arrival, most islanders rely solely on fires to provide night light. He also shows them that the toddy made from the sap of coconut trees can become sour, or fermented, if allowed to sit for a day or so. This sour toddy becomes a popular drink on Butaritari and spreads throughout the islands.

1840 Whalers ply area waters

Whaling ships begin to stop more frequently in the southern part of the Gilbert Island chain, particularly near the islands of Tamana and Nikunau. Sperm whales inhabit the waters near these islands, and over time, a plant is established to melt down whale fat.

The men aboard these ships often land in Kiribati in search for food and women. At first the landings are difficult because the islanders perceive the whalers, who carry guns, as enemies. Islanders often intercept ships and loot them, which gives them a reputation among the whalers for being dishonest. The islanders consider the ships' bounty to be their property since the boats had entered Kiribati waters.

However, the whalers eventually establish trade relationships. They offer islanders sticks of tobacco in exchange for women. The women often are taken aboard ships for several days while whales are being hunted. They often return with such highly desirable items as axes, knives, nails, and cooking pots, and with stories about the world beyond their islands.

1846 First European trading post is established

Two traders, known as Randell and Durant, arrive in the Gilbert Islands. Randell establishes a trading post on Butaritari, while Durant sets up shop on Makin. Both men encourage islanders to manufacture coconut oil, items made from turtle shells, and *beche-de-mer*, a dried fish.

The traders make big profits, largely because the islanders are ignorant of such things as money. Nevertheless, they are folded into the local societies by adapting certain aspects of the culture.

1852 Christian missionary activity begins

Members of the Protestant American Board of Commissioners for Foreign Missions visit the islands in 1852. Even though the king of Butaritari and Makin does not want missionaries on the island, the trader Randell serves as a guide and interpreter for the group. The London Missionary Society begins establishing missions in the southern part of present-day Kiribati in 1870, using teachers from such Polynesian islands as Samoa. Beginning in 1880, Catholics from Tahiti start converting residents on Nonouti.

Despite these efforts, many missionaries have a difficult time converting islanders to Christianity. As one historical account suggests, people would attend sermons because the missionaries would hand out gifts. But when asked to accept the new god, Jesus, many back away.

Missionaries are aghast over island lifestyles. Most residents do not wear clothing; many like to dance and drink sour toddy. Polygamy, prostitution, and sexual promiscuity are rife. Although anthropologists such as Arthur Grimble insist that the missionaries fail to understand the nuances of Gilbertese culture, the missionaries try to force islanders to stop going about naked and ban such activities as dance.

1860–90 Natives are pressed into labor

By the mid-nineteenth century, Europeans begin to seize control of much of Asia, Africa, and various islands throughout the world. The colonies become places where the Industrial Revolution is played out. Europeans use their possessions as a source of raw materials and a place where they can manufacture goods.

As part of this industrialization process, a system of indentured labor is established. After slavery is eliminated in the Americas, Europeans turn to places such as Kiribati and Tuvalu for cheap labor. Often, the islanders are carried onto ships against their will.

One incident from the island of Makin offers an example of how islanders become laborers. A ship pulls in, and the crew invites the people to dance on board. The islanders then are fed a grog, which causes them to fall asleep. The anchor is lifted, and the ship sails away. Islanders awaken, disoriented and without a clue as to where they are heading.

In this particular story, the islanders end up in Hawaii, where they work on plantations. In other cases, inhabitants are sent to Guatemala, Peru, or even as far north as the Pacific Northwest logging camps in the United States and Canada. Sometimes, the islanders remain in their new homes. Often, however, they return and are not recognized by their families because the journey away changes them so deeply. Historian David Chappell refers to these returnees as "double ghosts."

Not all of the labor ships were dishonest, and many accounts suggest that islanders often left their homes voluntarily. Unscrupulous practices occurred often enough, however, that an Office of British High Commissioners to the Western Pacific was established in 1877 to control abuses.

1864 Rev. Bingham translates Bible into Gilbertese

The Rev. Hiram Bingham (1831–1908) arrives in the islands from the United States in 1857, with his wife and two converts to Christianity from Hawaii. Bingham was born in 1831 in Hawaii and was the son of one of the first missionaries sent to Hawaii. After attending Yale and the Andover Seminary, he is sent to the Gilbert Islands.

Like many early missionaries, he has difficulty converting islanders because he cannot speak the local language. Although Bingham builds a church in 1859 on the island of Abaiang, he is best known for translating the Bible into Gilbertese. In doing this, he also creates a written form of the local language. After the Bible is published in 1864, a printing press is sent to the islands from the United States.

1892: May 27 British rule ends civil wars

Throughout the nineteenth century, a series of wars erupt in the islands of Abemama, Tabiteuea, Maiana, and Tarawa. Though the wars themselves are not unusual, the nature of the fighting changes with the introduction of deadly items brought by Europeans, such as guns, cannons, and axes. In addition, missionary activity becomes a source of many wars. Rulers who convert to Christianity often try to force the new faith upon the people, which leads to strife. Converts to one particular denomination also begin waging war with converts to another form of the Christian faith.

During this time period, European powers—particularly the British—begin patrolling the Pacific more vigilantly. These patrols are mainly designed to control the lawless behavior of Europeans, but gradually begin to take over governance of the islands.

On Tarawa, ten wars occur between chiefs of the various *kaainga* who are fighting for land and control of the island. The wars span about four generations and end in 1892 with a war at Nea between two chiefs known as Tekinaiti and Matang. During this war, Tekinaiti's people are said to have learned through their gods that visitors would be arriving on the island. Those visitors turn out to be the crew members of the *HMS Royalist,* a British ship commanded by Capt. E. H. M. Davis. Realizing that the only way to stop the fighting would be to establish a protectorate, Davis raises the British flag over present-day Kiribati on May 27, 1892.

1900 Phosphate is discovered

Banaba, an oyster-shell shaped island to the west of the Gilbert chain, was named Ocean Island by Europeans in 1804, in honor of a ship called *Ocean,* which had visited the island. It is ignored until Albert Ellis, a New Zealander and an employee of a British mining company, discovers that four-fifths of its land consists of rocks containing valuable phosphate deposits.

The British quickly annex the island, and through an unscrupulous (unmindful of right and wrong) deal, secure a 999-year lease. They then launch a full-scale phosphate industry.

The people of Banaba object to the mining of their land, but to no avail. As phosphate production intensifies, the British transport inhabitants of the Gilbert and Ellice islands to Banaba to work in the mines. They also recruit Japanese and Chinese plantation workers, which changes the character of Banaba considerably.

Eventually the Banabans lose their land and in 1946 are relocated to Rabi, Fiji. Although the lease eventually is rescinded, Banaba is stripped of its resources and remains only sparsely populated today.

1916 British colony is established

After establishing the Gilbert and Ellice chains as a protectorate, the British centralize their rule and establish a series of local governments. They declare Banaba, Fanning, Washington, the Gilbert and Ellice islands all to be one colony. Christmas Island is added to the colony in 1919 and the Phoenix Island group in 1937.

Although the British rule initially strives to preserve the individual autonomy of each island, the imperial administration is unable to satisfy that goal. An administrative headquarters is established on Banaba. Because phosphate becomes so profitable, the British run the colony by putting their interest in that industry at the top of their priority list. As a result, a traditional way of life deteriorates, and development of the other islands generally is neglected. Work in the phosphate mines becomes the primary means of subsistence among the islanders, and gradually such traditional foods as coconuts are replaced by canned imports.

British rule also transforms the islanders' way of life. Nudity becomes a public offense, and dancing in public is allowed only on Christmas, New Year's, and the Queen's birthday. In addition, the *kaainga* system is dismantled, and communal villages with local governing units are set up. All adults are required to spend at least four hours a week building and maintaining roads and community projects. The British also impose a 9 p.m. curfew and refuse to allow residents to leave their home island without a permit from the magistrate.

1937: July 2 Amelia Earhart disappears

American aviator Amelia Earhart (1897–1937) mysteriously vanishes from radar screens while attempting to fly to Howland Island, a U.S. possession in the Pacific Ocean. Earhart's glamorous piloting adventures were among the most closely-followed news events of the 1930s, and her mysterious disappearance has remained one of the twentieth century's great unsolved mysteries.

Neither her plane nor her body has been found. Some believe that her plane might have been shot down by Japanese fighters, but most concur that she and co-pilot Fred Noonan ran out of fuel and crashed into the Pacific.

Recent evidence indicates that Earhart may have died on Mikumaroro Island in Kiribati. In 1940 British soldiers turn up a skeleton which is believed to have been that of a human male. Records found in 1998 on the island of Tarawa, however, suggest that the skeleton actually was that of a white female. Even if the body is not that of Earhart, researchers believe a new analysis of the skeleton may shed additional light on her disappearance.

1941: December 8 Japanese bomb Banaba

Japan bombs Pearl Harbor, Hawaii, on Dec. 7, 1941, which draws the United States into World War II (1939–45). One day later, Japanese war planes drop six bombs on the British administrative headquarters on Banaba. A French cruiser evacuates the European and Chinese workers, and shortly afterwards Japan occupies the island.

Over the next few months, Japanese troops occupy most of the British colony. Most Europeans are evacuated, and the islands generally become isolated. Japanese administrators force the islanders to serve as sentries, looking out for enemy ships. They also ration food and order the islanders to serve at their command. Most obey out of fear.

1943 U.S. invades islands

After abandoning plans to move through Japanese territories in the Pacific and eventually arriving on the coast of Japan, Allied troops decide to make a drive through the central Pacific. The United States begins an amphibious training program in Hawaii and launches an invasion of Micronesia by landing in the Ellice Islands.

U.S. Marines invade Tarawa from the Ellice Islands. The Battle at Betio on Tarawa begins November 20. After a bloody, five-day assault, U.S. Marines capture Tarawa from the Japanese and take control of the Gilbert and Ellice Colony.

U.S. troops remain in the islands until 1946. During the U.S. occupation, many islanders are hired to rebuild homes, buildings and the British ruling apparatus. The American presence also brings more exports into the islands. Manufactured products and processed foods rapidly begin to replace handmade items and foods grown on the islands. On Tarawa, blasted bunkers and rusted pieces of artillery provide a reminder of the battle.

1965 Steps toward independence begin

After World War II, British rule is restored. The British begin to reestablish a system of local governments, and with assistance from the United States and groups such as Christian missionary societies, begin building schools and other community facilities.

The war drains the British treasury, however, and makes many of its colonies a financial liability. At the same time, the United Nations starts criticizing colonial rule, and many nations in Asia and Africa gain independence. On top of this, the island of Nauru gains independence from Australia, which causes the price of phosphate to rise. All of these changes make the Gilbert and Ellice Island Colony increasingly expensive for the British government to administer, so the country tries to rid itself of the colony.

A Native Magistrates' Conference begins meeting annually in 1952 to provide training to local governmental leaders. This is followed by the establishment of the biennial Colony Conference in 1965. The conference provides a forum where residents debate issues, air grievances, and propose ideas.

1974 Ellice Islanders choose to separate

When the British formed the Gilbert and Ellice Island Colony, they brought together dozens of islands with little in common. There were similarities among the islanders within each group, but the Gilbert and Ellice people shared no common ideological belief or language. The people of the Gilbert islands regarded themselves as Micronesian, while those of the Ellice chain saw themselves as Polynesian. The Gilbertese descend from Micronesian migrants who came from the group of islands to the northwest, while the Ellice Islanders are descentants of Samoan migrants. Each recognized separate leaders.

As a movement toward independence intensifies, these differences become more apparent. Ellice Islanders realize that they are a minority within a predominantly Gilbertese culture and fear that they will not be able to command power in a new nation. Many Ellice people also regard the Gilbertese as backward. At the same time, many Gilbert Islanders feel that residents of the Ellice chain have educational advantages over them and receive favorable treatment from foreigners.

Such differences lead to a referendum in which the Ellice Islands vote to separate. The group becomes the independent nation of Tuvalu, while the remaining islands are to become the nation of Kiribati.

1975 Banabans sue British government

After being relocated to Rabi, Fiji, many Banabans want to return to their home island. They sue the British government, seeking damages. They declare their wish to be an independent nation, separate not only from Great Britain, but also from the Gilbert and the Ellice island groups.

Phosphate production ends in 1979 when the deposits are exhausted. A court settlement in 1981 establishes a trust fund to redevelop Banaba, and a commission makes an inquiry into the issue in 1985. However, little is accomplished, and most Banabans continue to live on Rabi. In 1997 Kiribati president Teburoro Tito announces a plan in which Australia, Great Britain, and New Zealand will build a new airport, road, water, and port facilities on the island.

Kiribati eventually recognizes the minority rights of Banabans by reserving two seats for the group in Parliament. One Banaban representative is chosen from Banaba, while the other is chosen from Rabi. Banaba's desire for independence, however, remains unmet.

1979: July 12 Gilbert group wins independence

The islands are established as a democratic republic, with a unicameral legislature. Members of this body, the House of Assembly, are elected every four years. The legislature initially has thirty-five members, but this number increases to thirty-nine in 1987. Chief minister Ieremia Tabai becomes president, or *beretitenti.*

Under the constitution, the president is elected every four years and can serve up to three terms. Island councils administer local governments.

The new nation is known as Kiribati, a local way of saying Gilbert. The slogan "I Kiribati" is adapted to describe the inhabitants of the nation.

1982 National health plan is prepared

Kiribati develops a health program that is designed to serve the individual needs of people on each of its thirty-three islands. Because the population is so diverse and spread out, the program develops nationwide community self-help groups. Health staffers, nurses, and medical assistants on each island receive training from the national government. They then tailor the programs to meet the needs of individual villages.

The program places special emphasis on maternal and child health, and family planning services, and often encourages islands within the nation to compete for the title of "healthiest island."

1995: October Kiribati protests nuclear tests

An organization known as the South Pacific Forum cuts ties with France after the European power conducts a series of nuclear tests near Tahiti. Kiribati, along with Nauru, severs these ties well before the tests occur below the Fangataufa atoll, which is part of French Polynesia, a colony of France.

Ongoing nuclear tests after World War II have posed a grave concern to many island nations in the Pacific, largely because it is the people of these islands who must live with the consequences. Kiribati has insisted on a nuclear-free status since in 1984. Even though the nation has been independent for nearly two decades, its people have not forgotten how the British disregarded the feelings of the people when nuclear tests were conducted on Christmas Island in 1956–58 and in 1962. For many, the French tests represent a return of such behavior.

1997 Kiribati creates 'Millennium Island'

Kiribati changes the name of its easternmost island from Caroline Island to Millennium Island, claiming that it will be the first spot in the world to greet the beginning of the twenty-first century on January 1, 2000. The move is designed to generate tourism, but Millennium Island is uninhabited and unable to accommodate visitors. Kiribati officials hope visitors will stay on Christmas Island and travel to Millennium Island via cruise ship.

1998 Cruises from Hawaii to Kiritimati begin

Norwegian Cruise Line, a cruising company based in Miami, begins a regular round-trip cruising service between Honolulu, Hawaii, and Kiritimati, or Christmas Island. This marks the first time that a cruise has been offered from Hawaii to another Pacific Island other than Tahiti and reflects a belief that remote locales such as Kiribati represent attractive vacation spots for Americans and Europeans.

The cruise ships, however, carry more than passengers. Ham radio operators on the Big Island of Hawaii periodically collect books, bandages, stethoscopes, and bedpans to be transported to Fanning Island. These donations, in some ways, are needed more desperately than tourist dollars. Fanning Island, like many remote Pacific Island outposts, has no airplane landing strip. It is filled with cesspools that often drain into the island's water table, making the potential for infection strong. Even though there are medical services on Kiritimati, getting there requires traveling 160 miles by sea. The next nearest neighbor is Hawaii, 1,000 miles away.

1999: March Water supply hits low

An El Niño weather pattern in 1998 produces a drought in Kiribati that continues for almost a year and nearly depletes the islands' natural water supplies. Water comes from rain that people capture in tanks and from shallow wells that lie underneath the country's various atolls.

President Tito declares a national disaster, and the government announces plans to spend up to $1 million on a machine to convert sea water to drinking water. Tito also asks

Australia and other nations to help Kiribati finance a desalination plant.

Bibliography

Adams, Wanda. " 'Double Ghosts' Remembers Early Traders," *Honolulu Advertiser*, 29 March 1998.

"Bones Found in '40 May Have Been Hers," *Honolulu Star-Bulletin,* 3 December 1998.

Chappell, David. *Double Ghosts: Oceanic Voyagers on Euroamerican Ships.* Armonk, N.Y.: M. E. Sharpe Press, 1997.

Kaser, Tom. "Christmas Gift for Cruising," *Honolulu Advertiser,* 13 April 1998.

Kiribati: Aspects of History. Tarawa, Kiribati: Ministry of Education, Training and Culture, 1979.

"Kiribati's Banaba Island to Undergo Rebuilding." *Pacific Islands Report.* Pacific Islands Development Program/Center for Pacific Island Studies, 28 August 1997.

Krauss, Bob. "Heroes Heed Call From Sea," *Honolulu Advertiser*, 10 March 1999.

Kreifels, Susan. "Pacific Islands Vie to Greet Millennium," *Honolulu Star-Bulletin*, 1 January 1998.

MacDonald, Barrie. *Cinderellas of the Empire: Towards a History of Kiribati and Tuvalu.* Canberra, Australia: Australian National University Press, 1982.

McCreery, David and Doug Munro. "The Cargo of the Montserrat: Gilbertese Labor in Guatemalan Coffee, 1890–1908" *Americas* 49,3 (January 1993): 271–296.

Soetjahja, Iwan. "The Healthiest Island." *World Health* (March 1990): 12–14.

"South Pacific group cuts French ties," *Honolulu Star-Bulletin*, 3 October 1995.

Thompson, Rod. "Hilo Plans Gift for Christmas Isle Neighbors," *Honolulu Star-Bulletin,* 18 December 1998.

Troost, J. Maarten. "Taking Atoll," *Washington Post*, 28 June 1998.

North Korea

Introduction

The Democratic People's Republic of Korea (DPRK) is one of the world's few remaining Communist countries. An isolated and secretive nation, it was created after World War II when the Allied powers divided Korea along the thirty-eighth parallel. This division of the Korean peninsula into the DPRK in the north and the Republic of Korea in the south was later consolidated by the Korean War (1950–53). Following the collapse of its major ally, the Soviet Union, in 1991, there has been increased pressure on North Korea to adopt a greater degree of openness in its foreign relations, especially with the United States, and to make more meaningful progress toward a peaceful reunification with South Korea.

The Korean peninsula extends southward about 625 miles (1,000 kilometers) from the northeastern corner of China. To its east lies the Sea of Japan; to the west, the Yellow Sea (Koreans call these the East Sea and the West Sea). With an area of 46,541 square miles (120,540 square kilometers), the DPRK occupies slightly over half the peninsula and is composed mostly of mountainous terrain. The majority of its estimated 24 million people are concentrated in the low-lying portions of the country, which include the capital city of Pyongyang.

History

Korean civilization has a long history, and one in which China, its closest neighbor, has played an important part. Some rulers of Korea's ancient kingdoms were Chinese, and China's Han dynasty (202 B.C.–A.D. 220) invaded Korea (then known as Choson) in 108 B.C. to put an end to possible competition over valuable overland trade routes. The Chinese ruled a portion of Korea for nearly 300 years, bringing with them many aspects of their culture, such as Confucianism and a variety of technological advances. By the time the Chinese were driven out, three new kingdoms—Silla, Paekche, and Koguryo—had spread throughout the land. Between 668 and 935 all of Korea was unified under the Silla kingdom. This was an important period, both politically and culturally. It was the first time that the entire Korean peninsula had had a single language, culture, and social structure. However, much of this unified culture showed Chinese influences, such as the civil service system and the design of the capital city of Ch'angan.

Beginning in the tenth century, Korea became a target for foreign invaders, and in the thirteenth and fourteenth centuries it came under the domination of the Mongol Yuan empire (1260–1268). The Mongols were ousted by a Korean general, Yi Song-gye (1335–1408), who began one of the longest-lived dynasties in world history—the Yi (or Li) dynasty, which ruled Korea from 1392 to 1910. The first century of dynastic rule was an especially rich period culturally. A written version of the Korean language, called *Hangul*, was created, leading to the development of a vernacular (in the native tongue) literature. Important scholarly and administrative works were written, and technological advances, including the rain gauge, were perfected.

Repeated invasions by Japanese emperor Hideyoshi Toyotomi (1536–98) at the end of the sixteenth century left Korea in a weakened state, and it fell under the control of China's Manchu rulers in 1636. The seventeenth century also inaugurated a 300-year period of political, cultural, and economic isolation during which Korea came to be known as the Hermit Kingdom. Foreign influence during this period was extremely restricted, and the institutions of Korean society remained timelessly locked in place for nearly three centuries. The Kangwha Treaty of 1876, opening Korea to trade with Japan and the West, was the first major step toward reestablishing relations with the rest of the world.

Unfortunately, Korea's neighbors brought expansionist goals to these new relationships. In the Sino-Japanese War of 1894–95, the Japanese drove the Chinese out of Korea, but only so that they could later assume power there themselves. Japanese control of Korea was further increased by the Russo-Japanese War of 1904–05, and Japan annexed the peninsula outright in 1910, beginning a thirty-five-year period of oppressive foreign rule. The Japanese treated the Koreans as second-class citizens in their own country, confiscating large amounts of land and attempting to supplant the Koreans' language and culture with their own.

The Japanese occupation of Korea continued until the end of World War II (1939–45), when the United States and the Soviet Union divided the country into an American zone

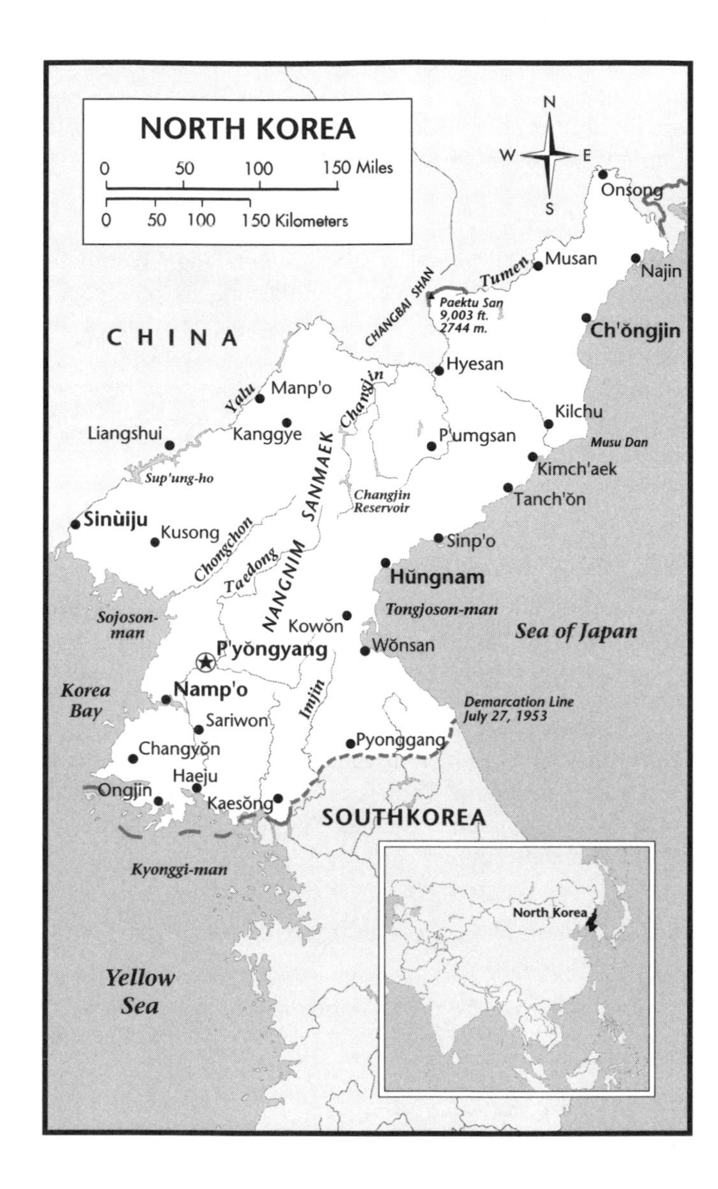

south of the thirty-eighth parallel and a Soviet zone to the north. The Democratic People's Republic of Korea, ruled by the Korean Workers' Party and supported by the U.S.S.R., was established in the north in May 1948, and the U.S.-supported Republic of Korea was formed in the south shortly afterward. In June 1950, troops from the north entered South Korea in an attempt to reunify the peninsula by force but were eventually driven back by U.S. and British troops. The battle lines were again stabilized back at the thirty-eighth parallel and an armistice agreement was signed in 1953.

Kim Il Sung (1912–94), the leader of the DPRK since its inception in 1948, remained in control until his death in 1994. A series of economic plans were implemented beginning in the 1950s, concentrating on heavy industry, with foreign aid coming mostly from the Eastern European Communist Bloc. From the 1950s to the mid-1970s, the country's economy was strong, led by labor-intensive export industries. After this point, however, it slowed due to competition from abroad and other factors. With the dissolution of the Soviet Union and the collapse of Communism in Eastern Europe, North Korea lost important trade partners and allies. In 1995 Russia declined to extend the mutual cooperation pact that had been in effect between the two nations since 1961.

North Korea was forced to adopt a more cooperative stance toward South Korea and the United States. In 1994, Kim Il Sung, the longest-serving world leader, died and was succeeded by his son, Kim Jong Il (1941–)(although the latter's succession to power was not made official until three years later). At the end of 1996, North Korea agreed to hold talks with the United States and South Korea on adopting a permanent peace agreement, formally ending the Korean War of 1950–53. This marked the first time since the end of the war that the DPRK had agreed to meet with representatives of South Korea.

Timeline

c. 3000 B.C. Arrival of Tungusic tribes

Neolithic tribes migrate to the Korean peninsula from northern China and Manchuria. Some combine farming with hunting, while others practice settled agriculture.

c. 1122 B.C. Kija becomes ruler of Choson

A nobleman fleeing China following the overthrow of the Shang dynasty (1766–1122) takes over the leadership of the ancient kingdom of Choson (present-day Korea). This country is said to have been founded in the third century B.C. by the legendary Tangun, fabled to be the offspring of a prince and a bear turned into a woman.

194–180 B.C. Wiman conquers Choson

Wiman, a military figure from China (possibly of Korean stock), gains power in the kingdom of Choson and overthrows its ruler, King Chun, said to be a descendent of Kija.

108 B.C. The Chinese invade Choson

The Chinese Han dynasty (202 B.C.–A.D. 220), perceiving a threat in the control of overland trade routes by the expanding Wiman Choson kingdom, seizes Choson after a year of bitter fighting. King Ugo (the grandson of Wiman) is assassinated, and the capital city is captured. The Chinese rule Choson, which they divide into four prefectures (regional district ruled by a government official) with their capital at Pyongyang.

37 B.C.–A.D. 313 Chinese rule

By 37 B.C. only the prefecture of Nangnang (Lolang in Chinese) is still under Chinese control, and it remains so for nearly three more centuries, during which many cultural and technological advances of the Chinese, including Confucianism, are absorbed.

Confucianism is a Chinese philosophy based upon the teachings of the Chinese philosopher Kung fuzi (Confucius, 551–479 B.C.). Confucianism is a prescription for ethical behavior in personal relations as well as governance.

A.D. 1st century–7th century Three kingdoms vie for dominance

Three new kingdoms emerge in the lands formerly controlled by the Chinese. Spread out among Korea's three natural topographical divisions, they are Silla in the southeast, Koguryo in the northwest, and Paekche in the southwest.

4th century Buddhism is introduced

Buddhism is introduced to Korea, from which it later travels to Japan. Buddhism originates in the Ganges valley in India during the fifth century B.C. The religion is based on the teachings of Siddhartha Gautama (c. 563–c. 483 B.C), a Hindu prince who comes to be called the Buddha (the Enlightened One or the Wise One). Buddhism, in its original form, has no gods, no priests, no social classes. Salvation is within the reach of anyone who accepts the Buddhist *Dharma* (Law) and the beliefs embodied in the Four Noble Truths. These four basic tenets are: (1) suffering exists, (2) suffering is caused by ignorance, (3) ignorance and desire must be eliminated to achieve Nirvana (the ultimate state), and (4) one can achieve Nirvana by following the eightfold path of behavior.

313 Koguryo drive the Chinese out of Nangnang

The aggressive and powerful Koguryo end Chinese rule in Korea by overrunning Nangnang.

372 First Korean school is founded

The *Taehak* (Great Learning) is established.

647 Astronomical observatory is built in Kyongju

An observatory, believed to be the oldest in the Far East, is built in the capital of the Silla kingdom.

660–68 Silla gains control of the Korean peninsula

Aided by an alliance with China, Silla conquers the kingdoms of Koguryo and Paekche.

668–935 Silla Unification

The kingdom of Silla rules over the whole of Korea, unifying its language, culture, and social structures for the first time. Many aspects of the Chinese Tang dynasty (618–906) culture also become widespread; the capital at Kyongju is modeled on the Tang capital of Ch'angan, and government administration emulates the Chinese civil service system, including its use of examinations to obtain government posts. A system, called *idu*, is developed for writing the Korean language using Chinese characters. Using this system, vernacular (commonly spoken) Korean poems called *hyangga* are first recorded.

A strict system of classes (called "bone-rank") is observed, with children of aristocrats receiving a Confucian education (based on the teachings of the Chinese philosopher Confucius). The economy flourishes through expanded trade, and Buddhism becomes increasingly important, inspiring the creation of many elaborate temples, shrines, and pagodas. Artworks include gigantic bronze bells (such as the Emille Bell weighing twenty-three tons), bronze statues, and delicate engravings.

682 First school for government workers is founded

Korea's first school for training civil servants, known as *kukhak*, is established.

857 Birth of poet Ch'oe Ch'i-won

Ch'oe Ch'i-won (b. 857) is an important early creator of *hansi* poetry, written in Chinese according to Chinese literary conventions. *Hansi* poetry remains popular for centuries and into the twentieth century.

918 Kingdom of Koryo is founded

Wang Kon (877–943) (later known as Taejo) establishes the kingdom of Koryo, from which the modern name "Korea" is derived.

936–1392 Koryo dynasty rules Korea

Defeating two other rebel leaders, Wang Kon gains control of Korea following the downfall of the Silla kingdom. To forge an alliance with the former Silla rulers, he allows many members of the ruling elite to retain their positions, including the Silla king who holds the highest government office. Wang Kon also marries a Silla princess. Buddhism remains the state religion. The Koryo period is one of warfare and invasion by rival empires, including the Khitan Liao and Chin states in northern China and Manchuria and, later, the Mongols, who carry out multiple invasions, eventually adding Koryo to their empire. The strain of continued warfare takes it toll on the Koryo government, which falls prey to intrigues and corruption.

957 State examination system is established

The Koryo kingdom sets up a system of state examinations for government workers based on the Chinese system. Scholars commonly apply for posts in the civil service, and the examination questions are based on the teachings of Confucius.

1231 First Mongol invasion of Korea

Mongols invade the kingdom of Koryo and relocate its capital to the Kanghwa Island.

1236–51 New set of Tripitaka woodblocks is carved

A set of woodblocks for printing Buddhist scriptures is created to replace the set destroyed in the Mongol invasion (see 1231). The set, consisting of over 81,000 blocks, represents the most sophisticated book production method until the invention of movable metal type in the fourteenth century. The blocks have been preserved up to the present and are still usable.

1254 First book is published using movable type

During the period when the Koryo court is in exile on Kanghwa Island, movable metal type is developed, a century before its invention by Johannes Gutenberg (1390–1468) in Germany. The first book published using movable type is a fifty-volume anthology of religious texts by Ch'oe Yun-ui.

1258 Koryo comes under Mongol domination

The Koryo leaders make peace with the Mongols, and their government comes under the control of the Mongol Yuan empire (1260–1368). Mongol officials take charge of government administration and exact heavy taxes from the population. Koryo crown princes are forced to marry Mongol princesses and live in the Mongol capital. One marries the daughter of Kublai Khan (1216–94).

1277 Glass tile is developed

The invention of glass tile, leading to the development of the arts of ceramics, is credited to a Korean Buddhist monk.

c. 1285 Oldest known folk collections appear

The earliest recorded collections of folk literature, *Samguk sagi* and *Samguk yusa*, are written down.

The Yi Dynasty

1392 Yi dynasty begins

A Koryo general, Yi Song-gye (1335–1408), ends Mongol control of the country and founds the Yi Dynasty (also called the Li dynasty), which will rule Korea until 1910. The capital is moved to Seoul. Yi Song-gye changes the name of the country back to Choson and inaugurates ties with China's new Ming dynasty (1368–1644), establishing an alliance known as *sadae*. In this system, modeled on the Confucian relationship known as "elder brother/ younger brother," the Yi king is subordinate to the Chinese emperor.

1392–1500 Yi era begins with period of peace and cultural development

The first century of the Yi dynasty was its high point, marked by peace and cultural advances, including the development of a written Korean language (see 1446), the invention of a rain gauge, and publication of many types of texts with government support. A set of administrative guidelines, the *Kyongguk taejon*, is issued. Also recorded are a history of the Koryo period (*Koryo sa*) and a definitive anthology of Korean music (*Akhak kwebom*). Painters such as An Kyon (1418–?) begin to develop a distinctly Korean style, in terms of their brushwork, composition, and treatment of space.

After 1500 the quality of Yi leadership and daily life begin a long decline due to political factionalism within and invasion from abroad.

15th century Koreans produce the first encyclopedia

Korean scholars compile what is believed to be the world's earliest encyclopedia, consisting of 112 volumes.

15th century First fiction is written down

Kim Sisup (1435–93) writes *New Stories of the Golden Turtle*, generally considered the first fictional literature in Korean.

1446 Hangul written language is developed

A written version of native Korean speech is created, hastening the development of a vernacular literature. Up to this point, all Korean written material has been in Chinese. The poetic form known as *sijo* flourishes with the introduction of Hangul.

1587 Birth of poet Yun Son-do

Yun Son-do (1587–1671) is considered the greatest master of the *sijo* poems (see 1728). In addition to writing poetry, he conducts an active political career. His most famous work is a cycle of forty poems, called *The Fisherman's Calendar*, that describe each of the four seasons as experienced in one of the poet's favorite settings. Altogether, seventy-six of Yun Son-do's *sijo* survive. The poet dies in 1671.

1592 Japanese emperor attempts to invade Korea

Japan's new emperor, Hideyoshi Toyotomi (1536–98), sends forces to Korea as part of a larger effort to invade China as well. They are turned back by troops under the leadership of Yi Sun-sin (1545–98), aided by forces from China's Ming rulers. The vessel—or "turtle ship"—carrying the Korean military leader is the first iron-clad armored ship to be used in battle.

1597 Japan tries to invade Korea for a second time

Hideyoshi again orders an invasion of Korea, but the Japanese emperor dies a year later, and the campaign ends without success. However, defending itself against the Japanese attacks weakens the Yi kingdom and hastens its decline and subsequent conquest by the Manchus of China. Much of Korea's agricultural land is ruined, and the kingdom suffers an early version of "brain drain," as many talented artists and other persons with specialized skills are forcibly transported to Japan to take advantage of their talents.

The Hermit Kingdom

1600–1900 Korea enters a period of isolation

From the seventeenth through much of the nineteenth century, Korea enters a period of isolation from foreign influence, becoming known as the Hermit Kingdom. Like the elite in Japan, which experiences a similar period under the Tokugawa shogunate (military dynasty), the entrenched aristocracy in Korea fears that its grip on power would be threatened by exposure to new ideas from abroad and the increased middle-class power that would result from growing trade with other nations. Thus, contact with outsiders is halted, and Korea's social structure and culture are virtually frozen in place for nearly 300 years.

1636 Manchus gain control of Korea

China's Manchu dynasty (1644–1912) wins control of the Yi kingdom, which is forced to cut its ties to the Ming rulers and becomes a vassal or subordinate state of the Manchus. Yi power is further curtailed by internal squabbling between political factions.

1671 Death of poet Yun Son-do

Acclaimed *sijo* poet Yun Son-do dies. (See 1587.)

1689 Publication of *The Nine Cloud Dream*

Authored by Kim Man-jung (1637–92), *The Nine Cloud Dream* is considered a milestone in the development of Korean fiction. It is noted for its use of symbolism and its examination of the contrasts between Buddhism and Confucianism and of human imperfection itself. Kim is also the author of *The Story of Lady Sa*, a satire on concubinage (the state of being a secondary wife in a polygamous society).

1724–1800 Two Yi monarchs arrest dynastic decline

Two effective rulers of the eighteenth century—Yongjo (see 1724) and Chongjo (see 1776)—prompt a resurgence in Yi power and a cultural renaissance known as the Practical Learning Movement (*Sirhak*). Increased value placed on empirical knowledge and practical matters brings with it a new appreciation for prose writing, including the composition of satirical, realistic fiction. A leader in this area is Pak Chiwon (1737–1805), whose works satirize aristocratic life.

Chinese and Korean poetry anthologies appear, and there is renewed respect for literature written in Hangul (the Korean alphabet), such as the novels *Chunhyang chon* ("Tale of Miss Spring-Fragrance") and *Kuun mong* ("Cloud Dream of the Nine").

1724–76 Reign of Yongjo

The Yi monarch Yongjo reigns. Yongjo succeeds in halting—albeit, briefly— the process of Korean decline.

1728 First anthology of *sijo* poetry is published

The first collection of *sijo*, *Ch'onggu yon-gon*, is published. This form of poetry has been written since the development of Hangul in the fifteenth century (see 1446) and may have existed in an oral form even earlier. In the most common form of *sijo*, called *p'yong-sijo*, the poem consists of three lines with fourteen to sixteen syllables in each line. The syllables are divided into four distinct sections, each introduced by the speaker (or singer) taking a breath, and the total number of syllables in the poem must not exceed forty-five.

1776–1800 Chongjo reigns

Reign of the second effective eighteenth-century Yi monarch, Chongjo. Along with his predecessor, Yongjo, he attempts to halt Korea's decline. He is followed by a succession of ineffective monarchs.

1785 Roman Catholicism is banned in Korea

Roman Catholicism, which has been introduced to the country by Koreans returning from China, is officially prohibited.

1860 Ch'ondogyo religion is founded

Ch'ondogyo (Religion of the Heavenly Way), a religion combining elements of Christianity and Buddhism, is established by Ch'oe Che-u. Originally called *Tonghak* (Eastern Learning), it emphasizes the present life over the afterlife and the elimination of injustice and corruption on earth. Its rapidly growing popularity among Korea's poor results in persecution of its followers and the death of its founder. It remains one of Korea's major religions throughout the twentieth century.

1864–73 Yi Ha-leng (1820–98) holds power as regent

In one of the final decades of the Yi dynasty, power resides with the Taewon'gun, or regent (the father of the king, Kojong), who restores some of the Yi power. However, he strongly upholds the policy of isolationism. Under his rule, thousands of Christians are persecuted.

1870s ***Kendo* is introduced to Korea**

The Japanese sport of *kendo* (fencing with bamboo swords) is introduced to Korea and later adopted as a training discipline by the military and the police.

1875: April 26 **Birth of Syngman Rhee**

Syngman Rhee (1875–1965), future president of the Republic of Korea, is born in Seoul to a family descended from the Yi dynasty. As a young adult, he is converted to Christianity and begins agitating for more democratic rule by the Korean monarchy. Between 1904 and 1910, Rhee lives in the United States, where he receives degrees from George Washington University, Harvard, and Princeton. Rhee spends the years of the Japanese occupation (see 1910–45) in Hawaii and Washington D.C., working for Korean independence and establishing a government-in-exile. He returns to Korea in triumph in October 1945 and begins a campaign opposing Communism.

In 1948 Rhee becomes the first president of the newly formed Republic of Korea (South Korea). He presides over early reforms, including land reform and the establishment of universal education. Rhee is reelected in 1952, 1956, and 1960, but his increasingly dictatorial methods arouse protest at home and censure (disapproval) abroad. Widespread fraud in the 1960 election sparks a rebellion that leads to Rhee's resignation on April 27. He goes into exile in Hawaii, remaining there until his death in July 1965.

1876 **Kanghwa Treaty is signed with Japan**

In the first move toward ending its centuries of isolation, Korea signs a treaty with Japan that opens it to trade with both Japan and the nations of the West.

1882 **Korea signs trade agreement with the U.S.**

Korea and the United States sign a commercial pact, which is followed by similar agreements with other Western nations.

1884 **Korea opens its ports to foreign trade**

Due to pressure by Japan and the West, Korea's ports are opened to foreign commerce.

1885 **Western music is introduced to Korea**

Hymns, the first form of Western music learned by Koreans, are brought to the country by the first U.S. missionaries, Horace G. Underwood (1859–1916) and Henry G. Appenzeller (1858–98).

1886 **First women's college is founded**

Methodist missionaries from the United States. establish Ehwa Women's University in Seoul.

1894–95 **First Sino-Japanese War**

The Japanese take advantage of peasant uprisings surrounding the 1894 Tonghak rebellion to rapidly crush Chinese power in Korea, forcing China to recognize Korean independence in the Treaty of Shimonoseki. The growing influence of the Japanese themselves, however, erodes this very independence.

Early twentieth century ***Sinmunhak* movement in literature**

Sinmunhak, or "new literature," written in Hangul, expresses a new national consciousness. It represents a reaction against the domination of the Chinese language and literary traditions in Korea. *Tears of Blood* (1906) is an example of the new novels associated with this movement.

1904–05 **Russo-Japanese War**

Russia and Japan go to war over Korea. The Japanese occupy the Korean peninsula, increasing their control of the country, and Korea is declared a Japanese protectorate.

1908–28 **Literary magazines serve as forum for new ideas**

A series of literary magazines, beginning with *Sonyon*, serve as vehicles for young writers to express and share their ideas.

Japanese Occupation

1910–45 **The Japanese rule Korea**

The Japanese maintain power in Korea for thirty-five years through an administration headed by a governor-general appointed from the ranks of its top military officials. Reinforced by two Japanese army divisions, Japan's rule is harsh. High-ranking government jobs all go to Japanese rather than Koreans, and Japan confiscates large amounts of royal land and land belonging to Buddhist temples. Some two thousand farmers also have their land appropriated. Korea is forced to trade exclusively with Japan.

Strict censorship is practiced, and schools are strictly controlled by the government. Dissent is swiftly crushed. Even misdemeanors are punishable by flogging. Formal instruction in Korean is virtually discontinued—and discontinued altogether in 1938, by which time Koreans are also forced to adopt Japanese names. They are also forced to accept the Japanese religion of Shintoism.

In spite of their repressive rule, the Japanese do achieve major economic and infrastructure improvements in Korea, modernizing roads and railways and constructing hydroelectric power plants and factories.

1910–20 Western painting is introduced to Korea

Western painting techniques are introduced to Korea, changing attitudes toward the status of painters, who have traditionally been regarded as craftsmen rather than creative artists.

1910: August 22 End of the Yi dynasty

After five centuries, the Yi dynasty comes to an end as Japan formally annexes Korea.

1912: April 15 Birth of Kim Il Sung

Kim Il Sung (1912–94), the first and longtime leader of North Korea, is born in Pyongyang and lives in Manchuria for much of his early life. There he is exposed to Communist ideas and joins the struggle to free Korea from Japanese occupation (see 1910–45). His birth name is King Song Ju, but in the 1930s he adopts the name Kim Il Sung to identify himself with previous Korean resistance fighters against the Japanese. Kim spends the years 1940–45 in the Soviet Union, returning to Korea as a major in the Red Army. He becomes the head of the Communist-oriented Korean Workers' Party, which becomes the dominant nationalist group in Korea.

When the Democratic People's Republic of Korea is founded in 1948 (see 1948), Kim becomes its first leader. In 1950 Kim orders the invasion of the Republic of Korea to the south, triggering the Korean War (see 1950), after which the two nations continue to be divided along the thirty-eighth parallel. Kim consolidates his power by brutally crushing all opposition and by encouraging a cult of personality to be created around the "Great Leader," whose portrait graces all public buildings and most private ones, as well as statues throughout the country. It is believed that hundreds of thousands of people die for opposing (or being suspected of opposing) Kim's government or its policies. Even more are interned in labor camps as political prisoners. Altogether, Kim leads his country for forty-six years, making him the longest-serving leader in the world at the time of his death.

1915 Birth of poet So Chong-ju

So Chong-ju (b. 1915) is regarded by many as the most significant Korean poet of the twentieth century. His early poetry reveals an integration of Western influences including Yeats and Baudelaire. His short, Zen-like lyrics are known for their evocation of the Buddhist ideals of the Silla kingdom (see 668–935) and the distinctive use of language characteristic of his home province of Cholla.

1915 Birth of fiction writer Hwang Sun-won

Considered Korea's foremost author of short stories, Hwang Sun-won (b. 1915) is associated with a group of writers who introduced a new emphasis on the Korean past into their short stories in the 1930s. He is also a respected novelist.

1917 Publication of novel, *The Heartless*

The Heartless, by Yi Kwang-su, is considered the first modern Korean novel. It deals with the themes of nationalism, love, and tradition, portraying conflicts between the values of the present and the past.

1917: November 14 Birth of Park Chung Hee

A military leader and president of South Korea, Park Chung Hee (1917–79) attends Japanese military training academies during Japan's occupation of Korea and later graduates from the newly formed Korean Military Academy in 1946. He attains the rank of brigadier general by 1953 and continues his rapid professional ascent throughout the 1950s. On May 16, 1961, Park leads a military coup against the South Korean civilian government that has replaced ousted president Syngman Rhee (see 1875: April 26). He heads a military government before becoming president when civilian government is restored in 1963. The South Korean economy achieves rapid growth during Park's years in office, but at the price of suppression of civil liberties. Park is reelected four times after 1963, the last time in 1978. In 1974 a bungled assassination attempt kills Park's wife, and repressive emergency measures are adopted to quell growing civil unrest. On October 26, 1979, Park is shot and killed by the head of South Korea's intelligence agency.

1918 Painters' association is established

The *Sohwa hyophoe* (Painting and Calligraphy Association) is established by a group of artists. Prominent among them is Ho Hu-dong.

1919: March 1 Peaceful demonstrations are crushed by the Japanese

A meeting of Korean leaders in Seoul to declare independence is followed by peaceful demonstrations in which an estimated seven thousand Koreans die and more than fifty thousand are jailed, resulting in a swift end to the *Samil* ("March 1") movement. Resistance activities continue to be carried out by Koreans abroad, especially in Shanghai, China, where future South Korean president Syngman Rhee (see 1875: April 26) organizes a provisional government in exile.

1922 Major annual art exhibition is inaugurated

The *Choson misul chollamhoe* (Korean Art Exhibition), popularly known as the *Sonjon*, has its debut.

1925 Publication of poetic manifesto *Sihon*

Poet Kim So-wol (1902–34) publishes an important essay on mysticism in modern Korean poetry. Kim is among the first twentieth-century Korean poets to attempt a synthesis of Western symbolism and other influences with the Korean tra-

Citizens of Pyongyang celebrate in the streets following elections. (EPD Photos/CSU Archives)

dition to create important new poetic forms written in Hangul (see 1446).

1937 Second Sino-Japanese War begins

Once the Japanese begin their second war with China (1937–45), the repression of Koreans becomes even worse than before, as they attempt to obliterate or wipe out the Korean sense of identity. Use of the Korean language itself is even banned.

World War II and the Division of Korea

1939–45 World War II

World War II brings new hardships to the Koreans. They are drafted into the Japanese army or conscripted to labor in factories to support the war effort.

1943: December 1 Cairo Declaration promises freedom for Korea

The Cairo Declaration, signed by the United States, the United Kingdom, and China, promises support for an independent Korea, giving Koreans hope that the end of the war will signal a return to full sovereignty over their own country.

1945: February The U.S. and U.S.S.R. agree to divide Korea

In spite of previous U.S. guarantees of Korean independence, the United States, Britain, and the U.S.S.R. secretly agree at the Yalta Conference to divide Korea once World War II is over and the Japanese are driven out. The part of the country south of the thirty-eighth parallel is occupied by the American army, while the northern section is under the jurisdiction of the Soviets. This arrangement is supposed to be temporary, with unification occurring within five years.

1945: October 10 Korean Workers' Party is formed

The Communist party that will rule the Democratic People's Republic of Korea is formed (see 1912: April 15).

1946 Kim Il Sung University is founded

Kim Il Sung University is established. Offering bachelors, masters, and doctoral degrees, it will serve as North Korea's only comprehensive postsecondary institution, enrolling 12,000 by the 1990s and offering programs including computer science, biology, chemistry, geology, physics, and mathematics.

1948–72 Kim Il Sung is premier of North Korea

Kim Il Sung becomes the premier of the newly formed Democratic People's Republic of Korea, an office he holds for twenty-four years. He strengthens his power throughout the 1950s, becoming a larger-than-life figure who is worshipped by the North Koreans in the same manner that Russians were encouraged to regard Josef Stalin (1879–1953).

1948: May 1 Democratic People's Republic of Korea is established in the north

Following U.S.S.R. opposition to Korean reunification, separate republics are created in the north and south. Neither one joins the United Nations. A constitution is adopted by the Democratic People's Republic of Korea (DPRK).

The Korean War

1950: June 25 Troops from the north invade the south

Troops from the northern, Soviet-occupied portion of Korea invade the south, capturing the city of Seoul and triggering the Korean War. President Truman (1884–1972) sends U.S. forces to drive back the North Koreans, and British troops arrive in the region as well. Chinese forces support the North Koreans.

1950: September 15 U.S. forces land at Inch'on

Led by General Douglas MacArthur (1880–1964), U.S. troops stage an amphibious landing from the Yellow Sea at Inch'on, driving back forces of the DPRK. When MacArthur leads his troops northward toward the Yalu River, China becomes involved to prevent the reunification of the Korean peninsula by the South Koreans and the United States, sending "volunteers" to aid the North Korean troops. MacArthur's forces are forced to retreat southward.

1951–53 Battle lines stabilize at the 38th parallel

By spring of 1951 the lines of combat stabilize at the thirty-eighth parallel, with the North Koreans above it and the South Koreans below. The situation remains static through two more years of heavy fighting and cease-fire negotiations.

1953: July 27 Armistice agreement is signed

The armistice agreement ending the Korean War is signed by all parties. In addition to ravaging the land on both sides, the war has killed over 1.5 million North Koreans and Chinese, and troop casualties among the United Nations command south of the line of demarcation number an estimated 350,000.

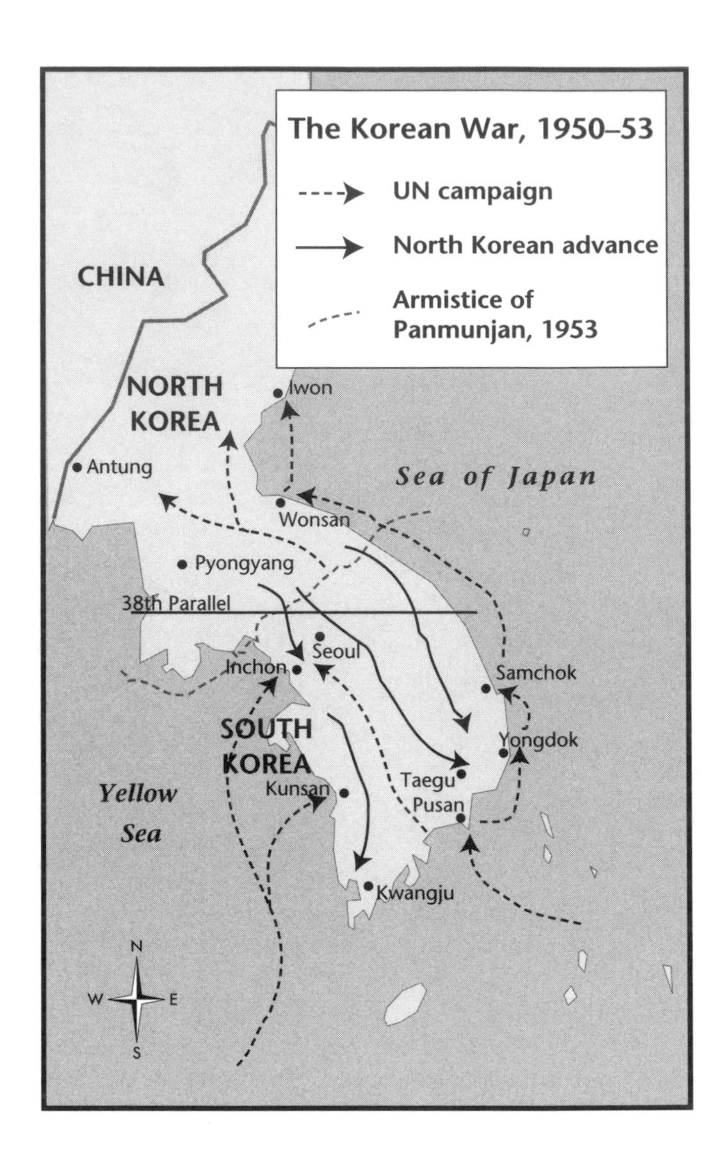

The Postwar Period

1954–58 Agriculture is collectivized

North Korean family farms are transformed into large cooperatives, most of which have areas averaging 1,200 acres and are run by as many as 300 families.

1954–57 DPRK adopts first economic plan

North Korea's Three-Year Plan is the first in a series of economic plans designed to restore its economy after the ravages of the Korean War. Its early plans concentrate on heavy industry, including chemicals, metals, machine building, and power plants. Most economic aid comes from the nations of the Eastern European Communist Bloc.

1954 Medical science academy is founded

The Academy of Medical Sciences, one of North Korea's main scientific institutions, is established.

1958 Major political purge

A major purge consolidates the power of Kim Il Sung by eliminating members of rival Communist groups from the government, either by removing them from their posts or executing them.

1964 Language reforms are announced

Kim Il Sung announces a new policy intended to purify the Korean language. Loan words from English and Japanese are to be replaced by Korean or Chinese terms.

1968: January 23 North Koreans seize the U.S.S. *Pueblo*

North Korean forces seize the U.S. intelligence-gathering ship the U.S.S. *Pueblo* at sea. The United States protests and eventually North Korea releases the ship's crew.

1971: September North and South Korean Red Cross societies hold talks

The Red Cross societies of North and South Korea meet to discuss an exchange of information on family members separated by the division between the two countries.

1972 New constitution is adopted

The DPRK replaces its 1948 constitution with a new one that creates the post of president.

1972 Reunification talks resume

North and South Korea resume reunification talks for the first time since 1954, sponsored by the international Red Cross. The DPRK proposes that the two nations form a unified political entity while retaining their contrasting political and economic systems, but the South rejects this idea.

1972: July 4 North-South Joint Communiqué is issued

The two Koreas issue a statement declaring their intentions to pursue efforts aimed at peaceful reunification of the two nations.

1972: December Kim Il Sung is elected president

Under the terms of the new constitution, Kim Il Sung, already the head of the government and of the Communist party, is elected to the newly created position of president.

1974 Tunnel is discovered under the DMZ

A North Korean tunnel is discovered underneath the demilitarized zone (DMZ) between the two Koreas. Two more tunnels are later found as well.

Mid-1970s North Korean economy begins to falter

After twenty years of strong growth, North Korea's economy begins a downturn that continues into the 1980s.

1977 North Koreans down U.S. helicopter

North Korea shoots down a U.S. helicopter that has mistakenly traveled into its territory.

1978 U.S. begins joint military exercises with South Korea

Apprehension is raised in the DPRK by the inauguration of the "Team Spirit" joint military exercise program by the United States and South Korea.

1979: January–March Reunification talks are held

Talks on reunifying North and South Korea are held, but little progress is made.

1984 DPRK provides flood relief to South Korea

The provision of flood relief by the DPRK to the Republic of Korea (ROK) is seen as a sign of improved relations between the two nations.

1986 Hydroelectric power plant at Taipingwan begins operations

A hydroelectric power project constructed jointly by North Korea and China goes on-line.

1988 Government plans new educational facilities

The government of the DPRK announces plans for new schools and colleges (currently, Kim Il Sung University is the country's only university). A new university, eight colleges, and other specialized facilities are planned.

1990: May Kim Il Sung is reelected president

Kim Il Sung is unanimously reelected to the presidency.

1991 Japan apologizes for abuses during occupation

Japan apologizes to North Korea for abuses inflicted during the occupation of 1910–45. A similar apology is issued to South Korea in 1990. Japan also offers to pay reparations.

1991 The U.S.S.R. dissolves

The collapse of the Soviet Union deprives the DPRK of a major source of political and economic support.

1991 The U.S. withdraws nuclear weapons from the ROK

The United States closes its nuclear weapons installations in the Republic of Korea. North and South Korea sign an agreement to halt the proliferation of nuclear weapons on the Korean peninsula.

1991: September Korea joins the United Nations

Both North and South Korea join the United Nations.

1992 New constitution mandates equal rights for women

Women's rights are guaranteed by the DPRK's new constitution.

1993 China demands cash trade instead of barter

The People's Republic of China changes the terms of its trade with the DPRK from barter to cash. The end of the Cold War leaves North Korea so desperate for trading partners that it agrees to the Chinese demands despite their unfavorable nature. Indeed, China becomes North Korea's largest trading partner with 1995 trade between them estimated at nearly $550 million. Smuggling from China also rises reaching a level of between $30 to $300 million.

1994: February North Korea agrees to IAEA inspections

Following demands by the United Nations, the DPRK agrees to open its nuclear program to international inspection by the International Atomic Energy Agency (IAEA).

1994: May IAEA inspects nuclear reprocessing plant

A team of inspectors from the IAEA examines a reprocessing plant for nuclear materials. Conflicts arise over the degree of access allowed by the North Korean government.

1994: July 8 Kim Il Sung dies on eve of talks between North and South Korea

North Korean president Kim Il Sung dies of a heart attack in Pyongyang shortly before a planned summit meeting between leaders of the two Koreas, the first such meeting to have taken place since 1945. The talks are canceled until after Kim's funeral. The announcement July 9 of Kim's death also disrupts high-level nuclear-development talks between Korean and U.S. officials begun the day before in Geneva. At the time of his death, Kim has served as head of state longer than any other leader in the world. (See 1912.)

1994: July 19 Funeral of Kim Il Sung

Two million attend the state funeral held for Kim Il Sung, North Korea's only leader since its formation in 1948. Mourners participate in a four-hour procession through Pyongyang, as a hearse carries the body of the deceased leader along a twenty-five mile (forty kilometer) route. The funeral also gives an opportunity for government officials to show their loyalty to Kim's son, Kim Jong Il (b. 1941), who is expected to succeed him in office.

1994: October 21 U.S. and North Korea reach nuclear inspection agreement

Officials from the United States and North Korea resolve an eighteen month diplomatic impasse over international inspection of Korea's nuclear-development sites. The Koreans had been suspected of diverting plutonium from nuclear reactors for military purposes, a charge they had repeatedly denied. The agreement specifies a ten-year timetable in which Korea will dismantle its nuclear-development program.

1994: December 17 U.S. helicopter downed by North Koreans

A U.S. Army helicopter engaged in reconnaissance activities is downed when it strays about three miles (five kilometers) past the demilitarized zone between North and South Korea and into North Korean territory. One of the two pilots is captured, and the other is killed in the crash. The United States claims the pilots had entered Korean territory accidentally.

1994: December 30 Captured U.S. pilot is returned safely

Bobby Hall, the helicopter pilot captured when his craft was shot down over North Korean territory (see 1994: December 17), is returned to the United States following talks between U.S. and Korean officials.

1995: August Disastrous flooding creates national emergency

Half a million North Koreans are made homeless by catastrophic flooding.

1995: September 10 Russia ends cooperation pact

Russia denies extension of the mutual cooperation pact with the DPRK that has been in effect since 1961.

1996: April 5–7 DPRK conducts military exercises in DMZ

North Korean troops take part in military exercises in the demilitarized zone separating the two Koreas. Analysts believe that North Korea is not planning a military offensive, but rather trying to pressure the United States into holding direct bilateral talks that exclude South Korea, something the United States has said it will not do.

1996: December 30 North and South Korea and U.S. to hold talks

North Korea agrees to hold talks with the United States and South Korea on adopting a permanent peace agreement formally ending the Korean War of 1950–53. This marks the first time since the end of the war that the DPRK has agreed to meet with representatives of South Korea.

1997: October 8 Kim Jong Il formally assumes leadership post

Three years after the death of his father, Kim Il Sung, Kim Jong Il formally takes over the post of general secretary of the Korean Workers' Party, the highest post in the DPRK. A traditional period of mourning had been observed in the time between the death of North Korea's longtime leader (see 1994: July 8) and the succession to power of his son.

Bibliography

Cumings, Bruce. *The Origins of the Korean War.* Princeton, N.J.: Princeton University Press, 1981.

Eckert, Carter J. *Old and New: A History.* Cambridge, Mass.: Harvard University Press, 1990.

Gills, Barry K. *Korea Versus Korea: A Case of Contested Legitimacy.* New York: Routledge, Inc., 1996.

Hoare, James. *Korea: An Introduction.* New York: Kegan Paul International, 1988.

Macdonald, Donald Stone. *The Koreans: Contemporary Politics and Society.* Boulder, Colo.: Westview Press, 1990.

Merrill, John. *Korea: The Peninsular Origins of the War.* Newark,: University of Delaware Press, 1989.

Oliver, Robert Tarbell. *A History of the Korean People in Modern Times: 1800 to the Present.* Newark,: University of Delaware Press, 1993.

Olsen, Edward A. *U. S. Policy and the Two Koreas.* Boulder, Colo.: Westview Press, 1988.

Smith, Hazel. *North Korea in the New World Order.* New York: St. Martin's Press, 1996.

Soh, Chung Hee. *Women in Korean Politics.* 2nd ed. Boulder, Colo.: Westview Press, 1993.

Suh, Dae Sook. *Kim Il Sung: The North Korean Leader.* New York: Oxford University Press, 1990.

Tennant, Roger. *A History of Korea.* London: Kegan Paul International, 1996.

South Korea

Introduction

The Republic of Korea (South Korea) occupies the southernmost half of the Korean peninsula, which has been divided along the thirty-eighth parallel since the end of World War II (1939–45), with the People's Democratic Republic of Korea (North Korea) occupying the northern portion. The two Koreas have declared their intention of reunifying their country since 1972, and talks have been held periodically, but little progress has been made. President Kim Dae Jung renewed his country's desire for unification at his inauguration in February 1998.

Geography

The Korean peninsula extends southward about 625 miles (1,000 kilometers) from the northeastern corner of China. To its east lies the Sea of Japan; to the west, the Yellow Sea (Koreans call these the East Sea and the West Sea). With an area of 38,315 square miles (99,237 square kilometers), South Korea is slightly larger than Hungary and occupies roughly forty-five percent of the Korean peninsula. Approximately two-thirds of South Korea is mountainous, with the remaining third consisting of lowlands and plains along the coast of the Yellow Sea. Twenty percent of the land is arable and rice is the chief agricultural product. The coastal areas provide South Korea with some of the best deep-sea fishing areas in the world. The country has a population of about 44.5 million.

Korean civilization has a long history, and one in which China, its closest neighbor, has played an important part. Some rulers of Korea's ancient kingdoms were Chinese, and China's Han dynasty (202 B.C.–A.D. 220) invaded Korea (then known as Choson) in 108 B.C. to put an end to possible competition over valuable overland trade routes. The Chinese ruled a portion of Korea for nearly 300 years, bringing with them many aspects of their culture, such as Confucianism and a variety of technological advances. By the time the Chinese were driven out, three new kingdoms—Silla, Paekche, and Koguryo—had spread throughout the land. Between 668 and 935 all of Korea was unified under the Silla kingdom. This was an important period, both politically and culturally. It was the first time that the entire Korean peninsula had had a single language, culture, and social structure. However, much of this unified culture showed Chinese influences, such as the civil service system and the design of the capital city of Ch'angan.

Beginning in the tenth century, Korea became a target for foreign invaders, and in the thirteenth and fourteenth centuries it came under the domination of the Mongol Yuan empire (1260–68). The Mongols were ousted by a Korean general, Yi Song-gye (1335–1408), who began one of the longest-lived dynasties in world history—the Yi (or Li) dynasty, which ruled Korea from 1392 to 1910. The first century of dynastic rule was an especially rich period culturally. A written version of the Korean language, called *Hangul*, was created, leading to the development of a vernacular (in the native tongue) literature. Important scholarly and administrative works were written, and technological advances, including the rain gauge, were perfected.

Repeated invasions by Japanese emperor Hideyoshi Toyotomi (1536–98) at the end of the sixteenth century left Korea in a weakened state, and it fell under the control of China's Manchu rulers in 1636. The seventeenth century also inaugurated a 300-year period of political, cultural, and economic isolation during which Korea came to be known as the Hermit Kingdom. Foreign influence during this period was extremely restricted, and the institutions of Korean society remained timelessly locked in place for nearly three centuries. The Kangwha Treaty of 1876, opening Korea to trade with Japan and the West, was the first major step toward reestablishing relations with the rest of the world.

Unfortunately, Korea's neighbors brought expansionist goals to these new relationships. In the Sino-Japanese War of 1894–95, the Japanese drove the Chinese out of Korea, but only so that they could later assume power there themselves. Japanese control of Korea was further increased by the Russo-Japanese War of 1904–05, and Japan annexed the peninsula outright in 1910, beginning a thirty-five-year period of oppressive foreign rule. The Japanese treated the Koreans as second-class citizens in their own country, confiscating large amounts of land and attempting to supplant the Koreans' language and culture with their own. The Japanese occupation of Korea continued until the end of World War II (1939–45), when the United States and the Soviet Union divided the

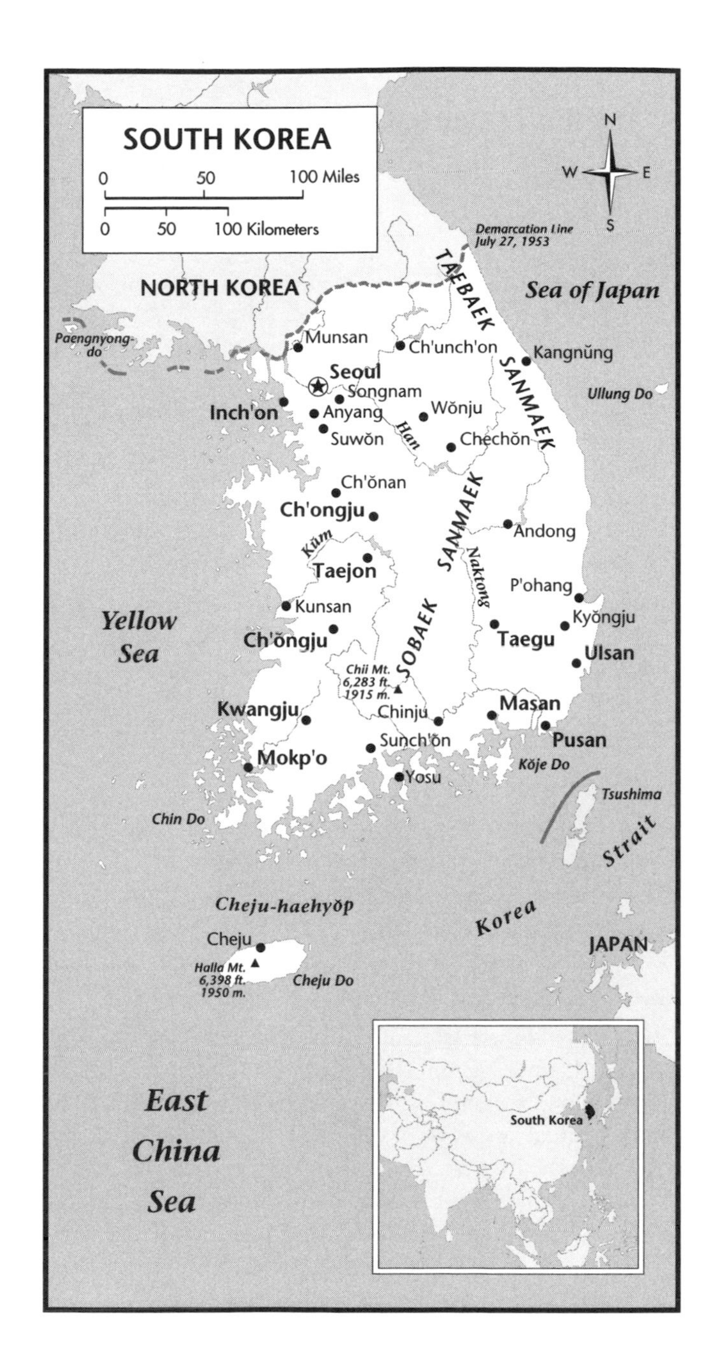

country into an American zone south of the thirty-eighth parallel and a Soviet zone to the north.

Division between North and South Korea

The People's Republic of Korea was established south of the thirty-eighth parallel in 1948, the same year that the Democratic People's Republic of Korea, ruled by the Korean Workers' Party and supported by the U.S.S.R., was created in the north. In June 1950, troops from the north entered South Korea in an attempt to reunify the peninsula by force but were eventually driven back by troops sent by the United States and Great Britain (Korean War: 1950–53.) An armistice agreement confirming the boundary at the thirty-eighth parallel was signed in 1953. In 1954 South Korea and the United States signed a mutual defense agreement.

Series of Military Rulers

South Korea's first president, originally elected in 1948, was Syngman Rhee, who had been educated in the United States and spent most of his life in exile fighting the Japanese occupation of Korea (1910–1945). Although he successfully rallied his country during the difficult days of the Korean War, Rhee's authoritarian mode of rule created growing discontent, and he was finally forced to resign in 1960 in response to mass demonstrations and riots.

Following a short period of democratic government, Major General Park Chung Hee came to power in a military coup in 1961 and made himself head of a new civilian government two years later. South Korea's economy improved under Park, whose policies resulted in increased exports and a heavy military buildup. His normalization of diplomatic ties with Japan in 1965 was an important step in the country's economic progress, although it drew heavy protests at home.

In the 1970s, North and South Korea took some steps toward a reconciliation. In 1972 the countries issued a joint communiqué declaring their intent to achieve peaceful reunification. Reunification talks were held at the end of the decade, but the two countries remained separate. Throughout the seventies, Park's rule grew increasingly repressive as opposition to his rule mounted. He was assassinated in 1979, and soon afterwards Major General Chun Doo Hwan came to power in a military coup. Under Chun, South Korea entered a period of rapid economic growth, becoming a world leader in the shipbuilding, automotive, and consumer electronics industries. Early in Chun's rule, however, his government's record was tarnished by its brutal response to student-led demonstrations in Kwangju, in which over 200 people were killed, an event that would return to haunt Chun later in his career.

With its new wealth, South Korea outbid all other contenders for the right to host the Twenty-fourth Olympic Games in 1988, and the games were held in Seoul in September and October of that year (and boycotted by North Korea because it was not named as co-host). They had been preceded by civil unrest and a political shake-up, as growing prodemocracy sentiments and opposition to President Chun, peaked in 1987. In June of the same year thousands had protested Chun's proposed postponement of constitutional reforms until after the Olympic Games. In July, 100,000 demonstrators marched in Seoul. Constitutional reforms were adopted in October, including direct election of the president.

Democratic Presidency

In December 1987, Roh Tae Woo became South Korea's first president to be elected by direct popular vote. More milestones in South Korea's presidential history occurred in 1993, when Kim Young Sam became the first civilian elected to the office in thirty years and 1997, when the election of opposition leader Kim Dae Jung to the presidency marked the first legal ouster of a ruling party in the nation's history (and the first peaceful transfer of power). However, not all news regarding South Korea's presidents in the 1990s was favorable. By 1996 two former presidents—Chun Doo Hwan and Roh Tae Woo—had been sentenced and imprisoned on charges ranging from bribery to sedition or stirring up rebellion (the latter for Chun's role in the 1980 massacre of demonstrators at Kwangju). In addition, other government officials were indicted on corruption charges involving the Hanbo Steel company.

Late in 1997, toward the end of his presidential term, Kim Young Sam pardoned both jailed former leaders, Chun and Roh. In his 1998 inaugural Kim Dae Jung, the victim of decades of persecution at the hands of South Korea's military dictators, expressed a similar spirit of national reconciliation in his call for the country's people to unite in facing its economic crisis, and for both Koreas to unite and once again become a single nation.

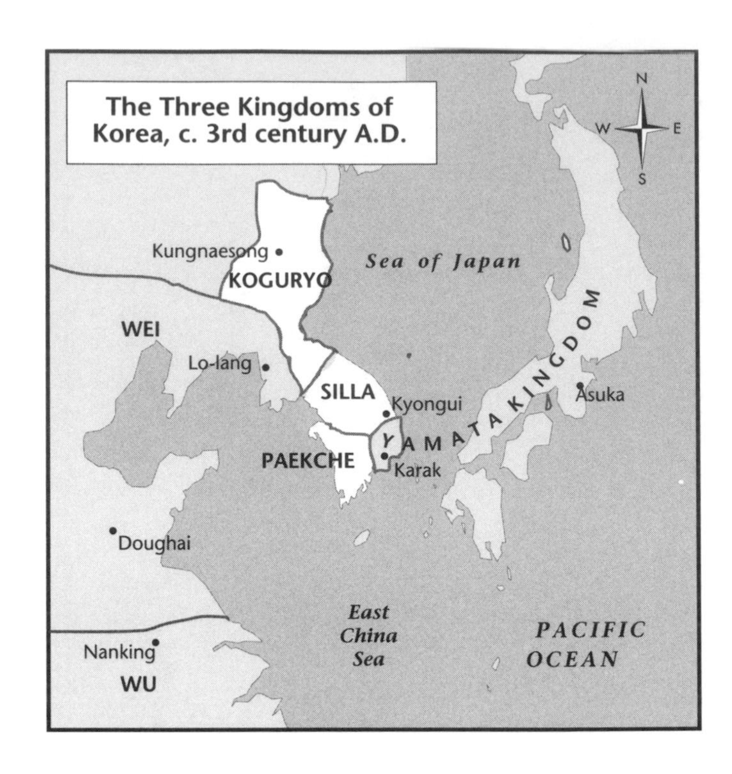

Timeline

c. 3000 B.C. Arrival of Tungusic tribes

Neolithic tribes migrate to the Korean peninsula from northern China and Manchuria. Some combine farming with hunting, while others practice settled agriculture.

c. 1122 B.C. Kija becomes ruler of Choson

A nobleman fleeing China following the overthrow of the Shang dynasty (1766–1122) takes over the leadership of the ancient kingdom of Choson (present-day Korea). This country is said to have been founded in the third century B.C. by the legendary Tangun, fabled to be the offspring of a prince and a bear turned into a woman.

194–180 B.C. Wiman conquers Choson

Wiman, a military figure from China (possibly of Korean stock), gains power in the kingdom of Choson and overthrows its ruler, King Chun, said to be a descendent of Kija.

108 B.C. The Chinese invade Choson

The Chinese Han dynasty (202 B.C.–A.D. 220), perceiving a threat in the control of overland trade routes by the expanding Wiman Choson kingdom, seizes Choson after a year of bitter fighting. King Ugo (the grandson of Wiman) is assassinated, and the capital city is captured. The Chinese rule Choson, which they divide into four prefectures (regional district ruled by a government official) with their capital at Pyongyang.

37 B.C.–A.D. 313 Chinese rule

By 37 B.C. only the prefecture of Nangnang (Lolang in Chinese) is still under Chinese control, and it remains so for nearly three more centuries, during which many cultural and technological advances of the Chinese, including Confucianism, are absorbed.

Confucianism is a Chinese philosophy based upon the teachings of the Chinese philosopher Kung Fuzi (Confucius, 551–479 B.C.). Confucianism is a prescription for ethical behavior in personal relations as well as governance.

A.D. 1st century–7th century Three kingdoms vie for dominance

Three new kingdoms emerge in the lands formerly controlled by the Chinese. Spread out among Korea's three natural topographical divisions, they are Silla in the southeast, Koguryo in the northwest, and Paekche in the southwest.

4th century Buddhism is introduced

Buddhism is introduced to Korea, from which it later travels to Japan. Buddhism originates in the Ganges valley in India during the fifth century B.C. The religion is based on the teachings of Siddhartha Gautama (c. 563–c. 483 B.C), a Hindu prince who comes to be called the Buddha (the Enlightened One or the Wise One). Buddhism, in its original form, has no gods, no priests, no social classes. Salvation is within the reach of anyone who accepts the Buddhist *Dharma* (Law) and the beliefs embodied in the Four Noble Truths. These four basic tenets are: (1) suffering exists, (2) suffering is caused by ignorance, (3) ignorance and desire must be eliminated to achieve Nirvana (the ultimate state), and (4) one can achieve Nirvana by following the eightfold path of behavior.

313 Koguryo drive the Chinese out of Nangnang

The aggressive and powerful Koguryo end Chinese rule in Korea by overrunning Nangnang.

372 First Korean school is founded

The *Taehak* (Great Learning) is established.

647 Astronomical observatory is built in Kyongju

An observatory, believed to be the oldest in the Far East, is built in the capital of the Silla kingdom.

660–68 Silla gains control of the Korean peninsula

Aided by an alliance with China, Silla conquers the kingdoms of Koguryo and Paekche.

668–935 Silla Unification

The kingdom of Silla rules over the whole of Korea, unifying its language, culture, and social structures for the first time. Many aspects of the Chinese Tang dynasty (618–906) culture also become widespread; the capital at Kyongju is modeled on the Tang capital of Ch'angan, and government administration emulates the Chinese civil service system, including its use of examinations to obtain government posts. A system, called *idu*, is developed for writing the Korean language using Chinese characters. Using this system, vernacular (commonly spoken) Korean poems called *hyangga* are first recorded.

A strict system of classes (called "bone-rank") is observed, with children of aristocrats receiving a Confucian education (based on the teachings of the Chinese philosopher Confucius (c. 551–479 B.C.). The economy flourishes through expanded trade, and Buddhism becomes increasingly important, inspiring the creation of many elaborate temples, shrines, and pagodas. Artworks include gigantic bronze bells (such as the Emille Bell weighing twenty-three tons), bronze statues, and delicate engravings.

682 First school for government workers is founded

Korea's first school for training civil servants, known as *kukhak*, is established.

857 Birth of poet Ch'oe Ch'i-won

Ch'oe Ch'i-won (b. 857) is an important early creator of *hansi* poetry, written in Chinese according to Chinese literary conventions. *Hansi* poetry remains popular for centuries and into the twentieth century.

918 Kingdom of Koryo is founded

Wang Kon (877–943) (later known as Taejo) establishes the kingdom of Koryo, from which the modern name "Korea" is derived.

936–1392 Koryo dynasty rules Korea

Defeating two other rebel leaders, Wang Kon gains control of Korea following the downfall of the Silla kingdom. To forge an alliance with the former Silla rulers, he allows many members of the ruling elite to retain their positions, including the Silla king who holds the highest government office. Wang Kon also marries a Silla princess. Buddhism remains the state religion. The Koryo period is one of warfare and invasion by rival empires, including the Khitan Liao and Chin states in northern China and Manchuria and, later, the Mongols, who carry out multiple invasions, eventually adding Koryo to their empire. The strain of continued warfare takes it toll on the Koryo government, which falls prey to intrigues and corruption.

957 State examination system is established

The Koryo kingdom sets up a system of state examinations for government workers based on the Chinese system. Scholars commonly apply for posts in the civil service, and the examination questions are based on the teachings of Confucius.

1231 First Mongol invasion of Korea

Mongols invade the kingdom of Koryo and relocate its capital to the Kanghwa Island.

1236–51 New set of Tripitaka woodblocks is carved

A set of woodblocks for printing Buddhist scriptures is created to replace the set destroyed in the Mongol invasion (see 1231). The set, consisting of over 81,000 blocks, represents the most sophisticated book production method until the invention of movable metal type in the fourteenth century. The blocks have been preserved up to the present and are still usable.

1254 First book is published using movable type

During the period when the Koryo court is in exile on Kanghwa Island, movable metal type is developed, a century before its invention by Johannes Gutenberg (1390–1468) in Germany. The first book published using movable type is a fifty-volume anthology of religious texts by Ch'oe Yun-ui.

The Chang Hee Moon Gate stands at the north enterance to Seoul, made the capital during the Yi era. (EPD Photos/CSU Archives)

1258 Koryo comes under Mongol domination

The Koryo leaders make peace with the Mongols, and their government comes under the control of the Mongol Yuan empire (1260–1368). Mongol officials take charge of government administration and exact heavy taxes from the population. Koryo crown princes are forced to marry Mongol princesses and live in the Mongol capital. One marries the daughter of Kublai Khan (1216–94).

1277 Glass tile is developed

The invention of glass tile, leading to the development of the arts of ceramics, is credited to a Korean Buddhist monk.

c. 1285 Oldest known folk collections appear

The earliest recorded collections of folk literature, *Samguk sagi* and *Samguk yusa*, are written down.

The Yi Dynasty

1392 Yi dynasty begins

A Koryo general, Yi Song-gye (1335–1408), ends Mongol control of the country and founds the Yi Dynasty (also called the Li dynasty), which will rule Korea until 1910. The capital is moved to Seoul. Yi Song-gye changes the name of the country back to Choson and inaugurates ties with China's new Ming dynasty (1368–1644), establishing an alliance known as *sadae*. In this system, modeled on the Confucian relationship known as "elder brother/younger brother," the Yi king is subordinate to the Chinese emperor.

1392–1500 Yi era begins with period of peace and cultural development

The first century of the Yi dynasty was its high point, marked by peace and cultural advances, including the development of a written Korean language (see 1446), the invention of a rain gauge, and publication of many types of texts with government support. A set of administrative guidelines, the *Kyongguk taejon*, is issued. Also recorded are a history of the Koryo period (*Koryo sa*) and a definitive anthology of Korean music (*Akhak kwebom*). Painters such as An Kyon (b. 1418) begin to develop a distinctly Korean style, in terms of their brushwork, composition, and treatment of space.

After 1500 the quality of Yi leadership and daily life begin a long decline due to political factionalism within and invasion from abroad.

15th century Koreans produce the first encyclopedia

Korean scholars compile what is believed to be the world's earliest encyclopedia, consisting of 112 volumes.

15th century First fiction is written down

Kim Sisup (1435–93) writes *New Stories of the Golden Turtle*, generally considered the first fictional literature in Korean.

1446 Hangul written language is developed

A written version of native Korean speech is created, hastening the development of a vernacular literature. Up to this point, all Korean written material has been in Chinese. The poetic form known as *sijo* flourishes with the introduction of Hangul.

1587 Birth of poet Yun Son-do

Yun Son-do (1587–1671) is considered the greatest master of the *sijo* poems (see 1728). In addition to writing poetry, he conducts an active political career. His most famous work is a cycle of forty poems, called *The Fisherman's Calendar*, that describe each of the four seasons as experienced in one of the poet's favorite settings. Altogether, seventy-six of Yun Son-do's *sijo* survive. The poet dies in 1671.

1592 Japanese emperor attempts to invade Korea

Japan's new emperor, Hideyoshi Toyotomi (1536–98), sends forces to Korea as part of a larger effort to invade China as well. They are turned back by troops under the leadership of Yi Sun-sin (1545–98), aided by forces from China's Ming rulers. The vessel—or "turtle ship"—carrying the Korean military leader is the first iron-clad armored ship to be used in battle.

1597 Japan tries to invade Korea for a second time

Hideyoshi again orders an invasion of Korea, but the Japanese emperor dies a year later, and the campaign ends without success. However, defending itself against the Japanese attacks weakens the Yi kingdom and hastens its decline and subsequent conquest by the Manchus of China. Much of Korea's agricultural land is ruined, and the kingdom suffers an early version of "brain drain," as many talented artists and other persons with specialized skills are forcibly transported to Japan to take advantage of their talents.

The Hermit Kingdom

1600–1900 Korea enters a period of isolation

From the seventeenth through much of the nineteenth century, Korea enters a period of isolation from foreign influence, becoming known as the Hermit Kingdom. Like the elite in Japan, which experiences a similar period under the Tokugawa shogunate (military dynasty), the entrenched aristocracy in Korea fears that its grip on power would be threatened by exposure to new ideas from abroad and the increased middle-class power that would result from growing trade with other nations. Thus, contact with outsiders is halted, and Korea's social structure and culture are virtually frozen in place for nearly 300 years.

1636 Manchus gain control of Korea

China's Manchu dynasty (1644–1912) wins control of the Yi kingdom, which is forced to cut its ties to the Ming rulers and becomes a vassal or subordinate state of the Manchus. Yi power is further curtailed by internal squabbling between political factions.

1671 Death of poet Yun Son-do

Acclaimed *sijo* poet Yun Son-do dies. (See 1587.)

1689 Publication of The Nine Cloud Dream

Authored byKim Man-jung (1637–92), *The Nine Cloud Dream* is considered a milestone in the development of Korean fiction. It is noted for its use of symbolism and its examination of the contrasts between Buddhism and Confucianism and of human imperfection itself. Kim is also the author of *The Story of Lady Sa*, a satire on concubinage (the state of being a secondary wife in a polygamous society).

1724–1800 Two Yi monarchs arrest dynastic decline

Two effective rulers of the eighteenth century—Yongjo (see 1724) and Chongjo (see 1776)—prompt a resurgence in Yi power and a cultural renaissance known as the Practical Learning Movement (*Sirhak*). Increased value placed on empirical knowledge and practical matters brings with it a new appreciation for prose writing, including the composition of satirical, realistic fiction. A leader in this area is Pak Chiwon (1737–1805), whose works satirize aristocratic life.

Chinese and Korean poetry anthologies appear, and there is renewed respect for literature written in Hangul (the Korean alphabet), such as the novels *Chunhyang chon* ("Tale of Miss Spring-Fragrance") and *Kuun mong* ("Cloud Dream of the Nine").

1724–76 Reign of Yongjo

The Yi monarch Yongjo reigns. Yongjo succeeds in halting—albeit, briefly— the process of Korean decline.

1728 First anthology of *sijo* poetry is published

The first collection of *sijo*, *Ch'onggu yon-gon*, is published. This form of poetry has been written since the development of Hangul in the fifteenth century (see 1446) and may have existed in an oral form even earlier. In the most common form of *sijo*, called *p'yong-sijo*, the poem consists of three lines with fourteen to sixteen syllables in each line. The syllables are divided into four distinct sections, each introduced by the speaker (or singer) taking a breath, and the total number of syllables in the poem must not exceed forty-five.

1776–1800 Chongjo reigns

Reign of the second effective eighteenth-century Yi monarch, Chongjo. Along with his predecessor, Yongjo, he attempts to halt Koreas decline. He is followed by a succession of ineffective monarchs.

1785 Roman Catholicism is banned in Korea

Roman Catholicism, which has been introduced to the country by Koreans returning from China, is officially prohibited.

1860 Ch'ondogyo religion is founded

Ch'ondogyo (Religion of the Heavenly Way), a religion combining elements of Christianity and Buddhism, is established by Ch'oe Che-u. Originally called *Tonghak* (Eastern Learning), it emphasizes the present life over the afterlife and the elimination of injustice and corruption on earth. Its rapidly growing popularity among Korea's poor results in persecution of its followers and the death of its founder. It remains one of Korea's major religions throughout the twentieth century.

1864–73 Yi Ha-leng (1820–98) holds power as regent

In one of the final decades of the Yi dynasty, power resides with the Taewon'gun, or regent (the father of the king, Kojong), who restores some of the Yi power. However, he strongly upholds the policy of isolationism. Under his rule, thousands of Christians are persecuted.

1870s *Kendo is introduced to Korea*

The Japanese sport of *kendo* (fencing with bamboo swords) is introduced to Korea and later adopted as a training discipline by the military and the police.

1875: April 26 Birth of Syngman Rhee

Syngman Rhee (1875–1965), future president of the Republic of Korea, is born in Seoul to a family descended from the Yi dynasty. As a young adult, he is converted to Christianity and begins agitating for more democratic rule by the Korean monarchy. Between 1904 and 1910, Rhee lives in the United States, where he receives degrees from George Washington University, Harvard, and Princeton. Rhee spends the years of the Japanese occupation (see 1910–45) in Hawaii and Washington D.C., working for Korean independence and establishing a government-in-exile. He returns to Korea in triumph in October 1945 and begins a campaign opposing Communism.

In 1948 Rhee becomes the first president of the newly formed Republic of Korea (South Korea). He presides over early reforms, including land reform and the establishment of universal education. Rhee is reelected in 1952, 1956, and 1960, but his increasingly dictatorial methods arouse protest at home and censure (disapproval) abroad. Widespread fraud in the 1960 election sparks a rebellion that leads to Rhee's resignation on April 27. He goes into exile in Hawaii, remaining there until his death in July 1965.

1876 Kanghwa Treaty is signed with Japan

In the first move toward ending its centuries of isolation, Korea signs a treaty with Japan that opens it to trade with both Japan and the nations of the West.

1882 Korea signs trade agreement with the U.S.

Korea and the United States sign a commercial pact, which is followed by similar agreements with other Western nations.

1884 Korea opens its ports to foreign trade

Due to pressure by Japan and the West, Korea's ports are opened to foreign commerce.

1885 Western classical music is introduced to Korea

Hymns, the first form of Western music learned by Koreans, are brought to the country by the first U.S. missionaries, Horace G. Underwood (1859–1916) and Henry G. Appenzeller (1858–98).

1886 First women's college is founded

Methodist missionaries from the United States. establish Ehwa Women's University in Seoul.

1894–95 First Sino-Japanese War

The Japanese take advantage of peasant uprisings surrounding the 1894 Tonghak rebellion to rapidly crush Chinese power in Korea, forcing China to recognize Korean independence in the Treaty of Shimonoseki. The growing influence of the Japanese themselves, however, erodes this very independence.

Early 20th century ***Sinmunhak*** **movement in literature**

Sinmunhak, or "new literature," written in Hangul, expresses a new national consciousness. It represents a reaction against the domination of the Chinese language and literary traditions in Korea. *Tears of Blood* (1906) is an example of the new novels associated with this movement.

1904–05 Russo-Japanese War

Russia and Japan go to war over Korea. The Japanese occupy the Korean peninsula, increasing their control of the country, and Korea is declared a Japanese protectorate.

1908–28 Literary magazines serve as forum for new ideas

A series of literary magazines, beginning with *Sonyon*, serve as vehicles for young writers to express and share their ideas.

Japanese Occupation

1910–45 The Japanese rule Korea

The Japanese maintain power in Korea for thirty-five years through an administration headed by a governor-general appointed from the ranks of its top military officials. Reinforced by two Japanese army divisions, Japan's rule is harsh. High-ranking government jobs all go to Japanese rather than Koreans, and Japan confiscates large amounts of royal land and land belonging to Buddhist temples. Some two thousand farmers also have their land appropriated. Korea is forced to trade exclusively with Japan.

Strict censorship is practiced, and schools are strictly controlled by the government. Dissent is swiftly crushed. Even misdemeanors are punishable by flogging. Formal instruction in Korean is virtually discontinued—and discontinued altogether in 1938, by which time Koreans are also forced to adopt Japanese names. They are also forced to accept the Japanese religion of Shintoism.

In spite of their repressive rule, the Japanese do achieve major economic and infrastructure improvements in Korea, modernizing roads and railways and constructing hydroelectric power plants and factories.

1910–20 Western painting is introduced to Korea

Western painting techniques are introduced to Korea, changing attitudes toward the status of painters, who have traditionally been regarded as craftsmen rather than creative artists.

1910: August 22 End of the Yi dynasty

After five centuries, the Yi dynasty comes to an end as Japan formally annexes Korea.

1912: April 15 Birth of Kim Il Sung

Kim Il Sung (1912–94), the first and longtime leader of North Korea, is born in Pyongyang and lives in Manchuria for much of his early life. There he is exposed to Communist ideas and joins the struggle to free Korea from Japanese occupation (see 1910–45). His birth name is King Song Ju, but in the 1930s he adopts the name Kim Il Sung to identify himself with previous Korean resistance fighters against the Japanese. Kim spends the years 1940–45 in the Soviet Union, returning to Korea as a major in the Red Army. He becomes the head of the Communist-oriented Korean Workers' Party, which becomes the dominant nationalist group in Korea.

When the Democratic People's Republic of Korea is founded in 1948 (see 1948), Kim becomes its first leader. In 1950 Kim orders the invasion of the Republic of Korea to the south, triggering the Korean War (see 1950), after which the two nations continue to be divided along the thirty-eighth parallel. Kim consolidates his power by brutally crushing all opposition and by encouraging a cult of personality to be created around the "Great Leader," whose portrait graces all public buildings and most private ones, as well as statues throughout the country. It is believed that hundreds of thousands of people die for opposing (or being suspected of opposing) Kim's government or its policies. Even more are interned in labor camps as political prisoners. Altogether, Kim leads his country for forty-six years, making him the longest-serving leader in the world at the time of his death.

1915 Birth of poet So Chong-ju

So Chong-ju (b. 1915) is regarded by many as the most significant Korean poet of the twentieth century. His early poetry reveals an integration of Western influences including Yeats and Baudelaire. His short, Zen-like lyrics are known for their evocation of the Buddhist ideals of the Silla kingdom (see 668–935) and the distinctive use of language characteristic of his home province of Cholla.

1915 Birth of fiction writer Hwang Sun-won

Considered Korea's foremost author of short stories, Hwang Sun-won (b. 1915) is associated with a group of writers who introduced a new emphasis on the Korean past into their short stories in the 1930s. He is also a respected novelist.

1917 Publication of novel, *The Heartless*

The Heartless, by Yi Kwang-su, is considered the first modern Korean novel. It deals with the themes of nationalism, love, and tradition, portraying conflicts between the values of the present and the past.

1917: November 14 Birth of Park Chung Hee

A military leader and president of South Korea, Park Chung Hee (1917–79) attends Japanese military training academies during Japan's occupation of Korea and later graduates from the newly formed Korean Military Academy in 1946. He attains the rank of brigadier general by 1953 and continues his rapid professional ascent throughout the 1950s. On May 16, 1961, Park leads a military coup against the South Korean civilian government that has replaced ousted president Syngman Rhee (see 1875: April 26). He heads a military government before becoming president when civilian government is restored in 1963. The South Korean economy achieves rapid growth during Park's years in office, but at the price of suppression of civil liberties. Park is reelected four times after 1963, the last time in 1978. In 1974 a bungled assassination attempt kills Park's wife, and repressive emergency measures are adopted to quell growing civil unrest. On October 26, 1979, Park is shot and killed by the head of South Korea's intelligence agency.

1918 Painters' association is established

The *Sohwa hyophoe* (Painting and Calligraphy Association) is established by a group of artists. Prominent among them is Ho Hu-dong.

1919: March 1 Peaceful demonstrations are crushed by the Japanese

A meeting of Korean leaders in Seoul to declare independence is followed by peaceful demonstrations in which an estimated seven thousand Koreans die and more than fifty thousand are jailed, resulting in a swift end to the *Samil* (March 1) movement. Resistance activities continue to be carried out by Koreans abroad, especially in Shanghai, China, where future South Korean president Syngman Rhee (see 1875: April 26) organizes a provisional government in exile.

1922 Major annual art exhibition is inaugurated

The *Choson misul chollamhoe* (Korean Art Exhibition), popularly known as the *Sonjon*, has its debut.

1925 Publication of poetic manifesto *Sihon*

Poet Kim So-wol (1902–34) publishes an important essay on mysticism in modern Korean poetry. Kim is among the first twentieth-century Korean poets to attempt a synthesis of Western symbolism and other influences with the Korean tradition to create important new poetic forms written in Hangul (see 1446).

1937 Second Sino-Japanese War begins

Once the Japanese begin their second war with China (1937–45), the repression of Koreans becomes even worse than before, as they attempt to obliterate or wipe out the Korean sense of identity. Use of the Korean language itself is even banned.

World War II and the Division of Korea

1939–45 World War II

World War II brings new hardships to the Koreans. They are drafted into the Japanese army or conscripted to labor in factories to support the war effort.

1943: December 1 Cairo Declaration promises freedom for Korea

The Cairo Declaration, signed by the United States, the United Kingdom, and China, promises support for an independent Korea, giving Koreans hope that the end of the war will signal a return to full sovereignty over their own country.

1945: February The U.S. and U.S.S.R. agree to divide Korea

In spite of previous U.S. guarantees of Korean independence, the United States, Britain, and the U.S.S.R. secretly agree at the Yalta Conference to divide Korea once World War II is over and the Japanese are driven out. The part of the country south of the thirty-eighth parallel is occupied by the American army, while the northern section is under the jurisdiction of the Soviets. This arrangement is supposed to be temporary, with unification occurring within five years.

1945: October 10 Korean Workers' Party is formed

The Communist party that will rule the Democratic People's Republic of Korea is formed (see 1912: April 15).

1946 Kim Il Sung University is founded

Kim Il Sung University is established. Offering bachelors, masters, and doctoral degrees, it will serve as North Korea's only comprehensive postsecondary institution, enrolling 12,000 by the 1990s and offering programs including computer science, biology, chemistry, geology, physics, and mathematics.

1948–72 Kim Il Sung is premier of North Korea

Kim Il Sung becomes the premier of the newly formed Democratic People's Republic of Korea, an office he holds for twenty-four years. He strengthens his power throughout the 1950s, becoming a larger-than-life figure who is worshipped by the North Koreans in the same manner that Russians were encouraged to regard Josef Stalin (1879–1953).

1948: May 1 Democratic People's Republic of Korea is established in the north

Following U.S.S.R. opposition to Korean reunification, separate republics are created in the north and south. Neither one joins the United Nations. A constitution is adopted by the Democratic People's Republic of Korea (DPRK).

1948: August 15 Republic of Korea is created

The Republic of Korea (ROK) is established in the southern portion of the Korean Peninsula, which has been occupied by U.S. forces since the end of World War II (1939–45). The Republic of Korea claims to be the legitimate government of all of Korea, including the northern portion where the Democratic People's Republic of Korea (DPRK) has been established with support from the Soviet Union. The first elected president of the Republic of Korea is Syngman Rhee (1875–1965).

The Korean War

1950: June 25 Troops from the north invade the south

Troops from the northern, Soviet-occupied portion of Korea invade the south, capturing the city of Seoul, triggering the Korean War. President Truman (1884–1972) sends U.S. forces to drive back the North Koreans, and British troops arrive in the region as well. Chinese forces support the North Koreans.

1950: September 15 U.S. forces land at Inch'on

Led by General Douglas MacArthur (1880–1964), U.S. troops stage an amphibious landing from the Yellow Sea at Inch'on, driving back forces of the DPRK. When MacArthur leads his troops northward toward the Yalu River, China becomes involved to prevent the reunification of the Korean peninsula by the South Koreans and the United States, sending "volunteers" to aid the North Korean troops. MacArthur's forces are forced to retreat southward.

1951–53 Battle lines stabilize at the 38th parallel

By spring of 1951, the lines of combat stabilize at the thirty-eighth parallel, with the North Koreans above it and the South Koreans below. The situation remains static through two more years of heavy fighting and cease-fire negotiations.

1953: July 27 Armistice agreement is signed

The armistice agreement ending the Korean War is signed by all parties. In addition to ravaging the land on both sides, the war has killed over 1.5 million North Koreans and Chinese, and troop casualties among the United Nations command south of the line of demarcation number an estimated 350,000.

South Korea under Rhee

1954 U.S. and ROK sign mutual defense treaty

The Republic of Korea and the United States sign a mutual defense pact authorizing the continuing presence of U.S. troops in South Korea.

1957 Major art exhibition is inaugurated

The daily newspaper *Chosun Ilbo* sponsors the Contemporary Korean Art Exhibition for Invited Artists, Korea's first independent exhibition. The creation of this exhibition, which is more innovative than its government-sponsored counterpart, the National Exhibition of Fine Art, encourages experimentation by contemporary artists.

1958 Composers form group to promote new music

Members of the composition faculty at Seoul National University form a group that plays a leading role in promoting the composition and performance of contemporary music in South Korea.

1960: April Rioters protest election results

South Koreans stage demonstrations at Masan to protest fraudulent March elections.

1960: April 27 Rhee is forced to resign

Popular unrest sparked by student demonstrations forces Syngman Rhee (see 1948: August 15) to resign as president.

1960: July Second Republic is established

Elections establish a parliamentary democracy with Yun Po Sun as president and John M. Chang as prime minister. Koreans enjoy expanded civil liberties, but continue to suffer the effects of a struggling economy.

Park Seizes Power

1961: May 16 Second Republic is overthrown in military coup

After only ten months in power, the Second Republic is overthrown in an armed coup. Major General Park Chung Hee (1917–79; see 1917: November 14) forms a military government and dissolves the National Assembly. Park declares martial law and rules by decree, promising an eventual return to civilian government.

1962: December New constitution is approved

Voters approve a new constitution in a referendum, paving the way for a transition to civilian rule.

1963: October 16 Park is elected to head civilian government

Heading the Democratic-Republican Party (DRP), Park is elected president under the terms of a new constitution. Park's government succeeds in improving the nation's economy, rapidly increasing growth, primarily through exports. Park also authorizes a heavy military buildup.

1965 Science and technology institute is established

The Korean Institute of Science and Technology is created with support from the United States.

1965 National Opera Group is formed

The National Opera Group is established. It debuts with a performance of *La Boheme* by Puccini.

1965: June Diplomatic ties with Japan are resumed

South Korea normalizes relations with Japan, which have been strained by the lengthy occupation of their country by the Japanese (see 1910–45). War reparations or compensations by Japan are to be replaced by economic aid. The Koreans, still resentful toward the Japanese for their forty years of harsh rule over the peninsula, protest and riot throughout South Korea.

1965: July 19 Death of former president Syngman Rhee

Former president Syngman Rhee dies in Hawaii. (See 1875: April 26)

1966 ROK sends troops to Vietnam

South Korea sends 45,000 combat troops to Vietnam to join the U.S.-led fight against the communist insurgency in South Vietnam, sparking a new round of anti-government protests.

1967: May Park is reelected

President Park wins his bid for reelection, defeating Yun Posun, and the Democratic-Republic Party wins a broad majority in parliament.

1969 Constitutional amendment authorizes third term for Park

The National Assembly approves a constitutional amendment allowing President Park to run for a third term.

1970s Monochrome movement becomes popular

The Monochrome movement, centered around Hong'ik University, plays an important role in Korean art. Works of this school are known for their meditative quality and feature traditionally neutral Korean hues such as gray, white, and brown.

1970 Publication of landmark novel *The Land*

The Land by Pak Kyong-ri is widely considered to be Korea's greatest contemporary novel. A sweeping historical novel, it follows the fortunes of one Korean family through the Japanese occupation and beyond.

1971: April Park wins third term as president

Park Chung Hee is reelected to a third term, defeating Kim Dae Jung (b. 1925, see 1987: July 8 and 1997: December 18) of the New Democratic Party (NDP).

1971: September North and South Korean Red Cross societies hold talks

The Red Cross societies of North and South Korea meet to discuss an exchange of information on family members separated by the division between the two countries.

1971: December 6 Park declares a national emergency

In response to student protests against his government, President Park declares a national emergency, but civil unrest continues to mount.

1972: July 4 North-South Joint Communiqué is issued

The two Koreas issue a statement declaring their intentions to pursue efforts aimed at peaceful reunification of the two nations.

1972: August 19 Hundreds die in heavy flooding

Torrential rains cause flooding and landslides in Kyonggi, Kangwon, and North Chungchong provinces, killing 638 people and making 144,000 homeless.

1972: October 17 Martial law is declared

President Park declares martial law to control growing civil unrest against the regime. The National Assembly is disbanded, and all political activities are outlawed.

1972: November New constitution is approved

A new constitution, approved by referendum, marks the beginning of the Fourth Republic. The president's powers are expanded, and he serves for six years.

1972: December Park is elected under new constitution

President Park is elected to a six-year term under the new constitution.

1974: August 15 Attempt to assassinate Park fails

In a failed attempt to kill President Park, an assassin murders Park's wife. The killer is a Korean carrying a Japanese passport with loyalties to North Korea. In response the his wife's assassination, Park strengthens internal security with a new series of emergency measures.

1975: May Harsh emergency measure drafted

Emergency Measure No. 9, one of a series of decrees tightening internal security, forbids students to take part in political protests and authorizes the arrest of anyone who criticizes the constitution.

1978: July Park is reelected

President Park is elected for another six-year term.

1979: January–March Reunification talks are held

Talks on reunifying North and South Korea are held, but little progress is made.

1979: October Government opponent is ousted from legislature

Kim Young Sam (b. 1927, see 1987: December 16, 1992: December 19, and 1997: December 22), leader of the opposition New Democratic Party (NDP) is thrown out of the legislature after demanding government reforms. This action causes rioting in Pusan and Masan, resulting in the imposition of martial law.

1979: October 26 Park is assassinated; prime minister takes over

Park Chung Hee is assassinated by the director of the government intelligence agency (the KCIA), Jim Jae-gyu. The prime minister, Choi Kyu-hah, takes over leadership of the government and promises new elections. Some of the emergency restrictions imposed by Park's government are rescinded. (See 1917: November 14.)

Chun Takes Over

1979: December Coup is led by Chun Doo Hwan

Major General Chun Doo Hwan (b. 1931) seizes control of the government in a military coup but retains a civilian as prime minister.

1980s Decade of rapid economic growth under Chun

South Korea enters a period of swift economic growth, especially in the shipbuilding, automotive, and consumer electronics sectors, becoming the twelfth largest trading nation in the world.

1980s Rise of Minjung misul movement in Korean art

Agitation for more democracy in government by students and others gives rise to the *Minjung misul* (people's art) movement, which is based on the social awareness and participation of artists. It emphasizes the daily lives of the common people and is concerned with Korea as a modern industrial society. Painters associated with this movement include O Yun (1948–86), Im Ok-sang (b. 1950), and Sin Hak-ch'ol (b. 1943).

1980 Office of Environment is established

South Korea's Office of Environment is created to monitor air, water, and land pollution and oversee solid waste management.

1980: May 18 Martial law is imposed

Chun Doo Hwan imposes martial law in response to civil unrest and arrests leading political figures.

1980: June 21 Student protests in Kwangju are crushed by government

The military government of Chun Doo Hwan ends the student-led siege of Kwangju, killing over 200 people and imprisoning thousands.

1980: July 9 Major government purge

A purge of government officials is undertaken with the announced goal of stemming corruption.

1980: September Chun becomes president of South Korea

Chun Doo Hwan assumes the presidency of South Korea.

1980: October 22 New constitution is approved

Voters approve a new constitution inaugurating the Fifth Republic and lifting the ban on political parties.

1981: January 24 Martial law is lifted

Martial law (see 1980: May 18) is lifted by President Chun.

1981: February Chun is reelected to seven-year term

President Chun is reelected to a seven-year term under the terms of the new constitution.

1983 Professional soccer league is formed

An eight-team professional soccer league is established.

1983: September 1 Korean Air Lines plane Is shot down over U.S.S.R.

A Korean Air Lines jet flying from New York to Seoul via Anchorage, Alaska, enters the airspace of the former Soviet Union and is downed by the military. The 269 persons on board the plane are all killed, provoking major international protests of the Soviet action.

1983: October 9 Rangoon assassination attempt kills seventeen

A bombing in Rangoon, Burma, fails to kill its target, President Chun, but seventeen government officials die in the blast, including four cabinet ministers. Chun blames North Korea, and Burma severs diplomatic ties with the DPRK.

1985 NKDP opposition party is formed

Groups opposing the government form the New Korea Democratic Party (NKDP), which mounts a strong electoral challenge to the ruling Democratic Justice Party (DJP), acquiring a strong minority position in the National Assembly and pushing for constitutional reforms.

1986: May Violent demonstrations take place against President Chun

Demonstrations against President Chun Doo Hwan at Inch'on end in violence.

1987: June Massive protests demand constitutional reform

Tens of thousands protest President Chun's order that constitutional reform be delayed until after the scheduled 1988 Summer Olympics in Seoul.

1987: July 8 100,000 protest in Seoul

In the most massive protest since the 1960 demonstrations that brought down Syngman Rhee (see 1875; 1948: August 15, and 1960: April 27), 100,00 demonstrators march in Seoul. The government restores political rights to Kim Dae Jung (see 1971: April) and 2,000 other opposition leaders.

South Korea Moves toward Greater Democracy

1987: October Constitutional reforms are enacted

Long-sought reforms authorize the direct election of the president, mandate a limit of one term, and place further curbs on presidential power. Members of the military are prohibited from serving as president. Further reforms guarantee greater freedom of expression to the population.

1987: December 16 Roh Tae Woo is elected president

DJP leader Roh Tae Woo (b. 1932) becomes South Korea's first president to be elected directly by popular vote, succeeding to the office to be vacated the following February by Chun Doo Hwan. The opposition vote is split between Kim Dae Jung and Kim Young Sam.

1988 Seoul Paralympics are held

The Seoul Paralympics involve over 4,000 disabled athletes from sixty countries and a budget of $26 million. The opening ceremonies are witnessed by a crowd of over 70,000.

1988: February Roh Tae Woo is inaugurated

Roh Tae Woo takes office as president of the ROK, as Chun Doo Hwan steps down.

1988: September–October Olympics are held in Seoul

The twenty-fourth annual Olympic Games are held in Seoul. North Korea boycotts the games.

1988: November Former president apologizes to the nation

Following government investigations into abuses of power, former president Chun apologizes to the nation on television. He announces that he is giving away all his money, including funds he obtained illegally, and retiring to a Buddhist temple.

1990 Japan apologizes to Korea for past abuses

The Japanese issue a formal apology to South Korea for abuses inflicted during the occupation of 1910–45 (see 1910–45).

1990 National Science Museum is completed

The National Science Museum in Daejon is completed.

1990: July 200,000 protest against the government

An antigovernment protest in Seoul is attended by 200,000 people.

1991 Seoul subway system opens

The Seoul subway system is inaugurated.

1991: September Korea joins the United Nations

Both North and South Korea join the United Nations.

1992 South Korea establishes diplomatic ties with China

A formal diplomatic relationship is established between the People's Republic of China and the Republic of Korea.

1992: December 19 Kim Young Sam is elected president

Kim Young Sam, head of the majority Democratic Liberal Party (DLP) party, is elected president with forty-two percent of the vote. The DLP is a new party that is born out of the 1990 merger of the DJP with two opposition parties.

1993: February Kim Young Sam is inaugurated

Kim takes office, becoming South Korea's first civilian president in thirty years. He institutes a wide-ranging anti-corruption program, carries out a purge of top military officials, and grants amnesty to 41,000 prisoners.

1995: November 16 Former president Roh Tae Woo is arrested for corruption

Former South Korean president Roh Tae Woo (see 1987: December 16) is arrested for accepting bribes during his term in office (1988–93). Roh, who has admitted the existence of a $654 million "slush fund" of campaign contributions, becomes the first South Korean leader charged with criminal activities while in office.

1995: December Former president Chun Doo Hwan is arrested for treason

Former president Chun Doo Hwan is arrested for his role in the 1979 military coup that brought him to power (see 1979: December) and the 1980 massacre of protesters in Kwangju (see 1980: June 21).

1996 Government rocked by corruption charges

Corruption charges connected with bribes by the Hanbo Steel Industry Company are leveled at government officials, including the son of President Kim Young Sam.

1996: January 23 Chun is indicted for role in massacre

Former president Chun Doo Hwan is indicted for treason for his role in the 1980 killing of over 200 pro-democracy protesters in the city of Kwangju. Government troops had opened fire on the demonstrators, who were protesting the martial law proclaimed by Chun's regime. Chun also faces bribery charges.

1996: August 26 Chun Doo Hwan is sentenced for treason

Former president Chun Doo Hwan is tried and found guilty of treason based on the 1980 military coup through which he gained power and the massacre of over 200 civilians demonstrators in Kwangju the same year. He receives a death sentence, which later reduced to life imprisonment.

1997: February Officials are indicted on bribery charges

Government officials are indicted on corruption charges in connection with loans made to the Hanbo Steel Industry Company.

1997: April 17 Supreme Court upholds sentences for Chun and Roh

South Korea's Supreme Court upholds the sentences given to former presidents Chun Doo Hwan and Roh Tae Woo for a variety of offenses leading to convictions of treason and corruption. Chun is sentenced to life imprisonment (a reduction from the original death sentence), and Roh is sentenced to seventeen years in prison (also a reduction from his earlier sentence).

1997: August 6 Korean plane crashes on approach to Guam airport

A Korean Air jet crashes as it prepares to land on the island of Guam, killing over 200 of its 254 passengers. The plane, attempting to land in a rainstorm, goes down in a jungle area and bursts into flames. U.S. Navy personnel stationed on the island mount a rapid rescue effort.

1997: December 3 Agreement reached on $57 IMF debt bailout

The South Korean government and the International Monetary Fund (IMF) reach agreement on the terms of a $57 million international bailout package, the largest ever put together by the IMF. South Korea's serious debt crisis will be alleviated by a combination of IMF credits, loans from the World Bank and the Asian Development Bank, and bilateral loans from individual countries including the U.S. and Japan. The money is intended to bolster Korea's dwindling foreign reserves, thus stabilizing its currency. Much of the country's debt problem is blamed on unrestrained expansion by industrial conglomerates.

1997: December 18 Opposition leader Kim Dae Jung is elected president

Kim Dae Jung (see 1971: April and 1987: July 8), leader of the National Congress for New Politics party, becomes South Korea's first opposition leader to be elected president. He will succeed Kim Young Sam, who is barred by law from seeking another term. A factor in Kim's election is the nation's economic crisis, which has necessitated a financial bailout by the IMF.

1997: December 22 Former presidents Chun and Roh receive pardons

Outgoing president Kim Young Sam grants an official pardon to former presidents Chun Doo Hwan and Roh Tae Woo, who have both been imprisoned since the end of 1996 on various charges of corruption and sedition. Both former officials are released from prison.

1998: February 25 Kim Dae Jung is inaugurated

Newly elected president Kim Dae Jung takes office in the first transfer of power to an opposition party in South Korea's history. The inaugural is attended by former presidents Chun and Roh, both of whose military regimes had inflicted persecutions ranging from imprisonment to assassination attempts on Kim, the nation's most prominent political dissident. In his address, Kim calls for national unity and reunification of North and South Korea.

Bibliography

Amsden, Alice H. *Asia's Next Giant: South Korea and Late Industrialization.* New York: Oxford University Press, 1989.

Cumings, Bruce. *The Origins of the Korean War.* Princeton, N.J.: Princeton University Press, 1981.

Gills, Barry K. *Korea Versus Korea: A Case of Contested Legitimacy.* New York: Routledge, Inc., 1996.

Grayson, James Huntley. *Korea: A Religious History.* New York: Oxford University Press, 1989.

Hahm, Sung Deuk. *After Development: Transformation of the Korean Presidency and Bureaucracy.* Washington, D.C.: Georgetown University Press, 1997.

Hoare, James. *Korea: An Introduction.* New York: Kegan Paul International, 1988.

Howe, Russell Warren. *The Koreans: Passion and Grace.* San Diego, Calif.: Harcourt Brace Jovanovich, 1988.

Hwang, Eui-Gak. *The Korean Economies: A Comparison of North and South.* New York: Oxford University Press, 1993.

Lone, Stewart. *Korea Since 1850.* New York: St. Martin's Press, 1993.

Macdonald, Donald Stone. *The Koreans: Contemporary Politics and Society.* Boulder, Colo.: Westview Press, 1990.

Merrill, John. *Korea: The Peninsular Origins of the War.* Newark: University of Delaware Press, 1989.

Oberdorfer, Don. *The Two Koreas: A Contemporary History.* Reading, Mass.: Addison-Wesley Pub. Co., 1997.

Oliver, Robert Tarbell. *A History of the Korean People in Modern Times: 1800 to the Present.* Newark: University of Delaware Press, 1993.

Olsen, Edward A. *U. S. Policy and the Two Koreas.* Boulder, Colo.: Westview Press, 1988.

Soh, Chung Hee. *Women in Korean Politics.* 2nd ed. Boulder, Colo.: Westview Press, 1993.

Tennant, Roger. *A History of Korea.* London: Kegan Paul International, 1996.

Kuwait

Introduction

Kuwait lies on the western head of the Persian Gulf (also known as the Arabian Gulf) with an area slightly smaller than the state of New Jersey. Because of its undefined borders, it is difficult to precisely estimate its size, but 6,880 square miles (17,820 square kilometers) is an accepted figure. Kuwait's boundaries with Saudi Arabia were settled by treaty in 1922. That treaty created a Neutral Zone, an area of approximately 2,500 square miles (6,475 square kilometers) in which each country would have an undefined half interest. In 1965, the two nations formally divided the Neutral Zone.

Nearly all Kuwait is a desert with summer temperatures that reach 120° F (49° C). Average rainfall is less than four inches, and the country has no rivers. Kuwait has invested heavily in building some of the most sophisticated desalination plants in the world to provide fresh water to its population of nearly 1.6 million people. A mere three-tenths of one percent of the total land area is cultivated. There's little vegetation but rains transform the desert between October and March. During this time, grass and foliage are plentiful, with a great variety of flowers and plants.

Ethnic Kuwaitis are mostly descendants of the tribe of Najd (Central Arabia), but some descend from Iraqi Arabs. A few are descendants of Iranians. Arabic is the official language, but English is widely spoken and used in business transactions. Kuwait is an Islamic country: Islam is the official state religion. About forty-five percent of its citizens are Sunni Muslims and thirty percent are Shiite Muslims. Since the seventh century, the two sects have been divided over the role of Ali, the cousin of the Prophet Mohammed. Sunni Muslims, who make up about eighty-five percent of all Muslims worldwide, regard Ali as the fourth successor of Mohammed, while the Shiites say he is first in the line of succession.

Discrimination is common against women and Bedouins. Kuwaiti women do not have equal rights and legal protection under the law. They are not allowed to vote, and their testimony in a court of law is not considered equal to that of men. Women must obtain permission from their husbands to apply for passports. Moreover, calls for equal rights have not been popular. Women who marry foreign men suffer legal discrimination and are not entitled to government housing subsidies. Sexual harassment, especially directed at women who wear Western clothing, is prevalent. Bedouin minorities also face discrimination. They are not entitled to citizenship and are unable to work or enroll their children in schools. The abuse of foreign domestic servants is also considered a serious ongoing social problem.

For centuries, Kuwait was little more than a desert outpost, a largely ignored corner of the Arab world. Inhospitable, unforgiving terrain on the western edge of the Persian Gulf kept most people away. Kuwait and the surrounding region didn't prosper until natives began to harvest pearls in the Persian Gulf in the mid-1800s. Europeans began to show interest in the Arabian peninsula in the eighteenth and nineteenth centuries, with the English, Ottomans, and Germans trying to get control. Fears of a takeover by the Ottoman Empire forced Kuwait to strike an alliance with England in 1899. Thanks to England's protection, the country remained relatively untouched by outside forces through 1961, when Kuwait declared its independence.

Discovery of Oil

The discovery of oil in the 1930s dramatically changed Kuwait, which is literally floating on top of oil. In the 1960s, the *emir,* the nation's leader, was earning $1 million per day from oil revenues. Foreigners were attracted by good wages, free education for their children, and free medical care. The numbers of foreign workers more than doubled during the 1970s. By the 1990s, the country had become used to a lavish lifestyle, hiring 2.4 foreign workers per every Kuwaiti, including more than 100,000 foreign maids.

Nonetheless, Kuwait remained just another one of the great oil producing countries with little else to distinguish it in the world community. Its relative anonymity ended in August 1990, when Iraq invaded the tiny nation in hopes of reclaiming territory. Iraq's border disputes with Kuwait have a long history. Within hours of declaring its independence back in 1961, Kuwait had faced its first crisis when Iraq reasserted its claims to the emirate. The British were forced to send a small force to deter the Iraqis. In March, 1973, Iraqi forces moved 3,000 troops into Kuwaiti territory, occupying one border post and bombing another one. Ultimately, however, Iraq finally withdrew under Arab League pressure.

Western nations quickly condemned the 1990 Iraqi invasion, with some Western leaders calling it an affront to democracy, even though Kuwait was anything but a democratic nation at the time. Iraq's leader, Saddam Hussein, refused to withdraw his troops and installed a provisional government to rule what he called the nineteenth province of Iraq. In response, the United States government sent more than 425,000 troops to the Gulf and twenty-seven other nations followed suit, sending another 265,000 troops. Hussein was warned to retreat or face attack. More than 150 ships in the Persian Gulf and 2,000 aircraft did not persuade him to relent.

Coalition forces began bombing Iraqi and Kuwaiti targets in the early morning of January 17, 1991. On February 28, coalition forces liberated Kuwait, pushing Iraqi troops deep into Iraq and inflicting massive damage. However, before leaving, the Iraqis devastated Kuwait, burning more than 700 of the country's oil wells and creating an environmental disaster of enormous proportions. While Iraq recognized its borders with Kuwait and the country's independence in November 1994, many Iraqi officials continued to assert claims to Kuwait.

After the war, Kuwaitis set about rebuilding the country, a job that was made easier with accumulated oil wealth. In many ways, the Kuwait that emerged after the war was slightly more democratic. Its citizens, who felt betrayed by their leaders—they left the country immediately after Iraqi tanks crossed the border—demanded more participation in their government. The emir was forced to resurrect the country's constitution and open up the political process. In free elections, citizens were allowed to vote for fifty of the Parliament's seventy seats. But under strict nationality laws, only ten percent of the country's citizens were eligible to vote. Women were not.

Before the Iraqi invasion, Kuwait had hired workers from neighboring Arab nations. But the invasion forced Kuwaitis to look elsewhere. They figured that a work force made up of Egyptians, Filipinos, Indians and others was less volatile politically than Palestinians, who were accused of supporting Iraq in the war. Kuwait also vowed never to let the number of foreigners surpass the native population, but it has been unable to do without them. By 1995, foreigners made up nearly fifty-eight percent of the country's population. However, the Kuwaiti government maintains tight control of the work force and carefully monitors union activities. The right to strike is severely limited, and foreign workers can only join a union after working in the country for five years. They are not allowed to run for office or vote in union elections. The abuse of foreign workers, especially domestic servants, remains a serious social problem.

Kuwaiti citizens receive free education, including free food, clothing, books, school supplies and transportation from kindergarten through the fourth year of college. Many of the country's top students study abroad at government expense. In 1995, about twenty-one percent of the population could not read or write.

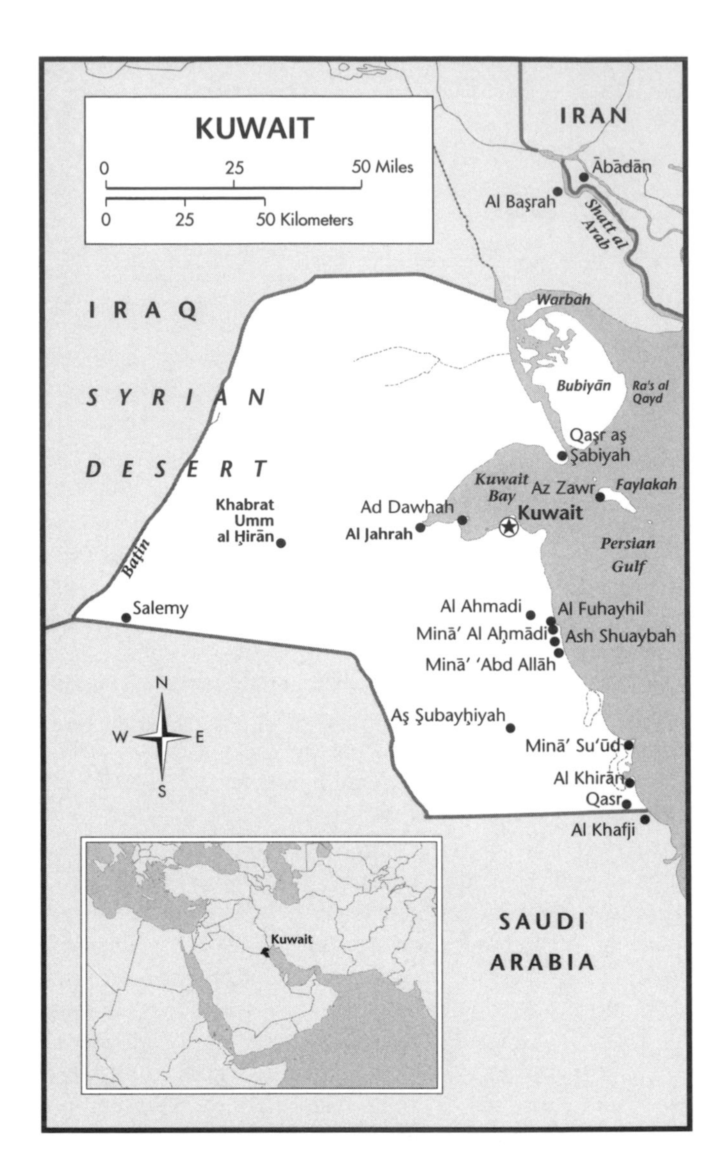

While oil remains the country's most important industry, Kuwait has attempted to diversify its economy. It is no easy task. This desert nation has negligible resources outside oil and natural gas. Much of its technology and those who run it have been imported. Clearly, the small nation faces a challenging future if it is to maintain its high standard of living.

Timeline

2800 B.C. Early contacts with Sumeria

An ancient civilization, which leaves archeological evidence, has trade links with the Sumarians. The Sumerians, who have one of the great ancient civilizations, settle in what is now Iraq and develop a writing system before 3000 B.C.

By the sixth century B.C., this part of the Persian Gulf becomes a major supply route for trade with India. Later evidence also shows early migrations to East Africa by seafaring inhabitants. During the reign of Alexander the Great (c. 323 B.C.), Greeks colonize the island of Failaka, which they call Ikaros. It becomes a center for trade, fishing and pearling. Yet the inhospitable area now called Kuwait has little importance for many centuries.

A.D. 632 Islam profoundly changes the Arab world

The emergence of Islam in present-day Saudi Arabia under the Prophet Mohammed (570–632) has a profound influence on the countries of the Middle East. Islam, like Christianity and Judaism, is a monotheistic religion that acknowledges the absolute sovereignty of God. Conquering Arabs quickly spread in all directions, taking Islam with them. By the end of the seventh century, the people of Kuwait have embraced Islam.

17th–18th centuries British establish a presence in the Persian Gulf

During the seventeenth and eighteenth centuries, the British drive French and Dutch competition out of the Gulf region and proceed to sign treaties with the small Arab states, today known as the nations of Kuwait, Oman, Qatar, Bahrain, and the United Arab Emirates, as well as present-day Saudi Arabia. Known as Exclusive Agreements, the documents give England control of foreign affairs, forcing the small Gulf states to seek British permission to make treaties with any nations or to cede land. In exchange, the British promise to protect them from any other foreign threat. The British administer the Gulf from India, and the Indian rupee becomes the common currency of the Arab states.

1716 Clans migrate from the interior of the Arabian Desert

Drought and famine force several clans of the tribe of Aniza to migrate from the Arabian Desert to a small region on the Persian Gulf, which later becomes Kuwait. The country's name is a diminutive of the word *kut*, meaning "fort."

1756 As-Sabah sheikhdom is founded

The tribesmen choose Sheikh Sabah 'Abd ar-Rahim as their leader, inaugurating the As-Sabah dynasty, which remains in power to the present day in Kuwait.

Late 18th century Kuwait becomes important trade center

After a period of pirate raids by land and sea, which are suppressed by the British, trade grows and shipbuilding thrives, and Kuwait becomes a major trade center. Kuwait's boat fleet grows to about 800 and camel caravans based in the city travel to Baghdad, Riyadh, and Damascus. The boats, called *dhows,* are unique to Arabia. The dhow is a lateen-rigged boat with a long overhang forward, an enclosed structure at the stern, and an open waist.

19th century Pearling industry grows

Kuwait's economy, as well as the economies of other Gulf states, depends heavily on the pearl industry. Pearling is a difficult and often brutal job, with boats constantly at sea from May to October. The work is divided between divers and pullers. The divers only carry a nose clip, leather guards to protect their hands, a knife to cut open the shells and a bag for the pearls. They usually tie rocks to their feet to sink to the bottom, where they stay for up to a minute. When they need air, they tug on the rope and they are pulled to the surface. Painful stings from jellyfish are a constant hazard.

While pearling makes the boat owners rich, divers and pullers barely manage on subsistence wages. They often borrow money from the boat owners before the pearling season starts and can never pay it back. If a diver dies in debt, his son must assume responsibility and work until it is paid off.

1866–92 Reign of Abdallah as-Sabah

Under the leadership of Sheikh 'Abdallah as-Sabah, Kuwait finds itself in the middle of a dynastic battle raging between the Arabian rival houses of Ar-Rashid and As-Sau'd. The Ottoman Turks, who want to extend their control over the coastal area south of Kuwait, support Ibn Rashid.

1870 Birth of Emir Abdallah al-Salim as-Sabah

Sir Abdallah al-Salim as-Sabah is one of Kuwait's most important emirs. Largely self-educated, he gains a reputation as a scholar and financial expert. He ascends the throne of Kuwait in 1950. Under his rule, previously underdeveloped Kuwait attains a prominent position among the great oil-producing nations of the world. With assistance from American and British advisers, his tiny nation adopts a social welfare program founded on a unique patriarchal system. The emir, who is revered as a man of simplicity, devotion and deep concern for his people, becomes the nation's first constitutional monarch in 1962.

Abdallah astutely handles foreign affairs so Kuwait can remain friendly with the West without alienating Arab neighbors. On one hand, he provides financial aid to Arab countries and maintains Kuwaiti support for the Arab cause against Israel. On the other, he calls for British help to keep Iraq from invading his country.

Reputed to earn as much as $1 million per day before his death in 1965, he leads a simple life with his wife and two sons, devoting most of his time to study and governing the country.

1896–1915 Mubarak as-Sabah rules Kuwait

Sheikh Mubarak as-Sabah overthrows and murders his brother, Mohammed, eliminates another brother, Jarrah, and installs himself as ruler of Kuwait. Mubarak opposes Turkey and fears annexation by them. The British, who want control of the sea routes between the Gulf and India, one of England's most profitable colonies, compete against Turkish, German, and Russian claims to Gulf territories. The Russians seek to set up a refueling station for vessels, while the Germans and the Turks want Kuwait to become a terminus for the Berlin-Baghdad railroad.

1899 Sheikh Mubarak sides with the British

Mubarak agrees not to give up any part of his territory to foreign interests or to receive representatives of any country without British consent. In return, the British offer protection as well as an annual subsidy to support the sheikh and his family.

1930s Decline of the pearling industry

Pearling in the Gulf collapses when the Japanese invent a method of culturing pearls.

1936 Kuwait modernizes its educational system

Beginning in 1936, Kuwait becomes the first country in the region to implement a modern school system and the first to grant scholarships in the arts.

1938 Discovery of oil transforms the country

Enormous oil reserves and massive quantities of natural gas turn tiny Kuwait into one of the richest countries in the world. In the next several decades, Kuwait becomes a fully developed, subsidized welfare state. Oil money builds ports, roads, an international airport, a seawater distillation plant, and modern government buildings, and it also supports free public services, including education through college. By 1996, Kuwait possesses oil reserves of about 96 million barrels, nearly ten percent of the world's known global resources of petroleum.

1946 Birth of artist Abdul Rasul Salman

One of Kuwait's artistic pioneers, Abdul Rasul Salman, is trained as a calligrapher. He uses a stylized type of script and rich colors.

1948 Gulf rupees introduced as new currency

With India's independence in 1948, the Indian rupee is replaced by the Gulf rupee, which remain in circulation until 1971 in the Arab states.

1950 Kuwait builds its first desalination plant

Until the twentieth century, Kuwait's population is quite small and a little rainfall and wells are enough to support its population. Before building its first desalination plant to obtain fresh water, Kuwait imported the precious liquid from Iraq, as desalination plants are extremely expensive to operate. Water imports by boat began in 1907 and hit a peak in 1947, when approximately 303,200 liters of water per day arrived in Kuwait by boat. By the 1990s, Kuwait's demand for fresh water is easily met by desalination.

1960 Art education programs are launched

In 1960, the government begins offering free art workshops for amateur artists, both Kuwaitis and foreign citizens. Classes are offered in painting, printmaking, and sculpture. At first only men are allowed to attend.

1961 Government provides support to artists

Among the major steps taken by the Kuwaiti government in the postwar years to encourage education and culture is the inauguration of a program offering support for full-time artists, giving them monthly salaries for two or more years. Most Kuwaiti art is figurative. It depicts local landscapes and still lifes.

1961: June 19 Kuwait declares sovereignty

By mutual consent, Kuwait and the United Kingdom end protective treaty relations. Kuwait declares itself fully sovereign and independent. This calculated move allows Kuwait to acquire controlling interests in the petroleum industry.

Refusing to recognize Kuwait's independence, Iraq asserts it has inherited Ottoman claims to the territory and threatens an invasion. The British dispatch troops to protect the country. Later, the Arab League, an organization of Arab countries that promotes friendship between member states and coordinates their political action, supports Kuwaiti independence.

Iraq appears to relent, but border issues remain unresolved. During the next two decades, Kuwait becomes a rich nation, supported mainly by a Palestinian and Egyptian foreign labor class.

1961: December Kuwait helps other Arab nations

Kuwait establishes the Kuwait Fund for Arab Economic Development and issues loans at low-interest rates. By 1985, its loans amount to $4.3 billion. Most of its aid goes to Egypt and Syria.

1965 Death of Emir Abdallah al-Salim as-Sabah

One of Kuwait's most notable emirs and the nation's first constitutional monarch dies. (See 1870.)

1970s Art diplomas are introduced

Kuwait's art education programs becomes more formal when diplomas are given to three-year participants. Foreigners are excluded. In the coming decade, neighboring countries copy Kuwait's workshop model.

1977: December Emir dies after 12-year reign

Emir Sabah as-Salim as-Sabah (1913–77) dies after a twelve-year reign and is succeeded by Emir Jabir al-Ahmad al-Jabir as-Sabah (b. 1928). *Emir* is the name given to the native ruler in the Middle East, Asia and Africa. The Emirate is his domain.

1977–85 Government provides homes for thousands

For centuries, housing in Kuwait consists of small cottages, mud huts, and some large homes built of coral and plastered with cement and limestone. Oil revenues allow Kuwaitis to make housing a national priority. Between 1977 and 1985, the National Housing Authority builds 50,000 housing units. It receives applications for another 25,000 between 1989 and 1994.

1980–88 Iraq and Iran wage war over territorial dispute

During the Iran-Iraq War, Kuwait claims neutrality but gives economic aid to Iraq. Goods and provisions amount to the equivalent of $6 billion. In response, members of Kuwait's Shiite minority and other dissidents wage a war of terrorism against the Kuwaiti government. Throughout the time of the Iran-Iraq war, Kuwait suffers bombings, assassination attempts, hijackings, and sabotage against oil facilities.

1982 Abortion in certain situations is legalized

Kuwait becomes the first country in the Persian Gulf region to legalize abortion for limited health reasons.

1983 Kuwait inaugurates National Museum

The country's new museum houses important works by leading Islamic artists and many Kuwaiti painters and sculptors. During the later Iraqi invasion of Kuwait, Iraqi troops burn and ransack the museum. The Islamic collection had been packed and crated and sent back to Iraq. The contemporary collections are lost, but Iraq returns the Islamic collection after United Nations intervention.

1987 Kuwait seeks help to keep Iraq-Iran War at bay

Iranian attacks on Persian Gulf shipping lead Kuwait to seek United States protection for its supertankers. Washington agrees and when a Kuwaiti vessel flying with a U.S. flag is attacked, Americans retaliate by bombing an Iranian offshore oil rig.

1990 Iraq-Kuwait relations deteriorate

At the end of the war, Kuwait and Iraq manage to maintain stable relations. But in 1990, Saddam Hussein accuses Kuwait of waging economic warfare by illegally drilling oil from the shared Rumailia field. Overproduction, Hussein says, drives oil prices down. Moreover, Kuwait is unfairly demanding repayment for war-time loans.

The Gulf War

1990: August 2 Iraq invades Kuwait

Negotiations and mediation do not defuse tensions and Iraq invades Kuwait, asserting the right to reclaim territory. The Kuwaiti defense forces crumble, and the emir and his cabinet flee to Saudi Arabia. Tanks cross the border at 2 a.m. and arrive in Kuwait City before dawn. By noon, Iraqi troops reach the Saudi border. On August 6 Iraq declares it has annexed Kuwait.

1990: August 9 United States offers to defend Saudi Arabia

King Fahd of Saudi Arabia accepts American troops to help defend the kingdom against a possible Iraqi invasion. More than 10,000 U.S. troops arrive by August 14. Hussein prohibits Western residents from leaving his country and begins to round up westerners in Kuwait.

1990: August 10 Hussein refuses to attend Arab meeting

The Arab League, in an emergency meeting in Cairo, Egypt, approves a resolution criticizing Iraq but remains deeply divided by Saudi Arabia's decision to allow American troops on its soil.

1990: August 18 Hussein cautions against U.S. intervention

U.S. President George Bush mobilizes reserve units, and Hussein announces that he will hold westerners as hostages at military and civilian installations to deter a U.S. attack. He threatens to use chemical and biological weapons.

1990: September 10 Americans and Soviets condemn invasion

President Bush meets with Soviet President Mikhail Gorbachev. Both condemn Iraq's occupation of Kuwait and call for a withdrawal of Iraqi troops. Instead, Iraqi troops in Kuwait increase to about 360,000.

Amnesty International, a human rights group, issues a report condemning Iraq for human rights violations in Kuwait. Atrocities include torture.

Kuwait's oil wells burn after bombing by Iraq during the Gulf War. (Corbis)

1990: November 8 Bush sends more troops to the Gulf

President Bush sends 200,000 troops to the Gulf, amassing a force of more than 425,000. Twenty-seven other nations send 265,000 additional troops, and the coalition commands more than 150 ships and 2000 aircraft.

1990: November 29 United Nations authorizes use of force

The United Nations Security Council authorizes the use of force if Iraq does not withdraw from Kuwait by January 15, 1991.

1991: January 9 Negotiations fail

Attempts to resolve the Gulf crisis are fruitless after U.S. Secretary of State James Baker and Iraqi Foreign Minister Tariq Aziz hold talks for more than six hours in Geneva.

1991: January 12 U.S. Congress authorizes use of force

A joint resolution by Congress authorizes President Bush to use force against Iraq and coalition forces begin bombing Iraq at 3 a.m. on January 17.

1991: January 20 Iraq creates massive environmental catastrophe

Iraqis pump millions of gallons of crude oil into the Gulf, apparently in an attempt to poison Saudi Arabia's drinking water supply. In all, they release between eight to ten million barrels and the oil slick affects more than 400 kilometers of coastline.

Experts from six nations try to control the slick after the war ends, and some oil companies manage to recover about one million barrels. The slick devastates marine life along the coast.

Inland, Iraqis begin to burn Kuwait's oil wells—more than 700 by some estimates—destroying millions of gallons of oil per day. Thick clouds of smoke turn day into night. It takes firefighting crews eight months to put out all the fires. Black rain and snow is reported as far as Pakistan and India. Oil lakes in the desert are several feet deep.

1991: February 24 Ground offensive gets under way

After Saddam Hussein ignores a February 22 deadline for removing Iraqi troops from Kuwait, forces from the U.S.-led multinational coalition launch a ground offensive against Iraq.

1991: February 27 Iraqi troops begin to withdraw from Kuwait

Coalition aircraft attack and massacre a retreating column of Iraqi forces. The same day coalition forces enter Kuwait.

1991: February 28 President Bush ends ground offensive

President George Bush ends the military ground offensive. Kuwait City is in chaos and martial law is declared. Many Kuwaitis turn their anger against Palestinian guest workers believed to have supported Iraq during the invasion. In special martial law courts, many Palestinians are charged and convicted of collaboration. Other Palestinians are exiled.

Kuwaitis resent the royal family and other leaders for leaving the country. Leaders are slow to return after liberation. The crown prince does not return to Kuwait until six days after the liberation, and the emir returns several days later. Kuwaitis also demand greater political participation.

1991: November Kuwait lifts press censorship

The constitution allows for freedom of speech and press; generally the country's citizens can criticize the government in all media.

1992: October Elections are held in Kuwait

Elections for a new National Assembly take place, with government opposition winning thirty of the fifty elected seats in the new parliament (twenty-five seats are appointed). The opposition secures six of the sixteen seats in the cabinet, but the as-Sabah royal family retains much of its power, with direct control over interior ministries, defense, and foreign affairs.

The new Assembly actively produces legislation and later begins to investigate allegations of government corruption. Political parties are prohibited, but opposition groups remain active.

Approximately ten percent of the country's population is allowed to vote. Kuwaiti laws restrict voting to about 70,000 adult literate Kuwaiti males holding "first class" citizenship, that is, whose ancestors resided in Kuwait prior to 1920. (See 1996.)

1993: May 27 United Nations support border demarcation

An international Boundary Demarcation Commission establishes the border between Iraq and Kuwait. The United Nations Security Council reaffirms the commission's decision. Relations between the two countries remain tense as some Iraqi officials continue to assert claims to Kuwait.

1994 Non-citizens remain majority in Kuwait

Foreign workers and their families make up nearly fifty-seven percent of the population even though Kuwait deports tens of thousands of workers after the Persian Gulf War. Only one-sixth of the estimated 400,000 Palestinians living in the country before the war are allowed to stay. Kuwait replaces its Arab workers with Filipinos, Pakistanis, and foreigners from countries outside of the Middle East.

1994: October Iraq troops move close to border

The Iraqis begin moving 60,000 troops within twenty miles of the Kuwait border. The UN Security Council condemns the action, and the United States and other countries offer assistance. Kuwait allows the United States to supply a squadron of twenty-four warplanes and a tank division to curb Iraq's military power.

1994: November 10 Iraq closer to accepting Kuwaiti sovereignty

Iraq agrees to recognize the current borders and independence of Kuwait. The decision is seen as a way to get the United Nations to lift some of its economic sanctions against Iraq.

1995: November The government sets up hot line for battered women

Kuwait sets up telephone hot line for battered women. The Islamic Affairs Ministry starts the service in cooperation with independent counselors for women suffering verbal, physical, sexual, or mental abuse.

1995: August Iraqi troop movements cause alarm

The United States sends ships and supplies to Kuwait after Iraq moves troops closer to the border.

1995 Military expenditures reach $3.1 billion

Kuwait rebuilds its defense and in 1995 has an army of 11,000 men and 200 tanks. The airforce has 2,500 men and 76 combat aircraft. Kuwait's military expenses on a per capita basis are the highest in the world. Yet the country continues to rely on foreign help to dissuade Iraq from invasion.

1996 Kuwaiti laws expand suffrage

Residents of Kuwait who have been naturalized citizens for at least thirty years become eligible to vote. Prior to this only a small percentage—about 70,000 adult Kuwaiti males holding "first class" citizenship—were eligible.

1996: April Joint military exercises

The United States, England, Russia, China, Italy, and Arab nations hold joint military exercises in Kuwait. The United Nations renews its multinational force of border observers across a nine-mile demilitarized zone that separates the two countries.

1997 Political prisoners remain behind bars

Amnesty International reports that more than 120 people, including prisoners of conscience, continue since 1991 to serve prison terms imposed after unfair trials.

1999: May 16 Emir orders full political rights for women

Emir Sheikh Jaber al-Ahmad al-Sabah orders full political rights for women, surprising political activists who have been pushing for several decades to allow women to vote. The order will allow women to vote in municipal and parliamentary elections and to run for office. Parliament must approve the emir's order.

1999: May Emir dissolves Parliament

The emir dissolves parliament after more than two years of clashes with his government. About 113,000 Kuwaiti men have the right to vote in elections for fifty parliamentary seats. The emir's order granting rights to women (see 1999: May 16) does not take effect until 2003. Because the order must be approved by parliament and the next parliament will be elected by men only, whether women will be able to exercise their new rights is not yet guaranteed.

Bibliography

Ali, Wijdan. *Modern Islamic Art: Development and Continuity.* Gainesville, FL: University Press of Florida, 1997.

Al-Moosa, Abdulrasool, and Keith McLachlan. *Immigrant Labour in Kuwait.* Dover, N.H.: Longwood, 1985.

Anscombe, Frederick F. *The Ottoman Gulf: The Creation of Kuwait, Saudi Arabia and Qatar.* New York: Columbia University Press, 1997.

Cordesman, Anthony H. *Kuwait: Recovery and Security after the Gulf War.* Boulder, CO: Westview Press, 1997.

Crystal, Jill. *Oil and Politics in the Gulf: Rulers and Merchants in Kuwait and Qatar.* New York: Cambridge University Press, 1990.

Ismael, J.S. *Kuwait: Social Change in Historical Perspective.* Syracuse, N.Y.: Syracuse University Press, 1982.

Khadduri, Majid. *War in the Gulf, 1990–1991: The Iraq-Kuwait Conflict and its Implications.* New York: Oxford University Press, 1997.

Robison, Gordon. *Arab Gulf States.* Hawthorn, Aus.: Lonely Planet Publications, 1996.

Kyrgyzstan

Introduction

Kyrgyzstan is a former Soviet republic located in the heart of the historical region of Central Asia. Independent since 1991, Kyrgyzstan is the nation state of the Kyrgyz people, a predominantly Muslim Turkic people who, until the beginning of the twentieth century, had led a largely pastoral nomadic way of life. Located in one of the most mountainous regions of the former Soviet Union, nearly all of Kyrgyzstan's territory lies above 1,500 feet in altitude.

The territory of Kyrgyzstan consists of 76,641 square miles (198,500 square kilometers) located within three major mountain ranges, the Alay Mountains (*Alayskiy Khrebet*) to the south, the Tian Shan Mountains to the east, and the Kyrgyzskiy Khrebet mountains in the north. All of these ranges reach heights well above 15,000 feet and are topped with permanent glaciers. The highest range is the Tian Shan, where the highest peak in Kyrgyzstan, Pik Pobedy, which is 24,406 feet (7439 meters), is located. The republic's main rivers are the Naryn, a tributary of the Syr Darya, the Chu, and the Talas, which empties into the southern Kazakh steppe. In addition to these major rivers, Kyrgyzstan has hundreds of smaller fast-flowing rivers that come down from the glaciers and other high elevations. Another important landmark of Kyrgyzstan is Lake Issyk-Kul, a huge 2,100-foot deep freshwater lake located high in the Tian Shan Mountains. Although much of Kyrgyzstan's territory is mountainous, the eastern, southern and northern fringes of the fertile Ferghana Valley lie within Kyrgyz territory. Kyrgyzstan's climate is largely determined by elevation. Lower elevations are mostly dry continental, while higher elevations are snowy in winter and cool in summer. Precipitation is highest in the western mountains and lowest in the north central highlands.

Kyrgyzstan is bordered to the north by Kazakstan, to the south and east by the People's Republic of China, with which it shares the Tian Shan as a border, to the south by Tajikistan, and to the west by Uzbekistan. Its capitol is Bishkek, which in the Soviet-era was known as Frunze, located in the extreme north of the country. Its other major cities are Osh and Jalal-Abad in the eastern Ferghana Valley. In 1994 the estimated population was just under 4.5 million. The largest ethnic group was Kyrgyz, with fifty-two percent; Russians, with twenty-two percent; Uzbeks, with thirteen percent; and Ukrainians with three percent. There are also substantial communities of Kyrgyz in the Tian Shan mountains in the People's Republic of China and in the Pamir mountains in southeastern Tajikistan.

The Kyrgyz are a Turkic people, and the Kyrgyz language is the country's national language. The literary language is based on the northern dialect, which is mutually intelligible with Kazakh. The southern dialect is mutually intelligible with Uzbek. Among Kyrgyz the urban population is for the most part fluent in Russian, and many Kyrgyz use Russian as their first language. Among Russians and Ukrainians, however, ignorance of the Kyrgyz language is very high. In urban areas Russian is still widely used, especially in commerce, as well as for official purposes.

History

The Kyrgyz are traditionally Muslim. After arriving in the Tian Shan region in the fifteenth century, ancestors of modern Kyrgyz began accepting Islam, their presence in Central Asia naturally integrating them into regional Islamic traditions. Once the Kyrgyz became Muslim, Islam and Islamic conceptions became an important element in virtually every aspect of their lives. In Soviet times (1917–91) Kyrgyz Islamic institutions were suppressed and effectively banned, and simultaneously Kyrgyz society became sovietized and secularized. However, certain Islamic customs, especially life-cycle rituals and pilgrimages to Muslim saints' tombs, continued to be important and widely observed among most Kyrgyz, especially in and around the Ferghana Valley. Since the independence of Kyrgyzstan, Islamic institutions have flourished.

Our understanding of Kyrgyz history is shaped by the fact that before the beginning of the twentieth century the Kyrgyz themselves left no historical accounts; in the absence of their own history, we have to rely on the accounts of their neighbors: Russians, Chinese, and sedentary Central Asians. According to these, the Kyrgyz came to this territory rather late, apparently in the fifteenth century, and it appears fairly certain that their ancestral homeland was the Altay and Sayan mountains of southern Siberia. As a result, much of the politi-

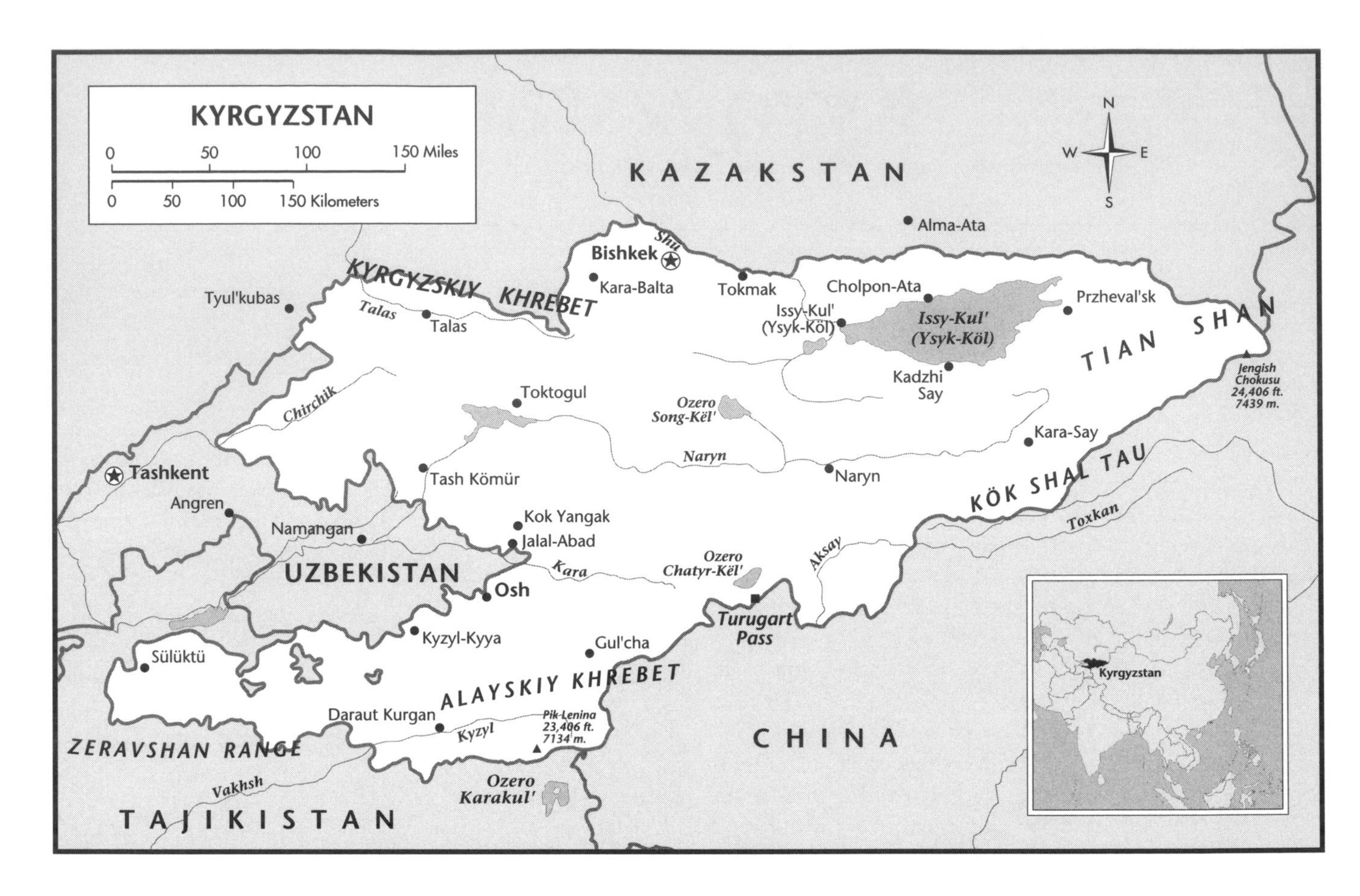

cal and ethnic history of the Kyrgyz unfolded in territories well beyond the borders of modern-day Kyrgyzstan. In any case, it was in South Siberia that the so-called Kyrgyz *Kaghanate* (an area in central Asia ruled by a prince) emerged onto the stage of history in A.D. 840 when Chinese sources speak of the overthrow of the Uyghur empire in Mongolia by an aggressive nomadic people who had lived to the northwest of the Uyghurs in the Sayan mountains. By 920 they had been driven back to their homeland along the Yenisei River by a new steppe dynasty, the Khitans. It appears that already at this time some Kyrgyz were present in the Tian Shan and remained there throughout the Middle Ages, although they were neither demographically nor politically dominant, and the center of Kyrgyz power remained in the upper Yenisei region. Then by the sixteenth century, the ancestors of modern Kyrgyz were present throughout the territory of Kyrgyzstan.

Kyrgyz society is characterized by an extensive tribal and clan system that remained strong through the Soviet period and down to the present day. The Kyrgyz tribal structure consists of two main federations: the first are the *Arkalïk* (Outer) Kyrgyz and the second are the *Ichkilik* (Inner) Kyrgyz. In Soviet times the Kyrgyz Communist Party was dominated by the Arkalïk Kyrgyz.

The traditional political structure of nomadic Kyrgyz differed substantially from that of their neighbors. Unlike the Kazakhs, their ethnic kinsmen to the north, the Kyrgyz never developed any tribal or supra-tribal aristocracy, nor did they develop or recognize any sort of charismatic dynasty like the *Chingisids* (descendants of Chingis Khan) who held sway among the Kazakhs into the nineteenth century. Authority was granted only to wealthy and authoritative nomadic leaders called *biys* or *manaps*. Traditionally political allegiances among the Kyrgyz tended to be based on kinship or tribal bonds, rather than on any sort of formalized charismatic authority.

The traditional economic pursuit of the Kyrgyz was pastoral nomadism. For centuries, until the Soviet period, the Kyrgyz would migrate with their flocks and derived their subsistence almost entirely from that activity. The animals most commonly raised by the Kyrgyz were sheep, cattle, horses, goats, Bactrian camels, and yaks. The cultivation of the Himalayan cattle, who were well suited to the topography and climate of the Tian Shan mountains, distinguished the Kyrgyz from the other pastoral nomadic peoples of Central Asia. Kyrgyz migration patterns were for the most part vertical. That is, winters were in the warmer sheltered valleys and summers in the alpine pastures. The traditional dwelling of the Kyrgyz was the *yurt,* the conical felt tent used by Inner Asian nomads from Mongolia to the Black Sea.

The nomadic diet consisted of meat and dairy products, especially yogurt and cheese. The Kyrgyz also traded with their sedentary neighbors for wheat, rice, and dried fruits. To some degree they supplemented their diets with hunting and fishing. In the Ferghana Valley irrigated agriculture had been practiced since antiquity, but in more recent times this activity was carried out primarily by Uzbeks and to a lesser extent by sedentary Kyrgyz communities. In Soviet times the Kyrgyz were permanently sedentarized, and the flatter steppe lands in the extreme north were plowed up for cereal cultivation.

Although the Kyrgyz developed a written culture relatively late, beginning in the late nineteenth and early twentieth centuries, they possess a particularly rich oral tradition. The centerpiece of Kyrgyz oral literature is their heroic epic, *Manas. Manas* was first recorded in the middle of the nineteenth century, and additional variations of the cycle were recorded throughout the Soviet period. In all up to a million verses have been recorded. The role of *Manas* in Kyrgyz culture is difficult to overstate. The traditional performer of the epic, a *manaschï*, was usually believed to have been elected for the role as a result of the intercession of the spirit of Manas himself. In fact, Manas became revered as a Muslim saint, and his tomb became an important pilgrimage site. In Soviet times *Manas* was aggressively promoted by both authorities and Kyrgyz intellectuals as the national epic of the Kyrgyz nation and as an example of the sort of literary heritage a small, recently literate people such as the Kyrgyz could have.

Kyrgyzstan has only been independent since 1991, following the fall of the Soviet Union. Until the nineteenth century the history of Kyrgyzstan was dominated by three recurring patterns. First, powerful, if fleeting, nomadic empires were established in the Semirechye region, that is, in the Chu and Talas River valleys, and extended their authority into the Tian Shan Mountains. Second, powerful sedentary states in the Ferghana Valley, extended their authority into surrounding mountains. Third, Chinese attempts to exert their authority into and beyond the Tian Shan Mountains recurred in the Ferghana Valley and the Chu River Valley. These patterns were evident until the middle of the nineteenth century when Central Asia, including the Kyrgyz lands, was divided between the Russian and Chinese empires.

Timeline

c. 12,000 B.C. Presence of Neolithic communities

Hunter-gatherer societies who produce stone tools live at Khodzha-Gora in the Ferghana Valley. Other Neolithic cultures on the territory of Kyrgyzstan leave numerous cave drawings.

c. 1400 B.C. Bronze-Age cultures appear

The so-called Andronovo Culture leaves numerous traces in northern Kyrgyzstan, including those along the shores of lake Issyk-Kul. The Andronovo Culture is a sedentary society that produces bronze tools and pottery. In the Ferghana Valley at this same time is the so-called Chust Culture, which also is a sedentary culture that survives until around 850 B.C. Kyrgyzstan's Neolithic cultures leave a wealth of cave and rock drawings.

750 B.C. Saka (Scythian) cultures appear

The Sakas are the first pastoral nomadic people in Inner Asia to develop an advanced political system and material culture. Sakas are established from Mongolia to the mouth of the Danube River on the Black Sea and are also present in Kyrgyzstan.

500 B.C. Sedentary Davan State in the Ferghana

The so-call Davan state emerges in the Ferghana Valley. Ethnically and culturally it is closely linked to the Sogdian states of Central Asia (Transoxiana). Unlike the sedentary Central Asian states of Sogdia and Bactria, the Ferghana Valley does not become politically dependent upon the Persian Achaemenian Empire but remains independent. According to Chinese sources the Davan state remains independent of the Greek state of Bactria after Alexander the Great's conquest of Central Asia in 329 B.C. The city states of the Ferghana Valley enjoy an already highly developed system of irrigation canals and practice intensive agriculture.

160 B.C. Wu-Sun Migrate to Issyk-Kul region

The Wu-Sun, as they are known in Chinese sources, a pastoral nomadic people related to the Sakas, are forced out of Mongolia by the Hsiung-nu nomadic confederation and settle in eastern Kazakstan and in northern Kyrgyzstan. Gradually the Wu-Sun are pushed further into the Tian Shan by more powerful groups.

104–101 B.C. Chinese campaigns against Davan State

Han Dynasty troops conduct two separate campaigns to conquer the Ferghana Valley, the first in 104–102 B. C. and the second in 101 B.C. Both are turned back.

c. A.D. 550 Sogdian settle Chu and Talas Valleys

Sogdian merchants and agriculturalists gradually settle the river valleys in the steppe regions of northern Kyrgyzstan and southeastern Kazakstan. These Sogdians are key middlemen in the trans-Asian silk trade between China and the Mediterranean, thereby linking these territories to major international trade routes.

552 Turk empire formed

The Turk empire is formed in Mongolia by the leader Bumin Kaghan. Up to this time the dominant ethnic group on the Eurasian steppes has been the Indo-European Sakas; with the expansion of the Turk empire the steppe regions begin to be populated by Turkic-speaking nomadic groups. Although the Turk empire is centered in central Mongolia, these Turks apparently originated in the mountains of South Siberia and include the ancestors of the Kyrgyz in the upper Yenisei Valley. Under Bumin Kaghan's younger brother, Ishtemi Turk, power extends rapidly across the steppes to Central Asia and as far as the Volga River and includes within it all of the nomadic communities inhabiting the territory of Kyrgyzstan.

603 Division of Turk empire

The Turk empire is divided between Eastern and Western Kaghanates. The Eastern Kaghanate remains centered in Mongolia, while the Western Kaghanate has its capitol in the territory of Kyrgyzstan on the Suyab River near the modern-day city of Tokmak. During the first half of the seventh century the Western Turks extend their power into Central Asian and Eastern Turkestan.

630 Western Turk Kaghanate collapses

The Karluk tribe, located to the north of Kyrgyzstan, rebels against the ruler in Tokmak, and the Kaghanate splits into two parts, the Dulu confederation and the Nushibi confederation. By 655 the Western Turk territories in Eastern Turkestan fall to the Chinese Tang Dynasty.

699 Turgesh Kaghanate founded

Under the leadership of Uch-Elig Kaghan (r. 699-706), the scattered Turkic tribes in and around the territory of Kyrgyzstan reunite to form the Turgesh Kaghanate and establish their capitol on the Suyab River in Kyrgyzstan. During the eighth century the Turgesh Kaghanate is involved in conflicts with Arabs, who are conquering Sogdia, and with the newly-formed Blue Turk Empire, based in Mongolia on the ruins of the Eastern Turk Kaghanate. Against the Arabs the Turgesh are especially active in Transoxiana in alliance with the Sogdians, although by 738 most of the Sogdian cities are in Arab hands. By 744 the Turgesh Kaghanate has collapsed under the Arabs and Blue Turks.

c. 700 Buddhist shrine built at Ak-Beshim

Buddhists build a temple at Ak-Beshim. Buddhist culture develops in northern Kyrgyzstan under the Turks, although Buddhism is not an official religion. Manichaeism, which combines elements of Zoroastrianism, Christianity, and Gnosticism, also is widespread in these areas.

716–76 Talas Runic Inscriptions

The earliest alphabet used by the Turks in Mongolia is the Runic alphabet. Between 716 and 776 the Turgesh leave a series of Runic funerary inscriptions in the Talas valley, which are among the oldest epigraphic monuments in Kyrgyzstan.

751: July Chinese defeat at battle of the Talas

After several decades of rivalry over Central Asia between the Arab Caliphate and the Chinese Tang empire, the armies of both sides meet in battle on the banks of the Talas River, just north of the modern-day border between Kyrgyzstan and Kazakstan. The Chinese are decisively beaten, and the western regions of modern-day Kyrgyzstan, including the Ferghana Valley, pass to Arab control. The cultural effect of this battle is that it limits Chinese political and cultural influence to the territories east of the Tian Shan mountains, while the process of Islamization of the local communities remains essentially uncontested from this date.

893 Karluk confederation disintegrates

The Karluk confederation, dominating northern Kyrgyzstan from the middle of the eighth century, is based in the town of Taraz, in the Talas Valley. The Karluks are among the first Turkic nomads in Central Asia to begin converting to Islam. In 893 their leader, Ogulchak Kadïr, is decisively defeated by the Samanid dynasty based in Bukhara. Ogulchak-Kadïr moves to Kashgar in Eastern Turkestan but is unable to defeat the Samanids.

c. 950 Karakhanids become first Turkic Muslim Dynasty in Central Asia

Under the leadership of Satuk Bughra Khan, the Karakhanid dynasty converts to Islam. This dynasty is based in Kashgar and fills the vacuum left by the collapse of the Karluk confederation. The Islamization of the Turkic nomads had been underway since the Arab conquest of Central Asia, but under the Karakhanids Islam becomes the official religion of the ruling dynasty and the nomads who constitute the Karakhanid state. One of the Karakhanid capitols is Balasagun, on the Chu River within the present-day territory of Kyrgyzstan.

c. 990 Karakhanids conquer Ferghana Valley

Under the successors of Satuk Bughra Khan, the Karakhanids extend their rule down the Syr Darya River in the north and into the Ferghana Valley, pushing out the weakening Samanid Dynasty. The center of Karakhanid administration in the Ferghana Valley is Uzgen.

995 Elected governing body convened by Manas

Manas, the national hero of the Kyrgyz people, convenes the first elected governing body according to *Manas,* the epic poem that recounts the exploits of the Kyrgyz. The poem is recounted from generation to generation by *manaschi,* members of certain families designated as the keepers of *Manas.* The *manaschi* are illiterate and rely on imagination and memory to tell the story. Manas is credited with initiating falconry, a revered activity for the Kyrgyz people. A small mausoleum near Talas in northern Kyrgyz is believed to be the place where Manas is buried.

999 Karakhanids occupy Bukhara

After taking the Ferghana Valley the Karakhanid armies defeat the Samanids and capture their capitol, ending Samanid rule in Central Asia and establishing a Karakhanid hegemony over eastern Central Asia. Following this victory the Karakhanid rulers in Bukhara, Balasagun, Uzgen, and Kashgar begin fighting among themselves.

1069 Yusuf of Balasagun writes the *Kutagdu Bilig*

Yusuf of Balasagun, a court scholar in the town of Balasagun, writes one of the earliest major works of Turkic prose, *Kutagdu Bilig,* a work of advice for his patron on the obligations and duties of a ruler.

1072 Mahmud Kashgari writes *Divan Lugh at Divan Lughat*

The court scholar Mahmud in Kashgar writes an Arabic-language dictionary of the Turkic languages spoken throughout Inner Asia. Known as *Divan Lugh at Divan Lughat,* it is the first Turkic dictionary.

c. 1120 Uzgen minaret built

One of the oldest monuments of Islamic architecture in Kyrgyzstan is the Uzgen minaret, built toward the end of Karakhanid period and based on Bukharan models.

1128–41 Kara-Khitay Conquest of Karakhanid realms

The Karakhanid city of Kashgar was the first city to fall to the Kara-Khitays, an ethnically Mongol Buddhist people who by 1141 had expelled the Karakhanid rulers from Balasagun, Bukhara, and Samarkand.

1211–19 Chingis Khan conquers Tian Shan

Under their leader the Mongols defeat the local rulers in the Tian Shan and annex the area to their empire. The local Turkic tribes are integrated into the Mongol Army. Large numbers of Mongol nomads migrate to the Tian Shan region. This region becomes a part of the heartland of the Mongol World empire. Urban civilization in the Chu and Talas Valleys is destroyed and is not reestablished for seven hundred years.

1227 Chingis Khan dies and his empire is divided

Following the death of Chingis Khan the Mongol empire is divided between his four sons. The Tian Shan and the Ferghana Valley, together with eastern Central Asia, become the *ulus* (realm) of Chingis Khan's son Chagatay and his descendants.

1269–1301 Qaydu rules the ulus of Chagatay

After the collapse of the Mongol empire into competing states, a Chingisid named Qaydu manages to unite the Chagatayids and rule independently. After Qaydu's death in 1301 the Tian Shan, Kashgar and Ferghana regions become known as Mogolistan and its rulers as Mongols.

1371–89 Timur-i-Lang invades Mogolistan

After establishing himself in Maverannahr, Timur-i-Lang (1336–1404, known in English as Tamerlane) conducts a series of military campaigns against Mogolistan, seeking to conquer the region. The Mongols remain in power, although Timur's repeated campaigns do much damage, especially in the Tian Shan region.

Timur-i-Lang is born in Kash near Samarkand. He earns the nickname of "Lame Timur" because of severe wounds he suffers during the wars of the 1370s and 1380s. He dies in a battle with the Chinese.

c. 1350 Tughluq Timur Khan converts to Islam

Ruler of the Chaghatayids of Mogolistan, Tughluq Timur converts to Islam and requires all nomads under his rule, including those in the Tian Shan and Semirechye, to become Muslims.

1400–40 Kyrgyz begin migrations from Altay region to Tian Shan region

Following attacks from the Tatars of the Blue Horde into the Altay Mountains region, the ancestors of the Kyrgyz begin migrating to the southwest into Mogolistan, that is, into the Tian Shan region. These poorly documented migrations also involve the assimilation into the Kyrgyz tribal system of various local Turkic groups in the Tian Shan area and in the fringes of the Ferghana Valley. The ethnic and linguistic peculiarities of the modern Kyrgyz are attributable to both the migration of the Kyrgyz and the assimilation of local groups. Over the course of the fifteenth and sixteenth centuries Kyrgyz continue to migrate to the south and west, into the Alay Mountains and into the Pamirs.

Modern Era

1512 Kyrgyz tribes submit to Uzbek Shībanid dynasty

The Uzbeks under Shaybani Khan conquer virtually all of Central Asia, expelling the Mongols from the Tian Shan and

Semirechye. The remaining Mongols establish a dynasty in Kashgar, in Eastern Turkestan, where local Kyrgyz communities remain under their authority. The bulk of the Kyrgyz tribes in the Tian Shan and Semirechye become vassals of the Uzbeks. During the sixteenth and seventeenth centuries the Kyrgyz will remain involved in the internal struggles of the Shïbanids in Tashkent and the Ferghana, of the Mongols in Kashgar, and in the conflicts between the Shïbanids and the Kazakh khanate.

1658 Establishment of Kalmyk dominance in Central Asia

Following the defeat of a combined Uzbek, Kazakh and Kyrgyz force along the Talas River by the Mongol Kalmyks in 1658 and the subsequent capture of Eastern Turkestan by the Kalmyks, the Kyrgyz are involved in frequent campaigns against the Kalmyks for nearly a hundred years. Strong Kalmyk pressure leads to large-scale migrations of Kyrgyz south into the Ferghana Valley.

1710 Khanate of Kokand founded

An Uzbek chieftain of the Ming tribe, Abdarrakhman Biy, founds a ruling dynasty in the Ferghana Valley in the city of Kokand.

1758 Manchu armies enter Tian Shan

After annihilating the Kalmyks and their armies, the victorious Manchu armies conquer Eastern Turkestan and begin bringing the Kyrgyz under their control. The ruler of Kokand, Irdane Biy, also recognizes Manchu suzerainty.

1760–1824 Systematic conquest of Kyrgyz by Kokand

Around 1760 the armies of the emirs of Kokand gradually conquer the Kyrgyz lands, beginning in the Ferghana Valley. The Kokandian conquest involves the establishment of fortresses with permanent garrisons in Kyrgyz territory. By 1824 the Kokandians have conquered the Issyk-Kul region and forced the northern Kyrgyz to submit.

1785 Russian caravans begin penetrating Northern Kyrgyzstan

Led by *Volga Tatars,* Muslim subjects of Russia, trade caravans begin trading with the Kyrgyz and collecting intelligence for the Russian authorities. Tatar merchants also begin trying to convince Kyrgyz leaders to submit to Russia.

1855: January 17 Bugu tribe become subjects of Russia

Under pressure from both the Manchus and the Kokandians, the northern Kyrgyz Bugu tribal confederation, located in the Issyk-Kul area, become Russian vassals.

c. 1850 First versions of *Manas* recorded

The oral epic poem, *Manas*, is first recorded in the middle of the nineteenth century, although it has existed in oral form since about A.D. 1000 (See 995 and 1995.)

1860–68 Russian conquest of Tian Shan region

Russian troops begin systematically destroying the Kokandian fortresses in the Tian Shan region and expelling their garrisons. The presence of Russian troops results in the formal submission of the Tian Shan Kyrgyz to Russia in 1863. By 1867 the Russians have completed their military occupation of the region, including much of the Ferghana Valley.

1876 Russia annexes all territories of the khanate of Kokand

Following a major anti-Russian rebellion that broke out in 1874, all territory of the former khanate of Kokand becomes annexed to Russia. Northern Kyrgyzstan becomes part of Semirechye *oblast'* (administrative district), and the Ferghana Valley becomes part of the Ferghana district.

1890–1910 Large scale Russian settlement in northern Kyrgyzstan

Large parts of the Semirechye, including northern Kyrgyzstan, is settled by Russian and Ukrainian peasants. This colonization reduces Kyrgyz pasture lands, impoverishes many Kyrgyz, and compels ever-larger numbers of nomads to abandon pastoral nomadism for sedentary agriculture.

1916: July–August Rebellion of Kyrgyz nomads against Russian rule

Facing enormous manpower losses on the western front during the First World War, Russian authorities seek to enroll Central Asians in labor units. Central Asians, including Kyrgyz, had heretofore been exempt from military conscription. Rebellion breaks out in the Ferghana Valley, over the rumored conscription, and quickly spreads throughout Central Asia. Russian reprisals are especially brutal against the nomads and thousands of Kyrgyz nomads perish.

1917: March 14 Czar Nicholas II abdicates

With the abdication of the Russian czar, central authority in Russia gradually weakens. There is little immediate reaction in Kyrgyzstan, and no local independence movement as such.

1917: November 7 Bolshevik seizure of power

The Provisional Government is overthrown by the Bolsheviks. The Bolshevik seizure of power reflects a larger current of radicalization in the Russian empire. There is little Bolshevik presence in Kyrgyzstan, and the center of Bolshevik power in Central Asia is Tashkent.

1918–20 Bolsheviks establish authority in Central Asia

After crushing the autonomous Muslim government in Kokand and defeating both White Cossacks in Semirechye, the Bolsheviks establish their authority in Kyrgyzstan.

1920: August 26 "Kirgiz" ASSR established

Soviet authorities create the "Kirgiz" Autonomous Soviet Socialist Republic (ASSR). This division combines the territories of the Kazakhs and the Kyrgyz. Until 1925 Russians refer to Kazakhs as *Kirgiz* and to Kyrgyz as *Kara Kirgiz.*

1924: October 14 Kara-Kirgiz Autonomous Region formed

The territory of Kyrgyzstan is separated from the Kirgiz ASSR and named Kara-Kirgiz Autonomous Region. It becomes a constituent part of the larger Russian Soviet Federated Socialist Republic (RSFSR). On May 25, 1925, it is renamed the Kyrgyz Autonomous Region. On February 1, 1926, it becomes the Kyrgyz Autonomous Republic. Despite the names of these territories, Kyrgyz political life remains firmly in the hands of the Communist Party in Moscow.

1929–31 Collectivization of Kyrgyz nomads

Under orders from Stalin, private land-holding is effectively banned throughout the Soviet Union; this order also applies to private ownership of livestock among nomads. Relatively wealthy Kyrgyz nomads are massacred or shipped en mass to prison camps, and poorer nomads are made to join collective farms (*kolkhozes*) or state farms (*sovkhozes*). These actions cause widespread social upheaval in the countryside, including armed resistance and the slaughter by the nomads of an enormous part of their herds.

Collectivization virtually puts an end to the traditional pastoral nomadic lifestyle among the Kyrgyz and also results in a widespread famine. Thousands of Kyrgyz nomads migrate to China. At this time all independent religious organization, including mosques and schools, are permanently closed and much of the Islamic clergy are exterminated, effectively ending traditional religious life in Kyrgyzstan.

1936: December 5 Kyrgyz Soviet Socialist Republic (SSR) founded

The Kyrgyz Autonomous Republic formally becomes the Kyrgyz SSR, a Union Republic of the USSR.

1937 Purges of Kyrgyz Communist Party

Although tens of thousands of Kyrgyz, if not more, have perished or been sent to labor camps (the Gulag) by 1937, in this year Stalin begins arresting and executing members of the Kyrgyz Communist Party, essentially turning over the membership and creating a new, more flexible organization. At this time the presence of ethnic Kyrgyz in the Party is gradually increased.

1941–45 Kyrgyzstan in the Second World War

Two years after the signing of the Nazi-Soviet Pact in 1939, Nazi Germany attacks the Soviet Union. Kyrgyz are drafted into the Red Army by the tens of thousands. The industrialization of Kyrgyzstan also begins as many factories are evacuated from the western regions of the Soviet Union and relocated in Kyrgyzstan. Together with the factories come tens of thousands of workers from Russia and the Ukraine. After the war much of this equipment and its workers remain in Kyrgyzstan. Thousands of ethnic Germans forcibly deported from central Russia are settled in Kyrgyzstan.

1956–64 Nikita Khrushchev implements Virgin Lands policy

Khrushchev brings millions of acres of steppe lands in Kazakstan and Kyrgyzstan under cultivation. This policy further increases Russian colonization of Kyrgyz lands and diminished pasture lands for stockbreeding.

1985 Mikhail Gorbachev comes to power

Mikhail Gorbachev becomes First Secretary of the CPSU and subsequently introduces the policies of *glasnost'* (openness) and *perestroika* (reorganization) which encourage more open debate and more criticism of Soviet political and life. The policies soon result in a growing demoralization of Communist Party elites throughout the Soviet Union and the emboldening of emerging nationalist forces.

1989: September Kyrgyz declared state language of Kyrgyzstan

Following the adoption of a language law Kyrgyz is declared the state language of the Kyrgyz SSR and Russian is declared the language of inter-ethnic communication.

1990: May–July Ethnic violence in Osh

Following disputes between Uzbeks and Kyrgyz in the city of Osh, large-scale rioting breaks out in the city. Soviet troops are called into the city to restore order.

Osh is among the oldest settlements of Central Asia. *Tash-Sulayman* (Solomon's throne) is a large, odd-shaped rock found in Osh. Historically it has been a place for Muslim pilgrimage.

1990: October 28 Askar Akaev elected president

Askar Akaev, president of the Kyrgyz Academy of Sciences, is elected president of the Kyrgyz SSR.

1991: March 17 Public votes to retain the Soviet Union

In an all-union referendum on maintaining the Soviet Union, ninety-five percent of the population of Kyrgyzstan votes to preserve the Soviet Union.

1991: August 31 Kyrgyzstan declares its independence

Following the unsuccessful coup attempt against Mikhail Gorbachev by the Communists, the Supreme Soviet of Kyrgyzstan declares its independence from the Soviet Union. The borders of the new Kyrgyz Republic correspond to those of the Kyrgyz SSR. The Supreme Soviet also outlaws the Communist Party.

1991: October 13 Akaev reelected president

In an uncontested presidential election Askar Akaev gets ninety-five percent of the vote.

1991: December 13 Kyrgyzstan joins the Commonwealth of Independent States (CIS)

Kyrgyzstan joins the CIS, thereby retaining economic, political, and security ties with the successor states of the now-defunct Soviet Union.

1992: May 15 Kyrgyzstan signs CIS collective security treaty

Kyrgyzstan signs CIS collective security treaty, allowing CIS troops, including Russian border troops, on Kyrgyz territory.

1992-93 Migration of non-Kyrgyz from Kyrgyzstan

Following the fall of the Soviet Union and the increase in nationalist legislation in the Kyrgyz parliament, thousands of non-Kyrgyz leave the country. This exodus includes primarily Russians returning to Russia, as well as ethnic Germans migrating to Germany.

1993: May 10 Kyrgyzstan introduces national currency

Kyrgyzstan votes to leave the ruble zone and introduces its own national currency, the *som*.

1994: April 27 Colorado Springs, Colorado and Bishkek become sister cities

The mayor of Colorado Springs, Colorado and the Kyrgyzstan ambassador to the United States and Canada sign an agreement to designate Colorado Springs and the Kyrgyz capitol, Bishkek, as sister cities.

Colorado Springs is the first city in the world to establish a sister-city link with Kyrgyzstan, through which the cities learn about each other's culture.

1994: October 22 Referendum allows bi-cameral legislature

In a referendum eighty-five percent of voters vote to amend the constitution to allow the creation of a bi-cameral legislature for Kyrgyzstan.

1995 "Traffic in Transit" agreement signed

China, Pakistan, Kazakstan, and Kyrgyzstan sign a five-year agreement, "Traffic in Transit," in Islamabad, Pakistan. According to the terms of the agreement, each country will provide facilities for traffic on mutually agreed routes for vehicles registered in the member states; in addition, freight rates and other charges will be consistent for all shipments. The four countries also promise to improve facilities and to standardize customs procedures for goods traveling through each other's territories.

The three land routes for transit traffic established by the agreement are Pakistan-China-Kyrgyzstan-Kazakstan, China-Kazakstan, and China-Kyrgyzstan.

1995: February 5 One thousand candidates for 105 seats in parliament

Names of one thousand candidates appear on the ballot for 105 parliamentary seats. Only thirteen win the necessary 50 percent of the vote. A second round of voting is held February 19 to decide the remaining ninety-two seats.

During the campaign period each candidate is given five minutes of television time and is limited to a specific number of campaign posters. All candidates' posters are produced in the same regulation format, making it difficult for those voting to see differences among the candidates. In the voting procedure, voters are required to cross off the names of the unwanted candidates, rather than to designate the name of the desired candidate.

Only one-fifth of all candidates belong to any political party. Just four of the thirteen winners belong to a political party. The Communist Party has just one winner.

1995: August Celebration of the millenium of the national poem

The millennium of the epic poem, *Manas* (see 995 and c. 1850), is celebrated. The United Nations (UN) recognizes 1995 as "the year of the celebration of the millennium of the Kyrgyz national epic, *Manas*." The poem recounts the life of the national hero, Manas, and takes from thirty-six hours to three weeks to recite, depending on the version.

UNESCO Director-General Federico Mayor comes to Kyrgyzstan from his offices in Paris, France, to join the celebration, which consists of poem recitations, traditional dances, and feasts including fermented mare's milk and boiled sheep eyeballs.

Guests travel to Kyrgyz from around the world to celebrate the Manas event. The government erects one thousand traditional yurts (circular domed tents) near the legendary site of Manas's burial near Talas. One thousand horsemen entertain festival attendees with wrestling matches.

1996: October **Oil refinery begins production**

In Jalal-Abad the first oil refinery in Kyrgyzstan produces petroleum products such as gasoline and diesel fuel.

1997 **Banking structure set**

The banking system consists of the National Bank of the Kyrgyz Republic and eighteen commercial banks, most of which are privately owned.

1997 **Production of antimony**

The state precious metals concern, Kyrgyzaltyn, says that antimony mines in the southwest are producing about 7600 metric tons annually. China and Kyrgyzstan plan joint sales of antimony in Europe. Antimony, an element found only in chemical mixtures is used in metallic alloys as well as medicines, pigments, and for fireproofing.

1998 ***Euromoney* magazine honors Kyrgyzstan Mercury Bank**

Euromoney mazagine's 1998 Awards for Excellence are presented to banks operating in Central Asia. Kyrgyzstan's Mercury Bank is awarded the Best Bank designation.

1998 **Jet fuel refinery project planned**

Manas Refinery Partners reveals plans to build and operate a $35 million refinery with capacity of 8,000 barrels per day. Located near Bishkek, the refinery will be a joint venture between Manas, based in San Antonio, Texas, and the government-owned Kyrgyzstan National Airlines. The complex is expected to begin operating in May 1999.

The refinery will produce jet fuel, diesel fuel, and high-octane gasoline. Kyrgyzstan National Airlines will take all of the jet fuel representing almost one-third of the refinery's planned production.

1998: May 21 **Cyanide spill pollutes river and lake**

A truck transporting twenty one-ton packages of cyanide overturns into a river on the way to a gold mine. Granular sodium cyanide spills into a river that feeds in Lake Issy-Kul. Environment Minister Kulubek Bokonbayev says in a news conference that the government will seek 160 million soms ($8.4 million) compensation for damage to the environment from Cameco Corporation based in Saskatoon, Saskatchewan, operator of the gold mine.

1998: October **Kazakh and Kyrgyz political leaders' children wed**

The eighteen-year-old daughter of Kazakh President Nursultan, Aliya, marries Aidar Akayev, the twenty-three-year-old son of Askar Akayev, the president of Kyrgyzstan. Both bride and groom are college students in the United States.

Bibliography

Attokurov, S. *Kïrgïz Sanjïrasï.* Bishkek: Kyrgyzstan, 1995.

Bennigsen, Alexandre & S. Enders Wimbush. *Muslims of the Soviet Empire: A Guide.* Bloomington & Indianapolis: Indiana Universiity Press, 1986.

Bernshtam, A. "The origins of the Kirgiz People," *Studies in Siberian Ethnogenesis*, edited by N. M. Michael, Toronto: Arctic Institute of North America, 1962, 119-128.

Bregel, Yuri. *Bibliography of Islamic Central Asia* vols. 1-3, Bloomington, Indiana: Research Institute for Inner Asian Studies, 1995.

Grousset, René. *The Empires of the Steppes: A History of Central Asia.* translated by Naomi Walford, New Brunswick, New Jersey: Rutgers University Press, 1970.

Huskey, Eugene. "Kyrgyzstan: The politics of demographic and economic frustration," in *New States, New Politics: Building the Post-Soviet Nations*, edited by Ian Bremmer and Ray Taras, Cambridge & New York: Cambridge University Press, 1997, 655-680.

Istoriia kirgizskoi SSR vols. 1-3, Frunze: Kyrgyzstan, 1984-1986.

Krader, Lawrence. *The Peoples of Central Asia.* Bloomington, Indiana: Uralic and Altaic Series, vol. 26, 1966.

Olcott, Martha B. "Kyrgyzstan," in *Kazakstan, Kyrgyzstan, Tajikistan, Turkmenistan and Uzbekistan*Area Handbook Series, Washington D.C.: Government Printing Office, 1997, 99-193.

Laos

Introduction

Overview

Laos is a landlocked nation located in the center of the Southeast Asian peninsula. Thailand borders Laos on the west and Vietnam on the east. Cambodia is to the south and China is on the north. Myanmar lies on the northwestern corner of Laos. The country is mountainous, with elevations commonly over 500 meters. Steep slopes and narrow river valleys characterize the country. Approximately four percent of the land is arable. The country has a monsoon climate, with a rainy season from May through October, a cool dry season from November through February, and a hot dry season in March and April.

Laos today is an underpopulated country of approximately four-and-a-half to five million people, the lowest population density of any nation in Southeast Asia. More than eighty-five percent of the population lives in small rural villages, usually smaller than 1,000 residents. The population of Laos is very young, with a life expectancy of less than fifty-two years. High birthrates are offset by one of the highest infant mortality rates in the region.

Lao is the ethnic term for some of the various ethnic groups, while Laotian refers to all the peoples of Laos. Laos is a multiethnic country with more than forty ethnic groups. Its mountainous topography has inhibited communication between groups, which has contributed to maintaining distinctions among them. The population of Laos is usually classified into three groups: the Lao Sung or upland Lao, the Lao Theung or midland Lao, and the Lao Loum, the lowland Lao.

The Lao Loum are the most numerous of the three groups, comprising about sixty-five percent of the population. The Lao Loum, or valley Lao, are descendants of the Tai peoples who migrated from China in the first millennium A.D. The lowlanders live as their name suggests, along the banks of the Mekong River and its tributaries. Many now live in the cities. The Lao Loum speak Laotian Tai, more closely related to the language of Thailand than those spoken by highlanders in Laos.

For most Laotians, the rhythm of life closely follows the cycles of season and crop. Slash-and-burn farmers in the middle and high uplands begin their work in January or February; for rice cultivators, the growing year begins with the rains. Despite the growth of urban areas, the increase in foreign aid and foreigners, and the influx of consumer goods, most Lao have remained in the villages where they can support themselves and their families.

Early History

According to legend, the first Lao, called the Lao Theung, came from a pumpkin. A nobleman named Khun Borom descended from the heavens on a pure white elephant with ebony eyes and a transparent tusk. He and his people, called the Tai, migrated from southern China into the mountains of northern Laos in the eighth century. One day, Khun Borom found a huge pumpkin that appeared suddenly on a vine. He jabbed the pumpkin with a hot poker and out of the hole fell ash-darkened seeds. These seeds became the Lao Theung, the first people to live in Laos. They were soon pushed south into Cambodia, where they still live today as Khmer. Lighter seeds also fell out of the pumpkin. These became the Lao Lum tribes of Laos. They then lived in Laos with the Tai who had come from southern China with the nobleman.

In the early centuries A.D., Laos was made up of fiefdoms where the primary activity was wet rice cultivation. Fiefdoms exerted control over their neighbors through *mandalas* (spheres of influence). One kingdom, Candapuri, was the earliest site of Vientiane (present-day capital of Laos). By the seventh century, new cultural influences were introduced by various peoples migrating into Laos. These outside influences actually served to strengthen the cultural identity of the Lao, because of the Lao's need to resist (or learn to survive under) foreign control. Through the 1200s, the Mongols from the north were among the most powerful political groups in the region of the middle Mekong Valley. In the 1300s, under the leadership of king Fa Ngum, a kingdom known as Lan Xang, meaning "kingdom of a million elephants," evolved in the region of present-day Laos. Fa Ngum was overthrown in 1373, but his kingdom continued for centuries although its degree of independence varied (at times it served as a tributary state of Myanmar).

Europeans first appeared in Laos in the seventeenth century as merchants and missionaries. European influence at this time was minimal. Not until the nineteenth century did Europeans conquer much of Southeast Asia. Nevertheless, the period between initial European penetration and eventual conquest proved traumatic for Laos. In 1690 the Kingdom of Lan Xang disintegrated due to factional struggles over succession. In the following century, the Lao states of Louangphrabang, Vientiane, and Champasak all fell to Thailand. Vientiane became a Thai vassal state and the Lao royal court moved to Thailand. Yet Lao resistance against the Vientiane continued until the arrival of the French in the nineteenth century.

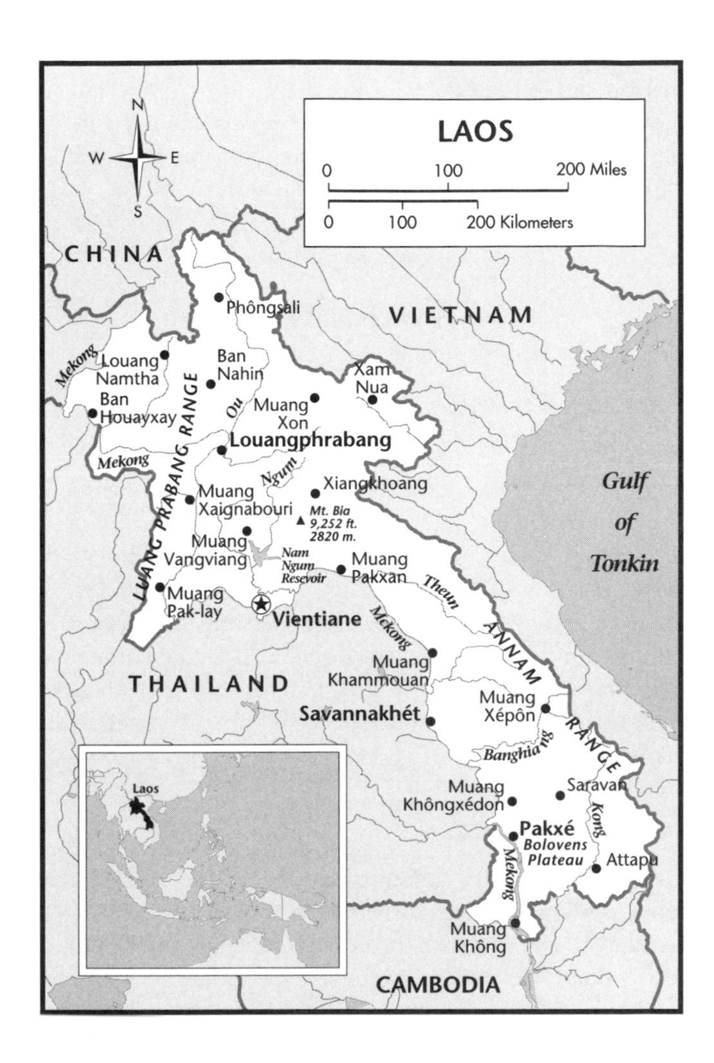

French Rule

After initial French exploration of the region in the 1880s, Laos passed to French control in 1890 and was administered as part of the colony of French Indochina. Under this system, the Laotian monarchy continued under the leadership of the king of Louangphrabang although real power rested in the hands of the French colonial administration. Overall, the French made little progress in modernizing the country. Most Lao remained uneducated and few benefited from contacts with French officials.

Laos remained under French control until Japan received occupation rights in 1940. Following the Japanese defeat in the Second World War (1939–45), France attempted to reassert its control over Southeast Asia but was opposed by its subject peoples, including the Lao. Although the French succeeded in regaining control over Laos in 1946, they quickly offered concessions to the Laotians that resulted initially in autonomy and, ultimately, in 1953, in full-fledged independence.

Independent Laos

Yet if most of the Laotian political leadership was united in opposition to renewed French control, they were hardly united in their visions of a future independent Laos. In 1949, Prince Souphanouvong founded what became known as the *Pathet Lao*, or People's Liberation Army, a Communist guerrilla movement. In foreign affairs, the Pathet Lao favored cooperation with the Soviet Union. The Royal Laotian Army, representing Prince Souvanna Phouma and supported by the United States opposed the Pathet Lao. Although the 1954 Geneva Conference called for a coalition government composed of the rival factions, the following two decades saw continued conflict between the parties during which the United States provided significant aid to the right-wing forces of Souvanna Phouma.

In 1975, after the withdrawal of U.S. forces from Southeast Asia, the People's Revolutionary Party came to power and established the Lao People's Democratic Republic. After the Lao People's Revolutionary Party took control of the government in the mid-1970s, many Laotians fled the country. Many were Hmong (a group of Tibeto-Burman origin) led by Vang Pao who had worked with the United States against the Pathet Lao. These refugees remained in southern China or in refugee camps in Thailand until resettled in third countries or repatriated to Laos. More than 300,000 Laotians have been resettled abroad.

The end of the Cold War has increased Laos' cooperation with countries outside the Soviet bloc. The limited ability of the former Soviet Union countries and Vietnam—now busy with their own problems—to provide assistance to Laos has been an incentive to Laos to look outward for aid and become more flexible in its foreign policies. Although living conditions have improved somewhat since the mid-1980s, Laos remains one of the poorest countries of Southeast Asia.

In the late 1990s, as in the mid-1970s, Laos is again ruled by a tiny group of elite: powerful, linked by family, involved in virtually all activities in the country. The majority of the population, however, for the first time has a sense of national unity. Prior to 1975 and the beginning of Communist rule, villagers had little contact with administrators from the central government: these officials from the capital depended on local authorities with almost feudal power to control the population and collect taxes. Since 1975, leftist governments have focussed their attention on uniting the country and giving all Laotians a sense of belonging to one country—at the expense, critics would argue, of political freedom.

Timeline

2000 B.C.–A.D. 500 Lao Bronze Age

First occupied by Austroasiatic peoples who originated in the region between Australia and Asia, these original residents live by hunting and gathering. They are skilled at navigating the rivers in canoes. Their traders use river routes through the mountains to trade with other groups. The Mekong River is the most important trade route, as traders travel its tributaries into the hinterland where they obtain food and cardamom, gum benzoin (a resin used especially in medicine and perfume), and sticklac (a substance used to make shellac and varnish).

100s B.C.–400s A.D. Early Mandala Period

Laos is comprised of fiefdoms ruled by princes with agriculture based on wet rice cultivation. These early occupants of Laos are associated with an early pottery and bronze culture called Ban Chiang that develops in the middle Mekong Valley. These fiefdoms control their neighbors through expanding and contracting spheres of influence called *mandalas*. In the mandala system of power, power is centered on a ruler, his court, and the related religious center from which he draws his legitimacy. Mandalas are extended through trade, marriage and war.

200s–300s Sikhottabong and Candapuri mandala

One mandala, Sikhottabong, is located on the left bank of the Mekong River, and moves west as it is pushed by the Kingdom of Champa, an early kingdom in Vietnam. The Mon kingdom of Candapuri, the earliest name of Vientiane (now the capital of Laos), is another mandala. These societies are strictly hierarchical with an aristocracy, commoners, and slaves.

600s Migrations continue

In the seventh century, migrations into Laos continue to expand cultural influences into the area. The Thai migrate northwest to create the Nan-chao kingdom in present-day Yunnan. This kingdom has enormous influence over populations occupying Lao territory, transmitting Indian Buddhism, political thought, and administrative organization to the area.

698 Khun Lo establishes a dynasty

A Thai prince, Khun Lo conquers Louangphrabang or Muang Sua, when Nan-chao administrators are engaged elsewhere. He establishes a dynasty that endures as an independent kingdom for 100 years.

Late 700s Nan-chao again occupies Muang Sua

The rulers of Nan-chao again conquer Muang Sua, replacing its Thai rulers with their own administrators. This reoccupation does not endure for more than a century. The local ruler of Xay Fong, near present-day Vientiane, then replaces the Nan-chao rulers of Muang Sua. The town becomes known as Xieng Dong Xieng Tong

700s–1100s Development of Lao culture

The rudimentary structures of a state comprised of numerous ethnic groups begin to emerge. Society is based in small confederate communities located in river valleys and among mountain people who find security away from well-traveled rivers and overland paths. In these localities, the institutions and customs of the Lao develop. Migration of various groups into the area and religious conflict and borrowing continue.

In the 1100s, the Candapuri Mandala in the Vientiane region is absorbed into the Khmer Empire. Lao's subjugation to regional powers, such as the Cham, Khmer, and the Sukhothai dynasty of Thailand (then called Siam), does not retard the process of cultural identification among the Lao. Influences from other powers, in fact, help shape Lao cultural identity.

1253 Mongolian influence in Laos

For more than a century, through the 1200s, the Mongols to the north exercise considerable political influence in the middle Mekong Valley. They destroy Nan-chao, making the area a province of their empire called Yunnan. Ramkhamhaeng, a Thai ruler, eliminates Khmer and Champa power in central Laos. He gains the allegiance of Muang Sua and the mountainous people to the northeast. But Mongol rule proves unpopular among the people of Muang Sua and causes internal disputes among the local ruling families.

1353 Fa Ngum becomes king

Mongol intervention and these family conflicts lead to the founding of the Kingdom of Lan Xang, meaning "kingdom of a million elephants." Lan Xang becomes the designation of what is now Laos. The beginning of recorded Laotian history begins with the twenty-second king of Laos, Fa Ngum. He is the first king of Lan Xang, and certainly one of the most important persons to reign in Laos. After twenty years of war, Fa Ngum unites Laos as the Kingdom of Lan Xang with thousands of troops mounted on hundreds of elephants.

Powerful and ambitious, King Fa Ngum is also cruel and capricious. His reign is troubled by religious disputes between his lamaistic Buddhism (an offshoot of the traditional religion) and the traditional Theravada Buddhism (original form of the religion) practiced by large numbers of the people. He destroys many Buddhist temples. He relents later in his reign, but not before many of his people have turned against him. In the next centuries, Fa Ngum's successors reign over a large indigenous kingdom with a hierarchical administration. Migration and religious crosscurrents continue. The kingdom endures with its original borders for nearly four centuries through a complex network of vassal relations with lesser princes, and Fa Ngum's descendants remain on the throne of Lan Xang for nearly 600 years. Small confederate communities begin taking shape.

1373 Fa Ngum loses his throne

Fa Ngum's generals finally overthrow him, placing his son, Oun Hueun, on the throne.

1373–1547 Kingdom of Lan Xang

For over a century, rulers expand the range of Lan Xang. By the mid-1400s, Lan Xang is the most powerful nation on the Southeast Asian peninsula. It includes portions of Myanmar (formerly Burma) Cambodia, Thailand, Vietnam and even southern China. In the next two centuries, however, the nation's size increases or decreases depending on the strength of its rulers and neighboring kings.

1400s Muang Phuan becomes dependent on Vietnamese

One kingdom, Muang Phuan, becomes semi-independent after being annexed by a Vietnamese army. This sets the precedent for a tributary relationship with the royal Vietnamese court at Hue.

1520–37 Conflict with Myanmar and Thailand

King Phetsarath involves the Kingdom of Lan Xang in battles with Myanmar and Thailand that continue for two centuries.

1574–78 Lan Xang becomes a vassal of Myanmar

Myanmar reduces Lan Xang to a vassal state.

1603 Lan Xang renounces vassal status

Lan Xang renounces its tributary ties to Myanmar.

1633–90 Kong Souligna Vongsa accedes to the throne

There are numerous succession struggles for the Lan Xang crown. Eventually, Kong Souligna Vongsa accedes to the throne. He unites the country, now ruled by only one king. This is often referred to as the golden age of Laos.

1641 First appearance of Europeans in Laos

European merchants come to trade, accompanied by missionaries. They describe a peaceful country ruled by King Souligna Vongsa. Emphasizing compromise rather than force during his 57-year reign, he presides over a peaceful nation. Once, he decides a border dispute between his country and Vietnam by suggesting that the border be determined by house construction: the Vietnamese build their houses on the ground while the Lao build theirs on stilts; let the border trace the difference between the two. Thus he avoids war with Vietnam.

1690 Kingdom of Lan Xang disappears

After 1690, the Kingdom of Lan Xang is split by power struggles. Leaving no successors, King Souligna Vongsa's nephews fight for control of the country, with Vientiane being captured by one pretender to the throne, the town of Louangphrabang by another. The country is now divided into three kingdoms: Louangphrabang, Vientiane, and Champasak. A fourth region, the tiny mountainous kingdom of Xieng Khousang, pays tribute to its neighboring and more powerful three Lao kingdoms and to Vietnam. The lowland and mountain Lao come very close to being absorbed by their powerful neighbors and rivals, Vietnam and Thailand.

1700s Lao struggles against Myanmar and Thailand

The Lao states of Louangphrabang, Vientiane, and Champasak try to maintain their independence from Myanmar and Thailand. They eventually come under the control of Thailand.

1771 Louangphrabang attacks Vientiane

The king of Louangphrabang attacks Vientiane, angered by its perceived compliance with an attack on Louangphrabang by the Burmese in 1765.

1778 Champassak and Vientiane conquered by Thailand

The country's internal disputes make it ripe for invasion, and Thailand, Vietnam, Myanmar and Cambodia all take their turns. The Lao states try to maintain independence from their neighbors, but eventually come under the control of Thailand. The Thai capture Vientiane for the first time, turning it into a vassal state. The invasion by Thailand forces thousands of Lao to resettle in areas under Thai control, and the Lao royal court is moved from Vientiane to the capital of Thailand in Bangkok.

Late 1770s–early 1880s Migration of Lao Sung from China

The Lao Sung, which includes the Hmong and the Man tribes, begin migrating south into Laos from China. Lao Sung is literally translated as the Laotian of the mountaintops. They speak their own Tibeto-Burmese, Miao-Yao languages and maintain identities separate from other groups in Laos. The Hmong, who during the period of monarchy are called the Meo, keep their tradition of having a king with subchiefs, with social organization centered on the village. Each village consists of several extended families, with a headman for each clan. The headman maintains authority over his clansmen in the village and is the link with regional and national authorities.

This is the last major migration into Laos in the 1800s. The Hmong avoid conflict with their lowland neighbors by staying in the mountainous areas, growing dryland rice and maize at high elevations as they did in southern China. Their major contact with lowland Lao is in selling them their chief cash crop, opium.

1824 Vientiane and Champassak attack Thailand

Early in the nineteenth century, Thailand controls much of Laos which includes Louangphrabang, Vientiane, and Champasak. The Lao ruler of Vientiane, Tiao Anou, persuades the Thai to make his son the ruler of Champassak. The two Lao kingdoms then betray Thailand, with the help of Vietnam. A Lao invasion of Thailand fails.

1828 Tiao Anou again attacks Thailand

The attack by Tiao Anou against Thailand so angers the Thai that they march to Vientiane and raze the city. Tiao Anou flees to the tiny mountainous Lao region of Xieng Khouang which has been keeping invaders out by alignments, gifts and promises. In order not to offend the Thai, the kingdom of Xieng Khouang turns Tiao Anou over to Thailand. This act enrages Vietnam which invades and occupies the tiny region.

At best these conflicts with the Thai and Vietnamese cause tremendous disruption in peoples' lives. At worst, they devastate families, their lands, and villages. Wars with the Thai in the early 1800s cause terror in the villages on the right bank of the Mekong River. People desert their villages, and officials flee to the nearest friendly town.

Mid-1880s Thai depopulation raids

On the left bank of the Mekong River, the Thai follow a policy of depopulating the country. The Thai carry off Lao villagers to use them as workers in the fields of the Thai upper classes. By removing populations from the area, Thai's rival, Vietnam, is deprived of food supplies, transport, and recruits—for both Thai and Vietnamese use Lao people to furnish food and manpower.

The Thai depopulate the area of Khamkeut, Khammouan, and the valley of the Xe Banghiang each year during the dry season. Entire villages are emptied. Depopulation raids cause the remaining population to migrate to the east and south where they form new villages. The Vietnamese, who see new areas of control and resources for themselves, encourage this expansion.

1860s France comes to Southeast Asia

Tired of purchasing Asian goods from Britain, their traditional rival, rather than obtaining them for themselves, France looks to Southeast Asia for colonies. After invading Vietnam, France turns its attention northeast to Laos. Now Thailand faces off with France for control over Lao territory.

1866: April French explorers visit Vientiane

France explores Southeast Asia primarily by its rivers, looking for access routes to China. An expedition by Frenchmen Ernest Doudart de Lagree (1823–68) and Francis Garnier (1839–73) visits the ruins at Vientiane. Other French expeditions do not yield them much information about the interior or its people.

1867: September 30 Frenchman Auguste Pavie comes to Laos

With a small party, French explorer Auguste Pavie (1847–1925) crosses into Laos on an exploring expedition. He hopes to meet King Oun Kham of Louangphrabang, one of the Lao kingdoms. Laos is still dependent on Thailand. In a struggle between Thai forces and Lao troops hostile to both French and Thai, the old king is trapped in his burning palace. Managing to be in the right place at the right time, Auguste Pavie and his men rescue the king and transport him by boat to safety. In gratitude, the old king signs papers transferring the loyalties of his Lao kingdom, Louangphrabang, to France.

Inspired by Pavie's journal writings, a group of wealthy businessmen and scholars in France form the French Syndicate of Laos. Their subsequent support of Pavie's scientific work, explorations, and contact with local Lao officials increases French interest in the wealth of the small Southeast Asian nation.

Thailand establishes a protectorate over Vietnam to extend its influence over Southeast Asia. The coming of the French has many ramifications, including preventing Laos from being devoured by its neighbors, Vietnam and Thailand.

1887: February Pavie becomes vice consul of Laos

France installs Auguste Pavie, its now-famous explorer and writer, as the first vice consul of Laos.

1890 France takes control of Laos

France begins sixty-three years of colonial rule over Laos. During the period France rules "the land of the Lao," they preserve local administrations and the royal house of Louangphrabang. French occupation also facilitates Vietnamese migration into the country, as the French bring Vietnamese in to staff the middle levels of the civil services and militia.

1893: May **France occupies area east of the Mekong**

France sends her troops into Lao territories east of the Mekong River.

1893: July **France swallows more land**

The "Paknam incident" gives France the excuse to demand that Laos give over land on the east banks of the Mekong River. Dispatching two warships to the Gulf of Siam, the French force their way into the river, training their guns on the royal palace. After a blockade on Bangkok, the Thai give in to the French demands for more land.

1893: October 3 **Treaty between France and Thailand**

The French government and the government of the king of Thailand conclude a treaty giving France occupation of the east bank territories in Laos. The Laotian government has no say in the matter. By the end of the nineteenth century, France rather than Thailand is the dominant power in Southeast Asia.

1896: January 15 **Southeast Asia is divided among Europeans powers**

In an effort to prevent conflict between the two European nations, the Anglo-French Convention is held. It defines spheres of influence for the British and the French in mainland Southeast Asia.

1900–40 **French rule**

Laos continues to have a monarchy, with the king of Louangphrabang serving as king over all, with the royalty of other kingdoms taking subordinate positions at the court. But this is a court of glamour, without any real power. Power remains with the French officials.

Laos never yields the wealth France hopes for. The Mekong River has too many rocks and rapids and the mountains are too steep to promote easy trade and exploration. Villagers trade along traditional lines with the Chinese, and are not interested in switching their allegiance to French traders. At any rate, most Lao villagers are primarily self-sufficient, growing their own food and making their own tools, clothes and houses, needing little from the outside world. During the entire occupation, Laos provides less than one percent of France's exports from Southeast Asia.

France continues her occupation not for wealth but for national pride, unwilling to yield her colony in the face of other European powers. In addition, occupation of Laos' mountainous area provides a buffer between French and English rule, with France occupying Vietnam, Cambodia and Laos and Britain occupying Burma. Occupation of Laos also keeps their mountains free of Vietnamese rebels, a threat to French occupation of the much more profitable Vietnam.

France's occupation of Laos produces mixed results. While Lao children learn French and sing the French national anthem in school, the vast majority of Lao do not go to formal schools. The French make few improvements in the country. Few roads, industries, or businesses are developed. While Lao villagers are expected to provide taxes and free labor to the French administrators, collection laws are not strictly enforced. France devotes far more attention to Vietnam, from which it draws far more profit. Nonetheless, Lao mountain tribes rebel occasionally against paying taxes and supplying *corvee* (forced) labor to the French. They are also inspired by outbreaks of religious fervor against the foreign invaders. The French are able to put down these rebellions and preserve internal peace.

With the higher administrative positions held by French or Vietnamese, French occupation leaves the Lao with little administrative experience. Lao hold only the lowest government positions, and these all by royalty or families of wealth. Their positions are low relative to French and Vietnamese in their country but high relative to their fellow countrymen. Most of the elite hold as many Western as Lao ideas and, with their wealth and position, have little in common with the vast majority of Lao villagers.

1902–07 **Laos acquires modern geographic shape**

French troops subdue rebellions on the Bolovens Plateau. Sisavang Vong becomes king of Laos. Laotian territories are lost to Thailand, so that Laos attains its present-day borders.

1907 **Laos is made a protectorate by France**

Weakened by war, Thailand relinquishes control of Laos to France. The people of Laos now live under French administration and law. Under the French, the country is unified for the first time since 1707. Each separate kingdom of Vientiane, Louangphrabang, Champassak and Xieng Khouang has had its own language, culture, and court. France unites these regions, calling their creation Laos. Its new borders are the Mekong River on the west, the Annamite Mountains on the east. Through its occupation, France reestablishes Laos as an independent country in the middle Mekong Valley, with northern borders with China and southern borders with Cambodia, dependent on neither Thailand nor Vietnam.

Auguste Pavie becomes the first French governor of Laos. He administers the country with seventy French officials. To assist the French and Vietnamese administrators, the French use Lao leaders, such as traditional chiefs, at the lowest administrative level.

1920 **Prince Phetsarath becomes viceroy**

The French make Prince Phetsarath viceroy. He establishes a system of ranks and titles for the civil service and modernizes the administration. He establishes a school of law and administration. He also reorganizes the administrative system of Buddhist monks, and creates a system of schools for educating the monks in Pali, the sacred language of Buddhism.

1925–26 French rule is solidified

Additional treaties and agreements finally resolve questions over the borders of Laos, and establish the permanent Franc-Siamese High Commission of the Mekong.

1930: June 5 Laos becomes official French colony

Laos is designated a French colony by the French Legislative Council.

1940: August 30 The Matsuoka – Henry Pact

The French Vichy government signs the Matsuoka-Henry Pact with Japan. This treaty grants Japan the right to station troops in Indochina and preserves France's sovereignty over Indochina.

1940 French cooperate with Japanese

Admiral Jean Decoux, governor-general of Laos, tries to maintain peace in Laos by compromising with the Japanese, who invade much of Asia at the beginning of World War II.

1941: May 9 Peace Convention

The Peace Convention between France and Thailand ends the Franco-Thai War and gives all Lao territories west of the Mekong River to Thailand. Mediated by the Japanese, it heavily favors the Thai.

1941: August 29 Treaty of Protectorate with France

In response to Lao anger over the Peace Convention with Thailand, the French sign the Treaty of Protectorate with the Kingdom of Louangphrabang, regularizing the protectorate and enlarging its domain by adding to it the provinces of Vientiane, Xiangkhoang, and Louang Namtha.

1945 Japanese drive the French from Indochina

Opposed to Admiral Decoux' compromise with the Japanese, French paratroopers drop into the Lao mountains and make guerrilla raids on Japanese troops stationed in Laos. Angered, the Japanese overturn their 1940 agreement with the French, and end French rule throughout Indochina. Japanese troops quickly move in, confiscating French property, and arresting French officials. After arresting the French colonial officials, Japanese replace them with Vietnamese administrators. Japan declares Laos an independent nation. Although the Japanese tell the king, Sisavang Vong, to declare Laos an independent country, they take no steps to prepare the Lao to rule themselves. Meanwhile, Franco-Laotian guerrillas fight to repel the Japanese from Laos.

In August 1945, the United States drops two atomic bombs on Japan. Japan surrenders, losing all her colonies and occupied land. When Japan surrenders to the Allies, Laos is left without occupiers for the first time in two hundred years. The occupying forces of France and Japan have been turned out, and neighboring countries such as China, Vietnam and Thailand are recoiling from the war. Into the void come new nationalistic Lao leaders.

The absence of an actively occupying French body encourages the elite to look elsewhere for support, for most feel their country too small, poor, and inexperienced to survive without it. Pro-Western elite, or rightists, look to France and the West for financial and military aid. The leftists look to China and the Soviet Union. The neutralists feel Laos should obtain support wherever it can, preferably from all sides. The Lao also fear its previous invaders, with some more fearful of France's return, others more afraid of Vietnam. At the end of World War II, Laos is populated by sixty clans and sizable urban populations of both Chinese and Vietnamese. All want security and prosperity for their own group.

Three brothers, Princes Phetsarath, Souvanna Phouma (1901–84), and Souphanouvong (1902–95) take control of the country. Working together sometimes, sometimes working against one another, these three brothers are Laos' most important leaders. Prince Phetsarath takes the lead in unifying Laos after the Japanese leave in 1945. When France informs Prince Phetsarath that they are returning to re-establish their administration, he tells them to stay away because Laos is now a free country.

His brother, Prince Souvanna Phouma, is a neutralist who believes in compromising with all sides in order to save Laos. His detracters say he "bends whichever way the wind is blowing," while his supporters claim his flexibility preserves the country in times of terrible conflict. Their half brother, Souphanouvong, is especially sympathetic to the common people of Laos, supposedly because his mother had not come from royalty. While studying in France, he becomes interested in Communism. While working in Vietnam, he marries a Vietnamese woman and becomes increasingly involved in nationalist and communist movements.

1945 Lao Issara seize power

The Lao Issara Movement is formed to resist any attempt by the French to return Laos to colonial status. Lao Issara, or Free Lao activists, seize power in Vientiane, Savannakhet, and other Lao towns, and establish a provisional government. In the few months when France's power in Laos is temporarily eclipsed, the consequences of earlier Vietnamese migration into the country nearly prove fatal to the new government, as they are accused of being dominated by the Vietnamese.

1946 French regain control of Laos

After a year of struggle, France regains control of the country. Thailand returns the former Lao territories of Xaignabouri and Champasak to Laos. The three princes and their "Lao Issara" government flee to Thailand, where they form a government in exile.

Souvavva Phouma (1901–84), prince and political leader, was an advocate of Lao neutrality. (Library of Congress)

Laos is given limited autonomy as the unified Kingdom of Laos within the French Union.

1946 Laos gains its first constitution

The three princely brothers head the "Free Laos" government in Thailand, and negotiate a constitution with France. Laos is made a constitutional monarchy and elections are held to elect members to the National Assembly. Prince Souvannarath forms the government of the Kingdom of Laos.

1949 Laos gains its independence within the French Union

Two years later, Laos gains internal autonomy from France. When the exiles are called to return, a number do so, but, of the three princes only Souvanna Phouma receives an invitation. Prince Souphanouvong goes to Vietnam to join anti-French Lao rebels. Distrusting both the French and the Vietnamese, Prince Phetsarath remains in Thailand.

1949 Prince Souphanouvong forms the Pathet Lao

In retaliation against French return to Southeast Asia, North Vietnam commences what becomes known as the French War in Indochina or the First Indochinese War.

In an effort to bring independence to Laos, Souphanouvong forms the Latsavong detachment, the first armed forces of the Pathet Lao, the beginning of the Lao People's Liberation Army, the communist-led guerrilla movement. The Army is often referred to as communist revolutionaries or guerrilla forces. These guerrilla troops begin attacking the Royal Laotian army. The Lao Issara government-in-exile dissolves and its members return to Laos or join the newly formed army on the Vietnam border. Two of the brothers are now arrayed against one another: Souphanouvong is leading the rebellion against his brother, Souvanna Phouma, the premier of a Laos still dependent on the French. The Pathet Lao are charged with being dominated by the Vietnamese when they proclaim themselves part of an Indochinese-wide revolutionary movement.

1950: February American recognition

The United States and Britain recognize Laos as an Associated State in the French Union.

1950: August Pathet Lao form government

The Pathet Lao formally establishes a resistance government.

1951: February Formation of individual Communist Parties

The Indochinese Communist Party dissolves and separate parties are formed in each of the Southeast Asian countries of Laos, Cambodia, and Vietnam.

1953: October 22 Independence for Laos

The Franco-Lao Treaty of Amity and Associations transfers remaining French powers to the Royal Lao Government, but retains control over Laotian military affairs. Laos resigns from the French Union.

1954 Geneva Conference

A conference is called in Geneva, Switzerland to discuss the fate of Southeast Asia, with France, the United States, China, the Soviet Union, Britain, Cambodia, and Vietnam participating. The Pathet Lao is not allowed to attend because they are considered part of the Vietnamese Communists. The Geneva Conference divides Vietnam into two countries: North Vietnam ruled by the Communists and South Vietnam with a pro-Western government. Laos is declared a neutral country, and thus becomes a buffer state between North Vietnam and Thailand.

The Royal Lao Government agrees to include the Pathet Lao, meaning Lao Nation, into the government coalition. Under an agreement signed by France and the Vietminh, the Vietminh agree to withdraw their troops from Laos. The two northern provinces of Phongsali and Houaphan are designated regrouping areas for the Pathet Lao until a political settlement is reached. The remainder of the Pathet Lao troops joins the Royal Lao Army. The International Control Commission is established to implement the Geneva agreements.

1954–75 Second Indochina War

The Second Indochina War begins in 1954. Between 1960 and 1973 especially, many Laotians are displaced by war, their villages threatened or destroyed.

1955 Beginning of the Lao People's Party

The Phak Pasason Lao, or Lao People's Party, is established as part of the Indochinese Communist Party.

1955: December 14 Laos joins the United Nations

Laos is admitted to the United Nations.

1955–57 Negotiations between Lao parties

The Royal Lao Government and the Pathet Lao continue negotiations over control of the government and the country.

1956 Beginning of the Lao Patriotic Front

The Lao Patriotic Front is established and secretly guided by the Lao People's Party. The Lao Patriotic Front serves as a political front for the Pathet Lao.

1956–60 Malaria eradication program

The first malaria eradication program occurs, and DDT is sprayed throughout much of the country to kill the mosquitoes that carry the disease.

1957 Souvanna Phouma leads first coalition government

Conflict between the Royal Lao Army and the Pathet Lao increasingly engulfs the country. Souvanna Phouma suggests a coalition government. This coalition retains Souvanna Phouma as premier and puts Souphanouvong in charge of economic planning and two Pathet Lao as cabinet ministers.

1957 First elections to include Pathet Lao

The first elections to include the Pathet Lao sees them winning a significant number of seats, much to the displeasure of the rightists and the Western powers. The United States withdraws financial aid to the country. Conflict occurs when Pathet Lao and Royal Government troops are combined. When the Pathet Lao flee, after perceiving they have been betrayed, their leader Prince Souphanouvong and a hundred other leftists are arrested. With support from the United States Central Intelligence Agency, Souvanna Phouma is removed from office for being too neutralist, and the rightists take control of the government under Phoui Sananikone.

Late 1950s Beginning of public education

The Pathet Lao tries to gain the support of Buddhist monks and, in turn, the support of the population. Before the 1950s, only lowland Laotians have any formal training because highland groups do not have written scripts. Formal education is based primarily in the Buddhist temple, where novice monks and other boys are taught to read Lao and Pali, the sacred Buddhist script. With little formal education available to the Laotians under the French, the Pathet Lao for the first time begin teaching Lao in schools.

1959 Laos gets a new king

King Sisavang Vong dies and is succeeded by Savang Vatthana. Meanwhile, fighting continues in northern Laos between the various factions

1960 Souvanna Phouma again prime minister

The 1960 elections are rigged so that rightists win all the seats. Disgusted with both sides, and with ties to each, a young paratroop captain in the Royalist army named Kong Le uses the army to overthrow the pro-Western government, demands a neutral government, and appoints Souvanna Phouma premier, the third time this brother prince has headed a Lao government. Kong Le hopes the neutralist government, led by neither rightists nor the communist Pathet Lao, will stop the fratricidal fighting. It does not. Souphanouvong escapes from jail and takes refuge in Vietnam with his followers. Laos is now in three parts: the rightists, the neutralists led by Souvanna Phouma, and the leftists led by his brother, Souphanouvong.

1961 Recognition of the various Lao governments

Rightist troops under General Phouumi Nosavan drive Souvanna Phouma's neutralist government from Vientiane. Souvanna Phouma's neutralists then naively ally themselves with the Pathet Lao. The Souvanna Phouma government is recognized by the Communist bloc and the Soviet Union sends airlift support.

Prince Boun Oum's Vientiane government is recognized by Western nations and receives considerable military and economic aid from the United States. There is heavy fighting between the forces of the two governments: the rightists and the neutralists.

For the first time, North Vietnamese troops intervene in Laos with regular troops, inflicting heavy military losses on the rightist government.

1961–62 The second Geneva Conference

In a second effort to ease tensions in Southeast Asia, France, Britain, the United States, and the Soviet Union again meet in Geneva. The conference concludes that Laos will be neutral, and sets up a coalition government that includes all three sides. The international agreement for an independent and neutral Laos occurs only on paper.

1962 Souvanna Phouma leads second coalition government

A second coalition government is formed. The rightist, Prince Boun Oum, who is the royal head of Champassak, joins the neutralist Souvanna Phouma and his leftist brother Souphanouvong.

Civil war soon resumes, continuing into the 1970s. Each side is backed by either the United States or Vietnam (itself supported by the Soviet Union). Each side trades accusations with the others of violating their agreements.

1963 The civil war resumes

After just a year and the assassination of several leftist leaders, the coalition dissolves. Again, Souphanouvong flees to Vietnam. Fighting resumes between the Pathet Lao and government forces. The Pathet Lao is supported by Chinese, Soviet and North Vietnamese troops.

1964 Americans are embroiled in Laos

Laos becomes increasingly linked with the situation in Vietnam. By 1964, the United States has more than three thousand Air Force personnel and seventy-five aircraft in Laos, all there to promote American efforts in South Vietnam against communist North Vietnam.

North Vietnamese troops refuse to leave Laos. The Ho Chi Minh Trail through Laos is expanded to enable North Vietnamese to send supplies down to their comrades in South Vietnam. The efforts of the International Control Commission to stop the fighting come to nothing. The United States begins bombing in Laos against leftist forces.

Late 1960s Development of formal curriculum

The Royal Lao Government begins developing a curriculum for the Lao language and other courses for the 200,000 children attending school, about thirty-six percent of the school-age population.

1968–74 Fighting increases

Fighting escalates between the Pathet Lao and the Royal Lao Army. The Hmong under General Vang Pao resist the Pathet Lao forces, fighting by the side of American troops.

1970s Pathet Lao struggle to make changes

By the 1970s, English begins to replace French as the language of urban dwellers and the elite. Buddhism is the religion of the majority, although the Communist government—instead of supporting the religion—tries to manipulate religious leaders into supporting its political goals. The Pathet Lao discourage traditional beliefs of the uplanders, such as the Hmong, in spirits and healing ceremonies. Frequent ceremonies are held to consult and beseech the khwan, each individual's thirty-two guardian spirits, for protection and assistance. Rituals to ensure the vitality of rice during the agricultural season decline in the face of communist disapproval.

1971 Laos is split in two

The Pathet Lao control eastern Laos, while Prince Souvanna Phouma's government troops control western Laos. By 1971, American jets bomb at a daily rate of seven hundred to one thousand missions. By this time, Prince Phetsarath is dead and Kong Le has been removed from command, eventually to take the side of the Pathet Lao. Some rightists, after attempting a coup, flee to Thailand. Souvanna Phouma is left, reluctantly encouraging American support of the rightists in his country.

1972 Factions begin negotiations

The Lao People's Party changes its name to the Lao People's Revolutionary Party. The Royal Lao Government and the Pathet Lao begin negotiations for a cease-fire.

Under the leadership of the Lao People's Revolutionary Party, allied closely with Vietnam, Vietnamese becomes the third language of the elite in Laos. Formerly, the elite was primarily urban, drawing income from renting lands and working in the cities. Increasingly, although many remain members of the nobility, the elite are members of the Communist Party.

1973: February Factions agree to a cease-fire

The Royal Lao Government and the Pathet Lao agree to a cease-fire in the Vientiane Agreement. The bombing by the United States ends.

The war goes better for the Communists than for the pro-Western government. Vast sums of money pour into the country from the United States, much going illegally into the pockets of the elite. A fourth of the population is homeless. The Pathet Lao controls over four-fifths of the land and two-fifths of the people. Many hill tribes work closely with American troops, feeling as alienated from the Lao government as from the Communists.

By 1973, conditions in Laos are such that Souvanna Phouma asks the Pathet Lao to form a coalition government. Half the power and half the government positions go to each side.

1974: April Third coalition government

A third coalition government is established with the participation of the Lao Patriotic Front. The new government takes office by royal decree under the name of the Provisional Government of National Union.

1974: August Increasing dissatisfaction with Communist rule

Fighting resumes among the various factions. Vang Pao flees to Thailand and senior rightist ministers and generals also leave for Thailand. The Lao People's Liberation Army frees the provincial capitals from the rightists and opens reeducation centers or "seminar camps" to hold political prisoners. The "Revolutionary Administration" takes power in Vientiane, and elections are held for local people's councils.

1974 Souvanna Phouma resigns

With the establishment of the Lao People's Democratic Republic, Souvanna Phouma resigns from office.

1975 End of third coalition government

The collapse of pro-Western South Vietnam and Cambodia in April 1975 helps the Communist Lao Patriotic Front, hastening the end of the third coalition government.

1975: December Lao People's Revolutionary Party assumes power

The Lao People's Revolutionary Party dissolves the Provisional Government of National Union amidst the dissatisfaction of the population, and persuades King Savang Vatthana to abdicate the throne. This ends the period of a conservative monarchy dominated by a few families. The Lao People's Revolutionary Party then establish the Lao People's Democratic Republic. Souphanouvong, known as the "red prince" because of his royal lineage and communist connections, becomes its first president. Kaysone Phomvihan (b. 1920) is the first prime minister. The government is ruthless in putting down internal opposition. The support of Buddhist monks helps the Pathet Lao gain governmental power.

The end of the Second Indochina War displaces about twenty-five percent of the entire Laotian population. Many return home. Others move to safer areas, and many others flee the country, including a number of elite and royal government officials.

1975: December 2 Pathet Lao establish People's Democratic Republic

After the United States withdraws from Southeast Asia, taking a number of Lao hill tribesmen with them, student demonstrators in Laos demand that the Pathet Lao take control of the country—which they do under the leadership of Prince Souphanouvong. Souvanna Phouma becomes advisor to the government. The six hundred-year-old monarchy is over.

The new Communist government blames the economic situation of Laos on the past and Western influences. Increased surveillance of the population, a weak economy, bad weather, reeducation programs and government bombing against resistance that killed thousands of Lao and hill tribespeople increases the population's disillusionment with the communist Pathet Lao government.

1975–90 Attempts to eradicate malaria

The government continues its attempts to wipe out malaria, including providing chemical prophylaxis to high-risk groups, eliminating mosquito breeding sites, and promoting individual protection.

1976: May Guidelines for Lao socialism

The Central Committee of the Lao People's Revolutionary Party passes the Third Resolution, which includes guidelines for establishing the socialist revolution in Laos.

1977: July Treaty with Vietnam

The Lao and the Vietnamese sign a Twenty-Five Year Lao-Vietnamese Treaty of Friendship and Cooperation.

1977–92 Laotians flee the country

A second refugee flight occurs between 1977 and 1981, with numerous villagers leaving the country for refugee camps in Thailand in response to economic hardship, poor weather for growing crops, and government incompetence. About 360,000 Laotians continue to flee through the 1980s—more than ten percent of the population. Over 300,000 are resettled in third countries. The remaining refugees are repatriated to Laos from the refugee camps in Thailand.

1979: February Establishment of the Lao Front

The Lao Front for National Construction is formed, replacing the Lao Patriotic Front.

1979 Government loosens its grip

Laos is facing collapse. Oppression, corruption, incompetent administration, and political factionalism turn much of the population against the government. Although Communist leaders see commonalties between communism and Buddhism (for example, that both teach that all men are equal and aim to end suffering), they try to get Buddhist leaders to conform more to their political ideology. Monks teach leftist ideas, leave the clergy, or flee to Thailand. People's participation in Buddhism declines rapidly.

In response, the leader of the country, Chairman Kaysone Phomvihan, announces that the government has been trying to turn the country toward socialism too rapidly. The government begins turning away from social goals. Taxes are decreased, trade increases and private ownership and foreign investment is increased. Concessions are made to Buddhist practice, and religious activity increases.

1981: January Five-Year Plan for Laos

Lao institutes its first five-year plan, aimed at increasing self-sufficiency in food production and the collectivization of agriculture, developing industrial activity within Laos, increasing trade with Thailand, improving the infrastructure, and increasing revenue from exports. All these goals are aimed at taking steps toward a market-oriented economy.

1982: April Third Congress is held

The third Congress of the Lao People's Revolutionary Party is held in Laos.

1983–84 Launch of adult literacy program

The government launches an intensive adult literacy program using educated villagers to teach their fellow adults. By 1985, the United Nations estimates that ninety-two percent of men and seventy-six percent of women between 15 and 45 can read, although many lose their new abilities because of disuse in village life. Books and reading materials are rare in villages.

Mid-1980s Conditions improve

In the early 1980s, Laos becomes self-sufficient in producing rice to feed its own people. As goods increase and political surveillance eases a bit, fewer Lao flee to Thailand. Increasing numbers of people are allowed to visit relatives in France and the United States. Laos agrees to help the United States search for American servicemen lost during the Vietnam War.

1985: February First Search for MIAs

Laos and the United States conduct their first search mission for American soldiers missing in action in Laos.

1985: March First Laos census

Laos takes its first national population census. According to the census, fifteen percent of the Lao population lives in urban areas, which includes the capital of Vientiane with over 250,000; three provincial capitals including Savannakhet with 109,000, Pakxe with 50,000, and Louangphrabang with 20,000; and district centers, with populations as low as 2 to 3,000. Eighty-five percent of the population remains in the rural areas.

1986 Laos and China normalize relations

Laos and China agree to normalize diplomatic relations which have been severed since the Vietnamese invasion of Cambodia in 1979.

The fourth Congress of the Lao People's Revolutionary Party is held. Kaysone Phomvihan is general secretary of the Lao People's Revolutionary Party and Phoumi Vongvichit becomes president of Laos.

Disappointed that growth during the First Five-Year Plan is less than anticipated, the government endorses a Second Five-Year Plan. This plan is called the New Economic Mechanism (NEM). The NEM formalizes reforms and begins opening the country to free market forces. It lifts numerous trade regulations and increases opportunities for foreign investment. As the economy improves, Laos turns its attention to instituting environmental protections, limiting the practice of slash-and-burn cultivation to protect its forests and encourage cash crops. Growth is not as rapid as hoped. The NEM also provides some, but not much in the opinion of most Western observers, journalist and diplomatic opportunity for observing human rights in Laos.

1987–88 Droughts occur

Droughts in Laos hinder rice production, and make it necessary to import rice into the country. The goal of achieving self-sufficiency for the country is not met.

1988 Laos holds its first national elections

National elections are held, and delegates are elected to the first Supreme People's Assembly. The opening session of the Assembly is held in May and June. The last Vietnamese troops leave Laos. Laos and Thailand make efforts to ease their border conflicts and lessen criticism of one another.

Late 1980s Increase in Buddhist activity

Buddhist practice increases because of both political liberalization and economic reform in the country. Ordination of monks increases and festivals become longer and more elaborate.

1990s Increased aid to Laos

Laos begins receiving increased assistance from Western countries, including Australia, France, Sweden and Japan, as well as increased support from regional and international organizations such as the International Monetary Fund, the World Bank, and the Asian Development Bank, devoted to assisting countries in developing economically.

1991 Laos and Thailand ease tensions

Laos and Thailand exchange military delegations in hopes of resolving border incidents. Both countries agree to withdraw troops from disputed areas.

1991: March Fifth Congress is held

Laos holds its fifth Lao People's Revolutionary Party Congress. The Secretariat is abolished. Kaysone Phomvihan becomes chairman of the Lao People's Revolutionary Party, and Souphanouvong retires.

1991: August Adoption of new constitution

A new constitution is approved and adopted by the Supreme People's Assembly. The new constitution affirms an individual's right to private ownership, and replaces the word "socialism" with "democracy and prosperity" in the national motto. Kaysone Phomvihan becomes president of the Lao People's Democratic Republic; Khamtai Siphandon (b. 1924) becomes prime minister.

1992 Kaysone Phomvihan dies

After Kaysone's death in November, Nouhak Phomsavan becomes president of the Lao People's Democratic Republic. Khamtai becomes chairman of the Lao People's Revolutionary Party and prime minister of the Lao People's Democratic Republic.

Laos gains observer status with the Association of Southeast Asian Nations (ASEAN), a first step in becoming a member. ASEAN was founded in 1967 to increase economic cooperation among its members.

1992 Conference on AIDS

The government holds a conference on Acquired Immune Deficiency Syndrome to discuss the likelihood of AIDS spreading in Laos.

1993: February National elections

Nouhak is elected president and Khamtai is elected prime minister of the Lao People's Democratic Republic. The Council of Ministers is reorganized. Laws are also passed to provide more protection for the environment and broaden economic reforms.

Although the goal to provide primary education to all eligible children is not reached in the early 1990s, twice as many students are in school as were in the 1950s. Still, less than half of the students entering complete the five years of elementary school. Secondary education remains limited in the number of students attending and curriculum available. The government sends numerous students abroad to compensate for the lack of trained teachers in Laos.

1993–2000 Socio-economic Development Plan

The government's Socio-economic Development Plan focuses on increasing the country's production of food and commercial products, improving infrastructure, and developing human and natural resources while protecting the environment. Specifically, the Plan calls for switching from slash-and-burn subsistence production toward sedentary market agriculture and a more diversified economy, building roads to improve public health and education, and irrigating unused lands. The Plan also calls for expanded economic relationships and cooperation with other nations.

1994 Lao-American cooperation

Laos and the United States have already conducted thirty-three joint searches and excavations for soldiers missing in action during the Vietnam War, and agree to carry out six more with additional American staff. The two countries also agree to measures for controlling the production and export of opium, heroin, and marijuana from Laos. Laos is the world's third largest producer of opium, making such cooperation necessary in controlling drug trafficking and consumption. Opium production has decreased in the 1990s.

1994: April First bridge across Mekong

Evidence of increasingly cordial relations between Laos and Thailand can be seen in the construction of the Friendship Bridge linking the two countries. The bridge has allowed for increased trade, tourism, transportation, and communications between Laos and Thailand.

1994: March Lao-Russian Trade Agreement

The Lao National Council of Trade and Industry and the Russian Council of Trade and Industry sign an agreement by which Laos will receive scientific and technical assistance from Russia and financial assistance from other countries, the International Monetary Fund, and businessmen. This assistance will provide aid for programs to protect the environment, restore the forests, increase agricultural productivity, and promote mining and exploration efforts.

1994: May Lao delegation visits China

A military delegation from the Lao People's Revolutionary Party makes an official visit to China to promote friendship and military solidarity between the two countries.

1994: July Lao-Thai Venture Agreement

Laos and Thailand sign a joint venture agreement allowing a Thai company to build and develop nine projects in the Vientiane Municipality.

1994: August Lao-Russian Trade Protocol

Laos signs a trade protocol with Russia by which Laos will purchase construction materials, electrical appliances, aircraft spare parts, and other products from Russia, and Russia will buy tin, coffee, wood products, and clothing from Laos.

1995 Return of Lao refugees to Laos

Laotian refugees in Thailand repatriate home to Laos from the refugee camps.

1995: January Death of Prince Souphanouvong

The death of Prince Souphanouvong ends the direct link between the monarchy established in the 1300s by Fa Ngun and the Communist government.

1996: March Sixth Congress focuses on health care

The Sixth Congress of the Lao People's Revolutionary Party is held and special attention is given to the state of health care in the country. Health care remains scarce for most Laotians. Vitamin and protein deficiencies, poor sanitation, and tropical diseases continue to plague the population. Western medicine is rare outside the cities, and its cost prohibits its use to all but the most comfortably well off. Most hospitals and clinics are understaffed with little equipment or supplies. The medical facilities, personnel, equipment, and supplies that are available are located in Vientiane. The vast majority of Laotians continue to rely on traditional healers to prevent and treat disease.

Bibliography

Cummings, Joe. *A Golden Souvenir of Laos*. New York: Asia Books. 1996.

Damrong, Tayanin and Kristina Lindell. *Hunting and Fishing in a Kammu Village*. London: Curzon Press. 1991.

Diamond, Judith. *Enchantment of the World. Laos*. Chicago: Children's Press. 1989.

Ellis, Dawn. *Laos*. New York: Pallas Athene. 1993.

Evans, Grant. *Lao Peasants under Socialism*. New Haven: Yale University Press. 1990.

Geddes, William Robert. *Migrants of the Mountains: The Cultural Ecology of the Blue Mieo (Hmong Njua) of Thailand.* Oxford: Clarendon Press. 1976.

Gunn, Geoffrey C. *Rebellion in Laos: Peasants and Politics in a Colonial Backwater.* Boulder, CO: Westview Press. 1988.

Halpern, Joel Martin. *Economy and Society of Laos. Southeast Asian Studies Monograph, No. 4*. New Haven: Yale University Press. 1964.

Izikowitz, Karl Gustav. *Hill Peasants in French Indochina*. New York: AMS Press, 1979.

Kremmer, Chistopher. *Stalking the Elephant King. In Search of Laos*. Honolulu, University of Hawai'i Press. 1997.

Lindell, Kristina, et. al. *The Kammu Year: Its Lore and Music. Studies on Asian Topics, No. 4*. London: Curzon Press. 1982.

Lewis, Paul and Elaine Lewis. *Peoples of the Golden Triangle*. London: Thames and Hudson. 1984.

Quincy, Keith. *Hmong: History of a People*. Cheney, WA: Eastern Washington University Press. 1988.

Scott, Joanna C. *Indochina's Refugees: Oral Histories from Laos, Cambodia, and Vietnam*. Jeferson, NC: McFarland. 1989.

Stieglitz, P. *In a Little Kingdom*. New York: M. E. Sharpe. 1990.

Stuart-Fox, Martin. *A History of Laos*. New York: Cambridge University Press. 1997.

Westermeyer, Joseph. *Poppies, Pipes, and People: Opium and Its Use in Laos*. Berkeley: University of California Press. 1982.

Lebanon

Introduction

The Republic of Lebanon, as it exists in the 1990s, was created by France in September 1920. Its area of approximately 4,015 square miles (10,400 square kilometers) includes the Lebanese Mountains, which traditionally were known as Mount Lebanon, coastal areas such as Beirut and Tripoli, and the Bekaa Valley in the east. Although historically Lebanon's distinctive geography protected its communities from outside invaders, more than two-thirds of this troubled country has been under foreign military occupation since January 1988. Syrian troops have controlled northern Lebanon and the Bekaa Valley since 1976. Syrians also gained control of West Beirut and the Beirut-Sidon coastal strip in February 1987. Meanwhile, Israel and a local militia group known as the South Lebanese Army have continued to control a 1,000-square-kilometer strip along the Israeli border in southern Lebanon known as the "security zone."

Lebanon lies on the eastern coast of the Mediterranean Sea. That body of water forms much of its western border. Israel lies south of Lebanon and Syria is located to its east and north. The heartland of the country is the rugged Mount Lebanon area. These mountains rise from sea level to a parallel range of 6,600 to 9,800 feet (2,000–3,000 meters) over less than twenty-five miles. Heavy rains in the winter have created many deep clefts and valleys in the region's soft rock. In historic times, the mountains allowed minority communities such as the Maronite Christians to thrive despite the spread of Islam and a series of invaders that swept the region beginning with the Roman Empire. More recently, this rugged terrain created natural fortresses for guerrilla activities. Lebanon's coastal strip also allowed for an early development of sea trade. Merchants exported cedar and spruce from the country's forests and traded extensively in copper and iron.

Although Lebanon is only about three-fourths the size of Connecticut, compared with most countries of similar size, the country is rich in natural resources. Olive and fig trees, grapevines, apples, and bananas are grown throughout its low-lying coastal areas. As the elevation climbs, cedar, maple, juniper, fig, cypress, valonia oak, and Aleppo pine trees take over. Lebanon has approximately 2,500 species of flora and many varieties of thrushes, nightingales and other songbirds. While hunting has killed off most wild animals, jackals, gazelles and rabbits continue to thrive. Because the country is small mountainous, and coastal, it is possible to both ski and swim within a forty-five minute drive.

Unlike other Middle Eastern countries, rain falls abundantly on Lebanon, particularly in the months of December, January, and February. The coastal area receives about 35 inches (90 centimeters) of rain each year, and western mountain slopes get about 50 inches (125 centimeters). The Bekaa area is much drier, receiving only 15 inches (38 centimeters) annually.

Most of Lebanon's 3.1 million people are Arabs, but several distinct religious and ethnic groups differentiate this population. Most Lebanese are either Christian or Muslim, but these two faiths are subdivided into several sects, most of which were created through the historic development of separate ethnic groups. While Muslims are mostly either Sunnis or Shiites, the Druzes, whose religion is derived from Islam, make up a significant minority. Christians are divided as Maronites, Greek Orthodox and Greek Catholics. Armenians, Jews, Syrians, Kurds, and other ethnic groups also reside in Lebanon. About 450,000 to 500,000 Palestinians live in Lebanon. In addition, about 180,000 are classified as stateless undocumented persons, many of whom live in disputed border areas.

History

Like Syria, its neighbor to the east and north, Lebanon was part of Phoenicia (c. 1800–c. 1600 B.C.) and later became part of the Roman Empire. Arabs conquered part of Lebanon in the seventh century A.D., and Islam gradually spread through the area as a result of conversions and Muslim migration to the region. Maronite Christians, however, long had been established around Mount Lebanon, and that part of the country remained predominantly Christian. Invasions by the Druzes (c. 1000), the Crusaders, the Mongols, and others followed.

After the area was brought under control of the Ottoman Empire, Fakhr ad-Din (1586–1635) of the Ma'an family worked to create an autonomous country. The Ma'an dynasty

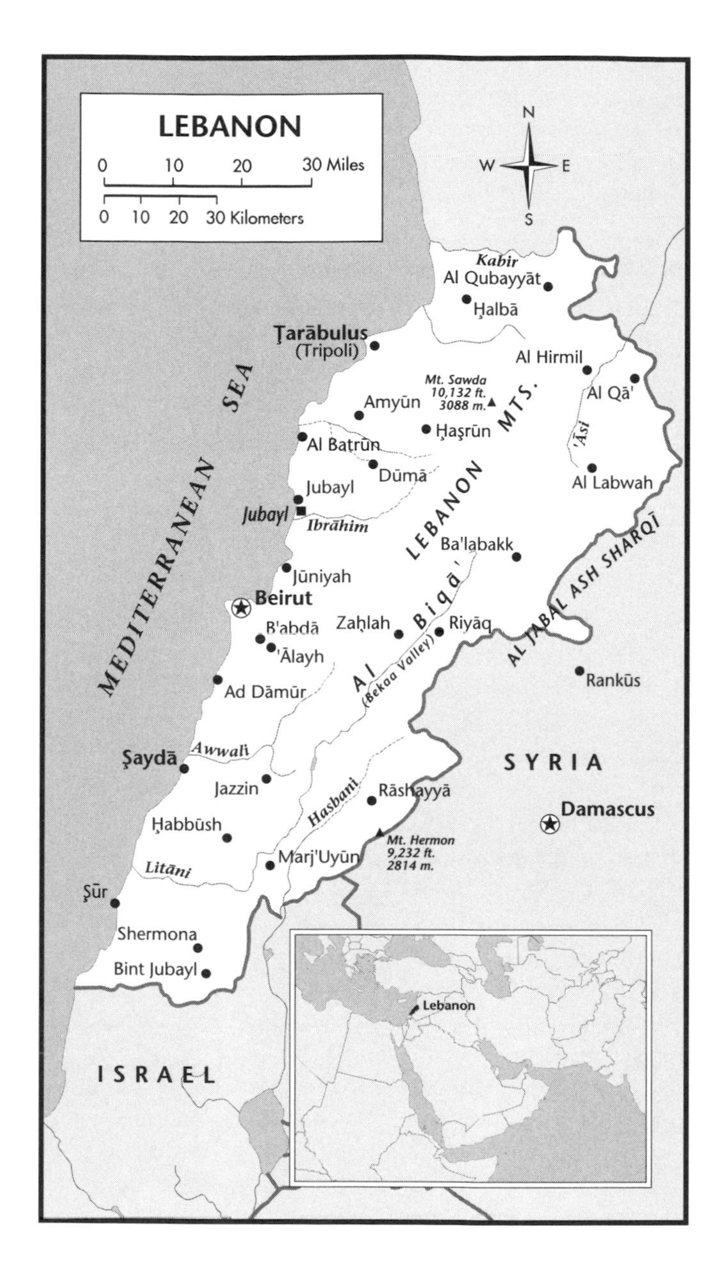

opened Lebanon to Western Europe through trade and military agreements, and these rulers encouraged Christian missionary activity. The strengthening of a Lebanese national identity continued under the Shihab family, who took control in 1697 and maintained authority until the British invasion of 1840–41. Lebanon was thrust into international affairs in the nineteenth century and eventually fell under control of the French, who re-drew its boundaries and governed the territory until it officially achieved independence in 1946.

Religious tensions have led to much of Lebanon's internal strife. An unwritten agreement made during the National Covenant of 1943 held that the president of Lebanon must be a Maronite Christian, and the prime minister must be a Sunni Muslim. The pact further required that the Legislature must consist of a ratio of six Christians to every five Muslims. This arrangement received a great deal of criticism because it did not accurately reflect the percentages of the varying groups within Lebanon's population. Christians, most of whom continue to be Maronite, comprised only about forty-three percent of Lebanon's population as of the early 1980s, while Muslims and Druzes made up fifty-seven percent. These ongoing tensions—coupled with large numbers of Palestinians who fled to the nation after World War II—helped fuel a bitter civil war that began in 1975 and did not conclude until 1990. Although the Taif Accord in 1989 modified the ratio to create a 50–50 balance of Christian and Muslim legislators, tensions between the two groups have continued.

Much of Lebanon's internal strife also was produced by ongoing tensions between Israeli Jews and Arabs. Even though Lebanon did not participate actively in the Arab-Israeli conflicts, its geographic locale meant it inadvertently was drawn into these disputes.

Despite the instability that characterizes Lebanon, its population is extremely well educated. Illiteracy rates in 1995 were estimated at 5.3 percent for males and 9.7 percent for females. However, these rates are not uniformly distributed among all ethnic groups. Shiite Muslims tend to be less well-educated and poorer than their Sunni Muslim and Christian counterparts. Lebanon has several well-known libraries and universities, including the American University in Beirut, St. Joseph University, the Lebanese (State) University, the University of the Holy Spirit, and the Arab University of Beirut.

In the 1990s, the sounds of construction, machinery and rebuilding pervaded Beirut. After Lebanon's long and costly civil war ended in 1990, its government set itself to the task of rehabilitating the nation, building up its industrial base and reviving its once romanticized image as the "Paris of the Middle East." While these efforts have looked promising, military tensions continued to haunt Beirut, with Israel supporting a militia and frequently launching attacks in southern Lebanon and Syria retaining 30,000 troops elsewhere in the country.

An air raid in April 1996, known widely as the "Grapes of Wrath" attack, has come to symbolize the ongoing unrest. International Studies scholar, Augustus Richard Norton, wrote in the January 1998 *Middle East Policy* journal that, following the attack, a memorial cemetery was established near a United Nations base where dozens of civilians had died. According to Norton, the cemetery has become a pilgrimage site for many Lebanon residents. They have made trips to the site to commemorate the dead and to express resentment against Israel. Banners that accuse Israel of terror and genocide decorate the cemetery. For many residents of Lebanon, the ongoing reality is a day-to-day life punctuated with war and violence.

Timeline

3500–800 B.C. **Phoenician settlers**

Phoenicians migrate from the Arabian Peninsula and settle an ancient land known as Phoenicia, which today includes Leba-

In a photo promoting tourism for Lebanon, the pillars of the Temple of Jupiter in Baalbek near Beirut are shown to survive to modern times. (EPD Photos/CSU Archives)

non, Syria and parts of Israel. They establish cities at Byblos, Tyre, Sidon, Baalbek, and Beirut, all of which become important as places of learning. They also introduce the twenty-two-letter Phoenician alphabet, a system of writing that evolved from Egyptian hieroglyphics, in which characters are used to represent certain words or ideas. This system becomes the basis for Hebrew, Aramaic, Arabic, Greek, and Roman scripts.

The Phoenicians have great commerce and navigational skills. After discovering that the North Star remains in a fixed position, they use it to guide journeys as far north as England. In Lebanon, they leave behind a heritage of pottery, purple dye and metal work.

Other cities also develop. The seaport of Jbail, for instance, known as Gebal in the Bible, is among the oldest continuously-inhabited cities in the world. Walls in this city date back to 2800 B.C. Remnants of a Greek temple, the Temple of Jupiter, a Roman theater and a citadel from the Crusaders offer reminders of the town's importance during successive periods of Lebanese history.

300 B.C. Papyrus trade develops

Much of Lebanon's pre-modern history is influenced by its nearness to the Mediterranean Sea and its reputation as a trading community. An example of this early commerce is the trade in papyrus which brings riches to Jbail, a seaport located about eighteen miles (twenty-nine kilometers) north of present-day Beirut. Papyrus is a tall, tufted marshy plant that thrives in the Nile Valley. When pressed together, it forms a writing material that closely resembles paper. Jbail serves as a transfer port for papyrus which is exported from Egypt to Aegean areas.

64 B.C. Roman conquest

Pompey the Great conquers territories that make up modern Lebanon. The region becomes part of the Roman Empire and is governed as part of the province of Syria.

Aramaic replaces Phoenician as the primary language. As Christianity becomes well established in the province of Syria, the rugged geography of Lebanon offers a haven for Christian sects fleeing persecution.

A.D. 517–694 Maronites set up community

Yuhanna Maroun establishes a Christian Maronite community in northern Lebanon and becomes its first elected patriarch. Maronites, who take their name from a monk known as St. Maroun, had established a settlement in Syria in the sixth century and had sent some disciples into present day Lebanon.

Over the next few centuries, the Maronites clash with other religious groups as well as other Christian sects over beliefs, for example, those concerning the divinity of Jesus.

Meanwhile, the Byzantine empire is established in the eastern half of the Roman Empire. Byzantine armies invade Syria in A.D. 694, destroy Maronite monasteries and kill 500 monks. As they move into Lebanon, Maronites in this area organize an army of about 12,000. This army defends its territory, and over time, becomes the primary stronghold of the religious sect. The Maronite church transfers its headquarters from northern Syria to the Mount Lebanon area in the tenth century and establishes its base at Qannubin, a monastery carved from Lebanon's soft rock.

Maronites are peasants who live in isolation from each other. Their religion, however, provides them with a sense of social cohesion.

A.D. 637 Muslims capture Jerusalem

Muslims capture Jerusalem from Roman rulers. Located in the land of Palestine south of Lebanon, the city becomes the most sacred shrine in Islam after Mecca, birthplace of the Prophet Muhammad (c. 570–632), founder of the faith.

After the death of the Prophet Muhammad, followers of Islam split into two main factions: Shi'ites and Sunni. Followers of both factions find their way to Lebanon and settle around Mount Lebanon as well as along coastal areas.

While Sunni Muslims gain a great deal of power and wealth in Lebanon, Shi'ites generally remain poor and are neglected by the ruling groups in the country until the early twentieth century. However, they remain the poorest and least educated community in Lebanon.

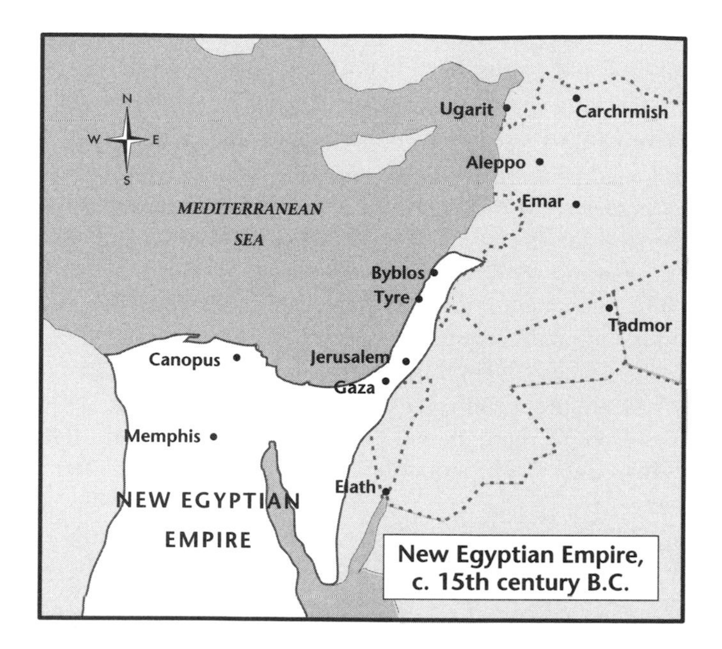

1043 Druzes secure toehold

The Druze faith originates in Egypt as a derivation of Islam. Followers believe that knowing the truth *(al-haqiqa)* is knowledge of the unity of God. Followers are persecuted heavily in Egypt but find a welcoming environment in Syria, Lebanon and Palestine. By 1043, a Druze community is well established in the southern part of Mount Lebanon. Like the Maronites, the community is mostly agricultural but remains fiercely loyal to its members. Its religious beliefs and practices provide strong social cohesion.

1100–1291 Crusaders arrive

As the influence of Islam spreads, Christians in Europe organize a series of assaults that attempt to regain control of Jerusalem, Palestine, and much of the Levant. These military campaigns last through the fourteenth century.

Crusaders initially arrive in Palestine in 1095 and retake Jerusalem from Muslim rulers. A delegation of 25,000 Maronites travels from the mountains to welcome these Christian warriors and provide them with guides and weapons. The Maronites see European Christians as a source of protection against Byzantine persecution and attacks from Muslims and other Christian sects. After the Crusaders establish the Latin Kingdom of Jerusalem in 1099, they reward the Maronites for their help by giving them a privileged position in the new kingdom. Among Christian denominations, the Maronites are considered second in social position only to Latins.

This honored position, however, does not last. Jerusalem falls to Muslims, led by the Egyptian sultan Saladin, in 1187. By 1291, the Mamlukes, a highly-trained military group, take control of Egypt and push the Crusaders out of the Middle East. Mamlukes, who are Sunni Muslims, invade Mount Lebanon and destroy Maronite villages. They also force Shi'ites to migrate to the south.

1516 Ottoman conquest

Ottoman Turks conquer the entire eastern Mediterranean Coast. Lebanon becomes part of the Ottoman Empire, an entity that a group of Turkish warriors established in northwestern Asia in the late thirteenth century and which came to dominate much of the Middle East and the Balkan regions until the end of World War I.

The Ottomans rule in Lebanon for nearly four centuries and generally allow local leaders within their vast empire a fair amount of local autonomy. The Ottoman system organizes people on the basis of their religious affiliation rather than geographically-defined communities, which means that each of the religious groups in Lebanon is allowed a great deal of freedom in setting up social policies that govern their communities. Through this system, Druzes, Maronite Christians, Sunni Muslims and Shi'ites operate largely as separate social, cultural, and economic entities.

The Ottoman system also establishes a hierarchy of communities in which Sunni Muslims are held to be the highest. However, the geography of Lebanon is such that the religious communities live separately and attain power as independent entities. Under Ottoman rule, two major chieftains emerge in Mount Lebanon: the Druzes and Maronite Christians. Meanwhile, the economic plight of Shi'ites, whom Ottoman rulers tend to regard as being loyal to Persia rather than to the Turks, worsen.

Under Ottoman rule, a town in the Lebanese mountains known as Beit Eddine (Bayt Ad Din) becomes the capital of the Mount Lebanon area.

Modern Chronology

1584 Maronite college opens in Rome

During their relationship with the Crusaders, the Maronites forge a lasting relationship with the papal authority in Rome. Their sect is folded into the Catholic Church, even though they are allowed to retain certain rites and customs, and are granted the freedom to have their own elected church patriarchs.

With help from The Vatican, the Maronites establish a college in Rome that is designed to educate clerics from Lebanon so they can return to the region to work as missionaries.

1586–1635 Westernization begins

A decentralized ruling structure fostered by the Ottoman Empire leaves districts in Mount Lebanon under control of local emirs and sheiks. Fakhr ad-Din of the Ma'an family tries to unify Lebanon and open it to European contact through commercial and military agreements. He also encourages the spread of Christian missionary activity.

1696 Monastery is founded

The St. John Monastery in Khonchara is established. Its library holds the first known printing press in the Middle East.

1697–1840 Shihab dominance

The Shihab family rises to power in 1697. Except for a few intervals, this family of Turkish emirs maintains control of Lebanon for much of the 18th and early 19th centuries. Bashir II of this family rules from 1788 to 1840. During this time, Lebanon's borders are extended, and Bashir II plays a critical role in strengthening its identity as a strong state.

Bashir also builds a palace in Beit Eddine that still stands in 1999. Located about thirty miles southeast of Beirut, the palace houses in one wing a folklore museum. The remainder serves as a summer home for Lebanon's president.

1832–40 Egypt occupies Levant

Egypt moves into Syria in 1832. This thrusts the Levant—the area that roughly includes Syria and Lebanon—into European politics and brings it under the influence of a steadily growing Arab nationalist movement.

Arab nationalism began to develop with the rise of Islam. It takes on a different character when Europeans, such as Napoleon Bonaparte, import French revolutionary ideas to Egypt. Nationalism emphasizes the idea that Arabs share a common land, language and culture with Islam serving as a binding force in the region. Many early Arab nationalists have their roots in Egypt.

Because Lebanon contains a strong Christian population, the religious communities there react with ambivalence to the rise of Arab nationalism. After Egypt invades Syria and Lebanon, it maintains control of the area by playing Christians and Muslims against each other.

1840–41 Britain invades Levant

The Egyptian occupation of Lebanon and Syria attracts interest from Europeans who gain additional power by pitting religious groups against each other. Russia, France and Great Britain all get involved in the politics of the area, and each assumes protection of a different religious group.

At this point, Druzes and Maronite Christians are the dominant communities on Mount Lebanon, while Sunni and Shi'ite groups live in coastal areas, around Beirut, and in the southern part of Lebanon that borders the area traditionally known as Palestine. France and Great Britain get involved in a series of disputes between Druzes and Maronites by each assuming protection for one of the groups. Britain becomes an advocate for the Druzes, while France acts as guardian of the Maronites.

British troops take control of the area in 1840 and drive out Egyptian forces. The British force the Turkish emir Bashir II into exile, and while the Ottoman Empire ostensibly retains authority, its power in the area erodes.

1842–64 Druzes-Christian tensions build

As feudal landlords, Druzes have traditionally held most of the power in Mount Lebanon, particularly in the south. Maronite farmers who migrated to the south in search of work often came under Druze power as peasants. This relationship begins to unravel in the mid-nineteenth century as European powers introduce a capitalist economy to the Levant.

Maronites thrive in the new money-based economy, and as they acquire wealth, the community becomes more cohesive. With encouragement from the Maronite church, they begin to fight Druze authority and demand a greater share of power.

Ottoman rulers, under pressure from Britain and France, try to resolve this tension by dividing Mount Lebanon into a northern district to be governed by a Maronite and a southern district to be ruled by the Druzes. This system, however, fails to quell the tensions between the two groups, and by 1860, a full scale war erupts.

Britain, France, Russia, Prussia, and Austria send naval fleets into Syrian waters, and France deploys armed troops in Mount Lebanon. When the insurrection ends, the European powers force the Ottoman government to establish an international commission that sets up a pro-Christian government. Mount Lebanon becomes an autonomous province with a Christian governor. Although this representative technically remains under Ottoman rule, the governor relies on European support in disputes with the Ottoman government.

With their new-found power, Maronites come to regard Mount Lebanon as a Christian homeland and begin to agitate for the creation of a nation-state that will protect their identity amid Arabs and Muslims.

1866 University is built

A cornerstone for the American University of Beirut is laid at what eventually becomes College Hall, the oldest building on the campus. This building remains intact until a bomb destroys its structure in 1991.

1867 *Rose Garden* is published

Poet Warda al-Yaziji publishes the first version of her *diwan,* or collected poetry, in Beirut. The collection, *Hardiqat al-Ward* (Rose Garden), is revised several times during her lifetime and speaks to a female presence within a broader, pan-Arab identity.

Born in 1838, in a Lebanese village, Al-Yaziji comes from a family of poets and scholars. She spends most of her formative years in Beirut, studying in both French and Arabic. Besides writing poetry, she publishes several articles on the status of Arab women in *Al-Diya*, a magazine based in Cairo, Egypt. She dies in 1924.

1883 Khalil Gibran is born

Khalil Gibran (d. 1931) is born in Bcharre, Lebanon, a town in the Lebanese mountains that is known for its giant cedar stands. A Maronite Christian, Gibran becomes a disciple of the French artist, Rodin, and studies in Paris for three years. He later moves to the United States where he lives until his death. He becomes one of Lebanon's greatest artists and achieves international renown for his paintings and literary works, notably his philosophical book, *The Prophet*.

While famous for being the birthplace of Gibran, Bcharre also is associated with another enduring symbol of Lebanon—its cedar trees. Much of Lebanon's early trade was in cedar, and Solomon, King of Israel, who ruled from about 1015 B.C. to 977 B.C., used Lebanese cedar to build the Great Temple of Jerusalem. While most of Lebanon's large trees have disappeared, one grove remains in modern Bcharre, preserved as a national monument known as the Grove of Giant Cedars. Gibran is buried near the Bsharre cedar grove.

1914 Ottoman Empire joins World War I

The Ottoman Empire enters World War I (1913–18) on the German side. This act leads to widespread hunger and poverty in Lebanon. First, Ottoman bureaucracy, along with a plague of locusts, produces a famine. Then, Allied troops establish a blockade in the Levant—the area that includes Lebanon and Syria—to cut off supplies to Ottoman Turk troops. About 120,000 people die in Lebanon during this period.

As victims of a war produced largely by outside forces, many Lebanese develop strong anti-foreign sentiment. The poet, Gibran, composes "Dead Are My People."

The war's hardships also produce a massive wave of emigration as thousands of Lebanese flee their country. As Robert Fisk notes in *Pity The Nation*, many Lebanese names are listed among the third-class passengers drowned aboard the Titanic. The war also brings a new minority community of immigrants into Lebanon—Armenians fleeing slaughter by Turkish forces.

1916: May 16 Sykes-Picot Agreement is signed

As World War I is being waged, a British leader promises to help Sharif Hussein create fully-independent Arab states if Hussein helps to overthrow the Ottoman Empire, a hazy incident that is known historically as the Hussein-McMahon correspondence. Hussein is the hereditary amir of Mecca, the King of the Hejaz (now largely Saudi Arabia), and grandfather of King Hussein of Jordan. He proclaims an Arab Revolt, which is led by his son, Faisal, who later becomes King of Iraq, and Thomas Edward Lawrence, a British scholar-adventurer whose life is depicted in the movie, *Lawrence of Arabia*. With help from the British, the revolt frees Palestine, Syria, and Lebanon from Turkish forces.

Meanwhile, two European officials—Mark Sykes of Britain and Georges Picot of France—negotiate secretly to carve up parts of the Ottoman Empire. Under this plan, Britain is to receive southern Mesopotamia, while France gets a protectorate over Syria (including Lebanon) and the southeastern part of Asia Minor. The agreement, signed in Petrograd, Russia, enrages Arab leaders who consider it a betrayal of the earlier agreement.

Statehood

1920 France draws Lebanese borders

When World War I ends, the Peace Conference at Versailles rejects British and French annexation of Arab nations. Instead, the League of Nations, which is set up after the war, develops a mandate system in which parts of the Middle East are divided into regions to be governed by the West.

Through this system, France receives a mandate over Syria and Lebanon. French rulers divide an area that they call Greater Lebanon (Grand Liban) from Syria in order to weaken Muslim nationalist aspirations in the larger country and to appease Maronites who see Lebanon as a Christian enclave that can protect their political power and identity in the Muslim-dominant Middle East.

The boundaries, however, transform the idea of Lebanon as a mountainous, agricultural area centered around Mount Lebanon, where Christians live, into a much larger entity that includes coastal areas and an almost equal number of Muslims and Christians. The new borders bring in a variety of urban communities that thrive on commerce, causing a rift in the Maronite ideal of Lebanon as a nation of rural villages. The boundaries established by France remain the official state boundaries of the Republic of Lebanon.

French rulers also establish Beirut as the Lebanese capitol, shifting the country's focal point from the mountains to the sea. The previous Ottoman capitol, Beit Eddine, becomes primarily a summer residence for Lebanese rulers of state.

French leaders build major roadways, enlarge the Beirut harbor, introduce a new judicial system, and upgrade schools and public health systems. However, much of this effort is designed to weaken Arab nationalist sentiments. Under the French mandate, Sunni Muslims lose the traditional ruling authority they had enjoyed under the Ottomans, and Maronites rise to prominence. Maronites speak and do business primarily in French, shrugging off their ties to the Arab world. Over the choice of Arabic as a national language, French becomes compulsory in schools.

1920–21 Cultural institutions open

As French leaders work to transform Beirut, the capitol city increasingly becomes a center for Lebanese national culture. The National Museum of Lebanon is founded in Beirut in 1920. Its collection consists of historical documents and antiquities such as the sarcophagus of King Ahiram on which are preserved some of the first-known alphabetical inscriptions. The National Library of Lebanon opens one year later.

1926 Constitution is established

A constitution is established which holds that Lebanon is an independent republic, with executive power vested in a president who is chosen by parliament for a single, six-year term. The constitution requires that the president be a Maronite Christian and the prime minister be a Sunni Muslim. Shi'ite Muslims make up the third largest community in Lebanon at this point, and it is determined over time that the position of Speaker of Parliament will be filled by a representative from their community.

1941–46 Lebanon gains independence

After France falls to the Nazis in World War II, Hitler establishes what is known as the Vichy government in which French loyalists continue to maintain French colonies but answer to German rulers. During this regime, an Allied force that includes British and Free French fighters moves into Lebanon and declares independence for the country. Fighting breaks out between Vichy and the Anglo-Free French troops.

After Free French troops emerge victorious, a Lebanese national government is set up, and Bishara al-Khuri is elected president. The government proves to be more nationalistic than the Free French would like, and on November 8, 1943, al-Khuri and several nationalist leaders are arrested. They are released on November 22, a date that Lebanese observe as a day of national celebration, and one month later, a Franco-Lebanese agreement that formally transfers power from France to the newly-independent Lebanon is signed and put into effect.

Lebanon establishes its national flag. It consists of two red horizontal stripes with a wider white stripe in between. A green- and brown-colored cedar tree is placed at the center of the white stripe.

The last Allied troops leave on December 31, 1946.

1945: March Arab League is formed

Representatives from Egypt, Iraq, Saudi Arabia, Syria, Lebanon, Jordan and Yemen establish the Arab League, an organization that works to strengthen friendship between member states and to coordinate political actions in the international sphere. Also known as the League of Arab States, it meets twice a year and comes to play a significant role in mediating regional disputes. Other Arab nations—Libya, Sudan, Tunisia, Morocco, Kuwait, Algeria, Yemen, Bahrain, Qatar, Oman and the United Arab Emirates—eventually become members.

1945: October 24 Lebanon joins U.N.

Lebanon becomes a charter member of the United Nations.

1948 Arab-Israeli war begins

Regarding the complicated Arab-Israeli war, Israel's official state history contends that Arab states—Lebanon, among them—opposed the creation of the state of Israel in Palestine and invaded the new nation immediately after it gained independence in 1948. Israelis say that the indigenous Palestinians who fled from Israeli-occupied areas did so out of a misguided terror that Arab leaders in the rest of the Middle East helped to encourage.

However, more recent historians present a different picture, one that shows massive numbers of Jews migrating into Palestine in the early twentieth century, generating fear among Palestinians that they would be pushed out of their homes. Zionist propaganda—which was trying to generate support for an Israeli Jewish nationalist movement—fed this fear by describing the land into which Jews from all over the world were moving as being "barren" and "empty" of people. Zionist groups also launched massive massacres in several Palestinian villages. In one notable incident, Zionists went from house to house in the village of Deir Yassin, killing Palestinians.

Israelis describe the 1948 war as a battle of independence. For Palestinians, it marks the beginning of decades of suffering. For neighboring nations such as Lebanon, the war marks the beginning of decades of regional hostilities and of a need to accommodate thousands of displaced Palestinian refugees.

1952–56 Economic prosperity grows

Increased U.S. aid flows into Lebanon and a variety of Western commercial enterprises give the young nation's economy a boost. Growing oil royalties also benefit the country; it remains calm, despite turmoil throughout the Middle East.

The Beirut Stock Exchange opens, and in 1956, a new law that prohibits banks from disclosing the details of their clients' business takes effect. These measures make Lebanon an attractive spot for foreign investment.

1958 Civil war breaks out

President Camille Chamoun reportedly tries to seek a second term. Because the Lebanese constitution prohibits this, Chamoun's action leads to a brief civil war. The United States sends troops into the country. The United States justifies its action with the Eisenhower Doctrine, which pledges to offer U.S. military and economic aid to any country that requests it in order to fight a communist threat.

General Fuad Chehab, commander-in-chief of the Lebanese army, is elected president in July, with support from governmental and opposition troops. A former member of the French army, he serves as president for six years. Under his regime, Lebanon gradually loosens its traditionally tight relationship with the West and cultivates its Arab ties.

1958–59 Diplomat serves as UN president

Charles Habib Malik (1906–87) serves as president of the 13th UN General Assembly. Malik is among Lebanon's most notable statesmen and is particularly known for his role in writing and adopting the United Nations' Universal Declaration of Human Rights, which defines and outlines the scope of basic individual rights. His service with the United Nations includes one year (1948) as president of the Economic and Social Council and one year (1953) as president of the Security Council.

He also serves as Lebanon's Minister for National Education and Fine Arts (1956–57) and as the country's Minister of Foreign Affairs between 1956 and 1958. While in these positions, he works with President Chamoun to develop a policy that encourages relationships with the West and keeps a distance from Egypt while cultivating a tie with the Arab world.

Born in Bitirram, Al-Khoura, Malik received degrees from the American University of Beirut in 1927 and Harvard University in 1937. He teaches mathematics, physics, and philosophy at AUB, and has written several books. A few titles include: *War and Peace* (1950), *Problem of Asia* (1951) and *Man in the Struggle for Peace* (1963).

1960 Arab poet settles in Lebanon

Known as one of the Arab world's most famous poets, Adonis chooses to become a citizen of Lebanon. Born in a village in Syria in 1930 as Ali Ahmed Said, he chose to move to Beirut in the mid-1950s.

Adonis is best known for his contributions to modern Arabic poetry. He started a poetry review in 1957 known as *Shi'r* and launches a literary journal known as *Mawaqif* in 1968. His work and his ideas are aimed at advocating liberty, creativity, and progress in Arab life.

1962 *September Birds* is published

Lebanese writer, Emily Nasrallah, publishes her novel about Lebanese village life while working and studying in Beirut. Lebanese high school children often read the book, known in Arabic as *Tuyur Ailul*. Nasrallah was born in 1938 in southern Lebanon and attended school at universities in Beirut. She has written several novels and short stories, many of which have been translated into English. She considers a novel she published in 1981 known as *Al-Iqla aks al-Zaman*, or *Flight Against Time*, to be a sequel to *September Birds*.

1963: July Green plan takes effect

Even though Lebanon is industrializing rapidly, agriculture remains an important part of its economy. Citrus fruits, bananas, and sugarcane grow in coastal areas, while the terraces on Mount Lebanon are used for cultivation of olives and cereals. In higher elevations, the country produces grapes, apples, peaches, pears and figs. The Beq'a Valley produces cereals, potatoes, chickpeas and beans, as well as wheat.

Despite the abundant rainfall that the coastal areas and mountains receive, some parts of the country—particularly the Beq'a Valley—suffer water shortages. The Green Plan is designed to alleviate these shortages by calling for construction of such projects as dams and irrigation canals.

1969 Government signs Cairo accord

Fighting in Israel between Jewish rulers and Palestinian guerrillas leads to clashes between guerrillas amassed at Lebanon's border and the Lebanese army. Hoping to avoid another civil war, the Lebanese government signs the Cairo Accord with the Palestinian Liberation Army (PLO) that allows Palestinian fighters to establish military bases within Lebanon and to launch cross-border raids into Israel, Lebanon's neighbor to the south.

1973: October Central bank's power is increased

The Code of Money and Credit is put into effect. Amendments to this law give the Bank of Lebanon (established April 1, 1964) a great deal of power to regulate and control commercial banks and other institutions in the country and to implement monetary policy. This measure follows a 1968 move to reduce the number of commercial banks in Lebanon from ninety-three to seventy-four, which also had increased the Bank of Lebanon's authority.

Lebanese Civil War (1975–90)

1975 War erupts

Like many conflicts that escalate into full-scale war, the fifteen-year civil war that engulfs Lebanon, claiming 150,000 lives, has origins that might seem obscure. Some scholars, such as Latif Abul-Husn, believe the fighting results from two unrelated conflicts.

First, the Lebanese army clashes with a rally in Sidon where a fishing monopoly was under protest. This conflict causes a former parliamentary deputy who was participating in a rally to be killed. Protesters retaliate against the army, and as these clashes intensify, the country slips toward war. Second, a group of right-wing Christian Phalangists ambushes a bus filled with Palestinians in east Beirut on April 13, killing twenty-seven. After this massacre, the PLO retaliates with attacks on Christians.

In the meantime, Israel attacks PLO bases in Lebanon, and Palestinians retaliate with strikes against Israel as well as several Maronite Christian-dominant villages. Over time, Muslim and Christian groups form militias, and the country slips toward civil war. By the early 1980s, more than thirty militias operate in Lebanon.

According to Abul-Husn, three major issues dominate the fighting: the political imbalance in the Lebanese structure of government that favors Christians, the identity of the nation, and the government's overall ability to hold the nation-state together. Maronites continue to regard Lebanon as their Christian homeland, while Muslims seek a greater Pan-Arab, pan-Islamic identity. Because the government's rule often is rendered ineffectual by the various militia, several nations often intervene in an attempt to maintain law and order. The most notable foreign mediator is Syria, which historically had been governed with Lebanon until France redrew the nation's borders in 1920. Israel, Iran, Libya, Iraq and the Palestinian Liberation Authority also play key roles in the war, sending in troops, launching attacks, offering support to militias, and negotiating for the release of hostages.

1977–82 Business networks grow

Despite continued outbreaks of violence, Lebanon remains a desirable area to foreign investors. After parliament passes legislation that allows for the establishment of a banking free zone, many overseas banks open headquarters or branches in the country.

Other Arab nations bolster Lebanon's economy by contributing $2 billion to help the country rebuild after previous fighting. In 1982, Arab shareholders decide to locate the headquarters for the Arab Stock Exchange Union in Beirut, reflecting Lebanon's importance in regional trade.

1982: June Israeli attack intensifies fighting

PLO rocket attacks on northern Israel and a Syrian installation of anti-aircraft missiles in the Bekaa Valley prompt Israel to launch a full-scale invasion of Lebanon. Israeli troops destroy PLO bases in the south, disable the Syrian missiles, and begin a two-month siege of West Beirut where many displaced Palestinians are encamped.

After this Israeli invasion, Arab nations who had agreed to contribute $2 billion to Lebanon's post-war reconstruction effort decide to withhold their money until the Israeli forces withdraw. At this point, only $381 million has been given to Lebanon.

After Israel, the PLO and Syria agree to a truce, more than 14,000 Palestinian and Syrian fighters are removed from Lebanon. British, French and Italian soldiers, along with U.S. Marines, establish a peacekeeping force around Beirut.

The assassination of Bashir Gemayel, a Christian Phalangist whom the Lebanese parliament had elected president, brings a shaky peace truce to an end. Israel quickly moves into West Beirut and kills tens of thousands of Palestinian resistance fighters.

After Phalangist forces are allowed into two refugee camps, hundreds of Palestinian civilians are massacred.

1983: April–October Militias attack Westerners

In April, the U.S. Embassy in Beirut is bombed. This event is one of several attacks on Westerners in and around Beirut. By this point, there are more than thirty militias in Lebanon. These private armies are fighting Israeli and Syrian troops, Lebanese government forces, and, often, each other. By the early 1980s, Muslim and Druzes forces increasingly turn their wrath toward Westerners.

Muslim and Druze forces begin to attack Israeli and Syrian troops which remain in Lebanon.

On October 23, a bomb explodes in a truck in a U.S. marine barracks at the Beirut International Airport, killing 241. A similar bombing in a French paratroop barracks leaves an additional 58 dead.

1984 AUB president is killed

Dr. Malcolm Kerr, president of the American University of Beirut is assassinated as militia groups and religious factions increasingly target Westerners. The university moves its operations to a New York office, which leads to massive administrative problems and considerably slows activity at the once thriving institution.

As the government deteriorates, France, the United States, United Kingdom, and Italy evacuate ground troops and nonessential personnel. A reconciliation conference ends without achieving any substantial agreement.

1986 National park is established

Ongoing fighting begins to take its toll on Lebanon's once lush natural environment. By the mid 1980s, four of its indigenous mammal species, fourteen of its bird species and five of its plant species are endangered. The National Preservation Park of Bte'nayel is established in the Byblos region in hopes of protecting some of the country's wildlife. Despite this move, industrialization and the war leave Lebanon with a legacy of water contamination and air pollution.

1987: January Foreigners are kidnapped

A series of kidnappings take place in and around Beirut. On January 13, a photographer covering hostage negotiations undertaken by Terry Waite, a special envoy of the Anglican Church, is kidnapped. Two kidnappings of West German nationals follow within a week. Three Americans and one Indian are kidnapped from Beirut University College on January 24. The four are teachers at the college. Two days later, gunmen abduct two more men in Beirut.

Special envoy Terry Waite is last seen on January 20. The Anglican Church's Archbishop of Canterbury in London tells reporters on February 1 that Waite left instructions that if he were taken hostage in Lebanon, no attempts be made to rescue him or exchange money for his release. This announcement is followed by one from Shi'ite Amal, leader Nabih Berri, and Druze leader, Walid Jumblat, both say Waite has been "arrested" by the same Shi'ite faction that is holding three Americans and other foreigners hostage.

1987: May 21 PLO agreement is revoked

As fighting continues, the Lebanese parliament repeals an agreement it had made with the PLO that allowed it to establish military bases in Lebanon. This so-called Cairo Accord had been in place since 1969.

1988 *House of Many Mansions* is published

As Lebanon continues to be battered by militias within the country while also being caught in the crossfire of ongoing Arab-Israeli fighting, historian Kemal Salibi receives a death threat from the Hezbollah, which forces him to take refuge in Amman, Jordan.

While in exile, Salibi writes *A House of Many Mansions: The History of Lebanon Reconsidered*, an inside look at the civil war as well as the strength and resilience of Lebanese nationalist sentiment. In an interview with William Dalrymple, Salibi explains that he wrote the book entirely from memory. He had no notes or reference books; all of these sources were destroyed when Maronite and Phalange forces bombed his home.

Salibi's experience illustrates the tangled sense of community that begins to evolve out of the incessant warfare. A professor at the American University of Beirut, Salibi is a Maronite Christian. Yet, during the war, he makes his home in a Muslim-dominant section of West Beirut. He had been in the basement as Christian militias began their bombing. The house, built in an Ottoman style by his grandfather, takes twenty-six hits and is reduced to rubble.

1988: September Succession crisis occurs

Amin Gemayel's term as president is set to expire on September 22. The Lebanese parliament is unable to decide on a successor so Christian and Muslim leaders meet in an emergency gathering in Syria.

Gemayel steps down after appointing a six-member Cabinet to rule the country until a new president is named. The group is headed by Major General Michel Aoun, a Maronite Christian, and includes three Muslim military leaders.

After the three Muslim leaders refuse to participate in the Cabinet named by Gemayel, Muslim leaders name a separate cabinet, headed by Prime Minister Selim al-Hoss.

Aoun remains head of the Cabinet appointed by Gemayel, and this group retains support among Lebanon's Christians. The two Cabinets rule in competition with each other.

1988: December 16–20 Red Cross pulls out of Lebanon

Peter Winkler, an official with the Swiss Red Cross, is freed December 16 after being held in captivity for nearly one month. After workers with the International Red Cross, based in Switzerland, receive death threats, the organization suspends its operations in Lebanon. This marks the first time in 125 years that the Red Cross pulls out of a country because of threats made to its workers.

1989: October Taif Accord is approved

Christian and Muslim militias as well as Syrian soldiers fight in and around Beirut through much of 1989. The Arab League attempts to impose cease fires, which fail to quiet the hostilities. Dozens of lives are lost and many injuries are suffered.

Despite a United Nations call for a cease fire, the fighting continues. Finally, Lebanese Army General Michel Aoun, the leader of the Lebanese Christian militia, agrees on September 22 to accept a cease-fire proposal from the Arab League.

The proposal goes into effect the following day, and by September 30, a peace conference begins in Ta'if, Saudi Arabia. This meeting includes representatives from Saudi Arabia, Algeria and Morocco as well as Lebanese legislators who represent both Christian and Muslim groups.

The conference produces the Ta'if Accord, which the Lebanese parliament approves. Among other things, the agreement allows for the presence of Syrian troops in Lebanon until full security is restored and blames Israel for much of the past fourteen years of violence. Most importantly, the pact affirms that Lebanon is an Arab state by identity and affiliation, an agreement that many observers say caused Maronite Christians to be the losers of the war but has allowed Lebanon to sustain a prolonged sense of peace. Aoun rejects the pact.

1989: November New government is formed

After Aoun orders that the Lebanese parliament be dissolved, members of Lebanon's parliament meet on November 5 in Qlailaat, a remote village in northern Lebanon, and elect Rene Moawad as president. The United States and Syria approve this election. After one week, Moawad appoints Selim al-Huss as prime minister and asks him to form a government.

This effort fails when Moawad is killed November 22 by a remote-control bomb that explodes in his motorcade. An organization known as the Christian Solidarity Front, which is loyal to Aoun, threatens that Lebanese legislators also will die if they attempt to elect a new president.

Despite this threat, Lebanese legislators meet in Shtaura, a town controlled by Syria, and choose Elias Hrawi, a Maronite Christian, to succeed Moawad as president. He also chooses al-Huss as his prime minister.

Aoun disputes the election and mobilizes his Christian militia troops. Thousands flee Beirut because they fear more fighting will ensue.

1990 Writer captures war experience

The civil war becomes such a formative aspect of Lebanese society that much of the country's cultural history deals with the fighting in one form or the other. Among many artists who try to capture the war experience is Jean Said Makdisi, a Palestinian-Lebanese woman who refuses to leave Beirut even as incessant bombings gut the very heart of the city. Makdisi's account, *Beirut Fragments: A War Memoir*, uses a variety of literary styles to describe her love for the city and her pain over what is happening to it.

1990: August 22 Parliament amends constitution

The Lebanese parliament alters the nation's constitution to give Muslims more political power. Under the new agreement, the President will continue to be Christian, but many of that position's powers are to be transferred to a Cabinet comprised equally of Christians and Muslims. The Prime Minister will continue to be Muslim and will countersign all presidential decrees.

1990: October Aoun surrenders his troops

Gunmen open fire October 1 on a rally in East Beirut, held by supporters of Aoun. Twenty-five people die.

After this event, President Elisa Hrawi asks the Syrian military to help force Aoun out. Syrian and Lebanese troops attack the house in East Beirut where Aoun is hiding on October 13 and force him to surrender. He takes refuge in the French Embassy in Beirut. Although Hrawi wants him to stand trial in Lebanon, France grants Aoun political asylum. At least 750 deaths are reported in the overthrow.

Aoun's surrender allows the terms of the Ta'if Accord to be put in place and brings about an end to the worst of the fighting.

1990: November 10 Militias pull out of Beirut

After Amal and the Party of God, the two major Shi'ite militias in Lebanon, sign a peace agreement, the Lebanese government sets a deadline to begin disarming all remaining militias in and around Beirut. The militias are given nine days to withdraw and are required to disband by March, 1991.

The militias begin withdrawing from Beirut on November 10.

Return to Peace: 1991– Present

1991 Refugees are allowed work permits

The Lebanese government agrees to allow Palestinians to receive work permits in Lebanon. About 300,000 Palestinian live in the country, and many of them are subject to widespread discrimination. They cannot vote, and they receive only limited health care and educational benefits. Many Palestinians also face human rights abuses and discrimination in the job market.

1991: February 6 Lebanon moves into security zone

Israeli troops and Palestinian guerrillas clash in the so-called security zone in southern Lebanon. After several days of fighting, about 1,000 Lebanese soldiers move into the area in an attempt to re-take it. This is the first time since 1978 that the Lebanese government has attempted to assert military control of the area.

The Lebanese government also arrests four Palestinian guerrilla officials, including Walid Khalid, spokesman for the PLO's Fatah Revolutionary Council.

1991: April 29–May 2 Lebanon opens coastal road

A powerful Christian militia known as the Lebanese Forces announces it will disarm and return territory it controls in the Kesrouan Mountains to the Lebanese army, in accordance with the 1989 Taif Accord. At the same time, Druze militia forces in the Shuf Mountains east of Beirut surrender their weapons to authorities.

With these militia forces finally subdued, the Lebanese army takes control of the Shuf Mountains on May 1, as well as Christian enclaves to the north and the east of Beirut. This action allows the government to remove barricades and reopen the coastal road connecting north and south Lebanon for the first time since 1975.

1991: May 22 Lebanon signs treaty with Syria

Lebanese President Hrawi and Syrian President Hafez Assad, meeting in Damascus, sign a Treaty of Brotherhood, Cooperation, and Coordination that lays the groundwork for joint defense, economic and foreign policy institutions and establishes a council that includes the presidents and prime ministers of both countries. This higher council is vested with the authority to make binding policy decisions.

Through this agreement, about 35,000 Syrian troops remain in Lebanon through the late 1990s. The Lebanese Parliament ratifies the treaty on May 27.

1991: July 1–7 Government disarms PLO groups

Palestinians initially refuse to meet a July 1 deadline set by the Lebanese government to relinquish control of the area east of Sidon. The government sends troops to the area on July 1 and subdues the guerrillas.

On July 5, Al Fatah, the main branch of the PLO, announces it will surrender its final bases in southern Lebanon and that two other Palestinian groups—the Popular Front for the Liberation of Palestine and the Democratic Front for the Liberation of Palestine—will relocate to the Bekaa Valley.

On July 7 the Lebanese army formally disarms the Fatah Revolutionary Council, which is led by Abu Nidal. By this point, most of the militias operating in Lebanon have either disbanded or left the country. The pro-Israeli South Lebanese Army, however, continues to launch attacks in the Security Zone.

1991: August 26 Amnesty agreement is approved

The Lebanese Parliament ratifies a measure that provides for general amnesty on war crimes committed over the past fifteen years. In a subsequent action, the parliament adds a paragraph to the Lebanese Constitution that emphasizes that the country guarantees freedom of opinion and religious belief, and social justice to all of its citizens.

The amnesty agreement directly affects General Michel Aoun, who leaves Lebanon after spending ten months in the French Embassy in Beirut. Hrawi promises him amnesty and a safe passage out of Lebanon if Aoun agrees not to take part in anti-government activities.

1992: August–October 22 Controversy mars elections

Lebanon holds its first elections in twenty years. Despite an assurance spelled out in the Taif Accord that Muslims and Christians would be equally represented in Parliament, Christians boycott the elections.

Three large Christian parties call for a general strike on August 21 to protest the presence of Syrian troops in central Lebanon, and many Christians refuse to participate in voting.

Two Muslim parties—Amal and the Party of God—win twenty-two of the twenty-three contested seats in Parliament. Because of boycotts, Christians win only fifty-nine seats in the 128-member Parliament. The elected Christians ran either as independent candidates or on Muslim party tickets.

On October 20, the newly-elected Parliament chooses former Shi'ite Amal militia leader Nabih Berri as its speaker, a move that reflects the political power that Shi'ites—historically an impoverished and overlooked group—command now in Lebanon. Two days later, President Hrawi appoints as Lebanon's prime minister Rafik al-Hariri, a Sunni Muslim and a billionaire.

1993 Archaeological excavations begin

Solidere, a quasi-public construction company, begins an $8 million project to rebuild central Beirut. Archaeologists join in by digging in the war rubble. They uncover remains of nearly 5,000 years of civilization.

These diggings suggest Beirut was founded before Athens, Greece, Damascus, or Jerusalem, making it one of the world's oldest cities. Archaeologists find traces of Canaanite, Phoenician, Hellenistic, Roman and Ottoman civilizations. Some key finds include a Canaanite hill, a Phoenician quarter, a Byzantine trading area, a Greek quarter and Roman ruins.

Archaeologists intend to preserve the ruins as part of the futuristic design that Solidere is developing for Beirut and to add such features as excavated mosaics to covered markets, or *souks*, being developed for the city.

1993: September 13 PLO-Israeli peace accord is signed

Representatives of the PLO and the Israeli government sign a historic peace agreement in Beirut. Despite this move, violence between factions loyal to the government and those who support the Palestinian movement continues. Lebanese troops fire on protesters of the peace accord, killing eight and wounding thirty-five. Most of those injured are members of the Party of God militia.

The Party of God reacts the following day with attacks against Israel in the security zone.

1994 Horizon 2000 plan is unveiled

Prime Minister al-Hariri initiates Horizon 2000, a plan to rehabilitate Lebanon's telecommunications, electrical, transportation, sewage, water, harbor, education and housing programs, renovate Beirut International Airport and turn the country into an industrial hub.

Hariri sets up a government agency, the Investment Development Authority of Lebanon, to oversee the plan and work to attract private investment. Among this agency's tasks is the job of upgrading Lebanon's forty-four industrial zones and overseeing the construction of fifteen new zones. Under the plan, the Lebanese government wants to use these areas of land to attract foreign investment by providing the necessary factories, warehouses, and offices that will appeal to businesses as well as the required roads, utilities and other infrastructure.

The plan envisions spending $17 billion over twelve years.

1994 Rule against women is loosened

Although Lebanon tends to confine women to traditional, subordinate roles, a new law nullifies a requirement that a wife must get her husband's permission to start a business or engage in commerce. Despite this change, many religious laws that govern status discriminate against women. Additionally, children born to Lebanese women cannot become citizens if their fathers are foreigners.

1994: March 22 Lebanon expands death penalty

President Hrawi signs a measure passed by the Lebanese Parliament that makes the death penalty mandatory for persons convicted of first-degree murder. The new law also allows for the death penalty in cases of political assassination.

Amnesty International, which monitors human-rights abuses throughout the world, sees the broadening of the death penalty in Lebanon and an increase in executions that follow from it as a negative trend. The organization notes that the Lebanese court system only sanctioned three executions from 1959 through the imposition of the new law, even though it acknowledges that Lebanon's government had little control over the country during its years of civil war. According to a 1998 report from Amnesty International, Lebanese courts have sentenced nearly twenty-seven people to death since the penalty's expansion.

1995: July Police beat demonstrators

The General Workers' Union calls for a general strike on July 19 to protest high prices, taxes, and many of the Lebanese government's post-war economic policies. Government security forces arrive at a protest rally and arrest about 200 participants. They are enforcing a ban on public demonstrations that has been in effect since 1993, but in doing so, they beat many of the protesters. Additionally, many of those arrested are Christians, who are increasingly becoming a minority group in a Muslim-dominant country.

Amnesty International sees the beatings as an ongoing concern of human-rights violations in Lebanon. While the organization applauds Lebanon's efforts to comply with international human-rights treaties, it suspects that the freedoms the country guarantees in its constitution may not be carried out in actual practice. Others who have studied Lebanon, including AUB historian Kemal Salibi, worry that Christians are being unduly marginalized in contemporary Lebanese society because, in part, of the role that Maronite-backed militias played in escalating the civil war.

1995: September Beirut Stock Exchange reopens

The Beirut Stock Exchange opens after being closed for twelve years. It will be several months before trading can actually take place, however (see 1996: January).

1996: January Trading begins

Trading begins on the Beirut Stock Exchange (see 1995: September) for the first time in twelve years. The three companies listed produce either cement or construction material. A fourth company joins the exchange later in the year.

1996: April Israel launches air raid

In an attack that has a code name, "Grapes of Wrath," Israel launches a seventeen-day air raid and ground attack into Lebanon. The nation's first full-scale attack on Lebanon in nearly fourteen years, the aggression conveys a message that despite Lebanon's attempts to rebuild, efforts to achieve peace remain elusive. The attack kills more than 200 civilians, including many who reside at a U.N. base in Qana.

Although Lebanon runs schools, courts, and police operations within the so-called security zone that separates it from Israel, Israeli fighters and their South Lebanese Army continue to exceed the nation in military might.

1996: December 4 Casino du Liban reopens

Operators of Lebanon's once famed Casino du Liban spend $50 million to reopen the gambling hall located about twenty kilometers north of Beirut. This casino, the largest in the Arab world, was closed in 1989 when fighting around the complex made it impossible to operate safely.

First opened in 1959 the casino attracted such notable celebrities as American singer, Frank Sinatra, and U.S. President Lyndon B. Johnson. Its new clientele is expected to include gamblers from the oil-rich Gulf States. However, it requires that gamblers earn at least $20,000 (U.S.) a year, which puts its gaming tables out of reach for more than two-thirds of the Lebanese populace.

1997 Horizon 2000 target is changed

The Lebanese government moves the target date for completion of the country's reconstruction from 2000 to 2007. Due to continued aggression from Israel and ongoing unrest in the country, Hariri is forced to limit the plan's spending budget from an initial $17 billion to $5 billion.

As the government pumps money into such projects, many protest that the average Lebanese citizen is being neglected. This frustration shows in a rally led by Sheik Soubhi Toufayli, a former Hezbollah chief. The rally, called "Revolution of the Hungry," protests the government's economic policies and wins support from labor unions.

1997: July U.S. lifts travel ban

The U.S. Department of State lifts its 1985 ban on U.S. travel to Lebanon. As institutions such as the American University of Beirut rebuild after the long civil war, Lebanese officials express optimism.

Bibliography

Abul-Husn, Latif. *The Lebanese Conflict: Looking Inward.* Boulder, CO: Lynne Rienner Publishers, 1998.

Adonis. *Transformations of the Lover.* Translated by Samuel Hazo. Athens, OH: Ohio University Press, 1982.

"Human Rights Development and Violations," *Amnesty International Country Report,* Lebanon, October 1997, www.amnesty.org

Badran, Margot and Cooke, Miriam. *Opening the Gates: A Century of Arab Feminist Writing.* Bloomington, IN: Indiana University Press, 1990.

Bleaney. C.H. *Lebanon: Revised Edition. World Bibliographical Series,* Volume 2, Oxford, Eng.: Clio Press, Ltd., 1991.

"Chronology," *Middle East Journal.* Vol. 52, No. 2, Spring 1998.

Dalrymple, William. *From the Holy Mountain: A Journey Among the Christians of the Middle East.* New York: Henry Holt & Co., 1997.

Doyle, Paul. "Betting on the Future," *The Middle East,* No. 269, July–August, 1997, p. 42.

Fisk, Robert. *Pity the Nation: Lebanon at War.* London: Andre Deutsch Limited, 1990.

Haravi, Mehdi, editor. *Concise History of the Middle East.* Washington, D.C.: Public Affairs Press, 1973.

King-Irani, Laurie. "War Gods Roar Again, Appear Unstoppable: Jet Streaks So Close She Could See Pilot," *National Catholic Reporter.* Vol. 34, No. 17, p. 9. February 27, 1998.

Makdisi, Jean Said. *Beirut Fragments: A War Memoir.* New York: Persa Books, 1990.

Norton, Augustus Richard. "Hizballah: From Radicalism to Pragmatism?" *Middle East Policy.* Vol. 5, No. 4, pp. 174–186. January 1998.

"Lebanon's Choreographed War-Dance," *The Economist.* Vol. 344, No. 8042, p. 52, November 8, 1997.

"Proud Traders: Lebanon," *The Economist.* Vol. 344, No. 8033, pp. 46–47, September 6, 1997.

Salibi, Kamal. *A House of Many Mansions: The History of Lebanon Reconsidered.* London: I.B. Tauris & Co. Ltd., 1988.

"Thinking the Holy Unthinkable: Lebanon," *The Economist.* Vol. 346, No. 8061, p. 45, March 28, 1998.

Trendle, Giles. "Lebanon Looks to Industry," *The Middle East.* No. 270, p. 26, September 1997.

Tuttle, Robert. "Americans return to AUB," *The Middle East.* No. 272, p. 42. November 1997.

"Understanding: Lebanon, Syria and Israel," *The Economist.* Vol. 344, No. 8031, p. 36, August 23, 1997.

Malaysia

Introduction

The modern nation of Malaysia is a group of states located on the Malay Peninsula, extending south from the mainland of Southeast Asia and located along the northern coast of the island of Borneo. After ages of past conflict and conquest, these states have found themselves linked by history and economics in the latter half of the twentieth century.

The people now known as the Malays migrated to the region from the north by land and the east by sea. Related by language and culture to Indonesians, they intermingled with many other ethnic groups and lent their name to a language, *Bahasa Malay*, and, eventually, a nation, Malaysia. The present government of Malaysia is prone to emphasizing distinctions between the Malays and later arrivals such as people whose ancestors immigrated from China and India during nineteenth–century British colonization. They term the Malays *Bumiputra* ("sons of the soil"), also granting that status to tribal people of Borneo and the inner mountains of the Peninsula.

These people have always mixed and traded with other natural seafarers. The location of Borneo on sea routes south of China and the setting of the Peninsula, defining the straits where ships go from Pacific to Indian Ocean waters, has created a history of cultural tides and flowing wealth. The monsoon winds and rains that sweep back and forth across Asia guided sailing ships into ports where they could exchange goods of value. The jungle interiors yielded exotica like delicate birds' nests for gourmet soup and the ivory of hornbill beaks and elephant tusks; now, centuries later, vast stands of plantation trees ooze rubber and palm oil. The earth held tin ore in great quantities, and the seabed gave forth high-quality crude oil. Control of trade lent immense power to Malay sultans, and disruption of it was a way of life for infamous pirates. The Islamic religion, practiced by the majority of Malaysians today, arrived via seafarers from the Middle East in ships brought along by the trade winds.

The Malay rulers, the sultans, acquired Hindu and Muslim trappings of divine power as they warred with each other and with neighbors like the Thais, Sumatrans, and the Sulus of the Philippines. Their pride and daring then came face to face with the Europeans, who also considered themselves divinely inspired. Rich, refined Malacca drew the Portuguese, the Dutch, and the British like moths to a flame. Britain managed the ultimate conquest, through the sleight of hand of indirect rule. Letting sultans be sultans, British administrators took actual charge of the Peninsula, and in Borneo adventurers ruled as "White Rajas" and cunning capitalists. Sublime arrogance coupled with flashes of appreciation for the special qualities of the locals kept the British ascendant until the next fortune-seekers, the Japanese, took over for a few years.

The Malays acquired a reputation for acquiescence to the authority of sultans and the self-proclaimed benevolence of British overlords. Stereotyped as easygoing and passive, the Malays nonetheless gave the world the phrase "running amok." Malay villagers would crack under societal pressure and go on murderous rampages, armed with knives or spears. This sporadic occurrence was considered to show the latent aggressive side of the local character, but more indicative of it are the periodic rebellions against British rule, not only by warlike headhunters of Borneo but by Malays of the Peninsula, who resented the imposition of outside authority as much as anyone in Europe or America ever did.

When the British colonials left following World War II, it was in an orderly process. Democratic institution were put rather firmly in place, but not without a full-scale communist jungle insurgency challenging lofty sultans and plantation capitalism. Eventually the "Red Menace" was replaced by a new enemy within—the Chinese and other ethnic groups—who might, through control of money sources and reputed hard labor, leave the Malays impoverished in their dust. Politicians would play the ethnic card relentlessly, district by district, neighborhood by neighborhood, getting elected by constituencies of "race." When this went to the disadvantage of the Malays, their ever-dominant party, UMNO, would resort time and again to suppression of "discussion" as their way to smooth over friction.

The first prime minister of Malaysia was Tungku Abdul Rahman (Tunku Abdul Rahman Putra bin Abdul Hamid Halimshah, b. 1903), an aristocratic liberal; he was followed by two more lawyers, British-trained as he was. These were Tun Abdul Razak (1922–76) and Dr. Mahathir bin Mohamed (Datuk Seri Mahathir bin Mohamed, b. 1925). Dr. Mahathir

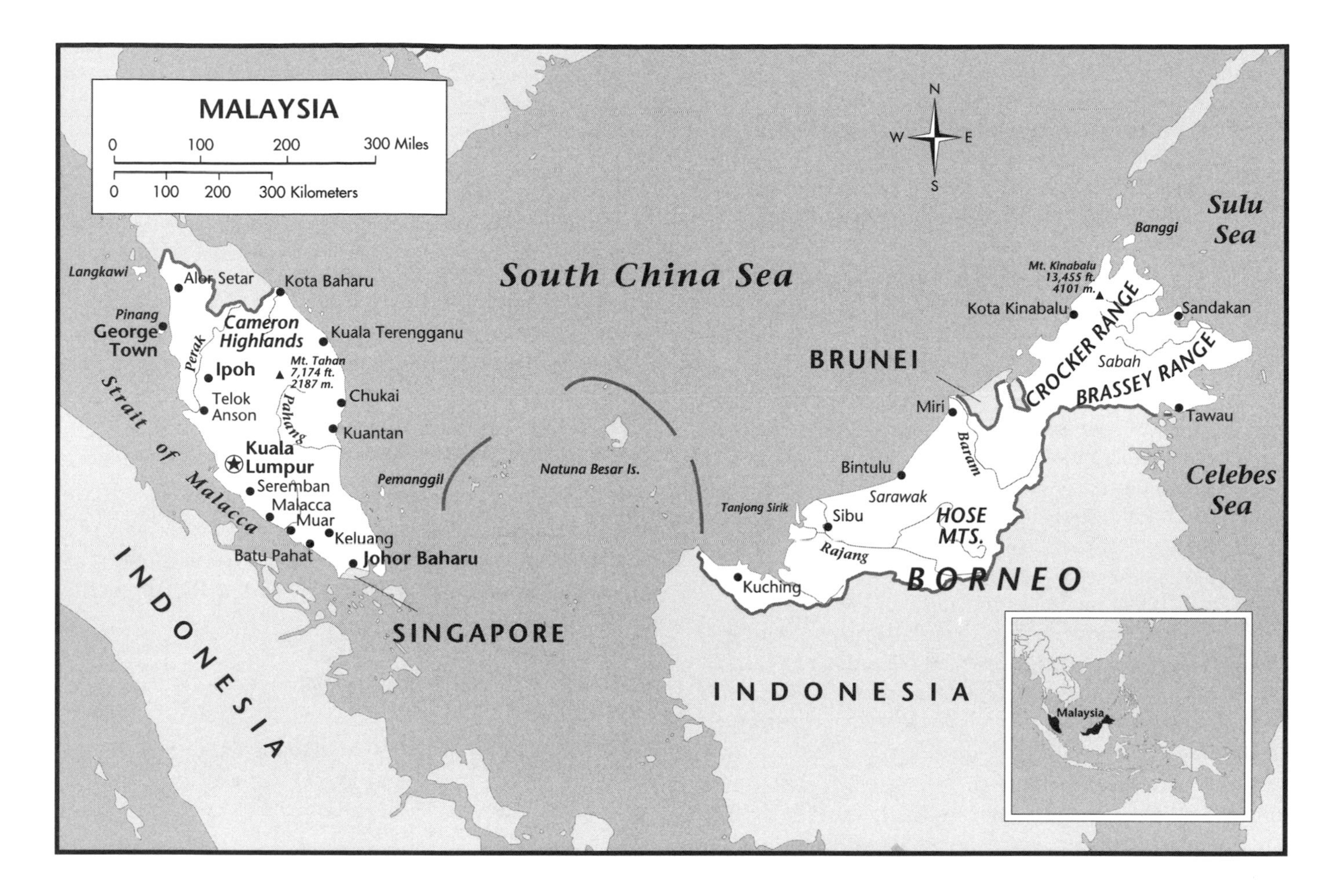

bin Mohamed was a general practitioner with a prescription of economic status for the nation. Throughout, Malaysia retained its own monarchy, with the Malay sultans of the various states waiting their turns for a five–year reign as Supreme Ruler over all. In contrast to other Asian nations, in Malaysia the military refrained from engaging in political and economic life.

Malaysia has a phenomenon of vibrant nongovernmental public interest organizations. People of all ethnicities work together in trade unions, community groups, and efforts to fight pollution and deforestation. As these activists go about their work, they are incessantly subjected to the government's use of arbitrary detention. While less horrific than imprisonment elsewhere in Asia, the days, weeks, or months of detention without trial are a serious form of harassment of people who actually may be Malaysia's greatest assets, including respected politicians like the Democratic Action Party's Lim Kit Siang and environmental activist Harrison Ngau.

Women have been nearly invisible in the political history of Malaysia, generally staying low-profile even in recent decades. But women's presence in commerce and social life is quite the opposite. Women of all classes have engaged in the buying and selling of trade goods. They have been investors, speculators, business owners, and market vendors as well as diligent farmers, tin miners, and factory workers.

Education in Malaysia is freely available for boys and girls in villages and cities. Until the recent economic downturn, many students had been going overseas for higher education, but Malaysian universities train new generations of technocrats and intellectuals back home as well. Increasingly, televisions, video-cassette recorders, and satellite dishes bring rural and urban areas closer together and enthrall Malaysians with images of the outside world. Malaysian urban culture is a whirlwind of traffic noise and air-conditioned mega-malls. Rural culture is catching up with the mood of the cities, as young people go off on journeys to offshore oil rigs or computer chip factories, returning with material goods and new notions. For the most part, Malaysians have enjoyed a very comfortable standard of living, but in 1997 the Asian economic crisis had a severe impact on the country. Gleaming concrete high-rises became hollow shells, luxury cars were repossessed, government annual growth projections were downsized, and thousands of people became suddenly unemployed, facing an insecure future. Immigrants from Indonesia and elsewhere who had provided a much-needed workforce for the preceding building boom found themselves shipped out, suddenly most unwelcome.

While Malaysia's health care is of a high standard, urban and rural pollution take their toll on the nation's health. The potential damage of the AIDS epidemic is highlighted by the efforts of Dr. Mahathir's daughter, Marina Mahathir, who is the head of the Malaysian AIDS Council. Population density is not high, but the noise and traffic jams of cars and motorcycles make Malaysia's cities seem more crowded than they really are.

The ravaging of Malaysia's forests has been the greatest tragedy of the post-independence years. Peninsular rain forests are now largely replaced with oil palm and rubber plantations and a slew of golf courses for local and Japanese players. In Sarawak and Sabah on Borneo, the world's oldest rain forests have been stripped down to embattled enclaves and encroached-on national parks. Their majestic trees have been dragged out on muddy roads and floated along silt-choked rivers to be sold cheaply as disposable plywood for Japan, the United States, and others. The tribal people of Borneo have protested the destruction with blockades, maps of their traditional lands, legal action, and world attention. But even while Borneo's rain forests vanish to relentless exploitation, Malaysian logging firms are turning to Cambodia, Africa, and South America for more raw material. International concern about global warming has not slowed their chain saws.

Known for huge exports of oil, minerals, and wood, Malaysia still strives for a cohesive national identity. The world's tallest buildings, twin spires, glitter above the skyscrapers of the capital, Kuala Lumpur. A car called the Proton Saga, made in Malaysia for Malaysians, graces traffic jams and country roads. But what Malaysia is remains elusive. Tourists are offered quaint kite-flying festivals and hotels with swooping wooden eaves and pricey pewter souvenirs crafted from local tin. Eco-tourists visit examples of Borneo's endangered tribes and then their mini-buses stop off at a center for orangutans orphaned by logging operations.

Perhaps the most attractive aspect of Malaysia is diversity experienced in everyday life. Malay *satay,* (grilled skewered meats with peanut sauce), is everywhere as are *roti,* (Indian pancakes) and Chinese Hainan Chicken Rice. And a full array of donut shops, burger franchises, and convenience chain stores abounds. Vivid Indonesian batik fashions, violent kung–fu movies, smashing punk rock, and sophisticated symphony orchestras are all available to Malaysians. If there is a "national culture," it has to take all of these things into account. Other places in the world are still coming to grips with the concept of multiculturalism, but in Malaysia it is inescapable reality. As acceptance replaces suspicion, Malaysians may find strength and comfort in their accidental wealth of ethnic juxtapositions. Asiabeat is a Malaysian "world beat" jazz-rock fusion group. It uses not only instruments associated with the peoples of Malaysia—gongs and drums—but instruments from other parts of Asia as well—a bamboo flute from Japan, for instance—along with saxophone and electric guitars. Perhaps Asiabeat could be a model for Malaysia's future as a society positioned to create a harmonious and adventurous culture out of many cultures.

Timeline

230 million B.C. The Malay Peninsula is formed

In the Paleozoic Era (230 to 600 million years ago), erosion from Southeast Asian land masses forms the Malay Peninsula, filling in the sea with sedimentary rock.

130–140 million B.C. Borneo's rain forest appears

The immense island of Borneo (288,150 square miles), formed by creases in the earth's crust, becomes largely covered by rain forests. The plant life has stayed much the same throughout history, and a variety of bird, mammal, reptile, and insect species thrive there.

120–250 B.C. Volcanic upheaval occurs on the Malay Pennisula

In the eastern part of the Malay Peninsula, during the late Jurassic and early Cretaceous Eras, volcanoes erupt, pushing mountain ridges up. The mountain chains contain layers of granite with tin embedded in it.

35,000–40,000 B.C. Modern humans live in Borneo

Homo sapiens (modern humans) inhabit Borneo, as evidenced by remains found in the Niah Caves in Sarawak.

5000 B.C. Australo-Melanesians arrive in Borneo from Taiwan

Arriving by outrigger canoe, Australo-Melanesian people reach Borneo from Taiwan and the Philippines. Their languages are related to those of present-day Pacific Islanders.

c. 2500–1500 B.C. Malays appear on the Peninsula

The ancestors of the Malays, migrating from the Asian mainland to the north, come to inhabit much of the Peninsula's coastal regions. Other inhabitants such as the Negrito people live in interior forest areas.

1st century A.D. Coastal society grows skilled in agriculture

As Malay people continue to migrate to the Peninsula and Borneo, their coastal settlements increase. The people develop the skills of rice-growing using irrigation, domesticating cattle and other animals, metal-working, and seagoing navigation.

1st centuries A.D. India influences Malay regions

As Indian seafaring traders visit, the people of the Peninsula are influenced by their culture, including Indian mythology and the religions; Hinduism, Buddhism, and, later, Islam. Coming to the Malays from India, all influence local art forms and win converts. Of these, Islam has the most widespread and lasting appeal for the Malays.

3rd Century A.D. The Chinese use the Peninsula as a trading stop

Twice each year, the wind and rain system called the Monsoon sweeps across Asia. Trading zones situated between the great power centers and markets of India and China can profit greatly. Indian traders bring their ships, sailing on monsoon winds, just as far as the Malay Peninsula, and stop there to exchange their cloth and other wares for the porcelain carried by Chinese sailing junks. Chinese traders on their way south or west use the Peninsula as a way station where they can replenish food and water or repair their vessels. For the Chinese, the Peninsula is also a convenient stopping place where they can exchange goods for those from India or the Indonesian Archipelago and then return home.

c. A.D. 350 Kedah becomes a city-state and port of call

Kedah, on the Peninsula's west coast, becomes a city-state, paying allegiance to the kingdom of Funan far to the north on the Mekong River Delta. By the seventh century Kedah is a stop on the routes of Buddhist pilgrims on their way from China to holy sites in India.

7th century A.D. A Sumatran kingdom takes over the Peninsula

A great kingdom called Srivijaya rises on Sumatra and begins campaigns of expansion in the region. As the Mekong Delta kingdom of Funan is conquered by the Khmers (Cambodians), Kedah, without its protector from the north, becomes the vassal of Srivijaya. Ships going through the passage between the Peninsula and Sumatra now must pay tariffs to Srivijaya. Control of that passage from the Pacific Ocean to the Indian Ocean, called the Straits of Malacca, is eagerly sought by the rulers of the region. By conquering an empire all around the Straits, Srivijaya becomes wealthy and powerful from taxes on the shipping trade.

13th–14th centuries The sea trade thrives but is attacked

The sea trade between China, the southern islands, India, Arabia, and beyond to Europe thrives, as spices and jungle products are exchanged for silks, ceramics, and other sophisticated goods. Pirates roam the shipping lanes in small, fast sailing crafts and attack the traders' fleets of ships, overwhelming them and looting their valuable cargoes.

1398–1400 The trading port of Malacca is founded

Sri Parameswara, a Sumatran aristocrat married to a Javanese princess, finds his way north to the Malay Peninsula, where, with a group of his followers, he takes charge of a west coast fishing village. Named Malacca after its river, the village is soon transformed into a trading port. As it is convenient for the Chinese, they prevent Malacca from being taken over by Thailand, a kingdom now in control of much of the Peninsula. Parameswara, his royal court, and the Malay people of Malacca eventually convert to Islam.

15th century Martial arts training becomes popular

Silat, a form of martial arts that probably originated in Northern Sumatra, becomes popular on the Peninsula; young Malay boys study it after their daily lessons in the Koran. According to legend, *silat* is based on observation of a flower floating in a waterfall, showing how to gracefully avoid the grasp of an attacker.

1400–1511 Malacca enjoys its "Golden Age"

Malacca flourishes as a very important trading stop, particularly in the exchange of cloth from India for spices from the Indonesian islands. The local Malay language is used by traders from all over the region to communicate with each other. The city-state of Malacca is set up as a conglomeration of enclaves where traders from different lands dwell. Each enclave has its representatives, and the local Malay sultan (an Islamic term for a ruler), reigns over them all, often using—or being used by—a powerful prime minister, who sets and impements state policies.

1407 A Chinese admiral drives out piracy

Zheng He (1371–1435), a eunuch and admiral of the Chinese Emperor, brings his fleet of massive junks to the Straits of Malacca and rids the area of pirates by destroying their hiding places. He is honored with a temple in Malacca.

1441–33 Malacca secures favorable relations with China and other lands

The sultans of Malacca journey to China to show allegiance to the Chinese Emperor. Sultan Mansur Shah of Malacca marries Hang Li Po, a Chinese princess. Also, Malacca pays tribute to Thai kings and befriends Sumatran rulers through marriage with their aristocracy and conversion to Islam.

Late 1400s Malacca becomes more powerful

Wealthy from trade, Malacca expands its territory. It has the power to demand tribute from central and southern regions of Sumatra; most of the states of the Peninsula are part of a Malaccan empire. In the royal court, arts like shadow puppet theater, dance, metalwork and woodcarving flourish. Only royalty are allowed to wear the color yellow. No one but the female aristocrats may wear golden ankle bracelets. Wearing *songket,* silk brocade woven with threads of gold, is another privilege of the royal court. A fifteenth–century romance called *Hikyat Hang Tuah* (The Tale of Hang Tuah), influenced by Indian literature, is written about the Malay aristocracy of Malacca. It recounts the adventures of Hang Tuah, a loyal bodyguard of Sultan Mansur Shah, who eventually becomes an admiral. Hang Tuah outwits the Javanese, is brave in battle, and is a devout Moslem.

Early 16th century A chronicle of Malay history glorifies Malacca

The *Sejarah Melayu* (Malay Annals) sets down the history, legends, and ceremonies of Malacca. Its tales of heroism and morality are written in the Malay language. In this book, Malacca's founder, Parameswara, is portrayed as a descendant of Alexander the Great. While romanticized, the *Sejarah Melayu* is a useful source of Malay history and the greatest classic of Malay literature. Poems, spoken aloud or sung, become popular at this time among royalty and commoners alike.

16th century Brunei rules Borneo

Brunei, a small Islamic state on the northern coast of Borneo, expands to rule the entire island and the Sulu islands of the southern Philippines. Its sultans gain great wealth by controlling trade between the rest of Borneo and China. The Chinese trade ceramic jars, glass beads, and other goods much prized by the people of Borneo for ivory, rattan, and birds' nests. These small white nests, glued together by the birds' saliva, are stuck to the walls of caves by birds called swiftlets. Chinese cooks use the fibrous nests in a very expensive soup.

1509 The Portuguese make contact with Malacca

Admiral De Sequeira brings his Portuguese fleet from Goa, on India's west coast, to Malacca. At first, the Portuguese are welcomed by curious crowds. Soon other traders, fearing Portuguese competition, instigate a military attack on the Europeans by the city's Malay administration. Portuguese are taken prisoner, and De Sequeira flees back to Goa with only three of his five ships.

1511 Malacca is conquered by the Portuguese

With the attack on De Sequeira as provocation, the governor of Goa, Affonso de Albuquerque, brings a large Portuguese fleet back to Malacca. First bombarding the city with cannons, then fighting a long battle on land against elephant-mounted Malay troops, the Portuguese capture Malacca. The new European overlords build a massive stone fort there, calling it "A Famosa" (The Famous One). The city takes on a Portuguese character with a Roman Catholic cathedral and other European-style buildings. Malacca, with its Portuguese occupiers and heavy fortifications, is attacked constantly by forces of other states, particularly Johor, a state located on the southernmost part of the Malay Peninsula.

1512 A Portuguese account of Malacca is written

Tome Pires, a Portuguese apothecary (dispenser of medicines) living in Malacca, writes a multivolume descriptive work on Asia called the *Suma Oriental.* The sixth book in his series describes the history of Malacca and its status as a thriving trading zone.

1540–1640 Malacca and Johor are attacked by Aceh

The Moslem state of Aceh in northern Sumatra attempts expansion by waging war against Malacca and Johor. The three states continue to be rivals without one gaining power over the other.

1586 Java besieges Malacca

While troops from Java attempt to starve Malacca into surrender, thousands die in the fortified city, but the Javanese are held back. Malacca faces constant military attacks by the forces of regional rulers, but manages to fend all of them off, using cannons placed atop the thick stone walls of its fort A Famosa (see 1511).

1619 The Dutch attempt a takeover of Malacca

The Dutch East India Company—backed by the Netherlands government—has taken over much of the Indonesian archipelago to the south of the Malayan Peninsula. Dutch traders attempt to undermine Portuguese-administered Malacca. The Dutch try to encourage attacks by Aceh and Johor on Malacca. Dutch ships blockade rice supplies to Malacca from Java and Sumatra.

1640: August A Dutch siege conquers Malacca

Many Dutch officers and troops die of disease as they besiege Malacca. The Dutch alliance with Johor against Malacca finally succeeds when the invasion force is able to burst through the walls of the fort. Malacca then becomes a military base for the Dutch. The commercial dominance of the Malay Peninsula's city-state ends, as the Dutch prefer routing trade through their capital in Indonesia, Batavia (their name for Jakarta), on Java.

1717 Peoples of Sumatra and Sulawesi become political forces on the Peninsula

Cut off from their legitimate trade routes by the Dutch, a fierce seafaring people called the Bugis, from Sulawesi in the Indonesian Archipelago, turn to piracy in large numbers. Many of the Bugis marauders are based on the Malay Peninsula in Selangor. The Minangkabau, a Sumatran people who have settled in the countryside close to Malacca, form an alliance with the Bugis in 1717, and the two groups are able to take control of Johor.

1721 The Bugis and Minangkabaus become rivals on the Peninsula

Having seized Johor with the Minangkabaus, the Bugis then push their allies out of power. The Bugis then take part in more conflict by taking sides in civil wars in the Peninsula state of Kedah and in Perak. In 1740, the Bugis become the rulers of Selangor. The more powerful the Bugis become, the more worried the Dutch are. Fighting them for decades, the Dutch finally overcome the Bugis and take over their islands and their seat of rule at Johor in 1783.

1771 The British attempt to gain trading posts of their own

The British East India Company tries but fails to set up trading centers on islands off the Borneo coast for their trade in Chinese tea. In 1771, Francis Light (1740–94), a British trader, is unsucessful in his negotiations with the sultan of Kedah, a state on the Thai border. Light's requests for a British trading post on the Peninsula are rejected by the sultan.

1786: August 11 Francis Light obtains a trading post on the island of Penang

Francis Light persists in his quest for British East India Company trading; and the sultan of Kedah grants him the use of the island of Penang, off the west coast of the Peninsula, but Light is not given a formal treaty for it.

1790 The Bugis drive out the Dutch on the Peninsula

The Bugis gather strength again and succeed in expelling the Dutch from Johor and Selangor. Then, the Bugis make an alliance with several states of the Peninsula and Sumatra for a concerted campaign against the Dutch. The Bugis also support the state of Kedah when it attacks the British base at Penang, but the British hold on to the island.

1790 Pepper growing is introduced

Pepper growing is introduced to the Peninsula in 1790, and later to northwest Borneo, as an export crop. The export of pepper grown in those locations rivals the spice trade from Indonesian islands to the south. The black seed-corns of the pepper vine are highly valued for food flavoring in Europe. In Asia, fiery chili peppers, from a different plant originating in South America, become the preferred spice for many dishes

1791 A treaty formally cedes Penang to the British

Cash payments win the formal cession of Penang to British rule. Francis Light becomes the island's administrator under the supervision of the British East India Company. The island's small population soon grows to 25,000, with many

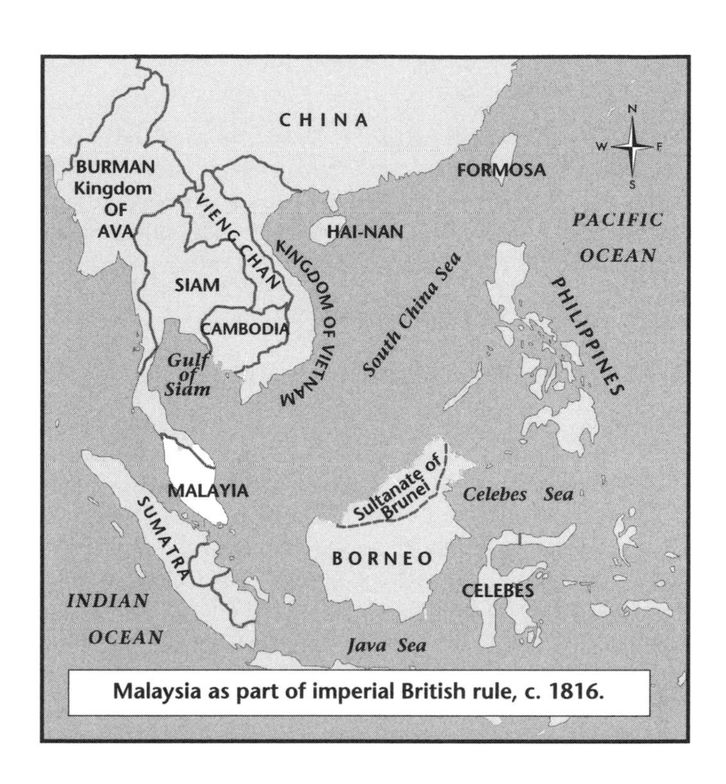

Malaysia as part of imperial British rule, c. 1816.

Chinese, Indians, Thais, and Burmese coming there. By 1788, the Chinese comprise two-fifths of the Penang residents. In 1800, the island is granted rice-producing land on the nearby mainland.

1795 War in Europe causes Britain to take over Malacca

While the Dutch government is in exile due to French conquest, Britain, an ally of the Dutch, takes over temporary administration of Malacca. The Bugis retake power over Johor.

1806 The Malay region's first newspaper is founded

The British start a newspaper, the *Government Gazette,* in Penang. The first in the Malay regions, the newspaper is written in English for thc British colony.

1807 British forces destroy the fort at Malacca

In order to prevent the Dutch from regaining their rival trading position at Malacca following the war in Europe, the Penang-based British decide to destroy Malacca's fort. After attempts to demolish it with sledgehammers fail, explosive charges are used to blow up the wall. Stones from the fort are then used for construction of new buildings in the city of Malacca.

1818: September 21 The Dutch return to Malacca

With the war in Europe over, the British allow the Dutch to reestablish their trade monopolies in the Indonesian islands and to reoccupy what is left of their fort at Malacca.

1819: January 29 The British decide on Singapore for a new trading post

Thomas Stamford Raffles (1811–16), an official of the British East India Company, visits the island of Singapore, just south of the Malay Peninsula, and decides it is a worthy site for a new British trading post. By arranging for a new sultan to rule at Singapore, Raffles wins permission to set up operations there.

1823 Thailand takes the state of Kedah

The state of Kedah borders the land of the Thais, whose king sends a military force to annex it in 1823. In 1826, the British East India Company makes a deal with the Thais, letting them control Kedah if they will leave the other northern Peninsular state of Perak alone. The Thais not only gain Kedah and the region called Perlis, but take control over the east coast states of Kelantan and Terengganu.

1824 The British and the Dutch carve out their territories

The Anglo-Dutch treaty of 1824 recognizes British colonial interests on the Peninsula and Dutch control of the Indonesian Archipelago. The British gain formal ownership of Malacca and relinquish claims south of Singapore. The two colonial powers set the division between modern Malaysia and Indonesia.

1826 British colonies become a unit called the Straits Settlements

The British-controlled territories of Singapore, Penang, and other areas of the Malay Peninsula become an administrative unit called the Straits Settlements. Singapore is their capital. The British colonists make money through "tax farms": franchises they sell for the charging of taxes on gambling, prostitution, or imported goods.

1831 Tax rebellion hits the British

Minangkabau settlers near Malacca rebel as their local chief refuses to pay taxes to benefit the British. The rebellion takes British forces a year to suppress.

1841 A British adventurer obtains territory in Borneo

The Sultan of Brunei cedes an area of northwestern Borneo known as Sarawak to a British adventurer, James Brooke (1803–68), who had helped him suppress a rebellion. Brooke is made the new raja (district chief) by the Sultan. Brooke fends off a rebellion led by the raja he replaced and by 1853 is head of a Sarawak state that is independent of Brunei. He extends his territory by suppressing raids by local tribes and pirates and then taking over their lands.

1845 The rise and fall of *guttapercha*

The *pertjah* trees of the Malay Peninsula yield a useful gum when their bark is cut. This gum, which has been used for waterproofing by Malays since the 1600s, is called *"guttapercha"* by the British. In 1845, demand for the gum grows for use as insulation in the undersea telegraph cables that are starting to make worldwide communications link-ups. Before this increase in demand, workers called tappers would cut the tree bark so that the gum could drip out over and over again without killing the tree (a process similar to that used on rubber trees). But, after 1845, the tappers cut the *pertjah* trees down to extract all the gum immediately. This nearly makes the *petrjah* trees extinct, foreshadowing the fate of many of the trees of the rainforests on the Peninsula and in Borneo.

1860s Tin discoveries bring more Chinese workers

Tin ore is discovered in large quantities in several areas of the Peninsula. Local Chinese business people bring immigrants from China, nearly all men, to work in the tin mines. With rebellions and civil wars disrupting China's countryside, many Chinese are eager to find work overseas and send money back home to their families In the tin-mining regions, Chinese clan and village secret societies, known as *kongsi,* become a form of administration, charging taxes, running company stores, and providing police forces. While the intent is to maintain order and social structure, violence often breaks out between rival *kongsi,* and they are used as troops by Malay aristocrats in wars of succession.

1867 British territories become crown colonies

The Straits Settlements, plus Labuan Island off the coast of Borneo, become direct possessions of Great Britain as crown colonies. Labuan has an excellent harbor and produces coal, making it a fine port for British ships. In 1869, the Suez Canal is completed, with its passage between Egypt and Arabia making sea journeys much faster from Europe to Asia. This increases British interest in the crown colonies and British hopes for more thorough control of the states on the Peninsula.

1874 A treaty establishes British rights in a Malay state

The Pangkor treaty sets out the method of British control of the Peninsula outside of the directly-ruled Straits Settlements. The treaty obligates Perak's sultan to defer to a British official, called a Resident, for policy on internal matters and to conduct all foreign affairs through the British. In return, the British are to provide military and commercial protection to the sultan.

Late 1870s A government school system is set up for Malays

Rural Malay-language primary schools become the Government Malay Schools, a British-approved system with public funding. A few years of basic education deemed appropriate for village children are provided. Reading and writing are taught in Malay using a Roman alphabet script. Higher education and English instruction are reserved for upper class Malay boys to train them for administrative jobs with the British.

1881 The British North Borneo Company is granted a charter

A consortium of British and Austrian speculators manage to receive grants for the same land in eastern Borneo from both the Sultan of Brunei and the Sultan of Sulu (in the Philippines); their territory is called Sabah. The business partners form the British North Borneo Company, which is granted a royal charter in London. They proceed to look for ways to develop their territory commercially, in spite of objections to their occupation of it by Spain, the Netherlands and Germany, which also claim the area. The British occupiers begin the large-scale logging of Sabah's rain forests, exporting the timber to China as railway ties for its vast new transport system.

1885 The first railway is built on the Peninsula

The first railway is built on the Malay Peninsula. Going from Taipang to Port Weld, it is followed by a line from Kuala Lumpur to Port Klang. The lines serve to connect tin mining areas to ports. Roads are built and improved for the same purpose. Eventually, the railway lines are extended to link most of the major towns on the Peninsula. City railway stations, notably the Kuala Lumpur terminal, are designed by the British in fanciful Moorish-Indian styles with arches and domed towers. Government buildings also take on this style, which is distinctive to the British Empire in Asia.

1888 Mt. Kinabalu, the highest mountain in the area, is climbed

Mt. Kinabalu, a 13,455 foot high peak in Sabah, is climbed for the first time on record by John Whitehead, a British zoologist. It is the highest mountain between the island of New Guinea and the Himalayas.

1891–95 The Pahang Rebellion challenges British authority

In the state of Pahang at the center of the Peninsula, a district chief initiates a campaign of civil disobedience against the British authorities. This develops into a campaign of guerrilla warfare led by a skilled fighter, Mat Kilau, and then a crusade led by a Moslem holy man, until British troops are able to suppress it.

1895–1905 Mat Salleh leads rebellion against British North Borneo Company

Mat Salleh (d. 1900), an indigenous leader of Bajau and Sulu ethnicity, launches a war against the British North Borneo Company in retaliation for the company's taking land from the local people and charging taxes in Sabah. He wins the loyalty of the worshipers of spirits and Moslems who consider him a *Mahdi* (Islamic savior). A skilled commander, he builds a fort that frustrates British attempts to capture it. Shot dead in January 1900, Mat Salleh becomes a legendary hero in Sabah for his resistance to colonization.

1896 The British centralize the Peninsula's government with a federal system

Sir Frank Swettenham, the British Resident in the state of Perak, devises a way to coordinate the policies of the Peninsula's states that have Residents through a federal system. In the Federated Malay states of Perak, Pahang, Selangor, and Negari Sembilan, the sultans are the sovereign powers, but the British Residents actually control the government and make the laws. Administration of the Federated Malay States is centralized in Kuala Lumpur, a tin-mining town, and a Malay civil service is established there. The Straits Settlements, under direct British rule, are not included in the Federation.

1897 Rubber cultivation is encouraged

Rubber trees are planted on the Peninsula from seeds brought from Brazil. In 1897, the British authorities enact land–use regulations to encourage the establishment of vast rubber plantations. Rain forests are replaced by rubber "estates" run by British businessmen. Immigrants are brought over from southern India for work as tappers (who cut the bark of the tall trees to cause the gum to seep out) and as processors preparing the raw material for export. Rubber becomes a major source of income for the British colonials, especially when motorcars become an industry requiring huge amounts of rubber for tires.

1900 Medical research is conducted

The Institute of Medical Research begins its work in Kuala Lumpur, and its scientists work to stop malaria, an often-deadly disease caused by a parasite spread through mosquito bites.

1906 Britain gains influence in Brunei

Rather than giving up territory to Sarawak, which is ruled by a British rajah, Charles Brooke (James Brooke's nephew), Brunei's Sultan agrees to accept a British Resident.

1909 The Malay Peninsula is changed by rubber production

As world demand for rubber for car tires grows, more of the Peninsula is covered with rubber plantations. Transportation is improved so that bales of rubber can be brought from plantations to ports. More communication between the states leads to more centralization, and a federal council is established in Kuala Lumpur in 1909 despite objections from the Malay sultans, whose authority is eroding.

1909 The Thai government gives up four Peninsula states

The Thai government cedes the states of Perlis, Kelantan, Kedah, and Terengganu to the British colonizers of the Malay Peninsula. The four newly–acquired states refuse to join the Federated Malay States, so they are termed the Unfederated Malay States and are given British "Advisers" instead of Residents for the appearance of less–direct British control. In 1914, Johor becomes another Unfederated Malay State, with a British Adviser.

1910: August 10 Oil is discovered in Borneo

The Shell Oil Company strikes oil near Miri in Sarawak. Oil fields are also found in Brunei. Sarawak and Brunei become petroleum exporters as demand grows due to the motorcar industry.

1911 Malaya's first census is conducted

The British conduct a census in their controlled states, now known as Malaya. The people are counted by ethnic group and include some 1.5 million Malays, more than 900,000 Chinese residents, and 267,000 Indian settlers.

1917 Palm oil cultivation takes hold in Malaya

Commercial palm oil cultivation takes place in Malaya in 1917. It becomes widespread in the 1920s as plantations are established along the west side of the Peninsula in order to take advantage of transportation for the rubber industry. Production of palm oil, used in the food and cosmetics industries, increases for export.

1923: September A causeway linking Singapore and Johor is completed

A 3,465 ft. long causeway, a roadway on a bridge with railway tracks and a water pipeline running alongside, is completed to link efficiently the thriving island city-state of Singapore with Johor on the Peninsula.

1928 Rebellion against the British breaks out in Terengganu

A peasant uprising in the state of Terengganu is led by To' Janggut, who organizes opposition to the Sultan's acceptance of a British Advisor. To' Janggut claims that British authority and land rents are not legitimate. He is killed in battle and the rebellion is suppressed by British troops sent from Kuala Lumpur.

1929 The worldwide Depression affects Malaya and Borneo

The international Depression damages the economies of Malaya and the Borneo states, particularly by lowering the prices for their export commodities such as rubber, tin, timber, petroleum, and spices. As workers become unemployed, immigrants are sent back to their countries of origin. In 1933, the British administration enacts the Aliens Ordinance, Malaya's first immigration law, which ends the importation of large numbers of workers from overseas.

1929 An association of women teachers is founded

The Union of Johor Women Teachers is founded. The first teachers' union in Malaya, its leader is Hajah Ibu Zain (1903-89). The Malacca-born supervisor of a girls' school in Johor, Ibu Zain is a champion of education for young Malay women. The Union's two hundred members travel throughout the Peninsula to encourage parents to send their girls to school. A journal is published by the Union, which urges girls to strive to succeed in life.

1930: April The Malayan Communist Party begins operating

The Malayan Communist Party (MCP) is founded at a secret meeting in Singapore. Led mainly by urban Chinese, the MCP receives directions from communists in Shanghai, China. Arrests wipe out the MCP in 1932, but it is revived by 1934 and becomes involved in the local labor movement.

1939 An enormous National Park is established

Teman Negara, a 4,343 square kilometer park, is founded by the British administration in 1939, following a fifteen–year campaign for its preservation by Theodore Hubback, a game warden. The park is located on three river systems and is in the states of Pahang, Kelantan, and Terengganu. Its ancient rain forest contains an estimated 14,000 species of plants, 250 species of birds, and 200 different species of mammals.

1941: December 8 Japan invades Malaya

The day after the Japanese attack the United States at Pearl Harbor, Japanese troops invade northern Malaya. British defenses are overwhelmed by the Japanese invasion. By Christmas Day, the Japanese take over Sarawak, and on February 15, 1942, the British surrender Singapore to the Japanese. Thousands of Chinese residents of Singapore and Malaya are killed by Japanese troops, and many flee from the cities to jungle areas.

1943 The Japanese occupiers of Malaya give states to Thailand

The Japanese authorities in charge of Malaya return control of the states of Kedah, Perlis, Kelantan, and Terengganu to Thailand. During the Japanese occupation, disruption of rice imports causes hardship in Malaya. Chinese members of the Malayan Communist Party form the Malayan People's Anti-Japanese Army, which is supported by the British as a guerrilla force under their orders. In Sabah, the Japanese respond to guerrilla attacks with executions, deadly prisoner–of–war camps, and forced labor. Allied bombing efforts to dislodge the Japanese destroy towns in Sabah.

1945: September The British return at the end of World War II

British forces return to Malaysia after the Japanese surrender. The communist resistance fighters of the Malayan People's Anti-Japanese Army are persuaded to give up their arms to the British, but they hide stocks of weapons for a future revolution. In 1949, Communist leader Chin Peng (b. 1922) is given the Order of the British Empire medal for his commando actions against the Japanese during the war.

1946 Malayan Communists rebel

The Malayan Communist Party goes into active rebellion against the British authorities and their Malay associates. The rebels raid police stations and sabotage commercial operations such as rubber plantations and tin mines.

1946: April Malayan Union is set forth as new government amid objections

A Malayan Union is planned by the British to take effect in April, 1946. It is to include Penang and Malacca, which contain considerable Chinese and Indian residents. These residents would become citizens equal with the Malays. Citizens would be subjects of the British crown rather than of the sultans. Malay opposition to the plan is strong enough to delay its actual implementation. A group founded in 1946 called United Malays National Organization (UMNO) organizes opposition to the Malayan Union, voicing fears that Malay interests would be undermined by a strong Chinese and Indian participation in government and by the diminished authority of the sultans.

1946: July 26 Sarawak is passed from the "White Rajahs" to Britain

Vyner Brooke, the third of the "White Rajahs" of Sarawak, cedes control of the state to Great Britain, and it becomes a crown colony, as does Sabah. Britain grants postwar relief aid to Sarawak and Sabah, while oil-rich Brunei is able to loan and donate funds to Malaya.

1947 Malaya's airline is founded

Malayan Airways is founded; it later becomes the international carrier Malaysian Airline System (MAS) and in 1981 embroils the Malaysian government in a trade dispute with Britain over landing rights in London.

1947: September 1 A women's political group is organized

A women's auxiliary of UMNO, Kaum Ibu (Women's Movement), is organized. The leaders promote women's rights in education, marriage and divorce, control of prostitution, education, and equal pay. They are instrumental in raising funds for UMNO and in encouraging women to vote for the party. Hajah Ibu Zain is an early leader of Kaum Ibu and advocates women's participation in politics.

1948: February 8 The Federation of Malaya is established

An alternative to the Malayan Union, the Federation of Malaya, is established by the British and is more acceptable to Malay interests. A British High Commissioner is in charge of the Federation, with an appointed council to advise him. Nine Malay states and the settlements of Penang and Malacca are included in the Federation.

1948: June 18 A State of Emergency is declared as communist rebellion flares

The British administration declares a State of Emergency in the Peninsula and Borneo as the communist rebellion gathers support from Chinese villagers. The British counter-insurgency effort uses the strategy of relocating civilians who are potential sources of food and other support for the rebels to "New Villages" under British security. The communists continue sabotage and raids and in 1951 their guerrillas assassinate the British High Commissioner in an ambush. The war with the communists, which is known as The Emergency, results in approximately 11,000 fatalities before its end in 1960.

1951 Municipal elections are held in the Federation

In preparation for independence and democracy, municipal elections are held in Malayan cities. Parties based on ethnic groups (Malay, Chinese, Indian) win most of the votes.

1952 An ethnic alliance wins local elections

Local elections are held, and an ethnic alliance is formed to successfully contest them. UMNO's leader, Tunku Abdul Rahman (1903–90), a Malay aristocrat, forges the Alliance Party with the Malayan Chinese Association.

1953 The Family Planning Association begins population effort

A nongovernmental group, the Family Planning Association, begins to offer birth-control advice. It opens offices in each state of the Peninsula and gains governmental support for its clinics and mobile family planning advisory teams

1955 The Alliance Party is victorious in elections

The Alliance Party is joined by the Malayan Indian Congress, and, with a goal of Independence from Great Britain in four years, the Party sweeps elections for seats on the Federal Legislative Council, an advisory board, in 1955.

1957: August 31 Independence for the Federation of Malaya is declared

A commission produces a constitution for the Federation of Malaya, in which Malays are guaranteed special rights. Malay citizens have privileges beyond those for Chinese or Indians in matters including civil service jobs, business licenses, and land use. On August 31, "Merdeka Day" (Freedom Day), Britain hands over power to the Federation of Malaya. Tunku Abdul Rahman (see 1952) becomes prime minister. A system under which the sultans of the various states serve as Supreme Ruler of Malaya—on a rotating basis according to seniority, for five year terms—is instituted. The Federation of Malaya is a member of the British Commonwealth, the association of former British colonies.

1959 A group is founded to promote the Malay language

The Language and Literature Commission is founded with government support. The group aims to promote and standardize the usage of the Malay language rather than English, Chinese, or other languages. Known as *Bahasa Malaysia,* the national language requires updated vocabulary as well as coordination with the related language, *Bahasa Indonesia,* which is the national language of Indonesia.

1960: May Tunku Abdul Rahman is a foe of South Africa's apartheid

At the Commonwealth conference in London, Malaysia's prime minister Tunku Abdul Rahman is an articulate and forceful foe of South Africa's apartheid system, and South Africa reacts to criticism by leaving the Commonwealth.

1960: July 11 The end of The Emergency is declared

Determining that the communist insurgency is essentially suppressed, the government of Malaya declares the end of The Emergency. Small groups of communist rebels hold out near the Thai border.

1962: February 19–April 18 Investigators in Sarawak and Sabah determine that people favor merger with Malaya

A five-man commission travels through Sarawak and Sabah to assess opinions on merger with Malaya. Interviews and meetings produce the determination that seventy percent of the people in the two states are in favor of becoming part of the nation of Malaya.

1962: September 1 Referendum held in Singapore regarding merger with Malaya

In the predominantly Chinese city-state of Singapore, a referendum is held to gauge the people's approval of independence from Britain in the form of merger with Malaya. The referendum offers three choices which are all forms of merger, so twenty-five percent of those polled cast a blank ballot to show their disapproval of any merger at all.

1962: December 7 Unrest hits Brunei

The North Kalimantan National Army, a nationalist faction in Brunei, wants to revive the old independent sultanate and its control of Sarawak and Sabah. On December 7, 1962, it attempts an armed revolt, aiming to capture the sultan and declare independence. The rebellion fails as Indonesian military aid promised by Indonesia's President Sukarno does not show up.

1963: June The Philippines and Indonesia object to the formation of Malaysia

The Philippines objects to the merger of the Borneo states with Malaya, because the Philippines still has claims on Sabah dating back to Spanish colonial times and the Sultan of Sulu. Indonesia, which controls the southern part of Borneo, called Kalimantan, is also strongly against the expansion of the Federation of Malaya. The expanded country, called Malaysia, is termed a neo-colonialist plot to keep Britain's interests dominant in the region by Indonesia's President Sukarno. Negotiations for a mutual defense alliance called MAPHILINDO to be formed of Malaysia, the Philippines and Indonesia, are a failure. In July, 1963, Sukarno threatens to use military force against Malaysia.

1963: September 16 Malaysia is formed, including Singapore, Sarawak, and Sabah

The Federation of Malaya combines with Singapore, Sarawak, and Sabah to form Malaysia. Brunei opts for complete independence, unwilling to share its oil revenues with other states or its sultan's supreme power with other sultans. Immediately after Malaysia's formal establishment, rioters attack Malaysian and British embassies in Indonesia. The Philippines and Indonesia cut their diplomatic ties with Malaysia.

1964: January Mediation fails to reconcile Malaysia and its neighbors

Talks held in Thailand and mediated by the United States in January, 1964, between Malaysia, the Philippines and Indonesia fail to bring about reconciliation, as does a meeting held in Tokyo in June, 1964. Indonesia makes attempts to infiltrate troops into Malaysia and attacks Malaysian ships in actions known as the *Konfrontasi* (Confrontation). Great Britain,

Australia, and New Zealand give military assistance to Malaysia.

1964: December Malaysia gains membership in the United Nations

Security Council Malaysia is elected to membership on the United Nations Security Council for a one–year term. Indonesia had objected strongly to this prestigious position for its neighbor, Malaysia, which Sukarno considers a puppet of the British.

1965: August 9 Singapore withdraws from Malaysia

The Singaporean political leader Lee Kwan Yew upsets the Malay elite, including Prime Minister Tunku Abdul Rahman, with speeches promoting equal rights for Malaysian citizens of all ethnic groups. Abdul Rahman holds meetings with Lee Kwan Yew, and it is decided that Singapore must cease being part of Malaysia. Singapore becomes independent on August 9, 1965.

1966: August 11 *Konfrontasi* between Indonesia and Malaysia is ended

With Sukarno replaced by Suharto as Indonesia's President, a peace treaty is signed between Indonesia and Malaysia, ending the hostile *Konfrontasi*. After a few territorial disputes, Malaysia and Indonesia seek tranquil relations with each other.

1967: August Association of Southeast Asian Nations is founded

A regional non-Communist trade support group, the Association of Southeast Asian Nations (ASEAN), is founded; its members include Malaysia, Thailand, Indonesia, Singapore, and the Philippines.

1967: November Riots occur in Penang

A general strike called by the Labour Party in protest of governmental economic policies turns into rioting, which pits the ethnic groups against each other in Penang.

1968: September The Philippines presses its claims to Sabah

Negotiations in Thailand between the Philippines and Malaysia fail to resolve the Philippines' claims to Sabah. In September, 1968, the official government map of the Philippines includes most of the state of Sabah. Malaysia suspends diplomatic relations with the Philippines. The following month, the Philippines call for the dispute to be taken to the World Court.

1969: May Inter-ethnic violence shakes Malaysia

In new elections, the Alliance Party wins its smallest majority so far. On May 13, the Malayan Chinese Association withdraws from the government, destroying the Alliance of the ethnic political parties. Demonstrations and parades in Kuala Lumpur are accompanied by rumors of ethnic violence, and rioting flares up for four days in the city and nearby Selangor until the police and military are able to quell it. At least 178 people die in the rioting, which largely targets ethnic Chinese. A State of Emergency is declared, with the constitution suspended; deputy prime minister Tun Abdul Razak (1922–76) is made head of a National Operations Council to temporarily replace the government.

1970: August 13 A National Ideology is proclaimed

Through the sultan serving as Supreme Ruler, the government proclaims a five-point National Ideology for Malaysia called the *Rukunegara*. The principles are: belief in God, loyalty to sovereign (Supreme Ruler) and country, respect for the Constitution (which includes special rights of Malays), the rule of law, and respectful behavior towards each other.

1970: September Tunku Abdul Rahman is replaced in office

Tunku Abdul Rahman resigns and deputy prime minister Tun Abdul Razak, an advocate of Malay rights from the state of Pahang, is named prime minister in his place.

1971 Economic and cultural policies are introduced by the government

The New Economic Policy, a long-term economic plan, is introduced. Through it the government seeks to end poverty by developing Malaysia and to boost the economic participation of Malays through affirmative action programs for them. A National Culture Policy is also announced in order to promote a conservative and sentimental version of "traditional" Malay culture.

1971: February 19 Malaysia returns to parliamentary government

The National Operations Council that had been in charge since the riots on May 13, 1969 is disbanded, and Parliament is convened again.

1971: March 3 The constitution is amended in hopes of preventing ethnic unrest

The parliament adopts amendments to the constitution that severely restrict public discussion of matters pertaining to ethnic rights and ethnic tensions. Such discussion, with the potential of inciting unrest, is considered seditious by the government, as is any criticism of the special Malay rights in the constitution.

1974: May 31 Malaysia recognizes communist China

Malaysia is the first member of ASEAN to agree to diplomatic recognition of communist China. Malaysia promotes neutrality for Southeast Asia and also seeks economic ties with as many Asian countries as possible. China ends support for communist rebels and gives up dual citizenship claims on Chinese Malaysians.

1974: December 3 Rural and student protests occur

Protests over rural poverty due to depressed rubber prices in Baling, Kedah state, spread into student unrest in Kuala Lumpur, Penang, and elsewhere on the Peninsula. A huge demonstration on December 3 in Kuala Lumpur leads to police occupying the campuses and arresting 1,100 students and professors.

1975 Boat people from Indochina seek refuge in Malaysia

Thousands of refugee "boat people," fleeing strife in Vietnam and Cambodia, begin to arrive on Malaysian shores. The military attempts to keep them from landing and tows their boats back to sea. Malaysian government policy firmly seeks to discourage the arrival of the boat people, most of whom are of Chinese ethnicity. In 1979, Prime Minister Mahathir threatens to have refugees shot on sight, which does not actually happen. The boat people are considered illegal immigrants in Malaysia and are handed over to other countries for resettlement as quickly as possible.

1975: August The Philippines formally suspends claims on Sabah

As a show of cooperative relations with Malaysia, the government of the Philippines announces that it will suspend pressing its claims of ownership of Sabah.

1975: August The Communists commit sabotage

In 1974 and 1975, Malaysian communist insurgents repeatedly sabotage the East-West Highway that is being built from coast to coast parallel to the Thai border. In August, 1975, they blow up the National Monument in Kuala Lumpur, which had been erected to honor the Malaysian government's victory over the communists in The Emergency.

1976: January 14 Prime Minister Tun Abdul Razak dies in office

Prime minister Tun Abdul Razak dies of leukemia and is succeeded as prime minister by Tun Hussein Onn (1922–90), a lawyer from Johor who has limited political support in UMNO.

1977 Aliran, a social activist group, is founded in Penang

A group of intellectual activists founds Aliran. The multiethnic group publishes a newsletter and hosts seminars and other events promoting social reform. Penang becomes a center for nongovernmental groups promoting workers' rights, environmentalism, fair treatment for women and ethnic minorities, and access to health care.

1978–81 The giant caves of Sarawak are explored

International expeditions explore the vast limestone cave systems of Gunung Mulu National Park in Borneo. They discover the largest known cave chamber in the world, the Sarawak Chamber, which is 600 meters long by 450 meters wide.

1981: July 17 Dr. Mahathir Mohammed becomes prime minister

The head of UMNO, a medical doctor named Dr. Mahathir bin Mohammed (b. 1925), becomes prime minister when P.M. Hussein Onn retires due to ill health. Mahathir is known as the author of *The Malay Dilemma,* a book promoting the idea that the Malays are the real owners of the nation and immigrant ethnic groups must shed their cultures and languages to become Malay. He also urges a tougher work ethic for the Malays so they can succeed economically. As prime minister, Dr. Mahathir promotes what he calls his "Look East" policy, calling for Malaysians to imitate the hard work and corporate loyalty of Japanese workers, and for the nation to seek investments from Japan, Korea, and other Asian countries as a way to develop without dependence on the West.

1982 Activists challenge loggers in Borneo

A village-schooled activist from the Kayan tribe, Harrison Ngau (b. 1960), establishes a branch of Sahabat Alam Malaysia (Friends of the Earth Malaysia) in Sarawak. From their small office, the group gathers and distributes information on Sarawak's deforestation and political corruption. Ngau is awarded the Goldman Award for environmental activism in 1990 and wins a seat in the Malaysian legislature that year.

1983: August–December UMNO is in conflict with hereditary Malay rulers

The Supreme Ruler refuses to assent to constitutional amendments which would limit the sultans' powers. Dr. Mahathir pressures the sultans to accept his amendments, but eventually compromises with them on the laws.

1984 The collapse of the international tin market affects Malaysia

An attempt by the Malaysian government to control tin prices and the International Tin Agreement's inability to hold a minimum tin price causes the price of tin to drop by half. Three hundred fourteen tin mines are closed down in Malaysia.

With prices for oil and rubber also low, Malaysia enters an economic recession.

1985 Severe penalties are enacted for narcotics

Viewing narcotics as a threat to society, the Malaysian government imposes severe penalties for drug possession and trafficking. The death penalty is made mandatory for anyone convicted of drug dealing, such as people arrested with more than half an ounce of heroin or seven ounces of marijuana. Foreigners as well as Malaysians are executed. In 1985, the Dangerous Drugs Act is amended to allow suspected narcotics traffickers to be detained without trial.

1985: April Kadazan politicians win elections in Sabah

Led by Dato Joseph Pairin Kitingan (b. 1940) of the Kadazan tribe, the Parti Bersatu Sabah (PBS) unseats the government-associated party in Sabah. Pairin becomes chief minister and the PBS enters conflicts with the federal government over oil and timber business revenues and immigration from the Philippines and Indonesia. By 1990, when PBS again wins elections, Dr. Mahathir and UMNO resort to having Pairin and other leaders accused of corruption charges. The PBS then loses the 1994 elections.

1985: September 1 Malaysia's national car is presented

Malaysia enters a joint venture with the Japanese corporation Mitsubishi to produce a "national car." The Proton Saga is presented to the public when Dr. Mahathir drives the first one over a newly completed bridge between the island of Penang and the mainland. The price of the car is heavily subsidized to ensure its success with Malaysian buyers.

1985: November Police attack a Moslem commune

Memali, a village commune of militant Moslems in Kedah's Baling district, is attacked by the police. The villagers, who are supporters of the Pan-Malaysia Islamic Party, an anti-UMNO opposition group, engage in a battle with the police. Fourteen of the villagers and four police are killed. One hundred sixty commune members, including children, are arrested.

1986: September Malaysia cracks down on heavy metal music

Following an exuberant "Battle of the Bands" in a Penang stadium at which the police are called in to stop the audience from dancing, the government issues a ban on open air rock concerts throughout Malaysia. It then bans heavy metal music on radio and television. Despite the censorship, local heavy metal bands such as Search, Bloodshed, and Rusty Blade remain popular, mainly with Malay teenage boys.

1987 Awareness is raised about effects of logging in Borneo

Evelyn Hong, a Malaysian journalist associated with the Consumers Association of Penang, publishes *Natives of Sarawak: Survival in Borneo's Vanishing Forest,* a book that raises international awareness of the despoiling of indigenous peoples' lands in Borneo by logging companies exporting timber overseas.

1987: March Indigenous people blockade logging roads in Sarawak

Protesting destruction of their forest homelands by timber companies, people from Sarawak tribes stage blockades of logging roads in 1985 and 1986. In March 1987, these protests gain momentum when Penan and Kelabit tribespeople blockade a Japanese-built road opposite a timber camp belonging to a company owned by Sarawak's Minister of Environment, a logging tycoon named James Wong. The blockade movement then spreads to numerous other sites in Sarawak. In June, 1987, leaders of the protests visit Kuala Lumpur to present their views to the Malaysian government. At the end of October, 1987, the police break down all of the blockades and put ninety-two people, including Harrison Ngau (see 1982), in detention under the Internal Security Act.

1987: October The government detains potential opponents

Tensions between UMNO and the Malaysian Chinese Association bring up fears of ethnic unrest in Kuala Lumpur. Using the Internal Security Act, which authorizes detention without trial or charges, the government launches Operation Lallang, in which 119 people are detained. The detainees, most of whom are released within months, include opposition politicians, labor leaders, environmentalists, Chinese education promoters, community organizers, and Christian activists.

1989: May UMNO splits into two parties

The powerful UMNO splits into UMNO Baru (New UMNO) and Semangat 46 (Spirit of 46). Dr. Mahathir Mohamad is head of UMNO Baru, and a rival, Tunku Razaleigh Hamzah, leads Semangat 46, which joins forces with an opposition coalition, the United Movement of Moslem People.

1989: December 2 Communist party signs peace accord

The remaining communist insurgents of Malaysia sign peace agreements with Malaysia and Thailand, at a meeting in Thailand, ending decades of armed operations along the frontiers of the two countries.

1990 An opposition coalition is formed

A large opposition coalition is formed, and develops proposals for policy changes. Named *Gagasan Rakyat* (People's

Might), it includes Semangat 46, the socialist-liberal Democratic Action Party, and others. Dr. Mahathir's UMNO Baru party campaigns hard against the new coalition and keeps a two–thirds majority in elections, but it loses control of the states of Kelantan and Sabah to the opposition.

1991 Tensions arise with neighboring countries over islands

Malaysia plans a deep-sea fishing resort with an airstrip on Terembu Layang Layang in the Spratly Islands, which are claimed by several Asian countries. Believing that the airstrip could be used militarily, Vietnam, China, and Taiwan make protests to Malaysia over the plan.

1992: December Sultan's attack calls for an end to their legal immunity

An outcry follows an assault on a school field-hockey coach by the sixty-year-old Sultan of Johor. The sultan had escaped jail for earlier assaults—even manslaughter—because the sultans, royal rulers of Malaysia's states, enjoy legal immunity. There are calls in the press and Parliament for an end to the immunity and even for the abolition of the system of monarchy. Ultimately, parliament revokes royal privileges not explicitly granted in the constitution. After expressing some reservations, all nine sultans relent, thus sanctioning the reform.

1994: March Scandal taints the national bank

The Bank Negara, Malaysia's national bank, admits to massive losses of foreign exchange in the past two years, which depleted its reserves and its paid-up capital. The government covers the bank's losses and installs new management.

1994: September The government cracks down on an Islamic movement

Interest in strict Islamic practices, such as the wearing of head-scarves called *hijab* by girls and women, increases among Malays in the 1980s and 1990s. The possible growth of Islamic fundamentalism sometimes worries the government, which wishes to promote a modern, economically powerful, secular image for Malaysia but still does not want to offend Malays' religious sensitivities. In 1994, the government cracks down on Al-Arqam, a well-funded Malay Islamic movement. With some 10,000 followers, Al-Arqam, led by the mystic Ashaari Mohamed, is accused of spreading heresy, and the government bans the organization. Ashaari Mohamed is arrested under the Internal Security Act.

1995: April 25 The ruling party wins election; Malaysia enjoys prosperity

The ruling National Front coalition, dominated by UMNO, wins sixty-five percent of the popular vote, its largest majority ever. Malaysia enjoys a phenomenal eight percent economic growth rate, making it one of the world's fastest growing economies. Huge construction projects are announced for continuing modernization and development into the 21st century.

1996: January High-tech development announced

Dr. Mahathir announces plans for a "Multimedia Super-Corridor," a fifteen-by-fifty kilometers special development zone for high-tech industry, to be located south of Kuala Lumpur. The corridor is to replicate California's Silicon Valley with a manufacturing area called "Cyberjaya" and to build a new national capital, "Putrajaya." The world's tallest buildings, the twin towers of Petronas (the national oil company), are also being built, rising 1,483 feet each over Kuala Lumpur. In May, a plan is announced for Malaysia to become fully developed by the year 2020 through infrastructure improvements and changes to the entire education system.

1997: July Dr. Mahathir blames international speculators for dramatic downturn in Asian economies

As the currencies of neighboring Thailand and Indonesia plunge on the world market, Malaysia begins to see a decline in investor confidence. Dr. Mahathir lashes out at international speculators, particularly U.S. billionaire George Soros, who he believes are responsible for manipulating the currency market. By September, the Malaysian stock exchange is often seemingly in free fall. On September 4, four large development projects, including the environmentally controversial Bakun Dam in Sarawak and the world's longest shopping mall, are put on hold.

1997: August A haze of smoke and pollution blankets Malaysia

Forest fires in Sumatra to the south create a haze which combines with air pollution from construction dust, cars, and motorcycles in Malaysia. The haze is especially severe in Kuala Lumpur, which is located in a valley that traps the smog. Many citizens wear cloth masks, and children and old people stay indoors. Sarawak also suffers from forest fires and smoke.

1997: December 5 A growth-slowing policy for Malaysian economy is announced

Deputy Prime Minister Anwar Ibrahim announces measures to control the overheating of Malaysia's economy. Under the new plan, government spending will be cut by eighteen percent, and huge development projects that have not yet been started will be delayed indefinitely.

1998: February Haze and drought affect Sarawak

Forest fires in Borneo and Sumatra renew a blanket of haze over Sarawak. Drought conditions and the effect on crops of the smoky air pollution cause food shortages in many villages.

1998: March The Malaysian government expels immigrants

Going from a labor shortage, particularly in the construction business, to unemployment due to the economic downturn, Malaysia takes steps to expel foreign workers and to turn away a new influx of impoverished Indonesians seeking work. Called "Operation Go Away," the crackdown allegedly includes beatings and other abuse of the Indonesians in Malaysian detention centers. Political refugees from Aceh in Sumatra are included in the repatriation.

1998: April Water shortages parch Kuala Lumpur

Water shortages in the Klang River Valley, where Kuala Lumpur is located, cause water to be turned off at homes and factories for days at a time. Drought, combined with river pollution, deforested watersheds, and industrial water demands produce the crisis. The normally restrained Malaysian press is critical of the government for promoting huge infrastructure projects but neglecting something so basic as water supply. A conference of citizens' groups regarding the water crisis is canceled by the government.

1998: September 2 Deputy Prime Minister Anwar Ibrahim ousted

Prime Minister Mohathir Mohammed sacks Anwar Ibrahim from his posts as deputy prime minister and finance minister. The following day, Mohathir removes Anwar from the ruling Unmo party. Anwar's ouster provokes a storm of protest in Malaysia, particularly among youth who clamor for reform. Anwar's sacking is widely viewed as an attempt by President Mohathir to destroy a potential rival.

1998: September 20 Anwar arrested

Police arrest Anwar Ibrahim. He is subsequently indicted and tried on six charges. Five of these charges deal with sexual improprieties while the sixth charge, corruption, alleges that Anwar used his position to prevent public exposure of his sexual misconduct.

1999: April Anwar guilty, sentenced to six years

A court finds Anwar guilty on all six counts and is sentenced to six years in jail. Although even Anwar, himself, predicted a guilty verdict, observers are shocked by the lengthy sentence. Outside the courthouse riot police hold protesters at bay.

Bibliography

Bellwood, Peter S. *Prehistory of the Indo-Malaysian Archipelago.* Honolulu: University of Hawaii Press, 1997.

Bevis, William W. *Borneo Log: The Struggle for Sarawak's Forests.* Seattle: University of Washington Press, 1995.

Broughton, Simon, ed. *World Music: The Rough Guide.* London: The Rough Guides Ltd., 1994.

Bunge, Frederica M., ed. *Malaysia: A Country Study.* Washington, DC: United States Government, 1984.

Dancz, Virginia H. *Women and Party Politics in Peninsular Malaysia.* Singapore: Oxford University Press, 1987.

Far Eastern Economic Review. Asia 1998 Yearbook: A Review of the Events of 1997. Hong Kong: Review Publishing Company, 1998.

Hong, Evelyn. *Natives of Sarawak: Survival in Borneo's Vanishing Forest.* Penang, Malaysia: Institut Mayarakat, 1987.

Jin-Bee, Ooi. *Land, People and Economy in Malaya.* Singapore: Asia Pacific Press, 1970.

Kahn, Joel S. and Loh Kok Wah, Francis, eds. *Fragmented Vision: Culture and Politics in Contemporary Malaysia.* Honolulu: University of Hawaii Press, 1992.

Kaur, Amarjit. *Historical Dictionary of Malaysia.* Metuchen, NJ: Scarecrow Press, 1993.

Means, Gordon. *Malaysian Politics.* London: University of London Press, 1970.

——— *Malaysian Politics: The Second Generation.* Singapore: Oxford University Press, 1991.

Paletz, Michael. *Reason and Passion: Representations of Gender in a Malay Society.* Berkeley: University of California Press, 1996.

Roseman, Marina. *Healing Sounds of the Malaysian Rainforest: Temiar Music and Medicine.* Berkeley: University of California Press, 1991.

Ryan, N. J. *The Making of Modern Malaysia and Singapore.* Singapore: Oxford University Press, 1969.

Selvanayagam, Grace Inpam. *Songket: Malaysia's Woven Treasure.* Singapore: Oxford University Press, 1990.

Spruit, Ruud. *The Land of the Sultans: An Illustrated History of Malaysia.* The Hague, Netherlands: Pepin Press, 1989.

Maldives

Introduction

Maldives is an independent nation whose territory comprises a group of small coral islands in the Indian Ocean. The northernmost islands lie 300 miles (480 kilometers) southwest of the southern tip of the Indian landmass. From this point, at latitude 7° N, the islands extend in a chain southwards to the Equator, covering a distance of roughly 500 miles (800 kilometers). At its widest point, the island group measures 70 miles from east to west. There are 1,190 islands in the Maldives with a total land area of 115 square miles (298 square kilometers), only twice the area of New York's Staten Island. Only two hundred of Maldives' islands are inhabited, with the country's population numbering 280,391 (1997 estimate).

The islands of the Maldives are grouped into twenty-six atolls, rings of coral islands surrounding a central lagoon. The word "atoll" is actually derived from the Maldivian word *atolhu*, because it was here that European sailors first came across these features of the tropical seas. Made entirely of coral resting on a submerged volcanic ridge, the islands rarely rise more than 6 feet (1.8 meters) above sea level. They are, however, protected from the ocean by fringing coral reefs. The largest of the islands, Fua Mulaku, lying just south of the Equator, is only three and a half miles long by one and a half miles wide. Most islands in the Maldives are less than one mile long.

The unique character and location of the Maldives have influenced every aspect of its culture, society, and economy. On the one hand, geographical isolation has meant that little is known of the country's early history. Much of our knowledge of the Maldives' past, for example, is based on descriptions by travelers or shipwrecked sailors who happened to land on its shores. Yet, at the same time, the location of the island chain astride the sea-routes across the Indian Ocean has exposed the Maldives to contacts not only from South Asia, but from many distant lands.

History

The Maldives were most probably settled around the fifth or fourth centuries B.C. by peoples from India and Sri Lanka. Although linguistic evidence indicates an underlying Tamil (i.e. South Indian) element, Maldives culture clearly originates in the Sinhalese culture of Sri Lanka. Recent archeological finds, however, may well require a new look at the region's prehistory. Pottery remains that possibly date to 2000 B.C., ancient sun-symbols, sculptures that suggest both a primitive religion and Hindu worship prior to the Buddhist period, mounds built by a mysterious people called the "Redin"—all raise intriguing questions as to the Maldives' past.

Little is known of the Maldives prior to the accounts of Arab travelers of the tenth to twelfth centuries. Buddhist mounds (*stupas*) and sculpture indicate that Buddhism, introduced from Sri Lanka, was probably well established by the beginning of the Christian Era. It was during the Buddhist period that the language of the Maldives evolved. Divehi is derived from an old form of Sinhala, the tongue of the Sinhalese peoples of Sri Lanka. It had split off from its parent language by the early centuries A.D. and, despite several centuries of Islamic rule in the Maldives, has remained relatively untouched by later Arabic influences (except for the addition of some Arabic words).

Islam reached the Maldives in the twelfth century following the conversion of Theemuge Maha Kalaminja, the ruler of the Maldives. Upon his conversion, he took the title and name of Sultan Mohammed Ibn Abdullah. Since that time—and despite contacts with non-Muslims through the centuries—the population has been staunchly Muslim. Islam is the state religion today, and no one who is not a Muslim can reside in the islands. There is no secular law, only the *Sharia*, Islamic Law as it is interpreted and administered by the authorities. The people of the Maldives are Sunni Muslims, and their religious life centers on the mosque, prayer, and festivals such as Ramadan and the Id celebrations. Muslim customs such as the ban on alcohol and avoidance of pork are strictly adhered to, and life cycle ceremonies follow traditional Islamic custom.

Although the Chinese navigator Ma Huan reached the Maldives in 1433, non-Muslim influence is not felt until the arrival of the Portuguese in the early sixteenth century. In return for help in securing the throne, Sultan Kalu Mohammed offered the Portuguese a trading post in 1517. In subsequent decades, the Portuguese occupied the Maldives and

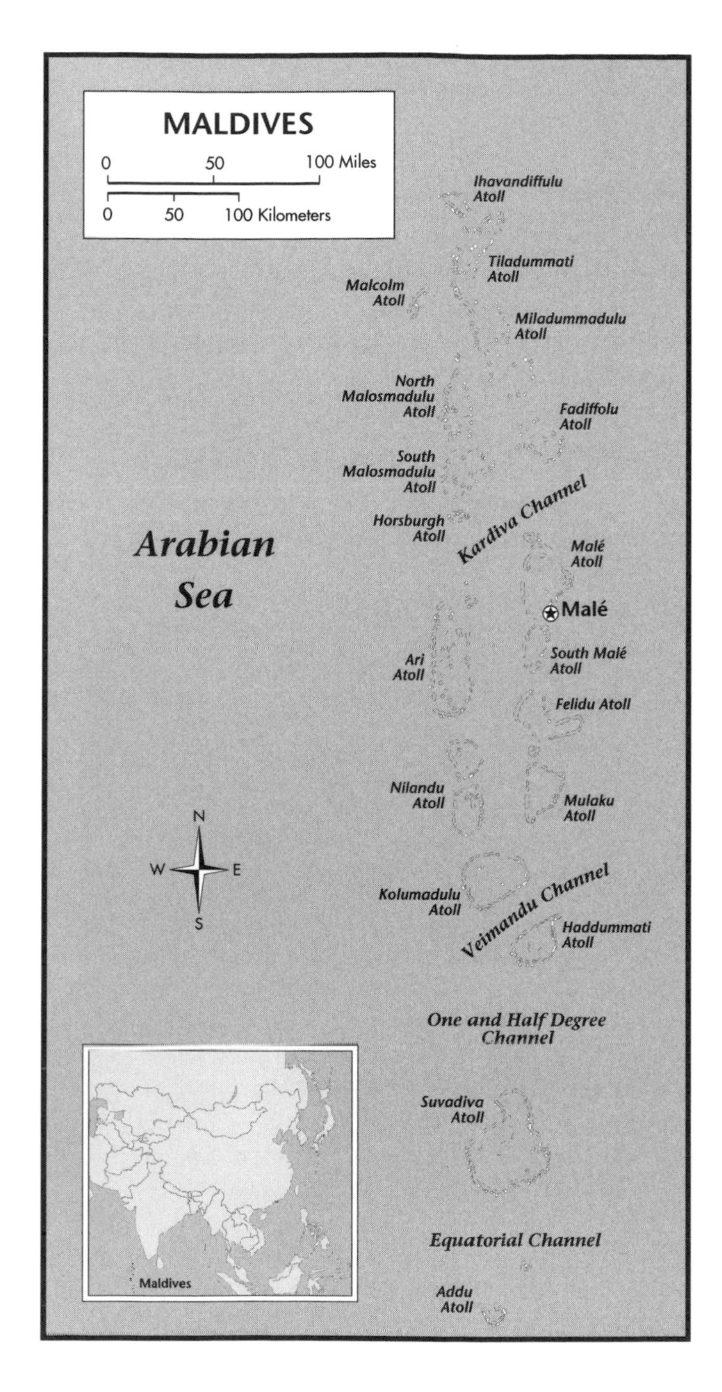

attempted to convert the islands' population to Christianity, often forcibly. Ultimately, the natives forced the Portuguese out in 1573. In the seventeenth century, the islands became a tributary state of Dutch-ruled Ceylon.

In 1796, Britain took Ceylon from the Netherlands but for nearly a century, the Maldives remained largely independent. Not until 1887 did Britain make the Maldives a protectorate. Although strategic interests dominated British policy toward the Maldives, they also introduced many reforms intended to modernize the islands' society. The most significant of these reforms were the establishment of a national education system along British lines and the creation of constitutional government while allowing the sultanate to continue as before.

In 1948, Britain granted the Maldives a measure of independence while remaining under the British crown. This arrangement allowed the islanders their first opportunity at self-government—a period of much turmoil. In January 1953, political leaders declared a republic only to restore the sultanate nine months later. The three southern atolls declared their independence in 1959 and pursued their rebellion for five years before the Maldives government suppressed it.

The Maldives received full independence in 1965. Since then, the sultanate has been abolished (1968) and the last British forces left their bases on the island (1976). The first decade of independence was characterized by the autocratic leadership style of President Ibrahim Nasir in which presidential authority outweighed that of parliament. Following Nasir's decision not to run for a third term in 1978, Maumoon Abdul Gayoom won election to the presidency on a platform that called for a decrease in presidential authority and a relaxation of press controls.

Gayoom remained popular and subsequently won election four times. Under his rule, the islands embarked upon a major development program that featured the construction of an international airport in 1981, a major hospital in 1983, and the hosting of an international global warming conference in 1989, and a South Asian Association for Regional Cooperation in 1997. In addition, the islands' democratic foundation has been solidified through membership in the British Commonwealth (1982) and the suppression of a military coup in 1988.

At century's end, economic development continues to serve as the main concern of the government. With its restricted agricultural and resource base, Maldives' options for economic growth and development are limited. Foreign investors have been induced by low operating costs to establish factories producing ready-made garments for export. The government has successfully developed a commercial shipping industry, with a fleet of ships carrying cargo to and from the Maldives. It has also taken advantage of the country's tropical climate, white sandy beaches, clear water, and coral formations to develop a rapidly-growing tourist industry. Tourism has now surpassed fishing as the Maldives largest source of foreign exchange.

Timeline

c. 2000–1500 B.C. Prehistory

Finds of pottery from Vadu lagoon during the 1980s suggest the Maldives are inhabited, or at least have contact with the outside world, at this time. The meaning of this archeological evidence is unclear, although the explorer Thor Heyerdahl argues it shows that seafarers from the Harappan civilization of the Indus Valley know of the islands and visit them frequently.

c. 500 B.C. Early settlement

The Maldives are settled during the fifth and fourth centuries B.C.

Once upon a time, so the Maldives' origin myth goes, a prince of royal birth named Koimala and his wife, the daughter of the king of Ceylon, make a voyage to the Maldives. They are becalmed, and rest awhile at Rasgetimu island. When the islanders learn the visitors are of royal blood, they invite them to remain and proclaim Koimala as king. The royal couple subsequently migrate to Giravaru Island in Malé Atoll. They send two ships to Lanka (Sri Lanka) to bring over other people of the "Lion Race" (the Sinhalese). Koimala and his queen have a son named Kalaminja, who reigns as a Buddhist before converting to Islam. Kalaminja's daughter, who rules as a Sultana, gives birth to a son also named Kalaminja. This prince marries a lady of the country, and all the subsequent rulers of the Maldives are their descendants.

This legend is clearly of relatively late origin (the conversion to Islam takes place in the twelfth century), but refers to many significant aspects of early Maldives' history. These include the presence of aboriginal peoples before settlers from Sri Lanka come to the islands, the cultural and historical ties with the Sinhalese (the dominant ethnic group in Sri Lanka), the significance of Buddhism, and the arrival of Islam in the islands.

c. 450 B.C. The *Jatakas*

The *Jatakas* mention islands that are possibly the Maldives.

A collection of tales that date as far back as the fifth century B.C, the Jatakas present Buddhist ethics and morals in the guise of stories woven around the life of an incarnation of Buddha (c. 563–c. 483 B.C.). An ancient commentary on the Bharu Jataka tells of exiles from the kingdom of Bharuch in Gujarat in India who "found standing room upon a thousand islands which are yet to be seen today about the island of Nalikera" (Maloney 1980: 38). Maloney claims these islands are the Maldives, and cites other references in the Jatakas as identifying the island group.

c. 450–300 B.C. Maldives in Sri Lankan legends

Legends in the Buddhist histories of Sri Lanka refer to islands that are possibly the Maldives.

Two Buddhist chronicles, the *Dipavamsa* (A.D. 350) and the *Mahavamsa* (A.D. 550), record the early history of Sri Lanka. They tell how the Sinhalese settle Sri Lanka and drive out its original inhabitants to live on the island of "Giri" or "Giridipa." *Giri* means "mountain," but there are few islands in the Indian Ocean that fit the description of the island found in the Buddhist texts. However, *giri* is also the Divehi word for "reef." Scholars cite this, and other supporting evidence, to identify the Maldives as the mythical Giri of the Buddhist histories. The events described in this story probably take place around the fifth or fourth centuries B.C.

c. A.D. 362 Roman histories mention the "Divi"

The Roman historian Ammianus Marcellinus (c. 330–395) writes that "...the Indian nations as far as the Divi and the Serendivi vied with one another in sending their leading men with gifts [to the Emperor Julian]..." (*Rerum gestarum libri*, trans. John Rolfe, vol. II: 213).

The Divi are the peoples of the Maldives (and possibly of Lakshadveep, the island chain immediately to the north), while the Serendivi are the Sri Lankans. The Maldivians call themselves Divehi, from *dive*-si, which means "islanders," and their country Divehi Rajje, the "island kingdom." The country's modern name is derived from the ancient Indian "Maladiv" (from the Sanskrit *mala* [garland] + *dvipa* [island]), which becomes Maldives in modern English usage.

629–635 Hsüan-tsang describes the "Coconut Island"

The Chinese traveler Hsüan-tsang, in his account of his travels in India, writes:

> Passing seawards to the south of this country (Ratnadvipa or Sri Lanka) some thousands of li [a Chinese unit of distance] we arrive at the island of Narakira (Na-lo-ki-lo). The people of this island are small of stature, about three feet high; their bodies are those of men, but they have the beaks of birds; they grown no grain, but live on cocoanuts....

c. 850–900 Sulaiman the Merchant

"Sulaiman the Merchant," an author whose identity is unknown, writes of the islands of Dibajat (Maldives) in the seas between Gujarat and the Bay of Bengal. He says:

> They say that their number goes up to 1900. These islands separate two seas. They are governed by a woman.... In these islands where a woman rules, coconut is cultivated. These islands are separated from one another by a distance of two, three, or four parasangs [a parasang is believed to be about five miles]....The wealth of the people is constituted by cowries....They build ships, houses and execute all sorts of work with consummate art.

Sulaiman never visited the Maldives, but his description of the islands (in *Akhbar as-Sinw'l-Hind*) is based on the accounts of other travelers. By the end of the ninth century, Arab traders have mastered the monsoon winds of the Indian Ocean. Arab sailing *dhows* (ships) ply the seas between Arabia, India, even voyaging as far as Southeast Asia and China. As the increasing frequency of Arab accounts of the Maldives indicates, these journeys bring them to the "garland of islands" strung across the sea routes to the Orient.

916 Al-Mas'udi

Al-Mas'udi visits the Maldives, and records that the people are subject to one government and ruled by a queen.

The mention of a reigning queen is of interest. The Koimala legend tells of a queen who ruled the Maldives, medieval Arab historians also relate this fact, and later historical record lists sultanas (queens) ruling in the islands. By the seventeenth century, this ancient matrilineal system has disappeared under prevailing Islamic influences.

c. 1000 Chola expedition to the Maldives

Rajaraja I (r. 985–1014), ruler of the powerful Chola kingdom based in the Tamil region of southeastern India, sends a naval expedition to the Maldives. An inscription dated A.D. 1017 tells he conquers Lanka and subdues "...the many ancient islands, 12,000 in number." The political influence of the Cholas in the Maldives is uncertain, but the islands clearly fall within the kingdom's economic sphere. The Divehi word for trader (*soliya*) is Tamil in origin.

c. 1030 Al-Biruni

The Arab traveler, Al-Biruni (973–1048), visits the Maldives and describes the country's physical geography, people, and economic activities. He mentions that the islands are divided into two classes according to their products. The *Diva-kanbar* are islands where ropes from coconut fiber are made, while the *Diva-kudha* produce cowrie shells.

The cowrie is a marine snail of the genus *Cypraea*. The shell of the species of cowrie found in the Indian Ocean (*C. moneta*), otherwise known as the money cowrie, is widely used as currency in the Old World—from Africa to China. Cowries clearly have value from earliest times, being found at the Harappan site of Lothal in Gujarat, India prior to 1500 B.C. Early Arab accounts stress the importance of cowries, both as currency and as a major export of the Maldives. Although the cowrie begins to loose its value as currency after the Europeans arrive in the Indian Ocean, as late as 1611 Pyrard writes that each year thirty to forty boats loaded with cowries leave the Maldives for Bengal in India.

1099–1168 Al-Idrisi

Al-Idrisi, another Arab traveler, visits the Dibajat Islands and describes its peoples, political system, and its products (e.g. cowries, tortoise-shell, coconuts). He leaves an account of the inhabitants' clothing and houses, noting they are skilled craftsmen and build wooden boats.

The maritime technology of the Maldives suggests a long tradition of seafaring (the ancestors of the Maldivians, after all, reach the islands by sea). There are several types of boats in the islands, the most common being the *dhoni*, the traditional sailing boat of the Maldives.

1153 Islam reaches the Maldives

Theemuge Maha Kalaminja, the Buddhist ruler of the Maldives, converts to Islam and assumes the title and name of Sultan Mohammed Ibn Abdulla.

A popular account of the conversion tells that every month a *djin* (demon), looking like a "ship full of lamps," comes from the sea and carries off a virgin to an idol temple by the sea shore. The djin dishonors the virgin, who then dies. A Muslim saint from North Africa named Abu-l' Barakat visits the Maldives and volunteers to exorcise the djin by reciting the *Qu'ran*, the sacred book of Islam. The djin disappears and the king receives the "true faith." The people of the island also convert to Islam, breaking idols, razing temples to the ground, and spreading the new religion to all the other islands.

Perhaps a more realistic interpretation of events is that the Maldives are gradually drawn into the new economic order of Arab trade in the Indian Ocean and the Islamic civilization that goes hand in hand with it.

1166 Sultan Mohammed Ibn Abdulla leaves for Mecca

The sultan sails on a pilgrimage to Mecca and is never heard of again.

Another version of the sultan's conversion and the events of his life appears in the *Tarikh*, or the State Chronicle of the sultans. This work, written in Arabic, and the *Radavali*, histories written in local scripts, provides details of Maldive history from the mid-twelfth century to 1821. The *Tarikh* for example, includes a list of all sultans who ruled during this period.

1195–96 Earliest copper-strip books

Divehi writings on copper plates (*loamaafaanu*) dating to the end of the twelfth century tell of the reigns of four Buddhist kings who rule the islands before the conversion to Islam. The text is written in *Eveyla akuru*, an early Maldives' script which gives way to the *Dives akuru* script by the mid-sixteenth century. Today, the Divehi language is written in the *Tana* script, which was developed during the sixteenth century. Unlike the earlier scripts, Tana reads from right to left, probably to accommodate the inclusion of Arabic words into the language. Copper plate texts are found until the mid-fourteenth century.

1343 Ibn Batutah arrives in the Maldives

The Moroccan traveler Ibn Batutah (1304–68 or 1369) lands in the Maldives where he serves the ruler as a *qazi* (Islamic judge) for eighteen months. He returns to the islands again in 1346.

Ibn Batutah provides the first detailed description of the Maldives, to which he devotes ten pages in the account of his travels around the world.

1433 Ma Huan

The Chinese sailor and navigator Ma Huan reaches the Maldives after a ten-day sail from the island of Sumatra, in Indonesia. On his return to China, he writes *The Overall Survey of the Ocean's Shores* in which he describes the Maldives.

1498 Vasco da Gama lands in South India

The Portuguese explorer Vasco da Gama sails into Calicut, becoming the first European to reach India by sea.

Portuguese exploration has as its aims trade and the spreading of Christianity. The Portuguese soon establish themselves along the southwest coast of India, from where they begin to wrest control of the trade routes across the Indian Ocean from the Arabs.

1507 Portuguese reach the Maldives

Dom Lourenço de Almeida "discovers" the Maldives for Portugal.

1512 Maldives Sultan seeks Portuguese help

Kalu Mohammed, twice deposed Sultan of the Maldives, seeks help from the Ali Raja of Kannaur on India's Malabar coast and the Portuguese in regaining his position as ruler of the Maldives. A combined Malabari-Portuguese fleet sails to the Maldives, attacks Malé, and reinstates Kalu Mohammed on the throne.

At this time, the Maldives begins paying tribute to the Malabaris and the Portuguese.

1517 Portuguese gain footing on the Maldives

Sultan Kalu Mohammed gives the Portuguese permission to build a factory, as European trading posts in Asia are called, in the Maldives.

The following year, a Portuguese expedition of four ships and 120 men under the command of João Gomes Cheiradinheiro lands at Malé. The Portuguese build a fort, and compel the Maldive islanders to channel all their trade through Portuguese hands. Within a short time, the local people rise up and kill all the Portuguese in the Maldives.

1552 Maldives sultan converts to Christianity

Sultan Hasan IX (r. 1550–52) declares his intention of renouncing Islam, causing turmoil in the country. He flees to Cochin in India where Francis Xavier baptizes him a Christian. He takes the name Dom Manoel and marries a Portuguese wife. On two separate occasions, the exiled ruler sends a Portuguese ship to the Maldives to bring his ministers and chiefs to Cochin. Both times, the vessels are seized and their crews killed.

Francis Xavier (1506–52), later canonized as Saint Francis Xavier and remembered as the greatest Roman Catholic missionary of modern times, is renowned for bringing Christianity to Asia.

1558 Portuguese occupy the Maldives

A large Portuguese expedition lands at Malé, kills Sultan Ali VI (r. 1557–58) in battle, and establishes Portuguese rule over the Maldives.

This military venture is in response to the unwillingness of the Maldive peoples to provide revenues for the exiled Sultan (Hasan IX) or to meet the demand that all trade pass through Portuguese hands. The leader of the expedition, Andreas Andre, is of part-Portuguese, part-Maldivian birth. He proclaims himself Sultan.

1558–73 The Portuguese period

The Portuguese set about exerting their monopoly over trade and violently converting the islanders to Christianity. Resentment at such treatment soon grows strong among Maldivians, who begin to organize resistance to the foreign occupiers. A group of young islanders from Utima island in the north commence guerrilla warfare against the Portuguese. Led by Mohammed Thakurufaanu, they build a boat which is used for hit-and-run attacks at night against Portuguese. Even though Mohammed's brother is captured and killed, the group succeeds in liberating most of the islands except Malé. Realizing that he does not have the strength to capture Malé on his own, he goes to Malabar in India to seek assistance in arms and men.

1573 Portuguese driven out

Assisted by four galleys (ships) from Malabar, Mohammed Thakurufaanu makes a surprise night attack on Malé. He succeeds in defeating the Portuguese garrison and kills Andreas Andre himself.

The Maldives are now free from Portuguese rule. The people unanimously elect Mohammed Thakurufaanu (r. 1573–85) to be Sultan.

1602–07 François Pyrard de Laval

A Frenchman, François Pyrard de Laval, is shipwrecked in the Maldives in 1602. The sultan holds him captive and it is five years before he manages to escape with an Indian fleet that attacks the islands. During his enforced stay, he learns the Divehi language, hears the history of the country, and observes the local people in the daily lives. On his return to France, Pyrard writes an account of his stay in the Maldives, a valuable work that provides the first substantial European study of the islands. Originally published in 1611, the work is reprinted in the Hakluyt Society's *The Voyage of François Pyrard of Laval to the East Indies, the Maldives, the Moluccas, and Brazil* (1888).

1625 Portuguese invade Malé

A Portuguese expedition to the Maldives lands on Malé but is driven off.

1645 Maldives sends tribute to the Dutch

The Maldives sultan begins sending annual tribute to the Dutch in Ceylon. This includes sweets, coir rope (made from coconut fiber), ambergris (a waxy substance found floating in the sea and thought to originate in the intestines of sperm whales—it is used in perfumes), 'sea coconuts,' Maldive mats, and cowries. Although the Dutch trade with the Maldives, they have no lasting cultural influence on the country.

1752 Malabaris occupy the islands

Malabaris temporarily occupy Malé and carry off the ruling sultan before being driven away. The Malabari threat leads the Maldives nobles to seek help from the French in Pondicherry (India). A French fleet arrives in the Maldives and defeats the Malabaris. A French garrison remains in Malé until 1754.

1796 British take Ceylon from the Dutch

The Dutch East India Company cedes Ceylon to Britain, which inherits the tributary relationship with Maldives. Contacts between Britain and the Maldives remain on an informal footing until the end of the nineteenth century.

1834 British survey expedition to the Maldives

Frequent shipwrecks in the Maldives lead the British to send an expedition to conduct a maritime survey of the islands. Two officers, Lieutenants Young and Christopher, are assigned to learn the language and observe the local people. Their account, *Memoir on the Inhabitants of the Maldives Islands*, published in 1844, is the most detailed description of Maldive society since Pyrard's in the early 1600s.

1879: November H. C. P. Bell is shipwrecked in the Maldives

H. C. P. Bell of the Ceylon Civil Service is shipwrecked on Gafaru Reef and spends a few days in the Maldives. Forty years later, the government of Ceylon sends Bell back to the Maldives to report on the islands antiquities. His subsequent report, *The Maldive Islands: Monograph on the History, Archaeology, and Epigraphy*, is the first serious study of the Maldives in modern times.

1887: December 16 Maldives becomes a British protectorate

An agreement signed by Sultan Mohammed Muinuddin II (r. 1886–87) and the Governor of Ceylon, Sir. H. A. Gordon, makes the Maldives a British protectorate. Britain guarantees the Maldives complete freedom in its internal affairs and undertakes to protect the country from external aggression. The Maldives continues to pay tribute to Britain.

Maldives in the Twentieth Century

1927 Secondary school education begins in Malé

Education at the secondary school level begins in Malé. The school initially accepts only boys, but later opens its doors to girls.

Prior to this time, the only educational institutions in the Maldives are traditional Islamic schools known as *kiyavages, mahktabs* and *madrasahs*. These are private institutions, focusing on religious instruction, arithmetic, and reading and writing in Divehi and Arabic. No national system of education exists in the country.

1932 Maldives becomes a constitutional monarchy

Under pressure from his people, Mohammad Shamsuddin III (r. 1893, 1903–35) adopts a written constitution, which provides for an elected sultan ruling through democratic institutions of government.

1935 Maldives elects Hassan Nuruddin sultan

The Maldives elects Hassan Nuruddin (r. 1935–43) sultan after charges of plotting against the constitution force Mohammad Shamsuddin III to leave office.

1939 Britain builds air strip on Gan Island

Britain constructs an air strip on Gan, an island in Addu Atoll at the southern end of the Maldives.

1941 Mohammad Amin Didi

Mohammad Amin Didi, who is to remain a central figure in Maldives' politics for over a decade, becomes prime minister.

Didi encourages islands to build primary schools, and makes education available for girls in Malé. Among his innovations are a power plant in Malé, telephones in offices, improved agricultural techniques, the writing of a history of the Maldives in Divehi, and promoting health ideas. At the same time, he angers people by some of his policies. He mandates a government monopoly of the trade in dried fish, and forces the local population to cut down trees and build roads on islands where there are no cars (only Malé and Addu Atoll have vehicles—bicycles are the main means of transportation for the people).

1943 Hassan Nuruddin resigns as sultan

Hassan Nuruddin resigns in the face of popular opposition. His elected successor, Abdul Majeed Didi, lives abroad and never assumes power. A Council of Regency, headed by

Prime Minister Mohammad Amin Didi assumes control of the government.

1948: April 23 New Anglo-Maldivian agreement

When Ceylon becomes independent from Britain, Maldives and Britain sign a new agreement in Malé. Maldives retains its protectorate status, but Mohammad Amin Didi stops payment of tribute to Britain. Britain continues to assume responsibility for the defense of the Maldives.

1952 Sultan-Designate dies

Abdul Majeed Didi, elected sultan in 1943, dies in Colombo, Ceylon.

1953: January The Maldives' First Republic

Maldives abolishes the sultanate and declares itself a republic, with Mohammad Amin Didi elected as its President. Policies such as a ban on smoking in the country (a strict adherence to Muslim law) and food shortages anger many people. Opposition to the president grows.

1953: September 4 The Republic falls

While the president is in India, the people in Malé gather to ask the vice-president to suspend the constitution and take over the reins of government. The new government arrests Mohammad Amin Didi on his return from India and banishes him from Malé.

1953: December 31 Mohammad Amin Didi escapes

The former president escapes from his place of exile and returns to Malé. He is attacked by angry mobs, and dies within a month.

1954: February New constitution restores sultanate

Maldives restores the sultanate after its brief flirtation with republicanism.

1954: March 7 Mohammad Fareed Didi becomes sultan

Mohammad Fareed Didi, the son of the former sultan-designate, assumes the office of sultan of the Maldives.

1956 Base on Gan established

Britain establishes a Royal Air Force (RAF) Base on Gan island.

1957 Ibrahim Nasir becomes prime minister

The sultan selects Ibrahim Nasir, a Maldivian educated in Colombo, as his prime minister.

1959–64 The Southern secession

The three southern atolls (Huvadu, Fua Mulaku, and Addu) declare themselves independent of the Maldives and form the Republic of Suvadiva. The revolt is eventually suppressed by the Maldives' government.

1965: July 26 Maldives fully independent

Britain agrees to relinquish all authority over the Maldives, which becomes a fully independent nation. The Maldives joins the United Nations the same year.

1966 Maldives Shipping Limited

The Maldives establishes its own shipping line, Maldives Shipping Limited (MSL). From two ships in 1966, MSL grows to a fleet of forty vessels. However, the world-wide shipping recession in the 1980s sees a reduction in the scale of the state-owned shipping line's operations.

1968: November 11 Republic of the Maldive Islands proclaimed

The Maldives becomes a republic again, with Ibrahim Nasir as its President. The former sultan is pensioned off and later goes to Colombo.

Under the 1968 constitution, the president is Head of State and is invested with full executive powers. The president appoints a cabinet, whose members are individually responsible to the Majlis (parliament). The Majlis has forty-eight members, forty of whom are elected by the people and eight appointed by the president. Every five years the Majlis elects a presidential candidate by secret ballot. A subsequent national referendum confirms the candidate by a simple majority vote. There are no political parties in the Maldives.

The constitution has no provisions for local government. For administrative purposes the government divides the country into nineteen atolls, with Malé forming the twentieth division. Each atoll has an atoll chief and every inhabited island has a *khatib* or island chief who are responsible for the functions of government. Each island also has a court, which administers justice in strict accordance with Islamic law.

1969: April Republic of the Maldive Islands renamed

The country is officially renamed Maldives.

1976: March 29 Last British forces leave the Maldives

The last British troops in the Maldives leave the islands. This date is subsequently celebrated as the country's Independence Day.

1977 Fish canning factory opens

The Maldive Nippon Corporation, a joint venture with Japan, opens a fish canning factory in Feliwaru in Fadiffolu Atoll.

The Maldives exports some ninety percent of its fish catch to Sri Lanka as dried fish known as "Maldive fish." In the 1970s, Sri Lanka limits imports of Maldive fish, and the emphasis in the industry switches to canned and, later, frozen fish. Government programs are put in place to modernize the fishing fleet.

1978 Television Maldives begins transmissions

Television Maldives begins operating in Malé, with its transmissions being received in a radius of twenty-five miles around the capital.

1978: November 11 Maumoon Abdul Gayoom becomes president

In the face of increasing public opposition, Ibrahim Nasir decides not to seek a third term as president. The Citizens' Majlis nominates Maumoon Abdul Gayoom to be president, and the appointment is confirmed by a national referendum.

Maumoon Abdul Gayoom is born to a middle class family in Malé in 1937. He studies at Al-Azhar University in Egypt, obtaining an MA in Islamic Studies. Maumoon Abdul Gayoom enters government service in Maldives in 1971, becoming Maldives' permanent representative to the United Nations in 1976. He returns to the Maldives in 1977 to become Ibrahim Nasir's minister of transport.

The new president commits himself to a more open and democratic style of government, to restoring freedom of the press, and to an increased role of the legislature and judiciary in running the country. On the economic front, the president makes development of rural regions a high priority.

1980 Garment production begins

The first factories producing ready-made garments for export open on Gan.

The port facilities, fuel storage areas, power plant, roads and buildings on the former RAF base prompts the government to promote its potential to foreign investors. By 1998, clothing and apparel rank second only to fish products in Maldives' exports.

1981: November 11 Maldives commissions international airport

Maldives enters the modern transportation age when it builds Maldives International Airport, which can handle wide-bodied jets. Malé has no space for the airport, so the government locates it on nearby Hulule Island, which has to be extended to accommodate the runway.

1982 Maldives joins the Commonwealth

Maldives joins the British Commonwealth.

1982 Central Hospital in Malé opens

Central Hospital, the first of four proposed regional hospitals, opens its doors to patients.

Health services in the Maldives is based on the primary care approach. Each village is served by family health workers, who provide preventative services and minor medical aid. Health workers refer more serious cases to the health centers that serve each of the atolls (administrative units) in the country. Each health center is equipped with a launch used for both routine services and emergency transportation. Cases that are beyond the capabilities of the health centers are referred to the regional hospitals.

1988: September Maldives competes in Olympic Games

The Maldives competes in its first Olympic Games, sending a 4 x 400 men's relay track team to Seoul, Korea.

1988: November 3 Mercenary coup attempt

A force of around eighty mercenaries, Sri Lankan Tamils allegedly recruited by a disaffected Maldivian business man, lands on Malé and attempts to seize control of the government. Although some members of the government are captured, the president escapes and goes into hiding. He appeals to the government of India for help and Indian troops are dispatched to the Maldives to restore the legitimate government.

1989: February Royal Dutch Shell prospects for oil

Maldives and the Royal Dutch Shell sign a contract allowing the Dutch company to prospect for oil in Maldives' waters. Energy is a critical sector of the Maldives economy. Petroleum ranks second among Maldives imports, and commercial and industrial development in the islands create a rapidly increasing demand for power. After a decade, however, no reserves of petroleum or natural gas have been discovered in the Maldives.

1989: November International conference on global warming

Maldives hosts an international conference to discuss the threat posed to low-lying island nations due to higher sea-levels that arise from global warming.

1993 Refrigeration plants under construction

Three refrigeration plants are under construction for the fishing industry. Two, being built with Japanese and Kuwaiti aid, are on Maamendhoo Island in Laamu Atoll, and the third in the Gaafu Alifu Atoll.

1993: October Maldivians re-elect President Gayoom

Maldivians elect the President to a fourth term in office by 92.8 percent of the popular vote in a national referendum.

1995 Fish exports

The value of marine products exported is 402.2 million rufiyaa (US $34.4 million at 1995 exchange rates). The main fish exported are canned tuna and frozen skipjack tuna.

1995: March Aid for educational development

The International Development Agency approves US $13.4 million credit to improve Maldives primary and secondary education. The main focus of development plans is to reduce teacher shortages and improve education in the atolls. English medium schools operate in Malé, preparing students for exams such as the London GCE (General Certificate of Education) 'O' and 'A' levels. Some higher education studies are possible in the Maldives, as for example at the vocational training center or the teacher training institute. For university education, however, students have to travel to foreign countries. Overall literacy levels in the Maldives stands at nearly 100 percent, unusually high for a South Asian country.

1997: May 12–13 Maldives hosts SAARC meeting

The leaders of seven Asian nations (Bangladesh, Bhutan, India, Maldives, Nepal, Pakistan, and Sri Lanka) gather in Malé for the Ninth Summit of SAARC, the South Asian Association for Regional Cooperation.

1997: September 13 Maldives takes Silver in SAFF Championships

The Maldives win the Silver Medal in the South Asian Football Federation (SAFF) Championships, losing 5–1 to India in the final. This makes up for the country's disappointing performance in World Cup 1998 qualifying in June, in which the Maldives went 0–6, losing by a record goal aggregate of 0–59. Football, i.e. soccer, is extremely popular in the Maldives.

Bibliography

Background Notes, Maldives. Washington, DC.: U.S. Department of State, Bureau of Public Affairs, Office of Public Communication, 1996.

Bell, H. C. P. *The Maldive Islands: Monograph on the History, Archaeology, and Epigraphy.* Colombo, Ceylon: Ceylon Government Press, 1940.

Chawla, Subash. *The New Maldives*. Colombo, Sri Lanka: Diana Agencies Limited, 1986.

Ibn Batuta, trans. H. A. R. Gibb., London, 1924.

Ellis, Kirsten. *The Maldives*. Hong Kong: Odyssey, 1993.

Heyerdahl, Thor. *The Maldive Mystery.* London: George Allen & Unwin, 1986.

Maldives: A Nation of Islands. Malé, Republic of Maldives: Media Tranasia, for Department of Tourism, 1983.

Maloney, Clarence. *People of the Maldive Islands*. Bombay: Orient Longman, 1980.

Phadnis, Urmila. *Maldives, Winds of Change in an Atoll State*. New Delhi: South Asian Publishers, 1985.

Pyrard. *The Voyage of François Pyrard of Laval to the East Indies, the Maldives, the Moluccas, and Brazil*. Trans. Albert Gray, 2 vols. London: Hakluyt Society, 1888.

Republic of Maldives, Office for Women's Affairs. *Status of Women, Maldives*. Bangkok: UNESCO Principal Regional Office for Asia and the Pacific, 1989.

Marshall Islands

Introduction

The Marshall Islands are composed of atolls and islands situated in two almost parallel chains in the easternmost region of Micronesia, midway between Hawaii and Australia. These two chains extend for approximately 800 kilometers each. One chain is referred to as the Ratak, or sunrise, group while the other is called the Ralik, or sunset, group. There are approximately 1,225 total islands, islets, and atolls that make up the Republic of the Marshall islands. In sum, these pieces of land cover a total of 181 square kilometers. Of this vast number there are only 29 coral atolls and five low-lying islands in the Republic of the Marshall Islands. The mean elevation of the land forms is only two meters above sea level. The highest point is Likiep Atoll which reaches a maximum elevation of six meters above sea level. The world's largest lagoon is located on Kwajalein.

The indigenous inhabitants of the Marshall Islands are called Marshallese. The Marshallese are a Micronesian culture group. Little is known about the prehistory of the Micronesian peoples or the Marshallese in particular. Most Micronesian groups did not engage in large-scale architectural endeavors so we do not have monumental remains as found in other parts of the world. Most of the items of material culture of the prehistoric Micronesians were made out of wood or other organic products which decay rapidly in damp environments of the Micronesian islands.

Traditional Marshallese society was organized according to principles of matrilineal descent. This means that an individual would trace his or her genealogy through his or her mother and his or her mother's relatives. Principles of land ownership and inheritance were also based on matrilineal relationships.

The complex system of classes that existed in pre-contact Marshallese society has been maintained almost completely intact to the present day. Basically, the Marshallese distinguish between commoners and nobles. Each matrilineage at each level of society had its own headman who would represent the group at larger meetings. In earlier times, marriages between nobles and commoners were not allowed.

Traditional practices of tattooing and ear lobe distending have long since disappeared from Marshallese culture. Tattooing among the Marshallese was accomplished through the use of a fish-bone lance that was struck by a small wooden mallet. The tip of the lance was dipped in coconut carbon to produce the pigmentation of the wound in the skin. Early explorers commented on the distended ear lobes of the Marshallese, some of which are reported to have reached the collarbone protrusions on the chest. While there have been attempts to invigorate traditional handicrafts for production and sale to tourists, there have not been any attempts to revive the traditional forms of bodily adornment.

Historically, the Marshallese language is closely related to the Kiribasi language of Kiribati (formerly known as the Gilbert Islands) as well as other Micronesian languages of the Caroline Islands group. The Ratak and Ralik island chains have separate dialects of Marshallese that are mutually intelligible. Today, English is the official language of the Marshall Islands and is almost universally spoken and understood.

Traders, beachcomers, and entrepreneurs began to take an interest in the Marshall Islands in the nineteenth century. The first European trader to settle in the Marshall Islands was a German merchant named Adolph Capelle. He worked initially as an agent for a German trading firm. Capelle was successful in catering to the diverse expectations of his hosts, the Marshallese, as well to those of the missionaries who had also become active in the area at the time. Capelle married a woman from Ebon, even though it was common for traders and beachcomers to take local mistresses, but not to marry them. In the 1860s, Capelle developed his own trading firm and began a coconut plantation. By 1875, he had relocated his operations to Jaluit and had his own fleet of schooners and agents representing him.

Japan seized control of the Marshall Islands in 1914, following the outbreak of World War I (1918–18). Japan placed many Micronesian islands, including Jaluit in the Marshalls, under naval administration. The Japanese seizure was not popular with the major Western European powers at the time and it was not until three years later that the Japanese occupation was formally recognized by Great Britain, France, and Russia. The League of Nations formally mandated Japan to administer Micronesia in 1920. During the Second World War

(1939–45), the islands were the scene of heavy fighting between American and Japanese forces.

Nuclear testing looms large in the history of the Marshall Islands since the end of World War II. Many of the islands within the region were adversely affected by fallout from repeated nuclear and hydrogen detonations. The inhabitants of Bikini Atoll were resettled after their island was contaminated by a series of nuclear tests that were conducted by the United States during the period between 1946 and 1958. A few Bikinians returned to their atolls surrounding Bikini Lagoon after the region was declared safe for human habitation in the late 1960s. However, soon after their repatriation, the United States government relocated them to Kili Island where most Bikinians still reside. About 500 Bikinians have moved to southern California.

The Marshall Islands became a self-governing republic on May 1, 1979. From 1947 until independence, the Marshall Islands had been part of the larger Trust Territory of the Pacific Islands (TTPI). The TTPI had been formed after the end of World War II (1939–45) and in response to recognized needs for development and military protection. The TTPI was classified by the United Nations, which set up the trusteeship, as a "strategic area." This designation allowed the United States, as the administrator of the trusteeship, to establish military bases, build fortifications and deploy armed forces as deemed necessary.

There were six district components of the TTPI: the Mariana Islands, the Marshall Islands, Ponape, Palau, Truk, and Yap. The administrative headquarters of the TTPI was located in Saipan. The entire area existed under what is referred to as a Trust Territory Government. The Trust Territory Government is democratic and consists of a judicial, legislative, and executive branch. The legislative authority was exercised by the Congress of Micronesia.

During the 1950s, the United States did little to aid the islands within TTPI. However, in the 1960s, the United States increased aid to the region dramatically. In the period between the early 1960s and the early 1970s, the United States increased the amount of financial aid that it gave the region tenfold. In 1970, a commission whose goal was to assess the political and economic future of the TTPI recommended self-government with "free association" with the United States. The Marshall Islands District became the first to declare self-government of the Federated States of Micronesia which also included Palau, Yap, Truk, Ponape, and Kosrae.

The Republic of the Marshall Islands adopted its constitution in 1979. It became independent on October 21, 1986. On that same date, the Republic of the Marshall Islands entered into a Compact of Free Association with the United States. Under this arrangement, the islands received full self-government and agreed to drop all future claims against the United States, while the United States agreed to continue to provide for the islands' defense. The president serves as the

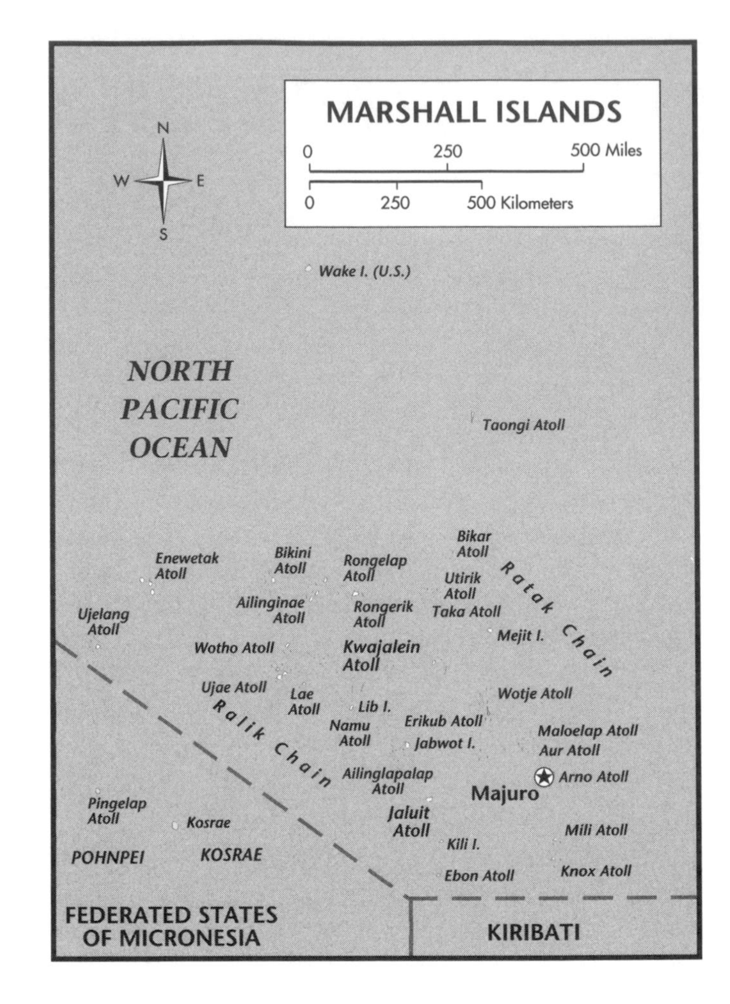

chief of state and the head of government. The president is elected to serve four-year terms without term limit. Suffrage is universal for all Marshallese eighteen years of age and older. The president appoints a cabinet selected from the members of parliament. The Republic of the Marshall Islands joined the United Nations in 1991.

The economic mainstays of the Republic of the Marshall Islands are agriculture and tourism, the most important commercial crops being coconuts, tomatoes, melons, and breadfruit. With few natural resources, the level of imports far exceeds the level of exports in the Marshallese economy.

Due to the efforts of the nineteenth century missionaries, Christianity is the dominant religion in the Republic of the Marshall Islands, and there is a church on almost every island. Protestants are in the majority in the Christian faith, although Roman Catholics and Seventh Day Adventists maintain long-established large missions in the islands. Baptists, Mormons, Jehovah's Witnesses, and other sects of Christianity operate missions out of Majuro. Private schools are also maintained by the Catholics and Seventh Day Adventists.

Migration has been a problem for the Marshallese government. The Republic of the Marshall Islands formally recognized the large immigrant Marshallese community in southern California by opening a consulate in Newport Beach in 1993. This immigrant community has developed strong economic, religious, and educational ties in the region and hopes to provide opportunities for future generations of Marshallese immigrants to the United States.

Timeline

2000 B.C. Earliest settlement of Marshall Islands

Archaeological evidence has produced a date of approximately 2000 B.C. for the earliest habitation of Bikini Atoll within the Marshall Islands. Controversy surrounds the dating of the material and other, more conservative, archaeologists adhere to date of only 500 B.C. for the earliest habitation.

1494 Treaty of Tordesillas cedes ownership of Micronesia to Spain

The Treaty of Tordesillas cedes ownership of all of the known Micronesian islands and atolls to Spain.

1529 First European records siting of Marshall Islands

Spaniard Alvaro de Saavedra is the first European to site the Marshall Islands.

1566 Sailors are marooned on Ujelang Atoll

A mutiny is thwarted on the *San Jeronimo*, a Spanish ship, and twenty-seven Spaniards are marooned on Ujelang Atoll.

1767 Captain Samuel Wallis lands on Rongerik

Since the Spanish siting of the Marshall Island chains in the sixteenth century, European activity in the islands has been non-existent. Captain Samuel Wallis (1728–95) who is credited with the European discovery of Tahiti, lands on Rongerik in 1767. His expedition is successful at circumnavigating the earth (1766–68).

1788 British naval captain makes additional discoveries in Marshall Islands

The islands are sighted by British naval captain, John Marshall. Marshall is sailing between Australia and China on his ship, the *Scarborough*. His name is later given to the islands by a Russian hydrographer (a person who charts bodies of water). Marshall frequently sails through the islands transporting convicts between Botany Bay and Cathay in southern China.

European explorers recount the elaborate tattoos that adorn the bodies of the inhabitants of the Marshall Islands. The tattoo ceremony extends over several weeks and is a rite of passage for both young men and women.

Tattoo patterns are based on designs from nature, especially from the ocean. Shell patterns, waves, and fish shapes are achieved through patterns of dots arranged in straight or wavy lines. For men, vertical designs, known as the "Mast," ascend from the navel. The lower chest is decorated to depict ocean waves. A man's back has what is referred to as Big Tattooing, consisting of heavy vertical bars, running from armpit to armpit. Women limit tattooing to arms, shoulders, and thighs.

1857 First mission is established

The Reverend Hiram Bingham, Jr. (1831–1908) of the American Board of Commissioners of Foreign Missions establishes the first mission on Ebon.

1878 Germany secures exclusive harbor use

Germany secures exclusive use of the Jaluit harbor and special trading privileges with islanders in the Ralik chain by entering into a treaty with a local ruler.

1885: October 13 The Marshall Islands become a protectorate of Germany

German chancellor Otto von Bismarck (1815–98) negotiates with Britain regarding colonial territories in the South Pacific. The Marshall Islands become an Imperial German Protectorate. This status is recognized internationally by Spain and Britain. The early German rule of the islands was left in the hands of the local Marshallese chiefs who are overseen by a small staff of Germans.

1898 Germany receives ownership of disputed territory

Ujelang and Enewetak Atolls which were disputed territory are transferred to German ownership at the end of the Spanish-American War.

1914: August Japan seizes territory from Germans

With the onset of World War I (1913–18), the Japanese government sends troops to the Marshall Islands, where they overcome German forces and assume occupancy of the islands.

1920 Japan formally assumes administration of Micronesia

When World War I ends, the League of Nations formally mandates Japan to administer the Micronesia which includes the Marshall Islands.

1933 Japan withdraws from League of Nations

Following international condemnation of its occupation of Manchuria, Japan withdraws from the League of Nations but maintains possession of the Marshall Islands. The Japanese initiate militarization of the Marshall Islands.

1938 Japan declares Micronesia a closed military area

Japan declares Micronesia as part of its empire and designates the entire region as a closed military region. Military build-up in the region culminates in a series of attacks on the United States (see 1941: December).

1941: December Japan uses Marshall Islands for attacks

Japan uses Marshall Islands as a base for attacks on the United States naval base at Pearl Harbor, Hawaii, and on other Allied targets to the south and east.

1943: December Allied invasion of Marshall Islands

U.S. troops begin invasion of Japanese stronghold in the Marshall Islands. They attack a Japanese stronghold on Kwajalein atoll and over 8,000 Japanese soldiers are killed. Fewer than 400 U.S. troops are lost in the attack.

1944 Allies take possession of Marshall Islands

Allied troops complete their invasion of the Marshall Islands and route the Japanese from their fortifications on the islands of Majuro and Kwajalein.

On Majuro, the Japanese government later erects the Peace Park Memorial, a small granite monument to honor soldiers who fought in World War II in the South Pacific.

1946: March Bikini Islanders are evacuated for nuclear testing

The first post-World War II nuclear testing to be undertaken by the United States is slated to begin in July under the name Operation Crossroads. To prepare for the upcoming tests, the 167 inhabitants of Bikini Atoll are relocated to Rongerik Atoll, 125 miles east.

1946: May Enewetak, Rongelap and Wotho Atolls are evacuated

Concerned about the effects of the nuclear tests, known as Operation Crossroads, that are set to commence in July, the United States government relocates the islanders of Enewetak, Rongelap, and Wotho Atolls.

1946: July "Able" and "Baker" nuclear tests on Bikini Atoll

Operation Crossroads commences with "Able" and "Baker" tests on Bikini Atoll. "Baker" test is an underwater test. Both nuclear detonations are nuclear bombs the size of the one that was dropped on Hiroshima during World War II.

1947: April United Nations approves trusteeship

The United Nations Security Council votes to designate Marshall Islands a United Nations Trust Territory, with the United States designated as administrator.

1947: December Enewetak Atoll is selected for second series of nuclear tests

A second series of nuclear tests, known as Operation Sandstone, is scheduled to take place in April 1948 on Enewetak Atoll. The inhabitants of Enewetak Atoll are relocated to Ujelang Atoll. This population and its descendants will not be repatriated to their homeland on Enewetak until 1980.

1948: March Bikinians moved to Kwajalein

The relocated Bikinian population in a state of near starvation is relocated to Kwajalein until a suitable new home can be found.

1948: April Operation Sandstone commences

A series of three nuclear tests are scheduled to take place during Operation Sandstone on Enewetak Atoll.

1951: April Operation Greenhouse commences on Enewetak Atoll

Four nuclear tests are scheduled to take place on Enewetak Atoll under the name Operation Greenhouse.

1952: November Operation Ivy commences on Enewetak Atoll

Operation Ivy, which includes the first testing of a hydrogen device, begins on Enewetak Atoll. The "Mike Test" vaporizes a small island with a force of 10.4 megatons, 750 times more powerful than the atomic bomb that was dropped on Hiroshima.

1954: March 1 Bravo hydrogen bomb test on Bikini Atoll

Despite unfavorable weather conditions, the United States proceeds with the detonation of the "Bravo" hydrogen bomb test on Bikini Atoll, part of the larger nuclear testing plan initiated earlier in the year under the name Operation Castle. The Bravo test is a fifteen megaton bomb. White ash fallout rains down on the islanders on Rongelap, Utrik, and Ailinginae Atolls, as well as on American weathermen stationed on Rongerik Atoll.

1954: March 3 Rongelap and Utrik islanders evacuated

The islanders from Rongelap and Utrik Atolls are evacuated to Kwajalein to receive medical treatment. Skin burns and hair loss are the results of the exposure to the fallout of the March 1st test.

1956: May Operation Redwing begins

A series of seventeen nuclear tests begin on Enewetak and Bikini Atolls.

1958: May Operation Hardtrack commences

Operation Hardtrack which plans thirty-two nuclear tests, begins on Bikini and Enewetak Atolls.

1958: August 18 Final nuclear test in Marshall Islands

The final nuclear detonation takes place as the culmination of Operation Hardtrack. A total of sixty-six nuclear tests have taken place in the Marshall Islands since the inception of nuclear testing by the United States in 1946.

1965 Congress of Micronesia is formed

The Congress of Micronesia convenes with representatives from all of the Trust Territory of the Pacific Islands (TTPI) islands. The Congress, established by the U.S. administration, will prepare Micronesia for self-government. Marshall Islands is among the TTPI entities to send representatives.

1966 Congress approves payments to Rongelap people

The U.S. Congress approves a plan of payments of $11,000 per person to the former inhabitants of Rongelap Atoll in compensation for hardship caused by nuclear testing.

1969: October U.S. government declares Bikini Atoll safe

United States government officials declare that Bikini Atoll is safe for rehabitation by the Bikinians. Based on information that radiation has dropped to normal levels, U.S. President Richard M. Nixon announces that it is safe for islanders to return to both Rongelap and Bikini.

Housing construction begins, and people return to their old way of life—fishing in the lagoons, harvesting coconuts, breadfruit, and taro. Although radiation readings seem to indicate that the environment is safe, the radioactive isotope cesium 137 is concentrated in the soil. It enters the food chain, and the islanders are affected.

1972: October Bikinian Council refutes U.S. claims of safety

The Bikinian Council votes not to return to Bikini Atoll as a group in spite of United States assurances that the levels of nuclear waste and radiation exposure are well within safe limits. Several families decide to return to the island anyway.

1973 Alele Museum is completed

The Alele Museum in Majuro opens. It houses historical documents and photographs, including hundreds of glass-plate negatives of photos taken by missionaries and traders from 1890 to 1930.

1974 Airport begins operations

The Majuro International Airport begins operations. The runways can accommodate jet aircraft.

1975 Bikinian Islanders file suit against United States

The Bikinian Islanders file a suit in federal court demanding the United States government initiate and complete a full scientific survey of the entire northern region of the Marshall Islands.

1976: July Congress approves 20 million dollars toward cleanup

The U.S. Congress approves $20 million for cleanup of nuclear waste on Enewetak Atoll, which begins in May, 1977.

1978 Contitutional Convention meets

Marshall Islands Constitutional Convention meets and adopts the nation's first constitution. (See 1979.)

1978: September Bikinians are evacuated

United States military forces evacuate the 139 Bikinians who have been residing on the atoll since 1969. At the same time, Congress establishes a trust fund for the future of the Bikinian Islanders in the amount of $6 million.

1979 Citizens of Marshall Islands approve a new constitution

The citizens of the Marshall Islands approve the new constitution (see 1978) and the republic becomes a self-governing territory. Amata Kabua (1928–96) is elected the first president. He leads the Ailin Kein ad Party. Kabua also writes the words and music for the new nation's national anthem.

1980 Airline of the Marshall Islands begins operation

The Airline of the Marshall Islands (AMI), offering air transportation between sixteen of the inhabited islands, begins operation. Service is also provided to neighboring Kiribati, Tuvalu, and Fiji.

1980: March U.S. Nuclear Agency declares Enewetak Atoll cleanup complete

The Nuclear Defense Agency of the United States declares that the cleanup of Enewetak Atoll is complete and that islanders can begin returning to their islands. The total cost of the cleanup is estimated to have been $218 million.

1982 Republic of the Marshall Islands becomes official name

The official name of the country becomes Republic of the Marshall Islands (RMI).

1982 Congress establishes a second trust fund for Bikini

The U.S. Congress establishes a second trust fund for Bikini Islanders in addition to the original trust fund established just four years earlier. Twenty million dollars is placed in the fund for compensation.

1983 Voters approve the Compact of Free Association with the United States

Voters in the Republic of Marshall Islands (RMI) approve the terms of the Compact of Free Association with the United States. The Compact takes effect three years later (see 1986: October 21).

1984 Environmental Protection Agency established

The Marshall Islands Environmental Protection Agency is established to monitor environmental quality and initiate protection programs. Initial programs deal with water quality and solid waste disposal.

1984 Court considers Traditional Rights cases

A Traditional Rights Court is created to consider legal issues related to land claims under traditional customs and similar issues. Marshall Island society is organized according to *bwij* (clan), and the *alap* (head of the clan) traditionally serves as spokesperson, negotiating with other clans in matters under dispute.

1984: May 2 Marshall Islands postal service inaugurated

Marshall Islands postal service is launched. Their first stamp series issued depicts maps of the islands and instruments used for ocean navigation.

1985 Plant to smoke tuna opens

A small plant, where tuna is smoked and dried, begins operation. Most of its product is exported to Japan.

1985: May Rongelap Islanders evacuate

Rongelap Islanders, concerned about medical conditions that continue to be present among the population, decide to relocate to Mejatto, a small island near Kwajalein Atoll.

1986 Government establishes dairy factory in Majuro

The government supports the establishment of a factory to produce milk and other dairy products in Majuro.

1986 College of Micronesia opens nursing and science college

The College of Micronesia establishes a School of Nursing and Science at Majuro.

1986: October 21 Compact of Free Association takes effect

The Compact of Free Association between the Marshall Islands and the United States, approved earlier by Congress, takes effect. The Compact includes a provision that prohibits Marshall Islanders from seeking future legal redress or compensation and dismisses all current claims against a trust fund totaling $150 million.

1991: September 17 U.N. admits Republic of the Marshall Islands

The Republic of the Marshall Islands (RMI) and the Federated States of Micronesia are both admitted to membership in the United Nations (UN).

1991: August Nuclear Claims Tribunal approves first compensation award

The Nuclear Claims Tribunal with a total fund of $45 million, approves the first compensation award for medical care resulting from nuclear testing in the Marshall Islands.

1995: October Advisory Committee on Human Radiation Experiments issues final report

President Clinton's Advisory Committee on Human Radiation Experiments issues its final report which includes testimony concerning the misrepresentation of radiation dangers to islanders as well as testimony concerning the intentional resettlement of islanders on contaminated atolls with the expressed goal of studying the long-term effects of radiation on human beings.

1995: November Amata Kabua is re-elected President

Having served as President of the Republic of the Marshall Islands since 1979, Amata Kabua was re-elected in November of 1995. The next scheduled presidential elections in the Marshall Islands will take place in November of 1999.

1995: November 30 Stamp honors Itzhak Rabin

The Republic of the Marshall Islands issues a postage stamp honoring Itzhak Rabin (1922–95), the prime minister of Israel and Nobel Peace Prize winner who was assassinated on November 4, 1995.

1996 Nuclear Claims Tribunal projects cost over-runs

The Nuclear Claims Tribunal projects that over $100 million in medical compensation claims will be filed before the termination of the Compact of Free Association in 2001. The total funding for the Tribunal is $45 million.

1996: June Bikini Lagoon is opened to divers

Bikini Lagoon is opened for scuba diving. Several battleships from World War II, including USS *Saratoga* and the submarines *Pilotfish* and *Apogon,* were sunk in Bikini Lagoon as part of the nuclear tests conducted during the 1940s and 1950s. One of the most famous of these, the Japanese flagship *Nagato* which served as the floating fortress for the Japanese attacks on Pearl Harbor, is a primary.

1996: December 12 President Kabua dies

President Amata Kabua, the Marshall Islands' only president since independence dies in office. He is sixty-eight years old. The government names the minister of transport and communications, Kunio Lemari, as Acting President.

1997: January 14 Imata Kabua elected President

Imata Kabua, a cousin of the late President, wins a special election and becomes President of the Marshall Islands.

2001 Compact of Free Association expires

The agreement between the United States and the Marshall Islands, which prohibits further legal claims against the government, expires.

Bibliography

Conrad, Robert. *A Twenty Year Review of Medical Findings in a Marshallese Population Accidentally Exposed to Radioactive Fallout.* Upton, NY: Brookhaven National Laboratory, 1975.

Denoon, Donald, ed. *The Cambridge History of the Pacific Islanders.* Cambridge: Cambridge University Press, 1977.

Grattan, C. Hartley. *The Southwest Pacific Since 1900: A Modern History.* Ann Arbor: University of Michigan Press, 1963.

Hines, Neal. *Proving Ground: An Account of the Radiobiological Studies in the Pacific, 1946–1961.* Seattle: University of Washington Press, 1962.

Johnson, Giff. *Collision Course at Kwajalein: Marshall Islanders in the Shadow of the Bomb.* Honolulu: Pacific Concerns Resource Center, 1984.

Langley, Jonathan, and Wanda Langley. "The Marshall Islands." *Skipping Stones,* Winter 1995, vol. 7, no. 5, p. 17+.

Lessa, William. *Drake's Island of Thieves: Ethnological Sleuthing.* Honolulu: University of Hawai'i Press, 1975.

Levesque, Rodrigu, ed. *History of Micronesia.* Honolulu: University of Hawaii Press, 1994.

Scarr, Deryck. *History of the Pacific Islands: Kingdom of the Reefs.* Sydney: Macmillan Australia, 1990.

Vayda, Andrew P., ed. *Peoples and Cultures of the Pacific: An Anthropological Reader.* Garden City, NY: *Natural History Press,* 1968.

Mongolia

Introduction

Modern Mongolia is the largest landlocked country in the world, with an estimated population of 2.5 million. Its people are scattered across 604,250 square miles (1.7 million square kilometers), giving it one of the lowest population densities of all nations. There are more Mongols living abroad than in Mongolia. Nearly 3.5 million Mongols live in China, mainly in Inner Mongolia, Xinjiang, and Höhnuur. Nearly one million Mongols live in the Buryatia, Irkutsk, Chita, and Volga regions of the Russian Federation.

Archeological evidence shows that people have lived in Mongolia since the Lower Paleolithic period, more than 130,000 years ago. A clan style of social organization based on horsemanship later emerged, with shamans functioning both as physicians and clerics. Driven westward during the Han dynasty in China (206 B.C.–A.D. 220), Turkic-speaking Huns created a nomadic empire in central Asia that nearly reached Rome during the leadership of Atilla (r. 433–456).

Under the leadership of the conqueror Temüjin, or Genghis Khan (Chinggis Khan, c. 1162–1227), the Mongol tribes united and rapidly conquered the steppes of Asia. The empire eventually stretched from the northern Siberian forest to Tibet and from the Caspian Sea to the Pacific. After his death in 1227, the empire was divided among his sons into several Mongol states known as *khanates.* Kublai (Khubilai, 1216–94), the grandson of Genghis Khan, later conquered China and became the founding emperor of the Yüan dynasty. The Mongol empire was at its height in the mid-1200s, stretching from the Korean peninsula and Vietnam all the way to Eastern Europe, making it the largest empire in history. By the mid-1300s, however, the great Mongol states had disintegrated. The Yüan dynasty in China collapsed in 1368, and thousands of Mongols returned to the Mongolian homeland. The other khanates also fell apart and were absorbed during that century into other empires. The Mongols' downfall stemmed from a dependence on their conquered peoples and infighting for power. Their period of dominance in Asia, however, permitted a significant amount of trade and cultural interchange. Modern cultures across Asia and Eastern Europe still have remnants of the Mongol influence in their languages, arts, and customs.

When the Mongols returned to their original homeland, they split into three major groups: the northern Khalka Mongols, the southern Chahar Mongols, and the western Oirat Mongols. The Khalka and Oirat groups fought a prolonged civil war. In the late 1400s, Mongols in the east unified under the leadership of Batmönh Dayan Khan (1460–1517), who split his khanate into six military districts: Tsahar, Halh, Urianghai, Ordos, Tümed, and Yongshebu. Mongolia's modern population is largely descended from the Halh, which has become the preferred ancestry of Mongolians. Altan Khan (1507–83) of Inner Mongolia later managed to reunite most of the Mongol clans with the vain hope of reestablishing the Mongol Yüan dynasty in China. During Altan Khan's reign, the Mongols and Chinese signed a peace treaty.

In either 1576 or 1578, Altan Khan bestowed the title *Dalai* (oceanic) upon Tibet's most senior *lama* (priest), Sonam Gyatso (1543–88), thus creating the office of the Dalai Lama. Lamaist Buddhism had been introduced to the Mongols from Tibet in the previous century. After the death of Altan Khan, however, Mongol unity fell apart. In either 1639 or 1641, Zanabazar (1635–1723) became the first *Bogd Gegeen* (Holy Enlightened One), leader of Mongolia's Buddhists, similar to the Dalai Lama of Tibet. A cultural rift developed between the Outer (northern) Mongols and the Inner (southern) Mongols. In 1636, all the southern Mongol *khans* (lords, or princes) came under the rule of China's Qing (Ch'ing) or Manchu dynasty, which held power until 1911. In 1691, the northern Mongols also accepted Manchu rule in order to wipe out the western Mongols. During the mid-1700s, border treaties between Russia and China gave the Chinese rule over both the southern and northern Mongol areas but assigned Buriat lands to Russia. The Buriats had previously been vassals of the northern Khalka Mongols.

The Modern Era

Following the overthrow of the Manchu dynasty in 1911 by the Chinese Revolution, northern Mongol princes proclaimed an autonomous Outer Mongolia under the rule of the Bogd Gegeen, the leader of Mongolia's Buddhists, with the capital

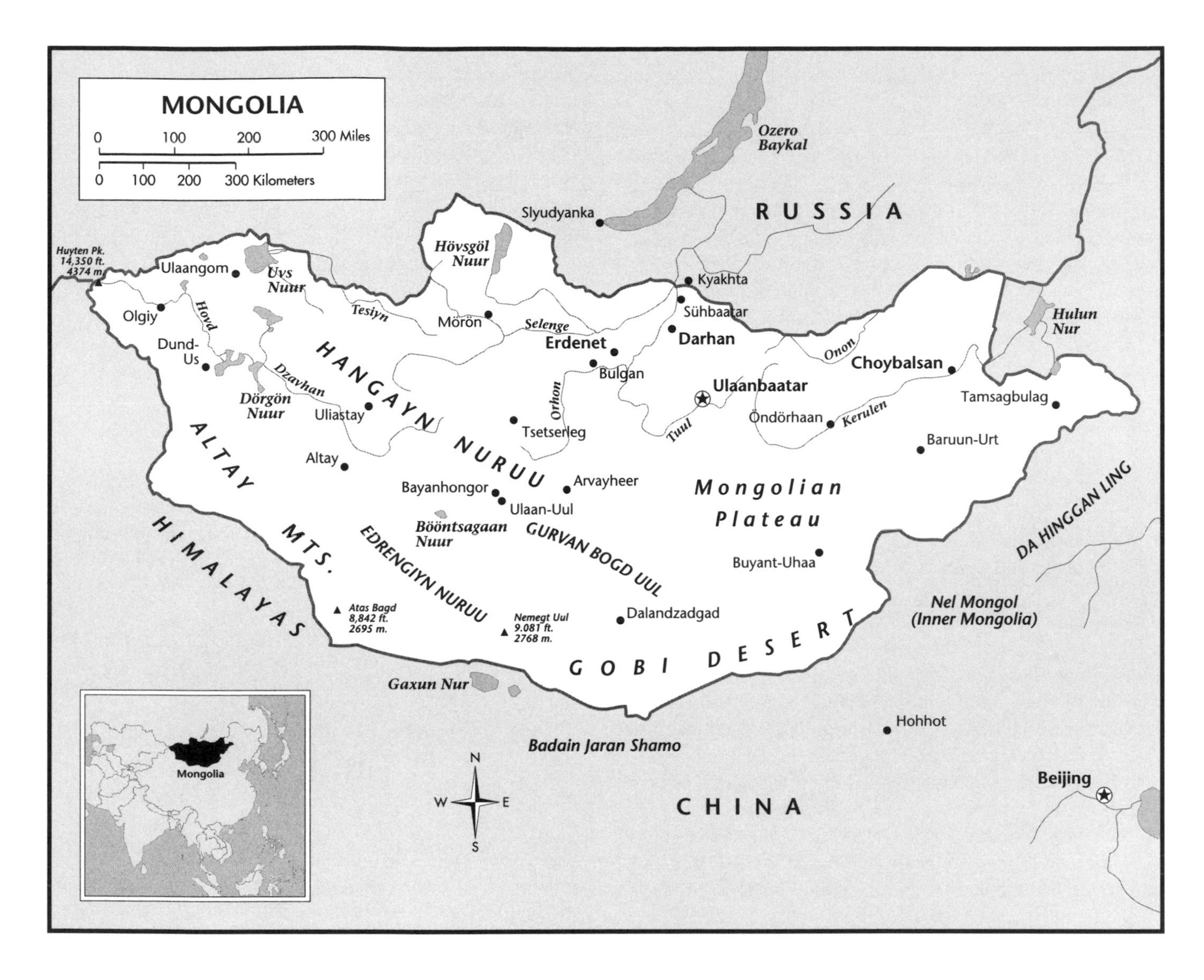

at Ugra (now Ulaanbaatar). In 1915, China and Russia signed a treaty that took away Mongolia's independence but made it an autonomous part of China. After the Bolshevik Revolution of 1917 in Russia, the Chinese exploited Russia's weakness by reoccupying Outer Mongolia in 1919 and making it part of the Chinese state in 1920. Counterrevolutionary Russian forces (Whites) drove out the Chinese in 1921, but the Whites were routed later that year by forces loyal to the Mongol revolutionary leaders, Sukhe Baatar (Damdiny Sükhbaatar, 1893–1923) and Khorloin Choibalsan (1895–1952), who were assisted by the Soviet Red Army.

Under Soviet influence, in July, 1921, a nominally independent Mongolian state was proclaimed. In a bizarre political alliance, the Bogd Gegeen was reinstated as its monarch, but many communist revolutionaries took key political positions in the new government. When the monarch died in May, 1924, the communist government prohibited Buddhist leaders from naming a successor.

In November, 1924, the Mongolian People's Republic was created as the world's second communist country (and the first Soviet satellite). Support from the USSR consolidated the Mongolian People's Revolutionary Party as the state's sole political party. The new government began confiscating property in 1929. The Soviet Union urged the Mongolian government to force people into collectivized agriculture. Since Mongolia was a primitive nomadic society, however, the scheme was hopelessly impractical. Order broke down in 1932 when popular rebellion arose against the policy. As a result of the national insurrection, the Soviet Union stationed troops and intelligence forces in Mongolia. The country became a police state, where political leaders, military officers, and civilians were controlled through fear and intimidation. Between 1937 and 1939, over 20,000 people were executed for allegedly conspiring or plotting against the state.

In the 1930s, the government also began promoting an athiest agenda to eliminate Mongolia's traditional religion, Lamaist Buddhism. By the late 1930s, virtually all the monasteries were closed and their livestock and lands confiscated. Thousands of lamas were either executed or died at labor camps in Siberia.

In the years leading to World War II, the Soviet Union signed a mutual assistance treaty with the Mongolian People's Republic. When Japan invaded Manchuria in August, 1939, Mongolia (with Soviet support) successfully fought invading Japanese forces along the border with Manchuria. Mongolia supplied the Soviet Union with equipment during the war. Toward the close of the war, the governments of Great Britain, the United States, and the Soviet Union agreed that Mongolia should remain independent. In the final days of the war, Mongolia formally declared war on Japan. Mongolian troops were utilized to finish off Japanese forces in Manchuria, the Gobi Desert, and Inner Mongolia.

Independence

After a unanimous plebiscite (binding vote on an issue decided by the citizenry) by the Mongolians in favor of independence, the Nationalist government of the Republic of China formally recognized the Mongolian People's Republic in early 1946. In 1950, the governments of the Mongolian People's Republic, the new People's Republic of China, and the USSR signed a treaty guaranteeing Mongolia's independence. Upon the death of Choibalsan in 1952, Yumjaagjiin Tsedenbal (1916–91) became Mongolia's leader. Tsedenbal's rule transformed the government into a highly centralized and ineffective bureaucracy, and the economy stagnated. His dominance extended from the 1950s until the mid-1980s. During the earlier years of his reign, Mongolia adopted a new socialist constitution and joined the United Nations. Mongolia also strengthened its ties with the Soviet Union as it distanced itself from China. In 1984, Tsedenbal was removed from power and replaced by Jambyn Batmunkh (b. 1926), supposedly because of his anti-Chinese rhetoric and policies.

As the Soviet Union adopted the policies of *glasnost* (openness) and *perestroika* (restructuring) in the mid-1980s, so also did Mongolia's communist leadership. The Mongolian government initiated its own policy of *il tod* (openness) and began reforms that served as transitional steps away from central planning and a collective economy and toward a market economy. After forty years of indifferent association, Mongolia and China established diplomatic relations. In late 1989, demonstrations in Ulaanbaatar prompted the ruling Mongolian People's Revolutionary Party (MPRP) to address political reforms in addition to the ongoing economic reforms. By early 1990, the first opposition party had formed, and a pro-democracy demonstration in Ulaanbaatar incited ruling government officials to resign. The vicious crackdown against pro-democracy demonstrators in China the previous year probably influenced officials to refrain from using violence to subdue protesters. The constitution was amended to permit multiparty elections, which were held for the first time in July, 1990. In 1991, the government issued vouchers to all citizens for the purchase of state-owned property as a step toward privatization. The collapse of the Soviet Union, however, slowed economic reforms. Russia dramatically cut its aid and insisted on cash-based trade rather than barter. The government rapidly sought to dismantle its socialist planned economy in favor of a market-oriented capitalist economy.

In February, 1992, the Communist-era legislature adopted a new constitution, the first one based on a democratic form of government. The MRRP won a majority in the first elections for the new parliament. Later that year, the last of the former Soviet Union's troops were withdrawn from Mongolia. In June, 1993, Mongolian voters chose incumbent Punsalmaagiyn Ochirbat as the country's first directly-elected president. Demonstrations against government graft and corruption in 1994 brought about legislative changes respecting freedom of the press and the right to peaceful demonstration. In the spring and early summer of 1996, wildfires across northern Mongolia rendered hundreds of people homeless and destroyed twenty percent of the country's coniferous forests. In June, 1996, the MPRP was defeated in parliamentary elections by a coalition called the Democratic Union. The election marked the first peaceful transfer of power in Mongolia's modern history.

Timeline

1000 B.C. Development of nomadic animal husbandry

Nomadic people (people who move, carrying their homes with them) raise animals for food and to work.

300 B.C. Development of an equestrian society

A clan style of social organization based on horsemanship emerges. The *tahi* (Przewalski's horse) is believed to be the ancestor of the Mongolian horse and several breeds of modern horses. It is believed that their strict social customs to avoid incest (such as the prohibition of intermarriage within nine generations) come from the observation of their horses, which instinctively avoid incest. Centuries later, these incest-avoidance customs result in an excellent knowledge of genealogy and family history among Mongols.

300 B.C. Housing is the ger

The ancient Mongols begin living in *gers* (yurts). The ger is a large round felt tent constructed with a wooden frame to withstand the high winds and extreme temperatures of the Asian steppe. Early rigid gers are drawn by horses intact from camp to camp, while later collapsible models are more easily transported. Many Mongol traditions, superstitions, and customs originate from the nomadic ger life-style.

300 B.C. Shamanism develops

Shamanism develops among the Mongolians. A shaman is usually born into the family of a shaman, and can be either male (a *boo*) or female (an *udgan*). The shaman functions as physician and cleric, treating illnesses and warding off evil spirits. One of the legacies of shamanism in Mongolia is the *ovoo* (or *oboo*), a ritual cairn consisting of wood, stones, animal parts, and trinkets erected as an offering to the gods. Ovoos are still erected and shown respect in Mongolia.

206 B.C.–A.D. 220 Huns rise to power

The Turkic-speaking Huns, driven westward during the Han dynasty in China (206 B.C.–A.D. 220) create a nomadic empire in central Asia that eventually extends into Europe by A.D. 370.

433–53 The rule of Atilla the Hun

Under the leadership of Atilla (406–53), the marauding Huns nearly reach Rome.

The Mongol Empire

1200s Some Mongol men have more than one wife

Some Mongol men, mostly those of higher social class, are polygamous (have more than one wife).

1206 Mongol tribes unite under Genghis Khan

Under the leadership of the conqueror Temüjin, or Genghis Khan (Chinggis Khan, c. 1162–1227), Mongol tribes unite and set up their capital at Karakorum. Genghis Khan receives his title in 1189 (which means "universal king" or "oceanic king"), and his army of up to 200,000 men rapidly captures the steppes of Asia. The Mongols' first move against China begins in 1211–16. The empire eventually extends from the northern Siberian forest to Tibet and from the Caspian Sea to the Pacific. The electrum (gold and silver alloy) *sükh* becomes the coin of the realm.

Among modern-day Mongolians, Genghis Khan is revered as an inspirational leader who united his people. The descendants of those whom he conquered, however, view Genghis Khan as a ruthless warrior.

1210–1949 Rapid delivery system operates

One of the world's first express delivery services is created during the reign of Genghis Khan. The system, known as *urton*, is a network that uses thousands of horses to deliver important messages across the vast empire. Through a system of relay stations, riders quickly change horses and transport messages some sixty miles per day. The urton system is steadily modernized over the centuries and continues in Mongolia until 1949. It later becomes the model for the U.S. pony Express.

1227 Genghis Khan dies

After the death of Genghis Khan, his empire is divided among his sons into Mongol states known as *khanates:* the Great Khanate of East Asia, the Khanate of Chaghadai in Turkestan, the Hulagid Khanate in Persia, and the Golden Horde in southern Russia.

The Great Khanate of East Asia gives rise to the Yüan dynasty in China and later reaches its peak under Kublai Khan (1216–94). The Golden Horde is founded by Batu Khan (d. 1255), who later invades Poland and Hungary in 1240. The Mongols' Golden Horde dominates Russia until 1502.

1228–40 *Secret History of the Mongols* is written

The *Secret History of the Mongols* (known as *Mongol-un Nigucha Tobchiyan* in Mongolian) is written allegedly by an anonymous Mongol author. The work is the only known substantial historical text of the period that is written in the Mongolian language. A phonetic Chinese script is used to represent Mongolian words, allowing the work to remain relatively unchanged over the centuries. The document, although not entirely factual, is nevertheless important because it chronicles some of the details of thirteenth-century Mongol society. The work is not translated into English until the twentieth century.

1229 Ögedei succeeds Genghis Khan

Genghis Khan's third son, Ögedei (Ögödey, Ogdai, 1186–1241), is made khan. Under his leadership, the Mongols begin construction of the fortified capital city, Karakorum, in 1235. Mongol soldiers invade Korea in 1231and take over the Crimea in 1239.

1241–46 Ögedei's widow rules

When Ögedei dies, his wife assumes power until a successor is selected. There is a five-year period characterized by power struggles among Ögedei's heirs.

1260 Mongol empire reaches its maximum size under Kublai Khan

Kublai (Khubilai, 1216–94), the grandson of Genghis Khan, is elected Great Khan. He conquers China and becomes the founding emperor of the Yüan dynasty, with the capital in Tatu (now known as Beijing). The Mongol empire stretches from the Korean peninsula and Vietnam (captured in 1257) all the way to Eastern Europe, making it the largest empire in history. Although the Mongols go on to invade Burma (1277) and India (1288), defeats in Egypt (1260), Java, and Japan (1274 and 1281) signal the turning point for the empire. Chabi, Kublai Khan's wife, assists him in governing China during the final years of his life.

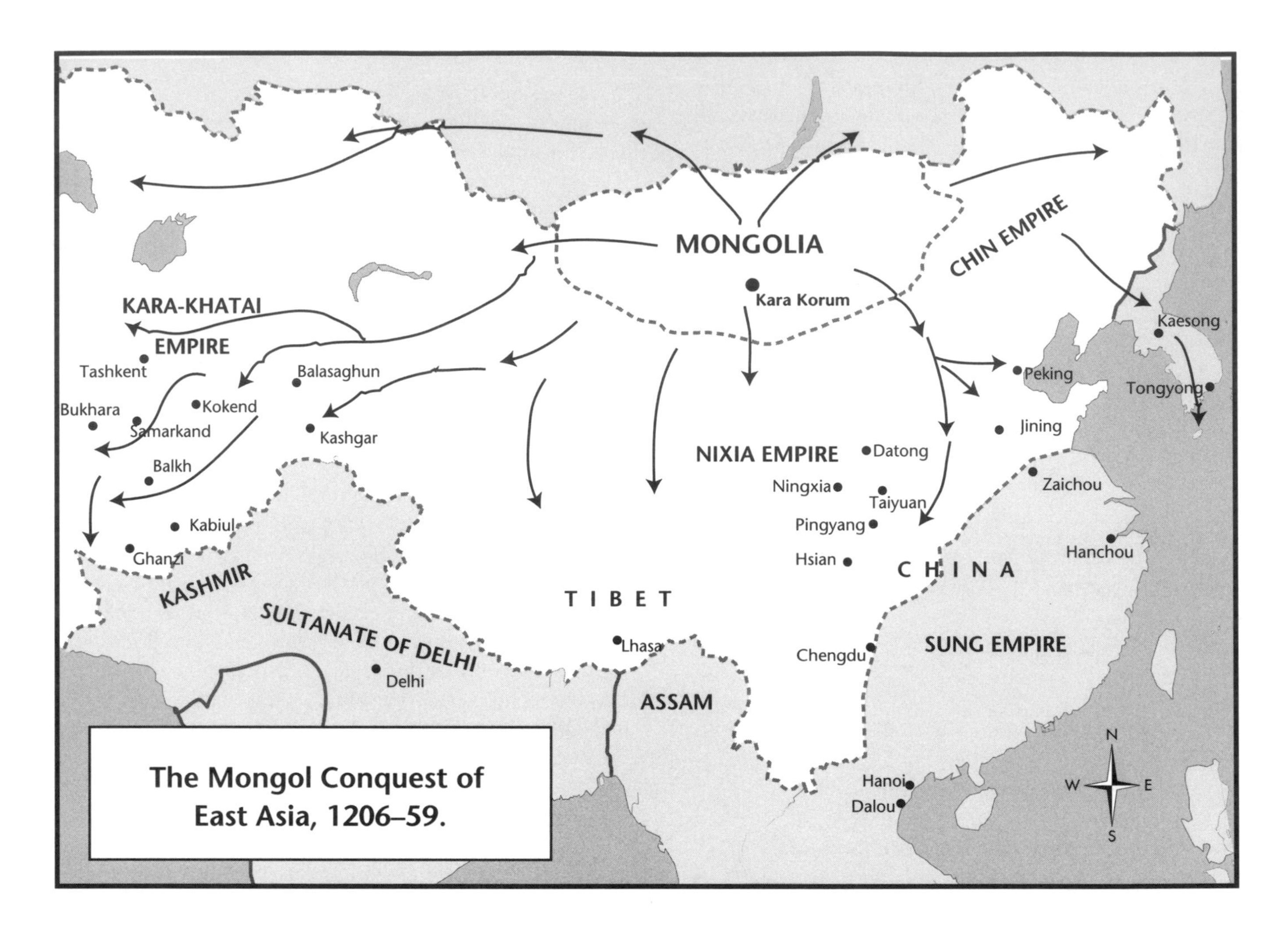

The Mongol Conquest of East Asia, 1206–59.

1346 Mongols spread bubonic plague to Europeans

Mongol troops in the Crimea infected with bubonic plague expose Europeans to the disease. The plague, known also as the Black Death, devastates Europe and Asia in the fourteenth century.

The Breakup of the Mongol Empire

1368 Mongol Yüan dynasty collapses in China

The great Mongol states disintegrate. The Yüan dynasty in China collapses in 1368, and over 60,000 Mongols return to Mongolia. It is replaced by the Ming dynasty; the western part of the Turkestan Khanate is incorporated into the empire of Timur in 1390; Hulagu's Persian empire disintegrates after 1335; and the Golden Horde is attacked and shaken by the forces of Prince Dmitry Donskoy in Russia in 1380 but rules in south Russia into the fifteenth century.

The Mongols' century of dominance in Asia allows for great trade and cultural interchange. The Mongol dynasty in China organizes a program to feed people during famine, expands the transportation system, and institutes paper currency as the sole legal exchange. The Mongols, however, are too reliant on their subjugated peoples and begin to squabble among themselves for power. During their dominance, Mongol rulers prohibit Mongols from learning Chinese culture and likewise forbid Chinese from adopting Mongol customs. Consequently, even after a century of domination there is little fusion between the two cultures, and Mongols are able to withdraw easily.

1400–54 Clan war breaks out among rival Mongol groups

When the Mongols return to their original homeland, they split into three major groups: the northern Khalka Mongols, north of the Gobi Desert; the southern Chahar Mongols, south of the Gobi; and the western Oirat Mongols. The Khalka and Oirat groups fight a prolonged civil war.

1479–80 Unification of the eastern Mongols

Eastern Mongols are unified under the leadership of Batmönh Dayan Khan (1460–1517), who splits his khanate into six *tümen* (military districts): Tsahar, Halh, Urianghai, Ordos, Tümed, and Yongshebu. Various leadership positions are split between the different districts. Mongolia's modern population is largely descended from the Halh tümen.

Mid-1500s Altan Khan revives Mongol power

Altan Khan (1507–83) of Inner Mongolia manages to reunite most of the clans and attempts to wage war against the Ming dynasty with the dream of reestablishing the Mongol Yüan dynasty in China. During Altan Khan's reign, China is extending the Great Wall (originally built in the third century B.C.) as a way to keep the Mongols out. In 1551 the Mongols and Chinese establish a common border and trade arrangements. In 1571 the two sides sign a peace treaty. After the death of Altan Khan, however, Mongol unity falls apart.

1576–78 Creation of the office of the Dalai Lama

In 1576–78, Altan Khan bestows the title *Dalai* ("oceanic") upon Tibet's most senior lama, Sonam Gyatso (Sodnomjamts, or Bsod nams rgya mtsho, 1543–88). Lamaist Buddhism had been introduced to the Mongols from Tibet in the previous century. The first two Dalai Lamas are named posthumously—Gedun Truppa (Gendündüv, 1391–1474) and Gedun Gyatso (Gendünjamts, 1475–1542). The third Dalai Lama proclaims himself the reincarnation of Pagspa Lama, Kublai Khan's spiritual advisor, and he announces that Altan Khan is the reincarnation of Kublai Khan. The Dalai Lama issues a decree prohibiting the ritual slaughter of women, slaves, and animals at shamanist funerals. It is believed that the Dalai Lama is an incarnation of *Avalokitesvara* ("glorious gentle one," known as *Chenresig* in Tibet), one of the three *bodhisattvas* (enlightened saviors) of Lamaist Buddhism. Avalokitesvara is the bodhisattva of compassion.

1586 Erdene Zuu lamasery is built

Construction of a large lamasery (monastery for lamas) begins at Erdene Zuu. The complex consists of several large buildings near the old imperial capital Karakorum. Lamaist Buddhism becomes institutionalized in the political structure and is often imposed on people by their leaders. The older shamanist traditions are not entirely replaced, however, and a syncretic form of Buddhism develops in Mongolia. The *tsam*, a theatrical dance using elaborate masks, is initially performed at the Erdene Zuu lamasery. The dance is performed to exorcise evil spirits and becomes a symbol of Mongol culture. The tsam dance is also possibly the origin of contortionism, which later evolves to become a form of art in Mongolia. A young girl trains for many years to become a skilled contortionist.

The construction of lamaseries across Mongolia changes the structure of the nomadic society by creating nodes of settlement. Such areas become centers of scholarly learning, as lamas study the arts, writing, and Tibetan medicine. Lamas create ornate manuscripts and books, incorporating inlaid metals, precious minerals, and embroidery. Lamas become highly respected members of society and are treated as nobility. In theory, lamas are celibate, but traveling lamas are believed to be responsible for the spread of venereal diseases throughout the countryside.

1636 Manchus subdue southern Mongols

A division develops between the northern (Outer) Mongols and the southern (Inner) Mongols. Ligdan Khan (r. 1604–34), the last emperor of the Mongols, fails to reunite the clans. When the Manchus conquer China, all the southern Mongol khans are also subdued and come under the rule of Abahai of China's Qing (Ch'ing) or Manchu dynasty (1644–1911).

The cultural division between Inner Mongols and Mongolians exists even into the twentieth cfentury when. Mongolians still often regard Inner Mongols (and other Mongols living outside Mongolia) as less than "pure" Mongols. Mongolians' prejudice against Inner Mongols is exacerbated by the Inner Mongolians' own nostalgic admiration for the "purity" of Mongol culture in Mongolia.

1639–41 Zanabazar made Mongolia's Buddhist leader

Zanabazar (1635–1723), a sculptor and the creator of the *soyombo* script, becomes the leader of Mongolia's Buddhists, akin to the Dalai Lama of Tibet. He takes the title of *Bogd Gegeen*—"Holy Enlightened One". At the age of three, Zanabazar is declared a possible saint, and he is sent to Tibet to study under the Dalai Lama. He returns to Mongolia to spread Tibetan Buddhism (also known as Yellow Buddhism or Lamaism) and creates many exquisite bronze statues of Buddha. Zanabazar resides in Örgöö (Urga, Urghe), which becomes the mobile capital. The settlement is located some 260 miles west of Ulaanbaatar's present location. Zanabazar is revered as a sculptor, statesman, Buddhist teacher, and envoy.

China and Russia Absorb Mongol Lands

1689 Nerchinsk border treaty is made between China and Russia

The Treaty of Nerchinsk divides Mongol territory between the two powers along the Ergun (Argun) and Geerbiqi (Gorbitsa) rivers and the Outer Hinggan mountains.

Late 1600s Traditional medicine develops

Luvsandanzanjantsan is considered the pioneer of traditional medicine in Mongolia. Medical diagnoses and treatments are a mix of Indian, Tibetan, and Chinese methods, based on five elements: earth, water, fire, wind, and wood. Herbs, plants, and animal organs are used in making medicines that are given according to the season and weather. Therapies also include bloodletting, acupuncture, and massage.

1691 Northern Mongols accept Chinese rule

The northern Mongols have been fighting with the western Mongols for supremacy. The three khans of Khalka (Zasagt Khan, Tüsheet Khan, and Setsen Khan), along with the head of the Buddhist church in Khalka, Zanabazar, the Jebtsundamba Khutukhtu (also known as the Living Buddha of Urga), accept Manchu rule in order to defeat the western Mongols. Under the Convention of Dolonnor, the northern Mongols give their allegience to the Manchus, who then occupy Mongolia. The Manchus, whose armies use cannon and firearms, eventually destroy the forces of prince Galdan, leader of the western Mongols, in 1758.

Under the agreement, the three khanates of Outer Mongolia (the lands occupied by the northern Mongols) are divided into thirty-four administrative regions. The official policy is to keep Outer Mongolia isolated from Chinese culture and society but to use Mongolian men for the military. China controls merchants' access to Outer Mongolia, discourages intermarriage, and limits economic development. Although China imposes trade restrictions with Outer Mongolia, Mongolians become increasingly dependent on Chinese trade.

1727 Kyakhta border treaty is made between China and Russia

The Russian-Chinese border treaties of Nerchinsk and Kyakhta confirm Chinese rule over both the southern and northern Mongol areas but assign Buriat lands to Russia.

1778 Urga (now Ulaanbaatar) becomes a permanent settlement

The town of Urga (Urghe, Örgöö), which had been a temporary town of tents, becomes a permanent village. Mongolia's capital has been moved about two dozen times over the years, fixed at various places along the Orkhon, Selenge, and Tuul rivers. The town later is called Ikhe Khuree and Niislel Khuree (1911–24). In 1924, the new communist government changes its name to Ulaanbataar, which means "red hero."

1837 Birth of Injanasi, Inner Mongolian novelist

Injanasi (1837–91), a member of the Inner Mongolian nobility, is a famous author of Sino-Mongolian romances, including *Negen dabkh*ar and *Ulaanaa uiakh tankhim*. He also writes *Khökh sudar* (The Blue Chronicle), a historical romance set in Inner Mongolia. His novels are still popular in China.

Mid-1800s Beginning of compulsory primary education

Prince To Wang introduces mandatory primary education for his subjects.

Early 1900s Marzan Sharav paints scenes of Mongol life

The traditional style of Mongol painting is called *Mongol zurag*, which uses bright colors with a lack of shadow and perspective. Marzan "Joker" Sharav (Balduugiyn Sharav, 1869–1920s), one of Mongolia's most famous traditional painters, portrays scenes of everyday Mongol life. He typically depicts many small mundane and bucolic scenes on large panels. His most famous painting is *One Day in Mongolia*.

1905 Birth of Rinchen, Mongolian folklorist and academician

B. Rinchen (1905–77) is one of Monglia's best-known avocational writers of the communist era. Rinchen becomes well-versed in Mongolian folklore and philology. His most famous novel is *Üüriyn Tuyyaa* (Rays of Light), set at the time of the revolution.

1906 Birth of Natsagdordz, Mongolia's most famous modern author

Dasdordziin Natsagdordz (1906–37) is considered the founder of modern Mongolian literature. In his early years, he works at the Leningrad Military Museum and studies journalism in eastern Germany. He produces a newspaper for the communist government. Natsagdordz is arrested in 1936 and disappears the following year during the purges. His works include poetry, short stories, plays, and even the text for an opera (*Three Fateful Hills*, or *The Story of Three Lives*). His most famous poem, *My Native Land*, describes the beautiful and idyllic vistas of Mongolia.

1908 Birth of Damdinsüren, scholar and writer

Tsendyn Damdinsüren (1908–87) is another famous communist-era writer whose literary contributions are created as a pastime activity. He writes poems, stories, travel adventures, and even a film script. His works express a concern about the assimilation of Mongolian culture.

Bogd Khan Mongolia, 1911–21

1911: December 1 Mongolia declares independence from China

Following the overthrow of the Manchu dynasty by the Chinese revolution, northern Mongol princes proclaim an autonomous Outer Mongolia under the rule of the Bogd Gegeen. The current eighth Bogd Gegeen (Agvaanluvsanchoyjindanzanvaanchigbalsambuu, 1870–1924) is proclaimed Bogd Khan (Holy King), the Living Buddha (*Jebtsun Damba Khutukhtu* or *Jebtsundamba Hutagt*), with the capital at Ugra (now Ulaanbaatar). The popular theocratic monarch is a relative of one of the Dalai Lama's officials who had been brought to Mongolia when he was four years old. The Bogdo

Khan essentially shares the throne with his wife because he is not celibate.

The Mongol forces at Urga unite with those of Dambijantsan (d. 1922) in the west. Dambijanstan, a Kalmyk rebel, is known for his vicious maiming and torture of prisoners. The government tries to convince princes in Inner Mongolia to join the new kingdom, which causes China to go to war with Mongolia.

A treaty with the Russian imperial government officially pledges Russian assistance for the autonomous state, but Russia does not want Mongolian independence to upset China's stability.

1912 Secondary school opens

The Russian Buriat scholar Jamtsarano opens one of the first secondary schools in Mongolia. The Russian and Mongol languages, as well as geography and mathematics, are taught.

1913 Treaty with Tibet is signed

The Treaty of Friendship is signed with Tibet.

1915: May 25 Treaty Kyakhta gives Mongolia some autonomy

Mongolia, China, and Russia sign a treaty which takes away Mongolia's independence but grants a limited amount of autonomy from China.

1919 China occupies Outer Mongolia

After the Bolshevik Revolution of 1917 in Russia, the Chinese exploit Russia's weakness and reoccupy Outer Mongolia to end its autonomy. The new government of the Soviet Union nullifies all treaties signed by the former imperial government, thus giving China the opportunity to claim Outer Mongolia as part of its territory. In actuality, however, the Soviet government supports an independent Mongolia under its control. In February of 1920 China incorporates Outer Mongolia into its territory.

Outer Mongolia Becomes a part of the Chinese State

1921: February 3 White Russian forces drive out Chinese

White (counterrevolutionary) Russian cavalry, led by the Baltic Baron Roman von Ungern-Sternberg, drive the Chinese from Outer Mongolia. Von Ungern-Sternberg, who is said to be insane, wants to create a Central Asian Empire and is already in Mongolia with several thousand soldiers because he is fleeing from Red (communist) forces. His forces consist largely of Buriats seeking revenge for oppression committed against them by invading Bolshevik soldiers. His troops round up and execute Russians in Urga whom they suspect are communist sympathizers, and they assault, abuse, and plunder the population.

The new government building at Urga, following the revolution led by White Russian forces. (EPD Photos/CSU Archives)

1921: March 13 First congress of the Mongolian People's Party is held

A delegation of revolutionaries meets in Russian Kiakhta, with the endorsement of the Living Buddha, and establishes a provisional government. Sukhbaatar begins organizing partisans to take power.

1921: July White Russians are defeated

Von Ungern-Sternberg's forces are overcome by the Mongol revolutionary leaders Sukhe Baatar (Damdiny Sükhbaatar, 1893–1923) and Khorloin Choibalsan (1895–1952), who are assisted by the Soviet Red Army. White Russian forces are totally defeated by January, 1922.

1921: July 11 Creation of the Mongolian constitutional theocracy

Under Soviet influence, a nominally independent state is proclaimed. The Living Buddha is reinstated as its monarch. Many revolutionaries occupy key political positions in the new government.

1921 Academy of Sciences is founded

The Academy of Sciences in Ulaanbaatar grows to include departments in agriculture, chemistry and biology, geography and geology, medicine, and technology. Jamtsarano founds the Mongolian Scientific Committee.

1923–28 Buyannemah contributes to Mongolian literature

Sodnombaljiryn Buyannemeh (1901–37) is a well-known writer and journalist of the early communist years. He works as a newspaper editor and is responsible for establishing a literary association and chairing the national arts council. During the 1920s, he writes several historical and literary works, including the plays *The Truth* and *Dark Rule*. In 1937, Buyannemeh is arrested, sentenced, and executed for suspicion of being a counterrevolutionary and spying for Japan.

1924: May The Living Buddha dies

The new Mongolian state only lasts until the death of Jebtsun Damba Khutukhtu. It is believed that he dies from complications of syphilis. The communist government prohibits Buddhist leaders from naming a ninth reincarnation, thus ending the possibility of theocratic succession. In 1939, Buddhist leaders in Japan gain permission from the Dalai Lama to name a ninth Jebstun Damba in Tibet, keeping his identity a secret to avoid assassination by Soviet secret police.

The Mongolian People's Republic, 1924–92

1924: November 26 Creation of the Mongolian People's Republic

The world's second communist country (and the first Soviet satellite country) is created. Support from the USSR consolidates the Mongolian People's Revolutionary Party as the state's only political party. Urga is renamed Ulaanbaatar (or Ulan Bator, "red hero") in honor of Sukhe Bator, who had died in February, 1923. (At the time, a lama is blamed for poisoning him, but it is now believed that the Soviets had him killed.)

Mongolia's first constitution nationalizes land, water, and mineral resources, and proclaims workers as the source of political power.

1925: December Mongolian People's Republic issues its own currency

The government issues the *tugrik* (or *tögrög*) as its legal currency. Traders have long used Russian gold as currency, or barter with commodities such as tea, silk, and furs.

1929: September Confiscation of property begins

Large landholdings from feudal lords are confiscated by the government. Many nomads settle down to raise livestock on state-owned farms. Foreign technical experts are deported, and Mongolia closes its borders.

1926 First air service to Ulaanbaatar

Mongolia's first air service begins operating between Ulaaanbaatar and Verkneudinsk in eastern Siberia.

1930s The government promotes an athiest campaign to eradicate religion

Before the rise of communism, Mongolia's traditional religion is Lamaist Buddhism, strongly influenced by indigenous shamanism. Before the government begins a campaign to eradicate religion, there are about 770 monasteries with approximately 100,000 lamas. The monasteries, as religious and cultural institutions, are also the focus of political and ecnomic power in Mongolia. The lamas, who are celibate and do not engage in labor, are seen as expendible and a drain on society by the communist government. The government starts executing lamas in 1932 but stops the policy amid public backlash and orders lamas to work in 1935. During 1936–39, virtually all the monasteries are closed and their livestock and lands are confiscated, while purges of lamas recommence. Some 17,000 senior lamas are tried for counterrevolutionary activities and are either executed or die at labor camps in Siberia. Lower lamas are persuaded to adopt a secular mode of life.

The Buriats, whose culture is similar to that of the Mongols, live to the north of Mongolia just inside the Soviet Union. All the Buriat lamaseries and temples are closed or destroyed and the lamas are sent to prison or executed.

1932 National rebellion against forced collectivization

The Soviet government pushes Mongolia to introduce forced collectivization of agriculture. Since Mongolia is a primitive nomadic society, however, the scheme is hopelessly impractical. The whole country rebels against the policy, and order breaks down. At Khorgo, a rebellion is organized by hundreds of lamas, whose supporters are outnumbered and crushed by the government.

As a result of the national insurrection, Mongolians abandon prior efforts at collectivization. The Soviet government quickly sends in troops to quell the rebellion, and stations a large garrison beginning in 1936.

1932 Japan occupies Manchuria

Japanese military forces move into Manchuria (in northeast China) and create the puppet state of Manchukuo. The Japanese occupy Inner Mongolia the following year.

1933: February Radio broadcasting begins

Mongolia broadcasts its first radio programming from Ulaanbaatar.

1936: March 12 Soviet Union signs mutual assistance treaty with Mongolian People's Republic

The Soviet Union and the Mongolian People's Republic formalize their close relations with the Treaty of Friendship and Mutual Defense. The ten-year agreement is renewed in 1946 for another ten years.

1937–38 Soviet purges extend to Mongolia

After the popular uprising in 1932, Soviet military and intelligence forces are sent to Mongolia. Secret police and paramilitary operations increase, coordinated by the NKVD (the Soviet secret police force that was the precursor to the KGB). Under Choibalsan's Special Commission, Mongolia becomes a police state, where political leaders, military officers, and civilians are controlled through fear and intimidation. Between 1937 and 1939, nearly eighty percent of the 25,824 people charged with espionage and counterrevolution are sentenced to execution. Only seven individuals are declared innocent, while the rest are sent to prison, typically for ten years. By the end of the 1940s, a total of 35,000 Mongolians have been falsely accused of conspiring or plotting against the state. Choibalsan's purges especially target Buriats—the adult male population of many Buriat enclaves virutally disappears.

One of the later victims of the purges is Anandyn Amar (1886–1941), who served as Mongolia's head of state during 1932–36 and was prime minister during 1928–30 and 1936–39. Amar, who is a founding member of the MPRP, is stripped of his political power in 1939 and arrested for suspicion of spying for Japan. Amar becomes one of nearly 200 Mongolians during the purges who are sent to the Soviet Union for sentence, trial, and execution.

1939 Choibalsan becomes prime minister

Choibalsan is made prime minister, although the country is already under his dictatorship. Choibalsan models his authoritarian leadership after his Soviet colleague, Stalin. His rise to power includes the condemnation and purging of many of his former associates. Authors, demoralized military officers, and many others perceived as threatening are executed.

1939: August Japan invades Manchuria

Mongolia, with Soviet support, fights invading Japanese forces at Nomonhan along the border with Manchuria. The invasion, which begins in the summer, is routed, and the Japanese are defeated by September. The event becomes known throughout Mongolia as the Victory of Khalkhin Gol. The battle pits Outer Mongolians (supported by the Soviet Union) against Inner Mongolians (supported by Japan).

1939 Ulaanbaatar is connected to the Trans-Siberian Railway

A railway linking Ulaanbaatar to Bayantümen is completed, connecting the capital to the Trans-Siberian Railway.

1940 Tsendebal rises to power

Yumjaagjiin Tsedenbal (1916–91) begins serving intermittently as General Secretary of the Central Committee of the MPRP. He later becomes Chairman of the Council of Ministers in 1952.

1941 Government introduces new phonetic alphabet, based on Cyrillic

The Mongolian language has used a variety of writing systems, including ones based on Chinese, Arabic, Tibetan, and some Indian-based scripts (including the square script of the thirteenth century). The Roman alphabet is introduced in the 1930s. Tsedenbal's advocacy for adopting a Cyrillic-based alphabet comes at a time when many ethnic groups in the Soviet Union are adopting similar scripts by order of Soviet dictator Stalin. The new alphabet is phonetic, making it much easier for illiterate Mongolians to learn to read and write in their own language.

1942 National University of Mongolia is founded

The National University of Mongolia (originally Choibalsan State University) at Ulaanbaatar has faculties in mathematics, natural sciences, physics, and biology. The university undertakes research in nuclear physics, biophysics, mineral resources, energy, and communications.

1944: July U.S. vice-president visits Mongolia

Vice-President Henry Wallace visits Ulaanbaatar and meets with Mongolia's leader, Marshal Choibalsan. It is the first meeting between a political leader of a major industrialized country and Mongolia's leader.

1945: February 11 Yalta agreement calls for continued independence for Mongolia

At the conference that would determine the fates of many postwar countries, the governments of Great Britain, the United States, and the Soviet Union agree that Mongolia should remain independent. Mongolian independence is one of the conditions stated by the Soviet Union for an agreement to declare war on Japan. Soviet recognition of independence is largely seen as a reward for Mongolia's supplying the Soviets with equipment during the war.

1945: August 10 Mongolia declares war on Japan

In the final days of the war, Mongolia formally declares war on Japan, just two days after the Soviet Union has made the same decision. Mongolian troops help secure Manchuria, the Gobi Desert, and Inner Mongolia.

1945: October 20 China votes to recognize Mongolia

After a unanimous plebiscite by the Mongolians in favor of independence, the Nationalist government of the Republic of China formally recognizes the Mongolian People's Republic starting on January 5, 1946.

1950: February 14 Mongolia signs treaty with China and Soviet Union

The governments of the Mongolian People's Republic, the People's Republic of China,and the USSR sign a treaty guaranteeing Mongolia's independence. In the early 1950s, the Soviet Union and China compete with each other in offering aid to Mongolia.

1952: January Choibalsan dies, Tsedenbal becomes Mongolia's leader

Upon the death of Choibalsan, Tsedenbal becomes Mongolia's leader. Tsedenbal's rule transforms the government into a highly centralized and ineffective bureaucracy, and the economy stagnates. His dominance at the apex of Mongolian politics extends from the 1950s until the 1980s.

1955 Completion of Trans-Mongolian railway

The Trans-Mongolian railway begins service, linking the Trans-Siberian railway with Chinese railroads through Ulaanbaatar.

1956 Soviet troops withdrawn

With the increased détente between the Soviet Union and China, the Soviet Union pulls its troops out of Mongolia. Some troops, however, begin returning in 1961.

1960: July 6 Mongolia adopts new constitution

The Mongolian People's Republic adopts a new constitution that establishes the country as a socialist state.

1961: October 28 Mongolia joins the United Nations

After seven attempts to join the UN (in 1946, 1947, 1955, and 1960), the Mongolian People's Republic is finally accepted into the organization.

1962: December 26 Boundary dispute with China settled

Mongolia and China sign a treaty ending conflicting boundary claims between the two countries.

1964 Government expels 2,000 Chinese

The government deports 2,000 ethnic Chinese who refuse to take part in an agricultural resettlement program.

1964: June 30 Mongolia strengthens ties with Soviet Union

The governments of the Mongolian People's Republic and the Soviet Union sign a twenty-year treaty of friendship, cooperation, and mutual assistance. Soviet influence increases in the 1970s, and Mongolia sends many students to the Soviet Union for training and education. Many Mongolians adapt to Russian culture through this exchange, and Russian becomes the country's defacto second language.

1969 Mongolian Technical University is founded

The Mongolian Technical University in Ulaanbaatar has several schools of engineering: power, mechanical, civil, and geology and mining.

1981 Tsedenbal named General Secretary of MPRP

Yumjaagjiin Tsedenbal (1916–91) is nominated General Secretary of the Mongolian People's Revolutionary Party. He had served as the chairman of the Presidium of the People's Great Hural in 1974 after rising through the ranks of the MPRP since 1940.

1981: February Completion of Erdenet mine

The government, with assistance from the Soviet Union, completes the development of a large copper-molybdenum at Erdenet, the new town founded in 1976 around the mine. The project began with a construction agreement in 1973.

1981: March First Mongolian in space

Jugderdemidiyn Gurragcha (b. 1947) becomes the first Mongolian in space aboard the Soviet Union's *Soyuz 39* and *Salyut 6*.

1983: March Government expels 1,700 Chinese

The government decides to resettle ethnic Chinese living in Ulaanbaatar.

1983–93 Mongolian government expels 7,000 ethnic Chinese

The Mongolian government deports thousands of ethnic Chinese who have been living in Mongolia for decades. The Chinese are expelled amid charges of engaging in illegal trade and spying.

1984: August Tsedenbal removed from power

After thirty-two years in power (either as president, prime minister, or MPRP leader), Mongolia's leader Tsedenbal is removed from power and replaced by Jambyn Batmunkh (b.

1926). Tsedenbal is removed supposedly because of his vituperative anti-Chinese rhetoric and plans. Batmunkh is made General Secretary of the Mongolian People's Revolutionary Party and in 1985 becomes Chairman of the Presidium of the People's Great Hural. Tsedenbal and his family go into exile in Moscow where he dies in 1991 before the new democratic government can bring charges against him for injustices committed during his rule.

1986: May 29 Mongolian Party Congress endorses Soviet restructuring

With their close ties to the Soviet Union, Mongolians are well aware of the Soviet policies of *glasnost* (openness) and *perestroika* (restructuring) and of the democratic movements in Eastern Europe after the mid-1980s. In a speech before the Mongolian Party Congress, Mongolian leader Batmunkh endorses the plan of the communist Party of the Soviet Union to open the economy and permit limited free market activity. The Mongolian government initiates its own policy of "openness"(*il tod*) and begins reforms that serve as transitional steps away from central planning and a collective economy and toward a market economy.

1989: March Mongolian foreign minister visits China

Mongolia and China establish full diplomatic relations. Tse Gombosuren, Mongolia's Foreign Minister, makes an official visit to Beijing, the first such meeting since the 1940s.

1989: December Demonstrations in Ulaanbaatar call for faster reforms

Demonstrations in Ulaanbaatar prompt the ruling Mongolian People's Revolutionary Party to begin political reforms in addition to the ongoing economic reforms.

1990: February 18 First opposition political party forms

The Mongolian Democratic Party (MDP) forms as the first political party to challenge the Mongolian People's Revolutionary Party.

1990: March 12–14 Political leaders resign

Prompted by a large pro-democracy demonstration in Ulaanbaatar and hunger strikes by ten MDP members, officials of the ruling Mongolian People's Revolutionary Party (MPRP) resign, including Batmunkh. By May the constitution (of 1960) is amended to permit new multiparty elections.

1990: July 29 Mongolia conducts its first multiparty elections

Candidates from six parties compete in the general elections. The MPRP wins eighty-five percent of the seats in the legislature.

1990 Trade union association is chartered

The Association of Free Trade Unions is chartered, consisting of about seventy unions.

1991 Political and economic transformation begins

The government issues vouchers to all citizens for the purchase of state-owned property as a step toward privatization. The collapse of the Soviet Union, however, complicates the pace of economic reforms. Russia dramatically cuts its aid and begins insisting on trade based on cash rather than barter.

1991 Education law brings sweeping changes to schools

With the passing of legislation, the traditional Mongolian script is introduced in first grade and teaching of English is made compulsary. There is a hesitancy, however, to shift away suddenly from using Cyrillic letters because Mongolia's books and signs use that alphabet, the only one that most adults know. In 1994 the government postpones the introduction of classical Mongolian script for official purposes and promotes use of the Cyrillic alphabet.

Democratic Mongolia

1992: February 12 Mongolia adopts new consititution

The Communist-era legislature, the People's Great Hural, adopts a new constitution, the first one based on a democratic form of government. According to the document, the existing legislature is to be replaced in June by the seventy-six elected representatives of the State Great Hural. The constitution completes Mongolia's transition from a single-party state to a multiparty parliamentary form of government and also establishes freedom of religion. At this time the country's name is officially changed from "Mongolian People's Republic" to "Mongolia."

1992: June 28 First elections held for the State Great Hural

The first election of the State Great Hural sees the MPRP win seventy-one of the seventy-six seats.

1992: September Soviet troops withdraw

Some 67,000 Soviet troops complete the process of withdrawal, which had started in 1990. Soviet troops had been stationed in Mongolia since 1966, when Sino-Soviet tensions were on the rise. Soviet forces were officially stationed in Mongolia during 1921–25, 1936–56, and 1966–92, although there has been an uninterrupted unofficial Soviet military presence since 1921.

1993: March **Snowstorms wipe out livestock**

Heavy snowfall in western Mongolia kills over one million head of livestock and a dozen people.

1993: June 6 **First direct presidential election is conducted**

Mongolian voters choose incumbent Punsalmaagiyn Ochirbat as the country's first directly elected president.

1994: April **Demonstrations against government**

Protesters, including some members of parliament from the opposition, stage a hunger strike outside government buildings in Ulaanbaatar. The protesters demonstrate against corruption in government and demand the withdrawal of legislation that would restrict public protest. Through the mediation of President Ochirbat, the government changes its position and proposes legislation respecting freedom of the press and the right to demonstrate.

1995 **Number of nomadic families on the rise**

The number of families formally registered as nomadic herders grows from 74,000 in 1990 to 170,000 in 1995. Mongolia is one of the only developing countries where internal migration to rural areas exceeds migration to cities. Early in this year, the government makes it legal for citizens to own land and for foreigners to lease it.

Rapid urbanization was the promoted policy during the communist years. The population living in urban areas increased from a few percent in the 1920s to nearly sixty percent by 1990. With the end of authoritarian rule, many Mongolians revert to a semi-nomadic way of life in order to make money from their herds. Many businesses, farms, offices, and institutions close or are sold to private owners as some cities become virtual ghost towns.

1995: August **Mongolia opens its own stock exchange**

The Mongolian Securities Exchange opens in Ulaanbaatar. In its first year of operation, the exchange sees more than 7.8 million shares from 400 companies trade, with an average daily trade volume of 60,000–80,000 shares.

1996: March–June **Wildfires spread across northern Mongolia**

After a winter of little snow, wildfires spread across the north for several months. The fires are the most extensive since records were first compiled in 1978. The fires cause 26 fatalities and nearly 800 injuries. About 700 are rendered homeless. An estimated twenty percent of Mongolia's coniferous forests are damaged in the fire. Damage to the economy is estimated at $1.9 billion.

1996: June 30 **MPRP defeated in parliamentary elections**

Discontent, especially among younger voters, leads to the defeat of the Mongolian People's Revolutionary Party. The MPRP wins only twenty-five seats, while the Democratic Union (a coalition of two smaller parties) wins fifty seats (the remaining seat is won by the Conservative party). The leaders of the Democratic Union, mostly political novices, promise to intensify market reforms. The election marks the first smooth transfer of power in Mongolia's modern history.

1996 **Cholera outbreak**

Mongolia reports its first outbreak of cholera, caused by tainted meat imported from Russia. The government orders the quarantine of infected regions and closes market stalls in Ulaanbaatar. Occurrence of the disease soon subsides.

1997: May **Government abolishes all trade tariffs**

Mongolia abolishes all tariffs, becoming the first country in the world to levy no taxes on trade.

1998: January **Labor code adopts five-day workweek**

The working week is changed from six to five days for the first time since the beginning of the communist era.

Bibliography

Akiner, Shirin. *Mongolia Today.* London: Kegan Paul International Ltd., 1991.

Bergholz, Fred W. *The Partition of the Steppe: The Struggle of the Russians, Manchus, and the Zunghar Mongols for Empire in Central Asia, 1619–1758.* New York: Peter Lang Publishing, Inc., 1993.

Bruun, Ole and Ole Odgaard, eds. *Mongolia in Transition.* Richmond, England: Curzon Press Ltd., 1996.

Bulag, Uradyn E. *Nationalism and Hybridity in Mongolia.* Oxford: Clarendon Press, 1998.

de Hartog, Leo. *Russia and the Mongol Yoke: The History of the Russian Principalities and the Golden Horde, 1221–1502.* London: British Academic Press, 1996.

Greenway, Paul. *Mongolia.* Hawthorn, Australia: Lonely Planet Publications, 1997.

Morgan, David. *The Mongols.* New York: Basil Blackwell Inc., 1986.

Sanders, Alan J. K. *Historical Dictionary of Mongolia.* Lanham, Md: Scarecrow Press, Inc., 1996.

Spuler, Bertold, translated by Helga and Stuart Drummond. *History of the Mongols, Based on Eastern and Western Accounts of the Thirteenth and Fourteenth Centuries.* New York: Dorset Press, 1988.

Myanmar

Introduction

Burma (whose official English name has been Myanmar since 1989) is a land of ancient civilizations stretching back over two thousand years. Its strong Buddhist tradition, of which evidence can be seen in its famous shrines and pagodas, has been a strong unifying force. The great ethnic diversity of this nation, however, which comprises some sixty different ethnic groups, has resulted in an often tumultuous political history and continues to pose serious challenges up to the present day.

Located on the Southeast Asian mainland, Burma is bordered by China to the north and east, Laos to the east, Thailand to the southeast, and Bangladesh and India to the west. To the south lies the Andaman Sea and to the west the Bay of Bengal. Burma's area of 261,970 square miles (678,500 square kilometers) makes it slightly smaller than the state of Texas. The country is divided into four major geographic regions: a mountainous area to the north and west that includes the Arakan coastal strip on the Bay of Bengal; the Shan Highlands, a steep plateau to the east; the central belt that lies between the Salween and Irrawaddy Rivers; and the delta and valleys of the Irrawaddy and Sittang Rivers to the south, a prime rice-growing region.

Early History

The earliest civilizations to people present-day Myanmar were the Mons, who settled there in the first millennium B.C., and the Pyu, who flourished in the first millennium A.D. Through their trade contacts with India, the Mons introduced Theravada Buddhism, which was to become one of the country's most important cultural influences. The Burmans—the ancestors of today's majority ethnic group—first entered the region from Tibet in the eighth century A.D. It was a Burman kingdom based in Pagan that first unified Upper and Lower Burma in the eleventh century, inaugurating a cultural golden age that saw the development of the Burmese alphabet and the construction of thousands of religious shrines by Buddhists from many lands, who voyaged to Pagan specifically to build these monuments to their faith. The Pagan empire reached its greatest point in the twelfth century. The following period saw the Mongol conquest in 1287 and centuries of fragmented rule by rival states. The Mons regained their former dominance in the south, while the Shans became the most powerful group in the north, establishing a capital at Ava.

In the fifteenth century the first European traders arrived on the scene, and by 1519 the Portuguese had established a trading center at Martaban. During this period, a single kingdom—this time based in Toungoo—once again gained control of Burma. The Toungoo dynasty lasted over two centuries and was followed by the establishment of Burma's last native dynasty, the Konbaung, founded by Alaungpaya in 1753. In the nineteenth century, the British became the dominant political force in Burma. Between 1824 and 1885 they won territorial concessions in three Anglo-Burman wars and finally annexed all of Burma in 1886, making it part of British India.

World War II and Foundation for Independence

The British ruled Burma as a province of India until 1937, when it became a separate colony. By this point, however, there was a strong nationalist movement agitating for complete independence. The advent of World War II (1939–45) complicated events, and Burma's nationalist army ended up helping Japan drive the British out of Rangoon in 1942. The treatment of the Burmese people under Japanese occupation turned the Burmese leaders against Japan, and they transferred their allegiance to the British toward the end of the war, helping the Allies retake their land in 1945.

By 1947 the Burmese had won a promise of independence within a year, and elections were held in April for a constituent assembly. In July Aung San, prime minister of the future nation, was assassinated together with six of his ministers when armed men opened fire on them in a council meeting. Responsibility for the attack was traced to a rival politician, U Saw, who was executed soon afterward. The thirty-two-year-old Aung San, who had led Burmese nationalist troops during World War II and laid the diplomatic foun-

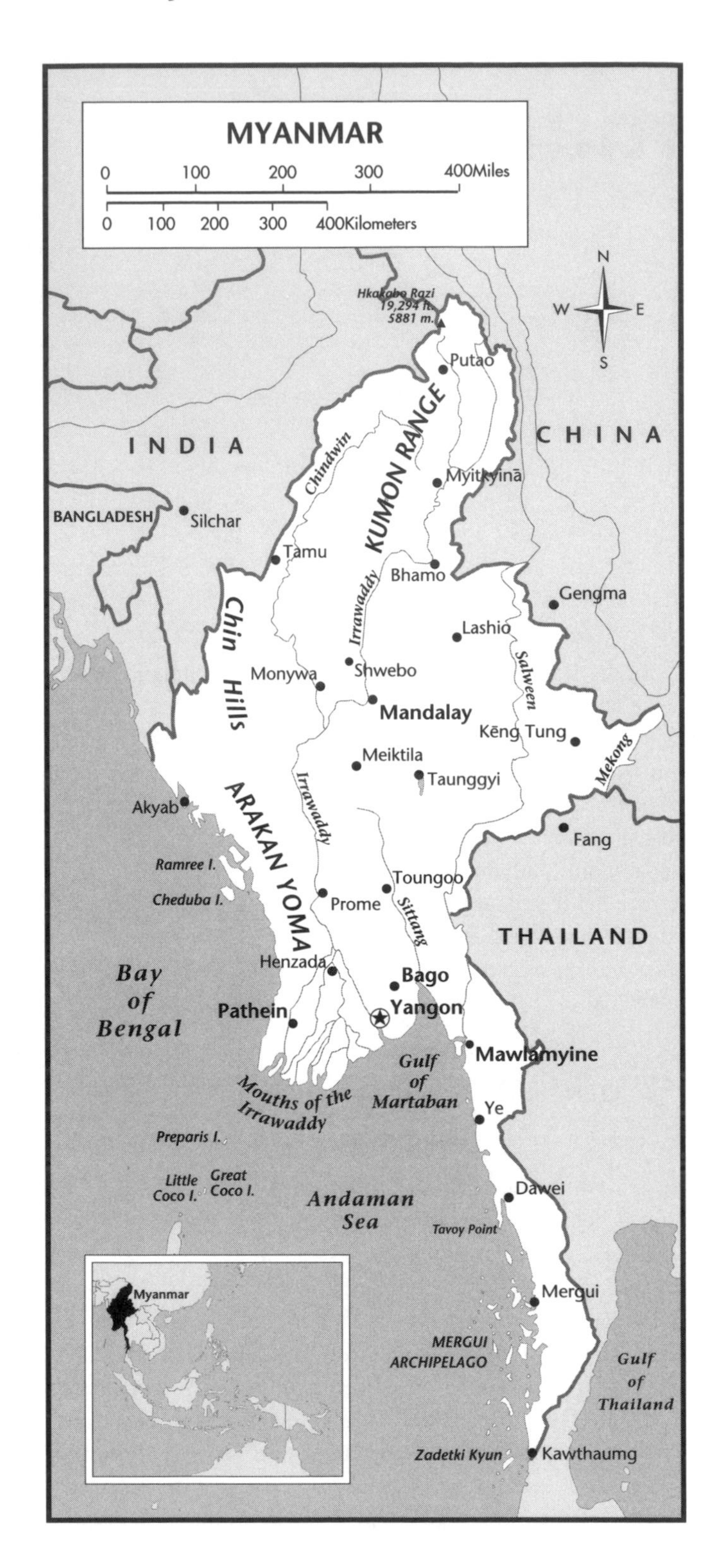

dation for independence, became Burma's foremost national hero of modern times.

Independence under Military Rule

The Union of Burma officially became a sovereign nation outside the British Commonwealth on January 4, 1948. U Nu became prime minister of a government that faced uprisings from both Communist and ethnic minority groups, as well as the troubled economy of a nation decimated by warfare and funding a new military buildup to counter the rebellions it faced. The Anti-Fascist People's Freedom League (AFPFL), founded by the murdered Aung San, won a wide majority in the new nation's first general elections, held in 1951. Although The AFPFL also won the 1956 elections, the party was hobbled by deep divisions that led to a 1958 crisis that saw U Nu resign and military leader Ne Win serve as prime minister of an interim government. U Nu returned to office in 1960, but in 1962 he was removed in a bloodless coup, and the military, under Ne Win, seized control of the country.

Ne Win instituted a socialist state governed by a Revolutionary Council made up entirely of military officials. The government controlled all sectors of the economy and nationalized foreign businesses. A policy of extreme isolationism, intended to end domination by foreign interests, was instituted with the expulsion of about 300,000 foreign residents, many Indian or Chinese. Opposition parties were banned, and civil liberties restricted. After twelve years of military rule, Ne Win proclaimed the Socialist Republic of Burma in 1974 and formed a civilian government. However, opposition parties were still forbidden, and the military still dominated the government.

In 1981 U San Yu succeeded Ne Win as president, although Ne Win continued as chairman of the official government party, the Burmese Socialist Program Party (BSPP). The 1980s saw an increase in separatist pressure by Burma's ethnic minorities living in border areas, culminating in warfare by a number of different rebel groups, twelve of whom banded together to form the National Democratic Front. Additional political turmoil was sparked by the government's 1987 withdrawal of the country's three largest currency notes from circulation in an attempt to thwart the black market activities that had become a deeply entrenched part of the Burmese economy. Protests sparked by the currency withdrawal were met with police brutality, and a protest movement was born, led at first by students and then by the National League for Democracy (NLD), one of whose leaders was Aung San Suu Kyi, daughter of the country's national hero, Aung San.

Pro-democracy Movement

Ne Win resigned as prime minister in July 1988. After a brief period of liberalization, another military dictatorship, led by Saw Maung and the State Law and Order Restoration Council (SLORC), took control in September following a coup. Aung San Suu Kyi was placed under house arrest in 1989 for her criticism of the government and her involvement in the pro-democracy movement. In May 1990 the government allowed multiparty elections to be held, and the NLD won a landslide victory. However, SLORC instituted a variety of delaying procedures to prevent the installation of the elected government and mounted a campaign of persecution against the elected NLD officials, many of whom were forced to leave the country. By the late 1990s the elected NLD officials had yet to take office. In 1991 Aung San Suu Kyi was awarded the Nobel Peace Prize for her role in Myanmar's pro-democracy movement. In spite of international pressure, she was detained until 1995. In 1996 she was placed under house arrest again and released in 1997.

In 1997 Myanmar won admission to ASEAN (the Association of Southeast Asian Nations).

Timeline

c. 1000– ca. 1 B.C. Mons settle in Lower Burma

The earliest civilization in present-day Myanmar is that of the Mons, who speak a language belonging to the Mon-Khmer family. They settle in Lower Burma, in an area they call *Suvannabhumi* ("the land of gold"), where they establish a capital at Thaton, using irrigation techniques to grow rice. As a result of trade links with India, the Mons bring Theravada Buddhism to Burma. Theravada is a form of Buddhism that believes one can improve one's karma through good works. It is closer to original Buddhism than Mahayana Buddhism.

A.D. 500–700 Pyus flourish in Upper Burma

Early in the Christian era, the Pyus enter Upper Burma from the north. Speakers of a Tibeto-Burman language, they build a capital at Srikshetra and establish trade and cultural ties with India and part of China.

A.D. 8th century Burmans enter the region

The Burmans migrate to the upper valley of the Irrawaddy River from Tibet, settling near present-day Mandalay and establishing a fortified capital called Pagan.

Burma is Unified under the Pagan Dynasty

A.D. 1044 Pagan Dynasty is founded

From his base in Pagan, King Anawrahta conquers the Mon capital of Thaton, becoming the first leader to unify the region that comprises present-day Myanmar and inaugurating a golden age in Burma's history. Anawrahta develops the Burmese alphabet and promotes the spread of Theravada Buddhism. With the arrival of scholars from a great Buddhist center in Bihar, India, Pagan gains a reputation as a center of Buddhist learning and architecture. King Anawrahta sponsors the construction of a plethora (overabundance) of temples, monasteries, and pagodas, and the development of religious art and crafts. Pagan becomes known for its religious architecture and also as a center of Buddhist learning.

Decline of the Pagan Dynasty

1287 Mongol conquest

Led by Kublai Khan (c. 1216–94), Mongol armies from China invade and conquer the kingdom of Pagan. The last ruler of the dynasty flees before the advancing enemy, and the capital is abandoned. A vast Chinese military force is stationed in the neighboring Chinese province of Yunnan, which becomes an important military and administrative headquarters and is considered a threat by the Burmese.

1300–1540 Period of turmoil and fragmentation

The fall of the Pagan empire is followed by a period of political fragmentation. The Mongols abandon Burma early in the fourteenth century, and the region falls under the control of various independent kingdoms. In Lower Burma, the Mons establish a new capital at Pegu, which becomes the center of a Buddhist religious revival, attracting monks from areas throughout Southeast Asia. In Upper Burma, the Shans found a kingdom whose center is the city of Ava. The Rakhines establish sovereignty in the west.

1385–1425 War between the Mons and the Shans

The Mon and Shan kingdoms go to war.

15th century Dance drama is developed

Burma's tradition of dance drama begins with the introduction of religious pageants (*nihat-khin*) performed by itinerant players who travel from village to village in carts. Their plays consist of historical scenes from the life of the Buddha and imaginative tales of his previous lives.

1435 Italian trader arrives at Pegu

The Venetian merchant Nicolo di Conti becomes one of the first Europeans to visit Burma when he completes a four-month sojourn at Pegu.

The Toungoo Dynasty

1486 Second unified Burmese kingdom is founded

King Minkyinyo of Toungoo, a Burman leader, establishes the second unified kingdom in the region. He occupies the Shan city of Ava after it is overrun by intruders early in the fifteenth century and later captures the Mon city of Pegu as well.

1505 Varthema describes Burma's mineral wealth

Ludovico di Varthema, a trader from Italy, reports on the quantities of precious stones to be found in Burma, including pearls, diamonds, and emeralds.

1519 Portuguese establish a port at Martaban

Portuguese merchant Antony Correa signs an agreement with the local ruler of Martaban to set up a port for trade with Siam (Thailand). The Portuguese begin trading in spices and other commodities.

1541 Tabinshwehti attacks Martaban

Toungoo ruler Tabinshwehti, disapproving of the local ruler's pact with the Portuguese (see 1519), attacks Martaban but does not succeed in ending the Portuguese presence there.

1600–13 Portuguese adventurer establishes a kingdom

Portuguese traveler Philip de Brito y Nicote, serving the ruler of the Rakhine kingdom in an official capacity as customs administrator of Syriam, takes control of the town and proclaims himself king of Lower Burma. He promotes Christianity in his "kingdom," gaining many converts, and destroys Buddhist holy sites. De Brito is finally defeated by King Anaukhpetlun of Toungoo and an invasion force of 12,000. De Brito is killed by impalement, and the remaining Portuguese in Syriam are driven out and settle in nearby villages.

1612 British establish a presence in the region

The British East India Company sets up trading stations near Rangoon.

The Konbaung Dynasty

1753 Alaungpaya founds Burma's last ruling dynasty

Maung Aung Zeya, the Burman chief of Moksobomyo, north of Ava, forms a power base among the surrounding villages and rebels against the forces of Binnya Dala, who has deposed the last Toungoo monarch. Taking the title Alaungpaya, he declares himself king of Burma. Defeating both the Shans in the north and the Mons in the south, he founds the Konbaung dynasty, Burma's last line of rulers. His kingdom gains territory in India and Thailand.

1767 Thai players introduce masked drama

Burma's masked drama (*zat gyi*) begins with performances of Thailand's dance dramas by Siamese (old name for Thai) court players following the conquest of the Thai capital of Ayutthaya.

The sophisticated art of the Siamese dancers wins great acclaim and is adopted by native performers, who start their own Burmese dance tradition.

1784 Arakan is added to Burma

During the reign of Bodawpaya (1782–1819), the Konbaung dynasty adds Arakan to its territory. This conquest creates tension with the British, as it gives Burma a common border with the Indian province of Bengal, and sets in motion the sequence of events that eventually results in the Anglo-Burmese wars of the nineteenth century and the British annexation of Burma.

Late 18th century Burmese puppet theater is developed

U Thaw, minister of fine arts during the reign of Singu Min, creates the Burmese puppet theater, or *yokthe pwe* as a means of getting around traditional taboos on having human actors perform intimate romantic scenes. A strictly regulated art form is developed, with standard settings, exactly twenty-eight marionettes in each play (a number significant in Buddhism), and scenes occurring in a predetermined order. The plot always portrays a romance between a prince and princess, named Mintha and Minthami. The puppets are jointed marionettes made from wood and cloth and manipulated by as many as forty separate strings.

1824–26 First Anglo-Burmese War

Warfare breaks out when the British East India Company seizes the regions of Arakan and Tenasserim on the coast. It ends when Burma cedes these and other territories and pays a sum of money to the British.

1852 Second Burma War

Complaints by two British sea captains about unfair treatment by a Burmese court lead to the arrival of a British expeditionary force and the second Anglo-Burmese war, at the end of which Britain controls Lower Burma.

1853–78 Rule of King Mindon

King Mindon is the first Burmese ruler to actively accept Western influences, allowing young Burmese men to be educated in Europe, increasing trade with the British, and introducing a modest degree of industrialization to the country.

1861 Court is moved to Mandalay

The court of King Mindon is transferred to Mandalay.

1872 Fifth Great Synod of Buddhism is held in Mandalay

King Mindon hosts this major Buddhist gathering to help unify the Burmese people within the Buddhist religion. The text of the Buddhist scripture, the *Tipitaka,* is written down permanently for the first time during this gathering.

1885–86 Third Burma War

With the Third Burma War, British annexation is complete, and the ruling Konbaung dynasty is dissolved. Burma becomes part of the British Empire and is ruled as a province of India.

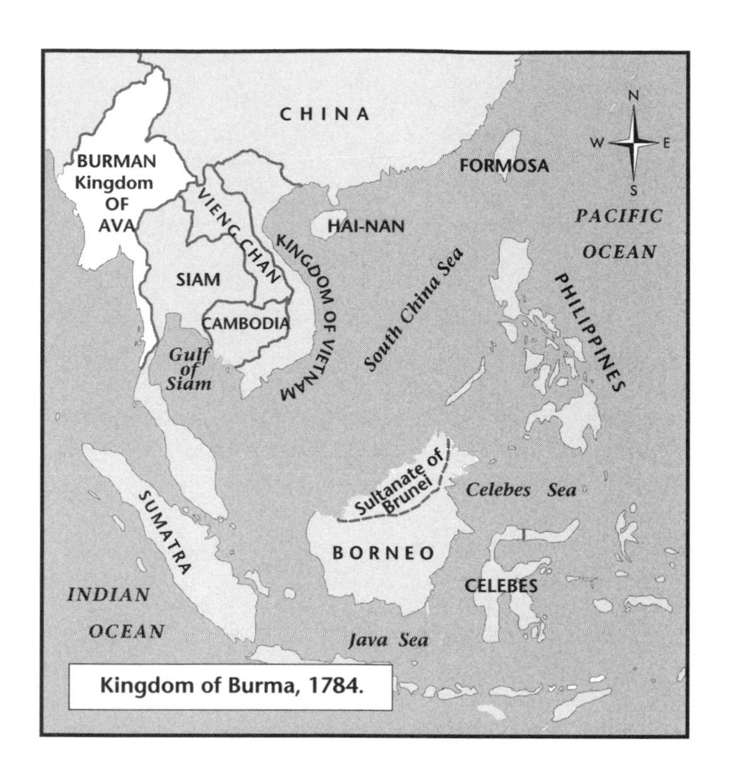

Kingdom of Burma, 1784.

British Rule

1886: January 1 British annexation is official

Burma is officially incorporated into British India.

1905 The National Museum of Mandalay is founded

The National Museum of Mandalay, housed in the Glass Palace and featuring cultural artifacts, is inaugurated.

1906 Young Men's Buddhist Association is formed

Burma's first twentieth-century nationalist group, the Young Men's Buddhist Association (YMBA), is established.

c. 1914 Birth of national hero Aung San

Aung San (c. 1914–47), the foremost nationalist leader associated with Burma's independence movement, attends Rangoon University, where he leads a student strike in 1936. After graduating, he becomes the leader of a nationalist organization lobbying for independence from Britain. On the eve of World War II (1939–45), he allies himself with Japan to obtain aid in fighting the British and organizes a Burmese army, which helps the Japanese invade Burma in 1942. Aung San serves in the puppet government of Ba Maw (1893–1977) during the Japanese occupation. By 1944, disillusioned with Japanese rule, he throws his support and that of his forces to the British.

After the war, Aung San becomes chairman of the Executive Council and negotiates an agreement for independence within one year with British prime minister Clement Atlee (1883–1967). The agreement is signed on January 27, 1947. Aung San's Anti-Fascist People's Freedom League (AFPFL) wins a landslide victory in April elections. On July 19, Aung San and six political colleagues, including his brother, are assassinated during a session of the Executive Council. U Saw, a political rival, is linked to the killings and later executed, while Aung San becomes a national hero. Aung San's daughter, Aung San Suu Kyi (see 1945: June 19), who is two years old when he is killed, later becomes a prominent human rights activist, protesting abuses by the government of Myanmar and winning the Nobel Peace Prize in 1991.

1919 Burma is excluded from "dyarchy" reform

India wins a greater degree of autonomy following World War I (1914–18), but Burma is excluded from this agreement, resulting in heated protests.

1922 Burma wins dyarchy privileges

Britain reverses its decision to exclude Burma from the dyarchy (government shared by two rulers; Montagu-Chelmsford) reforms, granting greater self-government.

1930–32 Anti-British revolt is staged north of Rangoon

An extended rebellion is carried out in the Tharrawaddy region, led by Saya San, a former monk.

1930 Student protest movement is launched

The All Burma Student Movement protesting British rule is begun at the University of Rangoon. It is both nationalist and Marxist, and its members address each other using a term of respect previously reserved for contact with the British: *Thakin*. Among its leaders is Aung San (see c. 1914–47), who will later be instrumental in winning Burma's independence.

1930: May 5 Earthquake rocks southern Burma

An earthquake with its epicenter at Bago shakes all of southern Burma, destroying ancient pagodas and other religious monuments. Restoration of some sites is still being completed five decades later.

1936 Student movement organizes school strike

Student protesters organize a boycott of high-school and university classes to call for an end to British rule.

1937: April 1 Burma becomes a separate colony

Burma is separated from India and becomes a Crown colony of the United Kingdom, with a new constitution and semi-autonomous status.

These women wearing decorative copper rings that elongate their necks live in the northern part of Burma.
(EPD Photos/CSU Archives)

1940 Nationalists are arrested

A group of nationalists, including Aung San, Ne Win, and the prime minister, Ba Maw, are arrested for treason by the British colonial government. Both Aung, Ne Win, and Ba Maw all favor Burmese independence.

1941: December Japan invades Burma

The Japanese invade Burma during World War II (1939–45). They are aided by the nationalist Burma Independence Army, trained in guerrilla warfare by the Japanese and led by Burmese nationalist leader Aung San.

1942: February Stilwell leads retreating Allied forces to India

U.S. general Joseph Stilwell ("Vinegar Joe") arrives in Burma to take charge of mostly Chinese troops resisting the Japanese advance. He leads 114 soldiers on a 930-mile (1,500-kilometer) retreat to safety in India.

1942: March 8 Rangoon is taken by the Japanese

The Japanese drive the British from Rangoon within four months; British and American forces are forced to retreat to India. The war inflicts great damage on Burma, resulting in heavy casualties, destroying buildings and infrastructure, and resulting in runaway inflation and shortages of consumer goods. The British withdraw from Mandalay, and the Burma Road is cut off.

1942: August Puppet government is formed

Former premier Ba Maw is named head of a puppet government that is controlled by the Japanese and has little real power.

1944 Resistance activities against the Japanese

In response to the impoverishment of their country and the harsh treatment of its people by the Japanese military government, Burmese nationalists become dissatisfied with the Japanese occupation and form the Anti-Fascist People's Freedom League, led by Aung San. The organization begins resistance activities against the Japanese, and the Burmese army's 10,000 troops transfer their allegiance to the British, aiding in the liberation of Burma.

1945 British retake Burma

Allied troops reopen the Burma Road and retake Myitkyina, and the Burma National Army revolts against the Japanese. Rangoon is captured, and Burma reverts to British rule.

1945: June 19 Birth of Aung San Suu Kyi

Acclaimed human rights activist Aung San Suu Kyi (1945–) is born in Rangoon, the daughter of nationalist leader Aung San, who is assassinated two years later by a political rival. She is raised in Burma by her mother, Khin Kyi, a prominent diplomat, and moves to India in 1960 when Khin Kyi is named ambassador to that country. Aung San Suu Kyi continues her studies in England, attending the University of Oxford and marrying an Englishman, Michael Aris, in 1972. She settles in England with her husband and raises two sons.

In 1988 Aung San Suu Kyi returns to Burma for an extended stay to nurse her mother, who is dying, and begins to speak out publicly against the silencing of dissent by Ne Win's (b. 1911) military regime. She is placed under house arrest in 1989. In 1990 the National League for Democracy (NLD), the political party with which she has become associated, wins a landslide victory in the country's first multiparty elections since 1962. However, the military government refuses to relinquish power to the democratically elected NLD. Aung San Suu Kyi is offered her freedom if she will agree to leave the country, but she refuses, and her continuing detention draws international attention to the military govern-

The Burma Road is a vital supply link to China during World War II. (EPD Photos/CSU Archives)

ment's continuing suppression of human rights and its refusal to let the elected government take office. In 1991 Aung San Suu Kyi is awarded the Nobel Peace Prize, which her son Alexander accepts in her stead. She is freed in 1995, arrested a second time in 1996, and released in 1997.

1946: September General strike spreads through Burma

A general strike, begun by the police and then joined by other groups, paralyzes Burma. Aung San's AFPFL and other nationalist groups help end the strike within one month and demand complete independence by 1948.

1947: January 27 Independence agreement is reached

In negotiations with nationalist leader Aung San, British prime minister Clement Atlee agrees to grant independence to Burma within one year. Elections for a governing assembly are to be held within four months. In an agreement worked out separately, Burma's minority groups are to come within the jurisdiction of the new country but will be permitted to form separate states after the first ten years of independence.

1947: April Elections are held

Elections are held for a constituent assembly to decide the fate of Burma. The AFPFL wins by a landslide. Thakin Nu (later known as U Nu, 1907–95) is chosen to be president, and Aung San is named prime minister.

1947: July 19 Aung San is assassinated

Aung San and six of his political ministers are killed in government chambers in Rangoon when a group of armed men burst in on a council session. The assassination is found to have been plotted by right-wing politician U Saw. U Saw and his accomplices are arrested and later convicted and executed for the crime. (See 1914.)

1947: September 24 New constitution is approved

The Constituent Assembly approves a constitution to be adopted upon independence. It mandates a parliamentary system of government with a bicameral (two legislative chambers) legislature for the new nation. Its political leader will be the prime minister, while the president will have formal powers as head of state. Freedom of religion and other basic civil rights, such as freedom of speech and assembly, are guaranteed, and provisions are made to guarantee social welfare in areas including education, health, and employment. The state claims the right to redistribute land, and large absentee landholdings are abolished.

1948–50 Challenges to government are overcome

The government of the new nation faces and overcomes serious challenges from Communist rebels and separatist leaders of the Karen ethnic minority, who attempt to set up their own state.

Independence

1948: January 4 The Union of Burma is proclaimed

The Union of Burma formally becomes a sovereign nation, also becoming the first former British colony to sever ties with the Commonwealth of Nations (an association of former British colonies united to encourage continued cooperation). U Nu becomes the first prime minister of the new nation, which faces immediate uprisings from Communist and ethnic groups, as well as factions of the military. Ne Win becomes commander in chief of the armed forces, which are granted increasing power in order to maintain government control of the country. The new nation faces difficult economic conditions, as it is simultaneously faced with heavy military expenditures, falling rice exports, and tax revenues.

1949–53 Chinese nationalists move into Shan province

Chinese Kuomindang nationalists fighting the Communists cross the border between the two countries and infiltrate Shan province, establishing a large illegal military base of as many as 12,000 troops and also engaging in the export of opium. Although resisted by Burma's army and ordered to leave by the United Nations, the Chinese establish a permanent presence in the province.

1951 General elections are held

The AFPFL wins a wide majority in Burma's first general elections. U Nu (formerly Thakin Nu) retains the post of prime minister.

1951 Karen state is created

A constitutional amendment creates the Karen state of Kawthule.

1952 The National Library is established

The National Library is founded in Rangoon. By the 1990s it houses more than 140,000 volumes.

1956 Social security program is begun

A newly inaugurated social security program provides for medical care, maternity leave, workmen's compensation, and survivors' benefits.

1956: June AFPFL wins general elections

The AFPFL returns to power, winning a legislative majority of over two-thirds. U Nu retains his post as prime minister.

1958: October Political shake-up in ruling party

Strong internal divisions threaten the ruling AFPFL, and Ne Win takes over as prime minister, forming an interim government. The military plays a large role in Ne Win's caretaker government, which restores order, streamlines the bureaucracy, and organizes the upcoming elections.

1960: February U Nu regains power

U Nu resumes the prime ministership following national elections that bolster his faction of the AFPFL. However, controversial campaign promises regarding establishment of Buddhism as the state religion and semi-autonomy for two ethnic groups create havoc once his new government is installed.

1961: August 26 Buddhism becomes the state religion

A constitutional amendment making Buddhism the state religion is passed and promulgated (widely spread). Soon afterward, another amendment is passed guaranteeing the rights of religious minorities.

Military Rule

1962: March 2 Military government installed in bloodless coup

The military, under Ne Win, removes U Nu's government in a largely bloodless coup. The constitution is suspended, and a Revolutionary Council composed entirely of military personnel is placed in power. Ne Win forms a new political party, the Burma Socialist Program Party (BSPP), and institutes "the Burmese Way of Socialism," nationalizing foreign businesses and taking control of all sectors of the economy, including banking (see 1963). More than 15,000 companies are reorganized as public enterprises.

In an attempt to end foreign domination of Burma, the new government also pursues isolationist policies, expropriating (to take for public use) many foreign assets and driving foreign capital out of the country. In addition, some 300,000 foreign residents, many of them Indians, Pakistanis, and Chinese, are expelled from the country, and tourist visas are limited to twenty-four hours.

1963 Government centralizes banks

The government centralizes the nation's banking system, making all banks affiliates of the Union of Burma Bank. Other state banks include the Burma Economic Bank, the Burma Foreign Trade Bank, and the Burma Agricultural Bank.

1964 Opposition parties are banned

All political parties except the BSPP (see 1962: March 2) are officially banned.

1968: November U Nu organizes political opposition

Former prime minister U Nu, following a period of imprisonment, comes out of political retirement to form the National Unity Advisory Board, consisting of himself and other politicians who had been jailed. The group makes recommendations to the ruling Revolutionary Council and demands a return to civilian, parliamentary government. The demands are refused, and U Nu leaves the country, traveling to gain international support for his political goals.

1970: May U Nu forms the National United Liberation Front

U Nu forms a new political group that brings together his followers and representatives of the Mon and Karen ethnic minorities. The group has an armed branch that launches raids on Burma from a base in Thailand.

1974 National flag is adopted

Burma's red, white, and blue national flag is adopted. It includes symbols of industry and agriculture, and its colors signify decisiveness (red), purity (white), and integrity (blue).

The Socialist Republic of Burma

1974: March 2 Return to civilian rule

After divesting himself of his military title, Ne Win forms a civilian government. Twelve years of military rule end as the Revolutionary Council is disbanded and the Socialist Republic of Burma is proclaimed. Ne Win becomes president, and a new constitution providing for a unicameral legislature is adopted. The only legal political party is the Burma Socialist Program Party (BSPP). Almost all top government posts are filled by members of the military.

1975 Mandatory school attendance is legislated

School attendance is made mandatory for children between the ages of five and nine.

1975 National Theatre tours the United States

Burma's National Theatre, a troupe of fourteen actors and dancers, undertakes an acclaimed concert tour of the United States.

1975: July 8 Pagan hit by earthquake

The ancient city of Pagan is hit by the worst earthquake in its history. Ancient religious structures suffer severe damage.

1977 Agricultural modernization program begins

The Whole Township Extension Program introduces modern cultivation methods and technology, including high-yielding varieties of rice seed, to improve the nation's agricultural production.

1979: September Burma gives up non-aligned position

Burma becomes the first of eighty-eight non-aligned nations to withdraw from the Non-Aligned Movement, giving political infighting within the movement as a reason for its action. The Non-Aligned Movement is an organization of countries that declare themselves neutral during the Cold War between the United States and the Soviet Union (c. 1947–c. 1991). The leading states of this movement are Yugoslavia, India, and Egypt.

1980 Political amnesty is declared

Ne Win declares an amnesty for all exiled former politicians. Aging former leader U Nu returns to Rangoon, but does not become involved in politics.

1981: November U San Yu becomes president

U San Yu succeeds Ne Win as president of Burma. Ne Win remains the chairman of the BSPP.

1983: October 9 North Korean bomb explodes at official ceremony

Burmese and South Korean government officials are killed when a bomb explodes at a ceremony, and South Korea's president narrowly escapes injury. Officers in the North Korean army are implicated in the bombing.

1983: November North Koreans convicted in bombing

North Korean officers are captured and convicted of plotting the bombing that killed Burmese and South Korean dignitaries (See 1983: October 9.)

1985–87 Prolonged warfare with ethnic minorities

The Burmese military carries on a protracted struggle against separatist ethnic minorities in border areas, including the Karen, Mon, Kachin, Shan, and Naga rebels, as well as Communist rebels near the Chinese border. Twelve rebel groups form the National Democratic Front with the goal of forming autonomous republics.

1985 U San Yu is reelected

U San Yu is reelected president. However, Ne Win still exercises most of the political power in virtually every area of government.

1986: November Government seeks "least developed" status in UN

Burma requests "least developed nation" status to qualify for emergency assistance from the United Nations. With agricultural and industrial production falling, the country suffers from a growing trade deficit, mounting inflation, shortages, and heavy foreign debt.

1987: September Currency withdrawn from circulation

In a move ordered by Ne Win and aimed at the country's black marketeers, the nation's three large currency denominations—the 25, 35, and 75 *kyat* notes—are withdrawn from circulation and declared worthless, wiping out eighty percent of the money in circulation, including personal savings, and creating economic turmoil. Student demonstrations and rioting follow.

1987: September 6 Universities are closed

In response to student protests, all universities in the country are closed indefinitely.

1988: March **Rioters protest police brutality**

Protesting an incident in which a student is clubbed to death by police and forty-one people are allowed to suffocate in an overcrowded van, rioters take to the streets of Rangoon. Protests spread across the nation, fueled by anger over Ne Win's economic policies. Thousands of demonstrators are killed by police, and thousands of students flee to border areas and join rebel minority forces.

1988: April **Aung San Suu Kyi returns to Burma**

Aung San Suu Kyi, daughter of slain national hero Aung San (see 1914), returns to Burma from exile in England.

1988: June 21 **Students march on Rangoon**

Following a political protest on the campus of Rangoon University, one thousand students converge on downtown Rangoon. Police crack down on political dissent, killing thousands in the next few weeks. A curfew is imposed in Rangoon.

1988: July **Ne Win resigns**

In response to mass protests of his handling of the economy, including the ill-advised withdrawal of large currency notes from circulation, Ne Win resigns his position at an emergency congress. The government adopts measures to introduce private enterprise and lessen controls on the economy. Sein Lwin, a longtime political associate of Ne Win, becomes president and chairman of the BSPP.

1988: August **Sein Lwin is replaced by Maung Maung**

The killing of peaceful demonstrators by government troops sparks mass protests and a general strike. Dr. Maung Maung, a lawyer and journalist, replaces General Sein Lwin as president. He lifts martial law, releases political prisoners, and vows to schedule free elections.

The S.L.O.R.C. Government

1988: September 18 **Military coup ousts Maung**

President Maung Maung is deposed in a military coup and succeeded by Saw Maung, who dissolves the legislature but promises free elections at a later time. The country's sole political party, the BSPP, is disbanded, and military rule is imposed under the authority of the State Law and Order Restoration Council (SLORC).

1988: September 24 **NLD political party is formed**

Aung San Suu Kyi and her associates form the National League for Democracy (NLD).

1989: June 18 **Burma's name is changed to Myanmar**

The English name for Burma is changed to Myanmar. The name of the capital, Rangoon, is changed to Yangon.

1989: July 5–7 **Mass protest rallies are held**

Mass demonstrations protesting government repression draw roughly 10,000 people.

1989: July 20 **Government cracks down on opposition**

The government arrests thousands of dissenters, including opposition leader Aung San Suu Kyi, daughter to Myanmar's national hero, General Aung San (see 1914). Suu Kyi, who has been criticizing the government in speeches and interviews, is placed under house arrest in Yangon.

1990 **Aung San Suu Kyi wins human rights award**

Prominent dissident Aung San Suu Kyi, under house arrest since 1989, is awarded the Sakharov Prize for Freedom of Thought by the European Parliament.

1990: May 27 **Multiparty elections are held**

Ninety-three parties field a total of 2,209 candidates in the multiparty elections promised by the SLORC government—the first since 1962. The NLD (National League for Democracy) wins an overwhelming eighty-eight percent of the votes cast. SLORC, however, claims that a national convention must draft a new constitution before the elected government can take office. The newly elected parliament is not allowed to convene, power is not transferred to the new government, and SLORC mounts a campaign of persecution against the leaders of the NLD, many of whom are forced to flee the country.

1990: November **NLD sets up a rival government**

The National League for Democracy, kept from power by the SLORC government, forms its own provisional government based in territory held by Karen rebels and led by Sein Win, a cousin of Aung San Suu Kyi. The rebel government wins support abroad, and foreign assistance to the SLORC government is reduced.

1991: October 15 **Aung San Suu Kyi wins Nobel Peace Prize**

The Nobel Peace Prize is awarded to Aung San Suu Kyi, under house arrest since 1989, for her opposition to the SLORC regime through nonviolent methods. Her son, Alexander, accepts the prize for her.

1992: April 23 **Than Schwe becomes president**

General Saw Maung (see 1988: September 18) resigns as president due to ill health and is succeeded by General Than

Schwe. The government ends its warfare with ethnic rebels along the Thai border, and other steps toward liberalization are taken.

1992: August **Universities reopen**

Myanmar's universities, closed since 1987 (see 1987: September 6), are allowed to reopen.

1993 **Constitutional convention meets**

A convention attended by 700 meets to draft a new constitution.

1993 **Kachin rebels sign cease-fire agreement**

The thirty-year insurrection by Kachin rebels in the north of the country is ended by a cease-fire agreement signed by the Kachin Independence Organization. In the next two years, fourteen other rebel groups, including the Karens and Mons, sign similar agreements. minority rebel groups.

1994: February 13–14 **U.S. congressman visits Aung San Suu Kyi**

Congressman William Richardson (D-NM) visits Aung San Suu Kyi, who is in her fifth year of house arrest. Richardson is the first non-family member to see the human rights activist in four-and-a-half years.

1995: June 19 **Red Cross withdraws from Myanmar**

The International Red Cross announces that it will close its office in Yangon and leave the country due to government restrictions on its ability to monitor prison conditions. Human rights abuses include arbitrary arrest and imprisonment, denying outside communication to prisoners, and denial of full citizenship for members of ethnic minorities. It is estimated that several hundred political prisoners are being held in jail.

1995: July 10 **Aung San Suu Kyi is freed**

Faced with mounting pressure from the United States and the ASEAN nations, the government releases dissident Aung San Suu Kyi from house arrest after six years. She vows to continue her struggle to bring democracy to Myanmar. The release of the 1991 Nobel laureate is lauded by human rights groups and political leaders worldwide.

1995: November 28 **NLD boycotts constitutional convention**

Unhappy with the direction taken by the government convention to draft a new constitution, the opposition National League for Democracy (NLD) announces that it will boycott future sessions. The announcement is made by Aung San Suu Kyi, released from a six-year period of house arrest in July. She calls for foreign leaders and potential foreign investors to pressure the country's military regime into improving its human rights record.

1996: January 1 **Drug lord surrenders to government troops**

Drug lord Khun Sa, head of the drug ring that controls heroin production in the "Golden Triangle," stretching across parts of Myanmar, Thailand, and Laos, surrenders to government troops, who seize his base in eastern Myanmar. The surrender is part of an amnesty agreement by which Khun Sa will avoid prosecution in exchange for ending his financing of a private rebel army fighting for an independent state for the Shan ethnic minority.

1996: May **NLD leaders arrested before conference**

The government arrests over 200 members of the NLD on the eve of a conference scheduled to be held at the home of Aung San Suu Kyi. Official rallies are held in Yangon to denounce the NLD and its leaders. Only eighteen of the delegates are able to attend the conference, but some 10,000 people gather outside Aung San Suu Kyi's home to show support for the beleaguered opposition party.

1996: July **Myanmar wins observer status within ASEAN**

Myanmar officials sign the Treaty of Amity and Cooperation, and ASEAN grants observer status to the country at its annual meeting, held in Brunei. The action is taken over the protests of European Union members and the United States, who condemn the government's support for international heroin traffickers and its record of human rights abuses.

1996: December **Aung San Suu Kyi placed under house arrest again**

Blaming Aung San Suu Kyi for inciting mass student demonstrations, the government places her under house arrest a year and a half after her release.

1997: January **Troops invade Thai refugee camps**

Government troops attack three Karen refugee camps in Thailand, leaving 35,000 homeless.

1997: May 31 **ASEAN admission is announced**

The Association of Southeast Asian Nations (ASEAN) formally announces that it will admit Myanmar as a member.

1997: July **Government releases Suu Kyi**

Bowing to international pressure once again, the government releases Aung San Suu Kyi from house arrest.

1997: November 15 Government reorganization is announced

Myanmar's military regime announces that it will be reorganized under a new name: the State Peace and Development Council (SPDC), composed of nineteen military leaders. Opposition leader Aung San Suu Kyi voices skepticism that the reorganization will produce any significant changes in the regime's policies.

1998: February U.S. lists Myanmar as uncooperative in drug war

The Clinton administration places Myanmar on a list of "decertified" nations that are deemed uncooperative in international efforts to halt the drug trade. The "decertified" status restricts foreign aid and U.S. approval of loans by international development banks.

Bibliography

Abbott, Gerry, ed. *Inroads into Burma: A Travellers' Anthology.* New York: Oxford University Press, 1997.

Aung San Suu Kyi. *The Voice of Hope: Conversations with Alan Clements.* New York : Seven Stories Press, 1997.

Aung-Thwin, Michael. *Pagan: The Origins of modern Burma.* Athens, Ohio: Ohio University Press, 1985.

Bunge, Fredericka J. *Burma: A Country Study.* Washington, D.C.: Government Printing Office, 1983.

Diran, Richard K. *The Vanishing Tribes of Burma.* New York: Amphoto Art, 1997.

Lintner, Bertil. *Burma in Revolt: Opium and Insurgency Since 1948.* Boulder, Colo.: Westview Press, 1994.

Maung Maung Gyi. *Burmese Political Values: The Socio-political Roots of Authoritarianism.* New York: Praeger, 1983.

Maung, Mya. *The Burma Road to Poverty.* New York: Praeger, 1991.

Parenteau, John. *Prisoner for Peace: Aung San Suu Kyi and Burma's Struggle for Democracy.* Greensboro, N.C. : Morgan Reynolds, 1994.

Renard, Ronald D. *The Burmese Connection: Illegal Drugs and the Making of the Golden Triangle.* Boulder, Colo.: L. Rienner Publishers, 1996.

Rotberg, Robert I., ed. *Burma: Prospects for a Democratic Future.* Brookings Institute Press, 1998.

Silverstein, Josef. *The Political Legacy of Aung San.* Ithaca, N.Y.: Cornell University Press, 1993.

Nauru

Introduction

The Republic of Nauru, covering only 8.1 square miles (21 square kilometers), is the smallest nation in the world. Nauru is approximately one-tenth the size of Washington, D.C. It is also one of the least populous countries with a population of around 10,500 people. The Republic of Nauru is one of the world's richest countries per capita.

Nauru is a single raised coral island whose nearest neighbor, Ocean Island, is 190 miles (305 kilometers) east. The island's highest point is only 216 feet (65 meters) above sea level. Lying at 167° east longitude, it is nearly at the midpoint of the Earth. The isolation of this small island has greatly impacted its historical development.

The climate of Nauru is tropical with a monsoon season that commences in November and ends in February. Since Nauru is fifty-three kilometers south of the equator, the monsoon season falls between late spring and early summer. Rainfall on the island is modest, with only an average of 18 inches (45 centimeters) per annum. Severe drought recurs regularly on Nauru.

Christianity is the majority religion of the Republic of Nauru. Ninety-two percent of the population is Christian, while the remaining eight percent is either Taoist or Confucionist. Protestants outnumber Catholics by a margin of two to one. Most Nauruans belong to the Nauruan Protestant Church. Most holidays on the Christian calendar are recognized and celebrated on the island. The major public holiday is Angam Day which occurs on October 26. This holiday recognizes the times during the history of Nauru when the population reached 1500, considered the minimal number of individuals needed for the society to survive.

History

The original settlers on Nauru remain unknown to archaeologists. As is the case with other Micronesian groups, Nauru oral history recounts that Nauruans were originally migrants to the island. The Nauruan language is classified as a Micronesian language within the larger Austronesian language family. The language contains archaic features which point to the long-term isolation of the original inhabitants.

The traditional Nauruans settled in the coastal fringe of the island where coconut trees are plentiful. Other groups settled in the land surrounding the interior Buada Lagoon. The majority of the land of the island of Nauru was unsuitable for habitation because it was composed of phosphate rock. Although phosphate in processed form is vital to modern agriculture, in its natural state it cannot support any significant agricultural pursuits.

The exact pre-European contact population is uncertain, however some scholars have speculated that the island supported as many as 1,400 individuals in the pre-contact era. Subsistence by the traditional islanders was confounded by frequent droughts and severe storms and high tides that made fishing beyond the reef extremely dangerous. In response to their needs for a reliable source of reserve food supply, the Nauruans developed a method of fish farming. The interior freshwater lagoon Buada Lagoon was utilized as a fish farm. The fry (recently hatched) of the *ibija* fish were collected from the reef. They were placed in a half coconut shell and their saline water was gradually replaced with freshwater. Once they were acclimated to the freshwater environment, they were released into Buada Lagoon or any smaller body of brackish, interior freshwater. The large Buada Lagoon was divided into sections which belonged to various social groups on the island. The sections were constructed from low coconut palms and fronds (palm leaves).

Traditional Nauruans were known for their activity of catching frigate birds by using boluses (lump of chewed food). They would then spend considerable time training the birds to eat on command. The birds' eating activities were part of ceremonial life and also formed a secular entertainment for the traditional Nauruans.

The traditional Nauruan society consisted of twelve clans. Membership in a clan was based on principles of matrilineal descent, a system where individuals trace their ancestry through their mothers. An individual was required to marry outside his/her mother's clan. The flag of the Republic of Nauru reflects the importance of this original social structure by having a twelve point yellow star on a field of blue.

Sustained European contact with Nauru began in 1830s. Beachcombers and castaways, two social types in the social structure of European expansion in the Pacific Basin, began to frequent the island due in part to its reputation for being receptive to foreigners. Most accounts by visiting European trading vessels indicate that the European expatriates lived without regard for life or law. Drunkenness was regularly reported and many traders did not want to drop anchor near Nauru. Nauruans themselves were not considered a threat to traders of European-descent until after the middle of the century. The Nauruans then took up arms against ships that stopped on the island, either attempting to destroy them or loot the cargo. Accounts from the end of the century describe the majority of Nauruans as suffering from venereal disease, speaking good English, and possessing an abundance of guns.

European beachcombers and castaways had a profound impact on Nauruan society. Many took Nauruan wives and raised children who interacted in two social worlds. Nauruan contact with Europeans was almost wholly with these beachcombers and castaways. Trade products that lured many other types of Europeans to other islands in Micronesia were not present on Nauru. Further, the island's natural defenses with no safe harbor and no obstructed views of the sea made raids by labor procurers impossible.

The influence of Australia on the social history of Nauru cannot be overestimated. The currency of the Republic of Nauru is the Australian dollar. Driving is on the left-hand side of the road, although there are only nineteen kilometers of sealed road that circle the island.

The system of local education has steadily increased over the years. An extension campus of The University of the South Pacific is located on the island. The literacy rate for Nauruans is ninety-nine percent.

The national economy of Nauru is completely dependent on the phosphate industry. The high standard of living coupled with the fact that the Republic of Nauru is tax-free insures an almost welfare state economy for the average Nauruan. Approximately forty percent of the population of Nauru comes from neighboring islands to work in the phosphate industry. External investment schemes have been enacted by the government to provide a permanent patrimony for the Nauruans.

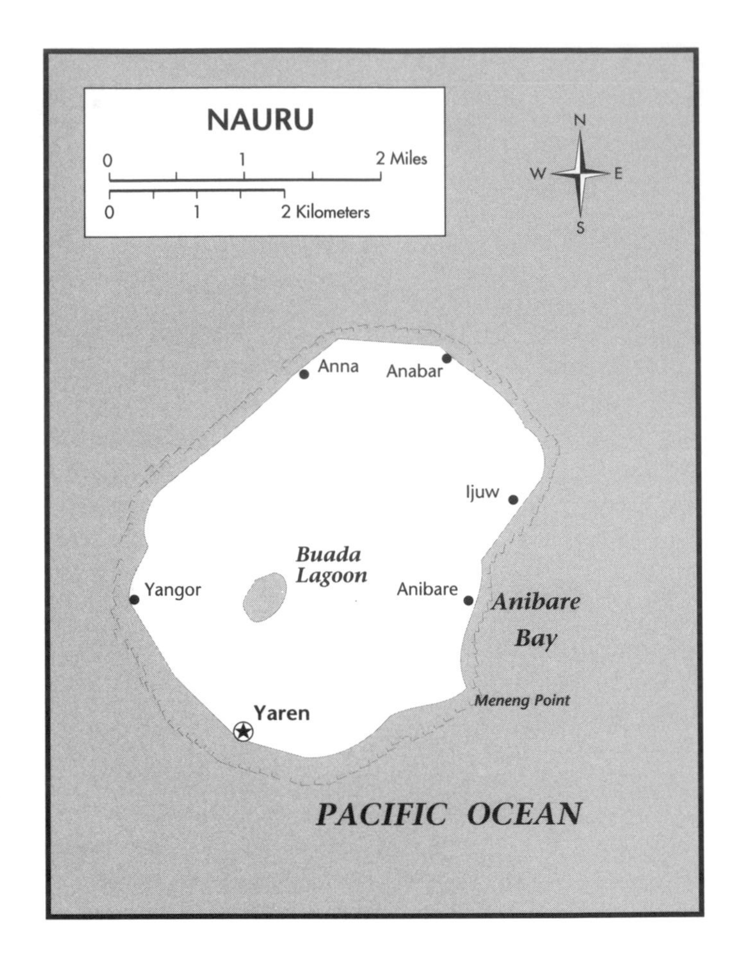

Timeline

Pre-1700s Nauru before European contact

Essentially nothing is known about the original inhabitants of Nauru. Early Nauruans probably originally migrate to the island. Since the Nauruan language contains archaic features, the original inhabitants are likely isolated from neighboring groups.

The first people on Nauru settle in the coastal fringe of the island where coconut trees are plentiful. Other groups settle in the land surrounding the interior Buada Lagoon. Most of Nauru is unsuitable for habitation because it is composed of phosphate rock. Although phosphate in processed form is vital to modern agriculture, in its natural state it cannot support any significant agricultural pursuits.

The island may support as many as 1,400 individuals in the era before contact with European explorers. Subsistence by the traditional islanders is confounded by frequent droughts and severe storms and high tides that make fishing beyond the reef extremely dangerous. In response to their needs for a reliable source of reserve food supply, the Nauruans develop a method of fish farming. The interior freshwater Buada Lagoon is utilized as a fish farm. The *ibija* fish are collected from the reef and are placed in a half coconut shell. The saline water is gradually replaced with freshwater. Once the fish are acclimated to the freshwater environment, they are released into Buada Lagoon or any smaller body of brackish, interior freshwater. The large Buada Lagoon is divided into sections which belong to various social groups on the island. The sections are constructed from low coconut palms and *fronds (palm leaves).*

1798: November 8 First European record of Nauru

Captain John Fearn, on a trip between New Zealand and southern China, sights Nauru, naming it Pleasant Island because of its beauty. Fearn's ship, *The Hunter,* does not drop anchor, although the Nauruans who come out to the ship in outrigger canoes plead with the Captain to come ashore.

From Fearn's account, the Nauruans do not show any aggression towards the Europeans and do not want to come aboard the trading vessel.

1830s European whaling ships visit Nauru for food

A number of brief accounts of Nauru appear during the 1830s. Deserters from ships take refuge on Nauru, often returning some years later to the European world only to describe an idyllic existence on the island.

1830s Escaped convict, John Jones, proclaims himself dictator

An Irish convict, John Jones, escapes from the penal colony on Norfolk Island, and becomes the self-imposed dictator of Nauru. European beachcombers who wind up on Nauru live in terror under Jones's reign.

1841: October John Jones is forced to leave Nauru

The Irish convict-turned-dictator, John Jones, kills eleven of his fellow beachcombers on Nauru. He attempts to blame the Nauruans for the act when confronted by other beachcombers and European sailors. The Nauruans force him to leave the island. Jones relocates temporarily to Ocean Island. He tries to return to Nauru to re-establish his reign of terror but is turned away by the Nauruans.

1878–88 Clan civil war

The population of Nauru is divided into rival clans. When Europeans introduce firearms to Nauru, the nature of the relationships that exist between rival clans on the island changes. This decade is plagued by constant fighting between members of rival clans with considerable loss of life.

1887 First Christian mission is established

The American Board of Commissioners for Foreign Missions, a Protestant group based in Boston, Massachusetts, sends three Gilbert Island missionary teachers to Nauru. They are sent back to the Gilbert Islands in the next year by German authorities.

1888: October 1 Nauru incorporated into Imperial German Protectorate

The decade of civil war on the island of Nauru ends with the incorporation of the island into the Imperial German Protectorate of the Marshall Islands. The Anglo-German Convention of 1886 has given Germany de facto control over Nauru due to the location of the demarcation line that separates German possessions from British possessions. Not until October 1888 does Germany show any interest in or exert any influence over Nauru.

1889 Census shows population imbalance

The effects of the ten year war on the Nauruan population are evidenced in the German census. Of a total population of 1,294 Nauruans, only a little over twenty percent are children. Adult women outnumber adult men by thirty percent. The population of young men has been severely reduced by the civil war.

1900 Albert Ellis discovers phosphate on Nauru

A young New Zealander, Albert Ellis, examines a doorstop in his Sydney geological office. Convinced that it is not 'fossilized wood' as the office workers claim, he runs tests and finds that the object is almost pure tricalcic phosphate. He learns the piece was brought to Sydney by a native of Nauru. Eliis concludes that Nauru as well as Ocean Island, which shares the same geological formation, may contain extensive phosphate deposits.

1902 First Roman Catholic missionary

Father Grundl comes from Germany to establish a Roman Catholic presence on Nauru. In the following year, an Alsatian, Father Kayser, also arrives. He has great success establishing Catholicism on the island.

1907 Phosphate mining begins

The Pacific Phosphate Company begins mining on Nauru.

1914: September 5 German administrator orders all British subjects deported

After the British declaration of war on Germany, the German representatives on Nauru ordered all British subjects deported to Ocean Island.

1914: November 6 German administrator surrenders

The German administrator of Nauru formally surrenders to British forces and is deported to Sydney, Australia.

1917 Australian becomes official administrator

Australia hopes to control Nauru's phosphate business. Australia has been the de facto administrator of the island since the deportation of the Germans three years before. An Australian, G.B. Smith-Rewse, becomes the official Administrator of the island in December.

1920: December 17 Mandate granted for British control of Nauru

The British Empire takes formal control of Nauru. In an agreement signed by Britain, Australia, and New Zealand, the Administrator of the island is to be chosen by the Australians.

1942: August 23 Japanese bomb Nauru

Japan has a vital interest in the phosphate that Nauru provides to Australia and New Zealand. After a lengthy blockade of the island, an air and sea bombardment of the island results in the surrender of the seven Europeans who remain on the island after the outbreak of World War II (1939–45) in the Pacific Theater.

1942: August 26 Japanese occupation

Three hundred Japanese soldiers take control of the island. Additional forces land shortly thereafter and begin construction of an airstrip.

1943: January Completion of first airstrip

Japanese forces with the help of Nauruan and Korean laborers complete the first airstrip on the island. Immediately, Japanese bombers begin arriving on Nauru.

1944 Allied blockade of Nauru

Allied naval and air forces mount a successful blockade of Nauru which prohibits the importation of food to the island. As a result, the Chinese, Korean, and Nauruan populations suffered malnutrition. Conditions for the Nauruans are very bad. Young Nauruan women are forced to work in brothels for Japanese officers.

1945: August 21 Japanese commander surrenders

One day after the peace between Japan and the United States is announced, the Japanese commander in charge of Nauru surrenders. Not until September 13 is an Allied occupation force led by Australian Brigadier J.R. Stevenson able to take possession of the island.

1945: September 14 Japanese are transported to Solomon Islands

Two Australian transports begin to remove all the Japanese from Nauru. The Japanese are relocated to Torokina on Bougainville Island.

1946: January 31 Final group of deported Nauruans return

The final group of Nauruans who have been deported as a result of the Japanese occupation of island return home from Truk. Among this group is the future President Hammer DeRoburt.

1948: June 7 Riot of Chinese indentured laborers

The constantly strained Chinese-Nauruan relations finally explode in a riot. A large group of Chinese laborers barricade themselves in a living compound and refuse entry to any non-Chinese. The Nauruan police attempt to arrest a Chinese man accused of threatening an interpreter. The Administration declares a state of emergency and sends a riot squad into the compound which results in the death of two Chinese men. Sixteen other Chinese were wounded.

1951: August 20 Nauru Local Government Council Ordinance enacted

The Nauruan Council of Chiefs, a traditional governing and judicial body, is given official status as a newly formed Local Government Council.

1951: December 15 First elections for Nauru Local Government Council

The first step towards self-governance on Nauru takes place with island-wide elections for nine councilor positions to represent the eight electoral areas.

1962 Discussions of resettlement of Nauruans

The Australian government makes several suggestions with regard to resettling the entire population of Nauru in anticipation of the depletion of the phosphate deposits on the island. The Nauruans have been steadily increasing their share of royalties from the extraction of the phosphate. All suggestions are rejected by the Nauruans who make a case for independence and self-government.

1968: January 31 Republic of Nauru established

On the anniversary of the return of Nauruans from Truk (see 1946: January 31), the Republic of Nauru becomes the smallest independent republic in the world.

1968 Radio Nauru begins public broadcast

Radio Nauru, broadcasting in English and Nauruan, is owned by the government. It does not generate original newscasts, but rather rebroadcasts reports from Radio Australia and the British Broadcasting Corporation (BBC).

1968 Hammer DeRoburt elected first President

Hammer DeRoburt (1923–92), a former teacher and head chief of the Nauru Local Government Council, is elected first President of the Republic of Nauru. DeRoburt is born in Nauru on September 25, 1923. After completing his secondary education on Nauru, DeRoburt studies at the Gordon Institue of Technology in Victoria, Australia. He has just begun teaching when World War II breaks out and he is deported to Truk. After returning to the island, he becomes active in politics on the island. He becomes Head Chief of Nauru in 1956 at the age of 33. DeRoburt is to become the dominant force in Nauruan politics, serving as President from 1968 until 1989. Only during the period of 1976–78 does he not fulfill the premier political leadership role of the country. In 1982, DeRoburt receives an honorary knighthood from Queen Elizabeth II. Sir Hammer DeRoburt dies in Melbourne, Australia on July 15, 1992, at the age of 69.

1970 Government takes complete control of phosphate industry

The government of the island takes complete administrative and financial control of phosphate production and exportation on Nauru. Half of the profits are set aside in Nauru Phosphate Royalties Trust established to insure the financial stability of the islanders after the depletion of the phosphate resources expected before the end of the century.

1971 DeRoburt is re-elected to presidency

Hammer DeRoburt re-elected president (see 1968).

1976 Bank of Nauru established

The government establishes the Bank of Nauru.

1980–87 DeRoburt is re-elected to presidency

In 1980, 1983, 1986, and 1987, Sir Hammer DeRoburt is re-elected to the position of President of the Republic. In Nauru, the president is elected by the members of Parliament who are themselves chosen through popular vote. All Naurans who are twenty and older are required to vote in popular elections.

1983 Yaren is established as capital

The only administrative and population center of the island, Yaren, is established as the capital of Nauru.

1987 Democratic Party of Nauru forms

Following the re-election of Hammer DeRoburt (see 1980–87), Kennan Adeang forms the Democratic Party of Nauru, recruiting eight members of parliament to join him. The goal of the Democratic Party, the only political party in existence on Nauru, is to limit the power of the president. The party fails to win much support and by 1995 ceases to exist.

Late 1980s University of the South Pacific opens campus

The University of the South Pacific, headquartered on Fiji, opens an extension campus on Nauru. It is the only local opportunity for higher education that Nauruans have. There are approximately 1,000 volumes in the university library.

1991 Television service begins on Nauru

Television service reaches Nauru although most programming is provided by Television New Zealand.

1991: September Nauru joins Asian Development Bank

Nauru becomes a member of the Asian Development Bank.

1993 Australia owes compensation to Nauru

The government of Australia loses its opposition to a law suit filed by the Republic of Nauru which seeks compensation for the extreme ecological damage inflicted by decades of phosphate mining. The International Court of Justice in The Hague, Netherlands finds in favor of Nauru and determines that Australia owes approximately 72 million dollars compensation to Nauru. Great Britain and New Zealand, although not specifically named in the suit, agree to provide 8 million dollars each toward the ecological revival of the island.

1994 Nauru announces plan to rehabilitate ecology

At the Small Island Conference on Sustainable Development, the Republic of Nauru announces its intentions to proceed with a plan of ecological rehabilitation. The estimated cost of the plan is set at 133 million dollars.

1995 France renews nuclear testing in South Pacific

Despite strong opposition by the leaders of many South Pacific nations including Nauru, French president Jacques Chirac allows the testing of nuclear weapons in the South Pacific following a lengthy ban. Nauran president Dowiyogo meets with French President Jacques Chirac to try to dissuade him from such action. When France resumes its testing, Nauru breaks off diplomatic relations with France.

1997: February 8 Kinza Clodumar elected President

Parliament elects Kinza Clodumar president. Nauruan presidents serve a three-year term unless they are required to leave office through a vote of 'no-confidence.' The next presidential elections are scheduled for the year 2000. The next parliamentary elections take place in November 1998.

1998 Census of population of capital

The population of the capital, Yaren, is 560.

1998: November Bernard Dowiyogo replaces Clodumar as president

Through a vote of 'no confidence,' the newly elected Parliament ousts Kinza Clodumar from the presidency and replaces him with Bernard Dowiyogo.

Bibliography

Bunge, Frederica and Melinda Cooke, eds. *Oceania: a regional study.* Washington, D.C.: U.S. Government Printing Office, 1984.

Hanlon, David. *Remaking Micronesia*. Honolulu: University of Hawai'i Press, 1998.

McKnight, Tom. *Oceania: the geography of Australia, New Zealand, and the Pacific Islands*. Englewood Cliffs: Prentice Hall, 1995.

Viviani, Nancy. *Nauru: phosphate and political progress*. Honolulu: University of Hawaii Press, 1970.

Nepal

Introduction

The Kingdom of Nepal is a small state set in the Himalayas, the mountain ranges that define the northern limits of Southern Asia. The country is landlocked, being surrounded on all sides by India and China. These basic geographical facts do much to explain the character of Nepal and its peoples, its historical and cultural evolution, and its modern political and economic problems.

Nepal has a population of 23 million people living in an area (56,139 square miles) roughly the size of the states of Iowa or Illinois. Nepal is truly a "mountain kingdom," with nearly a quarter of its land over 10,000 feet in elevation. The country falls into three distinct geographical areas. In the south is the *Terai,* a narrow belt of fertile, subtropical lowland along the plains of the River Ganges. The *Pahar*, or central hill zone, lies in the Himalayan foothills, a region of steep mountain ridges, swiftly flowing streams, and deep valleys. To the north rise the impressive peaks of the Great Himalayas, including Mount Everest, the world's highest mountain, Kanchenjunga, Dhaulagiri, and the Annapurna ranges.

The Nepali population shows considerable ethnic and cultural diversity. The inhabitants of the Terai resemble the peoples of the Indian plains, while groups such as the Sherpas and Bhotias of the mountainous north are physically and culturally related to the Tibetans. The central hill region is a transitional zone, where north Indian physical types represented by groups such as the Brahmans and the Chhetris live side by side with the Gurung, Magar, Tamang and Rai, whose appearance and culture clearly reflect their Central Asian origins.

Geographic isolation has meant that Nepal's various ethnic groups have preserved many individual customs and practices. Although the roots of Hinduism can be traced back over 3000 years, it emerged as a distinct religion only in the seventh to sixth centuries B.C. through a gradual reinterpretation of the *Vedas* (the traditional Indian religious sources) due to mass popular resentment toward the position of the Brahman (priestly) caste. Yet they have also developed a distinctive culture that blends South and Central Asian elements. Nepal is the world's only Hindu kingdom, and most Nepalis are Hindus by religion. Yet Hinduism as practiced in Nepal has been strongly influenced by both Tibetan Buddhism and local animistic beliefs (belief in the spiritual power of objects and places), and is quite distinct from the Hinduism of northern India. It is common for people to worship at both Hindu and Buddhist shrines. Nepalis in the Terai and Pahar regions have adopted the Hindu caste system, the social system that divides society into four hereditary, occupational classes. Caste influences whom one marries, the foods one eats, one's dress, and many other aspects of social life. However, it too is much less rigid and all-embracing than in India. Nearly sixty percent of the population speaks Nepali (also called Gorkhali), which is the official language of the country.

The historical, cultural, and economic core of Nepal is the Kathmandu Valley. The only extensive lowland area in Nepal outside of the Terai, this region has been a center of population since at least the eighth or seventh centuries B.C. The term 'Newar' describes the peoples who have settled here over the centuries, no matter what their ethnic origins. It is the Newaris who have been primarily responsible for the artistic achievements of Nepal, from the religious stone sculptures of the Licchavi period (fourth to ninth centuries A.D.), to the illuminated (painted) manuscripts of the eleventh century, and the wood carving and architecture of the fifteenth century. Even today, Newar culture is renowned for its music, its dances, and its colorful religious celebrations such as *Indrajatra* and the *Machhendranath* festivals.

History

Following the Muslim invasions of the India in the twelfth century, the area of present-day Nepal, came under the control of *Shas* (kings, descendents of the Rajputs, a warrior caste) who arrived from India in the sixteenth century. It was from the Kathmandu Valley, in the late eighteenth century, that Prithvi Narayan Shah's armies swept east and west along the Himalayan mountains, conquering lands that were to form the future state of Nepal. This expansion eventually led to conflict with Britain and China, and following a series of defeats, Nepal's rulers adopted a policy of self-imposed isolation. All foreigners were barred from the country during the nineteenth and early twentieth centuries. It was not until 1951, when a rising tide of agitation for democracy in the country over-

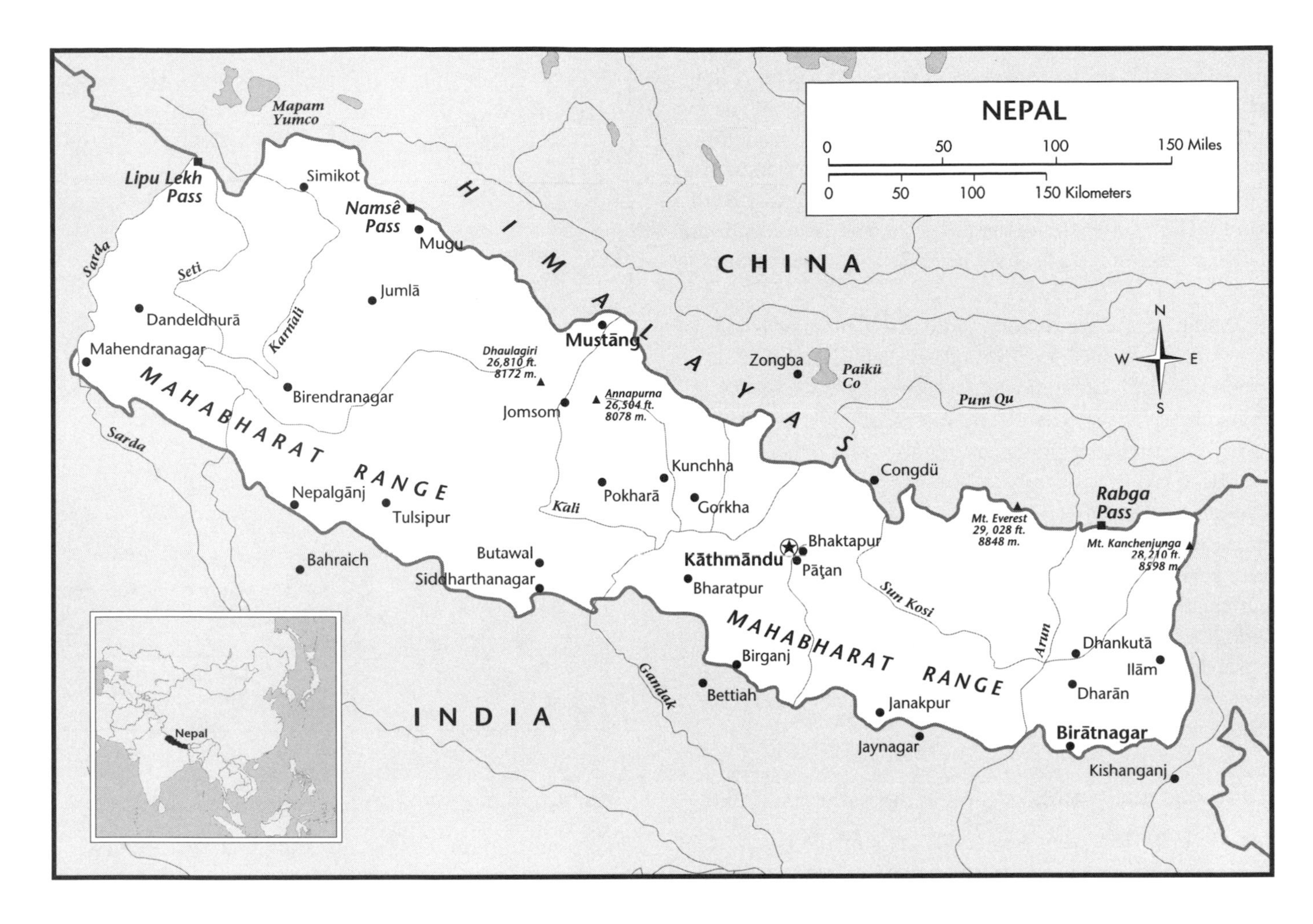

threw the old isolationist political order, that Nepal opened its borders to the outside world.

The decades following 1951 have been a time of political uncertainty and social unrest for Nepal. Democratic governmental institutions have slowly been evolving in the country, although with frequent setbacks. This process has involved a constant struggle for power between Nepal's king and popular political parties ranging from the Nepali National Congress to various Communist groups. A king, Birendra Bir Bikram Shah, remains on the throne today but rules as a constitutional monarch with limited powers.

Political instability has adversely affected the government's ability to deal with the country's social and economic problems. Nepal is among the poorest and least developed countries in the world, with an annual per capita income of around $200. Education is provided free in government schools, but literacy levels are low (around 38 percent for men twenty-four percent for women over fifteen years). The population is mostly agricultural, depending largely on rice, wheat, barley, lentils and vegetables for sustenance. However, only seventeen percent of Nepal's land can be farmed, and the country has to import food to feed its people. Rapid population growth is placing increasing pressure on agricultural resources, and poverty and malnutrition are widespread. Population pressure has also led to serious environmental problems such as deforestation, flooding, soil erosion, and pollution.

Nepal lacks the mineral resource base necessary for industrialization and economic growth. Poor transportation facilities impede development, and the economy faces severe shortages of skilled labor. Trade, formerly of great importance in Nepal's economy, has been curtailed by political problems. Tourism is a major source of foreign exchange, but by itself can have little influence on the standards of living of the population at large. The country remains heavily dependent on foreign aid. It remains to be seen if Nepal can overcome its locational and geographic disadvantages and achieve the levels of independence and prosperity to which it aspires.

Timeline

Pre-800 B.C. Nepal's legendary origins

Legend tells of a turquoise lake in the Himalayas at the very dawn of time. Upon this lake floats a lotus flower, which gives off a wondrous blue light and which represents the primeval Buddha. People come from many lands to pray on the shores of the lake and to worship the sacred flame. A holy man, Manjusri, comes from China to visit the lake and, desiring to approach the flame more closely, slices through the walls holding back the lake with his sword of wisdom. The waters of the lake drain away, forming the Valley of Kathmandu. The lotus settles to the valley floor at the site in Kathmandu where the Buddhist shrine of Swayambhunath stands today.

Much of the pre-history and early history of Nepal is rooted in legend (a Hindu myth relates that the god Krishna is responsible for the draining of the lake and the creation of the Kathmandu Valley). There is little archeological evidence to throw light on the region's past, and written materials date only from the fifth century A.D. Even then, a detailed and accurate account of Nepal's history is not possible until the latter half of the eighteenth century. Our understanding of the early period is based on accounts written down only in the fourteenth century or later, on references in early Hindu and Buddhist texts, and on descriptions left by travelers and pilgrims. The resulting picture is sketchy and incomplete, with most available information relating to the Kathmandu Valley rather than other areas of Nepal.

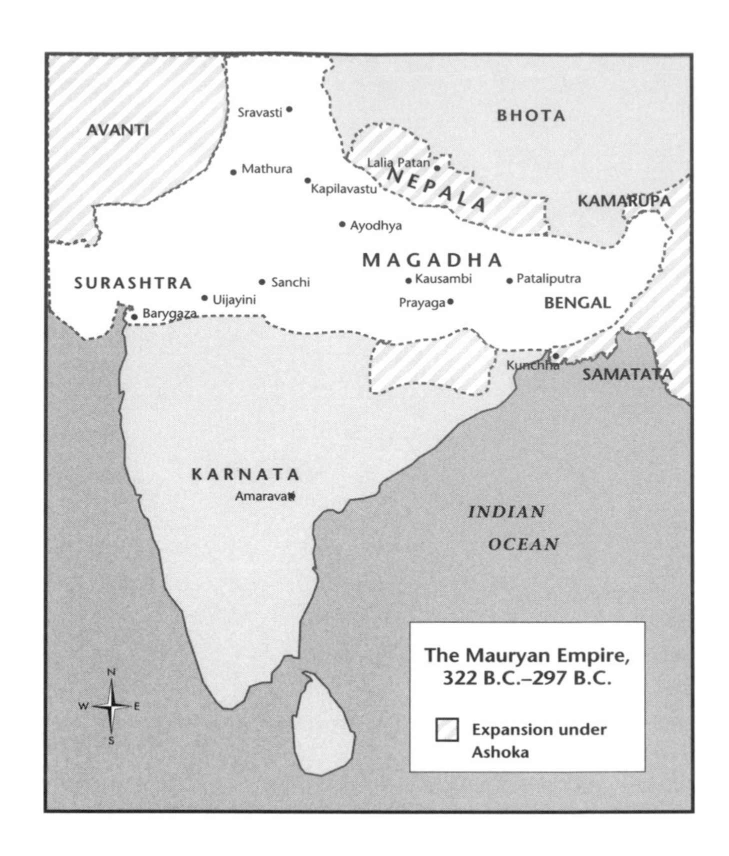

The Mauryan Empire, 322 B.C.–297 B.C.

c. 800–600 B.C. Early cultures in the Kathmandu Valley

Nepal's historical texts *vamsavalis* tell of the rule of the Gopalas, Abhiras, and Kiratas in the Kathmandu Valley. However, no independent evidence survives to confirm the existence of these legendary dynasties. The Kiratas, it is believed, migrate to the Kathmandu Valley from northeastern India or beyond in several waves ending in the seventh century B.C. They settle the area, conquer its inhabitants, and establish a dynasty that remains in place for 700 years. The Kiratas, who are mentioned in ancient Indian texts such as the epic *Mahabharata*, are a Mongoloid people, speak a Tibeto-Burman language, and live under a system of tribal government.

c. 563 B.C. Gautama Buddha is born

Siddhartha (c. 563–c. 483 B.C), the founder of Buddhism, is born in Lumbini, near the modern border with India. The son of a king and member of the Gautama clan of the Sakya tribe, Siddhartha belongs to the *ksatriya* or ruling warrior caste of Hindu society. At the age of 29 years, Siddhartha abandons his princely heritage and social standing, and wanders the countryside in search of supreme knowledge. He finds Enlightenment at Bodh Gaya, in what is now the Indian state of Bihar. From this time on, Siddhartha is honored as the Buddha, the Wise or Enlightened One. Over the centuries, Buddhism emerges as one of the world's great religions.

c. 520 B.C. Buddha visits the Kathmandu Valley

According to popular tradition, Buddha and his disciple Ananda visit Patan, a town in the Kathmandu Valley. The Valley remains a stronghold of Buddhism from this time forward.

c. 320 B.C. Nepala is mentioned in Kautiliya's *Arthasastra*

The kingdom of Nepala is mentioned in the *Arthasastra,* a treatise on economics and statecraft that describes India in the fourth century B.C. The text is attributed to Kautiliya, prime minister and chief counselor to Chandragupta Maurya, the first of the Mauryan emperors (dynasty from c. 322–185 B.C.). Nepala is cited for its woolen cloths, especially its high grade blankets, its carpets, skins and hides. The inhabitants of the region, the Kiratas, are described as sheep breeders and shifting cultivators with no knowledge of irrigation techniques.

c. 245 B.C. Asoka visits Nepal

Asoka (273–232 B.C.), the Mauryan emperor who converts to Buddhism, visits the birthplace of the Buddha and erects an engraved column there. Asoka also visits the Kathmandu Valley where, it is said, he builds *stupas* (hemispherical Buddhist monuments built to contain sacred relics) at the four cardinal points of the town of Patan.

Asoka (r. c. 265–238 B.C.) rules an empire that includes virtually all of the Indian subcontinent. His adoption of Buddhism as the state religion is responsible for its spread throughout Southern Asia.

c. A.D 300 Licchavi dynasty begins

The Licchavis, a tribal people from the plains of northern India, conquer the Kathmandu Valley and establish a dynasty that rules in Nepal for over five centuries. The invaders introduce the Hindu caste structure into the region, where Buddhism is already well-established. Under the Licchavis, artisans in the Kathmandu Valley create the first golden age of Nepalese classical art. The Licchavi period (fourth to ninth centuries) is particularly known for its stone sculpture focusing on religious themes. The 16-foot long statue of the reclining Vishnu, one of the main gods of Hinduism, at Budhanilantha near Kathmandu, is a classic example of the sculptures of this period.

467 The documented history of Nepal begins

A stone inscription at the Changu Narayan temple in the Kathmandu Valley is ascribed to the Licchavi king, Manadeva I (c. 464–505). This and inscribed coins from Manadeva's reign provide the earliest extant (existing) evidence of the Licchavi dynasty.

602 Amsuvarman becomes king

Amsuvarman becomes the first of several Thakuri kings to rule in Nepal. Like the Licchavis, the Thakuris are of Indian origin, are Hindu by religion, and claim high-caste princely *ksatriya* status.

During what is called the late Licchavi period (seventh to ninth centuries), Tibet rises to prominence and comes to dominate large areas of Central Asia, China, and the Himalayas. Nepal becomes a vassal (subordinate) state of Tibet, and is subject to Tibetan rather than Indian influences. Amshuvarman opens trade routes with Tibet and marries his daughter, Bhrikuti, to Tibet's ruler. This is also a period when Buddhism, along with Nepalese artistic and architectural forms, spread northwards across the Himalayas, changing forever the face of Tibetan culture.

c. 640 Hsüan-tsang describes Nepal

The Chinese traveler, Hsüan-tsang (c. 600–64), visits Nepal during his extensive journeys across northern India. He records the splendors of the court, describing the palace as seven stories tall and ornamented with pearls and precious gems. The king, he writes, sits on a lion throne surrounded by courtiers and officials, and guarded by hundreds of armed soldiers. The people of the Valley bathe several times a day, pierce and enlarge their earlobes as a a form of beauty, and love music and theatrical performances. Houses are carved of wood, and the stone sculptures are of exceptional beauty.

879 The Licchavi dynasty ends

The last Licchavi king is replaced by a Thakuri ruler. For over three centuries following the fall of the Licchavis, there are no written records of Nepal. This period, frequently called the Transition period or Nepal's "Dark Ages," is one of turmoil and strife, short-lived dynasties, and frequent foreign invasions. Throughout all this, however, Nepalese artistic traditions continue to evolve, with refinements in stone and bronze sculpture and the early development of architecture. Population grows, trade and commerce flourish, and religious monuments are built throughout the Valley.

1200 Mallas rise to power

King Ari-deva assumes the title of Malla, and founds a dynasty that rules in Nepal for several centuries. Periodic invasions during the early Malla period disrupt life in the Kathmandu Valley. An earthquake also devastates the region and kills many of its inhabitants.

1260 The architect Araniko visits China

Kublai Khan (1216–94), emperor of China, invites the Nepalese architect and artist Araniko to Peking to erect a stupa. Araniko introduces the pagoda style of architecture, with its distinctive multi-layered, upward-curving roofs, into East Asia. The modern highway linking Kathmandu to the Tibetan border is named after Araniko (d. 1306).

1346 Muslim invasions of Nepal

Ilyas Shah, the Muslim leader of Bengal, invades Nepal. The great Hindu shrine of Pashupatinath is destroyed and the Buddhist stupa at Swayambhunath is damaged. Other temples are burned, priests put to the sword, and many people killed. It takes decades for the Mallas to restore social and political order in the region.

1372 Third Malla dynasty established

Jaya Sthiti Malla captures Patan and establishes the third Malla dynasty. He unifies the Kathmandu Valley and initiates social and economic reforms such as new methods of land measurement and allocation. He introduces a new legal and social code based on Hindu principles, firmly establishing caste as the basis of Nepali society. Jaya Sthiti's short reign (c. 1382–95) ushers in the late Malla period, a span of over one hundred years that sees Malla political power and cultural achievements at their greatest.

c. 1429–82 Reign of Yaksha Malla

Yaksha Malla, Jaya Sthiti's grandson, extends his power beyond the confines of the Kathmandu Valley and by the mid-fifteenth century the Malla kingdom reaches its greatest territorial extent. The fifteenth century is also marked by a blossoming of Newari art and culture under the Mallas. Most of the great buildings in the Valley, the woodcarvings, and the sculptures for which Nepal is known date from this period. Newari is introduced as the court language.

c. 1484 The Malla Kingdom disintegrates

Following Yaksha Malla's death, the Valley is divided into three kingdoms (Kathmandu, Bhaktapur and Patan) ruled by Yaksha Malla's children. These small kingdoms are soon squabbling among themselves, and for nearly three hundred years remain in a constant state of war and turmoil. These centuries of conflict are so debilitating that the Malla kingdoms are able to offer little resistance to the Gorkha invasion of the late eighteenth century.

To the west of the Kathmandu Valley, there emerges during this period a series of minor states known as the Chaubisi (the twenty-four principalities), and to the west of these, the Baisi (the twenty-two principalities). Little is known of the histories of these kingdoms, though many of them are established by Rajputs (members of warrior caste) from the plains of northern India.

1559 Shahs rise to power in Gorkha

Drabya Shah, younger son of the Raja (ruler) of the Chaubisi kingdom of Lamjung, brings the principality of Gorkha (also Gurkha) under his control and establishes the line of kings that later becomes the royal house of Nepal. The Shahs are believed to be descendants of Rajputs, a warrior caste, who fled the plains of India to escape Muslim invaders.

1743 Prithvi Narayan Shah becomes Raja of Gorkha

Prithvi Narayan Shah (r. 1743–75) becomes Raja of Gorkha, a small principality in the Pahar zone to the west of the Kathmandu Valley. He embarks on a policy of conquest and expansion that eventually leads to the creation of the state of Nepal.

1744 Gorkhas march eastwards

Gorkha armies begin their march eastwards by capturing Nuwakot, a town between Gorkha and Kathmandu.

1753 Bhaktapur's "Golden Gate" is built

Bhaktapur's Sun Dhoka, or Golden Gate, is erected by Jaya Ranjit Malla. This gilded copper gate is regarded as the crowning achievement of Nepalese art in the Kathmandu Valley.

1756 Gorkhas surround the Kathmandu Valley

Prithvi Narayan occupies strategic positions in the hills surrounding the Kathmandu Valley, disrupting communications with the outside world. Gorkha gain control of the northern passes and stop all trade with Tibet. The defeat of British troops sent to help King Jaya Prakash Malla of Kathmandu in 1767 seals the fate of the Malla dynasty.

1768: September 25 Prithvi Narayan Shah captures Kathmandu

The Gorkha army marches into Kathmandu, where Prithvi Narayan is acclaimed king by the people. The subsequent fall of Patan and Bhaktapur marks the end of Malla rule in Nepal.

Modern History

1769 Prithvi Narayan Shah moves capital to Kathmandu

Prithvi Narayan now controls the entire Kathmandu Valley. He moves his capital from Gorkha to Kathmandu and establishes himself as the first king of Nepal.

1769–75 The beginnings of modern Nepal

Prithvi Narayan consolidates his position and lays the foundation of the modern state of Nepal. He ruthlessly executes potential rivals, thus avoiding the endless conspiracies and rivalries that often threaten the stability of many newly-formed states. Prithvi Narayan succeeds in bringing together people of differing ethnic and religious backgrounds and instilling in them a sense of national pride. He establishes a policy of excluding Europeans from the country. He initiates land and tax reforms, and his social and economic policies chart Nepal's course for the future. Prithvi Narayan describes his country, perched in the Himalayas between India and China, as "a root between two stones." This is an apt summation of Nepal's geo-political situation, with the country's relations with India and China dominating future foreign policy.

From his base in the Kathmandu Valley, Prithvi Narayan continues to push eastward and annexes the kingdoms of Chaudandi and Vijayanagar.

1775: January 10 King Prithvi Narayan dies

By the time of Prithvi Narayan's death, his kingdom encompasses almost one half the lands that make up modern Nepal. Gorkhas have conquered the Kirata lands to the east of the Kathmandu Valley and extended their territory as far as Darjiling (now in India). The western limits of the kingdom lie along the Marsyangdi River.

Prithvi Narayan is succeeded by his eldest son Pratap Singh Shah (r. 1775–77), and then by the latter's infant son Rana Bahadur Shah (r. 1777–99). A succession of minors, or child rulers, on Nepal's throne between 1777 and 1832 leads to power being concentrated in the hands of a series of regents (those who rule in place of a child king) and prime ministers during this period.

1777–85 Gorkhas expand westward

Bahadur Shah, the new king's uncle, is appointed regent. However, he is forced into exile by Rajendra Laxmi Shah, the queen mother, who assumes the regency herself. During her rule, Gorkhas push westward and annex most of the Chaubisi principalities.

1785 Bahadur Shah becomes Regent

Bahadur Shah returns from exile a few days before the queen mother's death and becomes regent, a post he holds until 1779. Under his leadership, the remaining Chaubisi principalities are conquered, the Baisi states are annexed, and Gorkha armies continue their march westwards along the Himalayas.

1788–92 Gorkhas invade Tibet

Conflict over issues such as the circulation of Nepalese coinage in Tibet and the taxation of goods moving from India to Tibet leads to war with China, which has suzerainty (political control) over Tibet. Gorkha troops invade Tibet in 1788 and again in 1791. A Chinese army enters Tibet and defeats the Gorkhas in 1792. Nepal surrenders territory taken earlier from Tibet, and agrees to send a tribute mission to the Chinese emperor in Peking every five years.

1806 Bhimsen Thapa becomes prime minister

Bhimsen Thapa becomes prime minister and holds this position during the reigns of two monarchs, Girvan Juddha Bikram (r. 1799–1816) and Rajendra Bikram Shah (r. 1816–47). Thapa supplants the monarch's authority and becomes the effective ruler of Nepal, wielding powers never before acquired by a minister. Nepal's kings are reduced to mere figureheads.

1809 Gorkhas overrun Kangra

With the capture of Kangra, a state in the western Himalayas, Gorkha expansion reaches its greatest extent. Nepal's territory at this time stretches out in a great arc along the Himalayas from the borders of Kashmir in the west to Sikkim in the east. Gorkhas at this time control territory that is double the area of modern Nepal.

1814–16 Anglo-Gorkha War

Territorial acquisitions by the British East India Company in northern India during the eighteenth century result in a common border with Nepal. During the early 1800s, frontier disputes along this border, repeated Nepalese incursions into British possessions, British concern at Gorkha expansion, and Nepal's persistent refusal to enter into trade or diplomatic relations leads to war. Britain invades Nepal and, although its troops meet determined resistance for almost two years, forces Nepal to sue for peace.

1816: March 4 The Treaty of Sagauli is signed

The Treaty of Sagauli ends the Anglo-Gorkha war on terms most unfavorable to Nepal. Nepal cedes (gives over) large areas of its territory, including most of its fertile lands in the Terai, to British India. The new boundaries that are drawn are essentially those of the modern state. Nepal agrees to accept a British political representative, to obtain British Government consent before employing any British subject, and to not employ Europeans or Americans. These last conditions are somewhat meaningless as Nepal subsequently closes its borders to all outsiders, leaving the British Resident as the only foreigner in the country.

One positive outcome for Nepal, however, results from the Anglo-Gorkha War. The British are so impressed with the fighting qualities of the Gorkhas that they begin recruiting them into the British Indian Army. Even today, Gorkhas serve with the armies of Britain and of India. The salaries and pensions earned by these soldiers are second only to tourism as a source of foreign exchange.

1837 Bhimsen Thapa is dismissed

The king's senior wife, with the support of the powerful Pande family, succeeds in bringing about the downfall of Prime Minister Bhimsen Thapa. Thapa is dismissed and dies in prison two years later. This creates a struggle for power between the king, members of the royal household, and families such as the Thapas, Pandes and Basnyats. The next nine years are marked by intrigue, rivalries, conflict, and assassinations. During this period, Nepal has eight governments and only one prime minister dies a natural death, the rest being killed in conflict.

1846: September 14 Ranas seize power in Kathmandu

General Gagan Singh Bhandari, the commander-in-chief of the military, a member of the government, and a favorite of the queen, is murdered. This triggers one of the most significant events in Nepal's modern history. An emergency meeting of the state council is called this night at the *Kot,* or Armory. Violence breaks out, and many of those present are killed. At the end of the 'Kot Massacre,' a young general, Jung Bahadur (1816–77), has seized power. He has himself designated prime minister, places family members in key positions of power, and confers on the family the name 'Rana.' Jung assumes virtually all the powers of the monarch, and establishes an oligarchy (rule by a powerful few) that effectively controls Nepal for the next hundred years. The Shahs are kings in name only, and are kept in close confinement in their palaces.

1847: May Jung Bahadur dethrones King Rajendra

Jung Bahadur removes Rajendra Bikram Shah and places the king's son, Surendra Bikram Shah (r. 1847–81), on the throne.

1850 Jung Bahadur visits Europe

Once he consolidates his power, Jung Bahadur visits England and France as ambassador of his new king. Impressed by what he sees of Britain's industrial and military might, he decides that Nepal's safest course is friendship with Britain as long as Nepal is allowed to continue its policy of isolation. On his return, he introduces a new legal code (the *Muluki Ain*) and restricts the use of capital punishment for certain crimes. He discourages, but cannot stop *sati.* Sati is the custom, practiced by the Hindu ksatriya castes, of widows willingly burning themselves alive on the funeral pyres of their dead husbands.

1854–56 War with Tibet

Mistreatment of Nepalis in Tibet and a desire to regain lands lost in 1792 lead Nepal to declare war on Tibet. Although the war ends in stalemate, Nepal gains certain trade concessions from Tibet.

1856 Ranas become hereditary prime ministers

Jung Bahadur Rana obtains a royal edict naming him Maharaja, with hereditary powers over both the king. The prime ministership is made a hereditary office within the Rana family, with the line of succession to Jung Bahadur clearly spelled out.

1857–1858 Nepal aids Britain during the Indian Mutiny

Nepal provides military assistance to the British during the uprising of *sepoys* (native soldiers) in India that is known as the Indian Mutiny. In return for this support, Britain gives back to Nepal areas of the Terai that were ceded (lost to the British) in 1816.

1877 Jung Bahadur Rana dies

At the death of Jung Bahadur Rana, his brothers persuade the king to appoint his designated successor as prime minister, Ranoddip Singh, to be Maharaja as well. From this time until the close of the Rana era, both positions are held by the same man.

1885: November The Shamsher faction seizes power

Bir Shamsher, Jung's nephew, and his supporters kill Ranoddip Singh and Bir is appointed Maharaja and Prime Minister by King Prithvi Bir Bikram Shah (r. 1881–1911). The Shamsher branch of the Rana family remains in control of Nepal until 1951.

1901 Chandra Shamsher comes to power

Prime Minister Chandra Shamsher (r. 1901–29) introduces limited social and economic reforms. His government sets up a college in Nepal, abolishes slavery and sati, brings electricity to Kathmandu, and builds a railroad in the Terai.

Nepal follows pro-British policies, assisting the 1904 Younghusband military expedition that forces Tibet to open trade relations with India. In recognition for loaning troops to fight in World War I (1914–18), Britain awards Nepal an annual payment of one million Indian rupees.

1911 Tribhuvan becomes king

Tribhuvan Bir Bikram Shah (r. 1911-55) becomes king of Nepal. Powerless like his predecessors, Tribhuvan later becomes a symbol for opponents of the Rana family.

1930–40 Political opposition to Ranas emerges

The first organized opposition to Rana rule from outside the Rana clan appears in the 1930s. Prime Minister Juddha Shamsher (r. 1932–45) uncovers a plot to overthrow the government in 1940. The leaders of the plot are imprisoned or executed although the king, who is implicated in the conspiracy, is allowed to remain, untouched.

1939–45 Gorkhas fight for Britain

Gorkha troops serve in the British Indian Army during World War II (1939–45).

1946: October 31 Nepali National Congress is founded

B. P. Koirala and other opponents of the Rana regime living in India found the Nepali National Congress, with the aim of bringing democracy to Nepal.

1948 Government of Nepal Act

In the face of agitation by the Nepali Congress and pressure from a now independent India, Prime Minister Padma Shamsher (r. 1945–48) passes the Government of Nepal Act. This allows for popular involvement in government through a system of village councils and a national *panchayat* (elected assembly of leaders). Rana hard-liners force Padma Shamsher to resign and his successor, Mohan Shamsher (r. 1948–51), postpones adopting the new constitution and declares the Nepali Congress illegal.

1950: November 6 King Tribhuvan escapes to India

Implicated in a Nepali Congress conspiracy against the Rana government and fearful for his safety, King Tribhuvan flees to New Delhi. At the same time a general uprising breaks out in Nepal, loosely coordinated by the Nepali Congress. By December, troops formerly loyal to the government begin to defect, and opposition to Mohan Shamsher within the Rana family increases.

1951: January Rana government accedes to India's demands

Faced with a deteriorating domestic situation and intense diplomatic pressure from India, Mohan Shamsher accepts India's demands that Tribhuvan be returned to the throne. He also agrees to an amnesty for the insurgents, popular representation in the government, a constituent assembly, and elections in 1952.

1951: February 18 King Tribhuvan returns from exile

King Tribhuvan returns to Kathmandu in triumph to assume his new role as a constitutional monarch. He rules with an interim cabinet made up of five Congress representatives and five Ranas, with Mohan Shamsher as prime minister. Nine months later, the Congress-Rana coalition collapses and Matrika Prasad Koirala, the Nepali Congress leader, becomes the first non-Rana prime minister of Nepal in over a hundred years.

1952 Elections postponed

The elections proposed for 1952 are delayed indefinitely. Several short-lived governments, unsuccessful coalitions, and periods of direct rule by the king (Tribhuvan in 1952–53, Mahendra in 1955–56 and 1957–59) contribute to political chaos in Nepal. Economic stagnation and failure of post-Rana governments to achieve political and economic reform leads to widespread discontent and sporadic outbreaks of violence across the kingdom.

1953: May 29 Everest is conquered

The New Zealander, Edmund Hillary (b. 1919), and his Sherpa companion, Tenzing Norgay (1914–86), successfully complete the first ascent of Mount Everest, the world's highest peak at 29,028 feet (8,848 meters).

1955: March 13 Mahendra ascends the throne

King Tribhuvan dies in Europe and is succeeded by his son, Mahendra Bir Bikram Shah (1920–72, r. 1955–72). Mahendra's policies bring him into conflict with Nepal's major political parties, whose leaders see a strong monarchy as a threat to their own political futures.

1958: June 25 Tribhuvan University is founded

The foundation stone of Tribhuvan University, Nepal's oldest university, is laid.

1959: February–March Nepal holds its first elections

Faced with the threat of civil disobedience by the major political parties, King Mahendra puts into effect a new constitution that provides for election of a legislative assembly. The Nepali Congress wins the election and forms a new government under B. P. Koirala.

1959 Hindu pilgrims bathe in Baghmati River

An estimated 50,000 Hindu pilgrims from many countries travel to the temple of Pashupatinath in Kathmandu to bathe in the Baghmati River and to celebrate Shiva, a Hindu deity, in a three-day ceremony.

1959 China closes Tibetan border

Following the Tibetan uprising against the Communist Chinese, China closes Tibet's border with Nepal. Tibetan refugees flee across the border into Nepal.

1960: December 15 King Mahendra takes control

Fearing the dominance of a single political party (as had happened with the Congress in India), Mahendra declares an Emergency, takes control of the government, and bans political parties.

1962 Mahendra establishes partyless government

The king establishes the "Partyless Panchayat System," in which village councils (panchayats) select representatives to district panchayats, which in turn send delegates to a national assembly. This body has very little real power, as the king selects the prime minister and his cabinet. Political parties continue to be outlawed.

c. 1967 Hippies come to Nepal

Nepal becomes a favorite destination for hippies. The climate, the tolerance of the local people, the mysticism of Hinduism and Buddhism, and the ready availability of drugs act as a magnet for young people seeking to drop out of Western society. Most of this group are gone from Nepal by the mid-1970s.

1972: January Birendra succeeds to the throne

King Mahendra dies and is succeeded by Birendra Bir Bikram Shah (b. 1945). The new king continues the political system established by his father, but also commits the government to improving the people's welfare and to promoting Nepal's economic development.

1979: April 23 Violence erupts in Kathmandu

Violence in the streets of Kathmandu climaxes a period of increasing discontent throughout the kingdom. In response to the situation, King Birendra announces a national referendum to determine whether the Nepali people wish to return to multi-party politics.

1980: May 2 Nepalis vote to keep partyless government

Nepalis vote, by a slim majority, to keep the existing system of government. The constitution is amended to allow for the direct election of the national legislature which, in turn, elects the prime minister.

Pilgrims wade in the waters of the Baghmati River outside the 5000-year-old temple of Pashupatinath. (EPD Photos/CSU Archives)

1985: December 28 Nepal Television commences service

Nepal Television begins broadcasting in Kathmandu. Both television and Radio Nepal, founded in 1951, are owned and operated by the government which controls programming content. Newspapers are published in Kathmandu in both Nepali and English, although the local reporting often leans towards the sensational. Periodic censorship is imposed by the government and King Birendra suspends publications of certain newspapers in 1985.

1987 SAARC locates its offices in Kathmandu

The South Asian Association for Regional Cooperation (SAARC) establishes its permanent secretariat in Kathmandu. SAARC was founded in 1985 by Nepal and the six other nations of South Asia (Bangladesh, Bhutan, India, the Maldives, Pakistan and Sri Lanka).

1989: March Trade treaties with India lapse

Nepal's trade treaties with India lapse, and India severely restricts the movement of goods and commodities, especially kerosene and staple foodstuffs, across its border with Nepal. The movement of goods from Nepal across Indian territory to the seaport of Calcutta and to Bangladesh remains an issue between Nepal and India.

1990: April 8 King lifts ban on political parties

Widespread popular agitation for a return to a multi-party system results in the lifting of the ban on political parties.

1991: May 12 Multi-party elections

A new constitution, adopted in 1990, paves the way for multi-party elections. The Nepali Congress Party secures a majority in the 205-seat House of Representatives and Girija Prasad Koirala becomes the new prime minister of Nepal. The Nepal

Communist Party (United Marxist-Leninist) forms the major opposition party.

1991 Refugees from Bhutan arrive in Nepal

Anti-Nepali sentiment in Bhutan leads to people of Nepali origin fleeing the country. Within five years, an estimated 100,000 Bhutanese refugees are living in camps in Nepal.

1995: August 3 Inferior status of women

The Nepal Supreme Court rules that the laws in the country's Civil Codes that prohibit women from inheriting property until they reach thirty-five years of age are unconstitutional and discriminatory. This highlights the growing status of women in a society in which they have traditionally occupied a subordinate role. Women's rights groups are working to expose problems ranging from physical abuse to trafficking (sale for prostitution) of women, especially to India.

1996: February Maoist insurrection begins

The Maoist United People's Front (UPF) launches a "people's war" against the government in mid-western Nepal. The insurrection is waged through torture, killings, and bombings of civilians and public officials. International human rights groups accuse government security forces of retaliation in kind.

1996: May 10 Disaster on Everest

A severe storm catches three climbing expeditions on Mount Everest. Eight climbers die in the blizzard in one of the worst disasters in Everest's history.

1998: January 1 Visit Nepal Year

King Birendra inaugurates "Visit Nepal '98", a year-long promotion designed to attract visitors to the country. Over ten percent of Nepal's land is designated as national parks and wildlife reserves. A popular destination is Royal Chitawan National Park, where visitors can see the endangered Bengal tiger and one-horned rhinoceros, and many other animals and birds.

1998: January–July Political instability continues

Surya Bahadur Thapa, the sixth prime minister since the multi-party system was introduced, resigns. The king asks Girija Prasad Koirala, of the Nepali Congress, to head a minority government. Koirala becomes prime minister and in July considers forming a coalition government with the Nepal Communist Party-Marxist Leninist Party (ML).

The democracy envisaged by the opponents of the Ranas remains in place in Nepal. However, the Maoist uprising, a series of weak governments, and limited progress in dealing with the country's social and economic problems, indicates an uncertain future for Nepal.

Bibliography

Anderson, Mary M. *Festivals of Nepal.* London: George Allen and Unwin, 1972.

Bista, Dor Bahadur. *People of Nepal.* Kathmandu: Ratna Pustak Bhandar, 1987.

Chauhan, R. S. *Society and State Building in Nepal: From Ancient Times to Mid-Twentieth Century.* New Delhi: Sterling, 1989.

Fisher, Margaret W. *The Political History of Nepal.* Berkeley: University of California, Institute of International Studies, 1960.

Hedrick, Basil C. and Anne K. Hedrick. *Historical and Cultural Dictionary of Nepal.* Metuchen, NJ.: Scarecrow Press, 1972.

Karan, Pradyumna P. *Nepal: A Cultural and Physical Geography.* Lexington, KY.: University of Kentucky Press, 1960.

Kramrisch, Stella. *The Art of Nepal.* New York: distributed by H. N. Abrams, 1964.

Matles, Andrea, ed. *Nepal and Bhutan: Country Studies.* Washington, D.C.: Federal Research Division, Library of Congress, 1993.

Rose, Leo E. and John T. Scholz. *Nepal: Profile of a Himalayan Kingdom.* Boulder, CO.: Westview Press, 1980.

Sanday, John. *The Kathmandu Valley: Jewel of the Kingdom of Nepal.* Lincolnwood, IL.: Passport Books, 1995.

Seddon, David. *Nepal, a State of Poverty.* New Delhi: Vikas, 1987.

Stiller, Ludwig. F. *The Rise of the House of Gorkha.* New Delhi: Manjusri, 1973.

New Zealand

Introduction

The island nation of New Zealand is known for the unspoiled natural beauty of its mountain scenery and for the two contrasting cultures that jointly make up its national identity: that of the white (or *pakeha*) descendants of its nineteenth-century British settlers, and that of the Maori, the native Polynesian people who arrived hundreds of years earlier and named the land Aotearoa ("land of the long white cloud"). This small, geographically remote nation is also known as a pioneer of progressive social legislation (such as its female suffrage law, enacted in 1893) and as one of the world's leading suppliers of wool, meat, and dairy products.

Located about 1,200 miles (1,930 kilometers) southeast of its much larger neighbor, Australia, New Zealand consists of two main islands, North Island and South Island, separated by the Cook Strait, and several smaller islands. Altogether, about 75 percent of New Zealand is mountainous or hilly. North Island, the smaller of the two main islands, is home to three-fourths of the country's population and to its two major cities: Wellington, the capital, and Auckland, the major commercial center. However, South Island, which is more mountainous, boasts the more spectacular scenery, as well as the world's southernmost city, Invercargill, at the furthest tip of the island. New Zealand has an estimated population of three and a half million people.

History

The Polynesian ancestors of today's Maori people are widely believed to have arrived in New Zealand in successive waves of immigration between the tenth and fourteenth centuries, traveling in outrigger canoes. They found no native mammals except bats on the islands, but ample aquatic life and many bird species. Farming of root crops, provided them with sustenance. They settled in villages organized around kinship communities and followed a religion that included ancestor and nature worship.

The first European to discover New Zealand was Dutchman Abel Tasman, sailing for the Dutch East India Company in 1642. Tasman did not settle, or even land on, the islands because of resistance by the Maori encountered by his men. In 1769 the Englishman James Cook circumnavigated and landed on both North and South Islands and claimed them for the British Crown. By the 1790s they were being unofficially settled by traders and renegade sailors. In 1814 Chaplain Samuel Marsden established the first Anglican mission, which was to play a significant role in the settlement of the islands. Another central force in the early colonization of New Zealand was the New Zealand Company, formed in England by Edward Gibbon Wakefield to promote land purchase and emigration to the islands.

One of the most significant historic dates in New Zealand history is the signing of the Treaty of Waitangi on February 6, 1840, by some fifty Maori chiefs, who agreed to grant Queen Victoria sovereignty over their land in exchange for British protection. However, differences in language and culture led to many subsequent disagreements about the provisions of the treaty, which were often interpreted differently by the British and the Maori. Some of the Maori land claims filed with New Zealand's government since the 1970s date back to the Waitangi Treaty.

Systematic colonization by the New Zealand Company began in earnest soon after the signing of the Waitangi Treaty, and the British government formally claimed New Zealand as a crown colony in 1841, giving it self-governing status in 1852. The discovery of gold on South Island in 1861 brought expanded immigration, tripling the population of the island within a decade. However, new immigration brought renewed tensions with the Maori, resulting in a series of land claim wars that ultimately resulted in the confiscation and sale of large areas of Maori land. Additional land was opened for colonization through the development of new transportation networks, and an agricultural boom led to the growth of exports, particularly of wool. As New Zealand grew and prospered, it began to develop its own distinctive culture. The year 1872 saw the publication of the colony's first major literary work, Alfred Dommet's epic poem *Ranolf and Amohia,* based on the life of the Maori. The following year Vincent Pike's *Wild Bill Enderby,* the first novel written and published in New Zealand, appeared. In 1877, New Zealand introduced free, compulsory education.

A new era began for New Zealand with the first successful shipment of a refrigerated cargo load to Britain in 1882. This innovation made it possible for the colony to become a

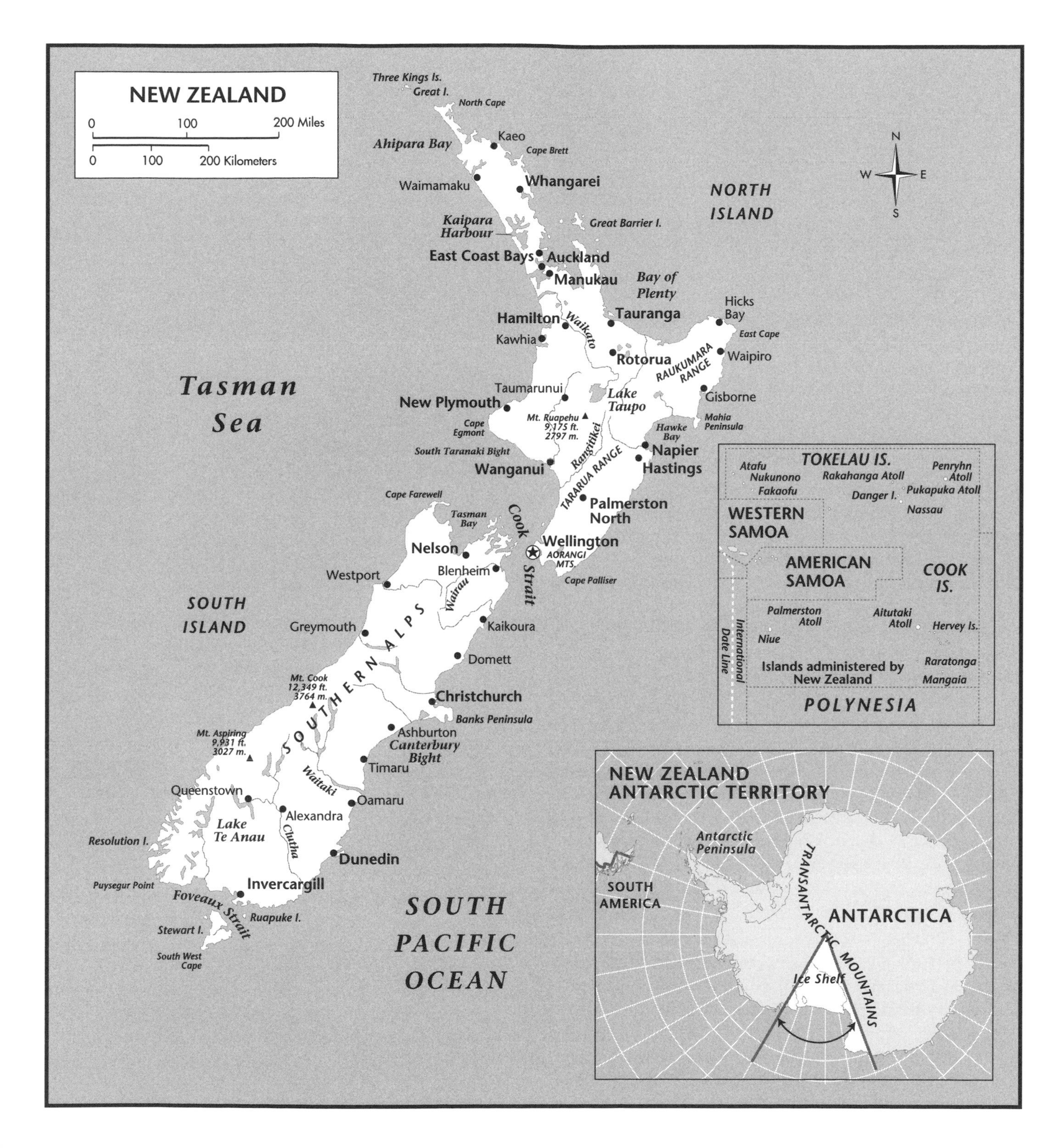

leading exporter of meat and dairy products in addition to wool. The resulting economic growth was accompanied by a growing awareness of social problems, and the Liberal government that came to power in 1890 made New Zealand a pioneer in the area of social reform. It became the world's first government to grant suffrage to women (1893), well in advance of both the United States and Great Britain. It also introduced compulsory arbitration of labor disputes, a minimum wage law, and legislation governing working conditions and pensions.

Over 100,000 New Zealanders volunteered to fight with British forces in World War I, and the colony's growing national awareness and pride was bolstered when its troops distinguished themselves in combat, most notably at Gallipoli. Some eighteen thousand were killed in action in the course of the war and many thousands more were wounded. In the 1930s, with New Zealand's economy hard hit by the global economic depression, the Labour Party came to power and re-established the government as a leader in the area of social reform, instituting social security, national health care, and a housing program. In World War II, New Zealanders once again mobilized to support Britain and its allies, with 200,000 troops sent into combat overseas. At home, as in many other nations, large numbers of women took their places in the work force. Wartime industry also employed an increased number of Maoris, contributing to the urbanization of the Maori population, an important social development of the postwar period.

Independence

In 1947 New Zealand attained independence when it formally accepted the 1931 Statute of Westminster which gave autonomy to British dominions. Following the war, the country developed ties with the United States, signing the ANZUS pact, a military alliance, with the U.S. and Australia in 1951. It also sent soldiers to aid the U.S. in both the Korean and Vietnam wars (although there was considerable domestic controversy over its involvement in the latter war, and its troops were withdrawn by 1971).

After a long period of prosperity, New Zealand's economy began to falter in the 1970s with the combined effects of rising oil prices and Britain's entry into the European Economic Community, which removed the traditional advantage New Zealand had enjoyed with its major trade partner. In response to this, New Zealand was forced to diversify its economy and find new export markets. In the 1980s, responding to rising inflation and other economic problems, a Labour government transformed New Zealand's economy with a series of free-market reforms, including extensive privatization and deregulation.

An important development in recent decades has been a revival of cultural awareness and political activism among the Maori, who began pursuing extensive land claims in the 1970s. In response to mass demonstrations, the government in 1975 established the Waitangi Tribunal to rule on land questions. In 1985 it agreed to recognize claims originating from the Waitangi Treaty of 1840, and four years later a cabinet-level committee was created to handle claims. In 1996 the Waikato Tainui tribe won a historic NZ$170 million (US$68 million) settlement for the confiscation of their lands in the nineteenth century.

An important political overhaul was initiated in 1993 when the New Zealand legislature adopted the "mixed member proportional" representation system, which allowed citizens to vote not only for individual candidates but also for the party of their choice. As predicted, the new system reduced the power of the dominant Labour and National parties and worked in favor of minor parties. The 1996 elections—the first following the reform—failed to produce a clear majority for either major party and the smaller New Zealand First party became the crucial factor in forming a new coalition government with the National Party.

In the 1990s, an international spotlight was turned on New Zealand through achievements in sports and in the arts. The entire country celebrated when team New Zealand won first place in the prestigious America's Cup yachting race—only the second time in history that a U.S. team had been defeated in this event. In the realm of popular culture, the country's burgeoning film industry produced internationally acclaimed motion pictures on New Zealand themes that both entertained and educated audiences worldwide. Peter Jackson's *Heavenly Creatures* (1994) dramatized a real-life scandal that shook the country in the 1950s. Lee Tomahori's *Once Were Warriors* (1994) presented a harsh picture of the lives of Maori living in urban poverty and torn from their cultural roots. Finally, Jane Campion, widely considered New Zealand's most distinguished filmmaker, directed both *An Angel at My Table* (1990), based on the memoirs of New Zealand novelist Janet Frame, and the Academy Award-winning film *The Piano* (1993), a tale of isolation and passion set in New Zealand in the colonial era.

Timeline

c. A.D. 950 First inhabitants arrive in New Zealand

A Polynesian people who are the ancestors of today's Maori begin arriving in New Zealand in canoes. Naming the land Aotearoa, "land of the long white cloud," they settle in communal villages composed of kinship groups and live by hunting, fishing, gathering, and the rudimentary farming of root crops. Each village centers around a sacred ground called a *marae.* These early settlers find no mammals on the islands other than bats, but there is abundant seafood and a variety of birds, most notably the huge, flightless *moa,* several times taller than an ostrich or emu.

European Discovery and Settlement

1642: December Tasman discovers New Zealand

Abel Tasman of the Dutch East India Co. becomes the first European to sight New Zealand. After some of his sailors are attacked by Maori, Tasman leaves the islands without landing on them. However, he does name them Niuew Zeeland, a name based on the Dutch province of Zeeland.

1769 Cook claims New Zealand for England

Captain James Cook circumnavigates New Zealand in the ship *Endeavor,* mapping the coastline. He lands on both islands and claims them for England. Cook makes three more trips to New Zealand in the 1770s.

1790s First settlements

Unofficial settlement of New Zealand is begun by runaway sailors, whalers, and sealskin traders.

1811 Birth of author Alfred Dommet

Dommet is widely regarded as the first major figure in New Zealand literature. A friend of British poet Robert Browning, Dommet authors the epic poem *Ranolf and Amohia* (see 1872). He later serves as prime minister of New Zealand.

1814 First mission is established

The first Anglican mission is established by Chaplain Samuel Marsden, who arrives from New South Wales, Australia, and plays a prominent role in the early settlement of the colony. In addition to proselytizing, he also advocates that the Maori be taught practical skills, including farming and carpentry. Wesleyan and Catholic missions are also established over the next twenty years.

1820s Written form of Maori is developed

A written form of the Maori language is devised by a scholar at Cambridge University.

1830s New Zealand Company is formed

The New Zealand Company is established in England by Edward Gibbon Wakefield. A joint stock company, it is intended to encourage and cover the costs of establishing settlements.

1832 Busby is appointed to represent Britain

James Busby is appointed British Resident, a post in which he represents British interests in New Zealand and attempts to institute a uniform basis on which to deal with the Maori chiefs, persuading them to establish the United Tribes of New Zealand.

1840 Settlement by New Zealand Company begins

The New Zealand Company, formed in England, plays an important role in the early settlement of the region. The company encourages settlement in New Zealand by selling land to speculators and giving free passage to prospective laborers to work on it.

1840: February 6 Treaty of Waitangi

About fifty Maori chiefs sign the Treaty of Waitangi giving sovereignty over their land to Queen Victoria of England in exchange for British protection. The chiefs also agree to sell land only to the British crown. Britain is represented by Captain William Hobson. The treaty proves to be a controversial document which still arouses disagreements today.

British Sovereignty

1840–47 Maori wars

Misunderstandings and disagreements over the provisions of the Treaty of Waitangi lead to intermittent warfare between the Maori and British settlers. Among the more notorious episodes are a massacre of New Zealand Company officials and the protracted conflict Bay of Islands conflict.

1840: May 21 British sovereignty is proclaimed

Britain formally proclaims its sovereignty over New Zealand, which is designated as a dependency of New South Wales (present-day Australia).

1841: May 3 New Zealand becomes a crown colony

The status of New Zealand is changed from a dependency of New South Wales to an English crown colony.

1845: June 22 Birth of statesman Richard Seddon

Richard Seddon (1845–1906), one of New Zealand's most famous and influential leaders, is born in Lancashire, England. He emigrates to New Zealand in 1863 during the gold rush era (see 1861). He works in the Victoria gold fields, where he becomes a leading spokesman for the miners in labor disputes.

Seddon is elected to Parliament in 1879. In 1891 he is appointed to the cabinet of liberal prime minister John Ballance and two years later becomes prime minister himself, holding that office for thirteen years during which New Zealand adopts some of the world's more progressive social legislation, including measures instituting female suffrage (1893), labor arbitration (1894), and old-age pensions (1898). He is nicknamed "King Dick" because of his powerful leadership role for over a decade in New Zealand politics. Seddon dies at sea in 1906 while returning from a trip to Australia.

1852 New Zealand is granted self-government

By the Constitution Act, the British Parliament grants autonomy to New Zealand's General Assembly, divides the colony into six provinces, and gives local governments the authority to buy and sell land.

1858–64 British author Samuel Butler in New Zealand

British essayist and novelist Samuel Butler, the future author of *Erewhon* and *The Way of All Flesh,* immigrates to New Zealand to set up a sheep farm on South Island with funds advanced by his father. During Butler's six years in New Zealand, he contributes articles to local newspapers and writes the memoir *A First Year in the Canterbury Settlement.* By the time Butler returns to England, he has doubled the value of his father's investment.

1858 The New Zealand Company is dissolved

After nearly going bankrupt and giving about one million acres (400,000 hectares) of land to the British government, the New Zealand Company is dissolved.

1860–72 Warfare between Maori and colonists

A period of warfare occurs between the British and Maori over government land purchases that the Maori claim are fraudulent. Many separate conflicts take place throughout the colony. The British are represented by both regular troops and local militias and police forces. Ultimately, the Maori are unable to withstand the superior force of the British, and large parcels of their land are confiscated and sold.

1861 Gold is discovered

The discovery of gold at Gabriel's Gully in Otago triggers a gold rush and a new wave of immigration to South Island. Within four months, the discovery brings four thousand prospectors to the area, and its population triples in ten years.

1865 Capital is moved to Wellington

The capital of New Zealand is moved from Auckland to Wellington.

1869 First art school is founded

New Zealand's first art school, the Otago School of Art, is established in Dunedin. It is headed by David Con Hutton.

1869: April 28 Birth of Frances Hodgkins

One of New Zealand's best-known artists, Hodgkins is born in Dunedin and begins watercolor painting under the tutelage of her father, William Mathew Hodgkins, a recognized painter in his own right. In 1901 she leaves for London and has a painting exhibited at the Royal Academy in 1903. Hodgkins lives in Paris between 1908 and 1914, where she establishes a reputation as a watercolorist and begins teaching classes in the art. While living in England during World War I, Hodgkins begins experimenting with oil painting. She becomes associated with English avant-garde artists in the 1920s and 1930s. Major critical recognition of Hodgkins's work does not come until shortly before her death in 1947. In 1969–70 a centenary exhibition of her work is held in New Zealand and also exhibited in Melbourne and London.

1870s and 1880s Artists' societies proliferate

Artists' societies formed in Auckland, Dunedin, Wellington, and other cities begin to hold regular exhibitions as an art scene becomes established in New Zealand.

1870s Public works bolster economy

Major public works projects, whose financing is arranged by colonial treasurer Sir Julius Vogel, trigger an economic boom and new immigration, including immigration from Germany and Scandinavia as well as Britain.

1872 Publication of epic poem *Ranolf and Amohia*

Alfred Dommet publishes *Ranolf and Amohia*, a poetic epic about Maori life. The poem, acclaimed by eminent British poets Robert Browning (a friend of Dommet's) and Alfred, Lord Tennyson, is considered New Zealand's first major literary work.

1873 First New Zealand novel is published

Wild Bill Enderby by Vincent Pike is the first novel to be written and published by a New Zealander.

1874: November 17 Ship fire kills hundreds of immigrants

Four hundred sixty-eight people—most of them immigrants from Britain—die when a fire aboard the *Cospatrick* rages out of control after the boat is steered into the wind. All but two lifeboats catch fire or sink from overcrowding. One of the two remaining boats sinks after four days at sea. Ultimately, only five people—four crew members and one passenger—survive the disaster.

1877 Free, compulsory education is introduced

The Education Act makes public school education free and mandatory.

1879 Male suffrage is expanded

All males over the age of 21, including Maoris, are granted the right to vote. Property qualifications are still retained, but they are reduced.

1882 Frozen meat shipping is introduced

The first cargo of refrigerated meat is shipped from New Zealand on the ship *Dunedin,* arriving in England three months later. The introduction of refrigeration makes New Zealand a leading exporter of meat and dairy products.

1888 Art gallery opens in Auckland

The Auckland City Art Gallery becomes New Zealand's first permanent venue for exhibiting art.

1888: October 14 Birth of Katherine Mansfield

Mansfield, one of the most important short story writers of the early twentieth century, is born Kathleen Mansfield Beauchamp in Wellington. After spending two years at school in London, she moves there at the age of nineteen and begins publishing her stories in periodicals, becoming friendly with other leading writers of her day, including T. S. Eliot, Virginia Woolf, and D. H. Lawrence. The collection *In a German Pension* appears in 1911. In 1918 she marries the essayist and critic John Middleton Murry. Subsequent story collections include *Prelude* (1918), *Bliss* (1920), and her masterpiece, *The Garden Party and Other Stories* (1922), whose title story and "The Daughters of the Late Colonel" are among Mansfield's most famous works. Katherine Mansfield dies of tuberculosis in France in 1923, at the age of thirty-four.

Liberal Social Policy

1890–1912 Liberal Party comes to power

The ascent of the Liberal Party, which is supported by both urban laborers and small farmers, represents the first time that a political party with a strong national agenda has come to power. The Liberal government enacts minimum-wage and pension legislation as well as laws regarding working conditions.

1890 Dock workers strike

Labor unrest stemming from industrial expansion culminates in an acrimonious dock strike.

1893–1906 Richard John Seddon serves as prime minister

Liberal Prime Minister Richard John Seddon presides over a notable series of reforms that make New Zealand a leader in enlightened social policy. Specific measures include the institution of progressive taxation, old-age pensions, child health services, land reform, programs to aid small farmers, labor reform (including the introduction of compulsory arbitration), and female suffrage. A highly popular leader, Seddon remains in office until his death.

1893 Female suffrage is enacted

After a petition signed by nearly 32,000 women (one-third of the women in New Zealand) is presented to the legislature, the government of New Zealand becomes the first in the world to enact female suffrage, some twenty-five years in advance of the United States and Britain. Women are granted equal suffrage rights as men five years later (see 1898).

1894 Arbitration is introduced

The Industrial Conciliation and Arbitration Act provides for compulsory arbitration to settle labor disputes. Minister of Labor William Pember Reeves oversees administration of the act, establishing a decentralized system of arbitration as well as an Arbitration Court empowered to rule on all aspects of labor law. Following the introduction of arbitration, there are no major strikes in New Zealand for over a decade.

1896 Cantata is composed using Maori themes

Based on a Maori legend, the cantata *Hinemoa* by Alfred Hill is the first significant musical composition by a New Zealand composer.

1898 Women are granted equal suffrage rights

After becoming the first to institute female suffrage, New Zealand's government also becomes the first to grant women voting rights identical to those of men.

1898 Social security is introduced

The Liberal government of prime minister Richard John Seddon passes legislation instituting old-age pensions.

1898 Birth of writer Henry Hector Bolitho

Henry Hector Bolitho (1898–1977), well-known author and historian, is born in Auckland. Bolitho's first published work, *With the Prince in New Zealand,* recounts his 1920 travels with England's Prince of Wales, who later becomes King Edward VIII. He works in both Australia and Europe during the 1920s, settling in London to become unofficial biographer of England's royal family. His books, favorably profiling royalty, include *Edward VIII: His Life and Reign,* and *King George VI: A Character Study,* both published in 1937. In 1946, he publishes *The Romance of Windsor Castle,* and in 1951, *A Century of British Monarchy.*

1899–1902 Boer War

New Zealanders fight for the British in the Anglo-Boer War in South Africa.

1901 The Cook Islands are annexed

New Zealand annexes the Cook Islands.

1906–07 First professional orchestra performs

New Zealand's first professional symphony orchestra is organized by composer and conductor Alfred Hill. The ensemble is enthusiastically received but disbands after one year due to lack of funding.

1906: June 10 Death of Richard John Seddon

Popular Liberal prime minister Richard John Seddon dies in a mishap at sea. (see 1845.)

1907 New Zealand becomes a dominion

New Zealand is granted the status of a dominion, or self-governing nation within the British Commonwealth.

1907 Rugby league is introduced

Rugby league, a form of rugby that is especially popular in working-class communities, is introduced to New Zealand. The same year, a team from New Zealand tours Britain under the leadership of A. H. Baskerville. A league is formed three years later.

1908: March 12 Birth of artist Rita Angus

Rita Angus, a leading twentieth-century painter, studies at the Canterbury School of Art in Christchurch. She first achieves professional success in the 1930s with works painted in a realist, regionalist style. Angus becomes known for her landscapes and portraits. Her works of the 1940s reflect her social and political concerns. Among her paintings are *Cass* (1936), the portrait *Betty Curnow* (1942), and the late work *Flight* (1968). Angus's work, most of which is collected in the National Art Gallery in Wellington, strongly influences younger artists. Angus dies in 1970.

1908: December 17 Birth of writer and educator Sylvia Ashton-Warner

Ashton-Warner, born in Stratford, is known for her work as an educator with Maori children and her support for better relations between New Zealand's Maori and white populations. Her novels, several of which become best-sellers, include *Spinster* (1958), *Incense to Idols* (1960), *Bell Call* (1964), *Greenstone* (1966), and *Three* (1970). She is also the author of several memoirs, including *Teacher* (1963), *Myself* (1967), and *I Passed This Way* (1979).

World War and Depression

1914–18 World War I

New Zealand troops fight alongside the British in World War I. Although a draft is instituted, most New Zealanders actively support the British cause and over 100,000 sign on as volunteers, serving with Australians in the ANZACS (Australia–New Zealand Army Corps) forces. They see combat at Gallipoli, in Palestine, and in France, and suffer heavy losses, with some eighteen thousand killed in action and thousands wounded.

1915: November 2 Birth of composer Douglas Lilburn

Douglas Gordon Lilburn (b. 1915), New Zealand's most acclaimed composer, is born in Wanganui. He is educated in New Zealand and England, where he studies with Ralph Vaughan Williams at the Royal College of Music in London. In 1936, his tone poem, *Forest,* wins a competition offered by Australian composer Percy Grainger. He returns to New Zealand in 1949, where he composes and serves as a professor at Victoria University of Wellington. Lilburn's early music is influenced by the romanticism of Jean Sibelius and Vaughn Williams. In the 1950s, he adopts a more atonal style, eventually becoming involved in electronic music and founding the first electronic music studio in the Pacific region (see 1964). Lilburn writes chamber music, symphonies, film scores, songs, guitar pieces, and other works. The titles of some of his compositions show their connection to the landscape and spirit of his native country: *Landfall in Unknown Seas, Aotearoa Overture, A Song of Islands,* and *Soundscape with Lake and River.* Through his compositions and as a teacher, Lilburn has been a major influence on succeeding generations of New Zealand composers. He is among those who establish the Archive of New Zealand music.

1916 Labour Party is formed

Several different labor organizations join to form the Labour Party.

1919 New Zealand receives mandate over Western Samoa

The League of Nations gives New Zealand a mandate to administer Western Samoa (formerly German Samoa), captured during World War I (1913–18).

1919 Birth of mountaineer Edmund Hillary

Edmund Hillary is born in Auckland. He becomes a professional apiarist (beekeeper) and amateur mountaineer. In 1953 he joins an expedition to Mount Everest, the world's tallest peak. With Sherpa guide Tenzing Norgey, Hillary reaches the summit of Mount Everest (29,030 feet or 8,850 meters) on May 29 of that year, the first to do so. This accomplishment earns him a knighthood.

In 1957–58, Hillary is co-leader of the British Commonwealth Antarctic Expedition, making the first overland trip to the South Pole. He returns to Mount Everest in 1960–61, and 1963–66, and to Antarctica in 1967 where he makes the first successful ascent of Mount Herschel (10,941 feet or 3,335 meters). Hillary wrote about his adventures in *High Adventure* (1955), *East of Everest* (1956), *No Latitude for Error* (1961), *Schoolhouse in the Clouds* (1965), and *Nothing Venture, Nothing Win* (1975).

1920s Cooperative export marketing is introduced

A network of export marketing agencies is set up, jointly run by agricultural producers and the government.

1923: January 9 Death of Katherine Mansfield

Katherine Mansfield, the distinguished New Zealand-born short story writer, dies in France. (See 1888.)

1924: August 28 Birth of author Janet Frame

Frame, a leading New Zealand writer of fiction and poetry, is born in Dunedin and grows up in poverty. As a young woman, she is misdiagnosed as a schizophrenic and spends years in and out of psychiatric hospitals, where she receives more than two hundred electric shock treatments. Her first published work is the short story collection *The Lagoon* (1951). Frame wins international acclaim for the experimental novel *Owls Do Cry,* which includes both poetry and prose and centers on the life of a small-town New Zealand family. Frame's other works include *Faces in the Water* (1961), *A State of Siege* (1966), *Daughter Buffalo* (1972), and *Living in the Maniototo* (1979). She is also the author of three autobiographical volumes. The second (*An Angel at My Table*; 1984) was made into a critically acclaimed motion picture by director Jane Campion in 1990.

1926: June 29 Birth of poet James Baxter

James K. Baxter, a leading literary figure of the postwar years and regarded by many as the country's finest poet, is born in Dunedin and educated both in New Zealand and England. His first book of poems, *Beyond the Palisade* (1944), is published when Baxter is only eighteen. In 1951 he publishes his first book of literary criticism, the well-received *Recent Trends in New Zealand Poetry.* Other poetry volumes include *Iron Breadboard* (1957) and *Pig Island Letters* (1966). Rejecting the values of mainstream New Zealand society, Baxter founds a commune in the 1960s. He dies in 1972.

1931 North Island earthquake kills 255

New Zealand's worst earthquake in modern times hits the town of Napier on the North Island, leveling the town and nearby villages and killing 255 people. International aid helps rebuild the town.

1931 Statute of Westminster

Through the Statute of Westminster, Britain grants autonomy to New Zealand. However, New Zealand does not accept the statute for a decade and a half. (See 1947.)

1932: April Economic crisis triggers rioting

Riots break out in Auckland as world prices (and British demand) for New Zealand's major exports—meat, dairy products, and wool—fall as a result of the global economic depression, and unemployment rises.

1932: June 4 Birth of author Maurice Shadbolt

Shadbolt is an acclaimed writer of novels and short fiction set in his native New Zealand. At the beginning of his career he works on documentary films and as a journalist. Turning to fiction in the late 1950s, he publishes two short story collections, *The New Zealanders* (1959) and *Summer Fires and Winter Country* (1963). Shadbolt's first novel, *Among the Cinders,* is published in 1965. Later works include *An Ear of the Dragon* (1971), *Danger Zone* (1975), *Season of the Jew* (1986), and *Monday's Warriors* (1990). A major theme in Shadbolt's fiction is the tension between urban life and traditional rural mores and values.

1935 Labour Party comes to power

The Labour Party wins its first election, coming to power under the leadership of Michael Savage. It once again makes New Zealand a pioneer in the area of social welfare legislation by establishing a full social security system, a national health service, and a state-subsidized housing program.

1936 Formation of the National Party

The National Party is formed by an alliance of groups opposed to the Labour Party. The most prominent groups are the Reform Party and the United Party. It takes thirteen years for the party to come to power, but it is the majority party for most of the period from 1949 to 1990.

1939–45 World War II

New Zealand declares war against Germany, serving with the British in North Africa and under U.S. command in the Pacific following the Japanese attack on Pearl Harbor. Nearly 200,000 troops are mobilized, and over 10,000 die. Major social changes occur at home during the war, notably the increased employment of Maoris in urban industrial jobs and the entry into the work force of large numbers of women.

1944: March 6 Birth of singer Kiri Te Kanawa

Acclaimed opera star Kiri Te Kanawa, born in Auckland of partly Maori ancestry, becomes one of the leading lyric sopranos of the postwar period. She receives her early musical training in New Zealand, winning prestigious music competitions and a scholarship to study in London. Since 1970 she has performed regularly at London's Covent Garden with the Royal Opera Company. In 1974 she makes her debut with the Metropolitan Opera Company in New York. Te Kanawa has also performed with the Paris Opera and other leading companies throughout the world. In 1981 she sings at the royal wedding of Prince Charles and Lady Diana Spencer. Although Te Kanawa has lived in London since the 1960s, she remains highly popular in her native New Zealand, returning periodically for concert engagements.

Independence

1947 New Zealand gains independence

Under the provisions of the 1931 Statute of Westminster, New Zealand claims full independence from Britain.

1947: March 6 Debut of national radio orchestra

The NZBS (New Zealand Broadcasting Service) National Orchestra gives its first concert, conducted by Andersen Tyler. The ensemble becomes an important force in the musical life of the country, with tours that sometimes cover over 16,000 kilometers a year. Guest conductors over the years include Josef Krips, Karel Ancerl, Sir William Walton, and Igor Stravinsky.

1947: March 9 Birth of writer Keri Hulme

Award-winning author of the novel *The Bone People*, Keri Hulme is born in Christchurch. She claims part-Maori ancestry and identifies strongly with the Maori people. Her first work is an innovative book of poetry entitled *The Silences Between: Moeraki Conversations* (1982). Hulme's first novel, *The Bone People* (1983), which incorporates Maori mythology, wins the 1985 Booker Prize, England's most prestigious literary award. Other works by Hulme include the short story collection *Kaihau/The Windeater* (1986) and two poetry collections, *Lost Possessions* (1985) and *Strands* (1992).

1947: May 13 Death of painter Frances Hodgkins

Critically esteemed painter Frances Hodgkins dies. (See 1869.)

1948: April 30 Mount Ngaurhoe spews ash and boulders

Mount Ngaurhoe, located in a remote area, begins pouring out ash, steam, and red-hot boulders, an eruption that lasts for several days. Geologist L.R. Allen heads an expedition to the crater, and reports no sign of molten rock (lava), only ash and hot boulders.

1949 First National Party electoral victory

The National Party, formed to represent rural and conservative interests, wins its first election. The new National Party government abolishes the upper house of parliament (the Legislative Council), making New Zealand one of the world's only democracies with a unicameral legislature.

1951 New Zealand signs ANZUS Pact

New Zealand joins with the U.S. and Australia in signing the ANZUS mutual security agreement.

1954 New Zealand joins SEATO

New Zealand becomes a founding member of the Southeast Asia Treaty Organization (SEATO). SEATO is a defensive alliance created by the United States and its Southeast Asian allies during the Cold War. It is the Asian equivalent of the North Atlantic Treaty Organization (NATO).

1954 Opera company is founded

The New Zealand Opera Company is established by singer Donald Munrow. Beginning with chamber operas, the group later undertakes full-length works including *Porgy and Bess* in 1965 and *Albert Herring* in 1966.

1954: April 30 Birth of filmmaker Jane Campion

New Zealand's best-known film director, Jane Campion is born in Wellington to a family involved in the arts. She is educated in New Zealand and London and begins making short films in the late 1970s. Her first theatrical feature, *Sweetie,* is released in 1989 and wins mixed reviews due to its eccentricity. Campion's next film, *An Angel at My Table* (1990), is based on the life of New Zealand author Janet Frame and depicts her struggle with mental illness and her evolution as a writer. Campion's best-known film is *The Piano* (1993), which wins multiple Academy Awards and a Best Director nomination for Campion. Campion's films are noted for their focus on the intellectual and emotional development of her female characters and on the dynamics of relationships between women.

1962 Ombudsman position is created

The government creates the office of ombudsman to investigate claims by citizens against government policies and rulings.

1964 Electronic music studio is founded

Composer Donald Lilburn establishes the first electronic music studio in the south Pacific at Victoria University of Wellington.

1965 Free trade pact is signed with Australia

The NAFTA (New Zealand Australia Free Trade Agreement) is signed to gradually eliminate trade barriers between the two countries.

1965 Troops sent to Vietnam

New Zealand sends troops to assist the United States in Vietnam, in a move that creates controversy at home.

Mount Ngaurhoe, dormant since 1934, pours out ash and red-hot rocks. (EPD Photos/CSU Archives)

1970s Economic malaise affects New Zealand

The global recession touches off economic difficulties in New Zealand, including rising inflation and unemployment and a growing trade deficit. In addition, the rise in oil prices forces New Zealand to incur a large foreign debt to cover the cost of importing the energy needed by its highly mechanized farming sector.

1970 Death of artist Rita Angus

Painter Rita Angus dies in Wellington. (See 1908.)

1972: October 22 Death of poet James Baxter

Baxter, one of New Zealand's most highly regarded and well-known poets, dies in Auckland at the age of forty-six. (See 1926.)

1973 U.K. entry into Common Market threatens New Zealand export market

Britain's entry into the European Economic Community (EEC, or Common Market) is a major blow to New Zealand's economy. It results in trade barriers for New Zealand's meat and dairy exports to the U.K., traditionally its major trade partner. New Zealand is forced to diversify its exports and seek new markets of its own, including the U.S. and Japan.

Mid-1970s Debut of popular cartoon, "Footrot Flats"

Murray Ball creates the comic strip "Footrot Flats," that becomes a staple of New Zealand popular culture. Spinoffs include a musical play, a feature-length cartoon film, and numerous books, all featuring the strip's main characters: a man a dog, and a sheep.

1974: November Voting age is lowered to eighteen

New Zealand's minimuum voting age is lowered from twenty-one to eighteen.

1975 Maoris march on parliament

Maoris march on New Zealand's parliament in a mass protest over loss of their tribal lands. They claim over two-thirds of the country's land (including the entire coastline) and half its fishing rights.

1975 Treaty of Waitangi Act creates tribunal

In response to massive protests, the government passes legislation establishing a seventeen-member tribunal to rule on Maori land claims.

1975 Strict immigration limits are adopted

Reacting to the country's economic downturn, New Zealand's government sharply lowers legal immigration levels.

1977 Government promotes film industry

The New Zealand government gives a boost to the country's film industry by forming a film commission and introducing tax incentives for film production. High-quality movies follow, by directors including Geoff Murphy, Vincent Ward, and Roger Donaldson.

1977–78 Naemi James is first woman to sail around the world solo

Naemi James of New Zealand becomes the first woman to circumnavigate alone in her yacht *Express Crusade.*

1982 Wage and price controls are imposed

The National Party government freezes wages, prices, and rents to deal with rising inflation. The freeze is in effect for two years.

1983 New trade pact signed with Australia

The CER (Closer Economic Relations) agreement, aimed at instituting free trade between Australia and New Zealand, replaces the 1965 NAFTA pact.

1984–90 Labour government reverses former economic policies

A newly elected Labour government restructures the economy along conservative lines, reducing its role in the economy through privatization and deregulation, with policies guided by finance minister Roger Douglas. The economic reforms presided over by Douglas (and popularly referred to as "Rogernomics") include the deregulation of financial markets and the transportation industry, the floating of New Zealand's currency, the virtual elimination of farm subsidies, and the removal of import controls. Douglas is eventually fired from his post by Prime Minister David Lange, but the Labour Party reinstates him, and Lange resigns.

1984 Government bans nuclear-armed vessels

New Zealand's government prohibits all vessels carrying nuclear weapons from entering its harbors, denying entry to a U.S. warship under the ban.

1985 Sinking of the *Rainbow Warrior*

In an incident that garners international attention, French agents bomb and sink the Greenpeace flagship *Rainbow Warrior* in Auckland harbor to halt a planned protest of French nuclear testing in Mururoa Atoll in the Pacific. One person is killed in the incident.

1985 Additional Maori claims are recognized

The Treaty of Waitangi Act is amended to recognize Maori claims dating back to the original signing of the treaty in 1840.

1985 First Maori governor-general is appointed

Anglican Archbishop Paul Reeves becomes the first person of Maori origin to become governor-general of New Zealand.

1986 U.S. suspends ANZUS obligations

In response to New Zealand's ban on vessels equipped with nuclear weapons in its harbors, the United States suspends its defense obligations under the ANZUS pact and terminates top-level diplomatic contacts between the two countries. They are later resumed (see 1990).

1986 Bungee jump off Eiffel Tower popularizes New Zealand sport

Alan John Hackett, a New Zealander credited by some as the inventor of bungee jumping, bungees off the Eiffel Tower in Paris. The sport is based on a ritual traditionally performed on the Pacific island of Vanuatu to guarantee a good yam harvest.

1987 Global stock collapse hits New Zealand speculators

The investment "bubble" resulting from the Labour government's free-market economic reforms bursts as world stock prices collapse.

1989: December Cabinet panet is formed to handle Maori claims

A new Cabinet-level committee is established to deal with Maori land claims.

1990 National Party comes to power

With the Labour Party in chaos following the resignation of Prime Minister David Lange, the National Party wins a majority in general elections and James Bolger becomes National Party leader.

Bolger is born in 1935 into a sheep and cattle-ranching family. He begins his political career as a National Party member of parliament in 1972. He becomes minister of agriculture in 1977, then minister of finance in 1981. He is elected prime minister in 1994, but defeated when he runs for reelection in 1996.

1990 U.S. restores normal diplomatic ties

The United States resumes normal diplomatic relations with New Zealand following a ban on meetings by top-level officials following a disagreement over access to New Zealand's harbors.

1990 Waitangi Treaty is commemorated

The 150th anniversary of the signing of the Waitangi Treaty (see 1840) between the British and the Maori is commemorated amid tensions over current Maori land claims.

1990 Free trade with Australia is achieved

The CER (Closer Economic Relationship) agreement (see 1983), providing for free trade with Australia, is fully implemented.

1993–94 New Zealand wins Whitbread yacht race

New Zealand wins the Whitbread round-the-world yacht race.

1993 *The Piano* wins international acclaim

Jane Campion's film *The Piano,* set in New Zealand's South Island in early colonial times, wins multiple Academy Awards and includes New Zealanders Sam Neill and Anna Paquin in its cast.

1993 Referendum on proportional representation passes

Voters pass a referendum providing for a major change in New Zealand's electoral system. In the past, all parliamentary seats have gone to candidates who won in their own districts. Under the new Mixed Member Proportional voting system, parties are guaranteed a certain number of seats based on their showing nationally, regardless of the outcomes of individual local contests. All voters will fill out two ballots, one for local candidates they favor, the other for the party they support. The electoral reform increases the size of New Zealand's unicameral parliament from 99 members to 120.

1994 Film dramatizes famous 1954 murder scandal

Director Peter Jackson releases his film *Heavenly Creatures,* based on a notorious 1954 murder committed by New Zealand teenagers Pauline Parker and Juliet Hulme, who together murdered Parker's mother, stoning her with a brick. Following her release from prison, Hulme establishes a career as a mystery writer under her new name, Anne Perry. The movie, which stars British actress Kate Winslet, is well received and joins the works of other film makers such as Jane Campion in gaining international recognition for New Zealand's film industry.

1994 *Once Were Warriors* depicts urban Maori plight

Part of a wave of films garnering international attention for New Zealand, the movie *Once Were Warriors,* directed by Lee Tomahori and based on a novel by Alan Duff, depicts the harsh life of the Maori under modern urban conditions.

1995: May New Zealand wins America's Cup

New Zealand becomes the only country other than Australia ever to defeat the U.S. in the prestigious America's Cup yacht race. Sailing the *Black Magic 1,* the New Zealand team, led by Peter Blake, dominates the event, winning forty-one out of the forty-two races in which it competes.

1995: September 23 Largest volcanic eruption in fifty years

Mount Ruapehu produces New Zealand's largest volcanic eruptions in fifty years. Sites near the volcano are evacuated and planes are not allowed to fly less than 25,000 feet above the mountain. The eruptions last over a week before dying down.

1996 Land settlement with Maoris is signed

A historic NZ$170 million (US$68 million) settlement agreement, approved and signed by Britain's Queen Elizabeth, compensates the Waikato Tainui tribe for nineteenth-century confiscation of its lands on the North Island.

1996: February 28 Coalition government is formed

The National Party and the newly established United Party form New Zealand's first coalition government in sixty years. The establishment of the coalition by the two conservative parties is linked to the 1993 referendum adopting a proportional voting system that gives increased influence to minor parties.

1996: October 12 Neither party wins clear majority in elections

In the first national elections held since the adoption of mixed-member proportional representation (see 1993), neither the Labour Party nor the conservative National Party wins a majority of seats in parliament. The New Zealand First Party is expected to form a coalition with one of the other two parties after protracted negotiations.

1996: December 10 New coalition government is formed

The National Party and the New Zealand First Party announce the formation of a coalition government after two months of negotiations following the failure of either of New Zealand's two leading parties—the National Party and the Labour Party—to win a decisive majority in the October elections. Jim Bolger, National Party prime minister since 1990, will retain his post, and Winston Peters (b. 1946), head of New Zealand First, will serve as deputy prime minister and treasurer.

Winston Peters, of Maori and Scots descent, enters politics as a member of the National Party in 1979. In 1990, he becomes minister of Maori affairs in the administration of Jim Bolger but he is expelled from the cabinet in 1991 and from the National Party in 1993 for his controversial views. He founds the New Zealand First party and serves in the coalition government with the National Party after the 1996 elections.

1997: November 3 Bolger resigns as prime minister

Jim Bolger, prime minister of New Zealand's coalition government, resigns due to lack of support nationwide for his government. Much popular criticism of the government stems from political scandals involving members of the New Zealand First Party, one of the two partners in the governing coalition.

1997: December 8 First woman prime minister is sworn in

Jenny Shipley, minister of transportation in the previous cabinet, is sworn into office as New Zealand's first female prime minister. Shipley has been the driving force behind the resignation of Jim Bolger as prime minister of the country's coalition government and head of the National Party.

1998: February 20–March 27 Auckland power supply fails

The electric power supply to the commercial district of Auckland, New Zealand's largest city, is shut off when the last of four underground cables serving the city fails, resulting in a five-week blackout. Thousands of inner-city residents are evacuated, and normal business operations grind to a halt or are relocated as the sharply reduced power supply is reserved for emergency services. Mercury Energy, the utility company serving the affected area, is widely blamed for failing to upgrade the cables adequately.

Bibliography

Alley, Roderick, ed. *New Zealand and the Pacific.* Boulder, Colo.: Westview Press, 1984.

Belich, James. *Making Peoples: A History of the New Zealanders from Polynesian Settlement to the End of the Nineteenth Century.* Honolulu: University of Hawaii Press, 1996.

Camilleri, Joseph A. *The Australia, New Zealand, US Alliance: Regional Security in the Nuclear Age.* Boulder, Colo.: Westview Press, 1987.

Hawke, G.R. *The Making of New Zealand.* Cambridge: Cambridge University Press, 1985.

Johnston, Carol Morton. *The Farthest Corner: New Zealand, a Twice Discovered Land.* Honolulu: University of Hawaii Press, 1988.

Mascarenhas, R.C. *Government and the Economy in Australia and New Zealand: The Politics of Economic Policy Making.* San Francisco: Austin & Winfield, 1996.

McKinnon, Malcolm. *Independence and Foreign Policy: New Zealand in the World since 1935.* Auckland: Oxford University Press, 1993.

Oddie, Graham, and Roy W. Perrett, ed. *Justice, Ethics, and New Zealand Society.* New York: Oxford University Press, 1992.

Rice, Geoffrey W., ed. *The Oxford History of New Zealand.* 2nd ed. New York: Oxford University Press, 1992.

Sharp, A. *Justice and the Maori: Maori Claims in New Zealand and Political Argument in the 1980s.* New York: Oxford University Press, 1990.

Oman

Introduction

Located in the northeastern part of the Arabian Peninsula, Oman occupies about 82,031 square miles (212,460 square kilometers), an area that is about the size of the state of Colorado. Its coastline stretches about 1,060 miles (1,700 kilometers) along the Strait of Hormuz and Gulf of Oman to the north, and the Arabian Sea to the east and south. It shares borders with Yemen, Sa'udi Arabia, and the United Arab Emirates, which separates Musandam, a rocky peninsula located at the tip of the Strait of Hormuz, from the rest of Oman. Oman's terrain includes pristine, white-sand beaches, fertile plains along the Batinah Coast, mountains in the Jabal al-Akhdar Range that soar more than 9,000 feet (3,000 meters) and vast stretches of desert.

Oman's population is diverse. Although most Omanis are Muslims and Arabs, peoples of Iranian, Baluchi and African descent populate the Batinah Coast. In addition, several cultural groups from India and Pakistan live in cities such as Muscat and Matrah. More than 200 tribal groups populate the interior. While Arabic is the official state language, it is not unusual to hear Urdu, Farsi and other South Asian languages being spoken, particularly in cities and along Oman's coasts. English is studied in schools.

Oman's official state religion is Islam, and most of its residents are followers of the Ibadi sect, which developed in the seventh century. Several tribal groups, however, are Sunni Muslims, and Oman enjoys a fairly strong level of religious tolerance. Qabus has encouraged such tolerance by setting aside land for Christian churches and Hindu temples, so that even non-Muslims can feel free to worship in their own faiths.

Early History

The earliest records indicate that the region of present-day Oman served as a supplier of copper to the Akkadian Empire c. 3000 B.C. Farming communities were established near oases and sustained themselves through the building of dams and, by the sixth century B.C., irrigation systems. However, the most significant event in the region's development occurred in the seventh century A.D. with the arrival of Arab armies which introduced Islam. Muslims in the area of Oman soon adhered to the Ibadi sect of Islam. Unlike other Muslims, the Ibadi believe that the *imam* (leader of a religious community) can be any male elected as such from his community; most Muslims believe that the *imam* can only be a direct descendant of the prophet Mohammed. The Ibadis in Oman managed to keep their independence at this time and soon developed a thriving state based upon extensive trade throughout the Indian Ocean. Trade with China and the Malay peninsula brought wealth to Oman's coastal cities.

Prosperity Amid Foreign Influence

The arrival of Portuguese traders in the early sixteenth century, put a damper on Oman's prosperity. However, this period of foreign domination lasted little more than a century as the Omanis ousted the Portuguese in 1622. Indeed, Oman followed up on its eviction of the Portuguese with attacks and conquests against its erstwhile occupiers. Thus, Oman captured Portuguese trading ports in East Africa and participated in the African slave trade.

Contact with England began in the seventeenth century when Oman offered the British East India Company trading concessions. Over the next two centuries, Oman remained a regional power. In the mid-eighteenth century, a sultanate was established under the Al-Busaid dynasty based in the city of Muscat. During this time, Oman-British friendship strengthened as the latter sought Oman's help in securing its supply route to its Indian colony.

At the same time, Oman's power diminished greatly when feuding over the succession to the throne led to a split in Oman's possessions. A separate Sultanate of Zanzibar was created for one of the claimants while the other held present-day Oman (called Muscat and Oman). The split of the empire brought economic hardship in its wake. The resulting unrest between Muscat and the interior of Oman eroded the sultan's power outside of the coastal areas and led to a formal split in 1920 between the sultan, who ruled over Muscat, and the *imam* (religious leader), who held sway over the interior. Not until the 1960s did the sultan regain control over the interior regions of Oman.

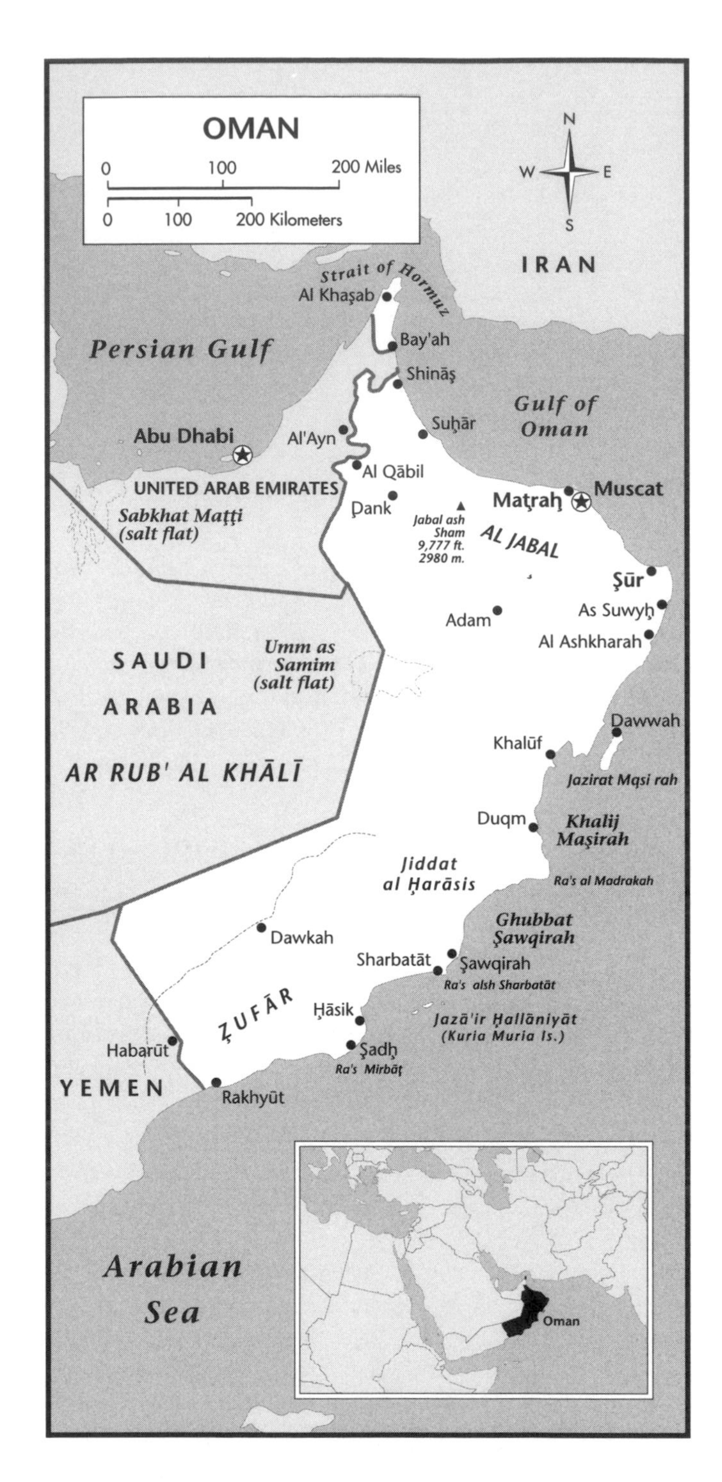

Oil-rich Oman

Most countries in the Persian Gulf prospered after World War II (1939–45). The discovery of oil made many of these nations rich. Most displayed their new wealth with flair. High rises made of concrete and glass dominated the skylines of major cities in Bahrain, Kuwait, Qatar, Sa'udi Arabia, and the United Arab Emirates. New roads were built, airlines fought for landing rights. Businesses vied to sign contracts.

Oil also was found in Oman. But this would be hard to believe from a brief examination of the country under the leadership of its eccentric leader, Sultan Said bin Taymur (1910–72). Under his rule, Oman had fallen further behind the West economically and socially, and the sultan preferred to keep it that way. He opposed education, internationalization, modernity, and even sunglasses. He forbade visitors from entering Oman and prohibited most of the country's residents from leaving. Under Said's orders, a cannon was fired each evening at sunset in the capital city of Muscat, and the gates to the ancient walled city officially were closed. The gates prevented anyone from leaving the city after dark, and served as a symbol of Oman's post-World War II repressive, inward-looking policy.

That isolationist stance ended in 1970 when Said's only son emerged from six years of virtual house arrest and overthrew his father in a bloodless coup. A group of soldiers entered the palace where Said was living and after a quick exchange of gunfire, one soldier was shot in the stomach and the deposed sultan was shot in the foot. Said was arrested and deported to London, where he lived in a hotel until his death in 1972. His son, Sultan Qabus bin Said, took control of the country and immediately repealed his father's repressive policies. It was time, he told Omani residents via a national radio broadcast, to modernize the country and enter the late twentieth century.

Qabus continues to rule Oman today as an absolute monarch. There is no constitution, no policy-creating legislature and no voting rights for its roughly 2 million residents. Yet, under Qabus, the country has thrived and in the late 1990s had begun to emerge as a cosmopolitan society that could provide its residents with the necessities of food, education, health care and, to an extent, freedom. The Basic Law by which Qabus rules Oman outlines a bill of rights ensuring press freedoms, religious tolerance and equality in accordance with Islamic *shari'ah*, or law. In a way, when Qabus promised to dramatically reform Oman, he was merely putting his country back in touch with its historic roots.

Qabus defended the need for a strong monarch when he seized power from his father in 1970. At the time, Oman had only six miles of paved roads, three schools, two hospitals, virtually no electricity grids and an average life-expectancy of thirty-eight years. Nearly 159 of every 1,000 babies died, giving the country one of the highest infant-mortality rates in the world.

Oman today has highly sophisticated medical facilities, a strong infrastructure, nearly 1,000 schools and a world-renowned university. All residents attend school, compared with a mere 3 percent before 1970. The country's gross-national product is nearly $15 billion compared with $15 million when Qabus came to power, and life-expectancy is 70 years. Despite the lack of a democracy, Omanis genuinely like and respect their sultan.

Timeline

3000 B.C. Copper trade flourishes

The early empire of Magan rises to power on the Batinah Coast, which arcs around the Gulf of Oman. Magan had been part of the Akkadian Empire centered in what is now southern Iraq but breaks off on its own and gains wealth through rich copper veins found in hills near Sohar.

Magan supplies copper to the kingdoms of Elam and Sumer, who use the metal for weapons. In their heyday, mines in inland Oman produce as much as 60 tons of copper a year.

Archaeological remains also show a sophisticated farming community thrives in the Western Hajar Mountains in inland Oman. Communities establish settlements along wadis, where farmers collect rainfall and trap water through dams. Settlers raise wheat, barley, sorghum, and dates. They use camels in their farmwork and live in stone dwellings.

563 B.C. Ancient water system is developed

As the copper trade begins to peter out, northern Oman is folded into the Achaemenid Empire of Persia, the first of three Persian empires to rule the area. During this period, Persians develop an irrigation system called *falaj*.

According to legends, Solomon, son of the Hebrew King David, built the system after traveling to Oman via a magic carpet. Regardless of how the system was built, it remains in use in present-day Oman. Through the system, water from the mountains flows to Oman's plains along man-made tunnels. A similar system is in place in Iran, China and in parts of the upper Rhone Valley.

Nomadic Arab tribes begin to arrive from what is now Yemen and Sa'udi Arabia. As the Arab tribes come into contact with Persians, some conflicts erupt. Generally, however, the Arabs and Persians remain separate. Persians tend to reside in towns while Arabs inhabit more pastoral areas.

A.D. 100–A.D. 300 Frankincense trade thrives

One gift that wise men of the Orient offered the baby Jesus was frankincense, an aromatic resin that once was more valuable than gold. One of the few places in the world where trees that produce frankincense can grow is the southern Omani region of Dhofar and by the first century A.D., production of the treasured product is thriving.

Producers obtain frankincense by making cuts in tree trunks. The resin has a natural oil content which allows it to burn well. It also possesses many medicinal and meditative qualities, and is used during this period in temples in Egypt, Jerusalem and Rome. Up to 3,000 tons of frankincense traverse trade routes between the south Arabian Sea and Greece and Rome. Near Salalah, the capital of Dhofar, lie ruins of the city of Sumhuram, where the trade had flourished.

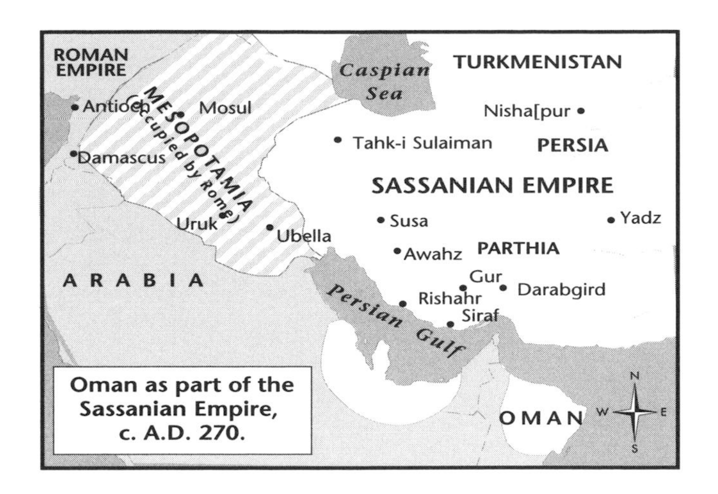

Oman as part of the Sassanian Empire, c. A.D. 270.

630 Oman accepts Islam

Soon after the Prophet Muhammad receives the message of God from the Angel Gabriel and establishes the Islamic religion, he sends envoys to the Persian rulers of Sohar. They and their peoples accept Islam, and soon become followers of the Ibadi sect.

Arab sheikhs in the interior also accept Islam. As the religion spreads through the region, use of Arabic increases.

Ibadi Muslims generally believe that the imam who leads a community need not be a descendent of the Prophet but can be any adult male who shows he is fit to lead the community They also hold that the position should be an elected one.

746 Ibadis rebel against Umayyad dynasty

After converting to Islam, the Omani people come under the rule of the Umayyad dynasty. Talib Al-Haqq leads Ibadis in a revolt against the Umayyad Caliphate based in Damascus, Syria, in A.D. 746, and manages to weaken its grip considerably. The Ibadis go on to conquer Medina in A.D. 748, only to lose it to remnants of the Umayyad force. Abbasids then overthrow the Umayyads and build a new capital in Baghdad.

The Ibadis in Oman remain independent of the Abbasid empire, and elect an imam, or prayer leader, in A.D. 749. Despite Ibadi tenets, the position of imamate becomes a hereditary one until the position is suppressed by the Abbasids in the ninth century. At this point, most tribal groups in Oman revert to choosing an elected imam.

A.D. 900–1506 Coastal areas thrive on trade

Maritime trade defines much of Oman's early history. Although inland copper mines and frankincense from Dhofar initially give Oman's interior more political and economic clout, the balance shifts by the early tenth century as Sohar becomes the country's largest city.

Sohar is the home port of Sinbad the Sailor, a fictional character. Located on the Batinah Coast, the city becomes quite wealthy as merchants trade with China and the Malay

Peninsula. Another coastal city, Qalhat, which is located at the opposite end of the Batinah Coast and is about 14 miles northwest of the modern city of Sur, later replaces Sohar in prominence.

As coastal communities reap commercial riches, clashes develop between them and tribal groups in the Omani interior. Despite the commercial riches that these coastal communities reap, Omani kings choose to live in Hormuz, an island that is separated from Oman proper by the Strait of Hormuz.

1507–1650 Portuguese occupy Oman

Portuguese traders, looking for a way to control the waterways between Europe and India, land in the Gulf in 1506 and occupy Oman one year later. They establish a base of operations in Hormuz, and build forts in Muscat as well as other cities along the Omani coast. Portugal, however, shows little interest in Oman beyond its use as a way station en route to India, and few traces of their influence remain in the country's cultural history.

The Portuguese are driven out of Hormuz in 1622, so they move their base to Muscat, Oman's present-day capital. Imam Sultan Bin Saif throws the Portuguese out of Muscat and the rest of Oman by 1650, and establishes the present-day Omani state.

After Sultan Bin Saif recaptures Muscat, Oman begins to take over trading ports throughout East Africa and participates in the African slave trade. By the end of the seventeenth century, an Omani empire extends from East Africa to Bahrain.

1646 Imam signs treaty with English

As Portuguese control weakens, the Omani imamate agrees to give the British East India Company trading rights. The Muslim country also sets up a separate judicial system for English subjects and allows them religious freedom.

1749 Al-Busaid dynasty begins

Ahmed Bin Said, the first ruler of Oman's current dynasty, is elected imam. His successor in 1783 also is named Ahmed Bin Said. The latter imam establishes Muscat as Oman's capital in 1786 and assumes the title of sultan.

This shifts the historical course of Omani society significantly. The Al-Busaid dynasty comes to power amid the early beginnings of an international capitalist economy. Commerce becomes increasingly important to the community's welfare, and the authority of a religiously oriented imam begins to diminish. Bin Said is elected imam after a civil war results from various tribal communities' inabilities to choose a successor to the previous leader. According to Carol J. Riphenburg, a scholar of Omani society and politics, this act transforms the imam from a religious leader to a political one. Al-Busaid rulers take the title of sultan because it presumes a less spiritual role than that of imam. It also allows them to sidestep the Ibadi objection to a hereditary leader.

Early Al-Busaid sultans are mercantilists who focus their energies on building what becomes a massive Omani empire. By the nineteenth century, Oman controls Mombasa and Zanzibar in Africa as well as ports further down the African coast and even establishes strongholds in what are now India and Pakistan. Oman retains its last colonial outpost at Gwadar, near the present-day border of Pakistan and Iran, until the British leave the Indian subcontinent in 1947.

1798–1800 Sultanate signs treaties with British

Two treaties signed with the British government ensure British protection of the sultanate. Unlike other agreements that Great Britain makes with Gulf States, these pacts are not designed to reduce Oman to a colony. Britain sees the treaties as a way to protect its own interests in India from threats by France and Egypt.

The treaties begin an era of friendship between Oman and Great Britain, and are supplemented by other agreements in 1891, 1939, and 1951. Over time, British civil servants play a larger role in the daily governance of Oman, serving as advisers to the sultan and as officers in the Omani army.

1850 Honey cultivation thrives

The grandfather of Mohammed bin Rashid Busaidi catches a queen bee in the Jabal Akhdar mountain range. More than a century and a half later, Mohammed continues to make honey in the village of Faiqan.

Honey production has roots in Islam. The Qur'an describes curative properties of honey, and many Omanis long have believed that a teaspoon of the bee nectar each day would enhance one's health. Omanis traditionally use honey as a source of defense as well; many forts have holes above doors through which boiling honey once was poured over invaders.

One bee that thrives in Omani mountains is the Bu Twiq, which may have been blown into Oman by southwest monsoon winds.

Apiarists—or bee keepers—enjoy a high status in traditional Omani society. The value of honey persists: Mohammed's honey sells in present-day Oman for $180 a liter.

1804–56 Oman's naval power grows

During the reign of Sultan Said bin Sultan, Oman's naval empire reaches its peak and ships among the sultanate's fleet call at ports throughout the world. Said bin Sultan rules from 1804 to 1856, and has an army of 6,500 men and a naval fleet of fifteen ships. He adds Dhofar to the sultanate and Oman's international power grows

However, unrest begins to build within Oman, most notably between the Arab tribes who inhabit inland areas and the mixed stock of traders who occupy coastal areas.

When Said bin Sultan dies, a quarrel develops over who to succeed him. British advisers intervene and the empire is divided between two sons. One becomes the Sultan of Zanzibar and commands Oman's African possessions, while the other assumes control of present-day country of Oman. It becomes known at this time as the Sultanate of Muscat and Oman.

1856–68 Sultanate is toppled

Arab tribes in the Omani interior had begun electing a separate imam at the end of the eighteenth century because they objected to the Al-Busaid family's establishment of a hereditary monarchy. The sultanate quietly recognized this rift and, realizing that much of the interior's political power rested with the Arabs, rarely extended much influence beyond the coastal areas in and around Muscat.

This changes after Said bin Sultan dies and the empire is divided. Without its African holdings, the sultanate loses a primary source of its income, and the country's difficulties mount when Great Britain begins pressuring the sultanate to end its role in the African slave trade.

As Oman stagnates economically, dissident groups in the interior launch a move to restore the religious imam as political leader. After a prolonged civil war, the Al-Busaid dynasty is toppled in 1868. The imamate is re-established and a leader known as Azzan bin Qays assumes office.

1868–1920 Split between Muscat and Oman intensifies

With British help, the sultanate regains control of Muscat and the coastal areas. Inland tribes elect imams as rulers over the next several decades. Neither the British nor the sultans recognize these leaders. The gap between the inland and coasts widens as cultural differences between the two become more apparent.

Unlike the interior communities who are largely indigenous Arabs, residents of the coastal areas represent a mixed stock. Former slaves of African descent populate the coasts, often speaking Swahili and English instead of Arabic. Hindu traders from India, who had migrated to Oman as early as the sixteenth century, play a large role in commerce. Baluchis from Iran and present-day Pakistan serve in the military.

One sultan, Faisal bin Turki, who rules from 1888 to 1913, illustrates the cultural diversity that has come to represent the coasts. Faisal's predecessors in the al-Busaid dynasty had intermarried with Africans, Abyssinians, Baluchis and Indians. Faisal speaks the Indian language of Gujarati better than Arabic and tends to regard India as a source of cultural inspiration rather than Arabia.

The division intensifies so greatly that by 1920, the sultan of Muscat and the imam of Oman formalize the split. They sign the Treaty of Seeb, in which the sultan recognizes the imam as a spiritual leader with power to rule autonomously in the interior and the imam recognizes the sultan as the ultimate sovereign.

1954–70: March 10 Sultan consolidates power

Said bin Taymur becomes sultan in 1931, and after the reigning imam dies in 1954, Said decides to assume the role for himself. He signs an oil-exploration agreement with Petroleum Development (Oman) Ltd., an oil company managed by the British. The corporation maintains a small army. This army occupies the Oman interior in 1955 and throws out a rebellious imam, who seeks to bring the interior into the Arab League as a state independent of Muscat. Said voids the imamate in 1959 and terminates the Treaty of Seeb.

Although other rebellions occur through the 1960s, Said quashes them and the sultanate retains absolute authority.

Said introduces a reactionary reign. He pays off debts to the British that his predecessors had incurred, and then literally closes Oman off to the outside world. He forbids most Omanis from leaving the country and restricts travel between the coasts and inland areas. He opposes education, and he secludes himself at a residence in Dhofar where only a few officials have access to him via radio. After the discovery of oil in 1964, money from oil exports flows into Oman. Said refuses to spend it, however, and the country's social and economic structures deteriorate considerably, and residents suffer a great deal of hardship.

1970: July 23 Coup brings new sultan to power

Qabus bin Said (b. 1940) is Said's only son. He grows up in a palace in Salalah in isolation until he is sent to England to attend school. He enrolls in the Royal Military Academy at Sandhurst in England and serves with British military units in West Germany. He also receives training in municipal administration and travels widely. He returns in 1964 and is placed under house arrest by his father.

While abroad, Qabus develops more progressive views and fears that his father's policies are causing undue suffering to the people. With help from a group of British officials and some elite Omanis, Qabus stages a coup and seizes power from his power. Said is arrested and spends the rest of his life in a London hotel, dying in 1972.

Qabus eases his father's restrictions considerably and embarks on a program to develop Oman. However, he retains power as an absolute monarch.

1979 Oman recognizes Israeli-Egyptian peace treaty

Oman becomes the only Arab state to accept the peace treaty that Egyptian president Anwar Sadat signs with Israel, a step that establishes an independent course that Qabus intends to follow. Despite widespread opposition against the United States that develops in the Middle East after the 1979 Islamic revolution in Iran, Qabus also agrees to give American ships and planes access to Omani military bases.

1981 Gulf Cooperation Council is formed

As a war between Iran and Iraq begins, Oman joins Bahrain, Kuwait, Qatar, Sa'udi Arabia and the United Arab Emirates in forming an alliance designed to protect regional security interests. Oman plays a leading role on this Gulf Cooperation Council in improving the region's hostile relationship with Israel.

1982 Oman launches environmental program

After Qabus issues a royal decree on the importance of environmental conservation, Oman launches several programs to prevent pollution. One particular worry is the presence of oil and tar on its beaches, a produce of discharge of oil by ships at sea. A report completed a decade later by the International Conservation Union finds that that this pollution is threatening many of Oman's migratory birds and that corals are being hurt by such things as fishing nets, anchors, plastic bags and empty paint tins.

Oman outlines plans to eliminate oil spills by building an area where tankers can get eliminate ballast. It also limits construction in coastal areas.

1987 University opens

Sultan Qabus University opens in Oman, a move that helps make higher education possible for a broader range of Omanis. Previously, education was limited to an elite class who could afford to leave the country to study abroad. Qabus' decision to open an institute for higher education particularly benefits women, and by the late 1990s, equal numbers of men and women attend Oman schools.

Besides the university, Oman also educates its youths in teacher training colleges, and technical and vocational institutes. Elementary and secondary schools are built throughout the country, which makes it easier for girls from traditional families to receive education. Traditional Muslim families often do not like their daughters to leave their communities to study.

1989 Muscat stock exchange opens

The Muscat Securities Market is established in 1988 and goes into operation the following year. It quickly becomes a dynamic stock exchange, and encourages investors from Gulf states as well as the West. A provision designed to encourage foreign investors allows companies that have up to 49 percent of their assets held by overseas owners to be treated as fully owned Omani companies for tax purposes. By late 1997, the market is described as being more "hot" than an Arabian summer, with a rate of return rising 123 percent from the previous year.

1990 Ancient forts are restored

Operation "Silk Route" gets under way in 1990 as part of an ongoing effort to preserve Oman's cultural heritage. Through a Ministry of National Heritage and Culture, Oman has put together an encyclopaedia of Omani heritage and is restoring several forts.

A few noteworthy projects include Bahia Fort, a massive city enclosed by about nine miles of walls; the fort of Nizwa, where the Arab-elected imam fought the sultan's authority until the Treaty of Seeb was signed; and a defense at Jabrin where a wood-paneled ceiling is covered with floral motifs and celestial illustrations decorate other rooms.

1994 Birth rate soars

Qabus encourages women to have fewer children and to slow the rate of their pregnancies. Omani women bear an average of 7.7 children, one of the world's highest birth rates.

In a *Middle East Policy* symposium, anthropologist Christine Eickelman attributes the high birth rate to ways in which women in Oman traditionally have attained status. "The roles of wife and mother belong to different categories in Oman—the role of wife belongs to the private world of family life, while the role of mother is one that a woman displays in public," Eickelman says. "A woman usually begins to play her second important role as provider of hospitality and networking in the community after the birth of her first child."

High fertility rates often strain health care systems, schools and water supplies. Because of this, Qabus tries to alter the traditional role women play in Omani society as child bearers by diversifying the opportunities available to them. As of 1993, women fill nearly 20 percent of Oman's government jobs.

1995 Vision 20/20 conference takes place

Qabus marks the twenty-fifth anniversary of his rule with a conference that emphasizes the importance of economic reforms. This comes at a time when Oman faces a high deficit and dwindling oil reserves. Unlike other Gulf states, Oman's oil supplies are relatively modest, and unlike his father, Sultan Qabus invests heavily in his country. While this policy improves roads, schools and social services in Oman, it threatens to bankrupt the government and force Oman's economy to stagnate.

The reform efforts aim at diversifying the country's economy and helping it reduce its deficit. As *Middle East* columnist Pat Lancaster describes it: "Vision 20/20 is not merely a development plan but a blueprint which will take Oman two decades into the new millennium." Its goal is to reduce Oman's deficit, cut back its reliance on oil revenue, diversify the economy, and make its private sector more competitive and efficient.

1995: June U.S. to help build desalination center

Oman and the United States each commit $3 million to build a Middle East Desalination Research Center in Oman. Water shortages are a pressing problem for Oman and other Gulf States, and many gain fresh water through desalination, the purifying of salt water.

Plans for this center come one year after the Ministry for Water Resources in Oman begins to worry about rising water consumption. Although desalinated sea water provides coastal areas with water, the agricultural areas of inland Oman use as much as 100 million cubic meters of water a month during Oman's hot summer months. The country builds dams to collect rainwater from mountains, and is looking for other ways to increase the sultanate's supply of water.

1996 Gallery promotes traditional cultures

A group of Omani expatriates opens the Oman Heritage Gallery in Muscat in hopes of helping artisans throughout the country create and market traditional crafts. Although the project is commercial, it provides hundreds of predominantly Arab nomadic craftspeople with a much-needed source of income.

The gallery features such items as incense burners, indigo pots and perfume from Dhofar, silks and handmade dishdasha robes from Sur, cushions from Ibra, camel saddle bags from the Wahiba Sands, palm mats from Musandam and woven rugs from villages in the Jebel Akhdar area.

A former Omani paratrooper buys products from villages throughout the Omani interior and transports them to the gallery in Muscat. Primary clients include business people and tourists.

1996: November Oman adopts Basic Law

Sultan Qabus announces the adoption of a constitution based on Islamic *shari'ah*, or law. It outlines a bill of rights that ensures press freedoms, tolerance for all religious faiths and equality for all, regardless of race, creed or sex. It also calls for the establishment of a court system that would interpret the law. Qabus hopes to put the Basic Law into full effect by 2000.

The Basic Law also codifies the process by which a successor to Qabus is to be chosen. After the sultan's death, members of the royal family will meet to select a candidate. If they cannot reach an agreement, Oman's Defense Council will name a successor based on a list of candidates suggested by the previous sultan. Qabus places his selections in sealed envelopes in two different parts of the country.

Bibliography

Arab Gulf States: A Travel Survival Kit. 2d edition. Melbourne Aus.: Lonely Planet Publications, 1996.

"A Well-Kept Secret: Omani Cuisine." *The Middle East*, April 1997, No. 266, pp. 38–39.

Bequette, France. "Environment-friendly Oman." *UNESCO Courier*, April 1995, pp. 39–41.

"Celebrating 25 Years of Achievement." *The Middle East*, November 1995, No. 250, pp. 24–27.

Gardner, Frank. "Pots of Money: The Oman Heritage Gallery." *The Middle East*, October 1997, No. 271, pp. 22–23.

Lancaster, Pat. "A New Optimism." *The Middle East*, November 1997, No. 272, pp. 19–21.

———. "Oman: Special Report." *The Middle East*, November 1996, No. 261, pp. 22–26.

Miller, Judith. "Creating Modern Oman: An Interview With Sultan Qabus." *Foreign Affairs*, May–June 1997, Vol. 76, No. 3, pp. 13–18.

Molavi, Afshin. "Oman's Economy: Back on Track." *Middle East Policy*, January 1998, Vol. 5, No. 4, pp. 1–10.

Osborne, Christine. "Omani Forts win Heritage Award." *The Middle East*, November 1995, No. 250, pp. 43–44.

Riphenburg, Carol. J. *Oman: Political Development in a Changing World*. Westport, CT: Praeger Publishers, 1998.

"Symposium: Contemporary Oman and U.S.-Oman Relations (Panel Discussion)." *Middle East Policy*, March 1996, Vol. 4, No. 3, pp. 1–28.

"The High Cost of a Sweet Tooth: Honey from Oman." *The Middle East*, October 1996, No. 260, pp. 38–39.

"U.S.-Oman Support for Middle East Desalination Research Center," U.S.-Oman Joint Communiqué. *U.S. Department of State Dispatch*, June 12, 1995, Vol. 6, No. 24.

"Where's our Sultan?" *The Economist*, August 9, 1997, vol. 344, No. 8029, pp. 38–39.

Pakistan

Introduction

Pakistan was created as a separate state for Muslims when Britain's Indian empire was partitioned in 1947. Areas of the subcontinent under British rule with Muslim majorities were assigned to the new state. This resulted in a country consisting of two blocks of territory, one located in the northwest of the Indian subcontinent and the other in the northeast, separated by 1,000 miles (1,600 kilometers) of territory belonging to the Republic of India. Although both wings of Pakistan were united by religion, serious cultural and political differences eventually led to the eastern region breaking away in 1971 to form Bangladesh. Modern Pakistan thus comprises the lands that made up West Pakistan at the time the states of India and Pakistan were established in 1947.

Northwestern India ("India" is used throughout this article in the geographical sense of the Indian subcontinent unless the context indicates otherwise) was among the first regions in the subcontinent to be exposed to Islam. Modern Pakistan is strongly Muslim, and its society and culture are deeply rooted in the Islamic faith. However, the country has historical antecedents that extend beyond a thousand years of Muslim rule to the very beginnings of civilization in the Indus valley some five millennia ago.

Geography and History

Now—as then—the Indus River is the lifeblood of the country. Some 1,900 miles (3,000 kilometers) in length, the Indus rises in southwest Tibet. It flows northwestwards for 500 miles (800 kilometers) before swinging around to the southwest, carving gorges and deep valleys through the mountain ranges of Kashmir. It continues in a general southwesterly direction, emerging onto the plains of the Punjab and flowing through Sind to its delta in the Arabian Sea. Most of Pakistan's 148 million people live on the Indus plains, supported by agriculture that is heavily dependent on the waters of the Indus and its tributaries for irrigation. This is of particular significance in a region that is mostly desert (Karachi, where summer temperatures may exceed 104ºF [40º C] for months at a time, receives only eight inches [204 mm] of rain a year).

To the west of the Indus plains lie the barren plateaus and hills of Baluchistan and the rugged mountains of the North West Frontier Province (N.W.F.P.). Reaching 10,000 feet (3,000 meters) in places, these areas have traditionally formed the boundary between South and West Asia. They are inhabited by tribes such as the Baluchis and Pathans who are culturally quite different from other Pakistanis and who display a fierce sense of independence from the central government in Islamabad. The northern limits of Pakistan are defined by the majestic ranges of the Hindu Kush and Karakoram Mountains, with numerous peaks exceeding 21,000 feet (6,500 meters).

The Indus River gives its name to the civilization that flourished in the region in the third millennium B.C. The Indus valley civilization, or the Harappan civilization (the preferred name because it extended over an area much more extensive than just the Indus valley), rivals Mesopotamia and ancient Egypt in its achievements. Archeological evidence reveals a people who farmed the Indus plains, growing wheat, barley, and cotton. They also had domesticated animals such as cattle, chickens, cats and dogs, and possibly even the elephant. They engaged in commerce, and planned and built cities and towns.

The Indus economy was dominated by two great cities, Harappa in the Punjab and Mohenjo-Daro in Sind. There is evidence of a complex society with an urban social structure that perhaps included an aristocracy and a thriving religion, complete with a priestly class. The Indus peoples had a system of writing, the Indus valley script, which has yet to be properly deciphered. It seems, however, that the Indus language was Dravidian (part of the group of people and languages indigenous to the Indian subcontinent) in origin and is related to the tongues now spoken in southern India. Although the Indus civilization disappeared rather suddenly around the middle of the second millennium B.C., many of its traits survive in the later cultures of the Indian subcontinent.

Invasions of peoples through the the northwest passes has been a recurrent theme in India's past. Some have had no more than a fleeting impact on the region, but others have changed the course of Indian history. The early years of the second millennium B.C., for example, saw the movement of Aryan peoples from Central Asia into northwestern India.

(Technically, "Aryan" is a linguistic rather than a racial term and refers to a family of languages [Indo-European] spoken by peoples who migrated out of Central Asia at this time.) The Aryans were a nomadic, tribal people who helped bring about the decline of the Indus civilization. But, over time, they also developed the beliefs, religious practices, and social system that underlie Hinduism.

By around 1000 B.C., Hindu civilization was well established in northern India. At times during the next two thousand years, the Indus valley and India's western borderlands were integrated into the great Indian empires (e.g. the Mauryan and Guptan empires) that grew up on the subcontinent. But wave after wave of invaders continued to sweep through the northwest passes into the plains of the Punjab, and during this period many peoples established themselves in the region. Persians, Greeks, Parthians, Scythians, Kushans, and White Huns—each ruled kingdoms that included some or all of the lands that now make up Pakistan.

Introduction of Islam

Arabs introduced Islam into Sind at the beginning of the eighth century, but it was not until after the Afghans conquered Delhi in 1193 that the religion came to have its full impact on the Indian subcontinent. Like all great religions,

Islam is not just a set of religious beliefs and practices. It is a complex civilization with specific social systems and its own cultural and artistic traditions. During six and a half centuries of Muslim rule in India, this civilization came into contact with and absorbed local traits to create a Muslim culture that is unique to India. This culture ranges from the Indo-Islamic style of architecture and Urdu poetry to the religious traditions of Sufism and Mughal miniature painting.

In addition, social organization in Pakistan represents an adaptation of Islam to local customs. Although Islam rejects the Hindu caste system of South Asia, Pakistanis are very much influenced by caste. For example, rural society in the agricultural lands of the Indus plains is organized into *jati (zat)* or caste. This is an occupational rather than a religious grouping and defines the role of people in the village economy. The *biradari* (patrilineage) also plays an important role in family relations and social life. Members of one's biradari are expected to help with ceremonies associated with birth, death, and marriage; to provide financial assistance to poorer families; and to rally around the family in times of need. Jati and biradari also define the marriage pool which provides potential spouses. Although details of the ceremony may vary according to community and region, Pakistanis follow the general customs of Islam in marriage (*nikah*). Purdah, the Islamic custom of keeping women in seclusion, is widely practiced in Pakistan.

The last significant foreign invasions of northwestern India came not from Central Asia but from the southeast. The British acquired Bengal in the mid-eighteenth century and pushed steadily northwestwards up the Ganges valley. Within a hundred years, Sind, the Punjab, and the North West Frontier Province were annexed, bringing virtually all of what is now Pakistan under British control. The British legacy can be seen in areas such as Pakistan's political, legal, and administrative systems; in its rail network and canals; and in the use of English for government and business. Many of Pakistan's problems, such as its hostile relations with India and the ongoing Kashmir dispute, also stem from Pakistan's colonial heritage.

If Islam and Islamic culture provide a sense of cohesion to Pakistan, they also overlay considerable ethnic and cultural diversity. In fact, most of the country's inhabitants identify with a region or tribal group before they see themselves as "Pakistani." Such regional and cultural affiliations often give rise to serious internal tensions. For example, the most numerous, and politically influential, group in the country are Punjabis, who provided the country's current prime minister, Mohammad Nawaz Sharif. The leader of the opposition in Pakistan—and Sharif's bitter political opponent—is Benazir Bhutto, whose family is from Sind. Baluchis are found in the southwest, numerous Pathan (also called Pushtuns or Pakhtuns) tribes dominate the northwestern region, and a bewildering array of tribal groups inhabit the mountains that stretch to the Chinese border. Ethnic complexity in Pakistan is further modified by presence of the *muhajirs*, Muslims who fled India for Pakistan at the time British India was partitioned. The various ethnic groups speak their own languages, have their own social patterns, and follow their own cultural and historical traditions. Indeed, creating a sense of "nationhood" among the diverse peoples who found themselves within the borders of Pakistan in 1947 has been a major task for post-independence governments.

Timeline

Pre-8000 B.C. Paleolithic cultures in the Soan valley

Stone tools from the valley of the Soan (Sohan) River, a tributary of the Indus in northern Pakistan, indicate Paleolithic ("Old Stone Age") peoples live in the area as far back as perhaps 500,000 years ago. The Pre-Soan and later Soan cultures are among the earliest to be identified on the Indian subcontinent. No human remains date this far back, but by approximately 50,000 years ago modern man (*Homo sapiens*) occupies the region. The population lives by hunting, fishing, and gathering.

c. 8000 B.C. Beginnings of agriculture

Climatic and related environmental changes at the end of the Pleistocene Ice Age initiate cultural processes leading to the development of agriculture and settled communities in India's western borderlands.

c. 6000–4000 B.C. Early settlements in Baluchistan

Settlements appear in the mountainous region of Baluchistan. These often lie in isolated valleys at heights of four to five thousand feet (c. 1200–1500 meters) above sea level. They are particularly common around Quetta and in the Zhob and Loralai River valleys. During the earliest stages of this period of settlement, the people are seminomadic, herding domestic sheep, goats, and cattle. By the end of the period, they build houses of mud-brick or clay, cultivate cereals, make pottery, and even use objects made of copper.

Towards the end of this period—perhaps as a result of population pressure or an awareness of the agricultural potential of the Indus plains—people move eastwards and settle in the Indus valley. Mehrgarh, at the foot of the Bolan Pass, is an important archeological site period, located in the transitional zone between the mountains and the Indus plains.

c. 4000 B.C. Pre-Harappan settlements

Agricultural settlements such as Amri and Kot Diji appear in the Indus valley. They and other sites show a convergence of cultural traits throughout the region that provides the foundation for the urban civilization that is to follow.

c. 3200 B.C. Beginnings of urbanization

The first urban settlements appear in the Indus valley.

c. 2500–2000 B.C. Harappan cities

Urban civilization reaches it peak in the Indus valley, as seen in the two great cities of Mohenjo-Daro and Harappa.

Mohenjo-Daro stands on the west bank of the Indus, south of Larkana in Sind. Roughly one mile square, Mohenjo-Daro has a layout typical of the more important Harappan settlements. A fortified "citadel" lies to the west, overlooking and guarding the lower city to the east. The city is planned, with streets laid out on a grid pattern and houses built of standard-sized fired or mud bricks. Houses have bathrooms and latrines connected to underground drains, and the city also has public baths and, apparently, a municipal water supply. Granaries indicate agricultural surpluses.

The people of the Indus valley cities engage in trade. They have a standardized system of weights and measures, and seals are apparently used on packaging or for identification of goods. The seals are small squares of steatite (soapstone) typically inscribed with pictures of animals and characters of the Indus script. Some scenes depicted on seals appear to be of ritual significance and suggest the existence of a mother-goddess cult.

c. 2000–1500 B.C. Aryans invasions

Several waves of nomadic tribes enter India from Central Asia through the passes of Afghanistan. The invaders speak Aryan languages, and are pastoralists (engaged in livestock grazing); their use of the horse and two-wheeled chariot gives them a military advantage over the local peoples.

c. 1750 Decline of Harappan civilization

Harappan civilization undergoes a rapid decline around 1750 B.C., with cities giving way to farming communities. Traditional explanations for this decline blame the Aryans, who sweep down from the northwestern passes to destroy the cities on the plains. A more likely scenario is that factors such as climatic change, environmental degradation, earthquakes, and changes in drainage patterns disrupt agricultural production and trade, undercutting the very basis of Harappan urban culture.

c. 1500–600 B.C. The Vedic Age

Aryans settle the Punjab, which emerges as the heartland of Aryan culture during the early Vedic period. In this land of rivers, the nomadic Aryan tribes make the transition from pastoralism to settled agriculture. After c. 1000 B.C., the Aryans push their settlements southeastwards into the upper Ganges valley.

This period sees the transformation of Aryan society, which is revealed in the Vedic literature, the songs and hymns composed by Aryan priests and handed down by word of mouth from generation to generation. The earliest of the Vedas, the *Rigveda*, probably dates to around 1200 B.C. The Vedas form the sacred literature of early Hinduism and describe religious beliefs, rituals, and the caste system that come to be integral elements of Hindu civilization.

c. 517 B.C. Achaemenid empire

Darius, ruler of Persia, conquers Gandhara, an ancient kingdom in northern Pakistan that includes the region around Peshawar and Taxila (Taksasila). In the following years, Darius incorporates neighboring lands west of the Indus into the eastern province of the Achaemenid empire.

327–325 B.C. Alexander invades India

Alexander the Great (356–323 B.C.), King of Macedonia, enters the Indian subcontinent from Afghanistan. He crosses the Indus River near Attock and proceeds to Taxila, which lies some sixteen miles or twenty-five kilometers northwest of Islamabad. Alexander continues southeast to the Jhelum River, where the Indian King Porus and a large army—including 200 war elephants—await the invaders. Alexander defeats Porus in at the Battle of Hydaspes (Jhelum) in July, 326 B.C, but the two kings become friends.

Alexander wishes to continue deeper into India, but his soldiers force him to turn back. The Greeks sail down the Indus to its mouth, where the army divides in two. Alexander sends one half of his army back to Persia by sea under his admiral Nearchus, while he leads the remainder overland through the Makran. Arrian's Anabasis (translated as The Campaigns of Alexander) provides a contemporary account of Alexander's military campaigns.

c. 322 B.C. Chandragupta conquers the Punjab

Chandragupta Maurya (r. 321–297 B.C.) conquers the Punjab.

Chandragupta establishes the Maurya dynasty (321–185 B.C.), which rules over the first great empire in India's history. In the following decades, he pushes the northwestern frontier of his domains beyond the Indus, conquering the lands formerly ruled by the Greeks and their vassals.

305 B.C. Chandragupta defeats the Seleucid Greeks

Chandragupta defeats Seleucus Nikator, the Greek general who rules Alexander's eastern empire, at a battle near Taxila. Seleucus cedes extensive territories to Chandragupta, whose empire now reaches the borders of Persia. Mauryan lands at this time include virtually all of modern Pakistan as well as large areas of Afghanistan.

c. 273 B.C. Asoka ascends throne

Asoka (reigned c. 265–238 B.C. or c. 273–232 B.C.), the third and perhaps greatest of the Mauryan emperors, succeeds to the throne. He continues to expand the empire, bringing all but the southernmost tip of the subcontinent under his rule.

261 B.C. **Asoka converts to Buddhism**

Asoka makes Buddhism the state religion and sets about promoting the religion and its precepts throughout his lands. He proclaims his beliefs and the laws of the land in edicts carved in stone and placed in strategic locations throughout the empire. The Rock Edicts are placed along the borders of the empire, and two are found in northern Pakistan—at Shahbazgarhi near Peshawar and at Mansehra (on the Karakoram Highway). Asoka founds a Buddhist university at Taxila.

c. 200 B.C. **The Indo-Greeks**

By the beginning of the second century B.C., the Mauryan empire is disintegrating. Bactrian Greeks, from what is now the area of Balkh in northern Afghanistan, push southeastwards into the subcontinent. They conquer the Punjab and establish themselves in Gandhara

The Greek king Menander (r. 155–130 B.C.) shifts his capital from Taxila to Sakala (the modern Sialkot), and the Indo-Greeks undertake raids down the Ganges valley. Menander even threatens Pataliputra (the modern Patna in Bihar), but turns back to face the Scythians (Sakas) who invade his northern territories.

Menander is known in Indian sources as Milinda and is a central figure in a Buddhist text known as *Milinda-pañha* or the "Questions of Milinda." This text is presented as a dialogue between Menander and a Buddhist monk that leads to Menander's conversion to Buddhism. Menander becomes a patron of the arts, supporting Buddhist learning, art, and architecture.

c. 100 B.C. **Scythian invasions**

Scythians, known as Sakas in India, invade the Indian subcontinent from Central Asia and conquer the Indo-Greek kingdoms of the northwestern region. They establish kingdoms at Taxila and the lower Punjab as well as in neighboring regions of India.

A.D. 20 **Parthians**

Parthians from east of the Caspian Sea conquer the Saka lands in what is now northwestern Pakistan and establish a short-lived kingdom. Gondophernes (reigned c. A.D. 20–46) is the most renowned of the Parthian rulers.

c. A.D. 48 **Kushans invade**

The Yüeh-Chih, known in India as the Kushans, are the next wave of invaders to enter the subcontinent through the northwestern passes. The Yüeh-Chih are a nomadic people originating in the modern Gansu province of China. During the second century B.C., they migrate westwards through the mountains and deserts of Central Asia, driving groups such as the Sakas before them. They conquer Bactria and establish themselves south of the Oxus River in northeastern Afghanistan. The Kushans push southeastwards into the Punjab in the middle of the first century.

The greatest Kushan leader is undoubtedly Kanishka (r. 120–162). He establishes his capital at Peshawar and rules an empire from there that stretches from the lower Ganges valley to the Aral Sea, from Persia to the western half of the Tarim Basin in China. All of modern Pakistan, except for Baluchistan, falls within the Kushan empire.

c. 150 **Gandharan art**

Kanishka becomes a Buddhist and extends his patronage to religion, art, and literature. He uses Greek and local artists to build temples and carve statues. During his reign, the Gandharan school of art flourishes. A unique feature of Gandharan art, which spans a period roughly from the first to the fifth centuries, is the blending of foreign (Greco-Roman or Kushana) influences with Indian themes. Thus, sculptures of the Buddha are common, but certain elements such as dress, and motifs such as vine scrolls, cherubs, and even facial features clearly reflect Western or Central Asian origins.

c. 300 **Sassanians**

Kanishka's successors are unable to hold the Kushan empire together. By the end of the third century, the Sassanian dynasty of Persia controls Sind and the former Kushan lands west of the Indus River.

320–510 **Gupta dynasty**

The Guptas rise to prominence on the Gangetic plain and unite much of northern India under their imperial power. Under Samudragupta (reigned c. 330–c. 380) and his son, Chandragupta II (reigned c. 380–415), the Guptas integrate the independent republics of the Indus plains into their empire. They bring Kashmir under their control and conquer much of northern Afghanistan.

455 **Huns invade Punjab**

During the fifth century, waves of tribesmen spread out from Central Asia, bringing death and destruction to neighboring lands. The Huns go westwards into Europe and, under Attila (406–453), lay waste to much of the Roman Empire. One branch, the White Huns, or Hunas as they are known in Indian sources, turn south and conquer the Kabul valley and Gandhara.

The Hunas invade Gupta domains in the Punjab in 455 but Skandagupta (r. 455–467) drives them back west of the Indus.

c. 500–c. 530 Hunas conquer Guptas

Under Toramana, the Hunas sweep southeastwards from their lands in the northwest frontier (their capital is probably Peshawar). They penetrate as far as the Gupta capital and make the Gupta emperor their vassal. The short-lived Huna empire extends from northern Afghanistan to the lower Ganges valley and includes Kashmir, Gandhara, and the Punjab. Although Huna chiefdoms and tribes survive for several centuries, the victory of King Yasodharman of Malwa over the Hunas in 528 effectively destroys Huna power in northern India.

c. 612 Harsha raids Sind

During the fifth century, Sind emerges as an important kingdom under the Hindu Rai dynasty. The emperor Harsha conquers Sind as well as Kashmir and other kingdoms along the Indus, but his presence in the region is short-lived.

632–712 The Chach dynasty of Sind

Under the Brahmin Chach rulers, Sind's power extends westwards along the Makran coast. Many small kingdoms in the uplands west of the lower Indus are tributary (they preserve their independence through the payment of a fee) to Sind.

711 Arab conquest of Sind

The inability of the Sind raja to deal with pirates along the Indus delta leads the Arab governor of Basra to send a punitive expedition to the region. Mohammad ibn-Qasim marches east along the Makran coast with six thousand picked cavalry. He defeats and kills Raja Dahir, and establishes Arab control over Sind and the southern Punjab. These lands now form the easternmost province of the Umayyad Caliphate, based in Damascus.

870 Hindu Shahis

Shahis establish a powerful Hindu kingdom that include areas of northern Afghanistan, Gandhara, and northwestern Punjab.

1001–26 Mahmud of Ghazna invades India

Mahmud of Ghazna (r. 997–1030) undertakes a series of devastating raids into northwestern India.

The Ghaznavid dynasty rises to power at Ghazna (the modern Ghazni in Afghanistan) in 962. The Ghaznavids expand eastwards, eventually destroying the Shahi kingdom and gaining control of the Punjab. Mahmud raids cities as far afield as the middle Ganges valley and sacks the famous Hindu temple at Somnath in Gujarat in 1024. In 1025, Mahmud invades Sind to punish Jats who attacked him on his retreat from Somnath.

1034–1337 Sind ruled by Sumras

The Sumra dynasty rises to power in Sind and rules successfully for three centuries.

1186 Ghurids take Lahore

Muhammad Ghuri (d. 1206), who captures Ghazna in 1173, takes Lahore and the Punjab, bringing Ghaznavid rule to an end.

1206 The Delhi sultanate

Qutb-ud-Din Aybak (d. 1210) founds the Delhi sultanate.

Following his defeat of the Hindu Rajputs (1192) and capture of Delhi (1193), Muhammad Ghuri returns to his capital in Afghanistan. He leaves his general Qutb-ud-Din as his viceroy in India. With his declaration of independence, Qutb-ud-Din establishes a Muslim power in northern India that is to extend its influence, to varying degrees, over the lands that make up Pakistan for the next six and a half centuries.

1221–1307 Mongol invasions

Genghis Khan (1155–1227) and his Mongols pursue a fugitive prince into the Punjab in 1221, finally catching him near the Indus at Attock. Iltutmish (r. 1211–36), the Delhi sultan, refuses the prince's plea for aid, fearing that the Mongols will attack Delhi. The Mongols control much of the Punjab during the next century and make periodic raids deep into India. Sultan Ala-ud-Din Muhammad Khalji (r. 1296–1316) successfully repulses several Mongol invasions in the latter years of his reign.

1337 Samma Rajputs in Sind

The Samma Rajputs come to power in Sind.

1345 Sufi mystic Lal Shabhaz Qalandar dies

Lal Shabhaz Qalandar, a Sufi mystic, dies and is buried in the village of Sehwan, near Lake Manochar in central Sind.

Sufis mystics or saints, known locally as *pirs*, play a major role in the conversion of Hindus to Islam following the Muslim conquest of India. Sufism is a mystical tradition within Islam that holds, unlike orthodox Muslim doctrine, that the average person needs a spiritual guide in his search for the truth. With its emphasis on love and personal devotion, Sufism has many similarities with Hindu devotional (*bhakti*) movements and finds a receptive audience among the common people. Prominent Sufis who spread Islam in northwestern India between the tenth and fourteenth centuries include Safiuddin-Gazruni, Data Ganj Bakksh, and Khwaja Moinuddin Chisti. The *urs* festivals held at the shrines of saints to celebrate the anniversaries of their deaths are an important aspect of popular religion in Pakistan today.

1362 Tughluqs conquer Sind

Firoz Shah, the Tughluq sultan of Delhi, subjugates Sind.

1388 Sind independent

Sind reestablishes its independence from the Delhi sultanate.

1398–99 Timur invades India

Timur (1336–1405), also known as Tamerlane, invades the Indian subcontinent from his base in Afghanistan. He conquers Multan and the Punjab and destroys Delhi. The Shah Mirs of Kashmir retain their independence by promising to pay Timur tribute.

1469 Birth of Nanak, founder of Sikhism

Nanak (1469–1538), founder of the Sikh religion, is born in the Punjab. At the very end of the fifteenth century, he begins his religious mission and comes to be regarded as the first in a line of ten Sikh *gurus* or "teachers." Sikhism is a branch of the Hindu devotional movement that combines aspects of Hindu religious thinking with elements of Islam, in particular Sufi mysticism. The followers of the Sikh religion are most numerous in the Punjab region.

1517 Babur raids Punjab

Zahir-ud-din Muhammad (1483–1530), commonly known as Babur, captures Kabul in Afghanistan in 1904. From this base, he mounts raids into the Punjab in 1517 and 1519.

1522–27 Arghuns conquer lower Indus valley

The Arghuns, descendants of the Mongol Il-Khans (subordinate *khans*) of Iran, bring Sind and Multan under their control.

1526: April 21 Babur establishes Mughal empire

Babur defeats and kills Ibrahim Lodi, the Sultan of Delhi, at the traditional Indian battle field of Panipat. He assumes the throne of Delhi and with it the lands controlled by the sultanate (including Punjab). Babur becomes the first in the line of Mughal (a corruption of "Mongol") emperors to rule northern India.

1556–1605 Mughal expansion under Akbar

Akbar (r. 1556–1605), the greatest of the Mughal emperors, conquers virtually all of the lands that now comprise Pakistan and integrates them into the empire. These include the Punjab (1557), Upper Sind (1574), Lower Sind (1591), Kashmir (1586–88), and Baluchistan and Makran (1595).

The territories fall with the Mughal provinces (*subahs*) of Multan and Lahore (which includes the subprovince of Kashmir).

1699 Sikhs organize as militaristic sect

Gobind Rai (1666–1708), the tenth Sikh Guru, forms the Sikhs into a fighting fraternity known as the *Khalsa.*

In its early years, the Sikh religion is pacifistic in nature. Sikhs face persecution by the Mughal government, however, and early in the seventeenth century they take up arms against the Muslims. Periodically the Sikhs rise against their overlords, and the early eighteenth century sees bitter conflict between the Khalsa and the Mughals in the Punjab.

1738–39 Nadir Shah of Iran invades subcontinent

The Shah of Iran overruns the Punjab and sacks Delhi, carrying off the famous Peacock Throne of the Mughal emperors.

1747–62 Afghan invasions

Afghans, under Ahmad Shah Abdali, mount a series of attacks on Lahore, and eventually gain control of the Mughal provinces of Lahore and Multan. By 1762, Afghans holds Baluchistan, Sind, Punjab and Kashmir.

1761: January 14 Third Battle of Panipat

An Afghan army, led by Ahmad Shah Abdali, defeats the Hindu Marathas at the third major battle to be fought at Panipat, near Delhi. The military power of both sides is severely weakened.

1762 Sikh confederacy in Punjab

A loose military confederacy of Sikh is in almost constant struggle with the Mughals. The Afghans have *de facto* control of the Punjab.

1799 Ranjit Singh seizes Lahore

Ranjit Singh (1780–1839) captures Lahore, the capital of the Punjab, from the Afghans. In 1801, he declares himself Maharaja, and creates a powerful Sikh state that extends from the Sutlej River to Kashmir.

1843 Britain annexes Sind

Britain conquers Sind, extending the British Indian empire to the borders of Baluchistan.

1849 Britain annexes Punjab

After Ranjit Singh's death, the Sikh kingdom in the Punjab rapidly disintegrates. The British fight two wars with the Sikhs (the First Sikh War [1845–46], the Second Sikh War [1848–49]) before adding the Punjab to their empire in India.

1857: May 10 Indian Mutiny

Indian troops kill their British officers and march to Delhi, where they proclaim Bahadur Shah emperor. The mutiny is put down by 1858 but leads to the British Crown taking over direct control of its Indian territories from the British East India Company.

c. 1858–95 The Great Game

The second half of the nineteenth century sees intense rivalry between Britain and imperial Russia for control of the mountains of Central Asia. Russia looks southwards, seeking the warm-water port and commerce that geography denies her. Britain, fearful for its Indian empire, establishes a tier of buffer states along Russia's border to contain the potential threat of invasion. The mountains of India's northwestern frontier (the Pamirs, the Karakorams, the Hindu Kush) become the scene of the "Great Game," with spies, explorers, and surveyors from both sides crossing the region—at considerable danger to themselves—to obtain information and intelligence. These events are fictionalized by Rudyard Kipling in works such as *Kim* and *The Man Who Would Be King*.

During this period, Britain extends its control over Baluchistan, the Khyber Pass, and the states and tribal regions of the northwestern frontier region.

1901 North West Frontier Province created

The British create the North West Frontier Province from parts of the Punjab, the Dir, Swat and Chitral Agency (a collection of tributary states), and frontier tribal areas.

1905–17 Canal Colonies

Although the British begin to build canals in the Punjab in the mid-nineteenth century, their greatest canal development occurs during this period. The canals channel waters from the Jhelum, Chenab, Ravi, and Sutlej Rivers to irrigate the *doab* or land between the rivers. Farmers settle in modern, planned agricultural communities or "canal colonies" along the new canals.

1906: December 30 Muslim League founded

Leading Muslims meet in Dhaka, in East Bengal, where they found the All-India Muslim League to promote the interests of Muslims in India.

1913 Jinnah joins Muslim League

Mohammed Ali Jinnah (1875–1948) is born and educated in Karachi. He studies law in London and subsequently sets up practice in Bombay. An ardent nationalist, he enters politics and joins the Indian National Congress. Finding himself in disagreement with the Congress's methods and increasing Hindu bias, he joins the Muslim League.

1914–18 World War I

Two divisions of the Indian Army, including the Lahore Division, fight for Britain in France and Mesopotamia.

1916: December Congress-Muslim League Lucknow Pact

Jinnah arranges an agreement by which Hindus agree to separate representation of Muslims in future legislative councils.

1919 The Khilafat movement

Muslims in India start the Khilafat movement to restore the Turkish sultan to power after Turkey's defeat in World War I. The Turkish sultan has been the Caliph, the leader of the Sunni Muslims, for over two centuries. Mohandas Karamchand Gandhi (1869–1948), the leader of the Indian nationalist movement, supports the aims of the Khilafat movement, and a period of Hindu-Muslim cooperation follows.

However, the Khilafat movement collapses in 1924 and introduces a period of bitter Hindi-Muslim antagonism. In the following years, communal violence between Muslim and Hindus reaches an unprecedented level.

1921 Discovery of Harappa

The discovery of Harappa on the west bank of a now dry channel of the Ravi River in the Punjab reveals the long-forgotten existence of the Harappan or Indus Valley civilization. Until this time, scholars believed that South Asian civilization began with the Aryans. Excavations at Harappa by Sir John Marshall pushed the beginning of Indian history back to around 3500 B.C. In 1922, archeologists uncover Mohenjo-Daro in Sind.

1930 Muhammad Iqbal proposes separate Muslim state

In his presidential address to the Muslim League at Allahabad, Muhammad Iqbal proposes that the Muslims of northwest India should demand a separate Muslim state.

Sir Muhammad Iqbal (1877-1938) is the leading Muslim poet and philosopher of his day (he is knighted by the British in 1922). He is known for his fiery and inspiring poetry, which focuses on the glories of Islam's past and the need for reform, renewal, and unity. His works include *Bang-e dara* (The Call of the Bell), written in Urdu, and the Persian *Asrar-e khudi* (Secrets of the Self). Iqbal is influenced by the views of Sir Syed Ahmad Khan (1817–98), who argues that Hindus and Muslims in India are separate nations. Iqbal promotes the "Two Nation" Theory that is eventually adopted as policy by the Muslim League, but he dies before the dream becomes a reality.

Choudhri Rahmat Ali and a group of students at Cambridge University in England create a name for the proposed Muslim state in northwest India in 1933. They write "Pakistan...is...composed of letters taken from the names of our homelands: that is Punjab, Afghana (NWFP), Kashmir, Iran, Sindh, Tukharistan, Afganistan, and Baluchistan. It means lands of the Paks, the spiritually pure and clean."

1935: May 31 **Quetta earthquake**

An estimated 26,000 people die in a violent earthquake that hits Quetta at 3:03 A.M.

1939–45 **World War II**

The Muslim League follows a policy of cooperation with the British government during the war, allowing the organization to consolidate its position as a political party in India. By contrast, when the British declare war for India without consulting Indians, Congress politicians resign from government positions.

1940: March 23 **Lahore Resolution**

At its annual session in Lahore, the Muslim League resolves that no plan for India's independence that does not include autonomous states for Muslims in the northwest and northeast of India is acceptable. The resolution gains Jinnah (see 1916) widespread popular support and he emerges as the undisputed leader of Muslim India.

1946: August 16 **Hindu-Muslim riots**

Jinnah calls for demonstrations to protest the omission of the Muslim League from an interim government of India formed by the British. On "Direct Action Day," communal violence between Muslims and Hindus breaks out in northeastern India, with bloody massacres of Muslims in Calcutta. Riots spread to other regions of the Indian subcontinent and continue up to the time of independence.

1947: June 3 **Partition announced**

Unable to persuade Jinnah to drop his demands for a separate Muslim state, the British government announces that India will be partitioned between Muslims and Hindus.

Pakistan after Independence

1947: August 14 **The birth of Pakistan**

Pakistan becomes an independent nation at midnight. The new country consists of regions of British India with Muslim majorities. In the northwest, this includes Sind, Baluchistan, and the Northwest Frontier Province. The western Punjab goes to Pakistan, the eastern regions of the province (including some Muslim majority areas) are assigned to India. In the northeast of the subcontinent, Pakistan's territory includes East Bengal and the Sylhet District of Assam.

In geopolitical terms (i.e. the relationships between geography and politics), the new state of Pakistan faces serious problems. The country consists of two territorial units 1,000 miles (1,600 kilometers) apart. Pakistan inherits British India's historical boundaries with Iran, Afghanistan, and Burma, but the new boundaries with India bisect long-standing historical and cultural regions. The partition of Punjab and Bengal leads to one of the greatest migrations in history. Up to fifteen million people abandon their homes and their land, carrying whatever they can with them as they flee to the safety of their new homelands. Hindus and Sikhs cross over into India, while Muslims head for Pakistan. Estimates of the loss of life during this period of social upheaval and sectarian violence run as high as a million people.

1947: August 15 **Kalat, in Baluchistan, declares its independence**

The Khan of Kalat, in Baluchistan, declares independence. Pakistan takes military action against the khan and other Baluchi chiefs to force accession.

Several hundred princely states, accepting British political control but governed by their own rulers, exist in the Indian subcontinent. The India Independence Act of 1947 leaves the princes free to accede to either Pakistan or India, a choice dictated largely by geography and the wishes of the local population. The frontier princely states (Dir, Chitral, Swat, Amb, and Hunza) opt for Pakistan while retaining substantial internal autonomy. Bahawalpur State, in the former Rajputana province, accedes to Pakistan. The Muslim Nawab (ruler) in Junagadh State in Kathiawar also accedes to Pakistan, but his actions are negated by an Indian police action.

Jammu and Kashmir is one of British India's largest princely states and is particularly important because of its strategic location on China's border. The state has a Muslim majority population but a Hindu ruler. Hoping to retain his independence, the Maharaja of Kashmir declines to accede to either Pakistan or India.

1947: October 26 **Kashmir accedes to India**

After Pathan tribesmen invade Kashmir in October, 1947, allegedly with Pakistan's assistance, the Hindu Maharaja of Kashmir accedes to India. The Indian government sends troops to defend Kashmir. Pakistan refuses to accept the accession, claiming that Kashmir's population is largely Muslim, and sends troops to fight with the Azad Kashmir ("Free Kashmir") forces facing the Indian regulars in the state.

1947 **Economic consequences of Partition**

The new boundaries of Pakistan disrupt traditional economic relationships and transportation systems. East Bengal is a major producer of jute, and cotton is an important crop in the Indus valley. The mills that process these crops, however, lie in territory assigned to India. Similarly, railroad networks are disrupted by the newly-drawn political boundaries. The resolving of such administrative problems is made difficult because the pre-independence government is largely in the hands of Hindus. With the migration of Hindus to India, Pakistan—particularly East Bengal—faces a serious shortage of experienced civil servants and administrators.

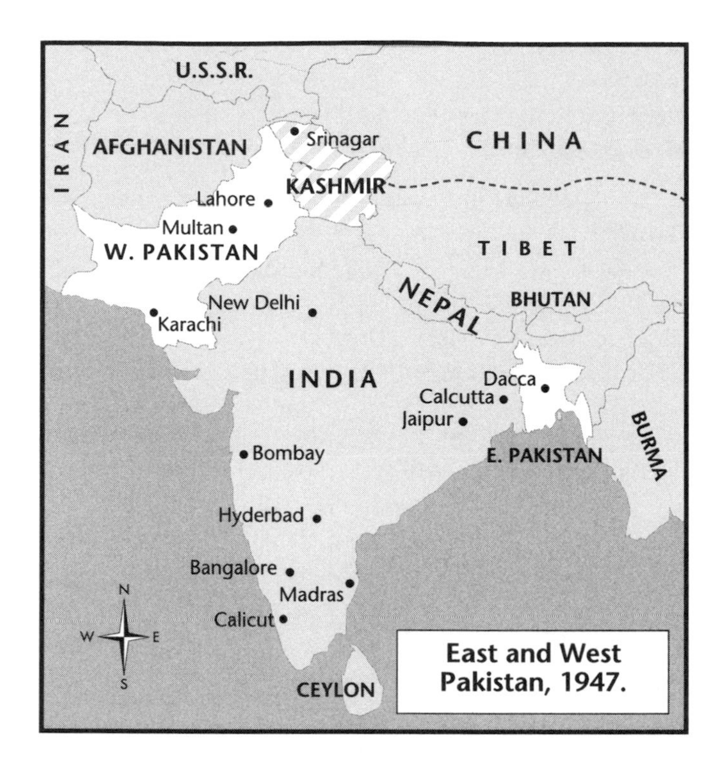

East and West Pakistan, 1947.

The boundaries of Pakistan and India also create problems over the use of the subcontinent's rivers. In the Punjab, the headwaters of the rivers that feed the Indus and the Punjab's canals fall into Indian territory. The Ganges and Brahmaputra rivers lie mainly in Indian territory, but their delta is in Pakistan's eastern wing (later Bangladesh). Disputes over the diversion of the waters of these rivers by India remain an issue in the relations between India and Pakistan for decades to come.

1948: September 11 Jinnah dies

Jinnah dies of tuberculosis in Karachi. He is known in Pakistan as *Quaid-e-Azam*, the "Father of the Nation."

1949: January 1 Hostilities in Kashmir cease

The United Nations arranges a cease-fire in Kashmir. The proposed plebiscite to decide Kashmir's fate never takes place. Kashmir remains divided, with about 30 percent of the state under Pakistan's control. Pakistani and Indian troops continue to face each other across the cease-fire line.

1951: October 16 Prime minister assassinated

Liaqat Ali Khan, Pakistan's first prime minister, is assassinated in Rawalpindi.

1952 Urdu chosen as the national language

Seeking to acquire a symbol of "nationhood," Pakistan chooses Urdu as the national language. Urdu, which is written in the Perso-Arabic script, evolves during the sixteenth and seventeenth centuries from the mix of languages spoken by Muslim soldiers (Persian, Turk, Arab, and Afghan) and the local speech in northern India.

However, Urdu is largely the language of the urban Muslims of the Ganges plains rather than the peoples of Pakistan. Bengalis in East Bengal riot over the imposition of Urdu as the national language. English continues to be used as the language of government and commerce.

1955 East Pakistan named

Pakistan's eastern territory, East Bengal, is renamed East Pakistan.

1956: March 23 Pakistan becomes a republic

A new constitution establishes the Islamic Republic of Pakistan, with a parliamentary form of government and Major-General Iskander Mirza as its president.

1958 New national capital

Pakistan decides to build a new national capital to replace Karachi.

The government enlists internationally known architects such as Konstantinos Doxiadis, Edward Durell Stone, and Gio Ponti to help design the new city. Islamabad, the "City of Islam," lies a few miles north of Rawalpindi in the Punjab and is administered as a federal district, much like Washington, D.C. in the United States. Construction on the capital begins in 1961.

1958: October 7 Martial law under Ayub Khan

The ineffectiveness of the parliamentary system, the splintering of political parties, corruption, and near bankruptcy leads the president to declare martial law. General Mohammad Ayub Khan (1908–74) becomes chief martial law administrator. Ayub Khan soon removes Mirza from the presidency and assumes the office himself.

1960 Signs of democracy

The country prepares for the first balloting under the country's new democracy. Candidates find visual ways, such as using familiar cultural items like the camel and the chair, to identify themselves with illiterate voters.

1960: September Indus Waters Treaty

Pakistan and India sign the Indus Waters Treaty after prolonged negotiations between the two countries mediated by the World Bank. The agreement allocates exclusive use of the Jhelum, Chenab, and Ravi Rivers to Pakistan.

1962: March 1 Constitution

A new constitution creates a presidential form of government in Pakistan. Although it establishes elected central and pro-

In Lahore, camels wear banners with political slogans in a parade to woo voters. Since much of the population is illiterate, political parties use pictures to represent themselves and their candidates. (EPD Photos/CSU Archives)

vincial legislatures, the ban on political parties imposed under martial law continues until late 1962. The president retains sweeping powers that give him virtual control of the government. Ayub Khan retains the office of president in elections held in 1965.

1965: September 6 Indo-Pakistan war

India invades the Punjab after serious border clashes in Kashmir and the Rann of Kutch. Neither India nor Pakistan can sustain a long war, and the U.N. arranges a cease-fire on September 23. The lack of military success and failure in the apparent objective of agitating an uprising in Kashmir leads to widespread dissatisfaction and antigovernment demonstrations in Pakistan.

1969: March 25 Martial law under Yahya Khan

Increasing popular opposition to the authoritarian government and spreading violence in both wings of Pakistan results in the proclamation of martial law. The army commander-in-chief, General Agha Mohammad Yahya Khan (1917–80), becomes chief martial-law administrator. He replaces Ayub Khan as president six days later. Yahya Khan promises early elections.

1970: December 7 General elections

The first elections in Pakistan based on one man, one vote take place.

The Awami League, under Sheikh Mujibur Rahman (1921–75), win 160 out of the 162 national assembly seats contested in East Pakistan. However, they gain none of the 138 seats at stake in West Pakistan. The Pakistan People's Party (PPP), led by Zulfikar Ali Bhutto (1928–79), wins a majority (81) of the seats in West Pakistan. Rahman, who is popularly known as "Mujib," claims the right to form a national government and draft a new constitution. This constitution is clearly to be based on the Awami League's platform of provincial autonomy. Bhutto refuses to accept the right of the Awami League to form a national government, arguing that its political base is confined to East Pakistan only.

1971: March 1 National Assembly convening postponed

Following the breakdown of talks with Mujib and Bhutto, Yahya Khan postpones the meeting of the national assembly.

The population in East Pakistan reacts to this violently, with strikes, demonstrations, and civil disobedience bringing business and government in the country to a standstill. Militant Bengali nationalists known as the Mukti Bahini, or "Freedom Fighters," stage terrorist attacks on central government installations.

1971: March 25–26 Civil War breaks out in East Pakistan

The government arrests Mujib; the army, made up mostly of troops from West Pakistan, begins operations against Bengali rebels. Bengali nationalists in East Pakistan proclaim the "independent, sovereign republic of Bangladesh" on March 26.

In the following months, Pakistani government forces and Bengalis fight a brutal and bloody war. Although exact numbers are uncertain, as many as three million civilians die in the conflict and another ten million refugees cross the borders into India.

India finally intervenes in the civil war. Following Pakistani preemptive air strikes on targets across northern India, Indian forces invade East Pakistan on December 4. India fights a holding action along the border with West Pakistan, aiming the main thrust of its attack at Dhaka in the east. Pakistani forces in East Pakistan lay down their arms on December 16 after twelve days of fierce fighting.

1971: December 20 Bhutto assumes power

The now discredited Yahya Khan resigns and Bhutto assumes the reins of government as president and chief martial law administrator of a dismembered and demoralized Pakistan.

1973: February Baluchi uprising

The Baluchi tribes rise against the central government and armed resistance continues for four years. The army is called in to the quell the uprising. The roots of the trouble lie in a deep-seated Baluchi nationalism and in the resentment of Punjabi domination of Pakistan's government.

1973: August 14 1973 Constitution adopted

Pakistan's leaders agree on a new constitution which puts a federal parliamentary system of government in place. Bhutto becomes the prime minister of Pakistan.

1974: February 22 Pakistan recognizes Bangladesh

Pakistan formally recognizes Bangladesh.

1974: February 22–24 OIC Summit in Lahore

Pakistan hosts the Second Summit of the Organization of the Islamic Conference (OIC), an association of Islamic countries committed to promoting close cooperation among themselves in the economic, political, cultural, and spiritual fields.

1974: February Shah Faisal Mosque

King Faisal of Saudi Arabia, who is attending the OIC Summit, donates funds for the construction of a mosque in Pakistan's new capital. Named after the king, the Shah Faisal Mosque is a striking modern structure that resembles an eight-sided Bedouin tent, flanked by four 300 foot-high (90 meters) minarets.

1975 Pakhthunistan

Tribesmen in the N.W.F.P. demonstrate for the creation of an independent state, Pakhtunistan. The Pakistan-Afghanistan border, fixed in 1893 along the Durand Line (at the time, it is the India-Afghanistan border), divides lands occupied by the Pathans, various tribal groups who speak Pashto. The idea of a homeland for the Pashto-speaking peoples of Afghanistan and Pakistan appears in the early twentieth century. The issue arises again in 1947 and in the decades following independence, but it is vigorously opposed by the Pakistani government.

1977: July 5 Martial law under Zia-ul-Haq

The armed forces intervene in Pakistan's deteriorating political situation, and General Mohammad Zia-al-Huq (1924–88) imposes martial law.

Zia takes this action to diffuse a crisis resulting from opposition to Bhutto and the PPP from the Pakistan National Alliance (PNA), a broad-based coalition of political parties. The PNA charges rigged elections and "undemocratic" and "authoritarian" government, and launches a country-wide civil disobedience campaign. This results in political arrests, a breakdown of law and order across the country, and widespread violence that virtually brings the country to a halt.

1978: September 16 Zia becomes president

Zia assumes the presidency with sweeping power over the government.

1979 Zia enforces Islamic laws

Zia announces measures to bring Pakistan's laws into conformity with Islamic (*Shari'a*) law. This includes the introduction of *zakat* (the obligatory Islamic religious tax to support the poor), of *ushr* (an agricultural tax), and other Muslim economic practices such as interest-free banking.

1979: April 4 Bhutto hanged

Zia's government executes Bhutto for instigating the murders of two political opponents. Zia brings charges against other former government members for misuse of public office and for using official position to amass personal wealth.

1979: December Karakoram Highway finished

Pakistan and China complete the 750-mile (1,200-kilometer) Karakoram Highway linking Pakistan with Kashgar in China's Xinjiang Province. The two-lane, all-weather road crosses some of the most difficult terrain in the world. The Khunjerab Pass, where the road crosses from Pakistan into China, lies at an elevation of 15,528 feet (4733 meters). Landslides, rockfalls, accidents, and severe weather conditions claim numerous lives during the road's construction.

The Karakoram Highway has strategic significance, providing a driveable land route between Pakistan and China. Pakistan develops close relations with China as a counterbalance to the India-Soviet Union alliance.

1979: December Afghan refugees

The Soviet invasion of Afghanistan results in a flood of refugees, estimated at around three and a half million people, crossing the border into Pakistan. Most settle in camps around Peshawar in the North West Frontier Province. International aid, particularly from the United States, helps Pakistan deal with the refugees. The camps in Pakistan provide bases from which Mujahideen guerrillas, with military aid from the United States, fight the Soviets in Afghanistan.

1985: December 30 Zia lifts martial law

Zia lifts martial law, following a national referendum that provides popular support for his policies and elections that return him as president for a five-year term. He appoints Mohammad Khan Junejo prime minister.

1986: May 1 Khunjerab Pass opens to tourism

The Khunjerab Pass opens to tourism, although the official traffic and trade across the pass began in August, 1982.

1988: August 17 Zia dies in air crash

Zia, several high ranking military officers, and the U.S. ambassador to Pakistan die in a plane crash at Bahawalpur.

1988: December 1 Benazir Bhutto becomes prime minister

Benazir Bhutto, leader of the PPP, becomes the first female to lead a Muslim country.

Benazir Bhutto (b. 1953) is the daughter of Zulfikar Ali Bhutto. Unlike most women in Pakistan who have little access to education, she is a graduate of Harvard University and the University of Oxford. Literacy among the female population of Pakistan is only 22.3 percent (1993 data). Girls tend to be kept home from school at an early age, especially in rural areas where children must work in the fields. Despite expansion of educational facilities since independence, only 35 percent of Pakistanis over fifteen years of age are literate.

1989: June Pakistan and the U.S. sign major arms deal

Pakistan and the U.S. sign an agreement by which Pakistan is to purchase sixty F-16 fighter aircraft. The following year, however, the U.S. withholds economic and military aid to Pakistan over concerns about Pakistan's association with terrorist groups and the country's developing nuclear weapons programs. Although the U.S. subsequently releases some military equipment to Pakistan, the F-16s remain under embargo.

1989 First Women's Bank

Pakistan sets up the First Women's Bank (FWB), a small public institution financed by the government and the country's other public-sector banks. FWB is run by women for women, and is an attempt to empower women in the male-dominated Islamic society.

1990: January Political strife in Sind

Political conflict breaks out in Sind between Bhutto's PPP and the Muhajir Quami Movement (MQM), the political party of the muhajirs. This strife continues for several years, especially in Karachi, where terrorism, murders, kidnappings, sectarian violence, and ethnic conflict are commonplace. Even Pakistan's security forces are unable to restore law and order.

1990: August–November Mohammad Nawaz Sharif replaces Bhutto

President Ghulam Ishaq Khan dismisses Bhutto's government on August 6. He charges the ousted government with corruption, nepotism, horse-trading, ineptitude, and failure to maintain law and order.

On November 6, Mohammad Nawaz Sharif, former chief minister of Punjab and leader of the Islamic Democratic Alliance (IDA), becomes prime minister of Pakistan.

1991: January Bumper opium harvest predicted

Narcotics experts predict an opium harvest of 600 tons, a threefold increase over the previous year. Drugs are a major social problem in the country, with drug-control programs having little effect.

1991 Sharia Bill

Nawaz Sharif's government enacts the Sharia Bill, perhaps the most important piece of legislation since the Constitution of 1973.

The bill declares Sharia to be the law of the land and provides for the Islamization of Pakistan's educational, judicial, and legal systems.

1992 Water Apportionment Accord signed

The Water Apportionment Accord assigns water resources to the various provinces of Pakistan. Debate continues on the proposed multipurpose Kalabagh Dam, to be built on the Indus 92 miles (150 kilometers) downstream from its confluence with the Kabul River. Environmentalists strongly oppose the dam's construction.

1992: March 25 World Cup victory

Pakistanis are jubilant over the national cricket team's victory over England in the final of the World Cup in Melbourne, Australia.

Few events serve to bring the Pakistani people together and create a sense of national belonging as this success in the international sports arena. This is especially true given Pakistan's past, one that contributes little to the process of nation-building—a country with deep regional divisions, that has fought and lost three wars with its neighbor, that has spent many years under martial law, and where no democratically elected government has completed its full term.

Sports play an important role in Pakistani life. In addition to cricket, Pakistani teams regularly win international (World and Olympic) field hockey championships and players such as Jehangir Khan and Jansheer Khan dominate the world of squash (a court game introduced by the British). Polo is popular in northern Pakistan.

1993: April–July Constitutional crisis

Continued ethnic violence in the country and a power struggle between President Ghulam Ishaq Khan and Nawaz Sharif result in the president dissolving Nawaz Sharif's government. The Supreme Court rules this action unconstitutional and reinstates Nawaz Sharif in May. Continued political turmoil in the country leads to the resignation of both the president and the prime minister.

1993: October 19 Bhutto returns to power

Following elections in early October, Benazir Bhutto returns to head a coalition government.

1994: December Sectarian violence

Shia gunmen assault the Masjid-e-Akbar in Karachi and kill eight Sunni worshippers.

Although Pakistan's people are overwhelmingly Muslim (minority religions account for only 3.3 percent of the population), sectarian violence is commonplace. Militants among the Shia Muslims, who account for some 25 percent of the population, are often in conflict with the Sunni majority. The Shia community itself is splintered into numerous sects. The Ismailis, a Shia sect that recognizes the Aga Khan as its leader, have a strong presence in the northern mountain region. The Ahmadiyas are a modern Islamic sect who face considerable discrimination and anti-Ahmadiya sentiment from other Pakistanis.

1995: February–May Violence continues in Karachi

A fresh wave of sectarian violence breaks out, with retaliatory attacks on mosques and other religious buildings. Lawlessness continues, and two U.S. consular officials are gunned down in broad daylight. Fighting breaks out involving various factions of the MQM and government security forces. The MQM uprising and acts of terrorism continue into 1996.

1995: March Central Asian trade routes reopened

Pakistan, the People's Republic of China, Kazakhstan, and Kyrgyzstan sign an agreement allowing Pakistan to trade with the Central Asian countries through China.

1995: June 10 Journalist arrested

Zafaryab Ahmed, a Pakistani journalist, is arrested on charges of harming the country's exports by writing about bonded labor by children. Workers who are bonded are kept in conditions very much like slave labor. Human rights groups charge that bonded labor is widespread in Pakistan, although it is technically illegal.

1995 World Bank assesses environmental problems

A World Bank team assessing Pakistan's economic problems identifies widespread water pollution, land degradation through water logging and salinity, and fast-increasing air pollution in cities as major environmental concerns.

1996: April 13 Family planning

At an Islamabad conference of the Economic Cooperation Organization (ECO), a group of ten Islamic nations, Pakistan announces plans to hire 12,000 new village-based family planning workers. Pakistan's annual rate of population increase (2.8 percent) is among the highest in the world and poses a major obstacle to economic development.

1996: November Bhutto government dismissed

President Sardar Farooq Ahmad Khan Leghari dismisses Benazir Bhutto and dissolves the federal and provincial assemblies. He cites his reasons for doing so as the deteriorating law-and-order situation, severe economic problems, disregard for judicial authority, widespread corruption, and various constitutional violations.

1997: February 17 Nawaz Sharif returned to power

Mohammad Nawaz Sharif is elected prime minister, replacing the interim government put in place after Bhutto's ouster.

1997: April 22–23 Pakistan obtains US $2.3 billion aid

At its meeting in Paris, the Aid to Pakistan Consortium awards Pakistan $2.3 billion (U.S.) in aid for 1997–98. The Consortium is an association of developed countries and multilateral agencies such as the World Bank that channels international aid to Pakistan. The Pakistani economy is heavily dependent on foreign aid.

1997: June 15 The "Developing Eight"

The first meeting of the Developing Eight countries (D-8) takes place in Istanbul, Turkey. The group consists of eight developing Muslim countries—Bangladesh, Egypt, Indonesia, Iran, Malaysia, Nigeria, Pakistan, and Turkey. These countries plan on mutual cooperation in various economic and cultural fields.

1997: July 23 Gas pipeline agreement

The governments of Pakistan and Turkmenistan, Unocal, and Saudi Arabia's Delta Oil sign an agreement to construct an 875-mile (1,400-kilometer) gas pipeline from Turkmenistan's Daulatabad field to Pakistan through Afghanistan.

1997: October 7 Queen Elizabeth visits Pakistan

Queen Elizabeth II of England and her husband Prince Philip arrive in Pakistan on a visit to the subcontinent to participate in the 50th anniversary celebrations of Pakistan (and India).

1997: November **U.S. oilmen murdered in Pakistan**

Four employees of Union Texas Petroleum and their Pakistani driver are killed in Karachi. The attack is apparently in retaliation for the guilty verdict handed down on Mir Aimal Kansi, a Pakistani accused of murdering two C.I.A. employees as they drove to work at C.I.A. headquarters at Langley, Virginia, in 1993.

1998: February 14 **Oil and gas found in Sind**

Pakistan's Oil and Gas Development Corporation reports the discovery of oil and gas reserves in Tando Allahyar in Sind Province. Energy shortages, a serious problem in Pakistan, are a major constraint on the country's economic development.

1998: May 28 **Pakistan tests nuclear devices**

Ignoring the pleas of world leaders for restraint, Pakistan explodes five nuclear devices (and a further one on May 30) at its test facility in the Chagai Hills in Baluchistan. This is in response to India's tests of nuclear weapons on May 11 and 13. The common people greet the tests with jubiliation, seeing them as enhancing the country's international standing and prestige as well as countering a perceived threat from India.

The potential for nuclear conflict in the Indian subcontinent is a source of widespread concern in the international community. Pakistan and India have fought three wars within the last fifty years, and still confront each other in Kashmir.

1998: May 28 **United States imposes sanctions**

President Bill Clinton announces sanctions against Pakistan, as required by United States law. The Glenn Amendment of 1994 bars American economic and military aid to countries developing nuclear weapons. Sanctions also include a ban on private banking assistance and require the U.S. to vote against funds for Pakistan in organizations such as the World Bank, International Monetary Fund, and Asian Development Bank.

The sanctions pose a serious threat to the stability of Pakistan. Financial analysts say that Pakistan—with a national debt of $50 billion, annual debt payments of $5.5 billion, and only about $1 billion in foreign exchange reserves—may default on its international loans. This could lead to spiraling inflation, currency devaluation, and social and political upheaval.

1998: July 26 **Bhutto faces corruption charges**

Benazir Bhutto returns to Pakistan from Dubai to faces charges of corruption during her tenure as prime minister. She also faces money-laundering charges in Switzerland.

1998: August 28 **Constitutional Amendment to adopt Islamic Law**

Prime Minister Nawaz Sharif introduces a constititional amendment to scrap the country's legal system, which is rooted in British common law, and replace it with one based on the *Qu'ran* (the holy book of Islam) and the *Sunnat* (the writings of the prophet Muhammad). Pakistan already has some Islamic laws in place, e.g. the death penalty by stoning for adultery, but, under the new legislation, the country will be governed entirely according to Islamic Law.

1998: September 4 **Artillery exchange in Kashmir**

Pakistani and Indian troops exchange artillery fire along the Line of Control in Kashmir. Regular border incidents such as this are a source of continuing tensions between Pakistan and India.

1998: September 23 **Pakistan willing to sign Comprehensive Test Ban Treaty**

Prime Minister Nawaz Sharif announces at the United Nations General Assembly in New York that Pakistan is willing to sign the 1996 international treaty banning nuclear tests (Comprehensive Test Ban Treaty) in return for the U.S. lifting its sanctions.

1998: November **IMF reschedules Pakistan's debt**

The International Monetary Fund (IMF) reschedules Pakistan's $5.5 billion debt and loans it $1.56 billion.

1999: February **Indian Prime Minister Vajpayee meets Sharif in Lahore**

Indian Prime Minister Atal Behari Vajpayee meets his Sharif in Lahore, Pakistan. The two prime ministers agree to continue their dialogue, consult each other prior to missile tests, and tacitly separate the nuclear weapons issue from the dispute over Kashmir. Vajpayee rode to Lahore on a bus that inaugurated the first bus service between the two countries since 1965.

Bibliography

Arrian. *The Campaigns of Alexander.* Translated by Aubrey de Selincourt. Harmondsworth, England, and Baltimore: Penguin Books, 1971.

Blood, Peter R., ed. *Pakistan, a Country Study.* 6th ed. Washington, D.C.: Federal Research Division, Library of Congress, 1995.

———. *Pakistan: A Nation in the Making.* Boulder, CO: Westview Press, 1986.

Burki, Shahid Javed. *Historical Dictionary of Pakistan.* Metuchen, NJ: Scarecrow Press, 1991.

Eglar, Zekiye. *A Punjabi Village in Pakistan.* New York: Columbia University Press, 1960.

Islamic Republic of Pakistan. U.S. Department of State Background Notes, November 1997. Available at http://infoweb3.newsbank.com/bin/gate.exe?f=doc&state=1t40pq.2.1

Kulke, Hermann and Dietmer Rothermund. *A History of India.* London and New York: Routledge, 1986.

Mahmud, S. F. *A Concise History of Indo-Pakistan.* Karachi: Oxford University Press, 1988.

Michel, Aloys Arthur. *The Indus Rivers: A Study of the Effects of Partition.* New Haven: Yale University Press, 1967.

Moorhouse, Geoffrey. *To the Frontier.* New York: Holt, Rinehart, and Winston, 1985.

National Institute of Folk Heritage. *Folk Heritage of Pakistan.* Islamabad: National Institute of Folk Heritage, 1977.

Quddus, Syed Abdul. *The Cultural Patterns of Pakistan.* Lahore: Ferozsons, 1989.

Robinson, Francis, ed. *The Cambridge Encyclopedia of India, Pakistan, Bangladesh, Sri Lanka, Nepal, Bhutan, and the Maldives.* Cambridge: Cambridge University Press, 1989.

Schwartzberg, Joseph E., ed. *A Historical Atlas of South Asia.* 2nd impression. Oxford and New York: Oxford University Press, 1992.

Shaw, Isobel. *Pakistan.* Lincolnwood, IL: Passport Books., 1988.

Taylor, David (revised by Asad Sayeed). "Pakistan: Economy," in *The Far East and Australasia 1997.* London: Europa Publications, 1996, pp. 873-79.

Wheeler, Sir Mortimer. *Civilizations of the Indus Valley and Beyond.* London: Thames and Hudson, 1966.

Palau

Introduction

The Republic of Palau is comprised of sixteen states that cover a total land area of 170.4 square miles (441 square kilometers), making the island nation about two and a half times the size of Washington, D.C. The Republic of Palau is located in the westernmost edge of the Caroline archipelago. The highest point is Mount Makelulu which rises to an elevation of 787 feet (240 meters) above sea level. The climate is hot and humid. There are two seasons: wet and dry. The wet season lasts from May through November. During the wet season, Palau is subject to damage from typhoons. Estimates in July 1997 placed its population at 17,000.

The economy of the Republic of Palau is based primarily on subsistence agriculture and fishing. Most people who work outside of these areas are employed by the government. The Republic of Palau depends heavily on financial assistance from the United States. With this aid, the per capita income of Palauans is higher than most other regions in Micronesia. The currency of the Republic is based on the U.S. dollar.

English is the official language of the islands—a legacy of nearly fifty years of United States rule. However, as a result of previous Japanese control of the island group, many older Palauans speak Japanese. In addition, several of the Palau states have indigenous official languages.

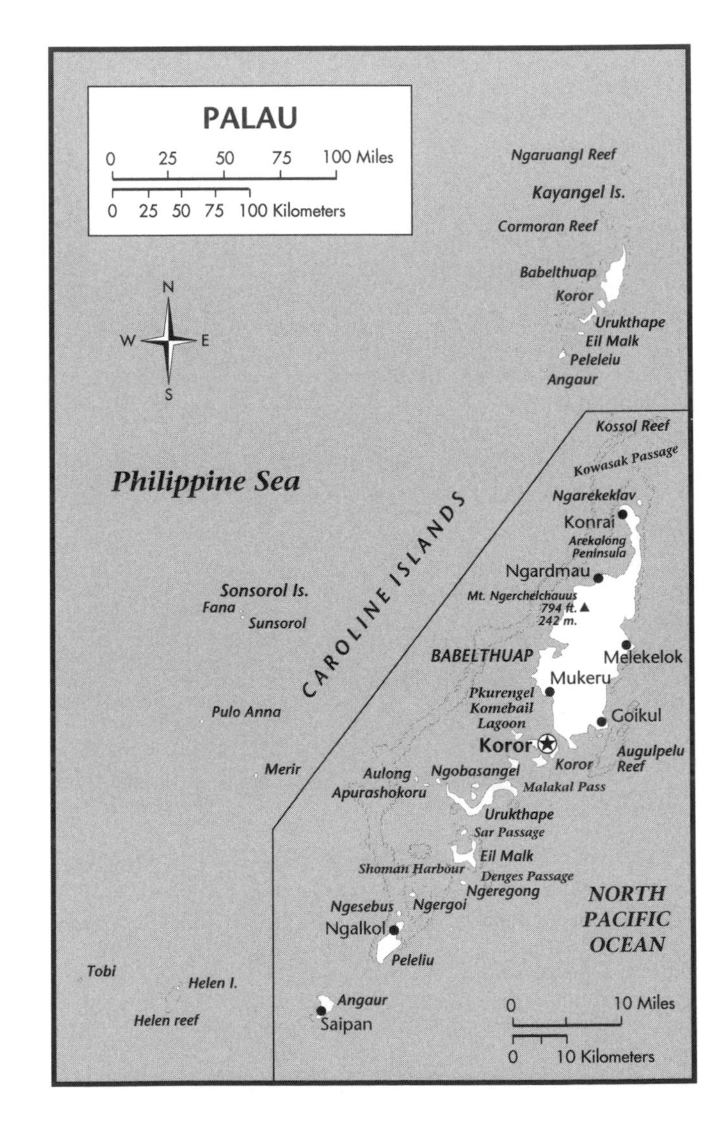

History

Palau belongs to the culture area of Micronesia in the South Pacific, north of the eastern half of the island of New Guinea. Carbon dating of shelters near Koror provide a date of approximately 1500 B.C. However, most archaeologists agree that settlement of Palau must have occurred nearly 1000 years before that. Throughout the island of Babeldaob, there are the remnants of terraces and huge earthen fortifications. There are also stone pavements and platforms for houses. Unusual remains include a series of upright monoliths with slotted tops. It seems that these fortifications were abandoned before the sixteenth century.

Contact with Europe had a profound impact on the islands' development. British and Spanish explorers sighted the island several times in the sixteenth through eighteenth centuries and published accounts of the islands appeared in the 1780s. Nevertheless, permanent foreign contact did not come to the Palau group until the 1880s when both Spain and Germany laid claim to the islands. By 1885, papal arbitration recognized the Spanish claim which lasted until 1899.

Although their rule lasted only fourteen years, the Spanish left a lasting legacy on the island through their establishment of Catholic missions on the island. As a result of the efforts of Christian missionaries, the majority of Palauans today are Christians. The major sects of Christianity are represented by Catholics, Seventh-Day Adventists, Jehovah's Witnesses, and the Church of Jesus Christ of Latter-Day Saints. By contrast, only about one-third of the population follows an indigenous belief system called Modekngei, which means "let us go forward together."

Spanish rule ended in 1899 when the German Empire bought the islands. German control ended when Japan conquered the islands during the First World War (1914–18). The Palau group played a significant part in Japanese defense, since they occupied the southeastern rim of the empire's defensive perimeter. American troops took heavy casualties when they invaded in September 1944.

World War II (1939–45) had a profound impact on the Palauan people. Heavily-fortified Koror had been bombed to rubble. As a result of the intense fighting on Peleliu and Angaur, these islands were completely denuded (stripped of vegetation). All of the Palauans on Peleliu were evacuated to Babeldaob. Here the indigenous Palauans retreated to the jungle and were only able to collect food and hunt animals in the periods between bombing raids. Older Palauans have strong memories of the famine during this time. The Japanese took everything the Palauans had while the Americans constantly dropped bombs on them from the air.

Following World War II, the Palau islands remained under United States administration as a United Nations Trust Territory. Palau adopted a constitution that allowed for self-government in 1980 and became fully independent in 1994.

Timeline

Prehistoric era Palau's topography evolves

Millions of years ago, volcanic forces thrust reefs in the Pacific Ocean upward above the ocean surface. Deep depressions in the reefs became landlocked saltwater lakes. One of the best-known of these lakes is found on Eil Malk Island, and is known as Jellyfish Lake. It is a saltwater lake filled with Mastigias jellyfish. Scientists believe that these jellyfish were trapped in the lake when the rocky reef was forced above the ocean surface to become an island. In the process of survival and evolution, the jellyfish lost their ability to sting. The school of jellyfish in Jellyfish Lake is estimated to number two million.

Early centuries A.D. Residents of Yap Island travel to Palau

People from the nearby island of Yap, 750 miles east of the islands of Palau, journey by canoe to a stone quarry. The Yapese use huge circular disks of stone as the main form of money. The rocks of Palau are well-suited for this purpose, so the Yapese travel to Palau to carve disks from the rocks and take them back to Yap. At the end of the twentieth century, visitors to Palau may see partially completed stone disks on Palau at an ancient quarry site, known as Babeldaob Jungle.

c. 1000 Cave used as burial site

Ngeruktabel is the site of an ancient burial cave. It is a huge cavern, whose mouth is open to the sea and has been eroded over the centuries by waves. Scattered everywhere are hundreds of human bones, left from centuries of ancient Palauan burials.

c. 1200 Tribal districts enact environmental laws

The chiefs in each of sixteen tribal districts establish controls, called *bul,* that are similar to modern-day conservation laws. The controls limit rights to lagoons to prevent environmental destruction and to avoid depletion to the food supply. It is believed that the controls include bans on fishing of species during their spawning periods.

1579: September 30 Sir Francis Drake's ship *The Golden Hind* sights Palau

Sir Francis Drake (c. 1540—96) sets out to circumnavigate the globe. On his voyage, his crew sights Palau and makes only limited contact.

1710: 11 December Captain Francisco de Padilla makes contact with Palau

Francisco de Padilla, a Spanish explorer, makes contact with the islands of Palau, but does not stay long.

1783 British East India Company ship *Antelope* runs aground in Koror

The British East India Company ship *Antelope,* commanded by Captain Henry Wilson, runs aground near Koror. The crew spends three months rebuilding the ship for their departure from the island. Communication between the Palauans and the sailors is accomplished by a member of each party who knows Malay. When Wilson departs, Lee Boo, a son of a local chief, accompanies him back to England. Lee Boo dies from smallpox six months after arriving in London.

1788 Publication of *An Account of the Pelew Islands,* written by George Keate.

An Account of the Pelew Islands by George Keate is the first published account of the local culture of the islands. Keate, who had never visited Palau, published his idealized account of the adventures of the crew of the British ship, *Antelope,* in Palau (see 1783). Keate used the notes and journals of Captain Henry Wilson to produce the book. *An Account of the Pelew Islands* was extremely popular in England and was soon translated into French for distribution throughout Europe.

1838–42 Explorer Charles Wilkes stops in Palau

U.S. explorer Charles Wilkes (1798–1877) leads an expedition to explore the South Pacific and Antarctica. They stop in Palau during their voyages. His exploration of Antarctica on this trip leads to a section of land being named for him. A commemorative stamp is issued in his honor by Palau. (See 1996: April 29.)

1840 Whaling vessels stop in Palau

European and American trading and whaling vessels begin making regular stops in Palau.

1885 Pope Leo XII accepts Spain's claim to Palau

Conflicting claims to Palau had been presented by Germany and Spain. Pope Leo XII (1810–1903) rules in favor of Spain's claim.

1885–99 Palau is under control of Spain

Palau was originally colonized by Spain which did little except send Catholic missionaries to the region.

1891 First permanent Roman Catholic mission established

Two Roman Catholic priests of the Capuchin order establish the first permanent Roman Catholic mission in the islands. Ibedul, a local chief of the southwestern portion of Palau that centers on Koror, provides an old *bai* for use as the permanent mission. *Bai* are large, elevated meeting houses which are the political heart of Palauan villages.

1899–1914 Palau is under the control of the German government

Germany purchases Palau from Spain. German rule is short-lived, ending with the outbreak of World War I (1914–18).

1914 Japanese forces occupy islands

During World War I (1914–18), Japan joins the Allies in declaring war against Germany. Japanese occupation of the islands of Koror and Anguar of Palau begins. Indigenous resistance to the imposition of Japanese culture and civilization takes the form of a religious movement known as Modekngei. Its founder, Temedad, promotes non-violent opposition to the Japanese oppression of the indigenous population.

1920–45 Palau is under Japanese administration

After World War I, the League of Nations mandates Japanese control over Palau and the rest of Micronesia. Japan views Palau as an extension of Japan. Japanese migration is promoted and acculturation of the Palauans to a Japanese way of life is encouraged. The Palauans become a minority within a larger migrant Japanese population. Koror becomes known as "Little Tokyo." Koror becomes the administrative center for Japanese rule over all of Micronesia.

1944: March Japanese cargo ship *Teshio Maru* sinks

During the World War II Operation Desecrate I, the Japanese cargo ship *Teshio Maru* is sunk. It is a 321-foot, 2,840-ton vessel. The wreckage becomes a popular site for scuba divers in the 1990s. Artifacts that are visible to divers viewing the wreckage include china, medical supplies, and personal belongings of the crew.

1944: September 15–17 United States armed forces invade Palau

The United States invades two of the southern Palauan islands. The Battle of Peleliu takes place on September 15. Angaur is invaded on September 17.

The battle for Palau is considered the most brutal of the Pacific Theater during World War II (1939–45). U.S. Marines come ashore at White Beach while Japanese troops fire at them from coral caves. Fighting continues for weeks. In late November, all but 93 of the remaining Japanese soldiers commit ritual suicide. The 93 who do not, surrender, and the island chain is secured by the U.S. Marines. More than 12,000 Japanese soldiers and about 1,800 U.S. Marines lose their lives in the battle.

After the war, the Peleliu State Museum is established. It is a one-room display of battle artifacts, including Japanese *sennibari* (good-luck belts) and photographs of U.S. Marines with their leader, Colonel Chesty Purer. Outside the museum, military vehicles and aircraft are visible.

1947 United States administers the Trust Territory of the Pacific Islands

The United States becomes the administering authority of the United Nations (UN) Trust Territory of the Pacific Islands, which includes Palau. The United States gains control over the 2,000 small islands that constitute Micronesia. The development plan for the region is as a strategic trust territory, in response to the Japanese strategy to use the region as a base from which they could strike and defend. The United States fortifies the islands, giving little attention to the fate of the indigenous populations who live there.

1955 Agriculture and fishing dominate Palauan economy

Most native Palauans are engaged in agriculture and fishing. Coconuts are the chief cash crop. A copra mill handles copra (dried coconut meat) from Palau and from other islands in Micronesia. Coconut oil and copra cake are the principal exports. These replace phosphate, which is an important export until reserves are depleted.

A fish-freezing plant in Palau is supplied by tuna fishing. Tourism, centered in Koror, the capital, is beginning to develop. The United States provides aid for harbor development and road construction on several Palauan islands.

1965 Meeting of the first Congress of Micronesia

Three Palauan representatives are elected at the meeting of the first Congress of Micronesia. This is the beginning of the independence movement for Palau and other Micronesian possessions.

1973 Micronesian Mariculture Demonstration Center founded

The Micronesian Mariculture Demonstration Center (MMDC) is founded by Palau and the United States to help preserve giant clams and other mollusks. Through the work conducted at MMDC, Palau becomes the world's leader in giant clam research and cultivation. Researchers there develop an efficient system of clam farming. About one million baby clams are cultivated each year at MMDC. These are distributed to coral reefs in an underwater farming project.

1979: April 30 Proposed Constitution for Republic of Palau is discussed

A joint session of the legislature and a Constitutional Convention meets to discuss the proposed Constitution of the Republic of Palau. Debate at the meeting centers on the nuclear ban that is included in the Constitution. United States opposition to this ban and the related loss of income by the fledgling nation become a focus of the movement towards independence.

1983 High Chief Ibedul Gibbons wins "Alternative Nobel Prize"

Palauan traditional leader, High Chief Ibedul Gibbons, is awarded the "Alternative Nobel Prize" by the Right Livelihood Foundation. He is recognized for his work to ensure peace in a nuclear-free world, especially in the Pacific region. This award is later commemorated on a Palau stamp issued on September 15, 1995.

1983 Postal service begins operation

An independent postal service is inaugurated. Palau issues commemorative stamps that become popular with stamp collectors worldwide. The stamp issues depict Palauan history, plants, and wildlife. They also commemorate significant world events and anniversaries.

1985: June 30 President Remeliik, the first elected President of Palau, is assassinated

The assassination of President Haruo Remeliik shatters the Palauans and the world. The island chain had been viewed as a sleepy backwater in Micronesia until this event changes the nature of Palauan politics and the image of Palau in the world's eyes. Vice-president Alfonso Oiterang serves as acting president until elections are held in August.

1985: August 28 Lazarus Salii is elected second president of Palau by special election

Vice-president Alfonso Oiterong, who served with Remeliik, runs against Lazarus Salii in the special election for president. Salii defeats Oiterong by a margin of 600 votes.

1988: 20 August Lazarus Salii, President of Palau, commits suicide

It is determined that the death of Lazarus Salii, the second president of Palau, is by his own hand. Early speculation suggested that another political assassination had occurred before the island chain had even become fully independent.

1989 Palau designated as "underwater wonder"

International marine experts, representing the Smithsonian Institution and other groups, designate the waters surrounding Palau as one of seven underwater "wonders of the world." Living in the seas around Palau are more than 1,000 species of fish, some 700 species of coral and anemones, and seven of the eight giant clam species.

1993 Poachers harvest giant clams

The giant clams inhabiting the seas around Palau are illegally destroyed by Taiwanese divers. Working from a fishing boat, the divers uproot nearly 15,000 clams from the coral reefs. The muscles that hold the two shells together, known as adductors, are considered a delicacy in many Asian countries and sell for up to $50 per pound. These muscles, only a fraction of the edible portion of the clam, are removed from the Palauan giant clams, and the remains of the clams left to rot on the ocean floor.

1994 United States gives financial aid to Palau

Under terms of an agreement negotiated in the 1970s—when the United States thought it might want to use Palau as a military base—the U.S. will give the 15,000 citizens of Palau about $500 million in grants over the next 15 years.

1994: October 1 Republic of Palau celebrates independence

Palau becomes the last of the U.S. Micronesian Trust Territories to become politically independent when it gains nationhood. The road to independence has been exceptionally long and arduous for the small island chain in Micronesia. After much internal strife and lengthy legal battles regarding the constitutionally mandated nuclear ban, Palau finally becomes independent. There is continuing debate over the spelling of the name of the nation, with many nationalists favoring Belau. In the Constitution, the nation is spelled Palau. Kuniwo Nakamura is the first president of the independent nation.

1994: November 2 Japan establishes diplomatic relations with Palau

Japan establishes diplomatic relations with the new nation of Palau, and discussions begin regarding the initiation of airline service between the two countries.

1994: November 5 Stamps featuring Disney characters issued

Eight stamps issued by Palau feature Disney characters Minnie and Mickey Mouse, Daisy, Donald, and Grandma Duck, Goofy, and Scrooge McDuck in Palau scenes.

1994: December 10 Stamps celebrate independence

Palau issues stamps commemorating the country's October 1 independence. The stamps feature national sites. These include the Henrik Starcke Peace statue at the United Nations; the official Presidential Seal; Palau president Kuniwo Nakamura meeting U.S. president Bill Clinton; the flags of Palau and the United States, and musical bars of the Palau national anthem.

1995 Palau issues stamps commemorating the end World War II

Paintings from United States military archives are used as the designs for twelve stamps called "Liberation and Victory." The U.S. navy unit that made air strikes against the Japanese at the Palau Islands had an artist-in-residence, Lt. William F. Draper, who made over twenty paintings. Eight of the twelve stamps reproduce Draper paintings, including "Preparing Tin-Fish," depicting getting torpedoes ready for action; "Hellcat's Takeoff into Palau's Rising Sun," revealing planes taking off from carrier ships; "Dauntless Dive Bombers Over Malakal Harbor;" "Planes Return From Palau;" illustrating planes returning to the carrier; "Communion Before Battle," showing church services held on the deck of an aircraft carrier before battle; "The Landing," featuring U.S. marines storming the beach; "First Task Ashore," showing marines clearing jungle and taking care of the wounded; and "Fire Fighters Save Flak-Torn Pilot."

Four stamps reproduce paintings by Tom Lea, an artist hired by U.S. magazine, *Life,* to cover World War II. These include "Young Marine Headed for Peleliu," "Last Rites for the Dead," and "The Thousand-yard Stare."

A second set of five stamps reproduces portraits by Albert K. Murray of the major military leaders, including "Admiral Chester W. Nimitz," who was the commander-in-chief of the U.S. navy's Pacific fleet; "Admiral William F. Halsey," commander of the South Pacific region and the Third Fleet; "Admiral Raymond A. Spruance," commander-in-chief of the Pacific Fleet and the naval administrator of Micronesian territories after the war; "Vice Admiral Marc A. Mitscher," commander of the carrier forces in the Pacific; and "Lt. General Holland M. (Howling Mad) Smith," the best known U.S. marine general and amphibious operations commander who led invasion ground forces.

1995: December 15 Palau becomes member of United Nations

The General Assembly unanimously votes to adopt a resolution admitting the Republic of Palau as the 185th member of the United Nations. Palau is the last of the United Nations (UN) Trust Territories to gain admission to membership in the UN.

1996: April 29 Palau releases stamps commemorating world exploration

Palau releases two sheets of nine stamps each featuring pictures of sailors and pilots who circumnavigated the globe. The sailors lived from 1500s to 1960, and include Ferdinand Magellan who was captain of the first sailing fleet to circumnavigate the world (1519–21), U.S. explorer Charles Wilkes (1798–1877) who led a sailing expedition from 1838–42 that stopped at Palau; U.S. sailor Joshua Slocum who in 1895–98 became the first person to circumnavigate alone; Australian Ben Carlin, who in 1951–58 circumnavigated in a vehicle that was half-jeep and half-boat; U.S. navy commander Edward L. Beach, who skippered the first underwater voyage around the world in 1960; Naemi James of New Zealand, who in 1977–78 was the first woman to circumnavigate alone; English explorer Sir Ranulf Fiennes, who in 1979–82 hiked 35,000 miles to circumnavigate via the North and South Poles; Canadian Rick Hansen, who in 1985–87 traversed the earth in his wheelchair, and Robin Knox-Johnson of Great Britain, who set a circumnavigation speed record.

The individuals who circumnavigated by piloting air or space vehicles include Lowell Smith, who in 1924 led a fleet on the first round-the-world flight; German Ernst Lehman, who in 1929, captained a dirigible; Wiley Post, who flew 15,474 miles across the northern hemisphere in 1931 in piloting the first flight around the world in a monoplane; Russian cosmonaut Yuri Gagarin, who in 1961 was the first to circumnavigate the earth from space; U.S. pilot Jerrie Mock, who in 1964 became the first woman to fly solo around the world; U.S. pilot H. Ross Perot, Jr., who in 1982 became the first to fly a helicopter around the world; U.S. pilot Brooke Knapp, who in 1984 set a circumnavigation speed record; U.S. pilots Jeana Yeager and Dick Rutan, who in 1986 made the first nonstop flight around the world without refueling; and 82-year-old U.S. pilot Fred Lasby who is the oldest pilot to complete a solo circumnavigation.

Bibliography

Hanlon, David. *Remaking Micronesia.* Honolulu: University of Hawai'i Press, 1996.

Hijikata, Hisakatsu. *Society and Life in Palau*. Tokyo: Sasakawa Peace Foundation, 1993.
Liebowitz, Arnold. *Embattled Island: Palau's Struggle for Independence.* Westport, CT: Praeger, 1996.
Morgan, William. *Prehistoric Architecture in Micronesia.* Austin: The University of Texas Press, 1988.
Parmentier, Richard J. *The Sacred Remains: Myth, History, and Polity in Belau.* Chicago: University of Chicago Press, 1987.
Roff, Sue Rabbitt. *Overreaching in Paradise: United States Policy in Palau Since 1945*. Juneau, Alaska: Denali Press, 1991.

Papua New Guinea

Introduction

The initial settlement of the region now known as Papua New Guinea occurred as far back as 40,000 years ago. At that time, the islands of New Guinea and Australia were joined together as a large land mass called Sahul. Approximately 10,000 years ago the two islands separated due to sea level changes creating the configurations we know today as New Guinea and Australia. Insular New Guinea remained a great unknown to the outside world until the early decades of this century. On September 16, 1975, the independent nation of Papua New Guinea was born.

Geographically, Papua New Guinea includes the eastern half of the island of New Guinea as well as the Bismarck Archipelago, Bougainville and Buka in the Western Solomon Islands, and the Woodlark, Trobriand, D'Entrecasteaux, and Louisiade Island groups which fan out from the southeastern tip of New Guinea. Papua New Guinea covers an area of 461,693 square kilometers. The area is equal to roughly the area of Oregon and Idaho combined. Eighty-five percent of the total land area of the country is made up by the main island. Papua New Guinea has nineteen provinces and the National Capital District.

An extensive and complex mountainous backbone bisects the northern and southern portions of the island. The highest point in Papua New Guinea is Mt. Wilhelm which rises to an elevation of 4,509 meters above sea level. Within this range there are expansive upland valleys at altitudes from 1,500 to 3,000 meters above sea level. These valleys provide extensive crop lands for the growing of sweet potatoes, yams, corn, and coffee.

The largest river is the Fly River which flows for more than 1,100 kilometers. It flows south and is navigable for approximately 800 kilometers but only by small vessels. Most traffic on the rivers is by dugout canoes, many of which are powered by outboard motors.

Although Papua New Guinea lies completely within the Tropics, its climate ranges from humid tropical to alpine in the heights of the Central Range. The country also falls within the monsoonal belt. As a result, the average yearly rainfall is extremely high; nearly 100 inches per year in most provinces, with some closer to 200 inches per year.

The generic word for the inhabitants of Papua New Guinea is Papuans. The word "papua" derives from the Malay word which means "frizzy hair." Malay-speakers from present-day Indonesia and Malaysia have a long history of contact and trade with the inhabitants of parts of insular New Guinea. Each one of the over 700 ethnolinguistic groups of the island has a unique name that it uses for self-reference. For example, the Itamul people of the Middle Sepik River refer to themselves as the "Nyara."

Archaeologists and anthropologists speculate that there were a number of waves of immigration into New Guinea from Southeast Asia beginning at least as far back as 40,000 years ago. These migrants reached the interior mountain valleys as far back as 26,000 years ago. The inhabitants of the mountain valleys radically changed the landscape by clearing the original forests with intentional burning. The grassy valleys which are now so characteristic of the high mountain valleys are anthropogenetic, or human-made. Around 5,000 years ago another type of people started to migrate to the island, but most of these people came via boats. These are the ancestors of the Austronesian speaking peoples who lived in the coastal regions of the country. These people are related culturally and linguistically to the other Austronesian speaking peoples of the Pacific such as the Maori, the Hawai'ians, the Samoans, and the Tahitians. They brought with them the technology to make pottery and navigational skills to cross large bodies of water.

Papua New Guinea has one of the highest concentrations of different languages of any nation in the world. Over 700 different languages are spoken by populations in Papua New Guinea. None of these had an indigenous writing system at the time of European contact. Each was transmitted orally and each has a complex body of oral history, folklore, and mythology. Anthropologists and linguists are working to record and analyze as many of these languages as they can before the languages are replaced by Tok Pisin (an English-based Creole). Since many languages have at most only a few hundred speakers, they are not seen as viable means of communication, so when people immigrate to towns and urban centers

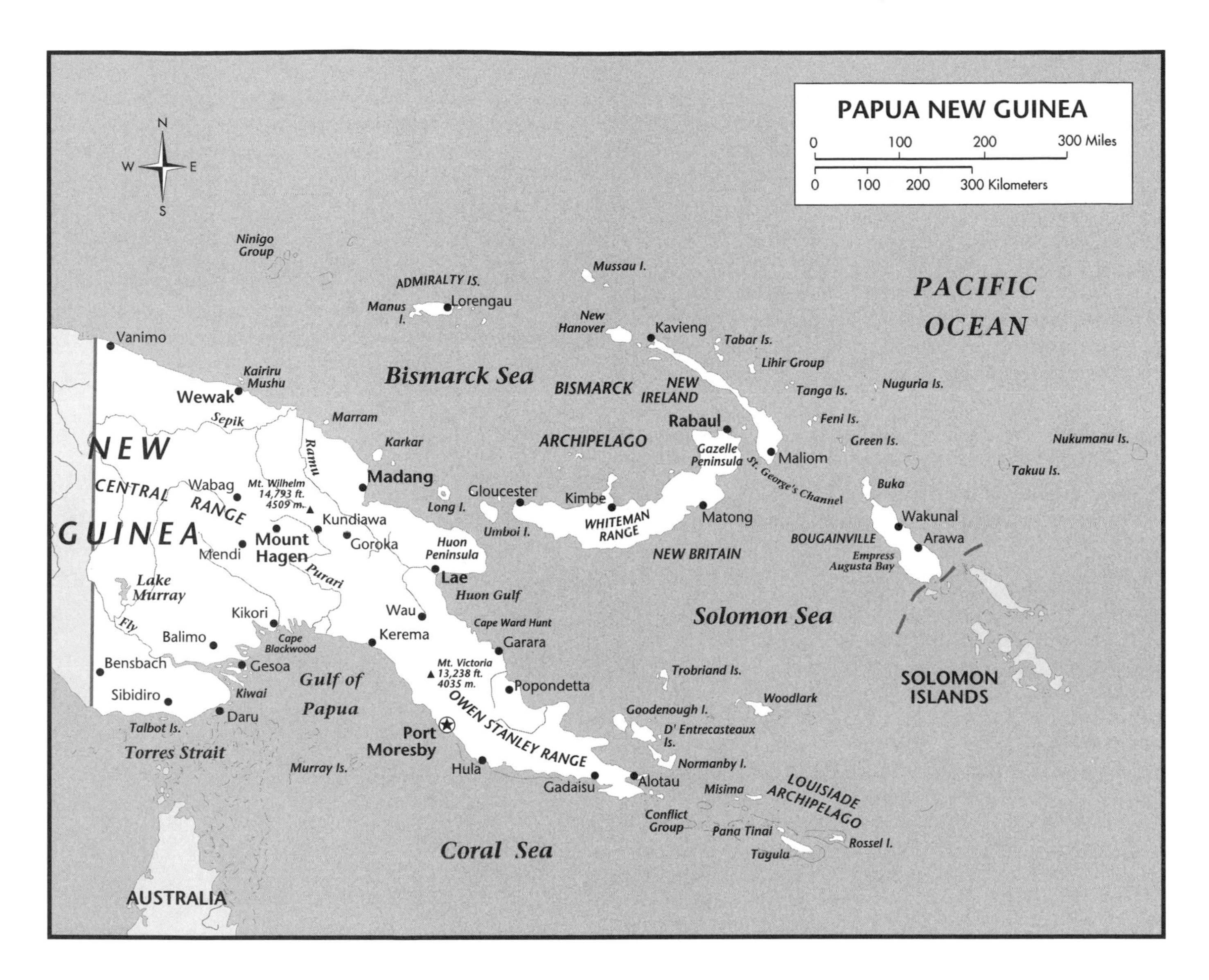

they begin to use Tok Pisin as their sole means of communication.

Papua New Guinea has three official, national languages: English, Hiri Motu, and Tok Pisin. English and Tok Pisin have wide currency throughout the urban centers and towns of the nation. Hiri Motu, however, is spoken and understood by only a limited population mostly living in the coastal areas near Port Moresby. Hiri Motu is a trade language that is spoken only as a second or third language. It developed out of the long distance, sea-going trade that the Motu engaged in with other groups in and around the Gulf of Papua. Hiri Motu was known as Police Motu since it was used by the post-colonial police force in Port Moresby to communicate with the immigrant Motu-speakers. Tok Pisin is also a language that developed out of trade and cultural contact. Tok Pisin is derived from a South Pacific English that was spoken by Europeans in the middle of the nineteenth century. Late in the nineteenth century plantations were formed to grow export crops for the world market and indigenous Papuans were transported, many unwittingly or even against their will, far from their homes to work there. Tok Pisin became a vehicle for intercultural communication on these plantations. It is now the fastest growing language in Papua New Guinea with radio and television broadcasts daily in the language as well as number of forms of print media solely in Tok Pisin.

Christianity has a strong foothold in Papua New Guinea with around two-thirds of the population claiming to be Christian. In many cases, Christian beliefs have melded with indigenous, pre-Christian beliefs in ancestor worship and ghosts. Catholics and Lutherans are the two largest Christian sects on the island. Seventh-Day Adventists and other evangelical sects are growing in popularity.

The government of Papua New Guinea has a legislative, executive, and judicial system. The legislature is a parliamentary system based on the Westminster system. Districts are represented by parliamentary members apportioned according to population, one provincial electorate per province, and

Carrying buckets of liquid latex on poles slung over their shoulders, Papaun workers must transport the latex from the collection point, the rubber tree plantation, to the processing factory. (EPD Photos/CSU Archives)

up to three nominated members. Parliamentary terms are five years.

The bulk of the population of Papua New Guinea exists in a subsistence economy, raising crops for internal consumption and limited trade. Cash cropping of coffee in the Highlands continues to expand, particularly with the growing coffee consumption in the United States. Arabica beans from Papua New Guinea are a moderate priced gourmet selection. Copra (dried coconut meat) and cocoa are grown in the lowland regions of the country.

Papua New Guinea has rich deposits of gold and copper. Rumors of extensive gold deposits in the Central Range is what brought the first non-Papuans into that area in the 1920s and 1930s. Major gold and copper deposits have been mined on Bougainville Island since the mid-1970s. There has been controversy and violence surrounding ownership of the mineral rights and over the distribution of profits from the mining endeavors. Another major gold and copper mining development was opened at Ok Tedi on the mainland in January of 1985.

The economy of Papua New Guinea is very stable. Exports and imports have been roughly equal since the mid-1980s. Australia, Japan, and the United States play major roles in investment, aid, and trade with the country.

Timeline

Pre-European contact **Society is agricultural**

The inhabitants of these islands use implements made of stone, bone, and wood. Most scholars believe that the island societies did not engage in metalworking. Agriculture was highly developed, and probably began around the same time as agriculture was beginning in settlements in Egypt and Mesopotamia

1511 The island of New Guinea is first sighted and recorded by Europeans

A Portuguese expedition lead by Antonio d'Abreau and Francisco Serrao sites and records the location of the island of New Guinea.

1526–27 Jorge de Meneses names the island of New Guinea "Papua"

Jorge de Meneses, a Portuguese naval explorer, accidentally comes across the island of New Guinea and calls it "Ilhos dos Papua." Speculation is that his naming derived from the Malay word for people which was 'papua,' meaning frizzy hair.

1545 The name "New Guinea" is bestowed upon the island

Ynigo Ortis de Retez, a Spaniard, gives the island the name "New Guinea" because its inhabitants resemble the Africans that he has seen in Guinea in West Africa.

1606 Two Spaniards land on Mainu Island

Luis Vaez de Torres and Diego de Prado land on Mainu Island. There is a skirmish between the local inhabitants and the Spanish sailors. The Spanish take fourteen boys and girls to Manila for conversion to Catholicism.

1871 London Missionary Society establishes missions

Teachers, representing the London Missionary Society, from the Loyalty Islands in New Caledonia set up mission schools in the Torres Straits.

1884 Germany takes possession of northeastern part of the island of New Guinea

Germany takes formal possession of the northeastern part of New Guinea, the Bismarck Archipelago, and New Britain.

1884: November 6 British protectorate proclaimed over southern New Guinea

The southern coast of eastern New Guinea which has been known as Papua is proclaimed a British protectorate and renamed British New Guinea. The northern half of what is now Papua New Guinea is a German possession and part of larger German New Guinea that includes New Britain, New Ireland, and Bougainville. The mainland portion is called Kaiser-Wilhelmsland.

1885 Rubber plantations established

Late in the nineteenth century, rubber plantations are formed to produce liquid latex for export. Indigenous Papauns are transported, many unwittingly or even against their will, far from their homes to work in the plantations.

1886 Germany takes possession of northern Solomon Islands

The German Empire annexes the northern Solomon Islands.

1888: September 4 Formal annexation of British New Guinea

Britain makes British New Guinea a crown colony and appoints William MacGregor the first lieutenant governor. Under his administration and guidance large parts of the unexplored interior are mapped and pacified. MacGregor also begins employing indigenous Papuans in the constabulary (police force).

1899 German New Guinea is made an imperial colony

In a final attempt to further control and develop their possession, the German government puts the possession under the administration of German government officials. Germany is attempting to create a German colony, owned and operated by immigrant Germans with the hopes of eventually establishing an overseas colony of locally-born Germans.

1902 British New Guinea placed under control of the Commonwealth of Australia

The Commonwealth of Australia receives administrative control over British New Guinea.

1905 Passage of Papua Act

With the passage of the Papua Act, British New Guinea becomes the Territory of Papua. Herbert Perry is appointed lieutenant governor of the Territory of Papua in 1908. He holds that position for the next thirty-two years.

1914: September German rule in New Guinea comes to an end

During World War I (1913–18), Australian troops seize Rabaul, the German administrative headquarters on the north coast at New Britain.

1920 League of Nations grants administration to Australia

Australia retains control of New Guinea according to the terms of a League of Nations mandate.

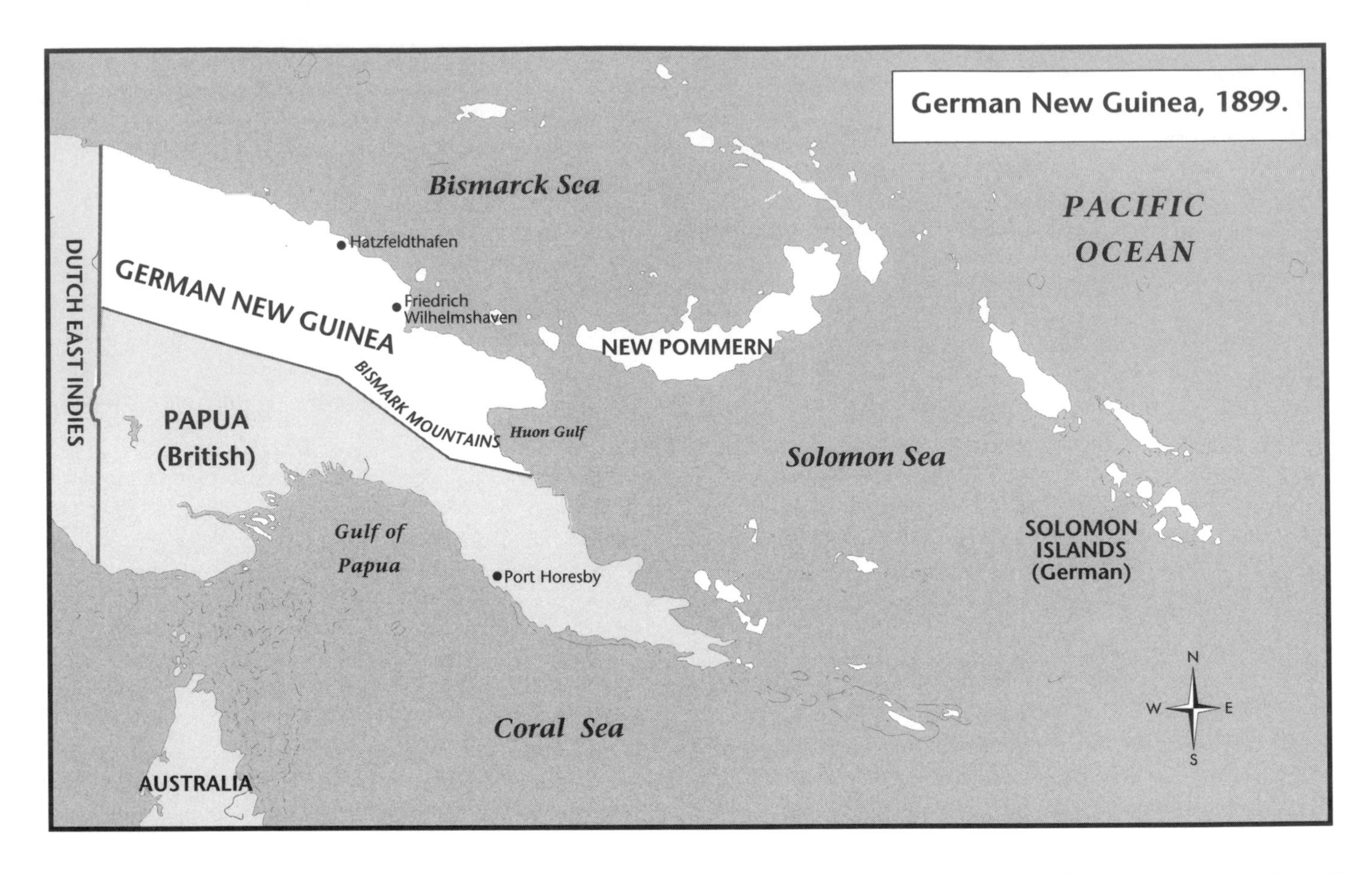

1920 British government assumes administration of former German New Guinea as Territory of New Guinea

The British government decides to administer former German New Guinea separately from neighboring Papua which it already administers. Australia takes charge of the rule of the Territory and continues with the German policy of exploitation. German private property is seized and then sold to Australian ex-servicemen.

1930 Leahy brothers are first whites to see Highlands

Three brothers from Australia make a series of expeditions in search of gold in the Central Range of the Territory of New Guinea. These are the first white people the indigenous Papuans have ever seen. Many believe that the brothers are the ghosts of their ancestors returned from the land of the dead.

1942: January 4 Japan bombs Rabaul

Following successful bombing raids in Malaya and Pearl Harbor, the Japanese bomb Rabaul on the island of New Britain. There is a small Australian garrison in the town that is taken by Japanese forces by January 23

1942: July–August Papua is invaded by Japan

After failing to land a naval force at Port Moresby, the Japanese mount land attacks at the Kokoda Trail and Milne Bay in Papua. These two sites are where the major concentrations of Japanese forces in Papua stay.

1949 Papua and New Guinea Act

Papua and New Guinea are joined together into one administrative unit, and placed under United Nations (UN) trusteeship. Australia is assigned the trust territory administration.

1951 Australia grants New Guinea home rule

As United Nations trustee, Australia is responsible for the internal affairs of Papua New Guinea. The Australian government grants limited home rule to New Guineans.

1960 Australia gives up more control

Australia allows the trust territory of New Guinea to have more extensive control over its own affairs.

1972 The name of the territory is changed to Papua New Guinea

Elections are held that result in the formation of a ministry headed by chief minister Michael Somare.

Michael Thomas Somare (b. 1936) is leader of the newly formed Pangu party: an acronym for Papua New Guinea Unity. Somare, whose training is in journalism, has been a strong advocate for changes in the political structure which legally allows discrimination against Papuans who work in the public service sector. His platform is based on self-governance. Michael Somare is from a village on the Sepik River in the East Sepik Province.

1973: December Papua New Guinea becomes self-governing

Papua New Guinea wins the right to self-government.

1975: September 16 Independence Day for Papua New Guinea

Australia grants Papua New Guinea complete independence.

Bougainville, an island of Papua New Guinea with a distinct culture, wants to assert its own independence when the Australians give up trusteeship. A conflict arises between Bougainville islanders and an Australian group over control of a Bougainville copper mine. Bougainville continues to campaign for independence from the central Papua New Guinea government. They want to be free of the control of a government that is headquartered in Port Moresby 500 miles (800 kilometers) away.

1977 Michael Somare elected Prime Minister

Michael Thomas Somare (see 1972) who had been in control of the independence movement earlier in the decade, now takes on the primary elected position within the newly formed nation. His government lasts only three years before it is ousted by a vote of no confidence in 1980.

1980 In vote of no confidence, Sir Julius Chan replaces Somare

After a vote of no confidence, the Somare cabinet is replaced by one headed by Sir Julius Chan.

1982 Somare and Pangu party win

Michael Somare, leader of the Pangu party, is restored to the position of prime minister following his Pangu party's successful reclamation of seats in Parliament.

1985: November Vote of no confidence

The cabinet of Michael Somare is once again toppled in November through a vote of no confidence. The parliament elects Paias Wingti, who heads a five party coalition, to prime minister.

1988 Wingti's government is toppled by vote of no confidence

After a vote of no confidence, Rabbie Namaliu replaces Wingti as Prime Minister. Namaliu has only recently taken over as leader of the Pangu Party.

1993 Forest depletion

An estimated 3 million trees are harvested and their logs exported each year. Since 1980, log exports have quadrupled. Experts estimate that if logging continues at this rate, wood of commercial value will be gone in ten to twenty years. Since PNG has no sawmills, its laws allow the unprocessed logs to be exported.

1993 Internal Security Act

The Internal Security Act is passed to strengthen anti-crime laws. International legal experts and human rights monitoring agencies observe that the laws may violate internationally accepted human rights. The act allows arrests to be made in certain situations with warrants. The laws may infringe on citizen's freedoms of expression, movement, and association. The increase in street crime in Port Moresby in the 1980s has encouraged the government to step up measures of control. The government considers establishing a national identification system and tightening regulation of the media.

1994 Wildlife Management Area created

One thousand acres are donated by tribal clans to establish a Wildlife Management Area. The area will preserve tropical fauna and flora of Papua New Guinea, and will be jointly managed by the New York Zoological Society, The Wildlife Conservation Society, the Papua Research and Conservation Foundation, the Foundation of the South Pacific, and local landowners. In return for donating their lands, clans will receive assistance with wildlife and agriculture management. A visitor center will attract tourists and revenue to the area.

1994: September Julius Chan is prime minister

Sir Julius Chan is sworn in as prime minister, and immediately tackles the problem posed by the continuing secessionist movement on Bougainville. Three days after taking office, Chan signs an agreement with Sam Kauona, leader of the Bougainville Revolutionary Army. Chan also plans to start production at the idle Panguna copper mine and to seek financing for a project to mine gold.

1994: August 2 Explosives plant accident

An explosion at a plant that produces explosions for gold mining kills eleven. The company that owns the plant, Australian-based Dyno Wesfarmers Ltd. Company, is named in a lawsuit brought by the Papua New Guinea government that cites the company's failure to maintain proper safety standards.

1994 Satellite launching station project planned

Russia launches a joint venture with an Australian firm to build a $1 billion space port in Papua New Guinea for commercial satellite launches. Papua New Guinea is near the equator, and the launch station can capitalize on the velocity of the Earth's rotation, which is highest at the equator.

1995: January   Pope John Paul II beatifies Peter To Rot

Pope Jahn Paul II visits Asia, and beatifies three people from the region. The one layman who is beatified is Peter To Rot (1912–45) of Papua New Guinea. To Rot converted to Christianity and became a martyr at the hands of Japanese troops occupying Papua New Guinea during World War II (1939–45).

1995: June 20   Government attacks Bougainville secessionists

The government of Papua New Guinea announces an attack on rebels on Bougainville island. Just a few weeks later, it is reported that the rebels are still in control. The government of Papua New Guinea has been battling with the Bougainville separatists for eight years with no resolution in sight.

1996   Cocoa bean exports

The cocoa beans used by the chocolate industry are a growing export, along with gold, copper, coffee, and copra (dried coconut meat).

1996: October 12   Assassination of Theodore Miriung, Bougainville

Theodore Miriung, legal advisor to the Bougainville Revolutionary Army, is murdered by two gunmen at his home. In 1995, Miriung had become convinced that Bougainville had achieved sufficient self-rule, and he accepted the post of provincial premier. As such, he led a transitional government for the island. It is believed that rebel secessionists are responsible for the murder which they commit to express their dissatisfaction with Miriung's peaceful cooperation with Papua New Guinea.

1998   Ceasefire declared between rebels and government

The rebel leaders on the island of Bougainville, engaged in a campaign to secede from Papua New Guinea since the country's independence (see 1975: September 16), sign a ceasefire with the PNG government. Almost 8,000 people (of a total population of 156,000) died in violent conflict over Bougainville secession since 1988.

The United Nations Security Council's observer mission in Bougainville, headed by the Australian army, are scheduled to leave Bougainville in March but will remain until at least September to help maintain peace.

1998: July 17   Tidal wave

An earthquake of magnitude 7.1 on the Richter scale rattles the villages of Arop, Warapu, Malol, and Sissano. The quake, which occurs fifteen miles off the shore of the island, creates ocean currents of over twenty miles per hour, and launches a forty-foot-high tidal wave. The tidal wave or tsunami kills almost 3,000 people, the majority of whom are children.

Bibliography

Campbell, I.C. *A History of the Pacific Islands.* Berkeley: University of California Press, 1989.

"Death of a Peacemaker." *The Economist,* October 19, 1996, vol. 341, no. 7988, p. 41.

Donohue, John W. "Of Many Things. (Beatification of Laypersons)," *America,* January 28, 1995, vol. 172, no. 2, p. 2.

Grattan, C. Hartley. *The Southwest Pacific since 1900.* Ann Arbor: University of Michigan Press, 1963.

Sinclair, James. *Papua New Guinea: the First 100 Years.* Bathurst: Robert Brown and Associates, 1985.

Spriggs, Matthew. *The Island Melanesians.* Cambridge, MA: Blackwell Publishers, 1997.

The Philippines

Introduction

As a Pacific nation colonized by Spain and later controlled by the United States, the Philippines has a unique history and cultural heritage. Among its distinctions, it is the only Catholic country in Asia (crowds greeting Pope John Paul II on his 1995 visit were the largest ever assembled during his tenure as pontiff). The resilience of its people has been tested by foreign domination and numerous natural disasters, especially massive earthquakes and volcanic eruptions. A little over a decade ago, authoritarian rule ended when dictator Ferdinand Marcos was ousted in a peaceful confrontation involving massive public demonstrations. Since then the Filipino government has worked to maintain democracy and meet serious economic challenges.

The Philippines is an archipelago of over seven thousand islands off the coast of Southeast Asia, running from Taiwan in the north to Malaysia and Borneo to the south. Its total land area of 115, 831 square miles (300,000 square kilometers) is slightly larger than the state of Arizona. Roughly two-thirds of this area is accounted for by the two largest islands: Mindanao and Luzon, where the capital city of Manila is located. These two islands also define two of the islands' major geographical divisions, with the third being the smaller and more numerous Visayan Islands (or Visayas), which lie between them. Nineteen ninety-eight estimates placed the Philippines' population at 77.7 million, of which over nine million lived in metropolitan Manila, the only large urban population center.

History

Malays, the ancestors of some ninety percent of Filipinos, began migrating to the islands in about 300 B.C. from the mainland of Asia and from Indonesia. Living in agricultural communities called *barangay*, they practiced animistic religions (faiths characterized by the belief that every object has a soul) and formed trade links with regions in East Asia, including China. The European presence on the islands began with the arrival of the explorer Ferdinand Magellan in 1521. Forty years later, the Spanish began colonizing the Philippines (named after their king Philip II). An important part of this effort was the successful conversion of the native populace to Christianity. The Philippines remains the only Catholic country in Asia, with about four-fifths of its population belonging to the Roman Catholic church. However, parts of Mindanao remained predominantly Muslim, and this religious division had political repercussions in the modern era. In the seventeenth century, Spain and the Netherlands competed for control of the region's spice trade, and at the end of the eighteenth century, the British occupied Manila for two years as part of the Seven Years' War between France and Britain.

Throughout its history the Philippines has been decimated by natural disasters, including volcanic eruptions, and earthquakes. In 1766 the Mayon volcano on the island of Luzon underwent its second-worst eruption, lasting four days and sweeping whole villages away in lava flows and mudslides. Mayon and Mount Pinatubo, whose 1991 eruption ravaged the rice crop and left over 100,000 people homeless, are only two of the islands' numerous volcanoes, many of them active. Mayon's worst eruption, in 1814, killed over 2,000 people, and a massive earthquake in 1863 nearly destroyed Manila. Up to 8,000 were killed by an earthquake on the island of Mindanao in 1976 that also triggered a giant tidal wave called a *tsunami*.

The Philippines enjoyed a trade boom in the nineteenth century as Spain relaxed its trade monopoly and the port of Manila was opened to other countries. The profits realized from exports of agricultural products, including sugar, coconuts, and hemp, drew entrepreneurs to the islands. In the latter part of the century, at the same time that a distinctly Philippine Spanish-language literary tradition was evolving, the islands' new self-awareness also found expression in nationalistic strivings toward independence from Spain. The foremost political figure of this period was José Rizal, a physician, author, and statesman who advocated peaceful reform rather than armed revolution. However, when members of the secret society, Katipunan, rose in rebellion in 1896, Rizal was arrested and executed by the Spanish, and his death provided a new rallying point for Philippine nationalists. The rebels continued their struggle, but it was ultimately the Spanish-American War of 1898 that freed the Philippines from Spanish rule, only to deliver them to U.S. control.

In the period of control by the United States (1898–1946), agricultural exports expanded, and the Philippines became a major supplier of sugar, tobacco, and other crops to the U.S. market. As a result large farming estates increased, but development of a modern industrial economy was not encouraged. Over a period of several decades, the U.S. gradually moved toward granting independence to the Philippines, which became a commonwealth of the United States in 1934 and an independent republic in 1946, after a three-year occupation by Japan during World War II.

Democratic government prevailed for over two decades after Philippine independence. Then, on September 21, 1972, Ferdinand Marcos, serving his second term as the country's elected president, declared martial law, dissolved Congress and emprisoned his political opponents. The following year he promulgated a new constitution allowing him to extend his presidency. Martial law ended in 1981, but Marcos continued his authoritarian rule. Ultimately, he was forced out by mass public demonstrations after fraudulently claiming victory in 1986 presidential elections over Corazon Aquino widow of slain opposition leader Benigno Aquino. Aquino became president and Marcos fled the country. In spite of the country's return to democracy, Aquino's presidency was troubled by multiple coup attempts, several natural disasters (including the 1991 eruption of Mount Pinatubo), tensions with the U.S. over renewal of the lease on the Subic Bay naval base, and growing economic problems.

Corazon Aquino declined to run for president when her six-year term in office ended and backed the candidacy of Fidel Ramos, her defense minister, who won the election and became president in 1993. It was the first orderly transition of power from one administration to another in over twenty years. Ramos was succeeded as president by Joseph Estrada in 1998.

Timeline

c. 30,000 B.C. Negritos migrate to the Philippines

The Philippines' earliest inhabitants, a short, dark-skinned people whose descendants are called "Negritos" by the Spanish, migrate to the present-day Philippines from Borneo and Sumatra over land bridges that are submerged during a subsequent ice age. They live as hunter-gatherers.

c. 300 B.C. Malay migration begins

Malays begin migrating to the Philippines from the Asian mainland and the Indonesian islands, forming agricultural communities called *barangay* based on kinship groups each consisting of roughly 30 to 100 families. Land is held in common and each village is ruled by the head of its kinship group. Animistic religions (spirit worship) are practiced under the leadership of medicine men and women, and ancestor worship is widespread. Trade relations are established with states in East Asia, including China.

14th century Islam is introduced

Arab traders bring Islam to the Sulu and Mindanao islands in the southern Philippines.

1380 Mount Pinatubo erupts

Mount Pinatubo, a volcano on Luzon island, erupts, causing great loss of life and damage to the surrounding land. Its next eruption occurs over 500 years later. (See 1991.)

1521: March 15 Magellan arrives in the Philippines

The Portuguese explorer Ferdinand Magellan, commissioned by Spain to circumnavigate the globe, arrives in the Philippines, where he is killed on the island of Mactan by a hostile chieftain named Lapulapu. The remaining members of Magellan's crew continue on to the Moluccas and eventually to Spain, thus succeeding in their goal of sailing around the world, albeit without their captain. The islands are claimed for Spain and later named in honor of King Philip II. By establishing a presence in the Philippines, the Spanish hope to participate in the spice trade in the Moluccas.

Spanish Conquest and Colonization

1565–71 Native population is converted to Christianity

During the late 1560s the Spanish convert almost all of the native populace to Christianity, which assists them in their conquest of the islands. The Spanish government funds conversions which are accomplished by members of religious orders, including the Dominicans, Franciscans, Jesuits, and Augustinians. The Spanish also view the islands as a potential base from which to spread Christianity to Asia and Japan. However, the Philippines remains the only Catholic state in Asia, and even there the inhabitants retain many of the beliefs and practices of their native religions, which they incorporate into their Christian observances. The inhabitants of eastern and southern Mindanao, Palawan, and the Sulu Archipelago retain their Islamic faith and are called Moros (Moors) by the Spanish.

1565 First Spanish settlement is established

An expedition headed by Miguel López de Legazpi founds the first Spanish settlement at Cebu.

1571 Manila becomes the new capitol

Manila is chosen as the capitol of the new colony, which is administered from Mexico.

1600–48 Spain and the Netherlands vie for the spice trade

The Spanish and Dutch navies compete for control of Philippine waters and with them the region's spice trade.

1762–63 British occupation of Manila

Following Spain's entrance into the Seven Years' War on the side of France, the British East India Company captures Manila and occupies it for two years. It is returned to Spain at the peace conference that ends the Seven Years' War. However, Spanish colonial rule weakens as Spain's prestige in the region declines due to the British occupation.

1766: October 23 Catastrophic eruption of the Mayon volcano on Luzon

Mayon, rising 8,000 feet on the island of Luzon, suffers its second-worst eruption to date. A river of lava 100 feet wide rolls down its eastern side for two days, burying or sweeping away dozens of villages. The volcano continues to erupt for four more days, sending mud cascading down its slopes and killing many more people. Altogether some 2,000 are thought to have perished in the disaster. Volcanic activity continues for two months.

1781 Government forms export production society

Governor José Basco y Vargas forms the Economic Society of Friends of the Country to encourage export production of products including hemp, tea, silk, indigo, and poppies for manufacturing opium.

1782 Spanish tobacco monopoly is established

The colonial government realizes a substantial income by creating a monopoly on growing, processing, and exporting tobacco.

1788 Birth of poet Francisco Balagtas

Like other writers of his era, Balagtas (1788–1862), commonly regarded as the Philippines' first major poet, writes narrative poems based on the Spanish chivalric ballad, the *corrido*, in the native Philippine language of Tagalog. His most famous work is *Florante at Laura.*

Nineteenth century Philippines undergoes a trade boom

With Spanish liberalization of commerce, a major trade expansion occurs, making the Philippines profitable for the first time and attracting entrepreneurs to the islands. Export products include coconut, sugar, and hemp.

1814: February 1 Mayon volcano's worst eruption

The most deadly eruption of Mayon, on the island of Luzon, pours burning stones, sand, and ash over the surrounding area, flattening the towns of Cagsauga and Badiao and killing more than 2,200 persons.

1823 First art school is opened

The Academia de Dibujo, the first formal artist training school in the Philippines, is established. Representational art is taught there.

1834 Manila is opened to trade

The Spanish allow the opening of Manila to international trade.

1861 Birth of national hero José Rizal

Rizal, a physician, scientist, and patriot, is born in Laguna Province to a *mestizo* (of mixed Spanish and native descent) family. He studies medicine in the Philippines and Spain and also gains prominence in Europe as a scholar in the natural sciences. He is particularly interested in anthropology for its potential to disprove racist beliefs about native Filipinos and other maligned groups. As a young man, he publishes two novels that portray the harsh conditions of Spanish rule in the Philippines: *Noli Me Tangere* (Touch Me Not) and *El Filibusterismo* (The Reign of Greed). Both become best-sellers in the Philippines although they are officially banned.

In June 1892 Rizal, who is dedicated to peaceful political reform, returns to the Philippines, where he starts the Liga Filipina, a nonviolent nationalist organization. Rather than full independence, he is in favor of making the Philippines a province of Spain with full political representation and equal treatment of Spaniards and Filipinos. However, he is soon arrested by the colonial government and exiled to a remote part of the islands. In 1896, he receives permission to go abroad and work as a doctor in Cuba but sets sail just as the radical nationalist group Katipunan launches an armed rebellion. Rizal's boat is intercepted, and he is unjustly arrested for complicity in the revolt. Rizal is executed by firing squad on December 30, 1896. Shortly before his execution he writes the poem "Ultimo adios" (Last Farewell). His death gives additional impetus to the nationalist movement, and he becomes the national hero of the Philippines.

1862 Death of poet Francisco Balagtas

Balagtas, generally considered the first major Philippine poet, dies. (See 1788.)

1863 Public education is inaugurated

The first public schools are established under the Spanish colonial government.

1863: July 3 Earthquake almost levels Manila

A massive earthquake kills approximately 1,000 people, nearly destroying Manila. All but a few of its historic structures are toppled, including cathedrals and three Dominican convents. Brief eruptions occur in nearby volcanoes. The Pampasinga River overflows, triggering mudslides that crush warehouses containing some $2 million worth of tobacco.

1872: January 20–22 Dockworkers revolt against Spanish officers

About 200 Filipino dockworkers in Cavite Province stage a revolt, killing commanding Spanish officers. The revolt is put down within three days.

1872: February 17 Priests are executed in connection with revolt

The Spanish hang three Filipino priests who have worked for government reform—José Burgos, Mariano Gomez, and Jacinto Zamora.

1884 Philippine painter is denied art prize

Juan Luna y Novicio (1857–99), one of the leading nineteenth-century Philippine painters, is awarded the First Gold Medal at the Madrid Art Exhibition for his painting *Spoliarum*, but the prize is ultimately denied him because of his native Filipino ancestry. However, both Luna and his Filipino colleague, Felix Resurrección Hidalgo (1853–1913), go on to win prestigious prizes in exhibitions in Paris, Rome, Berlin, and Barcelona.

Late 19th–early 20th centuries Development of Spanish-language literature

A Spanish-language literary tradition evolves in the Philippines. Major authors include novelist and national hero, José Rizal (1861–96); lyrical poets, Fernando Guerrero (1873–1929) and Cecilio Apóstal (1877–1938); and satirist, Jesús Balmori (1886–1948).

The Struggle for Independence from Spain

1892 Rizal founds the Liga Filipina

Patriot José Rizal, who supports a moderate reform agenda rather than armed revolution, founds the Liga Filipina, a nonviolent nationalist group, but it languishes after he is exiled to a town in northwestern Mindanao.

1892 Katipunan is formed

Following the arrest and exile of José Rizal, Katipunan, a secret society dedicated to winning Philippines independence, is founded by Andres Bonifacio. In many respects, it is based on a Masonic lodge (Bonifacio, Rizal, and a number of other Philippine nationalists are Masons). Members use secret passwords, wear hoods of different colors, and undergo an elaborate initiation. The group expands rapidly, gaining 30,000 members within four years.

1895 Birth of modernist painter Victorio Edades

Edades is popularly known as the "father of modernism" in Philippine art. After studying at the University of Washington in the United States, Edades shocks the Philippine art world with paintings featuring distortions of perspective and the rough brush strokes modeled on those of Cézanne and Gauguin. Edades's murals of the 1930s, such as *Bountiful Harvest* (1935), are known for their distinctively Filipino character. The School of Fine Arts at the University of Santo Tomás, founded by Edades in 1935, becomes a center for modernist artists to learn their craft. The controversy sparked by Edades and his colleagues between modernists and classicists, or conservatives, persists into the 1950s. Edades dies in 1985.

1896: August 29 Patriots launch revolt against Spain

On learning that the Spanish have discovered the existence of Katipunan, Bonifacio leads an attack on Spanish military installations. The rebellion has very limited success, and the rebels only gain control of Cavite Province. At the time of the revolt, José Rizal is leaving the country to work as an army doctor in Cuba, the only way he can end his detention without being considered an active rebel. His boat is intercepted, and he is arrested and falsely charged with cooperating in the rebellion.

1896: December 30 Rizal is executed

Convicted of conspiring with the Katipunan, physician and patriot José Rizal is executed by a firing squad. (See 1861.)

1897: March Katipunan leadership is divided

Aguinaldo is elected president of the secret nationalist group Katipunan, and its founder, Andres Bonifacio, withdraws and forms his own government.

1897: May 10 Bonifacio is executed by Katipunan leadership

The founder of the nationalist group Katipunan is executed by order of its new leader, Aguinaldo.

1897: December Government and rebels reach agreement

With the war between the colonial government and the Katipunan rebels at a military stalemate, the rebel leader, Aguinaldo, agrees to form a government in exile upon payment of $800,000 from the government, and the group relocates to Hong Kong.

1898: May 11 U.S. sails into Manila Bay

U.S. forces under the command of Admiral George Dewey (1837–1917) destroy the Spanish fleet anchored in Manila Bay.

1898: May 19 Aguinaldo returns to Manila

With the outbreak of the Spanish-American War, Aguinaldo returns from exile and organizes rebel forces to oppose the Spanish in league with the United States, believing that the U.S. will guarantee Filipino independence after the war. The Filipino rebels provide the U.S. with intelligence information, and their presence places additional pressure on an already beleaguered Spanish military.

1898: June 12 Aguinaldo declares independence

Philippine independence is declared by Aguinaldo's rebel government.

1898: August 13 U.S. forces take Manila

United States troops take control of Manila in a fake attack that has been prearranged with the Spanish governor to prevent the losses that would come from an actual battle and to keep him from losing face.

The Philippines under U.S. Control

1898: December 10 U.S. wins control of the Philippines

The Treaty of Paris, which ends the Spanish-American War, transfers control of the Philippines from Spain to the United States in exchange for a U.S. payment to Spain of $20 million.

1899: January 20 U.S. forms first Philippine Commission

U.S. President William McKinley appoints the First Philippine Commission, headed by Dr. Jacob Schurman. The commission recommends eventual independence for the islands but reports that they are not ready for it yet. Formation of a civilian government is recommended.

1899: February U.S. Congress ratifies the Treaty of Paris

The U.S. Congress gives final agreement to the document authorizing U.S. control of the Philippines.

1899: February 4 Philippine-American war begins

Filipino rebel forces inaugurate a new struggle for independence, this time against the United States, waging a guerrilla war in the mountains of northern Luzon. Sustained fighting also occurs in Mindanao and the Visayan Islands. As many as 200,000 Filipino civilians die, some in combat but most of famine and disease indirectly resulting from the war.

c. 1900 Red Cross services begin

Red Cross work is introduced in the Philippines under the leadership of Trinidad Tescon (1848–1928), a former freedom fighter who founds both a hospital and a nursing home.

1900: March 16 Second Philippine Commission is appointed

A second commission, headed by future U.S. president William Howard Taft (1857–1930), is appointed and given the power to issue not only recommendations but also laws governing U.S. policy in the Philippines. The Philippine civil service is reorganized, a judicial system is set up, and a system of municipal boards is established. Separation of church and state and freedom of worship are guaranteed. Taft serves as the first governor general of the islands.

1901: March 23 U.S. captures Aguinaldo

Rebel leader Aguinaldo is captured by forces loyal to the United States. He surrenders, calling on his troops to lay down their arms, and swears allegiance to the U.S. However, rebel activity continues in some areas as late as 1903.

1907 Nacionalista Party is formed

The Nacionalista Party, which will dominate Philippine politics until after World War II, is established.

1907: July First elections to the national assembly are held

Elections are held for the Philippine national assembly.

1908 University of the Philippines is founded

The state-supported University of the Philippines is established in Manila.

1909 Reciprocal free trade established with the U.S.

Reciprocal free trade is established between the Philippines and the United States. As a result, the Philippines concentrate heavily on producing agricultural exports, including sugar, tobacco, and copra (dried coconut meat), for U.S. consumption, while the U.S. provides most of its manufactured goods duty free. Thus industrial development on the islands lags.

1911: January 30 Major eruption of Lake Taal volcano

The volcano, on an island in Lake Taal, experiences its most lethal eruption in modern history. Following a series of earthquakes in the preceding week, there is a double eruption, which blasts away the floor of the volcano, spewing hot steam and white hot ash. Mudslides flatten thirteen villages and towns, and all vegetation and wildlife within a ten-mile radius of the eruption are destroyed. All human life and habitation is destroyed within a distance of six miles. The death toll reaches 1,335.

1915 Birth of composer Eliseo Pajaro

A well-known composer and conductor, Pajaro studies at the University of the Philippines before pursuing studies at the Eastman School of Music in the U.S. under Howard Hanson and others. He receives a doctorate from the University of Rochester in 1953. Two years later he founds the League of Philippine composers. Pajaro serves as a professor at the University of the Philippines and directs the conservatory there in the 1960s. He receives a Guggenheim Fellowship from the United States in 1969. He also receives two Republic Cultural Heritage awards, one in 1964 for his opera *Binhi ng kalayaan* (Seeds of Freedom) and one in 1970 for his ballet *Mi-re-nisa.* Pajaro is known for combining folk themes with advanced compositional techniques in his compositions.

1916 Jones Act promises political independence

The Jones Act, passed by the United States Congress, brings the Philippines closer to independence. The legislature is now under Philippine control, with the Philippine Commission replaced by the popularly elected Philippine Senate as the upper house of the legislature. However, the executive branch is still led by a governor general appointed by the U.S. president.

1917: May 4 Birth of author Nick Joaquin

Joaquin (b. 1917), a leading twentieth-century Filipino authors, is born in Manila. Joaquin writes in many genres, including history, drama, essays, short stories, novels, poetry, and biography. He also works as an editor of the Philippines *Free Press* and writes columns under the pen name, "Quijano de Manila" ("Manila Old-Timer").

Like most of his contemporaries, Joaquin writes in English. His novels include *The Woman Who Had Two Navels* (1961) and *Cave and Shadows* (1983), which is set during the period of martial law imposed by Ferdinand Marcos. One of his most acclaimed works is the play, *A Portrait of the Artist as a Filipino* (1966). Other works include a biography of slain opposition leader Benigno Aquino (*The Aquinos of Tarlac: An Essay on History as Three Generations*; 1983); the short-story collection *Stories for Groovy Kids* (1979); and the poetry collections *The Ballad of the Five Battles* (1981) and *Collected Verse* (1987).

1917: September 11 Birth of Ferdinand Marcos

Ferdinand Edralin Marcos, head of state of the Philippines for 20 years, is born in Sarrat and schooled in Manila, where he studies law at the University of the Philippines. He serves in the Philippine armed forces in World War II. After serving briefly as an assistant to the Philippines' first president, Manuel Roxas, Marcos has a sixteen-year career as a legislator, first in the House of Representatives and later in the Senate. In the 1965 presidential election, he defeats the incumbent, Diosdado Macapagal. Marcos's early years in office are

Because there is little flat terrain, most agriculture is carried out by terracing the hillsides, as seen in these rice fields. (EPD Photos/CSU Archives)

marked by educational reform and agricultural and industrial development, but opposition to his regime grows among students and other groups.

Allegedly in response to threats of Communist subversion, Marcos declares martial law on September 21, 1972, jailing his political opponents and drafting a new constitution permitting him to run for a third term. The Philippines remains under martial law until January 1981, and even afterwards Marcos continues his authoritarian rule. The assassination of opposition leader Benigno Aquino on his return to the country in 1983 fuels popular discontent, and in 1986 Aquino's widow, Corazon Aquino, challenges Marcos in presidential elections. Marcos is declared the winner, but fraud is widely suspected, and elements in the military turn against the dictator. Marcos is finally forced to flee the country on February 25, 1986, taking refuge in Hawaii.

After Marcos is ousted, he and his wife and cronies are found to have diverted billions of dollars in government funds for their own use. The Marcoses are tried in the United States for racketeering. Marcos dies on September 28, 1989, before his 1990 acquittal. However, Imelda Marcos, who returns to the Philippines in 1991, is convicted on corruption charges there in 1993.

1930s Rural poor are hard hit by global depression

Economic development in the Philippines under the United States has encouraged large landholdings with many tenant farmers, causing increased rural poverty. The situation of the rural poor worsens dramatically during the Great Depression, leading to some instances of violence.

1933: January 25 Birth of Corazon Aquino

Aquino, born Maria Corazon Cojuango, is the first female president of the Philippines. She is born to an influential and wealthy family and receives a Catholic education, both in the Philippines and the United States, where she earns her college degree in 1953. After returning to the Philippines, she begins work on a law degree but quits to marry Benigno Aquino, a young politician and scion of a prestigious political family. In

1972, when President Ferdinand Marcos declares martial law, Aquino, as a leading opposition leader, is jailed for eight years. In 1980 Marcos allows him to travel to the United States for heart surgery. After a three-year absence, Aquino returns to the Philippines, where military guards assassinate him when he disembarks from his plane at the airport in August 1983.

After her husband's assassination, Corazon Aquino returns to the Philippines to work for an end to the Marcos regime and is drafted to run against him in 1986. When Marcos is fraudulently declared the winner, popular protest mounts and Marcos is forced to flee the country. Aquino assumes the presidency, releases all political prisoners, and drafts a new constitution. Her years in office are hampered by multiple coup attempts and the effects on the economy of a string of natural disasters (including the 1991 eruption of Mount Pinatubo). Relations with the United States also become frayed when the legislature refuses to extend the U.S. lease on the Subic Bay naval base. Aquino chooses not to run again in 1992 and backs defense minister Fidel Ramos, who takes office in the nation's first peaceful transfer of power in twenty years, as Aquino retires to private life.

1934 Tydings-McDuffie Independence Law is passed

The Tydings-McDuffie law makes the Philippines an independent commonwealth of the United States and provides for complete Philippine independence by 1944.

1935: November 15 New Philippine government takes control

Under a newly adopted constitution, the Philippine commonwealth is established, under the leadership of elected president Manuel Quezon.

1941: December 8 Japanese invade the Philippines

The Japanese invade, driving U.S. personnel and the Quezon government off the islands by 1942.

1942–45 Japanese occupation

Following the battles at Bataan and Corregidor, the Japanese win complete control of the Philippines, which they occupy for the next three years. President Quezon establishes a government-in-exile in the United States, while the Japanese set up a puppet government on the Philippines, led by José P. Laurel. Many Filipinos, however, wage a guerrilla war against the Japanese.

1944: October MacArthur lands at Leyte

Forces under the leadership of U.S. General Douglas MacArthur land on the island of Leyte.

1945: February MacArthur liberates Manila

General Douglas MacArthur liberates Manila following his victory in the Battle of Leyte Gulf, one of the most bitter battles of World War II and the most extensive naval battle ever fought.

1946–54 Wartime Hukbalahap guerrilla rebellion continues

The Hukbalahaps, a Communist-led guerrilla group that fought against the Japanese occupation during World War II, oppose the U.S.-supported postwar government and continue to wage war until their leader, Taruc, is captured.

The Republic of the Philippines

1946: July 4 The Republic of the Philippines is formed

The Republic of the Philippines is established; Manuel A. Roxas y Acuna is elected to be its first president.

1948: February Roxas pardons Japanese collaborators

President Roxas pardons those who held posts in the Japanese-sponsored government of the Philippines during World War II. (Roxas had held such a post himself.)

1948: April President Roxas dies

President Roxas dies of a heart attack and is succeeded by Elpidio Quirino.

1949 Elpidio Quirino is elected president

Quirino, after having succeeded former president Roxas, is elected president in his own right.

1950s–60s Neo-realism dominates painting

Neo-realism becomes the dominant mode of painting, reflecting the post-World War II intellectual climate in its surrealist images and distortions. Prominent figures include Victor Oteyza (b. 1913), Romeo Tabuena (b. 1921), and Arturo Luz (b. 1926). Vicente Manansala (1911–81) is known for introducing Cubist influences.

1953–57 Magsaysay serves as president

Former defense minister Rámon Magsaysay, who has led the successful campaign to suppress the Hukbalahap guerrillas, is elected president as the Nacionalista Party candidate. He becomes one of the dominant political figures of the decade, instituting a program of land reforms. Magsaysay is killed in a plane crash in March 1957.

1954 The Philippines joins SEATO

The Philippines joins the Southeast Asia Treaty Organization (SEATO).

1957: March García succeeds Magsaysay in office

Vice President Carlos García becomes president when Rámon Magsaysay is killed in a plane crash.

1957: November Carlos García is elected president

The presidency of Carlos García is confirmed by election. García encourages the growth of industry by imposing import controls on goods manufactured abroad.

1961 Diosdado Macapagal is elected president

Diosdado Macapagal is elected to the presidency after an anti-corruption campaign. He lifts the import controls imposed by former president García, but economic growth lags.

The Marcos Era

1965 Ferdinand Marcos is elected president

Ferdinand Edralin Marcos is elected president, beginning his long period as leader of the Philippines.

1969 Cultural Center of the Philippines opens

Opened in Manila, the national Cultural Center of the Philippines includes both a theater for the performing arts and art galleries.

1969 Marcos is reelected

Ferdinand Marcos is reelected with a 62 percent majority. Criticism of his administration's corruption and human rights abuses and of his pro-U.S. policies increases.

1969 Guerrilla rebellion is resumed

The Hukbalahap left-wing guerrillas resume their insurgency (see 1946). They are brutally crushed by the Marcos government.

Early 1970s Militant groups oppose Marcos

Several militant rebel groups take up arms against President Ferdinand Marcos. They include the Moro National Liberation Front (MNLF), consisting of Muslims (Moros) on the island of Mindanao, and the Maoist New People's Army (NPA), a Communist-led group begun in Luzon.

1972 Educational reforms are instituted

Under new educational reforms, Pilipino, the native language of the Philippines, becomes an official language of instruction in the public schools (as does Arabic on the Muslim-dominated island of Mindanao). The curriculum is also updated to include vocational courses.

1972: September 21 Marcos imposes martial law

President Ferdinand Marcos declares martial law, allegedly to combat terrorism. Opponents of his regime are jailed, the media are strictly censored, and Congress is dissolved. The United States does not protest the martial law declaration and continues to provides military aid to the Marcos regime.

1973 New constitution is adopted

A new constitution is adopted extending the period of martial law and enabling Marcos, who would otherwise be barred from seeking a third term, to retain the presidency indefinitely.

1975 Martial arts federation is founded

The National Arnis Federation of the Philippines (NARAPHIL) is founded to promote *arnis,* a martial art that is the national sport of the Philippines. Arnis evolved from the ancient native martial art of kali, dating back to the thirteenth century, which survived the era of Spanish colonization by being disguised as a dance in stylized religious dramas.

It combines Western influences, such as fencing and boxing, as well as the Japanese martial arts, into a more modern form, known first as *eskrima* and later as arnis. Like such Asian counterparts as judo and *tae kwan do*, arnis is taught in formal, martial-style classes and divided into levels of difficulty ranked by different belt colors.

1976 Earthquake kills thousands on Mindanao

A earthquake strikes the Philippines, leaving 5,000 people dead, 3,000 missing, and 150,000 homeless. The heaviest damage and loss of life occur on the island of Mindanao. The quake also triggers an eighteen-foot-high *tsunami* (giant tidal wave) which wipes out entire fishing villages. Yet more deaths are caused by fires resulting from the disaster.

1976 Truce reached with Moro guerrilla group

The Philippine government reaches a truce with Mindanao-based Muslim guerrillas who have been conducting an insurgency since the early 1970s.

1977: November Aquino receives death sentence

The government sentences opposition leader Benigno Aquino to death. Marcos later grants a reprieve, and Aquino goes into exile outside the country.

1978 Marcos party wins a majority in suspect legislative elections

Kilusan Bagong Lipunan (New Society Movement), a new party formed by Ferdinand Marcos, wins three-quarters of the country's legislative seats in the first elections held since the imposition of martial law in 1972. The voting is marred by allegations of fraud.

1981: January Martial law is ended

President Marcos ends martial law, but many restrictions on personal liberties are retained, and Marcos still has wide-ranging emergency powers.

1981: June Marcos is re-elected

Ferdinand Marcos is elected to a new six-year term as president. Most of his political opponents boycott the election in protest of his authoritarian tactics.

1983 Government declares moratorium on debt

With debt levels approaching $25 billion, the Philippine government declares a moratorium on its payments.

1983: August 21 Aquino is assassinated

Exiled opposition leader Benigno Aquino is assassinated at the Manila airport when he returns to the Philippines after years of political exile. The blame is placed on government troops escorting him and on the chief of staff of the armed forces, General Fabian Ver. The murder further unites Filipinos in their opposition to the Marcos regime, which is now openly joined by the Roman Catholic Church. Mass protests take place in Manila.

1984 Opposition makes election gains

Opponents of the Marcos regime realize significant electoral gains in spite of electoral fraud on the part of the government and some opposition boycotts of the elections.

1984: September 2 Typhoon Ike kills over 1,000 people

Typhoon Ike, the worst storm to strike the Philippines in the twentieth century, hits the southern part of the islands with winds measuring up to 137 miles per hour. Estimated deaths total 1,363 and over a million people lose their homes. After first spurning foreign emergency aid, President Marcos accepts assistance from the United States and Japan.

1985 Death of modernist painter Victorio Edades

Pioneering modernist painter, Victorio Edades, dies. (See 1895.)

1985: February Aquino assassins are acquitted

General Fabian Ver and the other military personnel accused to murdering opposition leader Benigno Aquino in 1983 are acquitted by a specially appointed court, raising indignation at home and censure abroad.

1986: February Marcos is ousted in aftermath of electoral defeat

In an attempt to silence his critics, Ferdinand Marcos calls a special presidential election, naming a pro-U.S. running mate. He is decisively defeated by opposition candidate Corazon Aquino, widow of slain politician Benigno Aquino. When the national assembly still declares Marcos the winner, thousands of Filipinos take to the streets in massive demonstrations. Even some segments of the military turn against Marcos, and the dictator is finally ousted.

The Post-Marcos Period

1986: February 25 Marcos flees the country

With his family and associates, Ferdinand Marcos takes refuge on Guam and then Hawaii. Corazon Aquino becomes president and installs her own administration, replacing many Marcos-appointed officials and freeing persons imprisoned by Marcos on political grounds. She wins new loan commitments from the International Monetary Fund and foreign nations, including the United States and Japan.

1986: March 25 Interim constitution is adopted

An interim constitution replaces the 1973 constitution enacted under martial law declared by Marcos.

1986: July Military coup is crushed

An attempted coup by supporters of Marcos is suppressed. The military is critical of the Aquino government's tolerant policy toward Communist guerrillas.

1987: January New pro-Marcos coup is squashed

A coup involving some 500 members of the military tries to oust Aquino and bring Marcos back from exile, but the United States prevents him from returning.

1987: February 11 Plebiscite approves new constitution

A nationwide plebiscite (vote by the people) approves a new constitution by a wide majority. The document calls for the closing of U.S. military bases in the Philippines (the Subic Bay naval base and Clark air base).

1987: May 11 Aquino supporters win legislative elections

Over eighty percent of the electorate participates in a legislative contest that is the first free election in the Philippines in nearly twenty years. Political supporters of Corazon Aquino win by a wide margin.

1987: August 28 Military coup is suppressed

One of seven military coups to occur during Corazon Aquino's presidency is put down by loyal members of the military.

1987: December 20 Maritime collision kills 1,500

A passenger ship and an oil tanker collide off Mindoro Island, killing at least 1,500 people in one of the worst sea disasters in history.

1988 Vice President calls for Aquino's resignation

Corazon Aquino's vice president, Salvador Laurel, calls for her to resign.

1988: October Marcos is indicted in U.S.

Former ruler Ferdinand Marcos and his wife, Imelda, are indicted for illegally transferring $100 million out of the Philippines for their own use.

1989 U.S. military bases are closed

The Aquino administration order U.S. military bases in the Philippines closed.

1989: September 26 Ferdinand Marcos dies

Former Philippine president Ferdinand Marcos dies in Hawaii. The Aquino government denies his family permission to bury him in the Philippines. (See 1917.)

1989: December Most serious coup attempt against Aquino

The most serious of the seven military coups against Aquino during her term in office takes place. Over 3,000 troops take part, led by Colonel Gregorio Honasan, and U.S. military aid is required to suppress the rebels. Aquino is granted special emergency powers for six months by the Senate. Her government loses credibility abroad for its inability to deal with the coup without foreign intervention.

1990: May Land reform implementation draws protests

Protesters demonstrate for faster implementation of agrarian reforms authorized two years earlier.

1990: June Peace Corps personnel leave the Philippines

Following several incidences of anti-American violence attributed to Communist guerrillas, 261 U.S. Peace Corps volunteers depart from the Philippines.

1990: July Over 1,600 are killed in earthquake

An earthquake registering 7.7 on the Richter scale strikes the Philippines, killing more than 1,600 people. Baguio City, on the island of Luzon, suffers heavy damage.

1990: September Benigno Aquino's assassins are convicted and sentenced

Sixteen officers found guilty of assassinating political opposition leader Benigno Aquino at the Manila airport in 1983 receive life sentences.

1991 Philippine martial art is featured in the Southeast Asian Games

The native martial art of arnis (see 1975) is played as a demonstration sport at the Southeast Asian Games (SEA Games).

1991: June 10–15 Eruption of Mount Pinatubo

Located in Zambales province on the island of Luzon, Mount Pinatubo has been dormant for 500 years, during which time hundreds of thousands have settled, turning its slopes into lush farmland. The appearance of smoke, steam and ash beginning in April signal the beginnings of activity within the volcano, and the U.S. evacuates most of its military personnel from nearby Clark Air Force Base.

A moderate eruption on June 10 is followed by an explosive burst of ash and smoke on the morning of June 12, creating a mushroom cloud 15 miles high. Ash falls as far away as Singapore. Further eruptions occur on June 13 and 14 and residents begin to evacuate. The disaster is compounded then a typhoon strikes the area on June 15, turning fallen ash into mud, under which the roof of a bus station in the town of Angeles collapses, killing dozens of people.

Altogether, some 200 people die in the disaster, and 100,000 lose their homes. The area's rice crop is completely ruined, devastating its economy, and the once-green landscape is turned into a barren wasteland.

1991: September 16 Philippine Senate order U.S. military bases closed

Over the opposition of President Aquino, the Senate rejects the Philippine-American Cooperation Talks agreement (PACT) allowing U.S. military bases in the Philippines to remain open. The bases are declared a violation of Philippine sovereignty, and their leases are not renewed.

1991: November Imelda Marcos returns to face charges

Imelda Marcos, wife of former president Ferdinand Marcos, returns to the Philippines to face civil and criminal charges.

1991: November Typhoon hits the Visayas

The central Visaya islands suffer a massive typhoon, generating floods and landslides. Altogether, over 5,000 are killed.

1992: January 6 U.S. receives official notice to close Subic Bay station

The United States receives notice that it is to vacate Subic Bay Naval Station after nearly 100 years of military presence in the Philippines.

1992: May 11 Fidel Ramos is elected president

With the backing of Corazon Aquino, who decides not to run again, Fidel V. Ramos is elected president. Joseph E. Estrada is elected vice president. Ramos, a leader of the 1986 military opposition that ousted Ferdinand Marcos, is the first military man and first Protestant to be elected president of the Philippines. He pursues a conservative economic policy of privatization and deregulation that leads to strong growth during his first years in office.

1992: September 30 U.S. evacuates naval base

The United States turns the naval base at Subic Bay over to the Philippine government.

1993: September Marcos's body is returned to the Philippines

The embalmed remains of former president Ferdinand Marcos are returned to the Philippines and buried near his birthplace in northern Luzon.

1993: September 24 Imelda Marcos convicted on corruption charges

Imelda Marcos is convicted under the government's laws regarding graft and corruption, with a prison sentence of up to twenty-four years.

1994: January Cease-fire signed with Muslim guerrillas

The Ramos government signs a cease-fire agreement with the Moro National Liberation Front, a Muslim guerrilla group based in Mindanao, bringing to an end two decades of hostilities.

1994: June China protests Spratly Islands oil permit

The People's Republic of China protests Philippine's granting of an oil exploration license for the jointly claimed Spratly Islands to the U.S. firm Vaalco Energy.

1995 Weather-related rice crisis creates economic setback

The rice harvest is decimated by typhoons, leading to a rise in inflation and sparking public protests.

1995: January 6 Authorities uncover plot before papal visit

Philippine police foil a plot by Islamic extremists to assassinate Pope John Paul II on his approaching visit to the Philippines. Two Arabs are found with bomb-making materials and arrested.

1995: January 11–15 Pope John Paul II visits the Philippines

Pope John Paul II visits the Philippines as the first stop on an eleven-day tour of Asia. The papal celebration of mass draws a crowd estimated at up to four million people, topping even the turnout when the pope visited his native Krakow, Poland, in 1979. The climax of the visit is the World Youth Day celebration on January 15.

1995: February China occupies a portion of the Spratly Islands

The Chinese construction of shelters on *Panganiban* (Mischief) reef creates tension with the Philippines, which also claims the islands.

1995: March 17 Philippine domestic worker executed in Singapore

The execution of Flor Contemplacion, a Filipina working as a maid in Singapore, on charges of murder draws massive public protests in the Philippines. Witnesses claim that Contemplacion's confession to having killed another Filipina and a four-year-old Singaporean boy was coerced by Singaporean police and that the boy's father killed the maid and framed Contemplacion.

A stay of execution requested by President Ramos is rejected. In the week following the execution, Manila is the site of the largest public demonstrations to take place in the country since the ouster of Ferdinand Marcos in 1986.

1995: March 22 Ramos recalls ambassador to Singapore

The Philippine ambassador to Singapore is recalled in a diplomatic gesture protesting the execution of Flor Contemplacion (see March 17). He makes it illegal for women from the Philippines to continue working in Singapore and says that a pending inquiry into the Flor case may result in the Philippines severing diplomatic ties with Singapore. It is feared that the matter may have serious economic consequences, as some 75,000 Philippine women work as domestics in Singapore, and the money they send home is an important source of income. It is also feared that Singaporeans may withdraw large investments in the Philippines.

1995: May Government coalition is strong in midterm elections

The coalition of President Fidel Ramos has a strong showing in the midterms elections, strengthening their leader's political standing.

1995: August Agreement is reached with China over Spratly Islands

The Philippines and the People's Republic of China agree that neither party will occupy the disputed Spratly Islands and that they will resolve their conflicting claims to the islands peacefully.

1995: November 2–3 Powerful typhoon decimates Luzon

The Philippines are rocked by a "super typhoon" measuring 500 miles (800 kilometers) in diameter, with wind up to 155 miles (250 kilometers) per hour. It is the worst storm to affect the country in over a decade. Hardest hit is the main island of Luzon, location of the capital city of Manila. Altogether, 600 people die in the storm, and damage to crops and infrastructure is estimated at $77 million.

1996 Normal relations with Singapore are resumed

After the 1995 diplomatic crisis between the Philippines and Singapore over the death of Flor Contemplacion (see 1995), normal relations resume between the two nations.

1996: March 19 Disco fire kills 150 in suburb of Manila

A late-night fire in a nightclub outside Manila claims more lives than any other fire in the nation's history. The deaths are blamed on corruption in the building inspection system, which allowed the club's fire exit to be blocked by new construction on an adjacent property. In addition, the legal occupancy limit of the disco was only 35.

1996: April Terrorism by Muslim rebels continues

Despite a government cease-fire (see 1994), attacks by Muslim guerrillas continue. Fifty-seven people are killed in the town of Ipil and fire damages the town's commercial district. The rebels and the government eventually reach an agreement providing for limited autonomy in southern Mindanao.

1998: May 11 Estrada is elected president

Former film actor and current vice president Joseph Estrada is elected president, beating his nearest competitor, House Speaker José de Venecia, by a margin of 39.9 percent to 15.9 percent of the vote. About 27 million Filipinos, representing roughly 80 percent of eligible voters, participate in the election. Campaign-related violence results in fewer than thirty deaths, compared with eighty-three deaths in the 1995 midterm elections. Estrada promises to fight crime and corruption while working to relieve poverty through agricultural programs.

Bibliography

Abinales, Patricio N. *The Revolution Falters: The Left in Philippines Politics After 1986.* Ithaca: Cornell University, 1996.

Brands, H.W. *Bound to Empire: The United States and the Philippines.* New York: Oxford Univ. Press, 1992.

Bresnan, John, ed. *Crisis in the Philippines: The Marcos Era and Beyond.* Princeton, N.J.: Princeton Univ. Press, 1986.

Broad, Robin. *Plundering Paradise: The Struggle for the Environment in the Philippines.* Berkeley: University of California Press, 1993.

Casper, Gretchen. *Fragile Democracies: The Legacies of Authoritarian Rule.* Pittsburgh: University of Pittsburgh Press, 1995.

Davis, Leonard. *The Philippines: People, Poverty, and Politics.* New York: St. Martin's Press, 1987.

Dolan, Ronald E., ed. *Philippines: A Country Study.* 4th ed. Washington, D.C.: Library of Congress, 1993.

Karnow, Stanley. *In Our Image: America's Empire in the Philippines.* New York: Random House, 1989.

Kirk, Donald. *Looted : The Philippines After the Bases.* 1st ed. New York: St. Martin's Press, 1998.

Reid, Robert H.. *Corazon Aquino and the Brushfire Revolution.* Baton Rouge: Louisiana State University Press, 1995.

Steinberg, David Joel. *The Philippines, a Singular and a Plural Place.* 3rd ed. Boulder, Colo.: Westview, 1994.

Thompson, W. Scott. *The Philippines in Crisis: Development and Security in the Aquino Era 1986–92.* New York: St. Martin's 1992.

Qatar

Introduction

Qatar is located in the Middle East on a peninsula strategically located in the Persian Gulf and the Gulf of Bahrain, between latitude 25° 30' north and longitude 51° 15' east. The island of Bahrain is to the west. Sa'udi Arabia borders Qatar on the south. The United Arab Emirates (U.A.E.) is to the southwest. Directly across the Persian Gulf is Iran. Other countries bordering on the northern end of the Persian Gulf are Iraq and Kuwait. The Persian Gulf after a passage at the Straits of Hormuz opens into the Gulf of Oman, the access to the Indian Ocean.

Qatar is about the size of Connecticut and Rhode Island combined with an area of 4,427 square miles (11,437 square kilometers), and with about 350 miles (563 kilometers) of coastline. Its capital is Doha. Islam is the state religion and ninety-five percent of the population are Muslim. The official language is Arabic. In 1998 population estimates were 697,126. Qatar's major ethnic groups in 1998 were: Arab-forty percent, Pakistani-eighteen percent, Indian-eighteen percent, Iranian-ten percent, and other fourteen percent. Qataris have one of the highest standards of living in the world.

Qatar's terrain is mainly a flat and barren desert covered with loose sand and gravel. Its highest point, Jebel Dukhan, is only 250 feet above sea level. Only one percent of the land is arable. In 1993 irrigated land was eighty square kilometers. The climate is hot, up to 104°F in summer and very humid along the coast, about ninety percent. Annual rainfall is less than two inches. Its limited freshwater resources are overcome by desalination plants. The *shamal* is an unpleasant hot wind that blows from the north. Qatar contains the third largest natural gas reserves and the largest non-associated gas field in the world.

Society

The basis of social structure was the tribe. Two families were prominent in Qatar, the Al-Khalifah of the Utub tribe and the Al-Thani of the Tamim tribe. Within this tribal structure ruling families were in power for long periods. Ties with the ruling family and personal connections were the most significant allegiances. Male dominance was the rule at all age levels. Marriages were arranged by parents or grandparents.

Nomadic peoples, the Bedouin, or *badu* migrated from Arabia. They settled in the northern and eastern Qatar, pearling by summer and herding their camels, sheep, donkeys and goats in winter. Bedouin life was more traditional in southern Qatar. Bedouin life revolved around the camel as a source of milk and meat, and beast of burden.

Qatar's government is a traditional monarchy. There is neither a legislature, nor political parties. There is no suffrage at the national level. Sheikh Hamad bin Khalif Al-Thani proposes to extend the right to vote in municipal elections to women. *Shar'ia*, Islamic Law is the basis of the legal system. Traditionally women were veiled. The veil was an urban custom when women mingled with strangers. In the tribal community women did not necessarily cover their faces. The *batula*, a face mask that covers the whole face with two slits for eyes, was the traditional veil for Qatari women. Contemporary trends are to abandon the use of the *batula*, but to wear a head covering, a scarf, *hijab* or to wear the *'abaya*, as a shawl.

Traditional education involves teaching boys to recite the Holy Qur'an (Koran). Education is not compulsory. Schooling from first grade to university level is free. Medical care is free. Housing and furniture is provided free to poorer Qatari citizens.

History

The earliest records of human habitation in present-day Qatar date from c. 6000 B.C. However, the record of human settlement in the region is quite sketchy until the sixth century B.C. when the area came under the control of the Persian Empire. Following this state's conquest by Alexander the Great in the fourth century, present-day Qatar became linked to the rest of the Persian Gulf through trade. During this period, the region was known primarily for its shipbuilders, fishermen, and nomadic shepherds in constant search of water and grazing areas.

The most important cultural event in the history of Qatar came during the seventh century A.D. with the rise of Islam.

A monotheistic faith meaning "submission to God", Islam spread throughout the Arabian peninsula in the 620s and early 630s. Following the death of Mohammed, the faith's founder, Arabs spread the religion throughout the Middle East, North Africa, central Asia, and parts of southern Europe. As a result of the foundation of an Islamic empire, present-day Qatar became a part of one of the world's major states.

After the decline of the first Islamic empire, however, much of the Arabian coastline along the Persian Gulf faded into obscurity once more. Little is known about present-day Qatar until the arrival of Portuguese traders in the sixteenth century. The Portuguese, interested in securing trade routes to their Indian colony of Goa, establish trading posts along the Arabian peninsula, including the Persian Gulf coast. Subsequently Britain and Persia expelled the Portuguese; by the eighteenth century Britain had become the chief factor in Qatari foreign affairs, a position it would retain almost without interruption until 1971.

Collectively in the nineteenth century the region on the eastern part of the Arabian Peninsula and the Arabian Gulf was known by many descriptive names: Trucial Coast, Trucial Sheikhdoms, Pirate Coast, Coast of Oman and Trucial States. "Trucial" refers to the nineteenth century treaties between separate sheikhdoms and Oman, Bahrain, and Qatar with Great Britain. British rule ended—temporarily—in 1872 when Qatar came under the control of the Ottoman Empire. Not until after the outbreak of the First World War (1914–18) did Ottoman forces withdraw and British forces return. In 1916, Britain reached an agreement with the Thani family that recognized their authority over Qatar.

The discovery of oil in the 1930s vaulted Qatar into the modern age. Commercial production began in 1947 and soon thereafter the Qatari government began using revenues from its natural resources (deposits of natural gas were also discovered) on modernization programs. The first boys' school was founded in 1952, the first girls' school opened in 1956, and the first university was established in 1973. As a result of Britain's decision to withdraw all of its military forces east of Suez, the State of Qatar became independent on September 3, 1971. Qatar along with Bahrain (which also achieved independence) were invited to join the proposed United Arab Emirates (U.A.E.); both declined.

Since independence, Qatar has used its position as a leading oil supplier to become a leading player in international affairs. In 1973 Qatar joined the Arab oil embargo against the United States to protest American support for Israel in the Yom Kippur War against its Arab neighbors. In 1981, Qatar was a founding member of the Cooperation Council for the Arab States of the Gulf, a security grouping of the Arab Persian Gulf states.

Domestic stability in Qatar was shaken briefly in 1995–96 following a coup by Crown Prince Sheikh Hamad bin Khalifah Al-Thani against his father Emir Sheikh Khalifah bin Hamad Al-Thani while the latter was in Europe. Although the ousted emir initially formed a government in exile in the U.A.E., he ultimately reached a settlement with his son whereby the father renounced the throne and was allowed to return home. Since then, the Qatari government has occupied itself with continuing economic problems stemming from the worldwide drop in oil prices.

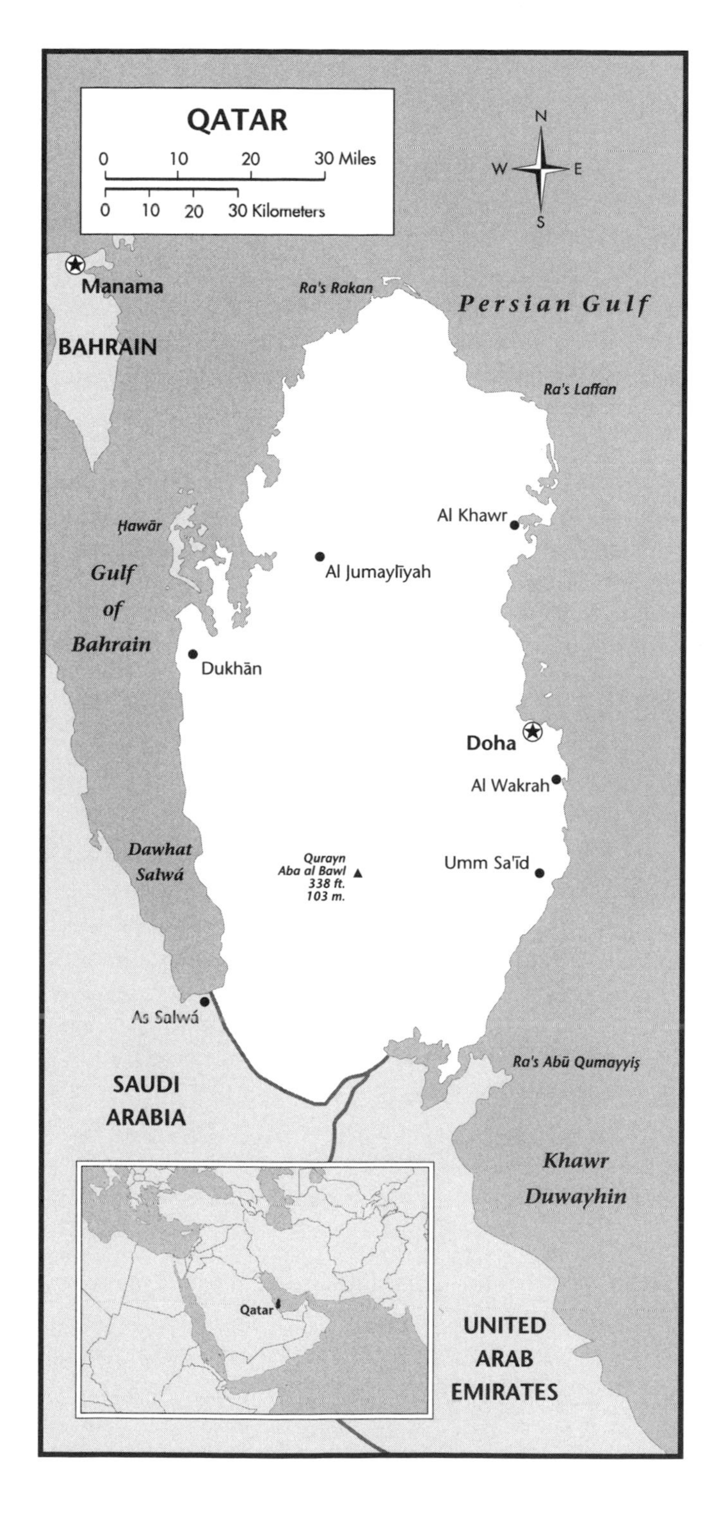

Timeline

Prehistoric Qatar

6th millennium B.C. **Early habitation**

An archaeological site at Shagra in southeast Qatar is designated as having the oldest structure in this area of the Arabian Gulf. Flints, arrowheads, remains of fish and marine mollusks, but no pottery is found in relation to a two-room structure constructed from small slabs of sandstone.

5th–4th millennium B.C. **Ubaid Period**

Al Ubaid pottery originating near Ur at the confluence of the Tigris and Euphrates Rivers is associated with the origins of agriculture. Pottery from this cultural period is found at Qatari sites in association with stone tools and domestic implements. These various sites indicate activities tied to fish-curing and cultivation or collection of grains.

3rd and 2nd millennia B.C. **Bronze Age and trade**

Evidence based on fragments of Barbar pottery suggest that Qatar is the site of seasonal migrations during this period.

2nd millennium B.C. **Kassite Period**

Kassites from the Zagros Mountains gain power with the decline of the Hittites. The Qatari site attributable to this period is on a small island in the bay of Khor. A shell midden (refuse heap) composed mainly of a single species of sea shell known for its purple die is found in association with Kassite pottery. Purple-dyed cloth is prized by the Kassites and later groups.

Historic Qatar

550 B.C. **Cyrus the Great founds the Persian Empire, the Achaemenid**

The Persian Empire unites western Asia and Egypt facilitating Persian Gulf trade as far as India.

5th century B.C. **A Greek historian associates the Canaanites with Qatar**

Herodotus attributes the original inhabitation of Qatar to the seafaring Canaanites.

331 B.C.: October 1 **Macedonian king, Alexander the Great, defeats King Darius III of Persia at Gaugamela**

Alexander the Great routes the Persians at Gaugamela and enters Babylon continuing on to Persepolis, the capital of Persia. As Darius flees again, he is killed by his officers. Alexander becomes King of Persia. Alexander continues into India and returns to Babylon in 325 B. C. He marries a Persian princess. In spite of his untimely death at the age of thirty-three his empire links Greek and Asian cultures and opens the Persian Empire to foreign traders. However, this trade remains under local control in the Persian Gulf.

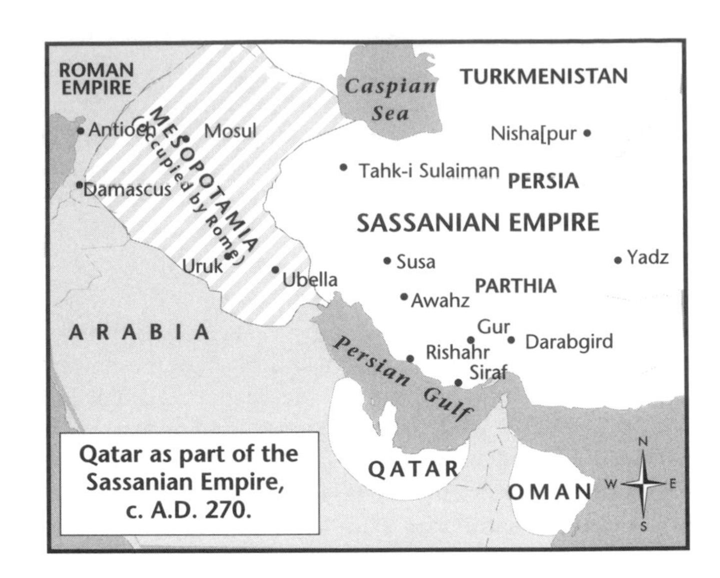

Qatar as part of the Sassanian Empire, c. A.D. 270.

300 B.C. **The region is part of a regional trade network**

Trade links the whole Arabian peninsula with the Mediterranean region and the Indian Ocean and the Indus civilization. Goods, such as, frankincense, amphorae (two-handled, narrow necked jars used by the Greeks and Romans), and alabaster containers, along with minted coins evidence the expanse and antiquity of this maritime trade.

A.D. 77 **Pliny the Elder writes of the rigors of nomadic existence in Qatar's interior**

Pliny's *Natural History* mentions the Arabian shore of the Gulf. He describes Persian Gulf shipbuilding techniques. Nomadic herders are in constant search for water and grazing for the herds in the barren interior. Coastal areas support trade in pearls and dried fish. Other classic writers, Ptolemy and Strabo, describe military and commercial use of the Persian Gulf by peoples of the area.

224–642 **The Sassanian Period; Persian seamen control; maritime trade**

A local dynasty, the Sassanid, triumphs in battle over the last Parthian ruler. A period of prosperity and governmental power ensues under over forty Sassanid rulers. The Sassanids extend their rule to eastern and southern Arabia and dominate trade. Struggles with Rome and Byzantium, and finally the Arabs end Sassanid reign as Islam is introduced.

630 Spread of Islam to Qatar

Qatar and environs are under the rule of the Al Mundhir Arabs. Their leader Al-Mundhir Ibn Sawi Al Tamimi converts to Islam.

632 The Prophet Mohammed dies

The Prophet Mohammed dies in Mecca.

8th–9th centuries Arab seafaring and trading are at their height

Arab Muslim rule unites both sides of the Persian Gulf. Arab and Persian merchants voyage to India and as far as China.

10th century The Ismaili tradition dominates the Gulf region

The Ismaili Shia faith is a powerful force. The Ismaili originate in Iraq but move to the gulf to escape Sunni authorities. In the ninth century a missionary, Hamdan Qarmat, sends a group from Iraq to Bahrain to start an Ismaili community. These followers become known as Qarmatians. In the tenth century these Ismailis of Bahrain control the coast of Oman collecting tribute from Baghdad and Cairo. Persian and Arab forces drive the Ismailis from Iraq and eastern Arabia by the end of the tenth century.

Modern History of Qatar

1498 Portuguese explorer Vasco Da Gama rounds the Cape of Good Hope

Da Gama opens maritime trade routes to India.

Sixteenth century The Portuguese dominate the Persian Gulf region

By 1521 Bahrain is a Portuguese possession as the Portuguese gain control of the silk and spice route to India. Portugal's expansion is held in check by the Ottomans and the Persians. In 1602 Portugal's control of Bahrain is lost to the Persians.

1507 Portuguese capture Muscat and then Affonso DeAlbuquerque captures Hormuz

In the 16th century the Portuguese dominate trade in the Gulf and the Indian Ocean. Control of Muscat (modern Oman), Hormuz an island off modern Iran, and the Indian coast allows the Portuguese to control trade routes and customs and duties. The Portuguese establish Goa as their capital in India.

Seventeenth century The Dutch, French and English vie for control of the region

The Dutch, French, and English East India Companies are rivals for trade in the Arabian Gulf. Tribes from Oman expel the Portuguese from Muscat, but fight the Persians into the next century as they take Muscat back. The founder of the dynasty ruling Oman, Imam Sultan bin Saif, finally expels the Persians from Oman. This dynasty rules Oman into the end of the twentieth century. The Persians remain in the region in Bahrain. Kuwait is settled by Utbi Arabs and governed by a branch of the Utub tribe, the al Sabah. It is the Utub who later migrate from Kuwait to Qatar.

1600 English East India Company forms

The English East India Company is a private trading company with land and naval forces to protect its economic interests.

1602 Dutch East India Company is established

The Dutch East India Company, a colonial trading organization, is established

1622 English and Persian forces drive the Portuguese from Hormuz

Portugal is driven from Hormuz by combined English and Persian forces.

1646 and 1659 Commercial treaties are signed between Oman's Imams and the British East India Company

The Imams (Islamic religious leaders) of Oman sign commercial treaties with the British East India Company.

1650 Imam Sultan bin Saif expels the Portuguese from Muscat

Second Omani ruler of the Ya'ruba dynasty, Imam Sultan bin Saif, drives the Portuguese out and works toward uniting Oman.

1742 The Persians under Nadir Shah establish control of Muscat

Within two years the Omani drive out the Persians.

1747 Ahmed bin Said drives the Persians from Oman and is elected Imam

After Ahmed bin Said drives Persian forces from Oman, he is elected Imam.

1747 Persian leader Nadir Shah dies and the Qawasim move into Persia

The Portuguese and Persians have been driven from Muscat. Following the death of Persian leader Nadir Shah the Qawasim of Ras al Khaimah seize the city of Lingah almost directly opposite on the Persian shore. The Qawasim now rival the imams (a leader considered a lawful successor of Mohammed by the Islamic community) of Omani and Muscat for control of the Gulf trade. The Qawasim (singular, Qasimi) are a migratory group of central Arabia called Hawala. In the mid-seventh century they settle in Persia, eventually dwelling on both shores of the Gulf. The Qawasim force the Omani to recognize Qasimi independence. During this time period Oman is encompassed in the territory now considered the U.A.E..

Eighteenth century The Al-Thani family comes to Qatar

The Al-Thani family arrives in Qatar in the early eighteenth century. They are a branch of the Arab tribe Tamim from the eastern parts of the Arabian peninsula.

1766 The Utub tribe migrates to Qatar from Kuwait

The Utub settle in the town of Zubara in northwest Qatar. The Al-Khalifas, the chief Utub family, formerly sailors and traders established Zubara as a successful port. The Al-Jalahima family arrives next. Unable to live together peacefully, the Al-Jalahima move on to Ruwais and then to Khor Hassan.

1783 The Persians and the Utubs are at war

The Persians are in control of Bahrain. The rise of Zubara is a threat to the Persians. The Persians realize that Oman would side with the Utub against them. Continuing hostilities with the Persians result in an Utub attack in Bahrain, and the Persians retaliate against Zubara. The Utub attack again and capture Bahrain. Upon winning the war with the Persians the Utub tribe settles in Bahrain. The dominant family is the Al-Khalifa who still govern Bahrain into the late twentieth century. Disputes break out between the Al-Khalifa and Al- Jalahima families leading to the return of the Al-Jalahima family to Qatar in Khor Hassan.

1795 Qatar is conquered by Sa'udi Arabia and the Wahhabis

The name *Wahhabi* is derived from the name of the founder of the movement, Muhammad ibn Abd al Wahhab, who died in 1792. Growing up in central Arabia he studies at the strictest of Islamic schools and considers himself a reformer. He wants to establish a new state based on Islamic principles. Joining with a local emir, Muhammad ibn Saud, they pledge to work together. Present-day Sa'udi Arabia results from their efforts.

Wahhabi thought is popular with the Al-Thani family. When the Al-Khalifa family attempts to take over Qatari territory before moving on to Bahrain the Al-Thani resist Al-Khalifa attempts because the Al-Khalifa reject Wahhabism. As a result Wahhabi beliefs are associated with Qatar, and not Bahrain.

1809 The Sultan of Oman loses Oman

This war succeeds under the combined forces of the Islamic sect, Wahabis, and the powerful maritime leader Rahmah ibn Jabir of the al Jalahima family. Zubara is Rahmah's reward.

1811 The Turks attack and Sa'udi Arabia is under Turkish rule.

The Sultan of Oman attacks Qatar in order to get Bahrain back. The Al-Khalifa defeat Rahmah ibn Jabir. They return to power in Bahrain and parts of Qatar.

1820 The first General Treaty of Peace is made between Britain and the coast of Oman (Trucial Sates)

The British campaign to quash the Qawasim, a maritime people of the Gulf, is ostensibly to safeguard the maritime routes of the British East India Company. However, as the British control the Persian Gulf, they are also establishing their supremacy in the region. Their increased involvement acts against the claims of other European powers.

1826 Ramah ibn Jabir dies

This fierce maritime warrior and foe of the Utubs and the Al-Khalifa dies by his own hand rather than surrender. Caught in a loosing battle with an Utub fleet he blows up his own ship with himself and his eight year old son aboard.

1830s The pearl trade is at its height

Special boats called *sambuqs* are used for pearling. They have both sails and oars. The upper hull is oiled with shark oil to prevent warping. The lower half of the hull is covered with a mixture of lime and sheep's fat. Pearling is dangerous work amongst sharks, barracuda, and swordfish. To prevent stings the divers' wear white cotton suits as protection against giant stingrays and jellyfish. Leather stalls are used to protect fingers and the big toe of the right foot used to push along the seabed is also leather covered. A bone or wooden clip over the nose, a knife to pry the oysters are the diver's tools. Divers can stay down for as long as two minutes and may make between 60–100 dives per day.

1832 First Maritime Truce

The British Resident in the Gulf, Captain Hennel, frames a truce outlawing warfare between May 21 and November 21 each year. This is the time of pearl-diving. The truce is so popular that it is renewed annually until 1843 when it is extended for another ten years. In 1853 the truce is made permanent. The regional ascription, Trucial Coast, designates the areas covered by the truces.

1853 The Perpetual Peace between Great Britain and the Trucial States is signed

The Trucial States agree to cease hostilities at sea and in return Britain safeguards navigation. The British guarantee to protect the Trucial States from external attack. This peace allows the sheikhs to concentrate on internal affairs and attend to border disputes.

1850–1878 Rule of Sheikh Mohammed Bin Thani

Settling in Doha Sheikh Mohammed aligns himself with the Islamic Caliphate, Midhat Pasha, the Turkish wali of Iraq. In an era of British domination of the region Sheikh Mohammed allows an Ottoman military force in Qatar. Mohammed Bin Thani is first of the hereditary rulers Qatar.

1868: September 12 Treaty of non-aggression between Bahrain and Qatar

Qasim bin Mohammed Al-Thani travels to Bahrain on Qatari official business and is detained. This controversy incites attacks on Doha and Wakra by Bahrain and Abu Dhabi. In retaliation the Qataris attack Bahrain. Colonel Lewis Pelly, British representative in the region, negotiates the Anglo-Qatari Treaty. Sheikh Muhammad is replaced by his brother Ali. A non-aggression treaty between Bahrain and Qatar compensating Qatar and ending the battle between Bahrain and Qatar by destroying the remaining fleets is implemented. The Al-Thani family remains in power.

1871: July Ottoman presence in Qatar

Turks occupy part of the eastern shores of Qatar. The Turkish garrison is supposedly in response to a request for protection by Qataris from the Bedouin forces harassing the Turkish forces in Arabia from the Qatar peninsula. Sheikh Muhammad does not recognize the occupation, but his son, Qasim, goes against his father's wishes. The British do not recognize Turkish rights to Qatar. The Turks remain in Qatar for forty years.

1872 Ottoman Turks occupy the Arabian peninsula

Ottoman Turks occupy Qatar and other countries until the World War I (1913–18).

1878 Turks appoint Qasim Governor of Qatar

He attacks Zubarah and routes the Bahrainis from the northwest coast.

1881 Sheikh Qasim expels British subjects and Indian traders from Doha

As part of a reaction to foreign influence, Sheik Qasim expels British subjects and Indian traders from Doha.

1888 Shipping attacks continue in the region

Already strained relations between Qatar and the British and the British and Turks are further strained by retaliatory attacks between the sheiks of Abu Dhabi and Doha. Qasim's son is killed.

1892 The Exclusive Agreement or Exclusive Treaties are signed

Great Britain signs separate one-sided treaties with each sheikdom. This prevents the rulers of the Trucial Coast from forming agreements with any other government. By the treaty terms Britain would control defense and foreign policy, but would not interfere in internal affairs. However, the British representatives in the region, the Political Resident in the Gulf and the Commissioners in the emirates, wield considerable influence as they act to control the slave trade, arms trade, the granting of pearl and sponge-hunting permits, and of oil exploration permits. These treaties are in force until 1971 with the formation of the separate states of Qatar, Bahrain and the United Arab Emirates.

1893: March 26 Qataris fight Turkish occupation

The Turks' surprise attack on Qasim at Wajbah fails. The Turks are driven to retreat to Doha and strengthen their defense. Qasim commandeers the water supply forcing negotiations. The defeat of the Turks returned the fighting forces to Arabia, but leaves the Turk's with nominal overlordship. Qasim remains as Governor, but friction remains between Qataris and the Turks.

1895 The al bin Ali tribe seeks permission from Qasim to relocate from Bahrain to Qatar

Qasim offers protection with the Turks help in rebuilding Zubara. The British react to the Turkish presence at Zubara by sending two war ships, the *Sphinx* and the *Pigeon*. Qasim's fleet is destroyed and the al bin Ali tribe returns to Bahrain.

1913 Sheikh Qasim Bin Mohammed Al-Thani dies

At eighty years old Sheikh Qasim dies and is succeeded by his son Abdullah bin Qasim Al-Thani.

1913–1949 Reign of Sheikh Abdullah bin Qasim Al-Thani

Sheik Abdullah bin Qasim Al-Thani succeeds his father Sheik Qasim and reigns for forty-six years.

1913: July 19 Anglo-Turk convention

The Ottoman Empire renounces all claims to Qatar.

1915: August Turks leave Qatar

During World War I (1914–18), Turkish forces leave Qatar. The Turkish departure ends forty years of partial occupation of Qatar.

1916: November Qatar is a member of the Trucial States

A treaty with the British is signed by Abdullah Al-Thani.

1916 Treaty between Britain and the Thani family

The Thani family are hereditary rulers of Qatar. Their sovereignty is recognized by the British, and a treaty, the Treaty of Protection, is negotiated between the two parties. The British agree to protect Qatar from foreign attack. In return, Qatar grants complete control of its foreign affairs to the British. (See also 1934, 1968, and 1971: September 3.)

1920s Qatari pearl trade disappears

The Japanese invention of the cultured pearl results in Qatar's loss of pearl trade. Financial depression ensues.

1930s The British discover oil in Qatar

British prospectors discover oil in Qatar. This discovery is to change radically Qatar's economy and its citizens' daily life.

1934 Treaty between Britain and Qatar extended

The treaty between the British and Qatar (see 1916) is extended. Under its terms, Qatar is protected by the British. (See also 1968 and 1971: September 3.)

1935 The Iraq Petroleum Company explores for oil and gas

Sheikh Abdullah grants permission to the Iraq Petroleum Company to explore for oil and gas.

1935 Great Britain grants Hawar Islands to Bahrain

This dispute festers until 1991 when Qatar refers it to the International Court of Justice in The Hague.

1940 The Iraq Petroleum Company discovers oil

The first successful oil well, Dukhan I, is discovered. The oilfields are located near the town of Dukhan near the west coast. The oil from Dukhan travels by pipeline to the port of Umm Said, 20 miles south of Doha.

1947 Commercial oil production begins in Qatar

Over a decade after the discovery of oil in Qatar, commercial production begins.

1948 Sheikh Hamad Bin Abdullah Al-Thani, proposed successor of Sheikh Abdullah, dies

Between 1940 and 1948 Sheikh Hamad, as heir apparent of his father, often represents his father as ruler.

1949–60 Sheikh Ali Bin Abdullah Al-Thani succeeds Sheikh Abdullah

Sheikh Abdullah abdicates due to old age in favor of his son, Sheikh Ahmed Bin Ali Abdullah Al-Thani. Sheikh Ali is the eldest brother of the now-deceased former heir apparent, Sheikh Hamad (see 1948).

1952 First boys' school opens

As part of an educational reform program, Qatar opens its first school for boys.

1952 Shell Petroleum of the Netherlands explores for offshore oil

Dutch Shell begins exploring for offshore oil.

1954 Desalination project

Qatar starts a desalination project intended to provide greater amounts of drinking water.

1956 Creation of the Ministry of Education.

The modern education system is designed. Three stages are outlined: primary (6 years), preparatory (3 years) and secondary (3 years). The education system is rapidly expanded to include all locales and all stages of education. Specialized and technical schools are included. Textbooks, stationary, transportation, sports, clothes and gear for all pupils at all levels of education are provided by the government. Programs on literacy, adult education and special classes for the handicapped are part of the expansion of the education system in the 1990s.

1956 First girls' school opens

Four years after the opening of a school for boys, Qatar opens its first school for girls.

1957 First stamps issued

Qatar issues its first stamps but they depict British themes. (See 1961.)

1960–72 Reign of Sheikh Ahmed Bin Ali Abdullah Al-Thani

Sheikh Ali Bin Abdullah Al-Thani abdicates turning the rule of Qatar over to his son Sheikh Ahmed.

1960s Three offshore oil fields are discovered

Shell Petroleum discovers oil fields 50 to 60 miles east of Doha in the gulf waters. Oil is piped to Halul Island for storage and export.

1961 Stamps with Qatari themes issued

The first stamps to be issued by Qatar depicting with local scenes are issued. (See also 1957.)

1963 Ras Abu Aboud power station generates electricity for Doha

Electricity for Doha is now generated by the Ras Abu Aboud power station.

1963 Ministry of Municipal Affairs and Agriculture establishes the Rawdat al Faras Agricultural Farm

Qatar's government establishes the Rawdat al-Faras Agricultural Farm.

1963 The Qatar Scouting Association is formed

The Association is registered in the World Scouting Bureau. Scouting programs and activities spread to the schools. Land is granted as the Permanent Khalifa Scouting Camp in 1979.

1968 Britain declares its intention to withdraw by ending its treaty relationships

Britain announces its intentions to sever its treaty relationships within the region (see 1916 and 1917: September 3). For Britain this would be a complete withdrawal of all its troops east of the Suez by 1971, including those based in the Gulf. This decision includes the seven Trucial Sheikhdom, together with Bahrain and Qatar, all under British protection.

1968 Federation is discussed among the leaders of Bahrain, Qatar, and the seven Trucial emirates

Bahrain, Qatar, and the seven Trucial emirates hold talks aimed at creating a federation of their states.

1968 Qatar's Umm Said Industrial District opens

The Umm Said Industrial District opens in Qatar.

1971: September 3 Provisional Constitution and Declaration of Independence

Qatar is independent with the abrogation of the 1916 Treaty of Protection with Great Britain. (See 1916.)

1971: September 21 Qatar joins the UN

Newly independent Qatar (see 1971: September 3) joins the United Nations (UN).

1971 Natural gas is discovered

North Field site of the single largest non-associated gas field in the world is discovered below the floor of the Persian Gulf.

1972: February 22 Sheikh Khalifa bin Hamad Al-Thani becomes Emir

Sheikh Ahmed turns over the rule to his cousin and heir apparent H. H. Sheikh Khalifa bin Hamad Al-Thani. Sheikh Khalifa is the seventh member of the Al-Thani family to rule Qatar since 1850. His eldest son, Hamad, the Minister of Defense and Commander-in- Chief of the Armed Forces is his future successor.

1973 The University of Qatar opens

The University opens with two Faculties of Education, separate faculties for men and women. Additional faculties are added: Science, Humanities, Islamic Studies, and Engineering.

1973 Qatar joins the Arab oil embargo against the United States

In retaliation for American support of Israel against its Arab neighbors in the Yom Kippur War (October 1973), Qatar and other Arab states embargo oil against the United States.

1974 A Special Education Section is established

Specialized training for the deaf and dumb and the mentally handicapped are begun. In 1981 and 1982, respectively, a Boys' Institute of Hope and a Girls' Institute of Hope are established. In 1985 the names are changed to Boys' School of Hope and Girls' School of Hope.

1977 Ras Abu Fontas power station opens

Upon opening Ras Abu Fontas is the largest gas-fueled power plant in the Middle East.

1978 Qatar's first steel mill opens

In yet another sign of its industrial development, Qatar opens its first steel mill.

1980–88 Qatar supports Iraq in the Iran-Iraq War

Fearful of the spread of Islamic fundamentalist Iran, Qatar supports Iraq in its war against Iran.

1981: May 25 Establishment of the Cooperation Council for the Arab States of the Gulf (CCASG)

The CCASG's aims are the comprehensive development of the member states, to ensure security, and to support the causes of the Arab and Islamic Nation. One result of these efforts is a Unified Economic Agreement that serves their common interests.

1986 Qatari helicopters remove and "kidnap" Bahraini workmen

Escalating a dispute with Bahrain over Hawar and adjacent islands Qatar removes Bahraini workmen constructing a coast guard station from a reef off the coast of Qatar. A truce is formed with Sa'udi intervention.

1990: October 5 Qatar cancels debts and interests of ten countries

To strengthen mutual cooperation with the Islamic States and Muslim peoples Qatar allocates a portion of its national income for assistance and support of less developed and underdeveloped countries. Qatar's representative to the United Nations announces the cancellation of all debts and interests owed to Qatar by ten friendly countries: Egypt, Syria, Morocco, Tunisia, Mauritania, Somalia, Guinea, Uganda, Cameron and Mali.

1990: December 25 The CCASG denounces Iraq's aggression against the State of Kuwait

The Doha Statement presented the joint strategy for security and defense including the total withdrawal of Iraqi forces from Kuwait, support of the Palestinian Intifada (the uprising), the Palestinian Cause as part of the Arab cause, joint economic policies, and the coordination of mass media.

1991 A new oil field is discovered

Elf Petroleum Qatar makes the first oil field discovery in 20 years. The new Al Khalij field is next to Iranian waters.

1991 Qatar refers dispute to the International Court of Justice in The Hague

Qatar's and Bahrain's long-running territorial dispute over the Hawar Islands is referred by Qatar to the International Court of Justice in The Hague. The Bahraini claim to Al Zubarah on the northwest coast of Qatar and to adjacent islands. The islands are close, only 3 kilometers from the Qatar mainland, but more than 20 kilometers from Bahrain. The Zubarah and Hawar claims had been settled in Qatar's favor, but Bahrain refuses to honor those settlements as based on western ideas and not local custom.

1992: QATARGAS sells liquefied natural gas (LNG)

A Japanese utility company agrees to purchase 4 million tons of LNG each year for twenty-five years as of 1997. Another 2 million tons of LNG per year for twenty-five years is sold to seven other Japanese utilities.

1992: September 30–October 3 Border dispute with Sa'udi Arabia turns violent

A 1965 border agreement with Riyadh that was never ratified leaves tensions simmering for over a year. Two Qatari guards and one Sa'udi national are dead with the circumstances of an armed clash unclarified.

1993: January Bahrain, Qatar, and U.A.E. agree to link communications networks

In order to improve telephone, data communication and television transmission systems among Bahrain, Qatar, and U.A.E., a fiber-optic marine cable will be installed. The partners will share the estimated $50 million cost.

1993: June Qatar bans ownership of satellite dishes

The ban on satellite dishes is a move to promote the so-far unsuccessful state-owned cable system.

1994: June A Qatar-to-Pakistan pipeline proposed

This 994 mile pipeline (the world's longest) would extend across the Persian Gulf and Arabian Sea from Qatar's North Field to a terminal near Karachi. The $3.2 billion project would be underway in 1996.

1994 A Qatari horse is named World Champion Mare at the Paris Horse Show

Qatar's horses achieve international recognition when a Qatari horse is named World Champion Mare at the Paris Horse Show.

1995: June 27 Qatar coup d' etat

Qatari Crown Prince and defense minister Sheikh Hamad bin Khalifah Al-Thani seizes control of the palace while his father, the Emir, is on a trip to Europe. Calling his father to announce the take-over, his father refuses to come to the phone. His father tours Arab capitals seeking support, promising to return to Qatar to retake his throne.

1996 Sheikh Hamad founds television service, al-Jezira

The emir, Sheikh Hamad, founds a television service, *al-Jezira* (The Peninsula), based in Doha. He recruits staff, including some from the British Broadcasting Service (BBC), to run the station. Al-Jezira is a satellite channel that broadcasts news and entertainment programs that some observers in the region find controversial. For example, one talk-show program, "Opposite Direction," featured a professor from Kuwait in a segment questioning the Persian Gulf region's economic organizations, and discussing the fitness of the region's monarchs to rule. Viewers across the Middle East have access to the broadcasts.

1996: February A countercoup is mounted

Hired assassins disclose the plot against Sheikh Hamad and senior officials. Hamad retaliates by freezing his father's bank accounts.

1996: October The Justice Ministry announces an out-of-court settlement

Sheikh Hamad bin Khalifah Al-Thani accuses his deposed father, Sheikh Khalifah bin Hamad Al-Thani, of making off with upto $7 billion. Traditional tribal practice is that revenues belong to the ruler and he distributes them as he sees fit. The issue is a question of ownership of this income: is this income a ruler's personal wealth, or is it state funds. Details of the settlement are not made public.

1996: November Sheikh Khalifah bin Hamad Al-Thani renounces his claim to the throne

His son, Sheikh Hamad bin Khalifah Al-Thani, allows him to return to Qatar from exile to act as an elder statesman. Sheikh Hamad quickly settles the question of succession by appointing his third son, eighteen-year-old Sheikh Jassim Bin Hamad Al- Khalifa Al-Thani as crown prince. The Emir also cedes power to his younger brother, Sheikh Abdulla Bin Khalifa Al-Thani, appointing him prime minister.

1997 Qatar tightens controls on foreign workers

Qatar increases pressure on foreign workers, most of whom are considered illegal immigrants. The interior ministry will allow those who have overstayed their visas just two months to prepare to leave the country. Severe punishment, including imprisonment, will be imposed on those who fail to comply.

1997: August The Emir of Qatar undergoes a kidney transplant

Sheikh Hamad bin Khalifah Al-Thani, forty-eight year old ruler of Qatar has a kidney transplant at the Cleveland Clinic, in Cleveland, Ohio. Sheikh Hamad suffers from diabetes.

1998

1998 Amnesty International (AI) raises issues relating to 1996 coup attempt

In response to the 1996 coup attempt Qatar charged 117 people, some *in absentia* and others are held incommunicado and without access to lawyers. Amnesty International reports that some detainees allege that they were tortured, and one person is reported to have died while in custody. AI is invited to attend the trial of the defendants.

1998: April Qatar fosters a more open business climate

Ruling Emir Sheik Hamad Bin Khalifa Al-Thani opens Qatar to investment by international companies.

1998: April Bahrain accuses Qatar of forging border dispute documents

Bahrain accuses Qatar of submitting forged documents to the World Court in the dispute over the Hawar islands. Qatar is given six months by the World Court to reply to these allegations.

1998: May Qatar holds the Gulf's first mixed athletics event

For the first time in the region women are allowed to participate in an international athletic competition. Women athletes agree to wear less revealing clothing and a special spectators section of the stadium is set aside for women.

1998: June Qatar is the first Gulf State to offer sovereign bonds

Oil prices drop 30 percent in 1997 precipitating an oil crisis for the exporting nations as prices fall and revenues decrease. Qatar's sovereign bond issue is postponed within two weeks of the initial offer. The lack of investor interest and unfavorable international conditions are cited.

1998: July Bahrain escalates the dispute over the Hawar islands

Bahrain escalates the two hundred year old dispute with Qatar over the Hawar islands by announcing development plans to construct housing units, a runway for light aircraft and a causeway to the mainland of Bahrain. The islands are uninhabited but may have oil and gas reserves.

1999: March The U.S. offers to share intelligence information with Qatar

The U.S. agrees to set up a telephone hotline between Washington and Qatar to share information on missile launches from Iran or Iraq.

1999: March Women candidates is Qatar are defeated

A national holiday is declared. Voter turnout is high with some voters arriving in limousines. In an election for a 29-member Municipality Council six women candidates are all defeated. This is the first time that women in Qatar are voting and running as candidates. It is also a first for the Gulf region. The poll is monitored by outside observers from nine countries who declare it is a fair election.

1999: March 10 Qatar opposes the U. S. air strikes against Iraq

Qatar's foreign minister, Sheikh Hamad bin Jassim Al-Thani protests the U. S. attacks in Iraq's no-fly zone. Arab critics propose that the U. S. attacks are provocations of the Iraqi. If the Iraqis fire, then retaliation would appear appropriate.

Bibliography

Abu Saud, Abeer. *Qatari Women Past and Present*. Essex, England: Longman Group Limited, 1984.

Anscombe, Frederick F. *The Ottoman Gulf: The Creation of Kuwait, Sa'udi Arabia, and Qatar.* New York: Columbia Press, 1997.

Crystal, Jill. *Oil and politics in the Gulf: Rulers and Merchants in Kuwait and Qatar*. New York: Cambridge University Press, 1995.

di Cardi, Beatrice, ed. *British Archaeological Expedition in Qatar (1973–1974)*. New York: Oxford University Press, 1978.

El Mallakh, Ragaei. *Qatar, Energy & Development*. London: Croom Helm, 1985.

Graham, Helga. *Arabian Time Machine: Self-Portrait of an Oil State*. New York: Holmes & Meier Publishers, Inc., 1978.

Nafi, Zuhair Ahmen. *Economic and Social Development in Qatar*. Dover, NH: F. Pinter, 1983.

Rickman, Maureen. *Qatar*. New York: Chelsea House Publishers, 1987.

Vine, Peter and Paula Casey. *The Heritage of Qatar*. London: IMMEL Publishing Limited, 1992.

Zahlan, Rosemarie Said. *The Creation of Qatar*. New York: Harper & Row Publishers, Inc., 1979.

Samoa

Introduction

Travel guides and history books typically depict Samoa as a quiet, gentle paradise where "time stands still." Photos capture scenes filled with palm trees, thatched huts, white-sand beaches, and smiling "native" women wearing flowers in their hair and clad in traditional *lavalava*, a brightly-colored piece of fabric wrapped around the body. The writing that accompanies such descriptions suggests that while the rest of the world rushes into the 21st century, life in Samoa remains primitive, innocent and content.

Like many such descriptions of island nations in the Pacific Ocean, this one is incomplete. For the past three centuries, Europeans and Americans have come to Samoa and exploited it. They came first as traders, then as Christian missionaries and eventually as colonizers who saw its islands' beaches as a strategic post for military and commercial operations in the Pacific. And these outsiders often defined and studied Samoans in a manner that differed from how the islanders thought of themselves.

When it suited their interests, many outsiders described Samoans as exotic and sexually seductive. In the late nineteenth and early twentieth centuries, for instance, photographers from Europe and the United States often used Samoans in posed portraits to convey an image of paradise. Samoan women often were photographed in the nude or lying in positions that invited dreams of erotic pleasure.

Samoa gained its independence in 1962, but the images that the colonial administrators used continue to haunt the nation as its leaders and its people struggle to survive amid poverty, lagging education, poor health care, and constant threats to the environment. By being painted as either exotic or child-like, Samoans ultimately were depicted as different—and usually inferior to—their Western counterparts. In order to succeed, Samoans were told that they had to surrender their culture and their way of life in order to become more like the West.

The Republic of Samoa is made up of several islands, which roughly equal the size of Rhode Island. Its two main islands are 'Upolu and Savai'i, and among the smaller islands, only Manono and Apolima are inhabited. The islands are located in the central basin of the Pacific Ocean, about 2,300 miles (3,680 kilometers) southwest of Hawaii and have a population of about 170,000. Although culturally part of Samoa, islands in the eastern part of the chain, the largest of which is Tutuila, have been separated politically from the western half of the country since the early twentieth century. The eastern part of Samoa is known as American Samoa, and is a U.S. territory.

Traditional Samoa

Most Samoans traditionally have worked in agriculture, and the country exports large quantities of coconut, cocoa, and taro, along with *'awa,* a medicinal root. However, two cyclones hit the country in 1991 and 1992, and destroyed much of its agricultural base. Since then, Samoa has relied heavily upon remittances—the money that islanders who migrate to the United States and elsewhere to work send home to their families—and tourism. That money made up as much as fifty percent of Samoa's national budget by the mid-1990s, and in the first eleven months of 1997, the government reported that remittances totaled $33 million and were the largest source of foreign revenue.

Samoans traditionally described their land as the "cradle of Polynesia." Like many Polynesian peoples of the Pacific, they drew their history from a rich oral, story-telling tradition that included a tale of creation. According to this story, a god known as Tagaloa once lived in empty space. He created a rock, which then split into clay, coral, cliffs, and stones. As the rock broke apart, the earth, sea, and sky came into being. According to the Samoan oral tradition, Tagaloa went on to create a man and a woman, Fatu and 'Ele'ele, words that mean "heart" and "earth." Tagaloa sent the pair to an area where fresh water existed and asked them to bring life into the area.

Over time, as the population grew, Tagaloa set up a government, led by Manu'a. The *tupu*, or kings, who descended from him were called "*Tu'i Manu'a tele ma Samoa atoa.*" This title means "king of Manu'a and all of Samoa." Tagaloa then divided his world into islands: Manu'a, Fiji, Tonga, and Savai'i. He later added 'Upolu and Tuitila. Before Tagaloa departed from the earthly world, he told his people always to respect Manu‘a.

The belief in this traditional story came into question when Europeans began to arrive in Samoa in the eighteenth century, bringing new technologies and Christianity. Before Europeans arrived, Samoans believed that the universe consisted of the islands in the central Pacific. They viewed the world as a sphere, in which spirits lived below the sea, humans lived on the islands above the water, and transparent domes of heaven, divided into nine layers, filled out the area above. Europeans who first sailed into Polynesia often were described as *papalagi*, or sky-busters. They appeared to have come from the heavens. Samoan religious beliefs could not explain the arrival of these white men in any other way, which eventually caused Samoan forms of knowledge and world perspectives to be undermined.

In the traditional Samoan system of government, the village or district where a person came from held a great deal of importance. Samoans expressed their origin in a variety of oral addresses, one of which was the *fa‘alupega*, a set of ceremonial greetings that were recited when the various district chiefs met. Historian Malama Meleisea observed that the *fa‘alupega* served as a spoken constitution, summarizing the origin and ranks of the chiefs who were present and the districts that each one served. This traditional system of ruling chiefs persists in Samoa's present-day government.

Missionaries had been active throughout Polynesia in the early nineteenth century. Before missionary John Williams (1796–1839) arrived, many Samoans had become exposed to Christianity through small cult groups that passed in and out of the islands, many from nearby Tonga. Williams, however, was the first missionary to gain legitimacy from a ruler. This caused the religion to spread more rapidly, and by the mid-1830s, missionaries managed to convert nearly all Samoans to Christianity. The religion continues to influence social and cultural life in the islands. Protestants, Catholics, and Mormons all maintain a presence, and some consider Samoa and Tonga to be like a Bible belt in the Pacific.

Because the missionaries had gained such a strong presence in Samoa, traders began to form alliances with them, and gradually established more permanent settlements in the islands. By the mid- to late nineteenth century, Germans had established a large plantation near the port of Apia. The British had set up several trading posts and settlements within the missionary communities, and Americans began to express a growing interest in taking over a port in Pago Pago on Tutuila. Competition for land and a desire to dominate trade in the Pacific led to frequent outbreaks of fighting among the European and American nations, and by the end of the nineteenth century, it was not uncommon to see gunboats from Germany, Britain, and the United States on the Samoan coast.

Since Samoans have regained their freedom, many scholars, writers, and artists have begun to reinterpret Samoa's culture. For many, this has meant linking traditional forms of storytelling with modern-day forms of communication. Like many Polynesian peoples, Samoans have a rich tradition of oral history. As Peggy Fairbairn-Dunlop has observed, "Samoans relish the spoken word. The love of word, building up into logical argument and woven into intricate plots, allusion and innuendo, are all here." However, Samoans traditionally viewed writing and books as the word of the *palagi*, or foreigner. The colonizers perpetuated this idea by limiting schools established in Samoa to *afakasis*, persons who were part-Samoan and part-European. Many teachers in these schools taught from books imported from Europe.

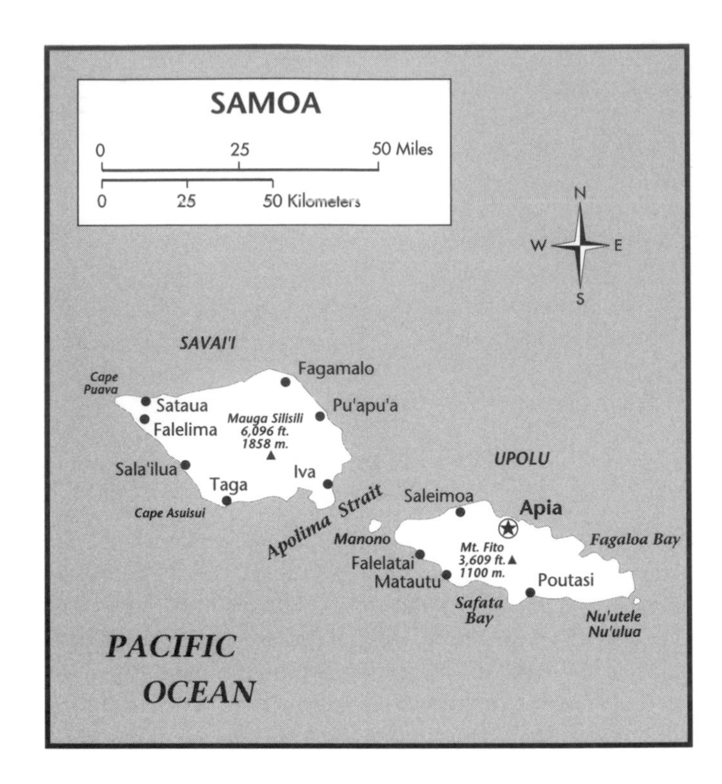

During the independence movement of the 1960s, students started to question the education they were receiving, and the values that the colonial schools conveyed. Through this questioning, many began to develop an interest in writing in Samoan and in mixing Samoan and English. Several writers such as Fanaafi Maiai, Apelu Aiavao, and most notably Albert Wendt nurtured their skills during this period.

With this literary revival has come a shift in how Pacific Islanders themselves imagine their places. Islands in the Pacific often are depicted as tiny dots in the ocean that are indistinguishable from one another. This has led many Westerners to see the islands as homogenous and their cultures as interchangeable. Lately, people such as Tongan intellectual Epeli Hau‘ofa have sought to change that perception. Hau‘ofa describes the Pacific as Oceania, as a sea of islands rather than islands in the sea. In this image, the islands are not dots in the ocean, but sections of a continent linked together by the ocean. While this description of Oceania creates a different idea of a continent, the shift in perception allows many Samoans and other Pacific Islanders to see their land as distinct entities within a larger geographic body, much as Americans would regard the South and the Midwest as culturally distinct.

For writers such as Wendt, this shift in perception is not only creative but also logical. "In the final instance, our countries, cultures, nations, planets are what we imagine them to be," he writes in his essay, "Towards a New Oceania." "One human being's reality is another's fiction."

Timeline

1500 B.C. Polynesians settle Samoan islands

Early archaeological evidence suggests Polynesian voyagers enter the central Pacific basin and establish settlements on several islands in present-day Samoa and Tonga. Much of this belief comes from traces of black lapita pottery found in Samoa.

Archaeologists have excavated traces of a village, partially submerged in a lagoon on 'Upolu, in which many structures were made of the clay material. In addition, platforms that appear to have been used to capture birds and to perform religious ceremonies have been found on 'Upolu, Savai'i, and Tutuila.

A.D. 500 Polynesian culture thrives

By the sixth century, a distinct Polynesian lifestyle has developed in Samoa, Tonga, and Fiji. Life in the islands focuses on the people's relationship to the land and surrounding ocean. Voyaging canoes carry as many as 250 people between the island chains, allowing residents of different islands to develop a shared culture. Polynesians plant yams, taro, breadfruit, bananas, and coconuts, and raise pigs and chickens. Many of the islanders use earth ovens to prepare large, festive meals shared by all people.

Women weave mats and make tapa cloth from bark, crafts which remain popular in Oceania today. Homes are built from woven coconut or sugar cane leaves. As a tradition of oral storytelling develops, islanders participate in group singing and dancing.

950–1250 Tongans rule in Samoa

In 950, a group of warriors from Tonga gain control of Savai'i, the island in the Samoan archipelago located closest to Tonga. After establishing a kingdom on Savai'i, Tongans rule for several generations and eventually move to conquer 'Upolu, one of the larger islands in Samoa. A group led by a Samoan chief known as Savea defeats the Tongan warriors in 1250. A peace treaty between the two island nations is created, and the Tongans leave Samoa permanently.

The war, however, leaves a lasting imprint on Samoan society in that it creates a new line of kings, known as Malietoa. Savea becomes the first to hold the position of Malietoa, and use of the name continues through the end of the nineteenth century.

1500 Salamasina begins new era in Samoan politics

By the early sixteenth century, a lively form of politics flourishes in Samoa. The islands are divided into eleven districts, each ruled by a chief or *matai*. Groups of orators, known as *tulafele*, then select four individuals from the leading districts to hold certain chiefly titles known as *papa*. These titles are known as the Tui A'ana, Tui Atua, Gato'aitele, and Tamasoali'i.

The practice of dividing the chiefly leadership among four individuals ends when Salamasina, an aristocratic woman who lived in the early sixteenth century, becomes the first person to hold all four titles collectively. She becomes known as *tupu*, and, according to many Samoan oral legends, holds special favor among the gods.

After Salamasina dies, her descendents form a royal dynasty that is centered in the districts of A'ana and Atua, both of which are on 'Upolu. Each successive *tupu* has several marriages and bears many children. After the death of a *tupu,* the *tulafele* continue to select a *tupu* from this pool of descendents, but competition for the position becomes increasingly fierce. District leaders often get involved in fights for power, forming what might loosely be described as political parties. Historian Malama Meleisea describes the battle as being between two factions: the *malo*, or victors, and the *vaivai*, or vanquished. Each time a *tupu* dies, a new contest for succession emerges.

1722 European ships near Samoa

Historical documents indicate that Jacob Roggeveen (1659–1729) is the first European to sight Samoa. Roggeveen, a Dutch explorer, spots Manu'a in 1722 while searching for *Terra Australis Incognita*, a mythical continent in the southern hemisphere. He gives the island a Dutch name but does not drop anchor.

French explorer Capt. Louis-Antoine de Bougainville (1729–1811) passes through the islands in 1768 and names them Les Iles des Navigateurs, or Navigator Islands, after observing inhabitants traveling about in voyaging canoes traditionally used in Samoa. He exchanges some goods with residents of Manu'a and records sightings of the more westerly islands.

1787 Massacre breaks out at Aasu

A French trader known as Jean-Francois de Galaup, Comte de La Perouse (1741–c.88), lands on the northern coast of Tutuila with an interest in trading glass beads. Europeans had introduced these beads to Tonga some time ago. Because Tongans and Samoans frequently came into contact with each other, Samoans are familiar with the beads by the time the Europeans arrive.

After La Perouse drops anchor, a group of Samoans board his ships and find items made of iron to be of more interest than beads. They take the products, which enrages La Perouse. He orders his sailors to punish them. This action causes residents of Aasu, located to the west, to organize an attack on the crew. While crew members are collecting water the next day, the Aasu warriors begin their ambush and wind up killing twelve French sailors. Thirty-nine Samoans also die in this encounter.

1791 British ship is attacked

A British ship known as the *HMS Pandora*, arrives in Samoan waters. Residents attack the ship as it nears Tutuila, and several Samoans are killed.

This attack causes many Europeans to regard Samoans as hostile and violent people. Although the reputation for savagery causes many passing ships to steer clear of the islands, a hunger for whale products, sandalwood, and other valued goods from them forces Europeans to keep their eyes on Samoa. In turn, Samoans begin befriending stray Europeans who arrive on the islands. Most of these newcomers are either escapees from convict ships or retired whalers. Both groups, however, offer knowledge of the *palagi* to Samoans, which they embrace readily.

1828 End of dynasty leads to war

After a *tupu* known as I'amafana dies, the royal dynastic line established by Salamasina dies out. *Matai* members from districts on Savai'i, Manono, and Apolima take advantage of the void by supporting a warrior known as Leiataua Tamafaiga as successor. Leiataua, who is the priest of the war goddess Nafanua, wages a war against the rulers on A'ana and wins. However, he is soon assassinated.

The death of Leiataua prompts his kinsman, Malietoa Vai'inupo, to launch a new war in 1828 to avenge Leiataua's death. Malietoa Vai'inupo is successful and claims all four *papa* titles as his own. He then drives most of the inhabitants of the district of A'ana off the island and seizes their land. The refugees retreat to Atua in defeat.

The *tulafale* then declare Malietoa Vai'inupo as *O Le Pule o Salafai*, or the authority of Savai'i, thus commemorating Savai'i's victory over the districts on 'Upolu.

1836 Malietoa Vai'inupo converts to Christianity

One Samoan oral history holds that the war goddess, Nafanua, had predicted one hundred years earlier that the heirs of Vai'inupo's ancestors would receive their *malo* from heaven. On the eve of Vai'inupo's victory over A'ana, a ship carrying the Rev. John Williams arrives in Vai'inupo's village of Sapapali'i.

This may have been one reason why Malietoa Vai'inupo was so willing to convert to Christianity. Williams represents the London Missionary Society, and he is guided to Vai'inupo by another Samoan chief who had traveled on Williams' ship from Tonga.

After Vai'inupo converts, he takes the name Tevita, after King David of Israel, and invites the people of A'ana to return to their lands. He also declares that the *malo* now belongs to God and that the four chiefly titles, or *papa*, should be returned to the districts with which they were traditionally associated. Before dying in 1841, he nominates four successors for the positions.

This move decentralizes the power base in Samoa and brings about an era of peace.

1861 Succession fight splits on religious lines

After one of Malietoa Tevita's sons dies, a new succession conflict breaks out. At this time, missionaries representing many Christian denominations are operating in Samoa and have managed to pit the followers of different denominations against each other. This provokes a war, which weakens the system that Malietoa Tevita had established.

By this point, several European powers have established business interests in Samoa and are quite busy acquiring land for agricultural plantations. The succession war prompts many of these countries to demand that Samoa establish a centralized government with one king with whom Western merchants could deal exclusively. Historians argues that this undermines the islands' traditional style of governance.

"The squabbling European settlers wanted a central authority and a Samoan king through whom they could legitimise their often dubious land acquisitions," writes Meleisea, in an introduction to a 1996 reprint of Robert Louis Stevenson's *A Footnote to History.* "But most failed to understand that the traditional political decision-makers were not the sacred chiefs of Samoa. ... Decisions of national importance were made, district by district, by councils constituted of the highest chiefs and their orators, who debated matters until there was a consensus of all leading polities. Without this consensus there could be no decision, unless one faction felt strong enough to force the issue upon those who disagreed by declaring war."

1888 Civil war breaks out

By the late nineteenth century, merchants from the United States, Great Britain, and Germany each establish sizeable business interests in the Samoan islands. Each Western country is eager to protect these investments from the tentacles of other Europeans, and as the era of European and American imperialism intensifies, frequently send warships to the Samoan coast to settle disputes.

The presence of these European warships gives the succession fights and other conflicts that crop up among Samoa's *matai* a more militaristic character. The Western powers play one potential successor against another in hopes of furthering their own commerical interests. This bickering leads to a civil war in 1888, which also pits Germany against the United States and Great Britain. Germany supports a leader known as Tupua Tamasese Titimaea. Another Samoan faction, whose leaders dislike the German presence on the islands, supports Tui Atua Mata'afa Iosefo. American and British troops, hoping to weaken the Germans, also rally behind Mata'afa.

The fighting brings seven foreign warships to the port of Apia. A hurricane hits the islands and six of the gunboats—three of which are American and three of which are German—sink. More than 150 people die in the fighting.

1889 Robert Louis Stevenson settles in Samoa

Robert Louis Stevenson (1850–94), a Scottish writer and poet, arrives in Apia via an ocean schooner that had taken him also to Tahiti, Australia, Kiribati, and Hawai'i. He falls in love with Samoa and builds a house known as Vailima, and settles in the islands with his wife Fanny de Grift Osbourne.

While in Samoa, he suffers from tuberculosis. But, perhaps feeling as if he is in a race against time, Stevenson writes prolifically, often about the injustices that Samoans suffer at the hands of Western powers. This love for a story endears him to many Samoans who nickname him Tusitala, or the teller of tales. His story "The Bottle Imp" becomes the first literary work to be translated into Samoan.

Stevenson dies December 3, 1894, after suffering a stroke. In his honor, Samoan chiefs order that a road be dug by hand from Vailima to the top of Mt. Vaea, where the storyteller wished to be buried. His coffin is placed on a base of coral and volcanic pebbles, and his grave is lined with black stones. In paying his respects, one Samoan chief, known as Tu'imaleali'ifano, declares: " 'Talofa e i lo matou Tusitala. Ua tagi le fatu ma le'ele'ele," or "Our beloved Tusitala. The stones and the earth weep."

1898 Americans and Germans colonize Samoa

Mata'afa wins the civil war, but he is not allowed to assume leadership in Samoa. Instead, in an effort to resolve their differences and protect their interests, the United States, Britain, and Germany sign the Berlin Treaty which installs another chief, Malietoa Laupepa, as *tupu*. Laupepa is designated as the leader of a neutral government, and Apia—a coastal city in which many of the Western businesses are concentrated—becomes a separate municipality, governed by Europeans.

After Malietoa Laupepa dies in 1898, a new succession dispute emerges. The Westerners intervene by abolishing the kingship and dividing up the islands. Americans receive the eastern portion of the archipelago, which allows the United States to maintain a naval harbor at Tutuila. The rest of the chain becomes a German protectorate known as German Samoa until 1915. In exchange for various concessions, Great Britain drops its claims to the islands.

1900: February 16 Treaty of Berlin is ratified

After the Treaty of Berlin is ratified, the United States formally gains possession of the eastern portion of Samoa. The United States annexes additional islands within the archipelago in 1904 and 1925. (See also 1929: February 20.)

1903 Chinese laborers arrive

After gaining control of Western Samoa, Germans want to make their planatations in the islands as profitable as possible. A desperate need for cheap labor arises, which the German administrators solve by bringing Chinese workers into the country.

Samoan chiefs had opposed the use of such labor in the late nineteenth century, fearing that such a massive influx of people would exhaust the islands' resources and endanger the Samoan way of life. Germans, however, overturn laws that prohibit Chinese immigration and bring nearly four thousand workers into the country between 1903 and 1913. New Zealand rulers who eventually occupy the islands bring an additional three thousands workers into the islands between World War I and World War II.

Chinese workers are recruited to Samoa through racist imagery that suggests Samoan men would serve their material needs and that exotic Samoan women would satisfy them sexually. Once in the islands, the workers are subjected to strenuous, back-breaking labor. A letter that a worker known as Cheng Wing sent to his uncle describes how they suffered:

"The monthly rate are only ten marks equal to $2.50," Cheng Wing wrote. "I do not mind receiving this low wage, but the tortures are hard to suffer. In the case of sickness. . . If you do not recover in a few days, you will be hit and kicked, and if you are still not recovered in a week or ten days, you will be carried out in the street and receive strokes with a cane."

Although most of the Chinese workers leave the islands after World War II, a few marry Samoans and their descendants remain in the islands today.

1914: 29 August New Zealand occupies Samoa

As World War I begins, troops from New Zealand move into Apia, and take over the administration of German Samoa, which then is renamed Western Samoa. The colony is placed under a military administration until 1920 when the League of Nations issues a mandate that gives New Zealand the authority to administer the islands.

German administrators had outlawed many traditional Samoan practices, calling them "communalistic" and "uncivilized". Germans had attempted to impose a centralized ruling system, which was undermined by the New Zealand invaders.

1927 Civil disobedience movement starts

The shift in rule gives Samoans a renewed belief in the value of their decentralized ruling system. This lays the groundwork for a nationalist movement known as Mau, in which Samoans, a handful of Chinese laborers, and even European settlers who disliked New Zealand's rule band together to oppose the administration.

For nearly a decade, activists in Mau refuse to participate in political life, attend school, or have any dealings with the government. Many focus on developing their villages and strengthening community ties, which had been lost during the past several decades of turmoil.

The Mau movement ends in 1936 when leaders agree to begin taking part in the colonial government. The power of the movement shows that European powers could not entirely reshape traditional Samoan society.

1928 Margaret Mead's book is published

Margaret Mead (1901–78), one of the twentieth century's most well-known and controversial anthropologists, spends two years researching the life of Samoans. Her research is published in 1928 in the book, *Coming of Age in Samoa,* which becomes one of the best-selling anthropological texts of all time.

Mead describes Samoa as an idyllic place in which nudity and promiscuity thrive. She describes the Samoan way of life as primitive, yet endearing, and recommends that Americans could learn a few things from the relatively casual ways in which Samoan teenagers are encouraged to experiment with sex.

In recent years, Mead has been accused of distorting her evidence and even of making up her research altogether. Although many scholars do acknowledge that her research was sound, most add that she overlooked many less positive aspects of Samoan life.

While the controversy about whether the book tells the truth or not isn't likely to be resolved, the impression that Mead conveys of Samoa as an innocent, primitive society has contributed to a stereotype that continues to haunt the country today.

1929: February 20 U.S. formalizes claim over American Samoa

U.S. Congress declares its sovereignty over the island group known as American Samoa and places the administration of the territory in the hands of the U.S. president. (See also 1900: February 16 and 1951: July 1.)

1940 U.S. military presence intensifies

As Japan becomes an increasingly hostile threat to the United States in the 1930s, a naval station established by the United States on Tutuila, American Samoa, gains new strategic importance. More U.S. troops are deployed to the naval base, and by 1940, the islands become a major training and staging area for the U.S. Marine Corps.

During World War II, U.S. military personnel on Tutuila outnumber native Samoans. This military presence deeply affects the development and daily life of American Samoa, and traces of its influence linger today. Even though many Samoans regard the two Samoas as culturally united, the U.S. presence has given American Samoa a more urban, commercial feel. Western Samoa, by contrast, remains much more agricultural and, according to some, more committed to the idea of *fa'a Samoa*, a Samoan way of life.

1951: July 1 U.S. control of American Samoa shifts from navy

The Department of the Navy governs American Samoa until July 1, 1951, when the jurisdiction is shifted to the Department of the Interior.

1960: April 27 Constitution enacted

The territory of American Samoa enacts a constitution. The constitution will be revised in 1967.

1961 United Nations sponsors vote in Western Samoa

A plebiscite is sponsored by the United Nations (UN). The Samoan people approve a constitution that establishes the Independent State of Western Samoa.

1962: January 1 Samoa becomes independent

After World War II, the United Nations and New Zealand establish a trusteeship agreement that is designed ultimately to make Western Samoa an independent nation. Samoa forms an Executive Council in 1957, and four years later, the New Zealand high commissioner removes himself from the colony's legislative assembly.

Samoa's constitution establishes a form of government that Pacific Studies scholar Michael Ogden describes as loosely resembling a constitutional monarchy. The position of head of state is largely ceremonial, and upon independence is occupied by the holders of two of the country's four traditional chiefly titles. The constitution specifies that after the remaining ruling chief dies, a legislative assembly will choose the new head of state. A prime minister is elected by Parliament, and that body consists of forty-nine members, forty-seven of whom are members of the traditional *matai* or ruling families.

This system, however, does not produce an effective government. Many of the politicians abuse their privileges and disenchantment with the day-to-day workings of government runs rampant in Samoan society.

1963 Albert Wendt begins writing career

Albert Wendt is one of the most prominent novelists and poets from Oceania. He is born in Samoa, but like many children of mixed European-Samoan marriages, goes to New Zealand for his education. He lives in New Zealand for thirteen years, and during that time gains not only a desire to radically change the subservient manner in which the islands of the Pacific are regarded, but also a profound sense of exile.

That sense of exile surfaces in many of his early works, such as "A Letter from Paradise," and his first novel, *Sons for the Return Home*.

Wendt returns to Samoa in 1965 to serve as principal of Samoa College. He later relocates to Suva, Fiji, where he teaches literature at the University of the South Pacific.

Much of Wendt's ongoing work deals with his experience of being an exile and with a growing sense of an almost spiritual relationship that he shares with his native soil. At the same time, he seeks an understanding of Samoa that places it within the context of a larger world. While from Samoa, he says he belongs to Oceania. His sense of reality is tied not just to one island in the Pacific, but to the Pacific and its many islands.

1979 Human Rights Political Party forms

Tofilau Eti Alesana (1924–99) and Va'ai Kolone form the Human Rights Political Party (HRPP) in 1979 with the intent to overthrow Prime Minister Tupuola Efi. (See 1982.)

1982 Tofilau comes to power

The Human Rights Political Party (HRPP), founded by Tofilau Eti Alesana and Va'ai Kolone (see 1979) wins power. Va'ai becomes prime minister. After charges of bribery and corruption are brought against Va'ai, Tupuola returns to power but is soon defeated by Tofilau.

Tofilau becomes the longest-serving prime minister in the Pacific. Although other members of the HRPP force Tofilau from power in the early 1980s, the political leader regains his position in 1988 and serves until 1998 when poor health forces him to retire. When he dies March 12, 1999, the HRPP commands a two-thirds majority in Samoa's legislature, and his nephew, Tauese Sunia, serves as governor of American Samoa.

The son of a Samoan missionary, Tofilau was born in 1924 as Aualamalefalelima Alesana. Tofilau notes that he was brought up in a "religious environment," and he becomes particularly devoted to his Christian faith in his final years of power.

1984 High chief pays huge phone bills

About two thousand people live in the village of Lano on Savai'i. All continue to regard Siaosi Talitimu as their high chief and ruler of their community, even though Siaosi lives not in Lano but in Denver, Colorado, where he works as an assembly technician for an aerospace company.

When villagers need advice, they call Siaosi long distance. Often, they call collect which means his phone bills run as high as $500 a month. From a distance of nearly 6,000 miles (9,600 kilometers), he gives advice on everything from land disputes to local politics.

Siaosi's situation illustrates some of the dilemmas that Samoans face today. He serves as high chief because he was the eldest son of his father, who held the title before him. Yet, because Samoa's economy is poor, he must seek work overseas. Although he admits in an interview with *People Weekly* than being a trans-Pacific chief can be a headache, he promises to continue to serve his people.

1991–92 Cyclones strike Samoa

Two cyclones named Ofa and Val, hit Samoa and destroy much of the island's agricultural base. As a result, remittances (money sent home by natives who emigrate abroad) and tourism become the chief sources of income for the island.

1994 Stamps honor Scottish writer

Samoa commemorates the one hundredth anniversary of the death of Robert Louis Stevenson by issuing four stamps in his memory. The stamps depict the ship on which he sailed to Samoa, his portrait, his tomb on Mt. Vaea, and his house in Vailima.

Samoa previously had issued stamps in Stevenson's honor in 1939 and 1969. The Virgin Islands also commemorated the author in 1969 with a set of stamps that depicted scenes from his book *Treasure Island* which is believed to have been set in those islands.

1997 Samoan diplomat warns of global warming

Tuiloma Neroni Slade, who is Samoa's ambassador to the United Nations, serves as chairman of an organization known as the Alliance of Small Island States. This group represents 36 island-nations in the Pacific, and is trying to pressure such countries as Australia, Japan, Great Britain, and the United States to make a conscious effort to cut back on the use of gasoline and other products that cause carbon dioxide to build up in the atmosphere.

Many environmental experts believe that this buildup of carbon dioxide is causing the temperature of the earth to rise. Higher temperatures will cause the polar ice caps to melt at a faster rate, which can cause the level of the ocean to rise. This puts many island-nations in the Pacific at risk because many rise only a few feet above sea level.

The Alliance of Small Island States wants industrialized nations to reduce their carbon dioxide emissions rates by twenty percent by 2005. Their members say that their very existence is at stake.

1997 Ethnobotanist helps save Samoan rainforest

American Paul Alan Cox (b. 1953) receives the Goldman Environmental Prize for his efforts to save the Samoan rainforest.

Cox is an internationally renowned ethnobotanist who first visits Samoa in 1973 as a missionary. He had earned a degree in botany at Brigham Young University, but gains a different understanding of how people use plants while in Samoa. He watches Samoans use plants to build huts known as *fales* and herbs and flowers to cure illnesses. After earning a doctorate in biology, Cox returns to Samoa in 1984 and spends nearly a decade studying healing practices. Many of these healing practices are passed down orally, from mother to daughter.

Over the years, Samoans have shown him and his students how they use *moso'oi* bark to treat asthma, a leaf from a plant known as *polo* to fight infection, and a root from *ulu ma'afala* for diarrhea, among other things. Through this work, the National Cancer Institute has found that a rainforest tree bark used to fight hepatitis may help treat AIDS.

Cox is working with Samoans to develop the rainforest as a source of income, in order to prevent logging by multinational corporations. Unlike many outsiders, he wins the respect of the Samoan people by showing a willingness to learn their language. Many Samoans call him Nafanua, which is the name of a protective deity.

1997 Western Samoa changes its name

In an effort to break from its colonial past, legislators in Western Samoa agree to drop the "Western" from the country's name, so that it officially becomes known as "Samoa." Germans had referred to the country as German Samoa and New Zealand later had described it as Western Samoa.

The change stirs up some opposition in American Samoa when some legislators and constituents complain that the independent nation is trying to bill itself as the "true" Samoa. Although American Samoa considers a bill that would refuse to recognize the name change, the measure fails. Residents of American Samoa are U.S. nationals but do not have citizenship. They elect their own governor and lieutenant governor, and choose legislators to represent them in their "Fono" or legislative branch. In addition, American Samoa has one nonvoting delegate in the U.S. House of Representatives.

1999: January New ferry begins service

The *Lady Naomi* begins passenger ferry service between the various islands of Samoa and American Samoa. Built with money provided by the Japanese government, the ship is handed over to the Samoan government in a ceremony whose participants include Japanese, Samoan, and U.S. Coast Guard officials. The *Lady Naomi* replaces a former ferry, known as the *MV Queen Salamasina.*

Bibliography

"American Samoa Concern Over Western Samoa's Name Change to Samoa," *Samoa News,* July 28, 1997.

"American Samoa's Faleomavaeaga Against Samoa Name Change Bill," *Samoa News*, January 26, 1998.

Brown, Richard P.C. "Do Migrants' Remittances Decline Over Time? Evidence from Tongans and Western Samoans in Australia," *The Contemporary Pacific*, 10:1 (Spring 1998) 107-152.

Fairbairn-Dunlop, Peggy. "Samoan Writing: Searching for the Written Fagogo," *Readings in Pacific Literature*, Paul Sharrad, ed., 136-160, (Wollongong: New Literatures Research Center, 1993).

Fialka, John J. "From Dots in the Pacific, Envoys Bring Fear, Fury to Global-Warming Talks," *Wall Street Journal*, September 31, 1997, A24.

Freidman, Jack and Mary Chandler. "Siaosi Talitimu, high chief of his Samoan tribe; wields power over a very long distance—from Denver," *People Weekly*, 24 (September 30, 1985), 65-66.

Ginzburg, Carlo. "Tusitala and his Polish Reader," *Raritan* 17:3 (Winter 1999), 85.

Hallowell, Christopher. "Rainforest Pharmacist," *Audubon* 101:1 (January 1999), 28.

Henry, Fred. *History of Samoa.* Apia, Western Samoa: Commercial Printers Limited, 1979.

Holmes, Lowell D., and Ellen Rhoads Holmes. *Samoan Village Then and Now.* Orlando, Fla.: Holt, Rinehart and Winston Inc., 1974, repr. 1992.

Liua'ana, Ben Featuna'i. "Dragons in Little Paradise: Chinese (Mis)fortunes in Samoa, 1900-1950," *The Journal of Pacific History* 32:1 (June 1997), 29-49.

Lynch, Colum F. "Heads Above Water: The Small Island States," *The Amicus Journal*, 18:1 (Spring 1996), 24.

Meleisea, Malama. *Change and Adaptation in Western Samoa.* Christchurch, New Zealand: MacMillan Brown Centre for Pacific Studies, 1992.

———. *The Making of Modern Samoa: Traditional Authority and Colonial Administration in the Modern History of Western Samoa.* Suva, Fiji: Institute of Pacific Studies, University of the South Pacific, 1987.

Owen, Elizabeth and Harry Benson, "Samoa," *Life*, 6 (May 1983), 32.

Paris, Sheldon. "Samoa Issues Stamps for Robert Louis Stevenson," *Stamps* 249:9 (Nov. 26, 1994), 5.

Samoa: A Travel Survival Kit (Lonely Planet Publications, 1996).

Stevenson, Robert Louis. *A Footnote to History: Eight Years of Trouble in Samoa*, 1892 (repr. Honolulu: University of Hawai'i Press, with introduction by Malama Meleisea, 1996)

Subramani. *South Pacific Literature: From Myth to Fabulation,* Suva, Fiji: Institute of Pacific Studies, 1985.

Taufe'ulungaki, 'Ana Maui. "The Cradle of Polynesian Culture," *UNESCO Courier*, October 1986, 15-19.

Wendt, Albert. "Towards a New Oceania," *Readings in Pacific Literature*, Paul Sharrad, ed. (Wollongong: New Literatures Research Centre, 1993), 9-19.

Williams, Ian. "Samoa through Western eyes ... the camera can lie," *Pacific Islands Monthly*, September 1996, 28-29.

Saudi Arabia

Introduction

Saudi Arabia is a modern state, having been created by royal decree in 1932. However, the first Saudi kingdom—i.e., a state ruled by the Al Saud family—dates to the mid-eighteenth century. At that time, the founder of the Saudi royal house, Muhammad ibn Saud ibn Muqrin ibn Markhan, was the *amir* (prince) of Dariya, a minor principality lying just west of Riyadh. In 1745, he accepted the teachings of Muhammad ibn Abd al-Wahhab, an Islamic reformer who founded a religious movement that came to be called *Wahhabism.* Embarking on a crusade to spread his new beliefs, Muhammad ibn Saud conquered several neighboring towns and so laid the foundations of the first Saudi state.

Religion and the fortunes of the ruling Al Saud family have thus been two dominant elements in the history of the Saudi kingdom. This history, however, can only be truly understood in the broader context of Saudi Arabia's physical geography and location, its resource base, the character of its peoples, and its role in modern geopolitics.

Saudi Arabia includes within its territory some four-fifths of the Arabian peninsula. Its area is approximately 865,000 square miles (1,960,582 square kilometers, or just under one quarter the size of the United States). The long coastline of the Red Sea forms the country's western border, while the Arabian (or Persian) Gulf lies to the east. In the north, Saudi Arabia shares land borders with Jordan, Iraq, and Kuwait. At the southern end of the Arabian peninsula, ill-defined borders separate Saudi Arabia from Yemen and Oman. The island state of Bahrain, Qatar, and the United Arab Emirates lie along the southern coast of the Arabian Gulf.

The Arabian peninsula is mostly desert, with annual rainfall averaging less than five inches and temperatures in summer reaching 120°F or 49°C. In the west, the rugged highlands of Hejaz and Asir parallel a narrow coastal plain along the Red Sea. These mountains stretch over 1300 miles from the Gulf of Aqaba in the north to the Yemen in the south, where they exceed 10,000 feet in elevation. The land slopes eastwards from this mountain fringe towards the interior deserts of the Najd. Further to the east lie the oil-rich lowlands bordering the Arabian Gulf. The southeast of the Arabian peninsula is an extensive sand desert called the *Rub al-Khali*, or the Empty Quarter.

Throughout history, physical geography has set Arabia apart from its surrounding areas. The peninsula's deserts, home to wandering Bedouin tribes and scattered oasis settlements, contrast dramatically with the fertile, more densely-populated lands that fringe the region. These lands, moreover—Mesopotamia, Egypt, and the *Levant* or eastern Mediterranean—have been persistent centers of economic and political power. A recurring theme in the history of the Arabian peninsula has been the relations between the Arab peoples and centers of civilization and empires that have arisen around them.

The population of Saudi Arabia is Arab by descent and culture. Ethnically, Saudis are descended from nomadic Semitic peoples who occupied the Arabian peninsula long before the coming of Islam. As with their ancestors, bloodlines are all important to the Saudis and the extended family (and, to a lesser extent, clan and tribal affiliations) dominates Saudi society. Saudis speak Arabic—a Semitic language that originated in Arabia—but perhaps even more important than language is the sense of cohesion religion brings to the Saudi people. Saudi Arabia is a Muslim state and governed in accordance with the *sharia*, the laws of Islam. Indeed, Islam originated in Arabia and the religion's most holy sites, the cities of Mecca and Medina, lie within the kingdom. Literally millions of Muslims flock to these cities each year at the time of the *hajj,* the pilgrimage to Mecca which is a Muslim's most sacred duty. As in most Muslim countries, the roles of men and women are well-defined, with women being confined to the home and playing virtually no role in public life.

The traditional economic systems of the Arabian peninsula were capable of supporting only a relatively sparse population. Although important cities existed and trade routes crossed the region, climate and geography were reflected in the dominance of pastoral nomadism. Fiercely independent tribes, paramount within their own tribal lands, roamed the peninsula grazing their herds of camels, goats, and sheep. Many a stereotype of Arabia is embodied in the image of the proud Bedouin, astride his camel, riding across the sandy wastes of the Rub al-Khali. Even today, Saudi Arabia stands out on a map of world population as a truly "empty quarter." Population densities average roughly 23 per square mile, with

the country's population totaling 20.1 million (1998 estimate). This figure, however, includes over 5 million foreigners, so the actual number of Saudi nationals is closer to 15 million. Saudi Arabia has a high rate of population increase, with annual growth estimated at 3.4% in 1998.

The early decades of the twentieth century brought many changes to the Arabian peninsula. But none was more dramatic in its impact on Saudi Arabia than the discovery of oil in the 1930s. Today, Saudi Arabia is the world's largest producer of petroleum and possesses more than one quarter of the world's proven petroleum reserves. Oil has raised Saudi Arabia to a place of prominence on the world economic and political stage. It has produced great wealth for its people. Saudis have one of the highest per capita incomes in the world, estimated at US$7,200 (US$1 = 3.75 Saudi riyals). Oil has supported the growth of towns and industries, of schools and hospitals, of roads and railways. It has also brought an influx of several million foreign workers into the labor force, along with ideas and attitudes that are alien to traditional Saudi society.

Saudi Arabia is coming under increasing pressure from global forces of modernization. How will the country and its people adapt to these forces? For some, the pace of progress towards a modern democratic state under the Al Sauds is too slow. For others, it is too rapid, raising the specter of a reactionary, and perhaps violent, turn towards Islamic fundamentalism.

Timeline

Prehistory to State Formation

4th–3rd millennia B.C. Early settlements along the Arabian Gulf

Pottery of the al-Ubaid period (c. 5500–4000 B.C. or later) of Mesopotamia occurs at sites along the Gulf coast of the Arabian peninsula north of Dammam. This suggests trade between the urban centers of Mesopotamia and the Indus civilization to the east.

Archeological remains dated to the thrid millennium B.C. on Tarut, a small offshore island just north of Damman, indicates trade continues to be of importance in Arabian Gulf. Some scholars associate the island with the ancient civilization of Dilmun (possibly Bahrain).

2nd millennium B.C. Walled towns in northwestern Arabia

Walled communities appear at several oases in northwestern Arabia. Their presence reflects the importance of trade as well as urban influences from Mesopotamia, Egypt, and the Levant.

1st millennium B.C. First Arab State

Al-Jwaf, also known as Dumat al-Jandal, rises to dominate much of northern and central Arabia. It is the first recorded independent Arab state in history.

c. 750 B.C.–A.D. 300 South Arabian civilizations

Advanced cultures emerge in the southern Arabian peninsula. The Sabaeans, based at *Saba* (the Biblical Sheba) in the Yemen, owe their prominence to control of the spice trade. Their influence extends along the caravan routes across Arabia, and they establish colonies at Mecca and Petra (in southern Jordan). Later, the Himyaritic culture (fl. c. 115 B.C.–A.D. 300) supplants the Sabaeans and Mineans, as the dominating element in South Arabian civilization.

4th century B.C. Nabataeans and Petra

The Nabataeans establish a kingdom in northwestern Arabia, with Petra as its capital. At its height, Nabataean power extends to Damascus in Syria and embraces much of what is now Saudi Arabia. The Nabataeans adopt the Aramaic alphabet for writing which evolves into the script used for the Qur'an, Islam's sacred book, and subsequently becomes the standard for modern Arabic.

24 B.C. Romans invade Arabia

Having conquered Egypt, the Romans invade Arabia from the Red Sea coast. Although this campaign fails, by the end of the first century A.D. Rome has diverted much of the spice trade between India and the Mediterranean to the sea routes skirting Arabia.

A.D. 50–60 Early description of Arabia

The *Periplus* of the Erythraean Sea, written by an anonymous Greek, provides one of the earliest written accounts of Arabia and Arabian civilization.

A.D. 106 Petra becomes a Roman province

Strategic interests lead the Romans to conquer Petra and incorporate the northern Nabataean territories into their empire as the province of Arabia. Most of the rock-cut monuments and tombs for which Petra is famous date from this period. Petra's inhabitants abandon the city around A.D. 551. It is eventually forgotten until its rediscovery in the early nineteenth century.

2nd–3rd centuries Rise of the Arabs

The decline of the Nabataeans, the waning of Parthian (Persian) influences in northeastern Arabia, and the decay of the South Arabian civilizations sees the expansion of the Bedouin tribes from the interior deserts. The populations of the peninsula become Arabized, adopting the language, social organization, and sometimes the nomadic economy of the desert Arabs or Bedouin.

c. 500 Quraysh tribe gains control of Mecca

Scattered Bedouin clans in central Hejaz come together to form the Quraysh tribe and seize Mecca, which soon emerges as a leading Arab commercial center.

c. 570 Birth of Muhammad, the founder of Islam

Muhammad is born in Mecca in A.D. 570.

Muhammad's full name is Abu al-Qasim Muhammad ibn Abd Allah ibn al-Muttalib ibn Hashim (*ibn* and also *bin* means "son of" in Arabic). He belongs to the Hashimi clan of the Quraysh tribe. Following tribal tradition, he becomes a merchant, accompanying caravans along the trades routes between Yemen and the Mediterranean.

c. 610 Muhammad's revelations begin

Muhammad sees a vision of the angel Gabriel who tells him he is chosen to be "the messenger of God." At frequent intervals until his death, he receives further revelations that he believes come directly from God, or Allah. Followers of

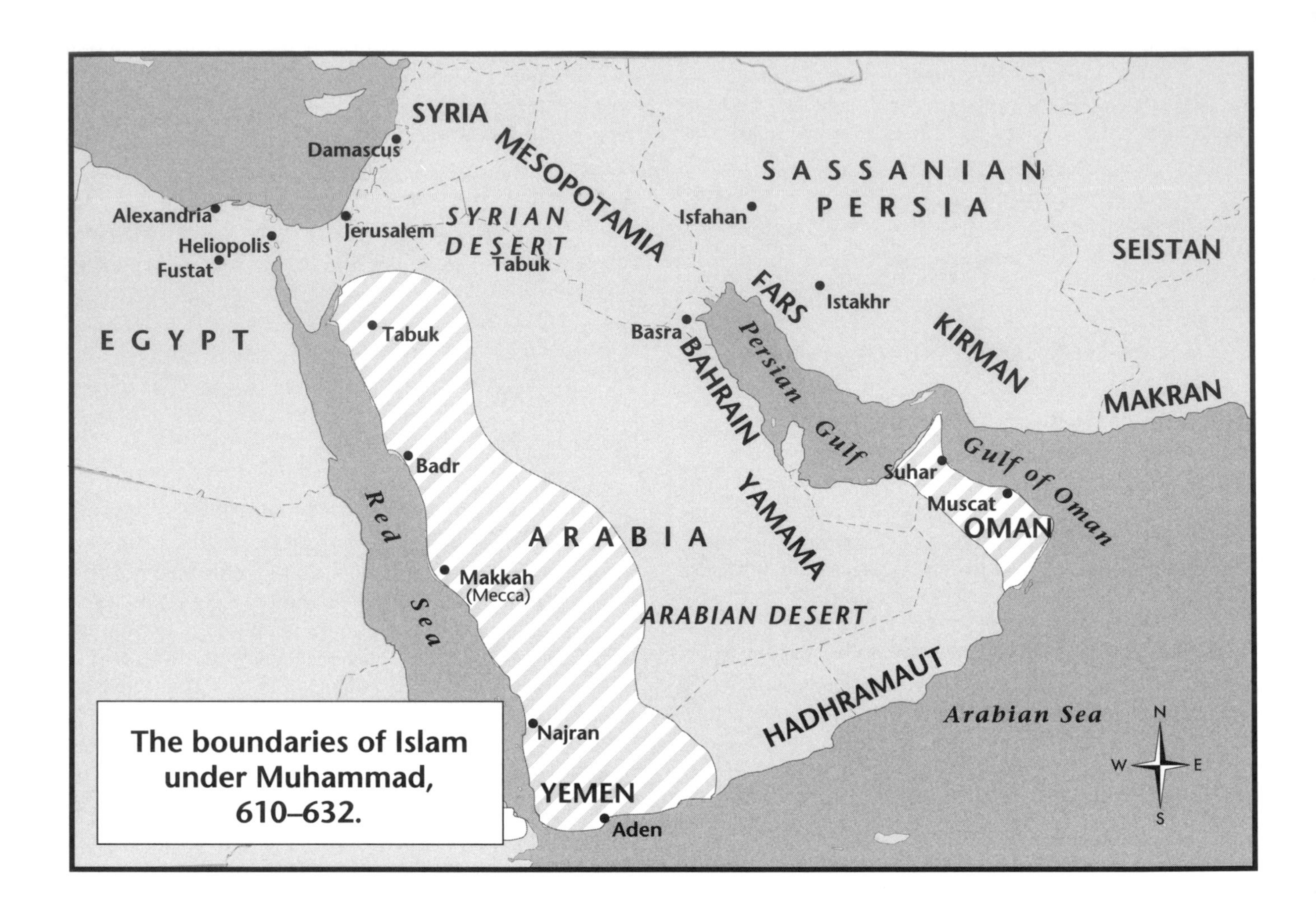

The boundaries of Islam under Muhammad, 610–632.

Muhammad later gather these divine messages and write them down in the Qur' an.

c. 613 Muhammad's public ministry

Muhammad begins preaching in public and attracts a small group of followers. The new religion eventually comes to be called *Islam* (surrender, i.e. to the will of Allah) and its followers known as *Muslims* (those who have surrendered).

615 Opposition to Muhammad begins

Although Muhammad's teachings are essentially religious in nature, they also embrace ideas such as the equality of all men and criticisms of the self-interested nature of the wealthy merchant classes in Mecca. The rich families of the city see the new religion as a threat to the political and social stability of the community. Opposition to Muhammad and his religion emerges in Mecca.

622: September 24 Muhammad reaches Medina

Fearing for his safety, Muhammad and some of his followers flee to Medina. This event is known as the *hijrah* (emigration) and marks the beginning of the Islamic calendar. The year in which the flight to Medina takes place (A.D. 622 according to the Western calendar) becomes the Year of the *Hegira* (hijrah) or A.H. (Anno Hegirae) 1 in Muslim history.

Medina is an oasis town some 220 miles (350 kilometers) to the north of Mecca. Muhammad and his companions from Mecca initially live in Medina as guests of the local Muslims. Eventually, however, they resort to the time-honored Arab custom of raiding passing caravans.

624: March 15 Victory at the Battle of Badr

Following an attempt to raid a wealthy caravan returning to Mecca, Muhammad and around 315 Muslims face an army of over 800 men from Mecca. Muhammad wins a decisive victory, which he sees as a sign of divine intervention.

625: March 23 Battle of Uhud

Mecca, increasingly aware of the growing threat from Muhammad, sends an army to attack Medina. Muhammad and his forces eventually drive the Meccans off but take heavy losses.

627: April **Medina under siege**

A confederacy of Arab clans opposed to Muhammad once again attacks Medina. The Muslims are well-prepared for the assault and successfully withstand a siege of two weeks. Disheartened by their failure, the Meccan forces disappear into the desert. Muhammad's position is greatly strengthened, and Medina is now clearly dominant in its power struggle with Mecca.

628 **Treaty of Al Hudaybiyah**

Muhammad leads a group of his followers on a pilgrimage to the *Kaba*, the sacred pagan site at Mecca. The Meccans are determined to stop the Muslims entering their city. Several days of negotiations result in the Treaty of Al Hudaybiyah, by which hostilities are ended, the Muslims withdraw, and the Meccans agree to allow the pilgrimage to occur in 629.

630 **Muhammad triumphs in Arabia**

Muhammad conquers Mecca without resistance and destroys the idols in the Kaba, the shrine Muslims thereafter consider to be the most sacred place on earth. Muslim forces also defeat a coalition of Bedouin tribes opposed to the Prophet at the battle of Hunayn. Virtually all of Arabia is now united under Muhammad's control and committed to Islam.

In the following centuries, the religion of Muhammad spreads far beyond the confines of Arabia. Yet, wherever Islam spreads, a few basic beliefs and practices serve to unite the Muslim community. These, the Five Pillars of Islam, are 1) the profession of faith; 2) frequent prayer; 3) the obligatory religious tax ; 4) fasting during the month of Ramadan; and 5) pilgrimage to Mecca.

632: June 8 **Death of Muhammad**

Muhammad dies after completing the Farewell Pilgrimage to Mecca.

Mid-7th century **Muhammad's successors**

The issue of succession divides the Arab tribes. Some argue that the successor (*caliph*) to the Prophet should be elected from the Muslim community at large, while others propose he come from the Quraysh tribe or even be descended from Muhammad. The tribes eventually accept Abu Bakr, a Qurayshi and father-in-law of Muhammad, as caliph.

Once the new caliph subdues the rebellious tribes who try to break away from Medina's control, he extends his empire far beyond the bounds of Arabia. Within two decades of Muhammad's death, Muslim armies from Arabia conquer Egypt, Syria, Iraq, and Persia.

656 **The caliphate leaves Medina**

The people of Medina refuse to recognize Ali (Muhammad's son-in-law) as the fourth caliph and force him flee from Medina. A rival from Damascus belonging to the Ummayyad clan claims the title and eventually secures recognition as caliph. From this time on (until 1924) the *caliphate*, the seat of the spiritual leader of the Muslim world, lies outside Arabia.

680 **Battle of Karbala**

Husayn, younger son of Ali and grandson of the Prophet through his daughter Fatima, asserts his claim to the caliphate. The Damascus caliph defeats and kills Husayn and his forces at the battle of Karbala in Iraq. Husayn becomes a martyr to the *Shias*, Muslims who believe the caliph should be a direct descendant of the Prophet.

8th century **Pilgrims' way built across Arabia**

Zubaydah, the wife of the caliph, builds the *Darb Zubaydah*, an ancient route taking pilgrims from Baghdad to Mecca.

Mid-9th century **Sunnism**

The Abbasid caliphs, starting with al-Mutawakkil (r. 847–861), establish a strict from of Muslim orthodoxy known as *Sunnism.* Most Saudi Arabians are Sunnis.

899 **Shias establish community at Al Ahsa**

A splinter sect of Muslims, the Carmathians (al-Qaramitah), establish a state at Al Ahsa in eastern Arabia. Originating in Iraq, the Carmathians, who are considered heretic by many Muslims, spread death and destruction across Arabia. They raid Mecca in 930 and briefly seize the sacred Black Stone of the Kaba.

c. 951 **Sharifate of Mecca established**

Jafar al-Musawi becomes sharif of Mecca, charged with watching over the sanctuaries of the Hejaz and the annual pilgrimage. The title *sharif* means nobleman, but in the Hejaz comes to be restricted to the descendants of Hasan, grandson of the Prophet and the eldest son of Ali.

10th–11th centuries **Fatimids of Egypt dominant in western Arabia**

The Fatimid dynasty of Cairo extends its influence into western Arabia. The Fatimids claim descent from the Prophet through his daughter Fatima, and their followers regard their leader as the true caliph. The Fatimid ruler is the head of the Ismaili sect of Shias and never recognizes the rival Abbasid caliphate. The Fatimids are in constant conflict with the Sunni caliph. Their eventual control of the Hejaz and its sacred sites bring them great prestige.

12th century **Crusaders in the Red Sea**

Sporadic attempts by Crusaders from Syria to gain a foothold in the Red Sea are unsuccessful.

c. 1200 Qatada bin Idris becomes Sharif of Mecca

Qatada bin Idris seizes power in Mecca, and his descendants retain the sharifate until the twentieth century. The new sharif succeeds in thwarting the designs of the Ayyubid dynasty of Cairo (successors to the Fatimids) on the Hejaz and Islam's holy cities.

Mid-13th century Rise of the Rasulids in Yemen

A dynasty founded by Ali ibn Rasul establishes a powerful state in Yemen and rules over extensive areas of southern Arabia for two centuries.

14th century Mamluk sovereignty over Hejaz

The Mamluk rulers of Egypt lay claim to the title *sultan* (supreme temporal authority) of the Muslim world. They bring the Hejaz and Mecca under their sovereignty.

c. 1446 Al Sauds' ancestors settle in Najd

The ancestors of the present Saudi royal family emigrate from Al Qatif in eastern Arabia and found the village of Dariya in the southern Najd. They are traditionally considered to belong to the Anazah, a group of Arab tribes viewed as being of pure lineage.

1453: May 29 Ottomans capture Constantinople

The Ottoman Turks capture Constantinople, the capital of the Byzantine (or Eastern Roman) empire and rename it Istanbul. The Ottoman empire survives until the early twentieth century.

1516–17 Ottomans defeat the Mamluks

Ottoman victories over the Mamluk sultans give them the former Mamluk lands in western Arabia. Their control, however, is limited mostly to the lands bordering the Red Sea.

1541 Portuguese attack Jiddah

The Portuguese unsuccessfully attack Jiddah, an important port on the Red Sea.

c. 1727 Muhammad ibn Saud becomes amir of Dariya

Muhammad ibn Saud (c. 1703–65) seizes power in Dariya and becomes its ruler.

1744–45 Muhammad ibn Saud accepts Wahhabism

Muhammad ibn Abd al-Wahhab (1703–92), an Islamic reformer, seeks refuge in Dariya. Muhammad ibn Saud grants him sanctuary and soon accepts his religious teachings. Ibn Abd al-Wahhab thus acquires the military backing to spread his beliefs, while the amir now has a religious cause to justify his military and political expansion. Muhammad ibn Saud subsequently acquires the title *Imam* (spiritual leader) of the Wahhabi Muslims. From this time on, Wahhabism and the fortunes of the Al Sauds are inextricably entwined.

The Wahhabi revival is a puritanical movement, based on the Hanbali school of Sunni teaching, that calls for a return to the original doctrines of Islam. Ibn al-Wahhab preaches against the loose practices of the townspeople and in particular against the widespread veneration of local saints and their graves. (Even today Saudis bury their dead in unmarked graves so they do not become shrines.) He also proclaims the necessity for the *jihad* or holy war against Muslims who do not accept his teachings.

1745–65 Rise of the First Saudi State

Wahhabism attracts followers among the local desert tribes, and Muhammad ibn Saud and his warriors conquer several neighboring towns and oases. This period is one of constant struggle with anti-Wahhabite forces and even occasional defeats. By the time of his death, however, Muhammad ibn Saud controls most of the southern Najd—except for the Riyadh amirate.

1765 Abd al-Aziz becomes amir

Muhammad ibn Saud dies in 1765 (1179 A.H, according to the Muslim calendar). His successor, Abd al-Aziz (r. 1765–1803), continues his father's policy of expansion, unifying most of central Arabia under Saudi control.

1773 Riyadh is captured

Saudi forces take Riyadh.

1801 Al Sauds sack Karbala

Saudi warriors raid northward into what is now southern Iraq and sack Karbala, a city sacred to the Shia sect. They destroy the tombs of several holy men, including Husayn, the grandson of the Prophet and one of the most revered of all the Shia saints.

1803 Al Sauds capture Mecca

Saudi forces take Mecca from the Hashimites, who are now vassals of the Ottomans.

1803 Abd al-Aziz assassinated

Imam Abd al-Aziz is assassinated and is succeeded by his son, Saud bin Abd al-Aziz (r. 1803–14).

1805 Saudis take Medina

Saudi forces capture Medina, the second holiest city in the Islamic world.

1811–14 Ottomans respond to Saudi expansion

Al Saud expansion within the Arabian peninsula attracts little attention. However, their incursions into Ottoman territory and particularly their capture of Islam's holy places evoke an Ottoman response. Muhammad Ali, the Ottoman viceroy and ruler of Egypt, sends an army to western Arabia. After several years of fighting, the Egyptians retake the Hejaz and its holy cities but never succeed completely in subduing the Saudis.

1814 Saud dies

Saud bin Abd al-Aziz dies. His son Abdullah (r. 1814–18) succeeds him as Saudi ruler. Following Saud's death, the Saudis retreat to their Najd heartland.

1818 End of the First Saudi state

Ibrahim Pasha, son of the Ottoman viceroy of Egypt, leads another army against the Saudis. After a two-year campaign, he captures and destroys Dariya. The Saudi amir, Abdullah ibn Saud, is taken to Istanbul and executed. The first Saudi state comes to an end.

1824–43 Rise of the Second Saudi state

Turki ibn Abdullah (r. 1824–34), an uncle of Saud, assembles an army of tribal warriors and drives the Egyptians from the Najd. He establishes himself at Riyadh in 1824, which becomes the capital of the second Saudi state. A rival assassinates Turki in 1834, but his son Faysal (r. 1834–38; 1843–65) defeats the killer and succeeds his father as the Saudi Imam. The Egyptians invade Najd in 1838, take Faysal captive, and install their own candidate as amir.

1843–65 Faysal's second reign

Faysal escapes from Cairo in 1843 and regains control of Najd, initiating the glory days of the second Saudi state. He restores order within the Saudi state and expands its territory northwards into the tribal lands of the Shammar and south to the borders of Oman.

1855–56 Burton visits Mecca and Medina

In disguise, Sir Richard Burton (1821–90) visits Islam's holy cities, from which non-Muslims are banned under pain of death. The English Orientalist and explorer describes his experiences in *Pilgrimage to El-Medinah and Mecca.*

1865–91 Decline of the Second Saudi state

Dynastic squabbles following Faysal's death in 1865 herald the downfall of the second Saudi state. Faysal's sons Abdallah (r. 1865–71; 1875–89), Saud (r. 1871–75) and Abd al-Rahman (r. 1875; 1889–91) are bitter rivals for the throne. They engage in constant intrigue and civil war, weakening the Saudi state. Meanwhile, Ottoman expansion extends Istanbul's influence into western Arabia and parts of the Gulf coast. The Al Rashids of Hail in northern Najd rise up against their Saudi masters and eventually wrest control of all the interior.

1891 Saudi Imam flees to Kuwait

After an unsuccessful attempt to regain power from the Al Rashids, Abd al-Rahman flees into exile in Kuwait.

The Kingdom of Saudi Arabia

1902: January Beginning of the modern Kingdom of Saudi Arabia

Abd al-Aziz ibn Abd al-Rahman Al Saud (1880–1953) leads a band of forty men from Kuwait into the interior. They cross the deserts of the Nadj and make a surprise attack on Riyadh, capturing the town. Abd al-Rahman is quite content to leave the new Saudi state in the competent hands of his son.

1908 Husayn becomes sharif of Mecca

Rival Hashimi clans select Husayn bin Ali al-Hashimi (1852–1931) as a compromise candidate to be sharif of Mecca.

1908 Hejaz railroad completed

The Ottomans complete construction of the Hejaz railroad linking Syria and Medina. The railroad is destroyed during the Arab rebellion against the Ottomans in 1917.

c. 1908 Saudi sovereignty extended over central Najd

The Saudis extend their control over central Najd. They repulse an army sent by Sharif Husayn of Mecca to recapture the oasis of Al Qasim in 1910.

1912 First Ikhwan settlement is founded

Realizing he needs more than the promise of plunder to keep the loyalty of the Najdi tribes, Abd al-Aziz rallies the tribesmen to the banner of religion. He actively encourages an Islamic revival movement, creating the *Ikhwan* or Brethren, a body of tough, dedicated Muslim warriors he settles at various oases in the Najd.

1912 Saudi state becomes a sultanate

Abd al-Aziz, better known in the West as Ibn Saud, renames his state the Sultanate of Najd and Its Dependencies.

The Ikhwan capture the oasis settlements of Al Ahsa and Al Qatif in eastern Arabia from the Ottomans and integrate the region into the Saudi kingdom.

1913: July 29 Anglo-Ottoman Convention

London and Istanbul establish the Blue Line, delineating the boundary of the Ottoman's eastern province (i.e. Najd) in Arabia. This cartographic line is the first attempt to define the boundary between what is to become the independent Kingdom of Saudi Arabia and the British-protected states of the Gulf region.

1914: May Treaty with Ottomans

Saudis formally accept Ottoman over-lordship.

1914: August–November Outbreak of World War I

Turkey signs a secret treaty to support Germany against Russia. This pits the Allies against the Ottoman empire, and Britain and France formally declare war against the Turks on November 5.

1915 Britain recognizes Ibn Saud

Britain recognizes Ibn Saud as Sultan of Najd and Al Ahsa.

1916: June 5 The Arab Revolt

Sharif Husayn of Mecca assumes the title King of Hejaz and, with British support, proclaims the Arab Revolt against the Ottomans. It is from the Hejaz that Colonel T. E. Lawrence launches his attacks against the railroad that runs from Damascus to Medina.

Thomas Edward Lawrence (1888–1935) is a British scholar and author who gains fame as a guerrilla leader commanding Arab forces behind enemy lines. Dressed in Bedouin robes and riding a camel, he leads his men against the Turks. He comes to be known as Lawrence of Arabia and writes about his wartime experiences in his *Seven Pillars of Wisdom.*

1916: October 29 Sharif Husayn takes the title King of Hejaz

Sharif Husayn of Mecca adopts the title of King of Hejaz

1916: November 5 Sharif Husayn becomes King of the Arabs

Sharif Husayn assumes the title King of the Arab Countries. The British (and the French also), who have interests in the region, reject Husayn's claim to pan-Arab leadership, as do the Al Saud family and other Arab leaders.

1918: May First armed clashes between Saudis and Hashimites

Both Husayn and Ibn Saud are allies of the British, who keep their natural rivalry in check during hostilities with the Turks. However, with the end of World War I approaching, their forces clash at Al Khurmah, an oasis on the border of the Hashimite and Saudi kingdoms.

1918: October 30 Armistice between Allies and Turks

The Mudros Armistice (signed aboard H.M.S. *Agamemnon* at anchor off Mudros on the island of Lemnos in the Aegean Sea) ends hostilities between the British and their Arab allies and the Ottoman Turks. The war in Europe ends twelve days later.

1920: April 19–26 San Remo talks

The victorious Allies (with the United States conspicuously absent) meet in Italy to draft a peace treaty with Turkey. The Ottoman provinces lying to the north of the Arabian peninsula are assigned to Britian as mandates of the newly created League of Nations (Transjordan, Iraq) and France (Syria, Lebanon).

1921: November 2 Ibn Saud annexes the Al Rashid emirate

Ikhwan warriors raid far north into Al Rashid territory, fighting the Shammar tribes, and eventually capturing Hail, the Al Rashid capital. Ibn Saud annexes the Rashidi lands.

1922 Abdullah seizes Transjordan

Abdullah, Sharif Husayn's son, seizes Transjordan and with British acquiescence proclaims himself king (the late king Husayn of Jordan was Abdullah's grandson). The British also make another brother, Faysal, King of Iraq.

Ibn Saud finds his northern borders ringed with what he sees as hostile Hashimite regimes.

1923: December 17 Kuwait Conference convenes

The Kuwait conference convenes to deal with continuing Saudi-Hashimi border problems.

1924–25 Saudis occupy Hejaz

Turkey abolishes the caliphate on March 3, and two days later Sharif (now King) Husayn declares himself caliph. This incenses the Ikhwan, and the Saudis invade Hejaz. They capture Mecca, and after a year-long siege, take Jiddah in December 1925.

1926: January Ibn Saud assumes title of king

Following his capture of Hejaz, Ibn Saud becomes King of Hejaz and Sultan of Najd and Its Dependencies.

1927: May 20 Treaty of Jiddah

Britain recognizes the new Saudi state.

1927–30 Ikhwan revolt

Ibn Saud has to suppress his unruly Ikhwan warriors. Tribal chiefs lead the Ikhwan on raids into Iraq and Kuwait, bringing British retaliation. Forces loyal to the king break the back of the Ikhwan movement at two battles, Sibilah (1929) and Shuayb Al Awjah (1930).

1929 First European crosses the Empty Quarter

Bertram Thomas becomes the first European to cross the desert wastes of the Rub al-Khali. Perhaps the best known European explorer of the Empty Quarter is Wilfred Thesiger.

The city of Mecca, holy to Muslims, earns much of its income from the thousands who travel there each year on the religious journey known as the hajj. (EPD Photos/CSU Archives)

1930: November Asir annexed

Ibn Saud annexes Asir in southwestern Arabia.

1932: September 23 Kingdom of Saudi Arabia

Ibn Saud unifies Najd and Hejaz, which are governed separately by members of the Al Saud family, into the Kingdom of Saudi Arabia and takes the title of king.

The new state differs little from the desert kingdoms that preceded it. Its stability is challenged by internal threats and external border conflicts. An autocratic monarch reigns over a country with limited resources and widespread poverty (the government's major source of revenues is its income from the annual hajj).

1932: October Idrisi uprising

The Idrisi amir of Asir leads an unsuccessful revolt against Ibn Saud.

1933 SOCAL gains oil concession

Ibn Saud grants Standard Oil of California (SOCAL) a concession to explore for oil in the Al Ahsa region of eastern Saudi Arabia.

1934: March–May Saudi-Yemeni war

Saudis defeat Yemen and impose a border agreement. The boundaries in the Empty Quarter between Saudi Arabia and the British-protected areas in southern Arabia remain undefined and continue to be disputed.

1938: March 3 Discovery of oil

Saudi Arabia's first commercial oil strike is made in the desert just south of Damman on the Gulf coast.

The modern city of Dhahran grows up at this location. The city is the site of ARAMCO's headquarters, and also has King Fahd University of Petroleum and Minerals, a Saudi Air Force base, and one of Saudi's three international airports. Dhahran, along with nearby Damman and the port terminal of Ras Tanurah, lies at the very heart of the Saudi petroleum industry.

1939–45 World War II

Saudi Arabia declares its neutrality at the outbreak of World War II, although in reality it is pro-British. Though Saudi Arabia is not involved in hostilities, the war is crippling to its economy. The country's major source of income, revenues from the hajj, is curtailed by the widespread conflict. Britain pays Saudi Arabia a war subsidy to replace its losses and stave off financial collapse.

1944 ARAMCO

The California Arabian Standard Oil Company is renamed the Arabian American Oil Company (ARAMCO).

1945: March 22 Arab League

Saudi Arabia joins with Egypt, Iraq, Lebanon, Syria, Transjordan (now Jordan), and Yemen to found the Arab League (also called the League of Arab States). The aim of the league is to strengthen political, economic, and cultural ties, although subsequently the League's members agree to cooperate on military defense issues.

1951 Railroad links Damman and Riyadh

A railroad is built between Saudi Arabia's main oil producing region, Damman, and the capital, Riyadh.

1953: November 9 Ibn Saud dies

Ibn Saud dies in Riyadh. His eldest surviving son, Saud bin Abd al Aziz (r. 1953–64), succeeds him as king.

1956: November Suez Crisis

Saudi Arabia breaks diplomatic relations with Britain and France after these countries, along with Israeli forces, attempt to seize the Suez Canal.

1958: March 24 Crown Prince Faysal becomes prime minister

King Saud proves unable to handle the complexities of administering the Saudi state. His reign is characterized by palace intrigue, corruption, profligate spending, and waste. Despite the large revenues generated by oil, the kingdom is virtually bankrupt.

Under pressure from senior princes and family members, Saud appoints his half-brother Faysal ibn Abd al-Aziz to be prime minister. Faysal assumes responsibility for all government matters, including foreign affairs, and immediately begins instituting administrative and fiscal reforms.

1960: September OPEC formed

Saudi Arabia becomes a charter member of the Organization of Petroleum Countries (OPEC).

Discoveries of new oil fields create a glut of petroleum on world markets in the immediate postwar years. Oil companies maintain prices by unilaterally cutting production, a policy that drastically reduces the revenues of the oil-producing countries. Saudi Arabia, along with Iran, Iraq, Kuwait, and Venezuela, band together in an organization that seeks to present a common front to the oil companies.

1960: December King Saud takes control of government

King Saud takes back control of the government from Faysal.

1960s Confrontation with Egypt

Saudi relations with republican Egypt under President Nasser deteriorate, particularly during the civil war in Yemen (1962–70). Egypt provides 70,000 troops to prop up the Yemen Arab Republic's government, while Saudi Arabia supports the Yemeni royalists. Egyptian planes even bomb Saudi border towns.

1962: October Faysal re-appointed prime minister

Dissatisfaction with King Saud, particularly over his relations with Egypt, leads the Al Saud family to force him to reappoint Crown Prince Faysal as prime minister. Khalid, another half brother, becomes deputy prime minister.

1964: November 3 King Saud abdicates

Religious leaders issue a *fatwa* (a legally binding Islamic proclamation) naming Faysal king and the Al Sauds, with widespread support, force the Saud to abdicate. He leaves Saudi Arabia and dies in Greece in 1969.

1964–75 Reign of Faysal, architect of modern Saudi Arabia

Faysal proves a strong and capable leader, and many regard him as the architect of modern Saudi Arabia. He uses rapidly expanding oil revenues to lay the foundation for Saudi Arabia's economic and social development. He establishes the direction of domestic and foreign policies for the next several decades. Faysal favors modernization but is nonetheless committed to preserving an Islamic society and state in Saudi Arabia.

King Faysal strongly supports education, seeing it as essential for social and economic progress. Four secular universities come into existence during his reign, to complement Saudi Arabia's religious universities. Education is free to all Saudis.

1973: October 6 Arab-Israeli war breaks out

Saudi Arabia sends token military units to support the Arab cause.

1973: October 16 Gulf producers raise oil prices

With a world shortage of oil, OPEC demands higher prices for its oil supplies. The oil companies meet with OPEC to negotiate prices but fail to come to an agreement. The oil-producing countries of the Persian Gulf unilaterally raise oil prices seventy percent on October 16.

1973: October 20 Oil embargo

Saudi Arabia and OPEC impose an oil embargo, cutting production and placing an absolute ban on the sale of oil to the U.S. and the Netherlands, two countries considered most friendly to Israel in its war with the Arab countries.

The oil embargo has a major impact on the industrialized world and especially on the United States. The American people, used to some of the lowest gasoline prices in the world, are outraged at the price increases and the shortages that lead to long lines at the gas pumps. One consequence of the embargo is that the U.S. and other affected countries seek to reduce their dependence on imported petroleum, implement conservation policies, and develop new sources of energy. Such policies contribute to the world oil glut of the 1980s.

1973: December OPEC raises prices again

OPEC raises its prices another 130 percent.

The increases in oil prices following the Arab oil embargo generate huge revenues for the oil producing countries. In Saudi Arabia, these funds help support the rapid economic growth and development pursued by King Faysal.

1974: March Oil embargo ends

Saudi Arabia calls off the oil embargo against the United States.

1975: March 25 King Faysal assassinated

King Faysal is killed by a deranged nephew, Khalid ibn Musaid. The assassin is beheaded according to Saudi custom. Faysal's half brother Khalid bin Abd al-Aziz (r. 1975–82) succeeds him as king.

1975 Various ministries established

Despite Faysal's death, his social and economic policies remain in place. The government establishes Ministries of Higher Education, Industry and Electricity, Municipalities and Rural Affairs, Planning, Public Works, and Housing, and Posts, Telephones, and Telegraphs.

1978 US sells F-15s to Saudis

U.S. Congress approves sales of F-15 combat aircraft to Saudi Arabia.

1979–80 Shia demonstrations

Anti-government Shia demonstrations, inspired by the revolution in Iran, occur in Saudi Arabia's eastern province.

1979: November 21 Great Mosque seized

Armed gunmen, under Juhayman bin Muhammad bin Sayf al-Utaybi, seize and occupy the Great Mosque of Mecca. The insurgents, mostly young theological students, oppose the royal family and its policies of modernization. Government forces recapture the mosque on December 5. Juhayman and most of his followers are subsequently executed.

1980 Saudi Arabia gains control of ARAMCO

The Saudi government, which has previously acquired a share of ARAMCO, negotiates one hundred percent ownership of the oil company.

1980–88 Iran-Iraq War

Saudi Arabia sees this prolonged and bloody war between its neighbors as a significant threat to its national security. Riyadh backs Iraq against the revolutionary government of the Ayatollah Khomeini in Teheran. Iran is a Shia country and attempts to ferment trouble among Saudi Arabia's Shia minority.

1981: May 25 Gulf Cooperation Council established

Saudi Arabia, Kuwait, Bahrain, Qatar, the United Arab Emirates, and Oman set up the Gulf Cooperation Council (GCC). Its aims are to provide for cooperation among the member states in the economic, social, cultural, political and security fields. The founding of the GCC is in part a reflection of Gulf concerns arising from the Iran-Iraq war.

1982: June Fahd becomes king

Fahd bin Abd al-Aziz al Saud, Khalid's half brother, succeeds to the throne when Khalid dies of a heart attack.

1985 Saudi Arabia signs arms deal with Britain

Saudi Arabia places orders for Britain's Tornado combat aircraft and other military equipment as part of an accord known as the Al Yamamah agreement.

In the post World War II era, Saudi Arabia relies primarily on the U.S. for its external security. Apart from the prestigious and well-equipped Royal Saudi Air Force, the Saudi professional military has limited capabilities. The U.S. Congress will not approve the sale of F-15s to Saudi Arabia at this time, so King Fahd turns to Britain. Three years later, Saudi Arabia and Britain sign an accord (Al Yamamah II) which is thought to be the largest arms deal ever (perhaps as much as $25 billion).

1986 King Fahd Causeway opens

The sixteen-mile long causeway linking Al Khobar in Saudi Arabia to Bahrain opens.

1986 End of guaranteed employment for graduates

University students are no longer guaranteed employment with the government on graduation. One consequence of this and a general lowering of educational standards is a growing number of unemployed—and often unemployable—graduates in the labor force.

1987: July 31 Mecca riots

Iranian government attempts to fan anti-Saudi feelings among the millions of pilgrims who converge on Saudi Arabia during the hajj lead to riots in which some 400 people are killed.

1990: July Pilgrims killed in stampede

Some 1,423 people making the hajj to Mecca are killed in a stampede that occurs in a pedestrian tunnel.

1990: August 2 Iraq invades Kuwait

Iraqi troops invade and quickly occupy Kuwait.

Fearing an Iraqi invasion, King Fahd asks for U.S. forces to be deployed in his country as part of a multi-national force to deter Iraqi aggression. Within a week, Operation Desert Shield is under way. Over the next five months some 540,000 U.S troops are sent to the Gulf region to join forces from over twenty nations, including several Arab countries.

1991: January 18 Operation Desert Storm air offensive

Failing to secure Iraq's withdrawal from Kuwait by diplomatic means, allied forces launch an air offensive on targets in Iraq and Kuwait.

1991: February 24 Ground campaign begins

Mounting a diversionary frontal attack on Kuwait, allied forces sweep through Iraq's southern deserts to the Euphrates River. Iraqi losses in the fighting are heavy. Prince Khalid bin Sultan Al Saud commands Saudi and Arab forces in Operation Desert Storm, while General H. Norman Schwarzkopf commands western troops.

1991: February 28 Cease-fire

Iraq accepts U.N. conditions for an end to the fighting and agrees to a cease-fire.

1991: November 6 Women protest driving ban

Forty-seven Saudi women drive through the streets of Riyadh to protest the ban on women driving vehicles.

Women continue to be subject to many restrictions in Saudi society. They must be veiled in public, cannot drive, cannot hold public office, and not are allowed to work (except in certain professions such as teaching or the health fields). At universities they are banned from studying subjects such as engineering, journalism, and law.

1992: March 1 Fahd announces Basic Law of Government

King Fahd issues a series of royal decrees that change the structure of Saudi government. He proclaims a Basic Law of Government, essentially a written constitution; he provides for the establishment of regional authorities; and he announces that a Consultative Council (*Majlis al-Shura*) is to be formed.

Such developments do little to alter Saudi Arabia's political system. The country remains an absolute monarchy, with no legislature or political parties. Power is concentrated in the hands of the King, who rules through a Council of Ministers, most of whom are senior princes of the Al Saud family.

1993: May 13 Opposition group declared illegal

The Saudi authorities disband the Committee for the Defense of Legitimate Rights (CDLR). This group is one of many opposing the Saudi government's policies of modernization. The "legitimate rights" in the Committee's name are rights granted by Islam rather than political or human rights as generally understood in the West.

Opponents of modernization call themselves *Salafiyin* (followers of the pious ancestors). In Saudi Arabia, these are mostly fundamentalist *ulama* or religious teachers, and there is little popular support for the movement. However, more young people are joining the movement. They are often graduates of Muhammad ibn Saud University in Riyadh, which offers its students a religious curriculum but no marketable skills. Many students volunteer for the Committee for Propagating Virtue and Suppressing Evil and act as religious police attempting to force the population to conform to strict Islamic norms. This disaffected element is increasingly resorting to violence as a means of religious (political) expression.

1995–2000 Sixth five year plan

Saudi Arabia's sixth five-year plan aims to reduce the country's dependence on oil (petroleum accounts for over ninety percent of the country's exports) as well as address other pressing economic problems. The plan provides support for the non-oil sector of the economy. It scales back spending on social development and economic infrastructure, a consequence of declining oil revenues. It also address another serious problem, labor. Saudis make up less than one-third of the Saudi labor force. Guest workers, from American and European petroleum engineers to domestic servants from South Asia and the Philippines, account for the rest. The plan promotes the Saudi-ization of the labor force. However, unemployment among the Saudis remains high. This reflects, in part, Saudi attitudes towards work (many Saudis see work as undignified), as well as an educational system that is not creating a trained labor force.

1995: November 19 Car bomb explodes in Riyadh

A car bomb explodes outside the offices of the Saudi Arabian National Guard. Seven foreign nationals, including five members of a U.S. training mission, are killed and some sixty people are injured. Four Saudis, with alleged links to fundamentalist Islamic groups outside the country, are arrested and later executed.

1996: January 1 Crown Prince assumes royal duties

Because of ailing health the seventy-four-year old King Fahd delegates the responsibility for government to his younger half-brother Crown Prince Abdullah bin Abd al-Aziz Al Saud.

1996: June 4 Military base bombed

A bomb attack on a military housing complex at Al Khobar near Dhahran kills 17 U.S servicemen and injures four hundred other, including Saudis and Bangladeshis as well as Americans.

1997 Foreign tourism begins

Saudi Arabian Airlines brings in the first groups of foreign tourists allowed to visit Saudi Arabia. Attractions include the mountainous Asir National Park near Abha and Madain Salah, a spectacular archeological site of the Nabataean period (c. 100 B.C.–A.D. 100) north of Medina.

1997: April Fires kill hajj pilgrims

Fires in a tent-camp housing pilgrims near Mecca kill 343 people.

1998: February Saudis deny U.S. use of its air bases against Iraq

As the conflict over U.N. weapons inspections in Iraq intensifies Saudi Arabia declares that it will not allow U.S. planes to stage strikes against Iraq from air bases in Saudi Arabia.

1999: April One million foreign pilgrims visit for the hajj

An estimated one million foreigners join over a million Saudis for the hajj pilgrimage to Mecca and Medina. The sacrifice of around a million animals (mainly sheep) accompanies the annual ritual.

1999: May 28 Drug smugglers executed

Saudi Arabia executes a Pakistani man and a Nigerian woman for smuggling drugs into the country. In addition to drugs, alcohol is banned in the kingdom.

Bibliography

Abir, Mordechai. *Saudi Arabia: Government, Society, and the Gulf Crisis.* London and New York: Routledge, 1993.

Caesar, Judith. *Crossing Borders: An American Woman in the Middle East.* Syracuse, NY: Syracuse University Press, 1997.

Lacey, Robert. *The Kingdom.* London: Hutchinson, 1981.

Lewis, Bernard. *The Arabs in History.* London: Hutchinson University Library, 1966.

Lipsky, George A. *Saudi Arabia: Its People, Its Society, Its Culture.* New Haven, CT: HRAF Press, 1959.

Long, David E. *The Kingdom of Saudi Arabia.* Gainesville, FL: University Press of Florida, 1997.

Peterson, J. J. *Historical Dictionary of Saudi Arabia.* Metuchen, NJ and London: Scarecrow Press, 1993.

Salibi, Kamal. *A History of Arabia.* Delmar, NY: Caravan Books, 1980.

Thesiger, Wilfred. *Arabian Sands.* New York: Dutton, 1959.

Wilson, Peter W. and Douglas F. Graham. *Saudi Arabia: The Coming Storm.* Armonk, NY: M. E. Sharpe, 1994.

Singapore

Introduction

Singapore has been an important trading center since the British first established a settlement there early in the nineteenth century. With a well-protected harbor and central location on the ocean passage that connects East Asia with the South Asian subcontinent and the lands to the west, Singapore is Southeast Asia's largest port. Since gaining its independence in 1959, the tiny island nation has thrived economically, becoming one of the world's leading trade and financial centers as well as developing its own high-tech industrial base.

Located south of the Malay Peninsula, the island of Singapore is only 26 miles (42 kilometers) long and 14 miles (23 kilometers) wide. The country also encompasses over fifty smaller nearby islets, which are mostly uninhabited. The main island consists largely of low-lying terrain with a central range of hills, as well as areas of rain forest. As of 1996, Singapore had an estimated population of 3.35 million people.

According to legend, the first trading center on the island of Singapore was founded in the twelfth or thirteenth century by an Indian prince and named *Singapura* ("lion city") to commemorate an encounter the prince had with a wild animal. The island is also mentioned in fourteenth- and fifteenth-century chronicles of the region in connection with pirate raids and other activities.

Colonialism and Sovereignty

Between the sixteenth and nineteenth centuries, the Western colonial powers of Portugal, the Netherlands, and Great Britain vied for control of the Malay Peninsula and the island of Singapore. The British presence on Singapore began in 1819 when Sir Thomas Stamford Raffles, then serving as governor of the settlement of Bengkulu, in Sumatra, established a trading post on the island, which was ceded to the British five years later. Over the next forty years the settlement flourished, becoming an important *entrepôt* center—a central shipping area for the re-export of raw materials from the surrounding region, especially Malayan rubber and tin. During this period, civil unrest in China brought many Chinese immigrants to the island, permanently changing its ethnic composition. (Today ethnic Chinese make up over seventy-five percent of the population. Malays, descendants of the island's original settlers, comprise the other major ethnic group.) By 1864, Singapore had become the foremost trading center in Southeast Asia, with annual trade valued at 13 million pounds. In 1867 the island became a British crown colony.

In 1922 the British established a major naval base on the island to protect its interests in the region. However, Singapore still fell to the Japanese during World War II, when it was occupied from 1942 to 1945. The 1950s saw the growth of both socialism and nationalism as political forces on the island, with the formation of the left-wing People's Action Party (PAP) and the inauguration of negotiations with the British on Singaporean self-rule. In 1959 the island became a self-governing state with its own legislature, prime minister, and cabinet. Lee Kuan Yew, the leader (and founder) of the PAP, became its first prime minister, a post he was to hold for the next thirty-one years. Under his leadership, Singapore joined with Malaya and other former British territories to form the Federation of Malaysia in 1963. However, tensions with Malaya ended this relationship by 1965, when Singapore became a sovereign republic—one of the smallest in all of Asia.

Lee presided over a period of unparalleled economic growth and industrialization, implementing policies that encouraged foreign investment and the development of labor-intensive industries. In the 1960s multinational electronics corporations began opening plants in Singapore. The seventies saw the establishment of a stock exchange and the introduction of offshore banking and other services. In the 1980s the industrial sector moved into the computer industry, which created thousands of jobs for skilled workers.

Lee's policies have been instrumental in producing the phenomenal economic success enjoyed by Singapore over the past decades, but Lee has also been criticized for suppressing some basic human rights in order to stay in power. The People's Action Party (PAP), founded by Lee in 1954, has ruled Singapore throughout its history as an independent republic, and opposition parties have held only a handful of the nation's legislative seats. Between 1966 and 1980, the major opposition party, the *Barisan Sosialis,* boycotted the political pro-

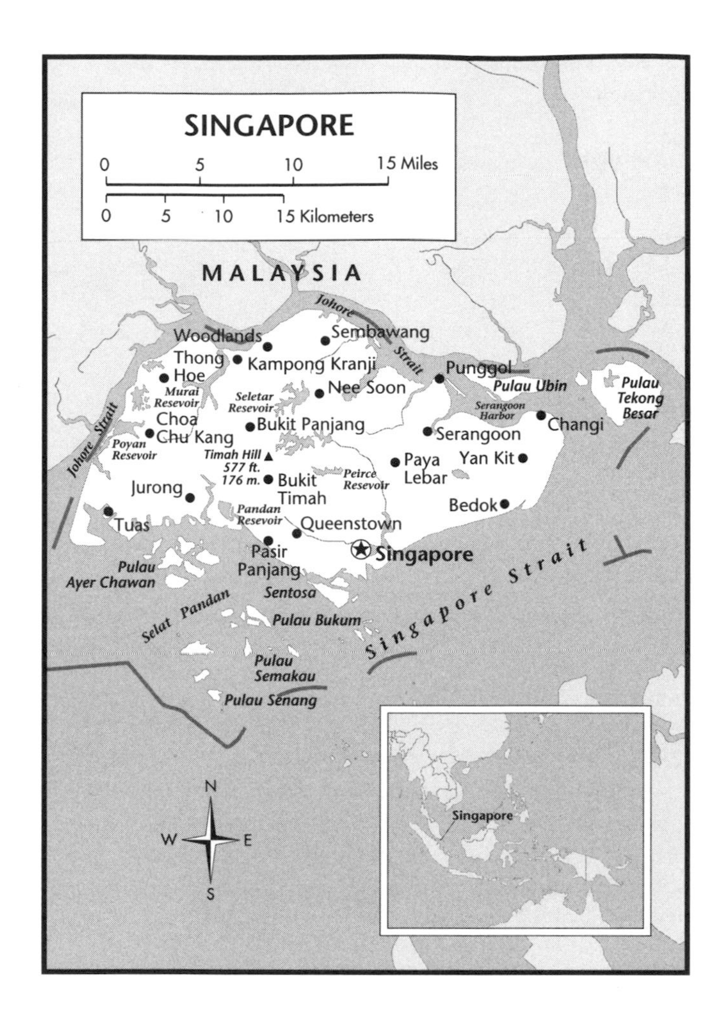

cess, and the PAP governed with no elected opposition whatsoever. In the 1990s harsh measures by Singapore's rulers attracted attention on the international stage with the 1993–94 imprisonment and government-mandated beating of an American teenager for spray painting cars, and the 1995 hanging of a young Filipino maid on double murder charges in spite of evidence that she had been framed. The government of Singapore has defended its actions in these and other situations and assailed liberal democratic freedoms as a cultural institution that the West has no right to impose on nations in other parts of the world.

Lee served as prime minister until 1990, when he was succeeded by Goh Chok Tong, but Lee remained the head of the PAP.

Timeline

A.D. 231 Chinese know of the island

A Chinese explorer makes reference to the island of Singapore.

1299 Traditional date for the founding of Singapore

According to legend, a trading city named Singapura ("lion city") is founded on the island currently called Singapore.

1349 Chinese traveler describes Singapore

Wang Dayuan mentions Singapore in a geographical handbook, calling it *Danmaxi* and noting that it is used by pirates as a base for attacking passing ships.

Late fourteenth century Javanese attack Singapore

Contemporary chronicles note destruction of the trading settlement on Singapore by the Javanese.

1511 Singapore becomes part of sultanate

The ruler of Malacca, fleeing the Portuguese, establishes a new sultanate in the Malay Peninsula, whose jurisdiction includes the island of Singapore.

1781 Birth of Sir Stamford Raffles

Sir Thomas Stamford Raffles, founder of the British port of Singapore, is born at sea to a captain in the merchant marine. Raised in poverty, he is forced to quit school at fourteen, when he is hired as a clerk by the British East India Company. He rises rapidly through the ranks, while studying science and languages on his own, and in 1804 wins an official post in the administration of Penang, an island in the Malayan archipelago. Seven years later, he uses his knowledge of the region to aid Lord Minto in annexing Java for the British and becomes its lieutenant governor.

After a three-year sojourn in England (1816–19), during which Raffles is knighted and inducted into the Royal Society, he returns east once again, receiving authority to establish a trading post on the sparsely inhabited island of Singapore, where he lands on January 29, 1819. Dividing his time between the new post and his official duties in Bengkulu, on the west coast of Sumatra, Raffles organizes the administration of Singapore. In ill health, he returns to England in 1824, where he participates in the founding of the London Zoo before his death in 1826 from a brain tumor.

1811 Sultanate of Johore annexes Singapore

Singapore is annexed by the Sultanate of Johore.

Arrival of the British

1819 British trading station established

A British East India Company trading post is founded by Sir Stamford Raffles, who leases the sparsely populated island of Singapore from the sultan of Johore for 3,000 pounds per year.

1824: March 17 Singapore is ceded to the British

The Sultan of Johoree cedes the island of Singapore to the British, and the Dutch relinquish all claims to the island. By the Anglo-Dutch Treaty, Malacca is also ceded to the British, and the Dutch, in turn, gain control of Sumatra.

1826 Formation of the Straits Settlements

Singapore becomes part of a colonial administrative unit, the Straits Settlements, that also encompasses Malacca and Penang.

1826: July 5 Death of Sir Stamford Raffles

Sir Stamford Raffles, the founder of the British port of Singapore, dies in London. (See 1871.)

A Trade Capital is Born

1830–67 Singapore flourishes as trading center

In its first decades under British management, Singapore becomes a busy seaport and *entrepôt* (re-export) center, enjoying a favorable location at the center of a trade network extending from Sumatra and Java to China. Within this period, the island's population quadruples to 85,000.

1832 Singapore becomes government headquarters

Singapore becomes the seat of government for the Straits Settlements. The island's location at the southern tip of the Malay peninsula makes it a strategic point.

1844 Raffles National Library is founded

Singapore's national library, the Raffles National Library, is established. It is later renamed the National Library of Singapore (See 1960).

1849 Raffles National Museum is established

A national museum is founded to display archaeological, ethnological, and natural history collections. It is later renamed the National Museum.

1850s Chinese migration grows

Chinese migration to Singapore increases due to civil wars in China.

1862 Victoria Theatre is built

The Victoria Theatre provides Singapore's first town hall. It later becomes a site for cultural events.

1864 Annual trade reaches 13 million pounds

Singapore, with its fine harbor, becomes the major collections and distribution point for the flourishing Malayan rubber and tin trade. The island becomes Southeast Asia's foremost trading center.

1867: April Straits Settlements become a crown colony

The Straits Settlements, including Singapore, becomes a crown colony when its administration is transferred from the India Office to the Colonial Office. British troops are now stationed there.

1869 Suez Canal opening boosts shipping

The inauguration of the Suez Canal, which opens a passage between the Mediterranean and the Red Sea, contributes to the growth of Singapore as a commercial center by boosting international shipping.

1887 First rubber plants are grown

Englishman H. N. Ridley grows Singapore's first rubber plants in its Botanical Gardens, laying the eventual foundation for the lucrative Malaysian rubber industry.

1887 National Museum opens

The first wing of the National Museum is completed. In addition to historical and ethnological collections, the Museum becomes home to a spectacular four-hundred-piece jade collection. It also presents dioramas depicting the island's social history.

1889 Secret societies are banned

Secret societies, a force for social cohesion common throughout Asia, are banned following violence between rival groups.

1905 Memorial hall is completed

Victoria Memorial Hall is built in honor of Britain's recently deceased monarch, Victoria. It later becomes the home of the Singapore Symphony Orchestra (see 1979).

1906 Electricity is introduced

The British install electricity in Singapore.

1922 Singapore becomes a defense base

The British fortify its military defenses on Singapore to shore up its interests in the region; eventually this site becomes Britain's principal naval base in the Far East.

1923: September 16 Birth of Lee Kuan Yew

Nationalist leader and Singapore's first prime minister, Lee is born to a wealthy Chinese family and attends Cambridge University, where he is introduced to socialism. He is admitted to the English bar and then returns to Singapore, becoming active in politics and founding the People's Action Party (PAP) with other prominent leftists in 1954. Lee plays a prominent role in negotiations with the British on Singaporean self-rule. When it is achieved in 1959, Lee becomes prime minister of Singapore as a self-governing state within the British Commonwealth.

Lee is instrumental in Singapore's entrance into the Federation of Malaysia in 1963. By 1965, growing tensions between ethnic Chinese and Malays force Singapore to secede from the federation and become a sovereign republic. Lee remains prime minister for the next twenty-five years, presiding over a period of phenomenal economic growth that finds Singapore with a standard of living second only to Japan's among the East Asian nations by the 1980s. Lee's moderate policies come to diverge from his initial socialist philosophy, and his government practices some authoritarian measures to silence left-wing opposition. Having assured himself of an acceptable political successor, Lee resigns the post of prime minister in 1990 but continues to head the PAP.

1926 First power station goes online

Singapore's first electric power generating station begins operation.

1927 Government housing program is started

The British administration establishes the Singapore Improvement Trust to oversee the construction of new urban housing and slum clearance.

1937 Fine arts academy is founded

The Nanyang Academy of Fine Arts is established, attracting gifted artists and teachers including Georgette Chen, Chen Chong Swee, Gheong Soo Pieng, and Lim Hak Tai, all of whom studied painting in China during the preceding decades, when Western techniques were first introduced.

1942: February 15 Japan captures Singapore

In the midst of World War II, the Japanese attack Singapore from the Malay Peninsula and occupy it until the end of the war. The British are caught totally by surprise by the Japanese move. British chief means of defense on the island fortress are a series of naval guns which point toward the sea. British military planners view the jungle-covered Malay peninsula as impenetrable and, therefore, discount the possibility of an invasion by land. This proves to be a costly mistake. The Japanese conquest of Singapore is the worst, most humiliating defeat in British military history.

The population of Singapore endures three years of brutal Japanese occupation.

1946 Separate colony is formed

With the dismantling of the Straits Settlements, Singapore becomes a separate colony. Penang and Malacca become part of Malaya.

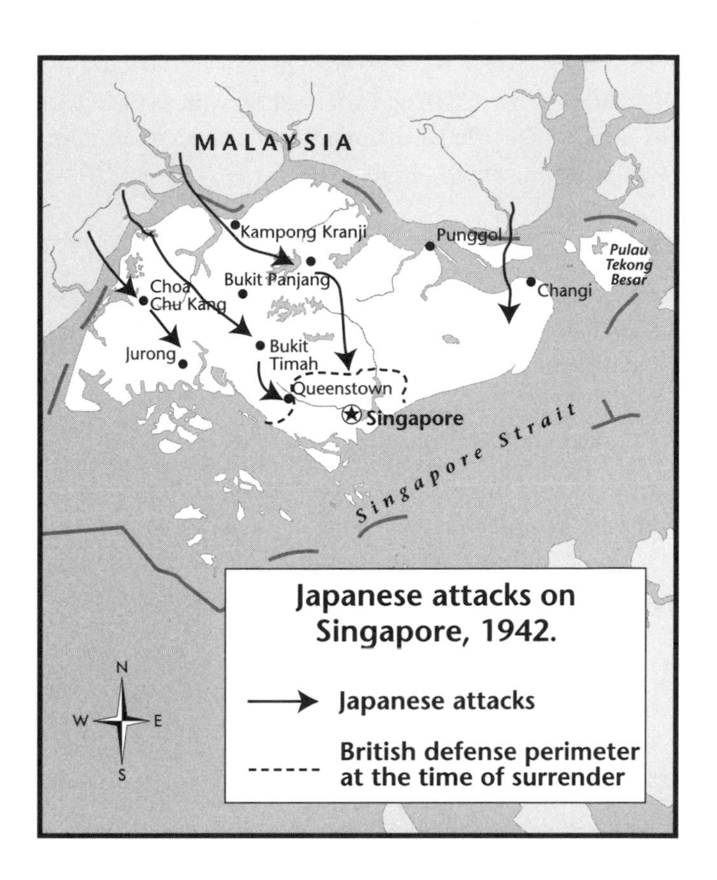

Move toward Independence

1954 People's Action Party is formed

The People's Action Party (PAP) is established, becoming Singapore's dominant political force. From the time of Singaporean independence (see 1959), it wins legislative majorities in every election. (Between 1968 and 1980 it wins every seat.) The PAP focuses on expanding economic growth and raising living standards. However, its continued power is also based on repression of all political opposition.

1955 New constitution is adopted

A new constitution increases the number of elective (rather than appointed seats) from 25 out of a total of 32.

1956–58 Talks on self-rule are held

Singaporean leaders hold talks in London to reach an agreement with the British for self-rule on the island.

1959: May Legislative Assembly is elected

With statehood pending, a new Legislative Assembly is elected. The majority of seats are won by the People's Action Party (PAP). Lee Kuan Yew, the head of the PAP, becomes prime minister and implements a five-year plan providing for the construction of new public housing, educational reform, increased rights for women, and industrial expansion.

1959: June 3 Singapore gains internal self-rule

Singapore becomes a self-governing state, with a constitution providing for a legislative assembly headed by a prime minister and a nine-member cabinet.

1960 National library is renamed

The Raffles National Library becomes the National Library of Singapore. It contains the government archives and all official publications printed since World War II.

1961 Economic Development Board is formed

The creation of the Economic Development Board heralds Singapore's transformation into a modern industrial nation. The country's first wave of industrialization is centered on import-substitute manufacturing requiring tariff protection, although many tariffs are later discontinued. Prime Minister Lee Kuan Yew also works to encourage foreign investment and support productive labor-management agreements.

1961: August Barisan Sosialis is formed

Dissatisfied with the moderate policies of the ruling PAP, its left-wing members form the opposition Barisan Sosialis (Socialist Front) party, which becomes Singapore's major opposition group.

1963: September 16 Singapore joins the Federation of Malaysia

Fearing a challenge from the Barisan Sosialis opposition party, the government of Singapore allies itself with Malaya, forming the Federation of Malaysia, which in addition to Malaya includes the former territories of Sabah and Sarawak in northern Borneo. Singapore retains some control over internal labor and education policies and participates in the Federation legislature together with the other members, although it does not have proportional representation.

1964–65 Malay-Chinese tensions produce riots

Ethnic rivalry between Chinese and Malays result in rioting.

The Republic of Singapore

1965: August 9 The Republic of Singapore is formed

Due to tensions between Singapore and Malaya, which are also exacerbating divisions between Singapore's ethnic Chinese and Malays, Singapore withdraws from the Federation of Malaysia to form an independent republic.

1966 Singapore joins international financial organizations

Singapore becomes a member of the International Monetary Fund (IMF), the World Bank, and the Asian Development Bank.

1966 Barisan Sosialis boycotts elections

The left-wing opposition party Barisan Sosialis begins a permanent boycott of national elections, leaving the PAP as the nation's sole political party until 1980.

1967 ASEAN is formed

The Association of South-East Nations (ASEAN) is formed by Singapore, Malaysia, Indonesia, Thailand, and the Philippines. This organization seeks to foster closer economic, cultural, and social cooperation among the region's non-communist states.

Growth of an International Financial Center

1967 Multinational firms open plants

Singapore's industrialization enters a new phase as multinational electronics corporations establish plants for labor-intensive manufacturing.

1967 Science Council is created

The Science Council is established to advise the government on research and development and on training workers in technological fields.

1971 British bases close

British military bases on Singapore are shut down, bringing an end to the century and a half of British presence on the island.

1972 Singapore Airlines is established

Singapore Airlines begins operations. Eventually, it becomes one of the world's major air carriers.

1973 Offshore banking is introduced

The establishment of an offshore banking system enhances Singapore's position as the major financial services center for the ASEAN region.

1973 Stock market is launched

The Stock Exchange of Singapore is inaugurated. It is regulated by a committee including both stockbrokers and non-brokers.

1976 National Art Gallery opens

The National Art Gallery, part of the National Museum (see 1849), exhibits the work of Southeast Asian artists.

1980 National University of Singapore is founded

The National University of Singapore is established by the merger of the University of Singapore and Nanyang University.

1980 Singapore Broadcast Corporation is created

The Singapore Broadcast Corporation operates three television networks and nine radio stations.

1980 Opposition parties win seats in Parliament

For the first time since 1966, parliamentary seats are won by members of opposition parties. However, the ruling PAP still holds the vast majority of political power.

1981 National Computer Board formed

With the formation of the National Computer Board, the government implements an economic strategy of moving the nation's labor-intensive industrial base away from products manufactured by low-wage, unskilled labor, and toward hi-tech products (especially in the area of information technology) produced by skilled workers.

1982 Video games are banned

Video games are banned by the Minister of Culture for their alleged negative effect on children.

1987 Mass transit system begins operations

A new billion-dollar Mass Rapid Transit (MRT) system is inaugurated. Upon full completion (slated for 1990) it will extend for sixty-seven kilometers and carry 800,000 passengers daily.

1987: May–June Detention of alleged Marxists draws international censure

Twenty-two people accused of participating in a Marxist conspiracy are detained by the government without trials and allegedly tortured. These actions draw criticism worldwide from human rights advocates.

1987: December Alleged conspirators are released

The government releases most of the persons arrested on conspiracy charges earlier in the year.

1988 Digital data network is installed

Singapore becomes the first ASEAN nation to set up a dedicated digital data network for voice communications and high-speed data transmission.

1988: April Eight detainees arrested again

After publicly describing their detention on conspiracy charges, eight of those arrested the preceding spring are taken into custody a second time. The last ones are released two years later.

1990: November 28 Prime Minister Lee resigns

Longtime Singaporean prime minister Lee Kuan Yew resigns after thirty-one years in office but remains a cabinet member. The new prime minister is Goh Chok Tong.

1993: August 28 First direct presidential election

Ong Teng Cheong becomes Singapore's first president to be elected directly by the people.

1993: October Foreign teens arrested for vandalizing cars

Nine foreign teenagers are taken into custody after spray painting about seventy automobiles. The alleged ringleader of the group, Michael Fay, is sentenced to pay a fine, spend four months in jail, and receive six lashes with a cane. In spite of an appeal by President Clinton, Singapore's government refuses to revoke the caning sentence, but it is reduced from six strokes to four.

1994: May 1994 Fay sentence is carried out

American teenager Michael Fay receives his punishment, which has been reduced from six cane strokes to four.

1994 Singapore sues newspaper for libel

The Singapore government sues the *International Herald Tribune* for libel over the contents of an editorial. The *Herald Tribune* is ordered to pay $667,000 in damages to the government leaders criticized in the editorial.

1995: March Hanging of Filipino maid creates diplomatic crisis

Relations between Singapore and the Philippines become strained following the execution of Flor Contemplacion, a Filipino working in Singapore as a maid, for a 1991 double murder. Growing evidence suggests that Contemplacion, accused of killing a four-year-old boy and another Filipino maid, was framed by the boy's father, who allegedly killed the other Filipino himself after his son, who had been left in her care, drowned in a bathtub. Mass public protests take place in the Philippines, which also cancels a planned visit by Singapore's prime minister, Goh Chok Tong, and recalls its ambassador to Singapore. In Davao, one of the Philippines' largest cities, the mayor leads a public burning of the Singaporean flag.

1996 Singapore Art Museum opens

This permanent collection houses over 3,000 works of art, including both paintings and sculpture, by artists from Southeast Asia.

1996: December Goh reelected by forfeit

Prime Minister Goh Chok Tong automatically wins a second term when no rival Parliamentary candidate is nominated in his district.

1997: January 2 PAP retains parliamentary majority

PAP keeps its parliamentary majority in national elections, winning 81 of the country's 83 legislative seats. Prime Minister Goh Chok Tong declares that the victory demonstrates popular approval of his government's authoritarian methods and rejection of Western democratic freedoms.

1997: September Haze covers Singapore

Singapore is part of an area—also including Malaysia, southern Thailand, and the Philippines—blanketed in a thick haze caused by forest fires in Indonesia. The haze, called the most severe ecological emergency in the history of Southeast Asia, produces respiratory ailments and prompts school closings and embassy evacuations. It is also thought to be a factor in an airplane crash over Indonesia and a tanker collision in the Strait of Malacca.

Bibliography

Bedlington, Stanley S. *Malaysia and Singapore: The Building of New States.* Ithaca, N.Y.: Cornell University Press, 1978.

Chew, Ernest and Edwin Chew, ed. *A History of Singapore.* New York: Oxford University Press, 1991.

Chiu, Stephen Wing-Kai. *City States in the Global Economy: Industrial Restructuring in Hong Kong and Singapore.* Boulder, Colo.: Westview Press, 1997.

Flower, Raymond. *Raffles: The Story of Singapore.* Singapore: Eastern Universities Press, 1984.

Lee, W.O. *Social Change and Educational Problems in Japan, Singapore, and Hong Kong.* New York: St. Martin's Press, 1991.

LePoer, Barbara Leitch, ed. *Singapore: A Country Study.* 2d ed. Washington, D.C.: Library of Congress, 1991.

Makepeace, Walter, ed. *One Hundred Years of Singapore.* New York: Oxford University Press, 1986.

Minchin, James. *No Man is an Island: A Portrait of Singapore's Lee Kuan Yew.* 2d ed. Sydney: Allen & Unwin, 1990.

Trocki, Carl A. *Opium and Empire: Chinese Society in Colonial Singapore.* Ithaca, N.Y.: Cornell University Press, 1990.

Turnbull, C.M. *A History of Singapore, 1819–1988.* 2d ed. New York: Oxford University Press, 1989.

Solomon Islands

Introduction

The Solomon Islands are a double chain of primarily volcanic islands located east of Papua New Guinea in the southwest Pacific Ocean. This double chain of ten high islands and a multitude of smaller ones extends for nearly 1,600 kilometers and covers a total land area of approximately 27,500 square kilometers. In terms of land area, the Solomon Islands are only slightly larger than the state of Maryland.

Over 900 islands make up the nation of the Solomon Islands. Of these, only around 350 islands are inhabited. There are six main islands, Malaita, Makira (formerly San Cristobal), New Georgia, Boghotu (formerly Santa Isabel), Choiseul, and Guadalcanal. Most of the communities are very small with sixty percent of the population living in villages of less than 200 people.

The largest and most important island within the Solomon Islands is Guadalcanal. The capital of the Solomon Islands, Honiara, is located on the northern coast of Guadalcanal. Honiara has an estimated population of 40,000. The entire island of Guadalcanal has only around 60,000. Malaita is the most populous island with an estimated total population of 100,000.

The climate of the Solomon Islands is tropical monsoon. The northeast monsoon dominates the climate from November to March, bringing rain and cooler temperatures. The average rainfall per annum in the Solomon Islands is between 84 to 119 feet (216 and 305 centimeters). The average daily temperature year-round in Honiara is eighty degrees Fahrenheit (twenty-seven degrees Celsius). Natural hazards included typhoons, volcanoes, and earthquakes. Due to the predominantly rural organization of the nation, these natural hazards rarely inflict major damage.

The terrain of the Solomon Islands is very rugged and mountainous. The highest point is Mount Makarakomburu which reaches a maximal elevation of 8,075 feet (2,447 meters) above sea level. The mountainous nature of the island nation makes communication and transportation extremely difficult. The Solomon Islands are characterized as a fragmented state due the geographical isolation of polity.

Culture

The Solomon Islands are part of the culture realm of Melanesia within Oceania. There is considerable variation in the cultures of the Solomon Islands. Groups on Malaita, Guadalcanal, Makira, and Choiseul trace descent according to patrilineal principles. On Santa Isabel and Santa Cruz Islands, descent is determined according to matrilineal principles. Although these societies trace descent and group membership through either the mother's or the father's line, they typically do not achieve status or inherit property according to these principles. Instead, most Solomon Islands indigenous groups rely on public feasts for the accumulation and eventual display of wealth and the acquisition of status.

The estimated population of the Solomon Islands for 1999 is 455,912. The estimated growth rate for the nation's population is 3.35 percent per annum. The average lifespan of an individual living in the Solomon Islands is seventy-one years. Of the total population, around ninety-three percent are Melanesian, four percent are Polynesian, 1.5 percent are Micronesian, 0.8 percent are European, 0.3 percent are Chinese, with the remaining 0.4 percent being of different ethnic groups including South Asians. More than eighty-five percent of the population lives in small, subsistence-based rural communities.

Christianity is the dominant religion of the islands. Anglicans make up thirty-four percent with Roman Catholics making up nineteen percent. Baptists are a significant group and make up nearly seventeen percent of the Christian adherents. Only four percent of the population claims to follow traditional, pre-Christian practices and beliefs. Seventh Day Adventists and United Methodist/Presbyterians have seen their numbers increase over the years.

There are over 120 different indigenous languages spoken in the Solomon Islands. Although English is the official language of the Solomon Islands, it is spoken and understood by less than three percent of the indigenous population. A form of pidginized English called "Solomon Islands Pijin" is widely used and understood.

History

The Solomon Islands were first settled by Papuan language speakers who migrated there approximately 10,000 years ago. About 4,000 years ago, another group began to settle in the islands. These peoples were Austronesian language speakers who brought with them technologies previously unknown by the original inhabitants. The Austronesian speakers were skilled agriculturalists who raised taro root employing slash-and-burn field rotation. They had also domesticated pigs, chickens, and dogs. They were accomplished navigators who could build ocean-going outrigger canoes. These Austronesian speakers also knew how to manufacture pottery.

The first contact with Europeans came in A.D. 1568 when the Spanish explorer Alvaro de Mendana de Niera encountered the islands. He named them Islas de Salomon, based on his belief that these islands provided King Solomon's jewelry. Following two unsuccessful Spanish attempts to establish settlements on the islands in 1593 and 1606, the islands received no European visitors until the British explorer Philip Carteret traveled there in 1767.

In the late 1800s, the Solomons became the scene of Australian labor recruiters' attemptd to find forced labor for their plantations. This activity led to British intervention (Australia was then a British colony) and Britain established a protectorate over most of the Solomons in 1893. The British faced opposition throughout the islands and did not succeed in gaining complete control over the main islands until 1941.

Shortly after the colonial administration stamped out the last native resistance to British authority, Japan embarked upon its Pacific conquests during World War II (1939–45). The light British forces, already reeling from defeat in Hong Kong, Malaya, and Singapore, evacuated all Europeans from the islands in February 1942, much to the resentment of the native islanders. Japanese forces occupied Tulagi that May and began construction of a major airfield on Guadalcanal in June. In response to the airfield, United States Marines invaded the island on August 7, 1942 and thus began one of the greatest battles of the war. After heavy fighting, Allied forces captured Guadalcanal in February 1943.

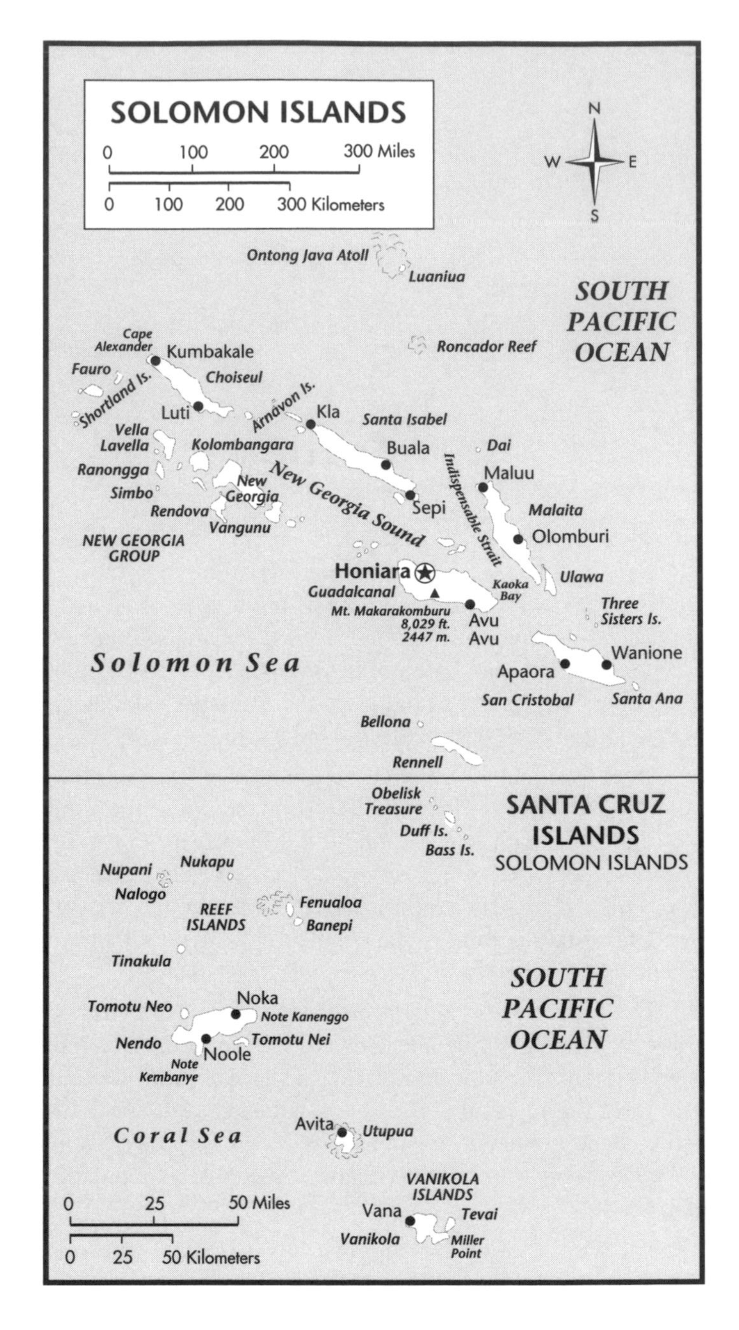

Solomon Islanders served as a critical portion of the labor force during World War II. During the Allied campaign and eventual occupation of the Islands, there were some 2,500 Solomon Islanders serving in the Solomon Islanders Labour Corps. Islanders also provided aid and shelter to wounded Allied soldiers that turned up in their villages or fell from the sky when their aircraft were shot down.

The Second World War devastated the minimal infrastructure that had been developed in the Solomon Islands. After the war ended, the British economy was not strong enough to provide immediate rebuilding in its colonial possessions and protectorates. The British government's *Colonial Development and Welfare Act* of 1940 which had been postponed due to the outbreak of war, was completed shelved after the war was over. It was not until the early 1950s that the British began rebuilding colonial infrastructure in the Solomon Islands. Beginning in the late 1940s, the Maasina Movement, a native movement demanding self-government, spread throughout the eastern Solomons and began challenging British colonial authority. The British reacted by dispatching recruits from the western Solomons and eventually defeated the rebels.

By the early 1960s the colonial authorities began introducing reforms intended to provide self-government and pave the way to independence. The administration introduced legislative and executive councils to be elected by Solomon citizens. In 1974 a parliament was formed and the islands received their independence on July 7, 1978. Since independence, the Solomons have concentrated their energies on economic development. Ironically, the numerous shipwrecks from World War II in the Solomon Islands' waters may be keys to the development of foreign tourism. Divers enjoy exploring submerged war vessels, and the Solomon Islands has the potential to develop its tourism industry to serve the scuba diving travelers.

Timeline

5100 B.C. Lapita culture complex spreads from Bismarcks to Solomons

Early Austronesian speakers who possess the Lapita cultural complex of a distinctive style of pottery spread from the islands of New Britain and New Ireland into the Solomon Islands.

4000 B.C. Earliest archaeological evidence of human habitation in Guadalcanal

The earliest traces of human settlement on Guadalcanal Island are dated to approximately 6000 years ago. These early settlers were likely Papuan language speakers who did not share the Lapita cultural complex of the early Austronesians.

A.D. 1568 First European to discover Solomon Islands

The Spanish explorer Alvaro de Mendana de Neira is the first European to discover the Solomon Islands. He names them Islas de Salomon, because he believes that the gold used in King Solomon's jewelry and other treasures came from these islands.

1593 Alvaro de Mendana de Neira attempts to establish settlement

Twenty-five years after his initial discovery, the Spanish explorer Alvaro de Mendana de Neira returns to the Santa Cruz group within the Solomon Islands and establishes a settlement. For reasons unknown, the settlement fails.

1606 Second Spanish attempt at permanent settlement

A second attempt by the Spanish to establish a permanent settlement in the Solomon Islands also meets with failure.

1767 Philip Carteret visits the Solomons

English explorer Philip Carteret travels to the Solomon Islands.

1845 First mission established on Makira

Catholic priests of the Marist sect establish a mission on Makira. The mission fails after several of the priests are killed by the local population.

1871 Massacre on board the *Carl*

Labor recruiters seeking workers for plantations in Australia and other neighboring islands travel on the *Carl*. They lure Malaitan canoes alongside and sink them. They then grapple with the survivors, pulling them on board the *Carl* and locking them below decks. A scuffle breaks out among the Malaitans; the crew of the *Carl* fires into the group, killing or wounding over seventy people. The dead and wounded are then thrown overboard.

1893 Britain declares protectorate over Solomon Islands

The practice of procuring indigenous Solomon Islanders for work on plantations in Australia, New Guinea, and Samoa called "blackbirding" causes violence and civil disruptions in the Solomons between indigenous peoples and Europeans. In an effort to control the level of civil disruption, Greta Britain establishes a protectorate over the Solomon Islands. Santa Isabel, Choiseul, Ontong Java, and the islands off of Bougainville remained under the control of Germany.

1896 Resident Commissioner Woodward arrives in Solomons

Britain's protectorate over the Solomon Islands (see 1893) is created on the condition that the resident commissioner is responsible for the protectorate's generation of its own revenue. Resident commissioner Woodward arrives in the Solomon Islands with a plan to establish plantations and to make the islanders recognize the authority of the state. Woodward's goal is to curtail the competing indigenous system of revenue generation that relies on barter, exchange, and the use of traditional items of wealth such as shells and feathers. Woodward also believes, like many other Europeans of the time, that the Melanesians will disappear: a version of Social Darwinism.

1899 Remaining Solomons become British protectorate

An Anglo-German agreement transfers control of Santa Isabel, Choiseul, Ontong Java, and the smaller islands off of Bougainville Island from Germany to Britain. These islands become part of the British protectorate of the Solomon Islands.

1902 Commissioner Woodward's militias continue campaigns

Resident commissioner Woodward believes that small forces of Fijian police are not sufficient to make the islanders recognize the authority of the state. He creates militias to undertake punitive expeditions against illegal traders. Woodward, who has moved to the Solomons from Fiji, hires his former police sergeant, William Buruku, of Fiji. Buruku is a native of Guadalcanal and returns to lead Woodward's militia on the island from Buruku's home village of Wanderer Bay.

News reports about the murder of whites on the island of Malaita in the Solomon Islands describe the hostile natives, and feature photographs of natives in their villages. (EPD Photos/CSU Archives)

1905 Solomons' protectorate budget is balanced

The financial success of Resident commissioner Woodward's plan is realized when the level of revenues matches the level of expenditures in the Solomon Islands. From 1905 forward, the Solomons budget remains balanced while under British protectorate status.

1927 Basiana murders on Malaita

Kwaio warriors, led by a man named Basiana, kill a British district officer, a cadet, and thirteen police who were in the area collecting taxes. The British respond with a large punitive expedition that results in the loss of many Kwaio lives.

1941 Administrative control of final main island in Solomon Islands

Full British administrative control of the main islands of the Solomon Islands is achieved with the cessation of headhunting and feuding on Choiseul Island. New Georgia and Malaita were notorious for headhunting and indigenous, local feuding.

1942: February Final evacuation of non-residents from Solomons

The last steamer leaves Tulagi with evacuees due to the battles of World War II (1939–45) between Japanese forces and Allied troops. The steamer is laden with Europeans—missionaries, planters, and colonial officials—who are fleeing before the imminent invasion of the island by Japanese forces. The Europeans leave behind their servants, their converts, and their employees. The islanders do not forget the European betrayal.

1942: May Japanese occupy Tulagi Island

Tulagi Island, which neighbors Guadalcanal Island, is occupied by Japanese forces during the Battle of the Coral Sea. It

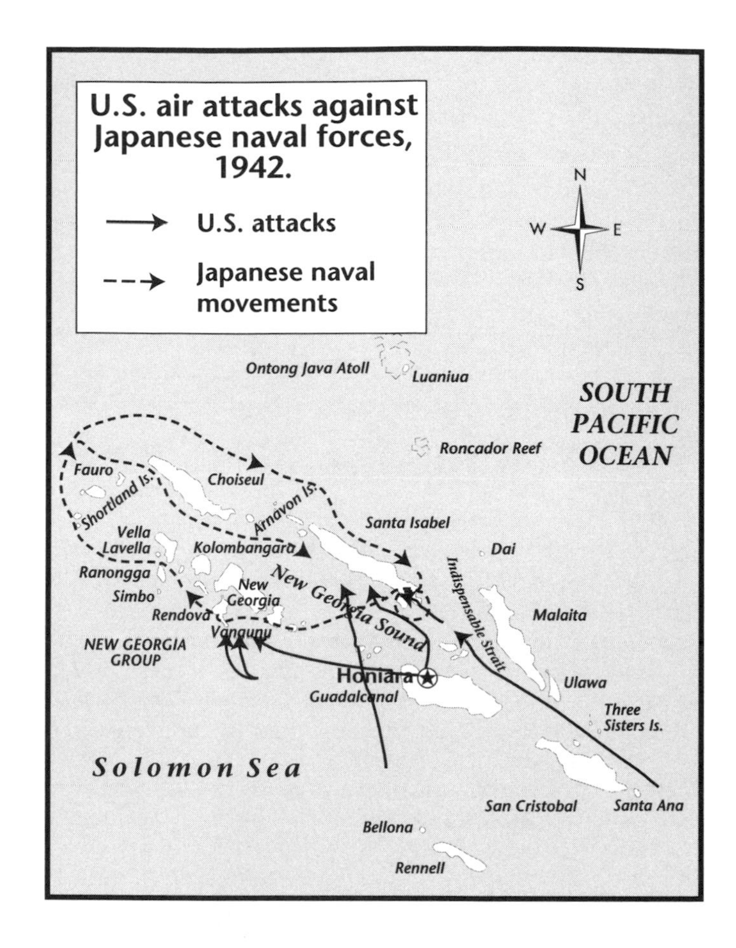

is from this location that the Japanese begin fortifications and military buildup on Guadalcanal.

1942: June Japanese begin military build-up on Guadalcanal

The Japanese military initiates a large-scale build-up of forces on Guadalcanal. The crowning piece of the military construction is the full-fledged airbase that serves as the attack point for most Japanese strikes against Allied forces in the Pacific.

1942: August 7 Battle of Guadalcanal begins

The First U.S. Marine Division lands on Guadalcanal and Tulagi Islands. Guadalcanal represents both the highest and lowest points of the World War II military campaign for the United States in the Pacific Theater. Guadalcanal is the most protracted naval battle that the United States experiences in the Pacific. In the late 1990s, scuba divers travel to the area to explore sunken World War II planes and ships. Among these sites is the SBD Dauntless Dive Bomber which crashed in 1943. The pilot, Lieutenant James W. Dougherty, escapes from his burning plane.

1943: February 8 Allied forces take control of Guadalcanal

Allied forces take full control of Guadalcanal Island. This Allied victory is the first complete defeat of the Japanese military in the Pacific campaign, with the Japanese suffering sea, land, and air defeats. With this victory, the momentum of the Pacific Theater shifts to the Allies for the first time in World War II.

1943–44 Maasina Rule movement develops in Malaita

Solomon Islanders who were working for the Allies are dissatisfied with their treatment and payment. As a result, they organize a movement to reinstitute customary law. The movement is centered among the Are'are, a group of native people in south Malaita. Leaders in the movement collect head taxes from movement supporters to finance their efforts.

1946 Maasina Rule spreads to eastern Solomons

Maasina Rule movement gains momentum as it spreads from Malaita to islands in the east.

1947 Maasina Rule spreads to central Solomons

With the continued spread of the Maasina Rule movement throughout the Solomons, the British government becomes uneasy. The British arrest leaders on Malaita, Guadalcanal, and San Cristobal. The British commence Operation De-Louse with mass arrests. The British create the "Black Army" with non-supporters from the western Solomons as well as Westerns of the Armed Constabulary. The "Black Army" is deployed against dissidents on a regular basis during 1946 and 1947.

1949 Rumors of American return incites Maasina Rule supporters

The Solomon Islanders are fond of Americans and the treatment that they had received from the American soldiers. When rumors of a return of American forces to the Solomon Islands begin to circulate among the Maasina Rule followers, actions are taken to aid the American return. Bonfires are consistently lit near the shorelines to aid ships, and warehouses are constructed to store the cargo that the Americans will bring with them.

1960 Councils established

The colonial administration establishes appointive legislative and executive councils to replace the advisory council that had existed until now.

1964 First elected members join government councils

Elected members are included in legislative and executive council membership for the first time in Solomon Islands history.

1970 Governing council is established by constitution

A new constitution for the Solomon Islands establishes a single governing council, with the majority of members being elected. This action set the stage for the development of a parliament comprised of elected officials (see 1973).

1973 All members of Governing Council are elected

A legislative change requires that all members of the governing council be elected officials.

1974 Parliament is formed

Through a new constitution, the Solomon Islands create a parliamentary body and a cabinet system. Chief minister Solomon Mamaloni is charged with heading the new legislative body.

1975 Official name change from British Solomon Islands to Solomon Islands

The territory of the British Solomon Islands Protectorate undergoes an official change of name to the Solomon Islands.

1978: July 7 Solomon Islands becomes an independent nation

The Solomon Islands receive independence from Great Britain.

1978: September 19 Solomon Islands admitted to United Nations

The Solomon Islands become a member of the United Nations.

1997: August 6 General election for parliament

In parliamentary elections, the Alliance for Change, led by Bartholomew Ulufa'ulu of the Solomon Islands Liberal Party, wins twenty-six of the fifty seats. The remaining seats go to the opposition Group for National Unity and Reconciliation.

1997 Guadalcanal government supports development of mine

The Guadalcanal government supports a mining project known as Gold Ridge Mine on Guadalcanal.

Bibliography

Bennett, Judith. *Wealth of the Solomons: A History of a Pacific Archipelago, 1800–1978.* Honolulu: University of Hawai'i Press, 1987.

Denoon, Donald, ed. *The Cambridge History of the Pacific Islanders.* Cambridge: Cambridge University Press, 1997.

Keesing, Roger. *Kwaio Religion: The Living and the Dead in a Solomon Islands Society.* New York: Columbia University Press, 1983.

Keesing, Roger and Corris, Peter. *Lightening Meets the West Wind: the Malaita Massacre.* Melbourne: Melbourne University Press, 1980.

Shineberg, Dorothy. *They Came for Sandalwood: A Study of the Sandalwood Trade in the Southwest Pacific, 1830–1865.* Carlton, Victoria: Melbourne University Press, 1967.

White, Geoffrey and Lindstrom, Lamont, eds. *The Pacific Theater: island representations of World War II.* Honolulu: University of Hawai'i Press, 1989.

Sri Lanka

Introduction

Sri Lanka is an island nation in the Indian Ocean that combines picturesque hills dotted with estates. The landscape features waterfalls, wildlife reserves where wild Asian elephants roam, tropical beaches indented by lagoons, and ruins of ancient civilizations.

The small country has had many names throughout history. The two-thousand-year-old Indian epic, the *Ramayana,* tells of Rama's beautiful wife, Sita, being abducted by the evil king Ravana and taken to a place called Lanka, meaning "resplendent land." The native name is Tambapani. Early Greek and Roman mariners called the island Taprobane because they could not pronounce Tambapani. Hindus from neighboring India called the island by a Sanskrit name, Sinhaladvipa, meaning "island dwelling place of the lions." Arab traders of the fourth century A.D. called the island Serendib (also spelled Serendip), mispronouncing this Sanskrit name. (It is interesting to note that this Arab name, Serendib, inspired a vocabulary word in the English language. Horace Walpole, an eighteenth-century British writer, coined the word "serendipity" after reading the Persian fairy tale "The Three Princes of Serendib." The heroes of the tale often made lucky discoveries of things for which they were not looking. Walpole's word is still used for happy, accidental discoveries.) The Portuguese who came to establish colonies in the region changed that Sanskrit name to their own version that was easier to pronounce: Ceilao. Subsequently, the Dutch colonists changed the spelling to Ceylan, and the British—who were the last foreign colonials—changed it to Ceylon.

The country was known as Ceylon until 1972, even after it gained independence in 1948. In 1972, when the nation was first made a republic, the ancient Pali (scholarly language of Buddhism) name of Lanka was taken, adding the honorific term "Sri."

The island is shaped like a teardrop and is located about fourteen miles (twenty-three kilometers) off the southeastern tip of India, separated by the Palk Strait. In area it is about the size of West Virginia (25,332 square miles, 65,610 square kilometers), and it measures 270 miles (435 kilometers) long with a breadth of 140 miles (225 kilometers) at its widest point. The coastline of sandy beaches gives way to rolling plains that become foothills followed by a massif (mountain mass) in the south central part of Sri Lanka with a top elevation of about 8000 feet (2500 meters). The most famous mountain is Adam's Peak (7359 feet or 2,243 meters), which is seen as a sacred site and is climbed by pilgrims of four religions (see Timeline: c. sixth century B.C.). Sri Lanka's longest river, the Mahaweli, originates close to Adam's Peak and runs north into the Bay of Bengal at Trincomalee.

As of the late 1990s, Sri Lanka included about 18.5 million people and was densely populated, with some 715 persons per square mile. However, the country's family planning programs have resulted in an average number of births per childbearing woman of 2.2. The ethnic profile of the nation includes a majority of Sinhalese (74 percent). Sri Lankan Tamils comprise 12.5 percent; Indian Tamils, 5.5 percent; and Sri Lankan Moors, 7.5 percent. Other groups, such as Malaysians; Burghers, who are descendants of Dutch colonials; and Veddhas, who are said to be descendants of the indigenous inhabitants, compise a half percent each. The Sinhalese majority are believed to be of Indo-Aryan origin, having first come from northern India about 500 B.C. when Vijaya, the legendary founding father of the Sinhalese, arrived. Vijaya was the son of King Sinhabahu, who was fathered by an ancestral lion (sinha). Thus, the Sinhalese people have a totemic (animal or plant as the emblem of ancestry) origin, being by legend descendants of the lion. The lion is now the central figure on the Sri Lankan flag and state emblem.

Adding to the ethnic diversity, the Sinhalese majority divide themselves into two groups, the Low-Country Sinhalese (about 7 million) and the Up-Country or Kandyan Sinhalese (about 5 million). The latter are a proud people, believing themselves to be the more pure because the Low-Country Sinhalese mingled and intermarried with the Portuguese and Dutch colonials from the early sixteenth century. The Sinhalese also have an ancient caste system based on occupations in the king's retinue (e.g. fishermen, potters, drummers, dancers, laundry washers, etc.). The highest caste (after the nobles) is the cultivator caste; this caste is also the most populous. Despite efforts by successive governments to eliminate caste distinctions, they still remain. Members of castes marry

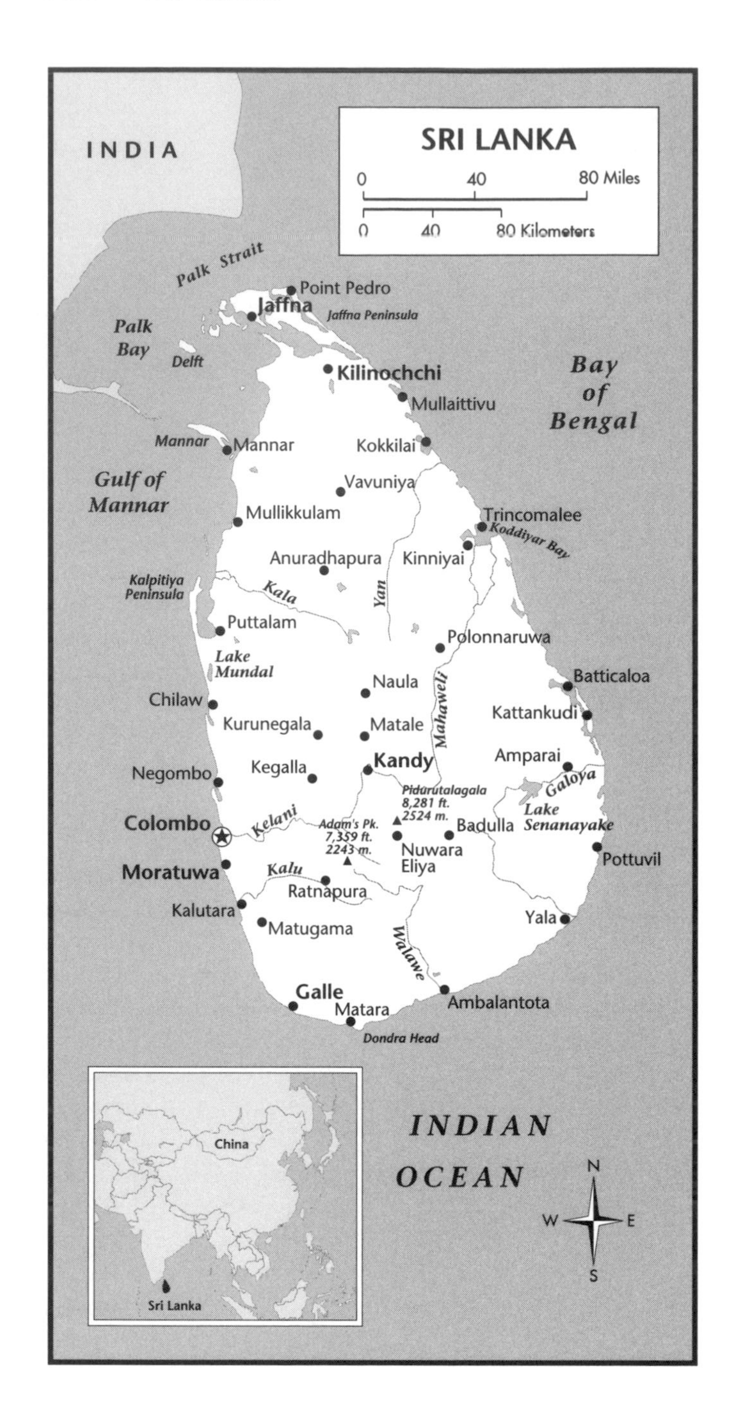

only among themselves, and caste membership may be a factor in occupational opportunities.

Early History

Little is known about the Paleolithic and Mesolithic cultures of Sri Lanka except that hunting/gathering societies inhabited the island as early as 10,000 B.C. The first recorded history began with the arrival of Indo-Aryan settlers from northern India around 500 B.C. The city of Anuradhapura, built by the Sinhalese King Pandukkabhaya, grew to greatness due to a sophisticated system of intensive rice irrigation. Early hydraulic engineering projects built dams across valleys to create reservoirs, along with canals and sluices to regulate water flow to vast areas of rice fields. Buddhism was introduced to the island in 247 B.C. by Prince Mahinda from India. The Sinhalese King Tissa converted to the new belief, which brought a rich culture with Buddhist art and architecture.

As a result of invasions from southern India, the glorious kingdom of Anuradhapura fell by the end of the eleventh century. The Indian Cholas people established their capital at Polonnaruwa. However, King Vijayabahu regained power from the Cholas, and his descendants restored the Sinhalese civilization to its former glory with Buddhist temples and immense irrigation reservoirs. Invasions from India continued intermittently. The capital was moved several times, being located in Kotte near present-day Colombo when the Portuguese arrived in 1505.

The Portuguese came to trade in spices but remained, gaining control of the coastal areas. The Dutch succeeded them and made the island part of their colonial empire from 1656 to 1796, when they were displaced by the British. Up through the early years of the British period, the Sinhalese kings of the Kandyan Kingdom in the central highlands retained sovereignty while surrounded by foreign rule. In 1815, however, the Kandyan kings surrendered to the British, who then had dominance over the whole island. Through a gradual peaceful process, Sri Lanka gained independence in 1948, ending four and a half centuries of colonial presence.

Modern Sri Lanka

The island nation is now a multiparty republic with membership in the Commonwealth of Nations. While the administrative capital of the country is Colombo, which has a population of about 800,000, the legislative capital is Sri Jayawardanapura at Kotte, just outside Colombo. The government has one legislative house, the National State Assembly, whose members are elected for a six-year term. The executive president is elected for six years and has the power to appoint or dismiss cabinet members, including the prime minister. The country is divided into districts, each with an appointed governor; there are also local authorities, from municipal councils to village councils. The judicial system resembles the British colonial system, with a Supreme Court and a series of lower courts.

During the 1980s and 1990s, Sri Lanka experienced flare-ups of an ongoing civil war. Ethnic violence between the Tamil minority and the Sinhalese majority had been simmering since 1958, although it was generally kept under control. In the summer of 1983, violence in the form of riots, burning, and looting exploded in earnest after a Sinhalese army patrol was ambushed and killed by Tamil extremists. The strife has continued intermittently with reciprocal retaliations up to the present as the Tamils have fought for their separatist cause.

Some 30,000 people have died in the civil war and about one million have been displaced.

The complex causes of the war are rooted in hundreds of years of cultural differences. It is not a religious war between the Tamil Hindus and the Sinhalese Buddhists, although many Sinhalese believe that Buddhist history shows the island is rightfully theirs. The problem seems to lie more in the economic sphere. After independence from Great Britain, the northern Sri Lankan Tamils felt discrimination in the areas of education and job opportunities, with the Sinhalese majority controlling the government, the military, and most large businesses. Under the British, the Tamils had received access to education and occupied relatively high positions in the colonial government. After independence, when the Sinhalese majority won control of the government, the Tamils grew to feel disadvantaged and persecuted. The extremists among them forged a separatist campaign for an independent Tamil nation. The Sinhalese considered the prospect of a separate Tamil nation dangerous and unnecessary on such a small island.

Sri Lanka is among the world's poorest nations, with a per capita income of $600. The costs of the ongoing civil war weigh heavily on the economy. During the British period, three agricultural products—tea, rubber, and coconuts—were the cash crops that boosted colonial interests. Since independence, these crops have remained important, especially tea, which is a major foreign exchange earner. Other important crops include spices such as cinnamon, cardamom, nutmeg, cloves, and pepper. The country is also a major exporter of precious and semi-precious stones. Since the 1990s, textile and garment manufacturing has grown in importance and has been encouraged by free trade zones. In peaceful times, tourism is another source of foreign currency. Additionally, many Sri Lankans go to the Middle East to work in such jobs as house maid, chauffeur, or engineer, and the wages they send home have been economically important for the country. Other nations continue to provide foreign aid, despite the criticism of the civil war and reports of human rights abuse.

Daily Life

Sri Lanka is primarily an agricultural nation, and the daily life of most people (about 80 percent) occurs in small villages where community values are strong. Village women spend much of the day cooking, which is a laborious task, often involving the fetching of water at the well in round pots and the collecting of firewood to cook the meal. Most meals consist of rice, the staple food, which is served with two or three curries—usually spicy vegetable dishes that are mixed with rice and eaten with the fingers. All family members share in the farm work, cultivating rice in irrigated paddy fields and other sorts of grain or vegetables in shifting highland fields. Village recreation centers around a variety of locally-celebrated festivities. Examples of celebrations include elaborate rite-of-passage events, such as a girl's "attaining age" party; religious holidays, such as the Wesak full-moon day in May, celebrating the birth, enlightenment, and passing away of the Buddha; or the Sinhalese and Tamil New Year occurring in mid-April, marking the entry of the sun into the first sign of the zodiac after passing through the twelve signs or months of the year. Family ties are strong, and respect for one's parents and elders is an important value. Hospitality is also important, and visitors are treated with generosity.

Despite their many household responsibilities, women have a relatively high status in Sri Lankan society. School enrollment figures and achievement scores are higher for girls than for boys, and more women than men attend universities. Women are found in politics and in all the professions, and more women have jobs than men. Sri Lanka boasts of having the world's first elected woman prime minister. In 1994, its fourth elected president was a woman, Mrs. Chandrika Bandaranaike Kumaratunga (b. 1946).

Religious practice roughly follows the ethnic composition of the population, with 69 percent Sinhalese Buddhists, 15.5 percent Tamil Hindus, 7.5 percent Sri Lankan Muslims, and 7.5 percent Christians. Buddhism in Sri Lanka is Theravada Buddhism or the "narrow path," which is considered close to the original form of the belief. While Buddhism has almost disappeared in its country of origin, India, it survives and thrives in neighboring Sri Lanka. It is a common sight to see saffron-robed monks—from young novices to highly respected elders—throughout Sri Lankan society. Buddhist temples, each with a pure white, bell-shaped *dagoba* or *reliquary* (shrines containing sacred texts), dot the countryside, often with large Buddha statues on the temple complex. Every Buddhist home—from the most humble wattle-and-daub house to the most extravagant mansion—has a small shrine with a Buddha figure or picture and a place to light oil lamps and to offer flowers and incense sticks. Hindu temples are also widespread. These are popular destinations for Buddhists and Hindus alike who seek the clairvoyant spiritual services of a Hindu priest. Most Sri Lankans believe in diverse forms of the supernatural—local deities, devils, evil spirits, and astrology. They frequently consult an astrologer to determine an auspicious moment for the beginning of an important event, such as breaking the first sod when starting to build a new house.

Sri Lanka has a life expectancy of seventy for males and seventy-five for females, and an infant mortality rate of seventeen per 1000 live births. An immunization program has eradicated a wide range of diseases. The country also has a network of rural clinics that make Western medicine free and available. Villages have practitioners of both indigenous medicine and ayurvedic medicine. Ayurvedic medicine is an ancient Indian practice that comes from the Vedas, the earliest Brahmanic sacred verses dating from the second millennium B.C. It is primarily based on homeopathic and herbal treatments. Ayurvedic medicine is highly regarded in Sri Lanka;

there is an ayurvedic hospital in Colombo where physicians are trained.

Education is free from kindergarten to university level. Almost all elementary–age children attend school, and a high proportion continue through high school. Nearly 90 percent of thc population is litcratc (can read and write). In early historical times, education was centered in the Buddhist monasteries, with the monks recording *jataka* stories from the Buddha's life as well as the *Mahavamsa* and the *Culavamsa* (A.D. fifth or sixth century)—great epics that chronicle the activities of kings and heroes from the time of Prince Vijaya's arrival from India. Poetry that dates from the eighth century includes the verse found at Siigiriya, a place famous for the secular paintings of the "Heavenly Maidens" on the walls of an overhanging cliff gallery on the side of the Siigiriya rock fortress. Sri Lankan writers have written throughout the ages in every literary genre—poetry, drama, short stories, and novels. Perhaps the best known internationally is Arthur C. Clarke (b. 1917), who has made Sri Lanka his permanent home since the 1950s. His work *2001:A Space Odyssey* (1968), became a major motion picture. Clarke continues to write science fiction. Canadian writer and Booker Prize winner Michael Ondaatje (b. 1942) was born in Ceylon.

The island's rich culture extends to sculpture, architecture, music, dance, and other artistic expressions. Images of the Buddha are the most prominent subject for sculpture. The Buddha is carved in three basic poses—sitting in meditation, standing, or reclining, with the hands arranged in diverse *mudras* (symbolic gestures) to convey messages such as protection or teaching. The most distinctive architecture is found in the Buddhist temple complexes. While orthodox Buddhism originally frowned on music and dance, these artistic forms have always been popular. Dance is divided into two traditional kinds: the colorful Up-Country Kandyan dancing, accompanied by drums played by the male dancers, and the Low-Country devil dancing based on folk rituals to drive out evil spirits. Here the dancers wear spectacular, grotesque masks with pointed teeth and bulging eyes, while drummers play the cylindrical demon drum. Such dances are still performed, especially in the southwestern coastal region, to cure a variety of afflictions.

Timeline

500,000–10,000 B.C. Paleolithic and Mesolithic cultures

Paleolithic and Mesolithic cultures inhabit the island, as evidenced by stone and bone implements, especially microliths (flint tools). The people are hunters and gatherers of southern Indian hill tribes.

15,000–10,000 B.C. Signs of early agriculture

Some evidence suggests that domestication of plants on the island, an early form of agriculture, takes place during this period.

10,000–500 B.C. Later Neolithic cultures

Later Neolithic (settled farming) cultures, such as the the Balangoda culture, are pottery-makers. These peoples are pushed into the rain forest after the early advance of settlers from India (see 500 B.C.) Historical documents recount that these early peoples belong to two tribes, the Yaksas and the Nagas. Legend has it that the people surviving into the twentieth century as hunters and gatherers—the Veddhas of Sri Lanka—are descendants of these two tribes.

c. 6th century B.C. The Buddha is said to visit Sri Lanka

Legends recount that the Buddha made three journeys to the island during his life. The first visit is said to have been to Mahiyangana, where he drove out the Yaksas and Nagas. The second visit was to Nagadipa, a small island near Jaffna, where he came to settle a quarrel between two Naga kings. The third visit is reported to have been to Kelaniya (near Colombo), at which time he also went to Ratnapura and to the present pilgrim's mountaintop, Sri Pada (Adam's Peak), where the Buddha's large footprint is said to be seen in a rock. (Hindus believe it to be the footstep of Siva, Hindu god of destruction and reproduction. Muslims claim it is Adam's footprint, while Christians are divided as to whether it is Adam's footprint or that of St. Thomas, who is believed to have preached in southern India 2,000 years ago.

c. 500 B.C. Indo-Aryan settlers from northern India arrive

The first historical records of the Sinhalese people are found in the *Mahavamsa* and its sequel, the *Culavamsa,* both literary epics compiled by Buddhist monks in the fifth or sixth century A.D. The *Mahavamsa* recounts that the legendary founding father of the Sinhalese, Prince Vijaya, is banished by his father due to his violence and misbehavior in destroying villages. The father, King Sinhabahu of North India, puts his son and 700 followers on a ship that arrives on the island in a region called Tambapanni shortly after the death of the Buddha (c. 543 B.C.). Because Vijaya's father, King Sinhabahu, is said to be the son of a lion *(sinha)* who married a runaway princess, the Sinhalese people are by legend descendants of the lion. The lion is the national symbol and the central figure on the Sri Lankan flag.

c. 3rd century B.C. Tamils arrive from southern India

The Tamil people are also mentioned in the *Mahavamsa* (see c. 500 B.C.). They establish numerous urban centers and trad-

Since ancient times Sri Lankans have cultivated wet rice paddies, a form of intensive agriculture that requires sophisticated knowledge of hydraulic engineering. (EPD Photos/Victoria J. Baker)

ing settlements in southern India, and probably come to the island of Sri Lanka as traders.

c. 275 B.C. Intensive agriculture is practiced

Sri Lankan engineers have invented canals and sluices to regulate the flow of water from artificial lakes or reservoirs (tanks) created by building dams across valleys. This sophisticated irrigation system permits intensive rice agriculture in the semi-dry and dry zones of the island. The third king in Vijaya's dynasty, Pandukkabhaya, builds the city of Anuradhapura, which flourishes due to rice cultivation.

247 B.C. King Tissa converts to Buddhism

Stories from the *Mahavamsa* relate that the Indian Emperor Asoka (c. 264–223 B.C.), a devout Buddhist who reigns over most of India, sends his son, Prince Mahinda, on a religious mission to Sri Lanka. He unexpectedly encouters King Devanampiya Tissa of Anuradhapura, who is deer hunting near Mihintale. Mahinda preaches a Buddhist sermon to King Tissa, thus converting him to Buddhism. Many of the king's subjects subsequently convert. The monastery of Maha Vihara is created, becoming the primary center of Buddhist worship on the island.

247 B.C. Sacred tree planted at Anuradhapura

Prince Mahinda sends to India for his sister, Sanghamittha. She arrives, bringing a sapling from the sacred Bo tree of Gaya in India, the tree under which the Buddha attained enlightenment. The sapling is planted in Anuradhapura. It continues to flourish, thriving at the end of the twentieth century. It is a sacred shrine and a destination for many pilgrims (travelers on a religious journey). It is considered the oldest historical tree in the world.

237 B.C. Cavalry officers revolt

Two officers of the cavalry overthrow the Sinahalese king. They then rule jointly as Sena I and Guptika. Power shifts back and forth between Sinhalese and Tamil rulers for centuries to come.

215 B.C. Sinhalese king takes power

The Sinhalese ruler, Asela, takes power from Sena I and Guptika.

205 B.C. Asela overthrown

Asela is overthrown by a Tamil warrior, Elala who rules until 161 B.C.

161 B.C.–A.D. 1070 The Anuradhapura period

The Sinhalese ruler Dutugemunu (r. 161–137 B.C.) expels the Tamil king Elala, kills him, and unifies the island under a single ruler. His subjects owe free labor *(rajakaria)* for a certain number of days each year in order to build irrigation projects, roads, and other public works. Dutugemunu rewards his senior officials with land, and a privileged class of landowners develops. He also gives land to monasteries, which become huge landholders of the state.

137 B.C. Death of Dutugemunu

When Dutugemunu dies, the country enters a period of instability and repeated Tamil invasions. The Brazen Palace at Anuradhapura is a memorial to Dutugemunu's support of Buddhism.

A.D. 276–303 Reign of King Mahasena

Vijaya's dynasty lasts until 65 A.D. when the Lambakanna dynasty is formed. A famous king of this dynasty, Mahasena (r. 276–303) builds the Kantalai Tank, covering 4,560 acres and fed by a twenty-five-mile long canal. Under his rule, the Buddhist monks preach in the Sinhalese language (not the

Among the most admired attractions in Sri Lanka are the frescoes of the "Heavenly Maidens" in a sheltered gallery on the rock face about halfway up the Siigiriya fortress. These are frescoes of beautiful women, the only ancient secular paintings in Sri Lanka. Twenty-two of the original paintings remain. (EPD Photos/Victoria J. Baker)

ancient Pali language); thus, the teachings are spread and the language becomes enriched. The Lambakanna dynasty ends with an invasion from southern India by the Pandyans.

4th century Relic of the Tooth is brought to Sri Lanka

The precious relic of the Buddha's eye tooth is brought to Sri Lanka. It is said to have been snatched from the flames of the Buddha's funeral pyre in 543 B.C. and smuggled into Sri Lanka, hidden in the hair of a princess. At first it is taken to Anuradhapura, then moved from place to place, ending in Kandy. (In 1283, it is carried back to India by an invading army, and later brought back to Sri Lanka by King Parapramabahu III. In the sixteenth century the Portuguese capture what they claim is the tooth; they burn it in Goa. However, the Sri Lankans say that what the Portuguese had was a replica, and the real tooth remains safe.)

459–77 Reign of King Dhatusena

King Dhatusena (or Datu Sen) drives out the Pandyans. He constructs the Kalawewa Tank covering seven square miles with a fifty-four-mile long canal. He commissions the sculpting of the Aukana Buddha statue near the tank; the statue stands twelve meters high. Dhatusena is murdered by his son (see 477–95).

477–95 Reign of King Kassapa I

Dhatusena's son, Kassapa (or Kasyapa), overthrows and murders his father. When the outraged citizens drive him from Anuradhapura because he murdered Dhatusena, Kassapa temporarily establishes a new capital, Siigiriya. A story recounts that Kassapa, whose mother is a palace consort, fears revenge by his half-brother, whose mother is the true queen. He builds an impregnable fortress and palace on the huge Siigiriya rock. In 491, when an invasion arrives, Kassapa rides out on an elephant at the head of his army. He takes a wrong turn trying to outflank his half-brother, gets bogged in a swamp, is deserted by his troops, and takes his own life.

495 Capital reestablished at Anuradhapura

After Kassapa takes his own life (see 477–95), the capital is reestablished in Anuradhapura.

684 Manavamma takes over the throne

The Sinhalese prince, Manavamma, is able to gain the throne with the help of the Tamils. His dynasty rules for almost three centuries, maintaining alliance with the Tamil Pallavas dynasty.

982 Rule of Mahinda V

Mahinda V ascends the throne as the last Sinhalese king to reign in Anuradhapura.

993–1070 Occupation by the Cholas

The Indian Chola people, led by Rajaraja the Great, invade and occupy the island from north to south. This period of Chola rule is the only time that Sri Lanka was ruled as a province of southern India. The Cholas sack Anuradhapura and destroy the city completely. They establish their capital in Polonnaruwa, farther to the southeast, making it possible to control the main route to the southern Sinhalese Kingdom of Ruhuna.

1070 Vijayabahu reestablishes Sinhalese rule

King Vijayabahu regains power from the Cholas and keeps Polonnaruwa as his capital.

1070–1220 The Polonnaruwa period

The capital remains at Polonnaruwa for about 150 years. The kings gain their income from irrigation dues and trade in surplus rice. They maintain strong military and political power and send expeditions abroad.

1153–86 Reign of Parakramabahu I

King Parakramabahu I restores Sinhalese civilization to its former glory. He plans beautiful parks, reconstructs many temples and stupas (domed Buddhist shrines) at Anuradhapura, and builds extensively in Polonnaruwa; his buildings include an immense tank, the Parakrama Samudra, which could irrigate 18,000 acres.

1187–96 Reign of King Nissankamalla.

King Nissankamalla ascends to the throne. During his reign, he continues Parakramabahu's generous support of the arts. He nearly bankrupts his kingdom trying to outdo his predecessors' accomplishments. He regulates the occupational caste system of Sri Lanka, with the *govigama* or cultivator caste being the highest caste after the king's nobles, who are able to own slaves.

1184–1235 Era of dispute and disease

The Sinhalese intermarry with the Kalinga dynasty of Indian origin. Dynastic disputes and wars destroy the irrigation system. It is likely that malaria, a disease transmitted by mosquitoes, appears in this period. The Sinhalese migrate to the malaria-free areas of the central highlands and to the southwestern and southeastern coasts. Those remaining in the dry zone revert to slash-and-burn agriculture. This early form of farming is typified by the slashing and burning of fields in order to clear land for planting crops.

1250–1500 Weakening and decline of the Sinhalese kings.

Parakramabahu III (r. 1287–93) is the last king to reign from Polonnaruwa. His successors move to Dambadeniya, Kurunegala, Gampola, and Kotte. None of the kings of this period have tight control of the island. The traditional rice irrigation system with tanks and canals, which had given power to the Dry Zone Polonnaruwa kings, collapses. Some tanks are deliberately destroyed by warring factions. People migrate to areas with greater rainfall in the central highlands, the south, and the southwest. Terraced rice fields in the mountainous areas and shifting highland slash-and-burn cultivation become important.

1344 Northern kingdom described by Ibn Battuta

The Muslim traveler and chronicler Ibn Battuta (1304–68) visits Sri Lanka and describes the northern kingdom and its trade system, which is supported by Arabs. Tamils control the pearl fishing in the north. A Tamil kingdom of Jaffnapatnam in the northern part of Sri Lanka begins to expand south. As a result, the island is weakened and divided and is vulnerable to foreign invasions. Sinhalese kings seek the aid of Arab traders and settlers.

1408 Revenge from China

A Chinese visitor is insulted during a visit to the island and the Chinese army retaliates. They capture the king and imprison him.

15th century Muslims secure a trade monopoly

The Muslims establish communities in the major coastal towns after securing a trade monopoly. Their presence as traders is still seen today. The rice agriculture in the dry zone has been lost, and consequently the island has to import rice. The coastal areas are planted with cinnamon, an export that is much in demand by Arab and Chinese traders.

1415 Parakramabahu regains control from the Tamils.

The Sinhalese king, Parakramabahu, takes over the throne with the help of Chinese traders. He establishes a fort at Kotte (near Colombo, the site of the modern governmental capital of Sri Lanka). He captures and holds Jaffna in the north for seventeen years.

1432 Jaffna reverts to Tamil control

Jaffna, an important trading center in the north, reverts back to the Tamil kings. They maintain a strong Hindu rule in the north, closely linked with the cultural traditions of southern India.

The Period of Colonial Rule

1505 Portuguese land

The Portuguese have their first contact with Sri Lanka when the fleet commanded by Lorenzo de Almeyda is blown off course and lands at Colombo. Almeyda finds the local king of Kotte hospitable. The Portuguese establish trading contact with the country.

1517 Portuguese build a fort

Portuguese traders are allowed to build a fort near Colombo. Due to disputes in the kingdom of Kotte, the Portuguese are able to gain a strong foothold in the coastal areas. The Portuguese East India Company dominates trade in the region.

1521 Portuguese aid king

The king of Kotte asks the Portuguese for aid against his brother, thereby making the Sri Lankan kings more dependent on the Portuguese.

c. 1530 Prince Dharmapala begins close Portuguese relations

Prince Dharmapala (d. 1597) is guaranteed Portuguese protection for ascending the throne in return for privileges and a large tribute of cinnamon.

1557 Dharmapala converts to Christianity

Portuguese Franciscans educate Prince Dharmapala and he converts to Christianity. He provides support to Christian missionaries, giving them land belonging to Buddhist and Hindu temples. Portuguese missionaries cross over from the north and convert many fishermen around Jaffna. In retaliation, the king of Jaffna kills several hundred of these Christian converts.

1580 King Dharmapala declares the Portuguese king as heir

Dharmapala declares the king of Portugal to be heir to Kotte Kingdom upon his death.

1591 Portuguese attempt to gain the Kandyan Kingdom

The Portuguese attempt unsuccessfully to capture the Kandyan Kingdom in the central highlands. They make another unsuccessful attempt three years later in 1594.

1597 Portuguese control southwest Sri Lanka

When Dharmapala dies, the Portuguese take possession of his southwestern Kotte Kingdom centered around Colombo.

1612 Kandyan kings seek aid

Portuguese threats continue. The Kandyan kings try to get help from the Dutch and the Danes in order to drive the Portuguese from the island.

1619 Portuguese control trade

The Portuguese succeed in annexing the kingdom of Jaffna in the north. They gradually gain control of Trincomalee and Batticaloa, principal ports on the east coast, thereby dominating the island's trade.

1619–38 Portuguese domination

The Portuguese fail to understand the Sinhalese traditional culture, exploiting the people with undue harshness in taxation as they monopolize the profitable trade in cinnamon, pepper, betel nuts, and elephants. They favor those who convert to Catholicism, giving them appointments as officials. Those who convert take on Portuguese surnames such as Fernando, Perera, da Silva, and Peiris. (In modern times, these Portuguese family names remain prevalent in Sri Lanka, although the families are likely no longer Christians.) The Portuguese establish Catholic churches and schools in diverse parts of the island. Many Portuguese words are found in the Sinhalese vocabulary today.

c. 1650s The Dutch recapture ports for the Kandyan king.

The Dutch assist the Kandyan king in recapturing land controlled by the Portuguese. The Dutch are successful in winning control of Batticaloa and Trincomalee ports from the Portuguese, but they refuse to give these cities to the Kandyan king, Rajasinha II, until all the war expenses are paid. The Dutch anger the Kandyan kings by making a truce with the Portuguese.

1652 Dutch-Portuguese truce expires

The truce between the Dutch and Portuguese expires, and the Dutch proceed to lay siege to ports controlled by the Portuguese. The Dutch already control Batticaloa and Trincomalee (see c. 1650s).

1655 The Dutch attack Colombo city

The Dutch attack Colombo by land and sea in their attempt to drive out the Portuguese.

1656 Portuguese in Colombo surrender

The Portuguese controlling Colombo surrender to the Dutch, but the Dutch do not give the city to King Rajasinha, the Kandyan king. In response, he destroys the land around Colombo and withdraws to his mountainous kingdom. The Dutch continue to expel the Portuguese, becoming rulers of the island except for the Kandyan Kingdom.

1656–1796 The Dutch Period

The Dutch rule is predominantly in the hands of the commercial Dutch East India Company, which first controls the coastal areas and slowly pushes inland. A Dutch governor resides in Colombo, and Dutch officials rule over the two additional administrative districts, Galle in the south and Jaffna in the north. The Dutch influence in the old walled fortress section of the town of Galle is still visible in the modern-day city.

1660–97 Robert Knox lives in the Kandyan Kingdom

Robert Knox, a nineteen-year old from London, England, is captured together with his father and several other members of an English ship. They are taken to the Kandyan highland and cared for by villagers. Knox spends eighteen years in the Kandyan Kingdom and writes a history of the island. His observations—though sometimes biased—give insight into the Up-Country Sinhalese culture of this period.

1687 Construction of the Temple of the Tooth begins

In Kandy, work is begun on the famous Dalada Maligawa temple, Temple of the Tooth. It houses a sacred relic—the Buddha's eye tooth. This temple is a major modern-day shrine and tourist attraction.

1766 Dutch and Kandyan agreement

The Dutch are unsuccessful in overthrowing the kings of Kandy. An agreement between the Dutch government and the Kandyan king gives the Dutch formal control of the coastal areas.

The Dutch interfere little in local government, but they maintain strict control of trade and commercialize agriculture to a great extent. The Dutch introduce the cash crops of coconuts, tobacco, coffee, sugar, and cotton. Dutch citizens, known as Burghers, are encouraged by land grants to settle in the countryside, but most remain in the urban centers. Their descendants make up about one percent of the population of modern Sri Lanka.

A big contribution of the Dutch is in the field of law. They try to apply folk law wherever possible, but eventually Roman-Dutch law is widely used and remains a big influence on the present Sri Lankan judicial system.

In the field of religion, the Dutch initially attempt to spread their Calvinist Protestant faith and to ban Catholicism. Ultimately, they relax the effort to convert the people—although they reserve the higher local offices for Christians. Their religion has little impact on the country.

1789 French revolution in Europe creates turmoil

The French Revolution puts the major European powers in turmoil and prepares the way for British takeover in Sri Lanka.

1795 Kandyan treaty with British

The Kandyan king negotiates a treaty with the British. Under the terms, the British are promised control of the coastal areas and the cinnamon trade in return for helping the Kandyan king push out the Dutch.

1796–1848 The British Period

The British capture the port of Trincomalee from the Dutch, thereby beginning their period of rule on the island.

1802 Ceylon become first British Crown Colony.

The European Treaty of Amiens provides that Ceylon—as it was then called by the British—becomes the first British Crown Colony. A Ceylon Civil Service is created to be controlled by a governor.

c. 1803 The first Kandyan War

The first Kandyan War is fought between the Kandyan king's followers and the British. The British try to gain control of the last Sinhalese kingdom to hold out against European colonizers. The British are successful in the beginning, but disease strikes during the return march, and the Sinhalese annihilate them. The British do not try to subdue Kandy for another ten years.

c. 1812 The Kandyan king alienates his subjects

The Kandyan king gradually angers and alienates his subjects by actions against Buddhist monks. Rebel groups form and undermine his ability to rule.

1815 Kandyan king is exiled

The Kandyan rebels invite the British to intervene against the king, who flees. Hardly a shot is fired. The king and his family are exiled to southern India. The fact that the Kandyan people of the central up-country were successful in remaining independent for 200 years after Europeans colonized the rest of the island is a source of pride among modern-day descendants.

1815: March 2 The Convention of 1815

The British government and the Sinhalese nobles sign the Convention of 1815. It provides sovereignty to the British Crown, rights and privileges for Sinhalese nobles, and protection of Buddhism by the British government.

1818 The British subdue a rebellion of nobles

A rebellion of the nobles is put down. Some of their privileges are reduced, the special position of Buddhism is modified, and the Kandyan provinces are integrated with the rest of the island under the British. Reforms are enacted, such as abolishing slavery and reducing the amount of compulsory service by the people for the nobles.

1820s British programs gain strength

A program for road building opens the highland area for development, and agriculture is encouraged. Restrictions on European land ownership are lifted. Christian missionaries start many schools with English as the medium of instruction. The Ceylon Civil Service, which is open to Ceylonese, emphasizes an English education. A Western-educated elite gradually appears; it is these people who will later be at the forefront of the independence movement in the twentieth century.

1821 Royal Botanical Garden established

A Royal Botanical Garden, founded on the English model of demonstration gardens in a park-like setting, is established at Perdeniya near Kandy.

1824 Coffee cultivation introduced

The first coffee plantation is opened in the Kandyan hills, introduced by the English governor Sir E. Barnes. There is a big demand for coffee but a shortage of labor, as the Sinhalese cultivator caste prefers to cultivate rice as a point of prestige.

1830s Indian Tamils are recruited

Many Tamils from southern India are recruited as inexpensive labor for the British coffee plantations.

1833 Colebrooke-Cameron reforms adopted

The British Colonial Office sends the Colebrooke-Cameron Commission to the island to evaluate the administration. Many of the proposals of the commission are adopted, among them uniform administrative and judicial systems for all of Ceylon. The plan opens the Ceylon Civil Service to Sri Lankans mainly educated in the English language and ultimately creates a Westernized elite.

1840s The estate system dominates the economy

Land belonging to Sinhalese farmers in the hill country (200,000 acres) is confiscated and sold as crown land (land owned by the British government). British entrepreneurs seeking a quick profit arrive from England. Financial and commercial power becomes concentrated in British hands. The estate (plantation) system dominates the island's economy.

1867 First tea seedlings are planted

James Taylor, a Scotsman, clears nineteen acres of forest to plant his first tea seedlings in the Loolecondera Estate.

1868 Coffee plants are struck by disease

A leaf disease strikes the coffee estates. A substitute is sought—for example, quinine—to take the place of the good cash crop, coffee. Tea eventually takes the place of coffee; by 1929, 500,000 acres of up-country are under tea cultivation, mostly British owned and managed.

1870s World depression brings economic decline, domestic unrest

World depression causes a big fall in coffee prices. The government imposes new taxes on everything from dogs to bullock carts.

Riots break out when the British suggest taking Buddhist temple lands for coffee plantations. The riots are harshly suppressed, causing an investigation. Many of the new taxes imposed by the British are discontinued.

1876 Rubber is introduced

Rubber is introduced as another estate crop. Additionally coconuts, which had always been part of Sinhalese gardens, become an estate crop for export. The gardens are mostly owned by Sinhalese entrepreneurs.

1915 Riots between Muslims and Kandyan Buddhists

Riots break out in the Kandyan highlands between the Muslim immigrants from southern India and the Kandyan Buddhists. The Muslims have constructed a mosque in Gampola and complain that the music and dancing of the Buddhist procession *(perehera)* violates their mosque. Some Sinhalese are imprisoned after the riots. Their representatives go to England to plead their cause.

1919 Ceylon National Congress

The Ceylon National Congress is formed. It unites Sinhalese and Tamil organizations and drafts proposals for constitutional reform and partial self-government.

1931 New constitution is adopted

A new constitution with reforms recommended by the Donoughmore Commission is adopted, giving Ceylonese leaders opportunities to gain experience in government with the intention of eventual self-government. The constitution provides for universal suffrage; thus, the Ceylonese people are able to gain experience in the democratic process.

1931 State Mortgage and Investment Bank established

To provide financial assistance to those engaged in agriculture, a new mortgage bank is established. Known as the State Mortgage and Investment Bank, its services are primarily in long-term mortgages and other long-term credit instruments.

1939–45 Trincomalee is a strategic harbor in World War II

The Japanese enter World War II (1939–45). Trincomalee becomes a strategic harbor. The British promise Ceylon self-government after the war.

1947 Soulbury Constitution

The British-drafted Soulbury Constitution is accepted, to become effective in stages. Full status in the Commonwealth is promised by 1948, at which time the country will have a parliamentary democracy.

The Period since Independence

1948: February 4 Ceylon gains independence

D. S. (Don Stephen) Senanayaka (1884–1952) of the United National Party (UNP) is elected as first prime minister of the newly independent Ceylon; under him, the people have majority rule in a parliamentary democracy. The country is in the hands of an English-educated elite that does little to preserve traditional culture. There is economic prosperity in Ceylon, due in part to the Korean conflict and the demand worldwide for rubber.

1952 Senanayaka dies in a riding accident

Prime minister D. S. Senanayaka is killed in a fall from a horse. His son, Dudley Senanayaka, is elected to succeed him.

1954 The rise of nationalism

The country faces economic difficulties due to the falling prices of rubber and tea. A new Sinhala (native Sinhalese language and culture) nationalism arises, embodied in the Sri Lanka Freedom Party (SLFP).

1955: December 14 Ceylon admitted to the UN

Ceylon is admitted to the United Nations.

1956 Election of Solomon Bandaranaike

Sinhalese politician Solomon West Rudgwat Dias (S.W.R.D.) Bandaranaike (1899–1959) becomes prime minister. Shortly before the elections, Bandaranaike forms the People's United Front (Mahajana Eksath Peramun, or MEP), a coalition of SLFP, the Lanka Sama Samaja Party (LSSP), and some indepedents. MEP wins fifty-one seats in partliament. Bandaranaike has a socialist platform, adopting national dress and standing up for Buddhist monks, local schoolteachers, and traditional (ayurvedic) doctors. The "Sinhala language only" policy is his rallying cry. His platform includes nationalizing estates, banks, and some essential industries, including the Ceylon Transport Board and Colombo Port, repatriation of Tamil estate workers to India, and nonalignment in foreign affairs. Tamils riot after his election when policy implementation begins.

1956 2500th anniversary of the Buddha's nirvana

The 2500th anniversary of the attainment of nirvana by the Buddha—Buddha Jayanthi Year—is celebrated. The All Ceylon Buddhist Congress is formed.

1956 2500th anniversary of Vijaya's arrival

The 2500th anniversary of the legendary landfall of the first Sinhalese king, Vijaya, is celebrated.

1958 Continued ethnic violence

Continuous riots occur due to the "Sinhala only" official language policy, which is opposed by the Tamils and the Christian community.

1959: September 25 Assassination of Solomon Bandaranaike

Prime Minister S.W.R.D. Bandaranaike is assassinated by a Buddhist monk with a personal grudge.

1960 Election of Sirimavo Bandaranaike

The widow of Solomon Bandaranaike, Sirimavo Bandaranaike (b. 1916), wins the next election for the SLFP party, becoming the world's first female prime minister. She continues to implement her husband's nationalistic policies, serving as prime minister until 1965 (see also 1970 and 1994).

1960 Association for the Advancement of Science founded

The Association for the Advancement of Science is founded with headquarters in Colombo. It is dedicated to providing a forum for the sharing of ideas among scientists of the country and with their colleagues internationally.

1965 UNP wins majority in elections

An economic crisis with much unemployment and a high cost of living causes unrest in the country. The UNP party headed by Dudley Senanayaka wins sixty-six of the seats in parliament, and he becomes prime minister. The new government encourages private enterprise but cannot solve the problems of many of the youth, who though educated are unemployed.

1970 The SLFP and the Marxist parties form an alliance.

With discontent in the country, the SLFP and the Marxist parties form an alliance and win a landslide victory in the election, with Sirimavo Bandaranaike (see 1960 and 1994) again becoming prime minister. She restricts private enterprise, nationalizes many private industries, and implements land reform, but her policies fail to get a grip on the economic crisis.

1971 Insurrection by Marxist youth

The Marxist youth, who want more radical change, rally under the JVP party (Janatha Vimukhti Peramuna or National Liberation Front), a militant Sinhalese party from south Ceylon. They rise in the first insurrection against the government and are put down by the army.

1972 Adoption of a new constitution and new name: Sri Lanka

The country adopts a new constitution and changes its name from Ceylon to Sri Lanka. A president is to become the formal head of state, with power to appoint a prime minister and cabinet. Sinhalese becomes the official language and Buddhism is given the most prominent place as religion. The economy continues to deteriorate, with high unemployment and long lines at shops.

Mid-1970s Tamil unrest

The number of Tamils admitted to universities is cut, and some youth begin to exercise violence, fighting for an independent state of Tamil Eelam.

1976 J.R. Jayawardana becomes president

Junius Richard Jayawardana of the UNP wins the presidential election and tries to revive the economy. He cuts subsidies, devalues the rupee, opens the country to tourism as well as foreign investment and imports, and sets up a Free Trade Zone near Colombo. He also accelerates the Mahaweli Project, damming the country's largest river to provide irrigation and hydroelectric power.

1978 Another new constitution

President Jayawardana introduces a new constitution that gives more power to the president.

1982–88 Reforms in President Jayawardana's new term

President Jayawardana is reelected. He provides for greater local control in government and allows Tamil to be used as a national language in Tamil-majority areas. Still, Tamil ethnic unrest and uprisings grow.

1983: October Unprecedented ethnic violence

The worst ethnic violence in the history of the country occurs between the Sinhalese and Tamils. (There had been smaller outbreaks of violence in 1958, 1977, and 1981.) A Sri Lankan army patrol is ambushed by a group of Tamil "Tigers," separatists who fight for an independent Tamil state. The ambush leads to unprecedented looting, burning, and killing. Shops in many Tamil business districts are completely destroyed, and atrocities are committed on both sides. Tens of thousands of Tamils flee to Tamil-majority areas in the north or northeast. Sinhalese move out of Tamil areas. Many Tamils go abroad as refugees.

1985: May Anuradhapura massacre

Approximately 150 Sinhalese Buddhist pilgrims are gunned down by Tamil separatists at the ancient sacred site of Anuradhapura.

Mid-1980s Tamils are given limited self-government

Although the Tamils are given limited self-government, ethnic violence continues, with atrocities perpetrated by both sides. The "Liberation Tigers of Tamil Eelam" (LTTE) attack Sinhalese villages. About 50,000 people are in refugee camps in Sri Lanka, and about 100,000 Tamils are in refugee camps in India and other countries. The economy suffers from the high military costs, the slump in tourism, and the reduction in foreign aid.

Mid–1980s Indian Peace Keeping Force (IPKF) is invited

President Jayawardana strikes an agreement with India. The Sri Lankan army discontinues fighting while the Indian Peace Keeping Force disarms Tamil rebels with the intention of keeping peace in the north and east. The LTTE attacks some Sinhalese villages in the east, provoking riots in Colombo. The IPKF takes Jaffna. Many are killed on both sides.

Mid–1980s The Marxist JVP party re-emerges

The Marxist National Liberation Front (JVP) reemerges in southern and central Sri Lanka under Rohana Wijeweera, and terrorizes these areas.

1989 Ranasinghe Premadasa is elected president

President Jayawardana retires at the end of 1988, and Ranasinghe Premadasa of the UNP becomes president in the new election. He promises to remove the Indian Peace Keeping Force in the north. He orders "death squads" to kill JVP suspects. Wijeweera and other JVP leaders are killed or captured. During the three-year insurrection period, about 17,000 people are killed, a large portion having "disappeared."

1990 Tamils declare an independent Tamil state

The LTTE agrees to a cease fire—while the Indians are withdrawing from the north. Later the Tamils declare an independent Tamil state in the north and northeast, provoking a renewal of the civil war.

Mid-1991 Civil war continues

War reaches a peak soon after the assassination of India's prime minister Rajiv Gandhi by a suspected Tamil Tiger. At times it looks like peace might be possible, but the Sinhalese are skeptical.

Mid-1992 New attacks on the north

The Sri Lankan army launches new attacks on the north. The LTTE begins using women fighters in their army. Tens of thousands of people are killed in 1992; 700,000 people are displaced as refugees.

1993: May 1 Premadasa is assassinated

President Premadasa is assassinated during the May Day rally. It is thought that the LTTE is responsible. Preparations are made for a new election, with Gamini Dissanayaka of the UNP running against Chandrika Bandaranaike Kumaratunga (the daughter of former Prime Minister Sirimavo Bandaranaike) of the People's Alliance party (PA).

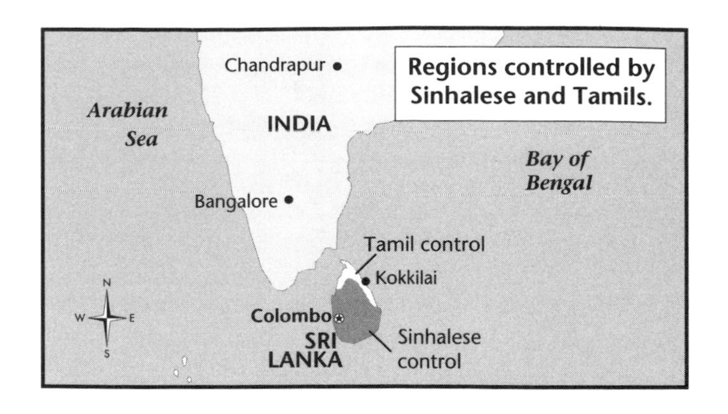

1994 Fishing on stilts

Fishers on stilts wade into the the waters of the Indian Ocean near Galle in an attempt to earn a living by fishing. Sri Lankan fishers do not catch enough to meet the country's needs, but a small amount of fish is exported. This fishing technique has been employed for centuries.

1994: November Bombing kills Dissanayaka; Kumaratunga wins election

Shortly before the November election, an LTTE suicide bomber kills Gamini Dissanayaka and many other senior UNP party members. Mrs. Chandrika Bandaranaike Kumaratunga (b. 1946) easily wins the election, becoming Sri Lanka's first woman president. She appoints her mother, Sirimavo, prime minister (see 1960). Her party promises to focus all efforts toward ending the civil war.

1995: January Peace talks fail

The peace talks fail. The war in the north and east continues, with sporadic violence.

1996: January Central Bank bombing

Eighty people are killed in a Central Bank bombing in Colombo. Terrorist acts continue, crippling the economy. The tourism industry declines. Over a million people continue to be displaced due to the ongoing ethnic strife.

1996–98 Attacks and counterattacks

The sixteen-year-old civil war persists. The government engages in major offenses on Tamil held areas, and the LTTE responds with counterattacks. Sporadic bombings occur in Colombo, undermining the morale of civilians. Attempts at peace talks become more difficult and agreements fail.

1997 Sprinter wins silver at World Athletic Championship

Sprinter Susanthika Jayasingha wins a silver medal in the women's 100 meter race at the World Athletic Championships held in Athens, Greece. Sri Lankans hope she will capture the country's first gold medal at the 2000 Olympic Games in Sydney, Australia.

1996: March Sri Lanka wins in cricket

Sri Lanka beats Australia in a World Cup cricket match. Cricket is a national passion, and the Sri Lankan team is a strong contender in international cricket competition. Cricket was brought to Sri Lanka (as it was to other nations colonized by the British in Asia, Africa, and the Caribbean) by English colonists in the late 1800s.

1997: August 6 Cricket team sets a record

The Sri Lankan cricket team sets a record for the most runs scored in a five-day test innings. The Sri Lankan team scored 952 for 6 in a test against India, a new test innings record. The previous record was set in 1938, when England scored 903 for 7 against Australia. The new record was achieved in 271 overs in three days of play.

1998: January Buddhist holy site bombed

The Temple of the Tooth, Sri Lanka's holiest Buddhist shrine is bombed, killing sixteen people. The government accuses the Tamil Tigers of committing the bombing.

1998: September 27 Tamils capture Kilinochchi

The Tamils score a major victory when they capture Kilinochchi, a city on the road to Jaffna in the north of the island. The Tamil capture of Kilinochchi cuts off Jaffna from all ground communication with the rest of the government area.

Bibliography

Anderson, John Gottberg, and Ravindral Anthonis, editors. *Sri Lanka* (Insight Guides series). Hong Kong: Apa Productions, 1993.

Baker, Victoria J. *A Sinhalese Village in Sri Lanka: Coping with Uncertainty.* Ft. Worth: Harcourt Brace College Publishers, 1998.

De Zoysa, Lucien. *Stories from the Mahavamsa.* Colombo: Marga Publications, 1995.

Ludowyk, E.F.C. *A Short History of Ceylon.* New York: Frederick A. Praeger, 1963.

Paranavitana, Senerat. *Sinhalayo.* Colombo: Lake House Publishers, 1967.

Robinson, Francis, editor. *The Cambridge Encyclopedia of Pakistan, Bangladesh, Sri Lanka, Nepal, Bhutan and the Maldives.* Cambridge: Cambridge University Press., 1989.

Ross, Russell R., et al. *Sri Lanka: A Country Study* (Area Handbook Series). Washington, D.C.: Library of Congress, 1990.

Vesilind, Pritt J. "Sri Lanka." *National Geographic,* January 1997, vol. 191, no. 1, pp. 110+.

Syria

Introduction

Africa, Asia, and Europe all converge at a small stretch of land, nestled in between the Mediterranean Sea, the Red Sea, and the Persian Gulf, about the size of Indiana and Illinois combined. This land has had many names but now officially is called the Syrian Arab Republic. This convergence of continents has given Syria not only a diverse geography, but also a diverse political and cultural history. Syria's land ranges from the fertile Euphrates Valley in the northeast, formed by the Euphrates River, to the arid desert in the southeastern area bordering Iraq.

Due to frequent invasions of its territory throughout thousands of years, Syria contains a rich variety of cultures inside its small borders. At the beginning of the twenty-first century, ninety percent of the 14.3 million people that lived in Syria were Arabs. Kurds made up nine percent of the population. The remainder of the people consisted of Armenians, Circassians, and Turkomans. This diversity makes the number of languages spoken in Syria and the number of religions practiced quite extensive. The languages include Arabic officially, English, French, Kurdish, Armenian, Aramaic, and Circassian. The religions range from Islam to Christianity to Judaism, not to mention all of the sects of these faiths. This cultural diversity began with Syria's rich past, which started what is the history of modern man and the concept of civilizations and empires.

Due to its location, the area now known as Syria was a major crossroads for trade, communications, and invading forces coming to and from war. The constant flow of people through Syria made it the base for the growth of civilizations. Many of the "firsts" of humankind can be traced back to this area. The first evidence of diplomatic texts and foreign relations were found in the city of Ebla as far back as 3000 B.C., and the first detailed record and administration systems were established at Tell Sabi Abyad, a site of several cities throughout Syria's history. The great empires such as the Assyrians and Akkadians, which spread the first recorded written languages of Phoenician and cuneiform were based in Damascus, the oldest continuously inhabited capital in the world. Many great philosophers and artists originated from this land as well, rivaling or sometimes even teaching famous philosophers like Aristotle and Cicero.

Two of the greatest religions of the present can trace their roots back to Syria as well. Christianity truly began to spread and take on popularity in Syria due to the work of religious figures like St. Paul, who was converted on the road to Damascus. The prophet Muhammad of the religion of Islam did much of his work in Syria as well.

Prize of the Middle East

Syria was considered the prize of the Middle East due to its convenient position at the center of many trade routes and communications lines. Many invaders attempted and often won reign over Syria, assimilating it into their own empires. Syria did not gain independence until 1946, after hundreds of years of occupation by forces such as the Romans, the Byzantines, the Ottomans, and the French. Even today, its territory is still in dispute with Israel's annexation of Syria's Golan Heights on December 14, 1981.

Syria was conquered by the Roman Empire around 64 B.C. The Syrian city of Damascus was named the Roman capital of its Syrian province, making it a central city of one of the most extensive and powerful empires in history. Coinciding with its occupation by Roman forces came an influx of new art, the Latin language, and new forms of government, deeply affecting the culture of Syria.

In A.D. 637, Syria was conquered by the Arab Umayyad Dynasty. With this new occupation came the introduction of the Islamic faith, which had a lasting effect on the people of Syria, who are now predominantly Muslim. Some of the greatest architectural feats of Syria were created in this time period such as the Umayyad Mosque, built in 706.

Shortly after the Umayyad Dynasty's reign came the invasion of the European Crusades in 1097. These Christian warriors, who came to rescue the "Holy Places" from "heathen rule," also had a lasting effect on the Syrian people. One of Syria's greatest historical sites, the Krak des Chevaliers, is one of the best testaments to the greatness of this time period. The effects of this temporary incursion of Christianity can still be seen today in the everyday lives of Syrians, who keep

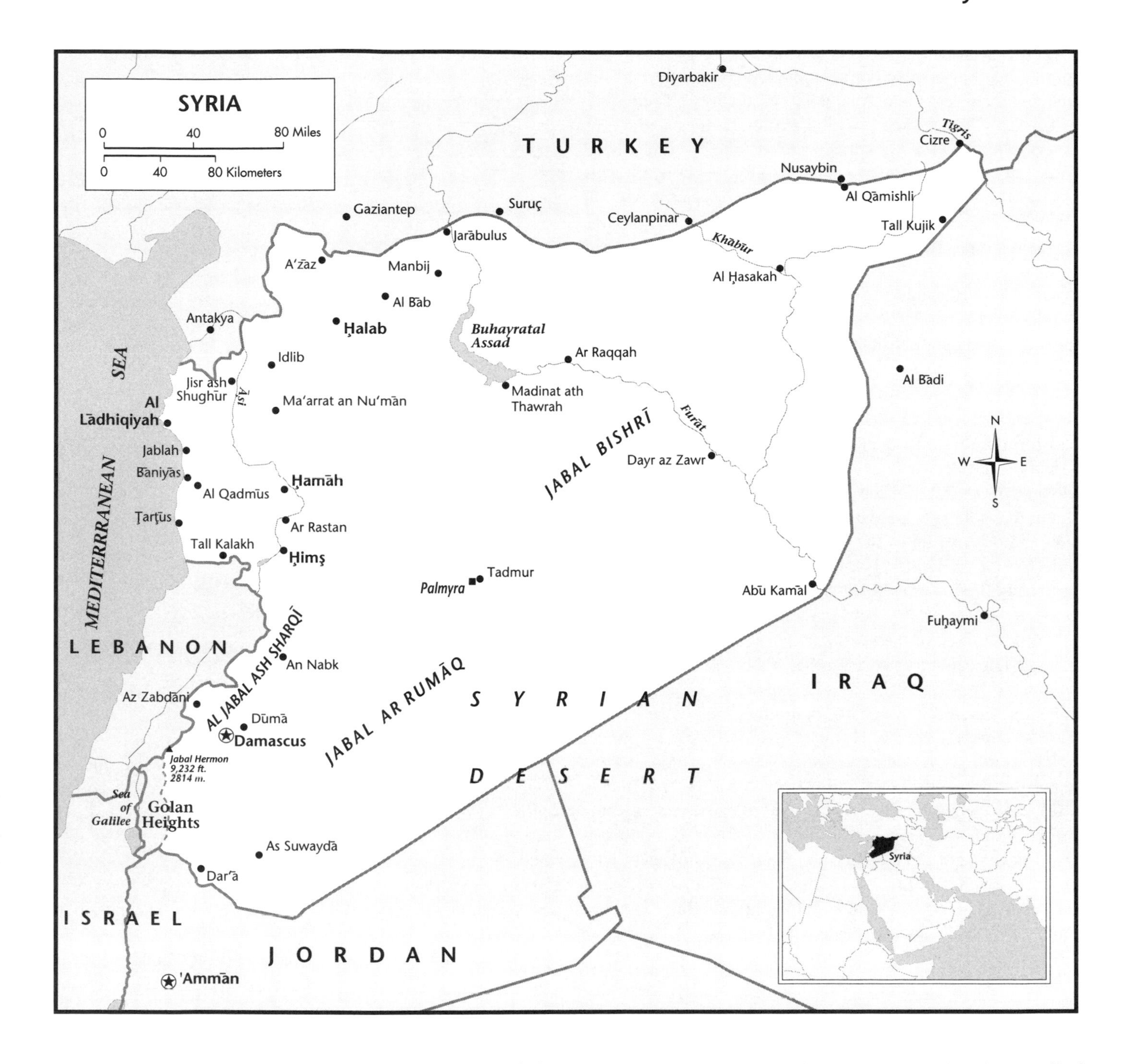

Christian names and practice Christian holidays despite their Islamic faith.

In 1516, Sultan Selim I of the Ottoman Empire invaded Syria, beginning a three hundred-year occupation. His reign brought temporary peace and prosperity to a politically tumultuous land. Selim and his successors also made significant changes in Syria's economy and political structure, many of which can still be seen in its present system of governance.

Syria enjoyed a brief period of independence due to the efforts of Sharif Hussein of Mecca during World War I but that independence ended when France invaded it on July 24, 1920, during the Battle of Maysaloun. Syria shifted between French and British rule until the French finally left in April of 1946 due to Syrian nationalist pressure.

The hundreds of years of invasions brought many cultures to Syria, affecting its arts, sciences, and ideologies. Syria's culture, however, also had an effect on the world. Its arts, inventions, and philosophies had all been spread to the far reaches of the globe. Hints of Syrian art can be seen in forms as far away as England and Germany. Some of its greatest myths have been tied to the parables in the Bible as well.

Unfortunately, the departure of the French did not bring immediate political stability to Syria. For thousands of years,

Syria had been under the control of many empires and rulers so it took a few years of political upheaval before Syria could find a solid base on which to stand politically and economically. Many leaders and political parties came in and out of power bringing with them drastic changes in government and constitutions. Many leaders lasted only for a couple of years, such as President Amin Hafiz, who gained power in May of 1963 and was imprisoned on February 23, 1966.

For several decades, Syria's political history was marked by strings of internal coups until President Hafiz al-Assad brought a long-awaited stability to the region when he was elected to his position on November 13, 1970. He has remained president until the present, bringing a stable constitution and major economic reform that has placed Syria competitively in the industrial age.

Despite President Assad's strong governance, some difficult foreign relations decisions on the part of Syria and its neighbors have caused some instability and political tension. Most of the difficulties are found in its relations with Israel, stemming from several clashes that began with Israel's invasion and occupation of the Golan Heights in 1967. Despite orders from the United Nations to do so, Israel has not yet returned all of Syria's territory. This fact remains a great cause of cultural tension between the two countries.

In spite of all of the historical hurdles to Syria's development, it has caught up to the other major industrial powers quite well. Due to reforms in its socialist system, many of its industries have been nationalized, or globalized in the case of its oil production. It also has made great strides in its human rights efforts by producing labor laws, social insurance, protection of religious freedom in its constitution (ratified in 1973), and the right of women to vote and participate in political parties. Critics, however, charge that these constitutional changes are superficial and do not restrict President Assad's power. Though it still has many areas in which to develop and improve, it will always be a country rich in culture and history that affects many other nations.

Timeline

3300 B.C. First cities are established

The first officially organized cities are built by man in Syria.

3000 B.C. Treaties are established between Ebla and Lebanon

Treaties are written between Ebla and Lebanon. These are the oldest recorded treaties in the history of man and are proof of a highly sophisticated civilization.

The city of Ebla is, for a time, a great metropolis. At one point in its history, it contains the world's largest library. It is also important economically for hundreds of years because of its participation in the silk trade. Ebla is famous for the intricate woodwork that comes from its artists as well as the delicate crafts containing inlays of ivory and mother of pearl.

2500 B.C. The city of Damascus is settled

The city of Damscus is settled. It is considered the oldest continuously inhabited capital in the history of man.

St. Paul, a leading figure in the history of Christianity, is converted from Judaism on the road to Damascus over two thousand years later. It is there that he establishes the first organized Christian church at Antioch. He leaves for many missions from Damascus that would inspire him to write his letters.

2300 B.C. Akkadians gain control

Syria is controlled by Akkadians after defeating the kingdom of Ebla (see 3000 B.C).

2000 B.C. Amorites gain control

Syria is controlled by Amorites after a brief attempt by Egyptians to rule all of its lands.

c. 1900 B.C. The city of Palmyra is built

The city of Palmyra is built. It serves as an oasis stop for trading caravans between the Euphrates River and the Mediterranean Sea. It is also a popular site to visit due to the supposed medicinal qualities of the cold, blue waters of the spring that flow through it.

The city is also important in Syria's history because it is one of the cities where monotheism (belief in one god) is introduced. Before the second century, the people of Palmyra worship Bel, a Babylonian god, Acolytes Yarshebal, Aglibol, and Belshemen. Christianity becomes more popular when a bishop from Palmyra takes a leading role in the Council of Nicea, an important establishment in the history of Christianity.

Palmyra is later ruled by Odenathus the Younger until he and his son are mysteriously murdered in 267. His wife, Queen Zenobia, takes power over Palmyra in 270 (see 270) and then continues on to conquer all of Egypt and Syria.

Palmyra never recovers from the loss of one of the greatest leaders and women in Syria's history. It remains somewhat important for a few hundred years as it is the only producer of turpentine, which is used for Egyptian mummification.

The city remains a great historical and architectural site today because of its well-known colonnade. It is the main street that stretches 1,000 meters in length, lined by towering columns that are capped by magnificent statues of public figures from the history of Palmyra.

1200 B.C. Sea Peoples attack Syria

The cities and cultures along the coastline of Syria are greatly altered by the continual attacks of an ethnicity specified only as Sea People and the consistent immigration of Semitic peoples.

1000 B.C. Aram of Damascus becomes king

Aram of Damascus becomes the first king to attempt to unite the people of Syria into a kingdom to resist King David, who is pushing in from the south of Syria. His attempt does not succeed.

700 B.C. Assyrians take control of Syria

The Assyrians take control of Syria.

323 B.C. The Battle of Issus occurs

The Battle of Issus occurs, giving the control of Syria to Alexander the Great from Macedonia. Seleucus I gains control after Alexander's death. Following Seleucus' death, a long conflict occurs between the Seleucids and the Ptolemids over dominion of Syria's central and southern territories. The Seleucids win the battle in the end, breaking up Alexander's grand empire.

100 B.C. The city of Maalula is formed

The city of Maalula is formed on the eastern slopes of the Al-Qalamoun Mountains, about 45 kilometers from Damascus. It belongs to the kingdom of Homs during this time period, the Aramaen era. It eventually comes under the control of the Romans and then the Byzantines. It later becomes an episcopate (church administrative center) in the late seventeenth century. (What makes it unique is that the people still speak the original Aramaic language, the language spoken by Jesus Christ. This language also provides the base for modern Arabic syntax and is still used in the vernacular outside of the city of Damascus.)

64 B.C. Syria is assimilated into the Roman Empire

Syria, Lebanon, Palestine, and Transjordan are all conquered by the Roman Empire. Syria is recognized as a province of the empire. Damascus is named the capital of this province. The Roman Empire's powerful presence has a deep impact on Syrian culture affecting every realm of life from architecture to literature. The official language of the Empire is Latin, but Arabic dialects are spoken among the native people.

A.D. 103 The tomb, Elhabel, is built

The tomb tower of Elhabel is built. This is onc example of the many famous tombs built at this time in this particular style of architecture. The other forms of tombs are the tomb tower, the house tomb, the hypogeum tower, a stairway linking underground tombs, and the hypogeum tomb. The Valley of Tombs, a site established around this time period, is well-known for its remarkable historical specimens of these forms of tombs.

106 The Roman Empire names Bosra as its capital

Bosra, a Roman city in Syria, is made the capital of the new province of the Roman Empire that includes Syria. The ruins of this once great city are a popular tourist attraction in the twentieth century and serve as a great historical documentation of the Roman Empire's occupation of Syria.

270 Queen Zenobia reigns over Palmyra and Syria

Queen Zenobia claims power over the city of Palmyra (see c. 1900 B.C.) after her husband, Odenathus the Younger and his son are mysteriously murdered in 267. After taking control of Palmyra, Queen Zenobia continues on to conquer all of Egypt and Syria. She insults the Roman Empire by sending her armies across Asia Minor and takes the name of August in competition with the great Roman Emperor Aurelian. Emperor Aurelian eventually raises a strong enough army to seize Palmyra in 274, dragging Queen Zenobia off in golden chains to Rome.

330 Byzantine gains control of Syria

The political allegiance of the area known as Syria changes from Rome to Byzantine. This change introduces a whole new influence of Byzantine culture on Syrian life.

c. 400 The great theologian Stylites lives

Stylites is a fifth century holy man who places himself in a cage on a forty-foot pillar to experience personal suffering as a way to gain enlightenment and to become closer to God and Jesus Christ. Pilgrims from as far away as England come to see this infamous Christian martyr. Qalaat Si is the shrine that is built for him many years later.

c. 400 Christianity becomes of the official religion of the Roman Empire

Constantine the Great recognizes Christianity as the official religion of the Roman Empire. The religion spreads quickly into regions of Syria making it one of the bases for its preliminary development. Many infamous religious figures originate from Syria.

c. 400 The Mar Sarkis Monastery is built

The Mar Sarkis Monastery is built. It is named after St. Sarkis, a Syrian horseman who fell during the reign of Marimanus in 297. This monastery remains an important religious site that both Muslim and Christian pilgrims visit.

570–632 The Prophet Muhammad lives

The Prophet Muhammad lives. In 622, he goes to the city of Yathrib and begins the Islamic era or *hegira*. His last pilgrimage to the holy city of Mecca is in 629.

634 Abu Bakr attempts to liberate Syria

Upright Caliph Abu Bakr launches the first Moslem-Arab excursion to liberate Syria but is not entirely successful.

637 Muslims conquer Syria

Arabs conquer Syria and Damascus comes under Muslim rule. It becomes the capital of the Arab world, making it the center of cultural, political, and economic activity. It is also the base for all Arabic conquests in the Middle Eastern area.

Most Syrians convert from Christianity to Islam at this time. Arabic also becomes the language of the area. The power and prestige of Damascus peaks at this time under the Umayyad Empire. The empire extends from Spain to India and lasts from 662–749 (see 662 and 749).

662 Mouaiwiya Bin Abi Sufyan establishes Umayyad rule

Mouaiwiya Bin Abi Sufyan, the most prominent mayor during the Upright Caliphs time period, establishes the rule of the Umayyad dynasty, seated in Damascus. He lends prosperity and power to the Arabs by Arabizing the administration and currency of Syria, leading to a vast Arab empire. He also makes it possible for the Arab style of architecture to be introduced, the most famous example of this being the great Umayyad Mosque located in Damascus (see 706).

706 Al-Walid builds the Umayyad Mosque

The great Umayyad Mosque is built by al-Walid. It is considered the jewel of the capital during the early Islamic Empire. In the middle of the mosque is a shrine, which is said to contain the head of Yahla or John the Baptist, who had helped to introduce and perpetuate Christianity from Syria.

The tomb of Saladin, the great military leader who takes Syria from the hands of the Crusaders, is directly outside of the mosque.

The Mosque is still intact today and is a popular tourist attraction because of its great beauty and historical value. It serves as an important religious site for people of the Islamic faith as well.

717 Umar bin 'Abd-Al'Aziz is born

Umar bin 'Abd-Al'Aziz is born. He is revered to this day as the great restorer of true Islam, the dominant religion in Syria.

749 The Abbasid dynasty succeeds the Umayyad Empire

The Abbasid dynasty succeeds the Umayyad Empire. The Abbasid caliphate is centered in Bagdhad, reducing Syria to the status of a province of the Abbasid Empire. Syria's name changes to Bilad esh Sham or "the Syrian lands."

Damascus is no longer the capital of the Arab world, and therefore it is no longer so highly protected. It begins to be the subject of a series of attacks by invaders due to its weakness. During this period, Byzantines, Crusaders, Seljuks, and Ayyubids, a Kurdish dynasty, all attack Syria.

915–965 Life of Abut at Fayyil Ahmad bin al-Husayn al Mutanabbi

Abut at Fayyil Ahmad bin al-Husayn al Mutanabbi lives. He is regarded as one of Syria's greatest poets.

c. 1000 Crusaders invade Bilad esh Sham

Crusaders invade the area known as Bilad esh Sham (see 749) claiming to be saving the Holy Places in Jerusalem from Moslem rule.

1000 A severe earthquake strikes the city of Apamea

The once great city of Apamea is completely buried due to a severe earthquake that strikes the area. Archeologists start to dig it up in 1965, finding a wealth of historical matter and information to understand more of Syria's complex history. The city was originally built in 300 B.C. by Seleucus, the Persian ruler of Alexander the Great's eastern empire (see 323 B.C.), and named after his wife. It was a major crossroad between the East and West and it was a stopping place for many invading leaders on their way to or from war, making it valuable territory.

1097–99 Crusaders invade Syria

The first Crusader excursions are successful in establishing four kingdoms in Bilad esh Sham or "Syrian Lands." The effects of the Crusades can still be seen on Syrian culture today. Many Muslims keep Christian names such as William and George. Peasants keep St. Barbara's Day, the saint who protects against sudden death, even today.

One of the most prominent pieces of evidence of the occupation of the Crusaders is the Krak des Chevaliers or Castle of Knights. It is run by the knights of the military and the Monastic Order of Hospitalities. It dominates a strategic point between the Mediterranean and the cities of Homs and Hama on the Orontes River. It is constructed over the base of the Arab fortress called Hisn-al-Ukrad.

1118–74 Nureddin annexes Aleppo

Nureddin, the ruler of Aleppo, annexes Damascus and brings Egypt under his control. He is an Umayyad caliph, and becomes famous for unifying Arab forces against the Crusaders. He is responsible for the victories of Saladin (see 1193), the sultan of Syria and Egypt.

1150 Saladin unites Egypt and Bilad esh Sham in preparation for war

Kurdish General Saladin (see 1193) unites Egypt with Bilad esh Sham in battle and wages a large-scale war against the

The Krak des Chevaliers or Castle of Knights stands on the top of a mountain. It was controlled by the Crusaders from 1097 until about 1271. (EPD Photos/CSU Archives)

Crusaders. They defeat the Crusaders in 1187 at the Battle of Hittin and successfully liberate the city of Jerusalem.

1191 Shibhab ad-Din Suhrawardi lives

Shibhab ad-Din Suhrawardi lives. He is a well-known mystic philosopher who is used by many of Syria's emperors to forsee military defeat or success.

1193 Saladin dies

Saladin, dies (see 1150). He is the founding father of the Ayyubid dynasty which lasts until 1250 (see 1250).

1250 The Mamluks defeat the Ayyubid dynasty

The Mamluks rise to power and defeat Saladin's Ayyubid dynasty (see 1193). They unite Egypt and Bilad esh Sham. The Mamluks divide their empire into six provinces and name Cairo as the capital of the entire empire.

The art of the Mamluk Empire reaches as far as Europe. Its styles can be seen on many European shields depicting the English lion, the German eagle, and French Fleur de Lys.

1260 Damascus loses its power

Damascus becomes a provincial capital of the Mamluk Empire for a period of about two hundred years.

c. 1300 Waterwheels for irrigation are developed

The first seventeen *norias* or waterwheels are built in the city of Hama. They bring water up from the Orontes River and transport it to irrigate lands and provide water for settlements, some of which are hundreds of miles away such as the Krak des Chevaliers. These waterwheels are still used today in some villages and represent a strong tie to the past and Syrian heritage.

1400 Mongols invade Syria

The Mongols of Tamerlane invade Syria frequently. They destroy settlements that had been left behind by other invasions, and kidnap many of Syria's prominent craftsmen including swordmakers, weavers, potters, and glassmakers, settling them in their city of Samarkand.

1516 Sultan Selin I defeats the Mamluks

Sultan Selin I of the Ottoman Empire defeats the Mamluks. Syria becomes a province of the Ottoman Empire until 1831 when the empire becomes weak and loses authority in the provinces (see 1831).

For a time, the Ottoman Empire brings peace and prosperity to the region. The Empire, under the direction of the sultan in Istanbul, stretches from the Balkans in eastern Europe to northern Africa. Each province is governed by a *wali* or governor.

The city of Aleppo prospers greatly under the empire when the sultan makes the bazaars or *suqs* and the caravanserais or *khans* more elaborate, adding to the culture of Syria at the time.

c. 1800 An intellectual revolution begins in Syria

Syrian thinkers turn their attention towards their heritage and *Nahda* or the Awakening. This philosophical movement stirs nationalistic feelings and pride. Political parties start to emerge as well.

1832–40 Ibrahim Paha conquers Damascus

Damascus is briefly occupied by Ibrahim Paha the son of Muhammad Ali, the governor of Egypt. In 1832, Muhammad is granted governorship of Syria for a lifetime from the sultan of Istanbul, and gives the position of commander-in-chief to his son. Ibrahim implements a regular tax system, security in the countryside, and equality for Muslims and non-Muslims alike. His efforts lead to flourishing trade and industries.

1866 Beirut becomes the center of cultural nationalism

Beirut, a city that used to be Syria's most important port, becomes the center of Arab cultural nationalism. It is known for its intellectual, cultural, and political sophistication. The University of Beirut is established at this time by American missionaries in order to promote this intellectual movement. From the university comes the first Arabic dictionary and entirely new literary forms.

1914–18 Sharif Hussein joins the Allied forces during WWI

Sharif Hussein of Mecca joins the Allied forces and revolts against the Ottoman Empire.

His son, Feisal leads the Arab forces under British control into Syria. They oppose French attacks and claim independence under terms of agreement determined between the British and King Hussein.

Damascus is liberated by Hussein the Sharif of Mecca and an Arab government is established on September 13, 1918.

Syria's newfound independence is soon revoked because of the Sykes-Picot Accord that France had made secretly with Britain on September 15, 1916, jointly constructing an Arab confederation with France governing Syria.

1919 The National Museum of Damascus is founded

The National Museum of Damascus is founded. It is the most important museum of the many located in Syria. It holds ancient Oriental, Greek, Roman, Byzantine, and Islamic collections.

1920 The Arab Kingdom is formed by King Feisal

The independent Arab Kingdom, including the area of Syria, is established under King Feisal of the Hashemite family. Soon after he is named king of Syria in March 1920, the French make the San Remo agreement with the League of Nations allotting France parts of Syria. Feisal's rule ends after the Battle of Maysaloun (see July 24, 1920). The conflict is a clash between the Syrian Arab forces and French forces that occupy Syria at this time.

1920: July 24 The Battle of Maysaloun occurs between the French and the Syrians

The French invade Syria during the Battle of Maysaloun. Yousef Azmeh, a military officer, leads encounters against hordes of invading colonialists but fails in the end to resist the French offensive.

The French divide Syria into four parts. France places Syria and Lebanon under its rule. Britain places Palestine and Transjordan under its own. Many revolts against the French occur almost immediately in Syria.

1928 Hafiz al-Assad, the future president of Syria, is born

Hafiz al-Assad is born. He rules Syria from 1970 to the present and is responsible for Syria's great development and for its revolution. Assad is a soldier in the Syrian forces that liberate Syria in 1946 from French rule (see 1946).

1930: October 6 Majlis al-Sha'ab is born

Sha'ab is born on October 6, 1930. He is jailed for participating in a student action group during French rule (see 1946). He joins the military in 1952. He is later exiled but then recalled in 1962. He is requested to become president in 1970 (see 1970) but his term can only be short-lived during a time of great political unrest.

1933 The first textile factory appears in Syria

The first mechanized plant appears in Syria, for spinning and weaving. The factory is established in the city of Aleppo. However, the ancient art of Syrian handicraft is kept alive, to Syria's slow entrance into the industrial age. Handicraft is Syria's most popular art and includes artistic forms such as the Damascus brocade and Syrian soap.

1940 Syria comes under the control of the Germans

France falls to Germany in World War II and Syria comes under the control of the Vichy government, the puppet administration run by Germany in France.

1941: July Syria becomes occupied by the British and Free French forces

British and Free French forces occupy Syria, liberating it from Vichy government control.

1944 The U.S. and United Kingdom recognize the independence of Syria

The United Kingdom and the United States recognize Syria's independence, despite France's continued occupation, granting it the ability to officially conduct political and economic relations with these countries. This recognition helps boost Syria's economic development for a short time.

1945 Syria joins in the formation of the League of Arab States

Syria is a charter member of the League of Arab States. Despite its membership, Syria's relations with fellow Arab states become strained at several points in the following decades due to several unfavorable alliances and political actions.

1945: October 24 Syria joins in the foundation of the United Nations

Syria becomes a founding member of the United Nations. This action helps its political and diplomatic standing with the West.

1946: April The French occupational forces evacuate

The French evacuate from Syria under pressure from Syrian nationalist groups. The departure of France leaves Syria in the hands of a republican mandate.

1946: April 17 Syria declares independence

Syria declares official independence as the last of the French occupation forces leave the country. It is the first Arab country to gain independence.

The independence leads to a short-term boom of economic development. Unfortunately, the period from independence to the 1960s is marked by upheaval, slowing the progress of the nation.

1947 The Ba'ath Party is founded

The Ba'ath Party is founded. Its goals are liberation, Arab unity, and socialism. This party remains the most powerful political party in Syria, holding the majority of the representative seats in Syria's legislature. This is due in no small part to the fact that the Ba'ath Party is that of President Assad who remains in power for over twenty years.

The Ba'ath Party is later combined with the National Progressive Front which consists of the Communist Party, the Syrian Arab Socialist Union, the Socialist Unionist Party, and the Arab Socialist Party.

c. 1950 Syria discovers oil

Syria discovers oil in commercial quantities in several areas of its territory. Oil will become Syria's largest source of national income. Western oil companies such as Shell will prospect for new deposits of oil through the 1990s.

1951 Adib Shishakli seizes power in Syria

Colonel Adib Shishakli seizes power in Syria. He is overthrown later in 1954 due to strong political maneuvering that brings the Arab nationalist and socialist forces to power. The instability at this time and the political parallelism with Egyptian policies appeals to the Egyptian President Gamal Abdel Nasser. His strong leadership after the 1956 Suez crisis creates support in Syria for a union of Egypt and Syria.

1956 France, Britain, and Israel attempt to halt Arab independence

France, Britain, and Israel join in a tripartite against Egypt and Syria in an attempt to stop Arab economic independence. Their aggression fails, however, resulting in the 1958 Egypt-Syrian unification (see 1958). The failure puts Zionist Israel in a poor political position with Syria, which will hold a diplomatic grudge against it for many years.

1956 The Syrian government forms the Central Bank of Syria

The Central Bank of Syria is formed. The bank issues the national currency and also acts as the financial agent of the government and as the cashier for the treasury department.

1958 Agrarian reform begins

The agrarian reform program begins. Legislation and execution of the program is completed in 1963. The program has three goals: to fix a maximum amount of irrigated land that a person can hold, to increase the amount of farmers employed, and to increase agricultural production.

1958: February 1 Syria and Egypt unite

Syria and Egypt join to form the United Arab Republic. All Syrian political parties cease their public activities because of the new Egyptian government.

1959 Social insurance is introduced

Syria's government introduces social insurance. It provides old-age pensions, disability pay, and death benefits to husbands and wives of the deceased, and to their children. Along with the newly introduced labor law (see 1959), these developments represent leaps forward in Syria's human rights and social service records.

1959 The first labor law is passed

Syria's legislature passes its first labor law. It establishes the right of workers to form unions. It also gives the government the power to regulate the hours of work that the laborers are required to fulfill, vacation time, sick leave, health and safety measures, and the minimum level of compensation that workers receive.

1960 Syria begins producing oil

Syria starts producing heavy grade oil that is discovered in fields in the northeastern area of the country. This production boosts Syria's economy, but represents a burden because Syria must import a great deal of light oil to mix with the heavy oil to produce the oil that is sold on the world market.

c. 1961 Syria begins to use the socialist five-year economic plan

Syria begins implementation of six five-year plans for its economic development following the five-year plan model of Russia. The plans deal largely with improving irrigation, land reclamation, and industrial projects geared towards promoting new industries and nationalizing others. Many enterprises are required to adopt nationalization policies to dispel regional and class differences.

1961: September 28 Syria secedes from the United Arab Republic

Resenting Egyptian domination, Syria secedes from the United Arab Republic and renames itself the Syrian Arab Republic.

1963 Syria is placed under a state of emergency

Syria is placed under a state of emergency by the government which requires Syria to be under martial law due to the ongoing war with Israel and the threats of attack from terrorist groups, including Islamic fundamentalists, and Iraqi and Lebanese factions.

1963: March 6 The NCRC is formed

Leftist Syrian army officers of the National Council of the Revolutionary Command (NCRC), are installed. This is a group of military and civilian officials who, after the installation, assume all control of executive and legislative affairs. This installation is conducted and planned by the Arab Socialist Resurrection Party (the Ba'ath Party) that has been active since the 1940s. This group's newfound power comes after the coup conducted by the Ba'ath Party in Iraq.

Seven years later, the Correctionist Revolution is led by Hafiz Assad to further the goals of this first revolution. His revolution is popular because people believe he is the embodiment of their aspirations.

1963: April 17 Plans are made to hold a referendum for Arab unity

An agreement is reached in Cairo to hold a referendum in September of 1963 on unity of the Arab countries. The referendum is never successfully held due to bureaucratic conflicts in several countries.

1963: May Syria's first constitution is written

President Amin Hafiz of the NCRC (see March 1963) writes a provisional constitution for Syria. This constitution provides for a legislative body composed of representatives of the popular political organizations. The legislature is to be called the National Council of the Revolution (NCR).

1964 All of Syria's oil rights go to the General Petroleum Authority

Despite many proposals from large international oil companies, Syria grants all oil concessions to the General Petroleum Authority of Syria.

1964–65 The Ba'ath Party begins socialization of the Syrian economy

The Ba'ath Party, the leading political party in Syria, assumes responsibility for socialization of the economy. In these two years 108 companies in the food and engineering industries are converted.

1965 The textile industry is nationalized

Syria's booming textile industry is nationalized and reorganized into thirteen state corporations to improve efficiency and promote growth. This action is in accordance with Syria's socialist five-year plans that promote the nationalization of industries.

1966: February 23 Army officers revolt

The army officers stage an internal coup and imprison President Amin. They dissolve the NCR and his cabinet (see May 1963).

1967 The General Women's Federation is formed

The General Women's Federation is formed. This group is a landmark in Syria's political history as it represents a granting of some equality to women in a country that is run by leaders who follow the Islamic faith which does not support equality of the sexes.

1967 Israel occupies the Golan Heights

Syria is an active protagonist in the Arab-Israeli war. Israel responds to this by occupying an area of Syria known as the Golan Heights and the city of Quneitra. The Golan Heights is an area of Syria that is 7,000 feet above sea level giving it a strategic military position from which Isreal can monitor Syria's actions.

1967 Syria occupies parts of Palestine

Syria aggressively occupies parts of Palestine for a short period.

1967: June War with Israel begins

The Syrians begin war with Israel. They are defeated in 1970 due to a conflict between the moderate military wing and the extreme wing of the Ba'ath Party. This conflict causes the retreat of the Syrian forces, who had been sent to aid the Palestine Liberation Offensive (P.L.O) during the Black September hostilities with Jordan.

1970 The Muslim Brotherhood is dissolved

The only major opposition to the Ba'ath Party is the Muslim Brotherhood, which claims that the Ba'ath Party is heretical. The Ba'ath Party allows the Brotherhood to be publicly active until 1982, when it is suppressed permanently.

1970 Majlis al-Sha'ab becomes president

Majlis al-Sha'ab becomes president of Syria for a short period before there is another internal coup bringing Hafiz al-Assad (see 1928) to power.

1970: November 30 Hafiz al-Assad becomes prime minister

Syria's Minister of Defense, Hafiz al-Assad, ousts the civilian party and assumes the role of Prime Minister. He quickly creates a solid structural make-up for the government that reduces the political instability of the nation. The provisional Regional Command of Assad's Arab Ba'ath Socialist Party nominates a 173 member legislature to the People's Council. The Ba'ath Party holds 87 of those seats.

1971 Syria's first phosphate plant begins production

Syria's first phosphate plant begins operating near the city of Homs. This event marks the beginning of Syria's chemical production industry. Chemical production develops into a major source of money in Syria's economy.

1972 Syria's pipelines are nationalized

Pipelines of the Iraq Petroleum Company in Syria are nationalized in conjunction with the nationalization of many other industries in Syria. Syria also joins OPEC, the Organization of Petroleum Exporting Countries at this time, increasing the amount of national income that it makes from oil production.

March 1972 Hafiz al-Assad becomes president

The Ba'ath Party holds its regional congress and elects a 21-member Regional Command to be headed by Hafiz al-Assad (see November 30, 1970). A national referendum is held to confirm Assad as president.

1972: March The National Progressive Front is formed

Assad forms the National Progressive Front, which is a coalition of parties, led by Assad's Ba'ath Party.

1973: March Syria's new constitution is written

A new national constitution is written, requiring that the president be Muslim. Islam, however, is not the official religion of the state.

A judicial system is established based on a varied mix of Ottoman, French and Islamic laws. All personal and family laws are dealt with in the religious courts. The Ba'ath Party establishes socialism and secular Arabism in the constitution.

1973: October 3 Syria attacks Israeli forces in the Golan Heights

The Syrian army performs a full-scale attack on Israeli forces in the Golan Heights (see 1967). Relations between the two countries become more strained.

1974 Henry Kissinger requires Israel to return Syria's land

U.S. Secretary of State Henry Kissinger achieves a disengagement agreement that permits Syria to recover the territory it lost to Israel (see 1967). Israel, however, does not completely comply with the terms of the agreement.

1975 Excavation of the city of Ebla begins

The city of Ebla is excavated. It is discovered that this city had once been part of a great Semitic empire that spread from the Red Sea to Turkey (2500–2400 B.C.). The population of the city at that time was 260,000, a number that was very large for that era. Ebla was later occupied by several groups of people: the Canaanites, the Phoenicians, the Hebrews, Arameans, the Assyrians, the Babylonians, the Persians, the Greeks, the Romans, the Nabalaens, the Byzantines, Crusaders, and finally Ottoman Turks. All of these ethnicities had a dramatic effect on its development through history.

1978: March 19 The United Nations commands Israel to leave Syria

The United Nations Security Council passes Resolution 425 during Israel's incursion into Lebanon. The resolution commands Israel to withdraw from Syria, which it does not do.

1978 Syria begins the Euphrates Dam project

The Euphrates Dam project begins. Syria attempts to increase irrigation to its farmlands by damming the Euphrates River, which at this time, serves only a small strip of Syrian land due to inefficient irrigation techniques. The project forms Lake Assad. The goal for this program is to allow farmers to move from rain dependent areas to irrigated ones in order to increase agricultural production. The government estimates the project will bring water to a new 652,000 hectacres of farmland but it only manages to reach 640,000 hectacres.

c. 1980s Oil exploration is revived

Syrian Petroleum Company, Marathon, Shell, and Coastal States begin to engage in renewed oil exploration in Syria.

c. 1980 Syria supports Iran in the Iran-Iraq war

Syria supports Iran during the Iran-Iraq war. This action isolates Syria from the rest of the nations in the Arab world, who support Iraq. This isolation hinders its economic and political development due to a decline in exports and political recognition by the other Arab countries. After 1988, it begins a slow reintegration with the rest of the Arab world.

c. 1980 The worldwide price of oil decreases

The global price of oil drops causing a decline in the economic growth of Syria from ten percent to a less than three percent per year. This rate of growth is less than the rate of population growth and the standard of living begins to rapidly decline.

c. 1980 Syria discovers light grade oil

Syria discovers light grade, low sulfur oil near Dayr as Zawr in Eastern Syria. This relieves pressure on Syria to import light grade oil to mix with its heavy grade oil to produce marketable oil.

1980: October Syria signs a pact with the USSR for weapons

Syria signs a twenty-year pact with the USSR agreeing that Syria will receive weapons such as missiles and new Russian technology. Their ties are based on the two countries' common socialist ideologies.

1981: December 14 Israel annexes the Golan Heights

Israel annexes the Golan Heights, which it had invaded in 1967 (see 1967). This action leads to years of conflict between the two nations. The hatred between the two cultures that is bred from this conflict causes irreparable damage to Israel-Syrian relations.

1982 The Syria-Iraq pipeline is completely closed

Syria finishes the construction of an oil pipeline to Iraq in a joint oil deal. This pipeline, however, is shut off after Syria's continued support for Iran in the Iran-Iraq conflict.

1982 Israel invades Lebanon

Israel invades Lebanon and provokes clashes between Syria and Israeli forces. Relations with Israel are soured even more due to yet another military offensive taken against Syria.

1986 Research on Tell Sabi Abyad begins

Heavy research of the archeological site of Tell Sabi Abyad or "Mound of the White Boy" begins. It is not a well-understood archeological site but it holds many important finds in the study of the history of man. The site contains several settlements that are on top of one another that date as far back as 5700 B.C.

1987 USSR technicians remain in Syria

Soviet military technicians still remain in Syria from the original pact that it signed in 1980. Their other primary research besides weapons development is with the ICARDA, the International Center for Agricultural Research in Dry Areas. Their research helps Syria greatly improve its agricultural industry, which had struggled under antiquated methods of farming and the inability to deal with their frequent droughts. This center eventually allows farmers to produce even during the harshest droughts, freeing up resources and money in Syria for other endeavors.

1989 Drought strikes Syria

A serious drought strikes Syria, causing worsened economic problems. Syria is forced to use its already small foreign exchange reserves to import large quantities of wheat, flour, and other food to keep the nation out of complete despair and chaos.

1989 Syria achieves its first trade surplus

The oil industry produces Syria's first trade surplus as it accounts for more than three-fourths of the export income.

c. 1989 The Arab world readmits Egypt into the Arab League of Nations

Syria joins the Arab world in readmitting Egypt into the 19th Arab League Summit at Casablanca. This marks the end of the Syrian-led opposition to Egypt following the split of their republic on September 28, 1961 (see September 28, 1961) which left many political tensions.

c. 1990 Syria's economic development remains hindered

Despite several development projects, economic progress is still hampered in Syria due to poorly reformed public sectors, low investment levels, and low industrial and agricultural productivity. Due to these issues, Syria has great difficulty in keeping up with the rapidly globalizing world trade, causing its economy to become depressed even more.

1989 Syria endorses the Charter of National Reconciliation

Syria endorses the Charter of National Reconciliation, or the "Taif Accord," a comprehensive plan for ending the Lebanese conflict.

c. 1990 Turkey begins construction of a dam of the Euphrates River

Turkey begins construction on a dam of the Euphrates River, doubling the effects of Syria's own dam (see 1978). This lowers the flow of water in the river, the main source feeding Syria's irrigation to its vital crops. The damming of the river also lowers Syria's ability to produce hydroelectric power, which makes the country place greater emphasis on its own oil, leaving less to be exported and causing more pollution of the environment in an already overpolluted area.

1990 Syria condemns Iraq's invasion of Kuwait

Syria is the first nation to condemn the Iraqi invasion of Kuwait. This puts Syria in better favor with the Western powers.

Syria participates in the multinational coalition against Iraq in the Gulf War. This ends years of Arab isolation and gives Syria access to European, Japanese, and Gulf financial resources. Total loans amount to $500 million from 1990–1992. Another $2 million is promised but is never delivered.

1990 Syria establishes an official parallel exchange rate

The government establishes an official parallel exchange rate with Arab countries of equal standing. This provides incentives for exports through official channels. This declaration also improves the supply of basic commodities and contains the rampant inflation in Syria by removing the risk premiums on investments, loans, and on smuggled commodities.

1990: August Socialism begins to be de-emphasized

The Ba'ath Party begins to de-emphasize its socialist views due to increasing pressure from the developing pan-Arab unity.

c. 1990 Central Europe and Russian markets de-socialize

The communist economies from Moscow and central Europe began to loosen, opening up new markets for Syria.

1991 The Cold War ends

The Soviet Union falls after the end of the Cold War, loosening the economic constraints once placed on Syria. Newspapers are more open and political criticism is once again possible.

1991 Syria joins a multilateral Middle East Peace Conference

Syria participates in a multilateral Middle East Peace Conference in Madrid aiding its relations with other Arabic countries. Its relations with the West remain strained due to the presence of terrorist groups, its human rights record, and Syria's heavy involvement in narcotics activity.

1991 Investment Law No. 10 is passed

Investment Law No. 10 is passed, allowing more foreign investors to seek opportunities in Syria. A rise in petroleum production combined with good harvests lead to an increase in the country's Gross Domestic Product (GDP, the measure of a nation's economy). The influx of foreign investment allows Syria to rehabilitate its declining infrastructure and public sector enterprises. The investment law permits enterprises to retain foreign exchange earned from exports so that these companies can finance certain imports of raw materials.

May 1991 Syria signs the Treaty of Brotherhood

Lebanon and Syria sign the Treaty of Brotherhood, Cooperation, and Coordination ending years of violence.

1997: November 3 Assad is sworn into office for another seven-year term

President Assad is sworn into office for his twenty-seventh year as president of Syria.

1998: December 28 A mob destroys the U.S. ambassador's home

A state-sponsored mob destroys the American ambassador's home in protest of the United States' renewed bombing of Iraq. Although Syria does not support the actions of Iraq's leader, Saddam Hussein, it is trying to attract the attention of the United States to show its growing power and importance. The United States, however, barely responds.

1999 The last remaining hakawati storyteller practices his art

Al-Nafinah is the last remaining storyteller who keeps alive the art of *hakawati* or the art of reading Arab epics. Some of the great stories in his repertoire are the romance of *Prince Antar,* and the adventures of Al-Zahir Rubn al-Din Baybou, the eminent thirteenth century sultan.

Bibliography

Blachowicz, James. "A Tour of the Ancient World in Syria." *The New York Times.* March 4, 1990.

Dourian, Kate. "City of Apamea, once lost in sand, partially restored." *Washington Times.* September 3, 1994.

Katler, Johannes. *The Arts and Crafts of Syria.* Thames and Hudson: London, 1992.

Kayal, Michele. "Ruins to Riches." *Washington Post.* February 13, 1994.

LaFranchl, Howard. "Ancient Syria's History Rivals That of Egypt, Mesopotamia." *The Christian Science Monitor.* February 17, 1994.

Parmelee, Jennifer. "Tracking Agatha Christie." *The Washington Post.* August 25, 1991.

"Syria." *Background Notes.* Central Intelligence Agency. November 1994.

"Syria." *Culturgram '97.* United States State Department.

"Syria: Country Report." The Economist Intelligence Unit. No. 21191.

"Syria: A Country Report." The Library of Congress Research Division. 1999

Theroux, Peter. "Syria: Behind the Mask." *National Geographic.* July, 1996.

Taiwan

Introduction

Located in the western Pacific Ocean and bisected by the Tropic of Cancer, Taiwan lies less than 100 miles (161 kilometers) from mainland China, and the histories of China and Taiwan have been closely linked for at least 2,000 years. Since 1949, both have claimed to be the seat of China's legitimate government. Both the Republic of China on Taiwan and the People's Republic of China on the mainland officially consider Taiwan a province of China. In practice, however, it is governed and operates as a separate country.

In addition to the main island, Taiwan comprises eighty-five smaller islands divided into two groups. The total area of both the large and small islands is 13,892 square miles (35,980 square kilometers). Mountains and foothills cover the eastern two-thirds of the main island, and the western third consists of a flat coastal plain. Taipei, the capital of Taiwan and its largest city, is located at the northern tip of the island. As of 1996, the total estimated population of Taiwan was 21.5 million.

The association between Taiwan and China goes back to the first millennium A.D. The Chinese launched expeditions to Taiwan in 239, 607, and 611, and migration from the Fujian and Guangdong provinces began in the seventh century. The first Westerners to arrive in Taiwan were the Portuguese, in the sixteenth century. Between 1624 and 1662, the island came under the control of the Dutch, who built two trading centers from which to export Chinese goods, such as silk and sugar, to the West. The Chinese population on Taiwan increased rapidly during this period, as immigrants fled from the mainland to escape the growing power of the Manchus. After two decades of rule by the Cheng family, who were Ming loyalists, Taiwan fell to the Manchus in 1683.

For the next two centuries it was administered as part of Fujian province. The opening of China to Western trade toward the end of this period, after centuries of isolation, affected Taiwan as well. Among the concessions exacted by the British under the Treaty of Tientsin (Tianjin)) was the opening of four Taiwanese ports. In 1885 Taiwan gained provincial status, becoming China's twenty-second province. Its first governor, Liu Ming-ch'üan (Liu Mingquan), presided over a wide-ranging modernization program that improved the island's defenses, transportation network, and educational system and introduced industrial production. However, only one decade later, Taiwan was ceded to Japan under the Treaty of Shimonoseki that ended the First Sino-Japanese War, and it remained under Japanese rule for the next fifty years. Modernization continued under the Japanese, who wanted to display Taiwan as an example of its success as a colonial power. Agricultural production was increased, harbors were expanded, and cities were modernized.

In 1911 the ruling Manchu dynasty on the mainland was deposed, and the Republic of China was formed, with Sun Yat-sen (1866–1965), leader of the nationalist Kuomintang (Guomindang) party, at its head. After 1915, rival warlords usurped much of the government's power, but by 1928 the Kuomintang, under the leadership of General Chiang Kai-shek (1887–1975), had reunified the country. Throughout the 1930s, Chinese nationalist forces battled both the growing communist movement and the Japanese, who invaded Manchuria in 1931 and had occupied the entire country by 1937. During World War II, the Chinese fought a war of resistance against the Japanese. At the same time, Taiwan was being used by Japan as a military base.

The Republic of China

Following the war, Chinese rule was restored on Taiwan. However, the native Taiwanese were dissatisfied with their treatment by the Chinese, and tensions between the two groups erupted into violence in early 1947 following a street altercation in which a customs officer killed a bystander. Troop reinforcements were sent in from the mainland, killing thousands and effectively crushing dissent on the island. Following the successful Communist Revolution of 1949, the nationalist government of the Republic of China (ROC), under the leadership of Chiang Kai-shek, fled the mainland and relocated to Taiwan, together with some two million Chinese emigrants, and the ROC government established martial law on the island.

In the 1950s the Republic of China received strong support from the United States as part of its worldwide campaign against communism. Economic assistance facilitated indus-

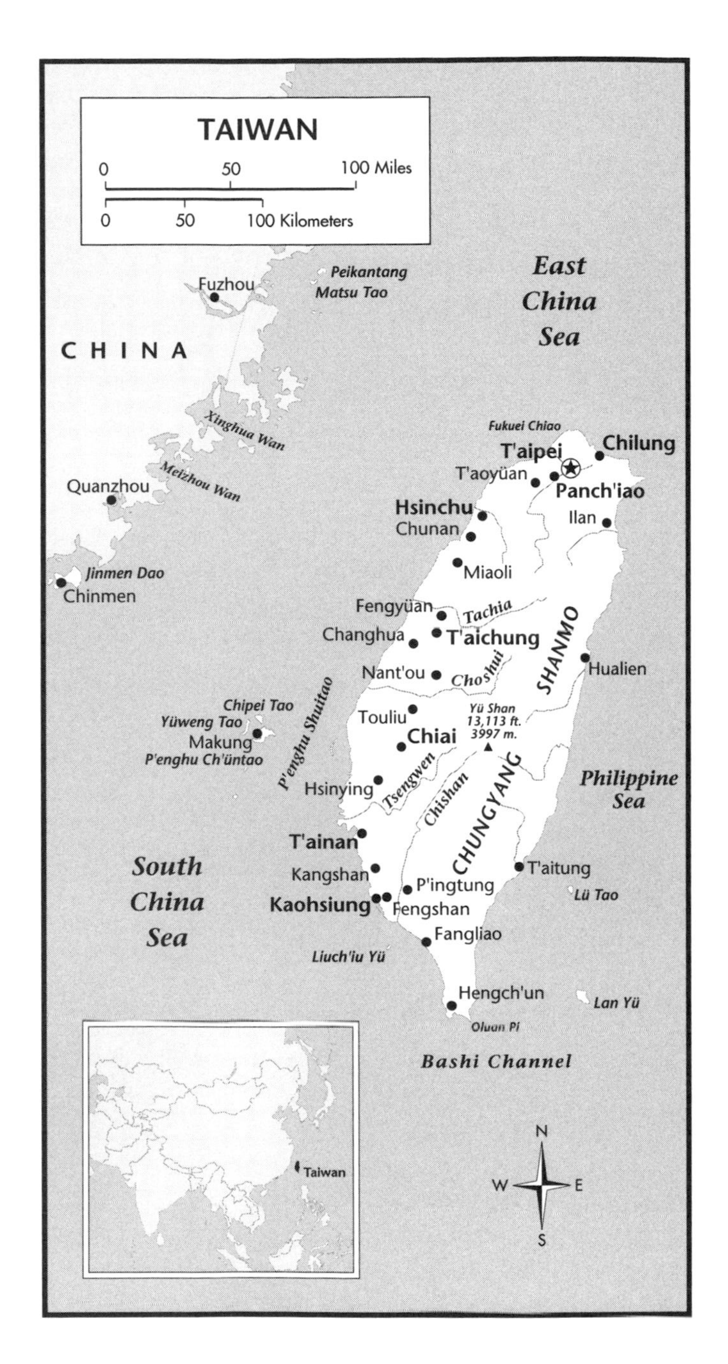

trial expansion and other types of development, while a 1954 mutual defense pact helped assure the security of this small island facing a mammoth potential adversary across the narrow Taiwan Strait. In the same decade, Taiwan, which had joined the United Nations following World War II, and the People's Republic began their long struggle over which government should represent the Chinese people in the world body.

With Taiwan's increasing economic success, U.S. foreign aid was terminated in 1965, and the U.S. ended its patrol of the Taiwan Strait, begun during the Korean War, in 1969. In the 1970s, as relations between the United States and the People's Republic began to thaw, the United States began to distance itself from Taiwan diplomatically. In 1972, the U.S. dramatically reversed its previous policies by joining the majority in the United Nations vote that transferred Chinese representation in the UN from Taiwan to mainland China. On January 1, 1979 the United States granted diplomatic recognition to the People's Republic of China. However, it maintained trade and cultural ties with Taiwan through non-diplomatic channels.

Chiang Kai-shek remained the head of the ROC government until his death in 1975 and was succeeded by his eldest son, Chiang Ching-kuo (Chiang Ching-kuo, 1910–88). Since the 1980s government on Taiwan has gradually been liberalized. In 1986 an opposition political party was officially recognized for the first time since ROC rule was instituted on Taiwan in 1949. Martial law, in effect for nearly forty years, was ended the following year. With the appointment of Lee Teng-hui (Li Denghui, b.1923) as president following the death of Chiang Ching-kuo in 1988, liberalization proceeded even more rapidly. The remaining curbs on basic civil rights were rescinded in 1991, and three years later constitutional changes mandated the direct election of the president. In 1996, Lee became Taiwan's first popularly elected president. In his 1997 inauguration, he offered to meet with leaders of the People's Republic of China.

Timeline

A.D. 239 Chinese arrive on Taiwan

The Wu kingdom sends a naval fleet with 10,000 men to explore Taiwan, as recorded in the *Sankuochi* (History of Three Kingdoms).

A.D. 607 and 611 China sends two expeditions

Two Chinese expeditions to Taiwan are launched by Emperor Yangdi during the Tang period (A.D. 618–907). Soon afterward, Chinese begin migrating to the island from the Fujian and Guangdong provinces.

960–1279 Trade develops during the Song period

Regular trade grows between Fujian province and a trading center on the west coast of Taiwan.

1517 Portuguese sailors sight Taiwan

The first Westerners to see Taiwan are the Portuguese, who name it *Ilha Formosa* (beautiful island).

Rule by the Dutch

1624–62 Taiwan under Dutch control

The Dutch and Chinese sign a treaty under which the Dutch withdraw from the Pescadore Islands and set up a trading station (Fort Zeelandia) on Taiwan for the transport of goods from China, such as sugar, silk, and ceramics, to Europe. The Dutch oust the Japanese settlers living in the southern part of the island. This period sees the first large-scale migration of Chinese to Taiwan. The growing power of the Manchus on the mainland unleashes a flood of refugees to Taiwan, and the Dutch encourage this migration because of the superior farming skills of the Chinese. The Chinese population on Taiwan quadruples from 25,000 to 100,000 between 1624 and 1650.

1626–42 Spanish presence on the island

A Spanish expeditionary force arriving from the Philippines gains a foothold on Taiwan, seizing part of Keelung (which they name Santissima Trinidad) and later Tanshui. The Spaniards are ultimately ousted by the Dutch.

1627 Missionaries arrive on Taiwan

The Dutch bring Christian missionaries to Taiwan. Most conversions are among the native aborigine population rather than the Chinese. All missionaries are later expelled by the Ching dynasty.

1630s Sugar exports begin

Sugar exports to Japan and Persia are inaugurated.

1644 Manchus come to power on the mainland

The Manchus take control of the Chinese mainland, inaugurating the Ching dynasty.

1650 Dutch build Fort Providencia

The Dutch build a second trading center on Taiwan, across the harbor from Fort Zeelandia, their first one.

1662 Cheng captures Taiwan from the Dutch

Following a six-month siege of Fort Zeelandia with a force of 30,000, Chinese military leader and Ming loyalist Cheng Ch'eng-kung (known as Koxinga in the West, 1623–63) captures Taiwan and ousts the Dutch, intending to use the island as a base from which to attack the Manchus on the mainland. On the site of the former Fort Zeelandia, he establishes a capital called An-p'ing (Calm Peace).

1662–83 Cheng family rules Taiwan

Control of Taiwan is transferred from Cheng Ch'eng-kung to succeeding generations of his family, who develop irrigation systems and introduce rice growing to southern Taiwan. Direct trade with Japan and with European colonies in Southeast Asia expands, and a cultural flowering occurs as scholars and other intellectuals flee the oppressive Ching dynasty rule by the Manchus on mainland China. Schools and Confucian temples are built on Taiwan, poetry flourishes, and woodcarving by the indigenous tribes reaches its highest point.

Manchu Rule

1683 Taiwan falls to the Manchus

The powerful Ching dynasty incorporates Taiwan into China's Fujian province, making Tainan the capital. Migration and settlement continue steadily, although Ching rule provokes seventy or more armed revolts during the two centuries of Ching administration as part of Fujian province.

1721 Founding of Taichung

The city of Taichung (at first named Tatun) is founded by immigrants from the mainland.

1739 Birth of artist Lin Chaoying

Lin Chaoying (1739–1816), a merchant-turned-artist, is known for his paintings and calligraphy, which are influenced by the Chinese Yangzhou School. His color paintings of native birds and bamboo are acclaimed.

1854 Pan-ch'iao garden built near Taipei

Modeled on the garden of Tang dynasty poet Bai Juyi, the Pan-ch'iao garden, designed by Lin Kuo-hua, is named for its many "moon bridges."

1854 U.S. forces arrive in Keelung

The U.S. East Indian Fleet commander, Commodore Matthew Perry (1794–1858), sends a detachment to assess water depth and mineral reserves at Keelung.

1858 Four Taiwanese ports are opened to trade

The Treaty of Tianjin (Tientsin) between China and the United Kingdom opens four Taiwanese ports (Anping, Tamsui, Kaohsiung, and Keelung) to foreign trade.

1866 U.S. naval forces attack southern Taiwan

U.S. warships bombard indigenous Taiwanese in the south as retribution for the murder of two shipwrecked American sailors.

1867 Earthquake batters Keelung

An earthquake levels much of the northern port of Keelung.

1879 Taipei is founded

The city of Taipei is established near the settlement of Mengjia.

1884 French occupy Keelung

During the Sino-French War (1884–85) the French attack the strategic city of Keelung, occupying it for eight months.

Provincial Status

1885 Taiwan becomes a separate province of China

Recognizing Taiwan's strategic importance, the Chinese government makes it a separate province (China's twenty-second province) and reorganizes its administration, dividing it into three prefectures. Rapid modernization accompanies the change in status, especially under the province's first governor, Liu Mingquan. Industry is introduced, education is reformed, the first railway is built, and the island's defenses are modernized.

1886 Birth of painter Shiozuki Toho

The work of Japanese-born painter Shiozuki Toho, who lives and works in Taiwan for much of his career, helps spread the popularity of Western painting styles. He lives in Taiwan for periods of time between 1921 and 1945, including the period between 1924 and 1932, when he teaches at the Taipei Normal School. Known for its brightly colored landscapes, Shiozuki's work resembles that of the later Impressionists and Fauvists.

1887: October 31 Chiang Kai-shek is born

Chiang Kai-shek, future leader of the Chinese Nationalist government, is born in the province of Zhejiang. He embarks on a military career at the age of eighteen, receiving his training at the Baoding Military Academy and a Japanese military academy in Tokyo. While in Japan, he meets Chinese nationalist Sun Yat-sen (1866–1925), who is gathering support for the overthrow of the Manchu emperors. As a sign of protest, Chiang cuts off his pigtail, a hairstyle the Ching dynasty has imposed on the Chinese throughout the period of Manchu rule. Chiang joins Sun Yat-sen's revolutionary group and takes part in a variety of revolts between 1911 and 1925, when he takes charge of the Kuomintang (Nationalist) forces in southern China following Sun's death.

After briefly allying his forces with the Communists, Chiang turns on them in 1927, carrying out a massacre of Communists in Shanghai and pursuing a civil war against them for two decades—through the 1930s, World War II, and beyond. By May 1949, the Communists are victorious, driving the Nationalists out of mainland China to Taiwan, where Chiang establishes the Republic of China, recognized by the United States as the official Chinese government until 1979. Throughout his period of rule on Taiwan, Chiang maintains the goal of retaking the mainland and unifying China, but it proves impossible to achieve. Chiang is elected president of the Republic of China for five successive terms, serving in that position until his death in 1975, when his son Chiang Ching-kuo, assumes the post.

1888 Postal service is inaugurated

Taiwan's postal service is launched, nine years before postal service is begun on the Chinese mainland.

1889 Electricity is introduced in Taipei

The first electricity is introduced in Taipei.

1893 Taiwan's first railroad is completed

Under the administration of Taiwan's first Chinese governor, Liu Mingquan, a railroad connecting Keelung and Hsinchu is completed.

Japanese Occupation

1895–1945 Taiwan annexed to Japanese Empire

Under the Treaty of Shimonoseki, which ends the First Sino-Japanese War, the Japanese annex Taiwan, together with the Pescadore and the Ryuku islands. Armed resistance on Taiwan flares briefly but is unsuccessful.

As part of the Japanese Empire, Taiwan undergoes a period of rapid development and modernization, as it is turned into a "model colony" to justify Japanese expansionism. Health care is modernized, and the educational system is reformed. Harbors, railways, and roads are expanded and modernized. Agricultural production is increased to meet demands for export to Japan, and Taiwan begins large-scale exports of rice and sugar. Widespread renovation is undertaken in urban areas.

However, nationalist and pro-Chinese sentiment is harshly suppressed, and the Japanese impose their language on the Taiwanese for all official purposes and in the schools. Nevertheless, the Taiwanese stage over a hundred revolts against Japanese rule.

1895: October 21 Japan ends Chinese resistance

Japanese soldiers enter Tainan, and the initial wave of organized resistance by the Chinese is halted.

1903 Hydroelectric power introduced

Taiwan's first hydroelectric dams help promote the development of industry.

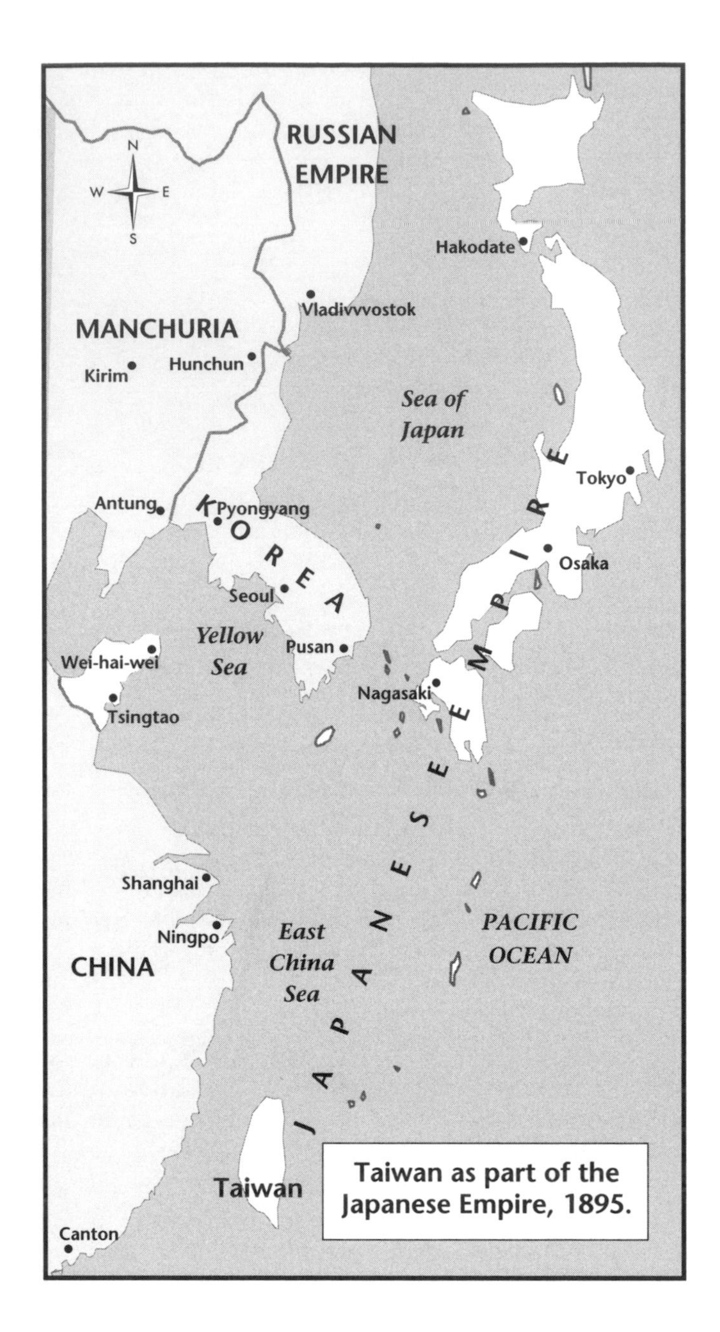

Taiwan as part of the Japanese Empire, 1895.

1905 First census is taken

Under Japanese rule, the first census taken on Taiwan records a Chinese population of nearly 3 million.

1911 Manchu dynasty ends

The last Manchu emperor of China is deposed, and the Republic of China is born. Sun Yat-sen, leader of the Kuomintang nationalist party, becomes the ROC's first president.

1915 Yuan Shih-kai declares himself emperor

Sun's successor, Yuan Shih-kai (1859–1916), declares himself emperor, and a chaotic period of rule by rival warlords follows.

1927 Annual Taiwan Fine Arts Exhibition is inaugurated

Sponsored by the Taiwan Education Committee, the first annual Taiwan Arts Exhibition is held. Encompassing both traditional and Western paintings, it is an important influence in the adoption of modern painting styles.

1928 Chiang Kai-shek defeats the warlords and unifies mainland China

Throughout the 1920s, the nationalists gradually regain power, and the country is reunified under the leadership of Sun Yat-sen's colleague, General Chiang Kai-shek. However, a growing communist movement now threatens the nationalists' control of the country.

1932 Birth of artist Liu Kuo-song

Liu Kuo-song, the most prominent painter of his generation, is born in Shandong Province of mainland China and emigrates to Taiwan in 1949. He studies art at the National Normal University and graduates in 1956. Liu's strongest influences early in his career are the works of twentieth century artists including Matisse (1869–1954), Cézanne (1839–1906), Picasso (1881–1973), and Klee (1879–1940). In the late 1950s, abstract expressionism becomes an important influence as well. Liu's works, like those of his contemporaries, merge the traditions of Chinese art with modern Western painting styles and techniques. However, unlike the preceding generation of artists, Liu and most of his peers do not actually study in the West but gain their exposure to Western art second-hand.

In 1965 Liu's first one-man show, at the National Taiwan Art Gallery in Taipei, attracts critical attention both at home and in the United States. The same year he publishes an influential collection of essays, *Zhongguo siandaihua de lu* (Whither Chinese Painting?), which helps to consolidate his reputation. Liu later serves on the faculties of art schools in Taiwan, the United States, and Hong Kong. He lectures at the Chinese University of Hong Kong beginning in 1971.

1937 Japanese occupy China

After initially invading Manchuria in the north, the Japanese, already occupying Taiwan, gradually gain control of the mainland.

1939–45 World War II

The mainland Chinese engage the Japanese in a war of resistance, aided by the Allied powers. Taiwan, used as a military base by the Japanese, becomes a target of Allied bombing attacks.

1945: September Chinese rule is restored

With the defeat of Japan in World War II, Chinese rule is restored to Taiwan under the Cairo Declaration and the Potsdam Proclamation.

1946 Provincial art exhibitions are instituted

The Taiwanese provinces begin to hold their own annual art exhibitions, supplanting the national exhibitions held before the war.

1947: February 2 "2-28 incident" sparks rioting and rebellion

In the process of a street arrest for black marketeering, a woman is injured, and when bystanders come to her aid, one is shot by a customs officer. The incident inflames native Taiwanese resentment of their treatment by the mainland Chinese, and violent demonstrations break out. Taiwan's Chinese governor, General Chen Yi (1901–72), calls for reinforcements from the mainland, and the protests are brutally suppressed, resulting in a death toll estimated by some to be as high as 28,000.

1947: May Chinese undertake extended suppression of dissenters

With new troop reinforcements from the mainland, the Chinese undertake a campaign to crush dissent on Taiwan, killing many of the island's political leaders.

The Republic of China

1949: December 8 Nationalist Chinese government flees to Taiwan

With the victory of the Communist Revolution on the mainland, Chiang Kai-shek's nationalist Kuomintang takes refuge on Taiwan. The entire Republic of China (ROC) government is moved to the island, together with a total of some 2 million mainland Chinese fleeing the Communists.

1950s Development carried out with U.S. aid

With foreign aid from the United States, Taiwan carries out agrarian reform and industrial development programs, as well as improvements in its education and transportation systems.

1950 U.S. protects Taiwan Strait during Korean War

During the Korean War, the United States Seventh Fleet patrols the Taiwan Strait to protect Taiwan against invasion by mainland China.

1950 UN debate over the two Chinas begins

The two-decades-long controversy over which government—the nationalist Republic of China on Taiwan or the communist People's Republic on the mainland—should represent China in the United Nations begins.

1951 Japan renounces its claim to Taiwan

Through the San Francisco Peace Treaty, the Japanese officially renounce all claims to Taiwan.

1954 Taiwan and the U.S. sign defense pact

Taiwan and the United States sign the Mutual Defense Treaty. Through a combination of political and military means, ROC leader Chiang Kai-shek hopes eventually to regain control of the mainland. The treaty is ratified by both nations the following year.

1955 Government limits the size of newspapers

Press censorship by the Kuomintang government applies not just to which papers can publish, but also how many pages. All newspapers are limited to eight pages.

1955 National Museum of History is founded

The National Museum of History opens in Taipei. It houses collections of artifacts from the Shang, Chou, Sui, Tang, and Han dynasties, including ritual vessels, earthenware, stone engravings, and jade carvings.

1957 Avant-garde art groups are formed

Taiwan's first *avant-garde* (experimental) art groups are formed, both in Taipei. They are the May Art Club, whose members include Liu Kuo-sung and Chuang Che, and the Eastern Art Club, whose members—all students of Li Chung-sheng—include Hsiao Ch'in and Hsia Yang. Both groups promote the adoption of Western abstract art together with the preservation and reinterpretation of traditional Chinese art.

1960s First color features produced

Taiwan is ahead of both mainland China and Hong Kong in its production of feature-length color movies. Movies are highly popular: Taiwan is one of the top filmgoing nations in the world. Movies featuring martial arts and swordfights are especially popular.

1960 Economy thrives

The economy of Taiwan is thriving, with exports doubled compared to last year. Industries, including textiles, pharmaceuticals, jewelry, and appliances, have all increased production.

1960 East-west highway is completed

The East-West Cross-Island Highway, costing almost $11 million, is completed after nearly four years of labor by 100,000 workers. It opens the Chungyang Shanmo in eastern

Businesses in Taipei are thriving. (EPD Photos/CSU Archives)

Taiwan to development by agricultural and logging interests and tourism.

1962 First television broadcasts

Television is introduced to Taiwan.

1965 The National Palace Museum opens its doors

The National Palace Museum, located in a northern suburb of Taipei, houses one of the world's most valuable collections of art objects, some dating back over 1,000 years to the Sung dynasty. The treasures in the collection were removed from imperial vaults in China's Forbidden City, initially to protect them during the Japanese invasion of China. They were later transported to Taiwan by the Kuomintang when it fled the mainland in 1949.

Altogether, the collections contains 620,000 artifacts, only 10,000 of which are on display at any one time. Included are bronze, jade, and porcelain objects, paintings, calligraphy, and rare books. Since the destruction of art objects on the mainland during the Cultural Revolution of the 1960s and 1970s, the collection of Chinese treasures in the National Palace Museum is thought to be the largest in the world.

1965 U.S. terminates financial aid

Financial assistance from the United States is discontinued. Nevertheless, Taiwan's economy continues to grow.

1966 First free trade zone is launched

The Kaohsiung Export Processing Zone is the first free trade zone established on Taiwan. This industrial park helps provide the country with additional foreign exchange and jobs.

1969 U.S. ends patrol of the Taiwan Straits

The United States terminates its patrol of the Taiwan Strait, begun during the Korean War.

1971 Taiwan loses its seat in the UN

The United Nations transfers its Chinese seats to the People's Republic of China, and Taiwan loses its membership in the world body. The United States reverses its previous policies and supports the UN decision.

1972: February U.S. issues Shanghai communiqué

In the "Shanghai communiqué" issued at the time of President Richard Nixon's (1913–94) trip to the People's Republic of China, the United States accepts the principle that there is only one China and that Taiwan is part of it. However, the U.S. does not sever diplomatic ties with Taiwan.

1975: April 5 Chiang Kai-shek dies

Chiang Kai-shek, longtime head of the nationalist Kuomintang and political leader of the Republic of China, dies. (See 1887.) Chiang's eldest son, Chiang Ching-kuo (Chiang Ching-kuo), takes over as head of the Kuomintang.

1978: March Chiang Ching-kuo is elected president

Chiang Ching-kuo, son of Chiang Kai-shek and head of the Kuomintang party, is elected to a six-year presidential term.

1979: January 1 U.S. grants official recognition to mainland China

The United States grants diplomatic recognition to the People's Republic of China and withdraws it from Taiwan. Trade and cultural ties with Taiwan are maintained through the American Institute and the Coordination Council for North American Affairs, and U.S. arms sales to Taiwan are continued.

1980s "Movie television" viewing gains popularity

Movie television, known as MTV (but not a cable channel like the American MTV), consists of stores where people watch movies in privately rented viewing rooms. Some fear that the popularity of rentals threatens the production of new films, but Taiwan's film industry continues to thrive, producing acclaimed films including *The Wedding Banquet, Farewell My Concubine,* and *The Puppetmaster.*

1980: January U.S. ends defense pact with Taiwan

The United States terminates the Mutual Defense Treaty in effect since 1954.

1984 Chiang Ching-kuo is elected to a second term

President Chiang Ching-kuo is elected to a second six-year term.

1984 Grass skiing is introduced to Taiwan

The Taiwanese first try grass skiing—skiing done on a grassy slope wearing boots attached to caterpillar treads with rollers. As a sport which can be done in most of Taiwan's mountainous areas when there is no snow, it quickly becomes popular. Apparel is the same as that worn for conventional downhill skiing: skintight outfits to decrease wind resistance. As in conventional skiing, poles are used.

1986 Opposition political party is recognized

For the first time an opposition group—the Democratic Progress Party—is permitted to function openly in the Republic of China, which has been under single-party rule by the Kuomintang since 1949.

1986 Taiwanese are permitted to visit relatives in mainland China

As part of a liberalization program begun by Taiwan's rulers, Taiwanese are allowed to visit relatives in the People's Republic of China.

1986 Trade is liberalized

Taiwan adopts trade liberalization policies to bring it into line with the General Agreement on Trade and Tariffs (GATT).

1986: November Demonstrators protest exile of dissident

Crowds estimated at between 5,000 and 10,000 protest the continuing exile of political dissident Hsu Hsin-liang (Xu Xinliang).

1987: July Martial law is revoked

Martial law—emergency rule—on Taiwan is lifted for the first time since 1949, marking an important step toward the achievement of democracy on the island. Freedom of the press and freedom of assembly are expanded, and opposition parties are legalized.

1988 Children's newspapers are launched

Two newspapers for children begin publication: *Children's Daily News* and *Mandarin Times.*

1988: January Chiang Ching-kuo dies

Chiang Ching-kuo, president of the ROC and son of Chiang Kai-shek, dies. He is succeeded as president and head of the Kuomintang by Lee Teng-hui (Li Denghui, b. 1923).

1990s Golf becomes popular

Golfing becomes one of the most popular pastimes on Taiwan. By the middle of the decade, there are over 400,000 golfers in the country, and new courses are being built rapidly.

1990 Lee is elected to six-year term as president

Lee Teng-hui, named to succeed Chiang Ching-kuo as Taiwan's leader, is elected to a six-year term in his own right—the first native Taiwanese to be elected president. He is also named chairman of the Kuomintang party.

1990: March Thousands march in pro-democracy demonstration

More than 10,000 Taiwanese demonstrate for direct presidential elections and a more democratic government.

1990: June–July President convenes policy forum

President Lee convenes the National Affairs Conference (NAC), a forum that includes academics and dissidents as well as government and business leaders, to offer recommendations on democratic liberalization, including possible constitutional revisions. The forum supports direct election of the president.

1991 National economic development plan is launched

The Taiwanese government launches the Six-Year National Development Plan. More than 600 projects are planned in the transportation, power, telecommunications, environmental, and other sectors. A total investment of $224 billion is planned.

1991: April 30 Government ends restrictions on civil rights

The government of President Lee announces that all remaining restrictions on basic civil rights, including freedom of the press and freedom of assembly, will be lifted.

1992 Taiwan gains GATT observer status

Taiwan is accorded observer status by the General Agreement on Trade and Tariffs (GATT).

1992 Baseball team wins Olympic silver

Taiwan's Olympic baseball team wins a silver medal at the Barcelona Olympics.

1994 Candidates convicted of vote buying in local elections

As part of the move toward a more open, democratic government, Taiwanese officials crack down on the widespread practice of vote buying, convicting more than one third of the candidates in city and county elections.

1994: March National Assembly meets to draft constitutional changes

A special session of the National Assembly is convened to revise or replace the original constitution of the Republic of Korea, drafted in 1947.

1995 Constitution is revised

Taiwan's constitution is rewritten to provide for direct popular election of the president.

1995: June 7–10 Lee visits the U.S.

President Lee Teng-hui, traveling to the United States to attend an alumni reunion at Cornell University, makes the first visit to the U.S. by a Taiwanese leader since the 1979 U.S. diplomatic recognition of the People's Republic of China. Mainland China, which regards Taiwan as a breakaway province, is angered by U.S. reception of the Taiwanese president and conducts military maneuvers in the East China Sea.

1995: August 26 Taiwanese team wins Little League honors

A team from Taiwan wins the Little League World Series held in Williamsport, Pennsylvania, defeating a team from Texas 17-3. In compliance with official regulations, the game is terminated after the fourth inning and ceded to the Taiwanese team because of its wide lead.

1996: March 23 Lee is reelected in first direct presidential election

Taiwan holds its first direct presidential election. Presidential incumbent Lee Teng-hui wins with a 54 percent majority, defeating three other candidates. Lee's running mate is current prime minister Lien Chan (b. 1923). Lee's victory is seen as a setback for the People's Republic of China, which fears that the elections will strengthen Taiwan's claim to legitimacy as an independent nation. Chinese military maneuvers in the waters off Taiwan are seen as an attempt to intimidate the Taiwanese electorate.

1997: May 4 50,000 demonstrators protest rising crime rate

Over 50,000 protesters lie down in the street outside the president's headquarters in Taipei to protest the government's inability to contain the nation's rising rate of violent crime, which has almost tripled in five years.

1997: May 14 Protests spur cabinet shake-up

President Lee Teng-hui changes the makeup of his cabinet in response to public protests against lack of government action to stem a serious rise in the nation's crime rate.

1997: May 20 Lee offers to visit the People's Republic

In his inaugural address, reelected Taiwanese president Lee Teng-hui offers to visit mainland China to confer with Chinese leaders. Reelected under the new constitutional provisions for direct presidential election, Lee becomes Taiwan's first popularly elected president.

1997: August 21 Prime minister resigns in response to government criticism

Prime minister Lien Chan resigns his post in response to popular discontent over Taiwan's rising rate of violent crime, which has included several high-profile cases involving top government officials and celebrities.

1998: February 16 Jet crash near Taipei kills 196

The crash of a China Airlines jetliner on its approach to Chiang Kai-shek International Airport kills all 196 passengers and crew aboard the plane, including the governor of Taiwan's central bank. After hitting the ground, the plane crashes through several houses, killing seven people on the ground. In the wake of the disaster, the head of Taiwan's civil aeronautics agency resigns.

Bibliography

Hood, Steven J. *The Kuomintang and the Democratization of Taiwan*. Boulder, Colo.: Westview, 1997.

Hsueh, Chi. *The Foreign Factor: The Multinational Corporation's Contribution to the Economic Modernization of the Republic of China.*

Li, Kuo-ting. *The Evolution of Policy Behind Taiwan's Development Success*. New Haven: Yale University Press, 1988.

Lin, Zhiling and Thomas W. Robinson, eds. *The Chinese and Their Future: Beijing, Taipei, and Hong Kong*. Washington, D.C.: AEI Press, 1994.

Long, Simon. *Taiwan: China's Last Frontier*. New York: St. Martin's Press, 1990.

Marsh, Robert. *The Great Transformation: Social Change in Taipei, Taiwan Since the 1960s*. Armonk, N.Y.: M.E. Sharpe, 1996.

Metraux, Daniel Alfred. *Taiwan's Political and Economic Growth in the Late Twentieth Century.* Lewiston, N.Y.: E. Mellen Press, 1991.

Shepherd, John Robert. *Statecraft and Political Economy on the Taiwan Frontier, 1600-1800.* Stanford, Calif.: Stanford University Press, 1993.

Tien, Hung-mao. *The Great Transition: Political and Social Change in the Republic of China.* Stanford, Calif.: Hoover Institution Press, 1989.

Yu, Bin. *Dynamics and Dilemma: Mainland, Taiwan and Hong Kong in a Changing World.* New York: Nova Science Publishers, 1996.

Tajikistan

Introduction

Tajikistan, with 5,945,000 inhabitants, was forged by the Soviets in the 1920s out of the Tajik-dominated region of Eastern Bukhara. The country assumed its present configuration in 1929 when the Leninabad *oblast'*, or Soviet administrative territory, was added to its territory. At that time, the capital Dushanbe, whose population had risen to 800,000 as of the late 1990s, had only 6,000 inhabitants.

Tajikistan is a landlocked country the size of the state of Wisconsin. Located in the center of Asia with an area of 55,300 square miles (143,100 square kilometers), Tajikistan borders China to the east, Afghanistan to the south, Uzbekistan to the west and northwest, and Kyrgyzstan to the northeast. The country is divided into four distinct regions. To the northwest is Leninabad, centered on the city of Khujand. This region, the industrial hub of the republic, has been its political hub as well. Khatlan, in the south, is gradually replacing Leninabad as Tajikistan's political center. Khatlan is composed of the two major, relatively independent, regions of Kulab and Qurqanteppe. The third or Central district is composed of Hissar, Dushanbe, and Gharategin. The fourth, Gorno-Badakhshan, is centered on the town of Khorog. The Islamic movement in Tajikistan is supported by Gorno-Badakhshan and Qarategin with nominal assistance from Hissar and Qurqanteppe.

About ninety-three percent of Tajikistan is mountainous, at least 10,000 feet (3,048 meters) above sea level some peaks in Gorno-Badakhshan reach 23,000 feet (7,010 meters). Most settlements, however, are located in the valleys at about 3,000 feet (1,000 meters). Rivers, whose sources begin in the Turkistan, Zarafshan, and Hissar Alai ranges, create favorable ground for agriculture, especially cotton.

The average temperature in winter ranges between 36°F (2°C) in the valleys and 4°F (20°C) in the highlands. In the summer the average readings are 86°F (30°C) in the valleys and 32°F (0°C) in the highlands.

History

The word *"Tajikistan"* is a compound of *-istan* meaning "place of" and *tajik,* a form with uncertain origins. It may have been used by medieval infidel Turks to refer to the Arab and Iranian Muslims of Central Asia. With the departure of the Arabs from the region, especially after the Mongol invasion, the designation only applied to the Iranians who remained.

Historically, after the breakup of the Indo-European family, the Aryan branch subdivided so that the Medes and the Pars migrated to the Iranian plateau where they created the Median and Persian Empires respectively; the Sughd and the Hind migrated to the Aral Sea region. Subsequently, the Hind migrated southeast and occupied the northwestern regions of the Indian subcontinent.

Sughdiana, settled between 1,000 and 500 B.C. by Iranian tribes, passed into the hands of the Persian Achaemenians in the sixth century, who lost it in turn to Alexander the Great in the fourth century B.C. The Arabs conquered Sughdiana in the early 600s. Under Muslim rule, especially with Samanid support, Sughdiana grew to encompass Maymurgh, Qabodian, Kushaniyya, Bukhara, Kish, Nasaf, Samarqand, and Panjekent, each a virtual kingdom.

During Mongol rule (1219–1370), agricultural development and urban expansion were halted, local traditions of kingship were dismissed, and the *Shari',* a system of Islamic laws, was replaced by the *Yasa* (Mongol Law). Indeed, the *Yasa* was used to enforce anti-Muslim policies, discouraging rebellion against the Chaghatai khans. Tajiks who could not tolerate the intensity of Mongol rule either migrated or lived in isolation in the highlands.

The fortunes of the Tajiks declined when the Golden Horde (Mongol empire) was dissolved and its tribes joined the Oguz Turks who had settled Transoxiana in the tenth century. Rather than settling on the fringes of the urban areas as they had on the Qipchak plain, the new invaders wrested the Tajiks' farms from them and became farmers. Leaving their cultural centers of Samarqand and Bukhara, the Tajiks continued to take refuge in the highlands. Thus, during the Shaibanid, Astarkhanid, and Manghit rule, Tajik cultural domination declined so that in 1920 the Tajiki language was

discontinued as the official language of the Emirate of Bukhara.

The Uzbeks (a Turkish people), however, were not the only intruders. Russians, after Muzaffar's defeat in 1868, dominated both the Turks and the Tajiks. Indeed, the Turks served as governors and tax collectors for the Russians. In 1924, the Soviets divided the Tajik population between the Autonomous Republic of Turkistan and the People's Republic of Bukhara. The Tajiks, however, continued their struggle to gain independence. Stalinabad became the capital of Tajikistan.

Between 1929 when Tajikistan Soviet Socialist Republic (SSR) came into existence and 1950, Tajikistan underwent intensive Sovietization which meant they were educated to value the political-economic system of collectivization and industrialization. Those with nationalistic tendencies were purged. However, this forced "modernization" was largely under the direction of Stalin, and when his administration ended the Soviet empire came to an end, an era of stagnation followed in the post-Stalin Soviet leadership, with leaders that were not as totalitarian and ruthless in their approach. This very stagnation eventually led to the dissolution of the Soviet Union and the emergence of the independent Republic of Tajikistan in 1991.

In 1992, the growing Muslim discontent against Soviet rule erupted in a bloody civil war in the south. The war, which resulted in 40,000 casualties, over 50,000 refugees, and 500,000 displaced people, ended after the UN intervened and after the Opposition members were allowed to participate in the governance of the republic.

The most recent history of Tajikistan consists of assassinations, perennial hostage taking, border conflicts, typhoid and plague epidemics, and food, water and fuel shortages. Nevertheless, in spite of clan wars, regional insurrections, and drug trafficking, the Tajiks have convened six rounds of talks under the auspices of the United Nations, achieving a semblance of unity.

Timeline

1000–500 B.C. Early settlement

The region known as Sughdiana is settled by Iranian peoples from the north.

600–400 B.C. The Achaemenians

Achaemenians rule Sughdiana.

329–328 B.C. Conquest by Alexander

Alexander the Great captures Sughdiana. Termez and Khujand are founded.

350–200 B.C. Bactrian kingdom

The Bactrian kingdom, covering a portion of present-day Tajikistan, is established out of Bactrian and Kushanid territories

A.D. 300–400 Hephthalite conquest

The region is captured by the Hephthalites.

400–600 Turkish invasions

Invasions by Turkish tribes culminate in Turkish control of the region which pays tribute to the Khaghan of the Turks.

600–700 Muslim domination

Central Asia is dominated by Muslim invaders from the west.

Tajik Identity Established

874–999 Samanid Empire

Drawing on the resources of the Arabs, Persians, and Turks, the Samanids establish and develop a major empire.

1020 Abual Qasim Firdowsi

Abu al-Qasim Firdowsi, author of the *Shahname,* dies.

1037 Ibn-i Sina

Physician and philosopher Ibn-i Sina dies.

1048 Abu Rayhan al Biruni

Abu Rayhan al-Biruni, scientist, philosopher, and scholar dies.

1122 Omar Khayyam

Omar Khayyam, Persian philosopher, mathematician, and poet dies.

1220 Chingiz Khan

Chingiz Khan invades Central Asia and in the process ravages Khujand.

1220 The poet Attar dies

1273 Marco Polo

Marco Polo crosses the Pamirs.

1373–88 Tamerlane

Toktamish's invasion of Transoxiana is repulsed by Tamerlane. Tamerlane (c. 1336-1405) is a Mongol conqueror who at the time of his death holds an immense empire from Delhi in the east (in modern day India) to the Black Sea in the west. Though Tamerlane's empire is responsible for making positive achievements in art, literature, science and vast public works, his historic reputation becomes that of a vicious conquerer. After conquering cities he is mercilious to the defenders, slaughtering thousands and grotesquely constructing large pyramids with their skulls.

1762 Manchus take over

Manchus annex eastern Turkistan and Tajikistan.

Modern History

1865 Glacier found

Alexis Fedchenko discovers the largest glacier in the world.

1868 Russia in Tajikistan

Russia annexes northern Tajikistan.

1894 Fortress at Termez

A fortress is built at Termez to guard against Afghan incursions.

1918: February–March 13 Bolsheviks in Kokand

Bolsheviks capture Kokand; land taken from individuals, are placed under the auspices of Sovnarkom, beginning of the

Bolshevik/feudal conflict in the region which leads to Basmachi uprisings and civil war.

1920s Soviet days

Soviet rule is established. Violence effects the government's willingness to spend funds on the region. Nevertheless, appropriate key personnel--social engineers, educators, mechanical engineers, agronomists, and psychologists--flood Tajikistan and transform it into a burgeoning agricultural enterprise in the south and a major industrial center in the north.

1924: October Creation of Tajik ASSR

Tajik ASSR, with its capital at Stalinabad, is established within Uzbekistan SSR. In January of 1925, the Gorno-Badakshan Autonomous region is created.

1929: October 15 Creation of Tajik SSR

Tajik Soviet Socialist Republic is created. Khujand Okrug, the future center of culture and politics of the republic, is added to Tajikistan; Dushanbe becomes the capital.

1930s Urbanization and industrialization

Tajikistan undergoes a profound transformation from a collection of medieval cities and rural towns, into a republic with a considerable industrial and agricultural economy. The society of many Tajiks begins to change too, as traditional agrarian life gives way to urbanization.

A number of purges (occuring in 1931 and 1937) take place within the Communist Party of Tajikistan (CPT) of the newly formed republic; Ibrahim Beg, the chief Basmachi leader, is executed in one.

1940s Muslims and Communists

People in Tavil Dara, Gharm, Vakhsh, and Qurqanteppe choose to respect their ancient traditions of land management and Islamic traditions. Known as Wahhabis, they institute their own schools, hire their own teachers, and organize their own society.

To safeguard against mass protest in the face of difficulty, moderate improvement is allowed in the lives of select groups of Tajiks. Over 233,000 houses, many schools, health centers, and recreation areas are built and given to deserving Sovietized Muslim families.

1940 Hissar Canal

To meet the needs of the nation at war, the completion of the Hissar Canal is expedited. The canal opens the door for large amounts of food and war supplies to be shipped into and across the Soviet Union.

1942 Factory development

The Dushanbe Textile Factory and a new cement factory become operational.

1948 Tajikistan State University

Tajikistan State University is completed. Access to knowledge, however, remains limited. Rather than educating Tajiks for the future, more factories and low-level jobs are created.

Early 1950s Continued development

Development of mining, fuel, textile, foodstuffs, and building materials are a priority. An improvement in lifestyle occurs with the death of Stalin. The expansion of industry outside of the production of war goods materially improves the lives of the Tajiks. In order to accommodate war veterans, two cotton processing factories are built in Qurqanteppe; a third hydroelectric station is built on the Varzob. These large factories not only serve to produce goods for the returning veterans, but also provides them with jobs. Psychologically the death of Stalin allows people more freedom from fear and oppression. The Soviet Union is still viewed by the west as a totalitarian dictatorship, the freedom of expression is still veiled, but is improving in contrast to rigid Stalinism.

Late 1950s Soviet contributions

Large tracts of marshland around Qurqanteppe, Kulab (lower Vakhsh), and Dushanbe (Varzob) are recovered and cultivated; the foundation of the Sarband hydroelectric stations (on the upper Vakhsh) is completed. This period, a watershed in the building of socialism, provides much of the infrastructure required by the agricultural sector; the pro-Islamic policies dictated by World War II are reversed.

Early 1960s Islam on the periphery

Jabbar Rasulov becomes the First Secretary of the CPT (1961). Light industry of lower Vakhsh is expanded and developed. The automotive industry responds to the immediate needs of engineers, managers, and farmers by producing more automobiles, not only for the transportation of goods, but for people as well. Alongside that, a gigantic hydroelectric station, with the capacity of 2.7 million kilowatts of energy, is built on the Vakhsh River. A combination of automation and the brigade system leads to the expansion of irrigation and development of some 13,615 hectares of land. The resulting agricultural surplus raises Tajikistan's purchasing power, benefits the economy, and the government. Illiteracy is eliminated, but complaints about the *mullahs'* (the Muslim teachers) methods and activities in Qurqanteppe and Kulab persist.

Late 1960s Era of prosperity

Production of automotive tools, electricity, and foodstuffs boosts both light industry and trade; naturally, it also tremendously improves Tajik lifestyle. Labor input and output increase twofold. The hydroelectric factory on the Vakhsh, the "Pamir" refrigerator plant, and the Hissar Hydrozal are added, and mining begins in Anzob. The production of cotton, milk, meat, and grain increases manifold, and 54,000 hectares of new land are reclaimed, the Yavan-Abkik Irrigation System is completed. The Dushanbe Textile Factory becomes fully operational. A major achievement of the Soviet Tajiks, the *kombinat* (a complex of textile factories and plants) is included among the textile production outlets of the USSR (1966). Kulabi youths are bused to Dushanbe to fight anti-Communist Pamiri youths (1967).

1971 Communist anti-religious agitation

Central Committee of the CPSU calls for an increase in atheistic work and vigorous enforcement of Soviet anti-religious laws.

1972 Soviet holidays

Soviet holidays replace Muslim holidays. The House of Scientific Atheism opens.

1973 Islam against communism

Planning stage of "Islam Against Communism," a program for Central Asia supported by the U.S., Britain, Saudi Arabia, and other countries, takes place.

1975 Wahhabi movement

The Wahhabi movement is established in Tajikistan. The Wahhabis are a fundamentalist Islamic sect that looks upon other sects as heretical.

1976 Anti-Communist protest

The first anti-Communist protest by a Muslim group is staged in Qurqanteppe by Mullah Abdullah Nuri.

1978 Islamic movement established

A covert Islamic movement, the Islamic Resurgence Party (IRP), is established in the USSR.

1982: April 2 Rahmon Nabiyev selected First Secretary

Rasulov, First Secretary of the CPT, dies. Rahmon Nabiyev, resolved to curb Islam promoted by the Pamiris and Gharmis, becomes First Party Secretary.

1985–86 Attempts at reform

Gorbachev asks First Secretary Rahmon Nabiyev to retire; Qahhor Mahkamov brings the question of increasing Islamic activities in Qurqanteppe and Kulab to the Plenum of the Central Committee and the Party's twentieth Congress; Nuri (b. 1947) urges his followers to petition the twenty-ninth Congress of the USSR to establish an Islamic republic.

1988 Campaign for *Qozikalon*

Akbar Turajonzoda and Haidar Sharifzoda compete for the position of *Qozikalon* (chief judge).

Turajonzoda, a graduate of the Islamic University of Jordan, is described, on the one hand, as an astute politician, a true democrat, and Tajikistan's pride and, on the other hand, as the ambitious leader of the Wahhabi movement. He is a major player in the 1990s peace negotiations.

Sharifzoda (b. 1946) is a graduate of the Bukhara Theological School. As leader of the Kulab mosque, he tries but fails to separate his mosque from the Qoziyyot of Tajikistan (1991).

1990: January Communism challenged

The Democratic Party of Tajikistan (DPT) is established amid intense opposition.

1990: February 12–15 Muslim-communist clash

First confrontation between Muslims and communists over the resettlement of Armenians ends in a number of deaths.

1990: February 23 73 years of Soviet rule

Tajikistan celebrates the seventy-third year of Soviet rule.

1990: March Supreme Soviet elections

Opposition candidates are barred from the Supreme Soviet elections. The communist-dominated parliament elects Mahkamov president.

1990: June Islamic party banned

The Islamic Resurgence Party (IRP) is formed and immediately banned by Tajik government.

1990: July 26–27 Agreement to normalize relations

An agreement is signed in Khorog among republic and regional leaders, governors, and armed groups to normalize relations; Presidium decrees that the government buy defensive weapons.

1991: March 9–13 Muslim agitation against teachers

Mullah Ajik Aliev incites popular resistance against teachers and Qiyomuddin Qoziev even issues a law making the killing of teachers legal. (See Late 1960s.)

1991: April 27 Demand for Turajonzoda to step down

Haidar Sharifzoda calls for Turajonzoda's resignation.

1991: August 29–31 Protest in Dushanbe

Mahkamov's support for the August 1991 failed coup in Moscow results in the first anti-government gathering in Dushanbe.

1991: September 7 Mahkamov replaced by Aslanov

Mahkamov resigns. Acting President Qadriddin Aslanov suspends the Communist Party and freezes its assets.

1991: September 9 Independence

Tajikistan declares its independence.

1991: September 21–23 Nabiyev reinstates CP

Lenin's statue is toppled; Aslanov is replaced by Rahmon Nabiyev who reinstates the Communist Party.

1991: October 2 Supreme Soviet loses power and property

The Communist Party of Tajikistan is put to rest by the Supreme Soviet which takes over its power and property.

1991: October 4–18 More agitation against teachers

In the south, Ishon Saidashraf, Ishon Qiyamuddin, and Mullah Muhammadjon Qufronov speak against teachers, call teachers *kafir* (infidel), refuse to perform prayers for deceased teachers and decree it lawful for the faithful to publicly ridicule teachers.

1991: October 6 Nabiyev quits presidency

Nabiyev is forced to step down temporarily from the presidency.

1991: November 24 Nabiyev elected president

Nabiyev is elected president by a popular vote (fifty-eight percent). Not receiving Moscow's assistance in time, Davlat Khudonazarov garners only thirty percent of the vote. The results delight the Leninabadis but enrage the Gharmis and the Pamiris. Safarali Kenjayev, a Yaghnabi, becomes the leader of the Supreme Soviet.

1991: December Badakhshan form autonomous republic

The Council of the People's Deputies in Badakhshan declares the formation of the Autonomous Republic of Badakhshan.

1991: December 21 CIS membership

Tajikistan joins the Commonwealth of Independent States (CIS).

1992: January 4 Prime minister resigns

Prime Minister I. Hayaev resigns. During his six years as prime minister, the economic and educational situation in Tajikistan has deteriorated.

1992: March 1–21 Mayor imprisoned

Attorney General Nurulla Huvaidullaev imprisons Mayor Maqsud Ikromov on corruption charges and refuses to release him even at the request of the dominant political voices in the city.

1992: March 2 Acceptance to UN

Tajikistan is accepted into the United Nations.

1992: March 25 Islam on the Offensive

Proceedings of the Presidium of the Supreme Soviet are televised. Kenjayev accuses Mamadayoz Novjavonov (a Pamiri and the Minister of Internal Affairs) of embezzlement and abusing his authority. Novjavonov resigns.

1992: March 26 Anti-government demonstrations

Three hundred Pamiri youths, gathering in front of the Central Committee headquarters, protest against Novjavonov's treatment; IRP and DPT join in the next day, creating an Opposition core demanding Kenjayev's resignation, a new constitution and parliament, and reversal of undemocratic legislation limiting the activities of newspapers.

1992: March 27 U.S. Embassy opens

Ikromov is moved to a prison in Khujand. U.S. embassy opens in Dushanbe.

1992: April 12–14 Protests continue

Turajonzoda and six Sufi shaykhs join the Shahidan gathering. Turajonzoda complains that the little freedom gained is being taken away rapidly; the number of protesters increases.

1992: April 20 Kenjayev becomes parliamentary leader

The Supreme Soviet reaffirms Kenjayev as the leader of the parliament. Opposition leaders leave the hall in protest.

1992: April 21 Kenjayev tenders resignation

Kenjayev requests inclusion of his resignation request in the agenda.

1992: April 22–23 Kenjayeve resigns

Imomali Rahmonov speaks out against the Opposition; Kenjayev's resignation is accepted; former head of Badakhshan Executive Committee, Akbarsho Iskandarov, a Pamiri, replaces Kenjayev.

1992: April 23 Turajonzoda elected to presidium; polarization continues

Turajonzoda and several other Opposition members are elected to the presidium of the Supreme Soviet; Turajonzoda appears on television. His sign for "victory" and his remarks about communist leaders and institutions rile the people of Kulab, a traditional communist stronghold.

1992: April 25 Gathering in Shahidan Square

Sharifzoda and Sangak Safarov arrive in Dushanbe with busloads of Kulabis. Sharifzoda opens the meeting of the Ozodi Square by protesting the newly reconfigurated government. Because of the appointment of Kenjayev as the Chief of National Security, the Shahidan Square, evacuated once to honor a "truce," is filled with people again.

1992: April 26 Protests mount

The Shahidan Square remains full and slogans become more aggressive. Reportedly 2 million rubles are gathered in the Ozodi Square of which 35,000 rubles is allocated to the people of Gharm for flood compensation. Apparently, 500,000 of the 2 million is contributed by the government.

1992: April 29 Government-opposition stand-off

The fourteenth Supreme Soviet reinstates Kenjayev; Opposition occupies the television studio and takes control of all the "city gates." The National Security and the Ministry of the Interior declare impartiality.

1992: April 30–May 2 Opposition issues demands, government responds

Ozodi Square demands include: calling a meeting of the Supreme Soviet, annulment of the Supreme Soviet that accepted Kenjayev's resignation, annulment of the decrees forced by the Shahidan on the Supreme Soviet, dismissal of those appointed to the Supreme Soviet at its last meeting, replacement of Turajonzoda with Sharifzoda, and a constitution whereby the head of the Supreme Soviet is elected by the people; Nabiyev issues an order to form a National Front for Tajikistan; freedom of conscience and religion in Tajikistan passes. Three hundred criminals set free from Dushanbe and other area prisons, don uniforms, pick up arms, and participate in the parade on Ozodi Square.

1992: May Kenjayev out

Kenjayev is dismissed again. The 50-days sit-in in Shahidan Square ends. Units of government and Opposition forces merge.

1992: May 5 Positions harden

Members of the National Front are armed; Tajik television is nationalized, and programming is changed to reflect current realities; Murodali Shiralizoda, a government paper editor, is killed.

1992: May 7–10 Government turmoil and fighting continue

After they sign the agreement to form a coalition government, Kenjayev and Vice-President Dustev are dismissed by the incoming Supreme Soviet. Iskandarov replaces Kenjayev. Prominent communists: Kenjayev, Huvaidullaev, Dustev, Abdullojonov, and Achilov leave Dushanbe.

In Dushanbe, 108 are killed, 233 are wounded, and 104 are unaccounted for.

1992: May 10–12 Nabiyev not recognized; war in Kulab

Nabiyev's coalition government is not recognized by Leninabad and Kulab and Alijan Salihbaev becomes the Head of the National Security Committee. War breaks out in Kulab.

1992: May 14–15 The Onset of the civil war

War moves from the squares to Qurqanteppe and Safarov and Rustam Abdurrahim form the Popular Front for the defense of Kulab. The Supreme Soviet of Khujand discusses secession from Tajikistan.

At this juncture begins the participation of people in a number of local forces to stem the tide of Islamic rule in the republic. People fear that the marches they witness in Dushanbe and the stories they hear about Muslim atrocities might become a reality.

1992: May 25 Coalition government constitutional

Constitutional Committee recognizes the legality of the coalition government.

1992: May 31 Muslims honor Khomeini

Opposition leaders participate in the anniversary memorial of the death of Ruhullah Khomeini in Tehran; Communist leaders are not invited.

1992: June 1 Opposition victories

Opposition defeats the Communists in the south and devastates the Vakhsh and Kirov *sovkhozes* (state-owned farms). Turkmenistani *kolkhoz,* (people from the Soviet collective farms) Kuybishev, Jilikul, and Kumsangir spearhead the invasion. Hundreds are killed.

1992: June 15 Iranian TV airs in Tajikistan

Regular broadcast of Iranian television programs into Tajikistan begins.

1992: June 16 –17 Hostages taken

In Lomonsov, Safarov takes twenty-two CIS members hostage but, fearing a major confrontation, releases them.

1992: June 20 Safarov recalls troops

Coalition government convinces Safarov to return his troops to Kulab.

1992: June 24 Dushanbe garrison declares allegiance to Russia

The officers of the Dushanbe garrison recognize the technology at their disposal to be the property of Russia and regard themselves as belonging to Russia.

1992: August 26 Government call on people to fight Opposition

Safarov, Abdurrahim, and Durbon Zardak visit Asht and encourage the villagers to rise against the Opposition.

1992: August 28–31 President's guards taken hostage

Between 150 and 180 young men try to meet with the President but fail, they invade the President's palace and hold its guards hostage asking for the President's resignation. In Tursunzoda, Kenjayev establishes his Peoples Front against the coalition government.

1992: September 2–27 Reign of Terror in the South

The Supreme Soviet gives Nabiyev a vote of no confidence. In Qurqanteppe, a meeting with Safarov is invaded by the Opposition. Safarov withdraws to the anti-fundamentalist Urgut Mahallah. At night, women, children, and old men of Urgut Mahallah are moved to Kalininabad.

A reign of terror begins in Qurqanteppe, a town which is cut off from the rest of Tajikistan and which is already suffering bread, water, and gas shortages. Those who try to leave the city are beheaded.

Opposition forces are kept at bay at Urgut Mahallah for three days. Thereafter, the Opposition loots Urgut Mahallah of all its savings of the past seven decades.

When Lomontosov Mahallah is captured, 200 of its youth are killed during the same night. The situation remains tense until the forces of Safarov, Saidov, and Langariev arrive.

1992: September 4–6 Supreme Soviet meeting cancelled; fighting spreads

The Supreme Soviet does not meet because the representatives of Qurqanteppe and Kulab are unable to attend. Intense fighting goes on in Qurqanteppe.

The Kumsangir region enters the war on the side of the Opposition. Hiding in the headquarters of the Russian 201st division, Nabiyev directs Safarov and Langariev to enter Qurqanteppe and Dushanbe.

1992: September 7 Nabiyev cornered

At the Dushanbe airport, the Opposition forces Nabiyev to resign at gunpoint. As president, Nabiyev fails to meet the needs of the parties that elect him; instead, he incites the government against the parties and the people who complain about shortages. Surrounded by sycophants, he is cut off from the academics, the media, and the traditional sources of his power.

Iskandarov becomes Acting President; Khudonazarov becomes his advisor.

The 201st division of the Russian army, stationed in Dushanbe for many years, controls the city and former Soviet industrial complexes as Russian property.

1992: September 20–21 Acting prime minister appointed

Supreme Soviet appoints Abdulmalik Abdullojonov as acting Prime Minister.

1992: September 20 Call for disarmament

Iskandarov urges the Interior Ministry to maintain security; an ultimatum is issued to the armed groups in Kulab and Qurqanteppe to disarm within four days; UN representatives visit Kulab and Qurqanteppe; the number of refugees is estimated at 126,000.

1992: September 22 Spiritual leader appointed

Imom Shohkarim Al-Husaini (Aga Khan IV) appoints Khudonazarov the spiritual *murid* (leader) of the Isma'ilis of Tajikistan.

1992: September 26–27 Fighting intesifies

Langariev and Abdurrahim are defeated at Narak. Russian troops rescue Iskandarov and Abdullojonov who are being held hostage in Kulab. Safarov and Sai'dov enter Qurqanteppe. Opposition moves into Kolkhozabad and Jilikul where it joins Mullah Umar.

Opposition's soldiers are village youths passed over by the Soviet army either because they had been declared dead or because they had been registered as girls. None is familiar with armaments.

1992: October 8 Opposition parties outlawed; barter for weapons

Parties that form the Opposition are disestablished. Iskandarov leaves for Bishkek. Regar reports that aluminum is being exchanged for weapons in Uzbekistan.

1992: October 20 Call on Russia to disengage

Turajonzoda urges Russia not to arm Kulab and to stop interference in Tajik affairs.

1992: October 22 More fighting

The Internal Affairs Office of Regar is looted of all its guns and ammunition.

Dushanbe-Kulab highway is placed under Russian protection. Fighting intensifies in Qurqanteppe; centers are set up in Kulab and Dushanbe to receive refugees; investigators examining the aluminum-for-weapons deal are themselves arrested and imprisoned; private cars are confiscated by hoodlums, occupants are executed.

1992: October 23–25 Evacuation of capital

High-ranking officials leave Dushanbe one day before Kenjayev and Abdurrahim shoot their way into Dushanbe, capture the President's Palace, and try to capture the radio and TV stations. Their plans fail because, for political reasons, Safarov, Sa'idov, and Langariev do not participate and carry out the tasks assigned to them.

The pro-Iskandarov forces retake the President's Palace. Abdurrahim, a Professor of English and a master guitar player, enters politics with Sangak and Kenjayev but, in the final assault on Dushanbe, sides with Kenjayev and is mysteriously abducted and killed.

Eight hundred armed Badakhshanis surround Dushanbe, tighten the circle around Kenjayev's forces. Discussion among Iskandarov, Kenjayev, and representatives of Russia results in Kenjayev pulling his forces back in exchange for setting a date for the Supreme Soviet to meet in Khujand.

1992: November Opposition on the run

After losing Qubadiyan and Shahrtuz, the Opposition moves to Kumsangir and Panj. A group occupies the "Bishe-yi Palangon" restricted zone in Jilikul.

Blockade of the railway from Uzbekistan creates a critical shortage of food in Dushanbe.

1992: November 3 Kenjayev seeks peace

Kenjayev claims that he has come to Dushanbe to offer peace; a curfew is imposed from the twenty-first hour on. The 201st division controls all entrance and exit points.

1992: November 7 Efforts to end the fighting

Safarov speaks in support of a constitutional government headed by Nabiyev and Kenjayev.

Andre Kozerov, visiting Kulab and Qurqanteppe, outlines Russia's concern for the safety of the Russians and Jews of Tajikistan and asks that the sixteenth Supreme Soviet meet by November 20.

Iran announces its readiness to bring peace to Tajikistan. Kazakhstan, Kyrgyzstan, Uzbekistan, and Russia express concern about the war spreading into neighboring countries. The 201st Russian division is assigned to provide security.

The needs of 7,000 refugees from Dushanbe and Qurqanteppe concern Badakhshani hosts so the Red Crescent and Red Cross announce assistance.

1992: November 9 Supreme Soviet to meet, body denounced by Opposition

Iskandarov announces November 16 as the date for the Khujand meeting of the Supreme Soviet; representatives of Tursunzoda, Hissar, and Shahr-i Nav denounce the Supreme Soviet as illegal, ask the commander of the 201st Russian division to take control until November 16.

1992: November 10 Government resigns, capital under Russian army protection

Dushanbe is under the protection of the Russian army which is not to interfere with the internal affairs of Tajikistan; Iskandarov and members of government resign; May-November Coalition government ends.

All eyes are on Leninabad and Khujand to save Dushanbe, and the country.

Eleven thousand of the 54,000 refugees in Dushanbe are accommodated in the Frunze district.

1992: November 12 Shartuz fighting

Conflict in Shahrtuz results in 10,000 deaths.

1992: November 16–21 Rahmonov takes charge

The sixteenth session of the Supreme Soviet in Khujand accepts Nabiyev's official resignation and Iskandarov's resignation. Imomali Rahmonov is elected Chairman of the Supreme Soviet; Abdulmalik Abdullojonov is elected Prime Minister.

Rahmonov (b. 1952, Danghara, Kulab) is an economics graduate of Tajikistan State University (1982). He serves as the director of the Workers Union of the Lenin kolkhoz and works for the Kulab Communist Party. In July 1988, he becomes the director of the Lenin sovkhoz in Danghara. He is elected Chair of the Executive Committee of the Kulab soviet on November 2, 1992; Chair of the Supreme Soviet of Tajikistan in November 19, 1992; and President of the Republic in November 6, 1994.

Abdullojonov (b. 1949, Khujand, Leninabad) is an engineering graduate of Technological College of Odessa (1971). He becomes the Minister of Agriculture in 1978 and prime minister in the coalition government in 1992; he is reelected to the position and is authorized to organize ministries, choose ministers, and suggest names to the Supreme Soviet. He is later dismissed in December 1993.

1992: November 22 Refugees go home armed, local guard units organize throughout country

Rahmonov meets the leaders of Shahrtuz, Kumsangir, and Panj regions in Shahrtuz. The refugees return, bringing arms with them.

City districts and various regions of the country organize their own local guards without regard for either the internal affairs officials or government troops. Some even invade the government troops and the 201st Russian division.

1992: November 24 Shortages widespread

Abdullojonov expresses concern for the 10,000-12,000 who gather in front of the bread factory daily; lack of fuel stops all transportation; schools are cold without heat.

1992: November 25 Opposition agree to give up arms

Leaders of the armed factions meet in Khujand with the people's representatives and Rahmonov. Safarov, Ya'qub Salimov, and Rahmanov pledge to put down their arms.

1992: November 30 Peace-keeping plan

Leaders of Kazakhstan, Kyrgyzstan, Uzbekistan, and Tajikistan meet in Termez, discuss the deployment of CIS peacekeeping troops in Tajikistan.

1992: December 1-6 Opposition defeated

201st Russian division takes over Dushanbe and closes all entrance points. Forces from Hissar, backed by Uzbekistan, attack the Opposition in Dushanbe. Opposition, defeated, moves to Faizabad, Kofarnihon, and the Romit Gorge.

1992: December 13 Essentials in Dushanbe

Wagons full of fuel, motor oil, wheat, and flour arrive in Dushanbe; a 39-day economic siege ends.

1993: January 21 Rahmonov-Yeltsin meeting

Rahmonov and Yeltsin meet to discuss problems and sign a number of documents.

1993: January 24–February 22 Renewed CIS offensive

Reinforced CIS forces launch offensive against Romit Gorge, capturing three strongholds, including Gharm.

1993: February New mufti

The Muftiyyot of Tajikistan is inaugurated. Mufti Fatkhullo Sharifzoda pledges to stay clear of politics.

1993: March UN membership

Tajikistan joins the UN and the Presidium of the Supreme Soviet grants amnesty to all participants of war.

1993: March 30 Tajik Patriots killed

Sangak Safarov and his deputy commander Faizali Saidov are killed.

Sangak (b. 1928) comes to prominence after his participation in the Ozodi meetings and appointment as National Front Commander. Between 1950 and 1978 he spends twenty-three years in prison for a variety of charges including murder. He works closely with Nabiyev to neutralize Turajonzoda and the Opposition forces in the south and install a constitutional government in Tajikistan.

1993: May 9 Clashes near Afghan border

Government accuses rebels of planning invasion from northern Afghanistan. In a battle with CIS forces on the Afghan border, eighty-six rebels are killed.

1993: June 26 Rahmanov seeks new democratic beginning

Rahmonov annuls all decrees issued by Iskandarov, pledges to build democracy.

1993: July 2 Russians clash

Russian troops kill fifty rebels at Panj.

1993: July 17 Border reinforced

Number of troops on the border is increased; Afghan villages are shelled. Pavel Grachev, the Russian Foreign Minister, visits Dushanbe.

1994: April–October Negotiations start

Inter-Tajik negotiations under the aegis of the UN begin in Moscow. Second round, in Tehran (June 18-27) results in a cease-fire effective October 20.

1994: August Renewed fighting

Armed conflict flares up in the regional centers of Gharm, Komsomolabad, Tajikabad, and Hait.

1994: November 6 Rahmonov elected

Rahmonov is elected president (with ninety percent of the vote). New Tajik constitution reconciling the former laws with new constitutional standards is adopted.

1994: December "Village Governance"

Reorganization of the ministries is finalized. The law of "Village Governance" is adopted.

1995: February 26 Elections

Local and parliamentary elections begin without the participation of radical Islamic or moderate parties.

1995: April 7-13 Afghan border fighting

The heaviest fighting of 1995 occurs along the Afghan border; nineteen soldiers are killed. Tajikistan appeals to the UN and CIS for assistance. In response CIS-owned Russian helicopters fly attack missions over the border. Twenty-nine guards and 170 rebels are killed and IRP headquarters in Afghanistan is bombed.

1995: May New currency

Tajik ruble is introduced to speed up privatization and boost foreign trade.

1995: May 17–21 Peace talks begin

Talks between Rahmonov and Nuri in Kabul, Afghanistan, result in extension of cease-fire agreement.

1995: October 8 Arrest of journalist

Journalist Abdulkayum Kayumov of the dissident journal *Charoq-i Ruz* is arrested.

1995: November 21 Electricity rationed

Electricity is rationed; residents receive six hours of electricity per day.

1995: November 30–1996: July 19 Successful peace talks

Fifth round of talks in Ashgabat results in an armistice (July 19) and a cease-fire (December 10-11).

1995: December 29 Defense treaty

Tajikistan signs defense agreement with Iran.

1996: January 21 Assassination

Mufti Sharifzoda, three members of his family, and three students are assassinated in Dushanbe.

1996: February 27 Government overture to Opposition

The government offers the Opposition the opportunity to speak at its March 11 session. Nuri claims that the Opposition controls seventy percent of Tajikistan.

1996: March 11 Parliament

Parliament opens without an Opposition representative.

1996: May 13–14 Violence at demonstrations

Six thousand people demonstrate in Urateppe asking for better living conditions; five die when troops fire on demonstrators.

1996: June 19 New mufti

In a Soviet-style election, Khoja Amanullo Negmatzoda becomes mufti.

1996: July 12–17 Action near Afghan border

Government regains the control of Tavil-Dara, closes border with Afghanistan.

1996: July 28 Problems with Uzbekistan

Uzbekistan halts gas delivery to Tajikistan. Tajikistan's debt to Uzbekistan is $140 million.

1996: July 29 Scholar/editor assassinated

Academician M. Osimi, Tajikistan's celebrated scholar and editor of *Payvand,* is gunned down.

1996: July 30 Riots

Food riots in Khorog kill five, injure eleven.

1996: August 10 New political group formed

"Congress of People's Unity" is registered. The movement includes many socio-political, ethnic, and economic organizations including the CPT and several ethnic minorities.

1996: August 20 Price controls lifted

Tajik authorities free prices for bread and provide subsidies for the poor.

1996: December 10–11 Agreement with Afghanistan

Meeting between Rahmonov and Nuri in Khosdekh, Afghanistan, results in a cease-fire agreement.

1997: March 8 Rebels to join Tajik army

In the sixth round of talks in Moscow, a multi-stage agreement is reached to integrate the rebels into the Tajik army.

1997: April 30 Rahmonov shot, lives

Rahmonov is wounded in an assassination attempt in Khujand.

1997: May 22–28 Talks in Tehran yield agreement

Government and UTO resume talks in Tehran and agree on political reforms, including recognition of opposition parties.

1997: June 27 Forced Compromise

A peace agreement is signed in Moscow.

1997: July 14 President signs on to reconciliation

Tajik president signs Mutual Forgiveness Act drawn up by the NRC.

1997: July 31 **Mufti and family taken**

Rezvon Sadirov kidnaps Mufti Negmatzoda and his family, holds them hostage to gain the release of his brother.

1997: August 1 **Amnesty**

Parliament approves a general amnesty.

1997: September 2 **Hostages freed**

Kidnapped mufti and family are released.

1997: September 25–November 15 **Refugee repatriation**

First stage of Tajik repatriation is completed; 6,300 refugees have returned.

1997: October 12 **Opposition members released**

One hundred and seventy out of 700 Opposition supporters are released.

1998: February 29 **Secularism reaffirmed**

Rahmonov says Tajik government will remain secular.

1998: March 27–28 **Tjikistan joins economic grouping**

Tajikistan is admitted to the Central Asian Economic Union. The government and the Opposition agree to exchange POWs.

1998: April 18 **Rahmanov becomes party leader**

Rahmanov is elected Chairman of the People's Democratic Party. The move paves his way to success in the 1999 elections.

1998: May 11 **Floods**

Heavy floods devastate Gharm and Khatlan.

1998: May 18 **Nuri speaks out**

Nuri denounces the anti-fundamentalist "troika"—Russia, Uzbekistan, Tajikistan—as a threat to the peace process.

1998: May 21 **Parliament rejects appointments**

Tajik parliament votes down presidential appointees Turajondoza and Davlat Usmon.

1998: May 23 **Parties banned**

Tajik parliament bans parties that receive either financing or "ideological guidance" from other countries. The ban affects the Communist Party and the IRP.

1998: May 26 **World Bank aid**

World Bank releases $50,000 to the Opposition to set up medical centers and vocational institutes.

1998: June 17 **Rahmanov-Nuri meeting**

Rahmonov and Nuri meet to discuss Opposition representatives in the coalition government and the ban on parties. The ban is opposed by Nuri.

1998: July 21 **UN members dead**

Four UN observers are found murdered 170 kilometers from Dushanbe.

1998: July 27 **Beards proscribed**

Rahmonov prohibits members of the armed forces from wearing beards—an unpopular measure among UTO fighters.

Bibliography

Atkin, Muriel. *The Subtlest Battle: Islam in Soviet Tajikistan,* Foreign Policy Research Institute, 1989.

Bashiri, Iraj. "Tajiks," *Worldmark Encyclopedia of Cultures and Daily Life,* Timothy Gall, ed., vol. 3, pp. 740–45.

Djalili, Mohammad-Reza, et al. *Tajikistan: the Trials of Independence.* New York: St. Martin's Press, 1997.

Famighetti, Robert (ed.). *The World Almanac and Book of Facts,* 1998.

Kenjayev, Safarali. *Tabaddulat-i Tajikistan,* Dushanbe-Tashkent, 1995.

Khatlani, Amirshah, et. al. *Tufan-i Sangak-i Safar va Faizali Sa'id,* Dushanbe, 1993.

Rakhimov, Rashid, et al. *Republic of Tajikistan: Human Development Report 1995,* Istanbul, 1995.

Rashid, Ahmed. *The Resurgence of Central Asia: Islam or Nationalism?* New Jersey: Zed Books, 1995.

Soviet Tajik Encyclopedia (vols. 1–8), Dushanbe, 1978–1988.

Tajikistan. Washington, D.C.: International Monetary Fund, 1994.

—I. Bashiri

Thailand

Introduction

Thailand, formerly known as Siam, extends almost two-thirds of the way down the Malay Peninsula. The peninsula, unlike most of Thailand, lies within the humid tropical forest. Thailand's heartland is the central valley, fronting the Gulf of Thailand and enclosed on three sides by hills and mountains. Thailand's main agricultural and population centers are on this plain, especially on its flat delta land bordering the Gulf. Along the Myanmar border from northern Thailand to the peninsula is a sparsely inhabited strip of rugged mountains, deep canyons, and restricted valleys. One of the few natural gaps in the mountains is the Three Pagodas Pass along the Thailand-Myanmar frontier, used by the Japanese during World War II for their "death railway" between Thailand and Myanmar.

Thailand has more than thirty ethnic groups, with the Thai comprising about eighty-four percent of the total population. The Thai may be divided into four major groups and two minor groups. Major groups are the Central Thai (Siamese) of the Central Valley; the Eastern Thai (Lao) of the northeast (Khorat); the Northern Thai (Lao) of north Thailand; and the southern Thai (Chao Pak Thai) of peninsular Thailand. Minor groups are the Phuthai of northeastern Khorat, the Shan of the far northwestern corner of northern Thailand, and the Lue in the northeastern section of northern Thailand. The several branches of the Thai are united by a common language. Major ethnic minorities are the Chinese, Malays, Khmers, and Vietnamese (or Annamese). Principal tribal groups include the Kui, Kaleung, Mons, and Karens. Hinayana Buddhism is the state religion; only Buddhists are employed by the government and the Thai monarch is legally required to be a Buddhist.

History

Artifacts have revealed the existence of an agrarian Bronze Age culture in Thailand as far back as 3600 B.C. Early kingdoms, notably the Funan, began to emerge around the beginning of the Christian Era. The rival Zhenla state became prominent in the seventh century A.D., and a confederation of Mon states, influenced by Indian culture and Theravada Buddhism, was formed between the ninth and eleventh centuries. It was taken over by the Khmer empire in the twelfth century in the course of its westward expansion, and Khmer culture spread throughout the region.

The ancestors of the present-day Thai people migrated to the area from China in the thirteenth century and established a number of states, of which the most powerful was Sukhothai, whose influence grew as that of the Khmers waned. By the late thirteenth century, they were free of Khmer rule and thus took the name *Thai,* meaning "free." Another national milestone during this period was the development of the Thai alphabet by Sukhothai's first king, Rama Khamheng, regarded as one of Thailand's national heroes. However, after Rama Khamheng, Sukhothai began to decline, and in the fourteenth century it was eclipsed by the Ayutthaya kingdom, which was to last four hundred years.

From their capital in the fertile Chao Phraya (or Menem) Basin, the first Ayutthaya kings expanded their territory northward and eastward. They attempted to centralize the power of government by issuing codes of civil and criminal law and declaring Theravada Buddhism the official state religion. Ayutthaya became a great trading center and, by the end of the fourteenth century, the dominant power in Southeast Asia. In the fifteenth century, King Trailok (r. 1448-88) developed a complex administrative system that eventually formed the basis of the modern bureaucratic system in Thailand. It mirrored the hierarchical structure of Thai society, in which the social rungs beneath that of the king were occupied by princes, nobles, free commoners, and slaves, with Buddhist monks forming the only classless social entity, as a group that could be joined by any member of society. The Ayutthaya kings gradually assumed greater and greater powers, consolidating authority in an absolute monarchy ruled by a king invested with divine power by the Buddhist religion.

The first contacts with Westerners were made in the sixteenth century, following the Portuguese annexation of Malacca, a trading center on the Malay peninsula, in 1511. By the end of the century, the kingdom had signed treaties with both the Portuguese and the Dutch. However, the Thais kept from being colonized by playing off the major colonial powers (Portugal, France, and the Netherlands) against each other. In

1605, in response to foreign pressure, the Ayutthayan kingdom did grant certain exclusive trade rights to the Dutch East India Company. Later in the century, the kingdom was obliged to fight off French attempts to subjugate their country and force Christianity on the population. During the seventeenth century, the kingdom's nobles gained power, turning Ayutthaya into a feudal state.

In the latter part of the eighteenth century, Burma invaded Thailand. The invaders were driven back within two years by the Thai military leader Phya Taksin (r. 1768-82), who formed the Kingdom of Siam (a term used since the twelfth century to refer to the Thai people) and installed himself as king. He rebuilt the country, expanded trade, and gained the support of his people. However, irrational behavior considered to portend madness drove Taksin from the throne by 1782 and he was succeeded by Chao Phraya Chakkri, who founded the dynasty that has ruled Thailand up to the present day, and made Bangkok the country's capital.

In the mid-nineteenth century, the Chakkri king Mongkut, the highly educated founder of an order of monks, opened the country, still known as Siam, to the West, signing a trade and friendship treaty with the British. Mongkut—the king portrayed in the book *Anna and the King of Siam* (and later the musical *The King and I*)—realized that modernizing the country would help strengthen it and thus avoid colonization by Western powers. Mongkut initiated many improvements in the country's infrastructure and economy, and European culture and customs spread among the upper classes. Mongkut's son, Chulalongkorn, who succeeded him as king (r. 1858-1910), continued the reforms and Westernization begun by his father, introducing such modern improvements as the railroad and the telegraph. In an important symbolic gesture, he ended the practice of having subjects prostrate themselves before the monarch. However, the Thai government remained an absolute monarchy.

Internal tensions bred by the Depression of the 1930s led to a bloodless coup in 1932, resulting in the creation of a constitutional monarchy and parliamentary government in Thailand. However, the king abdicated three years later, the prime minister was overthrown, and the military gained increasing power throughout the decade. The Thai government has remained highly volatile ever since, with a succession of constitutions, governments, and bloodless coups. Beginning with the government of Marshal Pibul Songgram, which lasted from 1935 to 1958, Thailand was led by a succession of military leaders until 1973, when students spearheaded an uprising that overthrew the government of Thanom Kittikachorn after two years of martial rule. Since that time, there has been a shifting balance of power between the military and civilian politicians.

In World War II, Thailand allied itself with Japan shortly after the bombing of Pearl Harbor and declared war on the United States. Later in the war, however, Thai sympathies shifted toward the U.S., and Bangkok became a center for

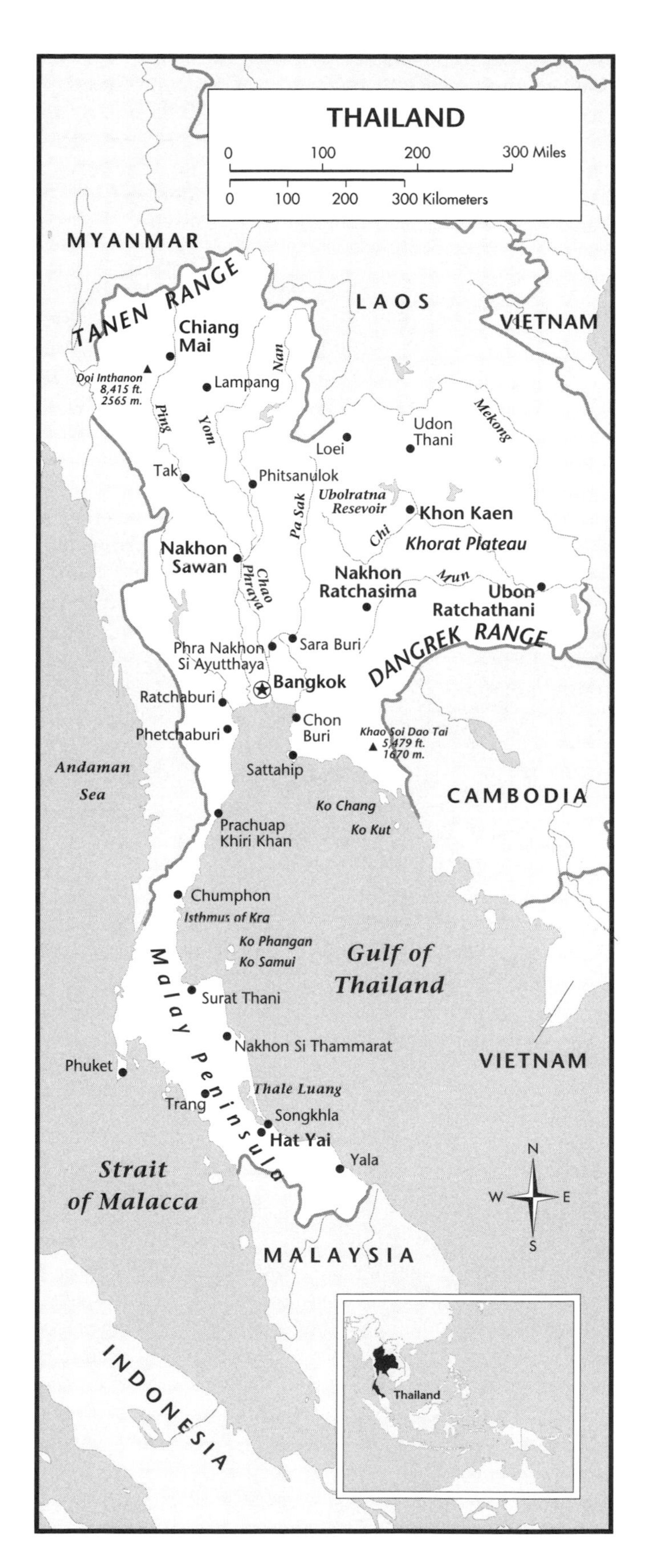

American intelligence operations. In the postwar decades, Thailand's leaders maintained an anti-Communist, pro-U.S. stance. Thailand sent troops to fight with UN forces in the Korean War, joined the Southeast Asia Treaty Organization (SEATO), and allowed the United States to use its air bases in the Vietnam War. When Vietnam invaded Cambodia in 1978, as many as 200,000 Cambodian refugees streamed into Thailand.

Government corruption was a major political issue in the 1990s. In 1991, the Thai military overthrew the elected government of Chatichai Choonhavan based on corruption charges and established its own regime. Public demonstrations protesting the appointment of coup leader Suchinda Kraprayoon as prime minister were brutally crushed by the military. Some 100 demonstrators were killed and thousands detained in Bangkok following civil unrest in that city. After intervention by King Bhumibol, Suchinda resigned, and a new government was elected, headed by reformist prime minister Chuan Leekpai. however, the volatility of Thai politics found Chuan out of office by 1995—by which time he was already the longest-serving civilian prime minister of modern Thailand. He was returned to office in 1997, in time to confront a mounting economic crisis. The Thai stock market plunged in May, leading to a currency crisis by July, as speculative trading devalued the country's currency (the baht). In August, Thailand accepted an International Monetary Fund (IMF) restructuring package that included $10-20 billion in stand-by credits. By the end of the year, the baht had lost half its value, and the crisis had spread to Singapore, the Philippines, Malaysia, and Indonesia.

Chuan Leekpai's government adhered strictly to the austerity program laid out by the IMF, and the worst of the crisis was over for Thailand by 1998.

Timeline

3600 B.C. Bronze Age culture develops

A Bronze Age people inhabit the Ban Chiang region of present-day northeastern Thailand. They are a settled agrarian people with an advanced knowledge of rice cultivation and bronze and iron metallurgy, thus predating Bronze cultures in China and the Middle East. Skills demonstrated in their pottery, housing, and printing of silk textiles are believed to be the product of at least 2,000 years of prior development. The most exquisite Ban Chiang pottery is made during 300 B.C.–A.D. 200.

1st–6th centuries A.D. Funan kingdom shapes cultural development in central plains

The Funan kingdom flourishes in the Mekong River Valley in the area between present-day Chaudoc and Phnom Penh. The Funan are an Indonesian people who speak an Austro-Asiatic language. Chinese tradition asserts that Funan culture is founded by Kaundinya, a Brahman Indian.

600 Khmer Zhenla state begins overtaking Funan society

Around 600, a vassal state of the Funan kingdom begins to usurp power. The Zhenla state is based in present-day southern Laos. Evidence of Zhenla inscriptions have been discovered in what is now northeast Thailand. By the 700s or 800s, a dynastic feud splits the Zhenla state into two parts.

802 *Devaraja* doctrine is established

The schism between the two Zhenla states is resolved when Jayavarman II declares himself to be the first *devaraja*, a king who is the earthly representative of a god (usually Shiva). An essential part of the devaraja cult is the concept of the temple-mountain (a symbolic representation of the sacred Mount Meru), which later dominates the religious architecture of the Khmers. The institution of the devaraja, a divinely sanctioned king, later shapes the development of ceremonies for the Thai monarchy.

The Khmers, centered around the Angkor region, later dominate present-day Cambodia, Thailand, and Laos during the reign of Yasovarman (r. 889–900).

9th–11th centuries Dvaravati Mon states emerge

A confederation of several Mon states develops in what later becomes Nakhon Pathom, Khu Bua, Phong Tuk, and Lawo (Lopburi). The Mons, speaking from the same language family as the Khmers, later settle in what is now southern Myanmar. Their earlier cities are typically surrounded by moats, while later they build earthen barriers and gates. Dvaravati culture is heavily influenced by India and Theravada Buddhism. By the twelfth century, however, Khmers to the east are expanding their empire into Mon areas.

The most important surviving Dvaravati landmark, the Phra Pathom Chedi, is at Nakhon Pathom, once the center of the Dvaravati kingdom. It was built sometime during 200 B.C.–A.D. 500, and remains one of Buddhism's most sacred monuments.

A *chedi* is the focal structure of a Buddhist monastery. It is a stepped spire that typically contains holy relics of the Buddha (or his possessions) in a relic chamber set in the middle of the spire. Two distinctive architectural styles of chedi develop: Sri Lankan, with a bell-shaped relic chamber topped with a tall spire; and Sukhothai, with a small relic chamber topped with a dome in the shape of a lotus bud.

12th century Khmer empire expands into Mon territory

The Khmer empire takes over the Mon states during its westward expansion. The Khmer rulers administer a highly centralized and organized society, whose culture spreads throughout its domain. Much of present-day Thailand is incorporated into the Khmer empire, as evidenced by the Khmer stone temples at Phimai and Phanom Ruang. Although the Khmers directly influence art and culture, their political control over the area is indirect and largely depends on a network of vassals administered by governors.

The Thai people are first mentioned by the Khmer in an inscription at the Khmer temple complex of Angkor Wat. They are referred to as *syan*, or "dark brown" people, the origin of the term "Siam." The term's early usage often refers to Thai slaves and mercenaries.

13th century Thai people establish Lan Na state

As the Thai people migrate south from China along the river valleys of southeast Asia, they mix with the Khmers, Mons, and Shans that they encounter along the way and begin establishing several small states. The Lan Na state is perhaps the largest of these new states, with its capitals at Chang Rai, Chang Mai (founded 1296), and Chiang Saen (founded 1327).

The Sukhothai Kingdom

13th–15th centuries First Thai Kingdom (Sukhothai)

Although there are other Thai states coexisting in what is now northern Thailand, Sukhothai is regarded as the predominant Thai state. Sukhothai develops during the wane of Khmer influence, as Thai rulers assert their authority and independence. Bang Klang Hao seizes power from the Khmers and is crowned Si Inthrathit (Sri Indraditya), independent Sukhothai's first king. The Sukhothai kingdom establishes its capital in Si Satchanalai, then Phitsanulok, and Kamphaeng Phet, all of which become cultural centers of Buddhist classical learning and civilization.

Although the Sukhothai kingdom is largely self-sufficient, Thais begin trading with nearby peoples. The Sukhothai kingdom becomes renowned for its *sangkhalok*, stylized glazed ceramics, which are exported throughout the South China Sea area.

Linguistic evidence indicates the Thai people's origin may have come from the Chinese provinces of Kwangtung, Kwangsi, and Yunnan. Another belief regarding their origin is that the Thai emerged from within Thailand already a mixture of different peoples and cultures.

1238 First declaration of independence from Khmer kingdom

A Thai chieftain declares independence from the Khmer kingdom, establishing a kingdom in the broad valley of the Mae Nam (Chao Phraya) River, at the center of present-day Thailand.

1253 Kublai Khan's Mongol forces destroy Nanchao kingdom

The true origin of the Thai as a distinct ethnicity is unknown. One hypothesis is that the Thai are racially related to the Chinese. While in the Sichuan province of south-central China, the "original" Thai create the powerful Nanchao kingdom, but continued pressure from the Chinese and the Tibetans and final destruction by Kublai Khan forces the Thai southward across mountain passes into southeast Asia. After entering the valley of the Chao Phraya River, they defeat the Khmer settlers, ancestors of the Cambodians, and establish the kingdom of Thailand. Thais, however, do not form a majority of the Nanchao kingdom's population and there is evidence that Thais are living in present-day Thailand before 1253.

1292 Rama Khamheng (the Great) chronicles Sukhothai history

Rama Khamheng (the Great, d. 1298) is the most famous monarch of the Sukhothai kingdom and the father of the Thai nation. The king (known also as Ramkhamhaeng) is traditionally regarded as the inventor of the Thai script which he based on the ancient Mon and Khmer writing systems. The king's own stone inscription serves as the most important account of Sukhothai history recording that the king imposes no sales taxes, inheritance taxes, or road tolls on his subjects. His rule is also marked by public accessibility—any subject may present a case to the king for his judgment.

The influence of Theravada Buddhism spreads during his reign, especially after the king invites monks from Ceylon (present-day Sri Lanka) to establish monasteries in Thailand. The monks' influence is not only religious, as elements of Ceylonese architecture and culture are imported as well.

Thai culture, however, retains many of its animist beliefs. One of these customs is the reverence for the *phi*, spirits that reside in the trees, hills, waters, and caves. Miniature houses on pedestals are erected and given offerings to appease the phi whenever ground is broken for a new building.

1298 King Rama Khamheng dies, kingdom falls apart

After the death of King Rama Khamheng, his son Lo Thai (r. 1298-1347) ascends to the throne, and the empire rapidly crumbles.

Mid-1300s Buddhism influences early Thai literature

King Mahathammarcha Lithai (1346–68 or 1374) compiles the *Tribhumikatha*, an early Thai book that explains the universe through Buddhism.

The Ayutthaya Kingdom

1350–1767 Kingdom of Ayutthaya

By the late 1300s, the Sukhothai kingdom is falling apart. Several smaller principalities in the north have become independent or allied themselves with emerging Lao states.

The Ayutthaya kingdom is officially founded by a lesser Thai prince known as U Thong. He is crowned King Ramathibodi I in 1350. The king is renowned as a warrior and lawgiver. As is typical among southeast Asian kingdoms of the era, it is founded on the belief in *dharma* (the Hindu and Buddhist doctrine of unchangeable natural law). However, the Ayutthayan kings also issue formalized codes of civil and criminal law. The kingdom rules over the lush Menem (Chao Phraya) Basin for over four centuries. The capital city, Ayutthaya, is strategically situated at the junction of three rivers (Chao Phraya, Pasak, and Lopburi). Surrounding rice fields flood during the rainy season, protecting it from attack.

During the reign of King Ramathibodi I (r. 1350–60) and immediately thereafter, the Ayutthaya kingdom's territory expands to the north and east, eventually overtaking the Sukhothai kingdom and the Khmer capital of Angkor. King Ramathibodi I promulgates the first-known Thai laws.

By the fifteenth century, the Ayutthaya kingdom is a centralized dominion with firm control over its territory.

Ayutthaya's proximity to the sea makes it possible to send and receive merchant vessels. The port later becomes one of the most active commercial centers in southeast Asia. Local products, along with goods from the Far East, the Malay/Indonesian archipelago, India, and Persia are traded at the port of Ayutthaya.

Ayutthaya produces a wide range of forestry products, including sapanwood (used to make a red dye), eaglewood and benzoin (used as an aromatic and incense), and gumlac (used as wax). Deerhides, elephant teeth, and rhinoceros horns are also exported.

1378 Ayutthaya kingdom conquers Sukhothai kingdom

King Borommaracha I of the Ayutthaya kingdom first takes over the Sukhothai city of Chakangrao (Kamphaengphet). The Sukhothai kingdom loses its independence and becomes a province of the Ayutthaya kingdom, which includes territory on the central plains of present-day Thailand. The Ayutthaya kingdom also assimilates other smaller states such as Suphannaphum and Lawo (Lopburi).

1400s Trailok institutes lasting governmental reforms

The smaller scale of government utilized during the Sukhothai kingdom is not practical under the larger Ayutthaya kingdom. King Trailok (Borommatrailokanat, 1448-88) creates an intricate administrative system that parallels the hierarchy of the society. This administrative system later becomes the integral part of modern Thailand's bureaucracy.

Thai society develops a class-based structure. The king is at the top of society, followed by princes (chao). Next come nobles *(khunnang)*, who serve as administrative officials. Free commoners (*phraj*) and slaves form the majority and are at the bottom of society. The fraternity of Buddhist monks (*sangha*) is the only classless social structure, because a man from any rank can become a monk.

During the consolidation of Ayutthayan power, kings assume more powers. The Buddhist kings rule as *devaraja* (demigods), whose powers are bestowed by the Hindu gods Indra and Vishnu. European merchants traveling to the kingdom during the seventeenth century write about the extreme reverence shown to the Ayutthayan monarch.

In order to profit from trade with China, the Ayutthayans willingly accept Chinese dominion and pay tribute to the Ming and Manchu rulers between the fourteenth and eighteenth centuries. In return, Chinese and Muslim merchants from India and Persia can sell their goods to the Thais. Eventually, small numbers of Chinese, Indians, Japanese, and Persians settle in Ayutthaya and are permitted complete religious freedom.

1511 Portuguese merchants arrive

Led by Afonso de Albuquerque, Portuguese arrive from Malacca, a Muslim commercial state on the west coast of the Malay peninsula that they had seized. The Portuguese are the first traders from Europe, and establish a treaty in 1516. The Portuguese are permitted to settle near the ports in exchange for guns and ammunition.

1569 Burmese capture Ayutthayan kingdom

During the heyday of the Ayutthayan kingdom, Burmese power rises and falls according to its military strength and the executive ability of its leaders, occasionally threatening the Ayutthayan kingdom. King Bayinnaung of Burma, seeking to expand his kingdom, captures Ayutthaya and controls it for a decade, installing Dhammaraja (a sympathetic Thai governor) as vassal king between 1569 and 1590. During traditional wars, combatants fight each other from the backs of elephants. Thai Prince Naresuan (who later serves as king between 1590 and 1605) arises to fight the Burmese conquerors, finally defeating the Burmese at Nong Sarai, when he kills the Burmese crown prince in battle.

Feudalism and Contact with the West

1600s The rise of the feudal state

Thailand becomes a feudal state with a powerful court of nobles. Trading contacts had been established during the previous century with the Dutch and Portuguese.

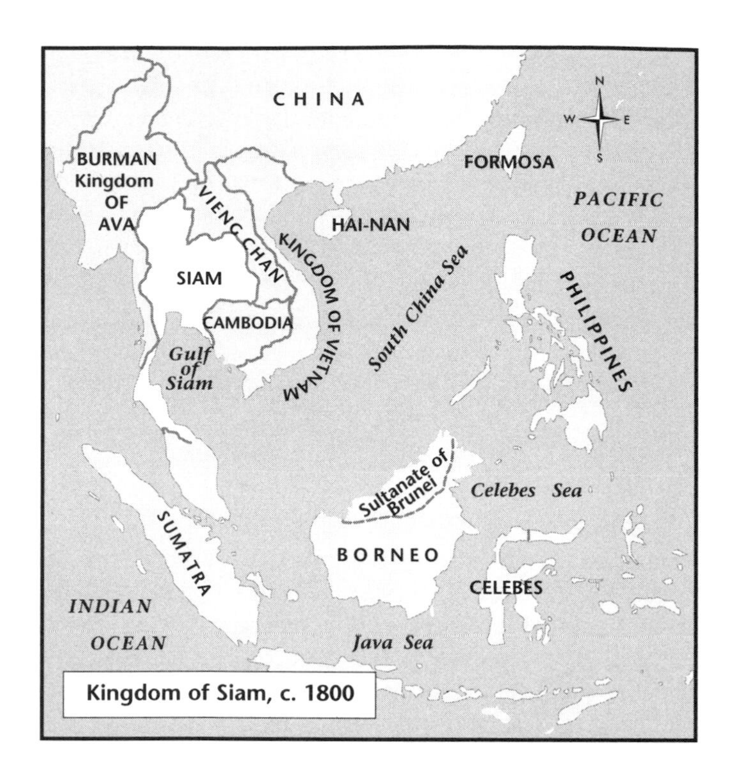

Kingdom of Siam, c. 1800

1605 Dutch East India Company begins trade deal with Ayutthayan kingdom

The Dutch East India Company (Dutch V.O.C.) enters into an exclusive trade deal with the Ayutthayan kingdom to buy its deerhide exports and tin from the Nakhon Si Thammarat (Ligor) mine. The Dutch V.O.C. plays a key role in developing Ayutthayan foreign trade until 1765.

1662 French merchants arrive

During the reign of King Narai (1632–88), who rules from 1656 until 1688, French merchants and missionaries arrive at Ayutthaya. European merchants consider the Ayutthayan kingdom to be the third wealthiest state in Asia (following China and India), calling it the "Venice of the East."

1687 French send troops to Ayutthaya

The French have been trying to coax King Narai (d. 1688) into accepting Christianity in order to form an alliance with King Louis XIV. France sends Jesuit missionaries and troops to Bangkok (which means "village of wild plums") and Mergui, which results in an anti-French backlash that coincides with a succession crisis within the kingdom. The French garrisons and priests are expelled, and trade with China increases.

1767: April Burmese subdue Ayutthayan kingdom

Although the Ayutthayan kingdom reaches its most opulent period during the reign of Borommakot (1733–58), the Burman leader Alauungpaya is recognized as king of Ava and begins attacking Siam in 1760.

During Burma's Alaunghpaya dynasty, expansionism into the Ayutthayan kingdom resumes. After fifteen months of siege, the Burmese military defeats the prosperous Thai kingdom. The Burmese ransack and loot the capital, ending the era of the Ayutthayan kingdom. Thousands of Thais are captured or killed, making it difficult to organize a counterattack.

The Kingdom of Siam

1767–1932 The Kingdom of Siam

Phya Taksin (r. 1768–82) a rebel general and former governor of Tak, consolidates enough power to drive out the Burmese invaders. Taksin sets himself up as the de facto Thai king, with his capital at Thonburi, across the river from Bangkok. He rebuilds war-defeated Siam. Trade with China resumes, thus reinvigorating the economy. The Thai nation rallies around the new king to reestablish its independence. King Taksin (the Great) wants to renew Thai religious activity, culture, literature, and the arts.

1782: April 6 Chao Phraya Chakri is appointed king

King Taksin's general, Chao Phraya Chakri, declares that the king is mad and unfit to rule, based on various episodes of religious fervor and random acts of malice. According to tradition, the deposed King Taksin is cudgeled to death in a velvet sack. Another tradition says that a substitute is pummeled to death in the sack, and that the real king lives until 1825 sequestered in a remote castle.

After the internal political struggle, Chao Phraya Chakri is made king, taking the name Phraphutthayotfa (Phra Phuttha Yofta) Chulalok, or Rama I, the founding king of the Chakri dynasty (the day of his accession is still celebrated in modern Thailand as Chakri Day). King Rama I is revered as a general, statesman, lawgiver, poet, and devout Buddhist. The king establishes the capital at Bangkok. Descended from an old Ayutthayan noble family, he models the capital after the Ayutthayan capital. On the artificial island of Ratanakosin along a bend of the Chao Phraya River, he oversees the construction of the Grand Palace and its royal temple, the Temple of the Emerald Buddha. The old Wat Photharam monastery is rebuilt in classical Thai style and renamed Wat Phra Chetuphon.

The practice of monastic Buddhism grows during the rule of King Rama I. It later becomes Thai tradition for every male to become a monk at some point in his life for a duration of at least a few months.

The reign of King Rama I marks a significant shift in the style of the monarch's rule, emphasizing a rational approach to reform. The forces of King Rama I rout the Burmese in 1785 and 1786, during their attempt to conquer Siam. The king's counterattack successfully takes Tavoy in Lower

Burma, thus adding Chang Mai to the Thai kingdom. Several Malay states begin paying tribute to King Rama I.

King Rama I continues the royal role of influencing Thai literature through his own writings. Later monarchs such as Rama II and Rama VI will also influence Thai literature through their personal writings.

1786 Birth of Sunthorn Phu, renowned poet

Sunthorn Phu (1786–1855) is one of Siam's most revered poets. King Rama II and King Rama III are both fond of his poems, which often have mythological characters and a moral. His most famous epic poem is "Phra Aphaimani," which tells the story of a prince held captive in an underwater realm who is aided by a mermaid.

1788–89 King's appointed council codifies Buddhist scriptures

A council of scholars, appointed by King Rama I, codifies and reedits the *Tripitaka*, Buddhist scriptures. A similar panel of experts revises the legal code, writing the *Kotmai tra samduang* (Three Seals Code).

1809 King Phra Phuttha Loetla, Rama II, ascends to the throne

King Rama I's son, Phra Phutthaloetlanaphalai (Phra Phuttha Loetla), ascends the throne as King Rama II. Rama II is revered for his poetry and artistic talent. The king's hand-carved door panels are regarded as masterpieces of Thai woodcarving.

1824–51 King Phra Nangklao, Rama III, reigns

Upon the death of King Rama II in 1824, Prince Chetsadabodin and Prince Mongkut (1804–68) vie for succession, with Chetsadabodin having greater experience in government and taking the regnant name Phra Nangklao (King Rama III). Prince Mongkut avoids creating a confrontation by remaining a monk throughout his brother's rule.

During his reign, Britain is at war with neighboring Burma. Fearing that the war will spill over into Siam, he orders the construction of a massive chain across the Chao Phraya River to prevent a naval incursion.

1827 Laotian leader seizes Nakhon Rachasima

The Laotian leader Chao Anu, who rules the city of Vientiane, attacks Nakhon Rachasima and Saraburi using the excuse of wanting to save Bangkok from the British. Siamese forces counterattack and totally destroy Vientiane. As a result, during the 1830s and 1840s the Siamese resettle many Lao on the west bank of the Mekong River and in the Khorat Plateau.

1830s Prince Mongkut establishes the Thammayutika order

Prince Mongkut travels throughout the kingdom as a priest and a prince to observe the living conditions of typical Thais. He gains valuable experience among the people and creates an order of monks known as the Thammayutika order (*dhammayutika nikaya*), an order closely allied with the royal family. Thammayutika is stricter than Mahanaikai, the other order of monastic Theravada Buddhism practiced in Siam.

Modernization and Opening to the West

1851–68 King Mongkut, Rama IV, reigns

Prince Mongkut becomes King Rama IV, King of Siam. As prince, Mongkut is eager to learn about foreign cultures, and his intellectual interests include learning Latin, French, English, the Buddhist scriptures, astronomy, mathematics, science, and geography. This knowledge proves useful when dealing with Western powers such as Britain and France. The king employs Western experts and consultants in his court and administration.

During the reigns of Mongkut (and his son Chulalongkorn), Siam emerges from feudalism and enters the modern world. As Rama IV, Mongkut focuses on the administrative and capital improvements of his kingdom. He authorizes the construction of roads and canals, improvements in the shipbuilding industry, reorganization of the military, and the minting of money. The aristocracy and upper classes begin adopting European customs and dress.

1855–56 Siam signs commercial treaties with Britain, France, and U.S.

The Bowring Treaty (named after Sir John Bowring) abolishes the government's monopoly in all industries except for opium, alcohol, and the lottery. Siam's rice cultivation increases as international trade opportunities open up. Duties on imports and exports are decreased, and British citizens residing in Siam are given special privileges. The treaty serves as an accommodation the to British, who are already present in Burma and the Malay peninsula. In 1856 Siam signs commercial treaties of friendship with France and the United States.

1868–1910 King Chulalongkorn, Rama V, reigns

King Rama V (Chulalongkorn, 1858–1910) continues his father's reforms, ascending to the throne when he is only sixteen years old. He is eager to continue learning about Western technology while simultaneously avoiding Western domination. He is the first Thai monarch to travel abroad, visiting Dutch and British colonies in Java, Malaya, Burma, and India. He also travels to Europe twice near the end of his reign. During one of these trips Tsar Nicholas II of Russia and Kaiser Wilhelm II of Germany agree to support Siam's efforts to resist Britain and France, who maintain a strong colonial presence in the area.

King Chulalongkorn often travels about his kingdom incognito in order to assess how his subjects live. He forbids his subjects to prostrate themselves in his presence and also does away with other royal taboos. He eventually does away with *corvée* labor (unpaid peasant labor done for a feudal lord) as well as the institution of slavery. The king becomes known as *Phra Piyamaharaj* (beloved great king) because of his close contact with the people of the country.

The kingdom's infrastructure is improved, with the construction of a telegraph system and a railway. The king also reorganizes the government, modeling his ministries after those of Western countries.

Kite flying, dating from the thirteenth century in Siam, reaches its peak popularity during the reign of King Chulalongkorn, who draws up the rules still used in kite-flying competitions.

1880s Secular schools established

Secular schools are established to create an educated elite capable of managing the administration of the government. Many students seek an education abroad not only to learn skills needed by the kingdom but also to establish contacts in Western countries.

1893 Siam cedes territory to France

To placate Western colonial powers, Siam agrees to give up some of its frontier territory in order to preserve its heartland. Lands on the east bank of the Mekong River are ceded to France. In 1904, all Thai territories on the west bank of the Mekong are given to France.

1904 Siam Society founded

Founded as one of Siam's first cultural organizations, the Siam Society issues studies on Thai art, literature, and science.

1905 National Library of Siam (now Thailand) founded

The National Library of Siam is founded. By the late twentieth century, it holds over 1.2 million books and over 180,000 manuscripts in its collection.

1907–09 Siamese government cedes external provinces

In order to maintain its independence, Siam must settle problems regarding land tax and trade duties from previous treaties signed with European powers. The Siamese government relinquishes its external territories. In 1907, the Thai government cedes its Khmer provinces (Siem Reap, Battambang, and Sisophon) to France. In 1909, the Malay states of Kedah, Perlis, Kelantan, and Trengganu are given away and become British protectorates. In total, Siam forfeits 176,000 square miles (456,000 square kilometers). Its size is only half as large as it was in 1851. However, Siam is only one of three Asian countries (Japan and China are the others) that are not dominated by a European power.

1910: October 23 King Chulalongkorn, Rama V, dies

King Chulalongkorn's forty-two-year drive towards rational modernization earns him a revered spot in Thai society. Upon his death, there is an extended mourning period and the date of his death becomes a national holiday.

1910–25 King Vajiravudh, Rama VI, reigns

King Chulalongkorn's son and successor is Vajiravudh, who becomes King Rama VI. He is the first Thai monarch to have received an education abroad (at Harrow School and Oxford University in England). He is renowned as a poet and dramatist, both in Thai and English. He also values Western technology but wants to maintain traditional Thai values. His policies advocate improving women's social status and monogamy. However, his poor management of the country's finances reduces the treasury to near bankruptcy (his lavish coronation ceremony in 1910 alone consumed eight percent of the country's annual budget). His foreign policy concentrates on undoing the inequitable treaties of the past. He also wants to rid the country of foreign economic domination, especially that of the Chinese merchants.

Interest in the military causes him to organize the Wild Tiger Corps *(Sua Pa)*, a nationalist paramilitary group formed out of the civil service and the king's favorite officers, and the Tiger Cubs, a separate junior form of the Boy Scout movement.

1912: February Disgruntled officers conspire to depose the king

A group of junior officers, stationed in the rugged northern part of Siam, plot to remove the king from power. The scheme is discovered before the plan is implemented and the conspirators are all sent to prison.

1913: March 22 Surnames legally mandated

King Vajiravudh makes the Western-style usage of surnames mandatory for his subjects. He gives all heads of families six months to choose and register a surname from a list of names he himself devises.

1917 Siamese government adopts a national flag

Siam adopts its national flag. The flag consists of five horizontal stripes. The outer two are red (symbolizing the Thai people), those adjacent are white (representing Buddhism), and there is a double-wide blue stripe in the center (symbolizing the monarchy).

1917 Chulalongkorn University founded

Bangkok's Chulalongkorn University opens, becoming Siam's first and most eminent university. Other universities in Bangkok later include the University of Thammasart (1933) and Kasesart University (1943).

1918: July 22 Siam officially enters World War I

Initially, Siam remains officially neutral during World War I but favors a German victory because of Siam's earlier loss of land to French and British colonies. When it becomes apparent that Germany is going to lose the war, King Rama VI supports the Allies and in September 1917 arranges to send 1,300 troops, a flying squadron, and an ambulance corps to join the Allies in France. The troops arrive at the front too late in the war to engage in any combat. After the war, his ceremonial military gesture is rewarded by international recognition. During the 1920s, Siam renegotiates its treaties with the Western powers and regains its power to collect trade duties.

1919–23 Siamese economy deteriorates

As the price of silver on the world market rises, Siam's silver coins become worth more than their face value. As a result, people begin to melt down the coins for export, even though it is illegal. In response, the government increases the currency's value. A simultaneous increase in the price of rice exacerbates the currency shortage, and Siam's own rice crop fails in 1920–21. The king prohibits rice exports. By 1923, Siam devalues its currency after its export trade has fallen by fifty percent.

1920s Chinese immigrants arrive in Siam

Some 70,000–140,000 Chinese immigrants arrive in Siam each year during the early 1920s. Strict immigration laws are imposed in 1927 to all but stop the legal flow of Chinese into the country.

1921 Introduction of compulsory education

The king issues a law to make primary education universal. Provisions call for compulsory school attendance starting at age seven through the fourth year of primary school, or through age fifteen.

1925–35 King Prajadhipok, Rama VII, reigns

After the death of King Vajiravudh in 1925, his younger brother Prince Prajadhipok becomes King Rama VII. During his reign, the global economic crisis of the late 1920s leads to domestic disputes concerning the pace of political reforms. Although he favors the creation of a constitutional form of government, he fears that politics will be dominated by wealthy Chinese. Ethnic Chinese account for over twelve percent of the country's population in the early 1930s. Other members of the royal family and the Supreme State Council do not support creating a parliamentary form of government.

1926 National Museum of Siam (now Thailand) founded

The National Museum in Bangkok features an extensive collection of Thai artifacts, including sculptures, textiles, ceramics, jewels, coins, weapons, and masks.

1928–29 First modern Siamese novels appear

As Siam has become more prosperous, a new group of writers and journalists arises. As the community of intellectuals in Siam increases in numbers, so does its criticism of the government. Novelists like Kulap Saipradit, Prince Akatdamkoeng, and M.L. Buppha Kunjura Nimmanhemin (a.k.a. Dokmai Sot). These early writers frequently use themes of modernization and the clash between Western and Siamese cultures. In 1940, Prince Akatdamkoeng's novel *Yellow Race, White Race* gains international notoriety. Kulap Saipradit also becomes famous in the 1940s and 1950s for his social protest fiction.

1932: April 6 Chakri dynasty marks 150th anniversary

Although recent kings in the Chakri dynasty have relaxed some formalities, the monarch still has absolute authority over his kingdom. Objections against the monarchy are on the rise, as public confidence in the current king falters. There are strict laws, however, that call for punishment of anyone who criticizes the king.

This time, too, many people are also remembering the prophecy made by Princess Narinthewi, sister of the first Chakri monarch, Rama I. She predicted the Chakri family would reign in Siam for only 150 years.

Constitutional Monarchy and the Rise of the Military

1932: June 24 Bloodless rebellion ends absolute monarchy

Siam's foreign reserves are dwindling, a drop in rice prices reduces farmers' incomes, and the king's indecisiveness over financial matters worsens the economy.

A group of French-educated Thai intellectuals known as the Promoters stage a coup d'état demanding an end to absolute monarchy. The king agrees to the change in power and pardons those responsible for the coup. The government, now controlled by the Khana Ratsadorn (People's Party), becomes a constitutional monarchy.

The military's bid for power has been growing since 1912, when a failed coup was launched against King Vajiravudh. After the revolution strips the king of his powers, Prajadhipok abdicates in 1935 and the monarchy disintegrates.

From 1932 until the 1940s, political power in Siam centers around Dr. Pridi Banomyong and Marshal Phibul Songgram.

1933 Siam fails to condemn Japan's invasion of Manchuria

Siam's pro-Japanese position is clarified when Japan invades Manchuria and Siam is the only League of Nations member to abstain from formally condemning the action. Siam later sends groups of army officers to Japan for training. Trade with Japan is on the rise—imports increase from three percent in 1928 to over twenty-five percent by 1938.

1933: April Banomyong advocates socialism

Dr. Pridi Banomyong proposes a radical plan to move the country toward socialism by nationalizing all land and capital and by making most workers government employees. The scheme, however, is rejected as too extreme by the People's Party and the nation's intellectual elites. The National Assembly is dissolved by royal decree and Banomyong goes into exile but returns in 1934 as Interior Minister.

1933: October 11 Counter-coup challenges new leaders

The Bowradet Rebellion, a royalist counter-coup, begins as rural soldiers march on Bangkok demanding that the government set up a true democracy. Prince Bowradet, a grandson of King Chulalongkorn, leads the coup, charging the new government with insulting the king and encouraging communism. Within two weeks the rebellion is suppressed. Although King Prajadhipok does not participate in the rebellion, the government increasingly distrusts him.

1933–38 Compulsory primary education expands

As part of the new government's plan, the People's Party promises to let a democratically-elected assembly take power when the country's level of education is high enough to allow an informed citizenry to conduct elections. Between 1933 and 1938, expenditures on education increase fourfold. Literacy rates for school-age children increase to over sixty percent.

1934 Phraya Phahol becomes prime minister

General Phraya Phahol Pholphayuhasena (1887–1947) serves as prime minister, trying to keep a balance of power in the government between the military and the civilian administration. He is the last of the senior officers that participated in the 1932 coup to still retain power. His policies advocate neutrality, rather than any inclination towards fascism or socialism. Upon his resignation in 1938, Luang Plaek Pibulsonggram (Phibunsongkhram, Phibul Songgram, or Phibun) becomes prime minister.

1935: March 2 King Rama VII abdicates

King Rama VII (Prajadhipok) abdicates while in England. His absolute power gone since 1932, the abdication comes as a protest against the increasing authoritarianism of the government and the steady erosion of constitutionally-protected royal privileges in the government. He goes into exile in England, where he dies in 1941, having been convicted of misappropriating six million baht. The ten-year-old Ananda Mahidol (b. Germany, 1925–46), a grandson of King Chulalongkorn, is crowned King Rama VIII and later goes off to study in Switzerland.

1938: December 16 Phibun becomes prime minister

Phibun becomes prime minister and rules during World War II. An admirer of Hitler and Mussolini, his authoritarian fascist policies likewise promote ultra-nationalism. He orders forty of his political opponents to be arrested, and has eighteen of them executed. He also initiates laws that discriminate against ethnic Chinese.

1939 Country's name is changed to Thailand

The nationalist regime of Prime Minister Phibun changes the name of the country from Siam to Muang Thai (land of the free), or Thailand. The name change endorses the preference for ethnic Thais over the economically dominant Chinese minority in the country. The name change is part of a package of government mandates designed to promote nationalism. By 1941, there are laws directing people how to dress, behave in public, and admire Thai culture. The government promotes nationalistic plays, music, and dance.

World War II

1940: November 28 Thailand invades Laos and Cambodia

Thailand wants to get back some of the territory lost to Indochina early in the century. Although the Allied governments do not favor granting land back to Thailand, Germany and Italy approve. Japan views Thailand's desire to expand its territory as a way to gain a dependent ally. After France falls to Germany, Phibun orders the invasion of Laos and Cambodia, which had been lost to France in 1893 and 1907. There had been reports of French aircraft flying across the Thai border for months, and after two bombs are dropped on the northeastern town of Nakhon Phanom, Thai forces launch a counterattack. A full-scale invasion occurs in January 1941. Thailand brokers an agreement with Japan that results in their gaining control of Cambodian territory, including the city of Angkor.

1941: December 21 Thailand collaborates with Japan

Thailand's dependence on its military for governance and development continues. On December 8, Japan sends troops to Thailand, which offers little resistance. Thailand signs an alliance with Japan. During the war, Thailand is the only state in Southeast Asia that Japan allows to remain independent.

1942: January 25 Thailand declares war on United States

Thailand declares war on the United States and the United Kingdom. This declaration, however, is never delivered or accepted in the United States. The Thai minister in Washington, Semi Primoj (the king's cousin), helps organize a Thai resistance movement known as *Seri Tha*i (Free Thai). The Office of Strategic Services, an American intelligence agency, helps coordinate the Free Thai movement in the United States and abroad.

1942 Bank of Thailand established

The Bank of Thailand is established and soon becomes a leading bank in the region.

1943–44 Japan begins construction of Burma-Siam Railroad

In 1942, Japan authorizes the construction of the Burma-Siam Railroad through south-central Thailand in order to circumvent the naval blockade of its cargo ships. During the construction, 60,000 Allied prisoners of war and 300,000 Asians work as slave laborers on the project under the brutal discipline of the Japanese military. About 60,000 die from abuse, cholera, malaria, and malnutrition. The project includes the construction of a railroad bridge over the Khwae Yai River at Kanchanaburi that is later chronicled in Pierre Boulle's novel *The Bridge on the River Kwai* and the famous 1957 film of the same name.

1944 Phibun resigns

Khuang Apaivongse (d. 1968) becomes prime minister upon Phibun's resignation. The government resigns because it is allied to Japan, which is losing the war.

1945: September 2 Japan surrenders

After Japan surrenders, British Indian troops occupy Thailand. Seni Pramoj (b. 1905) becomes interim prime minister in order to negotiate with the British. Thailand immediately tries to return British colonial lands granted by Japan during the war and cancels all other agreements made with Japan. Britain's demands will practically reduce Thailand to colonial status: reparations, preferential economic deals, the privilege to station British troops in Thailand indefinitely, and 1.5 million tons of free rice. By December, Thailand agrees to adhere to the strict terms of the peace treaty. However, after news of the treaty reaches the American public, Britain cancels the signing.

1946: March 24 Pridi Banomyong becomes prime minister

Seni Pramoj resigns as prime minister in January, and is later elected to parliament after forming the Prachatipat (Democratic) party with Khuang Apaivongse. Although Dr. Pridi Banomyong becomes prime minister, public sentiment against him increases after the unexpected death of King Rama VIII. He resigns in August and soon flees into exile to Singapore and then China.

1946: June 9 King Ananda Mahidol, Rama VIII, killed

King Rama VIII, having returned from Switzerland only six months earlier, is found slain in the royal palace with a gunshot wound to the head. Three royal servants are swiftly tried and executed, but the killing is never fully explained or solved. The public speculates that Pridi, well-known for his leftist anti-royalist beliefs, has some sort of role in the regicide.

1946: December 16 Thailand joins United Nations

After the war, Thailand becomes an ally of the United States through their common membership in the Southeast Asian Treaty Organization (SEATO).

1947 Immigration quotas imposed

Due to the large influx of refugees, the government limits immigration to one hundred persons per year from any country. During the Franco-Indochinese war some 45,000 Vietnamese refugees settle in Thailand. Thailand's 1947 census indicates that of the nearly eighteen million people living in the country, 2.7 percent are Chinese citizens.

Postwar Military Rule

1947: November 7–8 Field Marshal Phibun again assumes power

A military coup brings Phibun back to power. The Prachatipat Party organizes a government with Khuang Apaivongse as prime minister and Seni Pramoj as foreign minister.

1948 Thai Silk Company founded

After World War II, Jim Thompson, an American intelligence officer living in Thailand, starts the Thai Silk Company to rejuvenate the languishing traditional silk-weaving industry. Thompson becomes a celebrity in Thailand for his efforts in the silk industry until he mysteriously disappears while on vacation in Malaysia in 1967.

1948: April 8 Phibun becomes prime minister

Phibun is installed as prime minister, despite the democratic elections. Phibun soon begins consolidating his power by intimidating, harassing, and arresting his political rivals. Several are killed under the pretense of escaping from police.

1949: February 26–27 Pridi's attempted coup fails

Pridi Banomyong returns and, with the backing of the navy and marines, tries to wrest power from Phibun. After three days of violence in Bangkok, the attempted coup fails and Pridi Banomyong goes back into exile to France. Phibun intensifies his authoritarian rule by increasing the arrests of rivals and dissidents. Police conduct arbitrary arrests and even torture and kill prisoners.

1950: May 5 King Bhumibol Adulyadej, Rama IX, is crowned

Bhumibol (pronounced POOM-ee-pon) Adulyadej (b. United States, 1927), who had succeeded his brother in 1946, is formally invested as King Rama IX.

1950: June Thailand sends troops to Korea

Supporting the United Nations intervention force in Korea, Thailand sends a ground force of some 2,000 soldiers, along with a smaller naval and air force contingent. Thailand is the first country in Asia to send troops to fight with the United Nations in Korea.

The war in Korea helps Thailand's recovering economy. Exports of rice, rubber, and tin increase. Thailand also begins receiving economic and military aid from the United States. Between 1951 and 1957, the United States gives Thailand $149 million in economic assistance and $222 million in military aid.

1951: June 29 Attempted coup against Phibun fails

Phibun is taken prisoner by a group of naval officers while aboard the *Manhattan*, a dredge donated to Thailand by the United States. He is taken into custody on the flagship *Sri Ayudhya* during negotiations with his administration. Violence erupts in Bangkok between forces loyal to Phibun's government (police, army, and air force) and the navy and marines. The air force bombs the *Sri Ayudhya*, and Phibun swims to safety as it sinks. The attempted coup results in about 1,200 deaths, mostly of civilians.

1951: November 29 Phibun's government rescinds 1949 constitution

The military government decrees that the older 1932 constitution replaces the 1949 constitution. The older constitution is more restrictive and gives power to the monarchy. The decision comes while the new king is in Singapore. An appointed assembly replaces the elected legislature.

1952: July 28 Crown Prince is born

Crown Prince Wachiralongkon, heir to the throne, is born.

1955 Military government begins relaxing its control

The authoritarian government of Phibun starts easing some of the restrictions it had imposed earlier. Political parties are permitted to register, and local elections are allowed.

1955 Television station established

The first mainland Asian television station is established in Bangkok.

1957: February Miltary government conducts national elections

Phibun's Seri Manangkhalisa party wins over half the legislative seats in an election full of fraud, vote rigging, manipulation, and coercion. The Thai people are outraged by the election, which forces Phibun to declare a state of emergency. Phibun puts Sarit Thanarat in charge of maintaining public order, but Sarit undermines Phibun's authority by sympathizing with protesters. By August, Sarit and some other members of Phibun's cabinet resign, and appointed legislators resign from Phibun's political party.

1957 Chit Phumisak writes Marxist analysis of Thai history

Chit Phumisak (Chitr Phoumisak, 1930–65) writes *Chomnaa Sakdinaa Thai* (The Face of Thai Feudalism), a history of the Thai from a Marxist perspective. Chit is already renown in Thailand as a literary critic and essayist. Chit is later killed by Thai police.

Marxism becomes more popular in Thailand during the 1950s and 1960s, as anticolonialism swells throughout Southeast Asia and communism is viewed with high esteem by many. Many young intellectuals, disenchanted with current political conditions, are enamored of Marxist ideals.

1957: September 17 Field Marshal Sarit Thanarat made premier

Four days after demanding that Phibun step down from office, Sarit Thanarat (1900–63) replaces Phibun as Thailand's premier after seizing power. He focuses on cleanliness, orderliness, economic development, and national security through improving Thailand's infrastructure. As Sarit consolidates his power, his policies include street repair and maintenance and removing hoodlums from the streets. Prostitution is curbed and opium use is made illegal. Arsonists are subject to immediate public execution.

Sarit also increases the role of the monarchy in Thai society. Royal ceremonies, mostly ignored since 1932, are revived. Sarit wants the Thai people to revere their king as the model of Thai morality, society, and public order.

Late 1950s–1990s Resistance to anti-malarial drugs appears

The parasite that causes malaria first shows resistance to anti-malarial drugs in northern Thailand in the late 1950s. Resistance to the traditional anti-malarial drug, quinine, appears first. The synthetic form of quinine, chloroquine, and the next-generation drug, Fansidar, eventually lose most of their effectiveness in Thailand as well. By the 1990s, one of the last drugs in the line, mefloquine, is ineffective in fifty percent of the infected people along the Thai-Burmese border.

People who carry the resistant parasites in their blood transmit the resistance when they are rebitten. The first drug resistances are noticed among gem miners who travel from Cambodia across Thailand to Burma. The drug-resistant parasites are later transmitted through Bangladesh to India and finally, after thirty years, to Africa. The parasite develops resistance to all the anti-malarial drugs that have been developed, with the possible exception of a Chinese herbal drug.

Malaria is one of the most deadly diseases in the world, closely matched by tuberculosis. Some three hundred million people are infected with malaria every year, and more than two million die from the disease.

1959 Asian Institute of Technology is founded

The Asian Institute of Technology in Bangkok offers advanced degrees in agricultural engineering, human settlements, and computer applications.

1959 Opium production banned

Opium poppies have been cultivated in northern Thailand since their introduction from southern China in the late nineteenth century. The poppies are the main cash crop in northern Thailand because they thrive in poor soil at high altitudes. Despite the ban, opium production increases during the Second Indochina War (Vietnam War).

1960s New universities open in provincial areas

Chiang Mai University (1964), Khon Kaen University (1966), and Prince of Songkhla University (1968) open in provincial areas, making higher education available outside the Bangkok area.

1960–75 Thailand during the Second Indochina War

As a SEATO member, Thailand takes a direct role in the Vietnam war by supplying a small number of troops in support of the Republic of Vietnam (South Vietnam). Although Thailand is allied with the United States, it still pursues its own foreign policy. As the war in Vietnam resumes (from the First Indochina War, 1940–41) in the early 1960s, Laos and Cambodia are soon drawn into the conflict. As the situation in Indochina deteriorates, Thailand increasingly turns to the United States for protective assistance.

1960 Government endorses wildlife protection

The Wild Animals Preservation and Protection Act partially protects Thailand's natural fauna, but species have been depleted through poaching and trapping. Thailand's animals include bears, otters, civet cats, several species of monkeys, sheep, goats, oxen, rhinoceroses, deer, tapirs, wild cattle, wild hogs, snakes, crocodiles, lizards, and turtles. About 1,000 varieties of birds are indigenous. Elephants are used as draft animals in rural areas. Although logging is in decline, elephants are still used to move logs.

1963 Kittikachorn is appointed prime minister

Upon the sudden death of Field Marshal Sarit Thanarat, Field Marshal Thanom Kittikachorn is appointed prime minister. General Praphas Charusathien also becomes a prominent leader.

1965: January Thailand Patriotic Front forms

China announces the formation of the Thailand Patriotic Front, whose purpose is to "strive for the national independence" of Thailand. A limited insurgency subsequently develops in the north and northeast, growing in intensity in the 1960s and 1970s as the conflict in southeast Asia rages on Thailand's northern and northeastern borders.

1965 Thailand allows United States to use its air bases for bombing missions

Thailand gives the United States permission to station bombers at air bases in Thailand in order to conduct massive bombing sorties against the Democratic Republic of Vietnam (North Vietnam). By 1968, there are some 45,000 U.S. military personnel with about 600 aircraft stationed in Thailand. Thailand also receives U.S. aid to reduce the threat of insurgency within Thailand.

Between 1958 and 1967, the United States gives Thailand $358 million in economic aid and another $438 million in military assistance. After a power struggle in neighboring Laos in 1961, the United States had secretly agreed to defend Thailand in case it was invaded.

Thailand's economy becomes reliant on the presence of U.S. armed forces personnel, which has a profound effect on Thai culture and society. Urbanization intensifies, as young Thais go to the cities for service sector jobs. Many start working in restaurants, bars, hotels, brothels, and shops. The sex industry expands because of the large numbers of U.S. servicemen stationed in the area. Both residential and commercial construction grow rapidly to accommodate the economic surge. Thai society is directly exposed to Western culture for the first time.

1967 ASEAN established

The Association of Southeast Asian Nations (ASEAN) is established as a cooperative body, due primarily to the motivation of the Thai government.

1969 Royal Thai Army troops in South Vietnam number 11,000

The Royal Thai Army, originally agreeing to send 2,200 troops to South Vietnam in 1967, has 11,000 soldiers there by 1969 (about fourteen percent of its total strength).

1969: February Victory for Thanom's government party

Elections for the lower house produce a victory for Thanom's party.

1971: November Marshal Thanom declares martial law

Marshal Thanom Kittikachorn (b. 1911), who had been reconfirmed as prime minister in the 1969 general elections, leads a bloodless military coup that abrogates the constitution and imposes a state of martial law.

1972: December Interim constitution promulgated

The military government announces a new constitution in order to preserve its rule.

1973 Collection of popular Thai short stories published abroad

Khamsing Srinawk (b. 1930), Thailand's revered short story writer, has a collection of his short stories *The Politician and Other Stories* published by Oxford University Press. The collection introduces contemporary Thai literature to a foreign audience.

1973 National Housing Authority established

In order to house Bangkok residents living in shanties, the government forms the National Housing Authority to coordinate public and private housing programs.

1973: October 14 Demonstrations lead to rioting

In early October, demonstrations by student and labor groups, who have been protesting for a representative form of government, erupt into riots. Between 200,000 and 500,000 demonstrators are demanding the release of political prisoners and a new constitution. On October 14, Marshal Thanom resigns from office in favor of a professor, Sanya Dharmasakti. Many educated elites feel intimidated by the radical student demonstration, which they suspect has an ulterior communist motive. King Bhumibol Adulyadej steps into the vacuum and names a national legislative assembly to draft a new constitution and exiles Marshal Thanom Kittikachorn and General Praphas Charusathien. The 1973 revolt marks the end of Thailand's string of authoritarian personality-driven governments since the end of the absolute monarchy in 1932. Between 1973 and 1976, several short-lived governments are styled after the wishes of their leaders.

Military and Civilian Government

1974: October 7 Thailand adopts new constitution

Thailand's tenth constitution since 1932 is adopted.

1975: January 26 Parliamentary elections held

Thailand's first open parliamentary elections since 1957 are conducted. Some forty-two parties compete in the balloting, which produces a coalition government under Seni Pramoj, who had served as prime minister twenty-nine years earlier.

1975: March 14 Kukrit Pramoj becomes prime minister

A no-confidence vote forces Seni Pramoj's government to resign. A right-wing coalition government led by Kukrit Pramoj (Seni's brother, 1911–95) and the Social Action Party, then assumes control.

1975: April 30 Stock exchange opens in Bangkok

The Securities Exchange of Thailand opens as the country's first public stock exchange. All thirty of its members are Thai-owned securities firms.

1975 National Environment Board coordinates environmental protection programs

The Promotion and Enhancement of the Environmental Quality Act charges the National Environment Board with the coordination of the country's environmental programs. The nation's water supply is contaminated by industrial and farming activities, sewage, and salt water. Deforestation occurs in watershed regions, due to increased cultivation of upland areas. Land use in urban areas is regulated by the City Planning Act of 1975.

1976: March Thai government orders the removal of U.S. military forces.

In 1972, there are some 25,000 U.S. troops stationed in Thailand. With the withdrawal of U.S. troops from Vietnam in early 1973, the U.S. begins a gradual withdrawal of military personnel from Thailand as well. The Thai government orders the U.S. to close its remaining installations in March 1976, and calls for the removal of all but a few military aid personnel by July.

1976: April 20 **Seni Pramoj returns as prime minister**

Kukrit Pramoj's government resigns in January. The Prachitapat Party wins parliamentary elections (114 of 279 seats), returning Seni Pramoj as prime minister.

1976 **Queen Sirikit organizes revival of traditional textiles**

The queen founds Support, an organization dedicated to promote traditional hand-woven Thai textiles. With the help of designers and dress historians, she promotes a series of traditional silk costumes that gains popularity among Thai women.

1976: October 6 **Military overthrows elected government**

Prime Minister Seni Pramoj's government is overthrown by the military. During the mid-1970s, evidence and rumors of communist activities within Thailand cause the military to engage in assassinations of political leftists and police also routinely intimidate leftist parties. At Thammasat University in Bangkok, thousands of police force their way onto campus because two days earlier, some of the students had staged a demonstration that was pronounced *lèse-majeste*—treasonous because it defamed the monarchy (students were seen apparently burning Crown Prince Vajiralongkorn in effigy). Police terrorize students and slay some caught fleeing the campus. More than 3,000 people are arrested and 40 are bludgeoned to death, lynched, or burned alive during the turmoil. The hysteria serves as an excuse for the military to abolish constitutional government. Later that day, a mob gathers outside the Government House in Bangkok, and the prime minister personally makes an appeal to calm the crowd. The king declares a state of martial law. Labor strikes and political parties are banned.

1978: December 18 **New constitution eases martial law provisions**

Yet another constitution is put into effect, lifting the ban on political parties and easing some of the restrictions imposed by martial law in 1976.

1979 **Thai government estimates 10,000 communist insurgents reside in Thailand**

Activity of communist insurgents (rebels) in border areas, coupled with the presence of large numbers of refugees contributes to the nation's political instability. The insurgents are organized by the Communist Party of Thailand, whose active military arm is the Thai People's Liberation Army. Following the Vietnamese victory in Cambodia in January 1979, thousands of insurgents take advantage of a government offer of amnesty and surrender to Thai security forces. Others are apprehended. By the mid-1980s, there are fewer than 1,000 active communist insurgents.

1983 **Bangkok floods**

Much of Bangkok's land lies in a flood plain below sea level. An urban construction boom combined with an increase in the drilling of artesian wells causes some areas of the city to sink. During the 1983 flood, about 175 square miles (450 square kilometers) of the city lies under flood waters.

1985: September 9 **Military coup fails**

After several hours, an abortive military coup fails to take power from the elected government. It is the sixteenth coup, or attempted coup, since 1932.

1986 **Thai government forces repatriation of Laotian refugees**

During the 1970s, over four million refugees left Laos, Cambodia, and Vietnam, many of whom settled in Thailand. The government begins the forced repatriation of refugees from Laos. Cambodians living along the Thai-Cambodian border are repatriated in the early 1990s.

1987 **King receives award for encouraging conservation**

King Bhumibol Adulyedej receives a Magsaysay Award for International Understanding for his twenty years of effort to encourage crop diversity and to discourage swidden agriculture (slash-and-burn cultivation), which leads to deforestation.

1988: April 29 **King dissolves House of Representatives**

Upon the request of Prime Minister Prem, the king dissolves the House of Representatives.

1988: July **Chart Thai party wins parliamentary elections**

The Chart Thai gains the largest number of seats following the election. Although its leader, General Chatichai Choonhavan, declares his unsuitability to be prime minister, he is appointed nonetheless. He takes an active role in foreign affairs and makes bold initiatives to improve relations with Laos, Vietnam, and Cambodia.

1988: July 2 **King Bhumibol's reign becomes longest in dynasty**

King Bhumibol's reign exceeds in length that of his grandfather, King Chulalongkorn. He becomes the longest ruling Thai king of the Chakri dynasty.

1989 **Thai government imposes ban on logging government-owned timber**

Although teak and other tropical hardwoods have been forested in Thailand for centuries, overcutting in recent decades has left some areas deforested. The government bans commercial logging of teak.

1990 Social security system established

Since 1940, social welfare has been the domain of the government and only recently are private organizations involved in social welfare programs. A new law establishes a social security system that begins providing disability and death benefits in 1991.

1990: December General Chatichai resigns and is then reappointed

The House of Representatives rejected a motion of no-confidence in July. General Chatichai resigns as prime minister, only to be reappointed the next day, enabling him to form a new coalition government.

1991 Commission on Women's Affairs established

Although women have equal legal rights in most areas, inequities remain in divorce and child support areas. Also, many women are bonded laborers who are forced to work as prostitutes in brothels from which their parents borrowed money. Under Thailand's law, prostitutes are considered criminals, but brothel owners and clients are not.

1991: February 23 Bloodless military coup ousts Chatichai's government

The National Peace Keeping Council (NPKC) stages a bloodless four-hour military coup, which ousts Chatichai's government on the grounds of corruption. Martial law is declared, the constitution abrogated, and the government is dissolved. The king approves an interim constitution the following month. Anand Panyarachun, a business executive and former diplomat, is appointed prime minister.

1991: October 23 Cambodia's civil war ends, allowing refugees to return

After a thirteen-year civil war, Cambodia signs a peace treaty. Most of the Cambodian refugees in Thailand begin returning home.

1991: December 7 New constitution approved

The new constitution provides for a National Assembly comprised of elected representatives and a Senate appointed by the military, and a cabinet headed by an appointed prime minister.

1992: May 17–21 Thai government violently subdues demonstrators in Bangkok

General Suchinda had been appointed prime minister in March and public protest against his appointment has been growing. Retired Major General Chamlong has been calling for Suchinda's resignation and constitutional reform to prevent the appointment of an unelected prime minister.

A peaceful protest has been building support around the Parliament building in Bangkok for over a week. About 150,000 demonstrators meet at Sanam Luang parade ground in central Bangkok. Demonstrators break through police road blocks and torch nearby vehicles and a police station. In the early hours of the next morning, armored vehicles equipped with machine guns are brought to bear against the protesters, killing over 100. There are reports of government troops killing civilians and secretly burning the corpses in order to keep the official death toll artificially low. Four days of violence end with an intervention by the king. The violence in Bangkok rivals that of October 1973 and October 1976.

1992: May 24 Suchinda resigns

Suchinda resigns, after the military receives amnesty for firing upon the demonstrators.

1992: June 10 National Assembly approves constitutional changes

Constitutional changes mandate the election of a prime minister and remove the concentration of power from the Senate.

1992: September 13 General election brings Democratic Party to power

Chuan Leekpai of the Democratic Party becomes Thailand's new prime minister. His policies focus on four goals: to eradicate corrupt practices, to reduce the powers of the appointed Senate, to decentralize the government, and to enhance rural development.

1994: May Fire at toy factory kills 188 workers

A blaze at the Kader Industrial Toy Company near Bangkok brings national attention to the weak enforcement of labor laws designed to regulate hours and conditions of employment.

1995 Automotive industry booms

Thailand's automotive industry is the fastest growing in the world in terms of expansion; the country is the second largest producer of motorcycles and pickup trucks.

1995: January–April Thousands from Myanmar flee to northern Thailand

In late January, some 17,000 Karen refugees from Myanmar cross into northern Thailand, to escape the fighting in northeastern Myanmar between Burmese government forces and guerrillas. In April, 1,200 more flee to Thailand for refuge.

1995: May 19 Chuan dissolves parliament

Prior to a vote of no confidence, Chuan dissolves parliament and calls for new elections. Having served for two years (of a four-year term) as prime minister, Chuan becomes Thailand's longest serving civilian leader of the modern era.

1995: July 2 Chart Thai party wins elections

The Chart Thai party takes 92 of 391 parliamentary seats, selecting Banharn Silpa-archa (b. 1932), a wealthy Bangkok businessman, as the new prime minister. Banharn's cabinet appointees reflect a favoring of old corrupted elites. Even the king, who is revered by Thai society, is dissatisfied by the appointment of the new ministers. Banharn's government collapses before the end of 1996.

1995: December 27 Agreement reached for the return of last Vietnamese refugees

Over 5,000 Vietnamese living in a refugee camp in northern Thailand are scheduled to be returned to Vietnam by June 1996. This agreement will end twenty years of Vietnamese refugee presence in Thailand.

Mid-1990s Deaths from AIDS on the rise

HIV (human immunodeficiency virus) began spreading across Thailand in the late 1980s, spreading rapidly through the availability of commercial sex. By the mid-1990s, an estimated 860,000 Thais are infected with HIV and nearly 50,000 people in Thailand die of AIDS in 1995.

In response, Thailand develops an effective AIDS-prevention program. Thailand records a drop in the number of new HIV infections each year after 1991. According to government estimates supported by independent researchers, the number of new infections is expected to fall from 215,000 in 1990 to 90,000 in 2000.

Brothel patronage has long been considered a rite of passage for young Thai men. In Bangkok, brothels are found in virtually every neighborhood. There are an estimated 250,000 sex workers in Thailand, eighty percent of whom are female. A survey in 1990 finds that more than twenty percent of Thai men admit to paying for sex during the previous year. By 1994, however, the government estimates that only ten percent of Thai males between fifteen and forty-nine are paying for sex.

1996: November 17 New Aspiration Party coalition wins elections

The New Aspiration Party, led by coalition parties, wins the general election and names Minister of Defense Chavalit Yongchaiyudh as the new prime minister.

1997: February Last Vietnamese refugee camp in Thailand is closed

1997: May Thai stock market nose-dives

The Thai economy tumbles with its stock market. Speculative currency trading badly devalues the baht. Neighboring Asian countries scramble to protect their currencies from devaluation. By September, the crisis has spread to Singapore, the Philippines, Malaysia, and Indonesia.

1997: August IMF assists Thailand in economic restructuring

Thailand agrees to an economic restructuring package with the International Monetary Fund that includes $10–20 billion in stand-by credits.

Bibliography

Bello, Walden F., Shea Cunningham, and Li Kheng Poh. *A Siamese Tragedy: Development and Disintegration in Modern Thailand*. Oakland Calif.: Food First Books, 1998.

Dixon, C. J. (Chris J.) *The Thai Economy: Uneven Development and Internationalisation.* London: Routledge, 1999.

Kobkua Suwannathat-Pian. *Thailand's Durable Premier: Phibun through Three Decades, 1932–1957*. Kuala Lumpur, Malaysia: Oxford University Press, 1995.

Pattison, Gavin and John Villiers. *Thailand*. New York: Norton, 1997.

Phillips, Herbert P. *Modern Thai Literature*. Honolulu: University of Hawaii Press, 1987.

Stowe, Judith A. *Siam Becomes Thailand: A Story of Intrigue.* London: Hurst and Company, 1991.

Terwiel, B.J. *A History of Modern Thailand, 1767-1942*. St. Lucia, Queensland, Australia: University of Queensland Press, 1983.

Van Praagh, David. *Thailand's Struggle for Democracy: The Life and Times of M.R. Seni Parmoj*. New York: Holmes & Meier Publishers, Inc., 1996.

West, Richard. *Thailand, the Last Domino: Cultural and Political Travels.* London: Michael Joseph, 1991.

Wyatt, David K. *Thailand: A Short History.* New Haven, Conn.: Yale University Press, 1984.

Tonga

Introduction

The Kingdom of Tonga is an archipelago (a group or chain of islands) of 168 islands in the Pacific Ocean, although only 45 are inhabited. Tonga covers an area of 748 square kilometers; about four times the size of Washington, D.C. The Kingdom is divided into three main island groups: Tongatapu, which means 'sacred Tonga', Ha'apai Group north of Tongatapu, and Vava'u Group which is further north still. North of the Vava'u Group are three small islands, Niuafo'ou, Tafahi, and Niuatoputapu which are also part of the Kingdom of Tonga. Tongatapu accounts for approximately thirty-five percent of the total land area. It is also the major population center of Tonga.

The July 1997 census of the population of the Kingdom of Tonga records approximately 107,000 persons, making it one of the most densely populated countries within Oceania. Well over half of the population lives on less than half of the total land area of the nation. Approximately twenty-five percent of the population resides in the capital city of Nuku'alofa. There are nearly 20,000 expatriate Tongans living in the United States, especially in the western states of California, Washington, and Nevada.

Tonga's climate is tropical with warm and cool seasons. The warm season lasts from December to May while the cool season lasts from May until December. Cyclones arc a threat to the islands in the Tongan group between the months of October and April. Rainfall is variable between the northern and southern island groups. The northern islands of the Vava'u and Ha'apai Groups average around 2,600 millimeters per annum while Tongatapu in the south averages only around 1,700 millimeters per year. More than half of the annual rainfall is recorded in the months between January and April. Volcanoes and earthquakes pose threats to the inhabitants of the island of Fonuafo'ou which is formed from limestone deposits overlaying a volcanic base. Only a few islands in the Kingdom of Tonga have this type of geological configuration; the majority of islands have a limestone base formed from uplifted coral formations. The highest point in the Kingdom of Tonga is unnamed and is found on the island of Kao. This point rises to a maximal elevation of 1,033 meters above sea level.

History

To European explorers and sailors in the Pacific, Tonga was referred to as "The Friendly Islands," since the Tongans were not hostile and aggressive towards the Europeans in the region. The independent Kingdom of Tonga came into being on June 4, 1970, with the country's emancipation from Britain. The Tongan Constitution had been drafted nearly a century before independence. King Tupou I (1797–1893) requested Reverend Shirley W. Baker, a Weslyan clergyman, to draft a constitution in response to the European nations' interest in his country. King Tupou thought this would be a way to ward off European colonial advances.

Tonga has been ruled by dynasties in a manner akin to that of ancient Egypt. The Tui Tonga dynasty ruled the entire group of islands from approximately 1200 until A.D. 1500. Their dynastic control was centered in the district of Mu'a on Tongatapu. Tonga had one of the most highly stratified societies in Polynesia at the time of European contact in the eighteenth century.

Tonga is considered by many archaeologists and linguists to be the site of the earliest settlement in the Polynesian cultural area, dating back to at least 1300 B.C. By 1000 B.C., the early Polynesian settlers from Tonga had taken up residence in Samoa. Tongan chiefs colonized parts of Fiji, Samoa, and other central Pacific islands during the time between twelfth and the nineteenth centuries. This expansion was made possible by the voyaging canoes that the Tongans constructed. Canoes ranged in length from twenty to thirty-five meters. These canoes could carry between ninety and 250 people depending on length. By the time Europeans arrived in the eighteenth century, voyaging canoes had grown to their maximal size. The settlement history of gradual splits in the population is reflected in the contemporary linguistic differences between the various Polynesian languages.

Pre-European Tongan cosmology believed that their islands formed a complete universe of sea, land, and sky. The sky was divided into invisible layers, each one containing a place for a different set of gods. The entire region was cov-

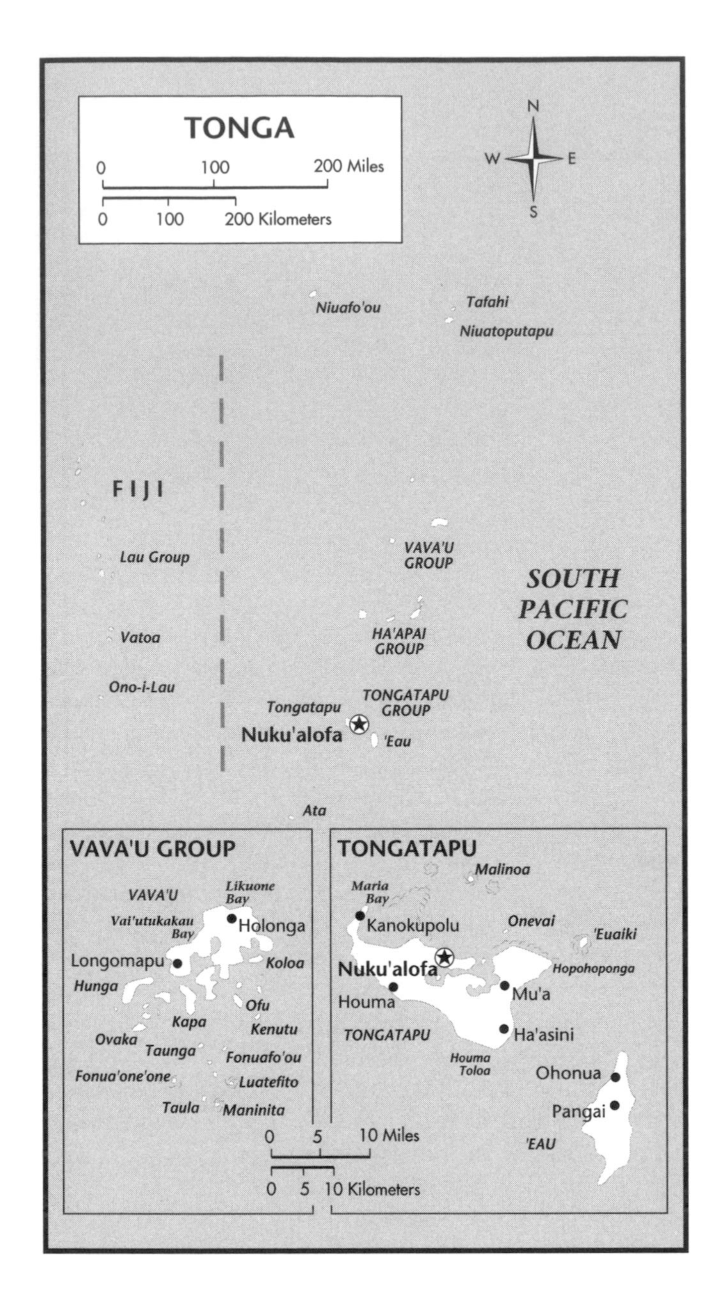

ered by a dome. The realm of Pulotu was an underwater region where the ghosts of the dead nobility resided. When Tongans first saw European ships on the horizon, they believed that the European vessels had broken through the dome of the heavens and invaded their universe.

Tonga was an important intermediary in the pre-European trade that took place between Samoa and Fiji. In this capacity, the Tongans were able to acquire the resources that they lacked on their islands. Lumber for constructing canoes, red feathers for the costumes for the nobility, adze blades, and pandanus cloth all came from neighboring Samoa and Fiji.

The Polynesian language group consists of approximately thirty different languages. These languages are all part of the larger Austronesian language family which is spread throughout the Pacific Basin. Polynesian languages are divided into two large groups: Tongic and Nuclear Polynesian. The Tongan language is the main member of the Tongic group reflecting the early settlement of the Tongan's Polynesian ancestors in the region as far back as 2300 years. Although Samoa was settled from Tonga, the language spoken by the early settlers diverged from Tongan, creating a separate branch within the Nuclear Polynesian branch of languages. The rest of Polynesia was settled from Samoa, and the languages of the other islands are more closely related to Samoan than they are to Tongan. Linguistically, Tongan is one of the most archaic Polynesian languages and therefore provides a number of linguistic clues as to the culture of the prehistoric Polynesians.

Prior to European contact, Tongan society emphasized the role of formal councils in making decisions for the society as a whole. These formal councils involved the ceremonial drinking of *kava*, a mild stimulant, which has now become popular as a means of relieving stress and anxiety among Westerners.

During this period the society was rigidly divided between chiefs and commoners. Social rules prohibited contact between the classes and intermarriage between the two was strictly forbidden. The chiefly class controlled all aspects of the society, and only the chiefly, or noble, class could own land. In Tonga's original constitution, King George attempted to initiate reforms to this system of land ownership.

Tonga was different from other stratified societies in Polynesia in that one ruling family was able to attain centralized control. In other parts of Polynesia such as Tahiti, several chiefly families controlled portions of the region and vied with each other for ultimate control.

Before Europeans arrived, yams were the main crop of Tongans. The diet of yams was supplemented by coconut, taro, and sweet potato. Since Tonga has no rivers or streams, irrigation was not possible for the pre-contact era Tongans. They had to rely on rainfall for crop irrigation and drought was a recurrent problem for Tongan subsistence.

The government of the Kingdom of Tonga is a hereditary constitutional monarchy. The king as of 1999, Taufa'aha Topou IV, had been reigning since December 16, 1965. He was born on July 4, 1918, and his birthday is celebrated as a national holiday in Tonga. The King of Tonga has three sons that stand in the line of succession to the throne. The King's eldest son Prince Tupouto'a who is Crown Prince, was born in 1948. The King appoints a Prime Minister and a Deputy Prime Minister as well as a Cabinet. The appointments of Prime Minister and Deputy Prime Minister are for life. The King and his appointed Cabinet comprise the executive body known as the Privy Council.

Tonga has a unicameral legislative assembly called Fale Alea in Tongan. The Fale Alea contains thirty seats of which twelve are reserved for Cabinet. Nine seats are for nobles selected by the country's thirty-three nobles. The other nine seats are elected by popular vote from the populace. Each member serves a three-year term. This division between nobles and commoners in the legislative assembly reflects the traditional social stratification between nobility and commoner classes in Tongan society.

The Kingdom of Tonga is divided into three administrative island units: Ha'apai, Tongatapu, where the capital Nuku'alofa is located, and Vava'u. Ha'apai and Vava'u each have their own governor. Both governors are part of the Privy Council; the highest executive body in the government. All persons aged twenty years and older have the right to vote in public elections: however, many Tongan positions are appointed or hereditary. There are two main political parties on Tonga: the Tongan People's Party and *Viliami Fukofuka*.

Tonga is an overwhelmingly Christian nation. Trough the efforts of missionaries, the Methodist Church was able to attain the first strong foothold in terms of the religious landscape of Tonga. The Free Wesleyan Church, an off-shoot of the Methodist Church, has the largest number of followers, claiming a total congregation of over 30,000. The Free Tongan Church which is in turn an off-shoot of the Free Wesleyan Church also has a strong following.

Tongan and English are the two most widely used languages in the Kingdom. Tongan is the official language of the nation. The Kingdom of Tonga boasts a literacy rate of 99 percent.

Local educational opportunities have been goals of Tongan kings and queens since the nineteenth century. The oldest operating secondary school in the Pacific Islands is located on Tonga. The Tongan branch of the University of the South Pacific, 'Atele campus, is located in Nuku'alofa. There are also two other universities on Tongatapu which provide higher education.

Tonga has been one of the poorest nations in Oceania. With limited natural resources and strict restrictions on foreign involvement in land holdings, the Tongan government has little opportunity to engage foreign investors in the local economy. King Tupou IV, the reigning King of Tonga, has sought opportunities to establish Tonga as a prosperous small nation not unlike Nauru, its neighbor in the Pacific. Tonga, as opposed to Nauru, has no natural resource that can be developed to pursue such an economic agenda. The King has courted many private foreign investors with little success. In the late 1970s, a banking scandal involving the King and his favorable treatment of controversial U.S. investor John Meier damaged Tonga's reputation internationally.

The Tongan economy is primarily based on agriculture. Over thirty percent of the nation's gross domestic product derives from agriculture. Industry accounts for only around ten percent of the gross domestic product (GDP). The main agricultural crops are squash, coconuts, bananas, and vanilla bean. Agricultural exports account for over two-thirds of the nations total exports. Japan receives the bulk of Tongan exports in the form of produce and fish. The United Kingdom receives the second largest amount of exports. Despite the large volume of agricultural production, population pressures have caused the Kingdom of Tonga must import a large portion of its food. Most of the imported food comes from New Zealand. Additional food products are imported from Australia, the United States, the United Kingdom, and Japan.

The currency of the Kingdom is the *pa'anga*. The value of the *pa'anga* is tied to the Australian dollar and the current value is approximately 98 Australian cents. Tourism is the primary source of hard currency earnings for the nation. In recent years the government has explored ways to enhance Tonga's image as a tourist destination.

Timeline

3000 B.C. Navigators in canoes reach Tonga

Seafarers known as Lapita leave their homes—probably in Southeast Asia or on the islands of the Pacific—travel east, moving from island to island, across about 12,000 kilometers (7,500 miles) of the Pacific Ocean. These people travel in outrigger and double canoes.

2000 B.C. Diet consists of birds

Early people rely on the natural resources of the island for food. Bats are the only native mammals, but birds are plentiful. Archaeologists believe that birds provide more than half of the diet of the early inhabitants of Tonga.

Archaeologists excavating on Tonga have uncovered bird bones in caves and at the bae of the rocky cliffs of of 'Eua (pronounced "ay-oo-ah"). Archaeologists compare the bone evidence in upper and lower layers of rock sediment in the caves. Based on their findings, these scientists theorize that humans, by hunting and eating, caused the elimination of over thirty species of birds.

A.D. 950 'Aho'eitu is born

The ancestor of all Tongan nobility is born out of the union of a Polynesian sky deity and a human female. 'Aho'eitu proclaims himself the first Tui Tonga, or King of Tonga in the year 1000. Unlike many other Polynesian cultures, the Tongans stress their native, or autochthonous, origins. Many other Polynesian groups have origin myths that trace their ancestry to other, established Polynesian groups.

1000 First Tui Tonga founds Tongan nobility

The first Tui Tonga, or king of Tonga, founds the Tongan nobility and settles the island group.

1500 Tui Ha'a Takalaua dynasty takes over rule in Tonga

The twenty-fourth king in the Tonga dynastic line transfers political power to his brother, and the dynastic sequence Ha'a Takalaua is founded.

1600 Tui Kanokupolu dynasty begins rule of Tonga

The Tui Kanokupolu dynasty gains control of Tonga in an act where the seventh king in the Ha'a Takalaua line transferred governing responsibility to his brother. Through this act, the Kanokupolu title is established. The reigning king feared for his life since many previous kings had been assassinated through Tongan history. The Kanokupolu line retains control of the Tongan islands until the present day.

1616 Dutch navigator Schouten encounters Tongan viceroy

A Tongan viceroy (representative of the king) is encountered by the Dutch navigator William C. Schouten (c. 1580–1625) on his landing at Niuatoputapu, now one of the far northern islands in the Tongan chain. Another European navigator, Le Maire, encounters another Tongan viceroy on Futuna. Tongan control over much of the Central Pacific was evident from these early encounters.

1773: October Captain James Cook lands on Tongatapu

Captain James Cook (1728–79) lands on the island of Tongatapu and records much of the traditional ways of life of the Tongan people. He was the first European to land on the Tongan chain of islands. This discovery was part of the larger second Cook voyage which commenced in 1772 and concluded in 1775. The purpose of the voyage was to confirm the existence of the Southern Continent. During the this voyage, Cook also landed on several Pacific Islands and circumnavigated Antarctica. The friendly reception that he and his crew receive from the Tongans results in the islands being called "The Friendly Islands." From the early accounts of Cook and his crew we learn about the social stratification of nobles, commoners, and slaves on the islands. In fact, Cook refers to the island of Tongatapu as the "Land of Chiefs" while all of the other islands in Tonga are referred to as the "Lands of Servants."

1777 Cook makes final visit to Tongatapu

Captain James Cook begins his third and final voyage on July 12, 1776. The purpose of the voyage is to find the Northwest Passage beginning from the Pacific side of the Passage. Cook and the crew of the *Resolution* return to several of the islands that they had visited on the first voyage including Tongatapu in the Tonga chain. Cook's third voyage is to be his final as he is killed in the Hawaiian Islands in 1779.

1797 First Christian missionaries arrive on Tonga

Missionaries from the London Missionary Society are the first Christian missionaries to take up residence in Tonga. The London Missionary Society had considerable success in converting the Tongans to Christianity.

1798 Taufa'ahau, King George, is born

Taufa'ahau, King George, the man who would dominate Tongan politics from 1803 until his death in 1893, is born.

1799–1852 Civil war

Rivalries between factions in the Tongan nobility lead to the eruption of civil war that lasts from 1799 until 1852. The missionaries from the London Missionary Society are forced to leave due to the conflict.

1822 First Wesleyan Missionaries arrive on Tonga

The Wesleyans, who felt a need to create a missionary society distinct from the London Missionary Society, establish their own Wesleyan Missionary Society and send their first missionaries to Tonga.

1826 John Thomas arrives on Tonga as Wesleyan missionary

John Thomas, a Wesleyan missionary, arrives on Tonga to bolster the failing Wesleyan Mission effort. Thomas spends nearly thirty years on Tonga and is generally considered to be solely responsible for the success of Christian conversion in the Kingdom of Tonga. Thomas teaches himself the Tongan language and translates almost the entire Bible into the Tongan language. He compiles extensive notes on the history and customs of the islands which are an invaluable resource on Tonga before the 1860s.

1831 Taufa'ahau is christened King George

A successful soldier and political leader, Taufa-ahau is christened King George, after the British king. Taufa'ahau is not yet King of Tonga at this point.

1839 Vava'u Code is drafted

King George drafts the first law of the Tongan government referred to as the Vava'u Code. This law commits Tonga to Christianity and prohibits murder, theft, adultery and the selling of liquor in the islands. The Code was drafted to show European powers that the Tongans were concerned with establishing social and moral order in their islands.

1845 King George becomes Tu'i Kanokupolu

King George He succeeds his uncle as the holder of the title Tu'i Kanokupolu. King George has spent the time since his christening developing alliances and converting local rulers to

Christianity. Later, he politically defeats all of his opponents and becomes the undisputed King of Tonga.

1850 Vava'u Code is extended

The Vava'u Code is extended under the direction of King George. In the revision, the sale of land to foreigners is forbidden. Two additional laws are added that direct the behavior of men and women to work and support their families, the causes of God, and the Chiefs of Tonga.

1862 Tongan Constitution is drafted

A complex set of revisions to the Vava'u Code are drafted to create what is referred to as the first version of the Tongan Constitution. This is the first step in Tonga's path to recognized independence. The Constitution shows a strong influence from Christianity. The sale of land to foreigners is reinstated and an attempt to create an equitable land tenure system is included.

1862 Emancipation Edict of King George I

King George I delivers a public Emancipation Edict to his subjects in which he states that inferior social classifications of serfs and vassals are abolished through the Constitution. The system of slavery that had existed in Tonga is also abolished.

1866 Tupou College is founded

Tupou College is founded as a Methodist boarding school for boys. It is the oldest secondary school still in operation in the Pacific Islands. As of 1999, the annual average enrollment is 1,000.

1870 First female students enroll in Tupou College

Four years after its founding, Tupou College enrolls its first female students. However, by 1873, female students are no longer enrolled in the College. Mr. S. Baker, all-around advisor to the King, decides that mixing males and females in a boarding school environment was not a wise social practice.

1874 Future King George Tupou II is born

The first son of King George Tupou I (1874–1918) is born. (See 1893: February 18.) King George Tupou II reigns from February 18, 1893 until his death on April 12, 1918. He is succeeded by his daughter Salote who takes the title "Tupou III" upon her accession to the throne.

1875 Tonga's new constitution is promulgated

The constitution of 1862 was heavily revised by King George and Reverend Baker in reaction to enhanced European interests in the Central Pacific Islands. The new constitution established a legislative assembly of forty members, although only half would be elected by popular vote. The constitution also calls for an autonomous Tongan church to supersede the Wesleyan mission. The constitution also calls for a treaty with the German Empire.

1876 German-Tongan treaty of friendship was enacted

The German-Tongan treaty of friendship allowed Germany to use the harbor at Vava'u as a coaling station in exchange for Germany's recognition of Tongan nationhood.

1879 British-Tongan treaty of friendship is entered into

The British are forced to develop a similar harbor use treaty with Tonga to secure trading interests in the Central Pacific. Tonga makes the treaty conditional on recognition of her nationhood, as had been done with Germany three years prior.

1881 Female students are permitted to enroll in Tupou College

After female students were not permitted to enroll in the only secondary school in the Tongan Kingdom in 1873, decisions were reversed in 1881, permitting female students the opportunity to pursue formal education.

1885 Free Church of Tonga is proclaimed

The Free Church of Tonga split from the Wesleyan Mission and established itself as an independent religious entity. The split was not based on religious ideology or doctrine but instead had to do with the King's desire to show that Tonga was an independent sovereign nation with its own church, not unlike England and Germany.

1893: February 18 King George Tupou II becomes ruler

The son of King George Tupou I becomes King George Tupou II (1874–1918). He reigns from February 18, 1893, until his death on April 12, 1918.

1900 Salote Tupou III is born

The only child of King George Tupou II, Salote Tupou (1900–65), is born. The exact date of her birth is not recorded. Salote is educated in New Zealand and ascends to the throne at the young age of eighteen (see 1918: April 12). Queen Salote Tupou III is remembered by Tongans as one of the finest monarchs in Tongan history; during her reign, the Tongan Free Church and the Weleyan church are united. In 1953, Queen Salote travels to England for the coronation of Queen Elizabeth II.

1901 Britain declares protectorate

In response to local and expatriot concerns over the lack of leadership shown by King George Tupou II, the government of Britain declares a protectorate over Tonga.

1905 Britain obtains right to review all government actions

The protectorate duties of Britain are extended so that the British have the right to review all internal government actions, specifically the right to review all official appointments and dismissals made by the King.

1909 Tonga Ma'a Tonga Kautaha is formed

In response to the price discrimination that the Tongans felt in response to foreign traders importing and exporting goods to and from their islands, they formed a cooperative society called *Tonga Ma'a Tonga Kautaha* (Tonga for the Tongans Association). The foreign traders protested to the British Counsel which found ways to impede the Association's activities.

1918: 12 April Queen Salote Tupou III ascends the throne

Salote Tupou III (see 1900) ascends the throne of the Kingdom of Tonga. Queen Salote Tupou III marries a man from a competing royal lineage to insure that her prodigy would have undisputed right to the throne upon her death.

1918: July 4 Queen Salote's son, Taufa'ahua, is born

In the same year that she ascends the throne, Queen Salote gives birth to her son, Taufa'ahua. During much of his mother's reign as queen, Taufa'ahua serves as prime minister. Taufa'ahua ascends to become king upon his mother's death (see 1965: September 4). King Taufa'ahua Tupou IV was the first Tongan to earn a college degree. He also established himself as an eminent scholar of the traditional Tongan calendar and traditional Tongan music. One of the King's primary goals was to establish independence for Tonga from Britain, which was achieved under his reign (see 1970). King Taufa'ahua Tupou IV is married to Queen Halaevalu Mataaho.

1918–19 Influenza epidemic hits Tonga

A widespread epidemic of influenza hits Tonga, affecting a sizable portion of the population.

1921 Girls branch of secondary school established

Queen Salote College breaks off from Tupou college to provide separate secondary education for girls. The girls' branch of the secondary school offers separate classes from the boys' school, although girls still take examinations with the boys.

1926 Queen Salote College achieves academic independence

Queen Salote College, which began as a branch of Tupou College (see 1921), achieves academic independence as female students take examinations independently from the male students enrolled in Tupou College.

1927 Mandatory education instituted

Queen Salote institutes mandatory primary education. As a strong advocate for the Tongan people both at home and abroad, Queen Salote expands education and health services in the Kingdom during her reign.

1929 Scholarships established for study abroad

Queen Salote (see 1900 and 1918: April 12) continues her support of education by creating scholarships for Tongan students to continue their education abroad.

1964 Stamps commemorate women's association meeting

A stamp is issued to commemorate the meeting of the Pan-Pacific Southeast Asia Women's Association Meeting in Nuku'alofa. The stamps are die-cut in the shape of a heart and the country of Tonga.

1944 Teachers college established

Queen Salote (see 1900 and 1918: April 12) established the first teachers college in the islands.

1965: September 4 Taufa'ahua Tupou IV becomes king

Queen Salote (see 1900 and 1918: April 12) dies. Her son, Taufa'ahua Tupou IV (see 1918: July 4 ascends the throne after her death.

1966 Prince Fatafehi Tu'ipelehake is appointed Prime Minister

The King's younger brother, Prince Fatafehi Tu'ipelehake, is appointed Prime Minister by the King. The position of prime minister is one of the components of the Privy Council within the Tongan political system. The Privy Council is the highest executive body. The King, his Cabinet, and the governors from the administrative districts of Ha'apai and Vava'u make up the Council. The cabinet is presided over by the prime minister. Each cabinet position is appointed by the King and the position is held until the member reaches retirement age.

1969: April Tonga issues unique stamps

Tonga's postal service issues stamps that are not in the traditional square of rectangular shape, but rather form the outline of a banana. Over the next decade, Tonga issues stamps in the shapes of coconuts, watermelons, and pineapples followed over the next decade.

1970: 4 June Tonga becomes completely independent

Over one hundred years since Tonga drafted a constitution, the Kingdom of Tonga becomes an entirely independent monarchy by shedding its protectorate from Britain.

1977–78 The 'Meier affair' disrupts Tongan economy

The "Meier affair" places Tonga in an unfriendly international spotlight. Accusations of corruption are directed toward the King of Tonga.

The Meier scandal involved U.S. investor John Meier. Meier was a former employee of U.S. millionaire businessman, film producer, aviator, and eccentric recluse Howard Hughes. In June 1977, Meier created the Bank of the South Pacific. The King of Tonga granted Meier and his bank a ninety-nine-year monopoly on merchant banking in the island nation. Meier was given a diplomatic passport by the King along with exceptional freedom from government regulations. Apparently unbeknownst to the King, Meier was wanted on criminal charges. Meier was arrested in Canada and extradited to the United States in 1978. Meier had a history of arrests for tax evasion and swindling. The series of financial risks that the King took left the nation with monetary losses and unfulfilled expectations.

1989 Tongan History Association formed

The Tongan History Association is formed to preserve, study, and discuss all aspects of Tongan history. The first president of the association is the Reverend Dr. Sione Latukefu, one of Tonga's foremost proponents of historical preservation and the undertaking of Tongan historical studies by Tongan scholars.

Dr. Sione Latukefu (1927–95) was born into a family of prominent commoners on the island of Tongatapu. He studied at Topou College, then at The University of Queensland, and earned his Ph.D. from The Australian National University in Canberra.

1991: August 22 Baron Vaea is appointed Prime Minister

King Tupou IV appoints a new Prime Minister, Baron Vaea, for life.

1998 Tonga establishes diplomatic relations with China

In an effort to promote the case for membership in the United Nations, the Kingdom of Tonga establishes diplomatic relations with China for the first time in its modern history. At the same time, Tonga cuts relations with Taiwan which it had maintained for the past twenty-six years. China's vote on the membership application of Tonga is cited as the primary reason for the shift in alliance.

1998 Royal School of Science for Distance Learning established

In an effort to allow its citizens more access to university-level education, the government of Tonga establishes the Royal School of Science in Distance Learning. Although the School does not offer any courses of its own design, it provides support in terms of machines and staff for students to take courses from international education providers offered over the internet.

1999: January 20–23 King Tupou IV visits Los Angeles

King Tupou IV spends four days in Los Angeles for high level meetings with business leaders from the United States as well as meeting with General Secretary of the United Nations Kofi Annan. The Kingdom of Tonga has been exploring membership in the United Nations for several years and was likely to be admitted as the 186th member later that year. If admitted to United Nations membership, the Kingdom of Tonga would be second-smallest member nation, after Monaco.

2000: January 1 First country to greet the millennium

The Kingdom of Tonga is thirteen hours ahead of Greenwich Mean Time (GMT) and one hour ahead of New Zealand. As a result of its unique location in relation to the International Date Line, Tonga will be the first nation to welcome in the twenty-first century.

Bibliography

Aswani, Shankae and Michael Graves. "The Tongan Maritime Expansion." *Asian -Perspectives* 1998. 37: 135–201.

Bellwood, Peter. *The Polynesians.* London: Thames and Hudson, 1978.

Campbell, I.C. *A History of the Pacific Islands.* Berkeley: University of California Press, 1989.

———. *Island Kingdom: Tonga Ancient and Modern.* Christchurch, NZ: Canterbury University Press, 1992.

Denoon, Donald, ed. *The Cambridge History of the Pacific Islanders.* Cambridge: Cambridge University Press, 1997.

Jennings, Jesse. D., ed. *The Prehistory of Polynesia.* Cambridge, MA: Harvard University Press, 1979.

Ledyard, Patricia. *The Tongan Past.* Nuku'alofa: Vava'u Press, Ltd., 1982.

Marcus, George E. Tonga's Contemporary Globalizing Strategies. In Lockwood, V., Harding, T., and B. Wallace, eds., *Contemporary Pacific Societies.* Englewood Cliffs: Prentice Hall, 1993, pp.21–33.

Morton, Helen. *Becoming Tongan: an Ethnography of Childhood.* Honolulu: University of Hawaii Press, 1996.

Perminow, Arne. *The Long Way Home: Dilemmas of Everyday Life in a Tongan Village.* Oslo: Scandinavian University Press, 1993.

Turkmenistan

Introduction

Situated in Central Asia, north of Iran and east of the Caspian Sea, Turkmenistan appears destined to emerge as one of the wealthiest of the former Soviet republics due to its vast amounts of untapped natural resources, particularly crude oil. However, such a transformation from an agrarian backwater to an up-and-coming economic power rests largely upon the fledgling state's continued stability in the midst of an inherently unstable region. Compared to its neighbors, Turkmenistan is small geographically, occupying an area of only 190,000 square miles (488,100 square kilometers). The 1995 census listed its population as 4,483,251, of which 77 percent were Turkmen. Other significant nationalities include Uzbeks (9.2 percent), Russians (6.7 percent), Kazakhs (2 percent), and Ukrainians (0.5 percent).

Despite its vast mineral resource wealth, Turkmenistan remains one of the poorest of the former Soviet republics. Most of the country is uninhabited desert; only three percent of the land is arable. Extensive irrigation in the southwestern and northern parts of the country allow for further cultivation but at great cost. The diversion of much of the water from Amu Darya River has resulted in the death of the Aral Sea. Despite the small amount of cultivable area, Turkmenistan is the world's tenth largest producer of cotton. Other crops include grain (the country's most valuable crop), vegetables, and grapes. Fish from the Caspian Sea are a major source of export income.

While the country has large expanses of heretofore unproductive dessert, the country's potential future wealth lies below the sand. Oil and gas reserves dot much of the Caspian coastline as well as the eastern part of the country from the Uzbekistan border down to Iran. The shores of the Caspian are also home to salt and coal mines while the eastern tip of the country between Afghanistan and Uzbekistan holds valuable sulfur mines. Attempts to profit on these natural resource deposits have been problematic, however. Part of this stems from Turkmenistan's inability to collect the large debt from natural gas exports owed by its neighbors. A weak currency compounds the problem.

History and Tradition

In many respects, Turkmenistan is still a traditional society. During the era of Soviet rule (1924–91), the Turkmen Soviet Socialist Republic (SSR) was the least integrated of all republics. This was primarily a result of the geography and culture. Located in the southern reaches of central Asia, the Turkmen SSR was far from Moscow and had only become a part of the Russian Empire late in the nineteenth century. Moreover, the republic's population of ethnic Russians remained below ten percent which prevented the infusion of Russian influence into a traditionally Islamic society. Finally, the relatively slow pace of urbanization in the region (as of the early 1990s, fifty-five percent of Turkmen still lived in rural areas) buttressed traditional cultural and societal values.

Tribal and clannish lines formed the traditional structure of Turkmen society. These traditional patterns of identity continue to play a role in Turkmen life, especially in rural areas. Extended families constitute the smallest unit of these clans. Indeed, the extended family is quite common even today among rural Turkmen. Significantly, the tribal system in Turkmenistan does not favor hereditary elites. Instead, the path to leadership was open to the mass of male members of that tribe. One peculiar feature of the tribal system is the significance accorded to six tribes known as the *övlat*. These tribes claim direct lineage from Muhammad. Within the tribal structure, the *övlat* tribes play a role as mediators in tribal disputes due to their continued veneration by the other tribes. Despite the strong cultural influence of Islam, Turkmen society never imposed the strict features of Islamic law on its females. Although females were kept out of tribal and clan authority positions, they contributed economically. Most leadership positions in the clan or tribe fell to male elders. To see this practice preserved in the present one need look no further than President Niyazov's oblique references to, and treatment of, the Turkmen parliament as an advisory council of elders.

A variety of rulers have held sway over what is now Turkmenistan. Early records indicate human settlement during the Stone Age. The region became a part of Alexander the Great's short-lived empire in the fourth century B.C. In the aftermath of Alexander's death, the Parthians made the region

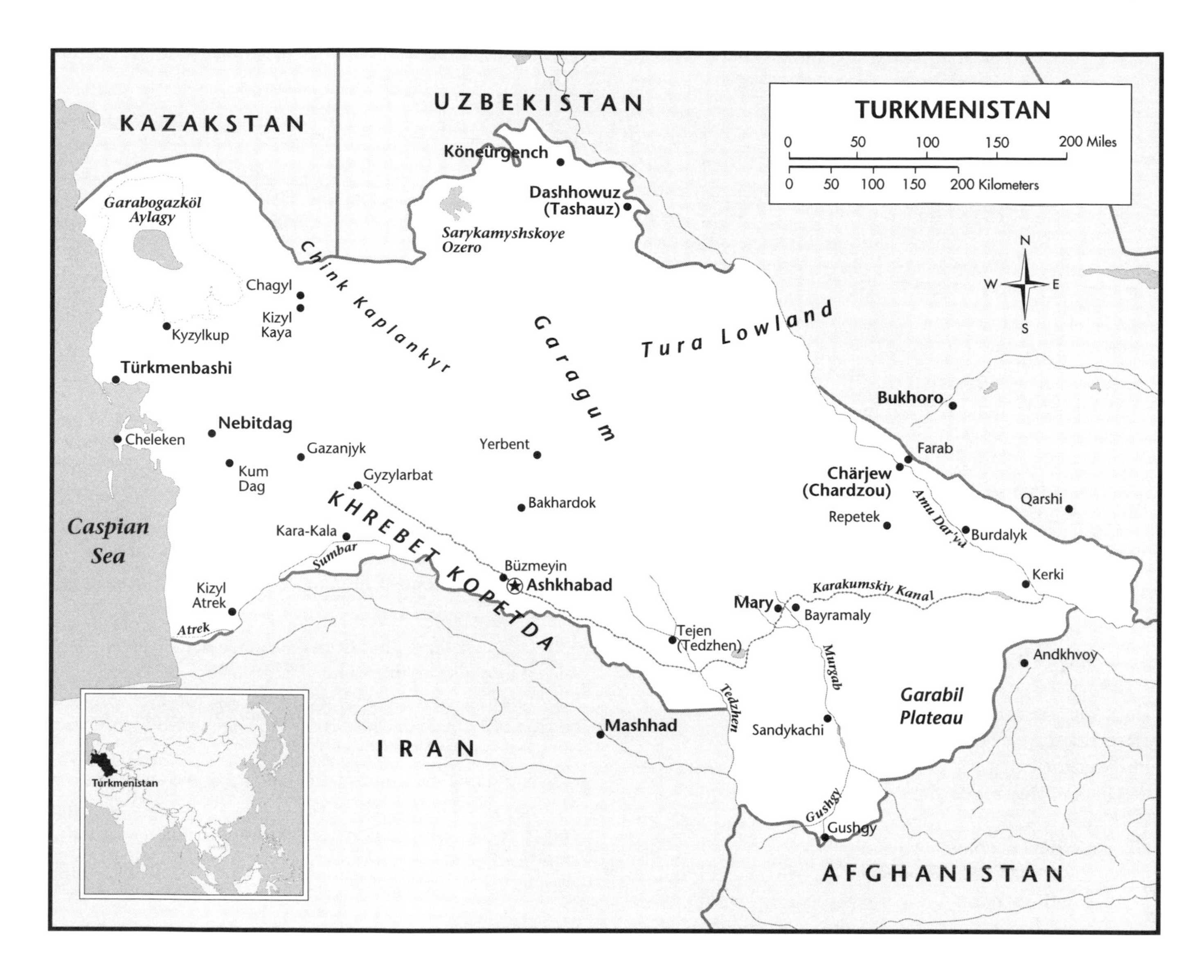

a province of their empire. By the fifth century A.D., however, the region came under the rule of the Sassanids, who were followed, in turn, by the Ephtalites. In the sixth century, the Turkic Tiu-chue nomads entered the area. In 716, the region fell to the Arab caliphate of the Umayyad dynasty which introduced Sunnite Islam to the inhabitants. Today's Turkmen trace their presence in the area to the Oghuz Turks who arrived in the ninth century. The rule of the Seljuk Turks, established in 1040, lasted for nearly two centuries before succumbing to new invading forces. The Mongols, under Chinggis Khan, conquered the region in 1219. Parts of modern Turkmenistan were incorporated into the Khanate of the Golden Horde, the Chagata Khanate, and the Hulaguid Khanate. In the late fourteenth century, Timur (Tamerlaine) captured the area.

The first contacts with Russia came in the sixteenth century. These contacts intensified in subsequent centuries as Turkmenistan became a transit center for trade between Europe and Asia. As trade increased, so too did the region's strategic importance for the great powers. During the nineteenth and twentieth centuries, Turkmenistan served as a major site of Anglo-Russian rivalry. Tsarist Russia annexed the area in 1865 as the Transcaspian District. However, Turkmen resistance, often with the support of Great Britain, thwarted Russian plans for an easy occupation; it took twenty years before tsarist officials finally suppressed resistance to their rule. In 1916 tensions flared once again as Turkmen rebelled against the tsarist government's efforts to conscript labor battalions for the war effort from among the empire's central Asians. The overthrow of autocracy in March 1917 and the subsequent Bolshevik takeover that November brought further chaos to the region. Not until 1924, when the Bolsheviks reorganized the region, did strong government return to Turkmenistan. The state's present borders were set with the establishment of the Turkmen SSR that same year.

The period of Soviet rule brought little economic and material change to the republic. In spite of its great natural wealth, Turkmenistan constantly lagged behind other Soviet republics in income. The dissolution of the USSR in 1991 ushered significant economic change that promises to continue. Under the leadership of president Sapamurat Niyazov (a former communist in power since 1985), Turkmenistan has opened its vast (and still mostly untapped resources) to foreign companies (albeit at a slow pace). The main source of the republic's untapped wealth lies in its huge deposits of oil and gas. These resources, in turn, have made Turkmenistan the focal point of U.S. efforts to build an oil pipeline through the Caspian Sea, Azerbaijan, and Turkey to an outlet on the Mediterranean Sea. A concerted effort to that end on the part of U.S. president Bill Clinton, which included a visit by president Niyazov to the White House in April 1998, appears to have borne fruit.

Critics, however, argue that the future for Turkmenistan is not so rosy. Human rights groups as well as the U.S. State Department have cited the country for countless human rights abuses. Elders declared Niyazov "president for life" (he declined the title supposedly on constitutional grounds) and he has canceled parliamentary and presidential elections. Opponents also contend that Niyazov has created a "cult of personality," and cite his adopted title of "Turkmenbashi" (Head of the Turkmen) as evidence of his autocratic tendencies.

In today's Turkmenistan school children recite "sayings of Turkmenbashi," downtown Ashgabat features a statue of "Hajji Turkmenbashi," his image is displayed on the currency, and the city of Krasnovodsk has been renamed after him. Nonetheless, the Turkmen president remains quite popular. Much of this support stems from the his desire to make Turkmenistan a "second Kuwait" by providing social benefits and increasing government subsidies with oil revenues. Niyazov has also sponsored a resurgence in Islam in order to foster stronger nationalism among his population. On the diplomatic front, Niyazov has excelled by cultivating good relations with all neighboring countries except Taliban-dominated Afghanistan. In large part, this is due to Turkmenistan's supply of natural resources. In 1997, for example, Turkmenistan opened a rail link with Iran and followed up the next year with a gas pipeline between the two countries. At the same time, however, Niyazov courted the United States and supported American efforts for an oil pipeline through the Caspian Sea, a project that would bypass Iran in favor of Washington's Middle East allies, particularly Turkey. Only time will tell if Niyazov's policy of neutrality will succeed over the long-term.

Timeline

Humans settle in present-day Turkmenistan during the Stone Age. These settlers converted their lifestyle from nomadic hunter-gathering to primitive farming. In the Neolithic era farmers organize in clans first settle in the Krasnovodsk (Turkmenbashi) region. Among the early tribes, the Anau are most significant. The Anau structure their society around agriculture and livestock raising.

1st millennium B.C. Emergence of Hyrcania

The first centers of development in what is now Turkmenistan occur in the Hyrcania along the southeastern coast of the Caspian Sea. Hyrcania comes under Median rule around the sixth century B.C. The Medes, are an Indo-European people from Persia. Under the Achaemenid dynasty, their rule spreads throughout much of Central Asia. This is the first of many periods of Persian rule over the country.

3rd–4th centuries B.C. Carpet-making

Carpet-making begins among the inhabitants of the region. Although this art has continued uninterrupted down to the present, it does not become popular until the nineteenth century. Turkmen carpets are known for their intricate designs and painstaking craftsmanship.

320s B.C. Conquest by Alexander the Great, Bucephalus

The greatest conqueror of his era, Alexander the Great (356–323) conquers the region from the Achaemenids and incorporates it into his empire. At his death at the age of thirty-three, his empire extends from Greece and Egypt in the west to the borders of India in the east, from the Persian Gulf in the south to Turkmenistan in the north.

While in Turkmenistan, the conqueror also adopts one of the region's Akhaltekin breed of horses, names it Bucephalus, and it becomes his favorite. Akhaltekin horses remain a source of pride and identity for the region down to the present.

c. 250 B.C. Parthian Empire

Modern-day Turkmenistan becomes a part of the Parthian Empire, essentially a revitalized Persian Empire. The Parthian Empire emerges after the collapse of the Seleucid Empire, one of the successor kingdoms from the empire of Alexander the Great. The Parthians reintroduce Persian culture and rule in Turkmenistan for nearly four centuries.

A.D. 224 Sassanid rule

The Sassanids, a Persian dynasty centered in the southern part of the Parthian empire, expand into the Caspian region. The Sassanids owe their rise to the erosion of Parthian power that results from a series of long wars with the Roman Empire. The Sassanid period brings stability to the southern part of central Asia for roughly three centuries.

6th century Turkish conquest

The Tiu-chue, a nomadic group of Turkish origin, enters the area by conquering the Ephtalites (also, possibly of Turkish origin). The Tiu-chue establish an empire following a revolt against Mongol rule. At this time, the various Turkish peoples emerge as the chief power in Central Asia and pose a formidable threat to the Chinese Sui and T'Ang dynasties.

716 Arab conquest

The Arab *caliphate* (area ruled by an Islamic leader) of the Umayyad dynasty conquers modern Turkmenistan. The Arabs introduce Sunnite Islam to the inhabitants. Sunnite Islam is the majority sect of the religion. Although Umayyad rule is relatively brief, its introduction of Islam is the most important religious and cultural imprint ever given to the region.

9th century Oghuz migration, rule of Samanids

The Oghuz Turks migrate to the region. Through their mixing with the existing population, the Oghuz become the direct ancestors of today's Turkmen. Indeed, the name Turkmen first emerges at this time.

The Samanids (874–999), yet another in a seemingly endless line of Persian dynasties, hold sway over Turkmenistan. At its height, the Samanid Empire includes all of Persia up to India, and reaches as far north as the Aral Sea region.

999 Rise of Ilak Khan

Samanid rule is overthrown by the first of the Ilak Khans, a dynasty centered in Bokhara that dominates most of the territory held by the Samanids.

1040 Arrival of the Seljuk Turks

The Seljuk Turks move into the region. Although the Seljuk Turks form the first powerful Turkish empire, they borrow heavily from the Persians. Thus, Persian influence continues in Turkmenistan, even under Turkic rule. The most important contributions of the Seljuks are their commitment to Sunnite Islam and the establishment of over one century of political stability.

1157 Fall of the Seljuks

Following a long period of upheaval over rival claims to the Seljuk throne, the Seljuk Empire of the East falls to the Khawarizm empire. In subsequent decades, the other Seljuk dynasties fall to rival forces.

Early 13th century Khawarizm rule

The forces of the Persian Khawarizm shah dominate the region. The Khawarizm dynasty, which had come under the domination of the Seljuks, now fills in the void left by the Seljuk collapse.

1219–21 Mongol conquest

The Mongols under Chinggis Khan (1167–1227) overrun the Khawarizm Empire. Chinggis Khan, one of the greatest conquerors in world history, founds a Mongol empire that eventually includes Mongolia, all of China, southern Siberia, central Asia, Persia, and much of European Russia (including the cities of Moscow and Kiev). The Mongols are arguably the fiercest warriors in all of history and leave a trail of destruction in their wake. During their wars for control of China, the Mongols rely upon gunpowder; over the centuries, cannons and muskets come to replace catapults and swords as the primary means of combat.

The period of Mongol rule is significant not only for its destruction. Along with devastation comes increased contact with the West, particularly Europe. Commercial ties increase during the thirteenth and fourteenth century. The events of one such European contact with the Mongols is provided in the story of the Venetian merchant Marco Polo in his *Description of the World*. The region of modern Turkmenistan becomes a crossroads for intercontinental communications and trade during this period. This role is best exemplified by the fact that Turkmenistan is the region where the Mongol Empire is divided into three *Khanates* (region of central Asia ruled by a *khan* or chieftain) for ease of rule (a fourth khanate, the Great Khanate, retained control over China and Mongolia, itself). Each of these Khanates is ruled by a descendant of Chinngis Khan. Soon, these three khanates begin to follow their separate courses, although they nominally maintained contact and recognize the Great Khan as their superior. The northern region comes under the control of the Khanate of the Golden Horde, a state that also includes most of European Russia, all of the Ukraine, and much of Eastern Europe. The southern portion of Turkmenistan falls under the domain of the Hulaguid Khanate of Persia. The Chagatai Khanate, situated in Central Asia, north of the Indian subcontinent, absorbs the eastern part of Turkmenistan.

1380s Tamerlaine

Tamerlaine (1336–1405), a descendant of Chinggis Khan, sweeps his armies through the region leaving further destruction in their wake. Tamerlaine reunites the central khanates and reconquers Persia as well as the Khanate of the Golden Horde in an effort to reunify the original Khanate. He also mounts attacks deep into India as far west as the Ottoman Empire. He dies suddenly while planning a campaign against China, and his empire disintegrates swiftly.

16th century Russian contacts

The first extensive contacts with Russia occur. These early contacts are primarily commercial. However, they set the stage for subsequent Russian military and political interest in the region. Ultimately, the Russian influence in Turkmen history is second only to that of the Arabs who introduced Islam.

Late 17th century **Turkmen migration to Russia**

Some Turkmen migrate to Russia and settle in the southern Caucusus region. These Turkmen facilitate Russian trading interests by acting as intermediaries. In addition, many clans send representatives to the Russian imperial court demanding that their homeland become part of the Russian Empire. The renewed Turkmen forays into commerce underscore the continuing importance of the region as a crossroads of trade.

18th century **Transit center, Chaghatai language**

Turkmenistan becomes a major transit center for trade between Europe and Asia. Its position as a trading post increases its strategic significance.

Turkmen writers use Chaghatai, the region's classical language, in the Arabic script. It proves troublesome for the average Turkmen since the language does not approximate Turkmen at all. Nevertheless, use of this language continues until the Russian Revolution.

18th–19th centuries **Point of great power contention**

Turkmenistan, like Afghanistan and Persia to the south, becomes a center of Anglo-Russian rivalry (i.e., the Great Game). The Russian empire looks south for expansion in order to gain access to a warm water port, while Britain covets central Asia and Persia in order to have a land route to India, the most important possession of the British Empire. In their rival endeavors, both great powers attempt to use the Turkmen to their advantage.

1865 **Russian annexation**

As part of the Anglo-Russian Great Game, Russia annexes Turkmenistan and its surrounding area as the Transcaspian District. However, Turkmen resistance to tsarist control is fierce and takes twenty years to subdue. This Russian expansion culminates the last great territorial drive by the tsarist empire.

1881 **Ashgabat founded**

The city of Ashgabat is founded. Originally built as a fortification, the city witnesses rapid growth over the course of the next century. By the mid-1990s, it has over 540,000 people. During the Soviet period, it becomes the center of industrial and commercial activity in the Turkmen SSR. The city's name means "city of love".

1903–04 **Social Democrats form rergional groups**

Russian Social Democrats form groups in the Turkmen towns of Kizyl-Arvat and Ashgabat. The outlawed Russian Social Democratic party later splits into two factions, the *Bolsheviks* (majority) and the *Mensheviks* (minority). In 1917, the Bolsheviks, under Vladimir Ilyich Ulyanov (Lenin), lead a revolution against the Russian government and, after a bloody civil war, establish their authority throughout most of the former Russian Empire. Shortly after their takeover, the Bolsheviks change their name to the Communist Party and institutionalize their hold on power. The Communist Party thus becomes the only legal political party of the country and retains a monopoly on power until 1991.

Support for the outlawed Bolsheviks is concentrated in the major cities of Russia where industry is concentrated. However, in rural, outlying, non-Russian regions of the empire, such as the Transcaspian District, support for the Bolsheviks is scant. Although the Bolsheviks promise equality for all nationalities in the empire and make appeals to peasants as well as workers, they are unsuccessful in their efforts to gain many converts to their cause among the Turkmen. This failure is largely attributable to the traditional clan-based, rural, Islamic traditions of the Turkmen, a group only recently added to a Russian Empire frantically trying to modernize and reach the same level of urban, industrial, secular Western Europe. The failure to integrate Turkmen into the efforts at modernization continue during the communist era when Turkmenistan becomes the least integrated of all of the Soviet Republics.

1905: October and November **Revolution**

The Revolution against the tsarist regime which follows the humiliating Russian defeat at the hands of Japan during the Russo-Japanese War (1904–05) does not leave Turkmenistan completely untouched. Some workers and soldiers stage political strikes in various parts of the Trascaspian District. The main protests, however, are in the capital of St. Petersburg where a peaceful demonstration in front of the Winter Palace (the residence of Tsar Nicholas II) results in a massacre of unarmed protesters. Among opponents to the tsar's regime, the day becomes immortalized as "Bloody Sunday" while Nicholas earns the nickname "Bloody Nicholas".

1914–18: **World War I and revolution**

Russia enters the First World War in August, 1914 as a member of the Triple Entente (Great Britain, France, and Russia) against the forces of Germany, and Austria-Hungary. Eventually, Russian forces are engaged against the Ottoman Turks as well. Initially, Russian forces manage an invasion of East Prussia in Germany, but are defeated in the fall of 1914 at Tannenberg. The years 1915–17 witness Russian reverses at the hands of the Germans, although the tsar's forces meet with some success against the Austro-Hungarians. By 1917, however, the army is exhausted and the Romanov dynasty is on the brink of collapse. Following a soldier's uprising in Petrograd (as St. Petersburg is called since 1911), the tsar abdicates in March 1917 and is replaced by a provisional government under the Socialist Revolutionary Alexander Kerensky.

Unable to provide order at home and reverse the tide at the front, Kerensky's regime totters until it is overthrown by the Bolsheviks that same November. The Bolsheviks, facing civil war in Russia and hoping to promote a global revolution, seek an immediate end to the war and issue peace feelers to the Germans. These attempts to end the fighting culminate in the Treaty of Brest-Litovsk in March 1918. Under the terms of this treaty, the Bolsheviks surrender vast tracts of Russian territory to German forces, including Byelorussia, Poland, Ukraine, and the Baltic coast. Finland becomes independent. From the outset, Germany shows no interest in upholding its part of the agreement while Bolshevik forces are engaged in a civil war whose outcome is uncertain. Thus, German forces continue their advance into the former Russian Empire and reach as far as the western edge of the Caspian Sea in the Caucusus before the war ends in a German defeat.

By all accounts, these years of war and revolution prove disastrous to the region. Normal lines of communication and trade are disrupted. Moreover, industrial, mineral, and agricultural production figures plummet as the economy slows down.

1916 Uprising

Although Turkmenistan is far from the front and does not see any fighting, it does not escape the effects of war. Turkmen stage a massive uprising against tsarist efforts at conscription into labor battalions. These types of demonstrations become commonplace throughout the empire as war drags on, and this wave of protests plunges the region into nearly a decade of anarchy as various groups vie for control.

1917: April 20 Emir of Bokhara declares independence

In the wake of the February Revolution, the Emir of Bokhara declares the independence of Turkmenistan. His action is similar to that taken by various other nationalist leaders in the non-Russian regions of the Empire who see in the tsar's overthrow the opportunity to rid themselves of Russian control by establishing independent states. Although the Bolshevik success during the Russian Civil War halts these efforts, most of these non-Russian regions do gain their independence after the collapse of communist rule.

1917: May All-Russian Moslem Congress

In an effort to gain support for their tottering grip on power, the Kerensky government convenes the All-Russian Moslem Congress. The creation of this body, however, has little effect in strengthening the Kerensky government among the Empire's Moslem population. Unlike the Christian European portions of Russia which had been conquered centuries earlier, most of the Empire's Moslem population come under Russian rule less than half a century before and feels no historic or religious ties to the empire. As a result, Kerensky's efforts to gain support among the empire's Islamic Central Asian inhabitants falls on deaf ears. Traditional clannish and religious leaders in the region favor an end to Petrograd's rule.

1917: November 14 Rebellion in Tashkent

A rebellion in Tashkent receives the support of radical workers in Turkmenistan. Although the size of the revolt is small and its effect is negligible, it demonstrates the cohesiveness of the revolutionary movement. One of the main reasons the Bolsheviks succeed in the revolution and civil war is their strict party discipline.

1917: November 30–December16 Soviet congress in Transcaspian Oblast

During the early stages of the revolution, radical workers and soldiers gain control of *soviets* (councils) that have organized more or less spontaneously throughout many regions of the empire. By the summer of 1917, it is apparent to the Bolsheviks that the majority of the members of the soviets are potential supporters and Lenin issues the call, "All power to the soviets." In November, 1918 the soviets of the Transcaspian Oblast hold a congress in support of the revolution. The soviets now claim to be the sole authority in the district and attempt to establish their control throughout the empire. However, they face strong opposition in their revolutionary efforts.

1918–21 War Communism

The Communists introduce the economic program of "War Communism" to the embattled former Russian Empire. The essential features of War Communism are nationalization of private enterprise and the institution of central planning for the entire economy. The plan proves disastrous and only exacerbates the preexisting economic upheaval. After three years, Lenin scraps War Communism and replaces it with the more liberalized New Economic Policy (NEP).

1918: April 30 Turkestan ASSR formed

The Fifth Congress of Soviets of Turkestan Krai meets in Tashkent. This body announces the formation of the Turkestan Autonomous Soviet Socialist Republic as a part of the Russian Soviet Federated Socialist Republic (RSFSR). The creation of an autonomous republic is in keeping with the Bolshevik desire to create separate republics for each of the major nationality groups of the empire.

1918: July 11–12 Soviets fall from power in Ashgabat

Bolshevik rule in Turkmenistan is overthrown by an uprising of Socialist Revolutionaries and Whites (those diverse groupings of forces opposed to the Bolshevik Red Army). This White success in Turkmenistan coincides with White successes throughout the former Russian Empire.

1919: May Red Army offensive into Turkmenistan

The Red Army begins its advance against its anti-Bolshevik opponents in Turkmenistan. Although it takes nearly a year for the Red Army to reestablish its control over the region, White power is on an steady decline.

1920: February 6 Krasnovodsk retaken

The Red Army retakes Krasnovodsk. This Bolshevik victory marks the end of significant resistance against communist rule in the Transcaspian region. Several factors cause the White defeat. Their forces are too fragmented and represented the gamut of anti-Bolshevik forces from leftist Socialist Revolutionaries to reactionary monarchists. In addition, the Whites have trouble cooperating and never produced a meaningful reform program to attract peasant support. The results of the war are devastating. Millions die and thousands emigrate. In addition, most of the industry that the Russian empire had built as part of its modernization efforts is destroyed.

1921 New Economic Policy introduced

Under the direction of Lenin, the Tenth Party Congress discards the program of War Communism in favor of a return to limited capitalism under a new program termed the New Economic Policy (NEP). NEP allows for the creation of small private enterprises. It also permits foreign investment in, thereby giving the new regime access to much needed foreign currency and opening the way for subsequent diplomatic recognition. NEP is ultimately discarded when Stalin embarks upon the first Five-Year Plan.

1922: December Tenth All-Russian Congress of Soviets forms USSR

The Tenth All-Russian Congress of Soviets (also known as the First All-Union Congress of Soviets) convenes and declares the formation of the Union of Soviet Socialist Republics (USSR). The new configuration replaces the RSFSR (Russian Soviet Federated Soviet Republic). The USSR consists of the Russian Socialist Federative Soviet Republic (RSFSR), the Transcaucasian Socialist Federative Soviet Republic (TSFSR), the Ukrainian Soviet Socialist Republic, and the White Russian Soviet Socialist Republic. Turkmenistan is divided between the Khorezm People's Soviet Republic and the Bukhara People's Soviet Republic, both of which are part of the TSFSR.

1923 New constitution for the USSR

The Second All-Union Congress of Soviets puts into action a constitution for the USSR which delineates between the powers of the national government and those reserved to the republics, including the right of secession from the USSR. In practice, however, the arrangement set down on paper is a fraud. For example, any attempt at secession is to be resisted with force. True power rests in the hands of the Communist Party.

1924: January 21 Lenin dies, power struggle ensues, rise of Stalin

Vladimir Ilyich Lenin dies in Gorky, a city east of Moscow. The death of the communist leader ushers in a power struggle among the leading communist leaders of the fledgling communist state. Initially, the leading contenders for the party leadership are Leon Trotsky (1879–1940), commander of the Red Army; Grigory Zinoviev (1883–1936), leader of the Communist International (Comintern); and Joseph Stalin (1879–1953), General Secretary of the Communist Party of the Soviet Union and Minister for Nationalities. Although a minor player in the Revolution and Civil War and overshadowed by his rivals, Stalin emerges, by 1929, as the leader of the Soviet Union. His rule is the most brutal in all of Russian history.

Stalin's system of rule is characterized by a "cult of personality" by which he glorifies himself as the leading, creative force in the country. He consolidates and maintains his force through the use of terror and repression. Opponents are repressed through coercion, arrest, exile, and execution at the hands of the secret police. The Stalinist system of terror reaches its height in the mid- to late-1930s.

His economic and foreign policies, in many respects, prove disastrous. Stalin discards NEP in favor of centralized state planning with the aim of industrializing the country through a series of Five-Year Plans. The ideology behind Stalin's development policy is the theory of "Socialism in One Country". This move away from revolution in favor of development in one state necessitates economic independence. Beginning in 1929 Stalin attempts to achieve this through a brutal collectivization of all agricultural lands. With state control of agriculture secure, Stalin then sells crops abroad for hard currency to subsidize his industrialization program. Although Stalin's methods succeed in industrializing the USSR, the cost is enormous. Collectivization decimates agriculture and millions of peasants starve. Indeed, the combined effects of the First World War, War Communism and civil war, and now collectivization render Soviet agriculture so ineffective that it fails to meet 1913 production levels late into the twentieth century.

Of all Soviet republics Turkmenistan is the least affected by these attempts at industrialization. Nonetheless, numerous new projects are begun in the republic under Stalin's rule. Given the republic's relative underdevelopment (even in comparison to other republics), a great emphasis is placed on education. The stress placed on education bears fruit. By 1990, nearly 98 percent of the population is literate.

1924: October 27 Bolsheviks establish Turkmen SSR

The region becomes the Turkmen Soviet Socialist Republic (SSR) USSR with the capital at Ashgabat (Ashgabat). Its borders coincide to those of today's independent Turkmenistan. That same year, the Uzbek SSR is created, while the Tajik SSR is formed in 1929, thereby completing the Soviet reorganization of Central Asia.

1925: January Turkmen State Publishing House formed

The Turkmen State Publishing House opens in Ashgabat. It is the first publishing house devoted solely to publishing works in the Turkmen language. During the Communist period, publishing grows tremendously in the Turkmen SSR. By 1989, the republic has 66 newspapers and 34 periodicals.

1925 First Turkmen film

B. Bash and A. Tseitlin shoot the first film about the fledgling Soviet republic, *The Proclamation of the Turkmen SSR.* Under the communist regime, films play a crucial role in propaganda.

1925–26 Expeditions discover untapped wealth

Two Soviet scientists, A. E. Fersman and D. I. Shcherbakov, venture in search of exploitable natural resources in the Turkmen SSR. Through their efforts, they discover sulfur deposits in the central Karakum region. These are first tapped during the Stalinist industrialization drive. The deposits are so immense, however, that they remain largely untapped for the remainder of the century.

1927 First radio broadcast

The first Turkmen language radio broadcasts begin. Under the Soviet system, radio not only serves as a form of entertainment, it also serves, most importantly, as a source of government propaganda.

1928 Adoption of Latin Script, institute founded

The Latin script is adopted as the alphabet for the Turkmen language after efforts in 1922 and 1925 to adapt Chatagai to Turkmen features failed. However, Latin serves as the script for Turkmen for only twelve years. It is readopted in the independent Turkmenistan of the 1990s.

The Institute of Turkmen Culture is founded.

1930s Research institutions formed

Several research institutions are formed in the Turkmen SSR: the M. I. Kalinin Turkmen Agricultural Institute, the Institute of Trachoma, the Institute of Skin and Venereal Diseases, and the Institute of Tropical Diseases.

1931 First Turkmen feature film

It Cannot Be Forgotten, the first Turkmen feature film, makes its debut.

1936–38 Height of the Stalinist terror

The mid to late-1930s witness the height of Stalinist terror in the Soviet Union. Millions lose their lives, and millions more are arrested and/or exiled. Although the terror begins as an effort to purge the Communist Party of *Old Bolsheviks* (i.e., communists whose rise in the party is not a result of Stalin's efforts), it soon takes on a dynamic of its own and extends to all levels of society. In addition to Communist Party members, others liable to arrest include proponents of nationality rights (who face charges of "bourgeois nationalism"), Jews, ordinary dissidents, anyone suspected of being an opponent of the regime, or anyone unfortunate enough to be perceived as suspicious (often for no reason at all) to the secret police. One of the many Turkmen who perishes in the purges isthe writer and political activist Abdulhekin Qulmukam Medoghli.

The results of the terror are catastrophic. Among the millions who perish are the leading scientists and industrial planners who provide the technical expertise behind the Stalinist effort to industrialize. In addition to the industrial expertise, the purges also claim scores of Red Army officers. Three of the five Marshals of the Soviet Union are shot as are most of the army, divisional,and brigade commanders. This leaves young, inexperienced officers in command of the largest army units.

1937: March 2 Constitution adopted

The Sixth Extraordinary All-Turkmen Congress of Soviets adopts a constitution for the Turkmen SSR. Like all Soviet-era constitutions which ostensibly guarantee basic human freedoms, the Turkmen document is a sham. Dissent is stifled by the prospect of arrest by the NKVD (secret police) and exile to the *gulag* (labor camp), or even execution. Indeed, the establishment of a constitution for the Turkmen SSR coincides with the height of the Stalinist terror during which millions of Soviet citizens are arrested, exiled, and/or executed.

1938 Turkmen Geological Administration founded

In keeping with the general trend toward an increase in scientific emphasis, the Turkmen Geological Administration is founded.

1940 Birth of Sapamurat Niyazov, Russian script instituted

Sapamurat Niyazov, future president of Turkmenistan, is born in Ashgabat. Beginning with his appointment as leader of the Turkmen communists in 1985, Niyazov establishes himself as arguably the most significant Turkmen leader ever.

In a major change, the Turkmen SSR drops the Latin script and adopts the Russian alphabet for the Turkmen language. Turkmen are taught their language in the Russian script for over half a century until independent Turkemenistan readopts the Latin script in a show of cultural independence from Moscow.

1941–45 The "Great Patriotic War"

On June 22, 1941, the forces of Nazi Germany with over three million men invade the Soviet Union and begin the greatest land battle in history. Initially, the Germans make extraordinary gains capturing all of Ukraine, Byelorussia, and the Baltic republics. The Red Army, felled by a purge of its leading officers in 1937–38, is in disarray and suffers defeat after defeat. By fall, both Moscow and Leningrad (as Petrograd has been named since 1924) are threatened. Not until early December are the Germans halted at the gates of Moscow. The Red Army counterattacks but German forces remain on Soviet soil for over three years. Not until 1945 does the war end, with the fall of Berlin to Soviet forces.

For the Soviet Union the cost is staggering. Over twenty million Soviet citizens lose their lives (including seven million soldiers). Industry and farmland are both destroyed. Nevertheless, the Soviet Union emerges from the war as a global superpower second only to the United States.

Although far from the fighting front, Turkmenistan, like the rest of the Soviet Union is affected by the Second World War. Many Turkmen work behind the lines, while others join the Red Army. Many, such as Niyazov's father, are killed. In 1942–43 the republic's rail lines and the port of Krasnovodsk prove crucial in providing supplies to beleaguered Red Army defenders at Stalingrad. Ultimately, these supplies enabled the Soviet forces to counterattack and wipe out the German Sixth Army and help turn the tide of the war.

1948 Earthquake rocks Ashgabat

A great earthquake strikes Turkmenistan and destroys much of Ashgabat. As a result, the city is largely rebuilt. New structures are built to withstand the shocks of earthquakes. One of those deeply affected by the devastation is future president Sapamurat Niyazov, who, at the age of eight, is orphaned when the rest of his family perishes.

1951 Academy founded

The Turkmen Academy of Sciences is established. It becomes the leading institute of higher learning in the republic. Originally, it concentrates its efforts on the exploitation of the republic's mineral resources. By the 1990s, it boasts eighteen different institutes.

1951 Compulsory eight-year education

Education is made compulsory for a minimum of eight years.

1956 Turkmen gift for a queen

Soviet Premier Nikita Khrushchev presents Great Britain's Queen Elizabeth II with a three-year-old Akhaltekin stallion, Melekush.

1959: November Television broadcasts begin

The first television broadcasts begin in the Turkmen SSR. The start of broadcasts brings the republic in closer contact to the rest of the USSR and the outside world.

1960 Olympic Gold

At the Summer Olympic games in Rome, Sergei Filatov wins the gold medal riding his stallion, Absent.

1972: April 13 New constitution

The Extraordinary Ninth Session of the Ninth Convocation of the Supreme Soviet of the Turkmen SSR sets into action a new constitution for the republic replacing the 1937 document. This constitution provides for a Supreme Soviet of 300 deputies. This body, in turn, elects the Council of Ministers, which serves as the government of the republic.

As was the case with the 1937 version, this Turkmen document embodies democratic values in theory, but is a sham in practice. The population "votes" from approved Communist Party lists and open opposition to the government is not tolerated.

1974: October Color television

Turkmen television broadcasts its first color programs.

1974: November 14 Republic awarded

The Turkmen SSR receives the Order of the October Revolution, in honor of the fiftieth anniversary of the founding of the Communist Party of Turkmenistan. Under the Soviet system it is quite common to "reward" republics or individuals for long-standing service to the development of the USSR. These ceremonies usually take place at Communist Party congresses.

1985 Niyazov becomes CP boss

New Soviet leader Mikhail Gorbachev ousts long-time Turkmen Communist Party boss M. Gapurov and replaces him with fellow reformer Sapamurat Niyazov. This change marks the beginning of the 45 year-old Turkmen leader's rise to prominence. Niyazov quickly becomes the popular and undisputed Turkmen leader both within the Communist Party of the Turkmen SSR and among the republic's population. Ironically, Niyazov, whom Gorbachev views as a reformer, uses the last years of Communist Party rule to delay the *glasnost* (openness) and *perestroika* (restructuring) reforms instituted by Gorbachev. Instead, Niyazov consolidates his own rule in the Turkmen SSR.

1986 Akhaltekin sells for record $50 million

Dancing Brave, the winner of the Parisian *Arc de Triomphe* race fetches a record $50 million. This record price indicates the continued reputation of and interest in Turkmenistan's Akhaltekin horses.

1990: October Niyazov President

Niyazov is elected president of the Turkmen SSR by the Supreme Soviet. With this new post, Niyazov becomes the official head of his republic's government and ruling party.

The chief characteristic of his rule is its personal nature. In a traditional society centered on clannish ties, Niyazov, who is unsure of his own roots, begins forging a personality cult. Pictures portraying Niyazov as a father figure multiply. Eventually, he discards his surname and adopts Turkmenbashi (Head Turkmen). School children memorize his sayings, and his image makes its way to currency and cities. In addition to the establishment of a personality cult, Niyazov also begins aligning himself with an Islamic revival that is sweeping the region. Indeed, the support of moderate Islam is interwoven with the president's cult of personality; Niyazov boasts of his *haj* (pilgrimage to Mecca), and in Ashgabat constructs the statue of *Hajji Turkmenbashi* to replace a toppled statue of Lenin. To make up for his lack of knowledge about his ancestors, he authorizes the state historical society to create a new division specifically for that purpose. In the words of one Western observer, "Niyazov's political style has nearly turned Turkmenistan into a caricature of one-man rule in a newly-independent state. Adulation of Niyazov has become an art form in a cult of personality that may even outdo Stalin's."

1991: October 27 Independence

In the wake of the failed August coup against Gorbachev and the failure to sign a new Union Treaty to preserve the USSR, the country declares independence as the Republic of Turkmenistan. However, Turkmenistan remains loosely tied with the former Soviet republics in the Commonwealth of Independent States (CIS).

1991: November 12 Law on Enterprises in Turkmenistan

The government establishes a guideline for the operation of public and private enterprises in Turkmenistan. This law is a radical departure from the system of state-run enterprises that existed under the old Soviet Union. Article 3 of this law governs foreign investment in the country.

However, Turkmenistan is among the slowest of the former Soviet republics to liberalize its economy. Niyazov's government keeps a tight grip on the transition from a command to a market economy. A year later, only twenty-three joint ventures exist. Even in the late 1990s as more efforts are made in the area of mineral exploitation, the state continues to have a heavy hand in economic affairs.

1992 Ten Years of Prosperity

President Niyazov embarks on an ambitious program to provide free heating, electricity, and water to his countrymen. In addition, he increases government expenditures on subsidies for rent and food. This effort copies social benefits offered by the oil-rich states of the Persian Gulf. Critics charge that the main purpose behind these gestures is to improve Niyazov's popular support; the granting of free utilities along with subsidies for other living expenses goes hand in hand with the creation of a cult of personality around the leader.

1992: February Ties with the United States

Turkmenistan establishes diplomatic relations with the United States. Washington wishes to secure access to Ashgabat's great mineral wealth and preventing others, such as Islamic fundamentalist Iran, from capitalizing on it.

1992: March 2 UN membership

Along with the other former Soviet republics, Turkmenistan joins the United Nations (UN).

1992: May Conference in Ashgabat, constitution adopted

Ashgabat hosts a conference of Asian states. Seven states attend. The conference is part of President Niyazov's efforts to cultivate good relations with his neighbors while keeping Turkmenistan free of any regional conflicts.

Later in the month, on May 18, Turkmenistan adopts a constitution.

1992: June 21 Niyazov elected president

Unopposed, Niyazov is elected president of Turkmenistan by popular vote. He receives 99.5 percent of all votes cast. His near-unanimous election is indication of both his genuine popularity among his people and the effectiveness of his restrictions on opposition to his regime. International human rights organizations are highly critical of Niyazov's methods and point out that his Democratic Party of Turkmenistan is merely the renamed Communist Party of Turkmenistan and is the only legal political party in the country.

With his domestic support buttressed by his personality cult and now sealed with the approval of the vast majority of his citizens, Niyazov embarks upon a cautious foreign policy. Of central importance is the continued maintenance of friendly relations with Russia and the CIS. Thus, Turkmenistan grants dual citizenship to ethnic Russians living in Turkmenistan and allows Russian control over its borders. With regard to its neighbors, Niyazov maintains cordial—but not close—relations with the other Central Asian republics. Militarily and economically weak, and fearful of potential upheaval in the region, Niyazov relies upon the Russian armed forces and his pledge of neutrality for continued security.

1992: September IMF and World Bank membership

Turkmenistan joins the International Monetary Fund (IMF) and the World Bank. With limited results both organizations put pressure on Turkmenistan to quicken the pace of its economic liberalization. Of all the former Soviet republics, Turkmenistan is the slowest to embark upon economic liberalization or modernization.

1993 Turkmenistan as "second Kuwait," change in script

As a result of Niyazov's efforts to make Turkmenistan a "second Kuwait," government spending on social programs skyrockets. The cost of these schemes is provided by the increased revenues garnered from oil and natural gas. Of all former Soviet republics Turkmenistan is arguably in the best position to provide such services; in 1992 it is the sole former republic with a balanced budget.

The government initiates a change in script from Cyrillic to Latin. Previously, Latin served as the script of the Turkmen language from 1928–40. This change in alphabet is done for cultural, political, and economic reasons. It signifies an effort to be rid of Russian cultural and political influence. At the same time, adoption of the Latin script may expedite trade relations with the West. The country expects the change to be complete by the late 1990s.

1993: November 1 New currency introduced

Turkmenistan introduces its own currency, the *manat*, to replace the Russian *ruble*. Initially, the *manat* is valued at 2 to $1 (US). However, by 1996, it takes 4,000 *manats* to equal $1 (US). The poor performance of the *manat* convinces experts that Turkmenistan's economy still needs significant reform.

1995 Pledge of neutrality

In an effort to foster close political and economic cooperation in the region, while steering his weak state clear of any regional rivalries, President Niyazov pledges Turkmenistan's "permanent neutrality." He hopes to make his country an "Asian Switzerland."

1996: May 13 Rail link to Iran

Turkmenistan and Iran inaugurate rail service between their two cities, Tedzhen and Mashad, respectively. This type of cooperation demonstrates Niyazov's efforts to act as a neutral state in the region. The United States favors the diplomatic isolation of Iran and fears that Turkmenistan's overtures will afford the Islamic fundamentalist government of Iran an opportunity to spread its influence in the region.

1997: May Ten-state summit in Ashgabat

Once again Niyazov's program of neutrality makes Ashgabat the host of a regional summit. This time the meeting is expanded to include the leaders of ten Asian states: Afghanistan, Azerbaijan, Iran, Kazakhstan, Kyrgyzstan, Pakistan, Tajikistan, Turkey, Uzbekistan, and Turkmenistan.

1998: January Pipeline to Iran

A gas pipeline opens from Turkmenistan to Iran. Even more than the rail link, this pipeline is a source of concern for the United States. As a result of this agreement, American diplomats double their efforts to build an oil pipeline from Turkmenistan across the Caspian Sea.

1998: April Niyazov meets Clinton

President Niyazov meets U.S. President Bill Clinton in Washington. Among the topics discussed are Washington's support for an oil pipeline through the Caspian Sea, Azerbaijan, and Turkey and American support for increased democratization in Turkmenistan. According to the *New York Times*, Turkmenistan agrees to "free and fair elections for parliament and the presidency."

Bibliography

Edwards-Jones, Imogen. *The Taming of Eagles: Exploring the New Russia*. London: Weidengeld and Nicolson, 1993.

Hunter, Shireen T. *Central Asia Since Independence*. Westport, CN: Praeger, 1996.

Kazakhstan, Kyrgyzstan, Tajikistan, Turkmenistan, and Uzbekistan, country studies. Washington: US Government Printing Office, 1997.

Maslow, Jonathan Evan. *Sacred Horses: The Memoirs of a Turkmen Cowboy*. New York: Random House, 1994.

Olcott, Martha Brill. *Central Asia's New States: Independence, Foreign Policy, and Regional Security*. Washington: United States Institute of Peace Press, 1996.

Tuvalu

Introduction

When Tuvalu became an independent nation in 1979, its first prime minister, Toaripi Lauti, observed that the new country of nine islands in the central Pacific had nothing but sunshine and a tiny part of the Pacific Ocean to claim as its own. It was so small, so remote, and such an insignificant part of the British Empire that on its independence day, Princess Margaret, the sister of Queen Elizabeth II, refused to go ashore to formally release the island chain from its status as colony. Although she had been sent to the islands to do the honors, she remained in her bunk on the New Zealand frigate *Otago*, feigning illness and sending an aid to perform the honors.

Even the United States questioned Tuvalu's ability to subsist as a nation; Sen. Jesse Helms, in fact, blocked Tuvalu's quest for independence for several years by insisting that the United States needed the islands for strategic purposes. Helms justified this contention by saying that the United States held a territorial claim on the island of Funafuti, where the capital of Funafuti is located, because its troops had occupied the area during World War II.

Despite this unenthusiastic beginning, Tuvalu has entered its third decade largely unscathed. With a population of barely 10,000, the island-nation is one of the smallest and poorest nations in the world, according to the United Nations, the World Bank, and many other international agencies. However, it is developing a strong sense of identity and maintains a proud tie to its Polynesian cultural past. These attributes have helped Tuvalu survive and perhaps fare better economically than many might have expected, according to Seve Paeniu, a native of Tuvalu whose doctoral research in economics at the University of Hawaii at Manoa focuses on Tuvalu.

As a nation with only sunshine and a small part of the ocean to its name, Tuvalu has learned to live within its means, according to S. Paeniu. By maintaining a concept of sharing and holding onto to a strong sense of Christian morality that was imported into the islands in the nineteenth century, S. Paeniu added that Tuvaluans have learned to cope with the challenges of modernity.

The country consists of nine low-lying coral atolls, essentially rings of land that encircle lagoons. It is located just below the equator, about 650 miles north of Fiji and 2,500 miles northeast of Australia. Scattered over 370 miles, Tuvalu's islands' total land-mass comprises about 10 square miles, about one-tenth the size of Washington, D.C. Although many of the islands are beautiful pieces of paradise complete with beaches, lagoons and palm trees, the nation is too remote to develop a significant tourism industry. As a result, the country depends heavily on outside investments for its income. Nearly one-tenth of its citizenry resides outside the country, and the money that these overseas Tuvaluans send home makes up a significant share of the nation's income.

History

The islands were first settled by Polynesian voyagers between the fourteenth and seventeenth centuries. Minimal contact also occurred with Micronesian migrants. An early European traveler referred to the islands as the Ellice Island chain in the early nineteenth century. Most official accounts referred to the islands in this manner until Tuvalu became independent in 1979. While Europeans plied the seas in and around the Ellice Chain, the islands themselves were largely ignored because of their small size and remote location. They became part of the British-run Gilbert and Ellice Island Colony after British sea captain E.H.M. Davis established the nearby Gilbert chain as a protectorate in 1892.

Because the Ellice Islands were part of the Gilbert chain (now part of the Republic of Kiribati), the histories of the two chains often are intertwined. However, the islands are quite different culturally: the Gilbert chain was settled primarily by Micronesians who migrated from Southeast Asia, while the Ellice Chain was inhabited mostly by Polynesians from Samoa. The manner in which they adapted first to the arrival of Europeans and later to British colonial rule differed dramatically: while the Gilbert chain often suffered from such corrupting influences as alcoholism, tobacco and prostitution, the whalers and traders who introduced these practices tended to bypass the Ellice Chain. When Christian missionaries began to arrive in both of the island chains, Gilbert residents often resisted the idea of a religion organized around the principle of one God. By contrast, Ellice Islanders were much more receptive to the idea and embraced Christianity quickly. Some Pacific Studies scholars have suggested that these contrasting attitudes developed at least partly because many of

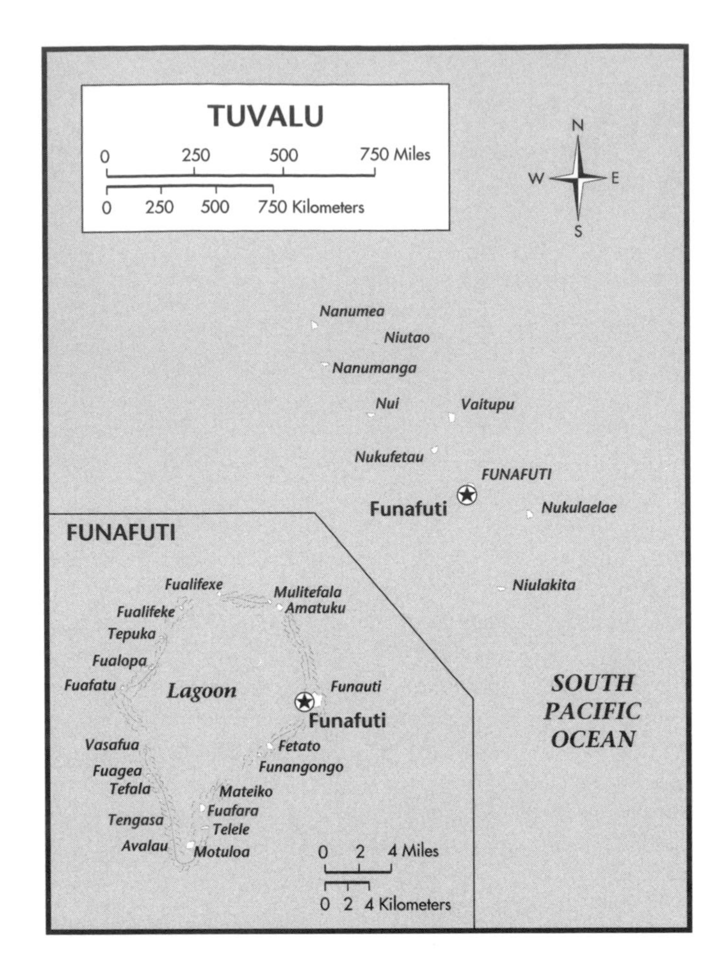

the missionaries who visited the Ellice Islands were Polynesian and, therefore, shared a cultural affinity with the islanders.

The differences between the two chains grew more pronounced during the nine decades that the islands were under British colonial rule. Gilbert residents often opposed colonial laws and educational practices; Ellice Islanders not only accepted the laws but also embraced the opportunity to have a Western education. This difference caused many Europeans to view the Ellice Islanders as more docile and obedient, and its people often were given more opportunities to advance in the colonial adminstration. As a result, many jealousies developed between the Gilbert and Ellice Islanders, with many Gilbertese complaining that the Ellice inhabitants were selling out to the "white man." Ellice Islanders, in turn, began to regard their Gilbert counterparts as lazy and unreceptive to change.

After World War II, political leaders in both island chains began to lay the groundwork for independence. As this work was being done, many felt that the cultural and social differences between the two chains were too great to be overcome, and Ellice Islanders voted in 1974 to separate from the Gilbert Islands. Two nations—Kiribati and Tuvalu—were established by the end of the 1970s.

Although Tuvalu is one of the nations of Oceania, which Tongan intellectual Epeli Hau'ofa has described as a "continent" of islands in the Pacific Ocean, it is a bit of an anomaly. Unlike many other such nations, Tuvalu's lack of natural resources has forced it not to try to develop by using what it has but by looking elsewhere for sources of income.

While the country does export small amounts of *copra* (dried coconut) that are grown on its islands, it derives much of its revenue from sales of fishing licenses and postage stamps. It also makes money through the Tuvalu Trust Fund, which was initially set up in 1987, and contained nearly $70 million as of late 1998, according to S. Paeniu. While some money that the fund earns is used to balance the Tuvalu budget, the rest is invested in overseas stocks and bonds.

In recent years, Tuvalu also has made money through less-orthodox means. The country used to rent out unused telephone numbers to companies that provide sex telephone lines, a venture that contributed nearly 10 percent of its $8.7 million budget for years. However, that practice was recently stopped when Prime Minister Bikenibeu Paeniu acknowledged that it contradicted the national motto of "Tuvalu for God." More recently, Tuvalu agreed in 1998 to sell its Internet domain name, "TV," to a Canadian company in a deal that was expected to net as much as $65 million a year. The company, TV Corp. of Toronto, expected to earn between $60 million and $100 million a year from the venture. Tuvalu was to receive sixty-five percent of the holdings.

While this deal may make the country considerably more economically self-sufficient, its long-term survival may face a different threat: global warming. Like many island-nations in the Pacific, Tuvalu's atolls rarely rise more than sixteen feet above sea level. This makes a possible rise in the ocean level a threat to their very existence. In recent years, the use of carbon dioxide has created a greenhouse effect in the earth's atmosphere, causing the polar ice caps to melt. This melting has caused many island nations in the Pacific to urge more industrialized societies to reduce their emission levels of carbon dioxide and other gases.

Prime Minister Paeniu has spoken to this issue by warning that his people would be among "the world's first victims of climate change." So far, very few industrialized nations have paid much attention to his words.

Timeline

A.D. 1300–1700 Samoan and Tongan migrants settle in Tuvalu

Although some oral histories suggest that Micronesians may have inhabited Tuvalu before the start of the Christian era, no archaeological evidence of such settlements exists. The first inhabitants are believed to be Polynesians who use prevailing winds to voyage west from Samoa and Tonga beginning in the early fourteenth century. These migrants establish settlements in both Tuvalu and the islands that comprise present-day Kiribati.

For these migrants, life is quite difficult. Unlike Samoa and Tonga, all of the islands in Tuvalu are atolls, which rise only a few feet above sea level. The soil is sandy and unfit for agriculture, and rainfall is sparse. Foods that thrive elsewhere in the Pacific, such as banana and the taro root, fare poorly in these islands.

Migrants who settle in present-day Tuvalu establish a hierarchical society in which a chief, or *tupu* or *ulu aliki*, heads a village district in which several extended families live. Each island has between four and eight of these districts, which are known as *puikaainga*, and each village contains a meeting house. The position of chief generally is hereditary, though some rise to the position through individual merit. The districts also establish a village council known as *fono o aliki*.

1537 Spanish explorers sight Christmas Island

Spanish explorers in the early sixteenth century begin sailing into the South Pacific in search of *Terra Australis Incognito*, a mythical southern continent that was supposed to be filled with wealth. On one such expedition, the crew of a ship kills its captain, Hernando de Grijalva, and sets sail for the East Indies. Only seven crew members survive this journey, but they record sighting the islands of Nonouti, and Kiritimati, or Christmas Island. Both are part of present-day Kiribati.

1568 Europeans discover Tuvalu

A captain known as Alvaro de Mendana sees canoes and fires lit on the shores of Nui, a palm tree-fringed island located about 160 miles from Funafuti, the largest of the country's islands. Nui is one of the few Micronesian communities in present-day Tuvalu.

Mendana does not drop anchor on this visit but returns in 1592. On this trip, he also sights Niulakita, the southernmost island in the chain.

1780–1820 More European sightings occur

After a penal colony is established at Botany Bay in Australia, more European ships travel through the central Pacific, many of them carrying convicts. Trade between Europe and the Americas, as well as China and India, also brings more foreign ships into Oceania.

Despite this traffic, Tuvalu remains relatively isolated. Ships captains and crew members record sighting various islands but rarely drop anchor.

1819 Islands are named

Arent de Peyser sights Nukufetau in present-day Tuvalu while commanding the ship *Rebecca*. He is traveling to Calcutta, India, and nearly has a shipwreck on Funafuti. He names the nine islands scattered around the lagoon where his ship ends up as the Ellice Group. The name refers to Edward Ellice, a British member of Parliament whose cargo de Peyser is carrying. An American explorer Charles Wilkes then uses that name on a map that he puts together of the area.

1840–60 Trade ties begin

Whalers and other European merchants intensify their activity in the Pacific. Traders such as J.C. Malcolm of Sydney, Australia, and Tom de Wolf of Liverpool, England, exchange items such as tobacco with the islanders. Generally, most of this trade activity occurs in the islands that comprise present-day Kiribati. The Ellice Islands are neglected because they are small and not heavily populated.

1861 Christian missionaries arrive

Elekana, a representative of the London Missionary Society, arrives on the island of Nukulaelae from the Cook Islands via canoe. After convincing some islanders to accept Christianity, he leaves. He returns in 1865 with the Rev. A.W. Murray and several teachers from Samoa. These missionaries convince virtually the entire Ellice population to convert to Christianity.

Ellice inhabitants are fairly receptive to Christianity, something that marks them as different from their counterparts in the nearby Gilbert Islands (which are part of present-day Kiribati). Some scholars believe that the chain's historical and cultural ties to Samoa may account for this. Samoans modify Christianity in such a way that pastors become influential political players in the local village councils. Through this, *fa'a Samoa*, or Samoan way of life, is encouraged. *Fa'a Samoa* encourages members of a village to support each other and to perform such activities as child rearing communally. This philosophy comes to be embraced by many in the Ellice chain.

Scholars also believe that Ellice inhabitants accept Christianity more readily because other Europeans generally bypass the islands in favor of the nearby Gilbert chain. Gilbert Islanders, as a result, are introduced to such temptations as tobacco, prostitution and the intoxicating beverage, sour toddy. All of these are less prevalent in the Ellice chain when missionaries arrive.

1860–90 Labor ships take natives

By the mid-nineteenth century, Europeans begin to seize control of much of Asia, Africa and various islands throughout the world. The colonies become places where the Industrial Revolution is played out. Europeans use their possessions as sources of raw materials and places where they can manufacture goods.

As part of this industrialization process, a system of indentured labor is established. After slavery is eliminated in the Americas, Europeans turn to places such as Kiribati and Tuvalu for cheap labor. Often, the islanders are carried onto ships against their will.

Inhabitants of Tuvalu end up on plantations in Australia and South America, as well as other Pacific Islands. This forced migration reduces the island chain's population from about 20,000 to 3,000.

An Office of British High Commissioners to the Western Pacific is established in 1877 to control such labor abuses.

1892: May 27 British rule begins

A series of wars erupts in the Gilbert Islands, spanning about four generations and ending in 1892 with a war at Nea. During this war, the people under one ruling chief from the Gilbert Islands are said to have learned through their gods that visitors would be arriving on the island. Those visitors turn out to be the crew members of the *HMS Royalist,* a British ship commanded by Capt. E.H.M. Davis. Believing the only way to stop the fighting is to establish a protectorate, Davis raises the British flag over the Gilbert Islands on May 27, 1892.

During this process, the Ellice Islanders also are taken over as a protectorate of the British. The British centralize their rule and establish a series of local governments. They declare Banaba, Fanning, Washington, the Gilbert and Ellice islands all to be one colony. Christmas Island is added to the colony in 1919 and the Phoenix Island group in 1937.

Although the British rule initially strives to preserve the autonomy of each island, the imperial administration is unable to satisfy that goal. An administrative headquarters is established on Banaba, which is located to the west of the Gilbert chain, after phosphate is discovered in 1900. Because phosphate becomes so profitable, the British run the colony by putting first their interest in that industry. As a result, development of the other islands is neglected, and the cultural differences between the Polynesians of the Ellice chain, the Micronesians of the Gilbert chain and the Banabans of Banaba are ignored.

Work in the phosphate mines becomes the primary means of subsistence among the islanders, and gradually such traditional foods as coconuts are replaced by canned imports.

1924 Ellice school opens

British educator David Kennedy opens the Ellice Islands School on Funafuti. It later is moved to Vaitupu, where it gains a reputation for academic excellence. Kennedy imposes a harsh, military style of discipline on his students and imposes a rigid schedule of lessons, sports and work in the school compound.

Even though he expects his students to behave properly, he drinks heavily and keeps a group of "house girls" around to serve his personal needs. Nevertheless, he gains the respect of many islanders because he encourages them to take pride in their culture and to preserve their native language. Many students from his school eventually become administrators in the colonial regime.

Historian Barrie MacDonald notes that Kennedy's legacy today is mixed. He says that while many of the students educated at the Ellice School eventually become involved in the colony's quest for independence, the encouragement of a singular culture may have led to the eventual partition of the colony into two separate nations.

1941: December 8 Japanese bomb Banaba

Japan bombs Pearl Harbor in Hawaii on Dec. 7, 1941, which draws the United States into World War II. One day later, Japanese war planes drop six bombs on the British administrative headquarters on Banaba in present-day Kiribati. A French cruiser evacuates the European and Chinese workers, and shortly afterwards Japan occupies the island.

Over the next few months, Japanese troops occupy the Gilbert Island chain. Most Europeans are evacuated, and the islands generally become isolated. Japanese administrators force the islanders to serve as sentries, looking out for enemy ships. They also ration food and order the islanders to serve at their command. Most obey out of fear.

1943: November 20 U.S. invades Tuvalu

After abandoning plans to move through Japanese territories in the Pacific and eventually arrive on the coast of Japan, Allied troops decide to make a drive through the central Pacific. The United States begins an amphibious training program in Hawaii and begins an invasion of Micronesia by landing in the Ellice chain.

The United States uses the Ellice Islands as a base from which to launch attacks against Japanese troops who have occupied the Gilbert chain. An assault begins on Tarawa in the Gilbert chain on November 20. After three days of heavy fighting, the Americans take control of the Gilbert group.

U.S. troops remain in the islands until 1946. During the occupation, many islanders are hired to rebuild homes and British administrative buildings. The American presence also brings more exports into the islands. Manufactured products and processed foods rapidly begin to replace handmade items and foods grown on the islands.

Reminders of the war remain. Several World War II landing craft and B-24 bombers remain on Nanumea. In addition, as part of the rebuilding effort, the United States digs an airstrip on Funafuti, which leaves about forty percent of this island permanently uninhabitable.

1965 Steps toward independence begin

After World War II, British rule is restored. The British begin to re-establish a system of local governments, and with assistance from the United States and groups such as Christian missionary societies, begin building schools and other community facilities.

However, the war drains the British treasury and makes many of its colonies a financial liability. At the same time, the United Nations starts criticizing colonial rule and many nations in Asia and Africa gain independence. On top of this, the island of Nauru gains independence from Australia, which causes the price of phosphate to rise. All of these changes make the Gilbert and Ellice Island Colony increasingly expensive for the British government to administer, so Britain tries to rid itself of the colony.

A Native Magistrates' Conference begins meeting annually in 1952 to provide training to local governmental leaders. This move is followed by the establishment of the biennial Colony Conference in 1965. The conference provides a forum where residents debate issues, air grievances, and propose ideas.

1974 Ellice Islanders choose to separate

When the British formed the Gilbert and Ellice Island Colony, they brought together dozens of islands with little in common. There were similarities among the islanders within each group, but the Gilbert and Ellice people shared no common ideology, culture, or language. Each recognizes separate leaders.

As a movement toward independence intensifies, these differences start to become more apparent. Ellice Islanders realize that they are a minority within a predominantly Gilbertese culture and fear that they will not be able to command power in a new nation. Many Ellice people also regard the Gilbertese as backward. At the same time, many Gilbert Islanders feel that residents of the Ellice Chain have educational advantages over them and receive favorable treatment from foreigners.

Such differences lead to a referendum in which the Ellice Islands vote to separate. The group becomes the independent nation of Tuvalu, while the remaining islands are to become the nation of Kiribati.

1979: October 1 Nation of Tuvalu is established

After the separation from the Gilbert group, the Ellice Islands take the name "Tuvalu," a traditional name for the islands, which means "eight standing together." Although there are nine islands in the nation, one is uninhabited, which is how political leaders justify the name.

A constitutional convention takes place in London in February 1978, and Tuvalu is declared an independent member of the British Commonwealth of Nations the following year. The new nation is a constitutional monarchy. The Queen of England is the monarch, and Sir Fiatau Penitala Teo is chosen as governor-general. The chief minister of the islands at the time of independence is Taoripi Lauti. He becomes Tuvalu's first prime minister.

Tuvalu's constitution, which is altered in 1986, establishes a Parliament of twelve elected members. Four islands send two members each, two vote together for one member, and the other three inhabited islands send one member each. Elections take place every four years.

1987 Tuvalu trust fund is established

Because Tuvalu is small, its economic resources are limited. The country receives some income from exports of *copra* (dried coconut), licenses it sells to foreign ships who wish to fish for tuna in its waters, and the sale of postage stamps. Otherwise, most of its income comes from remittances—the money that overseas Tuvaluans send back to families.

In hopes of developing an additional source of income, the government sets up the Tuvalu Trust Fund with help from Australia, New Zealand, and Great Britain. The fund starts with contributions from the three nations and, over time, receives money from Japan, South Korea, and the government of Tuvalu, as well. That money is placed in portfolio investments, and a certain amount is paid each year to the government so that it can balance its budget. The fund totals $36.8 million in mid-1994.

1989 Tuvalu elects first woman to Parliament

Naama Maheu Latasi, wife of Kamuta Latasi, becomes the first woman to hold office in Tuvalu. Although her father is from Nanumea, the island she represents, her mother is from Kiribati, and Latasi spends her childhood and early married years in that country. However, she and her husband move to Tuvalu shortly after the separation from Kiribati.

As a female member of parliament, Latasi devotes much of her energy to strengthening the role of women in the public sector. Cultural traditions in many Pacific island-nations assign traditional domestic roles to women. Latasi contends that these roles are changing as the islands modernize, and she encourages her constituents to adjust to a new era.

However, she believes that giving more opportunities to women should be a way of strengthening a society rather than disrupting it. "Socially, it's a very delicate transitional period through which Tuvalu is passing," she says in a 1990 interview with *Pacific Islands Monthly.* "We have to remain very firm and steadfast to keep the fabric of our society strong. We want to maintain our identity and dignity."

She adds: "I am a strong woman on principles, and I like to see our women take their rightful place in this changing society, but this should take place in harmony, rather than conflic with the men-folk of our nation."

The Mid 1990's Road Construction Begins

The main roads of Funafuti become paved through differing national works to improvetransportation.

1995: October 1 Prime minister changes flag

Prime minister Kamuta Latasi officially declares that the former flag of Tuvalu, which includes the symbol of the British Union Jack in its upper left-hand corner, no longer will be used. The new flag contains an insignia of a badge of arms that used to represent Tuvalu when it was part of the Gilbert and Ellice Island colony. In addition, the flag contains nine stars, one for each of the islands.

Latasi's move is both political and historical. A long-time politician, he was active first in the movement to gain independence from the British and later in the effort to separate Tuvalu from present-day Kiribati. He serves as Tuvalu's first High Commissioner to Fiji and is active in Parliament before becoming prime minister in December 1993. He wants Tuvalu to become a republic, so that its head of state no longer is the Queen of England.

In addition, Great Britain's support of France's nuclear tests near Tahiti in 1995 angers Latasi. Like many political leaders in the Pacific, he sees such tests as detrimental to the health of his people and contends that European decisions to go ahead with the experiments are acts of arrogance.

By removing the Union Jack from the national flag, he not only registers his opposition to such acts but also encourages Tuvaluans to sever their ties emotionally to their former colonizer. Unfortunately for Latasi, the people of the islands do not share his views and protest the changed the flag design. Latasi is voted out of office and the original flag is reinstated in April 1997.

1997: August Tuvalu and Kiribati sign friendship treaty

Tuvalu's decision to separate from Kiribati in the early 1970s represented a strong statement of independence, but relationships between the two island-nations always remained cordial. That spirit is exhibited in a formal Treaty of Friendship. According to Kaburoro Ruaia, Kiribati's Secretary for Foreign Affairs, the treaty deals mainly with transportation services, trade, and visitor arrangements.

1998: October 1st The 20th Anniversary of Tuvalu's Independence is celebrated

Early 1999 Telephone service quickly spreading

By 1999, Funafuti Atoll has a regular telephone service, and there are connections to all the outer islands through the Post Offices, with about 700 subscribers. A telephone directory is published for the first time, though communications with the outer islands is also available by radiophone.

Bibliography

Adams, Wanda. " 'Double Ghosts' remembers early traders," *Honolulu Advertiser*, March 29, 1998.

Chappell, David. *Double Ghosts: Oceanic Voyagers on Euroamerican Ships.* Armonk, NY: M.E. Sharpe Press, 1997.

Douglas, Norman. "Whither now, Tuvalu?" *Pacific Islands Monthly*, December 1993.

Field, Michael. "Tuvalu goes from phone sex to Internet riches," Agence France-Presse, Octoboer 20, 1998. (via *Pacific Islands Report,* Pacific Islands Development Program/Center for Pacific Islands Studies, University of Hawai'i at Manoa).

Ioane, Timeon. "Culture, A Key that Binds Tuvaluan Society," Pacific Island Development Program/Center for Pacific Island Studies, November 23, 1998. (via *Pacific Islands Report,* Pacific Islands Development Program/Center for Pacific Island Studies, University of Hawai'i at Manoa).

Kellner, Tomas. "Treasure Island.net (Tuvalu makes deal for use of .tv Internet domain," *Forbes*, September 21, 1998.

"Kiribati and Tuvalu Sign Treaty of Friendship," *PACNEWS*, August 8, 1997. (via *Pacific Islands Report*, Pacific Islands Development Program/Center for Pacific Island Studies, University of Hawai'i at Manoa).

Kiribati: Aspects of History. Tarawa, Kiribati: Ministry of Education, Training and Culture (1979).

Kristoff, Nicholas D. "In Pacific, Growing Fear of Paradise Engulfed," *New York Times*, March 2, 1997.

MacDonald, Barrie. *Cinderellas of the Empire: Towards a History of Kiribati and Tuvalu.* Canberra, Australia: Australian National University Press, 1982.

McManus, Diana. "A woman's champion," *Pacific Islands Monthly*, May 1990.

Nisbet, Irene. "Rumble of Republicanism," *Pacific Islands Monthly,* October 1991.

Paeniu, Seve. "First Tuvalu Governor General Rests in Peace," *Tuvalu Echoes,* December 1998. (via *Pacific Islands Report,* Pacific Islands Development Program/Center for Pacific Islands Studies, University of Hawai'i at Manoa)

United Arab Emirates

Introduction

The United Arab Emirates (UAE) is located on the Persian (Arabian) Gulf and the Gulf of Oman, between latitude 22° 50' and 26° 50' north and longitude 50° and 50° 25' east. Bordering countries are Oman (south), Saudi Arabia (south and southwest) and Qatar (west). Iran is directly across the Persian Gulf. Other countries bordering on the northern end of the Persian Gulf are Iraq, Kuwait, and the island of Bahrain. The Persian Gulf opens into the Gulf of Oman, the access to the Indian Ocean.

The UAE is about the size of Maine with an area of 29,182 square miles (75,581 square kilometers). With a coastline of 483 miles (777 kilometers) it is still a vast desert of rolling sand dunes with mountains in the east, and some agricultural areas. The lack of freshwater resources is surmounted by desalination plants.

The Federation of the United Arab Emirates, *Al Umarat al Arabiyah al Muthahidah* in Arabic, formed in 1971 from seven sheikhdoms or emirates. From east to west the sheikhdoms are Abu Dhabi (Father—or Fatherland—of the Gazelle), Dubai, Sharjah (Eastern), Ajman, Umm al Qaiwain, Ras al Khaimah (Point of Ten), and Fujairah.

Abu Dhabi comprises eighty-six percent of the land mass of the UAE with around 26,000 square miles (67,350 square kilometers). It is the richest of the seven emirates and important tribal alliances link its capital, Abu Dhabi, with the Rub al Khali (the Empty Quarter) edged by the fertile Liwa oasis. Dubai (or Dubayy) is the second largest of the emirates with 1,500 square miles (3,900 square kilometers). A natural harbor created by Dubai's creek created and sustains Dubai's centuries-long tradition as a major trading post. Sharjah on the Arabian Gulf is 1000 square miles (2,600 square kilometers) and has small enclaves toward the eastern coast. The smallest Emirate, Ajman, is 100 square miles (250 square kilometers). Ajman is nearly surrounded by the Emirate of Sharjah, with the small enclaves of Manama and Masfut located in the Hajar Mountains. Ajman is a traditional center for building *dhow* (ocean-going wooden craft), and a repair yard for modern steel ships. Fujairah (or Al-Fujayrah), at 450 square miles (1,150 square kilometers) lies on the east coast on the Gulf of Oman with direct access to the Indian Ocean. It has no coastline on the Arabian Gulf, nor any oil. Umm al Qaiwain (or Umm alQaywayn), the second smallest of the Emirates at 300 square miles (750 square kilometers), is a coastal town between Ajman and Ras al Khaimah. Its most important economic activity is fishing. Furthest north is fertile Ras al Khaimah (or Ras al-Khaymah), covering 650 square miles (1,700 square kilometersFrom ancient times the region was known for its seafaring and trade—in copper, pearls, and gold—and in the twentieth century for its oil and natural gas. The coastal areas have an ancient tradition of boat-building. Early canoelike craft, the *shasha* were made from palm fronds secured with palm fibers and waterproofed with fish oil. These were surpassed by the larger, timber-stitched dhows, and the galleon-like baghala. Baghala required a crew of twenty, and had a cargo capacity of five hundred tons.

History

From ancient times the region was known for its seafaring and trade—in copper, pearls, and gold—and in the twentieth century for its oil and natural gas. The coastal areas have an ancient tradition of boat-building. Early canoelike craft, the *shasha,* were made from palm fronds secured with palm fibers and waterproofed with fish oil. These were surpassed by the larger, timber-stitched dhows, and the galleon-like baghala. Baghala required a crew of twenty, and had a cargo capacity of five hundred tons.

Until the Japanese introduced cultured pearls on the world market in the 1930s, pearl diving was the primary occupation of coastal populations. Pearl traders did not openly discuss prices. Negotiations, based on an ancient ritual of finger language conducted under a cloth, allowed the seller to negotiate with several buyers simultaneously. No buyer would know the other bids; all deals were kept private.

Collectively in the nineteenth century the region on the eastern part of the Arabian Peninsula and the Arabian Gulf was known by many descriptive names: Trucial Coast, Trucial Sheikhdoms, Pirate Coast, Coast of Oman, and Trucial States. "Trucial" refers to the nineteenth century treaties between these separate sheikhdoms and Great Britain, including similar treaties between Great Britain and Oman, Bahrain, and Qatar.

The basis of social structure is the tribe. Within the patriarchal tribal structure the ruling families of the emirates remain in power for long periods. The Al Nahyan family of Abu Dhabi and the Al Maktoum family of Dubai, both of the Bani Yas tribe, have been in power since 1855 and 1886 respectively. The Al Qassimi in Sharjah and of Ras al Khaimah have ruled since 1883 and 1915 respectively. The remaining ruling families, the Al Mu'yam of Ajman, the Al Ali of Umm al Qaiwain, and the Al Sharqi of Fujeirah have ruled for long periods also. Qawasim, long the largest tribe, dominated the coast from Sharjah to Ras al Khaimah. Within this tribal structure ties with the ruling family and personal connections are the most significant allegiances.

In the late 1960s, prior to federation, an estimated twenty-seven to thirty-five tribes existed. Six of these were dominant tribes: Beni Yas, Manasir, Dhawawhir, Awamir, Na'im, Dubai, and Sharjah. In this same period the population of less than 200,000 people was spread through the seven emirates with one or two tribes dominant in each of the emirates. By 1995 the population had grown to 2.9 million.

A major distinction is made between nomadic peoples (the bedouin or *badu*) and the settled people (*hadar*) who reside in rural settlements and towns. The nomadic bedouins roam the inland regions with their camels, sheep, and goats between water sources in search of grazing lands. The land a tribe habitually ranges is its dirah. The *hadar* live at inland oases and coastal regions cultivating dates, fruit trees, barley, and vegetables. Coastal populations traditionally engaged in fishing and pearl-diving. Hunting with falcons is practiced in the region and bedu culture. Wild falcons— peregrine, saker, and lanner— are captured during their annual migration. Hunters lure the falcons with pigeons, and then trap the falcons in nets or pits. Larger, female birds are preferred. Training the falcons to hunt takes two to three weeks. Gradually tamed, the falcon becomes familiar with its owner and hunts the favored quarry, the *houbara* or MacQueen's bustard. Traditionally, all but the best birds were released back into the wild at the end of the season.

Islam is the state religion. Traditional education involves teaching boys to recite the Holy Koran. Since federation, government policy supports nationwide education for boys and girls, and includes literacy programs for adults. In addition, the government has a policy to provide free education for all children of Arab expatriates.

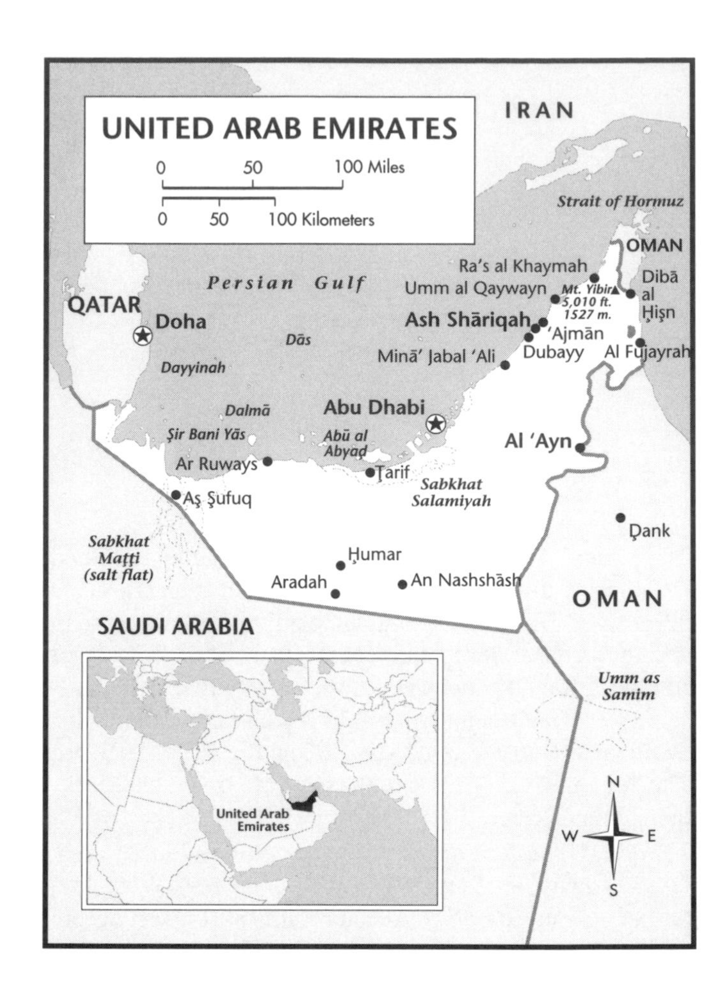

Timeline

5000–3000 B.C. Evidence of early contact within the Gulf region

Contact between Mesopotamia and the Gulf region is documented by archaeological evidence of coastal and inland settlements along the Arabian Gulf coast. Pottery found at these sites is similar to that of archaeological farming sites near Ur in Sumer. Labeled "Ubaid" after the Sumerian site where it is first found, this pottery documents contact with southern Iraq. Shell *middens* (a shell refuse heap) are found associated with settlement sites. Shells collected for food are used to make jewelry and decorative hooks.

3000 B.C. Trade and agriculture

Dilmun (now Bahrain) and Magan, an area roughly corresponding with the present-day United Arab Emirates (UAE) and Oman, supply the city states of southern Mesopotamia with a range of raw materials, including copper.

Tombs and stone cairns characteristic of this Bronze Age period are found at the entrance to *wadis* (stream beds that remain dry except during a rainy season), or along ridge lines. These tombs are also known as "Hafit tombs," named after the site where they are first excavated at Jebel Hafit near Al Ain. Another characteristic tomb of the same period is beehive-shaped. Signs of an oasis-based agricultural economy based on wheat, barley, and date palms and signs of the animal husbandry of sheep, goats and cattle are found at coastal sites.

2500–2000 B.C. Umm al Nar burial site

The small island of Umm al Nar near Abu Dhabi is used as a burial site. Circular tombs, about thirty feet (ten meters) in diameter, some with bas-relief depictions of animals, are used for complex burials. Multiple burials accompanied by sophisticated crafts are a result of the far-reaching trade networks. Later, the tomb style changes, becoming rectangular and standing above ground.

Tribes live in settlements with mud brick defensive towers, walls, and moats. They use fireplaces, ovens, water wells, and potsherds and remain continuously occupied for thousands of years. Copper mining, processing, smelting and export takes on increasing importance in the Magan region. Copper mines further to the north in the Magan region in Oman are supposed to be the site of King Solomon's Mines.

2000–1300 B.C. The Wadi Suq period is characterized by tomb sites

In the Wadi Suq period (named for the excavations sites in Wadi Suq of circular) below-ground tombs ten to thirteen feet (three to four meters) in diameter are used. Later rectangular, above-ground tombs are used in coastal settlements near Ras al Khaimah and Oman. Grave goods also change. The variety of goods indicate extensive maritime trade networks.

1300–300 B.C. The Iron Age

Agriculture develops and the number of settlements increases. The *falaj* system of irrigation is developed. A *falaj* sundial measures the amount of water used for irrigation by the minutes, hours and days of the week. A plot of land has a prescribed allocation of water regulated by the *falaj*. These water rights are sold with the property. The irrigation system is maintained by the owners or landlords.

The domestication of the camel and the development of a saddle permitting large loads of goods spurs overland transport. The rise of new land routes around 1000 B.C. complicates the competition between Persian Gulf and Red Sea trade routes. Overland trade routes also alter migration patterns by opening the coasts to inland nomads.

Historic United Arab Emirates

550 B.C. Cyrus the Great founds the Persian Empire, the Achaemenid

The Persian Empire unites western Asia and Egypt facilitating Persian Gulf trade as far as India.

331 B.C.: October 1 Alexander the Great defeats King Darius III

Macedonian king Alexander the Great defeats King Darius III of Persia at Gaugamela.

Alexander the Great routes the Persians at Gaugamela and enters Babylon continuing on to Persepolis, the capital of Persia. As Darius flees, he is killed by his officers. Alexander becomes King of Persia. Alexander continues into India and returns to Babylon in 325 B.C. He marries a Persian princess. In spite of his untimely death at the of age thirty-three his empire links Greek and Asian cultures and opens the Persian Empire to foreign traders. However, this trade remains under local Persian control.

300 B.C. Regional trade network develops

Trade links the whole Arabian peninsula with the Mediterranean region, the Indian Ocean, and the Indus civilization. Goods such as frankincense, amphorae (two-handled, narrow necked jars used by the Greeks and Romans), and alabaster containers, along with minted coins, provide evidence of the expanse and antiquity of maritime trade.

A.D. 77 Pliny's Natural History mentions the Arabian shore of the Gulf

Pliny describes Persian Gulf shipbuilding techniques. Other classic writers—Ptolemy and Strabo—describe military and commercial use of the Persian Gulf by peoples of the area.

1st century *Periplus of the Erythraean Sea,* is an account of Indian Ocean trade

A written account by a Greek hired in the Roman navy refers to the Arabian shore of the Gulf as Persian.

2nd–6th century Arab tribes migrate

Arab tribes migrate from the south Arabian coast to the Hajar Mountain area and the Al Ain Oasis. Later Arab migrations are from the north.

224–642 The Sassanian Period

A local dynasty, the Sassanid, triumphs in battle over the last Parthian ruler from Persia. A period of prosperity and governmental power ensues under more than forty Sassanid rulers. The Sassanids extend their rule to eastern and southern Arabia, absorbing both present-day United Arab Emirates and Oman, and dominate trade. Struggles with Rome, Byzantium, and finally the Arabs, end Sassanid reign as Islam is introduced.

632 The Prophet Muhammad dies

The Muslim prophet Muhammad dies. Muhammad founds a new religion, Islam, that is destined to become the second-largest religion in the world. A monotheistic faith, Islam is an offshoot of both Judaism and Christianity (although Muslims, followers of Islam, believe that Jesus is a prophet, rather than son, of God). Muhammad claims to be the last prophet. In the years following Muhammad's death, Islam spreads throughout the Arabian peninsula, into North Africa and the Middle East. Arabs spread the faith and incorporate newly converted territories into an Islamic Empire.

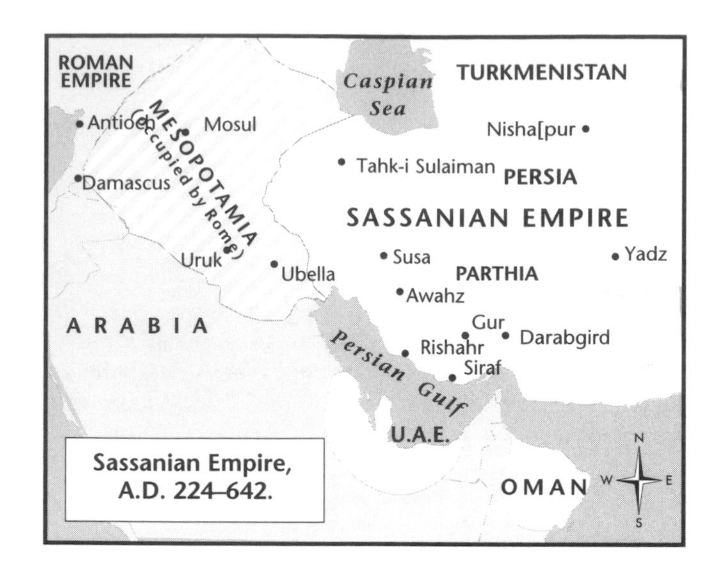

Sassanian Empire, A.D. 224–642.

8th–9th centuries Arab seafaring and trading at their height

Arab Muslim rule unites both sides of the Persian Gulf. Arab and Persian merchants voyage to India and as far as China. Julfar, a small town north of present day Ras al Khaimah, is an important port city and the terminal of a major trade route to what is now Al Ain. Ships sail with the northeast monsoon in September or October and await favorable winds to and from China. The dangerous route to China from Julfar can take eighteen months.

Ships are the pointed, double-ended *dhow* (term of Swahili origin). No iron nails are used. Instead boats are stitched together with coconut husk thread.

Tenth century The Ismaili tradition dominates the Gulf region

The Ismaili Shia faith is a powerful force. The Ismaili originate in Iraq but move to the gulf to escape Sunni authorities. In the ninth century a missionary, Hamdan Qarmat, had sent a group from Iraq to Bahrain to start an Ismaili community. These followers became known as Qarmatians. In the tenth century these Ismailis of Bahrain control the coast of Oman collecting tribute from Baghdad and Cairo. Persian and Arab forces drive the Ismailis from Iraq and eastern Arabia by the end of the tenth century.

Modern History

1498 Da Gama rounds the Cape of Good Hope

Portuguese explorer Vasco Da Gama opens maritime trade routes to India by sailing from Europe, around the Cape of Good Hope at the southern tip of Africa, to Asia.

1507 Portuguese capture Muscat and Hormuz

In the sixteenth century the Portuguese dominate trade in the Gulf and the Indian Ocean. They capture Muscat (modern-day Oman) and Hormuz. Control of Muscat, Hormuz (an island off modern Iran), and the Indian coast allows the Portuguese to control trade routes, customs, and duties. The Portuguese repress Islam. Goa is made their capital in India.

Seventeenth century The Dutch, French and English vie for control of the region

The Dutch, French, and English East India Companies are rivals for trade in the Arabian Gulf. Tribes from Oman expel the Portuguese from Muscat, but fight the Persians into the next century as they attempt to take Muscat back. The founder of the dynasty ruling Oman, Imam Sultan bin Saif, finally expels the Persians from Oman. This dynasty rules Oman into the end of the twentieth century. The Persians remain in the region in Bahrain.

1600 English East India Company forms

The English East India Company is a private trading company with naval forces to protect its economic interests.

1622 Portuguese driven out of Hormuz

English and Persian forces drive the Portuguese from Hormuz.

1650 Imam Sultan bin Saif expels the Portuguese from Muscat

Second Omani ruler of the Ya'ruba dynasty, Imam Sultan bin Saif, drives the Portuguese out and works toward uniting Oman.

1742 The Persians establish control of Muscat

The Persians, led by Nadir Shah (1688–1747), establish control of Muscat. Within two years the Omani drive out the Persians.

1747 Ahmed bin Said drives out the Persians

Ahmed bin Said drives the Persians from Oman and is elected Imam (leader claiming descent from Mohammed).

1747 Persian leader Nadir Shah dies and the Qawasim move into Persia

The Portuguese and Persians have been driven from Muscat. Following the death of Persian leader Nadir Shah, the Qawasim of Ras al Khaimah seize the city of Lingah almost directly opposite on the Persian shore. The Qawasim now rival the imams of Omani and Muscat for control of the Gulf trade.

The Qawasim (singular, Qasimi) are a migratory group of central Arabia called Hawala. In the mid-seventh century they settle in Persia, eventually dwelling on both shores of the Gulf. The Qawasim force the Omani to recognize Qasimi independence. During this time period Oman is encompassed in the territory now considered the UAE.

1761 Abu Dhabi is founded

The city, later the capital, Abu Dhabi, is founded.

1820 The first General Treaty of Peace

The British and the people along the coast of Oman (Trucial Sates) negotiate a treaty. The British campaign to quash the Qawasim, a maritime people of the Gulf, ostensibly to safeguard the maritime routes of the British East India Company. However, as the British control the Persian Gulf, they are also establishing their supremacy in the region. Their increased involvement acts against the claims of other European powers.

1830s Height of the pearl trade

Pearling is traditional in the coastal regions. India is a major market for pearls. Special boats called *sambuqs* are used for pearling. They have both sails and oars. The upper hull is treated with shark oil to prevent warping. The lower half of the hull is covered with a mixture of lime and sheep's fat.

Pearling is dangerous work that must be undertaken amongst sharks, barracuda, and swordfish. The divers wear white cotton suits to prevent stings by giant stingrays and jellyfish. Leather stalls are used to protect fingers. The big toe of the right foot, used to push the diver along the seabed, is also leather covered. A bone or wooden clip over the nose and a knife to pry open the oysters are the diver's tools. Divers can stay down for as long as two minutes and may make between sixty and one hundred dives per day.

1832 First Maritime Truce

The British Resident in the Gulf, Captain Hennel, frames a truce outlawing warfare between May 21 and November 21 each year. This is the time of pearl-diving. The truce is so popular that it is renewed annually until 1843 when it is extended for another ten years. In 1853 the truce is made permanent. The region becomes known as the Trucial Coast, designating the areas covered by the truces.

1833 Al-Maktoum family establishes dynasty in Dubai

A subtribe of the Bani Yas, the Bedu Al Bu Falasah, secede from Abu Dhabi and settle in Dubai. One of the leaders Maktoum bin Buti establishes the dynasty that continues to rule Dubai.

1853 The Perpetual Peace between Great Britain and the Trucial States is signed

The Trucial States agree to cease hostilities at sea and in return Britain will safeguard navigation, guaranteeing to protect the Trucial States from external attack. This peace allows the sheikhs to concentrate on internal affairs and attend to border disputes. It also allows the British to increase their dominance and control of the region.

1855–1909 Sheikh Zayid bin Khalifah of Abu Dhabi

Sheikh Zayid bin Khalifah, known as Zayid the Great, unites the Bani Yas tribe and influences neighboring tribes. In 1905 he calls a meeting of all the rulers of the Trucial states in order to resolve existing disagreements. These efforts are somewhat successful and this endeavor is seen as the forerunner of the federation promoted by his grandson and namesake.

Since no fixed order of succession exists after his death, rule passes amongst his seven sons with the eventual murder of his fifth son, Sheikh Hamdan bin Zayid (r. 1912–22) by his brother, Sultan bin Zayid (r. 1922–26). Sultan eventually incurs the wrath of his other brothers and is killed by Sheikh Saqr bin Zayid (r. 1926–28). Sheikh Saqr meets the same fate at the hands of the eldest brother, Sheikh Khalifah, who passes rule to Shakhbut bin Sultan, his nephew.

1874 The British East India Company dissolves

The British East India Company, which has been operating in the region for two centuries, dissolves.

1878–1958 Sheikh Said bin Maktoum al Juma

Sheikh Said bin Maktoum al Juma rules Dubai from 1912 to 1958.

1892 Exclusive Treaties are signed

Great Britain signs separate one-sided treaties with each sheikhdom. This prevents the rulers of the Trucial Coast from forming agreements with any other government. By the treaty terms Britain will control defense and foreign policy, but will not interfere in internal affairs. However, the British representatives in the region, the Political Resident in the Gulf, and the Commissioners in the emirates, wield considerable influence as they act to control the slave trade, arms trade, the granting of pearl and sponge-hunting permits, and of oil exploration permits. These treaties are in force until 1971 with the formation of the separate states of Qatar, Bahrain and the United Arab Emirates.

1898 Lord Curzon visits the Gulf

In response to increasing interest shown in the region by Russia and Germany, Britain reinforces its presence in the region by sending George Nathanial Curzon (1859–1925), known as

Lord Curzon, on a visit to the region. Lord Curzon is known for his expertise on Asia and the Middle East. His writings include *Asiatic Russia* (1889), *Persia* (1892), and *Problems of the Far East* (1894).

1903–89 Sheikh Shakbut bin Zayed

Sheikh Shakbut bin Zayed rules Abu Dhabi from 1928 to 1966.

1904–81 Sheikh Rashid bin Humaid al Nuaimi

Sheikh Rashid is Ruler of Ajman from 1928 to 1981. His son Sheikh Humaid succeeds him.

1904 The British India Steam Navigation Company calls at Dubai

Every two weeks, on the passage from Bombay to Basra, a ship delivering passengers and cargo calls at Dubai. The ships anchor a mile off the coast while passengers and cargo are transferred to barges that make way to the shore.

1908–75 Sheikh Mohammed bin Hamad al Sharqi

Sheikh Mohammed is the first Ruler of Independent Fujairah, 1952–75.

1908–81 Sheikh Ahmed bin Rashid al Mualla

Sheikh Ahmed is Ruler of Umm al Quwain from 1929 to 1981.

1912 Birth of Sheikh Rashid bin Said al Maktoum

Sheikh Rashid rules Dubai from 1958 and is the first Vice-President of the United Arab Emirates in 1971.

1918 Birth of Sheikh Zayed bin Sultan al Nahayyan

Sheikh Zayed bin Sultan al Nahayyan in born. He is named after his grandfather, known as Zayed the Great. His mother, Sheikha Salama, is the sister of the head of the Qubaisat subsection of the Bani Yas. He has three elder brothers, Shakhbut, Khaled, and Hazza. When his father, Sheikh Sultan, is killed, he and his brothers live in exile until his brother, Sheikh Shakhbut, is elected Ruler or Emir.

Sheikh Zayed bin Sultan al Nahayyan's upbringing is traditional as he spends his adolescence in Al Ain, village of the Buraimi Oasis. In 1966 Sheikh Zayed becomes the ruler of Abu Dhabi, and in 1969 he deposes his brother. In 1971 he becomes the first President of the newly federated United Arab Emirates.

1920 Birth of Sheikh Saqr bin Mohammed al Qawasimi

Saqr bin Mohammed al Qawasimi is born. His rule as Sheikh Saqr of Ras al Khaima begins in 1948.

1920s Post World War I is the height of British influence

Following the defeat of Germany in World War I and the collapse of the Ottoman Empire, Britain is at its most influential in the Gulf region.

1928–66 Reign of Abu Dhabi by Sheikh Shakhbut bin Sultan

Sheikh Shakhbut bin Sultan's mother promotes his long reign by pledging her sons to refrain from fratricide.

1930s The Gulf pearl industry declines

The pearl industry declines following the arrival of the Japanese cultured pearl. As the pearl market slumps, workers migrate as oil is discovered in Bahrain and Saudi Arabia and in Kuwait and Qatar in the 1940s. This abundant work in the oil sector accelerates social and economic change in the region.

1932 Britain's Imperial Airways establishes service to Sharjah

The establishment of an air route to Shajrah brings the region into the modern age of transportation.

1938: December 3 The first Koranic school opens

The Al Ahmadiyya school opens with nearly 200 students. The opening of two other schools follow.

1936–52 Trucial States sign agreement with Iraq Petroleum

The rulers of the Trucial States sign oil concessions with the Iraq Petroleum Company.

1946 First bank in the region opens

The British Bank of the Middle East opens an office in Dubai.

1951 Sheikh Mohammed bin Saqr Al-Qasimi is acting Ruler of Sharjah

Sheikh Mohammed bin Saqr Al Qasimi becomes the acting Ruler of Sharjah.

1951–65 Sheikh Saqr bin Sultan Al-Qasimi rules Sharjah

For fourteen years, Sheikh Saqr bin Sultan Al-Qasimi rules Sharjah.

1951 Causeway links Abu Dhabi and mainland

A causeway is built linking the island of Abu Dhabi to the mainland. This causeway is replaced by the Al Maqta Bridge in 1968.

1951 Formation of the Trucial Oman Levies

The Trucial Oman Levies (TOL) are established. These are local forces made up of nationals, commanded by British officers with the help of officers from Jordan and Aden. They are formed to maintain law and order throughout the territory. Relying less on British troops, the Levies operate with the sheikh's permission. TOL forces do not operate within towns where the local ruler's forces maintain security. (See also 1956.)

1952 The Council of Trucial States forms

The first council of Sheikhs, made up of the rulers of Abu Dhabi, Dubai, Sharjah, Ajman, Umm Al-Qaiwaon, Ras al Khaimah, and Fujairah, forms. The council meets twice a year and has no executive powers. Each emirate replaces British passports with its own.

1952 Disputes over the frontier occur

The Buraimi Oasis area is claimed by Saudi Arabia, Abu Dhabi, and Oman. Britain supports Oman's claim. The British occupy the Oasis and drive out the Saudi troops.

1953 School opens in Sharjah

The first modern school in the region opens in Sharjah. It offers education for boys, and teachers are recruited from Egypt. Recruiting teachers from Egypt is a politically sensitive issue with the British, since the 1952 Egyptian revolution advocates Arab liberation unity. Nevertheless, within a year a school for girls opens.

1956 The Trucial Oman Levies become the Trucial Oman Scouts

The Trucial Oman Levies (TOL—see 1951) become the Trucial Oman Scouts (TOS). These troops can be commanded by foreign nationals, e.g., British or Jordanian.

The Scouts offer training for the military and police of the separate emirates. The TOS also serve an educational function as they provide language and trade training before a school system is established. Upon the federation of the UAE the TOS provides the nucleus of the Union Defence Force.

1957 A school opens in Dubai

Dubai opens its first modern school, providing education for boys. In 1958, Britain opens a technical school in Sharjah.

1958 Company forms to supply electricity

A private joint-stock company forms in Dubai to supply electricity.

1959 Oil discovery in Abu Dhabi

Oil is discovered in commercial quantities in Abu Dhabi.

1959 Bank opens in Abu Dhabi

The British Bank of the Middle East opens an office in Abu Dhabi. This is the first bank to open in the emirate.

1960 Dubai's first airport open

An airport at Dubai opens. In 1965, a new paved runway will be opened. By the late 1990s Dubai International Airport handles six million passengers a year.

1963 The Maktoum Bridge opens

The Maktoum Bridge, the first bridge to be built over Dubais creek, opens. Prior to the opening of the bridge, the only way to cross the creek between Dubai and Deira is by *abras* (water taxi).

1966 The Dubai-Sharjah road is opened

A road connecting Dubai and Sharjah opens. In 1969 the road is extended to Ras Al Khaimah.

1966 Birth of Sheikh Yayid bin Sultan Al Nahyan

Sheikh Yayid bin Sultan Al Nahyan, Emir of Abu Dhabi and first president of the UAE (see 1971: December 2) is born. Of the tribe Bani Yas, and family Al bu Falah (or Nahyan), his people are predominantly *hadar* (settled people), but include a small bedouin (nomad) population. Prior to oil income, the Bani Yas made up about half the population of Abu Dhabi, and traditionally formed the power base of the ruler of Abu Dhabi.

1967 Dubai Petroleum Company begins offshore production

Dubai Petroleum Company begins offshore production from Fateh Field. A special oil storage container, a *khazzan* (Arabic for storage), is used for offshore oil storage. Shaped like an upturned champagne glass and as tall as a fifteen story building the *khazzan* stands on the sea bed. Because oil floats on water, when oil is pumped into the *khazzan* it displaces sea water and floats on top of the water. Tankers then withdraw oil from the top. *Khazzan* containers eliminate the need to pump crude oil to shore for storage.

1968 Britain declares its intention to withdraw by ending its treaty relationships

Britain announces its intentions to sever its treaty relationships within the region. For Britain this will be a complete withdrawal of all its troops east of the Suez by 1971, including those based in the Gulf. This decision includes the seven Trucial Sheikhdoms, together with Bahrain and Qatar, all under British protection.

1968: February Negotiations for federation begin

The seven emirates, Abu Dhabi, Dubai, Sharjah, Ajman, Umm al Quwain, Ras al Khaima and Fujairah, and Bahrain and Qatar begin negotiations toward federation. Sheikh Zayed bin Sultan Al Nahyan, Ruler of Abu Dhabi, and Sheikh Rashid bin Saed Al Maktoum, Ruler of Dubai, at an earlier meeting have agreed to unite Abu Dhabi and Dubai in a federation.

1968 Abu Dhabi International Airport

An international airport, the Abu Dhabi International Airport, opens at Bateen.

1971: December 2 The United Arab Emirates is formally inaugurated

The United Arab Emirates is formally inaugurated, with Sheikh Zayed of Abu Dhabi as President of the Federation. Sheikh Rasheed bin Said Al Maktoum is Vice-President and Prime Minister. Bahrain and Qatar established their own independence earlier this same year.

1972 Ras al Khaima joins the Federation

Ras al Khaima joins the United Arab Emirates.

1973 The UAE Currency Bank is established

The UAE Currency Bank is set up to institute the national currency.

1973 OPEC price hikes

The Organization of Oil Exporting Countries (OPEC) quadruples oil prices.

1975 Sheikh Hamad bin Mohammed al Sharqi becomes ruler of Fujairah

Sheikh Hamad bin Mohammed al Sharqi succeeds his father Sheikh Mohammed bin Hamad al Sharqi as ruler of Fujairah.

1976 The United Arab Emirates University is established

His Highness Sheikh Zayed Bin Sultan Al Nahayan establishes by federal law the United Arab Emirates University. A federal institution of Arab-Islamic identity its mission is to realize the ideals of the society and to contribute to the development of the country.

1977 Sharjah Airport opens

An international airport opens in the port city of Sharjah.

1977 The University of the UAE opens in Al Ain

The University of the United Arab Emirates opens with four colleges (faculties): Faculty of Arts (Humanities and Social Sciences); the Faculty of Sciences; the Faculty of Education; and the Faculty of Administrative and Political Science (College of Business and Economics). In the opening academic year 502 students enroll.

The University expands in 1978 with the addition of the Faculty of Shariah Law. In 1980 the Faculty of Agricultural Sciences and the College of Engineering are established. In 1986 the Faculty of Medicine and Health Sciences is added. In the 1996/1997 academic year more than 15,000 students are enrolled.

1980–88 Iran-Iraq War

The Iran-Iraq War threatens the security of the Persian Gulf. The two rival leaders—Iran's Ayatollah Sayyid Ruhollah Musavi Khomeini and Iraq's Saddam Hussein—battle over the Ayatollah's attempts to spread his Islamic fundamentalism to Iraq. Iraq attacks Iran's richest oil-producing area. Khomeini accepts a ceasefire in August 1988.

The UAE is ambivalent towards the attacks on international shipping. Vulnerable to Iranian attack of its offshore oil facilities neutrality is UAE policy. However, with substantial Iranian populations in Dubai and Ras al Khaimah, these emirates lean toward Iran. Abu Dhabi maintains a pro-Arab support of Iraq.

1981: May The first summit of the Arab Gulf Cooperating Council

The Arab Gulf Cooperating Council (GCC) forms. Saudi Arabia, Kuwait, Bahrain, the UAE, Qatar, and Oman are the members. They band together to protect their interests. If necessary, they will cooperate to defend themselves as Iran is considered a threat to their security. Kuwait turns to the United States for protection of their vessels in the Gulf. The U.S. interest is to protect the flow of oil. Kuwaiti tankers sail under U.S., Soviet, and British flags.

1985 The General Postal Authority is created

Based in Dubai, the General Postal Authority (GPA) becomes an autonomous federal institution with the power to issue its own stamps for the UAE.

1985: February 9 Jebel Ali Free Zone Authority is established

The Jebel Ali Free Zone Authority (JAFZ) is established in Dubai. Covering over thirty-eight square miles (100 square kilometers) as of 1998, the JAFZ contains the world's largest artificial port. Over eight hundred manufacturing and industrial companies from more than sixty-five countries are situated within the zone.

1988 Status of women

Women are 6.2 percent of the work force. The majority of these female workers who are UAE citizens work under either the Ministries of Education or Health. Women graduates outnumber men by a ratio of two to one at the United Arab Emirates University since the late 1980s.

1989 A network of Higher Colleges of Technology is set up

These are technical and vocational colleges. Tuition is free, as at the university and government schools. The curriculum is produced with the input of local industries, banks, airlines and oil.

1990: August 2 Iraq invades Kuwait

Saddam Hussein revives territorial claims against Kuwait and accuses Kuwait of illegally siphoning oil from Ar Rumaylah field. He also threatens to use force against other oil producers that exceed their oil quotas, including Kuwait and the UAE.

The UAE is the first gulf state to propose combined military action to deter Iraq. This reverses its earlier policy of the Iran-Iraq War of avoiding cooperation with foreign military powers. UAE and the United States cooperate in a combined military action. U.S. ships and aircraft carriers operate out of UAE ports and Iraqi positions are bombed from the UAE. The UAE also contributes troops and financially supports the war. Operation Desert Storm, the offensive land force coalition, gets underway on February 24, 1991. The city of Kuwait is recaptured on the third day of the offensive.

1993 Marriage Fund created to counteract high dowries

The Marriage Fund is set up by the government in an effort to reverse the practice of Emirati men marrying foreign brides because they cannot afford the exorbitant dowries (bride-price) expected of them. The Marriage Fund provides the couple with a wedding grant of $19,000. In order to qualify for the fund, the demand for dowry from the bride's father must not exceed $14,000.

1997 Iran names two vessels after disputed islands in the Gulf

UAE protests to the United Nations (UN) that Iran is making an illegitimate attempt to perpetuate Iranian occupation of the islands.

1997: December A Tarjik airliner crashes at the Sharjah airport

The aircraft, a Tupolev-154, en route from the former Soviet republic of Tajikistan to Sharjah, crashes eight miles from the airport. Of the eighty-six people aboard eighty die and at least two arc treated for critical injuries.

1998 Amnesty International (AI) reports alleged human rights violations

Cruel, inhuman, or degrading punishments continue to be reported, including flogging and amputation. Flogging is a punishment for a range of offenses, such as, theft and sexual offences such as adultery. In Ras al Khaimah the punishment of flogging is extended to traffic offenses and begging. Prisoners are detained without trial and possibly without charges. Increasing use if the death sentence is also reported.

1998: May Iran and UAE discuss their claims to small islands in the southern Gulf

Until 1992, both countires jointly administered the Abu Musa and the Greater and Lesser Tunbs when Iran took control. Iran rejects international arbitration. Both countries claim that the others refusal to recognize their rights is a hindrance to establishing confidence across the Gulf.

1998: December Asian workers stranded in UAE

More than 150 expatriate workers from south and east Asia are stranded as they are abandoned by their employers. They live on charity, some sleeping in buses. Suicide attempts and divorce are the results of their despair.

1998: December The Emirates Marriage Fund tries to curb lavish weddings

The Emirates Marriage Fund is passed to place limits on extravagant displays of wealth at weddings. The law makes it acceptable to slaughter no more than nine adult camels and twelve young camels. The common practice of slaughtering up to ninety camels for a wedding is made illegal. According to the new Marriage Law, those spending too much on their weddings will be jailed. The law does not specify a specific term for the jail sentence, but the fine can be as much as $140,000.

1999: January The Hamriyah Free Zone issues licenses

The Hamriyah port between Ajman and UmmAl-Quwain can accommodate large ships. The licensing process establishes local and international businesses within the zone.

1999: March 7 Low fuel prices threaten the economic stability of the Persian Gulf

The UAE derives sixty percent of its income from oil sales. Thus, it is known as a "petrodollar" economy deriving most of its foreign exchange from oil exports.

1999: March 13 Agreement to cut oil production

The Organization of Oil Exporting Countries (OPEC) includes Algeria, Indonesia, Iran, Iraq, Kuwait, Libya, Nigeria, Qatar, Saudi Arabia, the United Arab Emirates, and Venezuela. OPEC members, except Iraq, agree with non-OPEC members, Mexico and Oman, to cut oil production. These cutbacks increase the price of crude oil in 1999.

Bibliography

Ali Rashid, Noor. *The UAE Visions of Change*. Dubai: Motivate Publishing, 1997.

Butti, Obald A. *Imperialism, Tribal Structure, and the Development of Ruling Elites: A Socio-economic History of the Trucial States Between 1892 and 1939*. Georgetown University: Thesis (Ph. D.), 1992.

Codrai, Ronald. *The Seven Shaikhdoms: Life in the Trucial States Before the Federation of the United Arab Emirates*. London: Stacey International, 1990.

Forman, Werner. *Phoenix Rising: the United Arab Emirates, Past, Present & Future*. London: Harvill Press, 1996.

Kay, Shirley. *Emirates Archaeological Heritage*. Dubai, United Arab Emirates: Motivate Publishing, 1988.

Koury, Enver M. *The United Arab Emirates: Its Political System and Politics*. Hyattsville, Maryland: Institute of Middle Eastern and North American Affairs, Inc., 1980.

Johnson, Julia. *U.A.E.* New York: Chelsea House Publishers, 1987.

Mann, Major Clarence. *Abu Dhabi: Birth of an Oil Sheikhdom*. Beirut: Khayats, 1964.

O'Brien, Edna. *Arabian Days*. London: Quartet Books Limited, 1977.

Peck, Malcolm C. *The United Arab Emirates: A Venture in Unity*. Boulder, Colorado: Westview Press, 1986.

Taryam, Abdullah Omran. *The Establishment of the United Arab Emirates 1950–85*. London: Croom-Helm, 1987.

Zahlan, Rosemarie Said. *The Making of the Modern Gulf States*. London: Unwin Hyman Ltd., 1989

Uzbekistan

Introduction

After the Turks of Turkey, the Uzbeks are the largest ethnic group of Turkic people in the world. The Uzbeks were the third-largest ethnic group in the former Soviet Union when it collapsed in 1991, with a population of more than 16.7 million, of which 85% lived in what is now Uzbekistan. The population of Uzbekistan increased by seven million from 1970 to the mid-1980s, and is over twenty-three million. There are also some two million ethnic Uzbeks in northern Afghanistan (in the early 1920s, the Red Army expelled many people who were loyal to the Muslim independence movement). Tashkent, the capital, has a metropolitan area population of nearly 2.3 million. Historic Samarkand's population is 370,000. Modern Uzbekistan covers an area of more than 172,700 square miles (447,293 square kilometers), an area slightly larger than California. Uzbekistan is a land-locked country in Central Asia surrounded by other five other "-stan" countries: Kazakstan to the north, Kyrgyzstan and Tajikistan to the east, Afghanistan to the south, and Turkmenistan to the west. Uzbekistan and these neighboring countries with the exception of Afghanistan were all Soviet republics during the years of the Soviet Union (1922–91).

The country's topography is mostly flat to rolling sandy desert. The Ferghana Valley lies in the east and is surrounded by mountainous Tajikistan and Kyrgyzstan. There is semiarid grassland in the east, and the rapidly shrinking Aral Sea lies in the northwest. Uzbekistan's climate is that of a mid-latitude desert. Temperatures in the summer reach 90°Fahrenheit, (32°Celsius) while in the winter it can get as cold as –10°F (–23°C). Ecological damage since the 1950s has left much of the country devoid of animal life.

History

The Uzbeks are fairly new as a nationality, although historical accounts of a people called "Uzbeks" go back many centuries. The concept of Uzbek nationality was largely aided by the Soviet government of the 1920s, which set the geographical boundaries for their own homeland. Until then, the Uzbeks had been an amalgam of dozens of nomadic Turkic-Mongol clans.

Ancient Persians were the first recorded residents of the area now known as Uzbekistan, occupying it some 3,500 years ago. The Achaemenids of Persia consolidated power over the region in the sixth century B.C., and many of the region's oldest cities date back to that era. Alexander the Great (r. 336–323 B.C.) later seized the city of Samarkand and conquered the region. During the first millennium A.D., several regional empires came and went (Kushan, Ephthalite, Sassanian Persia). In the eighth century, Arabs seized the region's main urban areas, and in the ninth century, Bukhara became the seat of the Samanids, who controlled much of Persia. During the tenth century, Seljuq Turks began migrating and settling into the region. With the Mongol invasion of the early thirteenth century, the political climate and culture of Central Asia were suddenly changed. As the initial influence of the Mongol occupation wore off, one of the region's most famous leaders rose to power—Timur, also known as Tamerlane. Through shrewd political maneuvering, Timur went on to control the a huge portion of Central Asia by the mid-1300s.

In the early fifteenth century, the ancestors of the modern Uzbeks migrated to what is now Uzbekistan from the north. In the early 1500s, Uzbeks captured the cities of Samarkand and Bukhara and established the last great empire to arise from the region. In the 1700s, Russians were starting to explore Central Asia and the Uzbeks established khanates (independent empires each ruled by an autocratic khan) in Bukhara, Khiva, and Kokand.

During the nineteenth century, Britain and Russia competed for control over territories in the Caucasus region and Central Asia in what was called the Great Game. In the 1860s and 1870s, the Russians seized the cities of Tashkent and Samarkand, made the Bukhara and Khiva states into colonies, and annexed the Kokand state. These colonies became Russian Turkestan, and by the end of the nineteenth century there were revolts against the imperial government in towns across Central Asia. Uzbeks increasingly wanted independence from colonial control. World War I and the end of the imperial Russian government brought about political instability throughout Central Asia. From 1918 into the early 1920s, there were

several competing attempts to form independent governments, but by 1924 the Bolsheviks emerged victorious and created the Uzbek Soviet Socialist Republic.

During the 1920s and 1930s, the Soviet government promoted literacy but sternly suppressed religion, and constructed technological improvements but imprisoned and executed large segments of the population. During World War II, entire towns were evacuated from the European parts of the Soviet Union (along with their factories and workers) and moved to Central Asia. The large population of non-Uzbek immigrants made it possible to impose Russian language and culture. The primary leader of the Uzbek SSR in the post-war years was Sharif R. Rashidov, who was the top communist boss from 1959 until 1983. His government was conservative and went along with its orders from Moscow. He was ousted shortly before his death in a cotton industry scandal that uncovered rampant corruption of the government's leadership. By the 1980s, environmental mismanagement to meet cotton production quotas had caused the Aral Sea to lose much of its area and toxic chemical residue severely affected public health. As the Soviet Union began to loosen its control, ethnic tensions rose and riots broke out in the republic during the late 1980s.

After the failed Moscow coup of August 1991, Uzbekistan declared its independence from the Soviet Union. Initially, the Uzbek SSR had declared its loyalty to the anti-democratic coup plotters. The Soviet Union ceased to exist in December 1991. The new Republic of Uzbekistan adopted its constitution in 1992 and introduced its own currency in 1993. After a short period of democracy, President Islam Karimov stopped the democratic process and began to rule by decree.

Timeline

1500–1000 B.C. Earliest recorded inhabitants

Ancient Persian tribes are the first recorded inhabitants of what is now Uzbekistan, although archaeological evidence indicates that people have lived in the region for millennia. These early Persians migrate to the north, eventually settling between the Oxus (Amu Darya) and Jaxartes (Syr Darya) rivers.

Transoxiana refers to the region north of the Oxus River, the ancient name for the Amu Darya (the river that marks modern Uzbekistan's southern boundary).

6th century B.C. Transoxiana consolidated under the Achaemenids

The territory of what is now Uzbekistan is consolidated under the Achaemenids of Persia under the rulers Cyrus the Great (d. 530 BC) and Darius I (r. 522–486 BC) as part of outer Iran (known as Turan). The Achaemenids later divide the area into three administrative districts: Khwarazm, Bactria (covering present-day southern Uzbekistan, Tajikistan, and northeastern Afghanistan), and Zarafshan and Ferghana Valleys. During the Achaemenids' reign, cities are further developed and Zoroastrianism becomes the official religion.

329–327 B.C. Conquest of Transoxiana by Alexander the Great

The Greeks, under Alexander the Great (r. 336–323 B.C.), invade and conquer Transoxiana and capture the city of Samarkand, an oasis on the Zarafshan River. The Greeks rule from Bactria and Soghdiana.

Alexander stops in Maracanda, near Samarkand, and marries Roxana, daughter of a local Soghdiana leader. Over the centuries, the local legends surrounding Alexander (called Iskandar Zulqornai by the Uzbeks) grow until he becomes a larger-than-life heroic figure.

250 B.C. Greeks lose control over Soghdiana

Parthia (an ancient country located southeast of the Caspian Sea) gains control of Soghdiana from the Greeks.

A.D. 1–1000 Empires of the 1st millenium

The Parthian Empire, Bactria and Soghdiana, and Khwarazm become the first states established in what is now Uzbekistan.

1st century B.C. to 4th century A.D. Kushan Empire

In A.D. 50, the Kushan Empire is established by the ruler Kujula Kadphises, and it includes Persia, Transoxiana, and the Upper Indus. By the third century, however, both the Kushan and Parthian empires are in decline and are eventually taken over by Persia's Sassanian dynasty.

A.D. 440 Ephthalite Empire

The Ephthalites (also known as the Hephthalites, White Huns, or Avars) descend from the Altay region in the northeast and capture Transoxiana and much of eastern Persia.

550s–560s Fall of the Ephthalite Empire

The Ephthalites are conquered by an alliance between the Turks and Persians. During the 570s to 590s, the Turks then ally with Byzantium to defeat the Persians.

659 Chinese armed forces overrun West Turkic Khanate

After the fall of the Ephthalite Empire, the area then becomes a loose confederation of largely nomadic tribes organized into the West Turkic Khanate. The Chinese begin occupying the region.

661 Establishment of the Ummayid dynasty

The Arab Ummayid dynasty establishes a caliphate (a designation of Islamic sovereignty used by the successors of Mohammed—the caliph is the supreme ruler) in Damascus, marking the religious split of Islam into the Sunni and Shi'ite sects.

667 End of Persia's Sassanian dynasty

Perez, the last Sassanian shah, is defeated by the Arabs, who then enter Transoxiana.

673–704 Arab raids

Over the course of several decades, Arab raiders continually attack Transoxiana in order to seize Bukhara and Soghd.

705 Arabs declare war on Transoxiana

Arab forces begin a holy war against Transoxiana under the leadership of Qutayba Ibn Muslim. The Arabs use Merv (Mary, in modern Turkmenistan) as a base for the campaign. The Arabs introduce Islam into the region.

709 Arabs seize Bukhara and Samarkand

Arab forces capture and subsequently fortify Bukhara and Samarkand. This provides a strong bullwark against T'Ang Chinese forces to the east. The Arab seizures of Khiva in 711 and Khwarazm in 712 further strengthen their hold over the region.

715 Arabs conclude conquest of Transoxiana

Upon the death of Qutayba Ibn Muslim, governor of Khorasan, Arab forces cease hostilities in Transoxiana. Qutayba Ibn Muslim is assassinated by his own forces after he challenges the authority of the Ummayid caliphate, and his death brings more political instability to the region.

728 Revolt against Arabs

After the Arabs consolidate their power in Transoxiana, they try to make their new subjects accept Islam as their religion, resulting in widespread rebellion.

747–50 Abbasid dynasty replaces Ummayid dynasty

The Abbasid dynasty gains control of the caliphate from the Ummayid dynasty and moves the capital to Baghdad.

874–75 Samanids take control over Central Asia

The Samanid dynasty, which is Sunni, is granted control over Transoxiana by the caliph. The Samanid administration is based in the city of Bukhara. By 900, the Samanids gain control over the Shi'ite Saffarid dynasty and extend their dominion throughout Persia.

Turkic and Mongol Power

932 Turkic tribes form the Karakhanid state

Turkic tribes begin to push into the area from the east, and eventually form their own state. During the following years, the Karakhanid Turks convert from Buddhism to Islam.

985 Seljuq Turks settle at Bukhara

The Seljuqs are one of the Turkic tribes that originate from the western Syr Darya region in the middle of the eighth century. By the end of the tenth century, the Seljuqs have migrated to the area around Bukhara. The Seljuqs later go on in 1055 to seize Baghdad, establish their own sultanate, and control the caliphate.

999 Samanids lose power

The Karakhanid Turks gradually gain control over the Samanids and seize the city of Bukhara.

1073 Karakhanid state splits apart

The Karakhanid empire weakens during the eleventh century and finally falls apart in 1073, with its defeat by the Seljuqs. As a result, the Tarim Basin region becomes known as Eastern Turkestan and Transoxiana becomes West Turkestan.

1140–1210 Karakhitals hold power

The Karakhitals are originally Mongols who were driven out of China during the early twelfth century. They create their own state and overrun the Karakhanids, which opens up Transoxiana to their domination.

1194 Rise of the Khwarazm state

A lesser part of the Karakhanid state, Khwarazm in northern Persia grows more powerful. The control of Persia by the Seljuqs ends. The region's location along the Silk Road (a great trade route from China to Europe) makes it an important commercial center and the Khwarazm state goes on to dominate most of Central Asia. In 1212, the Khwarazm Shah Mohammed liberates Samarkand from the Karakhitals. However, the removal of the Karakhitals exposes Khwarazm to an even larger threat from the east, the Mongols.

1218 Khwarazm state executes Mongol emissaries

The ruling Khwarazm leader Shah Mohammed executes emissaries of Mongol leader Genghis Khan, which only provokes the Mongols to send armed forces into the area.

1219 Mongol invasion

Genghis Khan's powerful army crosses the Jaxartes River (Syr Darya) into Transoxiana and advances on the wealthy Muslim empire of Khwarazm. Genghis Khan and his warriors, with their lightweight armor, mangonels (devices used to throw heavy stones to break down walls), naphtha (a petroleum and sulfur mixture used to make firebombs), and clever tactics overrun the cities of Bukhara, Samarkand, and Urgench in present-day Uzbekistan, as well as Utrar (located in modern Kazakstan) and Merv (in Turkmenistan). The khan's army also conquers territory in what is now Pakistan, Afghanistan, Iran, Iraq, and southern Russia before returning to Mongolia.

1220 Mongols seize Bukhara and Samarkand

The Mongol conquest of Bukhara and Samarkand leaves much of Uzbekistan in ruins.

1227 Chagatai is made ruler

When Genghis Khan dies in 1227, his empire is split up among his sons. Chagatai (d. 1241) is made ruler of the area that becomes known as the Chagataid Khanate, which includes Transoxiana, the Tarim Basin, and Semirechye. The large region of the empire on the Russian steppes is given to Batu and becomes known as the Kipchak Khanate. In 1231, the Mongols decisively defeat the Khwarazm state. Khwarazm forces retreating from the Mongols later seize Jerusalem from the Crusaders.

1313–42 Uzbek Khan rules Golden Horde

Uzbek Khan (1282–1342) controls the Golden Horde, part of the Kipchak Khanate to the north that had split into two parts during the 1260s. During the khan's rule, the Golden Horde converts to Islam. The modern Uzbeks trace their genealogy back to Uzbek Khan, a grandson of Genghis Khan.

Mid-14th–15th centuries Timur and Samarkand Empire

As the influence of Genghis Khan's era wanes, Timur (or Tamerlane, 1336–1405) establishes an empire in Samarkand. Timur is born near Samarkand to an aristocratic Mongol family that that had adopted Turkic culture and Islam. Timur starts out as a sheep rustler and raider of caravans, and has intellectual interests in history, medicine, and astronomy. Due to arrow injuries in an arm and a leg, Timur becomes known as Timur-i-leng (Timur the Lame, corrupted into English as Tamerlane). Through political positioning and strategy, by 1369 Timur is the supreme leader over all of Transoxiana. The Chagataid Khanate splits into western and an eastern parts, with Transoxiana in the western half. Timur conquers parts of present-day India, Syria, and southern Russia. His archers go all the way to Moscow and Delhi, and Timur is the world's last great nomadic emperor.

An estimated seventeen million people are killed or die during Timur's conquests. Piles of skulls mark the trail of his troops. Timur enforces his authority through mass executions of thousands of prisoners, and he terrorizes his opposition with acts of cruelty such as burying people alive and using the heads of decapitated prisoners as cannon balls to fire upon his enemies.

1405 Death of Timur, Samarkand empire splits apart

Timur's exploits make Samarkand a prominent and wealthy city. Architects and builders from across Persia, India, Afghanistan, and Arabia are brought in to plan and construct the oasis city. Upon the death of Timur, the Samarkand empire splits into two parts: Khorasan, ruled by his son Shah Rukh (1337–1477); and Mawernahr, ruled by his grandson Ulgh Beg (or Ulughbek, 1394–1449).

1449 Ulgh Beg assassinated

Ulgh Beg is not only the ruler of Mawernahr, but also a prominent scholar, mathematician, and poet of his day. His astronomical observations plot the courses of nearly one thousand stars and he devises an accurate measurement for the length of the solar year. His astronomical charts are used by Chinese and Europeans afterwards for nearly four centuries, and many of his scientific discoveries are later embraced by European scholars during the Renaissance. Ulgh Beg is assassinated by his own son.

The Rise of the Uzbeks

Early 15th century Uzbeks migrate to Transoxiana

During the leadership of Abu al-Khayr (or Abul Kair, 1413–69), the Uzbeks move south into Transoxiana. Abu al-Khayr is a grandson of Genghis Khan who unified the various Turko-Mongol tribes inhabiting the steppes of present-day Kazakstan. He creates formal rituals and observes protocol, which helpes the new Uzbek confederation gain a sense of identity.

1441 Birth of Mir Alishar Navai

Mir Alishar Navai (1441–1501) is the premier poet of the Uzbeki language and a national hero. Navai creates the language's Turkic script and literature in order to replace Persian.

1497 Babur seizes Samarkand

Babur (1483–1530), the great-great-great grandson of Timur, is the ruler of Ferghana. His forces descend upon a weakened Samarkand and take the city.

1500 Uzbeks seize Samarkand

Although Timur is claimed to be the father of the modern Uzbeks, more likely candidates are the Sheibanid. The Sheibanid are named for their leader, Muhammad Sheibani Khan (1451–1510). These nomadic Uzbeks fight to take the area and capture Samarkand from Babur. After the khan is killed in battle, Samarkand becomes the capital, but political power is wielded from Bukhara.

1506 Uzbeks seize Bukhara

The Uzbeks make Bukhara their capital, and the old Bukhara state splits apart. Khwarazm, Balkh, and Khiva separate from Bukhara, becoming independent principalities.

1522 Babur founds Mogul Empire

Upon the capture of Delhi, Babur establishes the Mogul Empire in India.

1558 First commercial contact with Russians

Russians begin to trade in Transoxiana. By the early 1600s, Moscow and Bukhara establish diplomatic relations.

1557–98 Shaybanid Uzbek dynasty

Abdullah Khan rules Bukhara from 1557 until 1598. His Shaybanid Uzbek dynasty is the last great empire to arise from Transoxiana. During the 1600s, the Uzbeks come to dominate the area and absorb earlier Turkic and Persian populations into their culture.

1717 Russian expedition to Khiva

Russian armed troops sent on an initial expedition to Khiva are killed.

1740–47 Bukhara reconquered by the Persians

Persia invades Bukhara and goes on to dominate Transoxiana.

1747 Rise of the Uzbek Khanate of Bukhara

Although Bukhara is seized by the Persians, sovereignty is soon retaken by the Uzbek Mangyt dynasty of the Khanate of Bukhara, which rules until 1920.

1763 Rise of the Uzbek Khanate of Khiva

Shortly after the establishment of the dynasty in Bukhara, the Uzbek Kungrat dynasty consolidates its power in the Khanate of Khiva (Kharazm).

1798 Rise of the Uzbek Khanate of Kokand

The Kokand Khanate, in the eastern part of present-day Uzbekistan, grows powerful in the nineteenth century. During the eighteenth century, Russia begins trading with Bukhara, Khiva, and Kokand.

19th century Conquest by imperial Russia and the Great Game

During the nineteenth century, Britain and Russia compete for control over territories in the Caucasus region and Central Asia in what is called the Great Game. Each empire sends expeditions of geographers and scientists into its own territory, and these voyagers frequently engage in frontier-style espionage by exploring the other empire's territory. During the 1860s–1880s, concern about British expansion in India and Afghanistan leads eventually to the Russian conquest of Central Asia, including lands occupied by the Uzbeks. Rivalries between the various khanates hinder them from forming alliances to repel British and Russian forces.

1865: June Russians capture Tashkent

When the Russians, under General Konstantin von Kaufmann, seize the city of Tashkent, they make it the capital of Turkestan, a province created by the Russian Empire that encompasses its possessions in Central Asia.

1868: May Samarkand captured by Russians

By the 1880s, the Russian empire makes the region part of the Turkestan *guberniia* (administrative district), with Bukhara and Khiva administered as separate emirates under Russian protection.

1868: June Khanate of Bukhara becomes a colonial state

The Russian Empire officially makes Bukhara a protectorate. Many Uzbek leaders continue to resist, however, but their efforts are weak and only strengthen Russia's resolve to control the region.

1873: August Khanate of Khiva becomes a colonial state

The Russian Empire officially makes Khiva a protectorate. General von Kaufmann approves separate treaties for Bukhara and Khiva which make them both colonies and abolish slavery.

1876 Khanate of Kokand absorbed by Russia

The Russian Empire annexes Kokand.

1884 Russia introduces cotton to Turkestan

The Russians introduce American cotton to the region. During the Soviet years, cotton becomes the principal commodity and the Uzbeks liberally use irrigation, fertilizers, and pesticides to increase yields.

1885 Revolt against Russia

Muslims rebel against the tsar's troops in the towns of Osh, Margellan, and Andizan in the Ferghana Valley. The revolts are crushed by the Russians.

1887 Britain and Russia agree to Turkestan's border

Britain had long exercised its control over neighboring Afghanistan, and the two nations finally draw up a border marking the territorial boundaries of Turkestan. The Great Game is over, and the Anglo-Russian Convention of 1907 formally puts an end to the colonial rivalry.

1892 Riots in Tashkent

An epidemic of cholera causes widespread panic which leads to riots in Tashkent.

1894 Abdullah Quadiriy born

Abdullah Quadiriy (1894–1939), who will become the first novelist in the Uzbek language, is born. He is best known for his work *Days Gone By.*

1898: May 17 Riots in Andizan

Islamic militants in the Ferghana Valley town of Andizan kill twenty-two Russian soldiers in one of the most brazen uprisings of the imperial Russian colonial era. The rebellion spreads to neighboring towns before Russian reinforcements from Tashkent arrive. When the Russians regain control, over two hundred militants are executed and nearly eight hundred are imprisoned.

1906 Railroad completed in Turkestan

The Orenburg-Tashkent Railroad that is built in Turkestan ultimately connects the region to the European parts of Russia.

1909 Young Bukharans founded in Bukhara

The Young Bukharans are a social reform group that want to introduce socialism to the region. The Young Bukharans later come to power as the Soviet government consolidates its power in 1920. However, the Young Bukharans gradually have a falling out with the Bolsheviks.

1913–16 The height of the Jadid education movement

The Jadid (Reformist) education movement is an attempt by local moderates in Central Asia to improve the system of education from within the traditional Islamic framework. The Jadid movement gains popularity in the early years of the twentieth century in Bukhara, Khiva, Samarkand, and Tashkent, and begins spreading to other Muslim lands as well. The reformers want to reexamine issues such as ethics, religion, the judicial system, public health, and the treatment of women. One of the important Jadid scholars is Mahmud Khoja Behbudiy (1874–1919), who stresses the need to make an objective record of Central Asia's history.

1916 Conscription into labor battalions

During World War I (1914–18), Russia's Tsar Nicholas II issues a call for Central Asian males to be drafted into labor battalions. This sparks resistance throughout the region, which is violently repressed.

1917–20 Political upheaval

When the Russian empire falls apart in 1917, there are many challengers throughout its territory to fill the vacuum of power. Uzbekistan is the site of competing attempts to create governments; in 1918 the Bolsheviks announce a short-lived Autonomous Soviet Socialist Republic of Turkestan, while a Muslim Congress also attempts an Autonomous Government of Turkestan. Red Army forces intervene savagely, but armed resistance continues until 1924. An estimated one million people in Central Asia die during the chaos after nomadic peoples slaughter their herds in defiance of forced collectivization.

1917: April 16–23 First Central Asian Muslim Congress

The First Central Asian Muslim Congress convenes in Tashkent. The group demands that Russia stop colonizing Central Asia and return seized lands. A second congress held in September calls for the establishment of the Autonomous Federated Republic of Turkestan.

1917: November 7 Bolshevik Revolution in Russia

When the Bolsheviks seize power in Moscow, the Tashkent Soviet of Worker's and Peasant's Deputies takes control of the territory from the Tashkent Committee of the Provisional Government. Both groups had vied for control since the collapse of the Russian imperial government in February. The Bolsheviks then create the Council of People's Commissars in Tashkent to guarantee their supremacy.

1918: February 18 Tashkent Soviet defeats rival Kokand government

Immediately after the Bolshevik Revolution, the Fourth Central Asian Muslim Congress in Kokand creates the Muslim Provisional Government of Autonomous Turkestan. In January 1918, the Tashkent Soviet government declares war on the Kokand government. With the help of the Red Army, the Tashkent Soviet government defeats its rival and kills some fourteen thousand people.

1918–20 First Basmachi Rebellion

Although the Tashkent Soviet takes control, there are still several threats to its power. The region is cut off from Moscow by the Whites (forces loyal to the former Russian imperial government). Central Asia is the main scene of action between the White and Red armies during Russia's civil war. Thousands of European prisoners of war are also kept in Central Asia.

Muslims are also trying to win independence for the region. Under the leadership of Enver Pasha (1881–1921), a Muslim resistance called the Basmachi movement grows from the Ferghana Valley. The movement, however, lacks cohesion and clan rivalries weaken its effectiveness as a serious threat to the Bolsheviks.

1918: April Turkestan Autonomous Soviet Socialist Republic created

The Turkestan Autonomous Soviet Socialist Republic (ASSR) is a temporary construct of the early Soviet regime. Within two years, Lenin's Turkestan Commission proposes that the territory be split up into various smaller republics according to ethnicity.

1920: February 2 Soviets seize Khiva

When the Red Army takes control of Khiva, the Khanate of Khiva is abolished and the Kungrat dynasty comes to an end.

1920: September Soviets seize Bukhara

When the Red Army takes control of Bukhara, the Khanate of Bukhara is abolished and the Mangyt dynasty comes to an end. The following month, the People's Republic of Bukhara is established by the Young Bukharans and the Bukharan Communist Party. The leader of the Young Bukharans is Faizullah Khojaev (1896–1938), who is later executed by the Soviet government during the reign of Josef Stalin.

1922 The Uzbek Soviet Socialist Republic

The Union of Soviet Socialist Republics (USSR) is formally created in December 1922, with the Turkestan ASSR incorporated into the Russian Soviet Federated Socialist Republic (RSFSR). There are various smaller governmental units present including the Kharazmian Soviet Socialist Republic (SSR) and the Bukharan SSR. In October 1924, the USSR abolishes the Turkestan ASSR, the Bukharan SSR, and the Kharazmian SSR, and creates the Turkmen SSR and the Uzbek SSR (which includes the Tajik ASSR).

1924 Last Basmachi Rebellion

After Enver Pasha is killed, the Basmachi movement loses its morale. The attempt by local Muslim resistance fighters to resist the Red Army fails.

1924: October 27 Uzbek Soviet Socialist Republic created

The Soviet government creates the Uzbek Soviet Socialist Republic (Uzbek SSR) as part of its plan to redraw the political boundaries of Central Asia.

1925: March 6 Samarkand made capital of Uzbek SSR

The new Soviet government needs to select a capital. Tashkent is unpopular because it had been the seat of Russian imperial authority and has a large Russian population. Bukhara, Khiva, and Kokand had been the capitals of their own khanates. Samarkand is chosen because it is the center of the Jadid movement and had not recently been a political center.

1928 Anti-religious campaign intensifies

The Soviet government, officially atheist, begins to crack down on all religious institutions. As a result, Islamic courts and institutions are disbanded. Bukhara's remaining one hundred mosques are closed in 1924 and religious schools are all shut down.

Late 1920s Literacy program

During 1923–24, some ninety percent of the Uzbek SSR's population is illiterate, including the president of the Uzbekistan Communist Party and half of the organization's members. By 1932, only 40% of the population is illiterate. The Soviet government believes that literacy will improve the quality of life, and that the ability to read makes people more receptive to propaganda and more likely to renounce religion. During the late 1920s, Soviet authorities introduce the Latin alphabet to Central Asia in order to replace Arabic script, which is later replaced by the Cyrillic alphabet in 1940. The Latin alphabet is restored in 1993.

1929: October 15 Tajikistan removed from the Uzbek SSR

Tajikistan, which had been an administrative unit of the Uzbek SSR, is elevated to full republic level status, changing the boundaries.

1930: September 30 Tashkent made capital of Uzbek SSR

Samarkand falls out of favor as the political capital of the Uzbek SSR. With the return of the capital to Tashkent, the domination of the republic's administration by ethnic Russians resumes.

1936 Uzbek SSR's boundaries changed again

The Karakalpak ASSR, an area on the southern shores of the Aral Sea, is incorporated into the Uzbek SSR.

1936 Ilyas Malayev born

Popular Uzbek musician and poet Ilyas Malayev is born.

1939 Abdullah Qadiriy purged

Abdullah Qadiriy is the first Uzbek-language novelist, having written *Days Gone By,* a historical novel about Kokand. Many intellectuals throughout the Soviet Union are purged (executed or imprisoned) in the late 1930s by order of the Soviet leader Stalin. Once an intellectual is purged, no mention of the person is made on record. In 1956, Qadiriy is rehabilitated (returned to official history).

1939 Islam Karimov born

Future president Islam Karimov is born. Karimov will embark upon a political career in the Communist Party of the Soviet Union and will hold a variety of posts culminating in his appointment as First Secretary of the Uzbekistan Communist Party in 1989. The following year, he is elected president of the Uzbek SSR and in 1991 becomes the first president of the Republic of Uzbekistan.

1941–45 World War II

Although Nazi troops never make it all the way to Central Asia during World War II, that war has a profound effect on Uzbek culture. Entire towns are evacuated from the European parts of the Soviet Union (along with their factories and workers) and moved to Central Asia. The massive influx of people disturbs the native Uzbek culture and legitimizes the imposition of the Russian language and culture into the area that had already started at the turn of the twentieth century.

1942 Soviet suppression of Islam

By 1942, the number of mosques in Uzbekistan has declined from 25,000 in 1917 to just 1,700. As in other parts of the Soviet Union, religious buildings are typically razed or converted into museums and warehouses. Islam, however, is not obliterated in the Uzbek SSR but merely goes underground.

1954 Virgin Lands plan

Under the Soviet government, the steppes of the Uzbek SSR are selected by Soviet leader Nikita Krushchev to grow cotton and grain. At one point the republic grows seventy percent of the Soviet Union's cotton. Unfortunately the farmers have to heavily irrigate the crops to obtain meaningful results. Now the Amu Darya and Syr Darya rivers run dry in some places, half the Aral Sea is a dry lake bed, and the land is poisoned from the overuse of fertilizers.

1957 First ethnic Uzbek member of Politburo nominated

Nuriddun Akramovich Muhiddinov (b. 1917) becomes the first secretary of the Uzbekistan Communist Party in 1953, after the death of Soviet leader Josef Stalin. During the 1920s, Stalin orders the purging of over 90% of the Uzbekistan Communist Party's members. Ethnic Russians replace those leaders, and remain in control for decades. Muhiddinov supports the bid to make Nikita Krushchev the new Soviet leader, and afterwards he is nominated for the Politburo (then called the Presidium—the executive committee of the Communist Party of the Soviet Union).

1959 Rashidov made leader of Uzbekistan Communist Party

Sharif Rashidovich Rashidov (1917–83) succeeds Muhiddinov as the leader of the Uzbekistan Communist Party, a position he will hold for a quarter of a century. His years in power are marked with graft, corruption, and nepotism. Rashidov will later become infamous for perpetrating one of the most high-level government fraud scandals of the Soviet era.

1966: April 25 Earthquake demolishes Tashkent

An earthquake damages much of Tashkent. Afterwards, the city's housing is entirely rebuilt using technology that can resist any future earthquake damage.

1970s–early 1980s Conservative rule

Under the control of longtime leader Rashidov, the Uzbek SSR is politically conservative during the years (1964–82), when Leonid Brezhnev is the Soviet Union's leader. Political appointments and decisions often revert to competition among the Uzbeks' traditional system of three clans (Tashkent, Ferghana Valley, and Samarkand).

1979: December USSR invades Afghanistan

The Soviet Union invades Afghanistan, which borders the Uzbek SSR. Soviet troops make their first crossing from the military base at Termiz, just inside the Uzbek SSR border.

1983: October Corruption in government detected by KGB

Several top-level elites are removed from office during the mid-1980s, when considerable fraud in the state-run cotton industry is detected. The Uzbek SSR is designated as a cotton-growing region by the Moscow government, and it is even illegal for the Uzbeks to grow their own food. Under the centrally planned command economy, a system of production quotas is imposed on the cotton industry; these quotas are raised every year. Between 1940 and 1980, production increases by a factor of four. The cotton industry, however, increasingly relies on irrigation, fertilizers, and pesticides to increase production. These agricultural techniques eventually cause widespread environmental damage and many people die from exposure to toxic chemicals. The cotton industry cannot keep up with the demands of Moscow, and Rashidov fabricates records to show that the quotas are still being met. Between 1978 and 1983, Rashidov and his collaborators fraudulently record $2 billion in cotton receipts for a crop that is never even grown. Rashidov is dismissed from office, the minister for cotton production is sentenced to death, and some 2,600 officials are imprisoned. For many Russians, the Uzbek SSR then becomes synonymous with corruption. The Uzbeks, however, see the corruption as justified because of the food shortages and health problems.

1988 Aral Sea level is critical

The Aral Sea, a large inland sea lying partly in southwestern Kazakstan and partly in Uzbekistan separates into two smaller bodies of water. The two main rivers that feed into the sea, the Amu Darya and Syr Darya, are used so heavily for irrigation that almost no water flows into the sea between 1974 and 1986. The sea is only about 25% as large as it was in 1960. The sea's salinity increases threefold, and all twenty-four species of its fish die off. A residue of toxic chemicals used in agriculture is exposed by the drying seabed, creating a toxic dust that contaminates nearby towns. Infant mortality increases dramatically, as do birth defects, immune system damage, throat cancer, hepatitis, and respiratory diseases.

1989 Karimov appointed leader

Islam Karimov is first appointed to the position of first secretary of the Uzbekistan Communist Party by the Soviet government in 1989. Karimov is appointed chairman of the republic's State Planning Committee in 1986. Top leadership positions open up within the Communist Party after the corruption scandals, when several officials are removed from office.

1989: February–June Riots and violence in the Uzbek SSR

In February 1989, Uzbeks riot against the preferential treatment that ethnic Russians in the republic are receiving. On June 3, ethnic tensions between Uzbeks and Meshketian Turks erupt into violence, first in Tashkent and then in Kokand and several other towns. There are about 150 casualties. The Turks want to be repatriated to their homeland in Georgia; 15,000 are made homeless and many seek refuge in Kazakstan.

1990: March 24 Karimov elected president of the Uzbek SSR

Islam Karimov is elected to the newly-created post of president by the Uzbek Supreme Soviet.

1990: June **Uzbek-Kyrgyz riots**

In the border city of Osh, Kyrgyzstan, ethnic tensions erupt into violence, and hundreds of people are killed.

The Republic of Uzbekistan

Since its independence, international interests have been eager to win influence in Uzbekistan. After decades of Russian influence, the Uzbeks have become keenly interested in their own culture and Islam. International investors have also shown an interest in the area, reminiscent of the Great Game of the nineteenth century.

Ethnic Uzbeks living in northern Afghanistan have fought against the Islamist Taliban movement. Afghan Uzbeks have received assistance in their struggle from the government of Uzbekistan, which does not want the Taliban movement to gain any power in Uzbekistan.

1991: September 1 **Uzbekistan declares independence**

In the aftermath of the abortive Moscow coup of August 19–21, Uzbekistan declares its independence from the Soviet Union. President Karimov initially declares the Uzbek SSR's staunch loyalty to the anti-democratic coup in August. The Soviet Union ceases to exist in December 1991. Karimov quickly tries to repair the damage to his political career by getting the parliament to vote for independence and outlawing the Uzbekistan Communist Party (which is simply resurrected as the National Democratic Party of Uzbekistan).

1991: December **Karimov elected president of the Republic of Uzbekistan**

Karimov's presidency is reaffirmed in an election held only a few months after independence. After that, however, Karimov is hostile to even the most basic tenets of democracy and rules by decree.

1992: March 2 **Uzbekistan admitted into the United Nations**

Uzbekistan is one of ten new states admitted into the United Nations in 1992.

1992: June 2 **New labor code adopted**

Uzbekistan's labor code recognizes the right of all workers to voluntarily create and join unions. These unions may in turn voluntarily associate territorially or sectorially and can choose their own international affiliations.

1992 **Karimov bans opposition parties**

Parties opposed to Karimov's People's Democratic Party (PDP, the successor to the Communist Party) are outlawed. Political reformers are jailed or forced to flee the country.

1992: December 8 **Constitution adopted**

The Republic of Uzbekistan adopts a constitution that mandates a civil democratic society.

1993: November 15 **Uzbekistan introduces its own currency**

The *som* becomes the official currency, and is introduced in coupon form when Uzbekistan stops using the Russian ruble. National confidence in the coupons is initially low, and the black market exchange rate soars. On July 1, 1994, the permanent som is introduced.

1995: January 15 **Parliamentary elections**

Parliamentary elections are held on December 24, 1994 and January 15, 1995. The PDP wins 231 out of 250 seats. Following the elections, President Karimov holds a referendum that extends his presidency until 2000.

1995 **Mass Privatization Program**

The government begins the Mass Privatization Program, aimed at increasing the private sector's share of the domestic economy from 40% to 60%. During the next two years, 96% of small businesses and 11% of all farm lands are privatized. However, only 20% of Uzbekistan's medium and large enterprises are privatized.

1997 **IMF suspends loan program**

The International Monetary Fund (IMF), fearful of inflation from Uzbekistan's rapid monetary growth, suspends its $180 million loan program. Many small and medium-sized Western businesses begin to freeze their investments or pull out. Investors complain that once the required bribes are paid and an investment guaranteed, officials begin delaying, lengthening, and altering procedures so much that making a profit becomes impossible.

Bibliography

Alworth, Edward A. *The Modern Uzbeks: From the Fourteenth Century to the Present*. Stanford, Cal.: Hoover Institution Press, 1990.

Alworth, Edward A., ed. *Central Asia: 130 Years of Russian Dominance, A Historical Overview*. Durham, N.C.: Duke University Press, 1994.

Critchlow, James. *Nationalism in Uzbekistan: A Soviet Republic's Road to Sovereignty*. Boulder, Col.: Westview Press, 1991.

Rashid, Ahmed. *The Resurgence of Central Asia: Islam or Nationalism?* Atlantic Highlands, N.J.: Zed Books, 1994.

Undeland, Charles and Nicholas Platt. *The Central Asian Republics: Fragments of Empire*. New York: The Asia Society, 1994.

Vanuatu

Introduction

The New Hebrides became an independent nation in 1980 and with that act, changed its name to the Republic of Vanuatu. Prior to that, it had been governed jointly by Great Britain and France. Approximately eighty islands which extend over 800 miles (1,300 kilometers) from northwest to southeast between 13 and 20 degrees south latitude at approximately 168 degrees east longitude constitute the island nation. The islands of Vanuatu are mostly volcanic with dense vegetation. The climate is typical of an island chain within the tropical belt although the southernmost islands sometimes experience cool weather. The highest point in the Republic is Mount Tabwemasana which reaches a maximum elevation of 6,194 feet (1,877 meters) above sea level. The most recent approximation of the population of the island nation (July 1997) is 181,358 inhabitants.

Earthquakes are very common on Vanuatu. The island of Espirtu Santo experiences the most violent quakes. Mild earthquakes occur almost monthly throughout the island chain.

The indigenous population of the island nation of Vanuatu are called *ni-Vanuatu.* The vast majority of the *ni-Vanuatu* live in small, rural villages. Expatriate populations from Europe, Australia, China, Taiwan, and Vietnam also reside on Vanuatu but these populations are small and live mostly in the towns and the capital city, Port-Vila, located on the island of Efate. Port Vila has an estimated population of 19,000.

Epidemics of European diseases drastically reduced the population of indigenous peoples throughout the colonial history of the New Hebrides. Smallpox, measles, and dysentery were initiated by some groups of Europeans with the expressed intent of decimating the indigenous population.

Culture

Culturally, Vanuatu, along with Papua New Guinea, the Solomon Islands, Fiji, and New Caledonia, belongs to the area of Melanesia. This culture area is set apart from the other culture areas of the South Pacific, Polynesia and Micronesia, due to a configuration of societal and cultural patterns. In the times before European contact and immediately following, Melanesian societies lacked inherit social stratification. In other words, all men were created equal as were all women and children. There were differences between the tasks that men, women and children engaged in, but otherwise any person could aspire to do whatever she or he liked. Melanesian societies also had relatively small villages that were structured along lines of kinship. Important activities included pig raising, ceremonial exchanges with neighboring villages, warfare and in many cases head hunting, and ancestor cult worship. Most groups in Melanesia speak Papuan languages which are distinct from the better-known Polynesian languages such as Hawaiian, Tahitian and Maori. In the Republic of Vanuatu, there are around 115 different Papuan languages spoken, many with several dialects. According to the third edition of the *South Pacific Handbook,* the Republic of Vanuatu has the highest number of languages per capita of any country in the world. Despite its rich linguistic diversity, few studies of these unwritten languages have been conducted by anthropologists and linguists.

Vanuatu has three national languages: English, French and Bislama. Bislama is a form of English that derives from English colonization in the South Pacific in the eighteenth century. Some *ni-Vanuatu,* as the Melanesian inhabitants of the islands are known, speak Bislama as a second, third, or fourth language. The language facilitates communication across the diversity of indigenous Papuan languages spoken by the rural population. In towns and urban centers such as Port-Vila, Bislama is becoming the first language of migrants.

History

The earliest record of settlement on the islands of present-day Vanuatu extends to c. 2000–1000 B.C. with the migration of Melanesians from New Guinea. Melanesian culture was not very complex and lacked the ability to make pottery. Austronesians of the Lapita culture followed and settled the islands c. 1400 B.C. Unlike the Melanesians, with whom they intermingled, the Lapita had a more advanced culture who made pottery, lived in village houses, and engaged in trade with New Caledonia. Although subsequent migrations occurred, their impact was limited until European contact began in the seventeenth century A.D.

The colonial history of the Republic of Vanuatu is complex. The islands were first described by Spanish explorers in the very early seventeenth century, with the final explorations by the Spanish in the islands of present-day Vanuatu occurring in 1606. Despite the glowing accounts of the islands provided by Lieutenant Quiros of the Philippines, the Spanish did not choose to pursue settlement and colonization in the Vanuatu chain. The French explorer de Bougainville sighted two islands in the Vanuatu chain in 1768 naming them Ile de la Pentecôte and Ile Aurore. The islands retain these names to the present. British Captain James Cook sighted the islands in 1774 and named the island group the New Hebrides.

After a brief rush (1825–30) to exploit the islands' sandalwood trees, European missionaries expressed interest in converting the islanders. The first Christian missionary to visit the New Hebrides was killed by natives on the island of Erromanga in November 1839. Although most of the missionaries that followed were also killed, Christianity eventually took hold in the islands. Anglicans, Presbyterians, and Catholics all established missions in the New Hebrides and these have had a lasting effect on the religious development of the Republic of Vanuatu. The Presbyterians in particular were responsible for the development of modern medical care in the islands. Today, the majority of Christian *ni-Vanuatu* are Presbyterian.

By the mid-nineteenth century, with the islands' sandalwood exhausted, British traders in Australia began importing labor from the island by force, a practice that did not cease until 1901. The effects of this trade, together with the effects of disease, decreased the islands' population by over half in the nineteenth century. Politically, the colonial governance of the New Hebrides was uneasily shared between England and France until a condominium (joint administration) was proclaimed in Port Vila in December of 1907.

British-French administration of the New Hebrides continued until the islands' independence in 1980 as the Republic of Vanuatu. In the meantime, interaction with the outside world brought drastic changes to the islands' communities. Increased trade and contact with the United States resulted in the emergence of the John Prum cargo cult in the 1930s. This religious revitalization movement—the worship of industrial goods from the United States in an attempt to acquire them through magic—exists in opposition to the organized Christian church and most of its followers see participation as a direct method for increasing their economic and material wealth. As in other colonies, the post-1945 period saw the rise of anti-colonial movements. The first political party on the New Hebrides, the New Hebrides National Party, began its drive for independence in 1971.

Since independence, Vanuatu has attempted to maintain its own cultural identity while also attracting foreign investment and economic aid. As a result, Vanuatu passed laws in

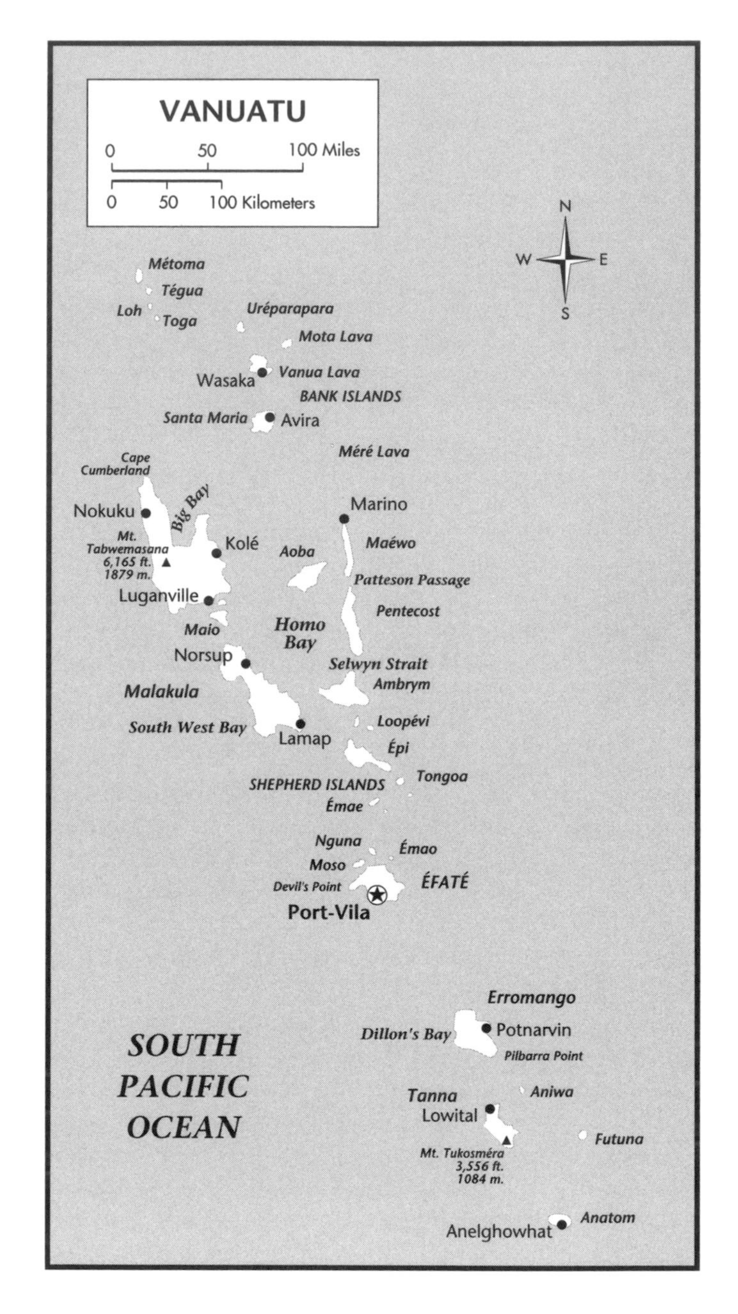

1982 and 1988 intended to restrict land ownership to native Vanuatans and to create a National Cultural Council for the preservation of *ni-Vanuatu* culture. In foreign affairs Vanuatu pursued a policy of neutrality in the last decade of the Cold War. Diplomatic relations were established with the United States, the Soviet Union, and China, and Vanuatu joined the Non-Aligned Movement.

The economy of Vanuatu is still heavily supported by foreign aid and loans. The principal foreign investors in the island nation are Australia, Great Britain and France. Recently, Vanuatu has developed into an international investment opportunity and tax haven. With no governmental control over currency exchange and few business taxes, Vanuatu has attracted over 1,200 international companies with registered official offices.

Timeline

c. 2000–1000 B.C. Melanesians settle islands in the New Hebrides

The archaeological record shows that an early culture settled various parts of the New Hebrides as far back as 4000 years ago. These migrants likely came in small waves of settlers, each speaking a distinct language. This pattern accounts for the complex linguistic diversity found in Vanuatu today. The early Melanesian settlers did not possess the technology to make pottery and they were not very accomplished seafarers.

c. 1400 B.C. The New Hebrides chain of islands is settled by Lapita culture

Seafaring Austronesians who possess a distinctive style of pottery, referred to as Lapita, settle on several of the islands in the New Hebrides chain. These newcomers intermingle with the Melanesians who had been migrating to the islands in small waves as far back as 2000 BC. The Lapita culture live in settled villages in houses that were elevated on posts. The Lapita are coastal dwellers who subsisted on fish and shellfish that they gathered from lagoons. There is evidence that shows that the Lapita groups in the New Hebrides participate in long-distance trade with Lapita groups in New Caledonia. The Lapita possess highly developed seafaring and navigational skills. The Lapita culture is the ancestral culture of many of the Polynesian groups of Oceania. There are, however, no indigenous Polynesian groups in present-day Vanuatu. The Lapita were likely absorbed into the migrant Melanesian cultures that continued to settle the islands throughout their prehistory.

c. 600 B.C.–A.D. 1300 Mangaasi culture prevails in central New Hebrides

A new cultural group appears in the archaeological record beginning about 600 B.C. They possess a style and manufacture of pottery that is distinctive from the co-existing Lapita style.

c. 1300 New migrant population settles in New Hebrides

A new archaeological complex appears at approximately 1300. These people did not possess pottery and were likely another wave of Melanesians who migrated from New Guinea.

1606 Quiros is first European to site Meralava, a small northern island in the New Hebrides

Lieutenant Quiros, exploring in the South Pacific for the Spanish court, sites an island that he calls Nostra Señora de la Luz. The island is now named Meralava. Quiros mistakenly believes that he has discovered a southern continent and names the island "Tierra Australia del Espirtu Santo," which it retains to this day.

1768: May 22 de Bougainville sights Ile de Pentecôte and Il Aurore

The French explorer de Bougainville sets sail from France in 1766 with the expressed goal of exploring the South Pacific. He encounters the chain of islands now referred to as Vanuatu in 1768, when on May 22 he sights two islands. Ile de Pentecôte is named for the day on which it was discovered while Ile de Aurore is named for the time of day of its discovery.

1774: July 16 Cook sights the New Hebrides

Captain James Cook sets sail from Deptford, England on April 9, 1772 with two ships, the "Resolution" and the "Adventure." In November of 1773, a violent storm separates the two ships with the "Adventure" returning to England. The "Resolution" captained by Cook sails on in search of the "Adventure." On 16 July 1774, he sights the New Hebrides.

1774: July 22 Cook sets foot on Malekula

After sailing through the chain of smaller islands, Cook finds what he believes to be the best anchorage in the area at Port Sandwich. He organizes and leads a landing party to the shore where an exchange of gifts takes place between Cook and an indigenous leader. Cook collects the firewood that was the purpose of the landing and returns to his ship without any violence.

1774: July 27 Cook records discovery of Erromanga

Cook sails throughout the New Hebrides charting and naming many islands within the group that had been previously unknown to Europeans. He also makes copious notes concerning the ways of the life of the indigenous peoples that he encounters, although his accounts often reflect the biases and beliefs of his own culture and society. As a result of his explorations, the New Hebrides become a relatively well-known group of islands in the South Pacific.

1825 Sandalwood rush on Erromanga

A South Pacific trader named Peter Dillon discovers huge stands of sandalwood trees in Erromanga, in the southern New Hebrides. A rush of loggers, traders, and shippers follow his discovery.

1830 Sandalwood boom is finished on Erromanga

As quickly as it started, the sandalwood boom on Erromanga is finished due to the depletion of the sandalwood stands.

1839: November First Christian missionary killed

The first Christian missionary to set foot on the islands is killed by natives on the island of Erromanga. Subsequently, most members of the first group of missionaries are killed. Despite the natives' original resistance to Christianity, however, the missionaries do not give up and eventually Christianize most of the population.

1863 First transportation of forced labor from the New Hebrides

The sandalwood trade established by the Europeans earlier in the century is the first large-scale exploitation of the islands. After the sandalwood is depleted, Europeans turn their sights to the indigenous population as a source of wealth. Laborers are transported against their will to the sugar plantations of Queensland, in northern Australia.

1871 Queensland Labour Act

The Australian government enacts control over the recruitment of indigenous peoples from the South Pacific islands for work on Queensland sugar plantations. The provisions do little of real significance to curb the atrocities of the labor trade in the New Hebrides and elsewhere.

1901 Pacific Islands Laborers Bill

The Australian government bans the importation of South Pacific laborers.

1907: December Condominium between England and France proclaimed

In order to secure the safety of the European inhabitants of the New Hebrides, the governments of England and France enters into a political agreement, called a condominium whereby a joint administration of the island chain is formally established and recognized.

1910: 9 November Earthquake

A strong earthquake strikes the New Hebrides. The main quake lasts over thirty seconds with smaller quakes following for a least twenty-four hours. Reports of sea captains indicate the southwest coastline of Espirtu Santo has risen approximately three feet as a result of the quake.

1930s John Prum cult emerges

The John Prum cult emerges on the island of Tanna. Little is known about the origins of the cult although its practices continue into the late twentieth century. The essence of the John Prum cult is the worship of the United States as the islanders' provider of modern goods through magic (hence, the characterization of the belief as a "cargo cult").

Early 1940s Vanuatu becomes important Allied base

Vanuatu becomes an important Allied base against the Japanese during World War II (1939–45). American forces occupy the islands in 1942 and build many roads and airstrips for military use.

1966 Radio Vanuatu is founded

Radio Vanuatu is founded. It transmits daily broadcasts in English, French, and Bislama.

1971 New Hebrides National Party is established

Father Walter Lini (b. 1943) establishes a nationalist political party whose overt goal is the independence of the New Hebrides from joint French and British rule. Father Lini, ordained as a priest in the Anglican Church in 1970, becomes the first prime minister of Vanuatu.

1974 New Hebrides National Party is renamed Vanua'aku Pati

The first political party established on the New Hebrides takes an indigenous name from the Bislama language. The party continues to push for independence.

1980: 30 July The New Hebrides becomes The Republic of Vanuatu

The New Hebrides officially becomes an independent nation with a parliamentary democracy. July 30 is celebrated annually with a national holiday.

1980: 29 August Nagriamel leader Jimmy Stephens is arrested

The founder and leader of a nationalist movement that seeks to repossess undeveloped lands under European ownership and establish full political control for the indigenous Melanesians of Vanuatu is arrested for asserting that the island of Espirtu Santo is independent of the newly formed Republic of Vanuatu. Within the next few months over 200 Nagriamel supporters are arrested by the government of Prime Minister Walter Lini. Nagriamel is formed in opposition to Prime Minister Lini's National Party.

1981: 15 September Vanuatu joins the United Nations

Vanuatu becomes a member of the United Nations (UN). Joining the UN the same year are Antigua and Barbuda (see Americas) and Belize (see Americas).

1982 Land ownership law passed

A law is passed that restricts eligibility for land ownership to natives of Vanuatu and their descendants. The law allows immigrants and other non-Vanuatuans to lease land only.

1982 New Hebrides Federal Party becomes Union of Moderate Parties

The New Hebrides Federal Party becomes the Union of Moderate Parties (UMP), led by Maxime Carlot Korman. UMP is a significant force in Vanuatu politics.

1987 National Advisory Committee on the Environment formed

The government forms a new agency, the National Advisory Committee on the Environment (NACE). NACE is charged with studying and planning for preservation of the environment, including the forests, which are being depleted through logging and the coral reefs around Vanuatu's coasts.

1987: February Cyclone causes extensive damage

A cyclone strike Vanuatu, damaging almost all the buildings in the capital, Port-Vila.

1988 Vanuatu National Cultural Council Act is adopted

The Vanuatu National Cultural Council Act of 1988 establishes the Vanuatu National Cultural Council for the preservation, protection, and development of various aspects of the rich cultural heritage of the *ni-Vanuatu*.

1991: December Maxim Carlot elected President of Vanuatu

The leader of the Union of Moderate Parties (UMP) was elected President of the Republic of Vanuatu in the general election. Quickly, a coalition was formed with two other rival parties to insure stability for the new government. Maxim Carlot Corman is prime minister.

1993: December Dam to be financed by China

Vanuatu and the People's Republic of China sign an agreement detailing the terms of a project to build a dam to generate hydroelectric power on the Vanuatu island, Malakula. Seventy-five percent of the project will be financed by China.

1994 Strike by Vanuatu Public Servants Union

The Vanuatu Public Servants Union stages a massive strike. The result is that hundreds of workers are fired, and overall union membership from 4,000 at the time of the strike to less than 1,000 in 1996.

1995 Parliamentary elections held

Parliamentary elections are held, with the Union of Moderate Parties (UMP) winning a slim majority of seats. Maxime Carlot Gorman thus remains prime minister, with Donald Kalpokas, leader of the Vanuatu Party (VP) as deputy prime minister.

1995 Royalties sought for bungee jumping

Many Vanuatans seek royalties for bungee jumping from operators throughout the world. The worldwide bungee jumping craze is based on a Vanuatan virility ritual in which men jump off of bamboo towers with vines around their legs.

1995: February Minimum wage increased

The minimum wage is increased by law to the equivalent of just over $140 per month.

Bibliography

Allen, Michael, ed. *Vanuatu: Politics, Economics, and Ritual in Island Melanesia.* Sydney, Aus.: Academic Press, 1981.

Bourne, Will. "The Gospel According to Prum. (Vanuatu's John Prum Movement), *Harper's*, January 1995.

Jennings, Jesse, ed. *The Prehistory of Polynesia.* Cambridge, MA: Harvard University Press, 1979.

Rice, E. *John Prum He Come.* New York: Doubleday, 1972.

Speiser, F. *Ethnology of Vanuatu.* translated by D. Stephenson. Hawaii: University of Hawaii Press, 1996 [1923].

Stanley, D. *South Pacific Handbook,* 3d ed. Chico, CA: Moon Publications, 1986.

"This Week's Sign that the Apocalypse Is upon Us," *Sports Illustrated*, 4 December 1995.

Vietnam

Introduction

An understanding of the geography of Vietnam makes clearer its history. Particularly, Vietnam's geographical position relative to China has had a tremendous influence on its history. Vietnam is located directly south of China which touches on Vietnam's northern border. Vietnam is over 1,000 miles in length, and shaped like an elongated S. In the northern loop of the S is the Red River delta, containing one center of population. The Mekong delta sits in the southern loop of the S and contains an even larger center of population as well as the most fertile land of Vietnam. The middle portion of the S borders the eastern coast. This cultivable strip connects the two main population centers and commands the South China Sea, thus occupying an internationally-important strategic position relative to India, Indonesia, Malaysia, Singapore, Thailand, the Philippines and southern China.

The Red River delta in northern Vietnam was settled before the Christian era; the expansion down to the Mekong Delta occurred primarily from the tenth to the eighteen centuries A.D. Vietnam was populated by clans of the Viet people, who lived in southern China, in the area from Shanghai to Hanoi in the Red River delta. They cultivated rice, which grew in rhythm with the monsoon rains falling between May and October and required extensive water system management. Viet people along the coast were also involved in fishing and shipping. Around 100 B.C., China's powerful Han dynasty conquered the Viet region. By A.D. 900, most of the Viets were incorporated into Chinese society, with the exception of those in the Red River delta. These Viets, or Vietnamese, although politically ruled by China, kept their own language and culture.

In the tenth century, the Vietnamese of the Red River Delta wrested independence from a China weakened by dynastic changes. They discarded Chinese provincial names and took over their own rule. The Vietnamese, however, preserved many of the traits learned from the Chinese, including an emperor who ruled a bureaucracy based on the mandarin system.

The Vietnamese continued their expansion, with farmer-soldier settlements pushing south from the Red River Delta, displacing locals as they went. There were several causes for this southern expansion, what the Vietnamese themselves called the "March to the South." Annual, sudden and uncontrollable flooding of the Red River drove farmers south in search of farmland. Also, after each successful Vietnamese military resistance of Chinese invasion, the government rewarded its soldiers with farmland in the south, thus the farmer-soldier settlements. These aggressive military migrants eventually defeated all the peoples already in their way. One of these was the Cham, an Indian-influenced group that was splintered and absorbed into the invading Vietnamese society or scattered to Laos and Cambodia. Another were the Khmer, many of whom continue to reside in the Mekong delta, a fact which gives rise to an on-going claim by Cambodia that southern Vietnam belongs to her. By the end of the eighteenth century, Vietnamese resided all the way to the southern sea, and in the nineteenth and twentieth centuries, the Mekong River delta became the rice bowl for all of Vietnam.

Vietnam's delta wealth, plus her strategic location on the South China Sea, has attracted numerous intrusions since the sixteenth century. Portugal, the Netherlands, France, Japan, the United States and China have invaded Vietnam. After the Chinese, the French remained the longest.

French Occupation

France governed the country in the late 1800s with French officials at the top assisted by Vietnamese in lower level positions. By the beginning of the twentieth century, French military forces had secured the country sufficiently for them to begin withdrawing its raw resources. Large tracks of land in southern Vietnam were given to French settlers and Vietnamese collaborators, who turned them into large, profitable rice plantations.

Except for a small number of Vietnamese collaborators and officials, French occupation did not benefit the Vietnamese people. Although Vietnam became a rice-exporting nation, the per capita consumption of rice declined. Taxes increased. Monopolies were placed on the production and sale of salt and alcohol, making them too expensive for most Vietnamese. Miners and plantation workers were fined if they left their jobs. Education declined. Moreover, Vietnamese were not allowed to protest their situation, for strict limits were placed on their participation in the political process.

The Vietnamese did not easily accept French occupation. Vietnamese mandarins, or government officials, refused to serve the French when they occupied southern Vietnam. When the French expanded into central and northern Vietnam, the educated elite led the peasants in a strong resistance consisting of pitched battles and guerrilla raids. The Vietnamese continued to struggle, even after the French exiled their emperor to Algeria in northern Africa, another country colonized by France.

During World War II, Vietnam was ruled by two foreign powers: the Japanese and the German-appointed French government. The Vietnamese continued to develop resistance groups who hoped to gain independence. Nguyen That Thank, later Ho Chi Minh (1892–1969), organized many young people into the League for Vietnamese Independence in 1941, which came to be known as the *Vietminh.* When the war ended, Ho Chi Minh declared Vietnam free and independent, hoping the nations he had cooperated with in fighting the Japanese would help him keep France from re-colonizing the country. Instead, Britain and the Nationalist Chinese helped France return, and the resistance began again fighting for the independence of Vietnam against the French.

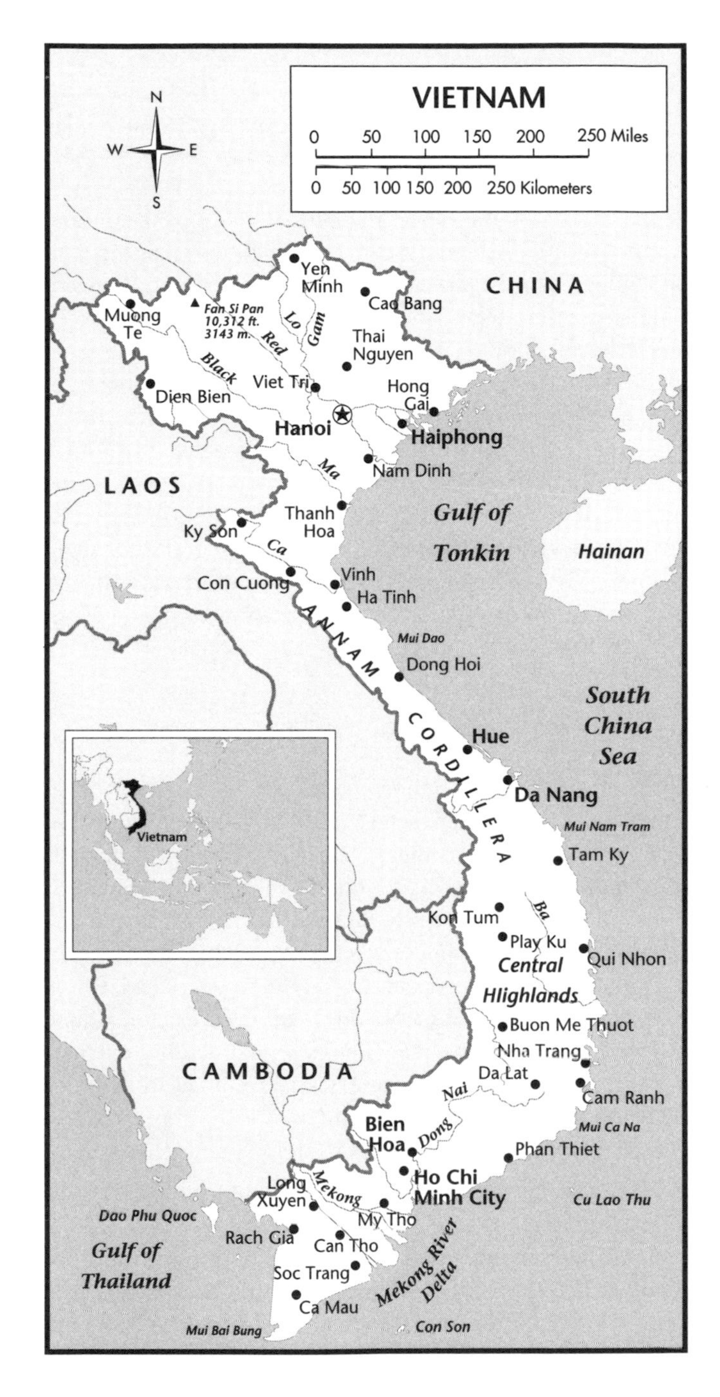

The Vietnam War

The United States disapproved of French occupation of Vietnam but, concerned with Communist growth in Asia and wanting to support an ally, America began cooperating with the French. By 1954, the United States was paying over seventy-five percent of the cost of the French war effort in Vietnam. From a reluctant ally, the United States soon became an aggressive supporter of the war against the Vietnamese Communists. The two sides locked in combat, with the Communist north supported by Communist China and the Soviet Union, and the south supported by Western allies, foremost among them the United States. The Geneva Conference of 1954 divided Vietnam along the 17th parallel into North Vietnam led by Ho Chi Minh and South Vietnam ruled by Ngo Dinh Diem (1901–63). Elections in 1956, it was hoped, would unify the country.

It did not happen. After the French suffered a humiliating defeat by the Vietnamese, America gave its support to Ngo Dinh Diem, a conservative nationalist from the elite class. A Catholic, he was supported by the large Catholic population of refugees, over 900,000, who migrated south from North Vietnam after the country was divided in two. Diem refused to hold elections, instead organizing a referendum that gave the Vietnamese a choice between him as president or Emperor Bao Dai (b. 1913) as monarch. Diem won, receiving more than 405,000 votes from the 605,000 registered voters in Saigon.

Diem's rule became increasingly harsh and authoritarian. The National Liberation Front was formed in 1960 in opposition to him. Later known as the *Viet Cong,* the National Liberation Front was comprised primarily of nationalist young people eager to see their country rid of foreigners and unite with North Vietnam. To prop up Diem, seen by Americans as the best of numerous bad options, President John Kennedy (1917–63) sent Green Beret advisors to provide increased military aid. By 1963, opposition to Diem's rule had Buddhist monks burning themselves in public protest. With American compliance, Diem was overthrown and assassinated in an army coup. The generals who overthrew him then took control of the country.

With full American support, Vietnam was now fully engaged in civil war. Worried about "losing" South Vietnam to Communism, President Lyndon Johnson (1908–73) in 1964 asked Congress to pass the Gulf of Tonkin Resolution, which gave the president broad powers to conduct the war. The United States began strategic bombing of North Vietnam, trying to pressure the North Vietnamese government to force the Viet Cong in South Vietnam to stop fighting. In the same year, 1964, the first American Marines landed in Vietnam. In 1965, another army coup put Nguyen Van Thieu (b. 1923) and Nguyen Cao Ky (b. 1930) in power. By 1968, the armies of South Vietnam and the United States were losing to the Vietnamese Communists. Soon after the Americans withdrew, North Vietnam conquered South Vietnam. Since then, the country has been united under a Communist government, which has faced considerable challenge in restoring an economy wracked by war and increasingly overpopulated.

Vietnam Today

The result of these foreign invasions has been the development of an intense national consciousness among Vietnamese of their own history and importance, a consciousness that has assisted them in defeating a number of powerful enemies. The word *Viet* stems from before the Christian era and has been used by the Vietnamese whenever possible. It was never used when the country was under foreign rule. Instead, China used Chinese provincial names, Portugal used Cochin China, and France added Annam and Tonkin to that. During much of the twentieth century, the country was referred to as Indo-China. As a consequence, the word Viet is an important one for the Vietnamese, signifying anti-foreign nationalism.

While nearly ninety percent of Vietnam's people are ethnic Vietnamese, over ten percent belong to other ethnic groups. These include two million Tai, over half a million Muong, and 200,000 of the Hmong Meo and Zao each. These groups live in the mountainous regions of northern Vietnam. More ethnic groups live in the mountains of central Vietnam, including 150,000 Jarai; 100,000 Ede; and 100,000 Bahnar. Most of these upland groups are slash-and-burn farmers: cutting down and burning trees to plant a crop, moving on to do the same elsewhere when the land becomes infertile and returning some years later to begin the cycle again. They speak their own languages, practice their own customs, and follow their own traditional beliefs, often different from even the other groups with whom they share the mountains. Their isolation and different ways made them natural enemies to lowland Vietnamese, who viewed them as inferior. Many uplanders joined foreign armies—the French and then the Americans—because of their traditional hatred for the lowlanders.

About 100,000 Cham also live in the highlands and almost a million Khmer in the south delta area, both groups left behind by kingdoms pushed aside by the advancing Vietnamese. Over a half million Chinese live in Vietnam's urban areas, often involved in trade, banking, and commerce. However, ethnic Vietnamese, the descendants of the early people of the Red River delta, dominate the country's industry, business, and government.

Approximately 75 million people live in Vietnam, a country that remains predominately agricultural. About twenty percent of the population lives in cities, the largest being Ho Chi Minh City (formerly and still frequently called Saigon), Hanoi, and Haiphong. With the influx of Western capital and business in the middle to late 1990s, these urban areas are seeing a revival. Most of Vietnam's population lives in the Red River and Mekong River deltas. The population is growing at an annual rate of 2.2 percent, a very fast rate indeed. At this rate, the population will double in thirty-one years.

The lives of most Vietnamese revolve around the agricultural cycle, the planting and harvesting of crops, and small village life. Most rural families live in houses built of wood, thatch and bamboo in the south that allow air circulation to cool people in the hot climate. In the north, many homes are made of stone to protect inhabitants from harsher weather. Farmers work either on private or on collectivized farms where the workers share tools, crops, and labor. Most families have a small garden. Most village and rural areas do not yet have electricity or public water supplies and sewer systems. Urban dwellers for the most part live in small apartments. Both rural and urban dwellers have little furniture.

Most Vietnamese live in extended families, with grandparents sharing the home with married children and grandchildren. It is not unusual for one's siblings to live nearby. The Vietnamese are proud of their close family ties, their history, language, literature, arts and cuisine.

Timeline

2800 B.C. Beginning of the Viet Nation

Legend has it that the Viet nation is formed when a Vietnamese prince named Lac Long Quan comes to northern Vietnam from his home in the sea. He marries a princess from the mountains and from their union are hatched one hundred eggs. When they hatch, a son emerges from each. The parents cannot stay together because they are of differing origin. Fifty sons travel with their mother across the mountains, to become ancestors of the Muong. Fifty sons accompany their father back to the sea to become ancestors of the Vietnamese. The bravest son becomes the first of eighteen kings of the Hung Vuong dynasty.

Local tradition says that a line of legendary kings who ruled the ancient kingdom of Van Lang for thousands of years found the small Vietnamese kingdom of Au Lac in the Red River Valley.

History has it that the first appearance of the Vietnamese is as one of many scattered peoples living in southern China and northern Vietnam just before the beginning of the Christian era. Archaeological evidence suggests that the early people of the Red River Delta may be among the first Asians to practice agriculture. By the first century B.C., they are practicing a Bronze Age level of civilization.

221 B.C. China completes its conquest of neighboring states

The Ch'in Dynasty in China completes its conquest of neighboring peoples, including the Viet people who live in southern China and northern Vietnam.

208 B.C. A Chinese general conquers Au Lac

A renegade general from China, Trieu Da, conquers Au Lac in the northern mountains of Vietnam. He establishes his capitol city and proclaims himself the emperor of Viet Nam.

111 B.C. The Han Dynasty conquers the Viet regions

The armies of the Han Dynasty migrate into northern Vietnam, incorporating the territory of Viet Nam as the province of Giao Chi within the Chinese Empire. Other Viet peoples of southern China are themselves incorporated into Chinese society, adopting Chinese language and custom.

In the area of what is now northern Vietnam, the Han Empire replaces chieftains from the local landed nobility with Chinese administrators. They impose political institutions modeled on the Chinese bureaucracy. Confucianism becomes the official ideology. Chinese language is introduced as the medium of official and literary expression, and Chinese ideographs are adopted as the written form of the verbal Vietnamese language. Chinese architecture, art, and music heavily influence the Vietnamese.

A.D. 39 Trung Sisters Revolt

The Vietnamese fiercely resist the Chinese, most famously when two aristocratic widows lead a revolt against Chinese rule. Successful for a time, the older sister, Trung Trac, becomes ruler of an independent state. By A.D. 43, however, Chinese armies put down the rebellion, and Vietnam's independence is over. The Trung revolt is only one of many intermittent uprisings that occur throughout the thousand years of Chinese rule over northern Vietnam.

The Viet of the Red River Delta in northern Vietnam, however, continue to speak their own language and practice their own ways. For over a thousand years, the Viet of northern Vietnam accept that they are a part of China and, despite occasional rebellions, they remain tied to China. They learn to harness water for cultivation. They adopt Chinese systems of education and bureaucracy for their own use and, while keeping their own language, learn to write Chinese. They also use Chinese *ideograms,* pictorial forms, to write Vietnamese words.

The Viet begin an expansion to the south that continues until the eighteenth century, with aggressive farmer-soldiers settlements taking the land from its owners, establishing villages, and taking over ricelands.

1–100 Founding of Funan

While the formation of the Vietnamese people in northern Vietnam is under the influence of the Chinese to their north, the entire Mekong delta in southern Vietnam and most of present-day Cambodia sees the founding of Funan. Funan is strongly influenced by India, not through military action but by Indian merchants, priests, and literati bringing economic and cultural innovations. Residents of the delta area adopt art, architecture, philosophy, religion, and the Sanskrit language.

100–200 Mahayana Buddhism comes to Vietnam

In the second century A.D., Buddhism spreads into Vietnam's Red River delta region from China. The Chinese form, Mahayana Buddhism, dominates northern Vietnam.

200 Kingdom of Champa

Champa is founded along the east coast, from the Mekong delta in the south to the mountain range of Hoanh Son in the north, near the eighteenth Parallel. Champa has a powerful fleet, which they use both for commerce and for incursions against their neighbors. Champa also borrows considerably from India, as can be seen in its art and architecture.

200s–500s Theravada Buddhism comes to Vietnam

Theravada Buddhism spreads into southern Vietnam from India. Mahayana believers say that the Buddha is only one of many enlightened beings, while Theravada believers think he was the only enlightened one. Mahayana adherents believe individuals can attain nirvana, or heaven, while Theravada believers think only select monks and nuns can do so.

500s Disappearance of Funan

After dominating Southeast Asia for several centuries, Funan loses power and is absorbed by the Khmer of Cambodia. Funan occupies part of Vietnam, but the Funan people probably have little or no direct contact with the Vietnamese who do not migrate south into Funan territory until 900 years later after Cambodia's conquest of Funan.

939 Defeat of the Chinese

Vietnamese troops led by General Ngo Quyen (d. 944) defeat Chinese occupation troops and declare independence for Vietnam.

939 Vietnam gains independence from China

The Vietnamese in the Red River delta claim their independence from China and begin a thousand-year struggle to retain their independence from China. Ngo Quyen founds the first authentic Vietnamese dynasty, the Ngo dynasty, locating his capitol at the site from which the country was ruled before being occupied by China.

944 Death of Ngo Quyen

Before Ngo can set up a strong central administration to control the numerous local chiefs constantly struggling to set up independent fiefdoms, he dies, leaving weak heirs. By 966, the country is in anarchy.

968 Dinh dynasty

Dinh Bo Linh, who founds the Dinh dynasty, overthrows the Ngo dynasty. He reunites the region as an independent country and pays tribute to the Chinese emperor. He calls his state Dai Co Viet, or Great Viet. He institutes universal military conscription, organizing a 100,000-man peasant militia comprised of ten districts, each defended by ten armies each composed of ten brigades. Each brigade is made up of ten companies, each with squads of ten.

980 Successive dynasties

The last ruler of the Dinh dynasty is succeeded by General Le Hoan, who makes himself ruler of Vietnam in the face of threats from China by overthrowing the six-year-old king. As founder of the Earlier Le dynasty, Le Hoan is able to keep the Chinese from invading his country. His is the third dynasty in the first seventy years of Vietnamese independence to be founded by strong capable leaders only to have their work destroyed by weak and less capable heirs.

A.D. 1010 The Ly Dynasty is founded

In the early eleventh century, the first of the great Vietnamese dynasties is established. The Ly Dynasty rules Vietnam for more than 200 years, from 1010 to 1225, under a series of strong dynamic leaders. The Ly Dynasty sees the emergence of a strong nationalist bent among the population but continues also to retain many Chinese political and social institutions. Confucianism provides the foundation for Vietnam's political institutions. Chinese civil service examinations are used for selecting government officials, although initially these are only given to nobility. The education system contains examinations based on the Confucian classics.

The Vietnamese also retain their own culture, admiring their past heroes and continuing village traditions. Some Vietnamese are agriculturists, owning their own land or working it for nobles. Others engage in commerce, manufacturing, and crafts.

The Ly rulers build roads, canals, and dikes, and provide the political stability necessary for economic progress. All this work is done against a backdrop of war and the constant threat of invasion: China from the north, Champa from the south, and Cambodians to the west. Cambodia and Champa attack in 1128, 1132, and 1138, and five times again between 1138 and 1216.

1100s Buddhism becomes official religion

Court officials and Buddhist monks compose the first major collection of poetry in the first centuries after Vietnam gains independence from China. Buddhism becomes the official religion, although its practice is restricted primarily to the royal court and scholars. Most Vietnamese continue to practice their traditional beliefs, including honoring their ancestors and respecting the powerful spirits who control the world around them.

A.D. 1225–1440 The Tran Dynasty rules Vietnam

The eventual decline of the Ly dynasty leads to a period of civil unrest and war which ends only with the victory of Tran Thai Thong in 1225.

In the 1200s the Tran dynasty takes over rule of Vietnam, bitterly resisting conquest by the armies of Kublai Khan. Tran rulers supervise extensive land reform, improve public administration, and encourage the study of Chinese language. They are best remembered for repelling the Mongols and Cham.

Historical writings become important during the Tran dynasty, the most influential being the *Historical Memoirs of the Great Viet,* a long work by Le Van Huu.

The greatest threat to the Vietnamese comes from the north. In 1257, a Mongol army sacks Hanoi and only heat and disease force them to leave. In 1284, a Mongolian army, with possibly half a million men, invades the Red River Delta but is defeated by Vietnamese armies who are helped by the malaria-spreading delta mosquitoes. The Vietnamese defeat a second Mongolian invasion three years later under the leadership of Tran Hung Dao, one of Vietnam's most famous heroes. After these fierce battles, the Vietnamese drive back the Mongol forces, retaining independence for Vietnam from China.

1371 Cham invasion

A Cham army invades the Red River valley and sacks Hanoi. Wars and widespread famine inspire numerous revolts. In a series of battles with the kingdom of Champa in central coastal Vietnam, Vietnam conquers this state, forcing the

Cham people to the south. The Vietnamese follow, settling where the Cham have lived, moving always south. By the end of the 1500s, the Champa Kingdom has lost its previous dominance over the region and is unable to prevent Vietnamese southward incursions.

1407 China again conquers Vietnam

Chinese troops again venture south and, after only seven years of reign by Ho Qui Ly, conquer Vietnam. For the next twenty years, the Ming Dynasty tries to reintegrate Vietnam into the Chinese Empire and denationalize the Vietnamese. Schools can teach only Chinese, local religions are suppressed, women are forced to wear Chinese dress, men are forced to wear long hair, and everyone is issued an identity card. These efforts strengthen nationalist sentiment and make Vietnamese determined to secure independence.

1428 Loi restores independence for Vietnam

Vietnamese troops led by Le Loi defeat the Chinese invaders. Le Loi then becomes the first emperor of the Le Dynasty. China recognizes Vietnam as an independent country by signing an accord with Le Loi. For over a hundred years, the Le family rules. The Le dynasty institutes a number of reforms, including redistribution of land, creation of new cultivable land, and appointment of dike inspectors to protect the land. Ly rulers create an advanced legal code, extend equality for women, reform education so that poor children of merit can be schooled, create hospitals, and build numerous temples. One ruler, Ho Qui Ly, introduces paper money 250 years before its use in Europe.

1471 Le dynasty expands its territory

The Le family also expands their territory to create more land for Vietnamese farmers, wresting land from the Champa until the Kingdom is almost wiped out. By 1471, the Champa Kingdom is defunct. Soldier-settlers move into the newly conquered lands, creating farms and villages. The Vietnamese then expand into Cambodian territory, settling the land by the same method.

1502 Beginning of national turmoil

The last able Le ruler, Hieu Tong, dies, which causes instability. During the next twenty-five years, Vietnam has eight kings, six of whom are assassinated by royal relatives or ambitious lords.

1500s Decline of the Le Dynasty

As the Le house declines, two families grow in influence: the Trinh and the Nguyen. When the Trinh family gains the edge, the Nguyen are granted a fiefdom in southern Vietnam. Now Vietnam is divided into two areas. The country is divided at the Gianh River in Quang Binh Province, with the north ruled by the Trinh and the south ruled by the Nguyen.

Civil war rages for over 150 years between rival warlords and between the Trinh and Nguyen families. Both families pay lip service to national unity for Vietnam, but each maintains its separate government over a portion of Vietnam. In their effort to keep out the invading Trinh who have 100,000 soldiers, 500 elephants, and 500 ships, the Nguyen build two great walls, six meters in height and twelve miles long on their frontier. The Nguyen are also better equipped because they have Portuguese weapons and gunpowder.

1500s First European influence in Vietnam

Conflict is intensified by manipulations of European powers who have just arrived in Southeast Asia and seek influence, wealth, and converts.

Europe's successful penetration of the East begins with Vasco Da Gama's (c. 1469–1525) discovery of a sea route via the southernmost African Cape to the Indian Ocean. This route opens up trade with Southeast Asia, which increases domestic and foreign economic activities in Vietnam. Trade relations extend down to the countryside, with trading villages springing up alongside agricultural-industrial-commercial villages and agricultural and handicraft villages. The village market network expands, with three or four villages sharing a market. New towns and commercial ports are created. Merchant ships from Portugal, Holland, France, and Great Britain come to trade, occasionally setting up business offices in Vietnam to increase their influence and ability to transact business. Chinese and Japanese come for extended periods to conduct business.

Europeans also bring new skills. For example, Portuguese help the Nguyen lords set up a guncasting workshop; a Vietnamese who spends two years in the Netherlands learning clock making returns to Vietnam to set up shop.

1516 Portuguese visitors

Dominican missionaries visit Vietnam.

1535 First European military visitor

Captain Antonio da Faria of Portugal visits Vietnam, becoming the first military man to do so. He establishes a Portuguese trading center at Faifo in central Vietnam.

1615 Famous missionary Rhodes

The Faifo mission becomes permanent. French Jesuits join the Portuguese, Spanish, and Italian missionaries already in residence. One Jesuit, Alexander of Rhodes from Avignon, becomes the famous father of French missionary activity in Vietnam and the first French author in Vietnam. He also develops *quoc ngu,* an alphabet that transcribes Vietnamese from ideograms into Latin characters and shows pronunciation with accent marks. Most Vietnamese authors continue to write in ideograms until the colonial period, when Vietnamese literature adopts quoc ngu, the transcribed Latin characters. The Trinh family, who resents the influence of foreigners on their subjects, expels Rhodes from Vietnam.

1672 Obstacles to European penetration

The Trinh and Nguyen administrators are interested in the Europeans as long as they can acquire modern weapons. When war between the north and the south ends, European trading centers in Vietnam do poorly.

1600s–1700s Vietnam remains divided

The division of the country continues until 1786. Although partition and civil war have been disastrous for the population, the country shows significant economic development in the seventeenth and 18th centuries. The Nguyen lords expand their land and control south to the Mekong delta. Wealthy landlords are encouraged to recruit poor farmers for land reclamation. Peasants migrate southward in large numbers, turning the marshy wasteland into productive rice fields and villages.

1765–1820 The poet, Nguyen Du

Nguyen Du writes one of the most famous Vietnamese works, *Kim Van Kieu* (Tale of Kieu), a 3,250-verse poem that tells the struggles of a beautiful girl named Thuy Kieu.

1772 Tay Son Revolt

Peasant uprisings against poverty and oppressive rule increase. In 1772, a local peasant rebellion spreads into a national revolt that overthrows the Trinh and Nguyen overlords of northern and southern Vietnam. The revolt, led by the Tay Son brothers, also dissolves the partition, reuniting Vietnam for the first time in two hundred years. In their revolt against the Nguyen dynasty, the Tay Son brothers kill the entire royal family except for one nephew, Nguyen Anh. While the Tay Son brothers fight the Trinh in the north, Nguyen Anh reoccupies Saigon in the south, aided by Cambodian mercenaries and Chinese pirates.

1783 The Tay Son brothers conquer the south

The Tay Son brothers take back the city of Ho Chi Minh from Nguyen Anh.

1788 Nguyen Anh conquers the south

China tries to exploit the Vietnamese crisis by invading Vietnam. Under the brilliant military leadership of the youngest Tay Son brother, Nguyen Van Hue, the Chinese are defeated. Meanwhile, Nguyen Anh, now with French assistance, takes over Ho Chi Minh City and the entire southern delta region.

1784–89 Ascendancy of the Tay Son brothers

The Tay Son Dynasty fights off invading Siamese in the south between 1784 and 1885 and the Qing in the north from 1788 to 1789. The leader of the Tay Son Revolt, the eldest brother Quang Trung Nguyen Hue, becomes a national hero for his military leadership.

1802 The Nguyen Dynasty takes over

Quang Trung Nguyen Hue reigns only a few years. After his death in 1792, the Tay Son Dynasty is weakened and defeated by Nguyen Anh in 1802. Nguyen Anh, now Emperor Gia Long (d. 1820), establishes a new dynasty. He moves the capitol to Hue and names the country Vietnam. The name of the country is later changed to Dai Nam. In the first half of the nineteenth century, the Nguyen dynasty institutes a number of reforms, such as land reclamation, to promote national unity.

But, on the whole, the dynasty is conservative, not interested in creating connections or contact with the developing world. Vietnam remains a centralized state headed by a monarch with absolute powers thought to come from heaven. The country is ruled by a hierarchy of civil and military mandarins recruited through civil service examinations taken after long years of study. Gia Long, like all Vietnamese rulers, is supreme lawmaker, chief justice, head of all administrative units, and religious leader.

Despite the military assistance of a French missionary, Pierre Pigneau de Behaine, who has raised a mercenary force to assist Nguyen Hue in seizing the throne of Vietnam, Gia Long is suspicious of the French.

1820–41 Rule of Minh Mang

Gia Long appoints Minh Mang as his successor, the eldest son of Gia Long's first concubine and an enemy to the French. He dismisses all French advisors, rejects French proposals to establish diplomatic and trade relations, and makes an official policy to persecute French missionaries and Vietnamese Catholics. Several are executed.

1840s Missionaries seek French assistance

French missionaries protest the persecution and execution of French and Vietnamese Catholics in Vietnam. Missionaries openly ask France to militarily intervene in Vietnam on behalf of the persecuted missionaries. Fearing an invasion of Vietnam, Minh Mang makes peace overtures to the French, who reject them on the advice of French Catholics and the powerful Society for Foreign Missions in France.

1841–47 Reign of Thieu Tri

Succeeding Minh Mang, Thieu Tri tries to curb missionary activities without antagonizing France. He expels missionaries from Vietnam but makes peace overtures to French diplomats arriving on warships in a Vietnamese harbor. French warships frequently threaten the Vietnamese government, sometimes on orders of the French government, sometimes on the initiative of the ship captains.

1847: April Naval bombardment of Tourane

The French clash with Vietnamese mandarins at Tourane, now Danang. A French warship bombards the harbor, killing hundreds of Vietnamese civilians. Tu Duc takes the throne and plans to eliminate Christianity in Vietnam.

1856 Second bombardment of Tourane

Another bombardment of Tourane occurs, after the ruler, Ta Duc, orders two French priests executed.

1857: July French naval force attacks Vietnam

French commercial and military interests in Southeast Asia increase. Emperor Napoleon III (1808–73) orders a French naval expedition launched against Vietnam in retaliation for Vietnamese aggression against the French. This aggression against Vietnam is more the result of missionary than commercial pressure.

1858: August 31 French begin their invasion of Vietnam

Fourteen French vessels carrying 2,500 men arrive in Tourane and take the city and harbor in one day. Resistance is heavy, however, and the French soon realize they cannot reach the capitol of Hue from Tourane.

1859: February The French take Ho Chi Minh City

The French attack Ho Chi Minh City and succeed in securing it but are unable to advance farther.

1859–83 France tries to conquer Vietnam

This shaky beginning at conquering the country is followed by eight years of struggle to conquer southern Vietnam and another sixteen years of diplomatic and military efforts to complete their conquest.

1862 Vietnam yields territory to France

Emperor Tu Duc signs a treaty with the French giving them wide religious, economic, and political concessions, including ceding to them several southern provinces in the Mekong Delta.

1883: August 25 Vietnam submits to French control

The French fleet is finally about to subdue the country, bombarding Hue in August 1883. The royal court submits, signing a Treaty of Protectorate with France that extends French authority over all Vietnam. France establishes a protectorate over the remaining territory of Vietnam and divides the country into three protectorates: Tonkin in the north, Cochin in the south, and Annam in the middle.

1885 Vietnamese rebel against the French

The Vietnamese resent French occupation, and anti-French sentiment soon emerges. The economic situation in the country exacerbates the situation. French occupation brings improvements in transportation, communications, commerce, and manufacturing, but most of the benefits from these changes go to the French residing in Southeast Asia. The Vietnamese population, by and large, remains poor. Those who do obtain jobs in factories, coal mines, and on the rubber plantations work in terrible conditions for very low wages. In addition, Vietnamese villagers are required to pay stiff taxes to their French overlords.

As a consequence, there are regional uprisings against the French. France subdues the northern Protectorate, Tonkin, by killing many Vietnamese. It takes another fourteen years of pacification before France establishes firm control over the Vietnamese.

1890 Birth of Ho Chi Minh

Ngyuen Ai Quec is born in central Vietnam. Better known as Ho Chi Minh (among a variety of names he adopts at various times), he is the George Washington of Vietnam, presiding over its struggle for independence from the French and the United States. He is president of Vietnam from 1945 until his death in 1969.

Nguyen leaves Vietnam and does not return for thirty years. He lives in Paris and joins the French Communist Party, lives in the Soviet Union, founds the Indochinese Communist Party in Hong Kong, and finally returns to Vietnam in 1941. The following year, Nguyen travels to China to ask Chiang Kai-Shek (1887–1975) for assistance in his fight against the Japanese, against whom Chiang Kai-Shek is also fighting. The Chinese arrest and imprison him. In jail for thirteen months, he composes his *Prison Diary,* a book of poetry.

The following year, Nguyen That Thank is released from Chinese jail, he returns to Vietnam, and changes his name to Ho Chi Minh, which means "Enlightened One." Ho leads the Vietminh against the Japanese in Vietnam with financial support from China and the United States. Vo Nguyen Giap (1912–75) forms the Vietminh army whose activities include sabotaging the Japanese army and rescuing American flyers.

1893 Phan Dinh Phung rebellion

Another armed rebellion breaks out in Annam, led by the famous scholar, Phan Dinh Phung. The French fight for more than two years before subduing the rebellion on Phan Dinh Phung's death in 1895.

1895 Creation of the Indochinese Union

Laos, Cambodia, and Vietnam, all French protectorates, become the Indochinese Union headed by a French governor-general residing in Hanoi in northern Vietnam.

1897 French exploitation of the Vietnamese

Governor-general Paul Doumer (1857–1932) arrives to rule Vietnam. He establishes direct and stern rule over all Vietnam. Staffing his administration with thousands of officials from France, Doumer uses Vietnamese officials only in minor positions. He oversees the building of roads, railroads, harbors, canals, bridges, and other works for the single purpose of drawing off Vietnam's raw resources and making Vietnam a tariff-protected market for French goods, all for the benefit of France. The French make no effort to develop local industries beyond local consumption. Their major goal is immediate profit. Little to none of the profit is reinvested in Vietnam. The benefits go to French private businessmen and a few Vietnamese landowners; virtually nothing goes to either the French or the Vietnamese people.

1880–1930 French rule

After fifty years of French occupation, most Vietnamese are worse off than they were before the French arrived. In these years, the individual peasant's rice consumption decreases (and is not increased by other foodstuffs) even though the surface of ricelands quadruples. Large landowners in southern Vietnam, 2.5 percent of all landowners, own 45 percent of the land. Peasants, 71 percent of all landowners, own 15 percent of the land. The situation is no better in central and north Vietnam. The peasants remain too poor to purchase French goods, most of which are consumed by the 6,000 to 7,000 Vietnamese landowners, a few Vietnamese officials, and the French residing in Vietnam.

Despite the Roman Catholic Church's rejection of ancestor worship, Catholicism spreads among the population. French rule brings Western education for some Vietnamese, the beginning of industrialization and urbanization, and the development of cash crops. Establishing a French-dominated ruling class depletes the power of the emperor and mandarins. Traditional Confucian schools, used to teaching in Chinese, begin teaching in Vietnamese and French.

1908 Vietnamese act against the French

Demonstrations against high taxes imposed by the French occur throughout Vietnam, with men cutting their long hair (a custom adopted from the Chinese). Resistance inspired by Phan Boi Chau (1875–1940) results in the establishment of the Association for the Modernization of Vietnam. Hoping for Japanese help in ousting the French, Phan and his followers, in addition to inspiring demonstrations, print materials and engage in guerrilla actions. He unites numerous nationalist groups in the League for the Restoration of Vietnam.

1911 Le Duc Tho

Le Duc Tho (1911–90) is born. He helps found the Viet Minh and serves as special advisor to the North Vietnamese delegation to the Paris Peace Conference from 1968 to 1973. He is nominated for the 1973 Nobel Peace Prize, with Henry Kissinger of the United States, but declines.

1913 Birth of Bao Dai

Bao Dai is born, the last emperor of Vietnam. He serves as a puppet ruler under the French, heading the government between 1945 and 1955.

1919 Paris Vietnamese write independence declaration

Vietnamese residing in Paris create the first document for Vietnamese independence to present to the Versailles Peace Conference. The document is based on a fourteen-point declaration written by American President Woodrow Wilson (1856–1924) that calls for independence for all peoples. Ho Chi Minh is one of the authors of this independence declaration. He tries but fails to meet President Wilson at the Versailles Peace Conference to ask his assistance in obtaining self-determination for Vietnam. The United States government never acknowledges the document.

1919 Development of Cao Dai religion

A low-level official, Ngo Van Chieu, has a series of revelations in the Mekong River Delta region, founding a religion that blends Buddhism, Christianity, Taoism, Confucianism and nineteenth century European romanticism. The religion grows to between one and two million adherents.

1920 Development of nationalist parties

By the early 1920s, a number of nationalist parties have formed increasingly demanding independence and reform from French colonialists. Their efforts are hampered, however, by several conditions of colonialism. First, the absence of civil liberties under the French prevents the rise of democratically-oriented groups and favors the rise of illegal organizations of radical bent. Second, Vietnamese under the French are excluded from the modern sector of the economy, preventing the rise of a property-owning middle class sympathetic to democratic and capitalist ideas.

1925 Bao Dai gains the throne

Twelve years of age, Emperor Bao Dai ascends the Vietnamese throne. Ho Chi Minh travels to China where he founds the Revolutionary Youth League of Vietnam, the first Marxist organization in Southeast Asia.

1930 The Communist Party of Vietnam is established

The unified Communist Party of Vietnam is founded in Kowloon, now Hong Kong, by a group led by Ho Chi Minh. Exploiting conditions of near starvation, the Vietnamese revolt against French rule at Yen Bay, northwest of Hanoi in northern Vietnam. The French in retaliation destroy the unified Communist Party of Vietnam.

1930s Beginning of Hoa Hao sect

The Hoa Hao sect, founded by Huynh Phu So in the Mekong River Delta, includes considerable Buddhist elements and stresses an individual's relationship to the Supreme Being, individual prayer, simplicity, and social justice over ritual and icons. It grows to include over a million followers.

1931–32 French retaliation

In these years, according to a French report, 699 Vietnamese are executed without trial, 83 are sentenced to death, and 546 are imprisoned for life, all in retaliation for activities against the French. Vietnamese nationalist groups report that over 10,000 Vietnamese are killed during this period. The French Governor-general, Pierre Pasquier, tells France that Communism has disappeared from Vietnam.

1932 Bai Dai returns to Vietnam

Emperor Bai Dai completes his studies in France. On his return to Vietnam, he attempts to convince the French to liberalize their rule of his homeland. When that effort fails, he loses interest in political matters.

1936 Reappearance of the Communist Party

The Communist Party reappears, fully reconstituted, and exploits every opportunity to create legal "front" organizations that conceal the true purpose of the group, which is resistance to colonialism.

1938 Creation of religious-political group

The progress of the Communist Party is retarded somewhat by the activities of two religious-political groups, the Hoa Hao established in 1938 and the Cao Dai established twelve years earlier.

1940 Communist Party foremost among nationalist groups

The Communist Party stands foremost among nationalist groups, the only group in command of indoctrinated, disciplined followers.

1940: September 22 Japanese troops arrive in Vietnam

The Japanese government demands the right to place Vietnam under its military control. The Japanese then restrict the power of the local French administrators to figureheads. French Vichy government allows Japan to station troops in Tonkin. Japanese forces cross the border into Vietnam from China and attack the French stronghold at Langson and Dong Dang. The French surrender to the Japan.

1941: May 10 Formation of the Vietminh

The Communists found the *Vietminh,* short for the League for the Independence of Vietnam and prepare for the end of World War II (1939–45) when they plan to take control of Vietnam. The major goal of the Vietminh is independence for the country.

1944: December 22 Vietminh attack French for first time

General Giap leads a small group of Vietminh against French outposts in northern Vietnam on this day, now celebrated as the beginning of Vietnam's struggle with France for independence.

1945: March Vietnam gains independence from Japan

The Japanese place Emperor Bao Dai back on the throne of Vietnam and grant independence to the country.

1945: July Accepting Japanese surrender in Vietnam

At the Potsdam Conference, the Allies assign the British to disarm the surrendering Japanese in southern Vietnam and the Nationalist Chinese to do so in northern Vietnam.

1945: August Ho Chi Minh seeks American support

After the Japanese surrender to the Allies, Vietminh forces throughout Vietnam take control of their area and declare the existence of an independent republic in Hanoi. The French do not favor this option, however, looking once again to Vietnam as a source of resources for the French.

1945: August 23 Emperor Bao Dai abdicates

Emperor Bao Dai abdicates the throne.

1945: September 2 Ho Chi Minh declares Vietnam's independence

In Hanoi, with American intelligence officers standing beside him, Ho Chi Minh announces Vietnam's new government: the Independent Democratic Republic of Vietnam. He quotes from the American Declaration of Independence and again seeks American aid in remaining an independent nation.

Other Vietnamese nationalist groups, French residents, and French officials creating chaos throughout the country challenge the Vietminh. Chinese troops invade northern Vietnam. The British, in the country to accept surrender from Japan, try to assert control with troops from Great Britain, India, France, and even Japan.

1945: September 24 France reasserts her claim over Vietnam

Vietminh leaders call for a general strike against the French after their offices in Saigon are destroyed. General Jacques Philippe Leclerc (1902–47) arrives from France to reassert French control over Vietnam.

1945: September 26 First American is killed in Vietnam War

Vietminh, who mistake him for a Frenchman, kills Lieutenant Colonel A. Peter Dewey, head of the American intelligence office in Vietnam.

1945: October French drive nationalist groups from southern Vietnam

Unwilling to concede independence, the French drive the Vietminh and other nationalist organizations out of southern Vietnam.

Ho Chi Minh asks American aid in gaining Vietnamese independence from France. The United States opposes French occupation of Vietnam but does not want a Communist government there either. America does not respond to Ho's requests for assistance.

1945: December War resumes between France and the Vietminh

The French and the Vietminh try to achieve a negotiated settlement to their dispute, but after a year of talks in France, war again breaks out. The war continues for almost eight years.

The Vietminh retreat to the hills in the west and north to organize and build up their troops. The French form a rival government to the Vietminh, headed by former Emperor Bao Dai, the last ruler of the Nguyen dynasty. Throughout the same period, approximately two million Vietnamese die of famine in northern Vietnam.

1946: March 6 France agrees to Vietnamese independence

France signs an agreement with Ho Chi Minh recognizing the Democratic Republic of Vietnam as a free state within the Indochinese Federation.

1946: June 1 Hostilities between Vietnam and France commence

At a conference in France to discuss the exact nature of the new state of Vietnam, France proclaims a separate French government for Cochin China or the southern portion of Vietnam. Ho Chi Minh walks out of the conference.

Despite attempts by both Vietnamese and French to resolve their differences, troops from each side engage in hostilities.

1946: December 19 Beginning of Indochina War

On this day, now considered the beginning of the Indochina War, troops of the Democratic Republic of Vietnam attack French troops in Hanoi. The Vietminh fight mainly as a guerrilla force because they are inferior to the French in troops, unable to challenge the French in the more populated centers on the Vietnamese eastern coast.

1947 War continues between Vietnam and France

Almost defeated in Tonkin and northern Annam by the French, the Vietminh take refuge in the mountains north of Hanoi. Attempts by the French to dislodge the Vietnamese from their strongholds in the mountains fail.

1948: June 5 French install new government

France declares Vietnam a free nation within the French Union and Emperor Bao Dai is made Chief and State. The Vietminh denounce the new government as a puppet of the French.

1949: March 8 Start of Western-supported anti-Communist force

French and Vietnam conclude the Elysee Agreement in which the French promise to help build a national, anti-Communist army.

1950: January 14 Soviet Union and China support Vietminh

Ho Chi Minh declares that the one true government of Vietnam is the Democratic Republic of Vietnam, and the Soviet Union and China agree. Both extend diplomatic recognition, and China supplies modern weapons to Vietminh troops.

1950: February 7 America and Britain support Bao Dai government

The Bao Dai government receives recognition from the United States and Great Britain, dividing the country into two: the Communist government of the north receives support from China and the Soviet Union, France, America and Great Britain support the anti-Communist government of the south.

1950: May Beginning of American support for southern Vietnam

The United States and France agree that America will supply weapons to the French Associated States of Indochina. Thus, America begins sending military aid to South Vietnam.

1950: June 27 First American military mission goes to Vietnam

President Truman (1884–1972) speeds up the amount of aid going to Vietnam through France. The first military mission arrives in Vietnam from the United States.

1950: August 3 First Military Assistance Advisory Group

The first U.S. Military Assistance Advisory Group of thirty-five advisors arrives in Vietnam to teach weapons use the southern Vietnamese troops.

1950: December 23 Mutual Defense Assistance Agreement signed

The United States signs a Mutual Defense Assistance Agreement with France, Vietnam, Laos, and Cambodia.

1951: September 7 United States sends direct aid to South Vietnam

The United States signs an agreement with South Vietnam to send assistance directly to the country. American civilians now join American military advisors in South Vietnam. American military aid to the south grows to over $500 million for 1951.

1953 Aid increases to both North and South Vietnam

Chinese aid to the Democratic Republic of Vietnam increases, while American military aid to South Vietnam grows to over $785 million.

1953: November 20 Border changes with China

General Vo Nguyen Giap defeats French troops at the last outpost in northwest Tonkin, leaving the entire border with China open to receive military supplies into North Vietnam.

1953: November French occupy Dien Bien Phu

After occupying and fortifying Dien Bien Phu, the French use it as a base from which to invade Laos.

1954–1975 Development of separate countries

North and South Vietnam develop into two different societies. In the North, the Communist government begins attempts to revolutionize the society and economy by eliminating classes and instituting a harsh land reform campaign. While the government improves education and health care services that existed before 1954, it is unable to effect economic transformations. In South Vietnam, the society remains much the same despite the change to rule by a Vietnamese elite. Increasing political instability causes enormous stress. Peasants are forced to relocate because of war, fear of the government abounds, and the economy is uneven in its distribution of goods.

1954: February 18 Big Four agree to a conference in April

France, the United States, Great Britain, and the Soviet Union agree to meet in Geneva to discuss the Vietnamese situation.

1954: March 13 Vietminh troops attack Dien Bien Phu

Troops numbering 40,000 of the Democratic Republic of Vietnam accompanied by heavy artillery surround the 15,000 French troops occupying Dien Bien Phu.

1954: March 18 Vietminh tighten siege of Dien Bien Phu

Vietminh troops tighten their hold on Dien Bien Phu. The Communist leaders of North Vietnam hope a victory there will influence the Big Four Geneva conference and encourage the French to negotiate for peace.

1954: March 20 American involvement intensifies

Hearing that the French are about to be defeated at Dien Bien Phu, Chairman of the Joint Chiefs of Staff in the United States, Arthur Radford, suggests that America help the French by launching a nuclear strike against the Vietminh. Instead, the United States launches a massive air strike, drops paratroopers into North Vietnam, and mines the harbor at Haiphong.

1954: March 25 Radford plan is approved

The National Security Council approves the Radford plan to give extensive aid to France in its struggles in Vietnam, a decision now commonly accepted as the date the United States committed to fighting in Indochina.

1954: April 7 Description of the "Domino Theory"

At a news conference, President Dwight Eisenhower (1890–1969) first introduces the Domino Theory in explaining the importance of assisting the French at Dien Bien Phu, saying: "You have a row of dominoes set up, and you knock over the first one and what will happen to the last one is the certainty that it will go over very quickly. So you have the beginning of a disintegration that will have the most profound influences!"

1954: April United States abandons air strike idea

Britain does not agree to the American plan for a massive air strike against North Vietnam. Several days later, President Eisenhower denies that the United States ever planned a massive air strike against North Vietnam.

1954: April 26 Beginning of Geneva Conference

The Big Four—France, Great Britain, the United States and the Soviet Union—meet in Geneva to discuss Vietnam (and Korea).

1954: May 7 French are defeated at Dien Bien Phu

Dien Bien Phu falls to the troops of the Democratic Republic of Vietnam. After engaging in fierce and continuous fighting for months and enduring heavy causalities, the Vietminh overrun the base in an important and decisive battle. The French army suffers from the siege, losing numerous men. More importantly, France decides to give up the battle for Vietnam, partly in response to French public opinion. Now, more than 35,000 French have been killed and 48,000 wounded in their war against the Vietminh. France's defeat weakens its position at the Geneva Conference.

1954: June 18 **France and Vietminh agree to negotiate end to war**

France and the Vietminh agree to hold negotiations for ending the war between them. Emperor Bao Dai selects Ngo Dinh Diem, a Catholic, as the Prime Minister of Vietnam. Under Diem, Catholicism spreads in the south. Roman Catholics receive advantages over non-Catholics in all occupations and educational opportunities. Growing opposition from Buddhists causes increasing difficulties for Diem's rule.

1954: August 11 **Cease-fire throughout Indochina**

After the conference in Geneva, a cease-fire takes effect in Southeast Asia. Approximately one million Vietnamese leave North Vietnam for South Vietnam. An unknown number migrate from South to North Vietnam.

1954: August 20 **French and Vietminh sign a Cease-fire Agreement**

The Final Declaration of the Geneva Conference is supported, but never signed, by the Big Four and China, Laos, Cambodia, and the Democratic Republic of Vietnam. The United States and South Vietnam oppose the Final Declaration.

The Vietminh occupy Vietnam north of the seventeenth parallel; France and its Vietnamese supporters occupy the region south of the seventeenth parallel. In order to prevent the country from remaining permanently divided, both sides agree to a political protocol calling for national elections in two years. It is hoped the outcome of these elections will reunite the country.

1954: September **Diem takes over South Vietnam**

In the south, the wealthy, conservative anti-Communist Ngo Dinh Diem soon shoves Emperor Bao Dai from office. Prime Minister Diem forms a coalition government, which includes leaders from two religious sects, the Hoa Hao and Cao Dai, and other groups hostile to Diem. When a majority of the cabinet resigns, he appoints new members loyal to him and his family.

1954: October 11 **Vietminh take over North Vietnam**

The Vietminh north of the seventeenth parallel begin constructing their Communist society. Although the North loses its traditional food source in the south, and its factories and public facilities are dismantled and destroyed by those departing for South Vietnam, Ho Chi Minh and his followers face no opposition as they take control of North Vietnam. Approximately 650,000 Catholics flee to the south.

1954: November 20 **France departs Vietnam**

France announces the end to its occupancy of Vietnam, leaving economic, commercial, and financial control to the South Vietnamese, and withdraws all its troops and military advisors from Vietnam. It turns over command of the National Army of Vietnam to the South Vietnamese government and turns over training of the National Army to the United States. In North Vietnam, French advisors are replaced by Chinese and Russian technicians.

With American support, President Diem in South Vietnam refuses to hold the elections mandated by the Geneva Conference. He sternly crushes opposition to his rule, trying especially to crush Communist activity in southern Vietnam.

1959 **Opposition to President Diem grows in South Vietnam**

His unwillingness to tolerate any opposition to his regime and the failure of his economic and social programs seriously erode public support for President Diem. His favoritism of Roman Catholics in a Buddhist and Confucian society is another strike against him. In the midst of social unrest and political resentment, the Communists decide to resume war for control of southern Vietnam.

1963 **Diem is assassinated**

In protest against Diem's regime, a Buddhist monk sets himself on fire. Diem's own generals overthrow Diem in a coup. The American government secretly welcomes and supports the coup against Diem. His immediate assassination is followed by considerable political uncertainty. The generals form a provisional government headed by Nguyen Ngoc Tho as Prime Minister. Washington, D.C. recognizes the new regime of South Vietnam. The Communists come close to assuming control of South Vietnam.

1964 **Beginning of the Second Indochina War**

Throughout 1964, the President of the United States, Lyndon Johnson, prepares two policies that lead to a sharp increase in fighting in 1965: the bombing of North Vietnam and the dispatch of American troops to Vietnam.

1964: January 6 **Establishment of military triumvirate**

A military triumvirate replaces the provisional government. General Duong Van Minh is Chief of State.

1964: January 30 **Nguyen Khanh leads a coup**

Less than a month after the establishment of a military triumvirate, the government is overthrown in a coup led by General Nguyen Khanh who becomes Prime Minister. Although he retains General Duong Van Minh as Chief of Staff, he takes over most of the power.

1964: August 4 **American air raid against North Vietnam**

The United States bombs North Vietnamese oil depots and patrol boat bases in retaliation for an attack by Vietnamese PT boats on the American destroyer *Maddox* in the Gulf of Tongking.

1965: February **First bombing of North Vietnam**

Dismayed that the Communists may wrest control of a South Vietnam that is rapidly disintegrating and convinced that the fall of Vietnam will lead to the Communist takeover of the rest of Southeast Asia, the United States begins bombing North Vietnam. These sustained bombing raids are undertaken in order to force Hanoi to stop supporting the Vietcong in South Vietnam and to agree to a negotiated end to the war. The U.S. uses napalm in raids against industries and civilians.

1965: February 21 **Nguyen Khanh is ousted**

General Nguyen Khanh is ousted after nine changes of government in a year. The Armed Forces Council takes over power and establishes a civilian government under Prime Minister Phan Huy Quat.

1965: March **First American troops dispatched to Vietnam**

President Johnson sends American regular troops into South Vietnam to fight the Communists. These consist of two battalions of Marines sent to the harbor city of Danang where their duties are limited to defensive security duties. The Marine troops, and advisory troops already in the country, now total 27,000. By October, the number has increased dramatically and swiftly to 148,000 troops. Moreover, the duties of American soldiers shift from advising and defense to active and aggressive combat in search-and-destroy missions.

The intense bombing and intrusion of American ground troops in Vietnam cause serious problems for the Communists. The North Vietnamese decide to send units of the North Vietnamese army into South Vietnam to assist guerrilla units of the *Vietcong,* South Vietnamese Communist troops. Despite severe bombing and heavy causalities among their troops, the North Vietnamese and Vietcong in South Vietnam continue the struggle.

1965: June 18 **A new government assumes power**

Air Vice Marshal Nguyen Cao Ky (b. 1930) assumes power as Prime Minister, with General Nguyen Van Thieu (b. 1923–) as Chief of State. Nguyen Cao Ky, a pilot, serves as vice-president of South Vietnam from 1965 to 1971. The instability of South Vietnam's government contrasts with that of North Vietnam, where the same men who began the revolution in 1945 remain in control until their deaths. South Vietnam's instability results in part from the unwillingness of the government to impose drastic political and social reforms that could gain popular support from the population. Land reform, control over landowners, and taxation are not addressed by those in power.

Tanks follow closely behind a family traveling by ox cart. (EPD Photos/CSU Archives)

1966: March **Opposition to the government**

Opposition to the government breaks out in violent demonstrations led by Buddhist spokesmen, especially in Hue and Danang. Approximately twenty Buddhist groups oppose the government, including the most vocal of all, the Unified Buddhist Church of Vietnam. The demonstrations are brutally put down by the police and army. Government corruption, estimated to consume forty percent of the aid being sent by the U.S. to South Vietnam, fuels continued opposition to the government.

1966: November **Ineffectiveness of bombing North Vietnam**

Secretary of Defense McNamara (b. 1916), who recognizes that bombing North Vietnam will have little effect on the war, writes in a memorandum to President Johnson that bombing the North has "had no significant impact on the war in South Vietnam."

1967: January **North Vietnamese bombing causalities**

The Central Intelligence Agency reports 36,000 North Vietnamese causalities as a result of American bombing raids since 1965, eighty percent of whom have been civilians.

1967: September **National elections**

Elections are held in a country with no freedom of the press, no freedom of assembly, and electioneering strictly limited by ordinance. Parties and candidates are barred from running for various reasons. Even with the assistance of considerable election fraud, President Thieu and Vice-President Ky receive only thirty-five percent of the vote. When Buddhists and students protest the fraudulent results, the election is annulled, but reinstated when the National Assembly votes, in the presence of the chief of police, to accept the election results. Nguyen Van Thieu is president of South Vietnam from 1967 to 1975.

1968: January Tet Offensive

The commander of the American troops in Vietnam, General William Westmoreland, expresses unrestrained optimism in his assessment of the military situation in Vietnam. There are now 500,000 American troops in Vietnam. Four days later, his words are shown false. North Vietnam launches a bloody offensive during the most important holiday season for both North and South Vietnam that demonstrates the incompetence and weakness of the South Vietnamese troops.

The Vietcong attack the South Vietnamese capital city, Saigon (today called Ho Chi Minh City); Danang; Hue; thirty-four provincial towns; and sixty-four district towns throughout South Vietnam; demonstrating that no area is safe from their incursion. It demonstrates also that the Vietcong are stronger in February 1968 than they had been at the beginning of the massive American military involvement in Vietnam that has included half a million men, 1.2 million tons of bombs a year, and billions of dollars.

Both Americans and South Vietnamese are badly shaken, and Nguyen Van Thieu's government is seriously threatened. President Johnson decides to seek a negotiated settlement to the war.

1968: May First peace negotiations

The first session of formal negotiations between the United States and North Vietnam is held in Paris.

1969: January First negotiations between four parties

The first session of the peace negotiations between all four parties, the United States, North Vietnam, South Vietnam, and the Vietcong (National Liberation Front), are held in Paris.

1969: June Americans begin troop withdrawals

There are now 541,500 American troops in Vietnam. The United States begins withdrawing its troops from Vietnam with an announcement that 25,000 men will leave by August. More withdrawals are announced, until by spring of 1971, 248,000 soldiers remain.

1969 Ho Chi Minh dies

The leadership of North Vietnam falls to another leader of Vietnam's revolution, Le Duan, after Ho Chi Minh's death at age 69. The new president of the United States, Richard Nixon (1913–94), gains office by promising to end the Vietnam War. Instead, he continues Johnson's policies while gradually beginning to withdraw American troops from Vietnam.

1971: October National elections

Elections are held, but there is only one candidate for president: President Nguyen Van Thieu. Thieu receives 91.5 percent of the vote.

1971: December Bombs on North Vietnam

The United States has dropped 3.6 million tons of bombs on North Vietnam, compared to a total tonnage of 2 million dropped during World War II and 1 million during the Korean War.

1973: January 27 War ends temporarily

The war ends when a peace agreement is signed in Paris. According to the peace settlement, all U.S. troops are to be removed from Vietnam, and North Vietnam agrees to accept the Thieu government in South Vietnam until elections can be held. Although the peace agreement does not endure long, the Vietnamese celebrate this day as the anniversary of the ending of the Vietnam War.

1973: March 29 American troops leave Vietnam

The last American troops leave Vietnam. Vietnam celebrates this anniversary.

1975 War resumes

The Communists launch a new military offensive. Within six weeks, the South Vietnamese government of Nguyen Van Thieu collapses.

1975: April 30 Communists overtake Saigon

The war ends as Communist troops enter Saigon and take over the government. President Thieu flees the country with a planeload of family, friends, furniture, and money.

The new government under Le Duan, who succeeds Ho Chi Minh as general secretary of the Communist Party, faces chaos. Much of the country is in ruins with much of the farmland, industry, and roads destroyed. The government tries to unify the economies of northern and southern Vietnam by combining small farms into large estates run by collectives rather than individual farmers, and takes over businesses in the south, substituting state ownership for private. Catholics are able to practice their religion as long as they support the government. There are almost two million Catholics in the south, one million in the north.

1976 Establishment of a united Vietnam

South Vietnam and North Vietnam are reunited as the Socialist Republic of Vietnam, but violence continues. Vietnam now faces conflict with Cambodia to the west over border disputes.

1976–1980 Second-Year Plan

The Communist Party develops a Second-Year Plan for Vietnam (the First-Year Plan applies only to northern Vietnam) designed to increase industry by sixteen to eighteen percent, agriculture by eight to ten percent, and the national income by thirteen to fourteen percent. In a series of such plans, the government hopes to promote Communism within the country and help integrate the North and the South. Vietnam requests the financial assistance of Western countries, Communist allies, and international aid organizations for its Plan.

1978: June Vietnam joins Comecon

Vietnam joins Comecon, the East European Economic Community.

1978: December Vietnam invades Cambodia

Vietnamese troops invade Cambodia, pushing the ruling Cambodian Communists to the west, to the Thai-Cambodian border. The Vietnamese install a puppet government of Vietnamese-friendly Cambodians. Thousands of Vietnamese "boat people" begin fleeing the country, including many persecuted Chinese Vietnamese.

1979 China attacks Vietnam

Vietnam and the Soviet Union sign a friendship pact, which China calls a threat to the security of Southeast Asia. China attacks Vietnam.

Late 1970s–1980s Failing economy

The government moves people from overpopulated areas of southern Vietnam to New Economic Zones, undeveloped areas where people are forced to farm. Many others are sent to re-education camps to learn Communist ideology. But the economy does not grow enough to provide sufficient food and basic consumer goods to a rapidly increasing population. By 1979, it is obvious that the Second-Year Plan is not addressing Vietnam's serious economic problems. The country continues with primarily small-scale production, low labor output, high unemployment, shortages in resources and technological knowledge, and insufficient food and consumer goods.

The government strictly controls religious organizations, although people's practice of religion does not change much as long as the leaders follow government policies. As a result, many flee the country in small boats.

1981–85 Third Five-Year Plan institutes economic reforms

With its Third-Year Plan, Vietnam hopes to solve its economic problems. The government proceeds carefully, compromising between ideology and practicality. Although efforts to collectivize the south continue, such as bringing separate farmers together into collective farms, the government allows some free enterprise.

An economic reformist, Nguyen Van Linh (b. 1913), becomes general secretary and with another reformist, introduces *doi moi,* meaning renovation, economic reforms to allow a return to private ownership. Farmers are granted long-term leases, the currency is devalued to control inflation, and Vietnam seeks more aid from Communist countries.

1988 Vietnam reduces troops in Laos

Vietnam reduces the number of its soldiers stationed in Laos.

1989: September Vietnam withdraws troops from Cambodia

Vietnam reduces the number of its soldiers stationed in Cambodia.

Early 1990s End of Cold War

Political instability and failing Communist economies shake the world, including Vietnam's standing with its long-time supporters. Imports into Vietnam of fertilizers, oil, and raw materials from the Soviet Union and other Communist countries fall sharply as the Soviet Union breaks into separate nations.

1990 Vietnam establishes diplomatic ties with Europe

The European Community (now the European Union) establishes diplomatic relations with Vietnam.

1991 Vietnam repairs international relations

Do Muoi becomes general secretary of the Communist Party, and Vo Van Kiet becomes Prime Minister. Both work to forge better ties with the Western world while keeping tight control at home.

Vietnam signs a peace agreement with Cambodia which leads to the restoration of diplomatic relations with China and to strengthened relations with members of the Association of Southeast Asian Nations (ASEAN). Contact with ASEAN leads to increased trade with other Southeast Asian nations.

1992 Vietnam strengthens ties with ASEAN

Vietnam signs an ASEAN agreement first concluded in 1976 on regional unity and cooperation regarded by members of ASEAN and others as the first step to Vietnam eventually joining the Association. Vietnam also establishes diplomatic ties with South Korea.

Early 1990s Vietnam improves ties with the United States

The United States removes a trade embargo with Vietnam, leading to improved relations and trade agreements. These changes come in part because Vietnam is assisting the United States in locating the remains of U.S. soldiers missing since the Vietnam War.

1995 Vietnam and the United States exchange diplomats

The United States and Vietnam agree to exchange low-level diplomats, with an eye to the eventual establishment of full diplomatic relations with an exchange of ambassadors and the establishment of embassies.

Bibliography

Buttinger, Joseph. *A Dragon Defiant: A Short History of Vietnam*. New York: Praeger. 1972.

Cima, Ronald J. ed. *Vietnam: A County Study*. Washington D.C.: U.S. Government Printing Office.

Crawford, Ann Caddell. *Customs and Culture of Vietnam*. Rutland, VT: Charles E. Tuttle. 1966.

Halberstam, David. *The Making of a Quagmire*. New York: Random House. 1964.

Hickey, Gerald Cannon. *Free in the Forest: Ethnohistory of the Vietnamese Central Highlands, 1954-1976*. New Haven: Yale University Press. 1982.

Kahin, George McT. *Intervention: How America Became Involved in Vietnam*. Garden City, NY: Anchor Books. 1987.

Karnow, Stanley. *Vietnam. A History. The First Complete Account of Vietnam at War.* New York: The Viking Press. 1983.

Newman, Bernard. *Background to Viet-Nam*. New York: Roy Publishers. 1965.

Sheehan, Neil. *A Bright Shining Lie*. New York: Random House. 1988.

Thuy, Vuong G. *Getting to the Know the Vietnamese and Their Culture*. New York: Frederick Ungar Publishing. 1975.

Wright, David K. *Enchantment of the World: Vietnam*. Chicago: Children's Press. 1989.

Yemen

Introduction

The present-day Republic of Yemen is less than a decade old. The territory it covers was formerly divided into the Yemen Arab Republic (YAR, or North Yemen) and the People's Democratic Republic of Yemen (PDRY, or South Yemen). The two regions had distinctly contrasting histories: North Yemen was ruled for generations by imams (religious leaders) of the Zaydi Muslim sect, while the area later called South Yemen comprised a group of British protectorates that grew out of Great Britain's occupation of the Port of Aden in 1839. The north won its independence from the Ottoman Empire at the close of World War I, becoming a pro-Western republic after a revolution in 1962; the south was governed by Britain until 1967, when it became a Marxist state. A border war between the two countries ensued, followed by gradual progress toward unification, which took place in 1990.

Occupying the southernmost part of the Arabian Peninsula, Yemen covers an area of 203,850 square miles (527,970 square kilometers). Its terrain includes an arid coastal plain (the Tihama) along the Red Sea, rising to mountains and cultivated highlands in the interior, and an extended desert plateau to the east, in what was formerly South Yemen. The December 1994 census for the unified country recorded a population of 14.5 million. Sana, the capital has a population estimated at roughly 700,000 to 800,000.

Historically, North and South Yemen, which had virtually no known natural resources, were among the world's poorest, least developed nations. Through much of the postwar period, earnings by Yemenis working abroad constituted the countries' main source of income. In the mid-1980s, however, crude oil was discovered in both the north and south, and living standards in the now-unified country have improved considerably since then. An oil refinery was completed in North Yemen in 1986, and a pipeline to deliver crude oil to the Red Sea opened the following year. In the south, a Soviet-backed joint venture found further oil reserves in the Shabwah region. Oil reserves for the Republic of Yemen were estimated at 3.75 billion barrels in 1990. By 1995, oil export revenues totaled $500 million. The nation's economy suffered a setback, however, when it took a position friendly to Iraq, a major trading partner, in the 1991 Gulf War. The Arab nations allied against the Iraqis discontinued foreign aid to Yemen, costing the country billions of dollars. In addition, roughly one million Yemeni "guest workers" were expelled from Saudi Arabia, resulting in a substantial loss of income.

History

Historical records for the area currently known as Yemen date back to ancient times. For most of the first millennium B.C., Yemen was dominated by the Sabaean civilization, which consisted of advanced city states that thrived on the region's trade in frankincense and myrrh and developed an extensive irrigation system that included a great dam at Marib. After A.D. 500, Ethiopians and Persians invaded the region, and Islam became the predominant religion in 628, when the reigning Persian ruler was converted to it. Control shifted to the Egyptian Fatimid and Rasulid dynasties between the twelfth and fifteenth centuries, and Ottoman rule began in 1517. In 1636 the Ottoman Turks were driven out by the imam of the Zaydi religious sect, which stretched back to the ninth century. The Zaydis ruled most of the region until the nineteenth century, when the Ottomans once again vied for control, as well as the Egyptians.

In 1839 the British established a presence in the area when they occupied the port of Aden. During the latter half of the nineteenth century, they established numerous protectorates to the north and in the area stretching east along the Gulf of Aden and the Arabian Sea. Following World War I, the victorious British maintained control over this region, which would come to be known as South Yemen, while the area to the north, freed from the vanquished Ottoman Empire, became fully independent under the rule of the Zaydi imams. After a border war with Saudi Arabia in 1934, this region won full recognition by both the Saudis and the British.

North Yemen remained under Zaydi rule until 1962, when the imam was deposed in the revolution that created the Yemen Arab Republic, headed at first by revolutionary leader 'Abdallah as-Sallal and, after 1967, by a Republican Council. Military rule was imposed in a 1974 coup, after which the next two leaders of the country were assassinated. In 1978

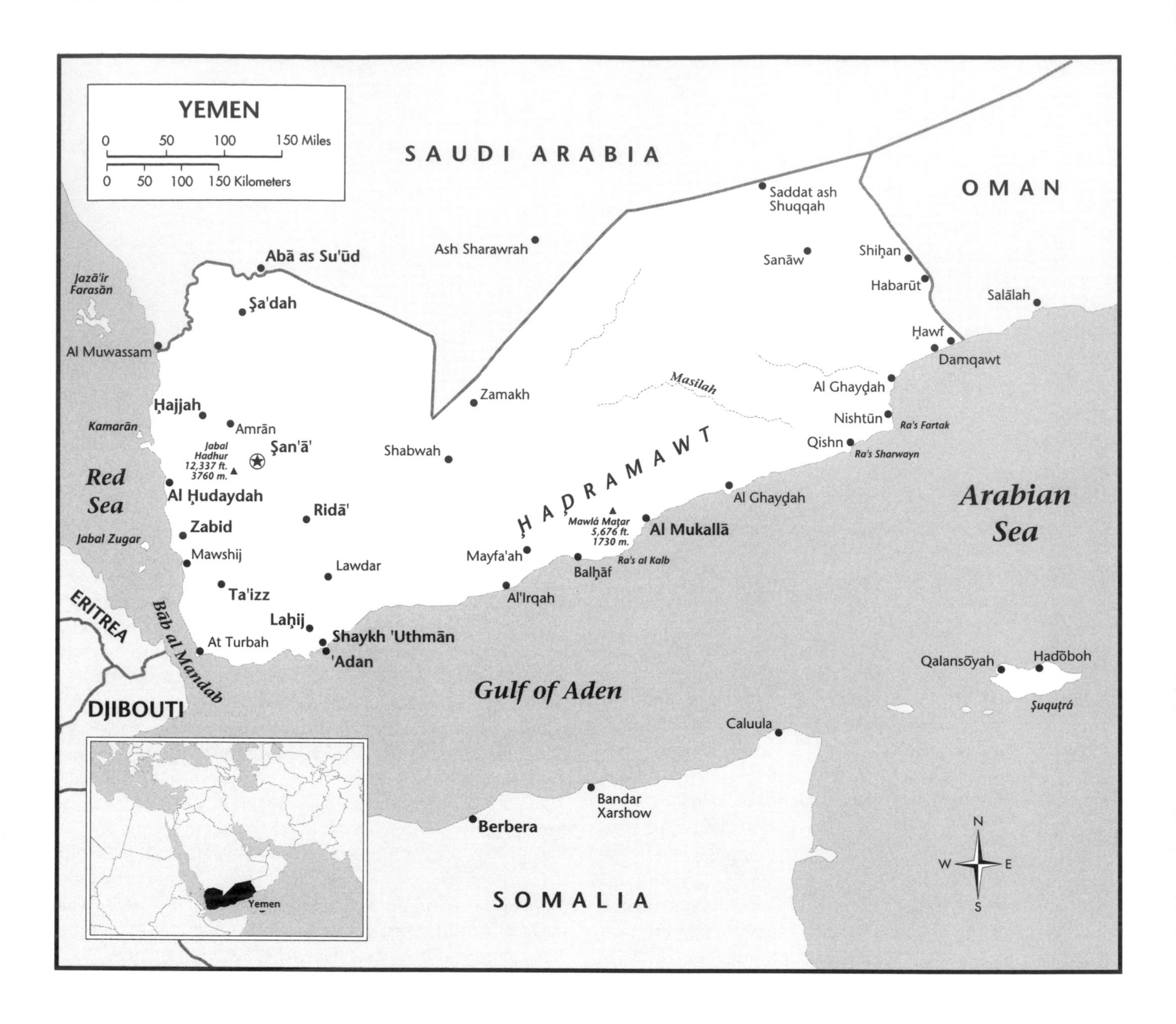

'Ali 'Abdallah Saleh came to power and ruled North Yemen until unification in 1990.

In the south, the British united Aden and the surrounding protectorates to form the Federation of South Arabia in 1963, promising independence by 1968. However, the end of British rule was hastened by the formation of Yemeni liberation groups. By 1967 the British withdrew and the National Liberation Front established the Republic of South Yemen, a Marxist state later renamed the People's Democratic Republic of Yemen. In the late 1970s and the 1980s, moderates and hardline Marxists vied for control of the country. In 1986 moderate president 'Ali Nasir Muhammad al-Hasani, challenged by the return of exiled leader Abd al-Fattah Isma'il, instituted repressive measures of his own, and an estimated 5,000 to 10,000 people were killed in the ensuing hostilities.

Unification of North and South Yemen was discussed as early as 1972, when an agreement was reached, but it was not carried out at that time. Following border clashes in 1979, a new unification agreement was arrived at, and throughout the 1980s the two Yemens gradually moved toward this goal. A draft constitution for the unified nation was approved in 1981, and a combined council of the two nations met for the first time in 1983. In 1988, economic as well as political ties were forged by the launching of a joint oil exploration project. Yemen was officially unified on May 22, 1990. The new nation was formed as a multiparty democracy to be ruled by a five-member presidential council. 'Ali 'Abdallah Saleh, the leader of the former Yemen Arab Republic, became the first president of unified Yemen.

In 1994 rebels in the former South Yemen attempted to secede from the Republic of Yemen, and civil war erupted. However, the secessionists, led by Vice President ‘Ali al-Baidh, were defeated in nine weeks. The following year the Yemeni government launched a five-year economic liberalization and privatization program that brought new assistance from the International Monetary Fund and the World Bank. In 1997 general elections increased the parliamentary majority of the ruling party (the General People's Congress).

Timeline

1st millennium B.C. Sabaean kingdom flourishes

The dominant civilization of the first millennium B.C. in present-day Yemen is that of Saba (or Sheba), whose capital is at Marib. The Sabaeans, who are known for developing a sophisticated irrigation system, grow wealthy by trading in spices and incense.

c. 500 B.C. Dam is constructed at Marib

The Sabaeans build a great dam at their capital city of Marib, with the capacity to irrigate approximately 1,600 hectares.

c. 100 B.C.—A.D. 525 Himyarites rule Yemen

The Himyarites rule and extend the former Sabaean territory in South Arabia. Ethiopian invaders occupy the region for a time during the fourth century A.D. but are driven out. Polytheism (belief in many gods) begins to give way to monotheism (belief in one god), both in the form of the well-established Jewish religion and the newer creed of Christianity.

341–46 Christianity is introduced

The first Christian missionary, said to have been sent by the patriarch Theophilus, arrives in Yemen.

Foreign Occupation

525 Second Ethiopian occupation ends Himyarite rule

The Ethiopians reoccupy Yemen.

575 Persians invade Yemen

Forces sent by the Sassanid emperor Chosros I (d. 579) begin a half century of Persian occupation. Chosros I (also known as Chosroes of Khosrow), known as “Immortal Soul,” ruled Persia from 531.

628 Persian ruler converts to Islam

Barely seventy years after the birth of the Prophet Muhammad, founder of Islam, the Persian leader Badhan adopts the new religion. Shiite and Sunni sects later vie for dominance of the region. Sunnis of the Shafi'i School control the coastal region (Tihama) and the south, while the Shiites hold the highlands.

Ninth century Zaydi dynasty is founded

Yahya al-Hadi ila'l Haqq, the leader of the Zaydis, a Shiite Muslim sect, founds a dynasty of theocratic (religious and political) leaders.

Eleventh–twelfth centuries Fatimids establish control

Allied with local leaders, Egypt's Fatimids control Yemen.

1173 Saladin conquers Yemen

The sultan Salah al-Din al-Ayyubi (1138–93), known as Saladin, conquers Yemen. Saladin rules Egypt and Syria, and later captures Jerusalem. He dies in Damascus in 1193.

1230–1400 Rasulid rule

Rasulid kings, of Turkish and Kurdish descent, rule Yemen.

The Ottomans and the British

1517 Ottoman rule

The Ottoman Turks formally take over control of eastern and southeastern Yemen, although local leaders continue to exercise most of the power.

1636 Ottomans are challenged by local ruler

The Zaydi imam (ruler) Qasim the Great (r. 1597–1620) organizes a revolt against Ottoman rule, and the Turks are driven out of the highlands and left only with control of the coastal regions. The highlands mostly remain under the control of imams.

Nineteenth century Different groups rule Yemen

Control of Yemen is divided among the Zaydi imams, Egypt, and the Ottoman Turks. The imams pay an annual tribute to the Ottomans but retain their autonomy in the northeastern highlands.

1839 British occupy Aden

The British occupy the port of Aden, using it as a refueling station along the trade route to India. It becomes an increasingly important trade and population center following the opening of the Suez Canal in 1869.

1854–1914 British protectorates are formed

Numerous small states north and east of Aden become British protectorates through purchase or treaty in the region that will later be known as South Yemen.

1918 World War I ends

With the fall of the Ottoman Empire, northern Yemen becomes fully independent. Britain, a victor in the war, retains control of its protectorates in the region.

1934–35 Saudi-Yemeni boundary dispute leads to war

A boundary dispute between Yemen and Saudi Arabia erupts into warfare, as Saudi forces invade northern Yemen.

1934 Treaty of Ta'if

The Saudi-Yemeni border war ends. Yemen loses some of its territory but wins official recognition by the Saudis and the British.

1937 Aden becomes a crown colony

As part of the Government of India Act, Aden becomes a British crown colony.

The British influence is seen in many aspects of life in Yemen. Some Yemenis dress in Western-style clothing, although many Yemenis favor traditional Arabic loose-fitting clothing. Many women observe the Muslim custom of fully covering their bodies, including their faces, whenever they leave their homes.

Many women in Yemen, like women in other Muslim societies, wear the heavy black veil covering their entire body when they venture into the street. (EPD Photos/CSU Archives)

1948 Imam Yahya is assassinated

Imam Yahya ibn Muhammad Hamid ad-Din (1869?–1948), of the ruling Zaydi sect, and two of his sons are assassinated as part of a failed coup by rival religious leader Sayyid Abdallah ibn-Ahmad el-Wazir. Emir Seif al-Islam Ahmed succeeds him as the religious and political leader of the country.

1949–50 Large-scale emigration of Jewish population

With the founding of the state of Israel, most of Yemen's Jewish population, numbering between 40,000 and 50,000, emigrates.

1956 Military alliance with Arab neighbors

The Jiddah Pact, signed by Yemen's imam Ahmad, Egyptian president Gamal Abdel Nasser, and Yemen forms a military alliance with Egypt, Syria, and Saudi Arabia.

1958–61 Yemen joins the United Arab Republic

North Yemen becomes part of the United Arab Republic (UAR), whose other members are Egypt and Syria. The loosely organized federation, an experiment in pan-Arab unity, is dissolved within three years.

1962–70 Civil war in newly declared republic

Following the ouster of Yemen's last ruling imam, republican forces wage civil war with royalists, who are supported by the Saudi Arabian and Jordanian monarchies. The republic's defenders receive aid from Egypt and the Soviet Union.

1962: September Imam is overthrown and a republic is proclaimed

Newly crowned imam Muhammad ad-Badr is deposed by members of the military, who declare Yemen a republic. Military leader 'Abdallah as-Sallal (1920–94) names himself president and commander-in-chief of the armed forces. Badr flees northward and organizes a royalist rebel force.

1967–72 Border war between North and South Yemen

North and South Yemen go to war over their disputed border.

1967: November **Sallal is ousted**

'Abdallah as-Sallal is removed as head of the Yemen Arab Republic, and a Republican Council takes over.

1970: March **Civil war ends**

Negotiations at Jiddah end the eight-year-long civil war between republicans and royalist followers of the imam in North Yemen. The republicans prevail, but royalists are given posts within the government.

1970 **Sana University opens**

Sana University is founded, offering programs in medicine, engineering, and agriculture.

1971 **A constitution is adopted**

North Yemen adopts a constitution.

1971 **The National Museum opens**

The National Museum, containing archaeological, Islamic, and ethnographic collections, opens in Sana.

1974–77 **Al-Hamdi seizes power**

Lieutenant Colonel Ibrahim Muhammad al-Hamdi comes to power in a bloodless coup. The constitution is suspended, normal political activity is outlawed, and al-Hamdi is named president.

1977: October **Al-Hamdi is assassinated**

Yemeni leader Muhammad al-Hamdi is assassinated and succeeded by Colonel Ahmad ibn Hussein al-Ghashmi, who forms a civilian government and institutes the Constituent People's Assembly.

1978: June **Al-Ghashmi is assassinated**

Ahmad ibn Hussein al-Ghashmi is assassinated in a bombing; South Yemeni involvement is suspected.

1978–90 **'Ali 'Abdallah Saleh serves as president**

Lieutenant Colonel 'Ali 'Abdallah Saleh (sometimes spelled Salih, b. 1942) succeeds Ahmad ibn Hussein al-Ghashmi as president of the Yemen Arab Republic. As his political opposition, supported by the People's Democratic Republic of Yemen to the south, is neutralized, his government becomes more democratic.

1980s **Farmers switch to qat**

In the early 1980s, many farmers switch from agricultural products and begin to grow qat. Qat is a woody shrub that is grown for its leaves, which contain a narcotic that produces a mild euphoria (feeling of well-being) when chewed. Qat, sometimes spelled khat, is widely chewed in the countries where it is grown, but it is not a significant export because it is quite perishable.

1981 **Oil exploration concession is granted**

The Yemen Arab Republic signs a concession agreement with U.S.-based Hunt Oil Company, which will form an international joint venture to explore for oil.

1982 **General People's Congress**

The YAR government becomes more representative with the formation of the General People's Congress, 70 percent of whose members are elected.

1984: July **First oil discovery is announced**

The Yemen Hunt Oil Company, an international joint venture, makes public the first discovery of high-quality crude oil reserves in the Marib/Al Jawf basin, at a location roughly forty-two miles (seventy kilometers) northeast of Marib. Reserves are estimated at one billion barrels. Plans are drawn up to construct a refinery to process the oil for domestic use and a pipeline to transport it to the coast for export.

1985 **Local elections are held**

Yemenis vote for members of local development councils in a preliminary step toward the holding of national elections.

1986 **Yemen's first art gallery is opened**

Gallery Number One in Sana, the first art gallery in Yemen, is launched by painter and sculptor Faud al-Futaih (b. 1948). Works by both Yemeni and foreign artists are shown.

1986 **Oil refinery is completed**

Yemen's first oil refinery is constructed at Marib by the Yemen Hunt Oil Company.

1987 **Oil pipeline is launched**

Construction of a 273-mile (440-kilometer) oil pipeline to transport oil to the Red Sea is completed.

1988 **National elections are held**

The YAR holds national elections for a consultative assembly. Opposition parties are not allowed to field candidates or to criticize the government of 'Ali 'Abdallah Saleh, who wins a third term as president.

1988 **Social programs enacted**

Social programs to cover all Yemenis—even those living abroad—are established to provide old age benefits, disability and workers' compensation benefits, and aid to surviving spouses. The old age benefits are paid as a lump sum, and are calculated from a formula that uses a percentage of the worker's final annual earnings and the total number of years he (or she) has worked.

1945–90: Southern Yemen

1947: September 30 Yemen joins the United Nations

Pakistan and Yemen (formerly Yemen Arab Republic or YAR) are admitted to membership in the United Nations (UN), almost two years after the UN was founded (October 24, 1945).

1959 Federation of Emirates is formed

The United Kingdom forms the Federation of Arab Emirates of the South, uniting six of its Aden protectorates.

1963–65 The Federation expands

The former colony of Aden is merged with the Federation of Arab Emirates of the South, now renamed the Federation of South Arabia. By 1965, seventeen small local states have been integrated into the new political entity.

Mid-1960s Factional violence erupts

Two liberation groups—the National Liberation Front (NLF) and the Front for the Liberation of South Yemen (FLOSY)—are formed to hasten the end of British rule and eventually vie with each other for dominance of the region. The FLOSY loses ground as its Egyptian support is withdrawn.

1966 British schedule withdrawal

The United Kingdom issues a white paper announcing that it will withdraw its forces from the region by 1968. (See 1967: November 30)

1967: November 30 Republic of South Yemen is created

The Federation of South Arabia dissolves, and the British withdraw from the region. The People's Republic of South Yemen, controlled by the National Liberation Front (NLF), is formed. Its first president is Qahtan al-Shaabi.

1969: June 22 South Yemen government is overthrown

The National Liberation Front, which controls the Republic of South Yemen, is overthrown by leftist rebels. Ruled by a five-member council, it allies itself with the USSR and the People's Republic of China. Salim Rubaya Ali becomes president.

1970 The republic is renamed

The Republic of South Yemen is renamed the People's Democratic Republic of Yemen (PDRY). A new constitution is adopted.

1975 NLF is merged into a new political party

The National Liberation Front becomes part of the United Political Organization—National Front (UPONF). It is later renamed the Yemen Socialist Party.

1975 University of Aden is established

The University of Aden is founded at Al-Mansoora. It offers programs in the sciences, agriculture, education, and the arts.

1978 President of South Yemen is overthrown and executed

President Salim Rubaya Ali is deposed by Abd al-Fattah Isma'il and executed, and Isma'il, a hard-line Marxist, becomes the new president. He oversees a treaty with the USSR, and Soviet troops are stationed in the country.

1980 Isma'il resigns as president

'Abd al-Fattah Isma'il resigns the presidency of the PDRY and goes into exile in the USSR He is succeeded by the more moderate 'Ali Nasir Muhammad al-Hasani (b. 1940).

1986: January Political violence erupts

Following the return of former president Abd al-Fattah Isma'il from political exile in the Soviet Union, President 'Ali Nasir Muhammad al-Hasani launches a campaign of political repression to silence his opponents. After Al-Hasani has four top government officials killed at a party meeting, fighting breaks out in Aden. Within two weeks, an estimated 5,000–10,000 people are killed.

1986: January 24 New president is named

After President al-Hasani flees to the Yemen Arab Republic, Prime Minister Haydar Abu Bakr al-'Attas (b. 1939) is named president.

1986: March General amnesty is proclaimed

The government of Haydar Abu Bakr declares a general amnesty for supporters of deposed president al-Hasani.

1987 Oil discovery is announced

A Soviet-led joint venture discovers oil reserves in the Shabwah region of South Yemen.

1989: July Government moves toward free market economy

The government of the PDRY enacts free-market economic reforms.

1972–present: Unification

1972 Unification agreement is reached between the two Yemens

North Yemen (the Yemen Arab Republic) and Southern Yemen (the People's Democratic Republic of Yemen) agree to unification of the two countries, but the agreement is not carried out.

1979: February–March Border war between North and South Yemen

Two weeks of border clashes occur between the two Yemens. The two sides meet for talks, and a new unification agreement is reached.

1981: December Constitution is drafted for unification

North and South Yemen agree on a draft constitution for a proposed unification of the country.

1983: August First meeting of North-South combined council

A combined council of North and South Yemen meets for the first time. Meetings continue on a regular basis.

1988: May Joint oil exploration agreement forges new ties

The two Yemens strengthen their relationship with a joint oil exploration agreement.

1989: November Unification is mandated within one year

North and South Yemen agree to unify within a year. Their leaders, fearing internal dissent on both sides, accelerate the process, and unification is achieved in close to six months.

1990: January Travel restrictions are lifted

Travel restriction between the two Yemens are relaxed in preparation for unification.

1990: May 22 Unified Republic of Yemen is established

The former Yemen Arab Republic (North Yemen) and People's Democratic Republic of Yemen (South Yemen) merge into the newly unified Republic of Yemen, a multiparty democracy ruled by an elected five-member presidential council. Former YAR leader 'Ali 'Abdallah Saleh becomes the first president of the new republic. The governments of both nations are to be integrated within thirty months. As specified in the constitution, Islamic religious law (*shari'ah*) will play a prominent role in determining the laws of the new country.

1991 Gulf War stance hurts Yemen's economy

Yemen, which has strong trade ties with Iraq, expresses disapproval of the Iraqi invasion of Kuwait but does not ally itself with the anti-Iraq coalition. The newly united country pays heavily for its support of Iraq, losing $2 billion in foreign aid from its neighbors in the Gulf States, as well as billions of dollars in earnings lost by about one million Yemeni workers expelled from Saudi Arabia.

1991 Pipeline to Gulf of Aden is opened

Soviet-supported Technoexport completes construction of a 115-mile (190-kilometer) pipeline to transport crude oil to Bir Ali on the Gulf of Aden.

1991 Central banks are merged

The central banks of the former North and South Yemen merge and prepare to issue a unified currency.

1992: April Parliamentary elections are held

President 'Ali 'Abdallah Saleh's General People's congress narrowly misses winning a majority in parliamentary elections, and a three-party governing coalition is formed.

1993: April Women vote in elections for the first time

Yemeni women vote in their first general election, which is also Yemen's first multiparty election.

1993: August Vice president boycotts parliament

Protesting the lack of political power accorded to oil-rich southern Yemen, Vice President 'Ali al-Baidh of the Yemen Socialist Party leaves the capital city of Sana and boycotts parliamentary sessions, endangering the survival of the governing coalition. The region of the former PDRY (South Yemen) charges that North Yemen holds a disproportionate amount of power in both the government and economy of the unified nation.

1994: May 5 Civil war breaks out

After a period of isolated skirmishes, full-scale warfare erupts between northern and southern Yemen, as secessionists attempt to reestablish a separate republic in the south. Tanks, artillery, rockets, and bombs are deployed in at least seven provinces, and thousands are killed. President Saleh declares a thirty-day state of emergency.

1994: May 21 Southern Yemen secedes

Vice president 'Ali al-Baidh proclaims a separate, breakaway state in the south and names himself president. A premier and cabinet are also appointed.

1994: July 1 UN calls for halt to fighting

The United Nations (UN) Security Council passes a resolution condemning the fighting in Yemen and calling for a cease-fire and peace negotiations.

1994: July 7 Aden is captured

Aden, the stronghold of the breakaway Yemeni state in the south, is captured, and the rebels are subdued, ending the nine-week civil war.

1995 Five-year economic program is launched

The government announces an economic program that includes privatization and removal of controls on prices and exchange rates. The measures meet with the approval of the International Monetary Fund (IMF) and the World Bank, which promise to provide financial assistance.

1997: April 27 General People's Congress retains parliamentary control

The first general elections since the nine-week civil war of 1994 extend the parliamentary majority of the ruling General People's Congress, which wins 187 of the nation's 301 legislative seats. The Yemen Socialist Party, which had ruled the former South Yemen, boycotts the elections. The party finishing in second place is the Yemeni Reform Group, which wins fifty-three seats. Approximately 4.6 million Yemenis vote in the election, and ten political parties sponsor some 2,000 candidates. The elections are deemed free and fair by international observers.

Bibliography

Al-Suwaidi, Jamal. *The Yemeni War of 1994: Causes and Consequences.* London: Saqi Books, 1995.
American University. *The Yemens: Country Studies.* Washington, D.C.: Government Printing Office, 1986.
Bidwell, Robin D. *The Two Yemens.* Boulder, Colo.: Westview, 1983.
Carapico, Sheila. *Civil Society in Yemen : The Political Economy of Activism in Modern Arabia.* New York: Cambridge University Press, 1998.
Chaudhry, Kiren Aziz. *The Price of Wealth: Economies and Institutions in the Middle East.* Ithaca, NY: Cornell University Press, 1997.
Crouch, Michael. *An Element of Luck: To South Arabia and Beyond.* New York: Radcliffe Press, 1993.
Dresch, Paul. *Tribes, Government, and History in Yemen.* New York: Oxford University Press, 1989.
Halliday, Fred. *Revolution and Foreign Policy: The Case of South Yemen, 1967–87.* New York: Cambridge University Press, 1990.
Ismael, Tareq Y., and Jacqueline S. Ismael. *The People's Democratic Republic of Yemen: Politics, Economics and Society.* Boulder, Colo.: Lynne Rienner, 1986.
Kostiner, Joseph. *Yemen: The Tortuous Quest for Unity, 1990–94.* London: The Royal Institute of International Affairs : Pinter, 1996.
Lackner, Helen. P.D.R. *Yemen: Outpost of Socialist Development in Arabia.* London: Ithaca Press, 1985.
Wenner, Manfred W. *The Yemen Arab Republic: Development and Change in an Ancient Land.* Boulder, Colo.: Westview Press, 1991.

Glossary

abdicate: To formally give up a claim to a throne; to give up the right to be king or queen.

aboriginal: The first known inhabitants of a country. A species of animals or plants which originated within a given area.

allies: Groups or persons who are united in a common purpose. Typically used to describe nations that have joined together to fight a common enemy in war.

In World War I, the term Allies described the nations that fought against Germany and its allies. In World War II, Allies described the United Kingdom, United States, the USSR and their allies, who fought against the Axis Powers of Germany, Italy, and Japan.

Altaic language family: A family of languages spoken in portions of northern and eastern Europe, and nearly the whole of northern and central Asia, together with some other regions. The family is divided into five branches: the Ugrian or Finno-Hungarian, Smoyed, Turkish, Mongolian, and Tunguse.

amendment: A change or addition to a document.

Amerindian: A contraction of the two words, American Indian. It describes native peoples of North, South, or Central America.

amnesty: An act of forgiveness or pardon, usually taken by a government, toward persons for crimes they may have committed.

animal husbandry: The branch of agriculture that involves raising animals.

Anglican: Pertaining to or connected with the Church of England.

animism: The belief that natural objects and phenomena have souls or innate spiritual powers.

annex: To incorporate land from one country into another country.

anti-Semitism: Agitation, persecution, or discrimination (physical, emotional, economic, political, or otherwise) directed against the Jews.

apartheid: The past governmental policy in the Republic of South Africa of separating the races in society.

appeasement: To bring to a state of peace.

arable land: Land that can be cultivated by plowing and used for growing crops.

archipelago: Any body of water abounding with islands, or the islands themselves collectively.

archives: A place where records or a collection of important documents are kept.

arctic climate: Cold, frigid weather similar to that experienced at or near the north pole.

aristocracy: A small minority that controls the government of a nation, typically on the basis of inherited wealth.

armistice: An agreement or truce which ends military conflict in anticipation of a peace treaty.

ASEAN *see* Association of Southeast Asian Nations

Association of Southeast Asian Nations: ASEAN was established in 1967 to promote political, economic, and social cooperation among its six member countries: Indonesia, Malaysia, the Philippines, Singapore, Thailand, and Brunei. ASEAN headquarters are in Jakarta, Indonesia. In January 1992, ASEAN agreed to create the ASEAN Free Trade Area (AFTA).

asylum: To give protection, security, or shelter to someone who is threatened by political or religious persecution.

atoll: A coral island, consisting of a strip or ring of coral surrounding a central lagoon.

atomic weapons: Weapons whose extremely violent explosive power comes from the splitting of the nuclei of atoms (usually uranium or plutonium) by neutrons in a rapid chain reaction. These weapons may be referred to as atom bombs, hydrogen bombs, or H-bombs.

austerity measures: Steps taken by a government to conserve money or resources during an economically difficult time, such as cutting back on federally funded programs.

Australoid: Pertains to the type of aborigines, or earliest inhabitants, of Australia.

Austronesian language: A family of languages which includes practically all the languages of the Pacific Islands—Indonesian, Melanesian, Polynesian, and Micronesian sub-families. Does not include Australian or Papuan languages.

authoritarianism: A form of government in which a person or group attempts to rule with absolute authority without the representation of the citizens.

autonomous state: A country which is completely self-governing, as opposed to being a dependency or part of another country.

autonomy: The state of existing as a self-governing entity. For instance, when a country gains its independence from another country, it gains autonomy.

Axis Powers: The countries aligned against the Allied Nations in World War II, originally applied to Nazi Germany and Fascist Italy (Rome-Berlin Axis), and later extended to include Japan.

Baha'i: The follower of a religious sect founded by Mirza Husayn Ali in Iran in 1863.

Baltic states: The three formerly communist countries of Estonia, Latvia, and Lithuania that border on the Baltic Sea.

Bantu language group: A name applied to the languages spoken in central and south Africa.

Baptist: A member of a Protestant denomination that practices adult baptism by complete immersion in water.

barren land: Unproductive land, partly or entirely treeless.

barter: Trade practice where merchandise is exchanged directly for other merchandise or services without use of money.

bicameral legislature: A legislative body consisting of two chambers, such as the U.S. House of Representatives and the U.S. Senate.

bill of rights: A written statement containing the list of privileges and powers to be granted to a body of people, usually introduced when a government or other organization is forming.

black market: A system of trade where goods are sold illegally, often for excessively inflated prices. This type of trade usually develops to avoid paying taxes or tariffs levied by the government, or to get around import or export restrictions on products.

bloodless coup: The sudden takeover of a country's government by hostile means but without killing anyone in the process.

boat people: Used to describe individuals (refugees) who attempt to flee their country by boat.
Bolshevik Revolution: A revolution in 1917 in Russia when a wing of the Russian Social Democratic party seized power. The Bolsheviks advocated the violent overthrow of capitalism.
bonded labor: Workers bound to service without pay; slaves.
border dispute: A disagreement between two countries as to the exact location or length of the dividing line between them.
Brahman: A member (by heredity) of the highest caste among the Hindus, usually assigned to the priesthood.
Buddhism: A religious system common in India and eastern Asia. Founded by and based upon the teachings of Siddhartha Gautama, Buddhism asserts that suffering is an inescapable part of life. Deliverance can only be achieved through the practice of charity, temperance, justice, honesty, and truth.
buffer state: A small country that lies between two larger, possibly hostile countries, considered to be a neutralizing force between them.
bureaucracy: A system of government that is characterized by division into bureaus of administration with their own divisional heads. Also refers to the inflexible procedures of such a system that often result in delay.
Byzantine Empire: An empire centered in the city of Constantinople, now Istanbul in present-day Turkey.
CACM *see* Central American Common Market.
canton: A territory or small division or state within a country.
capital punishment: The ultimate act of punishment for a crime, the death penalty.
capitalism: An economic system in which goods and services and the means to produce and sell them are privately owned, and prices and wages are determined by market forces.
Caribbean Community and Common Market (CARICOM): Founded in 1973 and with its headquarters in Georgetown, Guyana, CARICOM seeks the establishment of a common trade policy and increased cooperation in the Caribbean region. Includes 13 English-speaking Caribbean nations: Antigua and Barbuda, the Bahamas, Barbados, Belize, Dominica, Grenada, Guyana, Jamaica, Montserrat, Saint Kitts-Nevis, Saint Lucia, St. Vincent/Grenadines, and Trinidad and Tobago.
CARICOM *see* Caribbean Community and Common Market.
cartel: An organization of independent producers formed to regulate the production, pricing, or marketing practices of its members in order to limit competition and maximize their market power.
cash crop: A crop that is grown to be sold rather than kept for private use.
caste system: One of the artificial divisions or social classes into which the Hindus are rigidly separated according to the religious law of Brahmanism. Membership in a caste is hereditary, and the privileges and disabilities of each caste are transmitted by inheritance.
Caucasian or Caucasoid: The white race of human beings, as determined by genealogy and physical features.
ceasefire: An official declaration of the end to the use of military force or active hostilities, even if only temporary.
censorship: The practice of withholding certain items of news that may cast a country in an unfavorable light or give away secrets to the enemy.
census: An official counting of the inhabitants of a state or country with details of sex and age, family, occupation, possessions, etc.
Central American Common Market (CACM): Established in 1962, a trade alliance of five Central American nations. Participating are Costa Rica, El Salvador, Guatemala, Honduras, and Nicaragua.
Central Powers: In World War I, Germany and Austria-Hungary, and their allies, Turkey and Bulgaria.
centrist position: Refers to opinions held by members of a moderate political group; that is, views that are somewhere in the middle of popular thought between conservative and liberal.
cession: Withdrawal from or yielding to physical force.
chancellor: A high-ranking government official. In some countries it is the prime minister.
Christianity: The religion founded by Jesus Christ, based on the Bible as holy scripture.
Church of England: The national and established church in England. The Church of England claims continuity with the branch of the Catholic Church that existed in England before the Reformation. Under Henry VIII, the spiritual supremacy and jurisdiction of the Pope were abolished, and the sovereign (king or queen) was declared head of the church.
circuit court: A court that convenes in two or more locations within its appointed district.
CIS *see* Commonwealth of Independent States
city-state: An independent state consisting of a city and its surrounding territory.
civil court: A court whose proceedings include determinations of rights of individual citizens, in contrast to criminal proceedings regarding individuals or the public.
civil jurisdiction: The authority to enforce the laws in civil matters brought before the court.
civil law: The law developed by a nation or state for the conduct of daily life of its own people.
civil rights: The privileges of all individuals to be treated as equals under the laws of their country; specifically, the rights given by certain amendments to the U.S. Constitution.
civil unrest: The feeling of uneasiness due to an unstable political climate, or actions taken as a result of it.
civil war: A war between groups of citizens of the same country who have different opinions or agendas. The Civil War of the United States was the conflict between the states of the North and South from 1861 to 1865.
Club du Sahel: The Club du Sahel is an informal coalition which seeks to reverse the effects of drought and the desertification in the eight Sahelian zone countries: Burkina Faso, Chad, Gambia, Mali, Mauritania, Niger, Senegal, and the Cape Verde Islands. Headquarters are in Ouagadougou, Burkina Faso.
CMEA *see* Council for Mutual Economic Assistance.
coalition government: A government combining differing factions within a country, usually temporary.
Cold War: Refers to conflict over ideological differences that is carried on by words and diplomatic actions, not by military action. The term is usually used to refer to the tension that existed between the United States and the USSR from the 1950s until the breakup of the USSR in 1991.
collective bargaining: The negotiations between workers who are members of a union and their employer for the purpose of deciding work rules and policies regarding wages, hours, etc.
collective farm: A large farm formed from many small farms and supervised by the government; usually found in communist countries.
collective farming: The system of farming on a collective where all workers share in the income of the farm.
colonial period: The period of time when a country forms colonies in and extends control over a foreign area.
colonist: Any member of a colony or one who helps settle a new colony.
colony: A group of people who settle in a new area far from their original country, but still under the jurisdiction of that country. Also refers to the newly settled area itself.
COMECON *see* Council for Mutual Economic Assistance.
commerce: The trading of goods (buying and selling), especially on a large scale, between cities, states, and countries.
commission: A group of people designated to collectively do a job, including a government agency with certain law-making powers. Also, the power given to an individual or group to perform certain duties.

common law: A legal system based on custom and decisions and opinions of the law courts. The basic system of law of England and the United States.

common market: An economic union among countries that is formed to remove trade barriers (tariffs) among those countries, increasing economic cooperation. The European Community is a notable example of a common market.

commonwealth: A commonwealth is a free association of sovereign independent states that has no charter, treaty, or constitution. The association promotes cooperation, consultation, and mutual assistance among members.

Commonwealth of Independent States: The CIS was established in December 1991 as an association of 11 republics of the former Soviet Union. The members include: Russia, Ukraine, Belarus (formerly Byelorussia), Moldova (formerly Moldavia), Armenia, Azerbaijan, Uzbekistan, Turkmenistan, Tajikistan, Kazakhstan, and Kyrgyzstan (formerly Kirghiziya). The Baltic states—Estonia, Latvia, and Lithuania—did not join. Georgia maintained observer status before joining the CIS in November 1993.

Commonwealth of Nations: Voluntary association of the United Kingdom and its present dependencies and associated states, as well as certain former dependencies and their dependent territories. The term was first used officially in 1926 and is embodied in the Statute of Westminster (1931). Within the Commonwealth, whose secretariat (established in 1965) is located in London, England, are numerous subgroups devoted to economic and technical cooperation.

commune: An organization of people living together in a community who share the ownership and use of property. Also refers to a small governmental district of a country, especially in Europe.

communism: A form of government whose system requires common ownership of property for the use of all citizens. All profits are to be equally distributed and prices on goods and services are usually set by the state. Also, communism refers directly to the official doctrine of the former U.S.S.R.

compulsory: Required by law or other regulation.

compulsory education: The mandatory requirement for children to attend school until they have reached a certain age or grade level.

conciliation: A process of bringing together opposing sides of a disagreement for the purpose of compromise. Or, a way of settling an international dispute in which the disagreement is submitted to an independent committee that will examine the facts and advise the participants of a possible solution.

concordat: An agreement, compact, or convention, especially between church and state.

confederation: An alliance or league formed for the purpose of promoting the common interests of its members.

Confucianism: The system of ethics and politics taught by the Chinese philosopher Confucius.

conscription: To be required to join the military by law. Also known as the draft. Service personnel who join the military because of the legal requirement are called conscripts or draftees.

conservative party: A political group whose philosophy tends to be based on established traditions and not supportive of rapid change.

constituency: The registered voters in a governmental district, or a group of people that supports a position or a candidate.

constituent assembly: A group of people that has the power to determine the election of a political representative or create a constitution.

constitution: The written laws and basic rights of citizens of a country or members of an organized group.

constitutional monarchy: A system of government in which the hereditary sovereign (king or queen, usually) rules according to a written constitution.

constitutional republic: A system of government with an elected chief of state and elected representation, with a written constitution containing its governing principles. The United States is a constitutional republic.

Coptic Christians: Members of the Coptic Church of Egypt, formerly of Ethiopia.

Council for Mutual Economic Assistance (CMEA): Also known as Comecon, the alliance of socialist economies was established on 25 January 1949 and abolished 1 January 1991. It included Afghanistan*, Albania, Angola*, Bulgaria, Cuba, Czechoslovakia, Ethiopia*, East Germany, Hungary, Laos*, Mongolia, Mozambique*, Nicaragua*, Poland, Romania, USSR, Vietnam, Yemen*, and Yugoslavia. (Nations marked with an asterisk were observers only.)

counterinsurgency operations: Organized military activity designed to stop rebellion against an established government.

county: A territorial division or administrative unit within a state or country.

coup d'ètat or coup: A sudden, violent overthrow of a government or its leader.

criminal law: The branch of law that deals primarily with crimes and their punishments.

crown colony: A colony established by a commonwealth over which the monarch has some control, as in colonies established by the United Kingdom's Commonwealth of Nations.

Crusades: Military expeditions by European Christian armies in the eleventh, twelfth, and thirteenth centuries to win land controlled by the Muslims in the middle east.

cultivable land: Land that can be prepared for the production of crops.

Cultural Revolution: An extreme reform movement in China from 1966 to 1976; its goal was to combat liberalization by restoring the ideas of Mao Zedong.

customs union: An agreement between two or more countries to remove trade barriers with each other and to establish common tariff and nontariff policies with respect to imports from countries outside of the agreement.

cyclone: Any atmospheric movement, general or local, in which the wind blows spirally around and in towards a center. In the northern hemisphere, the cyclonic movement is usually counter-clockwise, and in the southern hemisphere, it is clockwise.

Cyrillic alphabet: An alphabet adopted by the Slavic people and invented by Cyril and Methodius in the ninth century as an alphabet that was easier for the copyist to write. The Russian alphabet is a slight modification of it.

decentralization: The redistribution of power in a government from one large central authority to a wider range of smaller local authorities.

declaration of independence: A formal written document stating the intent of a group of persons to become fully self-governing.

deficit: The amount of money that is in excess between spending and income.

deficit spending: The process in which a government spends money on goods and services in excess of its income.

deforestation: The removal or clearing of a forest.

deity: A being with the attributes, nature, and essence of a god; a divinity.

delta: Triangular-shaped deposits of soil formed at the mouths of large rivers.

demarcate: To mark off from adjoining land or territory; set the limits or boundaries of.

demilitarized zone (DMZ): An area surrounded by a combat zone that has had military troops and weapons removed.

demobilize: To disband or discharge military troops.

democracy: A form of government in which the power lies in the hands of the people, who can govern directly, or can be governed indirectly by representatives elected by its citizens.

denationalize: To remove from government ownership or control.

deportation: To carry away or remove from one country to another, or to a distant place.
depression: A hollow; a surface that has sunken or fallen in.
deregulation: The act of reversing controls and restrictions on prices of goods, bank interest, and the like.
desalinization plant: A facility that produces freshwater by removing the salt from saltwater.
desegregation: The act of removing restrictions on people of a particular race that keep them socially, economically, and, sometimes, physically, separate from other groups.
desertification: The process of becoming a desert as a result of climatic changes, land mismanagement, or both.
détente: The official lessening of tension between countries in conflict.
devaluation: The official lowering of the value of a country's currency in relation to the value of gold or the currencies of other countries.
developed countries: Countries which have a high standard of living and a well-developed industrial base.
dialect: One of a number of regional or related modes of speech regarded as descending from a common origin.
dictatorship: A form of government in which all the power is retained by an absolute leader or tyrant. There are no rights granted to the people to elect their own representatives.
dike: An artificial riverbank built up to control the flow of water.
diplomatic relations: The relationship between countries as conducted by representatives of each government.
direct election: The process of selecting a representative to the government by balloting of the voting public, in contrast to selection by an elected representative of the people.
disarmament: The reduction or depletion of the number of weapons or the size of armed forces.
dissident: A person whose political opinions differ from the majority to the point of rejection.
dogma: A principle, maxim, or tenet held as being firmly established.
dominion: A self-governing nation that recognizes the British monarch as chief of state.
dowry: The sum of the property or money that a bride brings to her groom at their marriage.
draft constitution: The preliminary written plans for the new constitution of a country forming a new government.
Druze: A member of a Muslim sect based in Syria, living chiefly in the mountain regions of Lebanon.
dual nationality: The status of an individual who can claim citizenship in two or more countries.
duchy: Any territory under the rule of a duke or duchess.
due process: In law, the application of the legal process to which every citizen has a right, which cannot be denied.
dynasty: A family line of sovereigns who rule in succession, and the time during which they reign.
Eastern Orthodox: The outgrowth of the original Eastern Church of the Eastern Roman Empire, consisting of eastern Europe, western Asia, and Egypt.
EC *see* European Community
ecclesiastical: Pertaining or relating to the church.
ecology: The branch of science that studies organisms in relationship to other organisms and to their environment.
economic depression: A prolonged period in which there is high unemployment, low production, falling prices, and general business failure.
elected assembly: The persons that comprise a legislative body of a government who received their positions by direct election.
electoral system: A system of choosing government officials by votes cast by qualified citizens.
electoral vote: The votes of the members of the electoral college.
electorate: The people who are qualified to vote in an election.
emancipation: The freeing of persons from any kind of bondage or slavery.
embargo: A legal restriction on commercial ships to enter a country's ports, or any legal restriction of trade.
emigration: Moving from one country or region to another for the purpose of residence.
empire: A group of territories ruled by one sovereign or supreme ruler. Also, the period of time under that rule.
enclave: A territory belonging to one nation that is surrounded by that of another nation.
encroachment: The act of intruding, trespassing, or entering on the rights or possessions of another.
endemic: Anything that is peculiar to and characteristic of a locality or region.
Enlightenment: An intellectual movement of the late seventeenth and eighteenth centuries in which scientific thinking gained a strong foothold and old beliefs were challenged. The idea of absolute monarchy was questioned and people were gradually given more individual rights.
epidemic: As applied to disease, any disease that is temporarily prevalent among people in one place at the same time.
Episcopal: Belonging to or vested in bishops or prelates; characteristic of or pertaining to a bishop or bishops.
ethnolinguistic group: A classification of related languages based on common ethnic origin.
EU *see* European Union
European Community: A regional organization created in 1958. Its purpose is to eliminate customs duties and other trade barriers in Europe. It promotes a common external tariff against other countries, a Common Agricultural Policy (CAP), and guarantees of free movement of labor and capital. The original six members were Belgium, France, West Germany, Italy, Luxembourg, and the Netherlands. Denmark, Ireland, and the United Kingdom became members in 1973; Greece joined in 1981; Spain and Portugal in 1986. Other nations continue to join.
European Union: The EU is an umbrella reference to the European Community (EC) and to two European integration efforts introduced by the Maastricht Treaty: Common Foreign and Security Policy (including defense) and Justice and Home Affairs (principally cooperation between police and other authorities on crime, terrorism, and immigration issues).
exports: Goods sold to foreign buyers.
external migration: The movement of people from their native country to another country, as opposed to internal migration, which is the movement of people from one area of a country to another in the same country.
faction: People with a specific set of interests or goals who form a subgroup within a larger organization.
Fascism: A political philosophy that holds the good of the nation as more important than the needs of the individual. Fascism also stands for a dictatorial leader and strong oppression of opposition or dissent.
federal: Pertaining to a union of states whose governments are subordinate to a central government.
federation: A union of states or other groups under the authority of a central government.
fetishism: The practice of worshipping a material object that is believed to have mysterious powers residing in it, or is the representation of a deity to which worship may be paid and from which supernatural aid is expected.
feudal society: In medieval times, an economic and social structure in which persons could hold land given to them by a lord (nobleman) in return for service to that lord.
final jurisdiction: The final authority in the decision of a legal matter. In the United States, the Supreme Court would have final jurisdiction.
Finno-Ugric language group: A subfamily of languages spoken in northeastern Europe, including Finnish, Hungarian, Estonian, and Lapp.

fiscal year: The twelve months between the settling of financial accounts, not necessarily corresponding to a calendar year beginning on January 1.

fjord: A deep indentation of the land forming a comparatively narrow arm of the sea with more or less steep slopes or cliffs on each side.

folk religion: A religion with origins and traditions among the common people of a nation or region that is relevant to their particular life-style.

foreign exchange: Foreign currency that allows foreign countries to conduct financial transactions or settle debts with one another.

foreign policy: The course of action that one government chooses to adopt in relation to a foreign country.

Former Soviet Union: The FSU is a collective reference to republics comprising the former Soviet Union. The term, which has been used as both including and excluding the Baltic republics (Estonia, Latvia, and Lithuania), includes the other 12 republics: Russia, Ukraine, Belarus, Moldova, Armenia, Azerbaijan, Uzbekistan, Turkmenistan, Tajikistan, Kazakhstan, Kyrgizstan, and Georgia.

free enterprise: The system of economics in which private business may be conducted with minimum interference by the government.

fundamentalist: A person who holds religious beliefs based on the complete acceptance of the words of the Bible or other holy scripture as the truth. For instance, a fundamentalist would believe the story of creation exactly as it is told in the Bible and would reject the idea of evolution.

GDP *see* gross domestic product

genocide: Planned and systematic killing of members of a particular ethnic, religious, or cultural group.

Germanic language group: A large branch of the Indo-European family of languages including German itself, the Scandinavian languages, Dutch, Yiddish, Modern English, Modern Scottish, Afrikaans, and others. The group also includes extinct languages such as Gothic, Old High German, Old Saxon, Old English, Middle English, and the like.

glasnost: President Mikhail Gorbachev's frank revelations in the 1980s about the state of the economy and politics in the Soviet Union; his policy of openness.

global warming: Also called the greenhouse effect. The theorized gradual warming of the earth's climate as a result of the burning of fossil fuels, the use of man-made chemicals, deforestation, etc.

GMT *see* Greenwich Mean Time

GNP *see* gross national product

grand duchy: A territory ruled by a nobleman, called a grand duke, who ranks just below a king.

Greek Catholic: A person who is a member of an Orthodox Eastern Church.

Greek Orthodox: The official church of Greece, a self-governing branch of the Orthodox Eastern Church.

Greenwich (Mean) Time: Mean solar time of the meridian at Greenwich, England, used as the basis for standard time throughout most of the world. The world is divided into 24 time zones, and all are related to the prime, or Greenwich mean, zone.

gross domestic product: A measure of the market value of all goods and services produced within the boundaries of a nation, regardless of asset ownership. Unlike gross national product, GDP excludes receipts from that nation's business operations in foreign countries.

gross national product: A measure of the market value of goods and services produced by the labor and property of a nation. Includes receipts from that nation's business operation in foreign countries

guerrilla: A member of a small radical military organization that uses unconventional tactics to take their enemies by surprise.

gymnasium: A secondary school, primarily in Europe, that prepares students for university.

harem: In a Muslim household, refers to the women (wives, concubines, and servants in ancient times) who live there and also to the area of the home they live in.

harmattan: An intensely dry, dusty wind felt along the coast of Africa between Cape Verde and Cape Lopez. It prevails at intervals during the months of December, January, and February.

heavy industry: Industries that use heavy or large machinery to produce goods, such as automobile manufacturing.

Holocaust: The mass slaughter of European civilians, the vast majority Jews, by the Nazis during World War II.

Holy Roman Empire: A kingdom consisting of a loose union of German and Italian territories that existed from around the ninth century until 1806.

home rule: The governing of a territory by the citizens who inhabit it.

homeland: A region or area set aside to be a state for a people of a particular national, cultural, or racial origin.

homogeneous: Of the same kind or nature, often used in reference to a whole.

Horn of Africa: The Horn of Africa comprises Djibouti, Eritrea, Ethiopia, Somalia, and Sudan.

human rights issues: Any matters involving people's basic rights which are in question or thought to be abused.

humanist: A person who centers on human needs and values, and stresses dignity of the individual.

humanitarian aid: Money or supplies given to a persecuted group or people of a country at war, or those devastated by a natural disaster, to provide for basic human needs.

hydroelectric power plant: A factory that produces electrical power through the application of waterpower.

IBRD *see* World Bank

immigration: The act or process of passing or entering into another country for the purpose of permanent residence.

imports: Goods purchased from foreign suppliers.

indigenous: Born or originating in a particular place or country; native to a particular region or area.

Indo-Aryan language group: The group that includes the languages of India; also called Indo-European language group.

Indo-European language family: The group that includes the languages of India and much of Europe and southwestern Asia.

infanticide: The act of murdering a baby.

infidel: One who is without faith or belief; particularly, one who rejects the distinctive doctrines of a particular religion.

inflation: The general rise of prices, as measured by a consumer price index. Results in a fall in value of currency.

insurgency: The state or condition in which one rises against lawful authority or established government; rebellion.

insurrectionist: One who participates in an unorganized revolt against an authority.

interim government: A temporary or provisional government.

interim president: One who is appointed to perform temporarily the duties of president during a transitional period in a government.

International Date Line: An arbitrary line at about the 180th meridian that designates where one day begins and another ends.

Islam: The religious system of Mohammed, practiced by Moslims and based on a belief in Allah as the supreme being and Mohammed as his prophet. The spelling variations, Muslim and Muhammad, are also used, primarily by Islamic people. Islam also refers to those nations in which it is the primary religion.

isthmus: A narrow strip of land bordered by water and connecting two larger bodies of land, such as two continents, a continent and a peninsula, or two parts of an island.

Judaism: The religious system of the Jews, based on the Old Testament as revealed to Moses and characterized by a belief in one God and adherence to the laws of scripture and rabbinic traditions.

Judeo-Christian: The dominant traditional religious makeup of the United States and other countries based on the worship of the Old and New Testaments of the Bible.

junta: A small military group in power in a country, especially after a coup.

khan: A sovereign, or ruler, in central Asia.

khanate: A kingdom ruled by a khan, or man of rank.

labor movement: A movement in the early to mid-1800s to organize workers in groups according to profession to give them certain rights as a group, including bargaining power for better wages, working conditions, and benefits.

land reforms: Steps taken to create a fair distribution of farmland, especially by governmental action.

landlocked country: A country that does not have direct access to the sea; it is completely surrounded by other countries.

least developed countries: A subgroup of the United Nations designation of "less developed countries;" these countries generally have no significant economic growth, low literacy rates, and per person gross national product of less than $500. Also known as undeveloped countries.

leftist: A person with a liberal or radical political affiliation.

legislative branch: The branch of government which makes or enacts the laws.

less developed countries (LDC): Designated by the United Nations to include countries with low levels of output, living standards, and per person gross national product generally below $5,000.

literacy: The ability to read and write.

Maastricht Treaty: The Maastricht Treaty (named for the Dutch town in which the treaty was signed) is also known as the Treaty of European Union. The treaty creates a European Union by: (a) committing the member states of the European Economic Community to both European Monetary Union (EMU) and political union; (b) introducing a single currency (European Currency Unit, ECU); (c) establishing a European System of Central Banks (ESCB); (d) creating a European Central Bank (ECB); and (e) broadening EC integration by including both a common foreign and security policy (CFSP) and cooperation in justice and home affairs (CJHA). The treaty entered into force on November 1, 1993.

Maghreb states: The Maghreb states include the three nations of Algeria, Morocco, and Tunisia; sometimes includes Libya and Mauritania.

majority party: The party with the largest number of votes and the controlling political party in a government.

Marshall Plan: Formally known as the European Recovery Program, a joint project between the United States and most Western European nations under which $12.5 billion in U.S. loans and grants was expended to aid European recovery after World War II.

Marxism *see* Marxist-Leninist principles

Marxist-Leninist principles: The doctrines of Karl Marx, built upon by Nikolai Lenin, on which communism was founded. They predicted the fall of capitalism, due to its own internal faults and the resulting oppression of workers.

Marxist: A follower of Karl Marx, a German socialist and revolutionary leader of the late 1800s, who contributed to Marxist-Leninist principles.

Mayan language family: The languages of the Central American Indians, further divided into two subgroups: the Maya and the Huastek.

Mecca (Mekkah): A city in Saudi Arabia; a destination of pilgrims in the Islamic world.

Mediterranean climate: A wet-winter, dry-summer climate with a moderate annual temperature range.

mestizo: The offspring of a person of mixed blood; especially, a person of mixed Spanish and American Indian parentage.

migratory workers: Usually agricultural workers who move from place to place for employment depending on the growing and harvesting seasons of various crops.

military coup: A sudden, violent overthrow of a government by military forces.

military junta: The small military group in power in a country, especially after a coup.

military regime: Government conducted by a military force.

militia: The group of citizens of a country who are either serving in the reserve military forces or are eligible to be called up in time of emergency.

minority party: The political group that comprises the smaller part of the large overall group it belongs to; the party that is not in control.

missionary: A person sent by authority of a church or religious organization to spread his religious faith in a community where his church has no self-supporting organization.

monarchy: Government by a sovereign, such as a king or queen.

Mongol: One of an Asiatic race chiefly resident in Mongolia, a region north of China proper and south of Siberia.

Mongoloid: Having physical characteristics like those of the typical Mongols (Chinese, Japanese, Turks, Eskimos, etc.).

Moors: One of the Arab tribes that conquered Spain in the eighth century.

mosque: An Islamic place of worship and the organization with which it is connected.

Muhammad (or Muhammed or Mahomet): An Arabian prophet, known as the "Prophet of Allah" who founded the religion of Islam in 622, and wrote *The Koran,* the scripture of Islam. Also commonly spelled Mohammed.

mujahideen (mujahedin or mujahedeen): Rebel fighters in Islamic countries, especially those supporting the cause of Islam.

mulatto: One who is the offspring of parents one of whom is white and the other is black.

municipality: A district such as a city or town having its own incorporated government.

Muslim: A follower of the prophet Muhammad, the founder of the religion of Islam.

Muslim New Year: A Muslim holiday. Although in some countries 1 Muharram, which is the first month of the Islamic year, is observed as a holiday, in other places the new year is observed on Sha'ban, the eighth month of the year. This practice apparently stems from pagan Arab times. Shab-i-Bharat, a national holiday in Bangladesh on this day, is held by many to be the occasion when God ordains all actions in the coming year.

NAFTA (North American Free Trade Agreement): NAFTA, which entered into force in January 1994, is a free trade agreement between Canada, the United States, and Mexico. The agreement progressively eliminates almost all U.S.-Mexico tariffs over a 10–15 year period.

nationalism: National spirit or aspirations; desire for national unity, independence, or prosperity.

nationalization: To transfer the control or ownership of land or industries to the nation from private owners.

NATO *see* North Atlantic Treaty Organization

naturalize: To confer the rights and privileges of a native-born subject or citizen upon someone who lives in the country by choice.

neutrality: The policy of not taking sides with any countries during a war or dispute among them.

Newly Independent States: The NIS is a collective reference to 12 republics of the former Soviet Union: Russia, Ukraine, Belarus (formerly Byelorussia), Moldova (formerly Moldavia), Armenia, Azerbaijan, Uzbekistan, Turkmenistan, Tajikistan, Kazakhstan, and Kirgizstan (formerly Kirghiziya), and Georgia. Following dissolution of the Soviet Union, the distinction between the NIS and the Commonwealth of Independent States

(CIS) was that Georgia was not a member of the CIS. That distinction dissolved when Georgia joined the CIS in November 1993.

Nonaligned Movement: The NAM is an alliance of third world states that aims to promote the political and economic interests of developing countries. NAM interests have included ending colonialism/neo-colonialism, supporting the integrity of independent countries, and seeking a new international economic order.

Nordic Council: The Nordic Council, established in 1952, is directed toward supporting cooperation among Nordic countries. Members include Denmark, Finland, Iceland, Norway, and Sweden. Headquarters are in Stockholm, Sweden.

North Atlantic Treaty Organization (NATO): A mutual defense organization. Members include Belgium, Canada, Denmark, France (which has only partial membership), Greece, Iceland, Italy, Luxembourg, Netherlands, Norway, Portugal, Spain, Turkey, United Kingdom, United States, and Germany.

nuclear power plant: A factory that produces electrical power through the application of the nuclear reaction known as nuclear fission.

OAPEC (Organization of Arab Petroleum Exporting countries): OAPEC was created in 1968; members include: Algeria, Bahrain, Egypt, Iraq, Kuwait, Libya, Qatar, Saudi Arabia, Syria, and the United Arab Emirates. Headquarters are in Cairo, Egypt.

OAS (Organization of American States): The OAS (Spanish: Organizaciûn de los Estados Americanos, OEA), or the Pan American Union, is a regional organization which promotes Latin American economic and social development. Members include the United States, Mexico, and most Central American, South American, and Caribbean nations.

OAS *see* Organization of American States

oasis: Originally, a fertile spot in the Libyan desert where there is a natural spring or well and vegetation; now refers to any fertile tract in the midst of a wasteland.

occupied territory: A territory that has an enemy's military forces present.

official language: The language in which the business of a country and its government is conducted.

oligarchy: A form of government in which a few people possess the power to rule as opposed to a monarchy which is ruled by one.

OPEC *see* OAPEC

open market: Open market operations are the actions of the central bank to influence or control the money supply by buying or selling government bonds.

opposition party: A minority political party that is opposed to the party in power.

Organization of Arab Petroleum Exporting Countries *see* OAPEC

organized labor: The body of workers who belong to labor unions.

Ottoman Empire: A Turkish empire founded by Osman I in the thirteenth century, that variously controlled large areas of land around the Mediterranean, Black, and Caspian Seas until it was dissolved in 1923.

overseas dependencies: A distant and physically separate territory that belongs to another country and is subject to its laws and government.

Pacific Rim: The Pacific Rim, referring to countries and economies bordering the Pacific Ocean.

pact: An international agreement.

panhandle: A long narrow strip of land projecting like the handle of a frying pan.

papyrus: The paper-reed or -rush which grows on marshy river banks in the southeastern area of the Mediterranean, but more notably in the Nile valley.

paramilitary group: A supplementary organization to the military.

parliamentary republic: A system of government in which a president and prime minister, plus other ministers of departments, constitute the executive branch of the government and the parliament constitutes the legislative branch.

parliamentary rule: Government by a legislative body similar to that of Great Britain, which is composed of two houses—one elected and one hereditary.

partisan politics: Rigid, unquestioning following of a specific party's or leader's goals.

patriarchal system: A social system in which the head of the family or tribe is the father or oldest male. Kinship is determined and traced through the male members of the tribe.

per capita: Literally, per person; for each person counted.

perestroika: The reorganization of the political and economic structures of the Soviet Union by president Mikhail Gorbachev.

periodical: A publication whose issues appear at regular intervals, such as weekly, monthly, or yearly.

political climate: The prevailing political attitude of a particular time or place.

political refugee: A person forced to flee his or her native country for political reasons.

potable water: Water that is safe for drinking.

pound sterling: The monetary unit of Great Britain, otherwise known as the pound.

prime meridian: Zero degrees in longitude that runs through Greenwich, England, site of the Royal Observatory. All other longitudes are measured from this point.

prime minister: The premier or chief administrative official in certain countries.

privatization: To change from public to private control or ownership.

protectorate: A state or territory controlled by a stronger state, or the relationship of the stronger country toward the lesser one it protects.

Protestant: A member or an adherent of one of those Christian bodies which descended from the Reformation of the sixteenth century. Originally applied to those who opposed or protested the Roman Catholic Church.

Protestant Reformation: In 1529, a Christian religious movement begun in Germany to deny the universal authority of the Pope, and to establish the Bible as the only source of truth. (*Also see* Protestant)

province: An administrative territory of a country.

provisional government: A temporary government set up during a time of unrest or transition in a country.

purge: The act of ridding a society of "undesirable" or unloyal persons by banishment or murder.

Rastafarian: A member of a Jamaican cult begun in 1930 as a semi-religious, semi-political movement.

referendum: The practice of submitting legislation directly to the people for a popular vote.

Reformation *see* Protestant Reformation

refugee: One who flees to a refuge or shelter or place of safety. One who in times of persecution or political commotion flees to a foreign country for safety.

revolution: A complete change in a government or society, such as in an overthrow of the government by the people.

right-wing party: The more conservative political party.

Roman alphabet: The alphabet of the ancient Romans from which the alphabets of most modern western European languages, including English, are derived.

Roman Catholic Church: The designation of the church of which the pope or Bishop of Rome is the head, and that holds him as the successor of St. Peter and heir of his spiritual authority, privileges, and gifts.

Roman Empire: A Mediterranean Empire, centered in the Italian peninsula, that was the most powerful state in the region in the first four centuries A.D. The empire helped spread Greek culture throughout its territory. After the fourth century, the empire served as a Christianizing influence. Although the western half of the area fell to barbarian invasions in the fifth century, the eastern half, based in Constantinople, continued until 1453.

romance language: The group of languages derived from Latin: French, Spanish, Italian, Portuguese, and other related languages.

runoff election: A deciding election put to the voters in case of a tie between candidates.

Russian Orthodox: The arm of the Orthodox Eastern Church that was the official church of Russia under the czars.

Sahelian zone: Eight countries make up this dry desert zone in Africa: Burkina Faso, Chad, Gambia, Mali, Mauritania, Niger, Senegal, and the Cape Verde Islands. (*Also see* Club du Sahel.)

savanna: A treeless or near treeless plain of a tropical or subtropical region dominated by drought-resistant grasses.

secession: The act of withdrawal, such as a state withdrawing from the Union in the Civil War in the United States.

sect: A religious denomination or group, often a dissenting one with extreme views.

segregation: The enforced separation of a racial or religious group from other groups, compelling them to live and go to school separately from the rest of society.

self-sufficient: Able to function alone without help.

separatism: The policy of dissenters withdrawing from a larger political or religious group.

serfdom: In the feudal system of the Middle Ages, the condition of being attached to the land owned by a lord and being transferable to a new owner.

Seventh-day Adventist: One who believes in the second coming of Christ to establish a personal reign upon the earth.

shamanism: A religion of some Asians and Amerindians in which shamans, who are priests or medicine men, are believed to influence good and evil spirits.

Shia Muslims: Members of one of two great sects of Islam. Shia Muslims believe that Ali and the Imams are the rightful successors of Mohammed (also commonly spelled Muhammad). They also believe that the last recognized Imam will return as a messiah. Also known as Shiites. (*Also see* Sunni Muslims.)

Shiites *see* Shia Muslims

Shintoism: The system of nature- and hero-worship which forms the indigenous religion of Japan.

Sikh: A member of a politico-religious community of India, founded as a sect around 1500 and based on the principles of monotheism (belief in one god) and human brotherhood.

Sino-Tibetan language family: The family of languages spoken in eastern Asia, including China, Thailand, Tibet, and Burma.

slash-and-burn agriculture: A hasty and sometimes temporary way of clearing land to make it available for agriculture by cutting down trees and burning them.

slave trade: The transportation of black Africans beginning in the 1700s to other countries to be sold as slaves—people owned as property and compelled to work for their owners at no pay.

Slavic languages: A major subgroup of the Indo-European language family. It is further subdivided into West Slavic (including Polish, Czech, Slovak and Serbian), South Slavic (including Bulgarian, Serbo-Croatian, Slovene, and Old Church Slavonic), and East Slavic (including Russian Ukrainian and Byelorussian).

socialism: An economic system in which ownership of land and other property is distributed among the community as a whole, and every member of the community shares in the work and products of the work.

socialist: A person who advocates socialism.

Southeast Asia: The region in Asia that consists of the Malay Archipelago, the Malay Peninsula, and Indochina.

state: The politically organized body of people living under one government or one of the territorial units that make up a federal government, such as in the United States.

subcontinent: A land mass of great size, but smaller than any of the continents; a large subdivision of a continent.

Sudanic language group: A related group of languages spoken in various areas of northern Africa, including Yoruba, Mandingo, and Tshi.

suffrage: The right to vote.

Sufi: A Muslim mystic who believes that God alone exists, there can be no real difference between good and evil, that the soul exists within the body as in a cage, so death should be the chief object of desire, and sufism is the only true philosophy.

sultan: A king of a Muslim state.

Sunni Muslims: Members of one of two major sects of the religion of Islam. Sunni Muslims adhere to strict orthodox traditions, and believe that the four caliphs are the rightful successors to Mohammed, founder of Islam. (Mohammed is commonly spelled Muhammad, especially by Islamic people.) (*Also see* Shia Muslims.)

Taoism: The doctrine of Lao-Tzu, an ancient Chinese philosopher (about 500 B.C.) as laid down by him in the *Tao-te-ching*.

tariff: A tax assessed by a government on goods as they enter (or leave) a country. May be imposed to protect domestic industries from imported goods and/or to generate revenue.

terrorism: Systematic acts of violence designed to frighten or intimidate.

Third World: A term used to describe less developed countries; as of the mid-1990s, it is being replaced by the United Nations designation Less Developed Countries, or LDC.

topography: The physical or natural features of the land.

totalitarian party: The single political party in complete authoritarian control of a government or state.

trade unionism: Labor union activity for workers who practice a specific trade, such as carpentry.

treaty: A negotiated agreement between two governments.

tribal system: A social community in which people are organized into groups or clans descended from common ancestors and sharing customs and languages.

undeveloped countries *see* least developed countries

unemployment rate: The overall unemployment rate is the percentage of the work force (both employed and unemployed) who claim to be unemployed.

UNICEF: An international fund set-up for children's emergency relief: United Nations Children's Fund (formerly United Nations International Children's Emergency Fund).

untouchables: In India, members of the lowest caste in the caste system, a hereditary social class system. They were considered unworthy to touch members of higher castes.

Warsaw Pact: Agreement made May 14, 1955 (and dissolved July 1, 1991) to promote mutual defense between Albania, Bulgaria, Czechoslovakia, East Germany, Hungary, Poland, Romania, and the USSR.

Western nations: Blanket term used to describe mostly democratic, capitalist countries, including the United States, Canada, and western European countries.

workers' compensation: A series of regular payments by an employer to a person injured on the job.

World Bank: The World Bank is a group of international institutions which provides financial and technical assistance to developing countries.

world oil crisis: The severe shortage of oil in the 1970s precipitated by the Arab oil embargo.

Zoroastrianism: The system of religious doctrine taught by Zoroaster and his followers in the Avesta; the religion prevalent in Persia until its overthrow by the Muslims in the seventh century.

Bibliography

Africa

Algeria

Lorcin, Patricia M. E. *Imperial Identities: Stereotyping, Prejudice, and Race in Colonial Algeria.* London: I. B. Tauris Publishers, 1995.

MacMaster, Neil. *Colonial Migrants and Racism: Algerians in France, 1900–62.* New York: St. Martin's Press Inc., 1997.

Metz, Helen Chapin. *Algeria: A Country Study. Area Handbook Studies.* Washington, D.C.: Federal Research Division, Library of Congress. U.S. Government, Department of the Army, 1994.

Sahnouni, Mohamed. *The Lower Paleolithic of the Maghreb: Excavations and Analyses at Ain Hanech, Algeria. Cambridge Monographs in African Archaeology 42.* BAR International Series 689. Oxford: Hadrian Books Ltd., 1998.

Angola

Bollig, M. *When War Came the Cattle Slept: Himba Oral Traditions.* Koln: R. Koppe, 1997.

Ciment, J. *Conflict and Crisis in the Post-Cold War World: Angola and Mozambique: Postcolonial Wars in Southern Africa.* New York: Facts on File, 1997.

Etienne Dostert , P., ed. "The Republic of Angola." *Africa.* Harpers Ferry, WV: Stryker-Post Publications, 1997.

Kaplan, I. *Angola: A Country Study.* Washington, D.C.: American University, Foreign Area Studies, 1979.

Oliver, R. and B. Fagan. *Africa in the Iron Age, c. 500 BC to AD 1400.* New York: Cambridge University Press, 1975.

Benin

Africa on File. New York: Facts on File, Inc., 1995.

Decalo, Samuel. *Historical Dictionary of Benin.* 3rd ed. Lanham, Md.: Scarecrow Press, 1995.

Miller, Susan Katz. "Sermon on the Farm." *International Wildlife.* March/April 1992: 4951.

Botswana

Country Profile of Botswana. McLean, Va.: SAIC, 1998.

Jackson, A. *Botswana, 1939–1945: An African Country at War.* Oxford: Clarendon Press, 1999.

Morton, R. F., and J. Ramsay, eds. *Birth of Botswana: The History of the Bechuanaland Protectorate, 1910–1966.* Gaborone: Longman Botswana, 1988.

Pickford, P. and Pickford, B. *Okavango: The Miracle Rivers.* London: New Holland, 1999.

Ramsay, J.; Morton, B. and Morton, F. *Historical Dictionary of Botswana.* 3rd ed. Lanham, Md.: Scarecrow Press, 1996.

Burkina Faso

McFarland, Daniel Miles, and Lawrence A. Rupley. *Historical Dictionary of Burkina Faso.* Second Ed. *African Historical Dictionaries,* No. 74. Lanham, Md., and London: The Scarecrow Press, 1998.

Sankara, Thomas. *Women's Liberation and the African Freedom Struggle.* London: Pathfinder, 1990.

Wilks, Ivor. *The Mossi and Akan States.* In *History of West Africa.* Third Ed. Ed. J.F.A. Ajayi and Michael Crowder. Harlow, U.K.: Longman, 1985.

Burundi

Lemarchand, René. *Burundi: Ethnic Conflict and Genocide.* Woodrow Wilson Center Press, Cambridge University Press, 1997.

Ramsay, F. Jeffress. *Burundi in Global Studies: Africa.* Connecticut: Dushkin Publishing Group/Brow, Benchmark Publishers, 1995.

Weinstein, Warren, Robert Schire. *Political Conflict and Ethnic Strategies: A Case Study of Burundi.* New York: Maxwell School of Citizenship and Public Affairs, 1976.

Cameroon

Bjornson, Richard. *The African Quest for Freedom and Identity: Cameroonian Writing and the National Experience.* Bloomington: Indiana University Press, 1991.

DeLancey, Mark. *Historical Dictionary of the Republic of Cameroon.* 2nd ed. Metuchen, N.J.: Scarecrow Press, 1990.

"Odd Man in: Cameroon. (African nation to become the 52nd member of the Commonwealth of Nations)." *The Economist (US),* October 7, 1995, vol. 337, no. 7935, p. 51.

Takougang, Joseph. *African State and Society in the 1990s: Cameroon's Political Crossroads.* Boulder, Colo.: Westview Press, 1998.

Cape Verde

Carreira, Antonio. *The People of the Cape Verde Islands: Exploitation and Emigration.* Hamden, Conn.: Archon Books, 1982

Chilcote, Ronald H. *Amilcar Cabral's Revolutionary Theory and Practice: A Critical Guide.* Boulder, Colo.: Lynne Rienner Publishers, 1991.

Foy, Colm. *Cape Verde: Politics, Economics and Society.* London: Pinter Publishers, 1988.

Lobban, Jr. Richard A. *Cape Verde: Crioulo Colony to Independent Nation.* Boulder, Colo.: Westview Press, 1995.

Russell-Wood, A.J.R. *A World on the Move: The Portuguese in Africa, Asia, and America, 1415–1808.* New York: St. Martin's Press, 1993.

Central African Republic

Decalo, Samuel. *Psychoses of Power: African Personal Dictatorships.* Boulder, Colo.: Westview Press, 1989.

Kalck, Pierre. *Central African Republic: A Failure in De-colonisation.* Trans. Barbara Thomson. New York: Praeger, 1971.

———. *Central African Republic.* Santa Barbara, Calif.: Clio Press, 1993.

———. *Historical Dictionary of the Central African Republic.* 2nd ed. Metuchen, N.J.: Scarecrow Press, 1992.

O'Toole, Thomas. *The Central African Republic: The Continent's Hidden Heart.* Boulder, Colo.: Westview Press, 1986.

Titley, Brian. *Dark Age: The Political Odyssey of Emperor Bokassa.* Montreal: McGill-Queen's University Press, 1997.

Chad

Azevedo, Mario Joaquim. *Chad : A Nation in Search of its Future.* Boulder, Colo.: Westview Press, 1998.

Collelo, Thomas, ed. *Chad: A Country Study,* 2nd ed. Washington, DC: Government Printing Office, 1990.

Decalo, Samuel. *Historical Dictionary of Chad.* 2nd ed. Metuchen, NJ: Scarecrow Press, 1987.

Nolutshungu, Sam C. *Limits of Anarchy: Intervention and State Formation in Chad.* Charlottesville: University Press of Virginia, 1996.

Wright, John L. *Libya, Chad, and the Central Sahara.* Totowa, NJ: Barnes & Noble Books, 1989.

Comoros

"Under the Volcano." *Time International,* March 23, 1998, vol. 150, no. 30, p. 35.

The Comoros: Current Economic Situation and Prospects. Washington, D.C.: World Bank, 1983.

Newitt, Malyn. *The Comoro Islands: Struggle Against Dependency in the Indian Ocean.* Boulder, Colo: Westview, 1984.

Ottenheimer, Martin. *Historical Dictionary of the Comoro Islands.* Metuchen, N.J: Scarecrow Press, 1994.

Weinberg, Samantha. *Last of the Pirates: The Search for Bob Denard.* New York: Pantheon Books, 1994.

Congo, Democratic Republic of the

Bobb, F. Scott. *Historical Dictionary of Zaire.* Metuchen, N.J.: Scarecrow Press, 1988.

Kanza, Thomas. *The Rise and Fall of Patrice Lumumba.* Cambridge, Mass.: Schenkman, 1979.

Leslie, Winsome J. *Zaire: Continuity and Political Change in an Oppressive State.* Boulder, Colo.: Westview Press, 1993.

Lumumba, Patrice. *Congo, My Country.* With a foreword and notes by Colin Legum. Transl. by Graham Heath. London: Pall Mall Press, 1962.

Mokoli, Mondonga M. *State Against Development: The Experience of Post-1965 Zaire.* Westport, Conn.: Greenwood Press, 1992.

Congo, Republic of the

Clark, John F. "Elections, Leadership, and Democracy in Congo." *Africa Today* 41, no. 3 (1994): 41–60.

Decalo, Samuel, Virginia Thompson, and Richard Adloff. *Historical Dictionary of Congo.* Lanham, Md.: The Scarecrow Press, Inc., 1996.

Fegley, Randall. *The Congo.* World Bibliographical Series, vol. 162, Oxford: Clio Press, 1993.

Hilton, Anne. *The Kingdom of the Kongo.* Oxford: Clarendon Press, 1985.

Vansina, Jan. *The Tio Kingdom of the Middle Congo: 1880–1892.* London: Oxford University Press, 1973.

Cote d'Ivoire

Ajayi, J.F. Ade, and Michael Crowder, eds. *History of West Africa, 2.* New York: Colombia University Press, 1974.

Clark, John F. and David E. Gardinier, eds. *Political Reform in Francophone Africa.* Boulder, Colo.: Westview Press, 1997.

EIU Country Reports. London: Economist Intelligence Unit, April 1999.

Handloff, Robert E. *Côte d'Ivoire: A Country Study. Area Handbook Series.* Third Edition. Washington, DC: Federal Research Division, Library of Congress, 1991.

Mundt, Robert. *Historical Dictionary of the Ivory Coast.* Metuchen, N.J.: Scarecrow Press, 1995.

Djibouti

Darch, Colin. *A Soviet View of Africa: An Annotated Bibliography on Ethiopia, Somalia, and Djibouti.* Boston: G. K. Hall, 1980.

Koburger, Charles W. *Naval Strategy East of Suez: The Role of Djibouti.* New York: Praeger, 1992.

Schrader, Peter J. *Djibouti.* Santa Barbara, Calif.: Clio Press, 1991.

Tholomier, Robert. *Djibouti: Pawn of the Horn of Africa.* Metuchen, N.J.: Scarecrow, 1981.

Woodward, Peter. *The Horn of Africa: State Politics and International Relations.* London: I. B. Tauris, 1996.

Egypt

Daly, M.W., ed. *The Cambridge History of Egypt*. New York: Cambridge University Press, 1998.

Metz, Helen Chapin. *Egypt, a Country Study,* 5th ed. Washington, D.C.: Library of Congress, 1991.

Rubin, Barry M. *Islamic Fundamentalism in Egyptian Politics.* New York: St. Martin's Press, 1990.

Shamir, Shimon, ed. *Egypt from Monarchy to Republic: A Reassessment of Revolution and Change*. Boulder, Colo.: Westview Press, 1995.

Equatorial Guinea

Fegley, Randall. *Equatorial Guinea: An African Tragedy.* New York: P. Lang, 1989.

Klitgaard, Robert E. *Tropical Gangsters.* New York: Basic Books, 1990.

Liniger-Goumaz, Max. *Historical Dictionary of Equatorial Guinea.* 2nd ed. Metuchen, NJ: Scarecrow Press, 1988.

———. *Small is Not Always Beautiful: The Story of Equatorial Guinea.* Translated from the French by John Wood. Totowa, N.J.: Barnes & Noble Books, 1989.

Sundiata, I.K. *Equatorial Guinea: Colonialism, State Terror, and the Search for Stability.* Boulder, Colo.: Westview Press, 1990.

Eritrea

Connell, Dan. *Against All Odds.* Trenton, N.J.: Red Sea Press, 1993.

Doombos, Martin, et al., eds. *Beyond the Conflict in the Horn.* Trenton, NJ: Red Sea Press, 1992.

Gebremedhin, Tesfa G. *Beyond Survival: The Economic Challenges of Agriculture and Development in Post-Independence Eritrea.* Trenton, NJ: Red Sea Press, 1997.

Okbazghi, Yohannes. *Eritrea: A Pawn in World Politics.* Gainesville: University of Florida Press, 1991.

Ethiopia

Bahru, Z. *A History of Modern Ethiopia, 1855–1974.* Athens: Ohio University Press, 1991.

Hassen, M. *The Oromo of Ethiopia: A History 1570–1860.* Cambridge: Cambridge University Press, 1990.

Kaplan, S. *The Beta Israel (Falasha) in Ethiopia: From Earliest Times to the Twentieth Century.* New York: New York University Press, 1992.

Marcus, H. G. *A History of Ethiopia.* Berkeley: University of California Press, 1994.

Prouty, C. and Rosenfeld, E. *Historical Dictionary of Ethiopia.* Metuchen, NJ: The Scarecrow Press, Inc., 1994.

Gabon

Alexander, Caroline. *One Dry Season*. New York: Alfred A. Knopf, 1989.

Gall, Timothy L., ed. *Worldmark Encyclopedia of the Nations.* 9th ed. Detroit: Gale Research, 1998.

Iliffe, John. *Africans: The History of a Continent.* Cambridge University Press, 1995.

Ungar, Sanford J. *Africa: The People and Politics of an Emerging Continent.* New York: Simon & Schuster, 1989.

The Gambia

Else, David. *The Gambia and Senegal.* Oakland, Calif.: Lonely Planet Publications, 1999.

Gailey, Harry A. *Historical Dictionary of The Gambia.* 2nd ed. African Historical Dictionaries, No. 4. Metuchen, N.J.: Scarecrow Press, 1987.

Gall, Timothy L., ed. *Worldmark Encyclopedia of Cultures and Daily Life*. vol. 1: Africa 1998.

Vollmer, Jurgen. *Black Genesis, African Roots: A voyage from Juffure, The Gambia, to Mandingo country to the slave port of Dakar, Senegal*. New York: St. Martin's Press, 1980.

Wright, Donald R. "The world and a very small place in Africa (history of Niumi)." *Sources and Studies in World History.* New York: M.E. Sharpe, Armonk, 1997.

Ghana

Ardayfio-Schandorf, Elizabeth and Kate Kwafo-Akoto. *Women in Ghana: An Annotated Bibliography.* Accra: Woeli Publishing Services, 1990.

Berry, LaVerle. *Ghana: A Country Study. Area Handbook Series.* Third Edition. Washington, DC: Federal Research Division, Library of Congress, 1995.

Davidson, Basil. *Black Star: A View of the Life and Times of Kwame Nkrumah.* Boulder: Westview Press, 1989.

Glickman, Harvey, ed. *Political Leaders of Contemporary Africa South of the Sahara: A Biographical Dictionary.* New York: Greenwood Press, 1992.

Guinea

Africa on File. New York: Facts on File, 1995.

Gall, Timothy L., ed. *Worldmark Encyclopedia of Cultures and Daily Life*. vol. 1: Africa. Detroit: Gale Research, 1998.

Nelson, Harold D. et al, eds. *Area Handbook for Guinea. Foreign Area Studies.* Washington, DC.: American University, 1975.

O'Toole, Thomas. *Historical Dictionary of Guinea.* Third Edition. Lanham, MD: London: The Scarecrow Press, 1994.

Guinea-Bissau

Africa on File. New York: Facts on File, 1997.

EIU Country Reports. London: Economist Intelligence Unit, Ltd., April 8, 1999.

Forrest, Joshua. *Guinea-Bissau: Power, Conflict, and Renewal in a West African Nation.* Boulder: Westview Press, 1992.

Lopes, Carlos. *Guinea-Bissau: From Liberation Struggle to Independent Statehood.* Boulder, Colo.: Westview Press, 1987.

Pedlar, Frederick. *Main Currents of West African History, 19401978.* London: The Macmillan Press, Ltd., 1979.

Kenya

Cohen, W. David, and E. S. *Atieno Odhiambo. Burying SM: The Politics of Knowledge and the Sociology of Power in Africa.* Portsmouth, NH: Heinemann, 1992.

Miller, Norman, and Roger Yeager. *Kenya: The Quest for Prosperity* (2nd edition). Boulder, Colo.: Westview Press, 1994.

Mwaniki, Nyaga. "The Consequences of Land Subdivision in Northern Embu, Kenya." *The Journal of African Policy Studies,* 2(1), 1996.

Ogot, A. Bethwell. *Historical Dictionary of Kenya.* Metuchen, NJ: The Scarecrow Press, Inc., 1981.

Lesotho

Eldredge, E. A. *A South African Kingdom: The Pursuit of Security in Nineteenth-Century Lesotho.* Cambridge: Cambridge University Press.

Haliburton, G. *Historical Dictionary of Lesotho.* African Historical Dictionaries, No. 10, Metuchen, N.J.: The Scarecrow Press, Inc., 1977.

Khaketla, B. M. *Lesotho 1970: An African Coup Under the Microscope.* London: Hurst, 1971.

Liberia

Africa on File. New York: 1995 Facts on File, Inc., 1995.

Africa South of the Sahara. London: Europa Publishers, 1998.

Dunn, D. Elwood, and Svend E. Holsoe. *Historical Dictionary of Liberia.* African Historical Dictionaries, No. 38. Metuchen, N.J., and London: The Scarecrow Press, Inc., 1985.

Dunn, D. Elwood and S. Byron Tarr. *Liberia: A National Polity in Transition.* Metuchen, N.J., and London: The Scarecrow Press, Inc., 1988.

Nelson, Harold D., ed. *Liberia: A Country Study.* Third Edition. Washington, DC: The American University, 1985.

Libya

Gall, Timothy, L., ed. *Worldmark Encyclopedia of Cultures and Daily Life.* vol. 1: Africa. Detroit: Gale Research, 1998.

Haley, P. Edward. *Qadhafi and the United States Since 1969.* New York: Praeger, 1984.

Simonis, Damien, *et al. North Africa.* Hawthorn, Aus.: Lonely Planet Publications, 1995.

Simons, Geoff. *Libya: The Struggle for Survival.* New York: St. Martin's Press, 1996.

Vanderwalle, Dirk ed. *Qadhafi's Libya: 19691994.* New York: St. Martin's Press, 1995.

Madagascar

Allen, Philip M. *Madagascar: Conflicts of Authority in the Great Island.* Boulder, Colo.: Westview Press, 1994.

Covell, Maureen. *Historical Dictionary of Madagascar.* Lanham, Md.: Scarecrow Press, 1995.

Kent, Raymond K. *Early Kingdoms in Madagascar 1500–1700.* New York: Holt, Rinehart and Winston, 1970.

Stratton, Arthur. *The Great Red Island.* New York: Charles Scribner's Sons, 1964.

Verin, Pierre. *The History of Civilisation in North Madagascar.* Rotterdam: A.A. Balkema, 1986.

Malawi

Baker, Colin. *State of Emergency: Crisis in Central Africa, Nyasaland 1959–1960.* London: I. B. Tauris Publishers, 1997.

Crosby, Cynthia A. *Historical Dictionary of Malawi.* Metuchen, N.J.: The Scarecrow Press, 2nd edition, 1993.

Phiri, D. D. *From Nguni to Ngoni: A History of the Ngoni Exodus from Zululand and Swaziland to Malawi, Tanzania and Zambia.* Limbe, Malawi: Popular Publications, 1982.

Mali

Economist Intelligence Unit. *Country Profile: Mali, 1998.* London: The Economist, 1998.

Historical Dictionary of Mali. Second edition. Metuchen, New Jersey: Scarecrow Press, 1986.

Mann, Kenny. *Ghana, Mali, Songhay: The Western Sudan.* Parsippany, NJ: Dillon Press, 1996.

McIntosh, Roderick J. *The Peoples of the Middle Niger: The Island of Gold.* Malden, MA: Blackwell Publishers, 1998.

Mauritania

Africa. Hawthorn, Aus.: Lonely Planet Publications, 1995.

Goodsmith, Lauren. *The Children of Mauritania.* Minneapolis: Carolrhoda Books, 1993.

McLachlan, Anne and Keith. *Morocco Handbook.* Bath, Eng.: Footprint Handbooks. 1993.

Pazzanita, Anthony G. *Historical Dictionary of Mauritania.* Lanham, Md.: Scarecrow Press, Inc., 1966.

Thompson, Virginia, and Adloff, Richard. *The Western Saharans.* Totowa, N.J.: Barnes and Noble Books, 1980.

Mauritius

Bunge, Frederica M., ed. *Indian Ocean. Five Island Countries.* Washington, D.C.: Dept of the Army, 1983.

Bunwaree, Sheila S. *Mauritian Education in a Global Economy.* Stanley, Rose Hill, Mauritius: Editions de l'Océan Indien, 1994.

Butlin, Ron (ed). *Mauritian Voices. New Writing in English.* Newcastle Upon Tyne: Flambard Press, 1997.

Selvon, Sydney. *Historical Dictionary of Mauritius.* Metuchen, N.J.: Scarecrow Press, 1991.

Morocco

Cook, Weston, F. *The Hundred Year War for Morocco: Gunpowder and the Military Revolution in the Early Modern Muslim World.* Boulder, Colo.: Westview Press, 1985.

Jereb, James F. *Arts and Crafts of Morocco.* New York: Chronicle Books, 1996.

Hermes, Jules M. *The Children of Morocco.* Minneapolis: Carolrhoda Books, 1995.

Hoisington, William A., Jr. *Lyautey and the French Conquest of Morocco.* New York: St. Martins Press, 1995.

Simonis, Damien, et al. *North Africa.* Hawthorn, Aus.: Lonely Planet Publications, 1995.

Mozambique

Azevedo, Mario Joaquim. *Historical Dictionary of Mozambique.* Metuchen, N.J.: Scarecrow Press, 1991.

Davidson, Basil. *Africa in History.* New York: Simon and Schuster, 1991.

Magnin, Andre; and Jacques Soulillou, eds., *Contemporary Art of Africa.* New York: Harry N. Abrams, Inc. Publishers, 1996.

Newitt, M. D. D. *A History of Mozambique.* Bloomington: Indiana University Press, 1995.

Slater, Mike. *Mozambique.* London: New Holland Publishers Ltd, 1997.

Namibia

Breytenbach, Cloete. *Namibia: Birth of a Nation.* South Africa: LUGA Publishers, 1989.

Cliffe, Lionel. *The Transition to Independence in Namibia.* Boulder, Colo.: Lynne Rienner Publishers, 1994.

Namibia: A Nation Is Born. Washington, D.C.: U.S. Dept. of State, 1990.

Swaney, Deanna. *Zimbabwe, Botswana, and Namibia: A Lonely Planet Travel Guide.* Hawthorn, Australia, Lonely Planet Publications, 1995.

Niger

Beckwith, Carol. *Nomads of Niger.* New York: Henry N. Abrams, 1983.

Charlick, Robert. *Niger. Personal Rule and Survival in the Sahel.* Boulder, Colo.: Westview, 1991.

Cooper, Barbara. *Marriage in Maradi: Gender and Culture in a Hausa Society in Niger 1900-1989.* Portsmouth, N.H.: Heinemann, 1997.

Decalo, Samuel. *Historical Dictionary of Niger.* Metuchen, N.J.: Scarecrow Press, 1979.

Gall, Timothy L., ed. *Worldmark Encyclopedia of Cultures and Daily Life.* vol. 1: Africa. Detroit: Gale Research, 1998.

Nigeria

Diamond, Larry, Anthony Kirk-Greene, and Oyeleye Oyediran, eds. *Transition Without End: Nigerian Politics and Civil Society under Babangida.* Boulder, Colo.: Lynne Rienner Publishers, 1997.

Graf, William D. *The Nigerian State: Political Economy, State Class, and Political System in the Post-Colonial Era.* Portsmouth, N.H.: Heineman, 1988.

Ihonvbere, Julius O., and Timothy Shaw. *Illusions of Power: Nigeria in Transition.* Trenton, NJ: Africa World Press, 1998.

Osaghae, Eghosa E. *Crippled Giant: Nigeria Since Independence.* Bloomington: Indiana University Press, 1998.

Wesler, Kit W. *Historical Archaeology in Nigeria.* Trenton, N.J.: Africa World Press, 1998.

Rwanda

Hodd, M. *East African Handbook,* Chicago: Passport Books, 1994.

Nyrop, Richard., *et al. Rwanda: A Country Study.* Washington, D.C.: U.S. Government Printing Office, 1984.

Pierce, Julian R. *Speak Rwanda.* New York: Picador USA, 1999.

Webster, J.B., B.A. Ogot, and J.P. Chretien. "The Great Lakes Region: 1500–1800." In *The General History of Africa,* Volume V, Calif.: Heinemann, UNESCO, 1998

Sao Tome and Principe

Carreira, Antonio. *The People of the Cape Verde Islands: Exploitation and Emigration.* Connecticut: Archon Books, 1982.

Davidson, Basil. *Black Mother: The Years of the African Slave Trade.* Boston: Little, Brown, and Co., 1961.

Denny, L.M. and Donald I. Ray. *São Tomé and Príncipe: Economics, Politics and Society.* London and New York: Pinter Publishers, 1989.

Garfield, Robert. *A History of São Tomé Island, 1470–1655: The Key to Guinea.* San Francisco: Mellen Research University Press, 1992.

Hodges, Tony and Malyn Hewitt. *São Tomé and Príncipe: From Plantation Colony to Microstate.* Boulder and London: Westview Press, 1988.

Thomas, Hugh. *The Slave Trade.* New York: Touchstone Books-Simon and Schuster, 1997.

Senegal

Clark, Andrew Francis, ed. *Historical Dictionary of Senegal.* 2nd ed. African Historical Dictionaries, No. 23. Metuchen, N.J.: The Scarecrow Press, Inc., 1994.

Else, David. *The Gambia and Senegal.* Oakland, Calif.: Lonely Planet Publications, 1999.

Gellar, Sheldon. *Senegal: An African Nation between Islam and the West.* Second Edition. Boulder, Colo.: Westview Press, 1995.

Sharp, Robin. *Senegal: A State of Change.* Oxford, UK: Oxfam, 1994.

Seychelles

Bennett, George. *Seychelles.* Oxford, England; Santa Barbara, CA: Clio Press, 1993.

Franda, Marcus. *The Seychelles, Unquiet Islands.* Boulder, Colo.: Westview 1982.

McAteer, William. *Rivals in Eden: A History of the French Settlement and British Conquest of the Seychelles Islands, 1724–1818.* Sussex, Eng.: Book Guild, 1990.

Vine, Peter. *Seychelles.* London, Eng.: Immel Pub. Co., 1989.

Sierra Leone

Africa South of the Sahara. "Sierra Leone." London: Europa Publishers, 1997.
Alie, Joe. *A New History of Sierra Leone.* New York: St. Martin's Press, 1990.
Foray, Cyril P. *Historical Dictionary of Sierra Leone.* African Historical Dictionaries, No. 12. Metuchen, N.J.: The Scarecrow Press, 1977.
Gall, Timothy L., ed. *Worldmark Encyclopedia of Cultures and Daily Life*. vol. 1: Africa. Detroit: Gale Research, 1998.

Somalia

Barnes, Virginia Lee. *Aman: The Story of a Somali Girl.* New York: Pantheon, 1994.
Clarke, Walter S., and Jeffrey Ira Herbst. *Learning from Somalia: The Lessons of Armed Humanitarian Intervention.* Boulder, Colo.: Westview Press, 1997.
DeLancey, Mark. *Blood and Bone: The Call of Kinship in Somali Society.* Lawrenceville, N.J.: Red Sea, 1994.
———, et al, eds. *Somalia.* Santa Barbara, Calif.: Clio, 1988.
DeLancey, Mark. *Blood and Bone: The Call of Kinship in Somali Society.* Lawrenceville, N.J.: Red Sea, 1994.
Metz, Helen Chapin, ed. *Somalia: A Country Study.* 4th ed. Washington, DC: Library of Congress, 1993.

South Africa

Brynes, Rita M. ed. *South Africa: A Country Study.* Washington D.C.: U.S. Government Printing Office, 1997.
Mostert, N. *Frontiers: The Epic of South Africa's Creation and the Tragedy of the Xhosa People*. New York: Knopf, 1992.
Riley, E. *Major Political Events in South Africa, 1948–1990.* New York: Facts on File, 1991.
Thompson, L. M. *A History of South Africa*. London: Yale University Press, 1990.

Sudan

Alier, Abel. *Southern Sudan.* Reading:Ithaca Press,1990.
Bovill, Edward William. *The Golden Trade of the Moors.*Princeton, New Jersey: Marcus Wiener Publishers,1995.
Metz, Helen Chapin, ed. *Sudan: A Country Study.* Federal Research Division/Library of Congress,1991.
Stewart, Judy. *A Family in Sudan.* Minneapolis: Lerner Publishing Co., 1988.
Voll, John Obert and Sarah Potts Voll. *The Sudan: Unity and Diversity in a Multicultural State.* Boulder, Col.: Westview Press, 1985.

Swaziland

Bonner, P. M. "Swati II, c. 1826–1865." In *Black Leaders in Southern African History.* Edited by Christopher Saunders. London: Heinemann, 1979: 61–74.
Davies, R. H.; O'Meara, D. and Dlamini, S. *The Kingdom of Swaziland: A Profile.* London: Zed Books, 1985.
Grotpeter, J. J. *Historical Dictionary of Swaziland.* Metuchen, N.J.: The Scarecrow Press, 1975.
Kuper, H. *Sobhuza II: Ngwenyama and King of Swaziland.* London: Duckworth, 1978.
Williams, G. and B. Hackland. *The Dictionary of Contemporary Politics of Southern Africa*. New York: Macmillan, 1988.

Tanzania

Hodd, M. *East African Handbook.* Chicago: Passport Books, 1994.
Hughes, A. J. *East Africa.* Baltimore: Penguin Books, 1969.
Iliffe, J. *A Modern History of Tanganyika.* Cambridge, Eng.: Cambridge University Press, 1979.
Maddox, Gregory, et al., eds. *Custodians of the Land: Ecology and Culture in the History of Tanzania.* Athens: Ohio University Press, 1996.

Togo

Ajayi, J.F.A. and Michael Crowder, eds. *History of West Africa,* Vol. 1. London,Eng.: Congman Group Limited, 1971.
———. *History of West Africa*, Vol. 2. Londond, Eng.: Congman Group Limited, 1971.
Decalo, Samuel. *Historical Dictionary of Togo.* 3rd ed. Metuchen, NJ: Scarecrow Press, 1996.
Packer, George, *The Village of Waiting*. New York: Vintage Books, 1988.

Tunisia

Ali, Wijdan. *Modern Islamic Art: Development and Continuity.* Gainesville, Fla.: University Press of Florida, 1997.
Lancel, Serge. Nevill, Antonia (translator). *Hannibal.* Oxford, Eng.: Blackwell Publications, 1998.
Ling, Dwight L. *Morocco and Tunisia: A Comparative History.* Washington DC: University Press of America, 1979.
Memmi, Albert. *Pillar of Salt.* Boston: Beacon Press, 1992.

Uganda

Byrnes, Rita. ed. *Uganda: A Country Study.* Washington, D.C.: U.S. Government Printing Office, 1992.
Gall, Timothy L., ed. *Worldmark Encyclopedia of Cultures and Daily Life*. vol. 1: Africa. Detroit: Gale Research, 1998.
Hodd, M. *East African Handbook.* Chicago: Passport Books, 1994.
Jorgenson, Jan Jelmett. *Uganda: A Modern History.* New York: St. Martin's Press, 1981.
Nzita, Richard. *Peoples and Cultures of Uganda.* Kampala: Fountain Publishers, 1993.

Zambia

Burdette, M. M. *Zambia Between Two Worlds*. Boulder, Colo.: Westview Press, 1988.
Dresang, E. *The Land and People of Zambia*. Philadelphia: Lippincott, 1975.
Grotpeter, J. J.; Siegel, B. V. and Pletcher, J. R. *Historical Dictionary of Zambia*. Lanham, Maryland: The Scarecrow Press, Inc., 1998.

Zimbabwe

Dewey, William Joseph. *Legacies of Stone: Zimbabwe Past and Present.* Tervuren: Royal Museum for Central Africa, 1997.

Nelson, H. D. *Zimbabwe: A Country Study.* Washington, D. C.: U.S. Government Printing Office, 1983.

Rasmussen, R. Kent and Rubert, Steven, C. *Historical Dictionary of Zimbabwe.* 2nd ed. Metuchen, N.J.: The Scarecrow Press, 1990.

Sheehan, Sean. *Zimbabwe.* New York: M. Cavendish, 1996.

Americas

Antigua and Barbuda

Coram, Robert. *Caribbean Time Bomb: the United States' Complicity in the Corruption of Antigua.* New York: William Morrow and Company, Inc., 1993.

Dyde, Brian. *Antigua and Barbuda: the Heart of the Caribbean.* London: Macmillan Publishers, 1990.

Kurlansky, Mark. *A Continent of Islands: Searching for the Caribbean Destiny.* Reading, Mass.: Addison-Wesley Publishing Co., 1992.

Argentina

American University. *Argentina: A Country Study,* 3rd ed. Washington, DC: Government Printing Office, 1985.

Sarlo Sabajanes, Beatriz. *Jorge Luis Borges: A Writer on the Edge.* New York: Verso, 1993.

Timerman, Jacobo. *Prisoner Without a Name, Cell Without a Number.* New York: Knopf, 1981.

Tulchin, Joseph S. *Argentina: The Challenges of Modernization.* Wilmington, Del.: Scholarly Resources, 1998.

Worldmark Press, Ltd. *Worldmark Encyclopedia of Cultures and Daily Life*. vol. 2: Americas 1998.

Bahamas

Hamshere, Cyril. *The British in the Caribbean.* Cambridge, Mass.: Harvard University Press, 1972.

Kurlansky, Mark. *A Continent of Islands: Searching for the Caribbean Destiny.* Reading, Mass.: Addison-Wesley Publishing Co., 1992.

Marx, Jenifer. *Pirates and Privateers of the Caribbean.* Malabar, Fla.: Krieger Publishing Company, 1992.

Barbados

Beckles, Hilary. *A History of Barbados: From Amerindian Settlement to Nation-State.* New York: Cambridge University Press, 1990.

Pariser, Harry S. *Adventure Guide to Barbados.* Edison, N.J.: Hunter Publishing, 1995.

Wilder, Rachel, ed. *Barbados.* Boston: Houghton Mifflin Co., 1993

Belize

Bolland, O. Nigel. *Belize: A New Nation in Central America.* Boulder, Colo.: Westview, 1986.

Edgell, Zee. *Beka Lamb.* London: Heinemann, 1982.

Fernandez, Julio A. *Belize: Case Study for Democracy in Central America.* Brookfield, Vt.: Avebury, 1989.

Gall, Timothy L., ed. *Worldmark Encyclopedia of Cultures and Daily Life*. vol. 2: Americas. Detroit: Gale Research, 1998.

Mallan, Chicki. *Belize Handbook.* Chico, Calif.: Moon Publications, 1991.

Bolivia

Blair, David Nelson. *The Land and People of Bolivia.* New York: J.B. Lippincott, 1990.

Hudson, Rex A. and Dennis M. Hanratty. *Bolivia, a Country Study.* 3rd ed. Washington, DC: Government Printing Office, 1991.

Morales, Waltrand Q. *Bolivia: Land of Struggle.* Boulder, Colo.: Westview Press, 1992.

Lindert, P. van. *Bolivia : A Guide to the People, Politics and Culture.* New York: Monthly Review Press, 1994.

Parker, Edward. *Ecuador, Peru, Bolivia. Country fact files.* Austin, TX: Raintree Steck-Vaughn, 1998.

Brazil

Carpenter, Mark L. *Brazil, an Awakening Giant*. Minneapolis, MN: Dillon Press, 1987.

Levine, Robert M. *Historical Dictionary of Brazil.* Metuchen, N.J.: Scarecrow Press, 1979.

——— and John J. Crocitti. *The Brazil Reader: History, Culture, Politics.* Durham, N.C.: Duke University Press, 1999.

Poppino, Rollie E. *Brazil: The Land and People.* New York: Oxford University Press, 1973.

Roop, Peter, and Connie Roop. *Brazil.* Des Plaines, Ill.: Heinemann Interactive Library, 1998.

Canada

Bothwell, Robert. *A Short History of Ontario.* Edmonton: Hurtig Publishers Ltd., 1986.

Dickinson, John A. and Brian Young. *A Short History of Quebec*. Toronto: Copp Clark Pitman Ltd., 1993.

McNaught, Kenneth. *The Penguin History of Canada.* London: Penguin, 1988.

Morton, Desmond. *A Short History of Canada.* 2nd revised ed. Toronto: McClelland & Stewart Inc., 1994.

Woodcock, George. *A Social History of Canada.* Markham, Ont.: Penguin Books Canada, 1989.

Chile

Arriagada Herrera, Genaro. Trans. Nancy Morris. *Pinochet: The Politics of Power.* Boston : Allen & Unwin, 1988.

Blakemore, Harold. *Chile.* Santa Barbara, Calif.: Clio Press, 1988.

Collier, Simon. *A History of Chile.* Cambridge: Cambridge University Press, 1996.

Falcoff, Mark. *Modern Chile, 1970–89: A Critical History.* New Jersey: Transaction Books, 1989.

Hudson, Rex A., ed. *Chile, a Country Study.* 3rd ed. Federal Research Division, Library of Congress. Washington, D.C., 1994.

Colombia

Bushnell, David. The Making of Modern Colombia: A Nation in Spite of Itself. Berkeley: University of California Press, 1993.

Davis, Robert H. *Colombia.* Santa Barbara, Calif.: Clio Press, 1990.

———. *Historical Dictionary of Colombia.* Metuchen, N.J.: Scarecrow Press, 1977.

Hanratty, Dennis M., and Sandra W. Meditz. *Colombia: A Country Study.* 4th ed. Washington, DC: Federal Research Division, Library of Congress, 1990.

Pearce, Jenny. *Colombia: The Drugs War.* New York: Gloucester Press, 1990.

Posada Carbs, Eduardo. *The Colombian Caribbean: A Regional History, 1870–1950.* New York: Clarendon Press, 1996.

Costa Rica

Biesanz, Richard; Biesanz, Karen Zubris; Biezanz, Mavis Hiltunen. *The Costa Ricans.* New Jersey: Prentice Hall, 1982.

Gall, Timothy L., ed. *Worldmark Encyclopedia of Cultures and Daily Life*. vol. 2: Americas. Detroit: Gale Research, 1998.

Stone, Doris. *Pre-Columbian Man in Costa Rica.* Cambridge: Peabody Museum Press, 1977.

Todorov, Tzvetan. *The Conquest of America.* New York: Harper & Row, 1984.

Cuba

Balfour, Sebastian. *Castro.* New York: Longman, 1990.

Brune, Lester H. *The Cuba-Caribbean Missile Crisis of October 1962.* Claremont, Calif.: Regina Books, 1996.

Gall, Timothy L., ed. *Worldmark Encyclopedia of Cultures and Daily Life*. vol. 2: Americas. Detroit: Gale Research, 1998.

Rudolph, James D., ed. *Cuba: A Country Study*, 3rd ed. Washington, D.C.: U.S. Government Printing Office, 1985.

Wyden, Peter. *Bay of Pigs: The Untold Story.* New York: Simon & Schuster, 1979.

Dominica

Baker, Patrick L. *Centering the Periphery: Chaos, Order, and the Ethnohistory of Dominica.* Montreal: McGill-Queen's University Press, 1994.

Philpott, Don. *Caribbean Sunseekers: Dominica*. Lincolnwood, Ill.: Passport Books, 1996.

Whitford, Gwenith. "Mining on 'Nature Island': the Dominican Government's Resource Extraction Plans Anger Conservationists." *Alternatives Journal,* Winter 1998, vol. 24, no. 1, p. 9+.

Dominican Republic

Cambeira, Alan. *Quisqueya la bella: the Dominican Republic in Historical and Cultural Pperspective.* Armonk, N.Y.: M.E. Sharpe, 1997.

Horowitz, Michael M. *Peoples and Cultures of the Caribbean: An Anthropological Reader.* New York: Natural History Press, 1971.

Logan, Rayford W. *Haiti and the Dominican Republic*, New York: Oxford University Press, 1968.

Plant, Roger. *Sugar and Modern Slavery: Haitian Migrant Labor and the Dominican Republic*. Totowa, N.J.: Biblio Dist., 1986.

Moya Pons, Frank. *The Dominican Republic: A National History.* New Rochelle, N.Y.: Hispaniola Books, 1995.

Ecuador

Bork, Albert William. *Historical Dictionary of Ecuador.* Metuchen, N.J.: Scarecrow Press, 1973.

Hemming, John. *The Conquest of the Incas*. San Diego: Harcourt Brace Jovanovich, 1970.

Rathbone, John Paul. *Ecuador, the Galápagos, and Colombia*. London: Cadogan Books, 1991.

Roos, Wilma, and Omer van Renterghem. *Ecuador in Focus: a Guide to the People, Politics and Culture*. New York: Interlink Books, 1997.

El Salvador

Browning, David, *El Salvador: Landscape and Society.* London: Clarendon Press, 1971.

Flemion, Philip, *Historical Dictionary of El Salvador* Metuchen, N.J.: The Scarecrow Press, 1972.

Haggerty, Richard, *El Salvador: A Country Study.* Washington, D.C.: Department of the Army, 1990.

Grenada

Gall, Timothy L., ed. *Worldmark Encyclopedia of the Nations.* 9th ed. Detroit: Gale Research, 1998.

Gunson, Phil, et. al., eds. *The Dictionary of Contemporary Politics of Central America and the Caribbean.* New York: Simon and Schuster, 1991.

Rouse, Irving. *The Tainos: Rise and Decline of the People Who Greeted Columbus.* New Haven: Yale University Press, 1992.

Weeks, John, and Ferbel, Peter. *Ancient Caribbean.* New York: Garland Publishing, 1994.

Guatemala

Gall, Timothy L., ed. *Worldmark Encyclopedia of the Nations.* 9th ed. Detroit: Gale Research, 1998.

Handy, Jim. *Gift of the Devil: A History of Guatemala*. Toronto: Between the Lines Press, 1984.

Nyrop, Richard, (ed.). *Guatemala: A Country Study.* (Washington DC: Department of the Army, 1983.

Schele, Linda. *A Forest of Kings: the Untold Story of the Ancient Maya*. New York: Morrow, 1990.

South America, *Central America and the Caribbean*, 6th ed. London: Europa Publication, 1997.

Guyana

Daly, Vere T. *A Short History of the Guyanese People.* London: Macmillan Education, 1975.

Gall, Timothy L., *Worldmark Encyclopedia of the Nations.* 9th ed. Detroit: Gale Research, 1998.

Mecklenburg, Kurt K. *Guyana Gold.* Carlton Press, 1990.

Singh, Chaitram. Guyana: *Politics in a Plantation Society.* New York: Praeger Publishers, 1988.

Haiti

Abbott, Elizabeth. *Haiti: The Duvaliers and their Legacy.* New York: Simon and Schuster. 1991.

Aristide, Jean-Bertrand. *Dignity.* Charlottesville and London: University Press of Virginia, 1996.

Gall, Timothy L., ed. *Worldmark Encyclopedia of the Nations.* 9th ed. Detroit: Gale Research, 1998.

McFadyen, Deidre; LaRamee, Pierre (editors). *Haiti: Dangerous Crossroads.* Boston: South End Press, 1995.

Honduras

Gall, Timothy L., ed. *Worldmark Encyclopedia of the Nations.* 9th ed. Detroit: Gale Research, 1998.

Schulz, Donald E. and Deborah S. Schulz. *The United States, Honduras and the Crisis in Central America.* Boulder, Colo.: Westview Press, 1994.

Todorov, Tzvetan (translated by Richard Howard) *The Conquest of America.* New York: Harper & Row, 1984.

Jamaica

Bayer, Marcel. *Jamaica: A Guide to the People, Politics, and Culture.* Trans. John Smith. London, Eng.: Latin American Bureau, 1993.

Davis, Stephen. *Reggae Bloodlines: In Search of the Music and Culture of Jamaica.* New York: Da Capo Press, 1992.

Gall, Timothy L., ed. *Worldmark Encyclopedia of the Nations.* 9th ed. Detroit: Gale Research, 1998.

Sherlock, Philip, and Hazel Bennett. *The Story of the Jamaican People.* Princeton, NJ: Markus Wiener Publishers, 1998.

Stone, Carl. *Class, State, and Democracy in Jamaica.* New York: Praeger, 1986.

Mexico

Burke, Michael E. *Mexico: An Illustrated History.* New York: Hippocrene Books, 1999.

Briggs, Donald C. *The Historical Dictionary of Mexico.* Metuchen, N.J.: Scarecrow Press, 1981.

Gall, Timothy L., ed. *Worldmark Encyclopedia of the Nations.* 9th ed. Detroit: Gale Research, 1998.

National Geographic. 190:2 (Aug. 1996) "Emerging Mexico". Entire isssue devoted to Mexico.

Randall, Laura, ed. *Changing Structure of Mexico.* Armonk, N.Y.: M.E. Sharpe, 1966.

Williamson, Edwin. *The Penguin History of Latin America.* New York: Penguin Books, 1992.

Nicaragua

Gall, Timothy L, ed. *Worldmark Encyclopedia of the Nations.* 9th ed. Detroit: Gale Research, 1998.

Kagan, Robert. *A Twilight Struggle: American Power and Nicaragua. 19771990.* New York: Free Press, 1996.

Rudolph, James D., ed. *Nicaragua: A Country Study.* 2nd ed. Washington, D.C.: Government Printing Office, 1994.

Walker, Thomas W., ed. *Revolution & Counterrevolution in Nicaragua.* Boulder, Colo.: Westview Press, 1991.

Panama

Flanagan, E. M. *Battle for Panama: Inside Operation Just Cause.* Washington, D.C.: Brassey's, Inc., 1993.

Guevara Mann, Carlos. *Panamanian Militarism: A Historical Interpretation.* Athens, Oh.: Ohio University Center for International Studies, 1996.

Hedrick, Basil C. and Anne K. *Historical Dictionary of Panama.* Metuchen, N.J.: Scarecrow Press, 1970.

Major, John. *Prize possession: The United States and the Panama Canal, 1903–1979.* New York: Cambridge University Press, 1993.

Noriega, Manuel Antonio. *America's Prisoner: The Memoirs of Manuel Noriega.* 1st ed. New York : Random House, 1997.

Pearcy, Thomas L. *We Answer Only to God: Politics and the Military in Panama, 1903–1947.* Albuquerque: University of New Mexico Press, 1998.

Paraguay

Gall, Timothy L., ed. *Worldmark Encyclopedia of the Nations.* 9th ed. Detroit: Gale Research, 1998.

Paraguay: A Country Study. Washington: U.S. Government Printing Office, 1990.

Wiarda, Howard J. and Harvey F. Kline, eds. *Latin American Politics and Development.* Boulder: Westview Press, 1996.

Peru

Gall, Timothy L., ed. *Worldmark Encyclopedia of the Nations.* 9th ed. Detroit: Gale Research, 1998.

Hemming, John. *The Conquest of the Incas.* New York: Harcourt Brace Jovanovich, 1970.

Holligan de Dmaz-Lmmaco, Jane. Peru: A Guide to the People, Politics and Culture. New York: Interlink Books, 1998.

Hudson, Rex A., ed. *Peru in Pictures.* Minneapolis: Lerner, 1987.

———. *Peru: A Country Study.* 4th ed. Washington, D.C.: Library of Congress, Federal Research Division, 1993.

Strong, Simon. *Shining Path : Terror and Revolution in Peru.* New York : Times Books, 1992.

Puerto Rico

Fernandez, Ronald. *The Disenchanted Island: Puerto Rico and the United States in the Twentieth Century.* 2d ed. Westport, Conn.: Praeger, 1996.

Figueroa, Loida. *History of Puerto Rico*. New York: Anaya, 1974.

Gall, Timothy L., ed. *Worldmark Encyclopedia of Cultures and Daily Life*. vol. 2: Americas. Detroit: Gale Research, 1998.

Morales, Carrion, Arturo. *Puerto Rico: A Political and Cultural History.* New York: Norton, 1983.

Morris, Nancy. *Puerto Rico: Culture, Politics, Indentity.* Westport, Conn.: Praeger, 1995.

St. Kitts and Nevis

Gall, Timothy L., ed. *Worldmark Encyclopedia of the Nations.* 9th ed. Detroit: Gale Research, 1998.

Hamshere, Cyril. *The British in the Caribbean.* Cambridge, MA: Harvard University Press, 1972.

Moll, V.P. *St. Kitts-Nevis.* Santa Barbara, CA: Clio, 1994.

Olwig, karen Fog. *Global Culture, Island Identity: Continuity and Change in the Afro-Caribbean Community of Nevis.* Philadelphia: Harwood, 1993.

St. Lucia

Claypole, William, and Robottom, John, *Caribbean Story, Book Two: The Inheritors.* Essex: Longman, 1986.

Craton, Michael, *Testing the Chains: Resistance to Slavery in the British West Indies.* Ithaca: Cornell University Press, 1982.

Gall, Timothy L., ed. *Worldmark Encyclopedia of the Nations.* 9th ed. Detroit: Gale Research, 1998.

Gunson, Phil, et. al., (eds.), *The Dictionary of Contemporary Politics of Central America and the Caribbean.* New York: Simon and Schuster, 1991.

Weeks, John, and Ferbel, Peter, *Ancient Caribbean.* New York: Garland Publishing, 1994.

St. Vincent and the Grenadines

Bobrow, Jill and Dana Jinkins. *St. Vincent and the Grenadines: Gems of the Caribbean*. Waitsfield, Vt.: Concepts Publishing Inc., 1993.

Hamshere, Cyril. *The British in the Caribbean*. Cambridge, Mass.: Harvard University Press, 1972.

Philpott, Don. *Caribbean Sunseekers: St. Vincent & Grenadines*. Lincolnwood, Ill.: Passport Books, 1996.

Suriname

Chin, Henk E. *Surinam: Politics, Economics, and Society.* New York: F. Pinter, 1987.

Cohen, Robert. *Jews in Another Environment: Surinam in the Second Half of the Eighteenth Century.* New York: E.J. Brill, 1991.

Dew, Edward M. *The Trouble in Suriname, 1975–1993.* Westport, CT: Praeger, 1994.

Hoogbergen, Wim S. M. *The Boni Maroon Wars in Suriname.* New York: Brill, 1990.

Sedoc-Dahlberg, Betty, ed. *The Dutch Caribbean: Prospects for Democracy.* New York: Bordon and Breach, 1990.

Trinidad and Tobago

Bereton, Bridget, *A History of Modern Trinidad* (Portsmouth: Heinemann, 1981).

Black, Jan, et. al., *Area Handbook for Trinidad and Tobago* (Washington D.C.: U.S. Government Printing Office, 1976).

Gunson, Phil, et. al., (eds.), *The Dictionary of Contemporary Politics of Central America and the Caribbean* (NY: Simon and Schuster, 1991).

Reddock, *Women Labour and Politics in Trinidad and Tobago: A History* (London: Zed Books 1994).

Weeks, John, and Ferbel, Peter, *Ancient Caribbean* (NY: Garland Publishing, 1994).

United States

Ayres, Stephen M. *Health Care in the Unites States: The Facts and the Choices.* Chicago: American Library Association, 1996.

Bacchi, Carol Lee. *The Politics of Affirmative Action: 'Women', Equality, and Category Politics.* London: Sage, 1996.

Barone, Michael. *The Almanac of American Politics.* Washington, D.C.: National Journal, 1992.

Bennett, Lerone. *Before the Mayflower: A History of Black America.* 6th ed. New York: Penguin, 1993.

Brinkley, Alan. *American History: A Survey.* 9th ed. New York: McGraw-Hill, 1995.

Brinkley, Douglas. *American Heritage History of the United States.* New York: Viking, 1998.

Carnes, Mark C., ed. *A History of American Life.* New York: Scribner, 1996.

Commager, Henry Steele (ed.). *Documents of American History.* Englewood Cliffs, N.J.: Prentice-Hall, 1988.

Davidson, James West. *Nation of Nations: A Narrative History of the American Republic.* 3rd ed. Boston, MA: McGraw-Hill, 1998.

Davies, Philip John. (ed.) *An American Quarter Century: US Politics from Vietnam to Clinton.* New York: Manchester University Press, 1995.

Donaldson, Gary. *America at War since 1945: Politics and Diplomacy in Korea, Vietnam, and the Gulf War.* Westport, Conn.: Praeger, 1996.

Foner, Eric. *The Story of American Freedom.* New York: Norton, 1998.

Garraty, John Arthur. *The American Nation: A History of the United States.* 9th ed. New York: Longman, 1998.

Goldfield, David, ed. *The American Journey: A History of the United States.* Upper Saddle River, NJ: Prentice Hall, 1998.

Hart, James David, ed.. *Oxford Companion to American Literature.* 6th ed. New York: Oxford University Press, 1995.

Hummel, Jeffrey Rogers. *Emancipating Slaves, Enslaving Free Men: A History of the Civil War.* Chicago: Open Court, 1996.

Jenkins, Philip. *A History of the United States.* New York: St. Martin's Press, 1997.

Kaplan, Edward S. *American Trade Policy, 1923–1995.* Westport, Conn.: Praeger, 1996.

Magill, Frank N., ed. *Great Events from History: North American Series.* rev. ed. Pasadena, CA: Salem Press, 1997.

Martis, Kenneth C. *The Historical Atlas of Political Parties in the United States Congress 1789–1989.* New York: Macmillan, 1989.

McNickle, D'Arcy. *Native American Tribalism: Indian Survivals and Renewals.* New York: Oxford University Press, 1993.

Mudd, Roger. *American Heritage Great Minds of History.* New York: Wiley, 1999.

Nash, Gary B. *The American People: Creating a Nation and a Society.* 4th ed., New York : Longman, 1998.

People Who Shaped the Century. Alexandria, VA: Time-Life Books, 1999.

Robinson, Cedric J. *Black Movements in America.* New York: Routledge, 1997.

Tindall, George Brown. *America: A Narrative History.* 5th ed. New York: Norton, 1999.

Virga, Vincent. *Eyes of the Nation: A Visual History of the United States.* New York: Knopf, 1997.

Woodward, C. Vann, ed. *The Comparative Approach to American History.* New York: Oxford University Press, 1997.

Uruguay

Gall, Timothy L., ed. *Worldmark Encyclopedia of the Nations.* 9th ed. Detroit: Gale Research, 1998.

Hudson, Rex A., and Sandra Meditz. *Uruguay: A Country Study.* Washington, DC : Federal Research Division, Library of Congress, 1992.

Weinstein, Martin. *Uruguay, Democracy at the Crossroads. Nations of Contemporary Latin America.* Boulder, Colo.: Westview Press, 1988.

Zlotchew, Clark M. Paul David Seldis, ed. *Voices of the River Plate : Interviews with Writers of Argentina and Uruguay.* San Bernadino, CA: Borgo Press, 1993.

Venezuela

Fox, Geoffrey. *The Land of People of Venezuela.* New York: HarperCollins, 1991.

Gall, Timothy L., ed. *Worldmark Encyclopedia of Cultures and Daily Life.* vol. 2: Americas. Detroit: Gale Research, 1998.

Haggerty, Richard A., ed. *Venezuela: A Country Study,* 4th ed. Washington, D.C.: Library of Congress, 1993.

Hellinger, Daniel. *Venezuela: Tarnished Democracy.* Boulder, Colo.: Westview Press, 1991.

Asia

Afghanistan

Adamec, Ludwig W. *Historical Dictionary of Afghanistan.* Metuchen, NJ: Scarecrow Press, 1991.

Gall, Timothy L., ed. *Worldmark Encyclopedia of the Nations.* 9th ed. Detroit: Gale Research, 1998.

Giradet, Edward. *Afghanistan: The Soviet War.* London: Croom Helm, 1985

Nyrop, Richard F. and Donald M. Seekins, eds. *Afghanistan: A Country Study.* 5th ed. Washington, DC: U.S. Government Printing Office, 1986.

Australia

Bassett, Jan. *The Oxford Illustrated Dictionary of Australian History.* New York: Oxford University Press, 1993.

Bolton, Geoffrey, ed. *The Oxford History of Australia.* New York: Oxford University Press, 1986–90.

Gunther, John. *John Gunther's Inside Australia.* New York: Harper & Row, 1972.

Heathcote, R.L. *Australia.* London: Longman, Scientific & Technical, 1994.

Rickard, John. *Australia, A Cultural History.* London: Longman, 1996.

Azerbaijan

Altstadt, Audrey. *The Azerbaijani Turks.* Stanford, Calif.: Hoover Institution Press, 1992.

Gall, Timothy L., ed. *Worldmark Encyclopedia of the Nations.* 9th ed. Detroit: Gale Research, 1998.

Nichol, James. "Azerbaijan." in *Armenia, Azerbaijan and Georgia.* Area Handbook Series, Washington, DC: Government Printing Office, 1995, pp. 81148.

Bahrain

Crawford, Harriet. *Dilmun and its Gulf Neighbors.* Cambridge: Cambridge University Press, 1998.

Gall, Timothy L., ed. *Worldmark Encyclopedia of the Nations.* 9th ed. Detroit: Gale Research, 1998.

Nugent, Jeffrey B., and Theodore Thomas, eds. *Bahrain and the Gulf: Past Perspectives and Alternate Futures.* New York: St. Martin's Press, 1985.

Robison, Gordon. *Arab Gulf States.* Hawthorn, Aus.: Lonely Planet Publications, 1996.

Bangladesh

Baxter, Craig. *Bangladesh: A New Nation in an Old Setting.* Boulder and London: Westview Press, 1984.

Bigelow, Elaine. *Bangladesh: the Guide.* Dhaka: AB Publishers, 1995.

O'Donnell, Charles Peter. *Bangladesh: Biography of a Muslim Nation.* Boulder and London: Westview Press, 1984.

Newton, Alex. *Bangladesh: a Lonely Planet Travel Survival Kit.* Hawthorne, Victoria, Australia: Lonely Planet, 1996.

Republic of Bangladesh. U.S. Department of State Background Notes, November 1997.

Bhutan

Aris, Michael. *Bhutan: The Early History of a Himalayan Kingdom.* Warminster, England: Aris & Phillips, 1979.

Crossette, Barbara. *So Close to Heaven: The Vanishing Buddhist Kingdoms of the Himalayas.* New York: A. A. Knopf, 1995.

Matles, Andrea, ed. *Nepal and Bhutan: Country Studies.* Washington, DC: Federal Research Division, Library of Congress, 1993.

Pommaret-Imaeda, Françoise. Lincolnwood, Ill.: Passport Books, 1991.

Strydonck, Guy van. *Bhutan: A Kingdom in the Eastern Himalayas.* Boston: Shambala, 1985.

Brunei Darussalam

Bartholomew, James. *The Richest Man in the World: The Sultan of Brunei.* London: Penguin Group, 1990.

Major, John S. *The Land and People of Malaysia & Brunei.* New York: HarperCollins 1991.

Pigafetta, Antonio. *Magellan's Voyage.* Trans. and ed. by R. A. Skelton. New Haven: Yale University Press, 1969.

Singh, D.S. Ranjit. *Brunei 1839–1983: The Problems of Poltical Survival.* London: Oxford University Press, 1984.

Vreeland, N. et al. *Malaysia: A Country Study.* Area Handbook Series. Fourth edition. Washington, D.C.: Department of the Army, 1984.

Cambodia

Barron, John and Anthony Paul. *Murder of a Gentle Land. The Untold Story of Communist Genocide in Cambodia*. New York: Reader's Digest Press. 1977.

Becker, Elizabeth. *When the War Was Over: The Voices of Cambodia's Revolution and Its People*. New York: Simon and Schuster. 1986.

Chandler, David P. *A History of Cambodia*. Boulder, Colo.: Westview Press. 1983.

Gall, Timothy L., ed. *Worldmark Encyclopedia of Cultures and Daily Life*. vol. 3: Asia. Detroit: Gale Research, 1998.

Ponchaud, Francois. *Cambodia Year Zero*. London: Allen Lane. 1978.

China

Bailey, Paul. *China in the Twentieth Century.* New York: B. Blackwell, 1988.

Cotterell, Arthur. *China: A Cultural History.* New York: New American Library, 1988.

Ebrey, Patricia B. *The Cambridge Illustrated History of China.* New York: Cambridge University Press, 1996.

Hsu, Immanuel Chung-yueh. *The Rise of Modern China.* 5th ed. New York: Oxford University Press, 1995.

Worden, Robert L., Andrea Matles Savada, and Ronald E. Dolan, eds. *China, a Country Study.* 4th ed. Washington, D.C.: Library of Congress, 1988.

Cyprus (Greek and Turkish zones)

Crawshaw, Nancy. *The Cyprus Revolt: The Origins, Development, and Aftermath of an International Dispute.* Winchester, Mass.: Allen & Unwin, 1978.

Gall, Timothy L., ed. *Worldmark Encyclopedia of Cultures and Daily Life*. vol. 3: Asia. Detroit: Gale Research, 1998.

Salem, Norma, ed. *Cyprus: A Regional Conflict and its Resolution.* New York: St. Martin's Press, 1992.

Solsten, Eric, ed. *Cyprus, a Country Study.* 4th ed. Washington, D.C. Government Printing Office, 1993.

Tatton-Brown, Veronica. *Ancient Cyprus.* Cambridge, Mass.: Harvard University Press, 1988.

Federated States of Micronesia

Denoon, Donald, ed. *The Cambridge History of the Pacific Islanders.* Cambridge: Cambridge University Press, 1997.

Gall, Timothy L., ed. *Worldmark Encyclopedia of the Nations.* 9th ed. Detroit: Gale Research, 1998.

Levesque, Rodrigu, ed. *History of Micronesia.* Honolulu: University of Hawaii Press, 1994.

Spate, Oskar. *The Pacific Since Magellan, vol. 1: The Spanish Lake.* Canberra: Australian National University Press, 1979.

Fiji

Gall, Timothy L., ed. *Worldmark Encyclopedia of the Nations.* 9th ed. Detroit: Gale Research, 1998.

Ogden, Michael R. "Republic of Fiji," *World Encyclopedia of Political Systems,* 3rd edition, New York: Facts on File, 1999.

Scarr, Deryck. *Fiji: A Short History.* Honolulu: The Institute for Polynesian Studies, 1984.

"Vaughn, Roger. "The Two Worlds of Fiji," *National Geographic,* October l995, vol. 88, no. 4, p. 114.

India

Heitzen, James and Robert L. Worden, eds. *India: A Country Study.* Washington, DC: Federal Research Division, Library of Congress, 1996.

Robinson, Francis, ed. *The Cambridge Encyclopedia of India, Pakistan, Bangladesh, Sri Lanka, Nepal, Bhutan and the Maldives.* Cambridge: Cambridge University Press. 1989.

Schwartzberg, Joseph E., ed. *A Historical Atlas of South Asia.* 2nd impression. New York and Oxford: Oxford University Press, 1992.

Wolpert, Stanley. *India.* Berkeley: University of California Press, 1991.

Indonesia

Bellwood, Peter S. *Prehistory of the Indo-Malaysian Archipelago.* Honolulu, Hawaii: University of Hawaii Press, 1997.

Broughton, Simon, ed. *World Music: The Rough Guide.* London: The Rough Guides Ltd., 1994.

Cribb, Robert. *Historical Dictionary of Indonesia.* Metuchen, N.J.: Scarecrow Press, 1992.

Frederick, William H., ed. *Indonesia: A Country Study.* Washington, D.C.: Library of Congress, 1993.

Schwarz, Adam. *A Nation in Waiting: Indonesia in the 1990s.* Boulder, Colo.: Westview Press, 1994.

Iran

Albert, David H. *Tell the American People: Perspectives on the Iranian Revolution.* Philadelphia: Movement for a New Society, 1980.

Bacharach, Jere L. *A Near East Studies Handbook, 570–1974.* Seattle: University of Washington Press, 1974.
Famighetti, Robert (ed.). *The World Almanac and Book of Facts*. New York: St. Martin's Press, 1998.
Nyrop, Richard F. (ed.). *Area Handbook of Iran: A Country Study.* Washington, D.C.: The American University, 1978.
Salinger, Pierre. *America Held Hostage: The Secret Negotiations*. Garden City, N.Y.: Doubleday & Company, 1981.

Iraq

Bulloch, John. *Saddam's War: The Origins of the Kuwait Conflict and the International Response.* Boston: Faber and Faber, 1991.
Chaliand, Gerard, ed. *A People without a Country: The Kurds and Kurdistan*. New York: Olive Branch Press, 1993.
Mansfield, Peter. *A History of the Middle East*. New York: Viking, 1991.
Metz, Helen Chapin, ed. *Iraq: A Country Study.* 4th ed. Washington, D.C.: Library of Congress, Federal Research Division, 1990.
Simons, Geoff L. *Iraq: From Sumer to Saddam*. New York: St. Martin's Press, 1994.

Israel

Blumberg, Arnold. *The History of Israel.* Westport, Conn.: Greenwood Press, 1998.
———. *Zion Before Zionism.* Syracuse, N.Y.: Syracuse University Press, 1985.
Grant, Michael. *The History of Ancient Israel.* New York: Charles Scribner's Sons, 1984.
Metz, Helen Chapin. *Israel: A Country Study.* Washington, DC : Library of Congress, 1990
Sachar, Abram Leon. *A History of the Jews.* New York: Alfred A. Knopf, 1955.
Shanks, Hershel. *Ancient Israel.* Englewood Cliffs, N.J.: Prentice-Hall, 1988.

Japan

Demente, Boye Lafayette. *Japan Encyclopedia.* Lincolnwood, Ill.: NTC Publishing, 1995.
Dolan, Ronald E., and Robert L. Dolan, eds. *Japan, a Country Study.* 5th ed. Washington, D.C.: Library of Congress, 1992.
Perkins, Dorothy. *Enclyclopedia of Japan: Japanese History and Culture, from Abacus to Zori.* New York: Facts on File, 1991.
Richardson, Bradley M. *Japanese Democracy: Power, Coordination, and Performance.* New Haven, Conn.: Yale University Press, 1997.
Thomas, J. E. *Modern Japan: A Social History Since 1868.* New York: Longman, 1996.

Jordan

Ali, Wijdan. *Modern Islamic Art: Development and Continuity.* University Press of Florida, Gainesville, Fla. 1997.
Contreras, Joseph and Christopher Dickey. "The Day After: King Hussein's Second Bout of Cancer Raises Questions About Jordan's Political Future." *Newsweek*, Vol. 132, No. 6, p. 38, August 10, 1998.
Shahin, Mariam. "Cracks Become Chasms," *The Middle East.* December 1997, No. 273, p. 13-14.

Kazakstan

Conflict in the Soviet Union: The Untold Story of the Clashes in Kazakhstan. New York: Human Rights Watch, 1990.
Edwards-Jones, Imogen. *The Taming of Eagles: Exploring the New Russia.* London: Weidenfeld & Nicolson, 1992.
Kalyuzhnova, Yelena. *Kazakstani Economy: Independence and Transition.* New York: St. Martin's Press, 1998.
Olcott, Martha Brill. *The Kazakhs.* Stanford, CA: Hoover Institution Press, Stanford University, 1987.
World Bank. *Kazakhstan: The Transition to a Market Economy.* Washington, DC: World Bank, 1993.

Kiribati

"Bones Found in '40 May Have Been Hers," *Honolulu Star-Bulletin,* December 3, 1998.
Krauss, Bob. "Heroes Heed Call From Sea," *Honolulu Advertiser*, March 10, 1999.
MacDonald, Barrie. *Cinderellas of the Empire: Towards a History of Kiribati and Tuvalu.* Canberra, Australia: Australian National University Press, 1982.
"New Meaning to the Term Down Under. The Tiny Islands of Kiribati, Tuvalu and Nauru Are Pressuring Australia to Reduce Greenhouse Gas Emissions.)" *The Economist (US),* September 27, 1997, vol. 344, no. 8036, p. 41.
"South Pacific Group Cuts French Ties," *Honolulu Star-Bulletin*, October 3, 1995.
Thompson, Rod. "Hilo Plans Gift for Christmas Isle Neighbors, " *Honolulu Star-Bulletin,* December 18, 1998.

Korea, Democratic People's Republic of

Gills, Barry K. *Korea Versus Korea: A Case of Contested Legitimacy.* New York: Routledge, Inc., 1996.
Oliver, Robert Tarbell. *A History of the Korean People in Modern Times: 1800 to the Present.* Newark,: University of Delaware Press, 1993.
Smith, Hazel. *North Korea in the New World Order.* New York: St. Martin's Press, 1996.
Soh, Chung Hee. *Women in Korean Politics.* 2nd ed. Boulder, Colo.: Westview Press, 1993.
Tennant, Roger. *A History of Korea.* London: Kegan Paul International, 1996.

Korea, Republic of (ROK)

Gills, Barry K. *Korea Versus Korea: A Case of Contested Legitimacy.* New York: Routledge, Inc., 1996.
Lone, Stewart. *Korea Since 1850.* New York: St. Martin's Press, 1993.
Oberdorfer, Don. *The Two Koreas: A Contemporary History.* Reading, Mass.: Addison-Wesley Pub. Co., 1997.
Oliver, Robert Tarbell. *A History of the Korean People in Modern Times: 1800 to the Present.* Newark: University of Delaware Press, 1993.

Tennant, Roger. *A History of Korea.* London: Kegan Paul International, 1996.

Kuwait

Ali, Wijdan. *Modern Islamic Art: Development and Continuity.* Gainesville, FL:

Anscombe, Frederick F. *The Ottoman Gulf: The Creation of Kuwait, Saudi Arabia and Qatar.* New York: Columbia University Press, 1997.

Cordesman, Anthony H. *Kuwait: Recovery and Security after the Gulf War.* Boulder, Colo.: Westview Press, 1997.

Crystal, Jill. *Oil and Politics in the Gulf: Rulers and Merchants in Kuwait and Qatar.* New York: Cambridge University Press, 1990.

Khadduri, Majid. *War in the Gulf, 1990–1991: The Iraq-Kuwait Conflict and its Implications.* New York: Oxford University Press, 1997.

Robison, Gordon. *Arab Gulf States.* Hawthorn, Aus.: Lonely Planet Publications, 1996.

Kyrgyzstan

Attokurov, S. *Kïrgïz Sanjïrasï.* Bishkek: Kyrgyzstan, 1995.

Bennigsen, Alexandre & S. Enders Wimbush. *Muslims of the Soviet Empire: A Guide.* Bloomington & Indianapolis: Indiana Universiity Press, 1986.

Huskey, Eugene. "Kyrgyzstan: The politics of demographic and economic frustration." in: *New States, New Politics: Building the Post-Soviet Nations*, edited by Ian Bremmer and Ray Taras, Cambridge & New York: Cambridge University Press, 1997, 655-680.

Krader, Lawrence. *The Peoples of Central Asia.* Bloomington, Indiana: Uralic and Altaic Series, vol. 26, 1966.

Olcott, Martha B. "Kyrgyzstan." in: *Kazakstan, Kyrgyzstan, Tajikistan, Turkmenistan and Uzbekistan*Area Handbook Series, Washington D.C.: Government Printing Office, 1997, 99-193.

Laos

Cummings, Joe. *A Golden Souvenir of Laos.* New York: Asia Books. 1996.

Kremmer, Chistopher. *Stalking the Elephant King. In Search of Laos.* Honolulu, University of Hawai'i Press. 1997.

Scott, Joanna C. *Indochina's Refugees: Oral Histories from Laos, Cambodia, and Vietnam.* Jeferson, NC: McFarland. 1989.

Stieglitz, P. *In a Little Kingdom.* New York: M. E. Sharpe. 1990.

Stuart-Fox, Martin. *A History of Laos.* New York: Cambridge University Press. 1997.

Lebanon

Abul-Husn, Latif. *The Lebanese Conflict: Looking Inward.* Boulder, Colo.: Lynne Rienner Publishers, 1998.

Bleaney. C.H. *Lebanon: Revised Edition. World Bibliographical Series,* Volume 2, Oxford, Eng.: Clio Press, Ltd., 1991.

King-Irani, Laurie. "War Gods Roar Again, Appear Unstoppable: Jet Streaks So Close She Could See Pilot," *National Catholic Reporter.* Vol. 34, No. 17, p. 9. February 27, 1998.

Norton, Augustus Richard. "Hizballah: From Radicalism to Pragmatism?" *Middle East Policy.* Vol. 5, No. 4, pp. 174-186. January 1998.

Tuttle, Robert. "Americans return to AUB," *The Middle East.* No. 272, p. 42. November 1997.

Malaysia

Bellwood, Peter S. *Prehistory of the Indo-Malaysian Archipelago.* Honolulu,: University of Hawaii Press, 1997.

Bevis, William W. *Borneo Log: The Struggle for Sarawak's Forests.* Seattle,: University of Washington Press, 1995.

Broughton, Simon, ed.*World Music: The Rough Guide.* London: The Rough Guides Ltd., 1994.

Kahn, Joel S. and Loh Kok Wah, Francis, eds. *Fragmented Vision: Culture and Politics in Contemporary Malaysia.* Honolulu,: University of Hawaii Press, 1992.

Kaur, Amarjit. *Historical Dictionary of Malaysia.* Metuchen, NJ: Scarecrow Press, 1993.

Maldives

Ellis, Kirsten. *The Maldives.* Hong Kong: Odyssey, 1993.

Republic of Maldives, Office for Women's Affairs. *Status of Women, Maldives.* Bangkok: UNESCO Principal Regional Office for Asia and the Pacific, 1989.

Marshall Islands

Johnson, Giff. *Collision Course at Kwajalein: Marshall Islanders in the Shadow of the Bomb.* Honolulu: Pacific Concerns Resource Center, 1984.

Langley, Jonathan, and Wanda Langley. "The Marshall Islands." *Skipping Stones,* Winter 1995, vol. 7, no. 5, p. 17+.

Levesque, Rodrigu, ed. *History of Micronesia.* Honolulu: University of Hawaii Press, 1994.

Scarr, Deryck. *History of the Pacific Islands: Kingdom of the Reefs.* Sydney: Macmillan Australia, 1990.

Mongolia

Bergholz, Fred W. *The Partition of the Steppe: The Struggle of the Russians, Manchus, and the Zunghar Mongols for Empire in Central Asia, 1619–1758.* New York: Peter Lang Publishing, Inc., 1993.

Bruun, Ole and Ole Odgaard, eds. *Mongolia in Transition.* Richmond, England: Curzon Press Ltd., 1996.

de Hartog, Leo. *Russia and the Mongol Yoke: The History of the Russian Principalities and the Golden Horde, 1221–1502.* London: British Academic Press, 1996.

Greenway, Paul. *Mongolia.* Hawthorn, Australia: Lonely Planet Publications, 1997.

Sanders, Alan J. K. *Historical Dictionary of Mongolia.* Lanham, Md: Scarecrow Press, Inc., 1996.

Myanmar (Burma)

Diran, Richard K. *The Vanishing Tribes of Burma*. New York: Amphoto Art, 1997.

Parenteau, John. *Prisoner for Peace: Aung San Suu Kyi and Burma's Struggle for Democracy*. Greensboro, N.C. : Morgan Reynolds, 1994.

Renard, Ronald D. *The Burmese Connection: Illegal Drugs and the Making of the Golden Triangle*. Boulder, Colo.: L. Rienner Publishers, 1996.

Rotberg, Robert I., ed. *Burma: Prospects for a Democratic Future*. Brookings Institute Press, 1998.

Silverstein, Josef. *The Political Legacy of Aung San*. Ithaca, N.Y.: Cornell University Press, 1993.

Nauru

Bunge, Frederica and Melinda Cooke, eds. *Oceania: a regional study*. Washington, D.C.: U.S. Government Printing Office, 1984.

Hanlon, David. *Remaking Micronesia*. Honolulu: University of Hawai'i Press, 1998.

McKnight, Tom. *Oceania: the geography of Australia, New Zealand, and the Pacific Islands*. Englewood Cliffs: Prentice Hall, 1995.

"New Meaning to the Term Down Under. The Tiny Islands of Kiribati, Tuvalu and Nauru Are Pressuring Australia to Reduce Greenhouse Gas Emissions.)" *The Economist (US)*, September 27, 1997, vol. 344, no. 8036, p. 41.

Viviani, Nancy. *Nauru: phosphate and political progress*. Honolulu: University of Hawaii Press, 1970.

Nepal

Bista, Dor Bahadur. *People of Nepal*. Kathmandu: Ratna Pustak Bhandar, 1987.

Chauhan, R. S. *Society and State Building in Nepal: From Ancient Times to Mid-Twentieth Century*. New Delhi: Sterling, 1989.

Matles, Andrea, ed. *Nepal and Bhutan: Country Studies*. Washington, D.C.: Federal Research Division, Library of Congress, 1993.

Sanday, John. *The Kathmandu Valley: Jewel of the Kingdom of Nepal*. Lincolnwood, IL.: Passport Books, 1995.

Seddon, David. *Nepal, a State of Poverty*. New Delhi: Vikas, 1987.

New Zealand

Belich, James. *Making Peoples: A History of the New Zealanders from Polynesian Settlement to the End of the Nineteenth Century*. Honolulu: University of Hawaii Press, 1996.

Mascarenhas, R.C. *Government and the Economy in Australia and New Zealand: The Politics of Economic Policy Making*. San Francisco: Austin & Winfield, 1996.

McKinnon, Malcolm. *Independence and Foreign Policy: New Zealand in the World since 1935*. Auckland: Oxford University Press, 1993.

Oddie, Graham, and Roy W. Perrett, ed. *Justice, Ethics, and New Zealand Society*. New York: Oxford University Press, 1992.

Rice, Geoffrey W., ed. *The Oxford History of New Zealand*. 2nd ed. New York: Oxford University Press, 1992.

Oman

Arab Gulf States: A Travel Survival Kit. 2d edition. Melbourne Aus.: Lonely Planet Publications, 1996.

Miller, Judith. "Creating Modern Oman: An Interview With Sultan Qabus." *Foreign Affairs*, May–June 1997, Vol. 76, No. 3, pp. 1318.

Molavi, Afshin. "Oman's Economy: Back on Track." *Middle East Policy*, January 1998, Vol. 5, No. 4, pp. 110.

Osborne, Christine. "Omani Forts win Heritage Award." *The Middle East*, November 1995, No. 250, pp. 4344.

Riphenburg, Carol. J. *Oman: Political Development in a Changing World*. Westport, CT: Praeger Publishers, 1998.

Pakistan

Blood, Peter R., ed. *Pakistan, a Country Study*. 6th ed. Washington, D.C.: Federal Research Division, Library of Congress, 1995.

Burki, Shahid Javed. *Historical Dictionary of Pakistan*. Metuchen, NJ: Scarecrow Press, 1991.

Mahmud, S. F. *A Concise History of Indo-Pakistan*. Karachi: Oxford University Press, 1988.

Schwartzberg, Joseph E., ed. *A Historical Atlas of South Asia*. 2nd impression. Oxford and New York: Oxford University Press, 1992.

Taylor, David (revised by Asad Sayeed). "Pakistan: Economy," in *The Far East and Australasia 1997*. London: Europa Publications, 1996, pp. 873-79.

Palau

Hanlon, David. *Remaking Micronesia*. Honolulu: University of Hawai'i Press, 1996.

Hijikata, Hisakatsu. *Society and Life in Palau*. Tokyo: Sasakawa Peace Foundation, 1993.

Liebowitz, Arnold. *Embattled Island: Palau's Struggle for Independence*. Westport, CT: Praeger, 1996.

Morgan, William. *Prehistoric Architecture in Micronesia*. Austin: The University of Texas Press, 1988.

Roff, Sue Rabbitt. *Overreaching in Paradise: United States Policy in Palau Since 1945*. Juneau, Alaska: Denali Press, 1991.

Papua New Guinea

Campbell, I.C. *A History of the Pacific Islands*. Berkeley: University of California Press, 1989.

"Death of a Peacemaker." *The Economist*, October 19, 1996, vol. 341, no. 7988, p. 41.

Grattan, C. Hartley. *The Southwest Pacific since 1900*. Ann Arbor: University of Michigan Press, 1963.

Sinclair, James. *Papua New Guinea: the First 100 Years*. Bathurst: Robert Brown and Associates, 1985.

Spriggs, Matthew. *The Island Melanesians.* Cambridge, Mass.: Blackwell Publishers, 1997.

Philippines

Brands, H.W. *Bound to Empire: The United States and the Philippines.* New York: Oxford Univ. Press, 1992.

Dolan, Ronald E., ed. *Philippines: A Country Study.* 4th ed. Washington, D.C.: Library of Congress, 1993.

Karnow, Stanley. *In Our Image: America's Empire in the Philippines.* New York: Random House, 1989.

Steinberg, David Joel. *The Philippines, a Singular and a Plural Place.* 3rd ed. Boulder, Colo.: Westview, 1994.

Thompson, W. Scott. *The Philippines in Crisis: Development and Security in the Aquino Era 198692.* New York: St. Martin's 1992.

Qatar

Abu Saud, Abeer. *Qatari Women Past and Present.* Essex, Eng.: Longman Group Limited, 1984.

Anscombe, Frederick F. *The Ottoman Gulf: The Creation of Kuwait, Sa'udi Arabia, and Qatar.* New York: Columbia Press, 1997.

Crystal, Jill. *Oil and Politics in the Gulf: Rulers and Merchants in Kuwait and Qatar.* New York: Cambridge University Press, 1995.

Samoa

Fialka, John J. "From Dots in the Pacific, Envoys Bring Fear, Fury to Global-Warming Talks," *Wall Street Journal*, September 31, 1997, A24.

Hallowell, Christopher. "Rainforest Pharmacist," *Audubon* 101:1 (January 1999), 28.

Holmes, Lowell D., and Ellen Rhoads Holmes. *Samoan Village Then and Now.* Orlando, Fla.: Holt, Rinehart and Winston Inc., 1974, repr. 1992.

Meleisea, Malama. *Change and Adaptation in Western Samoa.* Christchurch, New Zealand: MacMillan Brown Centre for Pacific Studies, 1992.

Samoa: A Travel Survival Kit. Sydney, Aus.: Lonely Planet Publications, 1996.

Sa'udi Arabia

Abir, Mordechai. *Saudi Arabia: Government, Society, and the Gulf Crisis.* London and New York: Routledge, 1993.

Caesar, Judith. *Crossing Borders: An American Woman in the Middle East.* Syracuse, NY: Syracuse University Press, 1997.

Long, David E. *The Kingdom of Saudi Arabia.* Gainesville, Fla.: University Press of Florida, 1997.

Peterson, J. J. *Historical Dictionary of Saudi Arabia.* Metuchen, N.J.: Scarecrow Press, 1993.

Wilson, Peter W. and Douglas F. Graham. *Saudi Arabia: The Coming Storm.* Armonk, N.Y.: M. E. Sharpe, 1994.

Singapore

Chew, Ernest and Edwin Chew, ed. *A History of Singapore.* New York: Oxford University Press, 1991.

Chiu, Stephen Wing-Kai. *City States in the Global Economy: Industrial Restructuring in Hong Kong and Singapore.* Boulder, Colo.: Westview Press, 1997.

Lee, W.O. *Social Change and Educational Problems in Japan, Singapore, and Hong Kong.* New York: St. Martin's Press, 1991.

LePoer, Barbara Leitch, ed. *Singapore: A Country Study.* 2nd ed. Washington, D.C.: Library of Congress, 1991.

Trocki, Carl A. *Opium and Empire: Chinese Society in Colonial Singapore.* Ithaca, N.Y.: Cornell University Press, 1990.

Solomon Islands

Bennett, Judith. *Wealth of the Solomons: A History of a Pacific Archipelago, 1800–1978.* Honolulu: University of Hawai'i Press, 1987.

Denoon, Donald, ed. *The Cambridge History of the Pacific Islanders.* Cambridge: Cambridge University Press, 1997.

White, Geoffrey and Lindstrom, Lamont, eds. *The Pacific Theater: island representations of World War II.* Honolulu: University of Hawai'i Press, 1989.

Sri Lanka

Anderson, John Gottberg, and Ravindral Anthonis, editors. *Sri Lanka.* Hong Kong: Apa Productions, 1993.

Baker, Victoria J. *A Sinhalese Village in Sri Lanka: Coping with Uncertainty.* Ft. Worth: Harcourt Brace College Publishers, 1998.

Robinson, Francis, editor. *The Cambridge Encyclopedia of Pakistan, Bangladesh, Sri Lanka, Nepal, Bhutan and the Maldives.* Cambridge: Cambridge University Press., 1989.

Ross, Russell R., et al. *Sri Lanka: A Country Study* (Area Handbook Series). Washington, D.C.: Library of Congress, 1990.

Vesilind, Pritt J. "Sri Lanka." *National Geographic,* January 1997, vol. 191, no. 1, pp. 110+.

Syria

Dourian, Kate. "City of Apamea, Once Lost in Sand, Partially Restored." *Washington Times.* September 3, 1994.

Katler, Johannes. *The Arts and Crafts of Syria.* Thames and Hudson: London, 1992.

Kayal, Michele. "Ruins to Riches." *Washington Post.* February 13, 1994.

LaFranchl, Howard. "Ancient Syria's History Rivals That of Egypt, Mesopotamia." *The Christian Science Monitor.* February 17, 1994.

Parmelee, Jennifer. "Tracking Agatha Christie." *The Washington Post.* August 25, 1991.

"Syria." *Background Notes.* Washington, D.C.: Central Intelligence Agency, November 1994.

Syria: A Country Report. Washinton, D.C.: Library of Congress Research Division, 1999.

Taiwan

Hood, Steven J. *The Kuomintang and the Democratization of Taiwan*. Boulder, Colo.: Westview, 1997.

Long, Simon. *Taiwan: China's Last Frontier*. New York: St. Martin's Press, 1990.

Marsh, Robert. *The Great Transformation: Social Change in Taipei, Taiwan Since the 1960s*. Armonk, N.Y.: M.E. Sharpe, 1996.

Shepherd, John Robert. *Statecraft and Political Economy on the Taiwan Frontier, 1600–1800*. Stanford, Calif.: Stanford University Press, 1993.

Tajikistan

Rakhimov, Rashid, et al. *Republic of Tajikistan: Human Development Report 1995,* Istanbul, 1995.

Rashid, Ahmed. *The Resurgence of Central Asia or Nationalism?,* Zed Books, 1995.

Thailand

Dixon, C. J. *The Thai Economy: Uneven Development and Internationalisation*. London: Routledge, 1999.

Pattison, Gavin and John Villiers. *Thailand*. New York: Norton, 1997.

West, Richard. *Thailand, the Last Domino: Cultural and Political Travels*. London: Michael Joseph, 1991.

Tonga

Aswani, Shankae and Michael Graves. "The Tongan Maritime Expansion." *Asian -Perspectives* 1998. 37: 135201.

———. *Island Kingdom: Tonga Ancient and Modern*. Christchurch, NZ: Canterbury University Press, 1992.

Denoon, Donald, ed. *The Cambridge History of the Pacific Islanders*. Cambridge: Cambridge University Press, 1997.

Perminow, Arne. *The Long Way Home: Dilemmas of Everyday Life in a Tongan Village*. Oslo: Scandinavian University Press, 1993.

Turkmenistan

Edwards-Jones, Imogen. *The Taming of Eagles: Exploring the New Russia*. London: Weidengeld and Nicolson, 1993.

Hunter, Shireen T. *Central Asia Since Independence*. Westport, CN: Praeger, 1996.

Kazakhstan, Kyrgyzstan, Tajikistan, Turkmenistan, and Uzbekistan, country studies. Washington: US Government Printing Office, 1997.

Maslow, Jonathan Evan. *Sacred Horses: The Memoirs of a Turkmen Cowboy*. New York: Random House, 1994.

Olcott, Martha Brill. *Central Asia's New States: Independence, Foreign Policy, and Regional Security*. Washington: United States Institute of Peace Press, 1996.

Tuvalu

Adams, Wanda. "'Double Ghosts' remembers early traders," *Honolulu Advertiser*, March 29, 1998.

Chappell, David. *Double Ghosts: Oceanic Voyagers on Euroamerican Ships*. Armonk, N.Y.: M.E. Sharpe Press, 1997.

Friendship and Territorial Sovereignty: Treaty Between the United States of America and Tuvalu, signed at Funafuti February 7, 1979. Washington, D.C.: U.S. Government Printing Office, 1985.

"New Meaning to the Term Down Under. The Tiny Islands of Kiribati, Tuvalu and Nauru Are Pressuring Australia to Reduce Greenhouse Gas Emissions.)" *The Economist (US),* September 27, 1997, vol. 344, no. 8036, p. 41.

Kristoff, Nicholas D. "In Pacific, Growing Fear of Paradise Engulfed," *New York Times*, March 2, 1997.

United Arab Emirates

Ali Rashid, Noor. *The UAE Visions of Change*. Dubai: Motivate Publishing, 1997.

Forman, Werner. *Phoenix Rising: the United Arab Emirates, Past, Present & Future*. London: Harvill Press, 1996.

Peck, Malcolm C. *The United Arab Emirates: A Venture in Unity*. Boulder, Colorado: Westview Press, 1986.

Taryam, Abdullah Omran. *The Establishment of the United Arab Emirates 1950–85*. London: Croom-Helm, 1987.

Zahlan, Rosemarie Said. *The Making of the Modern Gulf States*. London: Unwin Hyman Ltd., 1989.

Uzbekistan

Alworth, Edward A. *The Modern Uzbeks: From the Fourteenth Century to the Present*. Stanford, Calif.: Hoover Institution Press, 1990.

Alworth, Edward A., ed. *Central Asia: 130 Years of Russian Dominance, A Historical Overview*. Durham, N.C.: Duke University Press, 1994.

MacLeod, Calum and Bradley Mayhew. *Uzbekistan*. Lincolnwood, Ill.: Passport Books, 1997.

Rashid, Ahmed. *The Resurgence of Central Asia: Islam or Nationalism?* Atlantic Highlands, N.J.: Zed Books, 1994.

Undeland, Charles and Nicholas Platt. *The Central Asian Republics: Fragments of Empire*. New York: The Asia Society, 1994.

Vanuatu

Allen, Michael, ed. *Vanuatu: Politics, Economics, and Ritual in Island Melanesia*. Sydney, Aus.: Academic Press, 1981.

Jennings, Jesse, ed. *The Prehistory of Polynesia*. Cambridge, MA: Harvard University Press, 1979.

Speiser, F. *Ethnology of Vanuatu*. translated by D. Stephenson. Hawaii: University of Hawaii Press, 1996 [1923].

Stanley, D. *South Pacific Handbook,* 3d ed. Chico, Calif.: Moon Publications, 1986.

Viet Nam

Hickey, Gerald Cannon. *Free in the Forest: Ethnohistory of the Vietnamese Central Highlands, 1954–1976*. New Haven: Yale University Press, 1982.

Kahin, George McT. *Intervention: How America Became Involved in Vietnam*. Garden City, N.Y.: Anchor Books. 1987.

Karnow, Stanley. *Vietnam. A History*. New York: The Viking Press. 1983.

Sheehan, Neil. *A Bright Shining Lie*. New York: Random House. 1988.

Thuy, Vuong G. *Getting to the Know the Vietnamese and Their Culture*. New York: Frederick Ungar Publishing. 1975.

Yemen

Al-Suwaidi, Jamal. *The Yemeni War of 1994: Causes and Consequences*. London: Saqi Books, 1995.

Carapico, Sheila. *Civil Society in Yemen : The Political Economy of Activism in Modern Arabia*. New York: Cambridge University Press, 1998.

Chaudhry, Kiren Aziz. *The Price of Wealth: Economies and Institutions in the Middle East*. Ithaca, NY: Cornell University Press, 1997.

Crouch, Michael. *An Element of Luck: To South Arabia and Beyond*. New York: Radcliffe Press, 1993.

Halliday, Fred. *Revolution and Foreign Policy: The Case of South Yemen, 1967–87*. New York: Cambridge University Press, 1990.

Europe

Albania

Battiata, Mary. "Albania's Post-Communist Anarchy," *Washington Post,* March 21, 1998, A1–A18.

Biberaj, Ekiz. *Albania: A Sicuakust Maverick*. Boulder, Colo.: Westview Press, 1990.

Durham, M.E. *High Albania*. Boston: Beacon Press, 1985.

Pipa, Arshi. *The Politics of Language in Socialist Albania*. New York: Columbia University Press for Eastern European Monographs, 1989.

Andorra

Carter, Youngman. *On to Andorra*. New York: W. W. Norton, 1964.

Deane, Shirley. *The Road to Andorra*. New York: William Morrow, 1961.

Duursma, Jorri. *Fragmentation and the International Relations of Micro-States: Self-determination and Statehood*. Cambridge: Cambridge University Press, 1996.

Armenia

Batalden, Stephen K. and Sandra L. Batalden. *The Newly Independent States of Eurasia: Handbook of Former Soviet Republics*. Phoenix, AZ: The Oryx Press, 1993.

Croissant, Michael P. *The Armenia-Azerbaijani Conflict: Causes and Implications*. Westport, Conn: Praeger, 1998.

Curtis, Glenn E., ed. *Armenia, Azerbaijan, and Georgia: Country Studies*. Washington, D.C.: Federal Research Division, Library of Congress, 1995.

Economist Intelligence Unit, The. *Country Profile: Georgia, Armenia, 1998–1999*. London: The Economist Intelligence Unit, 1999.

Goldberg, Suzanne. *Pride of Small Nations: The Caucasus and Post-Soviet Disorder*. Atlantic Heights, N.J.: Zed, 1994.

Kaeger, Walter Emil. *Byzantium and the Early Islamic Conquests*. Cambridge: University Press, 1992.

Lang, David Marshall. *Armenia: Cradle of Civilization*. London, Boston: Allen and Unwin, 1978.

McEcedy, Colin. *The New Penguin Atlas of Medieval History*. London: Penguin, 1992.

Austria

Barkey, Karen and Mark von Hagen. *After Empire: Multiethnic Societies and Nation-Building: the Soviet Union and Russian, Ottoman, and Habsburg Empires*. Boulder, Colo.: Westview Press, 1997.

Brook-Shepherd, Gordon. *The Austrians: A Thousand-Year Odyssey*. London: Harper Collins, 1996.

Johnson, Lonnie. *Introducing Austria: A Short History*. Riverside, Calif.: Ariadne, 1989.

Steininger, Rolf, and Michael Gehler, eds. *Österreich im 20. Jahrhundert*. 2 vols. Vienna: Böhlau, 1997.

Belarus

Gross, Jan Tomasz. *Revolution from Abroad: The Soviet Conquest of Poland's Western Ukraine and Western Belorussia*. Princeton, N.J: Princeton University Press, 1988.

Marples, David R. *Belarus: From Soviet Rule to Nuclear Catastrophe*. London: Macmillan, 1996.

Sword, Keith, ed. *The Soviet Takeover of the Polish Eastern Provinces, 1939–41*. New York: St. Martin's Press, 1991.

Zaprudnik, I.A. *Belarus: At a Crossroads in History*. Boulder, Colo.: Westview Press, 1993.

Belgium

Fitzmaurice, John. *The Politics of Belgium: A Unique Feudalism*. London: Hurst, 1996.

Files, Yvonne. *The Quest for Freedom: The Life of a Belgian Resistance Fighter*. Santa Barbara, Calif.: Fithian Press, 1991.

Hilden, Patricia. *Women, Work, and Politics: Belgium 1830–1914*. Oxford: Clarendon Press, 1993.

Hooghe, Liesbet. *A Leap in the Dark: Nationalist Conflict ad Federal Reform in Belgium*. Ithaca: Cornell University Press, 1991.

Warmbrunn, Werner. *The German Occupation of Belgium: 1940–1944*. New York: P. Lang, 1993.

Wee, Herman van der. *The Low Countries in Early Modern Times*. Brookfield, Vt.: Variorum, 1993.

Bosnia and Herzegovina

Burg, Steven L., and Paul S. Shoup. *The War in Bosnia-Herzegovina: Ethnic Conflict and International Intervention*. Armonk, N.Y.: M.E. Sharpe, 1999.

Filipovic, Zlata. *Zlata's Diary: A Child's Life in Sarajevo.* New York: Viking, 1994.
Glenny, Misha. *The Fall of Yugoslavia: The Third Balkan War.* New York: Penguin, 1996.
Lampe, John R. *Yugoslavia as History: Twice There Was a Country.* Cambridge: Cambridge University Press, 1996.
Pinson, Mark, ed. *The Muslims of Bosnia-Herzegovina: Their Historic Development from the Middle Ages to the Dissolution of Yugoslavia.* 2nd ed. Cambridge, Mass.: Harvard University Press, 1996.
Prstojevic, Miroslav. *Sarajevo Survival Guide.* Trans. Aleksandra Wagner with Ellen Elias-Bursac. New York: Workman Publishing, 1993.
Rogel, Carole. *The Breakup of Yugoslavia and the War in Bosnia.* Westport, Conn.: Greenwood, 1998.
West, Rebecca. *Black Lamb and Grey Falcon.* Reprint. New York: Penguin Books, 1982

Bulgaria

Crampton, R.J. *A Concise History of Bulgaria.* Cambridge, New York: Cambridge University Press, 1997.
Curtis, Glenn E., ed. *Bulgaria, a Country Study.* 2nd ed. Federal Research Division, Library of Congress. Washington, D.C., 1993.
Melone, Albert P. *Creating Parliamentary Government: The Transition to Democracy in Bulgaria.* Columbus: Ohio State University Press, 1998.
Minaeva, Oksana. *From Paganism to Christianity: Formation of Medieval Bulgarian Art (681–972).* Frankfurt am Main, New York: P. Lang, 1996.
Paskaleva, Krassira, ed. *Bulgaria in Transition: Environmental Consequences of Political and Economic Transformation.* Brookfield, VT: Ashgate Publications, 1998.
Sedlar, Jean W. *East Central Europe in the Middle Ages, 1000–1500. A History of East Central Europe,* 3. Seattle and London: University of Washington Press, 1994.

Croatia

Cuvalo, Ante. *The Croatian National Movement, 1966–72.* New York: Columbia University Press, 1990.
Glenny, Michael. *The Fall of Yugoslavia: The Third Balkan War.* New York: Penguin, 1992.
Irvine, Jill A. *The Croat Question: Partisan Politics in the Formation of the Yugoslav Socialist State.* Boulder, Colo.: Westview Press, 1993.
Tanner, Marcus. *Croatia: A Nation Forged in War.* New Haven, Conn.: Yale University Press, 1997.

Czech Republic

Bradley, J. F. N. *Czechoslovakia's Velvet Revolution: A Political Analysis.* New York: Columbia University Press, 1992.
Kalvoda, Josef. *The Genesis of Czechoslovakia.* New York: Columbia University Press, 1986.
Kriseova, Eda. *Vaclav Havel: The Authorized Biography.* New York: St. Martin's Press, 1978.
Leff, Carol Skalnik. *The Czech and Slovak Republics: Nation Versus State.* Boulder, Colo.: Westview Press, 1997.

Denmark

Kjrgaard, Thorkild. *The Danish Revolution, 1500–1800: An Ecohistorical Interpretation.* Cambridge: Cambridge University Press, 1994.
Monrad, Kasper. *The Golden Age of Danish Painting.* New York: Hudson Hills Press, 1993.
Pundik, Herbert. *In Denmark It Could Not Happen: The Flight of the Jews to Sweden in 1943.* New York: Gefen Publishing House, 1998.

Estonia

Gerner, Kristian, and Stefan Hedlund. *The Baltic States and the End of the Soviet Empire.* New York: Routledge, 1993.
Hiden, John and Patrick Salmon. *The Baltic Nations and Europe.* London and New York: Longman, 1994.
Iwaskiw, Walter R. *Estonia, Latvia, and Lithuania: Country Studies.* Washington, DC: Federal Research Division, Library of Congress, 1996.
Lieven, Anatol. *The Baltic Revolution.* New Haven and London: Yale University Press, 1993.
Taagepera, Rein. *Estonia: Return to Independence.* Boulder, Colo.: Westview Press, 1993.

Finland

Lander, Patricia Slade. *The Land and People of Finland.* New York: HarperCollins, 1990.
Maude, George. Historical Dictionary of Finland. Metuchen, N.J.: Scarecrow Press, 1994.
.Rajanen, Aini. *Of Finnish Ways.* New York: Barnes & Noble Books, 1981.
Schoolfield, George C. *Helsinki of the Czars: Finland's Capital, 1808–1918.* Columbia, SC: Camden House, 1996.
Singleton, Fred. *A Short History of Finland,* 2nd ed. Cambridge, Eng.: Cambridge University Press, 1998.

France

Agulhon, Maurice. *The French Republic, 1879–1992.* Cambridge, Mass.: B. Blackwell, 1993.
Corbett, James. *Through French Windows: An Introduction to France in the Nineties.* Ann Arbor, Mich.: University of Michigan Press, 1994.
Gildea, Robert. *France Since 1945.* Oxford: Oxford University Press, 1996.
Gough, Hugh, and John Horne. *De Gaulle and Twentieth-Century France.* New York: Edward Arnold, 1994.
Hollifield, James F., and George Ross, eds. *Searching for the New France.* New York: Routledge, 1991.
Noiriel, Gérard. *The French Melting Pot: Immigration, Citizenship, and National Identity.* Minneapolis, Minn.: University of Minnesota Press, 1996.
Northcutt, Wayne. *The Regions of France: A Reference Guide to History and Culture.* Westport, Conn.: Greenwood Press, 1996.
Young, Robert J. *France and the Origins of the Second World War.* Basingstoke, England: Macmillan, 1996.

Georgia

Braund, David. *Georgia in Antiquity: A History of Colchis and Transcaucasian Iberia, 550 BC–AD 562.* New York: Oxford University Press, 1994.

Goldstein, Darra. *The Georgian Feast: The Vibrant Culture and Savory Food of the Republic of Georgia.* New York : HarperCollins, 1993.

Schwartz, Donald V., and Razmik Panossian. *Nationalism and History: The Politics of Nation Building in Post-Soviet Armenia, Azerbaijan and Georgia.* Toronto, Canada : University of Toronto Centre for Russian and East European Studies, 1994.

Suny, Ronald Grigor. *The Making of the Georgian Nation.* 2nd ed. Bloomington, Ind.: Indiana University Press, 1994.

———, ed. *Transcaucasia, Nationalism, and Social Change: Essays in the History of Armenia, Azerbaijan, and Georgia.* Ann Arbor: University of Michigan Press, 1996.

Germany

Davies, Norman. *A History of Europe.* New York: Oxford University Press, 1996.

Dülffer, Jost. *Nazi Germany 1933–1945: Faith and Annihilation.* London: Arnold, 1996.

Eley, Geoff, ed. *Society, Culture, and the State in Germany: 1870–1930.* Ann Arbor: The University of Michigan Press, 1996.

Friedländer, Saul. *Nazi Germany and the Jews.* Volume I - "The Years of Persecution, 1933–1939." New York: HarperCollins, 1997.

Fulbrook, Mary. *A Concise History of Germany.* Updated edition. New York: Cambridge University Press, 1994.

Gies, Frances and Joseph. *Cathedral, Forge, and Waterwheel: Technology and Invention in the Middle Ages.* New York: HarperCollins, 1994.

Kramer, Jane. *The Politics of Memory: Looking for Germany in the New Germany.* New York: Random House, 1996.

Greece

Costas, Dimitris, ed. *The Greek-Turkish Conflict in the 1990s.* New York: St. Martin's Press, 1991.

Jouganatos, George A. *The Development of the Greek Economy, 1950–1991.* Westport, Conn.: Greenwood Press, 1992.

Laisné, Claude. *Art of Ancient Greece: Sculpture, Painting, Architecture.* Paris: Terrail, 1995.

Lawrence, A.W. *Greek Architecture.* New Haven, Conn.: Yale University Press, 1996.

Legg, Kenneth R. *Modern Greece: A Civilization on the Periphery.* Boulder, Colo.: Westview Press, 1997.

Pettifer, James. *The Greeks: The Land and People Since the War.* New York: Viking, 1993.

Hungary

Bartlett, David L. *The Political Economy of Dual Transformations: Market Reform and Democratization in Hungary.* Ann Arbor, MI: University of Michigan Press, 1997.

Corrin, Chris. *Magyar Women: Hungarian Women's Lives, 1960s–1990s.* New York: St. Martin's, 1994.

Hoensch, Jorg K. *A History of Modern Hungary, 1867–1994.* 2nd ed. New York: Longman, 1996.

Litvan, Gyorgy, ed. *The Hungarian Revolution of 1956: Reform, Revolt, and Repression, 1953–1956.* Trans. Janos M. Bak and Lyman H. Legters. New York: Longman, 1996.

Szekely, Istvan P. and David M.G. Newberry, eds. *Hungary: An Economy in Transition.* Cambridge: Cambridge University Press, 1993.

Iceland

Durrenberger, E. Paul. *The Dynamics of Medieval Iceland: Political Economy and Literature.* Iowa City: University of Iowa Press, 1992.

Jochens, Jenny. *Women in Old Norse Society.* Ithaca: Cornell University Press, 1995.

Lacy, Terry G. *Ring of Seasons: Iceland: Its Culture and History.* Ann Arbor: University of Michigan Press, 1998.

Roberts. David. *Iceland.* New York: H.N. Abrams, 1990.

Ireland

Breen, Richard. *Understanding Contemporary Ireland: State, Class, and Development in the Republic of Ireland.* New York: St. Martin's Press, 1990.

Daly, Mary E. *Industrial Development and Irish National Identity, 1922–1939.* Syracuse, N.Y.: Syracuse University Press, 1992.

Hachey, Thomas E. *The Irish Experience: A Concise History.* Armonk, N.Y.: M. E. Sharpe, 1996.

Harkness, D. W. *Ireland in the Twentieth Century: Divided Island.* Hampshire, England: Macmillan Press, 1996.

MacDonagh, Oliver, et al. *Irish Culture and Nationalism, 1750–1950.* New York: St. Martin's Press, 1984.

Sawyer, Roger. *"We Are But Women": Women in Ireland's History.* New York: Routledge, 1993.

Italy

Baranski, Zygmunt G., and Robert Lumley, ed. *Culture and Conflict in Postwar Italy: Essays on Mass and Popular Culture.* New York: St. Martin Press, 1990.

Duggan, Christopher. *A Concise History of Italy.* New York: Cambridge University Press, 1994.

Furlong, Paul. *Modern Italy: Representation and Reform.* New York: Routledge, 1994.

Ginsberg, Paul. *A History of Contemporary Italy: Society and Politics, 1943–1988.* London: Penguin, 1990.

Hearder, Harry. *Italy: A Short History.* New York: Cambridge University Press, 1990.

Holmes, George. *The Oxford History of Italy.* New York: Oxford University Press, 1997.

Latvia

Gerner, Kristian, and Stefan Hedlund. *The Baltic States and the End of the Soviet Empire.* London and New York: Routledge, 1993.

Iwaskiw, Walter R. *Estonia, Latvia, and Lithuania: Country Studies.* Washington, D.C.: Federal Research Division, Library of Congress, 1996.

Lieven, Anatol. *The Baltic Revolution.* New Haven and London: Yale University Press, 1993.

Plakans, Andrejs. *The Latvians: A Short History.* Stanford, Calif: Hoover Institution Press, Stanford University, 1995.

Liechtenstein

Background Notes: Liechtenstein. Washington, D.C.: U.S. Department of State, Bureau of Public Affairs, Office of Public Communication, Editorial Division, USGPO, 1989.

Duursma, Jorri C. *Fragmentation and the International Relations of Micro-States: Self-determination and Statehood.* Cambridge: Cambridge University Press, 1996.

The Principality of Liechtenstein: A Documentary Handbook. Vaduz: Press and Information Office of the Government of the Principality of Liechtenstein, 1967.

Raton, Pierre. *Liechtenstein: History and Institutions of the Principality.* Vaduz: Liechtenstein-Verlag AG, 1970.

Lithuania

Gerner, Kristian and Stefan Hedlund. *The Baltic States and the End of the Soviet Empire.* London/New York: Routledge, 1993.

Hiden, John and Patrick Salmon. *The Baltic Nations and Europe.* London and New York: Longman, 1994.

Hiden, John. *The Baltic States and Weimar Ostpolitik.* Cambridge: Cambridge University Press, 1987.

———. *The Baltic States: Years of Dependence, 1940–1980.* Berkeley: University of California Press, 1983.

Vardys, Vytas Stanley. *Lithuania: The Rebel Nation.* Boulder, Colo.: Westview Press, 1997.

Luxembourg

Barteau, Harry C. *Historical Dictionary of Luxembourg.* Lanham, Md.: Scarecrow Press, 1996.

Clark, Peter. *Luxembourg.* New York: Routledge, 1994.

Dolibois, John. *Pattern of Circles: An Ambassador's Story.* Kent, Oh.: Kent State University Press, 1989.

Hury, Carlo. *Luxembourg.* Oxford, England: Clio Press, 1981.

Newcomer, James. *The Grand Duchy of Luxembourg: The Evolution of Nationhood.* Luxembourg: Editions Emile Borschette, 1995.

Macedonia

Billows, Richard A. *Kings and Colonists: Aspects of Macedonian Imperialism.* New York: E.J. Brill, 1995.

Danforth, Loring M. *The Macedonian Conflict: Ethnic Nationalism in a Transnational World.* Princeton, NJ: Princeton University Press, 1995.

Kofos, Evangelos. *Nationalism and Communism in Macedonia: Civil Conflict, Politics of Mutation, National Identity.* New Rochelle, N.Y.: A.D. Caratzas, 1993.

Poulton, Hugh. *Who Are the Macedonians?* Bloomington: Indiana University Press, 1995.

Shea, John. *Macedonia and Greece: The Struggle to Define a New Balkan Nation.* Jefferson, NC: McFarland, 1997.

Malta

Berg, Warren G., *Historical Dictionary of Malta.* Lanham, Md.: Scarecrow, 1995.

Caruana, Carmen M., *Education's Role in the Socioeconomic Development of Malta.* Westport Connecticut: Praeger, 1992.

Europa World Yearbook, Vol. 2, 39th Edition, 1998.

Evans, J.D., *The Prehistoric Antiquities of the Maltese Islands.* New York: Oxford University Press, 1971.

Moldova

Belarus and Moldova: Country studies. Washington: Department of the Army, 1996.

Bruchis, Michael. *The Republic of Moldavia: From the Collapse of the Soviet Empire to the Restoration of the Russian Empire.* Transl. by Laura Treptow. Boulder, Col.: East European Monographs, 1996.

Hitchins, Keith. *Rumania, 1866–1947.* Oxford: Clarendon Press, 1994.

Papacostea, Serban. *Stephen the Great: Prince of Moldavia, 1457–1504.* Transl. by Seriu Celac. Bucharest: Editura Enciclopedica, 1996.

Treptow, Kurt W. *Historical Dictionary of Romania.* Lanham, Md.: Scarecrow, 1996.

Monaco

Duursma, Jorri. *Self-determination, Statehood, and International Relations of Micro-states: The Cases of Liechtenstein, San Marino, Monaco, Andorra, and the Vatican City.* New York: Cambridge University Press, 1996.

Edwards, Anne. *The Grimaldis of Monaco.* New York: William Morrow, 1992.

Sakol, Jeannie and Caroline Latham. *About Grace: An Intimate Notebook.* Chicago: Contemporary Books, 1993.

Netherlands

Andeweg, R.B. *Dutch Government and Politics.* New York: St. Martin's, 1993.

Fuykschot, Cornelia. *Hunger in Holland: Life During the Nazi Occupation.* Amherst, N.Y.: Prometheus Books, 1995.

Israel, Jonathan Irvine. *Dutch Primacy in World Trade, 1585–1740.* New York: Oxford University Press, 1990.

Schilling, Heinz. *Religion, Political Culture, and the Emergence of Early Modern Society: Essays in German and Dutch History.* New York: E.J. Brill, 1992.

Slive, Seymour. *Dutch Painting 1600–1800.* New Haven, Conn.: Yale University Press, 1995.

Norway

Berdal, Mats R. *The United States, Norway and the Cold War 1954–60.* New York: St. Martin's, 1997.

Heide, Sigrid. *In the Hands of My Enemy: A Woman's Personal Story of World War II.* Trans. Norma Johansen. Middletown, Conn.: Southfarm Press, 1996.

Jochens, Jenny. *Women in Old Norse Society.* Ithaca, N.Y.: Cornell University Press, 1995.

Poland

Blazyca, George and Ryszard Rapacki, eds. *Poland into the 1990s: Economy and Society in Transition.* New York: St. Martin's, 1991.

Engel, David. *Facing a Holocaust: The Polish Government-in-Exile and the Jews, 1943–1945.* Chapel Hill: University of North Carolina Press, 1993.

Staar, Richard F., ed. *Transition to Democracy in Poland.* New York: St. Martin's, 1993.

Steinlauf, Michael. *Bondage to the Dead: Poland and the Memory of the Holocaust.* Syracuse, N.Y.: Syracuse University Press, 1997.

Tworzecki, Hubert. *Parties and Politics in Post-1989 Poland.* Boulder, Colo.: Westview, 1996.

Portugal

Birmingham, David, *A Concise History of Portugal*, Cambridge: Cambridge University Press, 1993.

Hermano Saraiva, Jóse, *Portugal: A Companion History.* Manchester: Carcanet Press, 1995.

Herr, Richard ed. *The New Portugal: Democracy and Europe*. Berkeley: University of California at Berkeley, 1992.

Russell-Wood, A.J.R. *A World on the Move: The Portuguese in Africa, Asia, and America, 1415–1808*, New York: St. Martin's Press, 1993.

Winius, George D. ed., *Portugal, The Pathfinder: Journeys from the Medieval toward the Modern World, 1300–ca.1600*, Madison: The Hispanic Seminary of Medieval Studies, 1995.

Romania

Bachman, Ronald D., ed. *Romania: A Country Study.* 2d ed. Washington, D.C.: Library of Congress, 1991.

Commission on Security and Cooperation in Europe. *Human Rights and Democratization in Romania.* Washington, D.C.: Commission on Security and Cooperation in Europe, 1994.

Economist Intelligence Unit. *Country Report: Romania.* London: The Economist Intelligence Unit, 1999.

Treptow, Kurt W. and Marcel Popa, eds. *Historical Dictionary of Romania.* Lanham, Md.: Scarecrow, 1996.

Russian Federation

Barner-Barry, Carol, and Cynthia A. Hody. *The Politics of Change: The Transformation of the Former Soviet Union.* New York: St. Martin's Press, 1995.

Channon, John, and Robert Hudson. *The Penguin Historical Atlas of Russia*. London: Viking, 1995.

Curtis, Glenn E., ed. *Russia: A Country Study.* Washington: Library of Congress, 1998.

Daniels, Robert V., ed. *The Stalin Revolution: Foundations of the Totalitarian Era*. 4th ed. Boston: Houghton Mifflin, 1997.

Fitzpatrick, Sheila. *The Russian Revolution*. New York: Oxford University Press, 1994.

MacKenzie, David, and Michael W. Curran. *A History of Russia, the Soviet Union, and Beyond*. 5th ed. Belmont, CA: West/Wadsworth, 1999.

Raymond, Boris, and Paul Duffy. *Historical Dictionary of Russia*. Lanham, Md.: Scarecrow, 1998.

San Marino

Catling, Christopher. *Umbria, the Marches and San Marino.* Lincolnwood, Ill.: Passport Books, 1994.

Duursma, Jorri C. *Fragmentation and the International Relations of Micro-States*. Cambridge: Cambridge University Press, 1996.

United States Bureau of Public Affairs; Background Notes: *San Marino*. Washington

World Reference Atlas: *San Marino*. London: Darling Kindersley Publishing, Inc., 1998.

Slovakia

Goldman, Minton F. *Slovakia Since Independence: A Struggle for Democracy.* Westport, Conn.: Praeger, 1999.

Jelinek, Yeshayahu A. *The Parish Republic: Hlinka's Slovak People's Party: 1939–1945.* New York: Columbia University Press, 1976.

Kirshbaum, Stanislav J. *A History of Slovakia: The Struggle for Survival.* New York: St. Martin's Press, 1995.

Leff, Carol Skalnik. *The Czech and Slovak Republics: Nation Versus State.* Boulder, Colo.:Westview Press, 1997.

Slovenia

Cohen, Lenard. Broken Bonds: *The Disintegration of Yugoslavia*. Boulder, Colo.: Westview Press, 1993.

Fink-Hafner, Danica, and John R. Robbins, eds. *Making a New Nation: The Formation of Slovenia*. Brookfield, Vt. Aldershot: Dartmouth Publishing, 1997.

Owen, David. *Balkan Odyssey.* New York: Harcourt Brace, 1995.

Rogel, Carole. *The Breakup of Yugoslavia and the War in Bosnia*. Westport, Conn.: Greenwood Press, 1998.

Silber, Laura, and Allan Little. *Yugoslavia: Death of a Nation*. New York: TV Books, 1996.

Zimmerman, Warren. *Origins of a Catastrophe: Yugoslavia and its destroyers—America's last ambassador tells what happened and why.* New York: Times Books, 1996.

Spain

Cantarino, Vicente. *Civilización y Cultura de España,* 3rd. ed., Englewood Cliffs, N.J.: Prentice Hall, 1995.

Hopper, John. *The New Spaniards.* Suffolk, Eng.: Penguin, 1995.

Jordan, Barry. *Writings and Politics in the Franco's Spain.* London, Eng.: Routledge, 1990.

Leahy, Philippa. *Discovering Spain.* New York: Crestwood House, 1993.

Wernick, Robert. "For Whom the Bell Tolled. (the Spanish Civil War)." *Smithsonian.* April 1998, vol. 28, no. 1, pp. 110+.

Sweden

Palmer, Alan. *Bernadotte: Napoleon's Marshal, Sweden's King.* London: Murray, 1990.

Roberts, Michael. *From Oxenstierna to Charles XII: Four Studies.* New York: Cambridge University Press, 1991.

Rothstein, Bo. *The Social Democratic State: The Swedish Model and the Bureaucratic Problem of Social Reforms.* Pittsburgh: University of Pittsburgh, 1996.

Scobbie, Irene. *Historical Dictionary of Sweden.* Metuchen, N.J.: Scarecrow Press, 1995.

Switzerland

Bacchetta, Philippe, and Walter Wasserfallen, ed. *Economic Policy in Switzerland.* New York: St. Martin's Press, 1997.

Eu-Wong, Shirley. *Culture Shock!: Switzerland.* Portland, Or.: Graphic Arts Center Pub. Co., 1996.

Hilowitz, Janet Eve, ed. *Switzerland in Perspective.* New York: Greenwood Press, 1990.

New, Mitya. *Switzerland Unwrapped: Exposing the Myths.* New York: I.B. Tauris, 1997.

Steinberg, Jonathan. *Why Switzerland?* 2nd ed. Cambridge: Cambridge University Press, 1996.

Turkey

Fodor's Turkey. New York: Fodor's Travel Publications, 1999.

Mitchell, Stephen. *Anatolia: Land, Men, and Gods in Asia Minor.* New York: Oxford University Press, 1993.

Stoneman, Richard. *A Traveller's History of Turkey.* New York: Interlink Books, 1998.

Time-Life Books. *Anatolia: Cauldron of Cultures.* Alexandria, Va.: Time-Life Books, 1995.

Ukraine

Hosking, Geoffrey A., ed. *Church, Nation and State in Russia and Ukraine.* New York: St. Martin's Press, 1991.

Kuzio, Taras. *Ukraine under Kuchma: Political Reform, Economic Transformation and Security Policy in Independent Ukraine.* New York: St. Martin's Press, 1997.

Shen, Raphael. *Ukraine's Economic Reform: Obstacles, Errors, Lessons.* Westport, Conn: Praeger, 1996.

Subtelny, Orest. *Ukraine: A History.* Toronto: University of Toronto Press, 1994.

Wilson, Andrew. *Ukrainian Nationalism in the 1990s: A Minority Faith.* New York: Cambridge University Press, 1997.

United Kingdom

Abrams, M. H., ed. *The Norton Anthology of English Literature.* 2 vols. 6th ed. New York : Norton, 1993.

Cook, Chris. *The Longman Handbook of Modern British History, 1714–1995.* 3rd ed. New York: Longman, 1996.

Delderfield, Eric F. *Kings & Queens of England & Great Britain.* New York: Facts on File, 1990.

Figes, Kate. *Because of Her Sex: The Myth of Equality for Women in Britain.* London: Macmillan, 1994.

Foster, R.F., ed. *The Oxford History of Ireland.* New York: Oxford University Press, 1992.

Glynn, Sean. *Modern Britain: An Economic and Social History.* New York: Routledge, 1996.

Jenkins, Philip. *A History of Modern Wales, 1536–1990.* New York: Longman, 1992.

Judd, Denis. *Empire: The British Imperial Experience from 1765 to the Present.* London : HarperCollins, 1996.

The Oxford History of Britain. New York: Oxford University Press, 1992.

Powell, David. *British Politics and the Labour Question, 1868–1990.* New York: St. Martin's, 1992.

Vatican

Accattoli, Luigi. *Life in the Vatican with John Paul II.* Trans. Marguerite Shore. New York : Universe, 1998.

Reese, Thomas J. *Inside the Vatican: The Politics and Organization of the Catholic Church.* Cambridge, Mass.: Harvard University Press, 1996.

Rosa, Peter de. *Vicars of Christ: The Dark Side of the Papacy.* New York: Crown Publishers, 1988.

Volpini, Valerio. *Vatican City: Art, Architecture, and History.* New York, Portland House: Distributed by Crown Publishers, 1986.

Yugoslavia

Cohen, Lenard J. *Broken Bonds: Yugoslavia's Disintegration and Balkan Politics in Transition.* Boulder, CO: Westview, 1995.

Denitch, Bogdan Denis. *Ethnic Nationalism: The Tragic Death of Yugoslavia.* Minneapolis: University of Minnesota Press, 1996.

Dyker, David A., and Ivan Vejvoda, ed. *Yugoslavia and After: A Study in Fragmentation, Despair and Rebirth.* New York: Longman, 1996.

Lampe, John R. *Yugoslavia as History: Twice There Was a Country.* New York: Cambridge University Press, 1996.

Pavkovic, Aleksandr. *The Fragmentation of Yugoslavia: Nationalism in a Multinational State.* New York: St. Martin's Press, 1997.

Ramet, Sbrina P. *Balkan Babel: The Disintegration of Yugoslavia from the Death of Tito to Ethnic War.* Boulder, Colo.: Westview, 1996.

Index

B

D

E

G

H

I

J

K

L

M

N

O

Q

R

S

T

U

V

W

X

Y

Z